ADVANCED WIRELESS COMMUNICATIONS & INTERNET

ADVANCED WIRELESS COMMUNICATIONS & INTERNET

FUTURE EVOLVING TECHNOLOGIES

Third Edition

Savo G Glisic
University of Oulu, Finland

A John Wiley and Sons, Ltd., Publication

This edition first published 2011
© 2011 John Wiley & Sons, Ltd

Registered office
John Wiley & Sons Ltd, The Atrium, Southern Gate, Chichester, West Sussex, PO19 8SQ, United Kingdom

For details of our global editorial offices, for customer services and for information about how to apply for permission to reuse the copyright material in this book please see our website at www.wiley.com.

Library of Congress Cataloging-in-Publication Data

Glisic, Savo G.
   Advanced wireless communications & Internet : future evolving technologies / Savo G. Glisic. – 3rd ed.
     p. cm.
   Summary: "The concept of the book will be organized within the framework of wireless access to future internet, so that a number of topics above physical layer will be added"– Provided by publisher.
Includes bibliographical references and index.
ISBN 978-0-470-71122-4 (hardback)
   1. Wireless communication systems–Technological innovations.   2. Internet. I. Title.
   TK5103.2.G548 2011
   621.384–dc22

                                        2010048265

A catalogue record for this book is available from the British Library.

Print ISBN: 9780470711224
ePDF ISBN: 9781119991625
oBook ISBN: 9781119991632
ePub ISBN: 9781119991755

Set in 9/11pt Times by Aptara Inc., New Delhi, India
Printed and bound in Singapore by Markono Print Media Pte Ltd

To my family

Contents

Preface to the Third Edition

At this stage of the evolution of wireless communications the dominant problems are to be found in the area of networking and the integration of wireless communications in the *Future Internet*. Even the classical concept of cellular networks is evolving and with multihop relaying its optimization includes not only the physical layer but also scheduling and routing in the network of passive or active relays. Contest awareness and cognitive solutions on all layers make the optimization process more challenging. Cloud computing and data centric services bring about new communication paradigms also. In order to reflect all of these trends in future evolving technologies the new edition of the book includes a number of new chapters.

In the introductory Chapter 1, there is a brief discussion on the next generation of the Internet, cloud computing, and network virtualization, the economics of utility computing, and wireless grids and clouds, which is intended to help with an initial understanding of the overall environment in which future wireless networks will be operating. In the rest of the book there are six new chapters.

Chapter 10 covers channel sampling, which is a basic problem in Cognitive Networks. Efficient spectrum utilization has attracted significant attention from researchers because most of the allocated spectrum is severely under-utilized. In order to improve spectrum utilization, a new spectrum allocation method, called cognitive radio, is proposed. While, in general, the terms cognitive radio and cognitive networks include a much broader scale of techniques, based on cognition, in this chapter we limit our interest to the network where users are classified into two groups: primary users (PUs) and secondary users (SUs). The PUs are licensed users for a given frequency band and have highest priority to access the allocated band, while the SUs share the bandwidth opportunistically with the PUs only when the bandwidth is not currently being used by PUs. Therefore, in order to avoid severe interference with the transmission from PUs, the SUs need first to sense channel availability and then carry out data transmission over idle channels.

Relay-assisted Wireless Networks are covered in Chapter 11. Cooperative communication is based on collaboration among several distributed terminals so as to transmit/receive their intended signals. This type of communication is based on the seminal works issued in the 1970s by van der Meulen, Cover, and El Gamal, where a new element is introduced in conventional point-to-point communication, the relay. The new network architecture exhibits some of the properties of MIMO systems, but in contrast to those systems, relay-assisted transmission is able to combat the channel impairments owing to shadowing and path-loss in the source-destination and relay-destination links. The chapter provides a comprehensive overview of the problems and solutions for such networks.

Chapter 12 covers the bio inspired paradigms in wireless networks. It discusses new paradigms in wireless networks inspired by existing biological concepts in the human body or other living organisms. It includes a biologically inspired model for securing hybrid mobile ad hoc networks, energy-aware routing algorithms in ad hoc networks, biological computations in the context of sensor networks, biologically inspired cooperative routing for wireless mobile sensor networks, as well as minimum power multicasting, genetic algorithm based topology control for wireless ad hoc networks, biologically inspired self-managing sensor networks, bio inspired mobility, immune mechanism based intrusion detection systems, artificial immune system, anomaly detection in TCP/IP networks using the immune systems paradigm, epidemic routing, and nano-networks.

Wireless Networks Connectivity is discussed in Chapter 14. Given a set of nodes and a set of commodities, the survivable network design problem involves designing the topology and dimensioning the links so that the network can carry all of the traffic demands and ensure full recovery from a range of link failures. The chapter reviews a range of methodologies for maintaining network connectivity by using tools ranging from genetic algorithms to stochastic geometry and random graphs theory, including discussion on percolation and connectivity.

Chapter 15 covers advanced routing and network coding including discussion of conventional routing versus network coding, a max-flow min-cut theorem, algebraic formulation of network coding, random network coding, gossip based protocol and network coding, multisource multicast network switching, the conventional route packing problem, multicast network switching as a matrix game, computation of the maximum achievable information rate for single-source multicast network switching, optimization of wireless multicast ad-hoc networks, matrix game formulation of joint routing and scheduling, extended fictitious playing and dominancy, optimization of multicast wireless ad-hoc network using soft graph coloring and non-linear cubic games, cubic matrix game modeling for joint optimum routing, network coding and scheduling, routing and network stability, time varying network with queuing, Lyapunov drift and network stability, Lagrangian decomposition of multicomodity flow optimization problem, flow optimization in heterogeneous networks, and dynamic resource allocation in computing clouds.

Finally, network formation games are discussed in Chapter 16. The chapter covers topics such as the general model of network formation games, knowledge based network formation games, and coalition games in wireless ad hoc networks.

The author would like to thank A. Agustin, J. Vidal, and O. Muñoz of Technical University of Catalonia, Barcelona, Spain, for putting together Chapter 11.

Savo G Glisic
Jacksonville, Florida

1

Fundamentals

1.1 4G and the Book Layout

The research community and industry in the field of telecommunications are considering continuously the possible evolution of wireless communications. This evolution is closely related to the future concept of the Internet. With the advances in multihop cellular networks (relaying) and the integrated elements of ad hoc and cellular networks the border between the Internet and wireless networks is disappearing rapidly. Instead of having *wireless access to Internet* we will see the extension of the Internet over wireless networks resulting in a *Wireless Internet*. For this reason an understanding of the future trends in the evolution of the Internet is necessary so as to be able to plan the necessary development of wireless networks in order to enable the closer integration of the two systems. This chapter will start with a generic 4G system concept that integrates the available advanced wireless technologies and will then focus on system adaptability and reconfigurability as a possible option to meet a variety of service requirements, available resources and channel conditions. The elements of such a concept can be found in [1–51]. This presentation will also try to offer a vision beyond the state of the art with an emphasis on how advanced technologies can be used for efficient 4G multiple access. The second part of the chapter will discuss the future evolution of the Internet, especially the concepts of the resource clouds and smart grids. Amongst a number of relevant issues the focus will be on:

- adaptive and reconfigurable coding and modulation including distributed source coding which is of interest for data aggregation in wireless sensor networks;
- adaptive and reconfigurable space time coding including a variety of turbo receivers;
- channel estimation and equalization and multiuser detection;
- orthogonal Frequency Division Multiple Access (OFDMA), Multicarrier CDMA (MC CDMA) and Ultra Wide Band (UWB) radio;
- antenna array signal processing;
- convex optimization based linear precoding for MIMO systems;
- channel sensing for cognitive radio;
- cooperative transmit diversity and relaying;
- biologically inspired paradigms in wireless networks;
- user location in 4G;
- reliability and redundancy design in communication networks;
- cross-layer optimization including adaptive and power efficient MAC layer design, adaptive and power efficient routing on IP and TCP layer including network coding and concept of green wireless network;
- cognitive networks modeling based on game theory.

Advanced Wireless Communications & Internet: Future Evolving Technologies, Third Edition. Savo Glisic.
© 2011 John Wiley & Sons, Ltd. Published 2011 by John Wiley & Sons, Ltd.

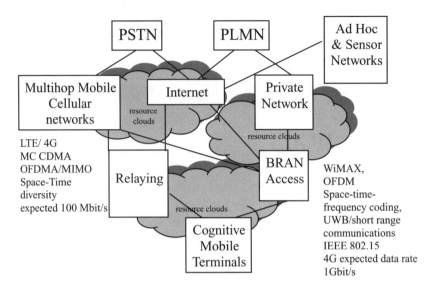

Figure 1.1   Cellular networks and WLAN convergence.

An important aspect of wireless system design is power consumption. This will be also incorporated in the optimization process of most of the problems considered throughout the book.

At this stage of the evolution of wireless communications there is a tendency towards even closer integration of mobile communications as specified by the International Mobile Telecommunications (IMT) standards and Wireless Local Area Networks (WLAN) or in the general Broadband Radio Access Networks (BRAN) specified by IEEE 802.xx. The core network will be based on Public Switched Telecommunications Network (PSTN) and Public Land Mobile Networks (PLMN) based on Internet Protocol (IP) [13, 16, 19, 24, 32, 41, 51]. This concept is summarized in Figure 1.1. Each of the segments of the system will be further enhanced in the future. The inter-technology roaming of the mobile terminal will be based on a reconfigurable cognitive radio concept presented in its generic form in Figure 1.2.

The material in this book is organized as follows:

Chapter 1 starts with the general structure of 4G signals, mainly Advanced Time Division Multiple Access – ATDMA, Code Division Multiple Access – CDMA, Orthogonal Frequency Division Multiplexing – OFDM, Multicarrier CDMA (MC CDMA) and Ultra Wide Band (UWB) signal. These

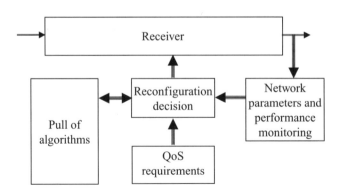

Figure 1.2   Reconfigurable cognitive radio concept intersystem roaming & QoS provisioning.

signals will be elaborated upon later in the book in more detail. In the second part of the chapter we discuss the future evolution of the Internet, especially the concepts of resource clouds and smart grids.

Chapter 2 introduces adaptive coding. The book is not intended to cover all the details of coding but rather to focus on those components that enable code adaptability and reconfigurability. Within this concept the chapter covers: adaptive and reconfigurable block and convolutional codes, punctured convolutional codes/code reconfigurability, maximum likelihood decoding/Viterbi algorithm, systematic recursive convolutional code, concatenated codes with interleaver, the iterative (turbo) decoding algorithm and a discussion on adaptive coding practice and prospects. The chapter also includes a presentation of distributed source coding which is of interest in data aggregation in wireless sensor networks.

Chapter 3 covers adaptive and reconfigurable modulation. This includes coded modulation, Trellis Coded Modulation (TCM) with examples of TCM schemes such as two, four and eight state trellis and QAM with 3 bits per symbol transmission. The chapter also discusses signal set partitioning, equivalent representation of TCM, TCM with multidimensional constellation, adaptive coded modulation for fading channels and adaptation to maintain a fixed distance in the constellation.

Chapter 4 introduces Space Time Coding. It starts with a discussion on diversity gain, the encoding and transmission sequence, the combining scheme and ML decision rule for a two-branch transmit diversity scheme with one and M receivers. Next, it introduces a general discussion on *space time coding* within a concept of space time trellis modulation. The discussion is then extended to introduce space-time block codes from *orthogonal design*, mainly linear processing orthogonal designs and generalized real orthogonal designs. The chapter also covers channel estimation imperfections. It continuous with quasy orthogonal space time block codes, space time convolutional codes and algebraic space time codes. It also includes differential space-time modulation with a number of examples.

Layered space – time coding and *concatenated* space time block coding are also discussed. Estimation of MIMO channel and space-time codes for frequency selective channels are discussed in detail. MIMO system optimization including gain optimization by singular value decomposition (svd) is also discussed. This chapter is extended to include a variety of turbo receivers.

Chapter 5 introduces multiuser detection starting with CDMA receivers and signal subspace-based channel estimation. It then extends this approach to iterative space time receivers. In Chapter 7 this approach is extended to OFDM receivers.

Chapter 6 deals with equalization, detection in a statistically known time-varying channel, adaptive MLSE equalization, adaptive joint channel identification and data demodulation, turbo-equalization Kalman filer based joint channel estimation and equalization using higher order signal statistics.

Chapter 7 covers orthogonal frequency division multiplexing (OFDM) and MC CDMA. The following topics are discussed: Timing and frequency offset in OFDM, fading channel estimation for OFDM systems, 64-DAPSK and 64-QAM modulated OFDM signals, space time coding with OFDM signals, layered space time coding for MIMO-OFDM, space time coded TDMA/OFDM reconfiguration efficiency, multicarrier CDMA system, multicarrier DS-CDMA broadcast systems, frame by frame adaptive rate coded multicarrier DS-CDMA system, intermodulation interference suppression in multicarrier DS-CDMA systems, successive interference cancellation in multicarrier DS-CDMA systems, MMSE detection of multicarrier CDMA, multiuser receiver for space-time coded multicarrier CDMA systems and peak to average power ratio (PAPR) problem mitigation.

Chapter 8 introduces *Ultra Wide Band Radio*. It covers topics such as: UWB multiple access in Gaussian channel, the UWB channel, UWB system with M-ary modulation, M-ary PPM UWB multiple access, coded UWB schemes, multiuser detection in UWB radio, UWB with space time processing and beamforming for UWB radio.

Chapter 9 covers linear precoding for MIMO Channels. This includes space–time precoders and equalizers for MIMO channels, linear precoding based on convex optimization theory and convex optimization-theory-based beamforming.

Chapter 10 discusses issues related to channel sensing for cognitive radio including optimal channel sensing in cognitive wireless networks, optimal sequential, parallel multiband channel and collaborative spectrum sensing and multichannel cognitive MAC.

Chapter 11: Introduces cooperative transmit diversity as a power efficient technology to increase the coverage in multihop wireless networks. It is expected that elements of this approach will be used in 4G cellular systems also, especially relaying which represents a simple case of this approach.

Chapter 12 covers a biologically inspired model for securing hybrid mobile ad hoc networks, biologically inspired routing in ad hoc networks, swarm intelligence based routing, analytical modeling of antnet as adaptive mobile agent based routing, biologically inspired algorithm for optimum multicasting, ant colony system (ACS) model, biologically inspired distributed topology control, optimization of mobile agent routing in sensor networks, epidemic routing, nanonetworks and genetic algorithm based dynamic topology reconfiguration in cellular multihop wireless networks.

Chapter 13 is modified significantly to include more detail on positioning. This is the result of a prediction that this technique will gain an increasing role in advanced wireless communications. This is also supported by activities within the Galileo program in Europe.

Chapter 14 covers survivable wireless networks design, survivability of wireless ad hoc networks, network dimensioning, genetic algorithm based network redundancy design, integer programming method, simulated annealing and survivable network design under general traffic.

Chapter 15 discusses conventional routing versus network coding, a max-flow min-cut theorem, algebraic formulation of network coding, random network coding, gossip based protocol and network coding, network coding with reduced complexity, multisource multicast network switching, conventional route packing problem, multicast network switching as matrix game, computation of maximum achievable information rate for single-source multicast network switching, optimization of wireless multicast ad-hoc networks, matrix game formulation of joint routing and scheduling, interference controlled scheduling, extended fictitious playing and dominancy theory for *AMG* games, optimization of multicast wireless ad-hoc network using soft graph coloring and non-linear cubic games, matrix game modeling for optimum scheduling, cubic matrix game modeling for joint optimum routing, network coding and scheduling and joint optimization of routing and medium contention in multihop multicast wireless network.

Chapter 16 covers network formation games including stability and efficiency of the game, traffic routing utility model, network formation game dynamics, knowledge based network formation games, network formation as a non-cooperative game, network formation as a cooperative game, dynamic network formation and topology control, greedy utility maximization, non-cooperative link utility maximization, cooperative link utility maximization via utility transfer, preferential attachment with knowledge of component sizes, preferential attachment with knowledge of neighbor degrees, link addition/deletion algorithm, coalition games in wireless ad hoc networks, stochastic model of coalition games for spectrum sharing in large scale wireless ad hoc networks and modelling coalition game dynamics.

The evolution of common air interface in wireless communications can be presented in general as in Table 1.1. The coding and modulation for 4G air interface are more or less defined. These problems are addressed in Chapters 1–9. The work on a new multiple access scheme still remains to be elaborated. In this segment the new generation of wireless networks will be significantly different from the solutions seen so far. A part of the solution is what we refer to as *intercell interference coordination (IIC) in MAC layer (IIC MAC)*, as a new multiple access scheme for 4G systems.

In multihop networks this problem will be addressed through different forms of joint optimization of scheduling, routing and relaying topology control. This is addressed in Chapters 10–16. Most of the material in Chapters 10–16 of the third edition of the book is new as compared with the previous edition.

Table 1.1   Evolution of common air interface in mobile communications

generation	2G	3G	4G
coding	convolutional coding	turbo coding	turbo coding & space time coding
modulation	GMSK, BPSK, QPSK	BPSK, QPSK, mQAM	NmQAM (OFDM)
multiple access	TDMA	CDMA	IIC MAC & joint optimization of scheduling/routing & topology control (new Chapters 10–16)

1.2 General Structure of 4G Signals

In this section we will summarize the signal formats used in the existing wireless systems and point out possible ways of evolution towards the 4G system. The focus will be on OFDMA, MC CDMA and UWB signals.

1.2.1 Advanced Time Division Multiple Access – ATDMA

In a TDMA system each user is using a dedicated time slot within a TDMA frame as in GSM (Global System of Mobile Communications) or in ADC (American Digital Cellular System). Additional data about the signal format and system capacity are given in [54]. The evolution of ADC system resulted in the TIA (Telecommunications Industry Association) Universal Wireless Communications (UWC) standard 136 [54]. The evolution of GSM resulted into a system known as Enhanced Data rates for GSM Evolution (EDGE) with parameters that are summarized in [54].

1.2.2 Code Division Multiple Access – CDMA

CDMA technique is based on spreading the spectra of the relatively narrow information signal S^n by a code c, generated by much higher clock (chip) rate. Different users are separated by using different uncorrelated codes. As an example the narrowband signal in this case can be a PSK signal of the form

$$S_n = b(t, T_m) \cos \omega t \tag{1.1}$$

where $1/T_m$ is the bit rate and $b = \pm 1$ is the information. The baseband equivalent of (1.1) is

$$S_n^b = b(t, T_m) \tag{1.1a}$$

Spreading operation, presented symbolically by operator $\varepsilon(\ )$, is obtained if we multiply narrowband signal by a pseudo noise (PN) sequence (code) $c(t, T_c) = \pm 1$. The bits of the sequence are called chips and the chip rate $1/T_c \gg 1/T_m$. The wideband signal can be represented as

$$S_w = \varepsilon(S_n) = cS_n = c(t, T_c)b(t, T_m) \cos \omega t \tag{1.2}$$

The baseband equivalent of (1.2) is

$$S_w^b = c(t, T_c)b(t, T_m) \tag{1.2a}$$

Despreading, represented by operator $D(\ )$, is performed if we use $\varepsilon(\ )$ once again and bandpass filtering, with the bandwidth proportional to $2/T_m$, represented by operator BPF( ) resulting into

$$D(S_w) = \text{BPF}(\varepsilon(S_w)) = \text{BPF}(cc\ b\ \cos \omega t) = \text{BPF}(c2\ b\ \cos \omega t) = b\ \cos \omega t \tag{1.3}$$

The baseband equivalent of (1.3) is

$$D(S_w^b) = \text{LPF}(\varepsilon(S_w^b)) = \text{LPF}(c(t, T_c)c(t, T_c)b(t, T_m)) = \text{LPF}(b(t, T_m)) = b(t, T_m) \tag{1.3a}$$

where LPF( ) stands for low pass filtering. This approximates the operation of correlating the input signal with the locally generated replica of the code $\text{Cor}(c, S_w)$. Nonsynchronized despreading would result in

$$D_\tau(\ ); \quad \text{Cor}(c_\tau, S_w) = \text{BPF}(\varepsilon_\tau(S_w)) = \text{BPF}(c_\tau\ c\ b\ \cos \omega t) = \rho(\tau)\ b\ \cos \omega t \tag{1.4}$$

In (1.4) BPF would average out the signal envelope $c_\tau c$ resulting in $E(c_\tau c) = \rho(\tau)$. The baseband equivalent of (1.4) is

$$D_\tau(); \quad \mathrm{Cor}\left(c_\tau, S_w^b\right) = \int_0^{T_m} c_\tau S_w^b dt = b(t, T_m) \int_0^{T_m} c_\tau c \, dt = b\rho(\tau) \tag{1.4a}$$

This operation would extract the useful signal b as long as $\tau \cong 0$ otherwise the signal will be suppressed because, $\rho(\tau) \cong 0$ for $\tau \geq T_c$. Separation of multipath components in a RAKE receiver is based on this effect. In other words if the received signal consists of two delayed replicas of the form

$$r = S_w^b(t) + S_w^b(t - \tau)$$

the despreading process defined by (1.4a) would result into

$$D_\tau(); \quad \mathrm{Cor}(c, r) = \int_0^{T_m} cr \, dt = b(t, T_m) \int_0^{T_m} c(c + c_\tau) dt = b\rho(0) + b\rho(\tau)$$

Now, if $\rho(\tau) \cong 0$ for $\tau \geq T_c$ all multipath component reaching the receiver with a delay larger than the chip interval will be suppressed.

If the signal transmitted by user y is despread in receiver x the result is

$$D_{xy}(); \quad \mathrm{BPF}(\varepsilon_{xy}(S_w)) = \mathrm{BPF}(c_x \, c_y \, b_y \cos \omega t) = \rho_{xy}(t) b_y \cos \omega t \tag{1.5}$$

So in order to suppress the signals belonging to other users (Multiple Access Interference – MAI), the crosscorrelation functions should be low. In other words if the received signal consists of the useful signal plus the interfering signal from the other user

$$r = S_{wx}^b(t) + S_{wy}^b(t) = b_x c_x + b_y c_y \tag{1.6}$$

despreading process at receiver of user x would produce

$$D_{xy}(); \quad \mathrm{Cor}(c_x, r) = \int_0^{T_m} c_x r \, dt = b_x \int_0^{T_m} c_x c_x dt + b_y \int_0^{T_m} c_x c_y dt = b_x \rho_x(0) + b_y \rho_{xy}(0) \tag{1.7}$$

When the system is synchronized properly $\rho_x(0) \cong 1$, and if $\rho_{xy}(0) \cong 0$ the second component representing MAI will be suppressed. This simple principle is elaborated in WCDMA standard resulting in a collection of transport and control channels. The system is based on 3.84 Mcips rate and up to 2 Mbits/s data rate. In a special downlink high data rate shared channel the data rate and signal format are adaptive. There shall be mandatory support for QPSK and 16 QAM and optional support for 64 QAM based on UE capability which will proportionally increase the data rate. For details see www.3gpp.com.

1.2.3 Orthogonal Frequency Division Multiplexing – OFDM

In wireless communications, the channel imposes the limit on data rates in the system. One way to increase the overall data rate is to split the data stream into a number of parallel channels and use

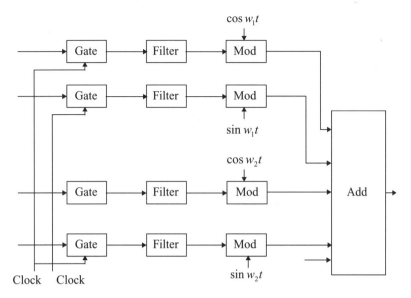

Figure 1.3 An early version of OFDM.

different subcarriers for each channel. The concept is presented in Figures 1.3 and 1.4 and represents the basic idea of OFDM system. The overall signal can be represented as

$$x(t) = \sum_{n=0}^{N-1} \left\{ D_n e^{j2\pi \frac{n}{N} f_s t} \right\}; \qquad -\frac{k_1}{f_s} < t < \frac{N + k_2}{f_s} \tag{1.8}$$

In other words complex data symbols $[D_0, D_1, \ldots, D_{N-1}]$ are mapped in OFDM symbols $[d_0, d_1, \ldots, d_{N-1}]$ such that

$$d_k = \sum_{n=0}^{N-1} D_n e^{j2\pi \frac{kn}{N}} \tag{1.9}$$

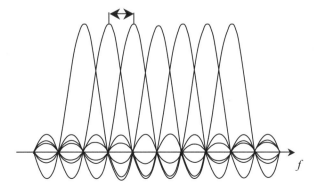

Figure 1.4 Spectrum overlap in OFDM.

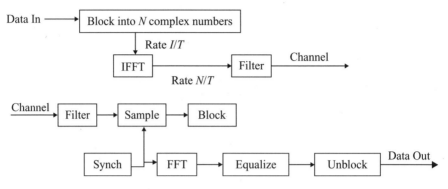

Figure 1.5   Basic OFDM system.

The output of the FFT block at the receiver produces data per channel. This can be represented as

$$\tilde{D}_m = \frac{1}{N} \sum_{k=0}^{N-1} r_k e^{-j2\pi m \frac{k}{2N}}$$

$$r_k = \sum_{n=0}^{N-1} H_n D_n e^{j2\pi \frac{n}{2N}k} + n(k) \tag{1.10}$$

$$\tilde{D}_m = \begin{cases} H_n D_n + N(n), & n = m \\ N(n), & n \neq m \end{cases}$$

The system block diagram is given in Figure 1.5.

In order to eliminate residual intersymbol interference a guard interval after each symbol is used as shown in Figure 1.6.

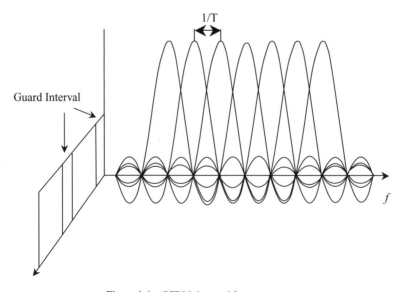

Figure 1.6   OFDM time and frequency span.

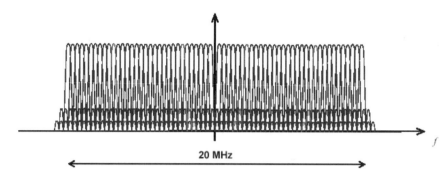

Figure 1.7 802.11a/HIPERLAN OFDM.

An example of an OFDM signal specified by the IEEE 802.11a standard is shown in Figure 1.7. The signal parameters are: 64 points FFT, 48 data subcarriers, 4 pilots, 12 virtual subcarriers, DC component 0, Guard interval 800 ns. A discussion on OFDM and an extensive list of references on the topic are included in Chapter 7.

1.2.4 Multicarrier CDMA (MC CDMA)

Good performance and the flexibility to accommodate multimedia traffic are incorporated in MC CDMA which is obtained by combining CDMA and OFDM signal formats.

Figure 1.8 shows the DS-CDMA transmitter of the j-th user for binary phase shift keying/coherent detection (CBPSK) scheme and the power spectrum of the transmitted signal, respectively, where $G_{DS} = T_m/T_c$ denotes the processing gain and $C^j(t) = [C_1^j C_2^j \cdots C_{G_{DS}}^j]$ the spreading code of the j-th user.

Figure 1.9 shows the MC-CDMA transmitter of the j-th user for CBPSK scheme and the power spectrum of the transmitted signal, respectively, where G_{MC}, denotes the processing gain, N_C the number of subcarriers, and $C^j(t) = [C_1^j C_2^j \cdots C_{G_{MC}}^j]$ the spreading code of the j-th user. The MC-CDMA scheme is discussed assuming that the number of subcarriers and the processing gain are all the same.

However, we do not have to choose $N_C = G_{MC}$, and actually, if the original symbol rate is high enough to become subject to frequency selective fading, the signal needs to be first S/P-converted before

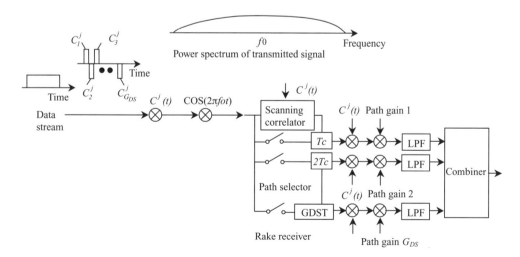

Figure 1.8 DS-CDMA scheme.

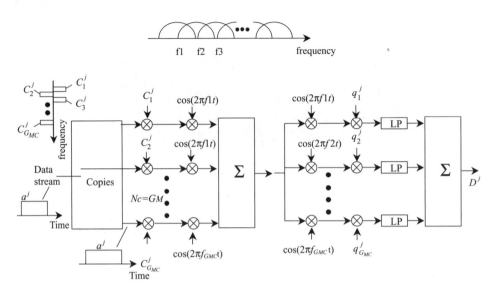

Figure 1.9   MC-CDMA scheme.

spreading over the frequency domain. This is because it is crucial for Multicarrier transmission to have frequency non-selective fading over each subcarrier.

Figure 1.10 shows the modification to ensure frequency non-selective fading, where T_S denotes the original symbol duration, and the original data sequence is first converted into P parallel sequences, and then each sequence is mapped onto G_{MC} subcarriers ($N_C = P \times G_{MC}$).

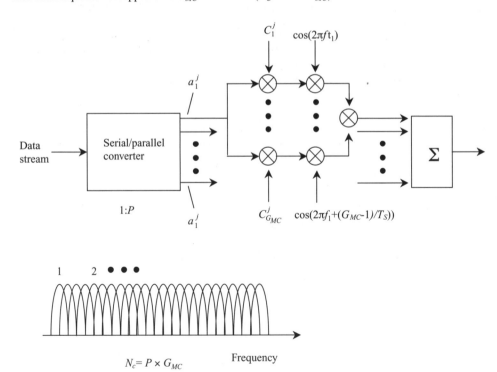

Figure 1.10   Modification of MC-CDMA scheme: spectrum of its transmitted signal.

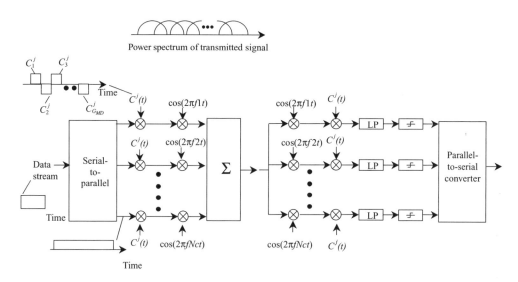

Figure 1.11   Multicarrier DS-CDMA scheme.

The Multicarrier DS-CDMA transmitter spreads the S/P-converted data streams using a given spreading code in the time domain so that the resulting spectrum of each subcarrier can satisfy the orthogonality condition with the minimum frequency separation. This scheme was proposed originally for a uplink communication channel, because the introduction of OFDM signaling into a DS-CDMA scheme is effective for the establishment of a quasi-synchronous channel.

Figure 1.11 shows the Multicarrier DS-CDMA transmitter of the j-th user and the power spectrum of the transmitted signal, respectively, where G_{MD} denotes the processing gain, N_C the number of subcarriers, and $C^j(t) = [C_1^j C_2^j \cdots C_{G_{MD}}^j]$ the spreading code of the j-th user.

The Multitone MT-CDMA transmitter spreads the S/P-converted data streams using a given spreading code in the time domain so that the spectrum of each subcarrier prior to the spreading operation can satisfy the orthogonality condition with the minimum frequency separation. Therefore, the resulting spectrum of each subcarrier no longer satisfies the orthogonality condition. The MT-CDMA scheme uses longer spreading codes in proportion to the number of subcarriers, as compared with a normal (single carrier) DS-CDMA scheme, therefore, the system can accommodate more users than the DS-CDMA scheme.

Figure 1.12 shows the MT-CDMA transmitter of the j-th user for CBPSK scheme and the power spectrum of the transmitted signal, respectively, where G_{MT} denotes the processing gain, N_C the number of subcarriers, and $C^j(t) = [C_1^j C_2^j \cdots C_{G_{MT}}^j]$ the spreading code of the j-th user.

All these schemes will be discussed in details in Chapter 7.

1.2.5   Ultra Wide Band (UWB) Signal

For the multipath resolution in indoor environments a chip interval of the order of few nanoseconds is needed. This results into a spread spectrum signal with the bandwidth in the order of few GHz. Such a signal can also be used with no carrier resulting in what is called impulse radio (IR) or Ultra Wide Band (UWB) radio. The typical form of the signal used in this case is shown in Figure 1.13. A collection of pulses received on different locations within the indoor environment is shown in Figure 1.14 and the corresponding delay profiles is presented in Figure 1.15. The ultra wideband radio will be discussed in detail in Chapter 8. In this section we will define initially only a possible signal format.

A typical time-hopping format used in this case can be represented as

$$s_{\text{tr}}^{(k)}(t^{(k)}) = \sum_{j=-\infty}^{\infty} \omega_{\text{tr}}(t^{(k)} - jT_f - c_j^{(k)}T_c - \delta d_{[j/N_s]}^{(k)}) \qquad (1.11)$$

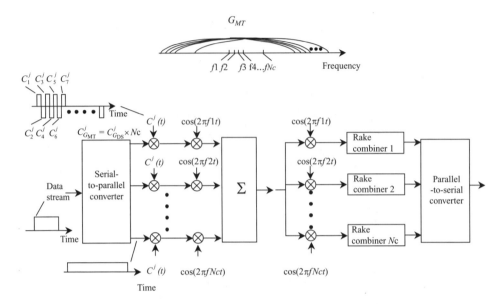

Figure 1.12   MT-CDMA scheme.

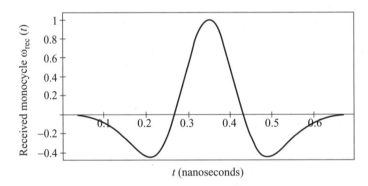

Figure 1.13   A typical ideal received monocycle $\omega_{rec}(t)$ at the output of the antenna subsystem as a function of time in nanoseconds.

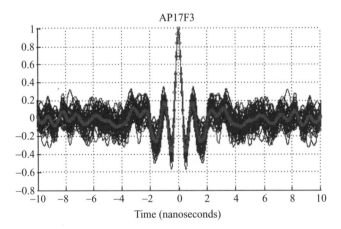

Figure 1.14   A collection of received pulses in different locations [53] © IEEE 2007.

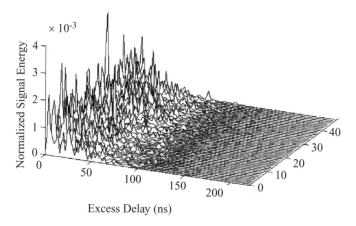

$\times 10^{-3}$

Normalized Signal Energy

Excess Delay (ns)

Figure 1.15 A collection of channel delay profiles [52] © IEEE 2002.

where $t^{(k)}$ is the kth transmitter's clock time and T_f is the *pulse repetition time*. The transmitted pulse waveform ω_{tr} is referred to as a *monocycle*. To eliminate collisions due to multiple access, each user (indexed by k) is assigned a distinctive time-shift pattern $\{c_j^{(k)}\}$ called a *time-hopping sequence*. This provides an additional time shift of $c_j^{(k)}T_c$ seconds to jth monocycle in the pulse train, where T_c is the duration of addressable time delay bins. For a fixed T_f the *symbol rate* R_s determines the number N_s of monocycles that are modulated by a given binary symbol as $R_s = (1/N_s T_f)s^{-1}$. The modulation index δ is chosen to optimize performance. For performance prediction purposes, most of the time the data sequence $\{d_j^{(k)}\}_{j=-\infty}^{\infty}$ is modeled as a wide-sense stationary random process composed of equally likely symbols. For data a pulse position data modulation is used.

When K users are active in the multiple-access system, the composite received signal at the output of the receiver's antenna is modeled as

$$r(t) = \sum_{k=1}^{K} A_k s_{\text{rec}}^{(k)}(t - \tau_k) + n(t) \tag{1.12}$$

The antenna/propagation system modifies the shape of the transmitted monocycle $\omega_{tr}(t)$ to $\omega_{\text{rec}}(t)$ on its output. An idealized received monocycle shape $\omega_{\text{rec}}(t)$ for a free-space channel model with no fading is shown in Figure 1.13.

The optimum receiver for a single bit of a binary modulated impulse radio signal in additive white Gaussian noise (AWGN) is a correlation receiver

"$decided_0^{(1)} = 0$" *if*

$$\sum_{j=0}^{N_s-1} \overbrace{\int_{\tau_1+jT_f}^{\tau_1+(j+1)T_f} r(u,t)\upsilon\left(t - \tau_1 - jT_f - c_j^{(1)}T_c\right)dt}^{\textit{pulse correlator output} \overset{\Delta}{=} \alpha_j(u)}$$

$$\underbrace{\phantom{\sum_{j=0}^{N_s-1} \int_{\tau_1+jT_f}^{\tau_1+(j+1)T_f} r(u,t)\upsilon\left(t\right)}}_{\textit{test statistic} \overset{\Delta}{=} \alpha(u)} \tag{1.13}$$

$$> 0$$

where $\upsilon(t) \overset{\Delta}{=} \omega_{\text{rec}}(t) - \omega_{\text{rec}}(t - \delta)$.

The spectra of a signal using TH is shown in Figure 1.16. If instead of TH a DS signal is used the signal spectra is shown in Figure 1.17(a) for pseudorandom code and Figure 1.17(b) for a random code.

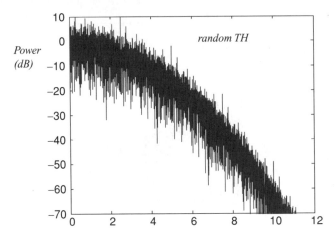

Figure 1.16   Spectra of a TH signal.

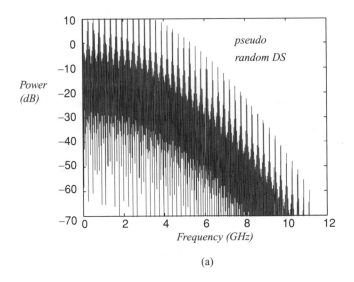

(a)

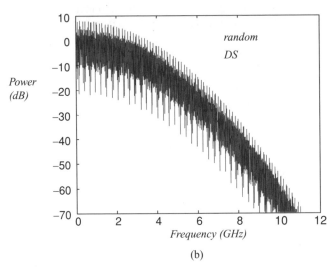

(b)

Figure 1.17   Spectra of pseudorandom DS and random DS signal.

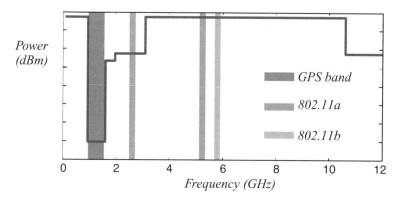

Figure 1.18 FCC frequency mask.

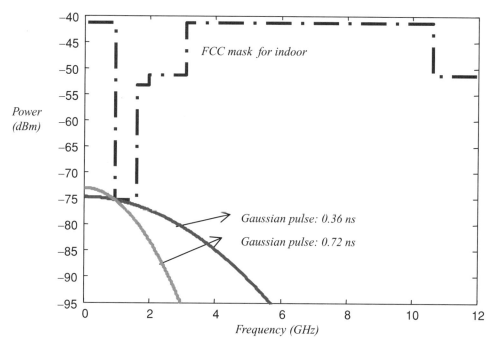

Figure 1.19 FCC mask and possible UWB signal spectra.

The FCC (Frequency Control Committee) mask for indoor communications is shown in Figure 1.18. Possible options for UWB signal spectra are given in Figures 1.19 and 1.20 for a single band and Figure 1.21 for a multiband signal format. For more detail see www.uwb.org and www.uwbmultiband.org.

The optimal detection in a multiuser environment, with knowledge of all time-hopping sequences, leads to complex parallel receiver designs [2]. However, if the number of users is large and no such multiuser detector is feasible, then it is reasonable to approximate the combined effect of the other users' dehopped interfering signals as a Gaussian random process. All of the detail regarding system performance will be discussed in Chapter 8.

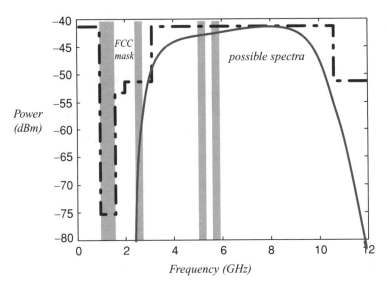

Figure 1.20   Single band UWB signal.

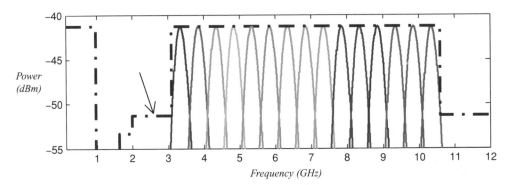

Figure 1.21   Multi band UWB signal.

1.3   Next Generation Internet

As already mentioned in the introduction to this chapter the evolution of wireless communications will be closely related to the evolution of the Internet. For this reason in this section we discuss this issue in more detail. The Internet architecture was developed almost 30 years ago and its basic framework has remained resistant to major changes. In order to predict future evolution within the Internet some authors [55] use the general theory of innovations [56–58] developed in economics.

Sustainable and Disruptive Innovation are the two important categories driving the market in two different ways along different economic dimensions. At some point in time in any industry, incumbent firms providing products and services in a market are competing by working to improve their offerings along a few narrowly defined dimensions. Typically, most products or services start out to be 'not good enough' in functionality, reliability and performance, and companies improve the products along one or more of these dimensions. The innovations involved in improving along these existing metrics are called *sustaining innovations,* even if such innovations from a technology standpoint may be quite radical.

Firms base their Sustaining innovation product improvement programs on the needs of their most demanding customers. The improvements at some point outstrip the needs of their low end customers.

These customers become 'overshot', while the needs of the high end customers remain 'undershot'. At this point, the low end customers begin to value *convenience, customizability*, and *price* more than functionality, reliability and performance.

When the number of low end, overshot customers becomes numerous enough, the industry becomes ripe for the entry of a collection of low-end innovators to take customers away from the incumbent firms. The entry firms are able to make good profits by providing a basic service or product that is unattractive to the high end customers but 'good enough' for low end customers due to lower price *and/or* much improved ease of use. If the entry firms are able to decouple their value chains completely from those of the incumbents, these entry firms can often force incumbent firms to abandon the low end customers and flee 'up market'. The innovations used by the entry firms in this kind of situation are called *disruptive*.

This entry and flight strategy only works for some time. Because the needs of the high end customers don't change fast enough, at some point, the entrant firms and the incumbent firms end up competing for the same pool of high end customers. At that point, the incumbent firms often go bankrupt or are merged with the entrants because the entrants' business models are honed to make money at a lower price point than the incumbents. On the other hand, disruptive entrants can fail if the incumbents are motivated to fight because the entrants' value chains overlap with theirs, or when an incumbent crams a technology successfully with disruptive potential into an existing business model.

There is also another, parallel disruptive innovation path, in which a firm offers a product or service that is not available within the existing market. The product or service often looks primitive or cheap because, initially, it does not provide the same level of performance as mainstream products. This strategy is referred to as *new market disruption* [56–58].

Conservation of Integration refers to the case when a market is in the initial stages of competing along the metrics of functionality, reliability and performance and companies need to control as many steps in the product architecture as possible in order to improve the product or service along the competition metrics. Firms that build their products or services around proprietary, integrated architectures have an advantage, because they can optimize along all metrics without having to compromise. Architectures that are modular – that is, which have well defined interfaces between components, allowing components to be independently developed – invariably fail to deliver along the competitive metrics since some aspects of functionality, reliability or performance must be compromised in order to achieve modularity.

However, once the market has flipped to competing along the metrics of convenience, customizability and price, firms which base their products on modular components have a competitive advantage. For an architecture to be modular, clients and implementers on both sides of the interface must agree on the *specifiability, verifiability* and *predictability* of the components. Modular interfaces allow flexibility in picking component suppliers, distributors and other participants in the ecosystem, while sacrificing some performance. Modular architectures therefore deliver convenience and customizability to customers much more quickly than integrated architectures, and at a lower price.

Internet architecture is a modular architecture with the Internet Protocol (IP) and the various IP transport protocols (such as TCP and UDP) as the modular interface between physical and data link layers and the upper layers. HTTP performs the same function for the application layer. The end-to-end principle, which is the fundamental architectural principle underlying the Internet architecture, is basically a modularity argument: the interface between applications and the network should be clearly defined and the functions specific to the communication system (i.e. transport and routing) are the only functions that should be within the network. IP and the base Internet transport protocol suite provide the specifiability, verifiability and predictability for applications and lower layer transport to agree on how packets get from one end to the other, and what the reliability characteristics of the transmission are. These characteristics are supported by the transparency of the Internet architecture: what goes in one end comes out the other.

The global communication networks replaced by the Internet (the circuit switched telephone network) were in contrast integrated, since applications embedded knowledge of the network operation within them and were deployed within the network. While the assumption among many in the technical community is that the end-to-end principle and the Internet superseded the circuit switched telephone network solely due to technical superiority, the Internet architecture would not have achieved widespread deployment over the earlier integrated architectures without a suitable economic driver. The economic driver was a change in the desired customer performance metrics. The basic network performance metrics that customers care about (namely bandwidth and latency) became optimized to the point that the performance losses that came from modularization by moving to IP no longer mattered, and customers began to value applications other than simple voice. As a result, the economic benefits to the customers of a modular architecture (many suppliers, price reduction, ease of customization, etc.) outweighed the costs.

It is *because* the Internet architecture has not changed much over the last 30 years that massive innovation has been possible above and below the network layer, and, more recently, above the HTTP layer. By *Conservation of Integration* principle, the data transport equipment (such as routers) below the IP layer should all exhibit integrated architectures, since they are not yet good enough, along their particular metrics of competition, for modularization to occur. Similarly, applications should exhibit the same characteristics above the HTTP layer, but we focus on IP and the lower layers in the rest of this section.

Routers are, in fact, highly integrated in the sense that they are complex integrated *hardware/software* products which include proprietary features and interfaces especially for management. The complexity and integration throw up high barriers to entrant firms. Routers also tightly integrate the control software with the data switching hardware, and both change in new versions. Thus operators are pressed to deploy control software upgrades (which reinforce the lock-in) when all they really need is additional switching capacity.

The evolution from an integrated to a modular architecture is not unidirectional, systems can evolve in the other direction. Something similar happened in the early 1990s when Microsoft integrated desktop office applications more tightly to the operating system in Windows in response to a change in the preferred customer performance metrics. Prior to that, MS-DOS provided a modular interface and desktop apps were not well integrated, either with each other or with the operating system.

This suggests that the most likely path to a major, incompatible shift in the Internet architecture is something that would cause the IP modular interface to lose its attractiveness to customers and cause an integrated architecture to come back into favor, since simply providing a slightly better modular interface is unlikely to win out against the enormous installed base of IP. A shift in the Internet architecture could occur if customers began to value some other metric than the current one of cheap, high bandwidth. The shift must cause the modular IP interface to deliver suboptimal performance along the new metric, requiring reintegration across the IP interface.

In the sequel, we discuss two technology areas that have the potential to foster radical innovation in the Internet in the short term.

Cloud Computing and network virtualization is a technology trend that is currently attracting a lot of attention. The *control/data* plane split is a technology trend that has yet to really develop, but holds potential. Although these innovations leave the basic Internet architecture untouched, they have the potential to catalyze radical change in the way Internet infrastructure and applications are deployed, following in the trend of many previous innovation waves in telecom and data networking.

1.4   Cloud Computing and Network Virtualization

In principle, enough unused computing capacity exists on the desktop so the computing needs handled by Cloud Computing could be handled by consolidating processing on desktops. The difficulty of software installation (that we can't locally execute the code we want) and of replicating large databases make providing cloud-like services difficult with the current service deployment model. By consolidating server capacity into large data centers, economies of scale allow companies hosting Web services to achieve much more cost effective hosting and data storage [59].

Servers in a Cloud Computing facility are virtualized, meaning that they can run multiple customer operating system images with different applications at the same time.

Consolidation of servers and virtualization simplifies management, both from the customers' viewpoint and from the cloud provider's viewpoint. While there are still many technical and business issues surrounding Cloud Computing, the ultimate vision is *utility computing*: providing resources for processing, bandwidth and storage in the same way that a utility provides electricity, water or telephone service.

Network virtualization is complementary technology. The idea is to divide the network into slices that are separate and run separate applications within the slices, with each slice allocated bandwidth and processing on network elements. The isolation between slices is used for privacy, security and guaranteed bandwidth. The goal is an on-demand network service that can provide a particular class of service (best effort, expedited forwarding, etc.) for reasonable cost.

Flexible connectivity into the cloud completes the end-to-end connection. The combination of Cloud Computing and network virtualization would allow a business to define a collection of applications and services in the compute cloud that could be accessed end-to-end within a network from virtual machines. The experience would be similar to a corporate *WAN/LAN* environment, except the compute and network resources would reside in the cloud and be rented by the business rather than owned.

For service providers, Cloud Computing is more likely to be a sustaining than a disruptive innovation. The large operators are likely to have the motivation and skills to master Cloud Computing and successfully offer their own cloud services. As an example, AT&T already has a Cloud Computing service [60] through their regional data centers. Cloud Computing is sustaining for incumbent service providers because any cloud provider must connect up to the network, so a disruptive entrant would have a hard time building an ecosystem independent from the incumbents. An incumbent service provider has an advantage, since it can also provide network virtualization as part of its cloud service.

The business models supported by incumbents and entrants are currently somewhat different. AT&T's service requires a contract including network SLAs to be negotiated between the business and cloud provider, whereas new entrant services such as Amazon's EC2 require a credit card for anonymous payment, and network bandwidth only is guaranteed to and from the data center. Initially, there may be room for both business models to grow but at some point the overlap in ecosystems may result in clashes.

For equipment vendors, the network equipment and server business seems to be realigning itself along a new performance metric, causing a re-integration across communication performance within data centers. Currently, standard L2 switches are used to build the switching fabric within data centers for communication between servers. Standard server blades are used to form the computational infrastructure. The cost advantage comes from using commercial off-the-self hardware optimized for deployment in individual chassis.

Optimization of communication between servers by integrating the communication more tightly with the server hardware seems a likely step.

The reintegration across the server backplane is enabled by a migration of the modular interface between the server and the network out to the virtualization layer. Newer releases of the virtualization systems, such as VMWare and Xen, include a virtualized switch that insulates running virtual machine images from the details of their location in the data center switching fabric. This allows the running images to be moved around the data center to different servers without changing the location of the running image in the IP address topology. The virtual machine images become independent of the switch hardware, opening a space for innovation to improve the performance of data center networking.

The basic idea behind the *control/data* plane split or split routing architecture is that IP subnet topologies, while quite useful in access networks, provide too much functionality in operator core networks at a high cost of configuration complexity. Most core network links are point to point, often switched, and therefore don't require the rich many-to-many provisioning capabilities offered by IP subnet topologies. In addition, basic IP routing protocols don't really take advantage of richly interconnected topologies anyway, since the shortest path first algorithm used by the interior gateway routing protocols concentrates traffic on one path.

MPLS [61], Carrier Ethernet [62] and GMPLS [63] are examples of split routing architecture technologies. Split routing architecture refers to splitting the router into two pieces in separate network nodes with a modular interface between: a route controller that handles the distribution of routing information and policy to control where packets go (the control plane) and a collection of routing switches where the actual forwarding decisions are made on individual traffic packets (the data plane). Another way to think of it is as a separation of policy-path setup, traffic engineering, etc. – from mechanism – the forwarding action. As they currently stand, split routing architecture technologies don't appear particularly threatening to the basic infrastructure providers. Expensive core routers running IP-MPLS make up a good chunk of their high margin business. MPLS still requires routers, and Carrier Ethernet requires the control plane software to be integrated with the switches.

Information-Centric Networking (ICN), or Content Centric Networking (CCN), is a relatively new research trend [14]. The basic idea is for networking to address the content, or information objects, and not the containers, or network nodes and connections. Receivers therefore get to choose what senders they get content from, rather than allowing senders to address receivers without permission, as in the current Internet architecture. From the innovation theory point of view, ICN requires integrating across the IP layer and creating a new modularity interface in the network stack. This new modularly interface creates a new 'waist' for the protocol stack, a role similar to the role played by IP protocol for the last 30 years. However unlike HTTP, which recently established an application 'waist' in the IP stack, ICN fundamentally changes the architecture to remove end node addressability. Three main metrics seem to be pushing the world towards ICN: need for reduced latency, dropping storage space/transmission price ratio, and ease of use. ICN, once developed to the full, is likely to address all of these metrics.

1.5 Economics of Utility Computing

Cloud Computing refers to both the applications delivered as services over the Internet and the hardware and systems software in the data centers that provide those services [59] (see Figure 1.22). The services themselves have long been referred to as Software as a Service (SaaS). The datacenter hardware and software is referred to as a *Cloud*. A Cloud, made available in a pay-as-you-go manner to the general public, is called Public Cloud; the service being sold is *Utility Computing*. The term Private Cloud is used to refer to internal data centers of a business or other organization, not made available to the general public. Thus, Cloud Computing is the sum of SaaS and Utility Computing, but does not include Private Clouds. From a hardware point of view, three aspects are new in Cloud Computing.

1. The illusion of infinite computing resources available on demand, thereby eliminating the need for Cloud Computing users to plan far ahead for provisioning.
2. The elimination of an up-front commitment by Cloud users, thereby allowing companies to start small and increase hardware resources only when there is an increase in their needs.

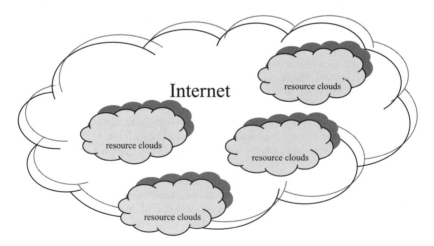

Figure 1.22   Utility computing concept.

3. The ability to pay for use of computing resources on a short-term basis as needed (e.g., processors by the hour and storage by the day) and release them as needed, thereby rewarding conservation by letting machines and storage go when they are no longer useful.

Any application needs a model of computation, a model of storage, and a model of communication. The statistical multiplexing necessary to achieve elasticity and the illusion of infinite capacity requires each of these resources to be virtualized to hide the implementation of how they are multiplexed and shared.

Regarding Cloud Computing economic models the following observations can be made:

a) Economic models enabled by Cloud Computing make tradeoff decisions about whether hosting a service in the cloud makes sense over the long term, more fluid, and in particular the elasticity offered by clouds serves to transfer risk.

b) Although hardware resource costs continue to decline, they do so at variable rates. Cloud Computing can track these changes and potentially pass them through to the customer more effectively than building one's own datacenter, resulting in a closer match of expenditure to actual resource usage.

c) In making the decision about whether to move an existing service to the cloud, one must additionally examine the expected average and peak resource utilization, especially if the application may have highly variable spikes in resource demand; the practical limits on real-world utilization of purchased equipment; and various operational costs that vary depending on the type of cloud environment being considered.

The economic appeal of Cloud Computing is based on converting capital expenses to operating expenses and the phrase 'pay as you go' is often used to capture directly the economic benefit to the buyer. Hours purchased via Cloud Computing can be distributed non-uniformly in time (e.g., use 80 server-hours today and no server-hours tomorrow, and still pay only for what you use); in the networking community, this way of selling bandwidth is already known as usage-based pricing. In addition, the absence of up-front capital expense allows capital to be redirected to core business investment.

Therefore, even though pay-as-you-go pricing could be more expensive than buying and depreciating a comparable server over the same period, it is argued that the cost is outweighed by the extremely important Cloud Computing economic benefits of elasticity and transference of risk, especially the risks of overprovisioning (underutilization) and underprovisioning (saturation).

The elasticity refers to the Cloud Computing's ability to add or remove resources at a fine grain (one server at a time) within minutes rather than weeks allowing matching resources to workload much more closely. Real world estimates of server utilization in datacenters range from 5% to 20% [65, 66]. This is consistent with the observation that for many services the peak workload exceeds the average by factors of 2 to 10. Few users deliberately provision for less than the expected peak, and therefore they must provision for the peak and allow the resources to remain idle at non peak times. The more pronounced the variation, the more the waste.

While the monetary effects of overprovisioning are easily measured, those of underprovisioning are harder to measure yet potentially equally serious since not only do rejected users generate zero revenue, they may never come back due to poor service.

In the concept of Cloud Computing the risk of mis-estimating workload is shifted from the service operator to the cloud vendor.

There are two additional benefits to the Cloud Computing user that result from being able to change their resource usage on the scale of hours rather than years.

First, unexpectedly scaling down (disposing of temporarily underutilized equipment), due to a business slowdown, or due to improved software efficiency, normally carries a financial penalty.

Second, technology trends suggest that over the useful lifetime of some purchased equipment, hardware costs will fall and new hardware and software technologies will become available. Cloud providers, who already enjoy economy-of-scale buying power, can potentially pass on some of these savings to their customers.

The previous discussion tried to quantify the economic value of specific Cloud Computing benefits such as elasticity. In the sequel we will extend our discussion to the equally important but larger question of whether or not it is more economical to move the existing data center-hosted service to the cloud, or to keep it in a data center?

This simple analysis includes several important factors.

First, most applications do not make equal use of computation, storage and network bandwidth; some are CPU-bound, others network-bound and so on, and may saturate one resource while underutilizing others. Pay-as-you-go Cloud Computing can charge the application separately for each type of resource, reducing the waste of underutilization. While the exact savings depends on the application, suppose the CPU is only 50% utilized while the network is at capacity; then in a data center you are effectively paying for double the number of CPU cycles actually being used.

The costs of power, cooling, and the amortized cost of the building are missing from our simple analyses so far. It is estimated that the costs of CPU, storage and bandwidth roughly double when those costs are amortized over the building's lifetime.

Today, hardware operations costs are very low. Rebooting servers is easy and minimally trained staff can replace broken components at the rack or server level. On one hand, since Utility Computing uses virtual machines instead of physical machines, from the cloud user's point of view these tasks are shifted to the cloud provider. On the other hand, depending on the level of virtualization, much of the software management costs may remain, e.g. upgrades, applying patches, and so on.

1.6   Drawbacks of Cloud Computing

In this section, we discuss the obstacles to the growth of Cloud Computing as classified in [59].

Availability of service is what organizations are concerned about. So far, existing SaaS products have set a high standard in this regard.

Just as large Internet service providers use multiple network providers so that failure by a single company will not take them off the air, it is believed that the only plausible solution to very high availability is multiple Cloud Computing providers.

Distributed Denial of Service (DDoS) attacks is another availability obstacle. Criminals threaten to cut off the incomes of SaaS providers by making their service unavailable. As with elasticity, Cloud Computing shifts the attack target from the SaaS provider to the Utility Computing provider, who can more readily absorb it and is also likely to have already DDoS protection as a core competency.

Data lock in is the next concern. Software stacks have improved interoperability among platforms, but the *application programming interfaces* (APIs) for Cloud Computing itself are still essentially proprietary, or at least have not been the subject of active standardization. Thus, customers cannot easily extract their data and programs from one site to run on another. Concern about the difficult of extracting data from the Cloud is preventing some organizations from adopting Cloud Computing. Customer lock-in may be attractive to Cloud Computing providers, but Cloud Computing users are vulnerable to price increases, to reliability problems, or even to providers going out of business.

Data Confidentiality and Auditability is also a concern. Current Cloud offerings are essentially public (rather than private) networks, exposing the system to more attacks. It is believed that there are no fundamental obstacles to making a Cloud-Computing environment as secure as the vast majority of in-house IT environments, and that many of the obstacles can be overcome immediately with well understood technologies such as encrypted storage, Virtual Local Area Networks, and network middleboxes (e.g. firewalls, packet filters). For example, encrypting data before placing it in a Cloud may be even more secure than unencrypted data in a local data center.

Data Transfer Bottlenecks are also a concern. Applications continue to become more data-intensive. If we assume applications may be 'pulled apart' across the boundaries of Clouds as indicated in Figure 1.22, this may complicate data placement and transport. The concern becomes of paramount importance in the case of *Wireless Internet*. This will be discussed in more details in the next section. One opportunity to overcome the high cost of Internet transfers is to *ship disks*. Jim Gray found that the cheapest way to send a lot of data is to physically send disks or even whole computers via overnight delivery services [67]. A second opportunity is to find other reasons to make it attractive to keep data in the Cloud, for once data is in the Cloud for any reason it may no longer be a bottleneck and may enable new services that could drive the purchase of Cloud Computing cycles.

A third, more radical opportunity is to try to reduce the cost of WAN bandwidth more quickly.

Performance Unpredictability is also a concern. The experience is that multiple Virtual Machines can share CPUs and main memory surprisingly well in Cloud Computing, but that I/O sharing is more problematic. One opportunity is to improve architectures and operating systems to virtualize interrupts and I/O channels efficiently.

Technologies such as PCIexpress (Peripheral Component Interconnect) are difficult to virtualize, although they are critical to the Cloud. Fortunately IBM mainframes and operating systems largely have overcome these problems, so we have successful examples from which to learn. Another possibility is that flash memory will decrease I/O interference. Flash is semiconductor memory that preserves information when powered off like mechanical hard disks, but since it has no moving parts, it is much faster to access (microseconds vs. milliseconds) and uses less energy. Flash memory can sustain many more I/Os per second per gigabyte of storage than disks, so multiple virtual machines with conflicting random I/O workloads could coexist better on the same physical computer without the interference we see with mechanical disks. Another unpredictability obstacle concerns the scheduling of virtual machines for some classes of batch processing programs, specifically for high performance computing.

Scalable Storage is one of the system parameters too. Early in this section, we identified three properties whose combination gives Cloud Computing its appeal: short-term usage (which implies scaling down as well as up when resources are no longer needed), no up-front cost, and infinite capacity on-demand. While it's straightforward what this means when applied to computation, it's less obvious how to apply it to persistent storage. It includes questions of the performance guarantees offered, and the complexity of data structures that are directly supported by the storage system. The opportunity, which is still an open research problem, is to create a storage system that would not only meet these needs but combine them with the Cloud advantages of scaling arbitrarily up and down on-demand, as well as meeting programmer expectations in regard to resource management for scalability, data durability and high availability.

Bugs in Large-Scale Distributed Systems is the next concern. One of the difficult challenges in Cloud Computing is removing errors in these very large scale distributed systems. A common occurrence is that these bugs cannot be reproduced in smaller configurations, so the debugging must occur at scale in the production data centers.

One opportunity may be the reliance on virtual machines (VM) in Cloud Computing.

Scaling Quickly is also important. Pay-as-you-go certainly applies to storage and to network bandwidth, both of which count bytes used. Computation is slightly different, depending on the virtualization level. Google AppEngine scales in response to load increases and decreases automatically, and users are charged by the cycles used. AWS (*Amazon Web Services*) charges by the hour for the number of instances you occupy, even if your machine is idle. The opportunity is then to scale quickly up and down automatically in response to load in order to save money, but without violating service level agreements.

Cloud Computing providers already perform careful and low overhead accounting of resource consumption. By imposing per-hour and per-byte costs, utility computing encourages programmers to pay attention to efficiency (i.e., releasing and acquiring resources only when necessary), and allows more direct measurement of operational and development inefficiencies.

Reputation Fate Sharing is a part of the overall system performance too. Reputations do not virtualize well. One customer's bad behavior can affect the reputation of the Cloud as a whole. An opportunity would be to create reputation-guarding services similar to the 'trusted email' services currently offered (for a fee) to services hosted on smaller ISP's, which experience a microcosm of this problem. Another legal issue is the question of transfer of legal liability. Cloud Computing providers would want legal liability to remain with the customer and not be transferred to them.

Software Licensing should be also considered. Current software licenses commonly restrict the computers on which the software can run. Users pay for the software and then pay an annual maintenance fee. Hence, many Cloud Computing providers originally relied on open source software in part because the licensing model for commercial software is not a good match to Utility Computing.

The primary opportunity is either for open source to remain popular or simply for commercial software companies to change their licensing structure to better fit Cloud Computing. For example, Microsoft and Amazon now offer pay-as-you-go software licensing for Windows Server and Windows SQL Server on EC2.

1.7   Wireless Grids and Clouds

In this section we continue our discussion on Cloud Computing with emphasis on wireless networks. *Wireless grid* is a wireless network based virtual system that consists of wireless-connected different types of electronic devices and computers [68–77]. It has broad application prospects in *e*-learning, mobile *e*-business, modern healthcare, smart home, wireless sensor networks and disaster management. Wireless grids are based on wireless networks but the communication infrastructure is not the only difference between wireless grids and traditional wired grids. Wireless grids have the following distinguishing features:

Ad hoc mode: Wireless networks, especially wireless ad hoc networks, facilitate the on-demand integration of heterogeneous devices. The ad hoc connected devices have no centric control. It makes the wireless grid more dynamic and brings more convenience to resource sharing and coordination. At the same time, the ad hoc mode also brings more challenges regarding security, resource discovery and trusted resource management to wireless grids.

Resource constrained devices like high performance servers, mass storage devices, are very common in computing grids, data grids, access grids and other traditional wired grids. But in wireless grids, the low-powered small devices, like sensors, smart phones, digital cameras, PDAs and laptops, are more popular. Most of them are battery-powered. Their storage and computing capabilities are also limited.

Heterogeneous components: Grid facilitates sharing of heterogeneous resources [78]. The evolution from PCs to smaller and mobile devices with computing capabilities brings more heterogeneous devices and applications into wireless grids. Wireless grids have to manage and support the inter-organizational applications that consist of more heterogeneous components from autonomic devices.

Dynamic resources and requirements: In wireless grids, the users, resources, requirements and computing environments are dynamic. The wireless networks are inherently volatile, due to their unstable wireless communication medium. In addition, the changes of natural environment also influence the wireless networks. In wireless grids, mobile users are common. And the users use mobile devices increasingly. Due to the ad hoc mode, the users and resources can join and quit the grid conveniently. The flexible resource sharing and coordination mode inspire the users to spontaneously raise their requirements for new applications.

Wireless grid architectures could be classified into the following categories according to the predominant devices and device mobility [70]: (a) Wireless sensor grid architecture; (b) Mobile wireless grid architecture; (c) Fixed wireless grid architecture.

The wireless sensor grid integrates wireless sensor networks and traditional grid computing technologies [71]. The applications are designed around the data from the sensors. And the sensors play an important role in this kind of wireless grid architecture. On the basis of wireless sensor network infrastructures, it adopts the technologies of data grids, computing grids and access grids to storing, processing, presenting and sharing the data collected from the sensors.

The mobile wireless grid has much more mobile devices, such as mobile phones and PDAs. In addition, the users are also mobile. The network infrastructure is a mobile ad hoc network. The devices participate in the mobile wireless grid as data collectors or client nodes. In most cases, these devices are mobile wireless access clients. They can freely enter and leave the wireless grid together with their mobile users.

The fixed wireless grid has no difference with traditional wired grid, except the communication medium is wireless.

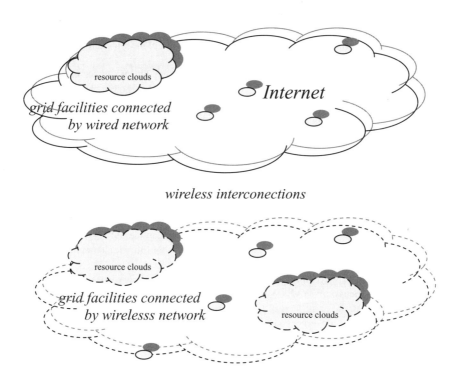

Figure 1.23   Hybrid wireless grid architecture.

In practice, besides the pure wireless network infrastructure, a hybrid network infrastructure is more commonly used in wireless grids, where the wireless networks are connected with wired networks.

Figure 1.23 depicts an architecture that integrates the wireless grid with wired network backbones.

In this architecture, the high resource consuming devices, such as high performance servers and mass storage devices, are connected by high speed wired networks; the small or mobile devices are integrated by wireless ad hoc networks. This hybrid architecture makes up the lack of resources in wireless grids.

Taking one with another, the architecture of a wireless grid can be divided into four layers: physical layer, communication layer, middleware layer and application layer.

The physical layer covers all the devices involved in the wireless grid. The communication layer chains the devices with wireless networks or connects the wireless-linked devices with wired networks. The middleware layer takes charge of resource description, resource discovery, coordination and trust mechanism, and supports development of wireless grid applications. On the basis of middleware layer, the application layer includes a variety of wireless grid applications, such as health care systems, environment surveillance systems, seismic monitoring systems and so on.

Many wireless communication technologies are used in wireless grids, to provide wireless interfaces to connect high-bandwidth powered PCs and low-bandwidth powered smart devices. These technologies include IEEE 802.llx, IEEE 802.16x, IEEE 802.15x, IEEE 802.20x, Zigbee, Bluetooth, wireless ad hoc networks, etc.

IEEE 802.11 is a set of protocol standards proposed by the IEEE *LAN/MAN* Standards Committee. The protocol family includes 802.11a~z. They support wireless local area network communications in the 2.4, 3.7 and 5 GHz frequency bands.

The 802.11 protocols are very similar to the 802.3 protocols. In fact, they are wireless Ethernet protocols. The 802.11 protocols cover the physical layer (PHY) and the media access control layer (MAC) of the data link layer. Different from the 802.3 protocols, the 802.11 protocols use *CSMA/CA*

(Carrier Sense Multiple Access with Collision Avoidance) technologies instead of *CSMA/CD* (Carrier Sense Multiple Access with Collision Detection) at the MAC layer, to meet the requirements of wireless communications.

Wi-Fi (Wireless Fidelity) is a trademark of the Wi-Fi Alliance. The Wi-Fi technologies cover the protocols of 802.11a, 802.11b, 802.11g and 802.11n, and can be used in mobile phones, home networks and other devices that are connected by wireless networks. With Wi-Fi, one can construct flexible wireless networks in infrastructure mode or ad hoc mode. These 802.11 based networks equip wireless grids with fundamental communication facilities.

The protocol family of IEEE 802.16 was proposed by the IEEE *LAN/MAN* Standards Committee also. It includes a group 802.16a–k, which are the standards of broadband wireless access. The IEEE 802.16 protocols support wireless communications in the frequency bands of 10–66 GHz.

The IEEE 802.16 defines the standards of protocols at the physical layer (PHY) and the media access control layer (MAC). These protocols support both mobile access and fixed access. At PHY layer, the 802.16 adopts the technologies of OFDM (Orthogonal Frequency Division Multiplex) and OFDMA (Orthogonal Frequency Division Multiple Access). OFDMA can adapt flexibly to requirements changes of bandwidth. With fixed sub-carrier frequency interval and symbol time, it also reduces the impact of changes at PHY layer on the MAC layer.

Different from the IEEE 802.11 protocols that support WLAN (Wireless Local Area Networks), the IEEE 802.16 protocols focus on WWAN (Wireless Wide Area Network). They use TDMA (Time Division Multiple Access) at MAC layer, instead of CSMA. In addition, the protocols have three sub-layers at MAC layer: service specific convergence sub-layer, common part sub-layer and privacy sub-layer.

In addition, the IEEE 802.16 protocols have more choices with regard to bandwidth and frequency and have higher transmission power. It makes them more suitable for the wireless grids that need wide area wireless networks.

WiMAX (Worldwide Interoperability for Microwave Access) is a technology based on the IEEE 802.16. It supports 802.11d and 802.11e. Similar to Wi-Fi, *WiMAX* can connect handsets, PC peripherals and embedded devices. As a backhaul technology for *3G* (Third Generation) and *4G* (Fourth Generation), *WiMAX* is a promising technology for constructing wireless grid.

The ad hoc is an organization mode of the wireless network nodes. A wireless ad hoc network consists of a group of autonomous nodes that communicate with each other in a decentralized manner. Each node of the network connects others with wireless links, and acts as both a host and a router. The nodes can join or quit the network freely. Therefore, the topology of wireless ad hoc network is dynamic. The decentralized nature makes the ad hoc network more suitable for wireless grid.

According to applications, wireless ad hoc networks are classified into three categories: (a) Mobile ad hoc networks where mobile devices connect with each other in a self-configuring way; (b) Wireless mesh networks where nodes are organized in mesh topology and in most cases, the nodes of wireless mesh networks are static; and (c) Wireless sensor networks where nodes are distributed autonomous devices that can monitor environmental conditions. Wireless sensor networks are typical tools for data collection in wireless grid.

Wireless ad hoc networks provide wireless grids with ubiquitous untethered communication, and allow wireless devices to discover and interact with each other in a *P2P* manner. With a special purpose gateway, an ad hoc network can bridge to wired LANs or to the Internet. It facilitates the integration of wireless grids and traditional grids.

Discovering and maintaining available routes is a key issue of wireless ad hoc networks. Although there are several routing protocols (such as AODV, DSR, ZIP, AOMDV et al.) [79], it still needs more efforts to provide adaptable routes for resource discovery in wireless grids. Traditional wired grids suffered the problems caused by heterogeneous resources and requirement changes. To solve the problems, many efforts have been made on traditional grid middleware. Regarding wireless grids, the problems are more serious, due to the dynamic characteristics. Meanwhile, the wireless grid middleware faces other challenges, for example, limited power, high-latency connectivity and unstable wireless connection.

In the following we review the typical work on wireless grid middleware.

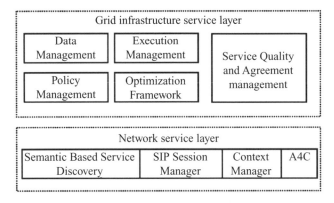

Figure 1.24 Akogrimo architecture.

Akogrimo, supported by the FP6-IST program, enables access to knowledge through wireless grids. Some systems of *e*-learning, disaster management or *e*-health used the middleware Akogrimo.

Akogrimo supports wireless grids with mobile grid clients. It provides fixed, nomadic and mobile users with mobile grid services. On the basis of Akogrimo, grid services can be composed dynamically in an ad hoc way. Figure 1.24 presents the architecture of Akogrimo.

The middleware consists of two logical layers: network service layer and grid infrastructure service layer. The network service layer is constructed on a mobile IPv6 based infrastructure. It provides network services like Authentication, Authorization, Accounting, Auditing and Charging (*A4C*). With a *QoS* broker, the network service layer enables context driven selection of different bandwidth bundles. On the basis of the network service layer, the grid infrastructure service layer provides service oriented facilities for wireless grid applications [80]. It includes the business partner management, service quality management, business process execution management and mobile virtual organization management.

Akogrimo adapts many traditional grid technologies (such as virtual organization management, grid workflow) to wireless mobile environments, and enables the awareness of the mobility of virtual organization members.

MORE is the middleware which adapts web services for embedded systems. The middleware facilitates construction of mobilized enterprise information systems.

It combines technologies of embedded systems and web services to facilitate the integration of heterogeneous devices. It hides the complexity of heterogeneous embedded systems with unified and simplified interfaces. It also supports scalable group communication, where the small devices such as sensors and smart phones can connect with each other through wireless networks. With *MORE*, application developers can deploy ubiquitous services on embedded devices, and share the services within a group.

The key blocks of *MORE* include core management service and application enabling services. The core management service adopts group management concept, and uses XML-based policies to create and configure groups to control the behavior of group participants. The enabling services implement the service functionalities and provide fundamental supports for the applications [81].

It uses connectors for the communication between the enabling services. The connectors include internal connectors, proprietary connectors, SOAP connectors and *μSOA* connectors. The proprietary connectors and *μSOA* connectors support the embedded device involved services. The *μSOA* connectors enhance the efficiency of mobile embedded devices. By the *μSOA* approaches, *MORE* reduces the resource consumption in wireless grids.

MoGrid is the middleware, which supports grid services in wireless ad-hoc networks. The middleware has much application potential in mobile collaborations. It orchestrates the distributed grid tasks among mobile devices in a *P2P* manner, and uses a resource discovery protocol referred to as *P2PDP* to coordinate tasks among the most resourceful and available mobile devices. *MoGrid* offers

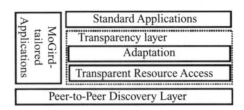

Figure 1.25   MoGrid architecture.

application-level mobility transparency, and supports the development of context-sensitive applications for mobile collaborations [72].

It leverages the resource coordination in wireless ad hoc networks by allocating the tasks for a group of mobile collaborators. The coordination consists of two phases. First, the *P2PDP* protocol discovers the resources that the coordination needs. Second, a collaborator submits the tasks to the participants according to the available resources they have [72].

As shown in Figure 1.25, the *MoGrid* architecture comprises a *P2P* discovery layer and a transparency layer. The applications are categorized into standard applications and *MoGrid*-tailored applications. The *P2P* discovery layer includes entities of collaborators, coordinators and initiators, and supports the resource registration and announcement, context definition, and resource discovery. After the devices register resources in *MoGrid*, they become collaborators. The registered resources are accessible to other devices in the *MoGrid*-enabled wireless grids. The initiators request the collaborators for task processing. The coordinators broadcast the request of initiators for resource discovery and coordinate the collaborators and initiators to complete the tasks. The transparency layer is responsible for the resource coordination among collaborators. It consists of two sub-layers: adaptation sub-layer and transparency resource access sub-layer (TRAS). The latter provides application-independent transparency for resource utilization. The adaptation sub-layer handles the mobility and connection-related events for each specific application.

MoGrid handles the issues of resource discovery and collaboration in wireless ad hoc networks. But its resource discovery protocol *P2PDP* should be improved further, in order to solve the problems caused by device mobility and transparency contexts [7].

MiPeG middleware enhances classic grid environments with wireless mobile mechanism. This middleware can be used in pervasive grid applications.

It enables integrating traditional grids and wireless devices. With a group of fundamental services, it enables deploying grid service clients on wireless mobile devices, and provides context-awareness for the applications. In addition, it can allocate user's tasks to different mobile devices also. In a *MiPeG*-powered grid, the mobile devices can be either active resources or grid service clients. When the *MiPeG*-powered mobile devices enter, the grid can automatically recognize the devices and coordinate them in applications.

MiPeG is a service-oriented middleware. As shown in Figure 1.26, it consists of a set of basic services which are compliant with the OGSA specifications.

In *MiPeG*, the asynchronous communication broker implements the WS-Brokered Notification specification, and provides an interaction mechanism based on the event publish-subscribe paradigm. It

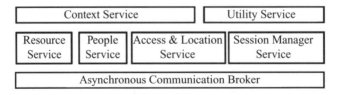

Figure 1.26   *MiPeG architecture.*

dispatches asynchronous messages in the grid. The resource service manages the grid resource registration, and enables the integration of wireless devices (such as PDAs and sensors) in the grid. The people service provides the mobile users with basic authentication mechanisms. The access and location service supports accessing to 802.11-enabled mobile devices. In addition, it can locate the position of the mobile devices. The session manager service handles the sessions for mobile users. The context service offers the context information including the state of resources, mobile user locations and users' profiles. The utility service implements the utility computing model and accepts the tasks submitted by users for execution. With the utility service, the users can complete a task without considering the issues of *reserving/releasing* resources et al.

MiPeG focuses mainly on the integration of mobile devices and traditional grids. It does not solve the problems of dynamic device changes well. In addition, *MiPeG* uses service-oriented technology, but its support for deploying the services on the resource-constrained mobile devices is still weak.

Applications of wireless grid technologies have broad prospects in many fields. This includes wireless sensor grids, environment data collection, *e*-health, e-learning, wireless-linked device integration and so on.

Wireless sensor grids combine wireless grids with sensor networks. It facilitates the transfer of information about the physical world around us to a plethora of web-based information utilities and computational services [76]. Wireless sensor grids can be used widely in disaster early warning and environment surveillance et al.

GridStix [82], Equator [83] and Floodnet [84] are three systems that use technologies of sensors, wireless networks and grids to monitor flood and provide early warning. In these systems, a variety of sensors equipped with wireless networking technology form a lightweight grid which is capable of collecting and transmitting data about flood disaster [82]. The lightweight sensor grid connects with traditional grid infrastructure which analyses the data gathered by the sensors for flood warning.

There are also wireless grid based supply-chain management system, the Smart Warehouse system. The Smart Warehouse is a prototype of a supply chain management system equipped with a wireless sensor grid [76]. With the smart sensor, it can track the items in the warehouse and identify problems in the supply chains.

e-Healthcare wireless grids can integrate a variety of medical devices (such as vital signs sensors) and provide a personalized healthcare service delivery paradigm which always requires a one-to-one connection between the patient and the medical expert [85]. Besides the personalized services, the wireless grid based *e*-healthcare systems can integrate the healthcare services into the patient's environment, forming a network of mobile and stationary monitoring and diagnosis facilities around the patient. By these means, medical professionals can establish an inter-organizational virtual hospital, and provide efficient detection of health emergencies and pervasive healthcare.

The Heart Monitoring and Emergency Service (HMES) system is an Akogrimo-based *e*-healthcare platform. It integrates the patient monitoring, emergency detection and subsequent rescue management, and enables an early recognition of heart attacks or apoplectic strokes and a proper treatment of patients as fast as possible [74].

Besides HMES, there are other wireless grid powered *e*-healthcare systems, for example, MobiHealth [86], Rural *e*-Healthcare Framework [87] and Hourglass [77].

e-Learning wireless grid technologies facilitates *e*-learning. Akogrimo enables almost universal availability of mobile devices to education and training. In an Ahogrimo-based *e*-learning system, the learners can use handheld devices to get learning contents from the wireless grid. And they can engage them in purposeful activity, problem solving, collaborations, interactions and conversations [88].

e-Campus system is another example of wireless grid powered *e*-learning. It adopts wireless grid technologies and supports mobile learning [89]. Based on grid service technology, the mobile learning combines traditional grids with mobile devices, and facilitates the sharing of learning resources distributed on different *e*-Learning platforms [75].

Wireless-linked device integration focuses on integrating wireless-linked small devices. *IMOGA* supports integrating mobile devices into a grid system, and sharing/producing information in a cooperative manner [76]. In IMOGA, the mobile devices, such as PDAs and smart phones, employ different kinds of sensors to collect data, such as Global Positioning System (GPS), temperature monitoring components, health monitoring systems and pollution monitoring components. The gathered data is shared between huge numbers of mobile devices and high capacity grid infrastructures where the developers can construct applications of *e*-healthcare, *e*-crisis management and *e*-government [76].

References

[1] Adachi, F. (2002) Evolution towards broadband wireless systems, *The 5th International Symposium on Wireless Personal Multimedia Communications*, **1**, 27–30 Oct. 2002, 19–26.

[2] Jun-Zhao Sun, Sauvola, J. and Howie, D. (2001) Features in future: 4G visions from a technical perspective, *IEEE Global Telecommunications Conference. GLOBECOM '01*. 25–29. Nov. 2001, **6**, 3533–3537.

[3] Axiotis, D. I., Lazarakis, F. I., Vlahodimitropoulos, C. and Chatzikonstantinou, A. (2002) 4G system level simulation parameters for evaluating the interoperability of MTMR in UMTS and HIPERLAN/2, *4th International Workshop on Mobile and Wireless Communications Network, 2002*, 9–11 Sept., 559–563.

[4] Mihovska, A., Wijting, C., Prasad, R., Ponnekanti, S., Awad, Y. and Nakamura, M. (2002) A novel flexible technology for intelligent base station architecture support for 4G systems, *The 5th International Symposium on Wireless Personal Multimedia Communications, 2002*, **2**, 27–30 Oct., 601–605.

[5] Kitazawa, D., Chen, L., Kayama, H. and Umeda, N. (2002) Downlink packet-scheduling considering transmission power and QoS in CDMA packet cellular systems, *4th International Workshop on Mobile and Wireless Communications Network*, 9–11 Sept., 183–187.

[6] Dell'Uomo, L. and Scarrone, E. (2002) An all-IP solution for QoS mobility management and AAA in the 4G mobile networks, *The 5th International Symposium on Wireless Personal Multimedia Communications*, **2**, 27–30 Oct, 591–595.

[7] Wallenius, E. R. (2002) End-to-end in-band protocol based service quality and transport QoS control framework for wireless 3/4G services, *The 5th International Symposium on Wireless Personal Multimedia Communications*, 2, 27–30 Oct, 531–533.

[8] Benzaid, M., Minet, P. and Al Agha, K. (2002) Integrating fast mobility in the OLSR routing protocol, *4th International Workshop on Mobile and Wireless Communications Network*, 9–11 Sept., 217–221.

[9] Kambourakis, G., Rouskas, A. and Gritzalis, S. (2002) Using SSL/TLS in authentication and key agreement procedures of future mobile networks, *4th International Workshop on Mobile and Wireless Communications Network*, 9–11 Sept., 152–156.

[10] van der Schaar, M. and Meehan, J. (2002) Robust transmission of MPEG-4 scalable video over 4G wireless networks, *2002 International Conference on Image Processing. 2002. Proceedings*, **3**, 24–28 June, 757–760.

[11] Qing-Hui Zeng, Jian-Ping Wu, Yi-Lin Zeng, Ji-Long Wang and Rong-Hua Qin (2002) Research on controlling congestion in wireless mobile Internet via satellite based on multi-information and fuzzy identification technologies, *Proceedings. 2002 International Conference on Machine Learning and Cybernetics*, **4**, 4–5 Nov., 1697–1701.

[12] 5th International Symposium on Wireless Personal Multimedia Communications. Proceedings (Cat. No. 02EX568), *The 5th International Symposium on Wireless Personal Multimedia Communications, 2002*, **1**, 27–30 Oct.

[13] Sukuvaara, T., Mahonen, P. and Saarinen, T. (1999) Wireless Internet and multimedia services support through two-layer LMDS system, *1999 IEEE International Workshop on Mobile Multimedia Communications, (MoMuC '99)* 15–17 Nov., 202–207.

[14] Martin, C. C., Winters, J. H. and Sollenberger, N. R. (2000) Multiple-input multiple-output (MIMO) radio channel measurements, *Proceedings of the 2000 IEEE Sensor Array and Multichannel Signal Processing Workshop 2000*, 16–17 March, 45–46.

[15] Pereira, J. M. (2000) Fourth generation: now, it is personal! *The 11th IEEE International Symposium on Personal, Indoor and Mobile Radio Communications, 2000. PIMRC 2000*, **2**, 18–21 Sept., 1009–1016.

[16] Otsu, T., Umeda, N. and Yamao, Y. (20010 System architecture for mobile communications systems beyond IMT-2000, *IEEE Global Telecommunications Conference, GLOBECOM '01*, **1**, 25–29 Nov., 538–542.

[17] Yi Han Zhang, Makrakis, D., Primak, S. and Yun Bo Huang (20020 Dynamic support of service differentiation in wireless networks *IEEE CCECE 2002, Canadian Conference on Electrical and Computer Engineering*, **3**, 12–15 May, 1325–1330.

[18] Jun-Zhao Sun and Sauvola, J. (2002) Mobility and mobility management: a conceptual framework, *10th IEEE International Conference on Networks, ICON 2002*, 27–30 Aug., 205–210.

[19] Vassiliou, V., Owen, H. L., Barlow, D. A., Grimminger, J., Huth, H.-P. and Sokol, J. (2002) A radio access network for next generation wireless networks based on multi-protocol label switching and hierarchical Mobile IP, *Proceedings of IEEE 56th Vehicular Technology Conference, 2002 VTC 2002-Fall*, **2**, 24–28 Sept., 782–786.

[20] Nicopolitidis, P., Papadimitriou, G. I., Obaidat, M. S. and Pomportsis, A. S. (2002) 3G wireless systems and beyond: a review, *9th International Conference on Electronics, Circuits and Systems, 2002*, **3**, 15–18 Sept., 1047–1050.

[21] 2002 IEEE Wireless Communications and Networking Conference Record, WCNC 2002 (Cat. No.02TH8609), *IEEE Wireless Communications and Networking Conference, WCNC2002*, **1**, 17–21 March.

[22] Borras-Chia, J. (2002) Video services over 4G wireless networks: not necessarily streaming, *2002 IEEE Wireless Communications and Networking Conference, WCNC2002*, **1**, 17–21 March, 18–22.

[23] Evans, B. G. and Baughan, K. (2000) Visions of 4G, *Electronics & Communication Engineering Journal*, **12**(6), 293–303.

[24] Kim, J. and Jamalipour, A. (2001) Traffic management and QoS provisioning in future wireless IP networks, *IEEE Personal Communications* [see also *IEEE Wireless Communications*], **8**(5), Oct., pp. 46–55.

[25] Aghvami, A. H., Le, T. H. and Olaziregi, N. (2001) Mode switching and QoS issues in software radio *IEEE Personal Communications*, [see also *IEEE Wireless Communications*], **8**(5), 38–44.

[26] Kanter, T. (2001) An open service architecture for adaptive personal mobile communication, *IEEE Personal Communications* [see also *IEEE Wireless Communications*], **8**(6), 8–17.

[27] Sampath, H., Talwar, S., Tellado, J., Erceg, V. and Paulraj, A. (2002) A fourth-generation MIMO-OFDM broadband wireless system: design, performance, and field trial results, *IEEE Communications Magazine*, **40**(9), 143–149.

[28] Kellerer, W. and Vogel, H.-J. (2002) A communication gateway for infrastructure-independent 4G wireless access, *IEEE Communications Magazine*, **40**(3), 126–131.

[29] Huang, V. and Weihua Zhuang (2002) QoS-oriented access control for 4G mobile multimedia CDMA communications, *IEEE Communications Magazine*, **40**(3), 118–125.

[30] Smulders, P. (2002) Exploiting the 60 GHz band for local wireless multimedia access: prospects and future directions, *IEEE Communications Magazine*, **40**(1), 140–147.

[31] Raivio, Y. (2001) 4G-hype or reality, *Second International Conference on 3G Mobile Communication Technologies, 2001*. (Conf. Publ. No. 477), 26–28 March, pp. 346–350.

[32] Becchetti, L., Delli Priscoli, F., Inzerilli, T., Mahonen, P. and Munoz, L. (2001) Enhancing IP service provision over heterogeneous wireless networks: a path toward 4G, *IEEE Communications Magazine*, **39**(8), 74–81.

[33] Abe, T., Fujii, H. and Tomisato, S. (2002) A hybrid MIMO system using spatial correlation, *The 5th International Symposium on Wireless Personal Multimedia Communications*, **3**, 27–30 Oct., 1346–1350.

[34] Lincke-Salecker, S. and Hood, C. S. (2003) A supernet: engineering traffic across network boundaries, *36th Annual Simulation Symposium*, 30 March–2 April, 117–124.

[35] Yamao, Y., Suda, H., Umeda, N. and Nakajima, N. (2000) Radio access network design concept for the fourth generation mobile communication system, *Proceedings of IEEE 51st Vehicular Technology Conference, VTC 2000-Spring Tokyo*, **3**, 15–18 May, 2285–2289.

[36] Ozturk, E. and Atkin, G. E. (2001) Multi-scale DS-CDMA for 4G wireless systems, *IEEE Global Telecommunications Conference, GLOBECOM '01*, **6**, 25–29, Nov., 3353–3357.

[37] Dell'Uomo, L. and Scarrone, E. (2001) The mobility management and authentication/authorization mechanisms in mobile networks beyond 3G, *12th IEEE International Symposium on Personal, Indoor and Mobile Radio Communications*, **1**, 30 Sept.–3 Oct., C-44–C-48.

[38] Kumar, K. J., Manoj, B. S., Murthy, C. S. R. (2002) On the use of multiple hops in next generation cellular architectures, *10th IEEE International Conference on Networks, 2002. ICON 2002*. 27–30 Aug, 283–288.

[39] Wang, S. S., Green, M. and Malkawi, M. (2002) Mobile positioning and location services, *IEEE Radio and Wireless Conference, RAWCON 2002*, 11–14 Aug., 9–12.

[40] Motegi, M., Kayama, H. and Umeda, N. (2002) Adaptive battery conservation management using packet QoS classifications for multimedia mobile packet communications, *Proceedings of IEEE 56th Vehicular Technology Conference, VTC 2002-Fall*, **2**, 24–28 Sept., 834–838.

[41] Ying Li, Shibua Zhu, Pinyi Ren and Gang Hu (2002) Path toward next generation wireless internet -cellular mobile 4G, WLAN/WPAN and IPv6 backbone, *Proceedings of 2002 IEEE Region 10 Conference on Computers, Communications, Control and Power Engineering TENCOM '02*, **2**, Oct. 28–31, 1146–1149.

[42] Qiu, R. C., Wenwu Zhu and Ya-Qin Zhang (2002) Third-generation and beyond (3.5G) wireless networks and its applications, *IEEE International Symposium on Circuits and Systems, ISCAS 2002*, **1**, 26–29 May, I-41–I-44.

[43] Bornholdt, C., Sartorius, B., Slovak, J., Mohrle, M., Eggemann, R., Rohde, D. and Grosskopf, G. (2002) 60 GHz millimeter-wave broadband wireless access demonstrator for the next-generation mobile internet, *Optical Fiber Communication Conference and Exhibit, OFC 2002*, 17–22 March, 148–149.

[44] Jianhua He, Zongkai Yang, Daiqin Yang, Zuoyin Tang and Chun Tung Chou (2002) Investigation of JPEG2000 image transmission over next generation wireless networks, *5th IEEE International Conference on High Speed Networks and Multimedia Communications*, 3–5 July, 71–7.

[45] Baccarelli, E. and Biagi, M. (2003) Error resistant space-time coding for emerging 4G-WLANs, *IEEE Wireless Communications and Networking, WCNC 2003*, **1**, 16–20 March, 72–77.

[46] Mohr, W. (2002) WWRF – the Wireless World Research Forum *Electronics & Communication Engineering Journal*, **14**(6), Dec., 283–291.

[47] Otsu, T., Okajima, I., Umeda, N. and Yamao, Y. (2001) Network architecture for mobile communications systems beyond IMT-2000, *IEEE Personal Communications*, [see also *IEEE Wireless Communications*], **8**(5), Oct., 31–37.

[48] Bria, A., Gessler, F., Queseth, O., Stridh, R., Unbehaun, M., Jiang Wu; Zander, J., and Flament, M. (2001) 4th-generation wireless infrastructures: scenarios and research challenges, *IEEE Personal Communications* [see also *IEEE Wireless Communications*], **8**(6), Dec., 25–31.

[49] Fitzek, F., Kopsel, A., Wolisz, A., Krishnam, M. and Reisslein, M. (2002) Providing application-level QoS in 3G/4G wireless systems: a comprehensive framework based on multirate CDMA, *IEEE Wireless Communications* [see also *IEEE Personal Communications*], **9**(2), April, 42–47.

[50] Classon, B., Blankenship, K., and Desai, V. (2002) Channel coding for 4G systems with adaptive modulation and coding, *IEEE Wireless Communications* [see also *IEEE Personal Communications*], **9**(2), April, 8–13.

[51] Yile Guo and Chaskar, H. (2002) Class-based quality of service over air interfaces in 4G mobile networks, *IEEE Communications Magazine*, **40**(3), 132–137.

[52] Cassioli, D., Win, M. Z. and Molisch, A. F. (2002) The ultra-wide bandwidth indoor channel: from statistical model to simulations, *IEEE Journal on Selected Areas in Communications*, **20**(6), 1247–1257.

[53] Win, M. Z. and Scholtz, R. A. (2002) Characterization of ultra-wide bandwidth wireless indoor channels: a communication-theoretic view, *IEEE Journal on Selected Areas in Communications*, **20**(9), 1613–1627.

[54] Glisic, S. (2004) *Advanced Wireless Communications, 4G Technology*, John Wiley and Sons, London.

[55] Kempf, J., Nikander, P. and Green, H. (2010) Innovation and the Next Generation Internet, *INFOCOM IEEE Conference on Computer Communications Workshops*, 1–6.

[56] Christensen, C. (2003) *The Innovator's Dilemma*, Harper Business, New York.

[57] Christensen, C. and Raynor, B. (2003) *The Innovator's Solution*, Harvard Business School Publishing Corp., Boston.

[58] Christensen, C., Anthony, S. and Roth, E. (2004) *Seeing What's Next*, Harvard Business School Publishing Corp., Boston.

[59] Armburst, M., Fox, A., Griffith, R., Joseph, A., Katz, R., Konwmski, A., Lee, G., Patterson, D., Rabkm, A., I. Stoica and Zaharia, M. (2009) Above the clouds: a Berkeley view of cloud computing, Technical Report No. UCB/EECS-2009-28, EE-CS Dept., University of California Berkeley, February 10.

[60] Synaptic Hostmg Enterprise, www.business.att.com/enterprise/Family/application-hosting-enterprise/synaptic-hosting-enterprise/.

[61] Rosen, E., Viswanathan, A. and Callon, R. (2001) Multiprotocol label switching architecture, RFC 3031, Internet Engineering Task Force, January.

[62] Green, H., Monette, S., Olsson, J., Saltsidis, P. and Takács, A. (2007) Carrier Ethemet: The native approach, *Ericsson Review, 3*.

[63] Farrel, A. and Bryskm, I. (2006) *GMPLS: Architecture and Applications*, Morgan Kaufman, San Francisco.

[64] Jacobson, V. (2006) in Stanford clean slate seminar, September, www.cleanstate.stanford.edu/seminars/jacobson.pdf.

[65] Rangan, K. (2008) The Cloud Wars: $100+ billion at stake. Tech. Rep., Merrill Lynch, May.

[66] Siegele, L (2008) Let It Rise: A Special Report on Corporate IT, *The Economist* (October).

[67] Gray, J., and Patterson, D. (2003) A conversation with Jim Gray, *ACM Queue* **1**(4), 8–17.

[68] Foster, I., Kesselman, C. and Tueck, S. (2001) The Anatomy of the Grid: Enabling Scalable Virtual Organizations, *International Journal of Supercomputer Applications*, **15**(3), 200–222.

[69] McKnight, L. W., Howison, J. and Bradner, S. (2004) Wireless Grids: Distributed Resource Sharing by Mobile, Nomadic, and Fixed Devices, *IEEE Internet Computing*, **8**(4), 24–31.

[70] Ahuja, S. P., and Myers, J. A. (2006) A Survey on Wireless Grid Computing, *Journal of Supercomputing*, **37**(1), 3–21.

[71] Gaynor, M., Moulton, S. L., Welsh, M., LaCombe, E., Rowan, A. and Wynne, J. (2004) Integrating Wireless Sensor Networks with the Grid, *IEEE. Internet Computing*, **8**(4), 32–39.

[72] Lima, L. M., Gomes, A. T. A., Ziviani, A., Endler, M., Soares, L. F. G. and Schulze, B. (2005) Peer-to-Peer Resource Discovery in Mobile Grid, *Proceedings of the 3rd International Workshop on Middleware for Grid Computing*, ACM, New York, 1–8.

[73] Coronato, A. and De Pietro, G. (2008) MiPeG: A middleware infrastructure for Pervasive grids, *Future Generation Computer Systems*, **24**(1), 17–29.

[74] Liao, C., Chi, Y., and Ou Yang, F. (2003) A Grid Service Oriented Platform for Mobile Learning of English, *Proceedings of World Conference on E-Learning in Corporate, Government, Healthcare, and Higher Education 2003*, 2265–2268.

[75] Ozturk, E. and Altilar, D. T. (2007) IMOGA: An Architecture for Integrating Mobile Devices into Grid Applications, *Proceedings of Fourth Annual International Conference on Mobile and Ubiquitous Systems: Networking& Services*, 1–8.

[76] Gaynor, M., Moulton, S. L., Welsh, M., LaCombe, E. and Rowan (2004) Integrating Wireless Sensor Networks with the Grid, *IEEE Internet Computing*, **8**(4), 32–39.

[77] McKnight, L. W. (2007) The future of the internet is not the internet: open communications policy and the future wireless grid(s), http//www.cecd.org/dataoecd/18/42/3 8057172 pdf.

[78] Li, G., Han, Y. B., Wang, J., Zhao, Z. F. and Wagner, R. M. (2006) Facilitating Dynamic Service Compositions by Adaptable Service Connectors, *International Journal of Web Services Research*, **3**(1), 68–84.

[79] Glisic, S. and Lorenzo, B. (2009) *Advanced Wireless Networks, 4G technology*, 2e John Wiley and Sons, London.

[80] Jähnert, J. M., Wesner, S. and Villagrá, V. A. (2008) The Akogrimo Mobile Grid Reference Architecture-Overview, Whitepaper, www.akogrimo.org/download/White_Papers_and_Publications/Akogrimo_WhitePaper_Architecture_v1-1.pdf.

[81] Wolff, A., Michaelis, S., Schmutzler, J. and Wietfeld, C. (2007) Network-centric Middleware for Service Oriented Architectures across Heterogeneous Embedded Systems, *Proceedings of the Eleventh International IEEE EDOC Conference Workshop*, IEEE Press, Washington, 105–108.

[82] Hughes, D., Greenwood, P., Coulson, G. and Blair, G. (2006) GridStix: Supporting Flood Prediction using Embedded Hardware and Next Generation Grid Middleware, *Proceedings of the 2006 International Symposium on a World of Wireless, Mobile and Multimedia Networks*, IEEE Press, Washington.

[83] Humble, J., Greenhalgh, C., Hampshire, A., Muller, H. L. and Egglestone, S. R. (2005) A Generic Architecture for Sensor Data Integration with the Grid, *Proceedings of Scientific Applications of Grid Computing 2004*, Springer-Verlag, Berlin.

[84] Roure, D. D. Improving Flood Warning times using Pervasive and Grid Computing, http://envisense.org/floodnet/ingenia/ingenia.htm.

[85] Daramola, J. O., Osamor, V. C. and Oluwagbemi, O. O. (2008) A Grid-Based Framework for Pervasive Healthcare Using Wireless Sensor Networks: a Case for Developing Nations, *Asia Journal of Information Technology*, **7**(6), 260–267.

[86] Mei, H., Widya, I. A., van Halteren, A. T. and Erfianto, B. (2006) A Flexible Vital Sign Representation Framework for Mobile Healthcare, *Proceedings of the 1st International Conference on Pervasive Computing Technologies for Healthcare*, IEEE Press, Washington, 1–9.

[87] Oludayo, O. Olugbara, Sunday, O., Ojo, Mathew, O., Adigun, Justice, O. (2007) Emuoyibofarhe and Sibusiso S. Xulu, An Architectural Framework for Rural e-Healthcare Information Infrastructure with Web Service-Enabled Middleware Support, *Proceedings of HELINA 2007*.

[88] Laria, G. (2008) Mobile and nomadic user in e-learning: the Akogrimo case, Whitepaper, http://www.akogrimo .org/download/White_Papers_and_Publications/Akogrimo_WhitePaper_eLearning.pdf.

[89] Chang, R. S. and Chang, J. H. (2006) E-campus system based on the pervasive grid, *Proceedings of TANET 2006*.

2

Adaptive Coding

Channel coding is a well established technical field that includes both strong theory and a variety of practical applications. Both theory and practice are well documented in open literature. In this chapter we provide a brief review of the basic principles and results in this field, with the emphasis on those parameters which are important for adaptability and reconfigurability of coding and decoding algorithms. This is an important characteristic for applications in wireless systems where a strong request for energy preservation suggests a system operation where quality of service (QoS) is met with minimum effort. In an environment with changing propagation conditions, this requires a possibility to adapt the complexity of the system. This is the focus of the presentation in this chapter and for the conventional details related to channel coding the reader is referred to the classical reference in the field.

2.1 Adaptive and Reconfigurable Block Coding

The simplest way to improve the probability of correct detection of a bit is to repeat the transmission of the same bit (repetition code) and base the detection of the bit on so-called majority logic. As an example, if each bit is repeated three times, the decoder will base the decision on the observation of the three bits. The error will now occur if two or more bits are received incorrectly. This is a simple solution but rather inefficient from the point of view of bandwidth utilization. The next option is the family of codes based on the parity check principle. An oversimplified example is given in Figure 2.1. For every two input bits $\mathbf{u} = (u_1, u_2)$, a parity check bit, $x_3 = u_1 + u_2$, is created so that the transmitted bits are $\mathbf{x} = (x_1, x_2, x_3) = (u_1, u_2, x_3)$ as indicated in the figure. This simple example can be further expanded to include a number of parity check bits. For a number of input bits k, $n - k$ parity check bits are generated, resulting in a code word of length n. For this we use the notation (n, k) block codes. The art of block code construction consists of finding such parity check rules that would provide the best error correction capabilities with the minimum amount of redundant bits.

The ratio $R_c = k/n$ is called coding rate. An example of such a code is the Hamming code $(7, 4)$ shown in Figure 2.2.

The Hamming code $(7, 4)$ is defined by the relations:

$$\begin{aligned}
x_i &= u_i, \ \ i = 1, \ 2, \ 3, \ 4 \\
x_5 &= u_1 + u_2 + u_3 \\
x_6 &= u_2 + u_3 + u_4 \\
x_7 &= u_1 + u_2 + u_4
\end{aligned} \tag{2.1}$$

The output *code words* $\mathbf{x}(x_1, x_2, x_3, x_4, x_5, x_6, x_7)$ for all possible *input words* $\mathbf{u}$ (u_1, u_2, u_3, u_4 are shown in Table 2.1.

Advanced Wireless Communications & Internet: Future Evolving Technologies, Third Edition. Savo Glisic.
© 2011 John Wiley & Sons, Ltd. Published 2011 by John Wiley & Sons, Ltd.

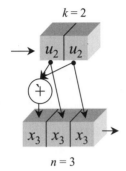

$$x_1 = u_1, x_2 = u_2, x_3 = u_1 + u_2$$

Data words	Code words
00	000
01	011
10	101
11	110

Figure 2.1   Encoder for the parity check code (3, 2).

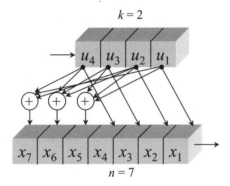

Figure 2.2   Encoder for the Hamming code (7, 4).

Table 2.1   Hamming code (7, 4)

Data words $\mathbf{u}(u_1, u_2, u_3, u_4)$	Code words $\mathbf{x}(x_1, x_2, x_3, x_4, x_5, x_6, x_7)$
0000	0000 000
0001	0001 011
0010	0010 110
0011	0011 101
0100	0100 111
0101	0101 100
0110	0110 001
0111	0111 010
1000	1000 101
1001	1001 110
1010	1010 011
1011	1011 000
1100	1100 010
1101	1101 001
1110	1110 100
1111	1111 111

One can see from Table 2.1 that out of 2^n possible code words, only 2^k are used in the encoder. A collection of these code words is called a code book. The decoder will decide in the favor of the code word from the code book that is the closest to the given received code word that may contain a certain number of errors. So, in order to minimize the bit error rate, the art of coding consists of choosing those code words for the code book that differ from each other as much as possible. This difference is quantified by Hamming distance d defined as the number of bit positions where the two words are different. A number of families of block codes are described in the literature. An example is cyclic codes, represented by Bose–Chaudhuri–Hocquengham (BCH) codes and its non-binary subclass known as Reed–Solomon (RS) codes. The RS codes operate with symbols created from m bits, which are elements in the extension Galois field GF(2^m) of GF(2). Now, an RS code is defined as a block of n m-ary symbols defined over GF(2^m), constructed from k input information symbols by adding an $n - k = 2t$ number of redundant symbols from the same extension field, giving an $n = k + 2t$ symbols code word. From now on we will use the notation for such codes: RS(n, k, t) over GF(2^m) or BCH(n, k, t) over GF(2). Such codes can correct t errors. For details of code construction and detection the reader is referred to the classical literature [1–18].

In the case of burst errors, bit interleaving, illustrated in Figure 2.3, is used. The figure represents a scheme for the interpretation of a (75, 25) interleaved code derived from a (15,5) BCH code. A burst of length $b = 15$ is spread into $t = 3$ error patterns in each of the five code words of the interleaved code.

In general, the decoding algorithms can be based on two different options. If a hard decision is performed for each bit separately and the detected word is compared with the possible candidates from the code book, the process is referred to as *hard decision* decoding. For such a decision the probability that a wrong code word is selected is given as [1–18]:

$$P_w(e) \leq (M - 1)\left[\sqrt{4p(1 - p)}\right]^{d_{\min}} \qquad (2.2)$$

where $M = 2^k$ indicates the number of code words in the code book, assumed to be equally likely, and $d_{\min}$ is the minimum Hamming distance between the code words. Parameter p represents the bit error rate. The second option is to create a sum of analog signal samples at bit positions and to compare this sum (metrics) with the possible values created from the code book. This is referred to in the literature as *soft decision* decoding. The code word error rate (WER) in this case is given as [1–18]:

$$P_w(e) \leq \frac{(M - 1)}{2}\mathrm{erfc}\left(\sqrt{\frac{d_{\min} R_c E_b}{N_0}}\right) \qquad (2.3)$$

The word error rate for Hamming code (7, 4) is given in Figure 2.4. One can see that soft decision decoding offers better performance.

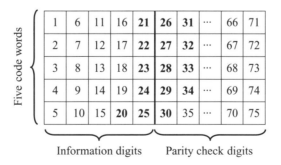

Figure 2.3   Interleaving.

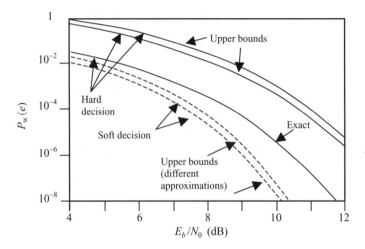

Figure 2.4   Word error probability for the (7, 4) Hamming code: hard decision and soft decision curves. Binary antipodal transmission.

If the word error occurs, then for a high signal to noise ratio, the decoder will choose the word with minimum distance from the correct one. It will make other choices with much lower probability. As a consequence, the decoded word will contain $2t + 1$ errors which can be anywhere in the n-bit word. So, the bit error probability can be approximated as:

$$P_b(e) \cong ((2t + 1)/n)P_w(e) \tag{2.4}$$

The bit error rate for some BCH codes is given in Figure 2.5. In general, from the figure, one can see that the longer the code the better the performance. One should be aware that this also means higher

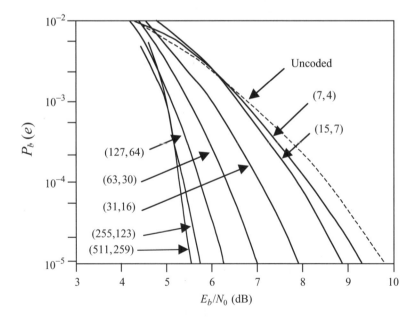

Figure 2.5   Bit error probability curves for some BCH codes, with rate R_c of about 0.5.

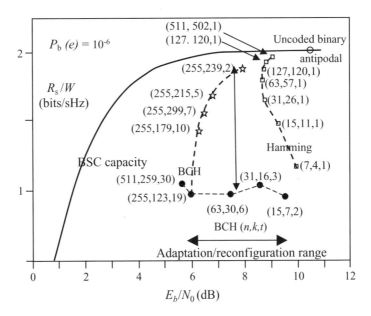

Figure 2.6 Performance of different BCH and Hamming codes. Each code is identified with three numbers: the block length n, the number of information bits k, and the number of corrected errors t.

complexity. This represents the basis for code adaptability and reconfigurability. Figure 2.6 summarizes these results for a broader class of codes.

For a Hamming code and available signal to noise ratio in the range $E_b/N_0 = 9$–10 dB, we can increase the data rate R_s/W approximately by a factor of two if we reconfigure the code from H(7, 4, 1) to H(511, 502, 1). A similar effect can be achieved by reconfiguration of the BCH codes in the range (255, 123, 19) to (255, 239, 2) if the available SNR changes in the range 6–8 dB. For a fixed data rate $R_s/W \cong 1$, we can reduce the required SNR for BER $= 10^{-6}$, from 9.5 to 5.5 dB (coding gain) by reconfiguring the BCH code from BCH(15, 7, 2) to BCH(255, 123, 19). The dependence of power consumption on the code reconfiguration range will be discussed in Chapter 10.

To relate these tradeoffs analytically we can use Relation (2.5), which represents a bound on word error probability as a function of coding rate, code length and required signal to noise ratio represented through bit error rate p.

$$P_w(e) \leq 2^{-n(R_0 - R_c)}, \quad R_c \leq R_0 \tag{2.5}$$

where, for the binary symmetric channel (BSC) with $p(0) = p(1) = 1/2$,

$$R_0 = 1 - \log_2 \left[1 + 2\sqrt{p(1-p)} \right] \tag{2.6}$$

These relations are illustrated in Figure 2.7.

Finally, adaptation/reconfiguration efficiency is defined as:

$$E_{ff} = \frac{-\Delta \text{SNR}}{\Delta_r \text{ complexity}}$$

$$= \frac{\text{coding gain}}{\text{relative increase in complexity}} = \frac{10^{g_c(\text{dB})/10}}{D_r} \tag{2.7}$$

This parameter will be used throughout the book to compare different schemes.

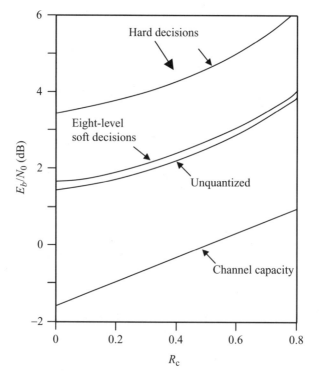

Figure 2.7　Cutoff rate-based bounds on the required signal to noise ratio as a function of the code rate R_c for hard and soft decisions and binary antipodal modulation over the AWGN channel.

2.2　Adaptive and Reconfigurable Convolutional Codes

In order to further increase the coding gain, a new class of codes, known as convolutional codes, is used [18–45]. A general block diagram of a convolutional encoder in serial form for an (n_0, k_0) code with constraint length N is given in Figure 2.8. N blocks, k_0 bits each, are used to produce a code word n_0 bits long. For every new input block a new code word is generated.

One can show that the encoder output can be represented as:

$$\mathbf{x} = \mathbf{u}\mathbf{G}_\infty \qquad (2.8)$$

where

$$\mathbf{G}_\infty = \begin{bmatrix} G_1 & G_2 & \cdots & G_N & & & \\ & G_1 & G_2 & \cdots & G_N & & \\ & & G_1 & G_2 & \cdots & G_N & \\ & & & G_1 & G_2 & \cdots & G_N \\ & & & & \cdots & \cdots & \cdots & \cdots & \cdots \end{bmatrix} \qquad (2.9)$$

Submatrices $\mathbf{G}_i$, containing k_0 rows and n_0 columns, define connectivity of the ith block of the input register with n_0 elements of the output register. In the notation, '1' means a connection and '0' no connection.

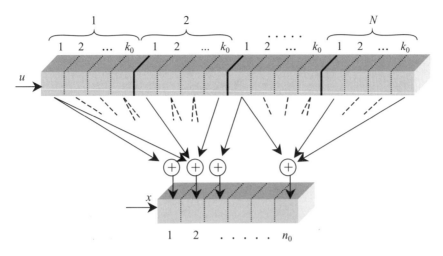

Figure 2.8   Convolutional encoder.

We will use the following notation: code rate $R_c = k_0/n_0$, memory $v = (N - 1)k_0$ and such a code will be denoted as an (n_0, k_0, N) convolutional code. An example of the encoder for $k_0 = 1$ is shown in Figure 2.9.

For the example in Fig. 2.9(a) we have:

$$\begin{aligned}
\mathbf{G}_1 &= \begin{bmatrix} 1 & 1 & 1 \end{bmatrix} \\
\mathbf{G}_2 &= \begin{bmatrix} 0 & 1 & 1 \end{bmatrix} \\
\mathbf{G}_3 &= \begin{bmatrix} 0 & 0 & 1 \end{bmatrix}
\end{aligned} \tag{2.10}$$

An equivalent representation is shown in Figure 2.9(b). For this representation we use notation that might be more convenient:

$$\mathbf{G} = \begin{bmatrix} g_{1,1} & \cdots & g_{1,n_0} \\ \vdots & & \\ g_{k_0,1} & \cdots & g_{k_0,n_0} \end{bmatrix} \tag{2.11}$$

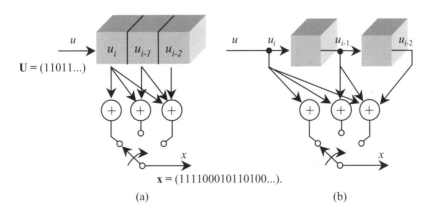

(a)   (b)

Figure 2.9   Two equivalent schemes for the convolutional encoder of the (3, 1, 3) code.

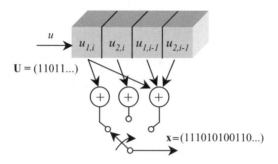

Figure 2.10   Convolutional encoder for the (3, 2, 2) code.

with

$$
\begin{aligned}
g_{1,1} &= (100) \\
g_{1,2} &= (110) \text{ octal numbers } (110 \rightarrow 6) \\
g_{1,3} &= (111)
\end{aligned}
\tag{2.12}
$$

An example for $k_0 = 2$ is shown in Figure 2.10. In this case we have:

$$
\mathbf{G}_1 = \begin{bmatrix} 1 & 0 & 1 \\ 0 & 1 & 0 \end{bmatrix}
$$

$$
\mathbf{G}_2 = \begin{bmatrix} 0 & 0 & 1 \\ 0 & 0 & 1 \end{bmatrix}
\tag{2.13}
$$

One can show that the same encoder can be implemented in parallel form, as shown in Figure 2.11. The N general equivalent parallel presentation is shown in Figure 2.12, and two more examples are shown in Figures 2.13 and 2.14.

For the encoder from Figure 2.9 one can verify, and then generalize for any encoder with arbitrary n_0 and $k_0 = 1$, that, for each input digit u_i, n_0 digits will be generated in accordance with:

$$
(x_{i1}, x_{i2}, x_{i3}, \ldots, x_{in_0}) = u_i \mathbf{G}_1 + u_{i-1} \mathbf{G}_2 + u_{i-3} \mathbf{G}_3 + \cdots + u_{i-N+1} \mathbf{G}_N
$$

$$
= \sum_{k=1}^{N} u_{i-N+1} \mathbf{G}_N
\tag{2.14}
$$

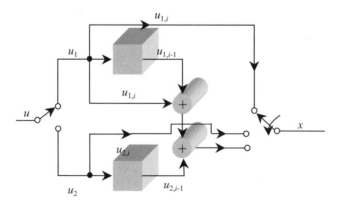

Figure 2.11   Parallel implementation of the same convolutional encoder shown in Figure 2.10.

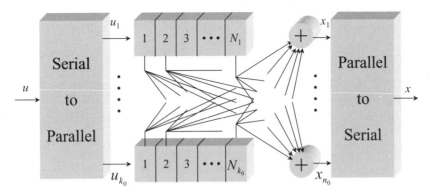

Figure 2.12   General block diagram of a convolutional encoder in parallel form for an (n_0, k_0, N) code.

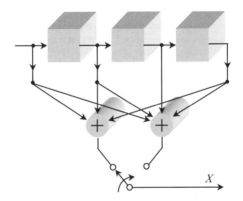

Figure 2.13   Encoder for the $(2, 1, 4)$ convolutional code.

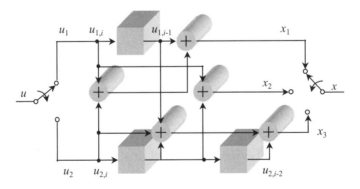

Figure 2.14   Encoder for the $(3, 2, 2)$ convolutional code.

This relation has the form of convolution, hence the name *convolutional coding*.

For an efficient insight into the operation of the encoder, a state diagram, illustrated in Figure 2.15, is used.

The $(3, 1, 3)$ encoder, from Figure 2.9, has memory $\nu = N - 1 = 2$. We define the state σ_l of the encoder at discrete time l as the content of its memory at the same time, $\sigma_l \triangleq (u_{l-1}, u_{l-2})$. There are $N_\sigma = 2^\nu = 4$ possible states. That is, $00, 01, 10$ and 11.

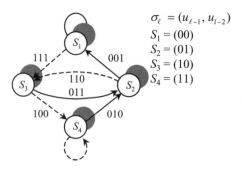

Figure 2.15   State diagram for the (3, 1, 3) convolutional code.

A solid arrow (edge) on Figure 2.15, represents a transition between two states forced by the input digit '0,' whereas a dashed arrow (edge) represents a transition forced by the input digit '1.' The label on each arrow represents the output digits corresponding to that transition. As an example we have:

$$\mathbf{u} = (\quad 1 \quad 1 \quad 0 \quad 1 \quad 1 \quad \dots \quad)$$
$$path \; S_1 \rightarrow S_3 \rightarrow S_4 \rightarrow S_2 \rightarrow S_3 \rightarrow S_4 \dots \qquad (2.15)$$
$$\mathbf{x} = (\quad 111 \quad 100 \quad 010 \quad 110 \quad 100 \quad \dots \quad)$$

The concept of a state diagram can be applied to any (n_0, k_0, N) code with memory v. The number of states is $N_\sigma = 2^v$. There are 2^{k_0} edges entering each state and 2^{k_0} edges leaving each state. The labels on each edge are sequences of length n_0. The equivalent representations using a trellis diagram or tree diagram are shown in Figure 2.16 and Figure 2.17 respectively.

An appropriate measure of the maximum likelihood decoder complexity for a convolutional code is the number of visited edges per decoded bit. Now, a rate k_0 / n_0 code has 2^{k_0} edges leaving and entering each trellis state and a number of states $N_\sigma = 2^v$, where v is the memory of the encoder. Thus, each trellis section, corresponding to k_0 input bits, has a total number of edges equal to 2^{k_0+v}. As a consequence, an

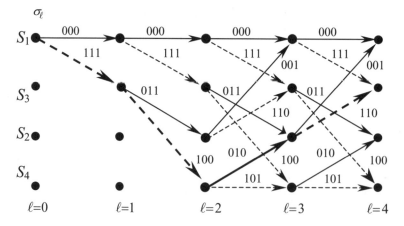

Figure 2.16   Trellis diagram for the (3, 1, 3) convolutional code. The boldface path corresponds to the input sequence 1101.

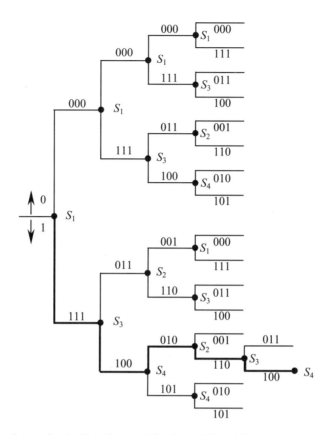

Figure 2.17 Tree diagram for the (3, 1, 3) convolutional code. The boldface path corresponds to the input sequence 11011.

(n_0, k_0, N) code has a decoding complexity:

$$D = \frac{2^{k_0+v}}{k_0} \tag{2.16}$$

Reconfiguration from code 1 to code 2 can offer the coding gain g_{12}. This process will, in general, increase the complexity from D_1 to D_2 resulting in:

$$D_r = D_2/D_1 = \frac{k_{01}}{k_{02}} 2^{(k_{02}-k_{01})+(v_2-v_1)}$$

$$= \frac{k_{01}}{k_{02}} 2^{\Delta k_0 + \Delta v} = \frac{k_{01}}{k_{02}} 2^{\Delta (k_0+v)} = \frac{k_{01}}{k_{02}} 2^{\Delta (k_0 N)} \tag{2.17}$$

where Δ should be used as an operator. Equation (2.7) defining the reconfiguration efficiency now becomes:

$$E_{ff} = 10^{g_{12}/10} \frac{k_{02}}{k_{01}} 2^{-\Delta (k_0+v)} \tag{2.18}$$

The maximum value of D_r will be referred to as the *reconfiguration range* and the maximum value of the gain g_{12} as the *adaptation range*.

2.2.1   Punctured Convolutional Codes/Code Reconfigurability

The increase of complexity inherent in passing from rate $1/n_0$ to rate k_0/n_0 codes can be mitigated using so-called *punctured convolutional codes*. A rate k_0/n_0 punctured convolutional code can be obtained by starting from a rate $1/n_0$ and deleting parity check symbols. An example is given in Figure 2.18.

Suppose now, that for every four parity check digits generated by the encoder, one (the last) is punctured, i.e. not transmitted. In this case, for every two input bits, three bits are generated by the encoder, thus producing a rate 2/3 code. The equivalent representation of the encoder is shown in Figure 2.19.

For more details, see [18, 46–48].

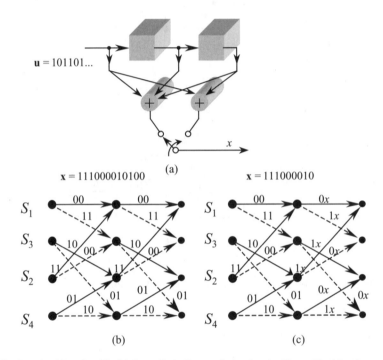

Figure 2.18   Encoder (a) and trellis (b) for a (2, 1, 3) convolutional code. The trellis (c) refers to the rate 2/3 punctured code.

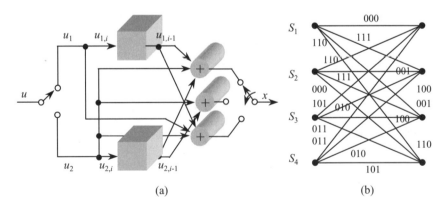

Figure 2.19   Encoder (a) and trellis (b) for the (3, 2, 3) convolutional code equivalent to the rate 2/3 punctured code described.

From the previous example, we can derive the conclusion that a rate k_0/n_0 convolutional code can be obtained considering k_0 trellis sections of a rate 1/2 mother code. Measuring the decoding complexity as before for the punctured code we have:

$$D_{\text{punc}} = \frac{k_0 2^{v+1}}{k_0}$$

so that the ratio between the case of the unpunctured to the punctured solution yields:

$$\Delta D_r = \frac{D}{D_{\text{punc}}} = \frac{2^{k_0}}{2k_0}$$

which shows that, for $k_0 > 2$, there is an increasing complexity reduction. Unfortunately, in general, this would also reduce coding gain with respect to the mother code so that the relative increase in the reconfiguration efficiency can be represented as:

$$\Delta E_{ff} = \Delta D_r 10^{\Delta g_{12}/10} \tag{2.19}$$

where Δg_{12} is negative.

2.2.2   Maximum Likelihood Decoding/Viterbi Algorithm

The maximum likelihood (ML) decoder will select the code word from the trellis, like the one in Figure 2.16, whose distance from the received sequence is minimal. Theoretically, the decoder must find the path through the trellis for which:

$$U(\sigma_{K-1}) \stackrel{\triangle}{=} \max_r \ U^{(r)}(\sigma_{K-1}) \stackrel{\triangle}{=} \max_r \ P\left(\mathbf{y}|\mathbf{x}^{(r)}\right)$$

$$\equiv \max_r \left[\ln \prod_{l=0}^{K-1} P\left(\mathbf{y}_l|\mathbf{x}_l^{(r)}\right) \right] = \max_r \left[\ln \prod_{l=0}^{K-1} \ln P\left(\mathbf{y}_l|\mathbf{x}_l^{(r)}\right) \right] \tag{2.20}$$

where $\mathbf{x}_l$ and $\mathbf{y}_l$ are n_0 binary transmitted and received digits respectively between discrete times l and $l + 1$. The branch metric

$$V_l^{(r)}(\sigma_{l-1}, \sigma_l) \stackrel{\triangle}{=} \ln P\left(\mathbf{y}_l \mid \mathbf{x}_l^{(r)}\right) \tag{2.21}$$

for antipodal modulation (with transmitted and received energy E) and assuming that the lth branch of the rth path has been transmitted, is:

$$y_{jl} = \sqrt{E}\left(2x_{jl}^{(r)} - 1\right) + v_j$$

So, we have:

$$P\left(\mathbf{y}_l \mid \mathbf{x}_l^{(r)}\right) = \prod_{j=1}^{n_0} P\left(y_{jl} \mid x_{jl}^{(r)}\right)$$

$$= \prod_{j=1}^{n_0} \frac{1}{\sqrt{\pi N_0}} \exp\left\{ -\frac{\left[y_{jl} - \sqrt{E}\left(2x_{jl}^{(r)} - 1\right)\right]^2}{N_0} \right\} \tag{2.22}$$

$$V_l^{(r)}(\sigma_{l-1}, \sigma_l) = \sum_{j=1}^{n_0} y_{jl}\left(2x_{jl}^{(r)} - 1\right) \tag{2.23}$$

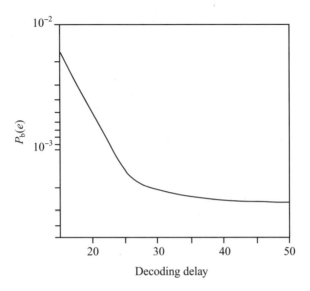

Figure 2.20   Simulated bit error probability versus the decoding delay for the decoding of the rate 1/2 (2, 1, 7) convolutional code. The signal to noise ratio E/N_0 is 3 dB.

In order to reduce the number of trajectories for which the metric is calculated, the Viterbi algorithm accumulates the result only for those branches with the highest metric (survivors) that have the best chances to be selected at the end as the final choice [1–18]. A maximum *a posteriori* probability (MAP) detector is defined in Appendix 2.1.

In the remainder of this section we provide several examples of the performance results.

Figure 2.20 demonstrates that increasing the decoding delay (length of the observed trajectories) over 30 bit intervals cannot reduce BER significantly. In other words, for the given code, only trajectories up to 30 bits should be observed. From Figure 2.21 one can see that soft decision decoding provides a roughly 3 dB coding gain compared with hard decision decoding. Figures 2.22 and 2.23 present BER performance for coding rates 1/2 and 1/3 respectively. For these examples, $k_0 = 1$ and reconfiguration

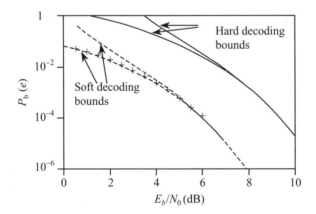

Figure 2.21   Performance bounds with hard and soft decision maximum likelihood decoding for the (3, 1, 3) convolutional code. The curve with '+' refers to simulation results obtained with the soft decision Viterbi algorithm.

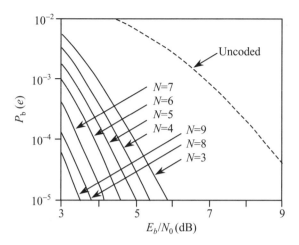

Figure 2.22   Upper bounds to the soft ML decoding bit error probability of different convolutional codes of rate 1/2.

efficiency, defined by Equation (2.18), becomes:

$$E_{ff} = 10^{g_{12}/10} 2^{-\Delta(1+(N-1))} = 10^{g_{12}/10} 2^{-\Delta N} \tag{2.24}$$

As an example, one can see from Figure 2.22 that for BER $= 10^{-6}$, the coding gain $g_{12} \cong 6.7 - 4 = 2.7$dB can be achieved by reconfiguration of the code from $N = 3$ to $N = 9$. So the reconfiguration efficiency is $E_{ff} (R_c = 1/2) = 10^{0.27} \ 2^{-6}$. On the other hand, from Figure 2.23, one can see that approximately the same coding gain can be achieved by reconfiguration of the rate $R_c = 1/3$ code from $N = 3$ to $N = 8$, giving $E_{ff}(R_c = 1/3) = 10^{0.27} 2^{-5}$. In other words, we find that the efficiency of the $R_c = 1/3$ code is higher by a factor of two.

Achievable coding gains for different codes are shown in Table 2.2. These results can be used for calculation of reconfiguration efficiency for different types of code.

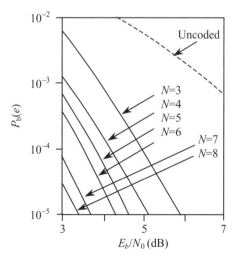

Figure 2.23   Upper bounds to the soft ML decoding bit error probability of different convolutional codes of rate 1/3.

Table 2.2 Achievable coding gains with some convolutional codes of rate R_c and constraint length N at different values of the bit error probability. Eight-level quantization soft decision Viterbi decoding is used. The last line gives the asymptotic upper bound $10 \log_{10} d_f R_c$

| | $R_c = 1/3$ | | $R_c = 1/2$ | | | $R_c = 2/3$ | | $R_c = 3/4$ | |
| | N | | N | | | N | | N | |
$P_b(e)$	7	8	5	6	7	6	8	6	9
10^{-3}	4.2	4.4	3.3	3.5	3.8	2.8	3.1	2.6	2.6
10^{-5}	5.7	5.9	4.3	4.6	5.1	4.2	4.6	3.6	4.2
10^{-7}	6.2	6.5	4.9	5.3	5.8	4.7	5.2	3.9	4.8
$\to 0$	7.0	7.3	5.4	6.0	7.0	5.2	6.7	4.8	5.7

2.2.3 Systematic Recursive Convolutional Codes

Systematic convolutional codes generated by feed-forward encoders yield, in general, lower free distances than non-systematic codes. In this section, we will show how to derive a systematic encoder from every rate $1/n_0$ non-systematic encoder, which generates a systematic code with the same *weight enumerating function* as the non-systematic one which is relevant for the free distance [18–45]. The systematic codes are used for turbo code construction, to be discussed later.

Consider, for simplicity, a rate 1/2 feed-forward encoder characterized by the two generators (in polynomial form) $g_{1,1}(Z)$ and $g_{1,2}(Z)$. Using the power series $u(Z)$ to denote the input sequence **u**, and $x_1(Z)$ and $x_2(Z)$ to denote the two sequences $\mathbf{x}_1$ and $\mathbf{x}_2$ forming the code **x**, we have the relationships:

$$\begin{aligned} x_1(Z) &= u(Z)g_{1,1}(Z) \\ x_2(Z) &= u(Z)g_{1,2}(Z) \end{aligned} \tag{2.25}$$

To obtain a systematic code we need to have either $x_1(Z) = u(Z)$ or $x_2(Z) = u(Z)$. To obtain the first equality, let us divide both equations by $g_{1,1}(Z)$, so that:

$$\begin{aligned} \tilde{x}_1(Z) &\triangleq \frac{x_1(Z)}{g_{1,1}(Z)} = u(Z) \\ \tilde{x}_2(Z) &\triangleq \frac{x_2(Z)}{g_{1,1}(Z)} = \frac{u(Z)}{g_{1,1}(Z)}g_{1,2}(Z) \end{aligned} \tag{2.26}$$

Defining now a new input sequence $\tilde{u}(Z)$ as:

$$\tilde{u}(Z) \triangleq \frac{u(Z)}{g_{1,1}(Z)} \tag{2.27}$$

the relations become:

$$\begin{aligned} \tilde{x}_1(Z) &\triangleq \tilde{u}(Z)g_{1,1}(Z) \\ \tilde{x}_2(Z) &\triangleq \tilde{u}(Z)g_{1,2}(Z) \end{aligned} \tag{2.28}$$

2.2.3.1 Example

Let us assume that the initial encoder is defined by the polynomials $g_1^A(Z) = 1 + Z^2$, $g_2^A(Z) = 1 + Z + Z^2$. Its equivalent recursive encoder is obtained as explained previously with generators $g_1^B(Z) = 1$, $g_2^B(Z) = (1 + Z + Z^2)/(1 + Z^2)$. These steps are illustrated in Figures 2.24 and 2.25.

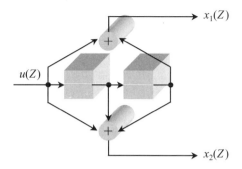

Figure 2.24   Rate 1/2, four-state feed-forward encoder generating the non-systematic code.

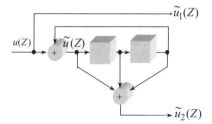

Figure 2.25   Rate 1/2, four-state recursive encoder generating the systematic code.

2.3   Concatenated Codes with Interleavers

Further improvements in performance can be achieved if two codes are combined (concatenated) to encode the same message, as illustrated in Figure 2.26 (parallel concatenation) and Figure 2.27 (serial concatenation).

For ML decoding we should consider an equivalent trellis characterized by $2^{v_1 + v_2}$ states. The complexity and reconfiguration efficiency are defined by the same equations as before, with $v \Rightarrow v_1 + v_2$. BER curves for the two types of concatenation are shown in Figures 2.28 and 2.29.

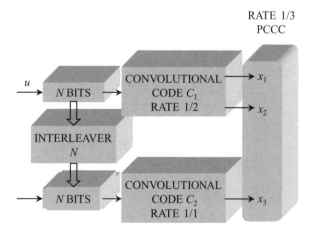

Figure 2.26   Block diagram of a rate 1/3 parallel concatenated convolutional code (PCCC). Rate 1/1 code could be obtained using the same rate 1/2 systematic encoder that generates C_1 and transmitting only the parity check bit.

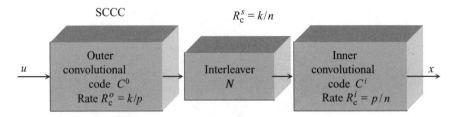

Figure 2.27   Block diagram of a rate k/n serially concatenated convolutional code (SCCC).

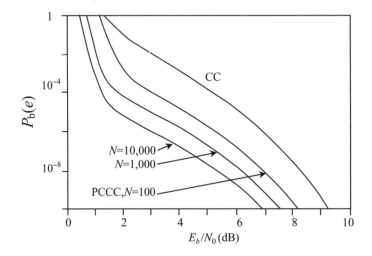

Figure 2.28   Average upper bounds to the bit error probability for a rate 1/3 PCCC obtained by concatenating two rate 1/2, four-state convolutional codes (CC), through a uniform interleaver of sizes $N = 100, 1000, 10000$.

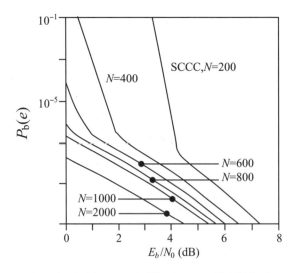

Figure 2.29   Average upper bound to the bit error probability for a rate 1/3 SCCC using as outer code a four-state, rate 1/2, non-recursive convolutional encoder, and as inner code a four-state, rate 2/3, recursive convolutional encoder with uniform interleavers of various sizes.

2.3.1   The Iterative Decoding Algorithm

Due to a significant increase in the number of states in the equivalent trellis of the concatenated codes, the complexity of the ML decoder might become unacceptable. A suboptimal, iterative algorithm, known as a *turbo decoder*, provides a sufficiently good approximation of the ML decoder and offers performance approaching Shannon's limit (channel capacity). An heuristic explanation for the iterative decoder is based on the assumption that Equation (2.20) can now be expressed in the form [18]:

$$\hat{u}_k \stackrel{\triangle}{=} \arg\ \max_i [APP(k, i)]$$

where the *a posteriori* probability APP (k, i) is defined as:

$$
\begin{aligned}
APP(k, i) &\triangleq p(u_k = i,\ \mathbf{y}_1, \mathbf{y}_2) \\
&= \sum_{\mathbf{u}:u_k=i} p(\mathbf{y}_1|c_1(\mathbf{u}))p(\mathbf{y}_2|c_2(\mathbf{u}))p_a(\mathbf{u})
\end{aligned}
\tag{2.29}
$$

$$p(\mathbf{y}_1|c_1(\mathbf{u})) = \prod_{j=1}^{N_1} p(y_{1j}|c_{1j}(\mathbf{u}))$$

$$p(\mathbf{y}_2|c_2(\mathbf{u})) = \prod_{m=1}^{N_2} p(y_{2m}|c_{2m}(\mathbf{u}))$$

$$p_a(\mathbf{u}) = \prod_{l=1}^{K} p_a(u_l)$$

In other words, it is assumed that the interleaver will randomize the data stream so that the output of the two encoders can be considered independent. Based on this assumption we can also write:

$$p(\mathbf{y}_1|c_1(\mathbf{u})) \cong \prod_l \tilde{P}_{1l}(u_l)$$

$$p(\mathbf{y}_2|c_2(\mathbf{u})) \cong \prod_l \tilde{P}_{2l}(u_l)$$

$$\tag{2.30}$$

and Equation (2.29) becomes:

$$APP(k, i) = \left[\sum_{\mathbf{u}:u_k=i} p(\mathbf{y}_2|c_2(\mathbf{u})) \prod_{l \neq k} \tilde{P}_{1l}(u_l)p_a(u_l) \right] \tilde{P}_{1k}(i)p_a(i)$$

$$APP(k, i) = \left[\sum_{\mathbf{u}:u_k=i} p(\mathbf{y}_1|c_1(\mathbf{u})) \prod_{l \neq k} \tilde{P}_{2l}(u_l)p_a(u_l) \right] \tilde{P}_{2k}(i)p_a(i)$$

$$\tag{2.31}$$

The solution to Equation (2.31) can be represented as:

$$\tilde{P}_{1k}(i) = \sum_{\mathbf{u}:u_k=i} p(\mathbf{y}_1|\ c_1(\mathbf{u})) \prod_{l \neq k} \tilde{P}_{2l}(u_l)p_a(u_l)$$

$$\tilde{P}_{2k}(i) = \sum_{\mathbf{u}:u_k=i} p(\mathbf{y}_2|\ c_2(\mathbf{u})) \prod_{l \neq k} \tilde{P}_{1l}(u_l)p_a(u_l)$$

$$\tag{2.32}$$

Finally, Equation (2.29) becomes:

$$APP(k, i) = \tilde{P}_{1k}(i)\tilde{P}_{2k}(i)p_a(u_l) \tag{2.33}$$

The maximization of Equation (2.33) can be performed in a number of iterations as:

$$\tilde{P}_{2k}^{(0)} = 1, \quad k = 1, \ldots, K$$

$$\vdots$$

$$\tilde{P}_{1k}^{(m)} = \sum_{\mathbf{u}:u_k=i} p(\mathbf{y}_1 | c_1(\mathbf{u})) \prod_{l \neq k} \tilde{P}_{2l}^{(m-1)}(u_l) p_a(u_l), \quad k = 1, \ldots, K \tag{2.34}$$

$$\tilde{P}_{2k}^{(0)} = \sum_{\mathbf{u}:u_k=i} p(\mathbf{y}_2 | c_2(\mathbf{u})) \prod_{l \neq k} \tilde{P}_{1l}^{(m)}(u_l) p_a(u_l), \quad k = 1, \ldots, K$$

For binary signaling, $u_k = 0, 1$ and the signal probabilities can be replaced by using log likelihood ratios (LLR) defined as:

$$\lambda_k(\text{APP}) \equiv \log \frac{\displaystyle\sum_{\mathbf{u}:u_k=0} p(\mathbf{y}_1 | c_1(\mathbf{u})) p(\mathbf{y}_2 | c_2(\mathbf{u})) p_a(\mathbf{u})}{\displaystyle\sum_{\mathbf{u}:u_k=1} p(\mathbf{y}_1 | c_1(\mathbf{u})) p(\mathbf{y}_2 | c_2(\mathbf{u})) p_a(\mathbf{u})} \tag{2.35}$$

By introducing the following notation

$$\lambda_k \equiv \log \frac{p(y_k | 0)}{p(y_k | 1)}$$

$$\lambda_{1j} \equiv \log \frac{p(y_{1j} | 0)}{p(y_{1j} | 1)} \qquad j = 1, \ldots N_1$$

$$\lambda_{2m} \equiv \log \frac{p(y_{2m} | 0)}{p(y_{2m} | 1)} \qquad m = 1, \ldots N_2$$

$$\lambda_a \equiv \log \frac{p_a(0)}{p_a(1)} \tag{2.36}$$

$$\pi_{1l} \equiv \log \frac{\tilde{P}_{1l}(0)}{\tilde{P}_{1l}(1)}$$

$$\pi_{2l} \equiv \log \frac{\tilde{P}_{2l}(0)}{\tilde{P}_{2l}(1)}$$

and

$$\lambda_k(\Lambda\text{PP}) = \log \frac{\displaystyle\sum_{\mathbf{u}:u_k=0} \frac{p(\mathbf{y}_1 | c_1(\mathbf{u}))}{p(\mathbf{y}_1 | 0)} \frac{p(\mathbf{y}_2 | c_2(\mathbf{u}))}{p(\mathbf{y}_2 | 0)} \frac{p_a(\mathbf{u})}{p_a(0)}}{\displaystyle\sum_{\mathbf{u}:u_k=1} \frac{p(\mathbf{y}_1 | c_1(\mathbf{u}))}{p(\mathbf{y}_1 | 0)} \frac{p(\mathbf{y}_2 | c_2(\mathbf{u}))}{p(\mathbf{y}_2 | 0)} \frac{p_a(\mathbf{u})}{p_a(0)}} \tag{2.37}$$

$$\frac{p(\mathbf{y}_1 | c_1(\mathbf{u}))}{p(\mathbf{y}_1 | 0)} = \prod_{j=1}^{N_1} \frac{p(y_{1j} | c_{1j}(\mathbf{u}))}{p(y_{1j} | 0)}$$

$$\frac{p(y_{1j} | c_{1j}(\mathbf{u}))}{p(y_{1j} | 0)} = \left\{ \begin{array}{ll} \exp(-\lambda_{1j}) & if \quad c_{1j}(\mathbf{u}) = 1 \\ 0 & if \quad c_{1j}(\mathbf{u}) = 0 \end{array} \right\} = \exp\left[-\lambda_{1j}\right)c_{1j}(\mathbf{u})\right]$$

$$\frac{p(y_{2m} | c_{2m}(\mathbf{u}))}{p(y_{2m} | 0)} = \exp\left[-\lambda_{2m}\right)c_{2m}(\mathbf{u})\right]$$

$$\frac{p_a(u_l)}{p_a(0)} = \exp(-u_l \lambda_a)$$

Equation (2.35) becomes:

$$
\lambda_k(\text{APP}) = \log \left\{ \sum_{\mathbf{u}:k_k=0} \exp - \left[\sum_{j=1}^{N_1} c_{1j}(\mathbf{u})\lambda_{1j} + \sum_{m=1}^{N_2} c_{2m}(\mathbf{u})\lambda_{2m} + \sum_{l=1}^{K} u_l \lambda_a \right] \right\}
$$
$$
- \log \left\{ \sum_{\mathbf{u}:k_k=1} \exp - \left[\sum_{j=1}^{N_1} c_{1j}(\mathbf{u})\lambda_{1j} + \sum_{m=1}^{N_2} c_{2m}(\mathbf{u})\lambda_{2m} + \sum_{l=1}^{K} u_l \lambda_a \right] \right\} \tag{2.38}
$$

and the iterative procedure (2.34) can be represented as:

$$
\pi_{2k}^{(0)} = 0
$$
$$
\vdots
$$
$$
\pi_{1k}^{(m)} = \log \left\{ \sum_{\mathbf{u}:k_k=0} \exp - \left[\sum_{j} c_{1j}(\mathbf{u})\lambda_{1j} + \sum_{l \neq k} u_l(\lambda_a + \pi_{2l}^{(m-1)}) \right] \right\}
$$
$$
- \log \left\{ \sum_{\mathbf{u}:k_k=1} \exp - \left[\sum_{j} c_{1j}(\mathbf{u})\lambda_{1j} + \sum_{l \neq k} u_l(\lambda_a + \pi_{2l}^{(m-1)}) \right] \right\} \tag{2.39}
$$
$$
\pi_{2k}^{(m)} = \log \left\{ \sum_{\mathbf{u}:k_k=0} \exp - \left[\sum_{m} c_{2m}(\mathbf{u})\lambda_{2m} + \sum_{l \neq k} u_l(\lambda_a + \pi_{1l}^{(m)}) \right] \right\}
$$
$$
- \log \left\{ \sum_{\mathbf{u}:k_k=1} \exp - \left[\sum_{m} c_{2m}(\mathbf{u})\lambda_{2m} + \sum_{l \neq k} u_l(\lambda_a + \pi_{1l}^{(m)}) \right] \right\}
$$

The final value of λ_k (APP) is calculated as

$$
\lambda_k(\text{APP}) = \pi_{1k} + \pi_{2k} + \lambda_a \tag{2.40}
$$

and the MAP decision is made according to the sign of λ_k. The details of the numerical evaluation of Equation (2.39) are available in standard literature [49–59], the most relevant being the original work by Berrou [60].

These iterations contain implicitly the trellis constraints imposed by the trellis structure. For these purposes forward and backward recursions are used [49–60].

In the block diagrams shown in Figures 2.30 and 2.31, these calculations are performed in soft input soft output (SISO) blocks [49–59]. Performance curves are given in Figures 2.32, 2.33 and 2.34. In order to avoid repetition, for details of iterative calculations, defined by Equation (2.39), the reader is referred to the classical literature [49–59]. Instead of going into these details we will get back, once again, to the issue of reconfiguration efficiency. As already mentioned, for two concatenated codes with $k_{01} = k_{02}$ and an ML decoder, Equation (2.18) becomes:

$$
E_{ff} = 10^{g_{12}/10} 2^{-\Delta(k_0 v)}
$$

So, for the two codes with $v_1 = v_2 = v$, the reconfiguration from a single convolutional code (CC) to parallel concatenated CC (PCCC) with $k_{01} = k_{02}$, gives:

$$
E_{ff} = 10^{g_{12}/10} 2^{-v}
$$

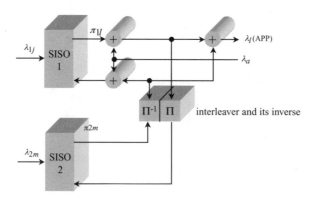

Figure 2.30 Block diagram of the iterative decoding scheme for binary convolutional codes.

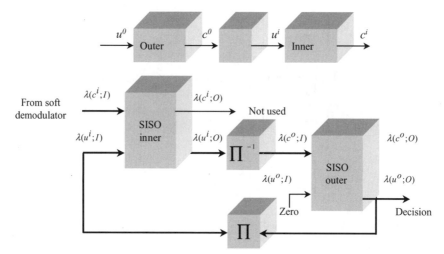

Figure 2.31 Block diagrams of the encoder and iterative decoder for serially concatenated convolutional codes.

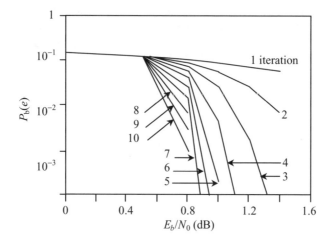

Figure 2.32 Bit error probability obtained by simulating the iterative decoding algorithm. Rate 1/2 PCCC based on 16-state rate 2/3 and 2/1 CCs, and interleaver with size $N = 8920$.

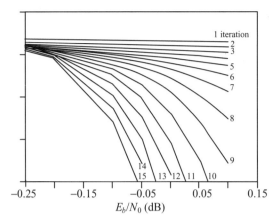

Figure 2.33 Simulated bit error probability performance of a rate 1/4 serially concatenated code obtained with two eight-state constituent codes and an interleaver yielding an input decoding delay equal to 16 384.

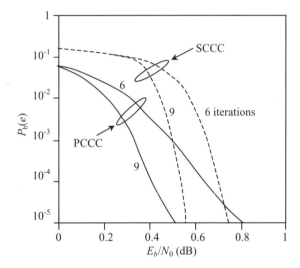

Figure 2.34 Comparison of rate 1/3 PCCC and SCCC. The PCCC is obtained concatenating two equal rate 1/2 four-state codes, whereas the SCCC concatenates two four-state rate 1/2 and rate 2/3 codes. The curves refer to six and nine iterations of the decoding algorithm and to an equal input decoding delay of 16 384.

If a turbo decoder is used, the relative complexity changes and Equation (2.19) gives:

$$\Delta E_{ff} = 10^{-|g_{12}|/10} \frac{2^v}{2I}$$

where I is the number of iterations.

2.4 Adaptive Coding, Practice and Prospects

Efficient error control on time varying channels can be performed by implementing an adaptive control system where the optimum code is selected according to the actual channel conditions.

There is a number of burst error correcting codes that could be used in these adaptive schemes. Three major classes of burst error correcting code are binary fire block codes, binary Iwadare–Massey convolutional codes [61], and non-binary Reed–Solomon block codes. In practical communication systems these are decoded by hard decision decoding methods. Performance evaluation based on experimental data from satellite mobile communication channels [62] shows that the convolutional codes with the soft decision decoding Viterbi algorithm are superior to all of the above burst error correcting codes.

Superior error probability performance and availability of a wide range of code rates without changing the basic coded structure, motivate the use of punctured convolutional codes [27–30] with the soft decision Viterbi decoding algorithm in the proposed adaptive scheme. To obtain the full benefit of the Viterbi algorithm on bursty channels, ideal interleaving is assumed.

An adaptive coding scheme using incremental redundancy in a hybrid automatic repeat request (ARQ) error control system is reported by Wu et al. [63]. The channel model used is BSC with time variable bit error probability. The system state is chosen according to the channel bit error rate. The error correction is performed by shortened cyclic codes with variable degrees of shortening. When the channel bit error rate increases, the system generates additional parity bits for error correction.

An FEC adaptive scheme for matching code to the prevailing channel conditions was reported by Chase [64]. The method is based on convolutional codes with Viterbi decoding and consists of combining noisy packets to obtain a packet with a code rate low enough (less than $^1/_2$) to achieve the specified error rate. Other schemes that use a form of adaptive decoding are reported in [65–70]. Hybrid ARQ schemes based on convolutional codes with sequential decoding on a memoryless channel were reported by Drukarev and Costello [71, 72] while a Type-II hybrid ARQ scheme formed by concatenation of convolutional codes with block codes was evaluated on a channel represented by two states [73].

In order to implement the adaptive coding scheme it is necessary again to use a return channel. The channel state estimator (CSE) determines the current channel state, based on counting the number of erroneous blocks. Once the channel state has been estimated, a decision is made by the 'reconfiguration block' whether to change the code, and the corresponding messages are sent to the encoder and locally to the decoder.

In FEC schemes only error correction is performed, while in hybrid ARQ schemes retransmission of erroneous blocks is requested whenever the decoded data is labeled as unreliable.

The adaptive error protection is obtained by changing the code rates. For practical purposes it is desirable to modify the code rates without changing the basic structure of the encoder and decoder. Punctured convolutional codes are ideally suited for this application. They allow almost continuous change of the code rates, while decoding is done by the same decoder.

The encoded digits at the output of the encoder are periodically deleted according to the deleting map specified for each code. Changing the number of deleted digits varies the code rate. At the receiver end, the Viterbi decoder operates on the trellis of the parent code and uses the same deleting map as in the encoder in computing path metrics [47].

The Viterbi algorithm based on this metric is a maximum likelihood algorithm on channels with Gaussian noise, since on these channels the most probable errors occur between signals that are closest together in terms of squared Euclidean distance. However, this metric is not optimal for non-Gaussian channels. The Viterbi algorithm allows use of channel state information for fading channels [74].

However, a disadvantage of punctured convolutional codes compared to other convolutional codes with the same rate and memory order is that error paths are typically long. This requires quite long decision depths of the Viterbi decoder.

A scheme with ARQ rate compatible convolutional codes was reported by Hagenauer [75]. In this scheme, rate compatible codes are applied. The rate compatibility constraint increases the system throughput, since in transition from a higher to a lower rate code, only incremental redundancy digits are retransmitted. The error detection is performed by a cyclic redundancy check which introduces additional redundancy.

2.5 Distributed Source Coding

In this section we discuss the problem of compressing correlated distributed sources, i.e. correlated sources that are not co-located or that cannot cooperate to directly exploit their correlation. The most interesting application is data aggregation in wireless sensor networks (WSN). This problem has been studied in the information theory literature under the name of the Slepian–Wolf source coding problem for the lossless coding case, and as 'rate-distortion with side information' for the lossy coding case [76–105].

In a WSN, sensors transmit their highly correlated information to a central processing unit (sink) that forms the best picture of the scene based on a fusion of the information collected by all of them.

Data aggregation is reduced to the problem of encoding of sources in the presence of side information available only at the decoder. Consider first the problem where X and Y are correlated discrete-alphabet memoryless sources, and we have to compress X losslessly, with Y (referred to as side information) being known at the decoder but *not* at the encoder. If Y were known at both ends (see Figure 2.35(a)), then the problem of compressing X is well understood: one can compress X at the theoretical rate of its conditional entropy [76] given Y, $H(X|Y)$. If Y were known only at the decoder for X and not at the encoder (see Figure 2.35(b)) one can still compress X using only $H(X|Y)$ bits, as is the case where the encoder *does* know Y. That is, by just knowing the joint distribution of X and Y, without explicitly knowing Y, the encoder of X can perform as well as an encoder which explicitly knows Y. This is known as the Slepian–Wolf coding theorem [78].

As an illustration, suppose X and Y are equiprobable 3-bit binary words that are correlated in the sense that the Hamming distance between X and Y is no more than one. If Y (side information) is available to both the encoder and the decoder, clearly we can describe X using 2 bits (there are only four possibilities for the modulo-two binary sum of X and Y: $\{000, 001, 010, 100\}$).

If Y were revealed *only* to the decoder but not the encoder, it is wasteful for X to spend any bits in differentiating between $\{X = 000$ and $X = 111\}$, since the Hamming distance between these two words is 3. Thus, if the decoder knows that either $X = 000$ or $X = 111$, it can resolve this uncertainty by checking *which of them is closer in Hamming distance to Y*, and declaring that as the value of X. Note that the set $\{000, 111\}$ is a 3-bit repetition code. Likewise, in addition to the set $\{000, 111\}$, each of the

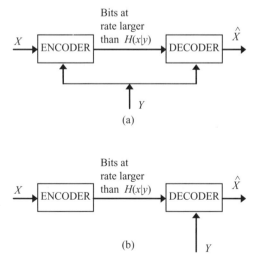

Figure 2.35 Communication system: (a) both encoder and decoder have access to the side information Y (which is correlated to X); (b) only decoder has access to the side information Y.

following three sets for X:$\{100, 011\}$, $\{010, 101\}$, and $\{001, 110\}$ is composed of pairs of words whose Hamming distance is 3. These are just simple variants *or cosets* of the 3-bit repetition code, and they cover the space of all binary 3-tuples that X can assume. Thus, instead of describing X by its 3-bit value, we encode *which coset X belongs to*, incurring a cost of 2 bits, just as in the case where Y is known to both encoder and decoder.

In general, by using terminology of algebraic channel codes, a linear block channel code [94] is specified by its 3-tuple (n, k, d), where n is the code length, k is the message length, and d is the minimum distance of the code. In the above example, we considered the cosets of the linear (3, 1, 3) repetition code. Every coset of a linear code is associated with a unique *syndrome* [95]. Recall that the syndrome s associated with a linear channel code is defined as $s^T = Hc^T$, where H is the parity-check matrix of the code, c is any valid codeword, T denotes transposition, and s, c are row vectors.

In general, for $X, Y \in \{0, 1\}^n$ and $d_H(X, Y) \le t$, the encoding of X is done in such a way that the encoder observes X and sends the index of the coset (say l) of an appropriately chosen binary linear code with parameters $(n, k, 2t + 1)$, in which it resides. The decoder recovers the value of X to be that vector in the coset l which is closest in Hamming distance to Y. The rate of transmission is $(n - k)$ bits per sample.

Implementation: Let $\mathbf{H}$ be the parity-check matrix of a binary linear $(n, k, 2t + 1)$ code in its systematic form, i.e., $\mathbf{H} = [\mathbf{A} \mid \mathbf{I}]$. Let $s^T = \mathbf{H}x^T$ be the syndrome when the outcome of X is x. The encoder transmits s to the decoder. Let $y' = y \oplus a$, where a is any vector in the coset whose syndrome is s ($\oplus$denotes addition modulo-2). Find a vector x' closest to y' in the coset whose syndrome is the all-zero vector. It can be seen that $x = x' \oplus a$. Since any vector in the coset with syndrome s will serve our purpose, we take the vector $[\mathbf{0}|s]$ to be a in all future discussions where $\mathbf{0}$ is the all-zero vector of length k.

Similarly to block codes, we can also define a syndrome sequence in convolutional codes. Let $\mathbf{H}(D) = [\mathbf{A}(D) \mid \mathbf{I}]$ be the systematic parity-check polynomial matrix of a convolutional code, and let $x(D)$ be the source sequence in polynomial representation: $x(D) = \sum_{i=1}^{L} x_i D^{i-1}$, where x_i denotes the ith element of x. The encoder sends the syndrome sequence $s(D)^T = \mathbf{H}(D)x(D)^T$ to the decoder. Let $a(D) = [\mathbf{0}| s(D)]$ and $y'(D) = y(D) \oplus a(D)$, where $y(D)$ is the side-information sequence. In the coset with the all-zero syndrome, let $x'(D)$ be the sequence closest to $y'(D)$, which can be found using the Viterbi algorithm [96] for $\mathbf{H}(D)$. Now the reconstruction $\hat{x}(D) = x'(D) \oplus a(D)$.

2.5.1 Continuous Valued Source

In this case X and Y are correlated memoryless processes characterized by independent and identically distributed (i.i.d.) sequences $\{X_i\}_{i=1}^{\infty}$ and $\{Y_i\}_{i=1}^{\infty}$, respectively. For illustration purposes we use a simple case where Y is a noisy version of X, i.e. $Y_i = X_i + N_i$, where $\{N_i\}_{i=1}^{\infty}$ is also continuously valued (defined on the real line R), i.i.d, and independent of the X_is. As before, *the decoder alone has access to the Y process (side information)*, and the task is to compress optimally the X process. In the sequel, we consider the case where the X_is and N_is are zero-mean Gaussian random variables with known variances, so as to benchmark our performance against the theoretical performance bounds.

The goal is to form the best approximation $\hat{X}$ to X given an encoding rate of R bits per sample. We assume encoding in blocks of length L. Let the distortion measure be $\rho(\cdot)$ over the L-sequence, defined as

$$\rho(x, \hat{x}) = \frac{1}{L} \sum_{i=1}^{L} \rho(x_i, \hat{x}_i), \qquad \rho : \text{R} \times \text{R} \to \text{R}^+$$

This problem can be posed as minimizing the rate of transmission R such that the reconstruction fidelity $E[\rho(X, \hat{X})]$ is less than a given value D, where $E(\cdot)$ is the expectation operator.

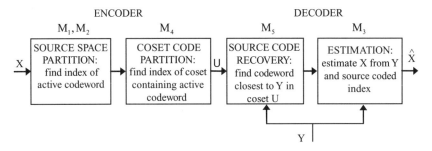

Figure 2.36   Encoder and decoder: X is quantized, and the index of the coset containing the quantized codeword is sent to the decoder. The decoder finds this codeword by decoding the side information in the given coset.

The encoder is a mapping from the input space to the index set: $R^L \rightarrow \{1, 2, \ldots, 2^{LR}\}$, and the decoder is a mapping from the product space of the encoded index set and the correlated L-sequence Y to the L-sequence reconstruction $\{1, 2, \ldots, 2^{LR}\} \times R^L \rightarrow R^L$. In the sequel, we focus on the mean-squared error (MSE) distortion: $\rho(x, \hat{x}) = (x - \hat{x})^2$.

The encoder and the decoder, as shown in Figure 2.36 consist of five mappings: $\{M_i\}_{i=1}^5$.

Source coding (M_1, M_2): The source X is quantized and a source codebook is constructed for a given reconstruction fidelity. The source space R^L is partitioned into 2^{LR_s} disjoint regions, where R_s is defined as the source rate. This is referred to as mapping M_1.

Let $\Gamma = \{\Gamma_1, \Gamma_2, \ldots, \Gamma_{2^{LR_s}}\}$ denote the set of 2^{LR_s} disjoint regions. Each region in the above partition is associated with a representation codeword. The set of representation codewords is referred to as the source codebook (S). This is a mapping M_2. We refer to the representation codeword to which X is quantized as the *active codeword*. Let the random variable characterizing the active codeword be denoted by W.

Estimation (M_3): The decoder gets the best estimate of X (minimizing the distortion) conditioned on the outcome of the side information and the element in Γ containing X. This is given by

$$\hat{x} = \arg \min_{a \in R^L} E\left[\rho(X, a) \left| \begin{array}{l} X \in \Gamma_i \\ Y = y \end{array} \right. \right] \qquad (2.41)$$

for the received message i and the side-information outcome y. It can be interpreted as a mapping M_3. The estimation error is a function of R_s, which is chosen to keep this error within the given fidelity criterion.

Channel coding (M_4, M_5): The random variable W characterizing the quantized source is correlated to X, and this in turn induces a correlation between W and the side information Y. This is characterized by a conditional distribution $P(Y/W)$ of the side information given W. With this conditional distribution we can associate a fictitious channel with W as input, which is observed at the encoder, and Y as output, which is observed at the decoder, whose information channel capacity is greater than 0 (due to this correlation). Actually to communicate W to the decoder in the absence of side information requires a transmission of R_s bits per sample. With the presence of Y at the decoder, we have this fictitious 'helper' channel carrying an amount of information $I(W; Y)$ about W. The remaining uncertainty in W after observing the side information Y is $H(W/Y) = H(W) - I(W; Y)$ and this is the desired final rate of transmission. The rebate in the rate of transmission is $I(W; Y)$.

The goal is to get a rebate as close to $I(W; Y)$ as possible by building a practical structured 'channel code' (C) for this fictitious channel on the space of W. Let 2^{LR_c} denote the number of codewords in the designed channel code where R_c is defined as the channel rate. Since, in general, any codeword in the source codebook can be a quantization outcome with a nonzero probability, we partition the source codebook space into cosets of this channel code. The channel code is designed in such a way that each of its cosets is also an equally good channel code for the channel $P(Y/W)$. Thus, each quantization outcome belongs to a coset of this channel code, and this alone has to be conveyed to the decoder, which can then proceed to use this coset of the channel code for finding the intended active codeword.

The encoder computes the index of the coset of the channel code containing the active codeword using a mapping $M_4 : \{1, 2, \ldots, 2^{LR_s}\} \rightarrow \{1, 2, \ldots, 2^{LR}\}$ and transmits this information with rate $R = R_s - R_c$ bits per sample to the decoder. The decoder recovers the active codeword in the signaled coset by finding (channel decoding) the most likely codeword given the observed side information. This is characterized by a mapping $M_5 : R^L \times \{1, 2, \ldots, 2^{LR}\} \rightarrow \{1, 2, \ldots, 2^{LR_s}\}$.

In this approach, there is always a nonzero probability of decoding error, where the side information is decoded to a wrong codeword, and this can be made arbitrarily small by designing efficient channel codes. For a given region in Γ, the choice of the representation codeword determines $I(W; Y)$, and hence R_c.

Example: For a fixed-rate ($\log_2 V$) scalar quantizer with $V = 8$ levels, designed for the distribution of X let $\nabla = \{r_0, r_1, \ldots, r_{V-1}\}$ be the set of reconstruction levels as shown in the Figure 2.37. ∇ partitions the real line into V intervals each associated with one of the reconstruction levels. Let $\Gamma = \{\Gamma_i\}_{i=0}^{V-1}$ be the partition of R, where Γ_i is the open interval $\left(\frac{r_{i-1}+r_i}{2}, \frac{r_i+r_{i+1}}{2}\right)$ and we take $r_{-1} = -\infty$ and $r_V = \infty$. We first quantize the source sample by sample using ∇. Thus, the source codebook S is given by ∇, and $R_s = 3$ bits/sample. Next we construct C by partitioning the set ∇ into $M(\leq V)$ cosets. For illustration, let $M = 2$. We group r_0, r_2, r_4, and r_6 into one coset. Similarly, r_1, r_3, r_5, and r_7 are grouped into another coset. This is done to keep the minimum distance between any two words in every coset as large as possible. Thus, the channel code is defined as $C = \{r_0, r_2, r_4, r_6\}$, making $R_c = 2$.

The information about the coset is transmitted by $R = 1$ bit/sample. The representation codeword r_i is the centroid of the disjoint region Γ_i for $0 \leq i \leq V - 1$.

The decoder deciphers (with a small probability of error) the active codeword by finding the codeword which is closest to Y in the coset whose index is sent by the encoder and the optimal estimate $\hat{x}$ is

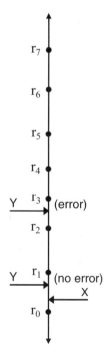

Figure 2.37   Reconstruction levels of scalar quantizer with eight levels.

computed as:

$$\hat{x} = \arg \min_{a \in R} E\left[\rho(X, a) \middle| \begin{array}{l} X \in \Gamma_i, \\ Y = y \end{array}\right] \tag{2.42}$$

where y is the outcome of Y, i is the index of the active codeword.

2.5.2 Scalar Quantization and Trellis-Based Coset Construction

In this section we use scalar (memoryless) quantization as before and a coset construction having memory L. In other words we still use fixed-length scalar quantizers for quantizing $\{X_i\}_{i=1}^{L}$, but the cosets are built on the space ∇^L. Thus, for the previous example, the source codebook is given by $S = \nabla^L$ and $R_s = 3$ bits/sample.

Source coding (M_1, M_2): For a general R_s, $M_1 : R^L \to \{1, 2, \dots, 2^{LR_s}\}$ is the L-product scalar quantizer and $M_2 : \{1, 2, \dots, 2^{LR_s}\} \to \nabla^L$ is the L-sequence extension of the mapping $\{1, 2, \dots, 2^{R_s}\} \to \nabla$.

For the previous example, the space ∇^L has 2^{3L} distinct sequences. The task is to partition this sequence space S into cosets of a set of sequences in such a way that the minimum distance between any two sequences in every coset is made as large as possible, while maintaining symmetry among the cosets. Let $R = 1$ bit/sample. In the following, a trellis-based partitioning with an algebraic structure is considered based on convolutional codes and set-partitioning rules of trellis-coded scalar-quantization (TCSQ) [99].

Consider a trellis code built on ∇ with block length L, having 2^{2L} sequences (code rate is $R_c = 2$ bits/sample), using a rate-2/3 convolutional code and a mapping $Q : \{0, 1\}^3 \to \nabla$. For this example, Ungerboeck's four-state trellis as given in [99] is used. The presentation is restricted to systematic trellis codes. ∇ is partitioned into two subsets as in Section 2.5.1. A coded bitstream selects points from one of these subsets depending on the state of the finite-state machine. The trellis for this code is shown in Figure 2.38(a), which we refer to as the *principal trellis coset*. For this example, let $Q(\zeta) = r_\eta$ where $\zeta \in \{0, 1\}^3$ is the binary representation of η. In the notation $Q(x) = Q(x_1, x_2, \dots, x_L) = [Q(x_1), Q(x_2), \dots, Q(x_L)]$, $x_k \in \{0, 1\}^3$ is a column vector for every $1 \le k \le L$. We take the set of all sequences that can be generated from the above machine (i.e., all the sequences of this trellis code) as a channel code C for the channel $P(Y/W)$. Clearly, $C \subset S$. If $H(D)$ be the parity-check matrix polynomial of the underlying convolutional code, and θ any sequence in ∇^L then we have $Q^{-1}(\theta) \in \{0, 1\}^{3L}$. In the sequel, we associate for any syndrome $s \in \{0, 1\}^L$ a coset of C, given by C(s), which consists of all the sequences $z \in \nabla^L$ such that $H(D)Q^{-1}(z) = s^T$. Each of 2^L cosets of C consists of 2^{2L} sequences. This has resulted in a partition of the space ∇^L into 2^L cosets of C. The encoder sends the index s of the coset C(s) containing the quantized source sequence.

Coset Indexing (M_4): For the case of general R_s, and for any $\theta \in \{1, 2, \dots, 2^{LR_s}\}$ we use $M_4(\theta) = D_R[H(D)\{Q^{-1}[M_2(\theta)]\}] + 1$, where D_R is the R-bit representation to decimal conversion mapping: $\{0, 1\}^{RL} \to \{0, 1, 2, \dots, 2^{RL} - 1\}$.

Decoder (M_5): The decoder receives L bits of syndrome s, and L samples of the process Y and searches through the list of codeword sequences in a given coset C(s) for the most likely codeword sequence, given

Figure 2.38 Trellis section for the trellis code built on an alphabet of size 8. The number of paths emanating from any state is four. (a) Principal trellis coset. (b) Complementary trellis coset.

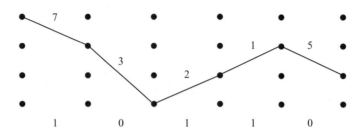

Figure 2.39   An example of computation of the syndrome $((R_s, R_c) = (3, 2)$ bits per sample): let the outcome of quantization be 7, 3, 2, 1, 5. The syndrome is given by 10110 for five samples. Sample numbers 0, 2, and 3 use complementary trellis coset and the rest use principal trellis coset.

the side-information sequence y. Consider the kth stage of the four-state trellis as shown in Figure 2.38(a). This is the trellis for the coset of C with the all-zero syndrome. If the kth bit of the syndrome sequence is 1 rather than 0, we need to modify the labels on each edge of the principal trellis coset at the kth stage. Let $a = [0|0|s^T]$, where $a_k = [0|0|s_k]^T$ for all $1 \leq k \leq L$, and $\mathbf{0}$ is the all-zero sequence. As discussed earlier, for the trellis code under consideration, the sequence $Q[[0|0|s^T]$ belongs to the coset whose syndrome is s. Thus, at the kth stage of decoding, if the kth bit of s is 1, we need to shift from the principal trellis coset to the other trellis coset (there are only two trellis cosets in the given example: see Figure 2.38(b)).

This is done by defining the trellis coset at the kth stage as having the edge labels $\xi = Q[Q^{-1}(r_l) \oplus a_k]$, where $r_l \in \nabla$ is the corresponding label in the principal trellis coset $a_k = [0|0|s_k]^T$ (this is possible due to the systematic trellis codes). The set of labels on the edges starting from any given state in the two trellis cosets forms a partition of ∇. For example, in the first state, the edge labels of the principal trellis coset $\{r_0, r_2, r_4, r_6\}$, and that of the complementary trellis coset $\{r_1, r_3, r_5, r_7\}$ form a partition of ∇. Thus, starting from $k = 1$ to $k = L$, at every stage we need to keep relabeling (shifting between the principal trellis coset and the other trellis coset) the edges in the trellis used in the Viterbi decoder. Let x' be the sequence that is closest to y obtained using this algorithm (characterizing the mapping M_5) with relabeled edges. This is illustrated in Figure 2.39 with an example. For a general R, there will be 2^R trellis cosets. It can be noted that the minimum distance between any two words in every coset of C has been increased from that in the memoryless coset construction.

After recovering the active codeword, the optimal estimate is given as follows:

$$\hat{x} = M_3(y, \theta) = \arg \min_{a \in R^L} \left[E \left\{ \rho(X, a) \middle| \begin{matrix} Y = y \\ X \in \Gamma \end{matrix} \right\} \right] \tag{2.43}$$

where θ is the index of the codeword x' in S, and Γ is the sequence of intervals associated with the elements of x'. Since S is memoryless, we can simplify the preceding expression as

$$\hat{x}_k = \arg \min_{a \in R} E \left\{ \rho(X_k, a) \middle| \begin{matrix} Y_k = y_k \\ X_k \in \Gamma_i \end{matrix} \right\} \tag{2.44}$$

where $\hat{x}_k$ is the kth sample of $\hat{x}$, for $1 \leq k \leq L$, and $x'_k = r_i \in \nabla$ (the estimation assumes correct decoding).

2.5.3   *Trellis-Based Quantization and Memoryless Coset Construction*

We now use a more efficient quantizers such as TCSQ as source codebook, and construct cosets on this space. Consider a TCSQ with rate R_s bits per sample built on a scalar quantizer alphabet ∇ of size 2^{R_s+1}. This has a finite-state machine with a rate-$R_s/(R_s + 1)$ convolutional code and a mapping

$Q : \{0, 1\}^{Rs+1} \rightarrow \nabla$. The set of 2^{LRs} 'valid' sequences of this TCSQ constitutes the source codebook S. The source is quantized using S by applying the Viterbi algorithm.

Source coding (M_1, M_2): M_1 corresponds to this trellis coded quantization, and M_2 corresponds to the rule of assigning the sequences in S to their corresponding regions. In example with $(R_s, R_c) = (2, 1)$ bits per sample and the trellis as shown in Figure 2.38(a) we consider a TCSQ with parallel transitions, and assign all the label points on a parallel transition to the same coset of a sequence set C, resulting in a reduction in rate by 1 bit/sample. In other words, we partition the scalar quantizer ∇ into four sets: $\{r_0, r_4\}$, $\{r_1, r_5\}$, $\{r_2, r_6\}$, and $\{r_3, r_7\}$, and the encoder does not expend bits to differentiate between the elements of a set (i.e., labels among parallel transitions between any two connected states in the trellis). Thus, the channel code is given by $C = \{r_0, r_4\}^L \subset S$. This amounts to discarding the uncoded bit in the 2-bit representation of the TCSQ codewords and sending only the coded bit to the decoder.

The set of codewords in the TCSQ can be viewed as a coset code [101]. ∇ is partitioned into four cosets of $\{r_0, r_4\}$, and a bit sequence set is used to index the valid set of L-dimensional coset sequences.

The extension of the above techniques to the case of trellis based quantization and trellis based coset construction is straightforward.

2.5.4 Performance Examples

The following model for X and Y: $Y = X + N$ is used in simulation, where X is i.i.d. Gaussian with zero mean and unit variance and N is i.i.d. Gaussian with zero mean and independent of X. For memoryless source and channel codes 4-, 8-, and 16-level scalar quantizers are used, each partitioned into two cosets, with each coset containing two, four, and eight codewords, respectively. Distortion during correct decoding only is plotted versus correlation-SNR (which is the ratio of the variance of X and N) for these three schemes in Figure 2.40(a). Figure 2.40(b) shows the probability of decoding error (P_e) for the same system. These are the results of Monte Carlo simulations. As can be noted, there is a tradeoff between

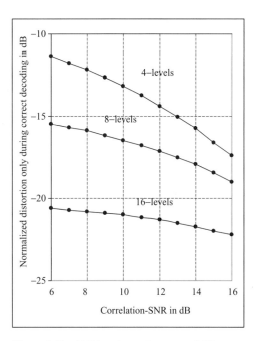
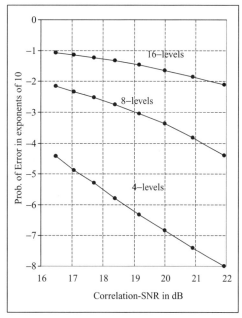

Figure 2.40 (a) Distortion performance of different quantizers during correct decoding for a Gaussian source. (b) Probability of error for $R = 1$ bit/sample for memoryless quantization and memoryless coset construction.

the distortion and probability of decoding error. For a given correlation-SNR, as the number of levels in the quantizer is increased, the distortion decreases and the probability of decoding error increases.

For the trellis-based coset construction we use 4- and 8-state trellis built on a 4-level scalar quantizer as shown in Figure 2.41(a). These trellises are designed based on the set partitioning rules of [100] and [101]. Figure 2.42(a) gives the probability of decoding error versus correlation-SNR. Note that by using trellis-based cosets, we get gains of around 3–4 dB in correlation-SNR over memoryless coset construction. Thus, without increasing the rate, at $P_e \leq 10^{-4}$, we can operate at correlation-SNRs no less than 12 dB (see Figure 2.42(a)) compared with 15.5 dB (see Figure 2.40 (b)). Similar coset constructions are done on an 8-level quantizer using 4-, 8-state trellises for $R = 1$ bit/sample as shown Figure 2.41(b). Figure 2.42(b) gives the performance in terms of P_e. Here again we get 3-dB gain over memoryless coset construction with four- and eight-state trellis at $P_e \leq 10^{-4}$.

We now construct trellis-based quantizers and coset construction. We construct such TCSQs and coset partitions on 8- and 16-level quantizers. The trellises for the source and the channel codes for the 8-level quantizer are shown in Figure 2.43. Recall that the encoder uses the trellis of Figure 2.43(a) in quantization, while the receiver uses the trellis of Figure 2.43 (b) in decoding the side information. The

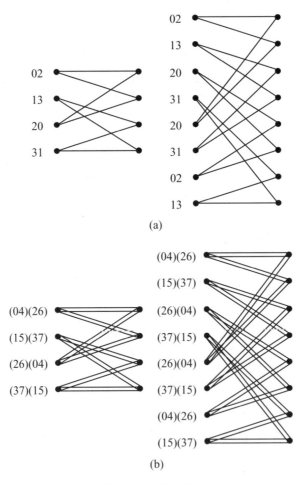

(a)

(b)

Figure 2.41   Trellises for the channel code C for $R = 1$ bit/sample using memoryless source codes (a) for 4-level quantizer; (b) for 8-level quantizer. Both (a) and (b) have the same structure for a given number of states.

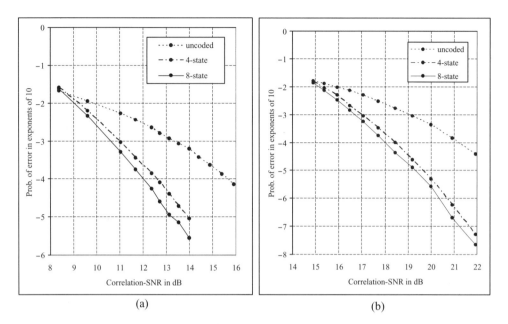

Figure 2.42 Probability of error for $R = 1$ bit/sample for memoryless quantization and trellis-based coset construction: (a) $R_s = 2$ bits/sample, S is 4-level quantizer; (b) $R_s = 3$ bits/sample, S is 8-level quantizer.

probability of decoding error as a function of correlation-SNR is plotted in Figure 2.44 with the number states ranging from 2 to 32. The performance of these systems is better than that of the corresponding constructions with memoryless cosets. Using these methods we can now operate at correlation-SNRs as low as 8.5 dB with 32 states $P_e \leq 10^{-4}$.

In order to analyze the system behavior for higher rates of transmission let us consider the constructions of memoryless source codes and trellis based coset construction for $R = 2$ bits/sample. For a trellis code (characterizing C with $R_c = 1$) built on an 8-level quantizer (characterizing S with $R_s = 3$), we need a rate-1/3 convolutional code, and a mapping rule Q to maximize the coding gain. In principle, for this trellis code, the set partition rule works by increasing the alphabet size four times, as compared to twice in most of the trellis designs. An alternative is to choose a subset (containing four elements) of the available alphabet (eight levels), and use only them in one coset. We partition the 8-level quantizer into two cosets (as in memoryless coset construction for $R = 1$), and construct trellis codes on each of these memoryless cosets. Let Coset-1 and Coset-2 specify the codeword sets $\{0, 2, 4, 6\}$ and $\{1, 3, 5, 7\}$, respectively. A rate-1/2 convolutional code with set partitioning on Coset-1 can be used to get trellis codes with a 4-state trellis as shown in Figure 2.45(a). The trellis uses an alphabet of size $2^{R_c + 2}$. There

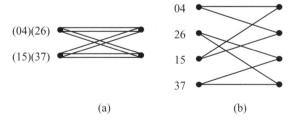

Figure 2.43 Code structure for $R = 1$ bit/sample on an 8-level quantizer. (a) Source code S: rate-2 trellis code. (b) Channel code C: rate-1 trellis code.

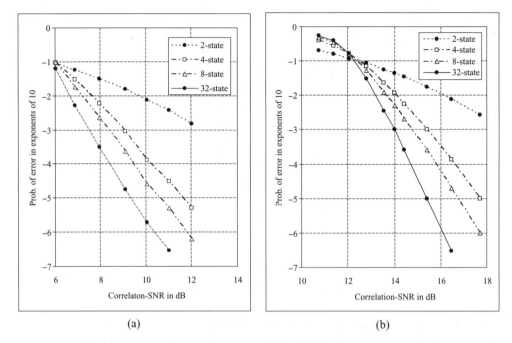

Figure 2.44 Probability of error for $R = 1$ bit/sample, trellis-based quantization and coset construction. (a) $R_s = 2$ bits/sample, S is TCSQ based on the 8-level quantizer. (b) $R_s = 3$ bits/sample, S is TCSQ based on the 16-level quantizer.

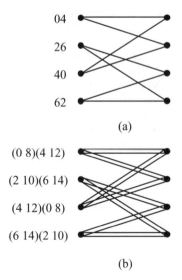

Figure 2.45 Codes obtained using the same convolutional code for memoryless quantization and trellis-based coset construction for $R = 2$ bits/sample. (a) 4-state trellis (of C) on the 8-level quantizer. (b) 4-state trellis (of C) on the 16-level quantizer.

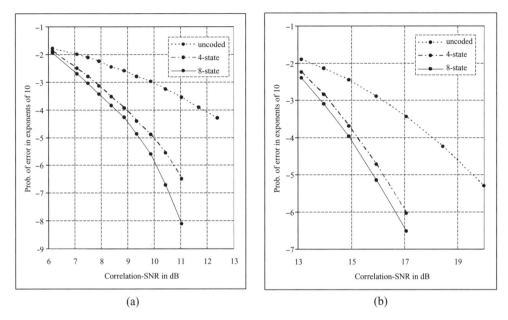

Figure 2.46   Probability of error for $R = 2$ bits/sample, memoryless quantization, and trellis-based coset construction. (a) $R_s = 3$ bits/sample, S is 8-level quantizer. (b) $R_s = 4$ bits/sample, S is 16-level quantizer.

are four trellis cosets, with two of them built on Coset-1 and two built on Coset-2. The index of one of the four trellis cosets containing the quantized source sample is sent to the decoder. Figure 2.46(a) shows the decoding error performance of this construction. Similarly to the case of $R = 1$, we get 3–4-dB gains over memoryless coset construction when $P_e \leq 10^{-4}$. As R is increased from 1 to 2 bits/sample, for the same trellis complexity and scalar quantizer, the probability of error is decreased, and the correlation-SNR that can be supported is decreased from 18 to 8.5 dB.

Similar codes are constructed on the 16-level quantizer using a rate-2/3 (instead of rate-2/4) convolutional code with memoryless partitioning of the alphabet into two cosets as shown in Figure 2.45(b). The probability of decoding error is shown in Figure 2.46(b). The operating range of the 16-level quantizer has been increased (from a correlation-SNR of ≥ 24 dB when $R = 1$ bit/sample) to ≥ 15 dB when $R = 2$ bits/sample corresponding to $P_e \leq 10^{-4}$.

Appendix 2.1 Maximum *a Posteriori* Detection

The BCJR Algorithm

The maximum likelihood decoder defined by Equation (2.20) minimizes the probability that the whole detected sequence is in error. In an alternative concept presented in this appendix we will be interested in minimizing the symbol error probability. The starting point is the average symbol *a posteriori* probability [18, 4–17]:

$$APP = E\{p(x_k/y)\} \tag{A2.1}$$

that should be maximized. In other words, the detector will decide in favor of the symbol for which the probability of correct detection is maximized. For simplicity **y** will now be replaced by y. For binary transmission, this means finding out which one of the two probabilities $P(x_k = 0 \mid y)$ and $P(x_k = 1 \mid y)$ is larger. For this we have to compare:

$$\Lambda_k = \frac{P(x_k = 1 \mid y)}{P(x_k = 0 \mid y)}$$

with a unit threshold. The transmitted symbol x_k is associated with one or more branches of the trellis stage at time k, and each one of these branches can be characterized by the pair of states, say (δ_k, δ_{k+1}), that it joins. Thus, we can write:

$$\Lambda_k = \frac{\displaystyle\sum_{(\delta_k,\delta_{k+1}):x_k=1} P(y, \delta_k, \delta_{k+1})}{\displaystyle\sum_{(\delta_k,\delta_{k+1}):x_k=0} P(y, \delta_k, \delta_{k+1})}$$

where the two summations are over those pairs of states for which $x_k = 1$ and $x_k = 0$, respectively, and the conditional probabilities from the previous equation are replaced by joint probabilities after using the Bayesian rule and canceling out the pdf of y, common to numerator and denominator.

Next, we need to compute the pdf $P(y, \delta_k, \delta_{k+1})$. By defining y_k^-, as components of the received vector before time k, and y_k^+, as components of the received vector after time k, we can write:

$$y = (y_k^-, y_k, y_k^+)$$

which results in:

$$
\begin{aligned}
p(y, \delta_k, \delta_{k+1}) &= p(y_k^-, y_k, y_k^+, \delta_k, \delta_{k+1}) \\
&= p(y_k^-, y_k, \delta_k, \delta_{k+1}) p(y_k^+ \mid y_k^-, y_k, \delta_k, \delta_{k+1}) \\
&= p(y_k^-, \delta_k) p(y_k, \delta_{k+1} \mid y_k^-, \delta_k) p(y_k^+ \mid y_k^-, y_k, \delta_k, \delta_{k+1})
\end{aligned}
$$

Due to the dependences among observed variables and trellis states, reflected by the trellis structure or, equivalently, by the Markov chain property of the trellis states, y_k^+ depends on $\delta_k, \delta_{k+1}, y_k^-$, and y_k only through δ_{k+1}, and, similarly, the pair y_k, δ_{k+1} depends on δ_k, y_k^-, only through δ_k. Thus, by defining the functions:

$$
\begin{aligned}
\alpha_k(\delta_k) &\equiv p(y_k^-, \delta_k) \\
\beta_{k+1}(\delta_{k+1}) &\equiv p(y_k^+ \mid \delta_{k+1}) \\
\gamma_{k,k+1}(\delta_k, \delta_{k+1}) &\equiv p(y_k, \delta_{k+1} \mid \delta_k) = p(y_k \mid \delta_k, \delta_{k+1}) p(\delta_{k+1} \mid \delta_k)
\end{aligned}
$$

we may write

$$p(y, \delta_k, \delta_{k+1}) = \alpha_k(\delta_k) \gamma_{k,k+1}(\delta_k, \delta_{k+1}) \beta_{k+1}(\delta_{k+1})$$

So, the *a posteriori* probability ratio can be rewritten in the form:

$$\Lambda_k = \frac{\displaystyle\sum_{\delta_k,\delta_{k+1}:x_k=1} \alpha_k(\delta_k) \gamma_{k,k+1}(\delta_k, \delta_{k+1}) \beta_{k+1}(\delta_{k+1})}{\displaystyle\sum_{\delta_k,\delta_{k+1}:x_k=0} \alpha_k(\delta_k) \gamma_{k,k+1}(\delta_k, \delta_{k+1}) \beta_{k+1}(\delta_{k+1})} \qquad (A2.2)$$

Finally, we now describe how the functions $\alpha_k(\delta_k)$ and $\beta_{k+1}(\delta_{k+1})$ can be evaluated recursively. We represent the forward recursion as:

$$
\begin{aligned}
\alpha_{k+1}(\delta_{k+1}) &= p(y_{k+1}^-, \delta_{k+1}) \\
&= p(y_k^-, y_k, \delta_{k+1}) \\
&= \sum_{\delta_k} p(y_k^-, y_k, \delta_k, \delta_{k+1}) \\
&= \sum_{\delta_k} p(y_k^-, \delta_k) p(y_k, \delta_{k+1} \mid \delta_k) \\
&= \sum_{\delta_k} \alpha_k(\delta_k) \gamma_{k,k+1}(\delta_k, \delta_{k+1})
\end{aligned}
$$

with the initial condition $\alpha_0(s_1) = 1$ (s_1 denotes the initial state of the trellis) and the backward recursion as:

$$
\begin{aligned}
\beta_k(\delta_k) &= p(y_{k-1}^+ \mid \delta_k) \\
&= \sum_{\delta_{k+1}} p(y_k, y_k^+, \delta_{k+1} \mid \delta_k) \\
&= \sum_{\delta_{k+1}} p(y_k, \delta_{k+1} \mid \delta_k) p(y_k^+ \mid \delta_{k+1}) \\
&= \sum_{\delta_{k+1}} \gamma_{k,k+1}(\delta_k, \delta_{k+1}) \beta_{k+1}(\delta_{k+1})
\end{aligned}
$$

with the final value $\beta_K(s_K) = 1$. The combination of the latter two recursions with Equation (A2.2) forms the BCJR algorithm, named after the authors who first derived it, Bahl, Cocke, Jelinek and Raviv [27]. Roughly speaking, we can state that the complexity of the BCJR algorithm is about three times that of the Viterbi algorithm.

References

[1] Hocquenghem, A. (1959) Codes correcteurs d'erreurs Chiffres (Paris), **2**, 147–156.

[2] Bose, R. and Ray-Chaudhuri, D. (1960) On a class of error correcting binary group codes. *Information and Control*, **3**, 68–79.

[3] Bose, R. and Ray-Chaudhuri, D. (1960) Further results on error correcting binary group codes. *Information and Control*, **3**, 279–290.

[4] Peterson, W. (1960) Encoding and error correction procedures for the Bose–Chaudhuri codes. *IEEE Transactions On Information Theory*, **IT-6**, 459–470.

[5] Gorenstein, D. and Zierler, N. (1961) A class of cyclic linear error-correcting codes in $p > n$ symbols. *Journal of the Society of Industrial and Applied Mathematics*, **9**, 107–214.

[6] Reed, I. and Solomon, G. (1960) Polynomial codes over certain finite fields. *Journal of the Society of Industrial and Applied Mathematics*, **8**, 300–304.

[7] Berlekamp, E. (1965) On decoding binary Bose–Chaudhuri–Hocquenghem codes. *IEEE Transactions On Information Theory*, **11**, 577–579.

[8] Berlekamp, E. (1968) *Algebraic Coding Theory*. McGraw-Hill, New York, USA.

[9] Massey, J. (1965) Step-by-step decoding of the Bose–Chaudhuri–Hocquenghem codes, *IEEE Transactions On Information Theory*, **11**, 580–585.

[10] Massey, J. (1969) Shift-register synthesis and BCH decoding. *IEEE Transactions On Information Theory*, **IT-15**, 122–127.

[11] Peterson, W. and Weldon Jr, E. (1972) *Error Correcting Codes*, 2nd edition, MIT Press, Cambridge, MA, USA.

[12] Sklar, B. (1988) *Digital Communications–Fundamentals and Applications*: Prentice Hall, Englewood Cliffs, NJ, USA.

[13] Clark Jr, G. and Cain, J. (1981) *Error Correction Coding for Digital Communications*, Plenum Press, New York, USA.

[14] Blahut, R. (1983) *Theory and Practice of Error Control Codes*. Addison Wesley, Reading, MA, USA.

[15] Lin, S. and Constello Jr, D. (1982) *Error Control Coding: Fundamentals and Applications*, Prentice Hall, Englewood Cliffs, NJ, USA.

[16] Michelson, A. and Levesque, A. (1985) *Error Control Techniques for Digital Communication*, John Wiley & Sons, New York, USA.

[17] Hanzo, L. *et al.* (2002) *Turbo coding, Turbo Equalization and Space–Time Coding*. John Wiley & Sons, Inc., New York.

[18] Benedetto, S. *et al.* (1999) *Principles of Digital Transmissions With Wireless Applications*, Kluwer, New York.

[19] Elias, P. (1955) Coding for noisy channels, *IRE Convention Record*, **4**, 37–47.

[20] Wozencraft, J. (1957) Sequential decoding for reliable communication, IRE Natl. *Convention rec.*, **5**(2), 11–25.

[21] Wozencraft, J. and Reiffen, B. (1961) *Sequential Decoding*, MIT Press, Cambridge, MA, USA.

[22] Fano, R. (1963) A heuristic discussion of probabilistic coding. *IEEE Transactions On Information Theory*, **IT-9**, 64–74.

[23] Massey, J. (1963) *Threshold Decoding*, MIT Press, Cambridge, MA, USA.

[24] Viterbi, A. (1967) Error bounds for convolutional codes and an asymptotically optimum decoding algorithm. *IEEE Transactions On Information Theory*, **IT-13**, 260–269.

[25] Forney, G. (1973) The Viterbi algorithm. *Proceedings of the IEEE*, **61**, 268–278.

[26] Heller, J. and Jacobs, I. (1971) Viterbi decoding for satellite and space communication. *IEEE Transactions on Communication Technology*, **COM-19**, 835–848.

[27] Bahl, L. R., Cocke, J., Jelinek, F. and Raviv, J. (1974) Optimal Decoding of Linear Codes for Minimising Symbol Error Rate. *IEEE Transactions On Information Theory*, **20**, 284–287.

[28] Viterbi, A. J. (1965) Optimum detection and signal selection for partially coherent binary communication, *IEEE Transactions On Information Theory*, **IT-11**, 239–246.

[29] Viterbi, A. J. (1966) *Principles of Coherent Communications*. McGraw Hill, New York.

[30] Viterbi, A. J. (1967) Error bounds for convolutional codes and an asymptotically optimum decoding algorithm. *IEEE Transactions On Information Theory*, **IT-13**, 260–269.

[31] Viterbi, A. J. and Omura, J. K. (1979) *Principles of Digital Communication and Coding*. McGraw Hill, New York.

[32] Clark, G. C. and Cain, J. B. (1981) *Error Correction Coding for Digital Communications*. Plenum Press, New York.

[33] Lin, S. and Costello, D. J. (1983) *Error Control Coding: Fundamentals and Applications*. Prentice Hall, Englewood Cliffs, NJ, USA.

[34] Forney, Jr, G. D. (1966) *Concatenated Codes*. MIT Press, Cambridge, MA, USA.

[35] Forney, Jr, G. D. (1970) Convolutional Codes I: Algebraic structure. *IEEE Transactions On Information Theory*, **IT-16**, 720–738.

[36] Forney, Jr, G. D. (1973) The Viterbi algorithm. *IEEE Proceedings*, **61**, 268–278.

[37] Forney, Jr, G. D. (1974a) Convolutional Codes II: Maximum-likelihood decoding, *Information and Control*, **25**, 223–265.

[38] Forney, Jr, G. D. (1974b) Convolutional Codes III: Sequential decoding, *Information and Control*, **25**, 267–297.

[39] Wozencraft, J. M. (1957) Sequential decoding for reliable communications, *Tech. Report 325*, RLE MIT, Cambridge, MA.

[40] Fano, R. M. (1963) A heuristic discussion of probabilistic decoding. *IEEE Transactions On Information Theory*, **IT-9**, 64–74.

[41] Jelinek, F. (1969) Fast sequential decoding algorithm using a stack. *IBM Journal of Research and Development*, **13**, 675–685.

[42] Zigangirov, K. S. (1966) Some sequential decoding procedures, *Problemy Peredachi Informacii*, **2**, 13–25.

[43] Wozencraft, J. M. and Jacobs, I. M. (1965) *Principles of Communication Engineering*. John Wiley & Sons, New York.

[44] Savage, J. E. (1966) Sequential decoding. The computational problem. *Bell Systems Technical Journal*, **45**, 149–176.

[45] Anderson, J. B. and Mohan, S. (1991) *Source and Channel Coding: An Algorithmic Approach*. Kluwer Academic Press, Boston, MA, USA.

[46] Cain, J. B., Clark, G. C. and Geist, J. M. (1979) Punctured convolutional codes of rate $(n - 1)/n$ and simplified maximum likelihood decoding. *IEEE Transactions On Information Theory*, **IT-25**, 97–100.

[47] Yasuda, Y., Hirata, Y., Nakamura, K. and Otani, S. (1983) Development of variable-rate Viterbi decoder and its performance characteristics, in *Proceedings Sixth International Conference on Digital Satellite Communications*, Phoenix, AZ, September 1983, pp. XII-24–XII-31.

[48] Yasuda, Y. Kashiki, K. and Hirata, Y. (1984) High rate punctured convolutional codes for soft decision Viterbi decoding. *IEEE Transactions On Communications*, **COM-32**, 315–319.

[49] Berrou, C. and Glavieux, A. (1996) Near optimum error-correcting coding and decoding: Turbo codes. *IEEE Transactions On Communications*, **COM-44**, 1261–1271.

[50] Benedetto, S. and Montorsi, G. (1996a) Unveiling turbocodes: Some results on parallel concatenated coding schemes. *IEEE Transactions On Information Theory*, **COM-44**, 591–600.

[51] Divsalar, D. and Pollara, F. (1995) Turbo codes for deep-space communications. *TDA Progress Report 42–120*, pp. 29–39, Jet Propulsion Laboratory, Pasadena, California.

[52] Divsalar, D. and McEliece, R. J. (1996) Effective free distance of turbo codes, *Electronics Letters*, **32**(5).

[53] Perez, L. C., Seghers, J. and Costello, D. J. (1996) A distance spectrum interpretation of turbo codes. *IEEE Transactions On Information Theory*, **IT-42**, 1698–1709.

[54] Benedetto, S. and Montorsi, G. (1996b) Design of parallel concatenated convolutional codes. *IEEE Transactions On Communications*, **IT-43**, 409–428.

[55] Bahl, L. R., Cocke, J., Jelinek, F. and Raviv, J. (1974) Optimal decoding of linear codes for minimizing symbol error rate. *IEEE Transactions On Information Theory*, **IT-20**, 284–287.

[56] Benedetto, S., Divsalar, D. and Hagenauer, J. (eds). (1998d) Concatenated coding techniques and iterative decoding: Sailing toward channel capacity. *IEEE Journal on Selected Areas in Communications*, **16**(2).

[57] Benedetto, S., Divsalar, D., Montorsi, G. and Pollara, F. (1998c) Soft-input soft-output modules for the construction and distributed iterative decoding of code networks. *European Transactions on TeleCommunications*, **9**, 155–172.

[58] Benedetto, S., Divsalar, D., Montorsi, G. and Pollara, F. (1998b) Serial concatenation of interleaved codes: Performance analysis, design, and iterative decoding. *IEEE Transactions On Information Theory*, **44**, 909–926.

[59] Hagenauer, J., Offer, E. and Papke, L. (1996) Iterative decoding of binary block and convolutional codes. *IEEE Transactions On Information Theory*, **IT-42**, 429–445.

[60] Berrou, C., Glavieux, A. and Thitimajshima, P. (1993) Near Shannon limit error correcting codes: Turbo codes. *IEEE International Conference on Communications* (ICC'93), Geneva, Switzerland, pp. 1064–1070.

[61] Lin, S. and Costello, D. (1982) *Error Control Coding: Fundamentals and Applications* Prentice Hall, Englewod Cliffs, NJ.

[62] Gordon, N., Vucetic, B., Musicki, D. and Du, J. *Joint error control and speech coding for 4.8 kbps digital voice transmission over satellite mobile channels*. Technical Report, Sydney University, Sydney, Australia.

[63] Wu, K., Lin, S. and Miller, M. (1982) A hybrid ARQ scheme using multiple shortened cyclic codes, in Proceedings GLOBECOM, Miami, FL, pp. C8.61–C8.65.

[64] Chase, D. (1985) Code combining – A maximum likelihood decoding approach for combining an arbitrary number of noisy packets. *IEEE Transactions On Communications*, **COM-33**, 385–393.

[65] Sovetov, B. and Stah, V. (1982) *Design of adaptive transmission systems*, Energoizdal, Leningrad, in Russian.

[66] Sullivan, D. (1971) A generalization of Gallagher's adaptive error control scheme. *IEEE Transactions On Information Theory*, **IT-17**, 727–735.

[67] Mandelbaum, D. (1974) An adaptive-feedback coding scheme using incremental redundancy. *IEEE Transactions On Information Theory*, **IT-20**, 388–389.

[68] Vucetic, B., Drajic, D. and Perisic, D. (1988) An algorithm for adaptive error control system synthesis, in *ISIT 1985*, Brighton, England; also in *Proceedings IEE*, Part F, 85–94.

[69] Mandelbaum, D. M. (1975) On forward error correction with adaptive decoding. *IEEE Transactions On Information Theory*, **IT-21**, 230–233.

[70] Kallel, S. and Haccoun, D. (1988) Sequential decoding with ARQ code combining: A robust hybrid FEC/ARQ system. *IEEE Transactions On Communications*, **26**, 773–780.

[71] Drukarev, A. and Costello, Jr, D. J. (1983) Hybrid ARQ control using sequential decoding. *IEEE Transactions On Information Theory*, **IT-29**, 521–535.

[72] Drukarev, A. and Costello, Jr, D. J. (1982) A comparison of block and convolutional codes in ARQ error control schemes. *IEEE Transactions On Communications*, **COM-30**, 2449–2455.

[73] Lugand, L. and Costello, Jr, D. J. (1982) A comparison of three hybrid ARQ schemes on a non-stationary channel, in *Proceedings GLOBECOM*, Miami, FL, pp. C8.4.1–C8.4.5.

[74] Hagenauer, J. and Lutz, E. (1987) Forward error correction coding for fading compensation in mobile satellite channels. *IEEE Journal on Selected Areas in Communications*, **SAC-5**, 215–225.

[75] Hagenauer, J. (1988) Rate-compatible punctured convolutional codes (RCPC codes) and their applications. *IEEE Transactions On Communications*, **36**, 389–400.

[76] Kahn, J. M., Katz, R. H. and Pister, K. S. J. (1999) Mobile networking for smart dust, in *Proceedings ACM/IEEE International Conference on Mobile Computing and Networking*, Seattle, WA Aug. 1990.

[77] Cover, T. M. and Thomas, J. A. (1991) *Elements of Information Theory*. Wiley, New York.

[78] Slepian, D. and Wolf, J. K. (1973) Noiseless coding of correlated information sources, *IEEE Transactions On Information Theory*, **19**, 471–480.

[79] Verdú, S. (1998) Fifty years of Shannon theory, *IEEE Transactions On Information Theory*, **44**, 2057–2078.

[80] Wyner, A. D. (1974) Recent results in the Shannon theory, *IEEE Transactions On Information Theory*, **20**, 2–10.

[81] Wyner, A. D. (1975) On source coding with side information at the decoder, *IEEE Transactions On Information Theory*, **21**, 294–300.

[82] Wyner, A. D. and Ziv, J. (1976) The rate-distortion function for source coding with side information at the decoder, *IEEE Transactions On Information Theory*, **22**, 1–10.

[83] Wyner, A. D. (1978) The rate-distortion function for source coding with side information at the decoder-II. General sources, *Information Contr.*, **38**, 60–80.

[84] Cover, T. M. (1975) A proof of data compression theorem of Slepian and Wolf for ergodic sources, *IEEE Transactions On Information Theory*, **21**, 226–228.

[85] Ahlswede, R. F. and Korner, J. (1975) Source coding with side information and a converse for degraded broadcast channels, *IEEE Transactions On Information Theory*, **21**, 629–637.

[86] Kaspi, A. H. and Berger, T. (1982) Rate-distortion for correlated sources with partially separated encoders, *IEEE Transactions On Information Theory*, **28**, 828–840.

[87] Berger, T. (1977) Multiterminal source coding, in *Information Theory Approach to Communications (CISM Courses and Lectures)*, Vol. **229**, G. Longo (Ed.), Springer-Verlag, Vienna/New York.

[88] Han, T. S. and Kobayashi, K. (1980) A unified achievable rate region for a general class of multiterminal source coding systems, *IEEE Transactions On Information Theory*, **26**, 277–288.

[89] Csiszár, I. and Körner, J. (1980) Toward a general theory of source networks, *IEEE Transactions On Information Theory*, **26**, 155–165.

[90] Zamir, R. and Shamai, S. (1998) *Nested linear/lattice codes for Wyner-Ziv encoding*, presented at the IEEE Information Theory Workshop, Killarney, Ireland.

[91] Flynn, T. J. and Gray, R. M. (1987) Encoding of correlated observations, *IEEE Transactions On Information Theory*, **33**, 773–787.

[92] Gersho, A. and Gray, R. M. (1992) *Vector Quantization and Signal Compression*. Kluwer, Norwell, MA.

[93] Shamai (Shitz), S., Verdú, S. and Zamir, R. (1998) Systematic lossy source/channel coding, *IEEE Transactions On Information Theory*, **44**, 564–579.

[94] MacWilliams, F. J. and Sloane, N. J. A. (1977) *The Theory of Error-Correcting Codes*. Van Nostrand, Amsterdam.

[95] Conway, J. H. and Sloane, N. J. A. (1988) *Sphere Packings, Lattices and Groups*. Springer-Verlag, New York.

[96] Forney, G. D. (1973) The viterbi algorithm, *Proceedings of IEEE*, **61**, 268–278.

[97] Zamir, R. and Feder, M. (1996) On lattice quantization noise, *IEEE Transactions On Information Theory*, **42**, 1152–1159.

[98] Proakis, J. G. (1995) *Digital Communications*. McGraw-Hill, New York.

[99] Marcellin, M. W. and Fischer, T. R. (1990) Trellis coded quantization of mem-oryless and Gauss-Markov sources, *IEEE Transactions on Communications*, **38**, 82–93.

[100] Ungerboeck, G. (1982) Channel coding with multilevel/phase signals, *IEEE Transactions On Information Theory*, **28**, 55–67.

[101] Forney, G. D. (1988) Coset codes, Part 1: Introduction and geometrical classification, *IEEE Transactions On Information Theory*, **34**, 1123–1151.

[102] Armstrong, M. A. (1988) *Groups and Symmetry*. Springer-Verlag, New York.

[103] Forney, G. D. (1991) Geometrically uniform codes, *IEEE Transactions On Information Theory*, **37**, 1241–1260.

[104] Wachsmann, U., Fischer, R. F. H. and Huber, J. B. (1999) Multilevel codes: Theoretical concepts and practical design rules, *IEEE Transactions On Information Theory*, **45**, 1361–1391.

[105] Garello, R. and Benedetto, S. (1995) Multilevel construction of block and trellis group codes, *IEEE Transactions On Information Theory*, **41**, 1257–1264.

3

Adaptive and Reconfigurable Modulation

3.1 Coded Modulation

In general we can use an $M = 2^b$ point constellation to transmit b bits of information. An example for $b = 2$ is shown in Figure 3.1(a). For this example the output symbol rate is $R_s = R_b/2$. If we use coding, for example a rate 2/3 convolutional encoder, and the same constellation as shown in Figure 3.1(b), the output symbol rate and the bandwidth required will be now higher, $R_s = (3/4)R_b$.

The third option is shown in Figure 3.1(c). Instead of 4PSK, 8PSK (8 points constellation) is used to transmit the encoded bits and the output symbol rate now remains the same. Because there are only $2^2 = 4$ possible code words and $2^3 = 8$ available constellation points, a proper choice of constellation points used in adjacent symbol intervals provides a way to encode the signal. This subset of signal trajectories, generated in K symbol intervals will again be referred to as a trellis in Euclidean space and the modulation is referred to as *Trellis Coded Modulation* (TCM) [1–38].

The above example illustrates a need to further elaborate the efficiency of reconfiguration in such a way as to explicitly incorporate constraints imposed by the limited available bandwidth. For these purposes, let us represent the coding gain g_{12} as the gain in energy per bit per noise density:

$$g_{12} = \Delta E / N_0 = \Delta (PT) / N_0$$

If there is no bandwidth limitation, the coding gain may be used in a number of ways.

1. Operate with reduced power and save the battery life.
2. Keep the same transmit power and data rate and increase the coverage of the network.
3. Increase bit rate (reduce the bit interval).

If the bandwidth is fixed and the coding gain is not available, we may have to reduce the data rate in order to maintain the required E_b/N_0 for a specified QoS. This suggests that the reconfiguration efficiency, defined by Equation (2.18), be further modified as follows:

$$E_{ff} = \frac{10^{g_{12}(\text{dB})/10} b_r}{D_r} = \left(\frac{k_{02}}{k_{01}} \right) \frac{10^{g_{12}(\text{dB})/10}}{D_r} \tag{3.1}$$

where $b_r = k_{02}/k_{01}$ is the relative change in the number of bits per symbol for the same symbol period T. One should notice, that in Chapter 2, coding gain was defined by taking into account this effect through

Advanced Wireless Communications & Internet: Future Evolving Technologies, Third Edition. Savo Glisic.
© 2011 John Wiley & Sons, Ltd. Published 2011 by John Wiley & Sons, Ltd.

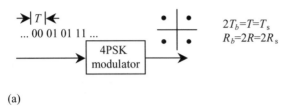

(a)

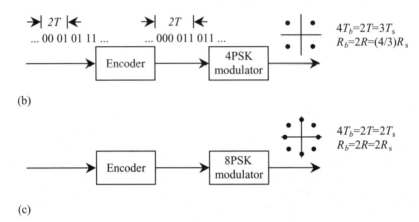

(b)

(c)

Figure 3.1   Three digital communication schemes transmitting two bits every T seconds. (a) Uncoded transmission with 4PSK; (b) 4PSK with a rate 2/3 encoder and bandwidth expansion; (c) 8PSK with a rate 2/3 encoder and no bandwidth expansion.

the coding rate R_c. In the next chapter, Equation (3.1) will be further modified to replace b_r by the relative change in the system capacity. This way we remove the portion of the gain due to bit rate reduction and take into account only those contributions due to increased efficiency of the decoding algorithm.

3.1.1   Euclidean Distance

The demodulator will decide in favor of the trajectory in the trellis which is closest to the received signal (*minimum* distance from the received trajectory). This is characterized by the Euclidean distance, defined as:

$$\delta^2 = \sum_{i=0}^{K-1} |\mathbf{r}_i - \mathbf{x}_i|^2 \tag{3.2}$$

where $\mathbf{r}_i$ is the received signal and $\mathbf{x}_i$ the possible transmitted signal. In other words, the Euclidean distance is minimized by taking $\mathbf{x}_i = \hat{\mathbf{x}}_i$, $i = 0, \ldots, K - 1$, if the received sequence is closer to $\hat{\mathbf{x}}_i, \ldots, \hat{\mathbf{x}}_{K-1}$ than to any other allowable signal sequence.

By increasing the constellation size M' to $M > M'$, and selecting M'^K sequences as a subset of S^K, we can have sequences which are less tightly packed and hence increase the minimum distance among them. We obtain a minimum distance, δ_{free}, between any two sequences, which turns out to be greater than the minimum distance, δ_{min}, between signals in S'. Hence, use of maximum likelihood sequence detection will yield a 'distance gain' of a factor of $\delta_{\text{free}}^2 / \delta_{\text{min}}^2$.

The free distance of a TCM scheme is the minimum Euclidean distance between two paths forming an error event.

3.1.2 Examples of TCM Schemes

In order to analyze Equation (3.1) in more detail, let us assume transmission with two bits per symbol. For such transmission a 4PSK modulation ($M' = 4$) would be enough. We can expand the constellation to $M = 8$, as shown in Figure 3.2, and use the trellis with $S = 2$ states, as shown in Figure 3.3.

The asymptotic coding gain of a TCM scheme is defined as:

$$\gamma = \frac{\delta_{free}^2(M)/\varepsilon}{\delta_{min}^2(M')/\varepsilon'}$$

For PSK signals, $M' = 4$ and $\delta_{min}^2/\varepsilon' = 2$. For a TCM scheme based on the 8PSK constellation whose signals we label $\{0, 1, 2, \ldots, 7\}$, as shown in Figure 3.2, we have:

$$\varepsilon' = \frac{\delta'^2}{4\sin^2\pi/8}$$

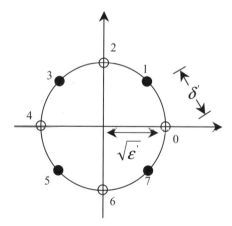

Figure 3.2 $M = 8$ point constellation.

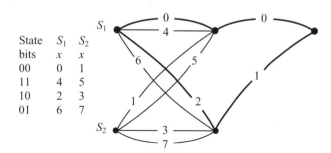

State	S_1	S_2
bits	x	x
00	0	1
11	4	5
10	2	3
01	6	7

Figure 3.3   A TCM scheme based on a two-state trellis, $M' = 4$ and $M = 8$.

3.1.2.1   Two-State Trellis

If the encoder is in state S_1, the subconstellation $\{0, 2, 4, 6\}$ is used. In state S_2, constellation $\{1, 3, 5, 7\}$ is used instead, as shown in Figure 3.3. The free distance of this TCM scheme is the smallest among the distances between signals associated with parallel transitions (error events of length 1) and the distances associated with a pair of paths in the trellis that originate from a common node and merge into a single node at a later time (error events of length greater than 1). The pair of paths yielding the free distance is shown in Figure 3.3. With $\delta(i, j)$ denoting the Euclidean distance between signals i and j, we have the following:

$$\frac{\delta_{\text{free}}^2}{\varepsilon} = \frac{1}{\varepsilon}[\delta^2(0, 2) + \delta^2(0, 1)] = 2 + 4\sin^2\frac{\pi}{8} = 2.586$$

The asymptotic coding gain over 4PSK is:

$$\gamma = \frac{2.586}{2} = 1.293 \Rightarrow 1.1 \text{ dB} \tag{3.3}$$

In this case, $\Delta k_0 = 0$ and $\Delta v = 1$ so that reconfiguration efficiency, defined by Equation (3.1), gives:

$$E_{ff} = \left(\frac{k_{02}}{k_{01}}\right)^2 \gamma 2^{-\Delta(k_0 + v)} = 1.293/2 = 0.646 \tag{3.4}$$

For a given symbol error probability, the bit error probability (BER) will depend on the mapping of the source bits onto the signals in the modulator's constellation (see Figure 3.3). To minimize the BER, this mapping should be chosen in such a way that, whenever a symbol error occurs, the signal erroneously chosen by the demodulator differs from the transmitted one by the least number of bits. For high signal to noise ratios, most of the errors occur by mistaking a signal for one of its nearest neighbors. So, a reasonable choice is a mapping where neighboring signal points in the constellation correspond to binary sequences that differ in only one digit. This is called *Gray mapping*. In this case, the bit and symbol error probabilities are related as $P_s(e)/b = P_b(e)$. For the evaluation of the symbol error probability the reader is referred to the classical references [39–42].

3.1.2.2   Four-State Trellis

In this case the trellis is as given in Figure 3.4.

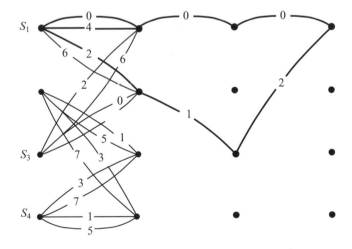

Figure 3.4   A TCM scheme based on a four-state trellis, $M' = 4$ and $M = 8$.

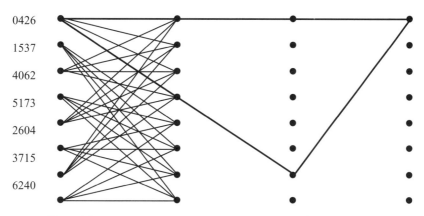

Figure 3.5   A TCM scheme based on an eight-state trellis, $M' = 4$ and $M = 8$.

We associate the constellation $\{0, 2, 4, 6\}$ with states S_1 and S_3, and $\{1, 3, 5, 7\}$ with S_2 and S_4. In this case, the error event leading to δ_{free} has length 1 (a parallel transition).

$$\frac{\delta_{\text{free}}^2}{\varepsilon} = \delta^2(0, 4) = 4$$

$$\gamma = \frac{4}{2} = 2 \Rightarrow 3 \text{ dB} \tag{3.5}$$

In this case, $\Delta v = 2$ and Equation (3.4) gives $E_{ff} = 2/4 = 0.5$, which is lower than Equation (3.4). This means that effort invested is larger than the gain obtained.

3.1.2.3   Eight-State Trellis

For the case of eight states, the trellis is shown in Figure 3.5. The four symbols associated with the branches emanating from each node are used as node labels. The first symbol in each node label is associated with the uppermost transition from the node, the second symbol with the transition immediately below it, etc.

The coding gain is calculated as:

$$\frac{\delta_{\text{free}}^2}{\varepsilon} = \frac{1}{\varepsilon}[\delta^2(0, 6) + \delta^2(0, 7) + \delta^2(0, 6)] = 2 + 4\sin^2\frac{\pi}{8} + 2 = 4.586$$

$$\gamma = \frac{4.586}{2} = 2.293 \Rightarrow 3.6 \text{ dB} \tag{3.6}$$

In this case, $\Delta v = 3$ and Equation (3.4) gives $E_{ff} = 2.293/8 = 0.286$, which is lower than in the previous case of the four-state trellis. This means again that effort invested is larger than the gain obtained.

3.1.2.4   QAM 3 bits per Symbol

In this case, the trellis is as given in Figure 3.6 and the signal constellation as in Figure 3.7.

In this case, we have two subsets of points $\{0, 2, 5, 7, 8, 10, 13, 15\}$ and $\{1, 3, 4, 6, 9, 11, 12, 14\}$. For the basic 8QPSK constellation we have $\delta_{\text{min}}^2/\varepsilon' = 0.8$, and coding gain can be represented as:

$$\frac{\delta_{\text{free}}^2}{\varepsilon} = \frac{1}{\varepsilon}[\delta^2(10, 13) + \delta^2(0, 1) + \delta^2(0, 5)] = \frac{1}{\varepsilon}[0.8\varepsilon + 0.4\varepsilon + 0.8\varepsilon] = 2$$

$$\gamma = \frac{2}{0.8} = 2.5 \Rightarrow 3.98 \text{ dB} \tag{3.7}$$

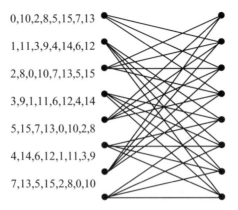

Figure 3.6   A TCM scheme based on an eight-state trellis, $M' = 8$ and $M = 16$.

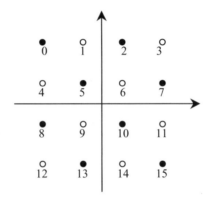

Figure 3.7   The 8 QAM constellation $\{0, 2, 5, 7, 8, 10, 13, 15\}$ and the 16 QAM constellation $\{0, 1, \ldots, 15.\}$.

In this case, $\Delta k_0 = 0$ and $\Delta v = 3$ so that reconfiguration efficiency, defined by Equation (3.1), gives:

$$E_{ff} = 2.5 \times 2^{-(0+3)} = 0.312 \tag{3.8}$$

which is higher than in the previous example. Additional results for free distances for a number of different schemes are given in Figure 3.8. These results can be used directly to evaluate the reconfiguration efficiency.

3.1.3   Set Partitioning

The M-ary constellation is successively partitioned into 2, 4, 8, ..., subsets with size $M/2$, $M/4$, $M/8, \ldots$, having progressively larger minimum Euclidean distances $\delta_{min}^{(1)}$, $\delta_{min}^{(2)}$, $\delta_{min}^{(3)}, \ldots$ as shown in Figure 3.9 and Figure 3.10.

Then, in accordance with Ungerboeck's rules, the following steps are taken:

1. Members of the same partition with the largest distance are assigned to parallel transitions.
2. Members of the next largest partition are assigned to 'adjacent' transitions, i.e. transitions stemming from, or merging into, the same node.

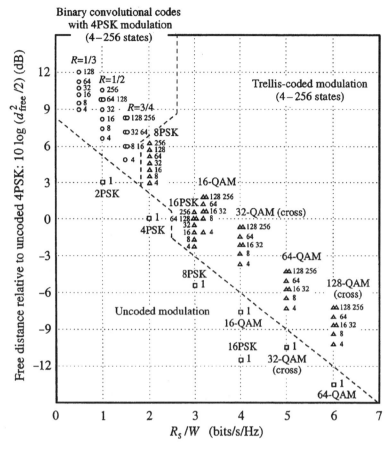

Figure 3.8 Free distance versus bandwidth efficiency of selected TCM schemes based on two-dimensional modulations. (Adapted from [2]) PSK and QAM © 1987, IEEE.

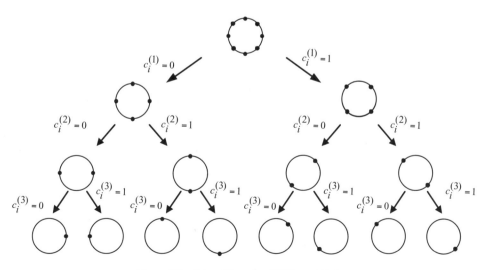

Figure 3.9 Set partition of an 8PSK constellation.

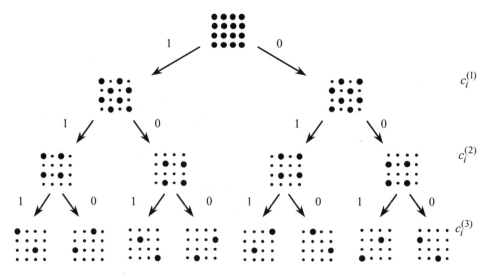

Figure 3.10   Set partition of a 16 QAM constellation.

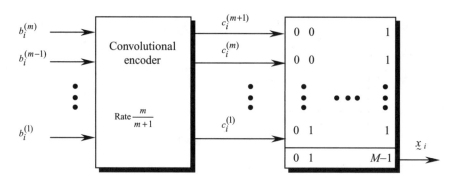

Figure 3.11   Representation of TCM.

3.1.4   Representation of TCM

A TCM encoder can be represented as a convolutional encoder encoding a block of m input bits $\mathbf{b}_i = (b_i^{(1)}, b_i^{(2)}, b_i^{(3)}, \dots, b_i^{(m)})$ into a block of $m + 1$ output bits $\mathbf{c}_i = (c_i^{(1)}, c_i^{(2)}, c_i^{(3)}, \dots, c_i^{(m)})$, followed by a memoryless mapper into points of the constellation of size $M = 2^{m+1}$ (Figure 3.11). In the case when there are parallel transitions, not all bits are encoded, which is represented explicitly in Figure 3.12.

Uncoded digits cause parallel transitions; a branch in the trellis diagram of the code is now associated with $2^{m-\tilde{m}}$ signals. An example for $m = 2$ and $\tilde{m} = 1$ is shown in Figure 3.13. The trellis nodes are connected by parallel transitions associated with two signals each. The trellis has four states, as does the rate 1/2 convolutional encoder, and its structure is determined by the latter.

3.1.5   TCM with Multidimensional Constellation

In general, we can use m channels for transmission and generate an m-dimensional trellis for the overall signal representation.

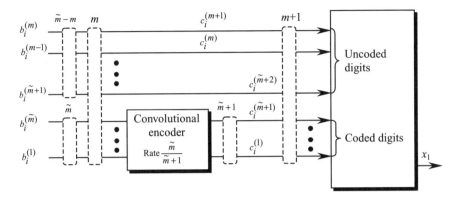

Figure 3.12   A TCM encoder where the bits that are left uncoded are shown explicitly.

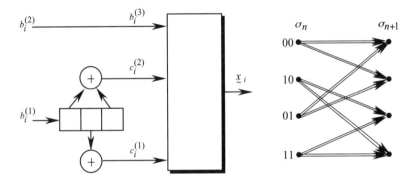

Figure 3.13   A TCM encoder with $m = 2$, $\tilde{m} = 1$ and the corresponding trellis.

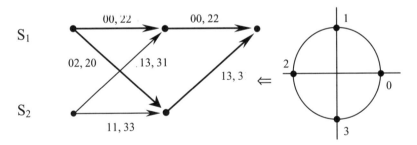

Figure 3.14   A two-state TCM scheme based on a $2 \times$ 4PSK constellation. The error event providing the free Euclidean distance is also shown.

As an example of a TCM scheme based on multidimensional signals, consider the four-dimensional constellation obtained by pairing $(m = 2)$ 4PSK signals. This is denoted as $2 \times$ 4PSK. With the signal labeling of Figure 3.14, the $4^2 = 16$ four-dimensional signals are:

$$\{00, 01, 02, 03, 10, 11, 12, 13, 20, 21, 22, 23, 30, 31, 32, 33\}$$

This constellation achieves the same minimum squared distance as two-dimensional 4PSK,

$$\delta^2_{\min} = \delta^2(00, 01) = \delta^2(0, 1) = 2$$

The following subconstellation has eight signals and a minimum squared distance four:

$$S = \{00, 02, 11, 13, 20, 22, 31, 33\}$$

With S partitioned into the four subsets:

$$\{00, 22\} \quad \{20, 02\} \quad \{13, 31\} \quad \{11, 33\}$$

the choice of a two-state trellis provides the TCM scheme shown in Figure 3.14. This has a squared free distance of 8.

If m channels are used independently, the overall data rate would be $k_{01} = mk_0$ and the complexity would be m times the complexity of the demodulation per trellis. So the normalized complexity per bit would be:

$$D_1 = \frac{m2^{k_0+v}}{mk_0} = \frac{2^{k_0+v}}{k_0}$$

For m dimensional trellises, only $k_{02} = k_0$ bits are transmitted:

$$D_2 = \frac{m2^{k_0+v}}{k_0} \quad \text{and} \quad D_r = \frac{D_2}{D_1} = m$$

Now we have:

$$E_{ff} = g_{12}\left(\frac{k_0}{mk_0}\right)\left(\frac{1}{m}\right) = \frac{g_{12}}{m^2} \tag{3.9}$$

For the previous example, $g_{12} = \gamma = 8/2 = 4$ and $m^2 = 4$ so that $E_{ff} = 1$.

For this reason, in the next chapter we will discuss multidimensional constellations obtained by using multiple antennas.

3.2   Adaptive Coded Modulation for Fading Channels

In this section we describe the system which uses reconfiguration to improve performance in time varying fading channels [43–66]. Basically, for a better signal to noise ratio $\hat{\gamma}$, estimated at the receiver side, the higher constellation M is used at the receiver, as shown in Figure 3.15.

Let $\bar{\tau}_j$ be the average time that the adaptive modulation scheme continuously uses the constellation M_j. Since the constellation size is adapted to an estimate of the channel fade level (instantaneous signal to noise ratio), several symbol times may be required to obtain a good estimate. In addition, hardware and pulse shaping considerations generally dictate that the constellation size must remain constant over tens to hundreds of symbols. This results in the requirement that $\bar{\tau}_j \gg T \; \forall j$, where T is the symbol time.

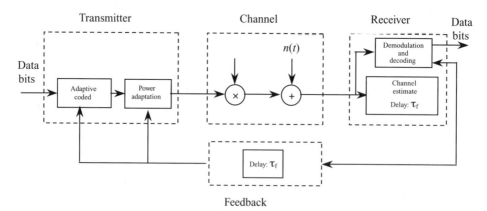

Figure 3.15   Block diagram of a system using adaptive modulation.

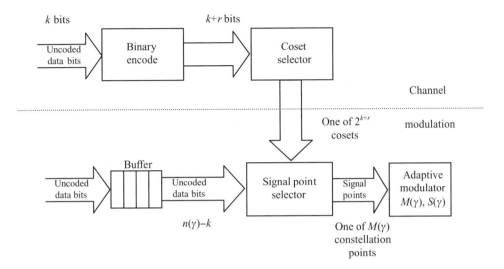

Figure 3.16   General structure for adaptive coded modulation.

Since each constellation M_j is associated with a range of fading values called the fading region R_j, $\bar{\tau}_j$ is the average time that the fading stays within the region R_j. The value of $\bar{\tau}_j$ is inversely proportional to the channel Doppler and also depends on the number and characteristics of the different fade regions. In Rayleigh fading with an average SNR of 20 dB and a channel Doppler of 100 Hz, $\bar{\tau}_j$ ranges from 0.7–3.9 ms, and thus, for a symbol rate of 100 ksymbols/s, the signal constellation remains constant over tens to hundreds of symbols. Similar results hold at other SNR values.

The flat fading assumption implies that the signal bandwidth B is much less than the channel coherence bandwidth $B_c = 1/T_M$, where T_M is the root mean square (rms) delay spread of the channel. For Nyquist pulses $B = 1/T$, so flat fading occurs when $T \gg T_M$. Combining $T \gg T_M$ and $\bar{\tau}_j \gg T$ we get $\bar{\tau}_j \gg T \gg T_M$.

Wireless channels have rms delay spreads less than 30 μs in outdoor urban areas and less than around 1 μs in indoor environments. Taking the minimum $\bar{\tau}_j = 0.8$ ms, rates of the order of tens of ksymbols/s in outdoor channels and hundreds of ksymbols/s in indoor channels are practical for an adaptive scheme.

Modulation uses ideal Nyquist data pulses with a fixed symbol period $T = 1/B$. We also restrict $M(\gamma)$ to square M-QAM constellations of size $M_0 = 0$ and $M_j = 2^{2(j-1)}, j = 2, \ldots, J$. Thus, at each symbol time a constellation from the set $\{M_j : j = 0, 2, \ldots, J\}$ is used – the choice of constellation depends on the fade level γ over that symbol time. Choosing the M_0 constellation corresponds to no data transmission. Since the constellation set is finite, there will be a range of γ values over which a particular constellation M_j is used. Within that range the power must also be adapted to maintain the desired distance d_0 between the trajectories in the trellis. Thus, for each constellation M_j, the power adaptation $S_j(\gamma)$ associated with that constellation is a continuous function of γ. The basic premise for using adaptive modulation is to keep these distances constant by varying the size $M(\gamma)$, transmit power $S(\gamma)$, and/or symbol time $T(\gamma)$ of the transmitted signal constellation relative to γ, subject to an average transmit power constraint $\bar{S}$ on $S(\gamma)$. By maintaining $d_{\min}(t) = d_{\min}$ constant, the adaptive coded modulation exhibits the same coding gain as coded modulation designed for an AWGN channel with minimum code word distance $d_{\min}$. The detailed system block diagram is given in Figure 3.16.

The details about the coset codes used in the scheme can be found in [67,68].

3.2.1   *Maintaining a Fixed Distance*

Define

$$M(\gamma) = \frac{\gamma}{\gamma_K^*} \tag{3.10}$$

where $\gamma_K^* \geq 0$ is a parameter which is optimized relative to the fade distribution to maximize spectral efficiency. For $\gamma < \gamma_K^* M_2$ the channel is not used. The constellation size M_j used for a given $\gamma \geq \gamma_K^* M_2$ is the largest M_j for which $M_j \leq M(\gamma)$. The range of γ values for which $M(\gamma) = M_j$ is thus $M_j \leq \gamma/\gamma_K^* M_{j+1}$, with $M_{J+1} \triangleq \infty$. We call this range of fading values the fading region R_j associated with constellation M_j.

3.2.2 Information Rate

For each γ, one redundant bit per symbol is used for the channel coding, so the number of information bits per symbol is $\log_2 M(\gamma) - 1$. Thus, the information rate for a single γ is $R_\gamma = [\log_2 M(\gamma) - 1]/T$ b/s, and the corresponding spectral efficiency is $R_\gamma/B = \log_2 M(\gamma) - 1$, since we use Nyquist pulses ($B = 1/T$).

Spectral efficiency is obtained by averaging the spectral efficiency for each γ weighted by its probability:

$$\frac{R}{B} = \sum_{j-2}^{J} (\log_2 M_j - 1) p(M_j \leq \gamma/\gamma_K < M_{j+1}) \tag{3.11}$$

where γ_K is picked to maximize Equation (3.11), subject to the average power constraint:

$$\sum_{j=2}^{J} \int_{\gamma_K M_j}^{\gamma_K M_{j+1}} S_j(\gamma) p(\gamma) \, d\gamma = \bar{S} \tag{3.12}$$

Similarly, the reconfiguration efficiency will now vary in time. Equation (3.1) can now be represented as:

$$E_{ff}(\gamma) = \left(\frac{k_{02}(\gamma)}{k_{01}}\right)^2 2^{-\Delta(k_0(\gamma)+v(\gamma))} g_{12}(\gamma) \tag{3.13}$$

and the average efficiency is obtained as:

$$E_{ff} = \int E_{ff}(\gamma) p(\gamma) \, d\gamma \tag{3.14}$$

Table 3.1 presents results of simulations for adaptive and non-adaptive systems with $M_j \in \{0, 4, 16, 64, 256\}$. One can see from the table that considerable SNR gains can be achieved with adaptive schemes.

Table 3.1 Comparison of adaptive and non-adaptive techniques

Spectral efficiency (bps/Hz)	BER	Trellis states	Average SNR (dB)	
			Adaptive	Non-adaptive
2	10^{-3}	4	10.5	18.5
		128	9.0	13.7
	10^{-6}	4	13	36.0
		128	11.3	21.0
3	10^{-3}	8	14.2	20.8
		128	13.1	16.8
	10^{-6}	8	16.4	36.5
		128	15.3	24.8

References

[1] Ungerboeck, G. (1982) Channel coding with multilevel/phase signals. *IEEE Transactions On Information Theory*, **IT-28**, 56–67.

[2] Ungerboeck, G. (1987) Trellis-coded modulation with redundant signal sets – Part I: Introduction. *IEEE Communications Magazine*, **25**, 5–11; and Trellis-coded modulation with redundant signal sets – Part 11: State of the art, *Ibidem*, pp. 12–21.

[3] Forney, Jr., G. D. (1989) Multidimensional constellation – Part II: Voronoi constellations. *IEEE Journal on Selected Areas in Communications*, **7**, 941–958.

[4] Forney, Jr., G. D., Gallagher, R. G., Lang, G. R., Longstaff, F. M. and Qureshi, S. H. (1984) Efficient modulation for band-limited channels. *IEEE Journal on Selected Areas in Communications*, **SAC-2**, 632–647.

[5] Forney, Jr., G. D. and Ungerboeck, G. (1998) Modulation and coding for linear Gaussian channels. *IEEE Transactions On Information Theory*, pp. 2384–2415.

[6] Forney, Jr., G. D. and Wei, L.-F. (1989) Multidimensional constellation – Part I: Introduction, figures of merit, and generalized cross constellations. *IEEE Journal On Selected Areas in Communications*, **7**, 877–892.

[7] Biglieri, E. (1984) High-level modulation and coding for nonlinear satellite channels. *IEEE Transactions On Communications*, **COM-32**, 616–626.

[8] Biglieri, E. (1992) Parallel demodulation of multidimensional signals. *IEEE Transactions on Communications*, **40**(10), 1581–1587.

[9] Biglieri, E., Divsalar, D., McLane, P. J. and Simon, M. K. (1991) *Introduction to Trellis-Coded Modulation with Applications*. Macmillan, New York.

[10] Imai, H. and Hirakawa, S. (1977) A new multilevel coding method using error-correcting codes. *IEEE Transactions On Information Theory*, **IT-23**(3), 371–377.

[11] Wei, L.-F. (1987) Trellis-coded modulation with multidimensional constellations. *IEEE Transactions On Information Theory*, **IT-33**, 483–501.

[12] Cavers, J. K. and Ho, P. (1992) Analysis of the error performance of trellis-coded modulations in Rayleigh-fading channels. *IEEE Transactions On Communications*, **40**(1), 74–83.

[13] Liu, Y. J., Oka, I. and Biglieri, E. (1990) Error probability for digital transmission over nonlinear channels with applications to TCM. *IEEE Transactions On Information Theory*, **IT 36**, 1101–1110.

[14] Zehavi, E. (1992) 8-PSK trellis-codes for a Rayleigh channel. *IEEE Transactions On Communications*, **40**, 873–884.

[15] Zehavi, E. and Wolf, J. K. (1987) On the performance evaluation of trellis codes. *IEEE Transactions On Information Theory*, **IT 33**(2), 196–201.

[16] Benedetto, S., Mondin, M. and Montorsi, G. (1994) Performance evaluation of trellis-coded modulation schemes. *IEEE Proceedings*, **82**, 833–855.

[17] Trott, M. D., Benedetto, S., Garello, R. and Mondin, M. (1996) Rotational invariance of trellis codes–Part I: Encoders and precoders. *IEEE Transactions On Information Theory*, **42**, 751–765.

[18] Sundberg, C.-E. W. and Seshadri, N. (1993) Coded modulation for fading channels: An overview. *European Transactions on Telecommunications*, **4**(3), 309–324.

[19] Pottie, G. J. and Taylor, D. P. (1989) Multilevel codes based on partitioning, *IEEE Transactions On Information Theory*, **IT-35**, 87–98.

[20] Pellizzoni, R., Sandri, A., Spalvieri, A. and Biglieri, E. (1997) Analysis and implementation of an adjustable-rate multilevel coded modulation system. *IEE Proceedings on Communications*, **144**, 1–5.

[21] Caire, G., Taricco, G. and Biglieri, E. (1998) Bit-interleaved coded modulation. *IEEE Transactions On Information Theory*, **44**, 927–946.

[22] Divsalar, D. and Simon, M. K. (1988) The design of trellis coded MPSK for fading channel: Performance criteria. *IEEE Transactions On Communications*, **36**, 1004–1012.

[23] Divsalar, D. and Simon, M. K. (1988) The design of trellis coded MPSK for fading channel: Set partitioning for optimum code design. *IEEE Transactions On Communications*, **36**, 1013–1021.

[24] Robertson, P., Worz, T. (1998) Bandwidth-efficient Turbo Trellis-coded Modulation Using Punctured Component Codes. *IEEE Journal on Selected Areas in Communications*, **16**, 206–218.

[25] Zehavi, E. (1992) 8-PSK trellis-codes for a Rayleigh fading channel. *IEEE Transactions on Communications*, **40**, 873–883.

[26] Li, X. and Ritcey, J. A. (1997) Bit-interleaved coded modulation with iterative decoding. *IEEE Communications Letters*, **1**.

[27] Li, X. and Ritcey, J. A. (1999) Trellis-coded Modulation with Bit Interleaving and Iterative Decoding. *IEEE Journal on Selected Areas in Communications*, **17**.

[28] Li, X. and Ritcey, J. A. (1998) Bit-interleaved coded modulation with iterative decoding – Approaching turbo-TCM performance without code concatenation, in *Proceedings of CISS 1998* (Princeton University, USA).

[29] Ng, S. X., Liew, T. H., Yang, L. L. and Hanzo, L. (2001) Comparative Study of TCM, TTCM, BICM and BICM-ID schemes. *IEEE Vehicular Technology Conference*, p. 265 (CDROM).

[30] Ng, S. X., Wong, C. H. and Hanzo, L. (2001) Burst-by-Burst Adaptive Decision Feedback Equalized TCM, TTCM, BICM and BICM-ID. *International Conference on Communications* (ICC), pp. 3031–3035.

[31] Ungerboeck, G. (1987) Trellis-coded modulation with redundant signal sets. Part 1 and 2. *IEEE Communications Magazine*, **25**, 5–21.

[32] Pietrobon, S. S., Ungerboeck, G., Perez, L. C. and Costello, D. J. (1994) Rotationally invariant nonlinear trellis codes for two-dimensional modulation. *IEEE Transactions On Information Theory*, **IT-40**, 1773–1791.

[33] Schlegel, C. (1997) Trellis Coded Modulation, in *Trellis Coding*, IEEE Press, New York, USA, pp. 43–89.

[34] Cavers, J. K. and Ho, P. (1992) Analysis of the Error Performance of Trellis-coded Modulations in Rayleigh-fading Channels. *IEEE Transactions On Communications*, **40**, 74–83.

[35] Pietrobon, S. S., Deng, R. H., Lafanechere, A., Ungerboeck, G. and Costello, D. J. (1990) Trellis-coded Multidimensional Phase Modulation. *IEEE Transactions on Information Theory*, **36**, 63–89.

[36] Wei, L. F. (1987) Trellis-coded modulation with multidimensional constellations. *IEEE Transactions On Information Theory*, **IT-33**, 483–501.

[37] Caire, G., Tariceo, G. and Biglieri, E. (1998) Bit-Interleaved Coded Modulation. *IEEE Transactions On Information Theory*, **44**, 927–946,

[38] Steele, R. and Webb, W. (1991) Variable rate QAM for data transmission over Rayleigh fading channels, in *Proceeedings of Wireless '91*, Calgary, Alberta, pp. 1–14.

[39] Hanzo, L., Wong, C. H. and Yee, M. S. (2002) *Adaptive Wireless Transceivers*. John Wiley & Sons, Inc., New York, USA and IEEE Press. (For detailed contents, please refer to http://www-mobile.ecs.soton.ac.uk/.).

[40] Benedetto, S. and Biglieri, E. (1999) *Principles of Digital Transmission with Wireless Applications*, Kluwer, New York.

[41] Proakis, G. (1995) *Digital Communications*, 3rd editon, McGraw Hill, New York.

[42] Simon, M. K. *et al.* (1995) *Digital Communication Techniques – Signal Design and Detection*, Prentice Hall, Englewood Cliffs, NJ.

[43] Sampei, S., Komaki, S. and Morinaga, N. (1994) Adaptive Modulation/TDMA Scheme for Large Capacity Personal Multi-Media Communication Systems. *IEIEE Transactions on Communications*, **E77-B**, 1096–1103.

[44] Goldsmith, A. J. and Chua, S. (1997) Variable-rate variable-power MQAM for fading channels. *IEEE Transactions On Communications*, **45**, 1218–1230.

[45] Wong, C. and Hanzo, L. (2000) Upper-bound performance of a wideband burst-by-burst adaptive modem. *IEEE Transactions On Communications*, **48**, 367–369.

[46] Matsuoka, H., Sampei, S., Morinaga, N. and Kamio, Y. (1996) Adaptive Modulation System with Variable Coding Rate Concatenated Code for High Quality Multi-Media Communications Systems, in *Proceedings of IEEE VTC'96*, Atlanta, USA, 28 April–1 May, **1**, 487–491.

[47] Lau, V. and Macleod, M. (1998) Variable rate adaptive trellis coded QAM for high bandwidth efficiency applications in Rayleigh fading channels, in *Proceedings of IEEE Vehicular Technology Conference* (VTC'98), Ottawa, Canada, 18–21 May, pp. 348–352.

[48] Goldsmith, A. J. and Chua, S. (1998) Adaptive Coded Modulation for Fading Channels. *IEEE Transactions On Communications*, **46**, 595–602.

[49] Won, C. H., Liew, T. H. and Hanzo, L. (1999) Burst-by-Burst Turbo Coded Wideband Adaptive Modulation with Blind Modem Mode Detection, in *Proceedings of 4th ACTS Mobile Communications Summit*, Sorrento, Italy, pp. 303–308.

[50] Goeckel, D. (1999) Adaptive Coding for Fading Channels using Outdated Fading Estimates. *IEEE Transactions On Communications*, **47**, 844–855.

[51] Choi, B. J., Munster, M., Yang, L. L. and Hanzo, L. (2001) Performance of Rake receiver assisted adaptive-modulation based CDMA over frequency selective slow Rayleigh fading channel. *Electronics Letters*, **37**, 247–249.

[52] Alamouti, S. M. and Kallel, S. (1994) Adaptive Trellis-coded Multiple-phase-shift Keying Rayleigh Fading Channels. *IEEE Transactions On Communications*, **42**, 2305–2341.

[53] A1-Semari, S. and Fuja, T. (1997) I-Q TCM: Reliable communication over the Rayleigh fading channel close to the cutoff rate. *IEEE Transactions On Information Theory*, **43**, 250–262.

[54] Webb, W. and Steele, R. (1995) Variable rate QANI for mobile radio. *IEEE Transactions on Communications*, **43**, 2223–2230.

[55] Torrance, J. and Hanzo, L. (1996) Performance upper bound of adaptive QAM in slow Rayleigh-fading environments, in *Proceedings of IEEE ICCS'96/ISPACS'96*, Singapore, 25–29 November, pp. 1653–1657.

[56] Torrance, J. M., Hanzo, L. and Keller, T. (1999) Interference aspects of adaptive modems over slow Rayleigh fading channels. *IEEE Transactions on Vehicular Technology*, **48**, 1527–1545.

[57] Matsuoka, H., Sampei, S., Morinaga, N. and Kamio, Y. (1996) Adaptive modulation systems with variable coding rate concatenated code for high quality multi-media communication systems, in *Proceedings of IEEE VTC'96*, Atlanta, USA, 28 April–1 May, pp. 487–491.

[58] Chua, S. G. and Goldsmith, A. J. (1996) Variable-rate variable-power mQAM for fading channels, in *Proceedings of IEEE VTC'96*, Atlanta, USA, 28 April–1 May, pp. 815–819.

[59] Torrance, J. and Hanzo, L. (1996) Optimisation of switching levels for adaptive modulation in a slow Rayleigh fading channel. *Electronics Letters*, **32**, 1167–1169.

[60] Torrance, J. and Hanzo, L. (1996) On the upper bound performance of adaptive QAM in a slow Rayleigh fading. *IEE Electronics Letters*, 169–171.

[61] Lau, V. and Macleod, M. (2001) Variable-rate adaptive trellis coded QAM for flat-fading channels. *IEEE Transactions On Communications*, **49**, 1550–1560.

[62] Lau, V. and Marie, S. (1999) Variable rate adaptive modulation for DS-CDMA. *IEEE Transactions On Communications*, **47**, 577–589.

[63] Chua, S. and Goldsmith, A. (1998) Adaptive Coded Modulation for Fading Channels. *IEEE Transactions On Communications*, **46**, 595–602.

[64] Liu, X., Ormeci, P., Wesel, R. and Goeckel, D. (2001) Bandwidth-efficient, low-latency adaptive coded modulation schemes for time-varying channels, in *Proceedings of IEEE International Conference on Communications*, Helsinki, Finland.

[65] Torrance, J. and Hanzo, L. (1996) Demodulation level selection in adaptive modulation. *Electronics Letters*, **32**, 1751–1752.

[66] Nanda, S., Balachandran, K. and Kumar, S. (2000) Adaptation techniques in wireless packet data services. *IEEE Communications Magazine*, **38**, 54–64.

[67] Forney, Jr., G. D. (1988a) Coset codes – Part I: Introduction and geometrical classification. *IEEE Transactions On Information Theory*, **34**, 1123–1151.

[68] Forney, Jr., G. D. (1988b) Coset codes – Part II: Binary lattices and related codes. *IEEE Transactions On Information Theory*, **34**, 1152–1187.

4

Space–Time Coding

4.1 Diversity Gain

In Chapters 2 and 3 the coding gain was used as a performance measure. Before we go into a detailed discussion on space–time coding, diversity gain will be defined and discussed. In Chapter 3 we briefly discussed the multidimensional trellis and pointed out the relatively high efficiency of such a concept. By using an additional dimension we provide a *diversity effect* which results in a considerable gain. These new dimensions may be additional frequency bands, different time slots or delayed replicas of the signal, or different antennas, resulting in frequency, time or space diversity respectively. In this section we elaborate the concept of diversity gain by using space diversity. In the subsequent sections of the chapter we will discuss space–time coding, where the concept of coding and diversity gain is combined into an integral performance measure.

A classical space diversity set-up with one transmitting and two receiving antennas is shown in Figure 4.1. The antenna diversity is realized in the receiver, hence the name *receiver diversity*. The following notation is used in the figure: the channel between the transmit antenna and the receiver antenna zero is denoted $\mathbf{h}_0$; that between the transmit antenna and the receiver antenna one is $\mathbf{h}_1$, where:

$$
\begin{aligned}
\mathbf{h}_0 &= \alpha_0 e^{j\theta_0} \\
\mathbf{h}_1 &= \alpha_1 e^{j\theta_1}
\end{aligned}
\tag{4.1}
$$

The resulting received baseband signals at antennas zero and one are:

$$
\begin{aligned}
\mathbf{r}_0 &= \mathbf{h}_0 s_0 + \mathbf{n}_0 \\
\mathbf{r}_1 &= \mathbf{h}_1 s_0 + \mathbf{n}_1
\end{aligned}
\tag{4.2}
$$

where $\mathbf{n}_0$ and $\mathbf{n}_1$ represent complex noise and interference.

In accordance with the discussion in Chapter 2, the ML decoder will choose signal s_i if and only if:

$$
d^2(\mathbf{r}_0, \mathbf{h}_0 s_i) + d^2(\mathbf{r}_1, \mathbf{h}_1 s_i) \le d^2(\mathbf{r}_0, \mathbf{h}_0 s_k) + d^2(\mathbf{r}_1, \mathbf{h}_1 s_k) \quad \forall i \ne k
\tag{4.3}
$$

where $d^2(\mathbf{x}, \mathbf{y})$ is the squared Euclidian distance between $\mathbf{x}$ and $\mathbf{y}$:

$$
d^2(\mathbf{x}, \mathbf{y}) = (\mathbf{x} - \mathbf{y})(\mathbf{x}^* - \mathbf{y}^*)
\tag{4.4}
$$

Advanced Wireless Communications & Internet: Future Evolving Technologies, Third Edition. Savo Glisic.
© 2011 John Wiley & Sons, Ltd. Published 2011 by John Wiley & Sons, Ltd.

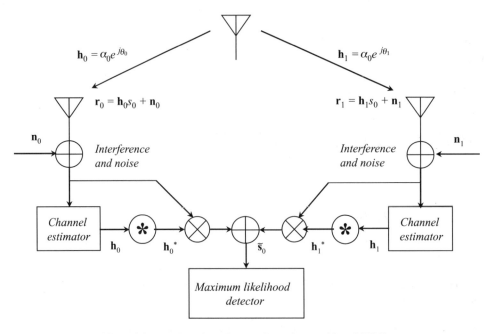

Figure 4.1   Two branch maximum ratio receiver combiner (MRRC).

A two branch *Maximum Ratio Receiver Combiner (MRRC)* would first create the signal:

$$\begin{aligned}
\tilde{s}_0 &= \mathbf{h}_0^* \mathbf{r}_0 + \mathbf{h}_1^* \mathbf{r}_1 \\
&= \mathbf{h}_0^* (\mathbf{h}_0 s_0 + \mathbf{n}_0) + \mathbf{h}_1^* (\mathbf{h}_1 s_0 + \mathbf{n}_1) \\
&= (\alpha_0^2 + \alpha_1^2) s_0 + \mathbf{h}_0^* \mathbf{n}_0 + \mathbf{h}_1^* \mathbf{n}_1
\end{aligned} \tag{4.5}$$

with an equivalent distance defined as:

$$\mathbf{d}_i^2 = (\tilde{s}_0 - \beta s_i)(\tilde{s}_0 - \beta s_i)^*; \quad \beta = \alpha_0^2 + \alpha_1^2 \tag{4.6}$$

and the ML detector would choose $\mathbf{s}_i$ if:

$$\left(\alpha_0^2 + \alpha_1^2\right) |\mathbf{s}_i|^2 - \tilde{s}_0 \mathbf{s}_i^* - \tilde{s}_0^* \mathbf{s}_i \leq (\alpha_0^2 + \alpha_1^2) |\mathbf{s}_k|^2 - \tilde{s}_0 \mathbf{s}_k^* - \tilde{s}_0^* \mathbf{s}_k \quad \forall i \neq k \tag{4.7}$$

If Equation (4.6) is used in (4.7), the latter can also be represented in the following form. Choose $\mathbf{s}_i$ if:

$$(\alpha_0^2 + \alpha_1^2 - 1) |\mathbf{s}_i|^2 + d^2 (\tilde{s}_0, \mathbf{s}_i) \leq (\alpha_0^2 + \alpha_1^2 - 1) |\mathbf{s}_k|^2 + d^2 (\tilde{s}_0, \mathbf{s}_k) \quad \forall i \neq k \tag{4.8}$$

For PSK signals (equal energy constellations):

$$|\mathbf{s}_i|^2 = |\mathbf{s}_k|^2 = E_s \quad \forall i, k \tag{4.9}$$

where E_s is the energy of the signal. So, for PSK signals, the decision rule (4.8) may be simplified to:
Choose $\mathbf{s}_i$ if

$$d^2 (\tilde{s}_0, \mathbf{s}_i) \leq d^2 (\tilde{s}_0, \mathbf{s}_k) \quad \forall i \neq k \tag{4.10}$$

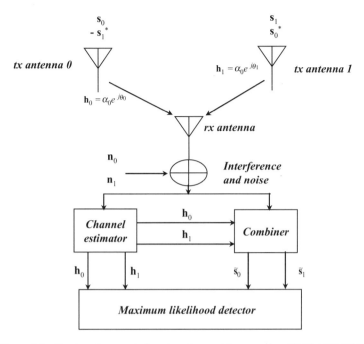

Figure 4.2   Two branch transmit diversity scheme with one receiver [1] © 1998, IEEE.

4.1.1   Two-Branch Transmit Diversity Scheme With One Receiver

Using two antennas in a mobile receiver might prove difficult in practice. For this reason, in this subsection we demonstrate how the same effect and diversity gain may be obtained by using two antennas at the transmitter and only one antenna at the receiver [1]. Implementing two antennas at the base station in mobile communication networks is a much simpler task. The system block diagram is shown in Figure 4.2.

4.1.1.1   The Encoding and Transmission Sequence

The encoding is done in space and time (space–time (ST) coding) as defined in Table 4.1. The encoding may also be done in space and frequency. Instead of two adjacent symbol periods, two adjacent frequency subbands may be used (space–frequency coding).

4.1.1.2   The Received Signal

Assuming that the fading is constant across two consecutive symbols we have:

$$\mathbf{h}_0(t) = \mathbf{h}_0(t + T) = \mathbf{h}_0 = \alpha_0 e^{j\theta_0}$$
$$\mathbf{h}_1(t) = \mathbf{h}_1(t + T) = \mathbf{h}_1 = \alpha_1 e^{j\theta_1}$$

(4.11)

Table 4.1   Space–time coding rules

	Antenna 0	Antenna 1
time t	s_0	s_1
time $t + T$	$-s_1^*$	s_0^*

where T is the symbol interval. The received signals in two adjacent symbol intervals can be represented as:

$$
\begin{aligned}
\mathbf{r}_0 &= \mathbf{r}(t) = \mathbf{h}_0 s_0 + \mathbf{h}_1 s_1 + \mathbf{n}_0 \\
\mathbf{r}_1 &= \mathbf{r}(t+T) = -\mathbf{h}_0 s_1^* + \mathbf{h}_1 s_0^* + \mathbf{n}_1
\end{aligned}
\tag{4.12}
$$

4.1.1.3 The Combining Scheme

The signals received in two adjacent symbol intervals are combined as follows:

$$
\begin{aligned}
\tilde{s}_0 &= \mathbf{h}_0^* \mathbf{r}_0 + \mathbf{h}_1 \mathbf{r}_1^* \\
\tilde{s}_1 &= \mathbf{h}_1^* \mathbf{r}_0 + \mathbf{h}_0 \mathbf{r}_1^*
\end{aligned}
\tag{4.13}
$$

$$
\begin{aligned}
\tilde{s}_0 &= (\alpha_0^2 + \alpha_1^2)s_0 + \mathbf{h}_0^* \mathbf{n}_0 + \mathbf{h}_1 \mathbf{n}_1^* \\
\tilde{s}_1 &= (\alpha_0^2 + \alpha_1^2)s_1 - \mathbf{h}_0 \mathbf{n}_1^* + \mathbf{h}_1^* \mathbf{n}_0
\end{aligned}
\tag{4.14}
$$

4.1.1.4 ML Decision Rule

For each of the signals s_0 and s_1 the decision rule is the same, Inequality (4.3) or (4.10). The resulting combined signals are equivalent to that obtained from two branch MRRC given by Equation (4.5). The only difference is phase rotations on the noise components which do not degrade the effective SNR. The resulting diversity order from the new branch transmit diversity scheme with one receiver is equal to that of two branch MRRC. This is known as Alamouti code [1].

4.1.2 Two Transmitters and M Receivers

There may be applications where a higher order of diversity is needed and multiple receive antennas are also feasible. In such cases, it is possible to provide a diversity order of $2M$, with two transmit and M receive antennas. Figure 4.3 represents an illustration of a special case of two transmit and two receive antennas. The generalization to N transmit and M receive antennas will be further elaborated in the subsequent sections. The notation is explained in Tables 4.2 and 4.3.

The space–time coding rule is given again by Table 4.1.

4.1.2.1 The Received Signal

The received signals in two adjacent symbol intervals with notation given in Table 4.3 are:

$$
\begin{aligned}
\mathbf{r}_0 &= \mathbf{h}_0 s_0 + \mathbf{h}_1 s_1 + \mathbf{n}_0 \\
\mathbf{r}_1 &= -\mathbf{h}_0 s_1^* + \mathbf{h}_1 s_0^* + \mathbf{n}_1 \\
\mathbf{r}_2 &= \mathbf{h}_2 s_0 + \mathbf{h}_3 s_1 + \mathbf{n}_2 \\
\mathbf{r}_3 &= -\mathbf{h}_2 s_1^* + \mathbf{h}_3 s_0^* + \mathbf{n}_3
\end{aligned}
\tag{4.15}
$$

where $\mathbf{n}_0$, $\mathbf{n}_1$, $\mathbf{n}_2$, and $\mathbf{n}_3$ are complex random variables representing receiver thermal noise and interference.

4.1.2.2 The Combiner

The combiner is defined by the following rule:

$$
\begin{aligned}
\tilde{s}_0 &= \mathbf{h}_0^* \mathbf{r}_0 + \mathbf{h}_1 \mathbf{r}_1^* + \mathbf{h}_2^* \mathbf{r}_2 + \mathbf{h}_3 \mathbf{r}_3^* \\
\tilde{s}_1 &= \mathbf{h}_1^* \mathbf{r}_0 - \mathbf{h}_0 \mathbf{r}_1^* + \mathbf{h}_3^* \mathbf{r}_2 - \mathbf{h}_2 \mathbf{r}_3^*
\end{aligned}
\tag{4.16}
$$

$$
\begin{aligned}
\tilde{s}_0 &= (\alpha_0^2 + \alpha_1^2 + \alpha_2^2 + \alpha_3^2)s_0 + \mathbf{h}_0^* \mathbf{n}_0 + \mathbf{h}_1 \mathbf{n}_1^* + \mathbf{h}_2^* \mathbf{n}_2 + \mathbf{h}_3 \mathbf{n}_3^* \\
\tilde{s}_1 &= (\alpha_0^2 + \alpha_1^2 + \alpha_2^2 + \alpha_3^2)s_1 - \mathbf{h}_0 \mathbf{n}_1^* + \mathbf{h}_1^* \mathbf{n}_0 - \mathbf{h}_2 \mathbf{n}_3^* + \mathbf{h}_3^* \mathbf{n}_2
\end{aligned}
\tag{4.17}
$$

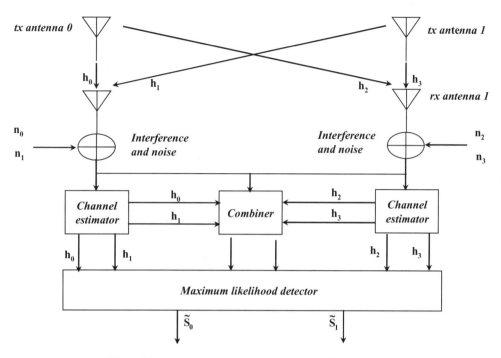

Figure 4.3   Two transmitters and two receivers [1] © 1998, IEEE.

So, the order of $2 \times 2 = 4$ diversity is achieved. For uncorrelated channel and noise, $\alpha_i^2 = \alpha^2$, and perfect channel estimation, the signal to noise ratio is given as:

$$(4\alpha^2)^2/4\alpha^2 N_0 = 4\alpha^2/N_0 \tag{4.18}$$

4.1.2.3   ML Decoder

The ML decoder will be operating as follows:
For $\mathbf{s}_0$, choose $\mathbf{s}_i$ if:

$$(\alpha_0^2 + \alpha_1^2 + \alpha_2^2 + \alpha_3^2 - 1)\,|\mathbf{s}_i|^2 + d^2(\tilde{\mathbf{s}}_0, \mathbf{s}_i) \le (\alpha_0^2 + \alpha_1^2 + \alpha_2^2 + \alpha_3^2 - 1)\,|\mathbf{s}_k|^2 + d^2(\tilde{\mathbf{s}}_0, \mathbf{s}_k) \tag{4.19}$$

Table 4.2   Channel notation

	rx antenna 0	rx antenna 1
tr antenna 0	$\mathbf{h}_0$	$\mathbf{h}_2$
tr antenna 1	$\mathbf{h}_1$	$\mathbf{h}_3$

Table 4.3   Signal notation

	rx antenna 0	rx antenna 1
time t	$\mathbf{r}_0$	$\mathbf{r}_2$
time $t + T$	$\mathbf{r}_1$	$\mathbf{r}_3$

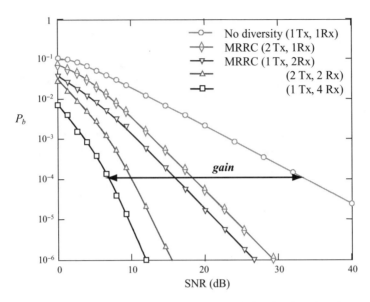

Figure 4.4 The BER performance comparison of coherent BPSK with MRRC and two branch transmit diversity in Rayleigh fading.

For $\mathbf{s}_1$, choose $\mathbf{s}_i$ if:

$$(\alpha_0^2 + \alpha_1^2 + \alpha_2^2 + \alpha_3^2 - 1)|\mathbf{s}_i|^2 + d^2(\tilde{\mathbf{s}}_1, \mathbf{s}_i) \leq (\alpha_0^2 + \alpha_1^2 + \alpha_2^2 + \alpha_3^2 - 1)|\mathbf{s}_k|^2 + d^2(\tilde{\mathbf{s}}_1, \mathbf{s}_k) \qquad (4.20)$$

4.1.2.4 The BER Performance

The BER results are shown in Figure 4.4. Significant diversity gain measured in SNR improvements is evident.

4.2 Space–Time Coding

In the previous section we introduced the concept of space–time coding. We are now going to look more closely at the general model and bring in more detail in the analysis of such a system. In general, we will consider a system where data is encoded by a channel code and the encoded data is split into n streams that are simultaneously transmitted using n transmit antennas. The received signal at each receive antenna is a linear superposition of the n transmitted signals plus noise.

4.2.1 The System Model

The system consists of n antennas in the base station and m antennas in the mobile. Data is encoded by the channel encoder, S/P converted, and divided into n streams of data. Each stream of data is used as the input to a pulse shaper. The output of each shaper is then modulated. At each time slot with index l, the output of modulator i is a signal c_l^i that is transmitted using transmit antenna (Tx antenna) i for $1 < i < n$.

The n signals are transmitted simultaneously, each from a different transmit antenna, and all of these signals have the same transmission period T. The signal at each receive antenna is a noisy superposition of the n transmitted signals corrupted by Rayleigh or Rician fading.

Elements of the signal constellation are contracted by a factor of $\sqrt{E_s}$ chosen so that the average energy of the constellation is 1.

At the receiver, the demodulator computes a decision statistic based on the received signals arriving at each receive antenna $1 < j < m$. The signal r_l^j received by antenna j at discrete time l is given by:

$$\mathbf{r}_l^j = \sum_{i=1}^{n} \alpha_{i,j} \mathbf{c}_l^j \sqrt{E_s} + \eta_l^j \tag{4.21}$$

where the noise η_l^j at discrete time l is modeled as independent samples of a zero mean complex Gaussian random variable with variance $N_0/2$ per dimension. The coefficient $\alpha_{i,j}$ is the path gain from transmit antenna i to receive antenna j. It is assumed that these path gains are constant during a frame of t symbol intervals and vary from one frame to another (quasistatic flat fading).

4.2.2 The Case of Independent Fade Coefficients

The ML receiver will decide erroneously in favor of a signal:

$$\mathbf{e} = e_1^1 e_1^2 \cdots e_1^n e_2^1 e_2^2 \cdots e_2^n \cdots e_l^1 e_l^2 \cdots e_l^n$$

assuming that

$$\mathbf{c} = c_1^1 c_1^2 \cdots c_1^n c_2^1 c_2^2 \cdots c_2^n \cdots c_l^1 c_l^2 \cdots c_l^n$$

was transmitted, with a probability that can be approximated by:

$$P(\mathbf{c} \to \mathbf{e} | \alpha_{i,j}, \; i = 1, 2, \ldots, n, \; j = 1, 2, \ldots, m) \le \exp\left(-d^2(\mathbf{c}, \mathbf{e}) E_s / 4 N_0\right) \tag{4.22}$$

where $N_0/2$ is the noise variance per dimension and

$$d^2(\mathbf{c}, \mathbf{e}) = \sum_{j=1}^{m} \sum_{l=1}^{t} \left| \sum_{i=1}^{n} \alpha_{i,j}(c_l^i - e_l^i) \right|^2 \tag{4.23}$$

is the distance between the two trajectories measured in a time frame of t symbol intervals. Setting $\Omega_j = (\alpha_{1,j}, \ldots, \alpha_{n,j})$, we rewrite Equation (4.23) as:

$$d^2(\mathbf{c}, \mathbf{e}) = \sum_{j=1}^{m} \sum_{l=1}^{t} \sum_{i'=1}^{n} \alpha_{i,j} \overline{\alpha_{i,j}} \sum_{t=1}^{l} (c_t^i - e_t^i)\overline{(c_t^{i'} - e_t^{i'})}.$$

where $\bar{x}$ stands for the complex conjugate of x. After simple manipulations, we observe that:

$$d^2(\mathbf{c}, \mathbf{e}) = \sum_{j=1}^{m} \Omega_j \mathbf{A} \Omega_j^* \tag{4.24}$$

where $\mathbf{x}^*$ stands for the transpose conjugate, and elements of matrix $\mathbf{A}$ are defined as $A_{pq} = \mathbf{x}_p \cdot \mathbf{x}_q$ and $\mathbf{x}_p = (c_1^p - e_1^p, c_2^p - e_2^p, \ldots, c_l^p - e_l^p)$ for $1 \le p, q \le n$. Thus:

$$P(\mathbf{c} \to \mathbf{e} | \alpha_{i,j}, \; i = 1, 2, \ldots, n, \; j = 1, 2, \ldots, m) \le \prod_{j=1}^{m} \exp\left(-\Omega_j \mathbf{A}(\mathbf{c}, \mathbf{e}) \Omega_j^* E_s / 4 N_0\right) \tag{4.25}$$

where $A_{pq} = \sum_{t=1}^{l} (c_t^p - e_t^p)\overline{(c_t^q - e_t^q)}$. This can be also represented as:

$$A = B^*B$$
$$\text{where} \quad B = \{b_{it}\} = \{e_t^i - c_t^i\}$$

(4.26)

Matrix B is given in an explicit form later by Equation (4.48). From now on we will express $d^2(c, e)$ in terms of the eigenvalues of the matrix $A(c, e)$ defined by $VA(c, e)V^* = D = \text{diag}\{\lambda_i\}$. For details of eigenvalue decomposition, see Appendix 5.1.

If we define vector $(\beta_{1,j}, \ldots, \beta_{n,j}) = \Omega_j V^*$, then we have:

$$\Omega_j A(c, e)\Omega_j^* = \sum_{i=1}^{n} \lambda_i |\beta_{i,j}|^2$$

(4.27)

At this point recall that $\alpha_{i,j}$ are samples of a complex Gaussian random variable with mean $\bar{\alpha}_{ij}$. Let

$$K^j = (\bar{\alpha}_{1j}, \bar{\alpha}_{2j}, \bar{\alpha}_{3j}, \cdots, \bar{\alpha}_{nj})$$

(4.28)

Since V is unitary, $\{v_1, v_2, \ldots, v_n\}$ is an orthonormal basis of C^n and β_{ij} are independent complex Gaussian random variables with variance 0.5 per dimension and mean $K^j \cdot v_i$. If $K_{i,j} = |\bar{\beta}_{i,j}|^2 = |K^j \cdot v_i|^2$, then $|\beta_{i,j}|$ are independent Rician distributions with pdf:

$$p(|\beta_{i,j}|) = 2|\beta_{i,j}| \exp(-|\beta_{i,j}|^2 - K_{i,j})I_0(2|\beta_{i,j}|\sqrt{K_{i,j}})$$

(4.29)

for $|\beta_{i,j}| \geq 0$, where $I_0(\cdot)$ is the zero order modified Bessel function of the first kind. To compute an upper bound on the average probability of error, we simply average:

$$\prod_{j=1}^{m} \exp\left(-(E_s/4N_0)\sum_{i=1}^{n} \lambda_i |\beta_{i,j}|^2\right)$$

(4.30)

with respect to independent Rician distributions of $|\beta_{i,j}|$ to arrive at:

$$P(c \to e) \leq \prod_{j=1}^{m}\left(\prod_{i=1}^{n} \frac{1}{1 + \frac{E_s}{4N_0}\lambda_i} \exp\left(-\frac{K_{i,j}\frac{E_s}{4N_0}\lambda_i}{1 + \frac{E_s}{4N_0}\lambda_i}\right)\right)$$

(4.31)

4.2.3 Rayleigh Fading

In this case $\bar{\alpha} = 0$, giving $K_{i,j} = 0$ for all i and j. Thus Inequality (4.31) can be written as:

$$P(c \to e) \leq \left(\frac{1}{\prod_{i=1}^{n} (1 + \lambda_i E_s/4N_0)}\right)^m$$

(4.32)

Let r denote the rank of matrix A, then the kernel of A has dimension $n - r$ and exactly $n - r$ eigenvalues of A are zero. Say the non-zero eigenvalues of A are $\lambda_1, \lambda_2, \ldots, \lambda_r$, then it follows from inequality (4.32) that:

$$P(\mathbf{c} \rightarrow \mathbf{e}) \leq \left(\prod_{i=1}^{r} \lambda_i \right)^{-m} (E_s/4N_0)^{-rm} \tag{4.33}$$

Thus, a diversity advantage of mr and a coding advantage of $(\lambda_1 \lambda_2 \cdots \lambda_r)^{1/r}$ is achieved. Recall that $\lambda_1 \lambda_2 \cdots \lambda_r$ is the absolute value of the sum of determinants of all the principal $r \times r$ cofactors of $\mathbf{A}$, Moreover, it is easy to see that the ranks of $\mathbf{A}(\mathbf{c}, \mathbf{e})$, and $\mathbf{B}(\mathbf{c}, \mathbf{e})$, defined as $\mathbf{A}(\mathbf{c}, \mathbf{e}) = \mathbf{B}(\mathbf{c}, \mathbf{e})^* \mathbf{B}(\mathbf{c}, \mathbf{e})$, are equal.

4.2.4   Design Criteria for Rayleigh Space–Time Codes

The rank criterion. In order to achieve the maximum diversity mn, the matrix $\mathbf{B}(\mathbf{c}, \mathbf{e})$ has to be full rank for any code words $\mathbf{c}$ and $\mathbf{e}$. If $\mathbf{B}(\mathbf{c}, \mathbf{e})$ has minimum rank r over the set of two tuples of distinct code words, then a diversity of rm is achieved.

The determinant criterion. Suppose that a diversity benefit of rm is our target. The minimum of r roots of the sum of determinants of all $r \times r$ principal cofactors of $\mathbf{A}(\mathbf{c}, \mathbf{e}) = \mathbf{B}(\mathbf{c}, \mathbf{e})^* \mathbf{B}(\mathbf{c}, \mathbf{e})$ taken over all pairs of distinct code words $\mathbf{e}$ and $\mathbf{c}$ corresponds to the coding advantage, where r is the rank of $\mathbf{A}(\mathbf{c}, \mathbf{e})$. Special attention in the design must be paid to this quantity for any code words $\mathbf{e}$ and $\mathbf{c}$. The design target is making this sum as large as possible. If a diversity of nm is the design target, then the minimum of the determinant of $\mathbf{A}(\mathbf{c}, \mathbf{e})$ taken over all pairs of distinct code words $\mathbf{e}$ and $\mathbf{c}$ must be maximized.

For large signal to noise ratios,

$$P(\mathbf{c} \rightarrow \mathbf{e}) \leq \left(\frac{E_s}{4N_0} \right)^{-rm} \left(\prod_{i=1}^{r} \lambda_i \right)^{-m} \left[\prod_{j=1}^{m} \prod_{i=1}^{r} \exp(-K_{i,j}) \right] \tag{4.34}$$

Thus, a diversity of rm and a coding advantage of:

$$(\lambda_1 \lambda_2 \cdots \lambda_r)^{-1/r} \left[\prod_{j=1}^{m} \prod_{i=1}^{r} \exp(-K_{i,j}) \right]^{1/rm} \tag{4.35}$$

is achievable. The derivation in this section is based on the original work presented in [2].

4.2.5   Code Construction

In the presence of one receive antenna, little can be gained in terms of capacity increase by using more than four transmit antennas. Similarly, if there are two receive antennas, almost all the capacity increase can be obtained using $n = 6$ transmit antennas.

As has been indicated earlier, we can use multidimensional trellis codes for a wireless communication system that employs n transmit antennas and (optional) receive antenna diversity where the channel is a quasistatic flat fading channel. The encoding for these trellis codes is obvious, with the exception that *at the beginning and the end of each frame, the encoder is required to be in the zero state.* At each time t, depending on the state of the encoder and the input bits, a transition branch is chosen. If the label of this branch is $q_t^1 q_t^2 \cdots q_t^n$, then transmit antenna i is used to send constellation symbols $q_t^i, i = 1, 2, \ldots, n$ and all of these transmissions are simultaneous. Let us consider the 4PSK and 8PSK constellations as shown in Figure 4.5.

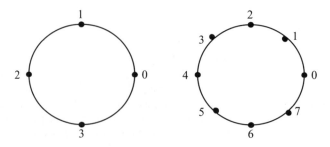

Figure 4.5   4PSK and 8PSK constellations.

4.2.5.1   Examples

A number of results have been published on space–time code design [1–46]. We will present several illustrations.

In Figure 4.5 the first signal constellation is 4PSK, where the signal points are labeled by the elements of $\mathbb{Z}_4$, the ring of integers modulo 4. We consider the four-state trellis code shown in Figure 4.6 [2]. The edge label $\mathbf{x}_1\mathbf{x}_2$ indicates that signal $\mathbf{x}_1$ is transmitted over the first antenna and that signal $\mathbf{x}_2$ is transmitted over the second antenna. This code has a very simple description in terms of a sequence (b_k, a_k) of binary inputs [2]. The output signal pair $\mathbf{x}_1^k\mathbf{x}_2^k$ at time k is given by:

$$(\mathbf{x}_1^k, \mathbf{x}_2^k) = b_{k-1}(2, 0) + a_{k-1}(1, 0) + b_k(0, 2) + a_k(0, 1) \tag{4.36}$$

where the addition takes place in $\mathbb{Z}_4$.

$$(\mathbf{x}_1^k, \mathbf{x}_2^k) = a_{k-2}(2, 2) + b_{k-1}(2, 0) + a_{k-1}(1, 0) + b_k(0, 2) + a_k(0, 1)$$
$$(\mathbf{x}_1^k, \mathbf{x}_2^k) = b_{k-2}(0, 2) + a_{k-2}(2, 0) + b_{k-1}(2, 0) + a_{k-1}(1, 2) + b_k(0, 2) + a_k(0, 1)$$

Figure 4.7 represents 2-space–time codes for the 4PSK constellation and 8 and 16 state encoders for 2 b/s/Hz. The two output bits $(\mathbf{x}_1^k, \mathbf{x}_2^k)$ as a function of the input bits are also shown in the figure. Several additional examples are given in Figures 4.8–4.11 [2].

$$\begin{aligned}
(\mathbf{x}_1^k, \mathbf{x}_2^k) = {} & a_{k-3}(2, 2) + b_{k-2}(3, 3) + a_{k-2}(2, 0) \\
& + b_{k-1}(2, 2) + a_{k-1}(1, 1) \\
& + b_k(0, 2) + a_k(0, 1)
\end{aligned} \tag{4.37}$$

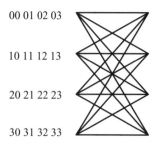

Figure 4.6   2-space–time code, 4PSK, four states, 2 b/s/Hz.

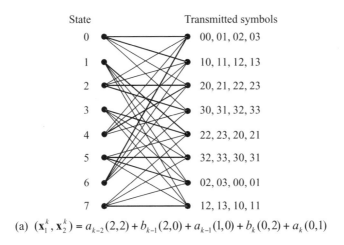

(a) $(\mathbf{x}_1^k, \mathbf{x}_2^k) = a_{k-2}(2,2) + b_{k-1}(2,0) + a_{k-1}(1,0) + b_k(0,2) + a_k(0,1)$

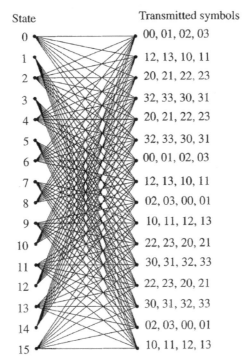

(b) $(\mathbf{x}_1^k, \mathbf{x}_2^k) = b_{k-2}(0,2) + a_{k-2}(2,0) + b_{k-1}(2,0) + a_{k-1}(1,2) + b_k(0,2) + a_k(0,1)$

Figure 4.7   2-space–time codes, 4PSK(a), 8 and (b)16 states, 2 b/s/Hz.

Assuming that the input to the encoder at time k is the three input bits (d_k, b_k, a_k), the output of the encoder at time k is:

$$(\mathbf{x}_1^k, \mathbf{x}_2^k) = d_{k-1}(4, 0) + b_{k-1}(2, 0) + a_{k-1}(5, 0)$$
$$+ d_k(0, 4) + b_k(0, 2) + a_k(0, 1) \qquad (4.38)$$

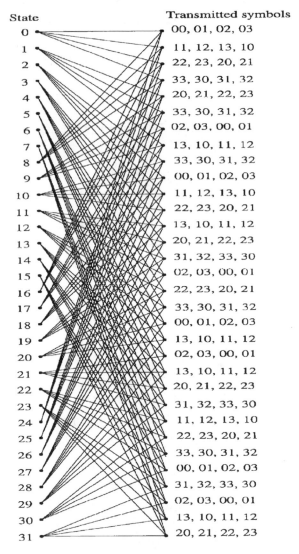

State	Transmitted symbols
0	00, 01, 02, 03
1	11, 12, 13, 10
2	22, 23, 20, 21
3	33, 30, 31, 32
4	20, 21, 22, 23
5	33, 30, 31, 32
6	02, 03, 00, 01
7	13, 10, 11, 12
8	33, 30, 31, 32
9	00, 01, 02, 03
10	11, 12, 13, 10
11	22, 23, 20, 21
12	13, 10, 11, 12
13	20, 21, 22, 23
14	31, 32, 33, 30
15	02, 03, 00, 01
16	22, 23, 20, 21
17	33, 30, 31, 32
18	00, 01, 02, 03
19	13, 10, 11, 12
20	02, 03, 00, 01
21	13, 10, 11, 12
22	20, 21, 22, 23
23	31, 32, 33, 30
24	11, 12, 13, 10
25	22, 23, 20, 21
26	33, 30, 31, 32
27	00, 01, 02, 03
28	31, 32, 33, 30
29	02, 03, 00, 01
30	13, 10, 11, 12
31	20, 21, 22, 23

Figure 4.8   2-space–time code, 4PSK, 32 states, 2 b/s/Hz.

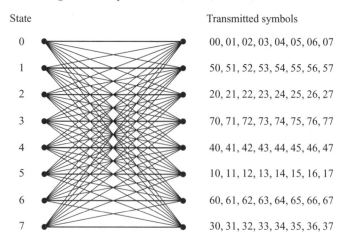

State	Transmitted symbols
0	00, 01, 02, 03, 04, 05, 06, 07
1	50, 51, 52, 53, 54, 55, 56, 57
2	20, 21, 22, 23, 24, 25, 26, 27
3	70, 71, 72, 73, 74, 75, 76, 77
4	40, 41, 42, 43, 44, 45, 46, 47
5	10, 11, 12, 13, 14, 15, 16, 17
6	60, 61, 62, 63, 64, 65, 66, 67
7	30, 31, 32, 33, 34, 35, 36, 37

Figure 4.9   2-space–time code, 8PSK, eight states, 3 b/s/Hz.

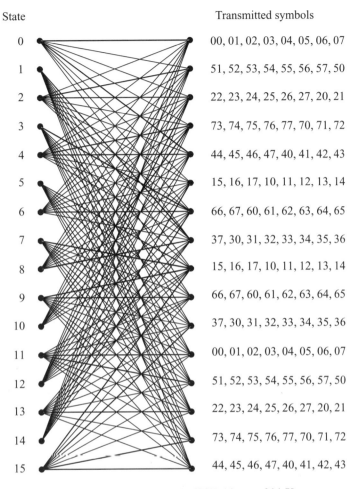

State   Transmitted symbols

0   00, 01, 02, 03, 04, 05, 06, 07

1   51, 52, 53, 54, 55, 56, 57, 50

2   22, 23, 24, 25, 26, 27, 20, 21

3   73, 74, 75, 76, 77, 70, 71, 72

4   44, 45, 46, 47, 40, 41, 42, 43

5   15, 16, 17, 10, 11, 12, 13, 14

6   66, 67, 60, 61, 62, 63, 64, 65

7   37, 30, 31, 32, 33, 34, 35, 36

8   15, 16, 17, 10, 11, 12, 13, 14

9   66, 67, 60, 61, 62, 63, 64, 65

10   37, 30, 31, 32, 33, 34, 35, 36

11   00, 01, 02, 03, 04, 05, 06, 07

12   51, 52, 53, 54, 55, 56, 57, 50

13   22, 23, 24, 25, 26, 27, 20, 21

14   73, 74, 75, 76, 77, 70, 71, 72

15   44, 45, 46, 47, 40, 41, 42, 43

Figure 4.10   2-space–time code, 8PSK, 16 states, 3 b/s/Hz.

where the computation is performed in $\mathbb{Z}_8$, the ring of integers modulo 8, and the elements of the 8PSK constellation have the labeling given in Figure 4.5.

If r_t^j is the received signal at receive antenna j at time t, the branch metric for a transition labeled $q_t^1 q_t^2 \cdots q_t^n$ is given by:

$$\sum_{j=1}^{m} \left| \mathbf{r}_t^j - \sum_{i=1}^{n} \alpha_{i,j} q_t^i \right|^2 \tag{4.39}$$

The Viterbi algorithm is then used to compute the path with the lowest accumulated metric. The frame error probability for four different examples of the coding is shown in Figures 4.12–4.14. The gain shown in these figures should be used in expressions for E_{ff} discussed in Chapters 2 and 3 to evaluate the overall reconfiguration efficiency for different schemes. In general, the expression for efficiency should be further modified as follows.

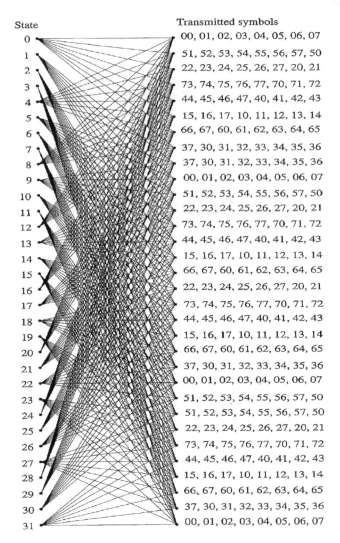

Figure 4.11   2-space–time code, 8PSK, 32 states, 3 b/s/Hz.

4.2.6   Reconfiguration Efficiency of Space–Time Coding

Let rm be the diversity advantage of the system with n transmit and m receive antennas. For a block of l symbols, with constellation Q of 2^b elements, the equivalent rate of transmission R in a system with ST coding satisfies:

$$R \le \frac{\log[A_{2^{bl}}(n, r)]}{l} \tag{4.40}$$

in bits per second per Hertz, where $A_{2^{bl}}(n, r)$ is the maximum size of a code length n and minimum Hamming distance r defined over an alphabet of size 2^{bl} [2].

On the other hand, if b is the transmission rate of a multiple antenna system employed in conjunction with an r-space–time trellis code, the trellis complexity of the space–time code is at least $2^{b(r-1)}$ [2].

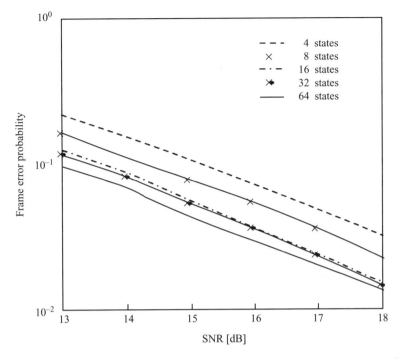

Figure 4.12   Codes for 4PSK with rate 2 b/s/Hz that achieve diversity 2 with one receive and two transmit antennas.

The reconfiguration gain is defined as the solution to:

$$P(\mathbf{c} \to \mathbf{e} \,|\, r, m, g(\text{ST})E_s/4T) = P(\mathbf{c} \to \mathbf{e} \,|\, r = 1, m = 1, E_s/4T) \tag{4.41}$$

Using Inequality (4.33) for Rayleigh fading, Equation (4.41) gives:

$$\left(\prod_{i=1}^{r} \lambda_i\right)^{-m} (g(\text{ST})E_s/4N_0)^{-rm} = \lambda_1^{-1}(E_s/4N_0)$$

resulting in

$$g(\text{ST}) = [\lambda_1^{-1}(E_s/4N_0)]^{rm} \left\{ E_s/4N_0 \left(\prod_{i=1}^{r} \lambda_i\right)^{-1/r} \right\} \tag{4.42}$$

The relative complexity over the signal constellation Q with 2^b elements is:

$$D_r(\text{ST}) > 2^{b(r-1)}/2^b = 2^{b(r-2)} \tag{4.43}$$

Now the reconfiguration efficiency defined by Equation (3.1) becomes:

$$E_{ff} = \left(\frac{R}{b}\right) \frac{g(\text{ST})}{D_r} = [\lambda_1^{-1}(E_s/4N_0)]^{rm} \left\{ E_s/4N_0 \left(\prod_{i=1}^{r} \lambda_i\right)^{-1/r} \right\} \left(\frac{R}{b}\right) 2^{-b(r-2)} \tag{4.44}$$

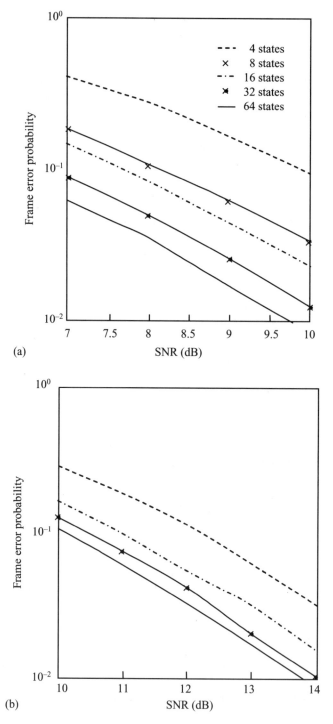

Figure 4.13 (a) Codes for 4PSK with rate 2 b/s/Hz that achieve diversity 4 with two receive and two transmit antennas; (b) codes for 8PSK with rate 3 b/s/Hz that achieve diversity 4 with two receive and two transmit antennas.

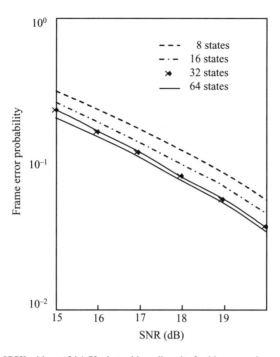

Figure 4.14   Codes for 8PSK with rate 3 b/s/Hz that achieve diversity 2 with one receive and two transmit antennas.

Now we continue with different examples of ST codes. Consider the 8PSK signal constellation, where the encoder maps a sequence of three bits $a_k b_k c_k$ at time k to ii with $i = 4a_k + 2b_k + c_k$. It is easy to show that the equivalent space–time code for this delay diversity code has the trellis representation given in Figure 4.15. The minimum determinant of this code is $(2 - \sqrt{2})^2$. As in the 4PSK case, one can improve the coding advantage of the above codes by constructing encoders with more states. An example is given in Figure 4.16 [2] with the constellation in Figure 4.17.

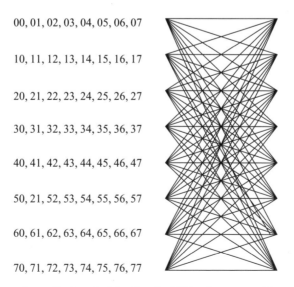

Figure 4.15   Space–time realization of a delay diversity 8PSK code constructed from a repetition code.

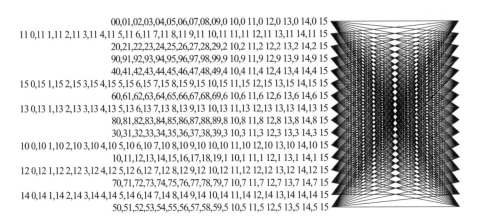

00,01,02,03,04,05,06,07,08,09,0 10,0 11,0 12,0 13,0 14,0 15

11 0,11 1,11 2,11 3,11 4,11 5,11 6,11 7,11 8,11 9,11 10,11 11,11 12,11 13,11 14,11 15

20,21,22,23,24,25,26,27,28,29,2 10,2 11,2 12,2 13,2 14,2 15

90,91,92,93,94,95,96,97,98,99,9 10,9 11,9 12,9 13,9 14,9 15

40,41,42,43,44,45,46,47,48,49,4 10,4 11,4 12,4 13,4 14,4 15

15 0,15 1,15 2,15 3,15 4,15 5,15 6,15 7,15 8,15 9,15 10,15 11,15 12,15 13,15 14,15 15

60,61,62,63,64,65,66,67,68,69,6 10,6 11,6 12,6 13,6 14,6 15

13 0,13 1,13 2,13 3,13 4,13 5,13 6,13 7,13 8,13 9,13 10,13 11,13 12,13 13,13 14,13 15

80,81,82,83,84,85,86,87,88,89,8 10,8 11,8 12,8 13,8 14,8 15

30,31,32,33,34,35,36,37,38,39,3 10,3 11,3 12,3 13,3 14,3 15

10 0,10 1,10 2,10 3,10 4,10 5,10 6,10 7,10 8,10 9,10 10,10 11,10 12,10 13,10 14,10 15

10,11,12,13,14,15,16,17,18,19,1 10,1 11,1 12,1 13,1 14,1 15

12 0,12 1,12 2,12 3,12 4,12 5,12 6,12 7,12 8,12 9,12 10,12 11,12 12,12 13,12 14,12 15

70,71,72,73,74,75,76,77,78,79,7 10,7 11,7 12,7 13,7 14,7 15

14 0,14 1,14 2,14 3,14 4,14 5,14 6,14 7,14 8,14 9,14 10,14 11,14 12,14 13,14 14,14 15

50,51,52,53,54,55,56,57,58,59,5 10,5 11,5 12,5 13,5 14,5 15

Figure 4.16   2-space–time 16 QAM code, 16 states, 4 b/s/Hz.

4.2.6.1   An r Space–Time Trellis Code For r > 2

As an example, a 4-space–time code for four transmit antennas is considered. The limit on the transmission rate is 2 b/s/Hz. Thus, the trellis complexity of the code is bounded below by 64. The input to the encoder is a block of length two of bits a_1, b_1, corresponding to an integer $i = 2a_1 + b_1 \in \mathbb{Z}_4$. The 64 states of the trellis correspond to the set of all three tuples (s_1, s_2, s_3) with $s_i \in \mathbb{Z}_4$ for $1 \le i \le 3$. At state (s_1, s_2, s_3) upon input data i, the encoder outputs (i, s_1, s_2, s_3) elements of the 4PSK constellation (see Figure 4.5) and moves to state (i, s_1, s_2). Given two code words $\mathbf{c}$ and $\mathbf{e}$, the associated paths in the trellis diverge at time t_1 from a state and remerge in another state at a later time $t_2 \le l$. It is easy to see that the t_1th, $(t_1 + l)$th, $(t_2 - 1)$th, and t_2th columns of the matrix $\mathbf{B}(\mathbf{c}, \mathbf{e})$ are independent. Thus, the above design gives a 4-space–time code.

4.2.7   Delay Diversity

The encoder block diagram of a delay diversity transmitter is given in Figure 4.18, with

$$c_t^1 = \tilde{c}_{t-1}^1$$
$$c_t^2 = \tilde{c}_t^2$$

(4.45)

0	1	2	3
○	○	○	○
7	6	5	4
○	○	○	○
8	9	10	11
○	○	○	○
15	14	13	12
○	○	○	○

Figure 4.17   The 16 QAM constellation.

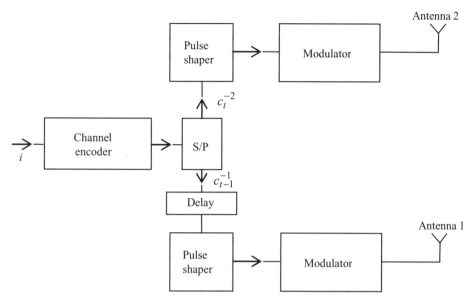

Figure 4.18   The block diagram of a delay diversity transmitter.

where c_t^1 and c_t^2 are the symbols of the equivalent space–time code at time t and $c_t^1 c_t^2$ is the output of the encoder at time t.

Next, we consider the block code [2]

$$\mathbf{C} = \{00, 15, 22, 37, 44, 51, 66, 73\}$$

of length two defined over the alphabet 8PSK instead of the repetition code. This block code is the best in the sense of product distance [10] among all the codes of cardinality eight and of length two defined over the alphabet 8PSK. This means that the minimum of the product distance $|c_1 - e_1||c_2 - e_2|$ between pairs of distinct code words $\mathbf{c} = c_1 c_2 \in C$ and $\mathbf{e} = e_1 e_2 \in C$ is maximal among all such codes. The delay diversity code constructed from this block code is identical to the space–time code given by the trellis diagram of Figure 4.9. The minimum determinant of this delay diversity code is thus 2.

The 16-state code for the 16 QAM constellation given in Figure 4.16, is obtained from the block code

$$\{0\ 0, 1\ 11, 2\ 2, 3\ 9, 4\ 4, 5\ 15, 6\ 6, 7\ 13, 8\ 8, 9\ 3, 10\ 10, 11\ 1, 12\ 12, 13\ 7, 14\ 14, 15\ 5\}$$

using the same delay diversity construction. Again, this block code is optimal in the sense of product distance.

The delay diversity code construction can also be generalized to systems having more than two transmit antennas. For instance, the 4PSK 4-space–time code given before is a delay diversity code. The corresponding block code is the repetition code. By applying the delay diversity construction to the 4PSK block code

$$\{0\ 0\ 0\ 0, 1\ 2\ 3\ 1, 2\ 1\ 2\ 3, 3\ 3\ 1\ 2\}$$

one can obtain a more powerful 4PSK 4-space–time code having the same trellis complexity.

4.3   Space–Time Block Codes from Orthogonal Designs

ML decoding of a multidimensional trellis requires a large complexity. In this section we present a special class of space–time codes for which maximum likelihood decoding is achieved in a simple way through decoupling of the signals transmitted from different antennas rather than joint detection. This uses the orthogonal structure of the space–time block code and gives a maximum likelihood decoding algorithm which is based only on linear processing at the receiver. The presentation is based on [5, 39–46].

Space–time block codes are designed to achieve the maximum diversity order for a given number of transmit and receive antennas subject to the constraint of having a simple decoding algorithm. Unfortunately, space–time block codes constructed in this way only exist for few sporadic values of n.

4.3.1   The Channel Model and the Diversity Criterion

At time t the signal r_t^j, received at antenna j, is given by Equation (4.21) which for $\sqrt{E_s} = 1$ becomes:

$$r_t^j = \sum_{i=1}^{n} \alpha_{i,j} c_t^i + \eta_t^j \tag{4.46}$$

Assuming perfect channel state information is available, the receiver computes the decision metric for l symbols and Expression (4.39) gives:

$$\sum_{t=1}^{l} \sum_{j=1}^{m} \left| r_t^j - \sum_{i=1}^{n} \alpha_{i,j} c_t^i \right|^2 \tag{4.47}$$

In order to achieve the maximum diversity mn, the matrix $\mathbf{A}\,(\mathbf{c},\mathbf{e}) = \mathbf{B}(\mathbf{c},\mathbf{e})^* \, \mathbf{B}(\mathbf{c},\mathbf{e})$ with

$$\mathbf{B(c,\, e)} = \begin{bmatrix} e_1^1 - c_1^1 & e_2^1 - c_2^1 & \cdots & \cdots & e_l^1 - c_l^1 \\ e_1^2 - c_1^2 & e_2^2 - c_2^2 & \cdots & \cdots & e_l^2 - c_l^2 \\ e_1^3 - c_1^3 & e_2^3 - c_2^3 & \ddots & \vdots & e_l^3 - c_l^3 \\ \vdots & \vdots & \ddots & \ddots & \vdots \\ e_1^n - c_1^n & e_2^n - c_2^n & \cdots & \cdots & e_l^n - c_l^n \end{bmatrix} \tag{4.48}$$

has to be full rank for any pair of distinct code words $\mathbf{c}$ and $\mathbf{e}$. If $\mathbf{B(c, e)}$ has minimum rank r over the set of pairs of distinct code words, then a diversity of rm is achieved.

4.3.2   Real Orthogonal Designs

For $n = 2, 4$ or 8, mathematics literature offers orthogonal sets of signals defined as:

$$\mathbf{O}_2(\mathbf{x}_1, \mathbf{x}_2) = \begin{bmatrix} x_1 & x_2 \\ -x_2 & x_1 \end{bmatrix} \tag{4.49}$$

The 4×4 design

$$\mathbf{O}_4(\mathbf{x}_1, \mathbf{x}_2, \mathbf{x}_3, \mathbf{x}_4) = \begin{bmatrix} x_1 & x_2 & x_3 & x_4 \\ -x_2 & x_1 & -x_4 & x_3 \\ -x_3 & x_4 & x_1 & -x_2 \\ -x_4 & -x_3 & x_2 & x_1 \end{bmatrix} \tag{4.50}$$

and 8×8 design

$$\mathbf{O}_8(\mathbf{x}_1, \mathbf{x}_2, \ldots \mathbf{x}_8) = \begin{bmatrix} x_1 & x_2 & x_3 & x_4 & x_5 & x_6 & x_7 & x_8 \\ -x_2 & x_1 & x_4 & -x_3 & x_6 & -x_5 & -x_8 & x_7 \\ -x_3 & -x_4 & x_1 & x_2 & x_7 & x_8 & -x_5 & -x_6 \\ -x_4 & x_3 & -x_2 & x_1 & x_8 & -x_7 & x_6 & -x_5 \\ -x_5 & -x_6 & -x_7 & -x_8 & x_1 & x_2 & x_3 & x_4 \\ -x_6 & x_5 & -x_8 & x_7 & -x_2 & x_1 & -x_4 & x_3 \\ -x_7 & x_8 & x_5 & -x_6 & -x_3 & x_4 & x_1 & -x_2 \\ -x_8 & -x_7 & x_6 & x_5 & -x_4 & -x_3 & x_2 & x_1 \end{bmatrix} \tag{4.51}$$

4.3.3   Space–Time Encoder

At time slot 1, nb bits arrive at the encoder and select constellation signals $s_1, \ldots, s_n$. Setting $x_i = s_i$ for $i = 1, 2, \ldots, n$, we arrive at a matrix $\mathbf{C} = \mathbf{O}(s_1, \ldots, s_n)$ with entries $\pm s_1, \ \pm s_2, \ldots, \pm s_n$. At each time slot $t = 1, 2, \ldots, n$, the entries $C_{ti}, i = 1, 2, \ldots, n$ are transmitted simultaneously from transmit antennas 1, 2, $\ldots, n$. The rate of transmission is b bits/s/Hz.

4.3.4   The Diversity Order

The rank criterion requires that the matrix $\mathbf{O}(\tilde{s}_1, \ldots, \tilde{s}_n) - \mathbf{O}(s_1, \ldots, s_n)$ be non-singular for any two distinct code sequences $(\tilde{s}_1, \ldots, \tilde{s}_n) \neq (s_1, \ldots, s_n)$. Clearly, $O(\tilde{s}_1 - s_1, \ldots, \tilde{s}_n - s_n) = O(\tilde{s}_1, \ldots, \tilde{s}_n) - \mathbf{O}(s_1, \ldots, s_n)$ where $\mathbf{O}(\tilde{s}_1 - s_1, \ldots, \tilde{s}_n - s_n)$ is the matrix constructed from $\mathbf{O}$ by replacing x_i with $\tilde{s}_i - s_i$ for all $i = 1, 2, \ldots, n$. The determinant of the orthogonal matrix $\mathbf{O}$ is easily seen to be:

$$\det(\mathbf{OO}^{\mathrm{T}})^{1/2} = \left[\sum_i x_i^2 \right]^{n/2} \tag{4.52}$$

where $\mathbf{O}^{\mathrm{T}}$ is the transpose of $\mathbf{O}$. Hence:

$$\det[\mathbf{O}(\tilde{s}_1 - s_1, \ldots, \tilde{s}_n - s_n)] = \left[\sum_i |\tilde{s}_i - s_i|^2 \right]^{n/2} \tag{4.53}$$

which is non-zero. It follows that $\mathbf{O}(\tilde{s}_1, \ldots, \tilde{s}_n) - \mathbf{O}(s_1, \ldots, s_n)$ is non-singular and the maximum diversity order nm is achieved.

4.3.5   The Decoding Algorithm

Rows of $\mathbf{O}$ are all permutations of the first row of $\mathbf{O}$ with possibly different signs. Let $e_1, \ldots, e_n$ denote the permutations corresponding to these rows and let $\delta_k(i)$ denote the sign of x_i in the kth row of $\mathbf{O}$. Then $e_k(p) = q$ means that x_p is up to a sign change the (k, q)th element of $\mathbf{O}$. Since the columns of $\mathbf{O}$ are pairwise-orthogonal, it turns out that minimizing the metric of Expression (4.47) amounts to minimizing:

$$\sum_{i=1}^{n} S_i \tag{4.54}$$

where

$$S_i = \left(\left| \left[\sum_{t=1}^{n} \sum_{j=1}^{m} \mathbf{r}_t^j \alpha_{e_t(i),j}^* \delta_t(i) \right] - s_i \right|^2 + \left(-1 + \sum_{k,l} |\alpha_{k,t}|^2 \right) |s_i|^2 \right) \tag{4.55}$$

The value of S_i only depends on the code symbol s_i, the received symbols $\{r_t^j\}$, the path coefficients $\{\alpha_{i,j}\}$, and the structure of the orthogonal design $\mathbf{O}$. It follows that minimizing the sum given in Expression (4.54) amounts to minimizing Equation (4.55) for all $1 \le i \le n$. Thus, the maximum likelihood detection rule is to form the decision variables:

$$R_i = \sum_{t=1}^{n} \sum_{j=1}^{m} \mathbf{r}_t^j \alpha_{e_t(i),j}^* \delta_t(i) \tag{4.56}$$

for all $i = 1, 2, \ldots, n$ and decide in favor of s_i among all the constellation symbols s if:

$$s_i = \arg \min_{s \in A} |R_i - s|^2 + \left(-1 + \sum_{k,l} |\alpha_{k,t}|^2 \right) |s_i|^2 \tag{4.57}$$

This is a very simple decoding strategy that provides diversity.

4.3.6 Linear Processing Orthogonal Designs

The above properties are preserved even if we allow linear processing at the transmitter. Therefore, we relax the definition of orthogonal designs to allow linear processing at the transmitter. Signals transmitted from different antennas will now be linear combinations of constellation symbols.

A linear processing orthogonal design in variables $x_1, x_2, \ldots, x_n$ is an $n \times n$ matrix $\mathbf{E}$ such that:

- the entries of $\mathbf{E}$ are real linear combinations of variables $x_1, x_2, \ldots, x_n$;
- $\mathbf{E}^T \mathbf{E} = \mathbf{D}$, where $\mathbf{D}$ is a diagonal matrix with (i, i)th diagonal element of the form $(l_1^i x_1^2 + l_2^i x_2^2 + \cdots l_n^i x_n^2)$, with the coefficients $(l_1^i, l_2^i, \ldots, l_n^i)$ strictly positive numbers.

It is easy to show that transmission using a linear processing orthogonal design provides full diversity and a simplified decoding algorithm as above.

4.3.7 Generalized Real Orthogonal Designs

Since the simple maximum likelihood decoding algorithm described above is achieved because of orthogonality of columns of the design matrix, we may generalize the definition of linear processing orthogonal designs. This creates new and simple transmission schemes for any number of transmit antennas.

A generalized orthogonal design $\mathbf{G}$ of size n is a $p \times n$ matrix with entries $0, \pm x_1, \pm x_2, \ldots, \pm x_k$, such that $\mathbf{G}^T \mathbf{G} = \mathbf{D}$ where $\mathbf{D}$ is a diagonal matrix with diagonal $\mathbf{D}_{ii}$, $i = 1, 2, \ldots, n$ of the form $(l_1^i x_1^2 + l_2^i x_2^2 + \cdots l_n^i x_n^2)$ and coefficients $(l_1^i, l_2^i, \ldots, l_n^i)$ strictly positive numbers. The rate of $\mathbf{G}$ is $R = k/p$.

A full-rate generalized orthogonal design has entries of the form $\pm x_1, \pm x_2, \ldots, \pm x_p$.

The generalized orthogonal signal sets are:

$$
\mathbf{G}_3 = \begin{bmatrix} x_1 & x_2 & x_3 \\ -x_2 & x_1 & -x_4 \\ -x_3 & x_4 & x_1 \\ -x_4 & -x_3 & x_2 \end{bmatrix}
\tag{4.58}
$$

$$
\mathbf{G}_5 = \begin{bmatrix}
x_1 & x_2 & x_3 & x_4 & x_5 \\
-x_2 & x_1 & x_4 & -x_3 & x_6 \\
-x_3 & -x_4 & x_1 & x_2 & x_7 \\
-x_4 & x_3 & -x_2 & x_1 & x_8 \\
-x_5 & -x_6 & -x_7 & -x_8 & x_1 \\
-x_6 & x_5 & -x_8 & x_7 & -x_2 \\
-x_7 & x_8 & x_5 & -x_6 & -x_3 \\
-x_8 & -x_7 & x_6 & x_5 & -x_4
\end{bmatrix}
\tag{4.59}
$$

$$
\mathbf{G}_6 = \begin{bmatrix}
x_1 & x_2 & x_3 & x_4 & x_5 & x_6 \\
-x_2 & x_1 & x_4 & -x_3 & x_6 & -x_5 \\
-x_3 & -x_4 & x_1 & x_2 & x_7 & x_8 \\
-x_4 & x_3 & -x_2 & x_1 & x_8 & -x_7 \\
-x_5 & -x_6 & -x_7 & -x_8 & x_1 & x_2 \\
-x_6 & x_5 & -x_8 & x_7 & -x_2 & x_1 \\
-x_7 & x_8 & x_5 & -x_6 & -x_3 & x_4 \\
-x_8 & -x_7 & x_6 & x_5 & -x_4 & -x_3
\end{bmatrix}
\tag{4.60}
$$

$$
\mathbf{G}_7 = \begin{bmatrix}
x_1 & x_2 & x_3 & x_4 & x_5 & x_6 & x_7 \\
-x_2 & x_1 & x_4 & -x_3 & x_6 & -x_5 & -x_8 \\
-x_3 & -x_4 & x_1 & x_2 & x_7 & x_8 & -x_5 \\
-x_4 & x_3 & -x_2 & x_1 & x_8 & -x_7 & x_6 \\
-x_5 & -x_6 & -x_7 & -x_8 & x_1 & x_2 & x_3 \\
-x_6 & x_5 & -x_8 & x_7 & -x_2 & x_1 & -x_4 \\
-x_7 & x_8 & x_5 & -x_6 & -x_3 & x_4 & x_1 \\
-x_8 & -x_7 & x_6 & x_5 & -x_4 & -x_3 & x_2
\end{bmatrix}
\tag{4.61}
$$

4.3.8   Encoding

At time slot 1, kb bits arrive at the encoder and select constellation symbols $s_1, s_2, \ldots, s_n$. The encoder populates the matrix by setting $x_i = s_i$ and at time $t = 1, 2, \ldots, p$ the signals $G_{t1}, \ldots, G_{tn}$ are transmitted simultaneously from antennas $1, 2, \ldots, n$. Thus, kb bits are sent during each p transmissions. It can be proved, as in Equation (4.53), that the diversity order is nm. It should be mentioned that the rate of a generalized orthogonal design is different from the throughput of the associated code. To motivate the definition of the rate, we note that the theory of space–time coding proves that for a diversity order of nm, it is possible to transmit b bits per time slot and this is the best possible. Therefore, the rate R of this coding scheme is defined to be kb/pb which is equal to k/p.

4.3.9   The Alamouti Scheme

The space–time block code, already used in Section 4.1 was proposed by Alamouti [1]. The code uses the complex orthogonal signal set:

$$
\begin{bmatrix} x_1 & x_2 \\ -x_2^* & x_1^* \end{bmatrix}
\tag{4.62}
$$

Suppose that there are 2^b signals in the constellation. At the first time slot, 2^b bits arrive at the encoder and select two complex symbols s_1 and s_2. These symbols are transmitted simultaneously from antennas one and two, respectively. At the second time slot, signals $-s_2^*$ and s_1^* are transmitted simultaneously from antennas one and two, respectively.

The ML detector will minimize the decision statistic:

$$\sum_{j=1}^{m} (|r_1^j - \alpha_{1,j}s_1 - \alpha_{2,j}s_2|^2 + |r_2^j - \alpha_{1,j}s_1^* - \alpha_{2,j}s_2^*|^2) \tag{4.63}$$

over all possible values of s_1 and s_2. The minimizing values are the receiver estimates of s_1 and s_2, respectively. This is equivalent to minimizing the decision statistics:

$$\left(\left| \left[\sum_{j=1}^{m} (r_1^j \alpha_{1,j}^* + (r_2^j)^* \alpha_{2,j}) \right] - s_1 \right|^2 + \left(-1 + \sum_{j=1}^{m} \sum_{i=1}^{2} |\alpha_{i,j}|^2 \right) |s_1|^2 \right) \tag{4.64}$$

for detecting s_1 and the decision statistics:

$$\left(\left| \left[\sum_{j=1}^{m} (r_1^j \alpha_{2,j}^* + (r_2^j)^* \alpha_{1,j}) \right] - s_2 \right|^2 + \left(-1 + \sum_{j=1}^{m} \sum_{i=1}^{2} |\alpha_{i,j}|^2 \right) |s_2|^2 \right) \tag{4.65}$$

for decoding s_2. The scheme provides full diversity of $2m$ using m receive antennas.

4.3.10 Complex Orthogonal Designs

Given a complex orthogonal signal set $\mathbf{O}_c$ of size n, we replace each complex variable $x_i = x_i^1 + x_i^2 \mathbf{i}$, $1 \le i \le n$ by the 2×2 real matrix:

$$\begin{bmatrix} x_i^1 & x_i^2 \\ -x_i^2 & x_i^1 \end{bmatrix} \tag{4.66}$$

In this way, x_i^* is represented by:

$$\begin{bmatrix} x_i^1 & -x_i^2 \\ x_i^2 & x_i^1 \end{bmatrix} \tag{4.67}$$

$\mathbf{i}x_i$ is represented by:

$$\begin{bmatrix} -x_i^2 & x_i^1 \\ -x_i^1 & -x_i^2 \end{bmatrix} \tag{4.68}$$

and so forth. It is easy to see that the $2n \times 2n$ matrix formed in this way is a real orthogonal design of size $2n$.

A complex orthogonal signal set of size n exists if and only if $n = 2$.

4.3.11 Generalized Complex Orthogonal Designs

Let $\mathbf{G}_c$ be a $p \times n$ matrix whose entries are $0, \pm x_1, \pm x_1^*, \pm x_2, \pm x_2^*, \ldots, \pm x_k, \pm x_k^*$ or their product with $\mathbf{i}$. If $\mathbf{G}_c^* \mathbf{G}_c = \mathbf{D}_c$ where $\mathbf{D}_c$ is a diagonal matrix with (i, i)th diagonal element of the form:

$$(l_1^i |x_1|^2 + l_2^i |x_2|^2 + \cdots l_k^i |x_k|^2)$$

and the coefficients $(l_1^i, l_2^i, \ldots, l_n^i)$ all strictly positive numbers, then $\mathbf{G_c}$ is referred to as a generalized orthogonal design of size n and rate $R = k/p$. For instance, rate 1/2 codes for transmission using three and four transmit antennas are given by:

$$\mathbf{G_c^3} = \begin{bmatrix} x_1 & x_2 & x_3 \\ -x_2 & x_1 & -x_4 \\ -x_3 & x_4 & x_1 \\ -x_4 & -x_3 & x_2 \\ x_1^* & x_2^* & x_3^* \\ -x_2^* & x_1^* & -x_4^* \\ -x_3^* & x_4^* & x_1^* \\ -x_4^* & -x_3^* & x_2^* \end{bmatrix} \tag{4.69}$$

and

$$\mathbf{G_c^4} = \begin{bmatrix} x_1 & x_2 & x_3 & x_4 \\ -x_2 & x_1 & -x_4 & x_3 \\ -x_3 & x_4 & x_1 & -x_2 \\ -x_4 & -x_3 & x_2 & x_1 \\ x_1^* & x_2^* & x_3^* & x_4^* \\ -x_2^* & x_1^* & -x_4^* & x_3^* \\ -x_3^* & x_4^* & x_1^* & -x_2^* \\ -x_4^* & -x_3^* & x_2^* & x_1^* \end{bmatrix} \tag{4.70}$$

These transmission schemes and their analogs for higher n give full diversity but lose half of the theoretical bandwidth efficiency.

4.3.12 Special Codes

It is natural to ask for higher rates than 1/2 when designing generalized complex linear processing orthogonal designs for transmission with n multiple antennas. For $n = 2$, Alamouti's scheme gives a rate one design. For $n = 3$ and 4, rate 3/4 generalized complex linear processing orthogonal designs are given by:

$$\mathbf{H_3} = \begin{bmatrix} x_1 & x_2 & \dfrac{x_3}{\sqrt{2}} \\[2mm] -x_2^* & x_1^* & \dfrac{x_3}{\sqrt{2}} \\[2mm] \dfrac{x_3^*}{\sqrt{2}} & \dfrac{x_3^*}{\sqrt{2}} & \dfrac{(-x_1 - x_1^* + x_2 - x_2^*)}{2} \\[2mm] \dfrac{x_3^*}{\sqrt{2}} & -\dfrac{x_3^*}{\sqrt{2}} & \dfrac{(x_1 - x_1^* + x_2 + x_2^*)}{2} \end{bmatrix}$$

$$\mathbf{H_4} = \begin{bmatrix} x_1 & x_2 & \dfrac{x_3}{\sqrt{2}} & \dfrac{x_3}{\sqrt{2}} \\[2mm] -x_2^* & x_1^* & \dfrac{x_3}{\sqrt{2}} & -\dfrac{x_3}{\sqrt{2}} \\[2mm] \dfrac{x_3^*}{\sqrt{2}} & \dfrac{x_3^*}{\sqrt{2}} & \dfrac{(-x_1 - x_1^* + x_2 - x_2^*)}{2} & \dfrac{(x_1 - x_1^* - x_2 - x_2^*)}{2} \\[2mm] \dfrac{x_3^*}{\sqrt{2}} & -\dfrac{x_3^*}{\sqrt{2}} & \dfrac{(x_1 - x_1^* + x_2 + x_2^*)}{2} & -\dfrac{(x_1 + x_1^* + x_2 - x_2^*)}{2} \end{bmatrix} \tag{4.71}$$

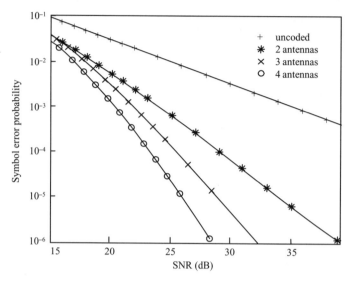

Figure 4.19   Symbol error probability versus SNR for space–time block codes at 3 bits/s/Hz; one receive antenna.

4.3.13   Performance Results

A collection of results is shown in Figures 4.19–4.26. The transmission using two transmit antennas employs the 8PSK constellation and the code G_2. For three and four transmit antennas, the 16 QAM constellation and the codes H_3 and H_4, respectively, are used. Since H_3 and H_4 are rate 3/4 codes, the

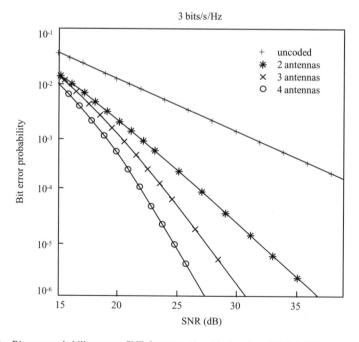

Figure 4.20   Bit error probability versus SNR for space–time block codes at 3 bits/s/Hz; one receive antenna.

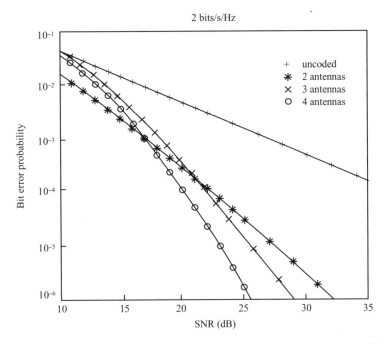

Figure 4.21   Bit error probability versus SNR for space–time block codes at 2 bits/s/Hz; one receive antenna.

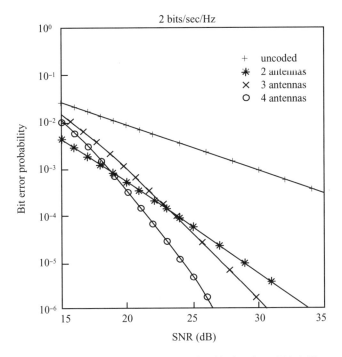

Figure 4.22   Symbol error probability versus SNR for space–time block codes at 2 bits/s/Hz; one receive antenna.

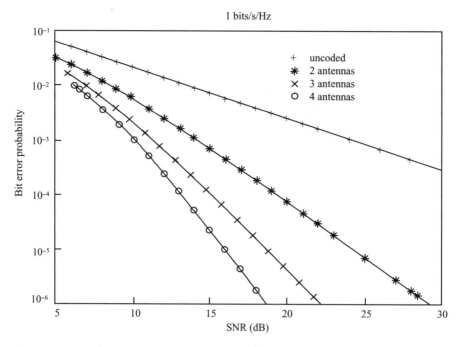

Figure 4.23 Bit error probability versus SNR for space–time block codes at 1 bit/s/Hz; one receive antenna.

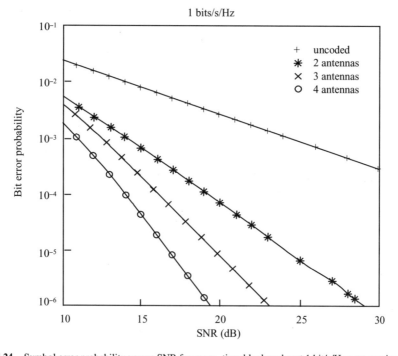

Figure 4.24 Symbol error probability versus SNR for space–time block codes at 1 bit/s/Hz; one receive antenna.

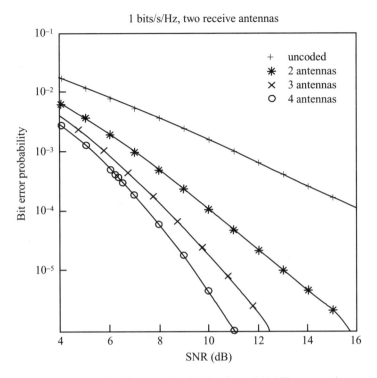

Figure 4.25   BER versus SNR for space–time block codes at. 1 bit/s/Hz; two receive antennas.

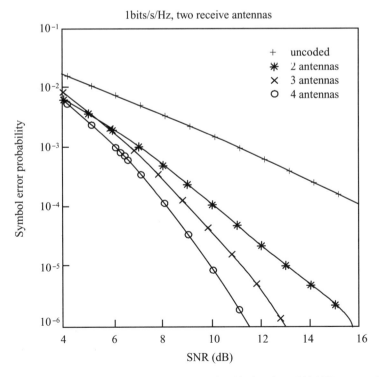

Figure 4.26   Symbol error probability versus SNR for space–time block codes at 1 bit/s/Hz; two receive antennas.

total transmission rate in each case is 3 bits/s/Hz. It is seen that at the bit error rate of 10^{-5} the rate 3/4 16 QAM code $\mathbf{H}_4$ gives about 7 dB gain over the use of an 8PSK $\mathbf{G}_2$ code.

Transmission using two transmit antennas employs the 4PSK constellation and the code $\mathbf{G}_2$. For three and four transmit antennas, the 16-QAM constellation and the codes $\mathbf{G}_3$ and $\mathbf{G}_4$, respectively, are used. Since $\mathbf{G}_3$ and $\mathbf{G}_4$ are rate 1/2 codes, the total transmission rate in each case is 2 bits/s/Hz. It is seen that at the bit error rate of 10^{-5} the rate 1/2 16-QAM code $\mathbf{G}_4$ gives about 5 dB gain over the use of a 4PSK $\mathbf{G}_2$ code.

The transmission using two transmit antennas employs the binary PSK (BPSK) constellation and the code $\mathbf{G}_2$. For three and four transmit antennas, the 4PSK constellation and the codes $\mathbf{G}_3$ and $\mathbf{G}_4$, respectively, are used. Since $\mathbf{G}_3$ and $\mathbf{G}_4$ are rate 1/2 codes, the total transmission rate in each case is 1 bit/s/Hz. It is seen that at the bit error rate of 10^{-5} the rate 1/2 4PSK code $\mathbf{G}_4$ gives about 7.5 dB gain over the use of a BPSK $\mathbf{G}_2$ code.

If the number of receive antennas is increased, this gain reduces to 3.5 dB. The reason is that much of the diversity gain is already achieved using two transmit and two receive antennas.

4.4   Channel Estimation Imperfections

So far we have been assuming that a perfect channel estimation is available for the operation of the ML decoder collecting statistics defined by Expression (4.39) or Expression (4.47). Let us look at this assumption more carefully. First of all, the errors in the channel estimation will depend on the estimator structure.

4.4.1   Channel Estimator

At the beginning of each frame of symbols to be transmitted from transmit antenna i, a sequence $\mathbf{W}_i$ of length k pilot symbols:

$$\mathbf{W}_i = (W_{i,1}, W_{i,2}, \ldots, W_{i,k})$$

is appended. The sequences $W_1, \mathbf{W}_2, \ldots, \mathbf{W}_n$ are designed to be orthogonal to each other:

$$\mathbf{W}_p \overline{\mathbf{W}}_q = \sum_{j=1}^{k} W_{p,j} \overline{W}_{q,j} = 0$$

whenever $p \neq q$. In the previous expression $\overline{W}$ stands for the conjugate of W and $\overline{\mathbf{W}}$ is the conjugate transpose of $\mathbf{W}$. Let $\mathbf{r}^j = (r_1^j, r_2^j, \ldots, r_k^j)$ be the observed sequence of received signals at antenna j during the training period. Then:

$$r_t^j = \sum_{i=1}^{n} \alpha_{i,j} W_{i,t} + \eta_{t,j}, \quad 1 \leq j \leq m, \quad 1 \leq t \leq k$$

where the channel coefficients $\alpha_{i,j}$ are independent samples of a complex Gaussian random variable with mean zero and variance 0.5 per dimension, and $\eta_{t,j}$ are independent samples of a zero mean complex Gaussian random variable with variance $N_0/2$ per dimension. Let $\eta^j = (\eta_{1,j}, \eta_{2,j}, \ldots, \eta_{k,j})$. Our goal is to estimate $\alpha_{i,j}$, $i = 1, 2, \ldots, n$, $j = 1, 2, \ldots, m$ using the statistic $\mathbf{r}^j$.

The unbiased estimator $\beta_{i,j}$ having the least variance is given by the ratio of inner products ($\mathbf{r}^j \cdot \overline{\mathbf{W}}^i$)/($\mathbf{W}^i \cdot \overline{\mathbf{W}}^i$). Indeed, since $\mathbf{W}^p \overline{\mathbf{W}}^q = 0$ it is easy to see that:

$$\mathbf{r}^j \cdot \overline{\mathbf{W}}^i = \alpha_{i,j}(\mathbf{W}^i \cdot \overline{\mathbf{W}}^i) + \eta^j \cdot \overline{\mathbf{W}}^i$$

thus

$$\alpha_{i,j} = \frac{\mathbf{r}^j \cdot \overline{\mathbf{W}}^i}{\mathbf{W}^i \cdot \overline{\mathbf{W}}^i} - \frac{\eta^j \cdot \overline{\mathbf{W}}^i}{\mathbf{W}^i \cdot \overline{\mathbf{W}}^i}$$

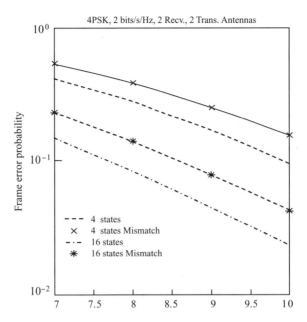

Figure 4.27   Performance of four and 16 state 4PSK codes in the presence of channel estimation error, 2 bits/s/Hz, two receive and two transmit antennas.

In other words,

$$\beta_{i,j} = \alpha_{i,j} + \frac{\eta^j \cdot \overline{\mathbf{W}}^i}{\mathbf{W}^i \cdot \overline{\mathbf{W}}^i}$$

The random variable $\beta_{i,j}$ has zero mean. The variance of the estimation error is $N_0/2kE_s$ per dimension which is the minimum given by the Cramer–Rao bound. Simulation results for imperfect channel estimation (mismatch) with $n = 2$, $k = 8$, and the frame length 130 symbols are shown in Figure 4.27.
   More details on system imperfections can be found in [7, 47, 48].

4.5   Quasi-Orthogonal Space–Time Block Codes

It was shown in Section 4.3 that a complex orthogonal design that provides full diversity and full transmission rate for a space–time block code is not possible for more than two antennas. Previous attempts have been concentrated in generalizing orthogonal designs which provide space–time block codes with full diversity and a high transmission rate. In this section we discuss rate one codes which are quasi-orthogonal and provide partial diversity. The decoder for these codes works with pairs of transmitted symbols instead of single symbols.
   An example of a full rate full diversity complex space–time block code is the Alamouti scheme already discussed in Section 4.3. In this section we will use the notation:

$$\mathcal{A}_{12} = \begin{bmatrix} x_1 & x_2 \\ -x_2^* & x_1^* \end{bmatrix} \tag{4.72}$$

Here we use the subscript 12 to represent the indeterminates x_1 and x_2 in the transmission matrix. Now, let us consider the following space–time block code with block length T symbol intervals where K

bits are transmitted over N transmit antennas and received with M receive antennas, for $N = T = K = 4$:

$$A = \begin{bmatrix} \mathcal{A}_{12} & \mathcal{A}_{34} \\ -\mathcal{A}_{34}^* & \mathcal{A}_{12}^* \end{bmatrix} = \begin{bmatrix} x_1 & x_2 & x_3 & x_4 \\ -x_2^* & x_1^* & -x_4^* & x_3^* \\ -x_3^* & -x_4^* & x_1^* & x_2^* \\ x_4 & -x_3 & -x_2 & x_1 \end{bmatrix} \tag{4.73}$$

It is easy to see that the minimum rank of matrix $A(s_1 - \bar{s}_1, s_2 - \bar{s}_2, s_3 - \bar{s}_3, s_4 - \bar{s}_4)$, the matrix constructed from A by replacing x_i with $s_i - \bar{s}_i$, is 2. Therefore, a diversity of $2M$ is achieved while the rate of the code is one. Note that the maximum diversity of $4M$ for a rate one code is impossible in this case.

4.5.1   Decoding

Assuming perfect channel state information is available, the ML receiver computes the decision metric:

$$\sum_{m=1}^{M} \sum_{t=1}^{T} \left| r_{t,m} - \sum_{n=1}^{N} \alpha_{n,m} A_{tn} \right|^2 \tag{4.74}$$

over all possible $x_k = s_k \in A$ and decides in favor of the constellation symbols $s_1, \ldots, s_K$ that minimize this sum.

4.5.2   Decision Metric

Now, if we define $\mathcal{V}_i$, $i = 1, 2, 3, 4$, as the ith column of A, it is easy to see that

$$\langle \mathcal{V}_1, \mathcal{V}_2 \rangle = \langle \mathcal{V}_1, \mathcal{V}_3 \rangle = \langle \mathcal{V}_2, \mathcal{V}_4 \rangle = \langle \mathcal{V}_3, \mathcal{V}_4 \rangle = 0 \tag{4.75}$$

where $\langle \mathcal{V}_i, \mathcal{V}_j \rangle = \sum_{l=1}^{4} (\mathcal{V}_i)_l (\mathcal{V}_j)_l^*$ is the inner product of vectors $\mathcal{V}_i$ and $\mathcal{V}_j$. Therefore, the subspace created by $\mathcal{V}_1$ and $\mathcal{V}_4$ is orthogonal to the subspace created by $\mathcal{V}_2$ and $\mathcal{V}_3$. Using this orthogonality, the maximum likelihood decision metric, Expression (4.74), can be calculated as the sum of two terms $f_{14}(x_1, x_4) + f_{23}(x_2, x_3)$, where f_{14} is independent of x_2 and x_3 and f_{23} is independent of x_1 and x_4. Thus, the minimization of Expression (4.74) is equivalent to minimizing these two terms independently. In other words, first the decoder finds the pair (s_1, s_4) that minimizes $f_{14}(x_1, x_4)$ among all possible (x_1, x_4) pairs. Then, or in parallel, the decoder selects the pair (s_2, s_3) which minimizes $f_{23}(x_2, x_3)$. This reduces the complexity of decoding without sacrificing the performance.

Simple manipulation of Expression (4.74) provides the following formulas for $f_{14}(.)$ and $f_{23}(.)$:

$$\begin{aligned}
f_{14}(x_1, x_4) = \sum_{m=1}^{M} \Bigg(& \left(\sum_{n=1}^{4} |\alpha_{n,m}|^2 \right) (|x_1|^2 + |x_4|^2) \\
& + 2\mathrm{Re}\{(-\alpha_{1,m} r_{1,m}^* - \alpha_{2,m}^* r_{2,m} - \alpha_{3,m}^* r_{3,m} - \alpha_{4,m} r_{4,m}^*) x_1 \\
& + (-\alpha_{4,m} r_{1,m}^* + \alpha_{3,m}^* r_{2,m} + \alpha_{2,m}^* r_{3,m} - \alpha_{1,m} r_{4,m}^*) x_4 \\
& + (\alpha_{1,m} \alpha_{4,m}^* - \alpha_{2,m}^* \alpha_{3,m} - \alpha_{2,m} \alpha_{3,m}^* + \alpha_{1,m}^* \alpha_{4,m}) x_1 x_4^* \} \Bigg)
\end{aligned} \tag{4.76}$$

$$f_{23}(x_2, x_3) = \sum_{m=1}^{M} \left(\left(\sum_{n=1}^{4} |\alpha_{n,m}|^2 \right) (|x_2|^2 + |x_3|^2) \right.$$

$$+2\mathrm{Re}\{(-\alpha_{2,m}r_{1,m}^* - \alpha_{1,m}^* r_{2,m} - \alpha_{4,m}^* r_{3,m} - \alpha_{3,m}r_{4,m}^*)x_2$$

$$+(-\alpha_{3,m}r_{1,m}^* + \alpha_{4,m}^* r_{2,m} + \alpha_{1,m}^* r_{3,m} - \alpha_{2,m}r_{4,m}^*)x_3 + (\alpha_{2,m}\alpha_{3,m}^*$$

$$\left. - \alpha_{1,m}^*\alpha_{4,m} - \alpha_{1,m}\alpha_{4,m}^* + \alpha_{2,m}^*\alpha_{3,m})x_2 x_3^*\} \right) \tag{4.77}$$

There are other structures which provide behaviors similar to those of Equation (4.73). A few examples are given below:

$$\begin{bmatrix} \mathcal{A}_{12} & \mathcal{A}_{34} \\ -\mathcal{A}_{34} & \mathcal{A}_{12} \end{bmatrix} \quad \begin{bmatrix} \mathcal{A}_{12} & \mathcal{A}_{34} \\ \mathcal{A}_{34} & -\mathcal{A}_{12} \end{bmatrix} \quad \begin{bmatrix} \mathcal{A}_{12} & \mathcal{A}_{34} \\ \mathcal{A}_{34}^* & -\mathcal{A}_{12}^* \end{bmatrix} \tag{4.78}$$

A similar idea can be used to combine two rate 3/4 transmission matrices (4×4) to build a rate 3/4 transmission matrix (8×8) and so on. An example of an 8×8 matrix which provides a rate 3/4 code is given below:

$$\begin{bmatrix}
x_1 & x_2 & x_3 & 0 & x_4 & x_5 & x_6 & 0 \\
-x_2^* & x_1^* & 0 & -x_3 & x_5^* & -x_4^* & 0 & x_6 \\
x_3^* & 0 & -x_1^* & -x_2 & -x_6^* & 0 & x_4^* & x_5 \\
0 & -x_3^* & x_2^* & -x_1 & 0 & x_6^* & -x_5^* & x_4 \\
-x_4 & -x_5 & -x_6 & 0 & x_1 & x_2 & x_3 & 0 \\
-x_5^* & x_4^* & 0 & x_6 & -x_2^* & x_1^* & 0 & x_3 \\
x_6^* & 0 & -x_4^* & x_5 & x_3^* & 0 & -x_1^* & x_2 \\
0 & x_6^* & -x_5^* & -x_4 & 0 & x_3^* & -x_2^* & -x_1
\end{bmatrix} \tag{4.79}$$

Here, $n = 8$ antennas, $k = 6$ symbols, $p = 8$ transmissions. In this code, if we define $\mathcal{V}_i$, $i = 1, 2, \ldots,$ 8, as the ith column, we have:

$$\begin{array}{ll}
\langle \mathcal{V}_1, \mathcal{V}_i \rangle = 0, \ i \neq 5 & \langle \mathcal{V}_2, \mathcal{V}_i \rangle = 0, \ i \neq 6 \\
\langle \mathcal{V}_3, \mathcal{V}_i \rangle = 0, \ i \neq 7 & \langle \mathcal{V}_4, \mathcal{V}_i \rangle = 0, \ i \neq 8 \\
\langle \mathcal{V}_5, \mathcal{V}_i \rangle = 0, \ i \neq 1 & \langle \mathcal{V}_6, \mathcal{V}_i \rangle = 0, \ i \neq 2 \\
\langle \mathcal{V}_7, \mathcal{V}_i \rangle = 0, \ i \neq 3 & \langle \mathcal{V}_8, \mathcal{V}_i \rangle = 0, \ i \neq 4
\end{array} \tag{4.80}$$

The maximum likelihood decision metric, Expression (4.74), can be calculated as the sum of three terms $f_{14}(x_1, x_4) + f_{25}(x_2, x_5) + f_{36}(x_3, x_6)$ and similarly the decoding can be done using pairs of constellation symbols.

Figure 4.28 illustrates the performance of transmission of 2 bits/s/Hz using four transmit antennas and the rate one quasi-orthogonal code, the rate 1/2 full diversity orthogonal code and the uncoded 4PSK. The appropriate modulation schemes to provide the desired transmission rate for the space–time block codes, are 4PSK for the rate one code and 16 QAM for the rate 1/2 code.

Figure 4.29 presents the performance for four transmit antennas for space–time block codes at 3 bits/s/Hz. The rate one code and the uncoded system use 8PSK and the rate 3/4 code uses 16 QAM. More details on the topic can be found in [42].

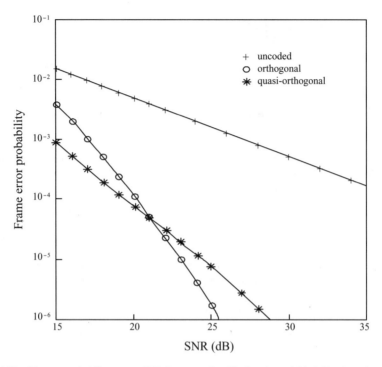

Figure 4.28   Bit error probability versus SNR for space–time block codes at 2 bits/s/Hz; 1 receive antenna.

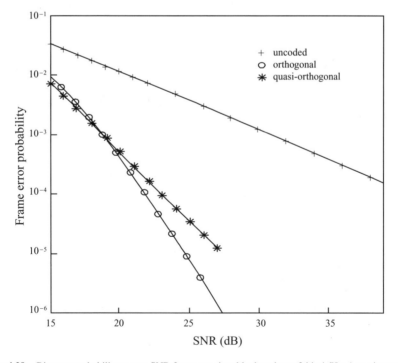

Figure 4.29   Bit error probability versus SNR for space–time block codes at 3 bits/s/Hz; 1 receive antenna.

4.6   Space–Time Convolutional Codes

In Section 4.2 the coding gain was defined as:

$$\eta = \min_{c,\,e} \left(\prod_{i=1}^{r} \lambda_i \right)^{1/r}$$

over all code word pairs. Due to the similarity of Inequality (4.33) to an error bound for trellis coded modulation, rm is called the diversity gain (the slope of the pairwise error probability on a log–log plot) and η is called the coding gain (η^{-rm} is an offset on a log–log plot).

Consider only codes that achieve maximum diversity gain. Of these, we search for codes that give the largest possible coding gain.

Similarly to Equations (2.8)–(2.13) in Chapter 2, consider a set of convolutional codes whose output at time k is:

$$[x_1(k),\ x_2(k), \ldots, x_n(k)] = \mathbf{b}_r(kR)G$$
$$= \mathbf{b}_r(kR)[G_1 G_2 \cdots G_n] \tag{4.81}$$

where G_i, $i = 1, \ldots, n$ is the ith column of the matrix $\mathbf{G}$ which is given by:

$$\mathbf{G} = \begin{bmatrix} g_{11} & g_{12} & \cdots & g_{1n} \\ g_{21} & g_{22} & \cdots & g_{2n} \\ g_{31} & g_{32} & \cdots & g_{3n} \\ g_{41} & g_{42} & \cdots & g_{4n} \\ \cdots & \cdots & \cdots & \cdots \\ g_{QR,1} & g_{QR,2} & \cdots & g_{QR,n} \end{bmatrix} \tag{4.82}$$

and $\mathbf{b}_r(kR) = [b_1, \ldots, b_{QR}]$ is a length QR binary row vector, $b_j \in \{0, 1\}$ for $i = 1, \ldots, n$ are taken to be in an alphabet of size s so that $g_{ij} \in \{0, 1, \ldots s - 1\}$. The arithmetic in Equation (4.81) is mod-s so $x_i(j) \in \{0, 1, \ldots s - 1\}$ also. Thus, the output code word is $(c_1(k), \ldots, c_n(k)) = (z[x_1(k)], \ldots, z[x_n(k)])$ where $z[x] = \exp(j(2\pi x/s + \phi))$ and $0 \le \phi \le 2\pi$ allows arbitrary rotation.

At each time slot, R bits are input into the convolutional encoder and the state is determined by $(Q-1)R$ bits. For space–time convolutional coding using Equation (4.81), the outputs $x_1(k), \ldots, x_n(k)$ are each mapped into a constellation of size s and transmitted simultaneously from n antennas. The input to the encoder at time slot k is $b_1, \ldots, b_R$. The state is given by $b_{R+1}, \ldots, b_{QR}$. For the next time slot, new input bits are received and the old input bits are shifted to the right into the state which explains the subscript on $\mathbf{b}_r(kR)$.

The results of the search for the good codes are presented in Tables 4.4–4.9 with the following notation:

η_t (4.2)   — coding gain for the trellis space–time codes defined in Section 4.2.
η_c (4.6)   — coding gain for the convolutional space–time codes defined in Section 4.6.
$e_{p\,\min}$ — minimum effective product distance.
L — total number of time slots considered in the bound calculation.
η_{AP} (L)   — considers all trellis paths describing error events of length L or longer.
η_{CP} (L)   — considers sets of code word pairs (continuous error paths) and searches for smallest upper bound.
$\tilde{Q}$ — the minimum length error event of $\tilde{Q}$ slots.

The performance results are shown in Figures 4.30–4.34.
More detail on the topic can be found in [35, 49–55].

Table 4.4　Optimum q-state 2 b/s/Hz 4PSK space–time codes [35] © 2002, IEEE

q	η_t (4.2)	η_c (4.6)	$\bar{\eta}_{AP}(\tilde{Q})$	$\bar{\eta}_{CP}(\tilde{Q})$	$\mathbf{G}^T$
4	2	$\sqrt{8}$	3.54	3.80	$\begin{bmatrix} 2 & 0 & 1 & 2 \\ 2 & 2 & 2 & 1 \end{bmatrix}$
8	$\sqrt{12}$	4	3.42	4.00	$\begin{bmatrix} 0 & 2 & 1 & 0 & 2 \\ 2 & 1 & 0 & 2 & 2 \end{bmatrix}$
16	$\sqrt{12}$	$\sqrt{32}$	5.76	6.24	$\begin{bmatrix} 0 & 2 & 1 & 1 & 2 & 0 \\ 2 & 2 & 1 & 2 & 0 & 2 \end{bmatrix}$
32	$\sqrt{12}$	6	5.63	6.33	$\begin{bmatrix} 2 & 0 & 1 & 2 & 1 & 2 & 2 \\ 2 & 2 & 0 & 1 & 2 & 0 & 2 \end{bmatrix}$

Table 4.5　Optimum q-state 2 b/s/Hz 4PSK space–time codes [35] © 2002, IEEE

q	η_t	η_c	$e_{p\,min}$	$\bar{\eta}_{AP}(L)$	$\bar{\eta}_{CP}(L)$	$\mathbf{G}^T$
4	2	$\sqrt{8}$	8	3.54	3.80	$\begin{bmatrix} 2 & 0 & 1 & 2 \\ 2 & 2 & 2 & 1 \end{bmatrix}$
8	$\sqrt{12}$	4	16	3.42	4.00	$\begin{bmatrix} 0 & 2 & 1 & 0 & 2 \\ 2 & 1 & 0 & 2 & 2 \end{bmatrix}$
16	$\sqrt{12}$	$\sqrt{32}$	32	5.76	6.24	$\begin{bmatrix} 0 & 2 & 1 & 1 & 2 & 0 \\ 2 & 2 & 1 & 2 & 0 & 2 \end{bmatrix}$
32	$\sqrt{12}$	6	36	5.63	6.33	$\begin{bmatrix} 2 & 0 & 1 & 2 & 1 & 2 & 2 \\ 2 & 2 & 0 & 1 & 2 & 0 & 2 \end{bmatrix}$

Table 4.6　Optimum q-state 1 b/s/Hz BSK space–time codes [35] © 2002, IEEE

q	η_c	$e_{p\,min}$	$\bar{\eta}_{AP}(L)$	$\bar{\eta}_{CP}(L)$	$\mathbf{G}^T$
2	4	4	4.00	4.00	$\begin{bmatrix} 0 & 1 \\ 1 & 1 \end{bmatrix}$
4	$\sqrt{48}$	12	7.33	6.93	$\begin{bmatrix} 0 & 1 & 1 \\ 1 & 0 & 1 \end{bmatrix}^{\dagger}$
4	$\sqrt{32}$	8	7.33	7.73	$\begin{bmatrix} 0 & 1 & 1 \\ 1 & 1 & 1 \end{bmatrix}$
8	$\sqrt{80}$	20	10.06	10.47	$\begin{bmatrix} 1 & 0 & 1 & 1 \\ 1 & 1 & 0 & 1 \end{bmatrix}$
16	$\sqrt{112}$	28	11.38	14.27	$\begin{bmatrix} 1 & 1 & 0 & 1 & 1 \\ 0 & 1 & 1 & 1 & 1 \end{bmatrix}$

†denotes that the code is catastrophic

4.7　Algebraic Space–Time Codes

Let C be a binary convolutional code of rate $1/L_t$ with the transfer function encoder $\mathbf{Y}(D) = X(D)\mathbf{G}(D)$, where $\mathbf{G}(D) = [G_1(D)\ G_2(D) \cdots G_{Lt}(D)]$. In the natural space–time formatting of C for BPSK transmission, the output sequence corresponding to $Y_i(D) = X(D)\ G_i(D)$ is assigned to the ith transmit antenna. The number of transmit antennas is L_t. The resulting space–time code C satisfies the binary rank criterion under relatively mild conditions on the connection polynomials $G_i(D)$.

Table 4.7   Optimum q-state 1 b/s/Hz BSK 3-space–time codes [35] © 2002, IEEE

q	η_c	$e_{p\,\min}$	$\bar{\eta}_{AP}(L)$	$\bar{\eta}_{CP}(L)$	$\mathbf{G}^T$
4	4	$\dfrac{64}{27}$	5.08	5.08	$\begin{bmatrix} 0 & 1 & 1 \\ 1 & 0 & 1 \\ 1 & 1 & 1 \end{bmatrix}$
8	$256^{1/3}$	$\dfrac{256}{27}$	7.65	7.05	$\begin{bmatrix} 1 & 0 & 0 & 1 \\ 1 & 0 & 1 & 0 \\ 1 & 1 & 1 & 1 \end{bmatrix}^{\dagger}$
8	$192^{1/3}$	$\dfrac{64}{9}$	7.65	8.32	$\begin{bmatrix} 1 & 0 & 1 & 1 \\ 1 & 1 & 0 & 1 \\ 1 & 0 & 1 & 1 \end{bmatrix}$
16	8	$\dfrac{512}{27}$	10.00	10.18	$\begin{bmatrix} 1 & 0 & 0 & 1 & 1 \\ 1 & 1 & 0 & 1 & 0 \\ 1 & 1 & 1 & 0 & 1 \end{bmatrix}$

†denotes that the code is catastrophic

Table 4.8   Optimum q-state 1b/s/Hz BSK 4-space–time codes [35] © 2002, IEEE

q	η_c	$e_{p\,\min}$	$\bar{\eta}_{AP}(L)$	$\bar{\eta}_{CP}(L)$	$\mathbf{G}^T$
8	4	1	5.97	5.97	$\begin{bmatrix} 0 & 1 & 0 & 1 \\ 0 & 1 & 1 & 1 \\ 1 & 0 & 1 & 0 \\ 1 & 1 & 1 & 0 \end{bmatrix}$
16	$1280^{1/4}$	5	8.11	8.37	$\begin{bmatrix} 1 & 0 & 0 & 0 & 1 \\ 1 & 0 & 1 & 1 & 1 \\ 1 & 1 & 0 & 1 & 1 \\ 1 & 1 & 1 & 1 & 0 \end{bmatrix}^{\dagger}$
16	$1024^{1/4}$	4	7.99	9.32	$\begin{bmatrix} 0 & 1 & 1 & 0 & 1 \\ 1 & 1 & 0 & 0 & 1 \\ 1 & 1 & 1 & 1 & 0 \\ 1 & 1 & 1 & 1 & 1 \end{bmatrix}$
32	$4352^{1/4}$	17	9.80	10.38	$\begin{bmatrix} 1 & 0 & 0 & 0 & 0 & 1 \\ 1 & 0 & 1 & 1 & 1 & 1 \\ 1 & 1 & 1 & 0 & 1 & 0 \\ 1 & 1 & 1 & 1 & 0 & 0 \end{bmatrix}$

†denotes that the code is catastrophic

Table 4.9   Optimum q-state 2 b/s/Hz 4 PSK space–time code using three transmit antennas [35] © 2002, IEEE

q	η_c	$e_{p\,\min}$	$\bar{\eta}_{AP}(L)$	$\bar{\eta}_{CP}(L)$	$\mathbf{G}^T$
16	$32^{1/3}$	$\dfrac{256}{27}$	3.90	4.72	$\begin{bmatrix} 0 & 2 & 1 & 2 & 2 & 0 \\ 1 & 2 & 2 & 0 & 0 & 2 \\ 2 & 2 & 0 & 2 & 1 & 2 \end{bmatrix}$

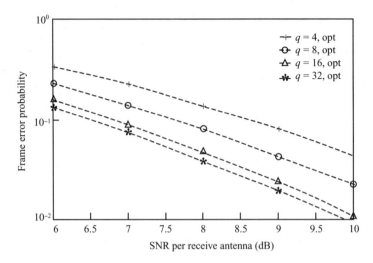

Figure 4.30 Performance comparison of some best 2 b/s/Hz, QPSK, q-state STCs with two transmit and two receive antennas (SNR per receive antenna = $n E_s/N_0$).

4.7.1   Full Spatial Diversity

Code C, associated with the $1/L_t$ convolutional code C, satisfies the binary rank criterion, and thus achieves full spatial diversity for BPSK transmission if and only if the transfer function matrix $\mathbf{G}(D)$ of C has full rank L_t as a matrix of coefficients over the binary field $\mathbb{F}$. This result follows directly from the stacking construction and can be easily generalized to recursive convolutional codes.

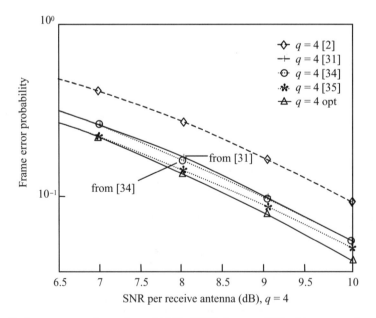

Figure 4.31 Performance comparisons of some 2 b/s/Hz, QPSK, four-state STCs with two transmit and two receive antennas.

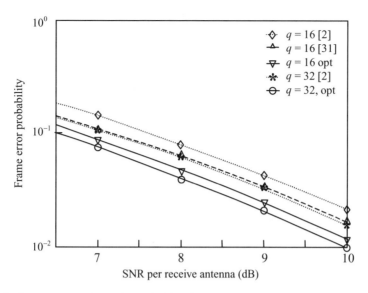

Figure 4.32 Performance comparisons of some 2 b/s/Hz, QPSK, 16- and 32-state STCs with two transmit and two receive antennas.

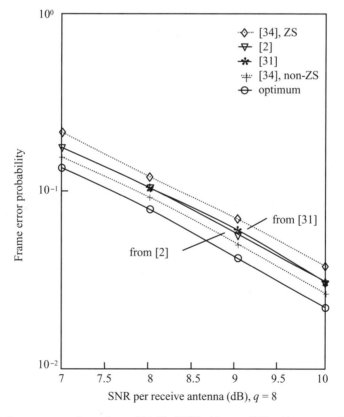

Figure 4.33 Performance comparisons of some 2 b/s/Hz, QPSK, eight-state STCs with two transmit and two receive antennas.

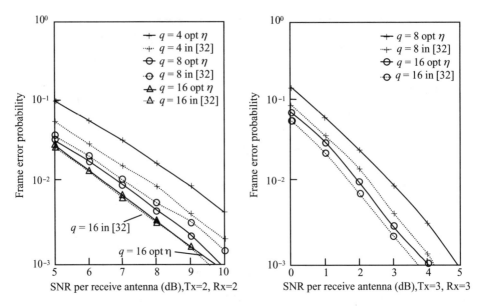

Figure 4.34   Performance of some 1 b/s/Hz, BPSK, q-state STCs with two transmit and two receive or three transmit and three receive antennas (performance of the best noncatastrophic codes similar).

It is straightforward to see that the zeros symmetry codes satisfy the stacking construction conditions. However, the set of binary rate $1/L_t$ convolutional codes with the optimal free distance d_{free} offers a richer class of space–time codes as their associated natural space–time codes usually achieve full spatial diversity. Furthermore, these codes outperform the zeros symmetry codes uniformly. A collection of these codes is given in Table 4.10.

4.7.2   QPSK Modulation

Experimentally, codes obtained by replacing each zero in the binary generator matrix by a two yield the best simulated frame error rate performance in most cases. These codes are reported in Table 4.11 for different numbers of transmit antennas and constraint lengths v and are used in generating the simulation results in the next section.

Table 4.10   Natural full diversity space–time convolutional codes with optimal free distance for BPSK modulation [56] © 2002, IEEE

L_t	v	Connection polynomials
2	2	5, 7
	3	64, 74
	4	46, 72
	5	65, 57
	6	554, 744
3	3	54, 64, 74
	4	52, 66, 76
	5	47, 53, 75
	6	554, 624, 764
4	4	52, 56, 66, 76
	5	53, 67, 71, 75
5	5	75, 71, 73, 65, 57

Table 4.11 Linear $\mathbb{Z}_4$ space–time codes for QPSK modulation [56] © 2002, IEEE

L	v	Connection polynomials
2	1	$1 + 2D, 2 + D.$
	2	$1 + 2D + D^2, 1 + D + D^2.$
	3	$1 + D + 2D^2 + D^3, 1 + D + D^2 + D^3.$
	4	$1 + 2D + 2D^2 + D^3 + D^4, 1 + D + D^2 + 2D^3 + D^4.$
	5	$1 + D + 2D^2 + D^3 + 2D^4 + D^5, 1 + 2D + D^2 + D^3 + D^4 + D^5.$
3	2	$1 + 2D + 2D^2, 2 + D + 2D^2, 2 + 2D + D^2.$
	3	$1 + 2D + D^2 + D^3, 1 + D + 2D^2 + D^3, 1 + D + D^2 + D^3.$
	4	$1 + 2D + D^2 + 2D^3 + D^4, 1 + D + 2D^2 + D^3 + D^4, 1 + D + D^2 + D^3 + D^4.$
	5	$1 + 2D + 2D^2 + D^3 + D^4 + D^5, 1 + 2D + D^2 + 2D^3 ++ D^4 + D^5,$
		$1 + D + D^2 + D^3 + 2D^4 + D^5.$
4	3	$1 + 2D + 2D^2 + 2D^3, 2 + D + 2D^2 + 2^3, 2 + 2D + D^2 + 2D^3, 2 + 2D + 2D^2 + D^3$
	4	$1 + 2D + D^2 + 2D^3 + D^4, 1 + 2D + D^2 + D^3 + D^4, 1 + D + 2D^2 + D^3 + D^4,$
		$1 + D + D^2 + D^3 + D^4.$
	5	$1 + 2D + D^2 + 2D^3 + D^4 + D^5, 1 + D + 2D^2 + D^3 + D^4 + D^5,$
		$1 + D + D^2 + 2D^3 + 2D^4 + D^5, 1 + D + D^2 + D^3 + 2D^4 + D^5.$
5	4	$1 + 2D + 2D^2 + 2D^3 + 2D^4, 2 + D + 2D^2 + 2D^3 + 2D^4, 2 + 2D + D^2 + 2D^3 + 2D^4,$
		$2 + 2D + 2D^2 + D^3 + 2D^4, 2 + 2D + 2D^2 + 2D^3 + D^4,$
	5	$1 + D + D^2 + D^3 + 2D^4 + D^5, 1 + D + D^2 + D^3 + 2D^4 + D^5,$
		$1 + D + 2D^2 + D^3 + D^4 + D^5, 1 + D + 2D^2 + D^3 + 2D^4 + D^5,$
		$1 + 2D + D^2 + D^3 + D^4 + D^5.$

A space–time code has zeros symmetry if every baseband code word difference $f(\mathbf{c}) - f(\mathbf{e})$ is upper and lower triangular and has appropriate non-zero entries to ensure full rank. The zeros symmetry property is sufficient for full rank but not necessary. Some simulation results for the system performance are shown in Figures 4.35–4.40.

4.8 Differential Space–Time Modulation

When channel estimation is not available, a differential modulation and detection might be a solution. We start with a simple transmission scheme [65] for exploiting diversity given by two transmit antennas when neither the transmitter nor the receiver has access to channel state information. At the receiver, decoding is achieved with low decoding complexity. The transmission provides full spatial diversity and requires no channel state side information at the receiver. The scheme can be considered as the extension of conventional differential detection schemes to two transmit antennas.

In traditional differential phase shift keying (DPSK) the data is encoded in the difference of the phase of two consecutive symbols. For b bits per symbol, the symbol would usually have the form:

$$\Delta(l) = e^{i\theta(l)} = e^{2\pi i l/L}, \quad l = 0, 1, \ldots L - 1 \tag{4.83}$$

where $i = \sqrt{-1}$ and $L = 2^b$. If the data sequence is $z_1, z_2, z_3, \ldots$, the transmitted symbol sequences would be $s_0, s_1, s_2, \ldots$, where, at time t,

$$s_t = \Delta(z_t)s_{t-1} \tag{4.84}$$

Bits are mapped into the symbols by using *Gray mapping*, e.g. for $b = 2$:

$$\begin{aligned}
M(z) \Rightarrow \Delta(z) &= \Delta(l) \\
\Delta(00) = \Delta(0) &= e^{2\pi i \times 0/4} = 1 \\
\Delta(01) = \Delta(1) &= e^{2\pi i \times 1/4} = e^{i\pi/2} \\
\Delta(10) = \Delta(3) &= e^{2\pi i \times 3/4} = e^{-i\pi/2} \\
\Delta(11) = \Delta(2) &= e^{2\pi i \times 2/4} = -1
\end{aligned} \tag{4.85}$$

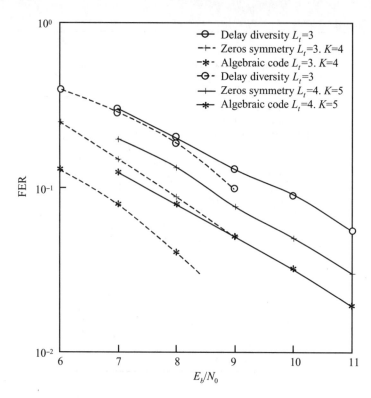

Figure 4.35 Performance of BPSK space–time codes with $L_r = 1$. The number of transmit antennas is represented by L_t, receive antennas by L_r, and constraint lengths by K. The number of bits per frame is 100.

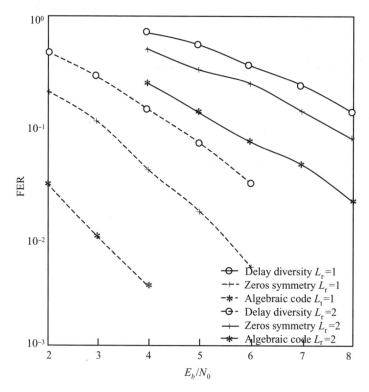

Figure 4.36 FER for BPSK space–time codes with $L_t = 5$ and $K = 6$.

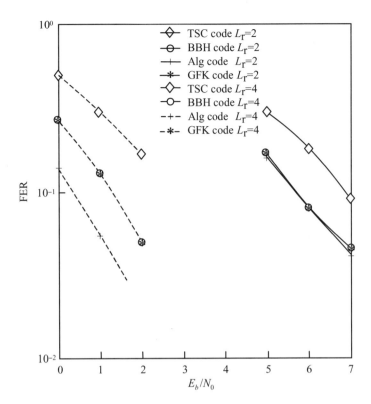

Figure 4.37 Performance of four-state QPSK space–time codes with $L_t = 2$. Four-state codes due to Tarokh, Seshadri, and Calderbank (TSC) [2], Baro, Bauch, and Hansmann (BBH) [31], Grimm, Fitz, and Krogmeier (GFK) [33], and the new linear $\mathbb{Z}_4$ code in Table 4.11 with two and four receive antennas.

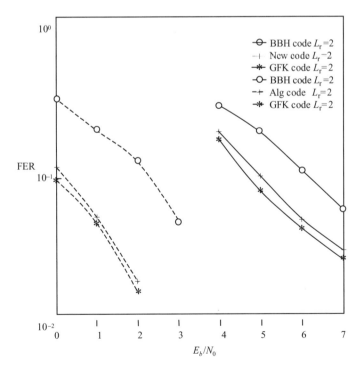

Figure 4.38 Performance of eight-state QPSK space–time codes with $L_t = 2$.

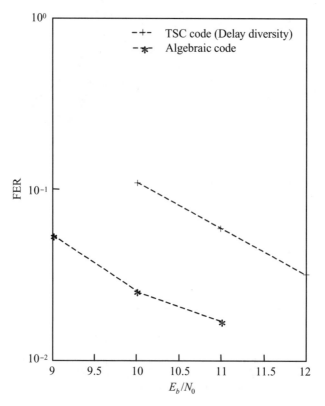

Figure 4.39   FER for QPSK space–time codes with $L_t = 4$ and $L_r = 1$. The new 64-state space–time code for four transmit antennas and the TSC 64-state code (i.e. delay diversity).

This ensures that the most probable symbol errors cause the minimum number of bit errors. Signal r_t from the received signal sequence $r_0, r_1, r_2, \ldots$ can be represented as:

$$r_t = h_t s_t + n_t \tag{4.86}$$

where h_t is the channel gain. The receiver would extract the information by processing the received signal samples as follows:

$$\hat{\theta}_t = \arg r_t r_{t-1}^* \tag{4.87}$$

which, for $h_t \cong h_{t-1}$, $|h_t| = 1$ and no noise, gives:

$$\hat{\theta}_t = \arg h_t s_t h_{t-1}^* s_{t-1}^* = \arg |h_t|^2 \Delta(z_t) |s_{t-1}|^2 = \arg \Delta(z_t) \tag{4.88}$$

By using analogy with the previous discussion we will now extend this principle to two-dimensional constellations. Later in the chapter we will generalize these schemes to multidimensional constellations.

Let us restrict the constellation A to 2^b-PSK for some $b = 1, 2, 3, \ldots$, but in reality only BPSK, QPSK, and 8PSK are of interest. Thus,

$$A = \left\{ \frac{e^{2\pi ki/2^b}}{\sqrt{2}} \,|\, k = 0, 1, \ldots, 2^b - 1 \right\} \tag{4.89}$$

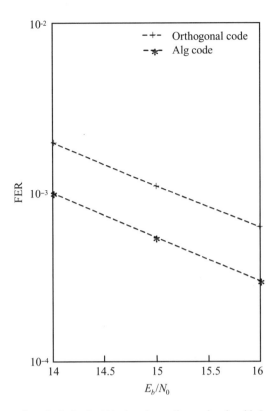

Figure 4.40 Performance of an algebraic short block and an orthogonal code with $L_t = 3$ and $Lr = 1$. The new algebraic code is used with QPSK modulation and the orthogonal code is used with 16 QAM. The increased size of the constellation in the case of the orthogonal code is necessary to support the same throughput [6]. More details on the topic can be found in [56–64].

In order to implement Equation (4.84) in a two-dimensional constellation, let us first consider the following relation between the vectors.

Given a pair of 2^b-PSK constellation symbols, x_1 and x_2, we first observe that the complex vectors $(x_1 x_2)$ and $(-x_2^* x_1^*)$ are orthogonal to each other and have unit lengths defined by $\sum_i x_i x_i^* = 1$. Any two-dimensional vector $\chi = (x_3 x_4)$ can be uniquely represented in the orthonormal basis given by these vectors. In other words, there exists a unique complex vector $\mathbf{P}_\chi = (\mathbf{A}_\chi \mathbf{B}_\chi)$ such that ($\mathbf{A}_\chi$ and $\mathbf{B}_\chi$ satisfy the vector equation:

$$(x_3 x_4) = \mathbf{A}_\chi (x_1 x_2) + \mathbf{B}_\chi (-x_2^* x_1^*) \tag{4.90}$$

or

$$x_3 = \mathbf{A}_\chi x_1 - \mathbf{B}_\chi x_2^*$$
$$x_4 = \mathbf{A}_\chi x_2 + \mathbf{B}_\chi x_1^*$$
$$x_3 x_1^* = \mathbf{A}_\chi x_1 x_1^* - \mathbf{B}_\chi x_2^* x_1^*$$
$$x_4 x_2^* = \mathbf{A}_\chi x_2 x_2^* + \mathbf{B}_\chi x_1^* x_2^*$$

giving $\mathbf{A}_\chi$ and $\mathbf{B}_\chi$ as:

$$\mathbf{A}_\chi = x_3 x_1^* + x_4 x_2^*$$
$$\mathbf{B}_\chi = -x_3 x_2 + x_4 x_1 \tag{4.91}$$

These relations will be used later to implement Equation (4.84). For an equivalent implementation of Equation (4.85) we will need a few additional definitions.

We define the set $\mathcal{V}_\chi$ to consist of all the vectors $\mathbf{P}_\chi$, $\chi \notin \mathcal{A} \times \mathcal{A}$. The set $\mathcal{V}_\chi$ has the following properties:

- *Property A*: It has 2^{2b} elements corresponding to the pairs $(x_3 x_4)$ of constellation symbols.
- *Property B*: All elements of $\mathcal{V}_\chi$ have unit length.
- *Property C*: For any two distinct elements $\chi = (x_1 x_2)$ and $\mathcal{Y} = (y_1 y_2)$ of $\mathcal{A} \times \mathcal{A}$

$$\|P_\chi - P_y\| = \|(x_1 x_2) - (y_1 y_2)\|.$$

- Property D: The minimum distance between any two distinct elements of $\mathcal{V}_x$ is equal to the minimum distance of the 2b-PSK constellation $\mathcal{A}$.

The above properties hold because the mapping $\chi \rightarrow \mathbf{P}_\chi$ is just a change of basis from the standard basis given by vectors $\{(1\,0), (0\,1)\}$ to the orthonormal basis given by $\{(x_1 x_2), (-x_2^* \, x_1^*)\}$, which preserves the distances between the points of the two-dimensional complex space.

The first ingredient of construction is the choice of an arbitrary set $\mathcal{V}$ having Properties $\mathcal{A}$ and $\mathcal{B}$. It is also handy if $\mathcal{V}$ has Properties C and D as well. As a natural choice for such a set $\mathcal{V}$, we may fix an arbitrary pair $\chi \in \mathcal{A} \times \mathcal{A}$ and let $\mathcal{V} = \mathcal{V}_\chi$. Because the 2^b-PSK constellation A always contains the signal point $1/\sqrt{2}$, we choose to fix $\chi = ((1/\sqrt{2})(1/\sqrt{2}))$.

We also need an arbitrary bijective mapping $\mathcal{M}$ of blocks of $2b$ bits onto $\mathcal{V}$. Among all the possibilities for $\mathcal{M}$, we choose the mapping analogous to Equation (4.85). Given a block $\mathcal{B}$ of $2b$ bits, the first b bits are mapped into a constellation symbol a_3 and the second b bits are mapped into a constellation symbol a_4 using *Gray mapping*.

Let $a_1 = a_2 = 1/\sqrt{2}$, then $\mathcal{M}(\mathcal{B}) = (\mathcal{A}(\mathcal{B}) \mathcal{B}(\mathcal{B}))$ is defined by Equation (4.91):

$$\begin{aligned} \mathcal{A}(\mathcal{B}) &= a_3 a_1^* + a_4 a_2^* \\ \mathcal{B}(\mathcal{B}) &= -a_3 a_2 + a_4 a_1 \end{aligned} \tag{4.92}$$

Clearly, $\mathcal{M}$ maps any $2b$ bits onto $\mathcal{V}$. Conversely, given $(\mathcal{A}(\mathcal{B}) \mathcal{B}(\mathcal{B}))$, the pair $(a_3 \, a_4)$ is recovered by Equation (4.90):

$$(a_3 \, a_4) = \mathcal{A}(\mathcal{B})(a_1 \, a_2) + \mathcal{B}(\mathcal{B})(-a_2^* \, a_1^*) \tag{4.93}$$

The block $\mathcal{B}$ is then constructed by inverse Gray mapping of a_3 and a_4.

4.8.1   The Encoding Algorithm

The transmitter begins the transmission by sending arbitrary symbols s_1 and s_2 from transmit antennas one and two respectively at time one, and symbols $-s_2^*$ and s_1^* at time two unknown to the receiver. These two transmissions do not convey any information. The transmitter then encodes the rest of the data in an inductive manner. Suppose that s_{2t-1} and s_{2t} are sent, respectively, from transmit antennas one and two at time $2t - 1$, and that $-s_{2t}^*$, s_{2t-1}^* are sent, respectively, from antennas one and two at time $2t$. At time $2t + 1$, a block of $2b$ bits $\mathcal{B}_{2t+1}$ arrives at the encoder. The transmitter uses the mapping $\mathcal{M}$ given by Equation (4.92) and computes $\mathcal{M}(\mathcal{B}_{2t+1}) = (\mathcal{A}(\mathcal{B}_{2t+1}) \mathcal{B}(\mathcal{B}_{2t+1}))$. Then, in accordance with Equation (4.92) it computes:

$$(s_{2t+1} \, s_{2t+2}) = \mathcal{A}(\mathcal{B}_{2t+1})(s_{2t-1} \, s_{2t}) + \mathcal{B}(\mathcal{B}_{2t+1})(-s_{2t}^* \, s_{2t-2}^*) \tag{4.94}$$

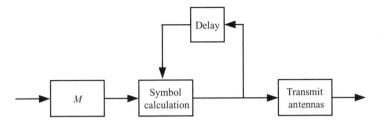

Figure 4.41 Transmitter block diagram.

The transmitter then sends s_{2t+1} and s_{2t+2}, respectively, from transmit antennas one and two at time $2t + 1$, and $-s_{2t+2}^*$, s_{2t+1}^* from antennas one and two at time $2t + 2$. This process is inductively repeated until the end of the frame (or end of the transmission). The block diagram of the encoder is given in Figure 4.41.

4.8.1.1 Example

We assume that the constellation is BPSK ($b=1$) consisting of the points $-1/\sqrt{2}$ and $1/\sqrt{2}$. Then the set $\mathcal{V} = \{(1\,0), (0\,1), (-1\,0), (0\,-1)\}$. Recall that the Gray mapping maps a bit $i = 0, 1$ to $(-1)^i/\sqrt{2}$. We set $a_1 = a_2 = 1/\sqrt{2}$. Then the mapping $\mathcal{M}$ maps two bits onto $\mathcal{V}$ and is given by:

$$\begin{aligned} \mathcal{M}(00) &= (1\,0) \\ \mathcal{M}(10) &= (0\,1) \\ \mathcal{M}(01) &= (0\,-1) \\ \mathcal{M}(11) &= (-1\,0) \end{aligned} \tag{4.95}$$

Now suppose that at time $2t - 1$, $s_{2t-1} = 1/\sqrt{2}$ and $s_{2t} = -1/\sqrt{2}$ are sent, respectively, from antennas one and two, and at time $2t$, $-s_{2t}^* = 1/\sqrt{2}$ and $s_{2t-1}^* = 1/\sqrt{2}$ are sent. Suppose that the input to the encoder at time $2t + 1$ is the block of bits 10. Since $\mathcal{M}(10) = (0\,1)$, we have $\mathcal{A}(10) = 0$ and $\mathcal{B}(10) = 1$. Then the values s_{2t+1} and s_{2t+2} corresponding to input bits 10 are computed as follows:

$$\begin{aligned} (s_{2t+1}\ s_{2t+2}) &= 0 \cdot \left(\frac{1}{\sqrt{2}}\ \frac{-1}{\sqrt{2}} \right) + 1 \cdot \left(\frac{1}{\sqrt{2}}\ \frac{1}{\sqrt{2}} \right) \\ &= \left(\frac{1}{\sqrt{2}}\ \frac{1}{\sqrt{2}} \right) \end{aligned} \tag{4.96}$$

Thus, at time $2t + 1$, $s_{2t+1} = 1/\sqrt{2}$ and $s_{2t+2} = 1/\sqrt{2}$ are sent, respectively, from antennas one and two, and at time $2t + 2$, $-s_{2t+2}^* = -1/\sqrt{2}$ and $s_{2t+1}^* = 1/\sqrt{2}$ are sent, respectively, from antennas one and two. Transmitted symbols at time $2t + 1$ and $2t + 2$ correspond to the input bits 00, 10, 01 and 11 for this scenario. The results are summarized in Tables 4.12 and 4.13.

Table 4.12 Transmitted symbols at time $2t + 1$ [65] © 2002, IEEE

Input bits at time $2t + 1$	Antenna 1	Antenna 2
00	$\dfrac{1}{\sqrt{2}}$	$\dfrac{-1}{\sqrt{2}}$
10	$\dfrac{1}{\sqrt{2}}$	$\dfrac{1}{\sqrt{2}}$
01	$\dfrac{-1}{\sqrt{2}}$	$\dfrac{-1}{\sqrt{2}}$
11	$\dfrac{-1}{\sqrt{2}}$	$\dfrac{1}{\sqrt{2}}$

Table 4.13   Transmitted symbols at time $2t + 2$ [65] © 2002, IEEE

Input bits at time $2t + 1$	Antenna 1	Antenna 2
00	$\dfrac{1}{\sqrt{2}}$	$\dfrac{1}{\sqrt{2}}$
10	$\dfrac{-1}{\sqrt{2}}$	$\dfrac{1}{\sqrt{2}}$
01	$\dfrac{1}{\sqrt{2}}$	$\dfrac{-1}{\sqrt{2}}$
11	$\dfrac{-1}{\sqrt{2}}$	$\dfrac{-1}{\sqrt{2}}$

4.8.2   Differential Decoding

For notational simplicity, we first present the results for one receive antenna. Write r_t for r_t^1, η_t for η_t^1 and α_1, α_2, respectively, for $\alpha_{1,1}, \alpha_{2,1}$, knowing that this can cause no confusion since there is only one receive antenna.

$r_{2t-1}, r_{2t}, r_{2t+1}$, and r_{2t+2} are received signal samples. Let

$$\Lambda\left(\alpha_1, \alpha_2\right) = \begin{bmatrix} \alpha_1 & \alpha_2^* \\ \alpha_2 & -\alpha_1^* \end{bmatrix} \tag{4.97}$$

and

$$N_{2t-1} = (\eta_{2t-1}\ \eta_{2t}^*) \tag{4.98}$$

Then the received signal can be represented as:

$$(r_{2t-1}\ r_{2t}^*) = (s_{2t-1}\ s_{2t})\,\Lambda\,(\alpha_1, \alpha_2) + N_{2t-1} \tag{4.99}$$

and

$$(r_{2t+1}\ r_{2t+2}^*) = (s_{2t+1}\ s_{2t+2})\Lambda(\alpha_1, \alpha_2) + N_{2t+1} \tag{4.100}$$

The receiver will create

$$\begin{aligned} (r_{2t+1}\ r_{2t+2}^*) \cdot (r_{2t-1}\ r_{2t}^*) ={}& (s_{2t+1}\ s_{2t+2})\Lambda(\alpha_1, \alpha_2)\Lambda^*(\alpha_1, \alpha_2)(s_{2t-1}^*\ s_{2t}^*) \\ &+ (s_{2t+1}\ s_{2t+2})\Lambda(\alpha_1, \alpha_2)N_{2t-1}^* \\ &+ N_{2t+1}\Lambda^*(\alpha_1, \alpha_2)(s_{2t-1}\ s_{2t})^* + N_{2t+1}N_{2t-1}^* \end{aligned} \tag{4.101}$$

which gives:

$$\begin{aligned} r_{2t+1}r_{2t-1}^* + r_{2t+2}^* r_{2t} ={}& (|\alpha_1|^2 + |\alpha_2|^2)(s_{2t+1}s_{2t-1}^* + s_{2t+1}s_{2t}^*) \\ &+ (s_{2t+1}\ s_{2t+2})\Lambda(\alpha_1, \alpha_2)N_{2t-1}^* \\ &+ N_{2t+1}\Lambda^*(\alpha_1, \alpha_2)(s_{2t-1}\ s_{2t})^* + N_{2t+1}N_{2t-1}^* \end{aligned}$$

For a more compact representation we use the following notation:

$$\mathcal{R}_1 = r_{2t+1}r_{2t-1}^* + r_{2t+2}^* r_{2t}$$
$$\mathcal{N}_1 = (s_{2t+1}\ s_{2t+2})\Lambda(\alpha_1, \alpha_2)N_{2t-1}^* + N_{2t+1}\Lambda^*(\alpha_1, \alpha_2)(s_{2t-1}\ s_{2t})^* + N_{2t+1}N_{2t-1}^*$$

which gives

$$\mathcal{R}_1 = (|\alpha_1|^2 + |\alpha_2|^2)\mathcal{A}(\mathcal{B}_{2t-1}) + \mathcal{N}_1 \tag{4.102}$$

For the second vector term in the right side of Equation (4.94) we have:

$$(r_{2t} - r^*_{2t-1}) = (-s^*_{2t}\ s^*_{2t-1})\Lambda(\alpha_1, \alpha_2) + N_{2t}$$

where

$$N_{2t} = (\eta_{2t} - \eta^*_{2t-1})$$

The receiver will now create

$$
\begin{aligned}
(r_{2t+1}\ r^*_{2t+2}) \cdot (r_{2t} - r^*_{2t-1}) &= (s_{2t+1}\ s_{2t+2})\Lambda(\alpha_1, \alpha_2)\Lambda^*(\alpha_1, \alpha_2)(-s_{2t}\ s_{2t-1}) \\
&\quad + (s_{2t+1}\ s_{2t+2})\Lambda(\alpha_1, \alpha_2)N^*_{2t} \\
&\quad + N_{2t+1}\Lambda^*(\alpha_1, \alpha_2)(-s^*_{2t}\ s^*_{2t-1})^* + N_{2t+1}N^*_{2t}
\end{aligned}
$$

giving:

$$
\begin{aligned}
r_{2t+1}r^*_{2t} - r^*_{2t+2}r_{2t-1} &= (|\alpha_1|^2 + |\alpha_2|^2)(-s_{2t+1}s^*_{2t} + s_{2t+2}s^*_{2t-1}) \\
&\quad + (s_{2t+1}\ s_{2t+2})\Lambda(\alpha_1, \alpha_2)N^*_{2t} \\
&\quad + N_{2t+1}\Lambda^*(\alpha_1, \alpha_2)(-s^*_{2t}\ s^*_{2t-1})^* + N_{2t+1}N^*_{2t}
\end{aligned}
$$

With the notation

$$
\begin{aligned}
\mathcal{R}_2 &= r_{2t+1}r^*_{2t} - r^*_{2t+2}r_{2t-1} \\
\mathcal{N}_2 &= +(s_{2t+1}\ s_{2t+2})\Lambda(\alpha_1, \alpha_2)N^*_{2t} + N_{2t+1}\Lambda^*(\alpha_1, \alpha_2)(-s^*_{2t}\ s^*_{2t-1})^* + N_{2t+1}N^*_{2t}
\end{aligned}
$$

we have

$$\mathcal{R}_2 = (|\alpha_1|^2 + |\alpha_2|^2)\mathcal{B}(\mathcal{B}_{2t-1}) + \mathcal{N}_2 \tag{4.103}$$

The final result can be represented as:

$$(\mathcal{R}_1\mathcal{R}_2) = \left(|\alpha_1|^2 + |\alpha_2|^2\right)(\mathcal{A}(\mathcal{B}_{2t-1})\ \mathcal{B}(\mathcal{B}_{2t-1})) + (\mathcal{N}_1\mathcal{N}_2) \tag{4.104}$$

Because the elements of $\mathcal{V}$ have equal length, to compute $(\mathcal{A}(\mathcal{B}_{2t-1})\mathcal{B}(\mathcal{B}_{2t-1}))$, the receiver now computes the closest vector of $\mathcal{V}$ to $(\mathcal{R}_1\mathcal{R}_2)$. Once this vector is computed, the inverse mapping of $\mathcal{M}$ is applied and the transmitted bits are recovered (see Table 4.12 and Table 4.13). The receiver block diagram and BER curves are given in Figures 4.42 and 4.43 respectively. As in the traditional, one dimensional DPSK system, there is loss in the performance (about 3 dB at BER = 0.001).

More results on the topic can be found in [65–76].

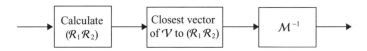

Figure 4.42   Receiver block diagram.

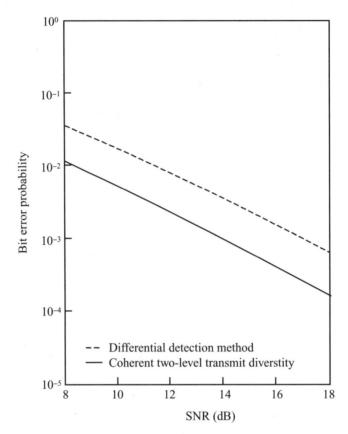

Figure 4.43   Performance of the differential detection and coherent detection: two-level transmit diversity scheme for BPSK constellation.

4.9   Multiple Transmit Antenna Differential Detection from Generalized Orthogonal Designs

In this section we explicitly construct multiple transmit antenna differential encoding/decoding schemes based on generalized orthogonal designs. These constructions generalize the two transmit antenna differential detection scheme discussed in the previous section. The presentation is based on [77, 78].

4.9.1   Differential Encoding

We consider the specific example code, Γ_{84}, and only one receive antenna, $M = 1$. The generalization to other codes is straightforward. The code set is defined as:

$$
\Gamma_{84} = \begin{bmatrix}
x_1 & x_2 & x_3 & x_4 \\
-x_2 & x_1 & x_4 & x_3 \\
-x_3 & x_4 & x_1 & -x_2 \\
-x_4 & -x_3 & x_2 & x_1 \\
x_1^* & x_2^* & x_3^* & x_4^* \\
-x_2^* & x_1^* & x_4^* & x_3^* \\
-x_3^* & x_4^* & x_1^* & -x_2^* \\
-x_4^* & -x_3^* & x_2^* & x_1^*
\end{bmatrix}
\tag{4.105}
$$

4.9.2   Received Signal

At each time, there is only one received signal, $r_{t,1}$, which can be noted by r_t and this is related to the constellation symbols s_1, s_2, s_3, s_4 by:

$$
\begin{aligned}
r_1 &= \alpha_1 s_1 + \alpha_2 s_2 + \alpha_3 s_3 + \alpha_4 s_4 + \eta_1 \\
r_2 &= -\alpha_1 s_2 + \alpha_2 s_1 - \alpha_3 s_4 + \alpha_4 s_3 + \eta_2 \\
r_3 &= -\alpha_1 s_3 + \alpha_2 s_4 + \alpha_3 s_1 - \alpha_4 s_2 + \eta_3 \\
r_4 &= -\alpha_1 s_4 - \alpha_2 s_3 + \alpha_3 s_2 + \alpha_4 s_1 + \eta_4 \\
r_1 &= \alpha_1 s_1 + \alpha_2 s_2 + \alpha_3 s_3 + \alpha_4 s_4 + \eta_1 \\
r_2 &= -\alpha_1 s_2 + \alpha_2 s_1 - \alpha_3 s_4 + \alpha_4 s_3 + \eta_2 \\
r_3 &= -\alpha_1 s_3 + \alpha_2 s_4 + \alpha_3 s_1 - \alpha_4 s_2 + \eta_3 \\
r_4 &= -\alpha_1 s_4 - \alpha_2 s_3 + \alpha_3 s_2 + \alpha_4 s_1 + \eta_4
\end{aligned}
\tag{4.106}
$$

By using Equation (4.165), one can show that the following relations are valid:

$$
(r_1, r_2, r_3, r_4, r_5^*, r_6^*, r_7^*, r_8^*) = (s_1, s_2, s_3, s_4)\Omega + (\eta_1, \eta_2, \eta_3, \eta_4, \eta_5^*, \eta_6^*, \eta_7^*, \eta_8^*)
\tag{4.107}
$$

where

$$
\Omega =
\begin{bmatrix}
\alpha_1 & \alpha_2 & \alpha_3 & \alpha_4 & \alpha_1^* & \alpha_2^* & \alpha_3^* & \alpha_4^* \\
\alpha_2 & -\alpha_1 & -\alpha_4 & \alpha_3 & \alpha_2^* & -\alpha_1^* & -\alpha_4^* & \alpha_3^* \\
\alpha_3 & \alpha_4 & -\alpha_1 & -\alpha_2 & \alpha_3^* & \alpha_4^* & -\alpha_1^* & -\alpha_2^* \\
\alpha_4 & -\alpha_3 & \alpha_2 & -\alpha_1 & \alpha_4^* & -\alpha_3^* & \alpha_2^* & -\alpha_1^*
\end{bmatrix}
\tag{4.108}
$$

If, instead of $S = (s_1, s_2, s_3, s_4)$, other orthogonal vectors are used, we have:

$$
\begin{aligned}
(-r_2, r_1, r_4, -r_3, -r_6^*, r_5^*, r_8^*, -r_7^*) &= (s_2, -s_1, s_4, -s_3)\Omega \\
&\quad + (-\eta_2, \eta_1, \eta_4, -\eta_3, -\eta_6^*, \eta_5^*, \eta_8^*, -\eta_7^*) \\
(-r_3, -r_4, r_1, r_2, -r_7^*, -r_8^*, r_5^*, r_6^*) &= (s_3, -s_4, -s_1, s_2)\Omega \\
&\quad + (-\eta_3, -\eta_4, \eta_1, \eta_2, -\eta_7^*, -\eta_8^*, \eta_5^*, \eta_6^*) \\
(-r_4, r_3, -r_2, r_1, -r_8^*, r_7^*, -r_6^*, r_5^*) &= (s_4, s_3, -s_2, -s_1)\Omega \\
&\quad + (-\eta_4, \eta_3, -\eta_2, \eta_1, -\eta_8^*, \eta_7^*, -\eta_6^*, \eta_5^*)
\end{aligned}
\tag{4.109}
$$

4.9.3   Orthogonality

Note that, in general, for $S = (s_1, s_2, s_3, s_4)^T$, vectors

$$
\begin{aligned}
V_1(S) &= (s_1, s_2, s_3, s_4, s_1^*, s_2^*, s_3^*, s_4^*)^T \\
V_2(S) &= (s_2, -s_1, s_4, -s_3, s_2^*, -s_1^*, s_4^*, -s_3^*)^T \\
V_3(S) &= (s_3, -s_4, -s_1, s_2, s_3^*, -s_4^*, -s_1^*, s_2^*)^T \\
V_4(S) &= (s_4, s_3, -s_2, -s_1, s_4^*, s_3^*, -s_2^*, -s_1^*)^T
\end{aligned}
\tag{4.110}
$$

are orthogonal to each other. Therefore, for specific constellation symbols S, vectors $V_1(S)$, $V_2(S)$, $V_3(S)$, $V_4(S)$ can create a basis for the four-dimensional subspace of any arbitrary four-dimensional constellation symbols and their conjugates in an eight-dimensional space. If the constellation symbols are real numbers, vectors

$$
\begin{aligned}
V_1'(S) &= (s_1, s_2, s_3, s_4)^T \\
V_2'(S) &= (s_2, -s_1, s_4, -s_3)^T \\
V_3'(S) &= (s_3, -s_4, -s_1, s_2)^T \\
V_4'(S) &= (s_4, s_3, -s_2, -s_1)^T
\end{aligned}
\tag{4.111}
$$

which only contain the first four elements of vectors $V_1(S)$, $V_2(S)$, $V_3(S)$, $V_4(S)$, create a basis for the space of any arbitrary four-dimensional real constellation symbols.

4.9.4   Encoding

Step 1. Let us assume that we use a signal constellation with 2^b elements. For each block of Kb bits, the encoding is done by first calculating the K-dimensional vector of symbols $\mathbf{S} = (s_1, s_2, s_K)^T$.

Step 2. Indeterminates $x_1, x_2, \ldots, x_K$ in Γ are replaced by $s_1, s_2, \ldots, s_K$ to establish the matrix $\mathbf{X}$ which is used for transmission in a manner similar to a regular space–time block code. The main issue here is how to calculate $\mathbf{S} = (s_1, s_2, \ldots, s_K)^T$ such that non-coherent detection is possible.

Step 3. Let us assume that $\mathbf{S}_v$ is the vector which is used for the vth block of Kb bits. Also, $\mathbf{X} \, (\mathbf{S}_v)$ defines what to transmit from each antenna during the transmission of the vth block. In other words, $\mathbf{X}_n$ $(\mathbf{S}_v)$, $n = 1, 2, \ldots, N$ is the nth column of $\mathbf{X} \, (\mathbf{S}_v)$ which contains T symbols which are transmitted from the nth antenna sequentially.

We fix a set V which consists of 2^{Kb} unit length vectors $\mathbf{P}_1, \mathbf{P}_2, \ldots, \mathbf{P}_{2Kb}$ where each vector $\mathbf{P}_l$ is a $K \times 1$ vector of real numbers $\mathbf{P}_l = (P_{l1}, P_{l2}, \ldots, P_{lK})^T$. We define an arbitrary one-to-one mapping β which maps Kb bits onto V. We start with an arbitrary vector $\mathbf{S}_1$. Then, let us assume that $\mathbf{S}_v$ is used for the vth block. For the $(v + 1)$th block, we use the Kb input bits to pick the corresponding vector $\mathbf{P}_l$ in V using the one-to-one mapping β. Then, we calculate:

$$\mathbf{S}_{v+1} = \sum_{k=1}^{K} P_{lk} \mathbf{V}'_k \, (\mathbf{S}_v) \tag{4.112}$$

where $\mathbf{V}'_k \, (\mathbf{S}_v)$ is a K-dimensional vector which includes the first K elements of $\mathbf{V}_k \, (\mathbf{S}_v)$. We use $\mathbf{X}(\mathbf{S}_{v+1})$ for transmission at the following T time slots.

4.9.4.1   Example

We pick four BPSK symbols, i.e. symbol $s_k = 1$ if the corresponding bit is zero and $s_k = -1$ otherwise. Then, $\mathbf{P}_l$ is defined as the projection of the vector (s_1, s_2, s_3, s_4) onto $\mathbf{V}'_1 \, (\mathbf{S}), \mathbf{V}'_2 \, (\mathbf{S}), \mathbf{V}'_3 \, (\mathbf{S}), \mathbf{V}'_4 \, (\mathbf{S})$, where $\mathbf{S} = (1,1,1,1)^T$. In fact, the elements of $\mathbf{P}_l$ can be calculated using the following equations:

$$\begin{aligned}
P_{l1} &= (s_1 + s_2 + s_3 + s_4)/4 \\
P_{l2} &= (s_1 - s_2 + s_3 - s_4)/4 \\
P_{l3} &= (s_1 - s_2 - s_3 + s_4)/4 \\
P_{l4} &= (s_1 + s_2 - s_3 - s_4)/4
\end{aligned} \tag{4.113}$$

This completes the one-to-one mapping β which is needed for encoding and decoding.

4.9.5   Differential Decoding

Let us recall that the received signal r_t is related to the transmitted signals by:

$$r_t = \sum_{n=1}^{N} \alpha_n \mathbf{X}_{t,n} + \eta_t \tag{4.114}$$

Assuming $T = 2K$, which results in a rate 1/2 code, and defining

$$\mathbf{R} = \left(r_1, r_2, \ldots r_K, r_{K+1}^*, r_{K+2}^*, \ldots, r_{2K}^* \right),$$

we have

$$\mathbf{R} = \mathbf{S}^T \Omega \, (\alpha_1, \alpha_2, \ldots, \alpha_N) + N \tag{4.115}$$

where $\mathbf{N} = (\eta_1, \eta_2, \ldots, \eta_K, \eta_{K+1}^*, \eta_{K+2}^*, \ldots, \eta_{2K}^*)$ and

$$\Omega(\alpha_1, \alpha_2, \ldots, \alpha_N) = (\Lambda(\alpha_1, \alpha_2, \ldots, \alpha_N)\Lambda(\alpha_1^*, \alpha_2^*, \ldots, \alpha_N^*)) \tag{4.116}$$

where $\Lambda(\alpha_1, \alpha_2, \ldots, \alpha_N)$ is the $K \times K$ matrix defined by

$$\Lambda_{i,j} = \delta(i, j)\alpha_{\varepsilon(i,j)}, \quad i = 1, 2, \ldots, K, \quad j = 1, 2, \ldots, K \tag{4.117}$$

where $\varepsilon(i, j) = l \Leftrightarrow B_{j,l} = s_i$ or $B_{j,l} = -s_i$, and

$$\delta(i, j) = \begin{cases} 1, & \text{if } B_{j,l} = s_i \\ -1, & \text{if } B_{j,l} = s_i \end{cases}$$

4.9.5.1 Example

For the rate 1/2 code defined by Γ_{84} in Equation (4.105), $\Lambda(\alpha_1, \alpha_2, \alpha_3, \alpha_4)$ is a 4×4 matrix defined by:

$$\Lambda = \begin{bmatrix} \alpha_1 & \alpha_2 & \alpha_3 & \alpha_4 \\ \alpha_2 & -\alpha_1 & -\alpha_4 & \alpha_3 \\ \alpha_3 & \alpha_4 & -\alpha_1 & -\alpha_2 \\ \alpha_4 & -\alpha_3 & \alpha_2 & -\alpha_1 \end{bmatrix} \tag{4.118}$$

Differential decoding is enabled by the fact that $\Omega\Omega^* = 2\sum_{n=1}^N |\alpha_n|^2 I_K$. To prove it, let us set $N = 0$ in Equations (4.114) and (4.115). Then, we have $\mathbf{R} = \mathbf{S}^T\Omega$ which results in:

$$\mathbf{RR}^* = \mathbf{S}^T \Omega\Omega^* S^{*T} \tag{4.119}$$

On the other hand

$$\mathbf{RR}^* = \sum_{t=1}^T |r_t|^2 = \sum_{t=1}^T \left(\sum_{n=1}^N \alpha_n X_{t,n}\right)\left(\sum_{n=1}^N \alpha_n X_{t,n}\right)^* \tag{4.120}$$

We rewrite the above formulas using matrix $\mathbf{C}$ as follows:

$$\begin{aligned} \mathbf{RR}^* &= (\alpha_1, \alpha_2, \ldots, \alpha_N)\mathbf{X}^T\mathbf{X}^{*T}(\alpha_1, \alpha_2, \ldots, \alpha_N)^* \\ &= (\alpha_1, \alpha_2, \ldots, \alpha_N)(\mathbf{X}^*\mathbf{X})^T(\alpha_1, \alpha_2, \ldots, \alpha_N)^* \\ &= (\alpha_1, \alpha_2, \ldots, \alpha_N)2\sum_{k=1}^K |s_k|^2 I_K(\alpha_1, \alpha_2, \ldots, \alpha_N)^* \\ &= 2\sum_{k=1}^K |s_k|^2 (\alpha_1, \alpha_2, \ldots, \alpha_N)(\alpha_1, \alpha_2, \ldots, \alpha_N)^* \end{aligned} \tag{4.121}$$

Therefore, we have:

$$\mathbf{RR}^* = \mathbf{S}^T\Omega\Omega^*\mathbf{S}^{*T} = 2\sum_{k=1}^K |s_k|^2 \sum_{n=1}^N |\alpha_n|^2 \tag{4.122}$$

Since Equation (4.122) is true for any $\mathbf{S} = (s_1, s_2, \ldots, s_K)^T$, we should have

$$\Omega\Omega^* = 2\sum_{n=1}^N |\alpha_n|^2 I_K$$

4.9.6   Received Signal

Let us recall that $\mathbf{S}_v$ and $\mathbf{S}_{v+1}$ are used for the vth and $(v + l)$th blocks of Kb bits, respectively. Using Γ_{84}, for each block of data we receive eight signals. To simplify the notation, we denote the received signals corresponding to the vth block by $r_1^v, r_2^v, \ldots, r_8^v$ and the received signals corresponding to the $(v + 1)$th block by $r_1^{v+1}, r_2^{v+1}, \ldots, r_8^{v+1}$. Let us construct the following vectors (see Equation (4.109)).

$$
\begin{aligned}
\mathbf{R}_v^1 &= (r_1^v, r_2^v, r_3^v, r_4^v, r_5^{v*}, r_6^{v*}, r_7^{v*}, r_8^{v*}) \\
\mathbf{R}_v^2 &= (-r_2^v, r_1^v, r_4^v, -r_3^v, -r_6^{v*}, r_5^{v*}, r_8^{v*}, -r_7^{v*}) \\
\mathbf{R}_v^3 &= (-r_3^v, -r_4^v, r_1^v, r_2^v, -r_7^{v*}, -r_8^{v*}, r_5^{v*}, r_6^{v*}) \\
\mathbf{R}_v^4 &= (-r_4^v, r_3^v, -r_2^v, r_1^v, -r_8^{v*}, r_7^{v*}, -r_6^{v*}, r_5^{v*}) \\
\mathbf{R}_{v+1} &= (r_1^{v+1}, r_2^{v+1}, r_3^{v+1}, r_4^{v+1}, r_5^{(v+1)*}, r_6^{(v+1)*}, r_7^{(v+1)*}, r_8^{(v+1)*})
\end{aligned}
\tag{4.123}
$$

By using Equations (4.107) and (4.109), we have:

$$
\begin{aligned}
\mathbf{R}_{v+1}\mathbf{R}_v^{k*} &= \mathbf{S}_{v+1}^{\mathrm{T}} \Omega\Omega^* \, \mathbf{V}_k' \, (\mathbf{S}_v)^{*\mathrm{T}} + \mathbf{N}_k \\
&= 2\sum_{n=1}^{4} |\alpha_n|^2 \, \mathbf{S}_{v+1}^{\mathrm{T}} \mathbf{V}_k' \, (\mathbf{S}_v)^{*\mathrm{T}} + \mathbf{N}_k \\
&= 2\sum_{n=1}^{4} |\alpha_n|^2 \, \mathbf{P}_{lk} + \mathbf{N}_k
\end{aligned}
\tag{4.124}
$$

Therefore, we can write:

$$
\begin{aligned}
\mathbf{R} &= (R_{v+1}R_v^{1*}, R_{v+1}R_v^{2*}, R_{v+1}R_v^{3*}, R_{v+1}R_v^{4*}) \\
&= \left(2\sum_{n=1}^{4} |\alpha_n|^2\right) \mathbf{P}_l + \mathbf{N}'
\end{aligned}
\tag{4.125}
$$

where $\mathbf{N}' = (\eta_1, \eta_2, \ldots, \eta_K)$.

4.9.7   Demodulation

Because the elements of $\mathbf{R}$ have equal lengths, to compute $\mathbf{P}_l$, the receiver can compute the closest vector of $\mathbf{V}$ to $\mathbf{R}$. Once this vector is computed, the inverse mapping of β is applied and the transmitted bits are recovered.

4.9.8   Multiple Receive Antennas

If there is more than one receive antenna, the same procedure can be used. In this case, first assuming that only the receive antenna m exists, we compute $\mathbf{P}^m$ using the same method for $\mathbf{P}$ given above. Then, after calculating M vectors $\mathbf{P}^m$, $m = 1, 2, \ldots, M$, the closest vector of $\mathbf{V}$ to $\sum_{m=1}^{M} P^m$ is computed. The inverse mapping of β is applied to the closest vector to compute the transmitted bits. It is easy to show that $4M$-level diversity is achieved by using this method.

4.9.9   The Number of Transmit Antennas Lower than the Number of Symbols

So far, we have considered an example where $N = K$, i.e. the real matrix which creates Γ is a square matrix. When the number of transmit antennas is less than the number of symbols, $N < K$, the same approach works. Let us use the following example:

$$\Gamma_{83} = \begin{bmatrix} x_1 & x_2 & x_3 \\ -x_2 & x_1 & -x_4 \\ -x_3 & x_4 & x_1 \\ -x_4 & -x_3 & x_2 \\ x_1^* & x_2^* & x_3^* \\ -x_2^* & x_1^* & -x_4^* \\ -x_3^* & x_4^* & x_1^* \\ -x_4^* & -x_3^* & x_2^* \end{bmatrix} \tag{4.126}$$

When there is only one receive antenna, $M = 1$, the received signals are related to the constellation symbols s_1, s_2, s_3, s_4 by:

$$\begin{aligned} r_1 &= \alpha_1 s_1 + \alpha_2 s_2 + \alpha_3 s_3 + \eta_1 \\ r_2 &= -\alpha_1 s_2 + \alpha_2 s_1 - \alpha_3 s_4 + \eta_2 \\ r_3 &= -\alpha_1 s_3 + \alpha_2 s_4 + \alpha_3 s_1 + \eta_3 \\ r_4 &= -\alpha_1 s_4 - \alpha_2 s_3 + \alpha_3 s_2 + \eta_4 \\ r_5 &= \alpha_1 s_1^* + \alpha_2 s_2^* + \alpha_3 s_3^* + \eta_5 \\ r_6 &= -\alpha_1 s_2^* + \alpha_2 s_1^* - \alpha_3 s_4 + \eta_6 \\ r_7 &= -\alpha_1 s_3^* + \alpha_2 s_4 + \alpha_3 s_1^* + \eta_7 \\ r_8 &= -\alpha_1 s_4 - \alpha_2 s_3^* + \alpha_3 s_2^* + \eta_8. \end{aligned} \tag{4.127}$$

Equation (4.107) now becomes:

$$(r_1, r_2, r_3, r_4, r_5^*, r_6^*, r_7^*, r_8^*) = (s_1, s_2, s_3, s_4)\,\Omega + (\eta_1, \eta_2, \eta_3, \eta_4, \eta_5^*, \eta_6^*, \eta_7^*, \eta_8^*) \tag{4.128}$$

where

$$\Omega = \begin{bmatrix} \alpha_1 & \alpha_2 & \alpha_3 & 0 & \alpha_1^* & \alpha_2^* & \alpha_3^* & 0 \\ \alpha_2 & -\alpha_1 & 0 & \alpha_3 & \alpha_2^* & -\alpha_1^* & 0 & \alpha_3^* \\ \alpha_3 & 0 & -\alpha_1 & -\alpha_2 & \alpha_3^* & 0 & -\alpha_1^* & -\alpha_2^* \\ 0 & -\alpha_3 & \alpha_2 & -\alpha_1 & 0 & -\alpha_3^* & \alpha_2^* & -\alpha_1^* \end{bmatrix} \tag{4.129}$$

The matrix Ω for Γ_{83} can be calculated from Ω for Γ_{84}, and Equation (4.108) by setting $\alpha_4 = 0$. Therefore, again for specific constellation symbols S, vectors $V_1(S)$, $V_2(S)$, $V_3(S)$, $V_4(S)$ can create a basis for the four-dimensional subspace of any arbitrary four-dimensional constellation symbols and their conjugates in an eight-dimensional space and the same encoding and decoding schemes are applicable. The only difference in the final result is that $\sum_{n=1}^{4} |\alpha_n|^2$ is replaced by $\sum_{n=1}^{3} |\alpha_n|^2$.

4.9.10   Final Result

Instead of Equation (4.125) we now have:

$$\begin{aligned} \mathbf{R} &= (\mathbf{R}_{v+1}\mathbf{R}_v^{1*}, \mathbf{R}_{v+1}\mathbf{R}_v^{2*}, \mathbf{R}_{v+1}\mathbf{R}_v^{3*}, \mathbf{R}_{v+1}\mathbf{R}_v^{4*}) \\ &= \left(2\sum_{n=1}^{3} |\alpha_n|^2 \right) \mathbf{P}_l + \mathbf{N}' \end{aligned} \tag{4.130}$$

4.9.11   Real Constellation Set

The above rate 1/2 space–time block codes can be applied to any complex constellation set. If the constellation set is real, rate 1 space–time block codes are available and the same approach works. For example, in the case of $T = K = 4$, the following space–time block code exists for $N = 4$:

$$
\Gamma = \begin{bmatrix}
x_1 & x_2 & x_3 & x_4 \\
-x_2 & x_1 & -x_4 & x_3 \\
-x_3 & x_4 & x_1 & -x_2 \\
-x_4 & -x_3 & x_2 & x_1
\end{bmatrix}
\tag{4.131}
$$

It is easy to show that:

$$
(r_1, r_2, r_3, r_4, r_1^*, r_2^*, r_3^*, r_4^*) = (s_1, s_2, s_3, s_4)\Omega + (\eta_1, \eta_2, \eta_3, \eta_4, \eta_1^*, \eta_2^*, \eta_3^*, \eta_4^*)
\tag{4.132}
$$

where Ω is defined by Equation (4.108). Similar differential encoding and decoding are possible if we use the following vectors for $\mathbf{R}_v^i$, $i = 1, 2, 3, 4$, and $\mathbf{R}_{v+1}$:

$$
\begin{aligned}
\mathbf{R}_v^1 &= (r_1^v, r_2^v, r_3^v, r_4^v, r_1^{v*}, r_2^{v*}, r_3^{v*}, r_4^{v*}) \\
\mathbf{R}_v^2 &= (-r_2^v, r_1^v, r_4^v, -r_3^v, -r_2^{v*}, r_1^{v*}, r_4^{v*}, -r_3^{v*}) \\
\mathbf{R}_v^3 &= (-r_3^v, -r_4^v, r_1^v, r_2^v, -r_3^{v*}, -r_4^{v*}, r_1^{v*}, r_2^{v*}) \\
\mathbf{R}_v^4 &= (-r_4^v, r_3^v, -r_2^v, r_1^v, -r_4^{v*}, r_3^{v*}, -r_2^{v*}, r_1^{v*}) \\
\mathbf{R}_{v+1} &= (r_1^{v+1}, r_2^{v+1}, r_3^{v+1}, r_4^{v+1}, r_1^{(v+1)*}, r_2^{(v+1)*}, r_3^{(v+1)*}, r_4^{(v+1)*})
\end{aligned}
\tag{4.133}
$$

This results in a full diversity, full rate scheme for differential detection. For example, as defined by Equation (4.133), performance results are shown in Figures 4.44 and 4.45 for four and three transmit antennas respectively.

More details on the topic can be found in [77, 78].

4.10   Layered Space–Time Coding

In this section we discuss a possibility of partitioning antennas at the transmitter into small groups, and using individual space–time codes, called component codes, to transmit information from each group of antennas. At the receiver, an individual space–time code is decoded by a linear processing technique that suppresses signals transmitted by other groups of antennas by treating them as interference. This receiver structure provides diversity and coding gain over uncoded systems. This combination of array processing at the receiver and coding techniques for multiple transmit antennas can provide reliable and very high data rate communication over narrowband wireless channels. A refinement of this basic structure gives rise to a multilayered space–time architecture that both generalizes and improves upon the layered space–time architecture proposed by Foschini, which is known in the literature as Bell Lab Layered space time (BLAST) coding [10, 77–88].

For $1 \leq j \leq m$ the signal r_t^j received by antenna j at time t is given by:

$$
r_t^j = \sum_{i=1}^n \alpha_{i,j} c_t^i + \eta_t^j
\tag{4.134}
$$

where c_t^i is the encoded signal transmitted from transmit antenna i.

For any vector $\mathbf{x}$, let $\mathbf{x}^T$ denote the transpose of $\mathbf{x}$. We can now write Equation (4.134) in the vector form given by:

$$
\mathbf{r}_t = \Omega \mathbf{c}_t + \eta_t
\tag{4.135}
$$

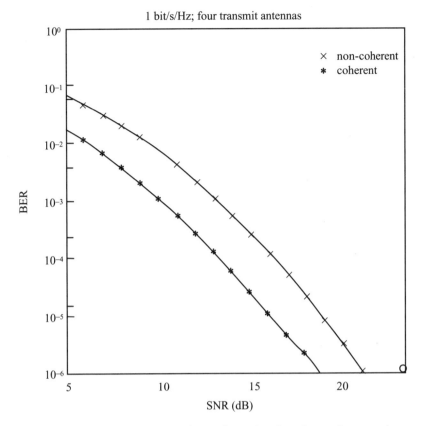

Figure 4.44 Performance results for coherent and non-coherent detection schemes; four transmit antennas, one receive antenna, BPSK.

where

$$c_t = (c_t^1, c_t^2, \ldots, c_t^n)^{\mathrm{T}}$$
$$r_t = (r_t^1, r_t^2, \ldots, r_t^m)^{\mathrm{T}} \qquad (4.136)$$
$$\eta_t = (\eta_t^1, \eta_t^2, \ldots, \eta_t^m)^{\mathrm{T}}$$

and

$$\Omega = \begin{bmatrix} \alpha_{1,1} & \alpha_{2,1} & \cdots & \cdots & \alpha_{n,1} \\ \alpha_{1,2} & \alpha_{2,2} & \cdots & \cdots & \alpha_{n,2} \\ \alpha_{1,3} & \alpha_{2,3} & \cdots & \cdots & \alpha_{n,3} \\ \vdots & \vdots & \ddots & \ddots & \vdots \\ \alpha_{1,m} & \alpha_{2,m} & \cdots & \cdots & \alpha_{n,m} \end{bmatrix} \qquad (4.137)$$

A *space–time product encoder* accepts a block of B input bits in each time slot t and these bits are divided into q strings of lengths $B_1, B_2, \ldots, B_q$ with $B_1 + B_2 + \cdots + B_q = B$. At the base station, n antennas are partitioned into q groups $G_1, G_2, \ldots, G_q$, respectively, comprising $n_1, n_2, \ldots, n_q$ antennas with $n_1 + n_2 + \cdots + n_q = n$. Each block B_j, $1 \leq j \leq q$ is then encoded by a space–time encoder $\mathbf{X}_j$. The output of $\mathbf{X}_j$ goes through a serial to parallel converter and provides n_j sequences of constellation

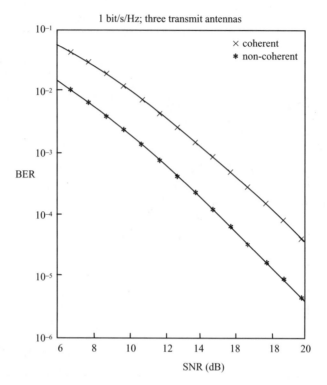

Figure 4.45 Performance results for coherent and non-coherent detection schemes; three transmit antennas, one receive antenna, BPSK.

symbols for $1 \leq j \leq q$ which are simultaneously transmitted from the antennas of the group G_j. This gives a total of n sequences of constellation symbols that are transmitted simultaneously from antennas $1, 2, \ldots, n$.

A space–time product encoder can be considered as a set of q space–time encoders, called the *component codes*, operating in parallel on the same wireless communication channel, with each encoder using n_j transmit and m receive antennas for $1 \leq j \leq q$. It will be denoted an $\mathbf{X}_1 \times \mathbf{X}_2 \times \cdots \mathbf{X}_q$ encoder.

4.10.1   Receiver Complexity

One approach to recovering the transmitted data at the receiver is to jointly decode the transmitted code word, but the difficulty here is decoding complexity. Indeed, if we require a diversity of rm, where $r \leq \min_j (n_j)$, then (see Section 4.2) the complexity of the trellis of $\mathbf{X}_j$ is at least $2^{B_j(r-1)}$ states and the complexity of the product code is at least $2^{B(r-1)}$ states. This means that if B is very large, the scheme may be too complex to implement.

4.10.2   Group Interference Suppression

The idea is to decode each code $\mathbf{X}_j$ separately while suppressing signals from other component codes. This approach has a much lower complexity but achieves a lower diversity order than the full diversity order nm, which is the product of the numbers of transmit and receive antennas.

4.10.3   Suppression Method

Without any loss of generality, we take $j = 1$ and look to decode $\mathbf{X}_1$. There are $n - n_1$ interfering signals. We assume that there are $m \geq n - n_1 + 1$ receive antennas and that the receiver knows the matrix Ω (channel state information). The matrix

$$\Lambda(\mathbf{X}_1) = \begin{bmatrix} \alpha_{n_1+1,1} & \alpha_{n_1+2,1} & \cdots & \cdots & \alpha_{n,1} \\ \alpha_{n_1+1,2} & \alpha_{n_1+2,2} & \cdots & \cdots & \alpha_{n,2} \\ \alpha_{n_1+1,3} & \alpha_{n_1+2,3} & \cdots & \cdots & \alpha_{n,3} \\ \vdots & \vdots & \ddots & \ddots & \vdots \\ \alpha_{n_1+1,m} & \alpha_{n_1+2,m} & \cdots & \cdots & \alpha_{n,m} \end{bmatrix} \tag{4.138}$$

has rank less than or equal to the number of its columns. Thus, rank $[\Lambda(\mathbf{X}_1)] \leq n - n_1$.

4.10.4   The Null Space

The *null space* N of this matrix is the set of all row vectors $\mathbf{x}$ such that $x\Lambda = (\mathbf{X}_1) = (0,0,\ldots,0)$. Furthermore,

$$\dim(N) + \text{rank}\,[\Lambda(\mathbf{X}_1)] = m$$

Since rank $[\Lambda(\mathbf{X}_1)] \leq n - n_1$, it follows that $\dim, (N) \geq m - n + n_1$. Hence we can compute a (not necessarily unique) *set of orthonormal vectors*:

$$\{\mathbf{v}_1, \mathbf{v}_2, \ldots, \mathbf{v}_{m-n+n_1}\}$$

in N. We let $\Theta(\mathbf{X}_1)$ denote the $(m - n + n_1) \times m$ matrix whose jth row is $\mathbf{v}_j$. Clearly, $\Theta(\mathbf{X}_1)\Theta(\mathbf{X}_1)^* = I_{n_1-n+m}$, where $\Theta(\mathbf{X}_1)^*$ is the Hermitian of $\Theta(\mathbf{X}_1)$ and $\mathbf{I}_{n_1-n+m}$ is the $(m - n + n_1) \times (m - n + n_1)$ identity matrix. We multiply both sides of Equation (4.135) by $\Theta(\mathbf{X}_1)$ to arrive at:

$$\Theta(\mathbf{X}_1)\mathbf{r}_t = \Theta(\mathbf{X}_1)\Omega\mathbf{c}_t + \Theta(\mathbf{X}_1)\eta_t \tag{4.139}$$

where

$$\Omega(\mathbf{X}_1) = \begin{bmatrix} \alpha_{1,1} & \alpha_{2,1} & \cdots & \cdots & \alpha_{n_1,1} \\ \alpha_{1,2} & \alpha_{2,2} & \cdots & \cdots & \alpha_{n_1,2} \\ \alpha_{1,3} & \alpha_{2,3} & \cdots & \cdots & \alpha_{n_1,3} \\ \vdots & \vdots & \ddots & \ddots & \vdots \\ \alpha_{1,m} & \alpha_{2,m} & \cdots & \cdots & \alpha_{n_1,m} \end{bmatrix} \tag{4.140}$$

Since $\Theta(\mathbf{X}_1)\Lambda(\mathbf{X}_1) = 0$ is the all zero matrix, Equation (4.139) can be written as

$$\Theta(\mathbf{X}_1)\mathbf{r}_t = \Theta(\mathbf{X}_1)\Omega(\mathbf{X}_1)\mathbf{c}_t^1 + \Theta(\mathbf{X}_1)\eta_t \tag{4.141}$$

where $\mathbf{c}_t^1 = (c_t^1, c_t^2, \ldots, c_t^{n_1})^{\mathrm{T}}$. Setting

$$\begin{aligned} \tilde{\mathbf{r}}_t &= \Theta(\mathbf{C}_1)\mathbf{r}_t \\ \tilde{\Omega} &= \Theta(\mathbf{C}_1)\Omega \\ \tilde{\eta}_t &= \Theta(\mathbf{C}_1)\eta_t \end{aligned} \tag{4.142}$$

we arrive at the equation

$$\tilde{\mathbf{r}}_t = \tilde{\Omega}\,\mathbf{c}_t^1 + \tilde{\eta}_t \tag{4.143}$$

This is an equation where all the signal streams out of antennas $n_1 + 1, n_1 + 2, \ldots, n$ are suppressed.

4.10.5   Receiver

Let us look at the structure of the decoder for the code $\mathbf{X}_1$, given that group interference suppression is performed to suppress all the signal streams out of antennas $n_1 + 1, n_1 + 2, \ldots, n$. To this end, suppose that $\Lambda(\mathbf{X}_1)$ is given. The receiver computes a set of orthonormal vectors

$$\{\mathbf{v}_1, \mathbf{v}_2, \ldots, \mathbf{v}_m - n + n_1\}$$

and the matrix $\Theta(\mathbf{X}_1)$ as described in the previous section. Let $\tilde{\Omega}_{ij}$ and $\tilde{\Omega}_{lk}$ denote the (i, j)th and (l,k)th elements of $\tilde{\Omega}$. By definition $\tilde{\Omega}_{ij} = \mathbf{v}_i\omega_j$ and $\tilde{\Omega}_{lk} = \mathbf{v}_l\omega_k$ where ω_j and ω_k are, respectively, the jth and kth columns of $\Omega(\mathbf{X}_1)$. The random variables $\tilde{\Omega}_{ij}$ and $\tilde{\Omega}_{lk}$ have zero means given $\Lambda(\mathbf{X}_1)$. Moreover,

$$\mathrm{E}\left[\tilde{\Omega}_{ij}\tilde{\Omega}_{lk}^*\right] = \mathrm{E}\left[\mathbf{v}_i\omega_j\omega_k^*\mathbf{v}_l^*\right] = \mathbf{v}_i\,\mathrm{E}\left[\omega_j\omega_k^*\right]\mathbf{v}_l^* = \delta_{jk}\mathbf{v}_i\mathbf{v}_l^* = \delta_{jk}\delta_{il} \tag{4.144}$$

where δ is the Kronecker delta function given by $\delta_{rs} = 0$ if $r \neq s$ and $\delta_{rs} = 1$ if $r = s$. We conclude that the elements of the $(m - n + n_1) \times n_1$ matrix $\tilde{\Omega}$ are independent complex Gaussian random variables of variance 0.5 per real dimension. Similarly, the components of the noise vector $\tilde{\eta}_t$, $t = 1, 2, \ldots l$ are independent Gaussian random variables of variance $N_0/2$ per real dimension.

4.10.6   Decision Metric

Assuming that all the code words of $\mathbf{X}_1$ are equiprobable, and given that group interference suppression is performed, the ML receiver for $\mathbf{X}_1$ decides in favor of the code word

$$c_1^1 c_1^2 \cdots c_1^{n_1} c_2^1 c_2^2 \cdots c_2^{n_1} \cdots c_l^1 c_l^2 \cdots c_l^{n_1}$$

if it minimizes the decision metric

$$\sum_{t=1}^{l}\left|\tilde{\mathbf{r}}_t - \tilde{\Omega}\mathbf{c}_t^1\right|^2 \tag{4.145}$$

4.10.7   Multilayered Space–Time Coded Modulation

The previous discussion reduces code design for a multiple antenna communication system with n transmit and m receive antennas to that of designing codes for communication systems with n_i transmit and $n_i + m - n$, $i = 1, 2, \ldots, q$, receive antennas where $\sum_{i=1}^{q} n_i = n$ and $n_i \geq n - m + 1$. Using this insight, we may design a multilayered space–time coded modulation scheme. The idea behind such a system is multistage detection and cancellation.

4.10.8   Diversity Gain

Suppose that $\mathbf{X}_1$ is decoded correctly using combined array processing and space–time coding. The space–time code $\mathbf{X}_1$ affords a diversity gain of $n_1 \times (n_1 + m - n)$. After decoding $\mathbf{X}_1$ we may subtract the contribution of these code words to signals received at different antennas. This gives a communication system with $n - n_1$ transmit and m receive antennas.

In the next step the receiver uses combined array processing and space–time coding to decode $\mathbf{X}_2$. The space–time code $\mathbf{X}_2$ affords a diversity gain of $n_2 \times (n_2 + n_1 + m - n)$. Proceeding in this manner, we observe that by subtracting the contribution of previously decoded code streams $\mathbf{X}_j, j \leq k - 1$ to the received signals at different antennas, the space–time code $\mathbf{X}_k$ affords a diversity gain of $n_k \times (n_1 + \cdots + n_k + m - n)$.

We can choose space–time codes $\mathbf{X}_i, 1 \leq i \leq q$ to provide these diversity gains and such that the sequence

$$n_1 (n_1 + m - n), n_2 (n_2 + n_1 + m - n), \ldots, n_k (n_1 + \cdots + n_k + m - n), \ldots, n_q m$$

is an increasing sequence. Assuming there was no decoding error in steps $1, 2, \ldots, k - 1$, then at decoding step k, the probability of error for the component code $\mathbf{X}_k$ is equal to the probability of error for $\mathbf{X}_k$ when employed in a communication system using n_k transmit and $(n_1 + \cdots + n_k + m - n)$ receive antennas.

4.10.9   Adaptive Reconfigurable Transmit Power Allocation

Since the diversity in each decoding stage k is more than that of the previous decoding stage $k - 1$, the transmit power out of each antenna at level k can be substantially less than that of the previous layer. Thus the transmitter should divide the available transmit power among different antennas in an unequal manner. Power allocation for this scenario is straightforward. In fact, powers at different levels could be allocated based on the diversity gains. In this way, the allocated powers may decrease geometrically in terms of the diversity gains. Other approaches are also possible.

4.10.9.1   Example 1

Transmitter. Here, four transmit and four receive antennas are used. The transmission rate is 4 bits/s/Hz. Let $\mathbf{X}$ denote the 32-state 4PSK (see Figure 4.46) space–time trellis code given in Figure 4.47. The product code $\mathbf{X}_1 \times \mathbf{X}_2$ where $\mathbf{X}_1 = \mathbf{X}_2 = \mathbf{X}$ will be used for transmission of 4 bits/s/Hz. At each time slot, upon the arrival of the four bits of the input data, the first two bits are used as the input to the encoder of $\mathbf{X}_1$ and the encoded symbols are transmitted by antennas one and two. The second two bits are used as the input to the encoder of $\mathbf{X}_2$ and the encoded signals are transmitted by antennas three and four. We assume that the average powers radiated from antennas one and two are equal but each is twice as much as the average power radiated from antennas three and four.

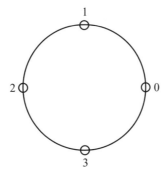

Figure 4.46   The 4PSK constellation.

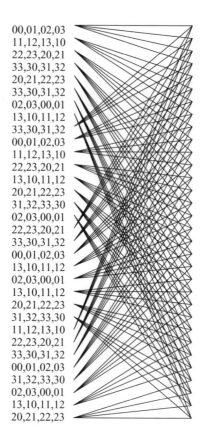

Figure 4.47   4PSK space–time code, 32 states 2 bits/s/Hz.

Receiver. Interference suppression is used to suppress $\mathbf{X}_2$ and decode $\mathbf{X}_1$. Upon decoding $\mathbf{X}_1$ the contributions of the code words transmitted from antennas one and two are subtracted from the received signals. Finally, $\mathbf{X}_2$ is decoded.

4.10.9.2   Simulation Results

Figure 4.48 demonstrates the performance of this multilayered space–time coded architecture. Each frame consists of 130 transmissions from each transmit antenna. It is assumed that the channel matrix is perfectly known at the receiver. The horizontal axis shows the receive signal to noise ratio per transmission time. Each transmission time corresponds to the transmission of four bits. Thus, the horizontal axis denotes the receive signal to noise ratio per four bits. For comparison, the graph of the outage capacity versus the signal to noise ratio for four transmit and four receive antennas is presented in Figure 4.49. The outage capacity is defined as the achievable capacity C_{out} for which the outage probability $P_{\text{out}} = P(C < C_{\text{out}}) < \varepsilon$. More details on MIMO channel capacity will be presented in Section 4.12. One can see that for a frame error probability of 10^{-1}, the system is about 6 dB away from the capacity.

4.10.9.3   Example 2

Here, eight transmit and eight receive antennas are used. The transmission rate is 8 bits/s/Hz. Let $\mathbf{X}$ denote the code given in Figure 4.47. We use the product code $\mathbf{X}_1 \times \mathbf{X}_2 \times \mathbf{X}_3 \times \mathbf{X}_4$ where $\mathbf{X}_1 = \mathbf{X}_2 = \mathbf{X}_3 = \mathbf{X}_4 = \mathbf{X}$ for transmission of 8 bits/s/Hz. At each time instance, upon the arrival of the eight bits of

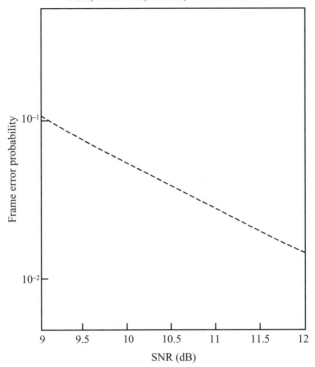

Figure 4.48 The performance of the scheme in Example 1.

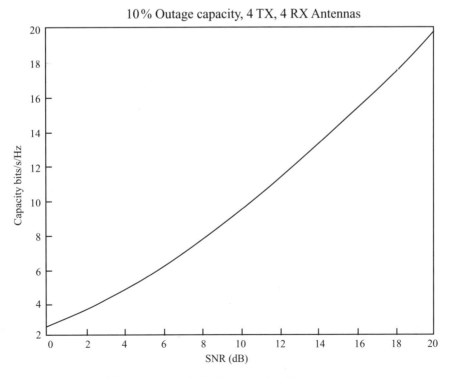

Figure 4.49 Outage capacity for four transmit and four receive antennas.

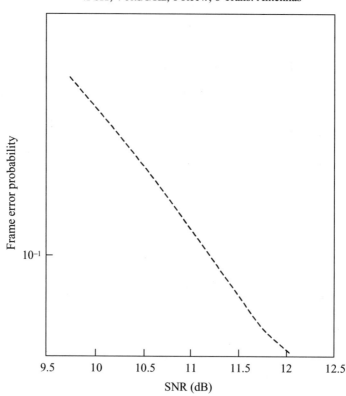

4PSK, 4 bits/s/Hz, 8 Recv., 8 Trans. Antennas

Figure 4.50   The performance of the scheme in Example 2.

the input data, the first, second, third, and fourth blocks of length two of the input bits are respectively used as the input to encoders of X_1, X_2, X_3 and X_4. The output of encoders of X_i, $1 \leq i \leq 4$ are, respectively, transmitted by antennas $2i - 1$ and $2i$. We assume that the average power radiated from antennas one and two is E_s, the average power radiated from antennas three and four is $E_s/2$, the average power radiated from antennas five and six is $E_s/4$, and the average power radiated from antennas seven and eight is $E_s/8$. Thus, the total signal to noise ratio at each receive antenna is $15\, E_s/4\, N_0$.

Decoder. Group interference suppression is used to decode X_1. Upon decoding X_1, the contributions of the code words transmitted from antennas one and two are subtracted from the received signals. Using this, X_2 is decoded next and so forth. In Figure 4.50, we provide simulation results to demonstrate the performance of this multilayered space–time coded architecture. Each frame consists of 130 transmissions from each transmit antenna. It is assumed that the channel matrix is perfectly known at the receiver. The horizontal axis shows the receive signal to noise ratio per transmission time. Each transmission time corresponds to the transmission of eight bits. Thus, the horizontal axis denotes the receive signal to noise ratio per eight bits.

For comparison, we provide in Figure 4.51 the graph of the outage capacity versus the signal to noise ratio for eight transmit and eight receive antennas, as computed by Foschini and Gans [89]. We observe that for a frame error probability of 10^{-1}, we are about 9 dB away from the capacity.

More details on this topic can be found in [10, 77–88].

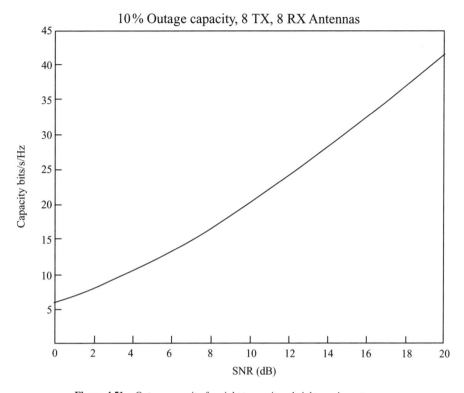

Figure 4.51   Outage capacity for eight transmit and eight receive antennas.

4.11   Concatenated Space–Time Block Coding

4.11.1   System Model

The system model is given in Figure 4.52.

We begin by performance analysis and design criteria in quasi-static fading, where the fading coefficients are constant during a frame of length $2L$ and vary from one frame to another.

4.11.2   Product Sum Distance

Let η be the set of all i for which $c_i \neq e_i$ or $c_{i+1} \neq e_{i+1}$. Denote the number of elements in η by l_η. Then, at high signal to noise ratios (SNRs) [90],

$$P\,(\mathbf{C} \to \mathbf{E}) \leq \frac{1}{\left[\left(\dfrac{E_s}{4N_0}\right)^{l_\eta} d_P\,(l_\eta)\right]^2} \tag{4.146}$$

where

$$d_P(l_\eta) = \prod_{i \in \eta} \left\lfloor |c_i - e_i|^2 + |c_{i+1} - e_{i+1}|^2 \right\rfloor$$

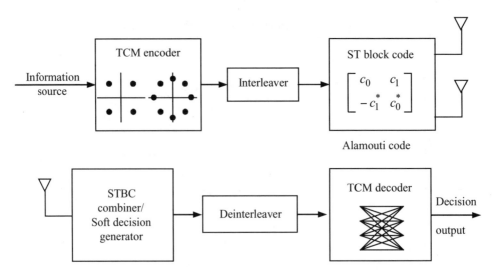

Figure 4.52   Block diagram of a space–time coded system concatenating STBC and TCM, $N = 2$, $M = 1$.

is the product of Euclidean distances associated with two consecutive symbols along the error event path ($\mathbf{C} \rightarrow \mathbf{E}$). Parameter $d_P(l_\eta)$ is referred to as the *product–sum distance* over span 2. In addition, l_η is referred to as the *effective length* of this error event over span 2.

Let P_e denote the error event probability, then by using the union bound, an upper bound can be obtained as:

$$P_e \leq \sum_{L=1}^{\infty} \sum_{C} \sum_{E \neq C} P(\mathbf{C}) P(\mathbf{C} \rightarrow \mathbf{E}) \tag{4.147}$$

where $P(\mathbf{C})$ is the *a priori* probability of transmitting the symbol sequence $\mathbf{C}$ with length $2L$. By summing over all possible l_η and all possible $d_P(l_\eta)$, the error event probability can be further written as:

$$P_e \leq \sum_{l_\eta} \sum_{d_P(l_\eta)} \Xi(l_\eta, d_P(l_\eta)) \left[\left(\frac{E_s}{4N_0} \right)^{l_\eta} d_P(l_\eta) \right]^{-2} \tag{4.148}$$

where $\Xi(l_\eta, d_P(l_\eta))$ is the average number of error events having the span 2 effective length l_η and the product–sum distance $d_P(l_\eta)$.

4.11.3   Error Rate Bound

The smallest l_η and the smallest $d_P(l_\eta)$ dominate the error event probability at high SNRs. Denoting $R = \min(l_\eta)$ and $d_{\min}(R) = \min(d_P(R))$, then the error event probability is asymptotically approximated as:

$$P_e \cong \Xi(R, d_{\min}(R)) \frac{1}{\left(\dfrac{E_s}{4N_0} \right)^{2R} [d_{\min}(R)]^2} \tag{4.149}$$

From Equation (4.149), we observe that the error event probability asymptotically varies with $2R$-power of SNR, so a diversity order of $2R$ is achieved. We further refer to R as the *built-in time diversity* or effective length of the concatenated space–time code.

The design criteria, in this case, involve the maximization of both the built-in-time diversity and the minimum product–sum distance of the trellis code at high SNRs for Rayleigh fading. This conclusion is, therefore, different from that of conventional TCM where the minimum product distance needs to be maximized. Thus, new optimal codes can be found, based upon these new criteria.

4.11.4   The Case of Low SNR

For low SNR we have [90]:

$$P\left(\mathbf{C} \rightarrow \mathbf{E}\right) \leq \left[1 + \frac{E_s}{4N_0} \sum_{i=0}^{2L-1} |c_i - e_i|^2 + o\left(\frac{E_s}{4N_0}\right)\right]^{-2} \tag{4.150}$$

where $o(E_s/4N_0)$ denotes the summation of all the terms which include higher order quantities of $(E_s/4N_0)$. Equation (4.150) indicates that the squared Euclidean distance becomes the main factor. Thus, the dominant factor affecting the performance of trellis coded modulation for use with space–time block coding at low SNRs is the free Euclidean distance rather than the product–sum distance and built-in time diversity.

4.11.5   Code Design

Here we explain the code design rules, by using the example with the four-state rate 2/3 8PSK trellis code for use with transmit diversity as it was presented in [90]. It is noted that the built-in time diversity (R) of a four-state code is equal to one and, therefore, optimized in this simplest case. In order to increase the product–sum distance of the code, parallel transitions in the trellis diagram are avoided. Thus, we can only consider the error events of actual length two to maximize the minimum product–sum distance of the underlying code. The signal transitions between states of consecutive stages can be represented by a 4×4 matrix $\mathbf{G}$, where the ijth element represents the signal transmitted from state i to state j between consecutive stages in the trellis diagram. Using set partitioning, the 8PSK signal set shown in Figure 4.53 is partitioned into two subsets $A_0 = (0, 2, 4, 6)$ and $A_1 = (1, 3, 5, 7)$. The design rules are given as follows.

Rule 1. Elements of each row of the matrix $\mathbf{G}$ are associated with signals from subsets A_0 or A_1. Specifically, signals with distance δ_1 and δ_3 are associated with branches diverging from one state to two adjacent states, with a state difference of two and one, respectively, where the state difference is defined as the number of bits in which two states differ (see Figure 4.54).

Rule 2. The distance between branches remerging at one state from two adjacent states with a state difference of two is δ_2. The pair of signals remerging from two states with a state difference of one is associated with distance δ_0 or δ_3.

According to Equation (4.149), the minimum product–sum distance should be maximized. Rule 1 associates distance δ_1 to signals diverging from one state to two adjacent states with a state difference of two (two states with state difference of two are always adjacent) and guarantees that the distance between any two signals diverging from one state is at least δ_1. Thus, if we assign δ_2 to signals remerging at one state from two states with a state difference of two (Rule 2), the minimum product–sum distance will be:

$$d_{\min}(R) = \min\left(\delta_1^2 + \delta_2^2, \delta_0^2 + \delta_3^2\right) = \delta_0^2 + \delta_3^2 = 4.586 E_s$$

which is greater than that of the optimal single antenna four-state 2/3 8PSK code.

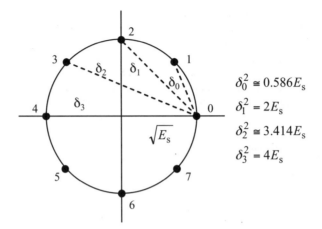

Figure 4.53 8 PSK signal set [90] © 2002, IEEE.

Using the above code design rules, the best four-state 2/3 8PSK trellis code for use with transmit diversity when perfect interleavers are assumed is shown in Figure 4.54. An equivalent code constructed using the design tools is also shown in the figure.

The eight-state trellis code can also be constructed. Due to the constraint of the trellis structure, the eight-state code becomes the Ungerboeck code [91]. Its minimum product–sum distance is $6E_S$. An eight-state code with a larger product–sum distance exists, but it is catastrophic. Obviously, the total diversity ($2R$) of both the four and eight-state codes is equal to two, but it will increase to four when the number of states is increased to 16, since R is increased from one to two. After experimentation with various trellis structures and signal assignments, *the 16-state code* is also found based on the design criteria, and is shown in Figure 4.55. The minimum product–sum distance of this code is equal to:

$$d_{\min}(R) = \left(\delta_0^2 \delta_0^1\right)\left(\delta_0^2 \delta_0^1\right) = (2 + 0.586)\,E_s \times (2 + 0.586)\,E_s = 6.69E_s^2$$

The constructed 16-state code has an Ungerboeck representation, which means that it can be generated by a feedback-free convolutional encoder followed by the natural mapping. Performance results for concatenated code obtained using Rules 1 and 2 (R1 and 2 code) and traditional trellis codes are compared in Figures 4.56–4.63 with transmission matrix Γ of the STBC by Alamouti.

More details on the topic can be found in [90–98].

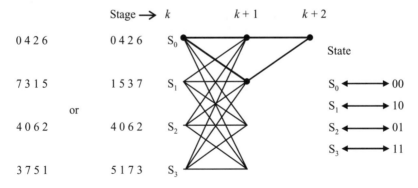

Figure 4.54 Trellis diagram of the four-state 2/3 8PSK trellis code [90] © 2002, IEEE.

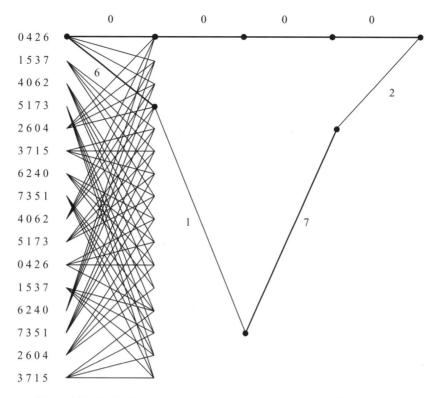

Figure 4.55 Trellis diagram of the 16-state 2/3 8PSK trellis code [90] © 2002, IEEE.

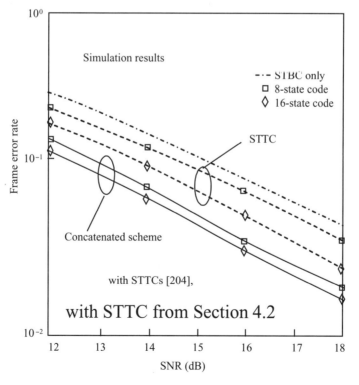

Figure 4.56 Quasi-static fading, $M = 1$.

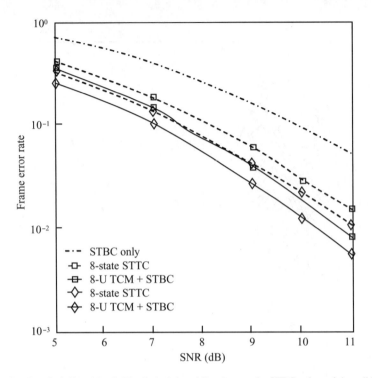

Figure 4.57 Quasi-static fading, $M = 2$. The dashed dotted line denotes the STBC only, and the solid lines denote the concatenated scheme.

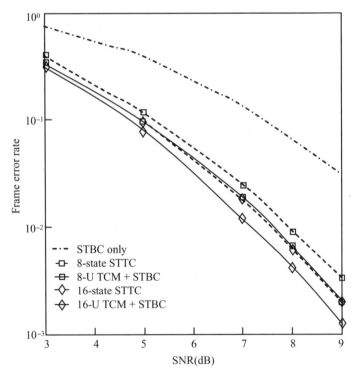

Figure 4.58 Quasi-static fading, $M = 3$. The dashed dotted line denotes the STBC only, and the solid lines denote the concatenated scheme.

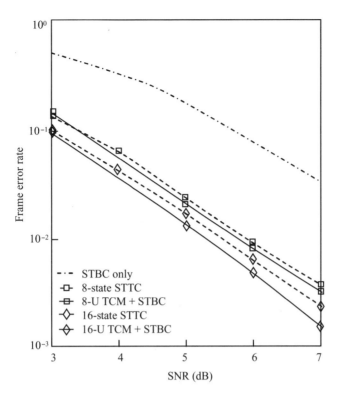

Figure 4.59 Quasi-static fading, $M = 4$. The dashed dotted line denotes the STBC only, and the solid lines denote the concatenated scheme.

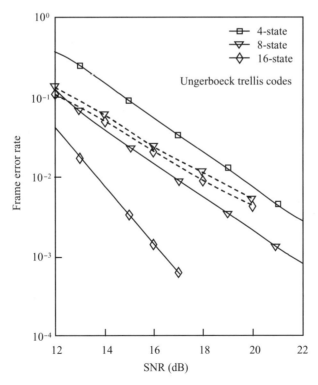

Figure 4.60 Perfect interleaving, $M = 1$. Performances without interleaving (dashed lines) are also shown for comparison.

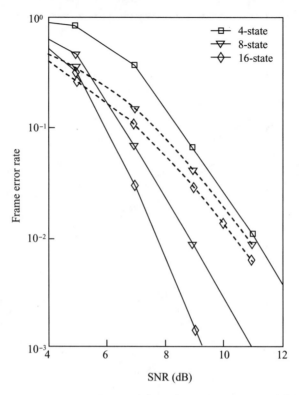

Figure 4.61 Perfect interleaving, $M = 2$. Performances without interleaving (dashed lines) are also shown for comparison.

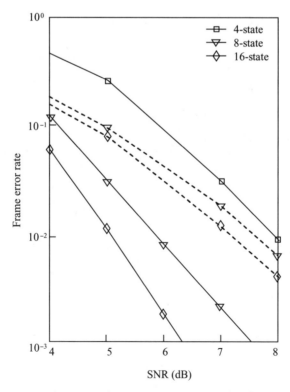

Figure 4.62 Perfect interleaving, $M = 3$. Performances without interleaving (dashed lines) are also shown for comparison.

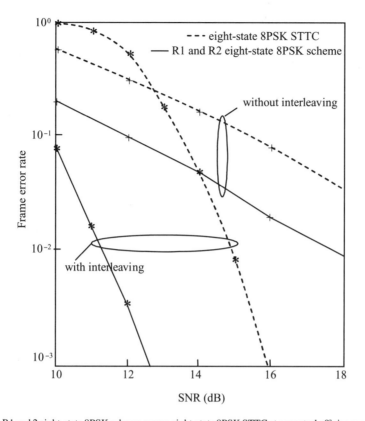

Figure 4.63   R1 and 2 eight-state 8PSK scheme versus eight-state 8PSK STTC at a spectral efficiency of 1.5 bit/s/Hz, $M = 1$.

4.12   Estimation of MIMO Channel

Channel estimation using training sequences is required for coherent detection in BLAST. In this section we present the maximum likelihood channel estimator and the optimal training sequences for block flat fading channels and analyze the estimation error. The optimal training length and training interval that maximize the throughput for a given target bit error rate are presented as functions of the Doppler frequency and the number of antennas.

4.12.1   System Model

The system consists of M transmitting antennas and N receiving antennas. The vector of signals at the output of N receive antennas can be represented as:

$$\mathbf{y}_i = \sqrt{\frac{\rho}{M}}\mathbf{H}_i\mathbf{s}_i + \mathbf{w}_i \tag{4.151}$$

where $\mathbf{H}_i$ is the $N \times M$ channel matrix, $\mathbf{s}_i$ is the $M \times 1$ transmitted signal vector and $\mathbf{w}_i$ is the $N \times 1$ vector of complex additive white Gaussian noise with zero mean and unit variance at time instant i. The average power of the components in $\mathbf{H}_i$ and $\mathbf{s}_i$ are normalized to unity, so the average signal to noise ratio (SNR) at each receiving antenna is ρ, independent of the number of transmitting antennas.

An alternative presentation without normalization gives:

$$\mathbf{y} = \mathbf{Hs} + \mathbf{w} \tag{4.151a}$$

In this case, the signal power is constrained by $E(\mathbf{ss}^*) \leq \mathbf{P}$ so that P/M is the maximum average power transmitted by each antenna, where $()^*$ stands for the Hermitian matrix/vector transpose conjugate. Starting from the definition of the mutual information exchanged in the MIMO channel, the capacity (maximum mutual information) $C(\mathbf{H})$ can be represented as [see Appendix 4.3, Equation (A4.3.21)]

$$C(\mathbf{H}) = \log \det \left(\mathbf{I}_N + \frac{P}{M} \mathbf{HH}^* \right) \tag{4.151b}$$

4.12.2   Training

During the training phase, training sequences of L_t symbols long are transmitted from all the transmitting antennas. An estimate of the channel, $\hat{\mathbf{H}}$, is obtained at the end. During the payload phase, data sequences of L_d symbols long are transmitted and jointly detected. We define L_t as the *training length* and $L = L_d + L_t$ as the *training interval*. The *duty cycle factor* $\eta = 1 - L_t / L$, is the fraction of time spent in data transmission.

4.12.3   Performance Measure

Since the channel is continuously fading, the actual channel will deviate progressively from the channel estimate obtained at time $i = L_t$. The BER performance will be dominated by the worst channel estimation error. Therefore, we consider $\hat{\mathbf{H}} - \mathbf{H}_L$, as a measure of the channel estimation error, where $\mathbf{H}_L$ is the channel at the end of the training period.

4.12.4   Definitions

Define the difference between the channel at time i and at time L as:

$$\Delta \mathbf{H}_i = \mathbf{H}_i - \mathbf{H}_L \tag{4.152}$$

then, we can rewrite Equation (4.151) as follows:

$$\mathbf{y}_i = \sqrt{\frac{\rho}{M}} \mathbf{H}_L \mathbf{s}_i + \sqrt{\frac{\rho}{M}} \Delta \mathbf{H}_i \mathbf{s}_i + \mathbf{w}_i \tag{4.153}$$

Let $\mathbf{S}$ be the matrix of training symbols, $\mathbf{S} = \begin{bmatrix} \mathbf{s}_1 \, \mathbf{s}_2 \, \cdots \, \mathbf{s}_{L_t} \end{bmatrix}$, where $\mathbf{s}_i$ for $1 \leq i \leq L_t$ is the $M \times 1$ training symbol vector at time i. Let the matrix of received signals be $\mathbf{Y} = \begin{bmatrix} \mathbf{y}_1 \, \mathbf{y}_2 \cdots \mathbf{y}_{L_t} \end{bmatrix}$, and the matrix of noise be $\mathbf{W} = \begin{bmatrix} \mathbf{w}_1 \, \mathbf{w}_2 \cdots \mathbf{w}_{L_t} \end{bmatrix}$. Then:

$$\mathbf{Y} = \sqrt{\frac{\rho}{M}} \mathbf{H}_L \mathbf{S} + \mathbf{W} \times \mathbf{S}^* + \sqrt{\frac{\rho}{M}} [\Delta \mathbf{H}_1 \mathbf{s}_1 \, \Delta \mathbf{H}_2 \mathbf{s}_2 \cdots \Delta \mathbf{H}_{L_t} \mathbf{s}_{L_t}] \tag{4.154}$$

4.12.5   Channel Estimation Error

For *block fading* channels where the channel realization remains constant within a block of certain length and then changes to an independent realization for the next block, the ML channel estimator is:

$$\hat{\mathbf{H}} = \sqrt{\frac{M}{\rho}} \mathbf{Y} \cdot \mathbf{S}^* \cdot (\mathbf{SS}^*)^{-1} \tag{4.155}$$

and the optimal training sequences which minimize the mean square estimation error are orthogonal across all transmitting antennas, i.e.

$$\mathbf{SS}^* = L_t \mathbf{I}_M \tag{4.156}$$

where $\mathbf{I}_M$ is the $M \times M$ identity matrix. A necessary condition for the matrix inversion $(\mathbf{SS}^*)^{-1}$ to exist is $L_t \geq M$.

Equations (4.155) and (4.156) are suboptimal but practically appealing for continuous fading channels too. By applying these to Equation (4.154), we obtain:

$$\hat{\mathbf{H}} = \mathbf{H}_L + \Delta \mathbf{H}_{\text{noise}} + \Delta \mathbf{H}_{\text{Doppler}} \tag{4.157}$$

where

$$\Delta \mathbf{H}_{\text{noise}} = \frac{1}{L_t} \sqrt{\frac{M}{\rho}} \mathbf{WS}^* \tag{4.158}$$

is the estimation error due to noise and

$$\Delta \mathbf{H}_{\text{Doppler}} = \frac{1}{L_t} \sum_{i=1}^{L_t} \Delta \mathbf{H}_i \cdot (\mathbf{s}_i \mathbf{s}_i^*) \tag{4.159}$$

is the estimation error due to the temporal variation of the channel.

4.12.6   Error Statistic

It is easy to show that $\Delta \mathbf{H}_{\text{noise}}$ has i.i.d. complex Gaussian entries of zero mean and variance of $M/(\rho L_t)$. We assume the components of $\mathbf{H}_i$ are uncorrelated with each other (rich scattering) and Rayleigh fading with respect to i. Let $\Delta \mathbf{h}_n^T$ represent the nth row of $\Delta \mathbf{H}_{\text{Doppler}}$.

$$E\{(\Delta \mathbf{h}_{n_1}^T)^* \Delta \mathbf{h}_{n_2}^T\} = \delta_{n_1 n_2} \cdot \frac{1}{L_t^2} \sum_{i_1=1}^{L_t} \sum_{i_2=1}^{L_t} \mathbf{s}_{i_1} \mathbf{s}_{i_1}^* \cdot [\xi(i_1 - i_2) - \xi(i_1 - L) - \xi(i_2 - L) + 1] \mathbf{s}_{i_2} \mathbf{s}_{i_2}^* \tag{4.160}$$

where δ_{jk} is the discrete Dirac delta function. $\xi(x) = J_0(2\pi f_{d\,\text{max}} T \cdot x)$, where $J_0(\cdot)$ is the zeroth order Bessel function of the first kind, $f_{d\,\text{max}}$ is the maximum Doppler frequency and T is the symbol period.

For channel estimation tracking, it is reasonable to assume that the phase change during one training period is small, i.e. $2\pi f_{d\,\text{max}} T L \ll 1$. Then Equation (4.160) can be simplified as:

$$E\{(\Delta \mathbf{h}_{n_1}^T)^* \Delta \mathbf{h}_{n_2}^T\} = \delta_{n_1 n_2} \cdot 2 \left(\frac{\pi f_{d\,\text{max}} T}{L_t}\right)^2 \cdot \left(\sum_{i=1}^{L_t} (L - i) \mathbf{s}_i \mathbf{s}_i^*\right)^2 \tag{4.161}$$

using $J_o(x) \approx 1 - x^2/4$ for small x.

4.12.7   Results

The result indicates that the estimation error due to temporal variation increases quadratically with the Doppler frequency. The error also depends on the training length L_t, the training interval L and the training sequences s_i.

In the simple case where there is only one transmitting antenna and one receiving antenna, i.e. $M = N = 1$, the variance of the estimation error in Equation (4.161) can be computed directly:

$$\sigma^2_{\text{Doppler}} = 2\left[\pi\, f_{d\max}T\left(L - \frac{L_t + 1}{2}\right)\right]^2 \tag{4.162}$$

We can see that if L is fixed and L_t increases, the error decreases. If L_t is fixed and L increases, the error increases. If both L_t and L increase at a fixed ratio L_t/L, the error increases. For $M, N > 1$, the expression for the mean square estimation error is generally very complicated and it depends on the exact training sequences. A possible choice of orthogonal training sequences is the FFT matrix, i.e.

$$S_{m,i} = e^{-j2\pi(m-1)(i-1)/L_t} \tag{4.163}$$

where $S_{m,i}$ is the (m, i)th component of the training matrix S, $1 \leq m \leq M$, $1 \leq i \leq L_t$. It can be shown that with such training sequences, the leading component of the variance is the same as Equation (4.162). Therefore, the earlier observations also apply to multiple antenna systems.

4.12.7.1   Example M, N = 4

In this example, $M = 4$ transmitting antennas and $N = 4$ receiving antennas are used. The training sequences are the fast Fourier transform (FFT) sequences in Equation (4.163). The average receiving SNR is $\rho = 15$ dB. The carrier frequency is $f_c = 2$ GHz and the maximum Doppler frequency is $f_{d\max} = 10$ Hz, which corresponds to a pedestrian speed. The symbol period is $T = 41$ μs, corresponding to the IS-136 standard (see Chapter 1). The channel coefficients are generated using the Jakes model and continuously fading. Figure 4.64 shows the mean square error (MSE) of the channel estimation as a function of the training length L_t. Both L_t and the training interval L increase at a fixed ratio, $L_t/L = 20\%$. The MSE due to noise decreases with L_t but the MSE due to temporal variation increases with L_t and L. As a result, the overall MSE first decreases and then increases.

Note here that as long as the flat fading model holds, the above results will apply to systems with different symbol period, T, if we scale the maximum Doppler frequency $f_{d\max}$ appropriately. This is valid because the estimation error due to temporal variation depends only on the product $f_{d\max}T$. Additional results are given in Figures 4.65 and 4.66.

More details on the topic can be found in [101–112].

4.13   Space–Time Codes for Frequency Selective Channels

The presentation in this section is based on [113]. If a frequency selective channel is modeled as a symbol spaced, tap delay line of length L, the sampled version of one frame ($K + L - 1$ time slots) of the received signal, at antenna r, after matched filtering can be represented as

$$y_k^r = \sum_{l=0}^{L-1}\sum_{t=1}^{M_T} h_t^r(l)c_{k-l}^t + n_k^r \quad k = 1, \ldots, K + L - 1 \tag{4.164}$$

In Equation (4.164) y_k^r is the received signal at antenna r and time slot k, n_k^r is a complex white Gaussian random noise sample at antenna r and time slot k with variance N_0 and $h_t^r(l)$ is a circularly

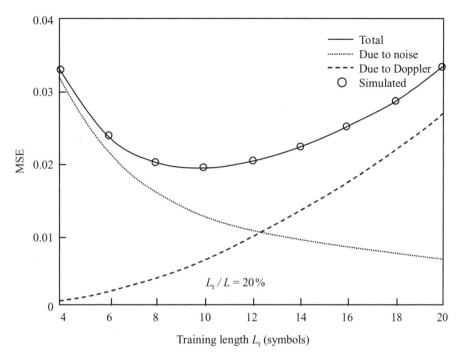

Figure 4.64 Channel estimation MSE versus the training length L_t. Four transmitting antennas and four receiving antennas. $L_t/L = 20\%$, ($\rho = 15$ dB, $f_{d\,max} = 10$ Hz).

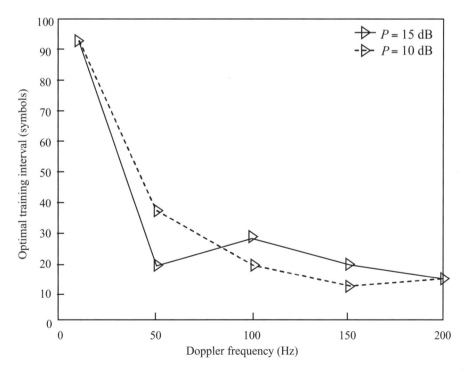

Figure 4.65 Optimal training interval versus Doppler frequency ($M, N = 4$, BER = 3%).

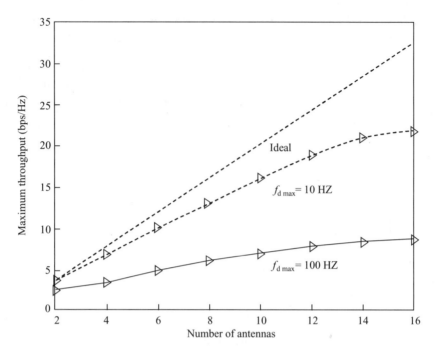

Figure 4.66 Maximum throughput versus the number of antennas resulting from the optimal training interval. 'Ideal' indicates the throughput with ideal channel estimation. ($\rho = 15$ dB, BER $= 3\%$).

symmetric complex Gaussian random variable with zero mean describing the lth tap gain. The variance of $h_t^r(l)$ is denoted as $\sigma^2(l)$ and $h_t^r(l)$ is normalized so that we have:

$$\sum_{l=0}^{L-1} \sigma^2(l) = 1 \tag{4.165}$$

No channel knowledge at the transmitter and a coherent receiver with perfect channel state information are assumed. The vector $v = (\sigma^2(0), \sigma^2(l), \ldots, \sigma^2(L-1))$ will be referred to as *the power delay profile vector* and is common for all subchannels. If the antennas are spaced sufficiently far apart, then $h_t^r(l)$ and $h_{t'}^{r'}(l')$ are independent if $t \neq t'$ or $r \neq r'$, which is referred to as *spatial independence*. The case when $h_t^r(l)$ and $h_t^r(l')$ are independent for $l \neq l'$, is referred to as the *uncorrelated tap* case. Channels are said to have *uniform power delay profiles* if all components in the power delay profile vector are equal. Otherwise we have a *non-uniform power delay profile* [111–114]. Equation (4.164) can be represented in vector form as:

$$\mathbf{y}_r = \sum_{l=0}^{L-1} \mathbf{h}^r(l)\mathbf{C}(l) + \mathbf{n}_r \tag{4.166}$$

where

$$
\begin{aligned}
\mathbf{y}_r &= (y_1^r, \ldots, y_{K+L-1}^r) \\
\mathbf{n}_r &= (n_1^r, \ldots, n_{K+L-1}^r) \\
\mathbf{h}^r(l) &= (h_1^r(l), \ldots, y_{M_T}^r(l)) \\
\mathbf{C}(l) &= (\mathbf{0}_{M_T \times l} \quad \mathbf{C} \quad \mathbf{0}_{M_T \times (L-1-l)})
\end{aligned}
\tag{4.167}
$$

with

$$\mathbf{C} = \begin{bmatrix} c_1^1 & c_2^1 & \cdots & c_K^1 \\ c_1^2 & c_2^2 & \cdots & c_K^2 \\ \vdots & \vdots & \vdots & \vdots \\ c_1^{M_T} & c_2^{M_T} & \cdots & c_K^{M_T} \end{bmatrix} \tag{4.168}$$

which is the traditional code word matrix from the flat fading channel analysis. In fact, the $l = 0$ term in Equation (4.166) is exactly the flat fading signal term when M_T transmit antennas are employed. Due to the similarity of the other terms, the lth term in Equation (4.166) for $l > 0$ can be thought of as coming from an lth set of M_T virtual transmit antennas.

The tth row of the matrix $\mathbf{C}(l)$ represents the modulated output symbols transmitted from the tth transmit antenna over $K + L - 1$ time periods. Combining these rows together for $l = 0, \ldots, L - 1$ in the order of increasing l gives:

$$\mathbf{C}_t = \begin{bmatrix} c_1^t & c_2^t & \cdots & c_K^t & 0 & 0 & \cdots & 0 \\ 0 & c_1^t & c_2^t & \cdots & c_K^t & 0 & \cdots & 0 \\ \vdots & \vdots & \vdots & \vdots & \vdots & \vdots & \vdots & \vdots \\ 0 & 0 & \cdots & 0 & c_1^t & c_2^t & \cdots & c_K^t \end{bmatrix} \tag{4.169}$$

Then, Equation (4.166) is also equal to:

$$\mathbf{y}_r = \sum_{l=1}^{M_T} \mathbf{h}_t^r \mathbf{C}_t + \mathbf{n}_r \tag{4.170}$$

where $\mathbf{h}_t^r = (h_t^r(0), h_t^r(1), \ldots, h_t^r(L - 1))$ is a subchannel impulse response vector between transmit antenna t and receive antenna r.

We stack up all the signals received by the M_R receive antennas to get:

$$\mathbf{Y} = \mathbf{H}\mathbf{C}_s + \mathbf{N} \tag{4.171}$$

where

$$\mathbf{Y} = \begin{bmatrix} \mathbf{y}_1 \\ \mathbf{y}_2 \\ \vdots \\ \mathbf{y}_{M_R} \end{bmatrix} \quad \mathbf{N} = \begin{bmatrix} \mathbf{n}_1 \\ \mathbf{n}_2 \\ \vdots \\ \mathbf{n}_{M_R} \end{bmatrix} \quad \mathbf{C}_s = \begin{bmatrix} \mathbf{C}_1 \\ \mathbf{C}_2 \\ \vdots \\ \mathbf{C}_{M_R} \end{bmatrix}$$

$$\mathbf{H} = \begin{bmatrix} \mathbf{h}_1^1 & \mathbf{h}_2^1 & \cdots & \mathbf{h}_{M_T}^1 \\ \mathbf{h}_1^2 & \mathbf{h}_2^2 & \cdots & \mathbf{h}_{M_T}^2 \\ \vdots & \vdots & \vdots & \vdots \\ \mathbf{h}_1^{M_R} & \mathbf{h}_2^{M_R} & \cdots & \mathbf{h}_{M_T}^{M_R} \end{bmatrix}$$

Assume that the transmitted code word is $\mathbf{C}_s$ and the erroneously decoded code word is $\mathbf{E}_s$. Define the codeword difference matrix as $\mathbf{B}_s = \mathbf{C}_s - \mathbf{E}_s$. Define the non-negative definite Hermitian matrix $\mathbf{A}_s = \mathbf{B}_s \mathbf{B}_s^H$, where H represents the conjugate transpose. Further, consider the $M_T L M_R \times M_T L M_R$ matrix $\mathbf{D}_s = \mathbf{I}_{M_R} \otimes \mathbf{A}_s$, where $\mathbf{I}_{M_R}$, is an $M_R \times M_R$ identity matrix and $\otimes$ is the Kronecker product. Now vectorize the channel matrix $\mathbf{H}^T$, where T represents the transpose operation, to define the channel vector $\mathbf{h} = \text{vec}(\mathbf{H}^T)^T$. Let $\mathbf{K} = E(\mathbf{h}^H \mathbf{h})$ denote the correlation matrix of $\mathbf{h}$. We only consider the case where $\mathbf{K}$ is full rank. Since $\mathbf{K}$ is a positive definite matrix, Cholesky factorization yields $\mathbf{K} = \mathbf{F}^H \mathbf{F}$, where

F is a lower triangular matrix. Using arguments from Section 4.2, the pairwise error probability is upper bounded by:

$$P(\mathbf{C}_s \rightarrow \mathbf{E}_s) \leq \frac{1}{\det(\mathbf{I} + \gamma \mathbf{F} \mathbf{D}_s \mathbf{F}^H)} \tag{4.172}$$

where $\gamma = E_s/4N_0$, E_s is the average energy per symbol at each transmit antenna. At high SNR, Inequality (4.172) reduces to:

$$P(\mathbf{C}_s \rightarrow \mathbf{E}_s) \leq \gamma^{-qM_R} \left(\prod_{i=1}^{qM_R} \lambda_i \right) \tag{4.173}$$

where q is the rank of the matrix $\mathbf{A}_s$ and λ_i, is the ith non-zero eigenvalue of $\mathbf{F} \mathbf{D}_s \mathbf{F}^H$. An equation similar to Inequality (4.172) can be developed by using Equation (4.166) instead of Equation (4.170).

The maximum rank of matrix $\mathbf{A}_s$ is $M_T L$. Thus, the maximum diversity gain of an STC employed in an frequency selective channel is $M_T M_R L$, L times greater than that of the same STC for flat fading channels, which is only $M_T M_R$.

4.13.1 Diversity Gain Properties

Similarly to arguments used in Section 4.2, let $\mathbf{C}$ and $\mathbf{E}$ be two code word matrices from Equation (4.168). Let $\mathbf{C}(l)$ and $\mathbf{E}(l)$ denote the corresponding matrices from Equation (4.167). Let $\mathbf{B}(l) = \mathbf{C}(l) - \mathbf{E}(l)$, then define $\mathbf{B}_{ds}$ as:

$$\mathbf{B}_{ds} = \begin{bmatrix} \mathbf{B}(0) \\ \mathbf{B}(1) \\ \vdots \\ \mathbf{B}(L-1) \end{bmatrix} \tag{4.174}$$

which can be easily derived from $\mathbf{B}_s$ and vice versa. Let $\mathbf{B} = \mathbf{C} - \mathbf{E}$ be the flat fading code word difference matrix and assume the rank of the matrix $\mathbf{B}$ is r_b. The matrix $\mathbf{B}$ is similar to an upper triangular matrix $\mathbf{T}$ whose first r_b diagonal elements are non-zero and other $M_T - r_b$ diagonal elements are zero. Likewise, the matrix $\mathbf{B}(l)$ is similar to $\mathbf{T}(l) = \mathbf{0}_{M_T \times l} \, \mathbf{T} \, \mathbf{0}_{M_T \times (L-1-l)}$. Thus, the difference matrix $\mathbf{B}_{ds}$ is similar to the following matrix:

$$\mathbf{T}_{ds} = \begin{bmatrix} \mathbf{T}(0) \\ \mathbf{T}(1) \\ \vdots \\ \mathbf{T}(L-1) \end{bmatrix} \tag{4.175}$$

It is apparent that the collection of vectors consisting of all r_b rows of the matrix $\mathbf{T}(0)$ and the r_bth row of each of the matrices $\mathbf{T}(l)$, $t = 1, \ldots, L-1$ will be linearly independent, so $\mathbf{T}_{ds}$ has rank $r_b + L - 1$ or larger. Therefore, the minimum rank of difference matrix $\mathbf{B}_{ds}$ is $r_b + L - 1$.

4.13.2 Coding Gain Properties

Inequality (4.172) shows that the coding gain depends not only on the matrix $\mathbf{A}_s$, but on the channel correlation matrix $\mathbf{K}$ as well. With the assumption that the STC provides maximum diversity gain, $\mathbf{K}$ is full rank, and SNR is large, Inequality (4.172) is well approximated by:

$$P(\mathbf{C}_s \rightarrow \mathbf{E}_s) \leq \frac{\gamma^{-M_T M_R L}}{\det(\mathbf{K}) \det(\mathbf{D}_s)} \tag{4.176}$$

Denote $\mathbf{K}$ as a partitioned matrix with $E(\mathbf{h}_{t_1}^{r_1\,H}\mathbf{h}_{t_2}^{r_2})$ as its $(M_T \times (r_1 - 1) + t_1, M_T \times (r_2 - 1) + t_2)$th block entry $(t_1, t_2 = 1, \ldots, M_T$ for each $r_1, r_2 = 1, \ldots, M_R)$, where $E(\mathbf{h}_{t_1}^{r_1\,H}\mathbf{h}_{t_2}^{r_2})$ is an $L \times L$ correlation matrix.

Channel 1. In this case the taps are uncorrelated with spatial independence. Based on spatial independence, $E(\mathbf{h}_{t_1}^{r_1\,H}\mathbf{h}_{t_2}^{r_2})$ reduces to the $\mathbf{0}$ matrix when $r_1 \neq r_2$ or $t_1 \neq t_2$. Furthermore, from the uncorrelated tap assumption, $E(h_t^{r\,H}h_t^r)$ simplifies to a diagonal matrix with $\sigma^2(0), \ldots, \sigma^2(L-1)$ along the diagonal. Then, in this case, the correlation matrix $\mathbf{K}$ becomes a diagonal matrix, whose determinant is:

$$\det(\mathbf{K}) = \left(\prod_{l=0}^{L-1} \sigma^2(l) \right)^{M_T M_R} \tag{4.177}$$

For a specific code and *Channel 1* we want to know conditions for the best coding gain. Due to the arithmetic mean and geometric mean inequality and $\sum_{l=0}^{L-1} \sigma^2(l) = 1$, $\prod_{l=0}^{L-1} \sigma^2(l)$ will achieve the maximum L^{-L} if and only if each $\sigma^2(l)$ is equal to each other, which implies the channel has uniform power delay profile.

Channel 2. In this case the taps are correlated with spatial independence. Again $E(\mathbf{h}_{t_1}^{r_1\,H}\mathbf{h}_{t_2}^{r_2})$ reduces to a $\mathbf{0}$ matrix when $r_1 \neq r_2$ or $t_1 \neq t_2$. However, $\mathbf{K}_t^r = E(\mathbf{h}_t^{r\,H}\mathbf{h}_t^r)$ cannot be simplified to a diagonal matrix as in the case of *Channel 1*, but $\mathbf{K}_t^r$ still has exactly the same entries along the diagonal as those in *Channel 1*. More precisely, $\mathbf{K}$ can be described as the matrix formed by arranging a set of non-diagonal matrices along the partition diagonal of a larger partitioned matrix with the appropriate zero padding. According to the property of the determinant of a partitioned matrix, we have:

$$\det(\mathbf{K}) = \prod_{r=1}^{M_R} \prod_{t=1}^{M_T} \det(\mathbf{K}_t^r) \tag{4.178}$$

According to Hadamard's inequality that the determinant of a non-negative definite square matrix is not greater than the product of all its diagonal elements, we have:

$$\det(\mathbf{K}_t^r) \leq \prod_{l=0}^{L-1} \sigma^2(l) \tag{4.179}$$

Combining Equation (4.178) and Inequality (4.179), results in:

$$\det(\mathbf{K}) \leq \left(\prod_{l=0}^{L-1} \sigma^2(l) \right)^{M_T M_R} \tag{4.180}$$

Comparing Equation (4.177) and Inequality (4.180), we can see that with the spatial independence assumption, the coding gain of a specific STC for the uncorrelated tap assumption is always larger than or equal to that under a correlated tap assumption. In other words, the tap correlation will generally degrade the coding gain.

Channel 3. In this case the STC is applied to a spatially correlated frequency selective fading channel with correlated taps. Now, in general, none of the submatrices $E(\mathbf{h}_{t_1}^{r_1\,H}\mathbf{h}_{t_2}^{r_2})$ can be reduced to a $\mathbf{0}$ matrix. In this case we will start from the following result in matrix calculus. Given three matrices $\mathbf{M}_1(m \times m)$, $\mathbf{M}_2(m \times n)$, $\mathbf{M}_3(n \times n)$, if

$$\mathbf{M} = \begin{bmatrix} \mathbf{M}_1 & \mathbf{M}_2 \\ \mathbf{M}_2^H & \mathbf{M}_3 \end{bmatrix}$$

is positive definite, then $\det(\mathbf{M}) \leq \det(\mathbf{M}_1)\det(\mathbf{M}_3)$. This is known as Fischer's inequality. The correlation matrix $\mathbf{K}$ is a positive definite Hermitian matrix. So, by successively using Fischer's inequality, we find:

$$\det(\mathbf{K}) \leq \prod_{r=1}^{M_R} \prod_{t=1}^{M_T} \det(\mathbf{K}_t^r) \tag{4.181}$$

4.13.3   Space–Time Trellis Code Design

A systematic design procedure for space–time trellis codes (STTCs), presented in Section 4.2 for flat fading channels, will now be modified to handle frequency selective channels. The relationship between the symbols transmitted by real antennas and those transmitted by the virtual antennas is used. Define an R-bit binary input vector $\mathbf{a}_k = (a_{k,1}, a_{k,2}, \ldots, a_{k,R})$ and concatenate this with a $(Q-1)R$-bit current state vector to get $\bar{\mathbf{a}}_k = (\mathbf{a}_k, \mathbf{a}_{k-1}, \ldots, \mathbf{a}_{k-Q+1})$. From Section 4.2 an STTC can be represented by:

$$\bar{\mathbf{x}}_k = (x_k^1, x_k^2, \ldots, x_k^{M_T}) = \bar{\mathbf{a}}_k \mathbf{G} \tag{4.182}$$

where

$$\mathbf{G} = \begin{bmatrix} g_{11} & g_{12} & \cdots & g_{1,M_T} \\ g_{21} & g_{22} & \cdots & g_{2,M_T} \\ \cdots & \cdots & \cdots & \cdots \\ g_{QR,1} & g_{QR,2} & \cdots & g_{QR,M_T} \end{bmatrix}$$

At each time slot, each component of the length M_T output vector from Equation (4.182) is mapped into a constellation symbol and these symbols are transmitted simultaneously from M_T antennas. In this case the $g_{ij}, i = 1, \ldots, QR, j = 1, \ldots, M_T$ can be taken from an alphabet whose size is equal to the constellation size s. The trellis encoder starts and ends in state zero at the beginning and end of each frame.

Consider the case with $M_T = 2$, $R = 1$, $Q = 5$, $s = 2$ (BPSK) and $L = 3$. This results in a 16-state 1 b/s/Hz BPSK STTC for a frequency selective channel with three taps. Let $\mathcal{L}$ denote the shortest length error event, as in Section 4.2, for all such codes. First, all codes with maximum diversity gain and maximum coding gain of $\eta = 5.532$ are found [113]. There are many such codes. Among them, those with larger $\bar{\eta}_{CP}(\mathcal{L})$ are chosen [113]. In this case, there are only two codes yielding maximum $\bar{\eta}_{CP}(\mathcal{L}) = 9.357$. They are $\mathbf{G}_{11} = (11101, 11011)^{\mathrm{T}}$ and $\mathbf{G}_{12} = (11011, 10111)^{\mathrm{T}}$ respectively,[1] as listed in Table 4.14. Monte Carlo simulation is used to evaluate the code performance. Figure 4.67 shows the frame error rate (FER) versus signal to noise ratio (SNR $= M_T E_s / N_0$) per receive antenna. The frame size is 130 symbol intervals. A simulated channel with spatial independence and uncorrelated taps is assumed. Maximum likelihood decoding is performed at the receiver. An additional selection of good codes is listed in Table 4.15 [113].

The latest results in the field are presented in [115–156].

4.14   Optimization of a MIMO System

4.14.1   The Channel Model

As before, the model consists of M transmit and N receive antenna elements. We assume that $M > N$. The antenna weights on the receive side are described as a column vector $\mathbf{U}$ with elements $U_1, U_2, \ldots,$ and the vector is normalized so the norm is unity ($\mathbf{U}'\mathbf{U} = 1$). Similarly, $\mathbf{V}$ denotes the transmit weight

---

[1] $(11101, 11011)^{\mathrm{T}}$ denotes $\begin{bmatrix} 1 & 1 & 1 & 0 & 1 \\ 1 & 1 & 0 & 1 & 1 \end{bmatrix}^{\mathrm{T}}$ for simplicity.

Table 4.14   16-state BPSK STTCs with maximum η and different $\bar{\eta}_{CP}(\mathcal{L})$ [113] © 2003, IEEE

No.	$\mathbf{G}^{\mathrm{T}}$	η	$\bar{\eta}_{\mathrm{CP}}(\mathcal{L})$
$\mathbf{G}_{11}$	$\begin{bmatrix} 1 & 1 & 1 & 0 & 1 \\ 1 & 1 & 0 & 1 & 1 \end{bmatrix}$	5.532	9.357
$\mathbf{G}_{12}$	$\begin{bmatrix} 1 & 1 & 0 & 1 & 1 \\ 1 & 0 & 1 & 1 & 1 \end{bmatrix}$	5.532	9.357
$\mathbf{G}_{13}$	$\begin{bmatrix} 1 & 1 & 1 & 1 & 0 \\ 1 & 1 & 1 & 0 & 1 \end{bmatrix}$	5.532	8.512
$\mathbf{G}_{14}$	$\begin{bmatrix} 1 & 1 & 1 & 0 & 1 \\ 1 & 0 & 0 & 1 & 0 \end{bmatrix}$	5.532	7.008
$\mathbf{G}_{15}$	$\begin{bmatrix} 1 & 1 & 1 & 0 & 1 \\ 0 & 0 & 1 & 0 & 1 \end{bmatrix}$	5.532	6.465
$\mathbf{G}_{16}$	$\begin{bmatrix} 0 & 0 & 0 & 1 & 1 \\ 1 & 0 & 0 & 0 & 1 \end{bmatrix}$	5.532	5.532

vector. For notation, * signifies complex conjugation, and $'$ transpose and conjugate. (M, N) refers to M antennas at the transmitter end and N antennas at the receiver end.

An example of a (3,2) system is shown in Figure 4.68. Similarly to Equation (4.151a) the transfer matrix from the transmit antennas to the receive antennas is described by transmission matrix $\mathbf{H}$ with elements H_{ik}. They are random complex Gaussian quantities. A normalization:

$$E\langle |H_{ik}|^2 \rangle = 1 \tag{4.183}$$

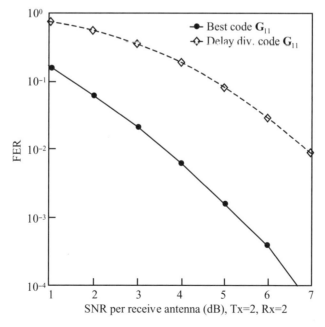

Figure 4.67   Performance comparison of best 16-state BPSK STTC and delay diversity code with two transmit and two receive antennas for a channel with uncorrelated power delay profile vector (1/3, 1/3, 1/3).

Table 4.15 q-state BPSK STTCs for channels with two taps [113] © 2003, IEEE

q	No.	$\mathbf{G}^{\mathrm{T}}$	η	$\bar{\eta}_{\mathrm{CP}}(\mathcal{L})$
4	$\mathbf{G}_{21}$	$\begin{bmatrix} 1 & 1 & 1 \\ 1 & 0 & 1 \end{bmatrix}$	4.000	5.968
8	$\mathbf{G}_{22}$	$\begin{bmatrix} 1 & 1 & 1 & 1 \\ 1 & 0 & 0 & 1 \end{bmatrix}$	5.981	7.825
16	$\mathbf{G}_{23}$	$\begin{bmatrix} 1 & 1 & 0 & 1 & 1 \\ 1 & 0 & 1 & 0 & 1 \end{bmatrix}$	7.445	9.973
32	$\mathbf{G}_{24}$	$\begin{bmatrix} 1 & 1 & 1 & 0 & 0 & 1 \\ 0 & 0 & 1 & 1 & 0 & 1 \end{bmatrix}$	9.514	10.942

is used. It is assumed that the angular spreads seen from both sides are so large that the antenna signals are spatially uncorrelated.

4.14.2   Gain Optimization By Singular Value Decomposition (SVD)

The matrix $\mathbf{H}$ will, in general, be rectangular with N rows and M columns. An SVD expansion of $\mathbf{H}$ can be represented as (see Appendix 5.1)

$$\mathbf{H} = \mathbf{U}_\lambda \cdot \mathbf{D} \cdot \mathbf{V}'_\lambda \qquad (4.184)$$

where $\mathbf{D}$ is a diagonal matrix of real, non-negative singular values, the square roots of the eigenvalues of $\mathbf{G}$, where $\mathbf{G} = \mathbf{H}' \cdot \mathbf{H}$ is an $M \times M$ Hermitian matrix. The columns of the unitary matrices $\mathbf{U}_\lambda$ and $\mathbf{V}_\lambda$ are the corresponding singular vectors. Thus, Equation (4.184) is just a compact way of writing the set of independent channels [157]:

$$\begin{aligned} \mathbf{HV}_1 &= \sqrt{\lambda_1}\mathbf{U}_1 \\ \mathbf{HV}_2 &= \sqrt{\lambda_2}\mathbf{U}_2 \\ &\vdots \\ \mathbf{HV}_N &= \sqrt{\lambda_N}\mathbf{U}_N \end{aligned} \qquad (4.185)$$

The SVD is particularly useful for interpretation in the antenna context. For one particular eigenvalue, one can see that $\mathbf{V}_i$ is the transmit weight factor for excitation of the singular value $\sqrt{\lambda_i}$.

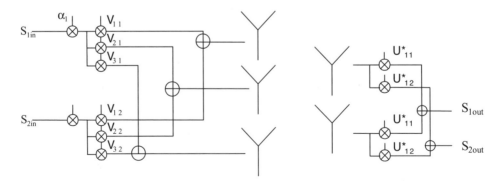

Figure 4.68   Transmission from three (M) transmit antennas to two (N) receive antennas [157].

A receive weight factor of $\mathbf{U}'_i$, a conjugate match, gives the receive voltage S_r, and the square of that the received power, P_r:

$$S_r = \mathbf{U}'_i \mathbf{U}_i \sqrt{\lambda_i} = \sqrt{\lambda_i}$$
$$P_r = |S_r|^2 = \lambda_i \tag{4.186}$$

This clearly shows that the matrix $\mathbf{H}$ of transmission coefficients may be diagonalized, leading to a number of independent orthogonal modes of excitation, where the power gain of the ith mode or channel is λ_i. The weights applied to the arrays are given directly from the columns of the $\mathbf{U}_\lambda$ and $\mathbf{V}_\lambda$ matrices. Thus, the eigenvalues and their distributions are important properties of the arrays and the medium, and the maximum gain is of course given by the maximum eigenvalue. The number of non-zero eigenvalues may be shown to be the minimum value of M and N. The situation is illustrated in Figure 4.68, where the total power is distributed among the N parallel channels by weight factors α. These coefficients are discussed later in more detail. An important parameter is the trace of $\mathbf{G}$, i.e. the sum of the eigenvalues

$$\text{Trace} = \sum_i \lambda_i \tag{4.187}$$

which may be shown to have a mean value of MN. We illustrate the above relations by an example [157].

4.14.2.1   Example: (2, 2) System

For the $(M, N) = (2, 2)$ example, matrix $\mathbf{G}$ is given by:

$$\mathbf{G} = \begin{bmatrix} |H_{11}|^2 + |H_{12}|^2 & H_{11} H_{21}^* + H_{12} H_{22}^* \\ H_{11}^* H_{21} + H_{12}^* H_{22} & |H_{22}|^2 + |H_{21}|^2 \end{bmatrix}$$
$$= \begin{bmatrix} a & c \\ c* & b \end{bmatrix} \tag{4.188}$$

and the two eigenvalues are (see Appendix 5.1)

$$\lambda_{\text{max}} = \frac{1}{2}(a + b + \sqrt{(a-b)^2 + 4|c|^2}) \tag{4.189}$$

and

$$\lambda_{\text{min}} = \frac{1}{2}(a + b - \sqrt{(a-b)^2 + 4|c|^2}) \tag{4.190}$$

Note that

$$\text{Trace} = \sum_i \lambda_i = a + b$$
$$= |H_{11}|^2 + |H_{12}|^2 + |H_{21}|^2 + |H_{22}|^2 \tag{4.191}$$

so the sum of the eigenvalues displays the full fourth-order diversity.

The distribution of ordered eigenvalues may be found in [9], from which the distributions for λ_{min} and λ_{max} may be derived:

$$p(\lambda_{\text{min}}) = 2e^{-2\lambda} \tag{4.192}$$
$$p(\lambda_{\text{max}}) = e^{-\lambda}(\lambda^2 - 2\lambda + 2) - 2e^{-2\lambda} \tag{4.193}$$

In this particular case, it may be shown that the mean values are:

$$E\langle\lambda_{max}\rangle = 3.5 \quad E\langle\lambda_{min}\rangle = 0.5 \tag{4.194}$$

The minimum eigenvalue is Rayleigh distributed with mean power 0.5.
The cumulative probability distribution for λ_{max} is:

$$\Pr(\lambda_{max} < x) = 1 - e^{-x}(x^2 + 2) + e^{-2x}$$
$$\approx x^4/12 \quad x \ll 1 \tag{4.195}$$

One can show that for the case of standard diversity $(M, N) = (1, 4)$:

$$\Pr(P < x) = 1 - e^{-x}(1 + x + x^2/2 + x^3/6)$$
$$\approx x^4/24 \quad x \ll, 1 \tag{4.196}$$

so the (2, 2) case displays full fourth-order diversity but with twice the cumulative probability for the same power level.

The cumulative probability distributions are shown in Figure 4.69, where the maximum eigenvalue (the array gain) follows the fourth-order maximum ratio diversity distribution quite closely.

One should be aware that in order to make full benefit of the maximum eigenvalue, the full knowledge of the channel at the transmitter is required, otherwise the eigenvectors cannot be found.

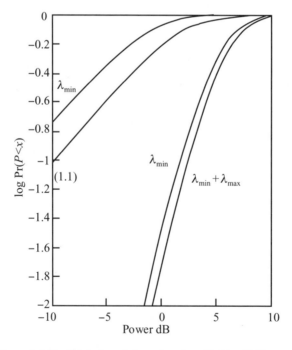

Figure 4.69 Cumulative probability distribution of eigenvalues for a $(T, R) = (2, 2)$ array with four uncorrelated paths. The maximum eigenvalue follows closely the fourth-order diversity with a shift of 0.75 dB.

4.14.3   The General (M, N) Case

For the (4, 4) case in Figure 4.70, the two arrays have 16 different uncorrelated transmission coefficients, so the diversity order is 16. In the asymptotic limit when N is large, it may be shown [158, 159] that the largest eigenvalue is bounded above by:

$$\lambda_{max} < (\sqrt{c} + 1)^2 N; \quad c = M/N \geq 1 \tag{4.197}$$

whereas the smallest eigenvalue is bounded below by:

$$\lambda_{min} > (\sqrt{c} - 1)^2 N \quad c \geq 1 \tag{4.198}$$

In the previous examples, $c = 1$, and the upper asymptotic bound for this case is $4N$. These bounds should not be understood as absolute bounds, but rather as limits approached as N tends to infinity for a fixed c.

The mean array gains (mean of the maximum eigenvalues) are shown in Figure 4.71, together with the upper bound and the gain for the correlated, free space case, N^2. For $N = 10$, the true mean gain is just 1 dB below the upper bound. For a partly correlated case, we can expect the gain to lie between the $\rho = 0$ and the $\rho = 1$ cases, where ρ is the spatial correlation coefficient between the elements.

In some situations, it might be advantageous to have more antennas on one side than on the other, especially for asymmetric situations with heavy downloading of data from a base station. Again, the asymptotic upper bound for the largest eigenvalue is useful, Inequality (4.197). Introducing $M = cN$ directly we find:

$$G_{\text{upper bound}} = (\sqrt{M} + \sqrt{N})^2 \tag{4.199}$$

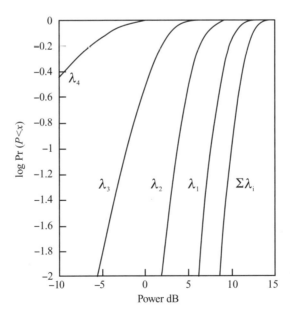

Figure 4.70   Cumulative probability distribution of eigenvalues (power) for two arrays of four elements each, including the sum of eigenvalues corresponding to a (1, 16) case.

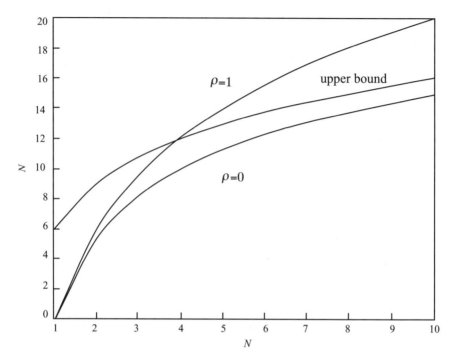

Figure 4.71   The gain relative to one element of (N, N) arrays in a correlated situation ($\rho = 1$), and in an uncorrelated case ($\rho = 0$). The upper bound equals $4N$, and is the asymptotic upper bound for the maximum eigenvalue for N tending to infinity.

which, asymptotically, will approach M for large values of M and fixed N. This clearly illustrates that the composite gain of the link cannot be factored into one belonging to the transmitter and one belonging to the receiver.

4.14.4   Gain Optimization By Iteration For a Reciprocal Channel

Since the channel is reciprocal, exactly the same weights may be used for transmission as for reception. Consider now the situation with transmission from M transmit antennas to N receive antennas (or vice versa). The iteration starts with an arbitrary $\mathbf{V}_1$, in the numerical calculations, chosen as a unit vector with equal elements. At the receive side, the weights are adjusted for maximum gain, and the same weights are then used for transmit since the channel is reciprocal. This may then be repeated a number of times. In principle, the process might converge to an eigenvalue different from the maximum one, but experience shows excellent performance [157].

An example of the convergence in the mean is shown for a (3, 3) case in Figure 4.72. After a few iterations, the gain has converged to the steady state. This might actually be a computationally efficient way of finding the maximum gain solution in practice without going through the trouble of finding the eigenvectors. For details regarding eigenvalue decomposition see Appendix 5.1.

4.14.5   Spectral Efficiency of Parallel Channels

From Figure 4.70 with four independent channels one can see that there are other options for using the eigenvalues than using the largest for maximum gain. Another option is to keep them as parallel channels with independent information, as discussed in the previous sections of this chapter. The knowledge about

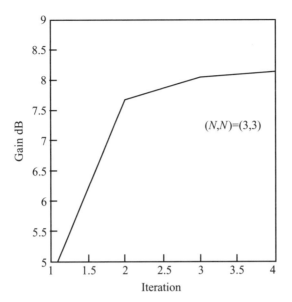

Figure 4.72 Convergence of gain by iterative transmissions between receiver and transmitter for a (3, 3) case. The gain values are mean values.

the distribution of the eigenvalues and the upper and lower bounds may now be used for evaluating bounds on the theoretical capacity of the link. Shannon's capacity measure gives an upper bound on the realizable information rates through parallel channels, and how the power should be distributed over the channels to achieve maximum capacity through 'water filling' [160]. From Equation (4.151b), the basic expression for the spectral efficiency measured in bits/s/Hz for one Gaussian channel is given by:

$$C = \log_2(1 + P) \quad \text{bits/s/Hz} \tag{4.200}$$

where P is the signal to noise ratio, SNR, for one channel.

Assuming all noise powers to be the same, the 'water filling' concept is the solution to the maximum capacity, where each channel is filled up to a common level D:

$$\frac{1}{\lambda_1} + P_1 = \frac{1}{\lambda_2} + P_2 = \frac{1}{\lambda_3} + P_3 = \cdots = D \tag{4.201}$$

Thus, the channel with the highest gain receives the largest share of the power. The constraint on the powers is that:

$$\sum_i P_i = P \tag{4.202}$$

The weight factors α_i in Figure 4.68 equal P_i/P. In the case where level D drops below a certain $1/\lambda_i$, that power is set to zero. In the limit where the SNR is small ($P < 1/\lambda 2 - 1/\lambda_1$), only one eigenvalue, the largest, is left, and we are back to the maximum gain solution of the previous section. For the case of $(M, N) = (2, 2)$ (Figure 4.69) for $P < 2 - 2/7 = 2.34$ dB, only the largest eigenvalue is active, using the mean values from Equation (4.192). The capacity equals

$$C = \sum_{N'} \log_2(1 + \lambda_i P_i) = \sum_{N'} \log_2(\lambda_i D) \tag{4.203}$$

where the summation is over all channels with non-zero powers. The water filling is of course dependent on the knowledge of the channels on the transmit side. In the case where the channel is unknown at the transmitter, the only reasonable division of power is a uniform distribution over the antennas, i.e.

$$P_i = \frac{P}{M} \tag{4.204}$$

M being the number of transmit antennas [100]. It may also be argued that the transmit antenna 'sees' M eigenvalues, not taking into account that there are only N non-zero eigenvalues. Thus, power is lost by allocating power to the zero-valued eigenvalues.

It also follows that when $M = N$, the difference between the capacity for known and unknown channels is small for large P.

4.14.6   Capacity of the (M, N) Array

It follows from Inequalities (4.197) and (4.198) that the instantaneous eigenvalues are limited by:

$$(\sqrt{M} - \sqrt{N})^2 < \lambda_i < (\sqrt{M} + \sqrt{N})^2 \tag{4.205}$$

So, for M much larger than N, all the eigenvalues tend to cluster around M. Furthermore, each of them will be non-fading due to the high MNth order diversity. Thus, the uncorrelated asymmetric channel with many antennas has a very large theoretical capacity of N equal, constant channels with high gains of M. The above illustrates in a mathematical sense the observation of Winters [161] that M should be of the order $2N$. In the limit of large M and N, with M much larger than N, the capacity is easily found to be:

$$C = N \log_2 \left(1 + \frac{P}{N}M\right) \tag{4.206}$$

with the result that the theoretical capacity grows linearly with the number of elements N [160, 161] for M/N fixed. The result may conveniently be interpreted as N parallel channels, each with $1/N$ of the power and each having a gain of M. Note that this capacity is higher than the one used in [160, 161], where the power is divided between the M antennas instead of the N channels, given as:

$$C_{\text{unknown}} = N \log_2(1 + P) \tag{4.207}$$

The numerical results shown in Figure 4.73 support this approximate analysis for the mean values. It should be remembered that the potential gains are higher when a certain outage probability is studied due to the high order diversity effects. This is illustrated in Figure 4.74, which shows the cumulative probability distribution (on a log scale) of the capacity for the case of four receiving elements, and four and twelve transmitting elements. The signal to noise ratio is 20 dB for the $(1, 1)$ case, and it is worth emphasizing that the total power radiated remains constant. The improvement going from four to twelve transmitting antennas is mainly due to the improved gain of the smallest eigenvalues as indicated by Inequality (4.205). Using Equation (4.206) in the $(12, 4)$ case gives 32.9 b/s/Hz.

4.15   MIMO Systems with Constellation Rotation

4.15.1   System Model

In this section we consider a system in which the base station transmitter has L antennas that transmit simultaneously. Each component of an L-dimensional signal point is transmitted on one antenna, and the receiver makes a decision based on the entire L-dimensional received vector. The performance of the system is optimized by rotating the baseline constellations in L dimensions.

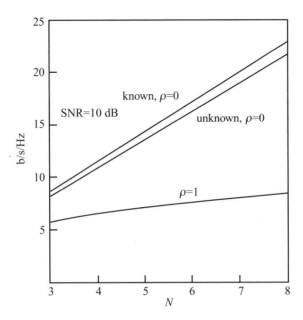

Figure 4.73 Mean capacity for two arrays of each *N* elements. The capacity grows linearly with the number of elements and is approximately the same for the known and the unknown channel. The total transmitted power is constant.

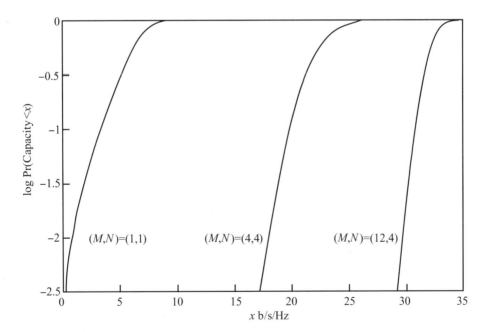

Figure 4.74 The cumulative probability distribution of capacity on a log scale for the (M, N) = (1, 1), (4, 4), and (12, 4) cases. The basic signal to noise ratio is 20 dB. The total radiated power is the same in all cases.

The received signal with one antenna receiver at the mobile is:

$$r(t) = \sum_{i=1}^{L} \alpha_i m_i s_i(t) \cos(w_c t + \theta_i) + n(t), \qquad 0 \le t \le T \tag{4.208}$$

The signal from the transmitter's ith antenna is a pulse-amplitude-modulated (PAM) signal with amplitude m_i, the pulse shape $s_i(t)$, the fading amplitude α_i and $n(t)$ is a white Gaussian noise process with power spectral density $N_0/2$. It is assumed that the fading amplitude for a given link is constant over the signalling interval $[0, T]$ and that the receiver uses coherent detection.

The signals $s_i(t), s_j(t), i \ne j$ are assumed to be orthogonal, and all of the energy of $s_i(t), 1 \le i \le L$, is contained in $[0, T]$. The optimum receiver consists of a bank of L correlators. The output of the ith correlator is:

$$y_i = 2 \int_0^T r(t) s_i(t) \cos(w_c t + \theta_i) \, dt = \alpha_i m_i E_s + \eta_i, \qquad 1 \le i \le L \tag{4.209}$$

The received vector $y = (y_1, y_2, \ldots, y_L)$ is fed into a decision device that estimates the transmitted vector $m = (m_1, m_2, \ldots m_L)$. To simplify notations in this section we do not use bold fonts for some vectors and matrices. It is assumed that the receiver can estimate the fading amplitudes $\alpha_i, 1 \le i \le L$. The receiver finds the L-dimensional constellation point in C, with coordinates suitably amplified, which has the closest Euclidean distance to the received vector y. That is, the receiver picks the symbol $\hat{m} = (\hat{m}_1, \hat{m}_2, \ldots, \hat{m}_L) \in C$ that minimizes $\sum_{i=1}^{L} (y_i/E_s - \alpha_i \hat{m}_i)^2$. A symbol detection error occurs when $\hat{m} \ne m$. The fading amplitudes α_i are modelled as independent and identically distributed Rayleigh random variables with the common probability density function.

4.15.2   Performance in a Rayleigh Fading Channel

Assuming that all points in an L-dimensional constellation C with $|C|$ points, are transmitted with equal probability, an upper bound on the average probability of symbol detection error can be obtained from the union bound:

$$P(\text{error}) \le \frac{1}{|C|} \sum_{m} \sum_{\substack{\hat{m} \\ \hat{m} \ne m}} P(m \to \hat{m}) \tag{4.210}$$

where $P(m \to \hat{m})$ is the probability that the received symbol is closer to the L-dimensional symbol $\hat{m} = (\hat{m}_1, \hat{m}_2, \ldots, \hat{m}_L)$ than to $m = (m_1, m_2, \ldots, m_L)$, given that m was transmitted.

If the receiver estimates the fading amplitudes in Equation (4.209) perfectly, and if a minimum distance decoding rule is used by the receiver, then the pairwise error probability, conditioned on the fading amplitude vector $(\alpha_1, \alpha_2, \ldots, \alpha_L)$, can be upper bounded using the Chernoff bound as follows:

$$P(m \to \hat{m} \,|\alpha_1, \ldots, \alpha_L) \le \prod_{i=1}^{L} \exp\left\{-\alpha_i^2 (m_i - \hat{m}_i)^2 E_s/(8N_0)\right\} \tag{4.211}$$

Assuming that the fading amplitudes are independent, averaging over the probability density function for α_1 gives

$$P(m \to \hat{m}) \le \prod_{i=1}^{L} \frac{1}{1 + (m_i - \hat{m}_i)^2 E_s/(8N_0)} \le \prod_{\substack{i=1 \\ m_i \ne \hat{m}_i}}^{L} \frac{8N_0}{E_s (m_i - \hat{m}_i)^2} \tag{4.212}$$

At sufficiently high SNR, Equation (4.210) is dominated by the largest $P(m \rightarrow \hat{m})$ term, and assuming $m_i \neq \hat{m}_i, 1 \leq i \leq L$, then $P(\text{error}) \propto 1/(E_s/N_0)^L$. From Equation (4.212), the following quantity, which we call the constellation gain in Rayleigh fading channel, gives an indication of the performance of a signal constellation at high SNR:

$$cg_{Rayleigh}(C) = \min_{\substack{m,\hat{m} \in C \\ m \neq \hat{m} \\ m_i \neq \hat{m}_i}} \prod_{i=1}^{L} (m_i - \hat{m}_i)^2 \tag{4.213}$$

In [162] this parameter is called the constellation figure of merit. In order not to degrade the performance of the constellation in the AWGN channel, we apply a transformation to the constellation that preserves the Euclidean distances between points, but improves the constellation's gain in fading. If we represent the original constellation as a $|C| \times L$ matrix $\mathbf{C}$, where each row of the matrix corresponds to a point in the L-dimensional constellation, one possible distance-preserving transformation is to multiply this matrix by an orthogonal $L \times L$ matrix $\mathbf{A}$. The optimal matrix $\mathbf{A}$ maximizes $cg_{Rayleigh}(\mathbf{CA})$. It is shown in Appendix 4.1 how an $L \times L$ orthogonal matrix $\mathbf{A}$ can be written as the product of $\binom{L}{2}$ rotation matrices and a reflection matrix. Such multiplication of the constellation matrix $\mathbf{C}$ by an arbitrary orthogonal matrix $\mathbf{A}$ has the following geometrical interpretation. The constellation is rotated with respect to the (i, j) plane by an amount θ_{ij}, $1 \leq i \leq L-1, i+1 \leq j \leq L$ and there are $\binom{L}{2}$ such rotations.

Then the constellation is reflected in the ith axis, where the matrix $\hat{I}$ has an (i, i) entry equal to -1 and the number of such reflections is equal to the number of -1 elements on the main diagonal of $\hat{I}$. Writing $\mathbf{A} = Q\hat{I}$ where Q is the product of $\binom{L}{2}$ rotation matrices in Equation (A4.1.1), we see that

$$cg_{Rayleigh}(\mathbf{CA}) = cg_{Rayleigh}(\mathbf{CQ\hat{I}}) = cg_{Rayleigh}(\mathbf{CQ}) \tag{4.214}$$

where the second identity in Equation (4.214) follows because the matrix $\mathbf{CQ\hat{I}}$ is just the matrix $\mathbf{CQ}$ with several of its columns negated, and negating the columns of a constellation matrix does not affect the $cg_{Rayleigh}$ of the constellation. Rather than look for an optimal constellation $\mathbf{CA}$ it is sufficient to look for an optimal constellation $\mathbf{CQ}$.

The method of obtaining an optimal constellation $\mathbf{C}_{opt} = \mathbf{CQ}$, one with maximum $cg_{Rayleigh}(\mathbf{CQ})$, given a starting constellation $\mathbf{C}$, is to vary $\binom{L}{2}$ rotation angles according to a numerical optimization or search algorithm. The constellation $\mathbf{C}$ is rotated with respect to the (i, j) plane, $1 \leq i \leq L-1, i+1 \leq j \leq L$.

For example, when $L = 2$ the baseline constellation matrix is

$$\mathbf{C} = \begin{bmatrix} 1 & 1 \\ 1 & -1 \\ -1 & 1 \\ -1 & -1 \end{bmatrix} \tag{4.215}$$

In this case, maximization of $cg_{Rayleigh}$ is done by varying only one rotation angle. The optimal angle of rotation for this constellation can be found using an exhaustive search to be $\theta_{opt} = 31.7°$, assuming a discretization interval of $0.1°$. The optimally rotated constellation is:

$$\mathbf{C}_{opt} = \begin{bmatrix} -0.325 & -1.376 \\ -1.376 & 0.325 \\ 1.376 & -0.325 \\ 0.325 & 1.376 \end{bmatrix} \tag{4.216}$$

These two constellations are shown in Figure 4.75. Each row in the constellation matrix corresponds to a point (m_1, m_2). With $L = 3$, the baseline constellation $\mathbf{C}$ consists of the vertices of a three-dimensional cube, and optimization is done over three rotation angles. Using a discretization interval of $1°$ our

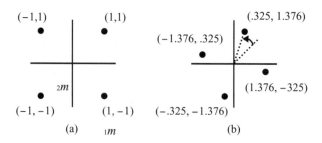

Figure 4.75   Examples of constellations for $L = 2$.

search procedure produced the rotation vector $\theta_{opt} = [\theta_{12}, \theta_{13}, \theta_{23}] = [24°, 36°, 66°]$. Both the baseline constellation and the rotated constellation are given as:

$$
\mathbf{C} = \begin{bmatrix}
1 & 1 & 1 \\
1 & 1 & -1 \\
1 & -1 & 1 \\
1 & -1 & -1 \\
-1 & 1 & 1 \\
-1 & 1 & -1 \\
-1 & -1 & 1 \\
-1 & -1 & -1
\end{bmatrix}
\qquad
\mathbf{C}_{opt} = \begin{bmatrix}
0.177 & 0.474 & -1.656 \\
-0.997 & -1.003 & -0.998 \\
-0.480 & 1.654 & -0.181 \\
-1.655 & 0.176 & 0.476 \\
1.655 & -0.176 & -0.476 \\
0.480 & -1.654 & 0.181 \\
0.997 & 1.003 & 0.998 \\
-0.177 & -0.474 & 1.656
\end{bmatrix}
\tag{4.217}
$$

For $L \geq 4$ an exhaustive search over the $\binom{L}{2}$ rotation angles in order to maximize $cg_{Rayleigh}$ for the L cube proved to be too time consuming. For these constellations, many rotation vectors could be picked at random, and a gradient-based approach can be used to vary the rotation angles so as to converge to a local maximum. For $L = 4$, there are six degrees of freedom, and the rotation vector found was $\theta_{opt} = [\theta_{12}, \theta_{13}, \theta_{14}, \theta_{23}, \theta_{24}, \theta_{34}] = [206°, 15°, 306°, 42°, 213°, 31°]$.

For $L = 5$ there are ten degrees of freedom, and the rotation vector was found as $\theta_{opt} = [\theta_{12}, \theta_{13}, \theta_{14}, \theta_{15}, \theta_{23}, \theta_{24}, \theta_{25}, \theta_{34}, \theta_{35}, \theta_{45}] = [294°, 349°, 18°, 340°, 103°, 184°, 114°, 275°, 212°, 25°]$. Some results are shown in Figures 4.76 and 4.77.

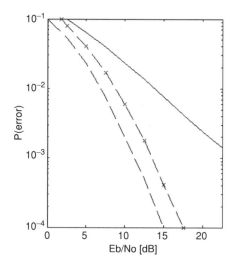

Figure 4.76   Probability of bit error for several transmission schemes. $L = 3$: (——) baseline, (-x-) optimal constellation, (– –) identical transmissions.

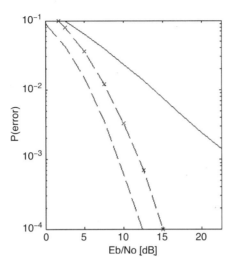

Figure 4.77 Probability of bit error for several transmission schemes. $L = 5$: (——) baseline, (- x -) optimal constellation, (– –) identical transmissions.

4.16 Diagonal Algebraic Space–Time Block Codes

In the previous section we established the basic concept of space time system improvements based on constellation rotation. In this section, we present a new family of linear ST block codes by the use of rotated constellations and the Hadamard transform that will be referred to as diagonal algebraic ST (DAST) block codes. These codes have a normalized rate of 1 symbol/s and achieve the full diversity over n transmit and m receive antennas. They maintain their diversity and coding gains over all real or complex constellations carved from the ring of complex integers Z[i], with $i = \sqrt{-1}$, such as pulse–amplitude modulation (PAM) or quadrature–amplitude modulation (QAM). Due to the lattice structure of these codes, the ML decoding can be implemented by the sphere decoder at moderate complexity independent of the transmission rate (see Appendix 4.2) or [171, 172]. The DAST block codes outperform the ST codes from orthogonal design (see Section 4.3) for $n > 2$.

4.16.1 System Model

Based on Equations (4.213) and (4.214), the minimum product distance of the constellation **Q** will be defined as

$$d_{d,\min} \equiv \min_{\substack{y=x_1-x_2 \\ x_1 \neq x_2 \in \mathbf{Q}}} \prod_{j=1}^{d} |y_j| \tag{4.218}$$

Some of the results for quasi-optimal rotations with the good values of the minimum product distance are reported in [171, 172]. In [171], the construction of rotations M_d in dimension d was done in an iterative manner in a 'Hadamard' way as follows:

$$\mathbf{M}_d = \begin{bmatrix} M_{d/2}^1 & -M_{d/2}^2 \\ M_{d/2}^2 & M_{d/2}^1 \end{bmatrix} \tag{4.219}$$

where $M_{d/2}^1$ is the optimal real rotation in dimension $d/2$ and $M_{d/2}^2$ is an orthogonal transformation in dimension $d/2$ depending only on one parameter [171]. Then, one varies this parameter in order to choose the rotation that maximizes the minimum product distance. This method works very well for

Table 4.16 First row of the optimal real rotation matrices in dimensions 2 and 4

Dimension	Column					$d_{d,\min}$
2	1–2	0.526	0.851			$1/\sqrt{5}$
4	1–4	0.201	0.325	−0.486	−0.786	1/40

the dimensions $d = 2, 3, 4$ and 6. It becomes less successful for $d \geq 8$, since too many parameters are excluded in the rotations in high dimensions. Table 4.16 presents the first row of the optimal real rotations found in [171] along with $d_{d,\min}$ for $d = 2, 4$ that is used to construct the DAST block codes in this section. The rest of the rotation matrix can easily be obtained from Equation (4.219). For the dimensions $d = 2^q$, q ≥ 3, the rotations given in [172] are used, which are constructed on the real part of the cyclotomic number field of degree $4d$: $\mathbf{Q}$ (cos $(2\pi/8d)$), which give relatively good values of the minimum product distance [172]:

$$d_{d,\min} = \sqrt{2}/ (2d)^{d/2} \tag{4.220}$$

Equation (4.221) shows the MATLAB program which generates the rotation matrix $\mathbf{M}_d$ of any dimension $d = 2^q$ constructed on the number fields $\mathbf{Q}(\cos (2\pi/8d))$. This method to generate full modulation diversity rotations is attractive, especially for large d. Note that the best rotations in [171] give better $d_{d,\min}$ for $d = 2, 4$ while starting from $d = 8$, the rotations given in (4.221) are better.

$$M = sqrt(2/d) * \cos(pi/4 * d) * (4 * [1 : d]' - 1) * (2 * [1 : d] - 1)) \tag{4.221}$$

4.16.2   The DAST Coding Algorithm

If M_n is a rotation of dimensions $n \times n$ (with, $n = 1, 2$ or n is a multiple of 4), which generates a full modulation diversity lattice, then DAST block code in dimensions $n \times n$ is constructed as:

$$\Xi_n \triangleq H_n \text{diag}(x_1, \ldots, x_n) \tag{4.222}$$

where $x = (x_1, \ldots, x_n)^T = M_n a$, and $a = (a_1, \ldots, a_n)^T$ is the information symbol vector. In the sequel, we denote the entries of the Hadamard matrix H_n by w_{ij} (Walsh) in order to differentiate them from the entries of the channel transfer matrix h_{ij}. The Hadamard transform is a real unitary transformation that exists for 1, 2, and all the dimensions multiple of 4. In dimension n, the Hadamard transform H_n satisfies $H_n H_n^T = n I_n$, with I_n the identity matrix in dimension n.

As an example for $n = 2$ the corresponding DAST block code is:

$$\Xi_2 \triangleq \begin{bmatrix} 1 & 1 \\ 1 & -1 \end{bmatrix} \begin{bmatrix} x_1 & 0 \\ 0 & x_2 \end{bmatrix} = \begin{bmatrix} x_1 & x_2 \\ x_1 & -x_2 \end{bmatrix} \tag{4.223}$$

where $x = (x_1, x_2)^T = M_2 a$, and M_2 is the two-dimensional rotation matrix given in Table 4.16. For $n = 4$, the corresponding DAST block code is given by

$$\Xi_2 \triangleq \begin{bmatrix} 1 & 1 & 1 & 1 \\ 1 & -1 & 1 & -1 \\ 1 & 1 & -1 & -1 \\ 1 & -1 & -1 & 1 \end{bmatrix} \begin{bmatrix} x_1 & 0 & 0 & 0 \\ 0 & x_2 & 0 & 0 \\ 0 & 0 & x_3 & 0 \\ 0 & 0 & 0 & x_4 \end{bmatrix} = \begin{bmatrix} x_1 & x_2 & x_3 & x_4 \\ x_1 & -x_2 & x_3 & -x_4 \\ x_1 & x_2 & -x_3 & -x_4 \\ x_1 & -x_2 & -x_3 & x_4 \end{bmatrix} \tag{4.224}$$

where $x = (x_1, x_2, x_3, x_4)^T = M_4 a$, and M_4 is the four-dimensional rotation matrix given in Table 4.16.

The DAST block code Ξ_n has a transmit diversity equal to n under quasi-static fading assumption. When n is a power of 2 and for the rotations given in Section 4.16.1, the coding gain of the DAST block code, equals:

$$\delta = \begin{cases} \dfrac{2^{2/n}}{\sqrt{5}}, & \text{for } n = 2, 4 \\[2mm] \dfrac{1}{2^{(n-1)/n}}, & \text{for } n \geq 8. \end{cases} \tag{4.225}$$

To prove it, let $y = x - e = M_n (a - b)$ such that $a \neq b$. We can write the DAST block code at y as Equation (4.222), $\Xi_n = H_n \text{diag}(y_1, \ldots, y_n)$. Since M_n generates a full modulation diversity lattice, one has $y_j \neq 0 \forall j = 1 \cdots n$ taken over all the vectors $a \neq b$ in the considered constellation. It follows that the matrix $\text{diag}(y_1, \ldots, y_n)$ is full rank, and also Ξ_n is full rank over all the differences of codewords. For the coding gain one computes (see Equations (4.25–4.30)):

$$\begin{aligned} \det\left(\Xi_n \Xi_n^H\right) &= \det\left(H_n \text{diag}(y_1, \ldots, y_n) \, diag\left(y_1^*, \ldots, y_n^*\right) H_n^T\right) \\ &= \det\left(n I_n \text{diag}\left(|y_1|^2, \ldots, |y_n|^2\right)\right) = n^n \prod_{j=1}^{n} |y_j|^2 \end{aligned} \tag{4.226}$$

The coding gain is defined as the minimum of $\det\left(\Xi_n \Xi_n^H\right)^{1/n}$, computed over all the differences between distinct codeword pair. By taking the minimum over y of the determinant above and then taking the nth root, one obtains the coding gain of the DAST block code:

$$\delta_n = n \left(d_{n,\min}\right)^{2/n} \tag{4.227}$$

From Table 4.16 one has $d_{2,\min} = 1/\sqrt{5}$, which gives $d_2 = 2/\sqrt{5}$, and $d_{4,\min} = 1/40$, which gives $\delta_4 = \sqrt{2}/\sqrt{5}$. For $n \geq 8$, and for the rotations given in Section 4.16.1, replacing Equation (4.220) in Equation (4.227) gives $\delta_n = 1/2^{(n-1)/n}$. Note that the coding gain given in Equation (4.225) is greater than 0.5 and it approaches this value when n increases. For example, $\delta_8 = 0.5453$ and $\delta_{32} = 0.5109$. In Equation (4.227) there is a multiplicative factor in the coding gain expression because in the model we normalize the radiated power at a given SNR by the number of transmit antennas n by multiplying the noise variance by n. If the normalization is done at the transmitter side then the coding gain will be $\delta_n = \left(d_{n,\min}\right)^{2/n}$.

4.16.3   The DAST Decoding Algorithm

During n periods of time, the received signal is given by an $m \times n$ matrix:

$$\mathbf{r} = H \left(H_n \text{diag}(x_1, \ldots, x_n)\right) + \mathbf{v} \tag{4.228}$$

where the $m \times n$ complex matrix $\mathbf{v}$ has independent Gaussian distributed random variables of variance per real dimension as entries. Equivalently, one has:

$$\mathbf{r} = (\mathbf{HH}_n) \text{diag}(x_1, \ldots, x_n) + \mathbf{v}$$

$$\text{vec}\left(\mathbf{r}^T\right) = \begin{bmatrix} \text{diag}(\mathbf{H}_1 \mathbf{H}_n) \\ \vdots \\ \text{diag}(\mathbf{H}_m \mathbf{H}_n) \end{bmatrix} x + \text{vec}\left(\mathbf{v}^T\right) \tag{4.229}$$

where H_j denotes the jth row of H(the MIMO channel matrix) representing the j_{th} receive antenna, and vec($\mathbf{r}$) arranges the matrix $\mathbf{r}$ in a one column vector by putting its columns one after the other. If at each receive antenna $1 \leq j \leq m$ we write diag $\left(H_j H_n\right) = \text{diag}\left(\alpha_{j1}, \ldots, \alpha_{jn}\right)$, then the received signal is given by:

$$\mathbf{r}_1 \triangleq \text{vec}\left(\mathbf{r}^T\right) = \begin{bmatrix} \alpha_{11} & 0 & \cdots & 0 \\ 0 & \alpha_{12} & \ddots & \vdots \\ \vdots & \ddots & \ddots & 0 \\ 0 & \cdots & 0 & \alpha_{1n} \\ & & \vdots & \\ \alpha_{m1} & 0 & \cdots & 0 \\ 0 & \alpha_{m2} & \ddots & \vdots \\ \vdots & \ddots & \ddots & 0 \\ 0 & \cdots & 0 & \alpha_{mn} \end{bmatrix} M_n a + \text{vec}\left(\mathbf{v}^T\right) \triangleq A M_n a + v_1. \tag{4.230}$$

Since the Hadamard transform is an orthogonal transformation, the variables $\alpha_{ij}, i = 1 \cdots m$, $j = 1 \cdots n$ are independent and identically distributed (i.i.d.) complex Gaussian variables with variance $n/2$ per real dimension.

From Equation (4.230) it is easier to understand how the DAST block codes exploit the transmit diversity. Using a DAST block code over n transmit antennas and m receive antennas is equivalent to sending the word $(x_1, \ldots, x_n)$ over one transmit antenna and m receive antennas during n periods of time, where the channel changes randomly every time instant (since the fading between each transmit–receive antenna pair is independent). The latter scheme has a diversity of mn since the lattice from which we transmit the words has a full modulation diversity [171].

A perfect CSI is assumed to be available at the receiver. First we perform maximum ratio combining of Equation (4.230). This yields:

$$r_2 = A^H r_1 = \text{diag}\left(\sum_{j=1}^m |\alpha_{j1}|^2, \ldots, \sum_{j=1}^m |\alpha_{jn}|^2\right) M_n a + v_2 \tag{4.231}$$

where v_2 is a colored Gaussian noise with covariance matrix $\mathbb{E}[v_2 v_2^H] = 2\sigma^2 A^H A$. In order to whiten the noise, we multiply the received signal in Equation (4.231) by $(A^H A)^{-1/2}$, giving:

$$r_3 = (A^H A)^{-1/2} r_2 = \text{diag}\left(\sqrt{\sum_{j=1}^m |\alpha_{j1}|^2}, \ldots, \sqrt{\sum_{j=1}^m |\alpha_{jn}|^2}\right) M_n a + v_3 \tag{4.232}$$

with v_3 an $n \times 1$ additive white Gaussian noise. Then we apply the sphere decoder [168, 169] on the real and imaginary parts of Equation (4.232). The sphere decoder takes advantage of the lattice structure of the received signals and searches the closest lattice points to the received signal, which are enclosed in a sphere of radius C_0 centered at the received signal. Each time a lattice point of a norm less than C_0 is found, we reduce the sphere radius accordingly and restart the search until an empty sphere is reached. The choice of C_0 depends on the considered lattice, which is generated by $\text{diag}(\sqrt{\sum_{j=1}^m |\alpha_{j1}|^2}, \ldots, \sqrt{\sum_{j=1}^m |\alpha_{jn}|^2}) M_n$ in Equation (4.232), as well as on the additive noise level. Some results are shown in Figures 4.78–4.81. In Figure 4.78, we compare the Alamouti code $\mathbf{G}_2$ with the code Ξ_2 for one and two receive antennas with the 4-QAM modulation. At the same spectral efficiency of 2 b/s/Hz, the Alamouti scheme shows almost 1 dB of gain over the code Ξ_2. For $n = 2$ transmit antennas it seems difficult to outperform the Alamouti scheme since it is the unique complex orthogonal design transmitting at a normalized rate of 1 symbol/s. However, when n increases (Figures 4.79–4.81), the DAST block codes give better performances.

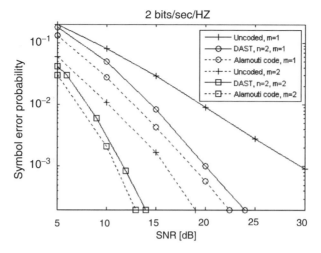

Figure 4.78 Average BER as a function of SNR, two transmit, one and two receive antennas.

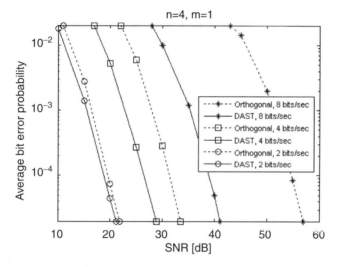

Figure 4.79 Average BER as a function of SNR, four transmit and one receive antennas.

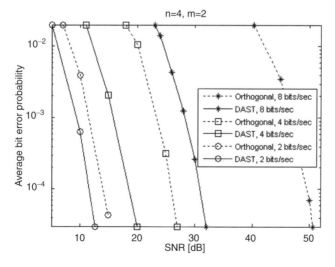

Figure 4.80 Average BER as a function of SNR, four transmit and two receive antennas.

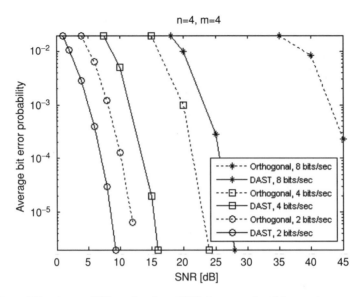

Figure 4.81 Average BER as a function of SNR, four transmit and four receive antennas.

Appendix 4.1 QR Factorization

Orthogonal matrix triangularization (*QR* decomposition) reduces a real (*m*,*n*) matrix **A** with $m \geq n$ and full rank to a much simpler form. A suitably chosen orthogonal matrix **Q** with triangularize the given matrix.

$$A = Q \begin{bmatrix} R \\ 0 \end{bmatrix}$$

with the (*n*, *n*) upper triangular matrix **R**. One only has then to solve the triangular system $Rx = Pb$, where **P** consists of the first *n* rows of **Q**. The least squares problem $Ax \approx b$ is easy to solve with $A = QR$ and $Q^T Q = I$. The solution $x = (A^T A)^{-1} A^T b$ becomes $x = (R^T Q^T QR)^{-1} R^T Q^T b = (R^T R)^{-1} R^T Q^Y b = R^{-1} Q^T b$.

This is a matrix-vector multiplication $Q^T b$, followed by the solution of the triangular system $Rx = Q^T b$ by back-substitution.

Many different methods exist for the **QR** decomposition, e.g. the *Householder transformation*, the *Givens rotation*, or the *Fram–Schmidt decomposition*.

Householder Transformation

The most frequently applied algorithm for **QR** decomposition uses the Householder transformation $u = Hv$, where the Householder matrix **H** is a symmetric and orthogonal matrix of the form $H = I - 2xx^T$, with the identity matrix **I** and any normalized vector **x** with $||x||_2^2 = x^T x = 1$. Householder transformations zero the $m - 1$ elements of a column vector **v** below the first element.

$$\begin{bmatrix} v_1 \\ v_2 \\ \vdots \\ v_n \end{bmatrix} \rightarrow \begin{bmatrix} c \\ 0 \\ \vdots \\ 0 \end{bmatrix} \quad \text{with} \quad c = \pm \left(\sum_{i=1}^{m} v_i^2 \right)^{1/2}$$

One can verify that:

$$\mathbf{x} = f \begin{bmatrix} v_1 - c \\ v_2 \\ \vdots \\ v_m \end{bmatrix} \quad \text{with} \quad f = \frac{1}{\sqrt{2c(c - v_1)}}$$

fulfils $\mathbf{x}^T \mathbf{x} = 1$ and that with $\mathbf{H} = \mathbf{I} - 2\mathbf{x}\mathbf{x}^2$ one obtains the vector $[c, 0 \cdots 0]^T$. To perform the decomposition of the (m,n) matrix $\mathbf{A} = \mathbf{QR}$ (with $m \geq n$) we construct in this way an (m,m) matrix $\mathbf{H}^{(1)}$ to zero the $m - 1$ elements of the first column. An $(m - 1, m - 1)$ matrix $\mathbf{G}^{(2)}$ will zero the $m - 2$ elements of the second column. With $\mathbf{G}^{(2)}$ we produce the (m,m) matrix:

$$\mathbf{H}^{(2)} = \begin{bmatrix} 1 & 0 & \cdots & 0 \\ 0 & & & \\ \vdots & & \mathbf{G}^{(2)} & \\ 0 & & & \end{bmatrix}, \text{ etc}$$

After n ($n - 1$ for $m = n$) such orthogonal transforms $\mathbf{H}^{(i)}$ we obtain:

$$\mathbf{R} = \mathbf{H}^{(n)} \ldots \mathbf{H}^{(2)} \, \mathbf{H}^{(1)} \mathbf{A}.$$

$\mathbf{R}$ is upper triangular all the orthogonal matrix $\mathbf{Q}$ becomes:

$$\mathbf{Q} = \mathbf{H}^{(1)} \mathbf{H}^{(2)} \ldots \mathbf{H}^{(n)}.$$

In practise the $\mathbf{H}^{(i)}$ are never explicitly computed.

Givens Transformations

Given an $n \times m$ matrix $\mathbf{A}$ with $n \geq m$ there is an $n \times m$ matrix $\mathbf{Q}$ with orthonormal columns and an upper triangular $m \times m$ matrix $\mathbf{R}$ such that $\mathbf{A} = \mathbf{QR}$. If $m = n$ $\mathbf{Q}$ is orthogonal (e.g., see [164, p. 112]).

The method based on Givens transformations is of particular interest to us (e.g., see [165, p. 214]). If $\mathbf{A}$ is an $n \times n$ matrix, then $\mathbf{A} = \mathbf{QR} \Rightarrow \mathbf{Q}^T \mathbf{A} = \mathbf{R}$.

The $\mathbf{QR}$ factorization method based on Givens transformations gives us a method for writing $\mathbf{Q}^T$ as the product of $\binom{n}{2} = n(n - 1)/2$ Givens matrices:

$$\mathbf{G}(i, k, \theta) = \begin{bmatrix} 1 & & & & & & \\ & \ddots & & & & & \\ & & c & \cdots & s & & \\ & & \vdots & & \vdots & & \\ & & -s & \cdots & c & & \\ & & & & & \ddots & \\ & & & & & & 1 \end{bmatrix}$$

is an $n \times n$ matrix called the Givens (or Jacobi) matrix, with $c = \cos \theta$, $s = \sin \theta$. It consists of 1's on the main diagonal, except for the two elements c in rows (and columns) i and k. All off-diagonal elements are zero, except the two elements s and $-s$ in rows (and columns) i and k. Postmultiplication of a vector by G rotates the vector counterclockwise by θ degrees with respect to the (i, k) plane.

Proposition 1. *If an* n × n *matrix* A *is orthogonal, then a QR factorization algorithm yields* $\mathbf{Q} = \mathbf{A}\hat{\mathbf{I}}$ *where* $\hat{\mathbf{I}}$ *denotes a matrix in which each diagonal element is either 1 or −1 and all off-diagonal elements are zero.*

Proof. $\mathbf{Q}^T$ is orthogonal because $\mathbf{Q}$ is orthogonal. If $\mathbf{A}$ is orthogonal, then $\mathbf{Q}^T\mathbf{A} = \mathbf{R}$ is also orthogonal since the product of two orthogonal matrices is an orthogonal matrix. Since $\mathbf{R}$ is an upper triangular orthogonal matrix, each main diagonal element must be either 1 or −1 and all off-diagonal elements must be zero. We have $\mathbf{A} = \mathbf{Q}\hat{\mathbf{I}} \Rightarrow \mathbf{Q} = \mathbf{A}\hat{\mathbf{I}}^{-1} = \mathbf{A}\hat{\mathbf{I}}$.

Proposition 2. *Any orthogonal* n × n *matrix* A *can be written as the product of* n(n − 1)/2 n × n*-Givens matrices and an* n × n $\hat{\mathbf{I}}$*-matrix.*

Proof. The proof follows directly from the proof of the previous proposition. Given an arbitrary orthogonal $n \times n$ matrix $\mathbf{A}$ we have a method for constructing Q, the product of $n(n − 1)/2$ Givens matrices, such that $\mathbf{A} = \mathbf{Q}\hat{\mathbf{I}}$. Note that the identity matrix is a Givens matrix as well as an $\hat{\mathbf{I}}$ matrix.

For example, 2 × 2 all orthogonal matrices have the form:

$$\mathbf{A} = \begin{bmatrix} \cos\theta & \sin\theta \\ -\sin\theta & \cos\theta \end{bmatrix}$$

or

$$\mathbf{A} = \begin{bmatrix} \cos\theta & \sin\theta \\ \sin\theta & -\cos\theta \end{bmatrix}$$

This is a well-known fact (for example, see [166, p. 282]). The first form corresponds to $\mathbf{G}(1, 2, \theta)\mathbf{I}_{2\times 2}$ and the second form corresponds to $\mathbf{G}(1, 2, \theta)(-\mathbf{I}_{2\times 2})$. Proposition 2, together with the **QR** factorization method using Givens transformations, implies that all 3 × 3 orthogonal matrices have the form:

$$\mathbf{A} = \mathbf{G}(1, 2, \theta_{12})\mathbf{G}(1, 3, \theta_{13})\mathbf{G}(2, 3, \theta_{23})\hat{\mathbf{I}}_{3\times 3} =$$

$$= \begin{bmatrix} \cos\theta_{12} & \sin\theta_{12} & 0 \\ -\sin\theta_{12} & \cos\theta_{12} & 0 \\ 0 & 0 & 1 \end{bmatrix} \cdot \begin{bmatrix} \cos\theta_{13} & 0 & \sin\theta_{13} \\ 0 & 1 & 0 \\ -\sin\theta_{13} & 0 & \cos\theta_{13} \end{bmatrix}$$

$$\cdot \begin{bmatrix} 1 & 0 & 0 \\ 0 & \cos\theta_{23} & \sin\theta_{23} \\ 0 & -\sin\theta_{23} & \cos\theta_{23} \end{bmatrix} \cdot \begin{bmatrix} \pm 1 & 0 & 0 \\ 0 & \pm 1 & 0 \\ 0 & \pm 0 & 1 \end{bmatrix}$$

In general, any $L \times L$ orthogonal matrix $\mathbf{A}$ can be factored into $L(L − 1)/2 = \binom{L}{2}$ rotation matrices and an $\hat{\mathbf{I}}$ matrix as follows:

$$\mathbf{A} = \left(\prod_{\substack{1 \le i \le L-1 \\ i+1 \le j \le L}} \mathbf{G}(i, j, \theta_{i,j}) \right) \cdot \hat{\mathbf{I}}_{L\times L} \tag{A4.1.1}$$

for suitable θ_{ij}.

Appendix 4.2 Lattice Code Decoder for Space–Time Codes

Consider the system of M transmit and N receive antennas, the single data stream in the input is demultiplexed into M substreams, and each substream is modulated independently then transmitted by its dedicated antenna. It is assumed that the same constellation is used for all the substreams. The transmission is done by burst of length l over a quasi-static Rayleigh fading channel changing randomly every l symbol durations. The power launched by each transmitter is proportional to $1/M$ so that the total radiated power is constant and independent of M. The proximity of antennas presupposes the synchronization of the system.

Notation $\mathbb{Z}$ is the ring of integers. $\mathbb{R}$ is the field of real numbers. $\mathbb{Z}_{(i)}$ is the set of numbers $a + ib$ with a, $b \in \mathbb{Z}$, and $i = \sqrt{-1}$.

The received signal at each instant time is given by

$$\mathbf{r} = \mathbf{H}\mathbf{a}^T + \mathbf{v} \qquad (A4.2.1)$$

where $\mathbf{a} = (a_1, a_2, \ldots, a_M)$ is the transmitted vector which belongs to the constellation QAM carved from $\mathbb{Z}_{(i)}$, and $\mathbf{v}$ is a $N \times 1$ complex vector AWGN component-wise independent with a variance σ^2 per dimension. Moreover, $\mathbf{H}$ is an $N \times M$ transfer matrix of the channel with entries h_{kj}, where h_{kj} is the fading between transmitter j and receiver k. In the sequel we set $M = N$. One can write the system (A4.2. 1) as

$$\mathbf{r}' \triangleq \left[\Re(\mathbf{r}^T)\Im(\mathbf{r}^T)\right] = \mathbf{u} \begin{pmatrix} \Re(\mathbf{H}^T) & \Im(\mathbf{H}^T) \\ -\Im(\mathbf{H}^T) & \Re(\mathbf{H}^T) \end{pmatrix} + \mathbf{v}' = \mathbf{u}\mathbf{M}_H + \mathbf{v}' \qquad (A4.2.2)$$

where $\mathbf{u} = [\Re(\mathbf{a})\,\Im(\mathbf{a})] \in^{2M}$, and $\mathbf{v}' = \left[\Re(\mathbf{v}^T)\,\Im(\mathbf{v}^T)\right] \in \mathbb{R}^{2M}$.

$\Re(\mathbf{a})$, $\Im(\mathbf{a})$ denote the real and imaginary part of $\mathbf{a}$, respectively. Note that the rank of $\mathbf{M}_H$ is $2M$ almost always, and its Gram matrix $\mathbf{G}_M = \mathbf{M}_H\mathbf{M}_H^\dagger$ is positive definite. Hence, we can represent the multi-antenna environment by a lattice sphere packing [173, 174], and one can apply the universal lattice decoder in a multi-antenna system. The principle of the algorithm is to search the closest lattice point to the received signal within a sphere of radius $\sqrt{C}$ cantered at the received signal (see [174] and references therein). The choice of C is very crucial to the speed of the algorithm. In practice, C can be adjusted according to the noise (and eventually the fading) variance. When a failure is detected, one can either declare an erasure on the detected symbol, or increase C. The complexity of the algorithm is independent of the lattice constellation size, which is very useful for high data rate transmission. In [175], Fincke and Phost showed that if d^{-1} is a lower bound for the eigenvalues of the Gram matrix $\mathbf{G}_M$, then the number of arithmetical operations is

$$O\left(n^2 \times \left(1 + \frac{\eta - 1}{4dC}\right)^{4dC}\right) \qquad (A4.2.3)$$

which, for a judicious choice of the radius $C = d^{-1}$, is approximated by $O(n^6)$ arithmetical operations, where n is the lattice dimension.

Appendix 4.3 MIMO Channel Capacity

The transmitted signals in each symbol period are represented by an $n_T \times 1$ column matrix $\mathbf{x}$, where the ith component x_i, refers to the transmitted signal from antenna i. We consider a Gaussian channel, for which, according to information theory, the optimum distribution of transmitted signals is also Gaussian. Thus, the elements of x are considered to be zero mean independent identically distributed (i.i.d.) Gaussian variables. The covariance matrix of the transmitted signal is given by $\mathbf{R}_{xx} = E\{\mathbf{x}\mathbf{x}^H\}$ where $E\{\cdot\}$ stands for expectation and $\mathbf{A}^H$ denotes the Hermitian of matrix $\mathbf{A}$ (transpose and component-wise complex conjugate). The total transmitted power is constrained to P, regardless of the number of transmit antennas n_T. It can be represented as $P = \text{tr}(\mathbf{R}_{xx})$, where $\text{tr}(\mathbf{A})$ denotes the trace of matrix $\mathbf{A}$ (sum of the diagonal elements of $\mathbf{A}$). If the channel is unknown at the transmitter, we assume that the signals transmitted from individual antenna elements have equal powers of P/n_T. The covariance matrix of the transmitted signal is given by

$$\mathbf{R}_{xx} = \frac{P}{n_T}\mathbf{I}_{n_T} \tag{A4.3.1}$$

where $\mathbf{I}_{n_T}$ is the $n_T \times n_T$ identity matrix. The transmitted signal bandwidth is narrow enough, so its frequency response can be considered as flat. In other words, we assume that the channel is memoryless. The channel is described by an $n_R \times n_T$ complex matrix, denoted by $\mathbf{H}$. The ijth component of the matrix $\mathbf{H}$, denoted by h_{ij}, represents the channel fading coefficient from the jth transmit to the ith receive antenna. For normalization purposes we assume that the received power for each of n_R receive branches is equal to the total transmitted power. Physically, it means that we ignore signal attenuations and amplifications in the propagation process, including shadowing, antenna gains, etc. Thus we obtain the normalization constraint for the elements of $\mathbf{H}$, on a channel with fixed coefficients, as:

$$\sum_{j=1}^{n_T}\left|h_{ij}\right|^2 = n_T, \qquad i = 1, 2, \ldots, n_R \tag{A4.3.2}$$

The noise at the receiver is described by an $n_R \times 1$ column matrix, denoted by $\mathbf{n}$. Its components are statistically independent complex zero-mean Gaussian variables, with independent and equal variance real and imaginary parts. The covariance matrix of the receiver noise is given by $\mathbf{R}_{nn} = \mathrm{E}\{\mathbf{nn}^H\}$. If there is no correlation between components of $\mathbf{n}$, the covariance matrix is obtained as $\mathbf{R}_{nn} = \sigma^2\,\mathbf{I}_{n_R}$. Each of n_R receive branches has identical noise power of σ^2.

The receiver is based on a maximum likelihood principle operating jointly over n_R receive antennas. The received signals are represented by an $n_R \times 1$ column matrix, denoted by $\mathbf{r}$, where each complex component refers to a receive antenna. We denote the average power at the output of each receive antenna by P_r. The average signal-to-noise ratio (SNR) at each receive antenna is defined as:

$$\gamma = \frac{P_r}{\sigma^2} \tag{A4.3.3}$$

As it was assumed that the total received power per antenna is equal to the total transmitted power, the SNR is equal to the ratio of the total transmitted power and the noise power per receive antenna and it is independent of n_T and can be written as:

$$\gamma = \frac{P}{\sigma^2} \tag{A4.3.4}$$

By using the linear model the received vector can be represented as $\mathbf{r} = \mathbf{H}x + \mathbf{n}$. The received signal covariance matrix, $\mathrm{E}\{\mathbf{rr}^H\}$, is given by $\mathbf{R}_{rr} = \mathbf{H}\mathbf{R}_{xx}\,\mathbf{H}^H$ while the total received signal power can be expressed as tr $(\mathbf{R}_{rr})$.

MIMO system capacity: The system capacity is defined as the maximum possible transmission rate such that the probability of error is arbitrarily small. Initially, we assume that the channel matrix is not known at the transmitter, while it is perfectly known at the receiver.

By the singular value decomposition (SVD) theorem [Appendix 5.1] any $n_R \times n_T$ matrix $\mathbf{H}$ can be written as $\mathbf{H} = \mathbf{U}\mathbf{D}\mathbf{V}^H$, where $\mathbf{D}$ is an $n_R \times n_T$ non-negative and diagonal matrix, $\mathbf{U}$ and $\mathbf{V}$ are $n_R \times n_R$ and $n_T \times n_T$ unitary matrices, respectively. That is, $\mathbf{U}\mathbf{U}^H = \mathbf{I}_{n_R}$ and $\mathbf{V}\mathbf{V}^H = \mathbf{I}_{n_T}$, where $\mathbf{I}_{n_R}$ and $\mathbf{I}_{n_T}$ are $n_R \times n_R$ and $n_T \times n_T$ identity matrices, respectively. The diagonal entries of $\mathbf{D}$ are the non-negative square roots of the eigenvalues of matrix $\mathbf{H}\mathbf{H}^H$. The eigenvalues of $\mathbf{H}\mathbf{H}^H$, denoted by λ, are defined here as:

$$\mathbf{H}\mathbf{H}^H\mathbf{y} = \lambda\mathbf{y}, \quad \mathbf{y} \neq 0 \tag{A4.3.5}$$

where $\mathbf{y}$ is an $n_R \times 1$ vector associated with λ, called an eigenvector.

The nonnegative square roots of the eigenvalues are also referred to as the singular values of $\mathbf{H}$. Furthermore, the columns of $\mathbf{U}$ are the eigenvectors of $\mathbf{HH}^H$ and the columns of $\mathbf{V}$ are the eigenvectors of $\mathbf{H}^H \mathbf{H}$. We can write for the received vector $\mathbf{r}$:

$$\mathbf{r} = \mathbf{UDV}^H \mathbf{x} + \mathbf{n} \qquad\qquad (A4.3.6)$$

In the sequel we will need the following transformations:

$$\begin{aligned} \mathbf{r}' &= \mathbf{U}^H \mathbf{r} \\ \mathbf{x}' &= \mathbf{V}^H \mathbf{x} \\ \mathbf{n}' &= \mathbf{U}^H \mathbf{n} \end{aligned} \qquad\qquad (A4.3.7)$$

as $\mathbf{U}$ and $\mathbf{V}$ are invertible. Clearly, multiplication of vectors $\mathbf{r}$, $\mathbf{x}$ and $\mathbf{n}$ by the corresponding matrices as defined in (A4.3.6) has only a scaling effect. Vector $\mathbf{n}'$ is a zero mean Gaussian random variable with i.i.d real and imaginary parts. Thus, the original channel is equivalent to the channel represented as:

$$\mathbf{r}' = \mathbf{D}x' + \mathbf{n}' \qquad\qquad (A4.3.8)$$

The number of nonzero eigenvalues of matrix $\mathbf{HH}^H$ is equal to the rank of matrix $\mathbf{H}$, denoted by r. For the $n_R \times n_T$ matrix $\mathbf{H}$, the rank is at most $m = (n_R, n_T)$, which means that at most m of its singular values are nonzero. Let us denote the singular values of $\mathbf{H}$ by $\sqrt{\lambda_i}$, $i = 1, 2, \ldots, r$. By substituting the entries $\sqrt{\lambda_i}$ in (A4.3.7), we get for the received signal components:

$$\begin{aligned} r_i' &= \sqrt{\lambda_i} x_i' + n_i', \quad i = 1, 2, \ldots, r \\ r_i' &= n_i', \quad i = r+1, r+2, \ldots, n_R \end{aligned} \qquad\qquad (A4.3.9)$$

As (A4.3.8) indicates, received components, r_i', $i = r+1, r+2, \ldots, n_R$, do not depend on the transmitted signal, i.e. the channel gain is zero. On the other hand, received components r_i', for $i = 1, 2, \ldots, r$ depend only on the transmitted component x_i'. Thus the equivalent MIMO channel from (A4.3.7) can be considered as consisting of r uncoupled parallel subchannels. Each sub-channel is assigned to a singular value of matrix $\mathbf{H}$, which corresponds to the amplitude channel gain. The channel power gain is thus equal to the eigenvalue of matrix $\mathbf{HH}^H$. For example, if $n_T > n_R$, as the rank of $\mathbf{H}$ cannot be higher than n_R.

Equation (A4.3.8) shows that there will be at most n_R nonzero gain subchannels in the equivalent MIMO channel, as indicated in Section 4.14. On the other hand if $n_R > n_T$, there will be at most n_T nonzero gain subchannels in the equivalent MIMO channel. The eigenvalue spectrum is a MIMO channel representation, which is suitable for evaluation of the best transmission paths.

The covariance matrices and their traces for signals $\mathbf{r}'$, $\mathbf{x}'$ and $\mathbf{n}'$ can be derived from (A4.3.6) as:

$$\begin{aligned} \mathbf{R}_{r'r'} &= \mathbf{U}^H \mathbf{R}_{rr} \mathbf{U} & \mathrm{tr}(\mathbf{R}_{r'r'}) &= \mathrm{tr}(\mathbf{R}_{rr}) \\ \mathbf{R}_{x'x'} &= \mathbf{V}^H \mathbf{R}_{xx} \mathbf{V} & \mathrm{tr}(\mathbf{R}_{x'x'}) &= \mathrm{tr}(\mathbf{R}_{xx}) \\ \mathbf{R}_{n'n'} &= \mathbf{U}^H \mathbf{R}_{nn} \mathbf{U} & \mathrm{tr}(\mathbf{R}_{n'n'}) &= \mathrm{tr}(\mathbf{R}_{nn}) \end{aligned}$$

The above relationships show that the covariance matrices of $\mathbf{r}'$, $\mathbf{x}'$ and $\mathbf{n}'$, have the same sum of the diagonal elements, and thus the same powers, as for the original signals, $\mathbf{r}$, $\mathbf{x}$ and $\mathbf{n}$, respectively.

Note that in the equivalent MIMO channel model described by (A4.3.8), the subchannels are uncoupled and thus their capacities add up. Assuming that the transmit power from each antenna in the equivalent MIMO channel model is P/n_T, we can estimate the overall channel capacity, denoted by C, by using the Shannon capacity formula as in (4.203):

$$C = W \sum_{i=1}^{r} \log_2 \left(1 + \frac{P_{ri}}{\sigma^2} \right) \qquad\qquad (A4.3.10)$$

where W is the bandwidth of each sub-channel and P_{ri} is the received signal power in the ith subchannel given by

$$P_{ri} = \frac{\lambda_i P}{n_T} \qquad (A4.3.11)$$

In (A4.3.11) $\sqrt{\lambda_i}$ is the singular value of channel matrix $\mathbf{H}$. Thus the channel capacity can be written as

$$C = W \sum_{i=1}^{r} \log_2 \left(1 + \frac{\lambda_i P}{n_T \sigma^2} \right) = W \log_2 \prod_{i=1}^{r} \left(1 + \frac{\lambda_i P}{n_T \sigma^2} \right) \qquad (A4.3.12)$$

Now we will show how the channel capacity is related to the channel matrix $\mathbf{H}$. Assuming that $m = \min(n_R, n_T)$, Equation (A4.3.5), defining the eigenvalue-eigenvector relationship, can be rewritten as:

$$(\lambda \mathbf{I}_m - \mathbf{Q}) \mathbf{y} = 0, \quad \mathbf{y} \neq 0 \qquad (A4.3.13)$$

where $\mathbf{Q}$ is the Wishart matrix defined as

$$\mathbf{Q} = \begin{cases} \mathbf{H}\mathbf{H}^H, & n_R < n_T \\ \mathbf{H}^H\mathbf{H}, & n_R < n_T \end{cases} \qquad (A4.3.14)$$

That is, λ is an eigenvalue of $\mathbf{Q}$, if and only if $\lambda \mathbf{I}_m - \mathbf{Q}$ is a singular matrix. Thus the determinant of $\lambda \mathbf{I}_m - \mathbf{Q}$ must be zero

$$\det(\lambda \mathbf{I}_m - \mathbf{Q}) = 0 \qquad (A4.3.15)$$

The singular values λ of the channel matrix can be calculated by finding the roots of Equation (A4.3.15). We consider the characteristic polynomial $p(\lambda)$ from the left-hand side in Equation (A4.3.15):

$$p(\lambda) = \det(\lambda \mathbf{I}_m - \mathbf{Q}) \qquad (A4.3.16)$$

It has degree equal to m, as each row of $\lambda \mathbf{I}_m - \mathbf{Q}$ contributes one and only one power of λ in the Laplace expansion of $\det(\lambda \mathbf{I}_m - \mathbf{Q})$ by minors. As a polynomial of degree m with complex coefficients has exactly m zeros, by counting multiplicities, we can write for the characteristic polynomial:

$$p(\lambda) = \prod_{i=1}^{m} (\lambda - \lambda_i) \qquad (A4.3.17)$$

where λ_i are the roots of the characteristic polynomial $p(\lambda)$, equal to the channel matrix singular values and we can write Equation (A4.3.15) as

$$\prod_{i=1}^{m} (\lambda - \lambda_i) = 0 \qquad (A4.3.18)$$

Further we can equate the left-hand sides of (A4.3.15) and (A4.3.18):

$$\prod_{i=1}^{m} (\lambda - \lambda_i) = \det(\lambda \mathbf{I}_m - \mathbf{Q}) \qquad (A4.3.19)$$

Substituting $-\frac{n_T \sigma^2}{P}$ for λ in (1.28) we get:

$$\prod_{i=1}^{m} \left(1 + \frac{\lambda_i P}{n_T \sigma^2} \right) = \det \left(I_m + \frac{P}{n_T \sigma^2} \mathbf{Q} \right) \qquad (A4.3.20)$$

and Equation (A4.3.12) can be written as:

$$C = W \log_2 \det \left(\mathbf{I}_m + \frac{P}{n_T \sigma^2} \mathbf{Q} \right) \qquad \text{(A4.3.21)}$$

As the nonzero eigenvalues of $\mathbf{HH}^H$ and $\mathbf{H}^H\mathbf{H}$ are the same, the capacities of the channels with matrices $\mathbf{H}$ and $\mathbf{H}^H$ are the same. Note that if the channel coefficients are random variables, formulas (A4.3.12) and (A4.3.21), represent instantaneous capacities or mutual information. The mean channel capacity can be obtained by averaging over all realizations of the channel coefficients as it was used in Chapter 14 (see Equation (14.54)).

Water-filling principle: Let us consider a MIMO channel where the channel parameters are known at the transmitter. The allocation of power to various transmitter antennas can be obtained by a 'water-filling' principle used in Equation (4.201). The 'water-filling principle' can be derived by maximizing the MIMO channel capacity under the power constraint:

$$\sum_{i=1}^{n_T} P_i = P \quad i = 1, 2, \ldots, n_T \qquad \text{(A4.3.21)}$$

where P_i is the power allocated to antenna i and P is the total power, which is kept constant. The normalized capacity of the MIMO channel is determined as:

$$C/W = \sum_{i=1}^{n_T} \log_2 \left[1 + \frac{P_i \lambda_i}{\sigma^2} \right] \qquad \text{(A4.3.23)}$$

Following the method of Lagrange multipliers, we introduce the function:

$$Z = \sum_{i=1}^{n_T} \log_2 \left[1 + \frac{P_i \lambda_i}{\sigma^2} \right] + L \left(P - \sum_{i-1}^{n_T} P_i \right) \qquad \text{(A4.3.24)}$$

where L is the Lagrange multiplier, λ_i is the ith channel matrix singular value and σ^2 is the noise variance. The unknown transmit powers P_i are determined by setting the partial derivatives of Z to zero:

$$\frac{\delta Z}{\delta P_i} = 0$$

$$\frac{\delta Z}{\delta P_i} = \frac{1}{\ln 2} \frac{\lambda_i/\sigma^2}{1 + P_i \lambda_i/\sigma^2} - L = 0 \qquad \text{(A4.3.25)}$$

which gives for P_i

$$P_i = \mu - \frac{\sigma^2}{\lambda_i} \qquad \text{(A4.3.26)}$$

where μ is a constant, given by $1/L \ln 2$. It can be determined from the power constraint of Equation (A4.3.22).

MIMO channel capacity for adaptive transmit power allocation: When the channel parameters are known at the transmitter, the capacity given by Equation (A4.3.21) can be increased by assigning the transmitted power to various antennas according to the 'water-filling' rule. It allocates more power when the channel is in good condition and less when the channel state gets worse. The power allocated to channel i is given by:

$$P_i = \left(\mu - \frac{\sigma^2}{\lambda_i}\right)^+, \quad i = 1, 2, \ldots, r \tag{A4.3.27}$$

where a^+ denotes max $(a, 0)$ and μ is determined so that $\sum_{i=1}^{r} P_i = P$.

We consider the singular value decomposition of channel matrix $\mathbf{H}$, then, the received power at subchannel i in the equivalent MIMO channel model is given by:

$$P_{ri} = \left(\lambda_i \mu - \sigma^2\right)^+ \tag{A4.3.28}$$

The MIMO channel capacity is then:

$$C = W \sum_{i=1}^{r} \log_2 \left(1 + \frac{P_{ri}}{\sigma^2}\right) \tag{A4.3.29}$$

Substituting the received signal power from Equation (A4.3.28) into Equation (A4.3.29) we get:

$$C = W \sum_{i=1}^{r} \log_2 \left[1 + \frac{1}{\sigma^2} \left(\lambda_i \mu - \sigma^2\right)^+\right] \tag{A4.3.30}$$

The covariance matrix of the transmitted signal is given by:

$$\mathbf{R}_{xx} = \mathbf{V} \operatorname{diag}\left(P_1, P_2, \ldots, P_{n_T}\right) V^H \tag{A4.3.31}$$

References

[1] Alamouti, S. (1998) A simple transmit diversity technique for wireless communications, *IEEE Journal on Selected Areas in Communications*, **18**(7), 1451–1458.
[2] Tarokh, V., Seshadri, N. and Calderbank, A. R. (1998) Space–time codes for high data rate wireless communications: Performance criterion and code construction, *IEEE Transactions On Information Theory*, **44**, 744–765.
[3] Furuskar, A., Mazur, S., Muller, F. and Olofsson, H. (1999) EDGE: Enhanced data rates for GSM and TDMA/136 evolution, *IEEE Personal Communications Magazine*, **6**, 56–66.
[4] Naguib, A., Seshadri, N. and Calderbank, A. R. (2000) Increasing data rate over wireless channels, *IEEE Signal Processing Magazine*, **17**, 76–92.
[5] Tarokh, V., Jafarkhani, H. and Calderbank, A. R. (1999) Space–time block codes from orthogonal designs, *IEEE Transactions On Information Theory*, **45**, 1456–1467.
[6] Stoica, P. and Lindskog, E. (2001) *Space–time block coding for channels with intersymbol interference*, presented at the thirty-fifth Asilomar Conference on Signals, Systems and Computing, October 2001.
[7] Tarokh, V., Naguib, A., Seshadri, N. and Calderbank, A. R. (1999) Space–time codes for high data rate wireless communication: Performance criteria in the presence of channel estimation errors, mobility and multiple paths, *IEEE Transactions On Communications*, **48**, 199–207.
[8] Naguib, A., Tarokh, V., Seshadri, N. and Calderbank, A. R. (1998) A space–time coding modem for high-data-rate wireless communications, *IEEE Journal on Selected Areas in Communications*, **16**, 1459–1477.
[9] Seshadri, N. and Winters, J. (1993) Two signaling schemes for improving the error performance of frequency-division-duplex (FDD) transmission systems using transmitter antenna diversity, in *Proceedings of Vehicular Technology Conference*, pp. 508–511.
[10] Foschini, G. J. (1996) Layered Space–time architecture for wireless communication in a fading environment when using multi-element antennas, *Bell Labs Technical Journal*, **1**, 41–59.

[11] Tarokh, V., Jafarkhani, H. and Calderbank, A. R. (1999) Space–time block coding for wireless communications: Performance results, *IEEE Journal on Selected Areas in Communications*, **17**, 451–460.

[12] Franz, V. and Anderson, J. (1998) Concatenated decoding with a reduced-search BCJR algorithm, *IEEE Journal on Selected Areas in Communication*, **16**, 186–195.

[13] Bahl, L., Cocke J., Jelinek F. and Raviv, J. (1974) Optimal decoding of linear codes for minimizing symbol error rate, *IEEE Transaction On Information Theory*, **IT-20**, 284–287.

[14] Fragouli, C., Al-Dhahir, N., Diggavi, S. and Turin, W. (2002) Prefiltered space–time M-BCJR equalizer for frequency-selective channels, *IEEE Transactions On Communications*, **50**, 742–753.

[15] Van Etten, W. (1976) Maximum likelihood receiver for multiple channel transmission systems, *IEEE Transactions On Communications*, **COM-24**, 276–283.

[16] Naguib, A. and Seshadri, N. (2000) MLSE and equalization of space–time coded signals, in *Proceedings of Vehicular Technology Conference*, May 2000, 1688–1693.

[17] Al-Dhahir, N. (2001) FIR channel-shortening equalizers for MIMO ISI channels, *IEEE Transactions On Communications*, **50**, 213–218.

[18] Bauch, G. and Al-Dhahir, N. (2000) Iterative equalization and decoding with channel shortening filters for space–time coded modulation, in *Proceedings of Vehicular Techology Conference*, 2000, pp. 1575–1582.

[19] Al-Dhahir, N., Naguib, A. and Calderbank, A. R. (2001) Finite-length MIMO decision-feedback equalization for space–time block-coded signals over multipath fading channels, *IEEE Transactions on Vehicular Technology*, **50**, 1176–1181.

[20] Eyuboglu, M. and Qureshi, S. (1999) Reduced-state sequence estimation for coded modulation of inter-symbol interference channels, *IEEE Journal on Selected Areas in Communications*, **17**, 989–999.

[21] Weinstein, S. and Ebert, P. (1971) Data transmission by frequency-division multiplexing using the discrete Fourier transform, *IEEE Transactions On Communications*, **COM-19**, 628–634.

[22] Zhou, S. and Giannakis, G. (2001) Space–time coded transmissions with maximum diversity gains over frequency-selective multipath fading channels, in *Proceedings of Globecom*, November 2001, pp. 440–444.

[23] Sari, H., Karam G. and Jeanclaude, I. (1995) Transmission techniques for digital terrestrial TV broadcasting, *IEEE Communications Magazine*, **33**, 100–109.

[24] Clark, M. V. (1998) Adaptive frequency-domain equalization and diversity combining for broadband wireless communications, *IEEE Journal on Selected Areas in Communications*, **16**, 1385–1395.

[25] Al-Dhahir, N. (2001) Single-carrier frequency-domain equalization for space–time block-coded transmissions over frequency-selective fading channels, *IEEE Communications Letters*, **5**, 304–306.

[26] Kaleh, G. (1995) Channel equalization for block transmission systems, *IEEE Journal on Selected Areas in Communications*, **13**, 110–121.

[27] SC FDE PHY Layer System Proposal for Sub 11 GHz BWA [Online]. Available: http://www.ieee802. org/16/tg3/contrib/802 163p–0131r2.pdf

[28] Diggavi, S., Al-Dhahir, N., Stamoulis A. and Calderbank, A. R. (2002) Differential space–time transmission for frequency-selective channels, *IEEE Communication Letters*, **6**(6), 253–255.

[29] Fragouli, C., Al-Dhahir, N. and Turin, W. (2002) Reduced-complexity training schemes for multiple-antenna broadband transmissions, in *Proceedings of WCNC*, **1**, 78–83.

[30] Fragouli, C., Al-Dhahir, N. and Turin, W. (2002) Finite-alphabet constant-amplitude training sequences for multiple-antenna broadband transmissions, *International Control Conference*, **1**, 6–10.

[31] Baro, S., Bauchs, G. and Hansmann, A. (2000) Improved codes for space–time trellis coded modulation, *IEEE Communications Letters*, **4**, 20–22.

[32] Hammons, R. and Gammal, H. E. (2000) On the theory of space–time codes for PSK modulation, *IEEE Transactions On Information Theory*, **46**, 524–542.

[33] Grimm, J., Fitz, M. P. and Krogmeier, J. V. (1998) Further results in space–time coding for Rayleigh fading in *Proceedings of 1998 Allerton Conference*, 1998, pp. 391–400.

[34] Grimm, J. (1998) *Transmitter diversity code design for achieving full diversity on Rayleigh channels*, Ph.D. dissertation, Purdue University, West Lafayette, IN.

[35] Blum, R. S. (2002) Some analytical tools for designing space–time convolutional codes, *IEEE Transactions On Communications*, **50**, 1593–1599.

[36] Guey, J.-C., Fitz, M. P., Bell, M. R. and Kuo, W.-Y. (1999) Signal design for transmitter diversity wireless communication systems over Rayleigh fading channels, *IEEE Transactions On Communications*, **47**, 527–537.

[37] Seshadri, N. and Winters, J. H. (1994) Two signaling schemes for improving the error performance of frequency division duplex (FDD) transmission system using antenna diversity, *International Journal of Wireless Infomation Networks*, **1**, 49–60.

[38] Raleigh, G. and Cioffi, J. M. (1998) Spatio-temporal coding for wireless communication, *IEEE Transactions On Communications*, **46**, 357–366.

[39] Jafarkhani, H. and Seshadri, N. (2003) Super-orthogonal space–time trellis codes, *IEEE Transactions On Information Theory*, **49**(4), 937–950.

[40] Genyuan Wang and Xiang-Gen Xia (2002) An orthogonal Space–time coding for CPM systems, *Proceedings of 2002 IEEE International Symposium On Information Theory*, pp. 107–112.

[41] Agrawal, A., Ginis, G. and Cioffi, J. M. (2002) Channel diagonalization through orthogonal space–time coding, *IEEE International Conference On Communications* (ICC 2002), 28 April–2 May 2002, **3**, 1621–1624.

[42] Sharma, N. and Papadias, C. B. (2002) Improved quasi-orthogonal codes, *IEEE Wireless Communications and Networking Conference* (WCNC2002) 17–21 March 2002, **1**, 169–171.

[43] Rouquette, S., Merigeault, S. and Gosse, K. (2002) Orthogonal full diversity space–time block coding based on transmit channel state information for 4 Tx antennas, *IEEE International Conference On Communications* (ICC 2002), 28 April–2 May 2002, **1**, 558–562.

[44] Jongren, G., Skoglund, M. and Ottersten, B. (2002) Combining beamforming and orthogonal space–time block coding, *IEEE Transactions On Information Theory*, **48**(3), 611–627.

[45] Larsson, E. G., Ganesan, G., Stoica, P. and Wing-Hin Wong (2002) On the performance of orthogonal space–time block coding with quantized feedback, *IEEE Communications Letters*, **6**(11), 487–489.

[46] Hsiao-feng, Lu, Kumar, P. V. and Habong, Chung (2002) On orthogonal designs and space–time codes, *Proceedings of 2002 IEEE International Symposium On Information Theory*, 30 June–5 July 2002, pp. 418–423.

[47] Gozali, R. and Woerner, B. D. (2002) The impact of channel estimation errors on space–time trellis codes paired with iterative equalization/decoding, *IEEE 55th Vehicular Technology Conference* (VTC Spring 2002), 6–9 May 2002, **2**, 826–831.

[48] Garg, P., Mallik, R. K. and Gupta, H. M. (2002) Performance analysis of space–time coding with imperfect channel estimation, *IEEE International Conference on Personal Wireless Communications*, December 15–17 2002, pp. 71–76.

[49] Yan, Q. and Blum, R. S. (2002) Improved space–time convolutional codes for quasi-static slow fading channels, *IEEE Transactions on Wireless Communications*, **1**(4), 563–571.

[50] Yan, Q. and Blum, R. S. (2000) Optimum space–time convolutional codes, *2000 IEEE Wireless Communications and Networking Conference* (WCNC), 23–28 September 2000, **3**, 1351–1355.

[51] Sellathurai, M. and Haykin, S. (2001) Random space–time codes with iterative decoders for BLAST architectures, *Proceedings of 2001 IEEE International Symposium On Information Theory*, 24–29 June 2001, pp. 105–110.

[52] Zhou, G., Wang, Y., Zhang, Z. and Chugg, K. M. (2001) On space–time convolutional codes for PSK modulation, *IEEE International Conference On Communications, 2001* (ICC 2001), 11–14 June 2001, **4**, 1122–1126.

[53] Youjian Liu, Fitz, M. P. and Takeshita, O. Y. (2000) QPSK space–time turbo codes, *2000 IEEE International Conference On Communications* (ICC 2000), 18–22 June 2000, **1**, 292–296.

[54] Jayaweera, S. K. and Poor, H. V. (2002) Turbo (iterative) decoding of a unitary space–time code with a convolutional code, *IEEE 55th Vehicular Technology Conference* (VTC Spring 2002) 6–9 May 2002, **2**, 1020–1024.

[55] McCloud, M. L., Brehler, M. and Varanasi, M. K. (2002) Signal design and convolutional coding for noncoherent space–time communication on the block-Rayleigh-fading channel, *IEEE Transactions On Information Theory*, **48**(5), 1186–1194.

[56] El Gamal, H. and Hammons, A. R. Jr. (2002) On the Design and Performance of Algebraic space–time codes for BPSK and QPSK Modulation, *IEEE Transactions On Communications*, **50**(6).

[57] Damen, M. O. and Beaulieu, N. C. (2003) On two high-rate algebraic space–time codes, *IEEE Transactions On Information Theory*, **49**(4), 1059–1063.

[58] El Gamal, H. and Hammons, A. R., Jr. (2000) 3 Algebraic designs for coherent and differentially coherent space–time codes, *2000 IEEE Wireless Communications and Networking Conference* (WCNC), 23–28 September 2000, **1**, 30–35.

[59] El Gamal, H., Hammons, A. R., Jr. and Stefanov, A. (2001) Algebraic space–time overlays for convolutionally coded systems, *12th IEEE InternationalSymposium on Personal, Indoor and Mobile Radio Communications*, 30 September–3 October 2001, **1**, C-149–C-154.

[60] Damen, M. O., Abed-Meraim, K. and Belfiore, J.-C. (2002) Diagonal algebraic space–time block codes, *IEEE Transactions On Information Theory*, **48**(3), 628–636.

[61] Damen, M. O., Tewfik, A. and Belflore, J. C. (2002) A construction of a space–time code based on number theory, *IEEE Transactions On Information Theory*, **48**(3), 753–760.

[62] Gamal, H. E. and Hammons, A. R., Jr. (2003) On the design of algebraic space–time codes for MIMO block-fading channels, *IEEE Transactions On Information Theory*, **49**(1), 151–163.

[63] Hammons, A. R., Jr. and El Gamal, H. (2000) Further results on the algebraic design of space–time codes, *Proceedings of IEEE International Symposium On Information Theory*, 25–30 June 2000, p. 339.

[64] El Gamal, H. and Damen, M. O. (2002) An algebraic number theoretic framework for space–time coding, Proceedings of the 2002 IEEE International Symposium On Information Theory, pp. 132–137.

[65] Tarokh, V. and Jafarkani, H (2000) A differential detection scheme for transmit diversity, *IEEE Journal on Selected Areas in Communications*, **18**(7).

[66] Yi Yao and Howlader, M. (2002) Multiple symbol double differential space–time coded OFDM, *IEEE 55th Vehicular Technology Conference* (VTC Spring 2002) 6–9 May 2002, **2**, 1050–1054.

[67] Diggavi, S. N., Al-Dhahir, N., Stamoulis, A. and Calderbank, A. R. (2002) Differential space–time coding for frequency-selective channels, *IEEE Communications Letters*, **6**(6), 253–255.

[68] Jianhua Liu, Jian Li, Hongbin Li and Larsson, E. G. (2001) Differential space-code modulation for interference suppression, *IEEE Transactions on Signal Processing* (see also *IEEE Transactions on Acoustics, Speech, and Signal Processing*), **49**(8), 1786–1795.

[69] Zhiqiang Liu, Giannakis, G. B. and Hughes, B. L. (2001) Double differential space–time block coding for time-selective fading channels, *IEEE Transactions On Communications*, **49**(9), 1529–1539.

[70] Ingram, M. A., Kuo-Hui Li, van Nguyen, A. and Pratt, T. (2000) Beamforming, Doppler compensation, and differential space–time coding, *Proceedings of the 2000 IEEE Sensor Array and Multichannel Signal Processing Workshop*, 16–17 March 2000, pp. 158–162.

[71] Hughes, B. L. (2000) Differential space–time modulation, *IEEE Transactions On Information Theory*, **46**(7), 2567–2578.

[72] Lampe, L. H.-J., Schober, R. and Fischer, R. F. H. (2003) Coded differential space–time modulation for flat fading channels, *IEEE Transactions on Wireless Communications*, **2**(3), 582–590.

[73] Steiner, A., Peleg, M. and Shamai, S. (2002) Iterative decoding of space–time differentially coded unitary matrix modulation, *IEEE Transactions on Signal Processing* (see also *IEEE Transactions on Acoustics, Speech, and Signal Processing*), **50**(10), 2385–2395.

[74] Zhiqiang Liu and Giannakis, G. B. (2003) Block differentially encoded OFDM with maximum multipath diversity, *IEEE Transactions on Wireless Communications*, **2**(3), 420–423.

[75] Steiner, A., Peleg, M. and Shamai, S. (2003) SVD iterative detection of turbo-coded multiantenna unitary differential modulation, *IEEE Transactions On Communications*, **51**(3), 441–452.

[76] Meixia Tao and Cheng, R. S. (2003) Trellis-coded differential unitary space–time modulation over flat fading channels, *IEEE Transactions On Communications*, **51**(4), 587–596.

[77] Jafarkhani, H. and Tarokh, V. (2001) Multiple transmit antenna differential detection from generalized orthogonal designs, *IEEE Transactions On Information Theory*, **47**(6), 2626–2631.

[78] Kun Wang and Hongya Ge (2001) New differential transmission scheme with transmit diversity for DS-CDMA systems, *IEEE 54th Vehicular Technology Conference* (VTC 2001 Fall), **1**, 232–236.

[79] Sellathurai, M. and Haykin, S. (2001) Further results on diagonal-layered space–time architecture, *IEEE 53rd Vehicular Technology Conference* (VTC 2001 Spring), 6–9 May 2001, **3**, 1958–1962.

[80] Zhiqiang Liu and Giannakis, G. B. (2001) Layered space–time coding for high data rate transmissions, *IEEE Military Communications Conference* (MILCOM 2001), 28–31 October 2001, **2**, 1295–1299.

[81] Matache, A., Wesel, R. D. and Jun Shi (2002) Trellis coding for diagonally layered space–time systems, *IEEE International Conference on Communications* (ICC 2002), 28 April–2 May 2002, **3**.

[82] Gamal, H. E. and Hammons, A. R., Jr. (2001) A new approach to layered space–time coding and signal processing, *IEEE Transactions On Information Theory*, **47**(6), 2321–2334.

[83] El Gamal, H. (2002) On the design of layered space–time systems for autocoding, *IEEE Transactions On Communications*, **50**(9), 1451–1461.

[84] Sellathurai, M. and Haykin, S. (2002) Turbo-BLAST for wireless communications: theory and experiments, *IEEE Transactions on Signal Processing* (see also *IEEE Transactions on Acoustics, Speech, and Signal, Processing*) **50**(10), 2538–2546.

[85] Wubben, D., Bohnke, R., Rinas, J., Kuhn, V. and Kammeyer, K. D. (2001) Efficient algorithm for decoding layered space–time codes, *Electronics Letters*, **37**(22), 1348–1350.

[86] Yan Xin and Giannakis, G. B. (2002) High-rate space–time layered OFDM, *IEEE Communications Letters*, **6**(5), 187–189.

[87] Sellathurai, M. and Haykin, S. (2003) T-BLAST for wireless communications: first experimental results, *IEEE Transactions on Vehicular Technology*, **52**(3), 530–535.

[88] Tarokh, V., Naguib, A., Seshadri, N. and Calderbank, A. R. (1999) Combined array processing and space–time coding, *IEEE Transactions On Information Theory*, **45**(4), 1121–1128.

[89] Foschini, G. J., Chizhik, D., Gans, M. J., Papadias, C. and Valenzuela, R. A. (2003) Analysis and performance of some basic space–time architectures, *IEEE Journal on Selected Areas in Communications*, **21**(3), 303–320.

[90] Yi Gong and Ben Letaief, K. (2002) Concatenated space–time block coding with trellis coded modulation in fading channels, *IEEE Transactions on Wireless Communications*, **1**(4), 580–590.

[91] Ungerboeck, G. (1982) Channel Coding with Multilevel Phase Signal, *IEEE Transactions On Information Theory*, **28**, 55–67.

[92] Liew, T. H. and Hanzo, L. (2002) Space–time codes and concatenated channel codes for wireless communications, *Proceedings of the IEEE*, **90**(2), 187–219.

[93] Dongzhe Cui and Haimovich, A. M. (2001) Performance of parallel concatenated space–time codes, *IEEE Communications Letters*, **5**(6), 236–238.

[94] Goulet, L. and Leib, H. (2003) Serially concatenated space–time codes with iterative decoding and performance limits of block-fading channels, *IEEE Journal on Selected Areas in Communications*, **21**(5), 765–773.

[95] Schlegel, C. and Grant, A. (2001) Concatenated space–time coding, 12th IEEE International Symposium on Personal, Indoor and Mobile Radio Communications, 30 September–3 October 2001, 1, C-139–C-143.

[96] Ying Li, Jun-hong Hui and Xin-mei Wang (2002) Non-full rank space–time trellis codes for serially concatenated system, *IEEE Communications Letters*, **6**(9), 397–399.

[97] Xiaotong Lin and Blum, R. S. (2000) Improved space–time codes using serial concatenation, *IEEE Communications Letters*, **4**(7), 221–223.

[98] Steiner, A., Peleg, M. and Shamai, S. (2003) SVD iterative detection of turbo-coded multiantenna unitary differential modulation, *IEEE Transactions On Communications*, **51**(3), 441–452.

[99] Foschini, G. J. and Gans, M. J. (1998) On limits of wireless communications in a fading environment when using multiple antennas, *Wireless Personal Communications*, **6**, 311–335.

[100] Telatar, I. E. (1999) Capacity of multi-antenna Gaussian channels, *European Transactions on Telecommunications*, **10**(6), 585–595.

[101] Stege, M., Zillmann, P. and Fettweis, G. (2002) MIMO channel estimation with dimension reduction, The 5th International Symposium on Wireless Personal Multimedia Communications, 27–30 October 2002, **2**, 417–421.

[102] Qinfang Sun, Cox, D. C., Huang, H. C. and Lozano, A. (2002) Estimation of continuous flat fading MIMO channels, *IEEE Transactions on Wireless Communications*, **1**(4), 549–553.

[103] Tugnait, J. K. (1997) Blind spatio-temporal equalization and impulse response estimation for MIMO channels using a Godard cost function, *IEEE Transactions on Signal Processing* (see also *IEEE Transactions on Acoustics, Speech, and Signal Processing*), **45**(1), 268–271.

[104] Bai, W., He, C., Jiang, L. G. and Li, X. X. (2003) Robust channel estimation in MIMO-OFDM systems, *Electronics Letters*, **39**(2), 242–244.

[105] Tugnait, J. K. (2001) Blind estimation and equalization of MIMO channels via multidelay whitening, *IEEE Journal on Selected Areas in Communications*, **19**(8), 1507–1519.

[106] Zhi Ding and Li Qiu (2003) Blind MIMO channel identification from second order statistics using rank deficient channel convolution matrix, *IEEE Transactions on Signal Processing* (see also *IEEE Transactions on Acoustics, Speech and Signal Processing*), **51**(2), 535–544.

[107] Haidong Zhu, Farhang-Boroujeny, B. and Schlegel, C. (2003) Pilot embedding for joint channel estimation and data detection in MIMO communication systems, *IEEE Communications Letters*, **7**(1), 30–32.

[108] Komninakis, C., Fragouli, C., Sayed, A. H. and Wesel, R. D. (2002) Multi-input multi-output fading channel tracking and equalization using Kalman estimation, *IEEE Transactions on Signal Processing* (see also *IEEE Transactions on Acoustics, Speech, and Signal Processing*), **50**(5), 1065–1076.

[109] Xavier, J. M. F., Barroso, V. A. N. and Moura, J. M. F. (1998) Closed-form blind channel identification and source separation in SDMA systems through correlative coding, *IEEE Journal on Selected Areas in Communications*, **16**(8), 1506–1517.

[110] Tugnait, J. K. and Bin Huang (2000) Multistep linear predictors-based blind identification and equalization of multiple-input multiple-output channels, *IEEE Transactions on Signal Processing* (see also *IEEE Transactions on Acoustics, Speech, and Signal Processing*), **48**(1), 26–38.

[111] Yang, H., Yuan, F. and Vucetic, B. (2002) Performance of space–time trellis codes in frequency selective WCDMA systems, *Proceedings of IEEE 56th Vehicular Technology Conference* (VTC 2002-Fall), 24–28 September 2002, **1**, 233–237.

[112] Sun, Q., Cox, D. C., Huang, H. C. and Lozano, A. (2002) Estimation of Continuous Flat Fading MIMO Channels, *IEEE Transactions on Wireless Communications*, **1**(4).

[113] Mu Qin and Blum, R. S. (2003) Properties of space–time codes for frequency selective channels and trellis code designs, *IEEE International Conference On Communications* (ICC' 03), **4**, 2286–2290.

[114] Shengli Zhou and Giannakis, G. B. (2003) Single-carrier space–time block-coded transmissions over frequency-selective fading channels, *IEEE Transactions On Information Theory*, **49**(1), 164–179.

[115] Bahceci, I. and Duman, T. M. (2002) Combined turbo coding and unitary space–time modulation, *IEEE Transactions On Communications*, **50**(8), 1244–1249.

[116] Rouquette-Leveil, S. and Gosse, K. (2002) Space–time coding options for OFDM-based WLANs, *IEEE 55th Vehicular Technology Conference* (VTC Spring 2002), 6–9 May 2002, **2**, 904–908.

[117] Yue, J. and Gibson, J. D. (2002) Performance of OFDM systems with space–time coding, *2002 IEEE Wireless Communications and Networking Conference* (WCNC2002), 17–21 March 2002, **1**, 280–284.

[118] Ben Lu and Xiaodong Wang (2000) Iterative receivers for multiuser space–time coding systems, *IEEE Journal on Selected Areas in Communications*, **18**(11), 2322–2335.

[119] Larsson, E. G., Stoica, P. and Li, J. (2002) On maximum-likelihood detection and decoding for space–time coding systems, *IEEE Transactions on Signal Processing* (see also *IEEE Transactions on IEEE Transactions on Acoustics, Speech, and Signal Processing*), **50**(4), 937–944.

[120] Youjian Liu, Fitz, M. P. and Takeshita, O. Y. (2002) A rank criterion for QAM space–time codes, *IEEE Transactions On Information Theory*, **48**(12), 3062–3079.

[121] Gore, D. A. and Paulraj, A. J. (2002) MIMO antenna subset selection with space–time coding, IEEE Transactions on Signal Processing *(see also* IEEE Transactions on Acoustics, Speech, and Signal Processing*)*, **50**(10), 2580–2588.

[122] Shengli Zhou, Muquet, B. and Giannakis, G. B. (2002) Subspace-based (semi-) blind channel estimation for block precoded space–time OFDM, *IEEE Transactions on Signal Processing* (see also *IEEE Transactions on Acoustics, Speech, and Signal Processing*), **50**(5), 1215–1228.

[123] Goulet, L. and Leib, H. (2003) Serially concatenated space–time codes with iterative decoding and performance limits of block-fading channels, *IEEE Journal on Selected Areas in Communications*, **21**(5), 765–773.

[124] Wing Hin Wong and Larsson, E. G. (2003) Orthogonal space–time block coding with antenna selection and power allocation, *Electronics Letters*, **39**(4), 379–381.

[125] Caire, G. and Colavolpe, G. (2003) On low-complexity space–time coding for quasi-static channels, *IEEE Transactions On Information Theory*, **49**(6), 1400–1416.

[126] Byoungjo Choi and Hanzo, L. (2003) Optimum mode-switching-assisted constant-power single- and multi-carrier adaptive modulation, *IEEE Transactions on Vehicular Technology*, **52**(3), 536–560.

[127] Ghrayeb, A. and Duman, T. M. (2003) Performance analysis of MIMO systems with antenna selection over quasi-static fading channels, *IEEE Transactions on Vehicular Technology*, **52**(2), 281–288.

[128] El Gamal, H. and Damen, M. O. (2003) Universal space–time coding, *IEEE Transactions On Information Theory*, **49**(5), 1097–1119.

[129] Xiaoxia Zhang and Fitz, M. P. (2003) Space–time code design with continuous phase modulation, *IEEE Journal on Selected Areas in Communications*, **21**(5), 783–792.

[130] Lampe, L. H.-J., Schober, R. and Fischer, R. F. H. (2003) Coded differential space–time modulation for flat fading channels, *IEEE Transactions on Wireless Communications*, **2**(3), 582–590.

[131] Jafarkhani, H. and Seshadri, N. (2003) Super-orthogonal space–time trellis codes, *IEEE Transactions On Information Theory*, **49**(4), 937–950.

[132] Il-Min Kim and Tarokh, V. (2003) Variable-rate space–time block codes in M-ary PSK systems, *IEEE Journal on Selected Areas in Communications*, **21**(3), 362–373.

[133] Gesbert, D., Shafi, M., Da-shan Shiu, Smith, P. J. and Naguib, A. (2003) From theory to practice: an overview of MIMO space–time coded wireless systems, *IEEE Journal on Selected Areas in Communications*, **21**(3), 281–302.

[134] Banister, B. C. and Zeidler, J. R. (2003) Feedback assisted transmission subspace tracking for MIMO systems, *IEEE Journal on Selected Areas in Communications*, **21**(3), 452–463.

[135] Petre, F., Leus, G., Deneire, L., Engels, M., Moonen, M. and De Man, H. (2003) Space–time block coding for single-carrier block transmission DS-CDMA downlink, *IEEE Journal on Selected Areas in Communications*, **21**(3), 350–361.

[136] Yan Xin, Zhengdao Wang and Giannakis, G. B. (2003) Space–time diversity systems based on linear constellation precoding, *IEEE Transactions on Wireless Communications*, **2**(2), 294–309.

[137] Wallace, J. W., Jensen, M. A., Swindlehurst, A. L. and Jeffs, B. D. (2003) Experimental characterization of the MIMO wireless channel: data acquisition and analysis, *IEEE Transactions on Wireless Communications*, **2**(2), 335–343.

[138] Gamal, H. E. and Hammons, A. R., Jr. (2003) On the design of algebraic space–time codes for MIMO block-fading channels, *IEEE Transactions On Information Theory*, **49**(1), 151–163.

[139] Steiner, A., Peleg, M. and Shamai, S. (2003) SVD iterative detection of turbo-coded multiantenna unitary differential modulation, *IEEE Transactions On Communications*, **51**(3), 441–452.

[140] Larsson, E. G. (2003) Unitary nonuniform space–time constellations for the broadcast channel, *IEEE Communications Letters*, **7**(1), 21–23.

[141] Meixia Tao and Cheng, R. S. (2003) Trellis-coded differential unitary space–time modulation over flat fading channels, *IEEE Transactions On Communications*, **51**(4), 587–596.

[142] Young-Hak Kim and Kaveh, M. (2003) Coordinate-interleaved space–time coding with rotated constellation, *The 57th IEEE Vehicular Technology Conference* (VTC 2003-Spring), April 22–25 2003, **1**, 732–735.

[143] Le Nir, V., Mard, M. and Le Gouable, R. (2003) Space–time block coding applied to turbo coded multicarrier CDMA, *The 57th IEEE Vehicular Technology Conference* (VTC 2003-Spring), April 22–25 2003, **1**, 577–581.

[144] Yang, J., Sun, Y., Senior, J. M. and Pem, N. (2003) Channel estimation for wireless communications using space–time block coding techniques, *Proceedings of the 2003 International Symposium on Circuits and Systems* (ISCAS '03), May 25–28 2003, **2**, 220–223.

[145] Blum, R. S. MIMO capacity with interference, *IEEE Journal on Selected Areas in Communications*, **21**(5), 793–801.

[146] Mietzner, J., Hoeher, P. A. and Sandell, M. (2003) Compatible improvement of the GSM/EDGE system by means of space–time coding techniques, *IEEE Transactions on Wireless Communications*, **24**(5), 690–702.

[147] Sellathurai, M. and Haykin, S. (2003) T-BLAST for wireless communications: first experimental results, *IEEE Transactions on Vehicular Technology*, **52**(3), 530–535.

[148] Shengli Zhou and Giannakis, G. B. (2003) Optimal transmitter eigen-beamforming and space–time block coding based on channel correlations, *IEEE Transactions On Information Theory*, **49**(7), 1673–1690.

[149] Debbah, M., Hachem, W., Loubaton, P. and de Courville, M. (2003). MMSE analysis of certain large isometric random precoded systems, *IEEE Transactions On Information Theory*, **49**(5), 1293–1311.

[150] Uysal, M. and Georghiades, C. N. (2003) An efficient implementation of a maximum-likelihood detector for space–time block coded systems, *IEEE Transactions On Communications*, **51**(4), 521–524.

[151] Baccarelli, E. and Biagi, M. (2003) Error resistant space–time coding for emerging 4G-WLANs, *2003 IEEE Wireless Communications and Networking* (WCNC 2003), 16–20 March 2003, **1**, 72–77.

[152] Hochwald, B. M. and ten Brink, S. (2003) Achieving near-capacity on a multiple-antenna channel, *IEEE Transactions On Communications*, **51**(3), 389–399.

[153] Banister, B. C. and Zeidler, J. R. (2003) A simple gradient sign algorithm for transmit antenna weight adaptation with feedback, *IEEE Transactions on Signal Processing* (see also *IEEE Transactions on Acoustics, Speech, and Signal Processing*), **51**(5), 1156–1171.

[154] Larsson, E. G., Stoica, P. and Jian Li (2003) Orthogonal space–time block codes: maximum likelihood detection for unknown channels and unstructured interferences, *IEEE Transactions on Signal Processing* (see also *IEEE Transactions on Acoustics, Speech, and Signal Processing*), **51**(2), 362–372.

[155] Hyundong Shin and Jae Hong Lee (2003) Effect of keyholes on the symbol error rate of space–time block codes, *IEEE Communications Letters*, **7**(1), 27–29.

[156] Shengli Zhou and Giannakis, G. B. (2003) Single-carrier space–time block-coded transmissions over frequency-selective fading channels, *IEEE Transactions On Information Theory*, **49**(1), 164–179.

[157] Andersen, J. B. (2000) Array gain and capacity for known random channels with multiple element arrays at both ends, *IEEE Journal on Selected Areas in Communications*, **18**(11), 2172–2178.

[158] Geman, S. (1980) A limit theorem for the norm of random matrices, *Annals of Probability*, **8**, 252–261.

[159] Silverstein, J. W. (1985) The smallest eigenvalue of a large dimensional Wishart matrix, *Annals of Probability*, **13**, 1364–1368.

[160] Raleigh, G. G. and Cioffi, J. M. (1998) Spatio-temporal coding for wireless communication, *IEEE Transactions On Communications*, **46**, 357–366.

[161] Winters, J. H. (1987) On the capacity of radio communication systems with diversity in a rayleigh fading environment, *IEEE Journal on Selected Areas in Communications*, **5**, 871–878.

[162] DaSilva, V. M. and Sousa, E. S. (1997) Fading-resistant modulation using several transmitter antennas, *IEEE Transactions On Communications*, **45**, 1236–146.

[163] Divsalar, D. and Simon, M. K. (1988) The design of trellis coded MPSK for fading channels: performance criteria, *IEEE Transaction Communications*, **36**, 1004–1012.

[164] Horn, R. A. and Johnson, C. R. (1985) *Matrix Analysis*. Cambridge University Press, Cambridge.

[165] Golub, G. H. and Van Loan, C. F. (1989) *Matrix Computations*, 2nd ed. The Johns Hopkins University Press, Baltimore, MD.

[166] Nicholson, W. K. (1986) *Elementary Linear Algebra with Applications*. PWS-Kent, Boston, MA.

[167] Conway, J. H. and Sloane, N. J. A. (1993) *Sphere Packings, Lattices and Groups*. Springer-Verlag, New York.

[168] Viterbo, E. and Boutros, J. (1999) A universal lattice code decoder for fading channel, *IEEE Transactions On Information Theory*, **45**, 1639–1642.

[169] Damen, M. O., Chkeif, A. and Belfiore, J.-C. (2000) Lattice codes decoder for space–time codes, *IEEE Communications Letters*, **4**, 161–163.

[170] Damen, M. O., Meraim, K. A. and Belfiore, J.-C. (2002) Diagonal algebraic space–time block codes, *IEEE Transactions On Information Theory*, **48**, 628–637

[171] Boutros, J. and Viterbo, E. (1998) Signal space diversity: a power and bandwidth efficient diversity technique for the Rayleigh fading channel, *IEEE Transactions On Information Theory*, **44**, 1453–1467.

[172] Belfiore, J.-C., Giraud, X. and Rodriguez, J. (2000) Linear labeling for joint source channel coding, in *Proceedings International Symposium On Information Theory (ISIT'2000)*, Sorrento, Italy, June 2000.

[173] Conway, J. and Sloane, N. (1998) *Sphere Packings, Lattices and Groups*, 3rd ed. Springer-Verlag, New York.

[174] Viterbo, E. and Boutros, J. (1999) A universal lattice code decoder for fading channel, *IEEE Transations On Information Theory*, **45**, 1639–1642.

[175] Fincke, U. and Phost, M. (1985) Improved methods for calculating vectors of short length in a lattice, including a complexity analysis, *Math. Comput.*, **44**, 463–471.

5

Multiuser Communication

The basic principles of CDMA were discussed in Chapter 1. In this chapter, after brief discussion of code generation, we focus on multiuser detection. More details can be found in the recent book on WCDMA [1].

5.1 Pseudorandom Sequences

5.1.1 Binary Shift Register Sequences

Let us define a polynomial

$$h(x) = h_0 x^n + h_1 x^{n-1} + \cdots + h_{n-1} x + h_n \tag{5.1}$$

in the discrete field with two elements $h_i \in (0, 1)$ and $h_0 = h_n = 1$. An example polynomial could be $x^4 + x + 1$ or $x^5 + x^2 + 1$. The coefficients h_i of the polynomial can be represented by binary vectors 10011 and 100101, or in octal notation 23 and 45 (every group of three bits is represented by a number between 0 and 7). A binary sequence u is said to be a *sequence generated by* $h(x)$ if, for all integers j,

$$h_0 u_j \oplus h_1 u_{j-1} \oplus h_2 u_{j-2} \oplus \cdots \oplus h_n u_{j-n} = 0 \tag{5.2}$$

where $\oplus$ = addition modulo 2.

If we formally change the variables

$$j \rightarrow j + n, \quad \text{and} \quad h_0 = 1 \tag{5.3}$$

then Equation (5.2) becomes:

$$u_{j+n} = h_n u_j \oplus h_{n-1} u_{j+1} \oplus \cdots h_1 u_{j+n-1} \tag{5.4}$$

In this notation, u_j is the jth bit (called chip) of the sequence u. Equation (5.4) suggests that the sequence u can be generated by an n-stage binary linear feedback shift register which has a feedback tap connected to the ith cell if $h_i = 1, 0 < i \leq n$. As an example, for $n = 5$, Equation (5.4) becomes:

$$u_{j+5} = h_5 u_j \oplus h_4 u_{j+1} \oplus h_3 u_{j+2} \oplus h_2 u_{j+3} \oplus h_1 u_{j+4} \tag{5.5}$$

Advanced Wireless Communications & Internet: Future Evolving Technologies, Third Edition. Savo Glisic.
© 2011 John Wiley & Sons, Ltd. Published 2011 by John Wiley & Sons, Ltd.

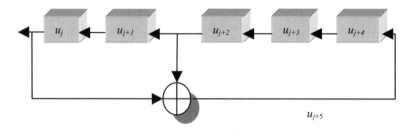

Figure 5.1   Sequence generator for the polynomial (45).

For $x^5 + x^2 + 1$, octal representation (45), the coefficients h_i are:

$$
\begin{array}{cccccc}
h_0 & h_1 & h_2 & h_3 & h_4 & h_5 \\
1 & 0 & 0 & 1 & 0 & 1
\end{array}
\quad \text{(Octal representation 45)}
$$

and the block diagram of the circuit is shown in Figure 5.1.

Similarly, for the polynomial $x^5 + x^4 + x^3 + x^2 + 1$, the coefficients h_i are given as:

$$
\begin{array}{cccccc}
h_0 & h_1 & h_2 & h_3 & h_4 & h_5 \\
1 & 1 & 1 & 1 & 0 & 1
\end{array}
\quad \text{(Octal representation 75)}
$$

and by using Equation (5.4) one can get the generator shown in Figure 5.2.

Some of the properties of these sequences and definitions are listed below. Details can be found in standard literature listed at the end of the book, especially [2–22].

If u and v are generated by $h(x)$, then so is $u \oplus v$, where $u \oplus v$ denotes the sequence whose ith element is $u_i \oplus v_i$. The all zero state of the shift register is not allowed because, for this initial state, Equation (5.5) would continue to generate zero chips. For this reason, the period of u is at most $2^n - 1$, where n is the number of cells in the shift register, or equivalently, the degree of $h(x)$. If u denotes an arbitrary $\{0, 1\}$ valued sequence, then $x(u)$ denotes the corresponding $\{+1, -1\}$ valued sequence, where the ith element of $x(u)$ is just $x(u_i)$:

$$
x(u_i) = (-1)^{u_i} \tag{5.6}
$$

If T^i is a delay operator (delay for i chip periods) then we have:

$$
\begin{aligned}
T^i(x(u)) &= x(T^i u) \ \text{ and } \ \sum x(u) = x(u_0) + x(u_1) + \cdots + x(u_{N-1}) \\
&= N^+ - N^- = (N - N^-) - N^- \\
&= N - 2N^- = N - 2wt(u)
\end{aligned} \tag{5.7}
$$

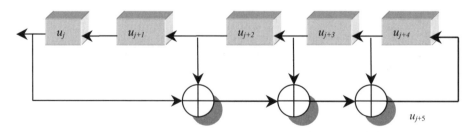

Figure 5.2   Sequence generator for polynomial (75).

where $wt(u)$ denotes the Hamming weight of unipolar sequence u, that is, the number of ones in u, N is the sequence period and N^+ and N^- are the number of positive and negative chips in bipolar sequence $x(u)$. The crosscorrelation function between two bipolar sequences can be represented as:

$$\theta_{u,v} \equiv \theta_{x(u),x(v)}(l) = \sum_{i=0}^{N-1} x(u_i)x(v_{i+l})$$

$$= \sum_{i=0}^{N-1}(-1)^{u_i}(-1)^{v_{i+l}} = \sum_{i=0}^{N-1}(-1)^{u_i \oplus v_{i+l}} = \sum_{i=0}^{N-1} x(u_i \oplus v_{i+l}) \tag{5.8}$$

By using Equation (5.7) we have:

$$\theta_{u,v}(l) = N - 2wt(u \oplus T^l v) \tag{5.9}$$

The periodic autocorrelation function $\theta_u(\cdot)$ is just $\theta_{u,u}(\cdot)$

$$\theta_u(l) = N - 2wt(u \oplus T^l u)$$
$$= N^+ - N^- = (N - N^-) - N^- = N - 2N^- \tag{5.10}$$

5.1.2   Properties of Binary Maximal Length Sequences

As was mentioned earlier the all zero state of the shift register is not allowed because, based on Equation (5.4), the generator could not get out of this state. Bear in mind that the number of possible states of the shift register is 2^n. So, the period of a sequence u generated by the polynomial $h(x)$ cannot exceed $2^n - 1$, where n is the degree of $h(x)$. If u has this maximal period, $N = 2^n - 1$, it is called a maximal length sequence or m-sequence. To get such a sequence, $h(x)$ should be a primitive binary polynomial of degree n. There are exactly N non-zero sequences generated by $h(x)$, and they are just the N different phases of u, $T u$, $T^2 u, \ldots, T^{N-1}u$. Given distinct integers i and j, $0 \le i, j < N$, there is a unique integer, k, distinct from both i and j, such that $0 \le k < N$ and

$$T^i u \oplus T^j u = T^k u \tag{5.11}$$

From the above discussion on the number of ones and zeros, $wt(u) = 2^{n-1} = 1/2 (N + 1)$, so that from Equation (5.9) we have:

$$\theta_u(l) = \begin{cases} N, & \text{if } l \equiv 0 \mod N \\ -1, & \text{if } l \neq 0 \mod N \end{cases} \tag{5.12}$$

From here on, $\tilde{u}$ will be called a characteristic m-sequence, or the characteristic phase of the m-sequence u if $\tilde{u}_i = \tilde{u}_{2i}$ for all $i \in \mathbb{Z}$.

Let q denote a positive integer, and consider the sequence v formed by taking every qth bit of u (i.e. $v_i) = u_{qi}$ for all $i \in \mathbb{Z}$). The sequence v is said to be a *decimation* by q of u, and will be denoted by $u[q]$.

Assume that $u[q]$ is not identically zero. Then, $u[q]$ has period $N/\gcd(N, q)$, and is generated by the polynomial whose roots are the qth powers of the roots of $h(x)$, where $\gcd(N, q)$ is the greatest common divisor of the integers N and q. The tables of primitive polynomials are available in any book on coding theory. The reciprocal m-sequence v is generated by the reciprocal polynomial of $h(x)$, that is,

$$\hat{h}(x) = x^n h(x^{-1}) = h_n x^n + h_{n-1}x^{n-1} + \cdots + h_0 \tag{5.13}$$

5.1.3   Crosscorrelation Spectra

Frequently, we do not need to know more than the set of crosscorrelation values together with the number of integers $l (0 \le l < N)$ for which $\theta_{u,v}(l) = c$ for each c in this set. Let u and v denote m-sequences of period $2^n - 1$. If $v = u[q]$, where either $q = 2^k + 1$ or $q = 2^{2k} - 2^k + 1$, and if $e = \gcd(n, k)$ is such that n/e is odd, then the spectrum of $\theta_{u,v}$ is three-valued [23–26] as:

$$
\begin{array}{lll}
-1 + 2^{(n+e)/2} & \text{occurs } 2^{n-e-1} + 2^{(n-e-2)/2} & \text{times} \\
-1 & \text{occurs } 2^n - 2^{n-e} - 1 & \text{times} \\
-1 - 2^{(n+e)/2} & \text{occurs } 2^{n-e-1} - 2^{(n-e-2)/2} & \text{times}
\end{array}
\tag{5.14}
$$

The same spectrum is obtained if instead of $v = u[q]$, we let $u = v[q]$. Notice that if e is large, $\theta_{u,v}(l)$ takes on large values but only very few times, while if e is small, $\theta_{u,v}(l)$ takes on smaller values more frequently. In most instances, small values of e are desirable. If we wish to have $e = 1$, then clearly n must be odd in order that n/e be odd. When n is odd, we can take $k = 1$ or $k = 2$ (and possibly other values of k as well), and obtain $\theta(u, u[3])$, $\theta(u, u[5])$ and $\theta(u, u[13])$ all having the three-valued spectrum given by Expressions (5.14) (with $e = 1$). Suppose next that $n \equiv 2 \bmod 4$. Then, n/e is odd if e is even and a divisor of n. Letting $k = 2$, we obtain that $\theta(u, u[5])$ and $\theta(u, u[13])$ both have the three-valued spectrum given by Expressions (5.14) (with $e = 2$).

Let us define $t(n)$ as:

$$
t(n) = 1 + 2[(n + 2)/2]
\tag{5.15}
$$

where $[\alpha]$ denotes the integer part of the real number α. Then if $n \ne 0 \bmod 4$, there exist pairs of m-sequences with three-valued crosscorrelation functions, where the three values are -1, $-t(n)$ and $t(n) - 2$. A crosscorrelation function taking on these values is called a *preferred three-valued crosscorrelation function* and the corresponding pair of m-sequences (polynomials) is called a *preferred pair of m-sequences* (polynomials).

Let u and v denote m-sequences of period $2^n - 1$, where n is a multiple of 4. If $v = u[-1 + 2^{(n+2)/2}] = u[t(n) - 2]$, then $\theta_{u,v}$ has a four-valued spectrum represented as:

$$
\begin{array}{lll}
-1 + 2^{(n+2)/2} & \text{occurs } (2^{n-1} - 2^{(n-2)/2})/3 & \text{times} \\
-1 + 2^{n/2} & \text{occurs } 2^{n/2} & \text{times} \\
-1 & \text{occurs } 2^{n-1} - 2^{(n-2)/2} - 1 & \text{times} \\
-1 - 2^{n/2} & \text{occurs } (2^n - 2^{n/2})/3 & \text{times}
\end{array}
\tag{5.16}
$$

5.1.4   Maximal Connected Sets of m-sequences

The preferred pair of m-sequences is a pair of m-sequences of period $N = 2^n - 1$, which has the preferred three-valued crosscorrelation function. The values taken on by the preferred three-valued crosscorrelation functions are -1, $-t(n)$ and $t(n) - 2$, where $t(n)$ is given by Equation (5.15). The pair of primitive polynomials that generate a preferred pair of m-sequences is called a preferred pair of polynomials. A connected set of m-sequences is a collection of m-sequences which has the property that each pair in the collection is a preferred pair. A largest possible connected set is called a *maximal connected set*, and the size of such a set is denoted by M_n. Some examples are given in Table 5.1.

5.1.5   Gold Sequences

A set of Gold sequences of period $N = 2^n - 1$, consists of $N + 2$ sequences for which $\theta_c = \theta_a = t(n)$. A set of Gold sequences can be constructed from appropriately selected m-sequences as described below.

Table 5.1 Set sizes and crosscorrelation bounds for the sets of all m-sequences and for maximal connected sets [2] © 1980, IEEE

n	$N = 2^n - 1$	Number of m-sequences	θ_c for set of all m-sequences	M_n	$t(n)$
3	7	2	5	2	5
4	15	2	9	0	9
5	31	6	11	3	9
6	63	6	23	2	17
7	127	18	41	6	17
8	255	16	95	0	33
9	511	48	113	2	33
10	1023	60	383	3	65
11	2047	176	287	4	65
12	4095	144	1407	0	129
13	8191	630	≥ 703	4	129
14	16383	756	≥ 5631	3	257
15	32767	1800	≥ 2047	2	257
16	65535	2048	≥ 4095	0	513

Suppose $f(x) = h(x)\hat{h}(x)$, where $h(x)$ and $\hat{h}(x)$ have no factors in common. The set of all sequences generated by $f(x)$ is of the form $a \oplus b$, where a is some sequence generated by $h(x)$, b is some sequence generated by $\hat{h}(x)$ and we do not make the usual restriction that a and b are non-zero sequences. We represent such a set by:

$$G(u, v) \triangleq \{u, v, u \oplus v, u \oplus Tv, u \oplus T^2 v, \ldots, u \oplus T^{N-1} v\} \tag{5.17}$$

$G(u, v)$ contains $N + 2 = 2^n + 1$ sequences of period N. Let $\{u, v\}$ denote a preferred pair of m-sequences of period $N = 2^n - 1$ generated by the primitive binary polynomials $h(x)$ and $\hat{h}(x)$ respectively. Then set $G(u, v)$ is called a set of Gold sequences. For $y, z \in G(u, v)$, $\theta_{y,z}(l) \in \{-1, -t(n), t(n) - 2\}$ for all integers l, and $\theta_y(l) \in \{-1, -t(n), t(n) - 2\}$ for all $l \neq 0 \bmod N$. Every sequence in $G(u, v)$ can be generated by the polynomial $f(x) = h(x)\hat{h}(x)$. Note that the non-maximal length sequences belonging to $G(u, v)$ also can be generated by adding together (term by term, modulo 2) the outputs of the shift registers corresponding to $h(x)$ and $\hat{h}(x)$. The maximal length sequences belonging to $G(u, v)$ are, of course, the outputs of the individual shift registers. Let us compare the parameter $\theta_{\max} = \max\{\theta_a, \theta_c\}$ for a set of Gold sequences to a bound due to Sidelnikov, which states that for any set of N or more *binary* sequences of period N,

$$\theta_{\max} > (2N - 2)^{1/2} \tag{5.18}$$

For Gold sequences, they form an *optimal* set with respect to the bounds when n is odd. When n is even, Gold sequences are not optimal.

5.1.6 Gold-Like and Dual-BCH Sequences

Let n be even and let q be an integer such that $\gcd(q, 2^n - 1) = 3$. Let u denote an m-sequence of period $N = 2^n - 1$ generated by $h(x)$, and let $v^{(k)}$, $k = 0, 1, 2$, denote the result of decimating $T^k u$ by q. The $v^{(k)}$ are sequences of period $N' = N/3$ which are generated by the polynomial $\hat{h}(x)$ whose roots are qth powers of the roots of $h(x)$. Gold-like sequences are defined as:

$$H_q(u) = \{u, u \oplus v^{(0)}, u \oplus Tv^{(0)}, \ldots, u \oplus T^{N'-1}v^{(0)},$$
$$u \oplus v^{(1)}, u \oplus Tv^{(1)}, \ldots, u \oplus T^{N'-1}v^{(1)}, \qquad (5.19)$$
$$u \oplus v^{(2)}, u \oplus Tv^{(2)}, \ldots, u \oplus T^{N'-1}v^{(2)}\}$$

Note that $H_q(u)$ contains $N + 1 = 2^n$ sequences of period N.

For $n \equiv 0 \bmod 4$, $\gcd(t(n), 2^n - 1) = 3$ vectors $v^{(k)}$ are taken to be of length N rather than $N/3$. Consequently, it can be shown that for the set $H_{t(n)}(u)$, $\theta_{max} = t(n)$. We call $H_{t(n)}(u)$ a set of Gold-like sequences. The correlation functions for the sequences belonging to $H_{t(n)}(u)$ take on values in the set $-1, -t(n), t(n) - 2, -s(n), s(n) - 2\}$ where $s(n)$ is defined (for even n only) by:

$$s(n) = 1 + 2^{n/2} = \frac{1}{2}(t(n) + 1) \qquad (5.20)$$

5.1.7 Kasami Sequences

Let n be even and let u denote an m-sequence of period $N = 2^n - 1$ generated by $h(x)$. Consider the sequence $w = u[s(n)] = u[2^{n/2} + 1]$. w is a sequence of period $2^{n/2} + 1$ which is generated by the polynomial $h'(x)$, whose roots are the $s(n)$th powers of the roots of $h(x)$. Furthermore, since $h'(x)$ can be shown to be a polynomial of degree $n/2$, w is an m-sequence of period $2^{n/2} - 1$. Consider the sequences generated by $h(x)h'(x)$ of degree $3n/2$. Any such sequence must be of one of the forms $T^i u, T^j w, T^i u \oplus T^j w, 0 \leq i < 2^n - 1, 0 \leq j < 2^{n/2} - 1$. Thus, any sequence y of period $2^n - 1$ generated by $h(x)h'(x)$ is some phase of some sequence in the set $K_s(u)$ defined by:

$$K_s(u) \stackrel{\Delta}{=} \{u, u \oplus w, u \oplus Tw, \ldots, u \oplus T^{2^{n/2}-2}w\} \qquad (5.21)$$

This set of sequences is called a small set of Kasami sequences with:

$$\theta = \{-1, -s(n), s(n) - 2\}$$
$$\theta_{max} = s(n) = 1 + 2^{n/2} \qquad (5.22)$$

θ_{max} for the set $K_s(u)$ is approximately one half of the value of θ_{max} achieved by the sets of sequences discussed previously. $K_s(u)$ contains only $2^{n/2} = (N + 1)1/2$ sequences, while the sets discussed previously contain $N + 1$ or $N + 2$ sequences.

Let n be even and let $h(x)$ denote a primitive binary polynomial of degree n that generates the m-sequence u. Let $w = u[s(n)]$ denote an m-sequence of period $2^{n/2} - 1$ generated by the primitive polynomial $h'(x)$ of degree $n/2$, and let $\hat{h}(x)$ denote the polynomial of degree n that generates $u[t(n)]$. Then, the set of sequences of period N generated by $h(x)\hat{h}(x)h'(x)$, called the *large set of Kasami sequences* and denoted by $K_L(u)$, is defined as follows:

1. If $n \equiv 2 \bmod 4$, then:

$$K_L(u) = G(u, v) \bigcup \left[\bigcup_{i=0}^{2^{n/2}-2} \{T^i w \oplus G(u, v)\} \right] \qquad (5.23)$$

where $v = u[t(n)]$, and $G(u,v)$ is defined in Equation (5.17).

Table 5.2 Polynomials generating various classes of sequences of periods 31, 63, 65, 127, and 255 [2]
© 1980, IEEE

N	Polynomial	Construction	No.	Values taken on by the correlation functions
31	3551	G	33	7 −1 −9
	2373	G	33	11 7 3 −1 −5 −9
63	14551	G	65	15 −1 −17
	14343	G	65	15 11 7 3 −1 −5 −9 −13
	12471	H_3	64	15 7 −1 −9 −17
	1527	K_s	8	7 −1 −9
	133605	K_L	520	15 7 −1 −9 −17
65	10761		63	15 11 7 3 −1 −5 −9 −13
127	41567	G	129	15 −1 −17
255	231441	G	257 31	15 −1 −17
	264455	G	257 31,...	15 11 7 3 −1 −5 −9 −13 −17,... −29
	326161	H_{33}	256 31	15 −1 −17 −33
	267543	H_3	256 31	15 −1 −17 −33
	11367	K_s	16	15 −1 −17
	6031603	K_L	4111 31	15 −1 −17 −33

2. If $n \equiv 0 \bmod 4$, then:

$$K_L(u) = H_{t(n)}(u) \bigcup \left[\bigcup_{i=0}^{2^{n/2}-2} \{T^i w \oplus H_{t(n)}(u)\} \right]$$

$$\times \bigcup \{v^{(j)} \oplus T^k w : 0 \le j \le 2, \ 0 \le k < (2^{n/2} - 1)/3\}$$

(5.24)

where $v^{(j)}$ is the result of decimating $T^j u$ by $t(n)$ and $H_{t(n)}(u)$ is defined earlier by Equation (5.19). In either case, the correlation functions for $K_L(u)$ take on values in the set $\{-1, -t(n), t(n) - 2, -s(n), s(n) - 2\}$ and $\theta_{max} = t(n)$. If $n \equiv 2 \bmod 4$, $K_L(u)$ contains $2^{n/2}(2^n + 1)$ sequences, while if $n \equiv 0 \bmod 4$, $K_L(u)$ contains $2^{n/2}(2^n + 1) - 1$ sequences. The large set of Kasami sequences contains both the small set of Kasami sequences and a set of Gold (or Gold-like) sequences as subsets. More interestingly, the correlation bound $\theta_{max} = t(n)$ is the *same* as that for the latter subsets. The previous discussion is summarized in Table 5.2 for some examples of codes.

5.1.8   JPL Sequences

These sequences are constructed by combining sequence $S_1(t, T_c)$ $S_1(t, T_c)$ of length L_1 and $S_2(t, T_c)$ of length L_2 with L_1, L_2 prime, as $S = S(t, T_c) = S_1(t, T_c) \oplus S_2(t, T_c)$ of length $L = L_1 \times L_2$. If the composite sequence is delayed for L_1 chips

$$\begin{aligned} S(t - L_1 T_c, T_c) &= S_1(t - L_1 T_c, T_c) \oplus S_2(t - L_1 T_c, T_c) \\ &= S_1(t, T_c) \oplus S_2(t - L_1 T_c, T_c) \end{aligned}$$

(5.25)

and, summed up with its original version,

$$\begin{aligned} S(t, T_c) \oplus S(t - L_1 T_c, T_c) &= S_1(t, T_c) \oplus S_2(t, T_c) \oplus S_1(t - L_1 T_c, T_c) \oplus S_2(t - L_1 T_c, T_c) \\ &= S_1(t, T_c) \oplus S_1(t, T_c) \oplus S_2(t - L_1 T_c, T_c) \oplus S_2(t, T_c) \\ &= S_2(t - L_3 T_c, T_c) \end{aligned}$$

(5.26)

The result is only a component sequence S_2. In a similar way, by delaying the composite sequence for L_2 chips, a component sequence S_1 will be obtained. This can be used to synchronize sequence S of length $L_1 \times L_2$ by synchronizing separately component sequences S_1 and S_2 of length L_1 and L_2, which can be done much faster. The acquisition time is proportional to $T_{acq}(S) \sim \max[T_{acq}(S_1), T_{acq}(S_2)] \sim \max[L_1, L_2]$.

5.1.9 Kronecker Sequences

In this case the component sequences $S_1(t, T_{c1})$ of length L_1 and chip interval T_{c1}, and $S_2(t, T_{c2})$ with $L_2, T_{c2} = L_1 T_{c1}$ are combined as:

$$S(t, T_{c1}, T_{c2}) = S_1(t, T_{c1}) \oplus S_2(t, T_{c2}) \tag{5.27}$$

The composite sequence, S, synchronization is now performed in cascade, first S_1 with a much faster chip rate and then S_2. Correlation of S by S_1 gives:

$$F_2(S_1 \cdot S) = \rho_1 S_2 \tag{5.28}$$

and after that, this result is correlated with sequence S_2. The acquisition time is proportional to $T_{acq}(S) \sim T_{acq}(S_1) + T_{acq}(S_2) \sim L_1 + L_2$.

5.1.10 Walsh Functions

A Walsh function of order n can be defined recursively as follows:

$$\mathbf{W}(n) = \begin{bmatrix} \mathbf{W}(n/2), & \mathbf{W}(n/2) \\ \mathbf{W}(n/2), & \mathbf{W}'(n/2) \end{bmatrix} \tag{5.29}$$

$\mathbf{W}'$ denotes the logical complement of $\mathbf{W}$, and $\mathbf{W}(1) = [0]$. Thus,

$$\mathbf{W}(2) = \begin{bmatrix} 0, & 0 \\ 0, & 1 \end{bmatrix} \quad \text{and} \quad \mathbf{W}(4) = \begin{bmatrix} 0, & 0, & 0, & 0 \\ 0, & 1, & 0, & 1 \\ 0, & 0, & 1, & 1 \\ 0, & 1, & 1, & 0 \end{bmatrix} \tag{5.30}$$

$\mathbf{W}(8)$ is as follows:

$$\mathbf{W}(8) = \begin{bmatrix} 0, & 0, & 0, & 0, & 0, & 0, & 0, & 0 \\ 0, & 1, & 0, & 1, & 0, & 1, & 0, & 1 \\ 0, & 0, & 1, & 1, & 0, & 0, & 1, & 1 \\ 0, & 1, & 1, & 0, & 0, & 1, & 1, & 0 \\ 0, & 0, & 0, & 0, & 1, & 1, & 1, & 1 \\ 0, & 1, & 0, & 1, & 1, & 0, & 1, & 0 \\ 0, & 0, & 1, & 1, & 1, & 1, & 0, & 0 \\ 0, & 1, & 1, & 0, & 1, & 0, & 0, & 1 \end{bmatrix} \tag{5.31}$$

One can see that any two rows from the matrix

$$w_k(n) = \{w_{k,j}(n)\}; j = 1, \ldots n$$
$$w_m(n) = \{w_{m,j}(n)\}$$

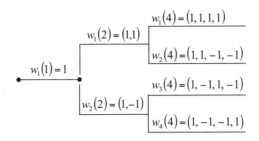

Figure 5.3   Flow graph generating OVSF codes of length 4.

represent the sequences whose bipolar versions have crosscorrelation equal to zero (orthogonal codes). This is valid as long as the codes are aligned as in the matrix.

A modification of the previous construction rule is shown in Figure 5.3, producing orthogonal variable speeding factor (OVSF) sequences. At each node of the graph, a code $w_k(n/2)$ of length $n/2$ is producing two new codes of length n by a rule

$$w_k(n/2) \rightarrow w_{2k-1}(n) = \{w_k(n/2), w_k(n/2)\}$$
$$\rightarrow w_{2k}(n) = \{w_k(n/2), -w_k(n/2)\}$$

5.1.11   Optimum PN Sequences

If we represent the information bit stream as

$$\{b_n\} = \cdots, b_{-1}, b_0, b_1, b_2, \ldots; \quad b_k = \pm 1 \tag{5.32}$$

and the sequence as a vector of chips

$$\mathbf{y} = (y_0, y_1, \ldots, y_{N-1}) y_k = \pm 1 \tag{5.33}$$

then the product of these two streams would create

$$\hat{\mathbf{y}}_i = \cdots; \quad b_{-1}\mathbf{y}; \quad b_0\mathbf{y}; \quad b_1\mathbf{y}; \quad \cdots \tag{5.34}$$

In other words, $\hat{\mathbf{y}}$ is the DSSS baseband signal which has as its ith element $\hat{y}_i = b_n y_k$ for all i such that $i = nN + k$ for k in the range $0 \le k \le N - 1$. A synchronous correlation receiver forms the inner product

$$\langle \hat{\mathbf{y}}_n, \mathbf{y} \rangle = b_n \langle \hat{\mathbf{y}}, \mathbf{y} \rangle = b_n \theta_y(0) \tag{5.35}$$

If the other signal is $\hat{\mathbf{x}}$ which is formed from the data sequence $\{b'_n\}$ and the signature sequence x (generated by a binary vector $\mathbf{x} = (x_0, x_1, \ldots, x_{N-1})$ in exactly the same manner as $\hat{\mathbf{y}}$ was formed from $\{b_n\}$ and $\mathbf{y}$, then we have for the overall received signal:

$$\hat{\mathbf{y}} + T^{-l}\hat{\mathbf{x}} \quad \text{where}$$
$$\hat{\mathbf{x}} = \cdots; \quad b'_-\mathbf{x}; \quad b'_0\mathbf{x}; \quad b'_1\mathbf{x}; \quad \cdots \tag{5.36}$$

The output of a correlation receiver which is in synchronism with $\mathbf{y}$ is given by:

$$\mathbf{z}_n = \langle \hat{\mathbf{y}}_n, \mathbf{y} \rangle + \left[b'_{n-1} \sum_{i=0}^{l-1} x_{N-l+i} y_i + b'_n \sum_{i=l}^{N-1} x_{i-l} y_i \right] \tag{5.37}$$

Having in mind the following relations:

$$\sum_{i=0}^{l-1} x_{N-l+i} y_i = \sum_{i=0}^{N-1+m} x_{i-m} y_i$$
$$\sum_{i=l}^{N-1} x_{i-l} y_i = \sum_{j=0}^{N-1-l} x_j y_{j+l} \tag{5.38}$$

and the definition of aperiodic crosscorrelation function $C_{x,y}$

$$C_{x,y}(l) = \begin{cases} \displaystyle\sum_{j=0}^{N-1-l} x_j y^*_{j+l}, & 0 \le l \le N-1 \\[2mm] \displaystyle\sum_{j=0}^{N-1+l} x_{j-l} y^*_j, & 1-N \le l < 0 \\[2mm] 0, & |l| \ge N \end{cases} \tag{5.39}$$

Equation (5.37) becomes

$$\mathbf{z}_n = b_n \theta_y(0) + [b'_{n-1} C_{x,y}(l-N) + b'_n C_{x,y}(l)] \tag{5.40}$$

The optimum sequences should minimize the interfering term for all values of l. Further details may be found in [14, 15, 27–40].

5.1.12 Golay Code

As an appendix to this section, we will present Golay code [27] which is used in the WUMTS syncho channel due to its good aperiodic correlation properties. Additional information on the construction and implementation of these sequences can be found in [27–40].

In general for the two complementary sequences a and b of length N, with aperiodic autocorrelations $A(k)$ and $B(k)$ (delay k) we have:

$$A(k) + B(k) = 0, k \ne 0 \quad \text{and} \quad A(0) + B(0) = 2N$$

The primary synchronization code (PSC), C_{psc} is constructed as a so-called generalized hierarchical Golay sequence. Define:

$$a = \langle x_1, x_2, x_3, \ldots, x_{16} \rangle = \langle 1, 1, 1, 1, 1, 1, -1, -1, 1, -1, 1, -1, 1, -1, -1, 1 \rangle$$

The PSC is generated by repeating the sequence modulated by a Golay complementary sequence, and creating a complex-valued sequence with identical real and imaginary components. The PSC C_{psc} is defined as:

$$C_{psc} = (1+j) \times \langle a, a, a, -a, -a, a, -a, -a, a, a, a, -a, a, -a, a, a \rangle;$$

where the leftmost chip in the sequence corresponds to the chip transmitted first in time. The 16 secondary synchronization codes (SSCs), $\{C_{ssc,1}, \ldots, C_{ssc,16}\}$, are complex-valued with identical real and imaginary components, and are constructed from position wise multiplication of a Hadamard sequence and a sequence z, defined as:

$$z = \langle b, b, b, -b, b, b, -b, -b, b, -b, b, -b, -b, -b, -b, -b \rangle, \quad \text{where}$$

$$b = \langle x_1, x_2, x_3, x_4, x_5, x_6, x_7, x_8, -x_9, -x_{10}, -x_{11}, -x_{12}, -x_{13}, -x_{14}, -x_{15,} - x_{16} \rangle$$

and $x_1, x_2, \ldots, x_{15}, x_{16,}$ are same as in the definition of the sequence a above.

The Hadamard sequences (see also Equations (5.29–5.31)) are obtained as the rows in a matrix $\mathbf{H}_8$ constructed recursively by:

$$\mathbf{H}_0 = [1]$$

$$\mathbf{H}_k = \begin{bmatrix} \mathbf{H}_{k-1} & \mathbf{H}_{k-1} \\ \mathbf{H}_{k-1} & -\mathbf{H}_{k-1} \end{bmatrix}, \quad k \geq 1$$

The rows are numbered from the top starting with row 0 (the all ones sequence). Denote the nth Hadamard sequence as a row of $\mathbf{H}_8$ numbered from the top, $n = 0, 1, 2, \ldots, 255$, subsequently. Furthermore, let $h_n(i)$ and $z(i)$ denote the ith symbol of the sequence h_n and z, respectively where $i = 0, 1, 2, \ldots, 255$ and $i = 0$ corresponds to the leftmost symbol. The kth SSC, $C_{ssc,k}, k = 1, 2, 3, \ldots, 16$ is then defined as:

$$-C_{ssc,k} = (1 + j) \times \langle h_m(0) \times z(0), h_m(1) \times z(1), h_m(2)$$
$$\times z(2), \ldots, h_m(255) \times z(255) \rangle;$$

where $m = 16 \times (k-1)$ and the leftmost chip in the sequence corresponds to the chip transmitted first in time.

5.1.12.1 Alternative Generation

The generalized hierarchical Golay sequences for the PSC described above may also be viewed as being generated (in real-valued representation) by the following methods:

Method 1: The sequence y is constructed from two constituent sequences x_1 and x_2 of length n_1 and n_2 respectively, using the following formula:

$$y(i) = x_2(i \bmod n_2) \, x_1(i \text{ div } n_2), i = 0 \cdots (n_1 n_2) - 1$$

The constituent sequences x_1 and x_2 are chosen to be the length 16 (i.e. $n_1 = n_2 = 16$) sequences:

- $x1$ is defined to be the length 16 ($N^{(1)} = 4$) Golay complementary sequence [27] obtained by the delay matrix $\mathbf{D}^{(1)} = [8, 4, 1, 2]$ and weight matrix

$$\mathbf{W}^{(1)} = [1, \quad -1, \quad 1, \quad 1].$$

- x_2 is a generalized hierarchical sequence using the following formula, selecting $s = 2$ and using the two Golay complementary sequences x_3 and x_4 as constituent sequences. The lengths of the sequences x_3 and x_4 are called n_3 and n_4 respectively.
- $x_2(i) = x_4(i \bmod s + s(i \text{ div } sn_3)) \, x_3((i \text{ div } s) \bmod n_3), i = 0 \cdots (n_3 \, n_4) - 1.$
- x_3 and x_4 are defined to be identical and the length 4 ($N^{(3)} = N^{(4)} = 2$) Golay complementary sequence obtained by the delay matrix $\mathbf{D}^{(3)} = \mathbf{D}^{(4)} = [1, 2]$ and weight matrix $\mathbf{W}^{(3)} = \mathbf{W}^{(4)} = [1, 1]$.

The Golay complementary sequences x_1, x_3 and x_4 are defined using the following recursive relation:

$$a_0(k) = \delta(k) \quad \text{and} \quad b_0(k) = \delta(k);$$
$$a_n(k) = a_{n-1}(k) + W_n^{(j)} \cdot b_{n-1}(k - D_n^{(j)});$$
$$b_n(k) = a_{n-1}(k) - W_n^{(j)} \cdot b_{n-1}(k - D_n^{(j)});$$
$$k = 0, 1, 2, \ldots, 2^{**}N^{(j)} - 1; \; n = 1, 2, \ldots, N^{(j)}$$

The desired Golay complementary sequence x_j is defined by a_n assuming $n = N^{(j)}$. The Kronecker delta function is described by δ; k, j and n are integers.

Method 2: The sequence y can be viewed as a pruned Golay complementary sequence, generated using the following parameters which apply to the generator equations for a and b above:

- Let $j = 0$, $N^{(0)} = 8$.
- $[D_1^0, D_2^0, D_3^0, D_4^0, D_5^0, D_6^0, D_7^0, D_8^0] = [128, 64, 16, 32, 8, 1, 4, 2]$.
- $[W_1^0, W_2^0, W_3^0, W_4^0, W_5^0, W_6^0, W_7^0, W_8^0] = [1, -1, 1, 1, 1, 1, 1, 1]$.
- For $n = 4, 6$, set $b_4(k) = a_4(k)$, $b_6(k) = a_6(k)$.

5.2 Multiuser CDMA Receivers

In this section we present a number of methods for CDMA multiple access interference cancellation. Multiple access interference is produced by the presence of the other users in the network which are located on the same bandwidth as our own signal. The common characteristic of all these schemes is some form of joint signal and parameter estimation for all signals present on the same bandwidth. It makes sense to implement this in a base station of a cellular system because all these signals are available there anyway. At the same time this concept will considerably increase the complexity of the receiver. Although very complex, these schemes are being standardized already because they offer significantly better performance.

If user k transmits bit stream b_k, with bit interval T, using spreading sequence s_k, then the low pass equivalent of the overall signal received in the base station can be represented as [41, 42]

$$r_t = S_t(\mathbf{b}) + \sigma n(t) \tag{5.41}$$

$$S_t(\mathbf{b}) = \sum_{i=-M}^{M} \sum_{k=1}^{K} b_k(i) s_k(t - iT - \tau_k) \tag{5.42}$$

where K is the number of users, $\mathbf{b} = (b_1, b_2, \ldots b_K)^T$ is the vector of bits of all users and the signal is observed in time interval [–MT, MT]. The noise component is represented by the second term of Equation (5.41) and τ_k is the delay of the signal from user k.

5.2.1 Synchronous CDMA Channels

If the signals from different users are received synchronously, Equation (5.41) becomes:

$$r(t) = \sum_{k=1}^{K} b_k(j) s_k(t - jT) + \sigma n(t), \quad t \in [jT, jT + T] \tag{5.43}$$

If we use notation y_k for the output of the matched filter of user k then we have

$$y_k = \int_0^T r(t) s_k(t) \mathrm{d}t, \quad k = 1, \ldots, K \tag{5.44}$$

and we can write

$$y_1 = \sum_j b_k R_{1j} + n,$$

$$y_2 = \sum_j b_k R_{2j} + n_2 \tag{5.45}$$

$$\vdots$$

$$y_k = \sum_j b_k R_{kj} + n_k$$

The vector form of these outputs can be presented as:

$$\mathbf{y} = \mathbf{Rb} + \mathbf{n} \tag{5.46}$$

where $\mathbf{R}$ is the non-negative definite matrix of crosscorrelations between the assigned waveforms:

$$R_{ij} = \int_0^T s_i(t) s_j(t) \, dt \tag{5.47}$$

Conventional single user detection can be represented as:

$$\hat{b}_k^c = \text{sgn } y_k \tag{5.48}$$

The optimum multiuser detector becomes:

$$\hat{b} \in \arg \min_{b \in \{1,1\}^K} \int_0^T \left[r(t) - \sum_{k=1}^K b_k s_k(t) \right]^2 dt \tag{5.49}$$

$$= \arg \max_{b \in \{-1,1\}^K} 2\mathbf{y}^T \mathbf{b} - \mathbf{b}^T \mathbf{Rb}$$

5.2.2   The Decorrelating Detector

In the absence of noise, the matched filter output vector is $\mathbf{y} = \mathbf{Rb}$. This suggests that the detector should perform the following operation $\hat{b} = \text{sgn } \mathbf{R}^{-1}\mathbf{y}$. Note that the noise components in $\mathbf{R}^{-1}\mathbf{y}$ are correlated, and therefore sgn $\mathbf{R}^{-1}\mathbf{y}$ does not result in optimum decisions. It is interesting to point out that this detector does not require knowledge of the energies of any of the active users.

5.2.3   The Optimum Linear Multiuser Detector

The linear detector which minimizes the probability of bit error will be referred to as the optimum linear multiuser detector. Its operation can be represented as:

$$\hat{\mathbf{b}} = \text{sgn}\,(\mathbf{Ty}) = \text{sgn}\,(\mathbf{TRb} + \mathbf{Tn}) \tag{5.50}$$

We will consider the set $I(\mathbf{R})$ of generalized inverses of the crosscorrelation matrix $\mathbf{R}$ and analyze the properties of the detector:

$$\hat{\mathbf{b}} = \text{sgn}\,\mathbf{R}^I\mathbf{y} \tag{5.51}$$

in Section 5.3. The special case $I(\mathbf{R}) = \mathbf{R}^{-1}$ is referred to as a decorrelating detector.

5.2.4   Multistage Detection in Asynchronous CDMA [43]

If the indexing of users is arranged in increasing order of their delays, then the output of the correlator of user k can be represented as:

$$z_k^{(i)}(0) = \int_{-\infty}^{\infty} r(t)s_k(t + iT - t_k)\,dt$$

$$= \eta_k^{(i)} + \sum_{l=k+1}^{K} R_{kl}(1)b_l^{(i-1)} + \sum_{l=1}^{K} R_{kl}(0)b_l^{(i)} + \sum_{l=1}^{k-1} R_{kl}(-1)b_l^{(i+1)} \qquad (5.52)$$

where $\eta_k^{(i)}$ is the component of the statistic due to the additive channel noise. In vector notation, letting $z^{(i)}(0) = [z_1^{(i)}(0), z_2^{(i)}(0), \ldots, z_K^{(k)}(0)]^{\mathsf{T}}$, we have:

$$z^{(i)}(0) = \eta^{(i)} + \mathbf{R}(1)\mathbf{b}^{(i-1)} + \mathbf{R}(0)\mathbf{b}^{(i)} + \mathbf{R}(-1)\mathbf{b}^{(i+1)} \qquad (5.53)$$

The multistage detector recreates the interfering term for each user based on bit estimations in the previous stage (iteration), subtracts the estimated MAI and then makes the new estimate of data which can be represented as:

$$\hat{b}_k^{(i)}(m + 1) = \mathrm{sgn}[z_k^{(i)}(m)] \qquad (5.54)$$

where

$$z_k^{(i)}(m) = z_k^{(i)}(0) - \sum_{l=k+1}^{K} h_{kl}(1)\hat{b}_l^{(i-1)}(m) - \sum_{l\neq k} h_{kl}(0)\hat{b}_l^{(i)}(m) - \sum_{l=1}^{k-1} h_{kl}(-1)\hat{b}_l^{(i+1)}(m) \qquad (5.55)$$

Examples of probability of error curves are shown in Figure 5.4. All parameters are shown in the figure itself. One can see that even a two-stage detector may significantly improve the system performance. In

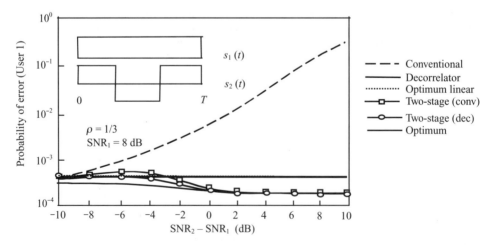

Figure 5.4   Error probability comparison of the linear, two-stage and optimum detectors for a two-user channel with $r_{12} = 1/3$ and signal to noise ratio of user 1 fixed at 8 dB [44] © 1991, IEEE.

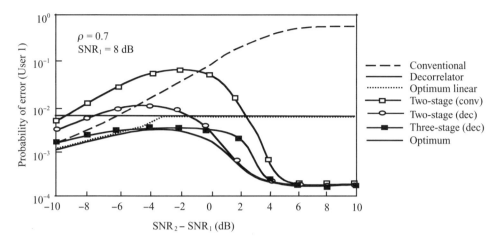

Figure 5.5   Error probability comparison for a two-user channel with $r_{12} = 0.7$ and signal to noise ratio of user 1 fixed at 8 dB [44] © 1991, IEEE.

order to further emphasize the role of MUD in the presence of the near–far effect, Figure 5.4 presents the BER for the case when the crosscorrelation is very high $r_{12} = 1/3$ (three chips long sequences). One can see that when the second user becomes stronger and stronger, the improvement compared with a conventional detector is more significant.

This conclusion becomes more and more relevant if either r_{12} is increased, as in Figure 5.5, or signal to noise ratio is increased, as in Figure 5.6. Figure 5.7 demonstrates the same results for five users in the network.

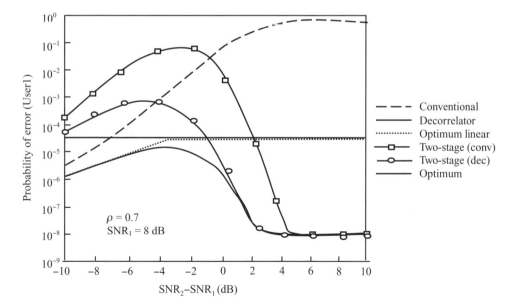

Figure 5.6   Error probability comparison of the linear, two-stage and optimum detectors for a two-user channel with $r_{12} = 0.7$ and signal to noise ratio of user 1 fixed at 12 dB © 1991, IEEE.

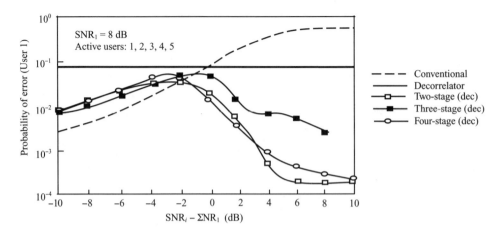

Figure 5.7   Probability of error, five users in the network [44] © 1991, IEEE.

5.2.5   Non-Coherent Detector

A conventional detector for differential phase keying signals is defined by the following equation:

$$\hat{b}_m = \text{sgn}[\text{Re}\{\overline{z_m(-1)}z_m(0)\}]$$

$$z_m(i) = \frac{1}{2}\int_{iT}^{(i+1)T} r(t)\overline{f_m(t-iT)}dt \tag{5.56}$$

where $f_m(t)$ is the signal matched filter function. In the trivial case it is the signal spreading code only.

In general, a non-coherent linear multiuser detector for the mth user, denoted by a non-zero transformation $\mathbf{h}^{(m)} \in C^K$, is defined by the decision:

$$\hat{b}_m = \text{sgn}\left[\text{Re}\left\{\overline{\sum_{k=1}^{K}\bar{h}_k^{(m)}z_k(-1)}\sum_{l=1}^{K}\bar{h}_l^{(m)}z_l(0)\right\}\right] \tag{5.57}$$

where K is the length of the code. A non-coherent decorrelating detector for user m is defined by the decision with the linear transformation $\mathbf{h} = \mathbf{d}$, where $\mathbf{d}$ denotes the complex conjugate of the mth column of a generalized inverse $\mathbf{R}^I$ of $\mathbf{R}$. If the mth user is linearly independent, it can be shown that $\mathbf{R}\bar{d} = \mathbf{u}_m$, the mth unit vector. If all the signature signals are linearly independent, the $\mathbf{R}^{-1}$ exists and the decorrelating transformation $\mathbf{d}$ is uniquely characterized as the complex conjugate of the mth column of the inverse of $\mathbf{R}$. The receiver block diagram is shown in Figure 5.8.

5.2.6   Non-Coherent Detection in Asynchronous Multiuser Channels [45]

The z-transform of Equation (5.53) gives:

$$\mathbf{Z}(z) = \mathbf{S}(z) \cdot \hat{\mathbf{D}}(z) + \mathbf{N}(z) \tag{5.58}$$

where

$$\mathbf{S}(z) = \mathbf{R}(-1)z + \mathbf{R}(0) + \mathbf{R}(1)z^{-1} \tag{5.59}$$

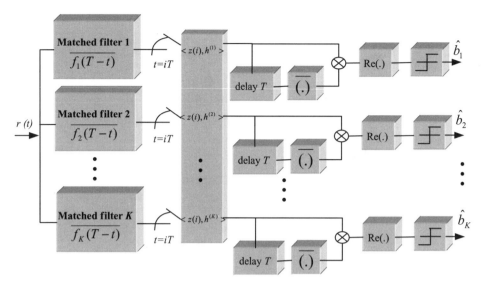

Figure 5.8 Linear multiuser DPSK detector.

and $\mathbf{Z}(z)$, $\hat{\mathbf{D}}(z)$ and $\mathbf{N}(z)$ are the vector-valued z-transforms of the matched-filter output sequence, the sequence $\{\hat{d}(l) = A(l)d(l)\} = A(l)d(l)$ and the noise sequence $\{n(l)\}$ at the output of the matched filters. If we define

$$\mathbf{G}(z) = [\mathbf{S}(z)]^{-1} = \frac{\text{adj}\,\mathbf{S}(z)}{\det \mathbf{S}(z)} \tag{5.60}$$

then we have

$$\hat{\mathbf{d}}(z) = \mathbf{G}(z)\mathbf{Z}(z) \tag{5.61}$$

and

$$\hat{\mathbf{b}}(i) = \text{sgn}\,\text{Re}[\tilde{\mathbf{d}}(i-1) \otimes \tilde{\mathbf{d}}(i)] \tag{5.62}$$

The system block diagram is shown in Figure 5.9.

5.2.7 Multiuser Detection in Frequency Non-Selective Rayleigh Fading Channels

Topics covered in the previous chapter are now repeated for the fading channel. Previously described algorithms are extended to the fading channel by using as much analogy as possible in the process of deriving the system transfer functions. In frequency selective channels, decorrelators are combined with the RAKE-type receiver in order to further improve the system performance. A number of simulation results are presented in order to illustrate the effectiveness of these schemes. The concept of this chapter is based on proper understanding of the channel model, covered in Chapter 14. The overall system model, including the channel model for frequency non-selective fading, is shown in Figure 5.10.

Parameters $c_k(i)$ are, for fixed i, independent, zero mean, complex-valued Gaussian random variables, with variances $\overline{|c_k|^2}$ with independent quadrature components. The time-varying nature of the channel is

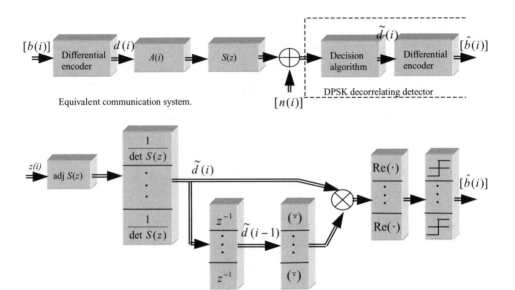

Figure 5.9   Non-coherent decorrelating detector.

described via the spaced time correlation function of the kth channel $\Phi_k(\Delta t)$:

$$E\{c_k^*(i)c_k(j)\} = \Phi_k((j-i)T) \qquad (5.63)$$

The received signal at the central receiver can be expressed as:

$$r(t) = S(t, \mathbf{b}) + n(t)$$

$$S(t, \mathbf{b}) = \sum_{i=-M}^{M}\sum_{k=1}^{K} b_k(i)c_k(i)u_k(t - iT - \tau_k) \qquad (5.64)$$

$$u_k(t) = \sqrt{E_k}s_k(t)e^{j\phi_k}$$

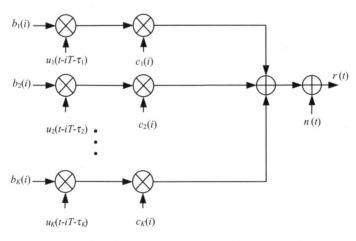

Figure 5.10   Asynchronous CDMA flat Rayleigh fading channel model.

where $u_k(t)$ is referred to as the user k signature sequence, and includes the signal amplitude (square root of signal energy), the code itself and the signal phase. By using proper notation, $r(t)$ can be represented as:

$$r(t) = \mathbf{b}^T \mathbf{C} \mathbf{u}_t + n(t) \tag{5.65}$$

where

$$\begin{aligned}
&\mathbf{b}^T [b_1(-M) b_2(-M) \cdots b_K(-M) \cdots b_1(M) b_2(M) \cdots b_K(M)] \\
&\mathbf{u}_t = [\mathbf{u}^T(t+MT) \cdots \mathbf{u}^T(t-MT)]^T \\
&\mathbf{u}(t) = [u_1(t-\tau_1) \cdots u_K(t-\tau_K)]^T \\
&\mathbf{C} = \mathrm{diag}(\mathbf{C}(-M) \cdots \mathbf{C}(M)) \\
&\mathbf{C}(i) = \mathrm{diag}(c_1(i) \cdots c_K(i))
\end{aligned} \tag{5.66}$$

5.2.7.1 Multiuser Maximum Likelihood Sequence Detection

By using analogy from the previous section, the likelihood function in this case can be represented as:

$$L(\mathbf{b}) = 2\,\mathrm{Re}\{\mathbf{b}^H \mathbf{y}\} - \mathbf{b}^H \mathbf{C}^H \mathbf{R}_u \mathbf{C} \mathbf{b} \tag{5.67}$$

Upper index $(\ )^H$ denotes the conjugate transpose and

$$\mathbf{y} = \int_{-\infty}^{+\infty} r(t) \mathbf{C}^H \mathbf{u}_t^* \, dt \tag{5.68}$$

represents the vector of matched filter outputs. The correlation matrix $\mathbf{R}_u$ can be represented as

$$\mathbf{R}_u = \int_{-\infty}^{+\infty} \mathbf{u}_t^* \mathbf{u}_t^T dt = \begin{bmatrix} \mathbf{R}_u(0) & \mathbf{R}_u(-1) & 0 & \cdots \\ \mathbf{R}_u(1) & \mathbf{R}_u(0) & \mathbf{R}_u(-1) & \cdots \\ & \ddots & \ddots & \\ \cdots & \mathbf{R}_u(1) & \mathbf{R}_u(0) & \mathbf{R}_u(-1) \\ \cdots & 0 & \mathbf{R}_u(1) & \mathbf{R}_u(0) \end{bmatrix} \tag{5.69}$$

with block elements of dimension $K \times K$

$$\mathbf{R}_u(i-j) = \int_{-\infty}^{+\infty} \mathbf{u}^*(t-iT) u^T(t-jT) \, dt \tag{5.70}$$

and scalar elements

$$[\mathbf{R}_u(i-j)]_{mn} = \int_{-\infty}^{+\infty} u_m^*(t-iT-\tau_m) u_n(t-jT-\tau_n) \, dt \tag{5.71}$$

5.2.7.2 Decorrelating Detector

If we slightly modify the vector notation, Equation (5.65) becomes:

$$r(t) = \sum_{i=-M}^{M} \mathbf{s}^T(t-iT) \mathbf{E} \boldsymbol{\Phi} \mathbf{C}(i) \mathbf{b}(i) + n(t) \tag{5.72}$$

with normalized signature waveform vector:

$$\mathbf{s}(t) = [s_1(t - \tau_1)s_2(t - \tau_2) \cdots s_K(t - \tau_K)]^\mathrm{T} \tag{5.73}$$

$K \times K$ multichannel matrix

$$\begin{aligned} \mathbf{C}(i) &= \mathrm{diag}\ (c_1(i)c_2(i) \cdots c_K(i)) \\ \mathbf{E} &= \mathrm{diag}(\sqrt{E_1}\sqrt{E_2} \cdots \sqrt{E_K}) \end{aligned} \tag{5.74}$$

and matrix of carrier phases

$$\mathbf{\Phi} = \mathrm{diag}\ (e^{j\phi_1} e^{j\phi_2} \cdots e^{j\phi_K}) \tag{5.75}$$

The $K \times K$ crosscorrelation matrix of normalized signature waveforms becomes

$$\mathbf{R}(\ell) = \int_{-\infty}^{+\infty} \mathbf{s}^*(t)\mathbf{s}^\mathrm{T}(t + \ell T)\, dt \tag{5.76}$$

The asynchronous nature of the channel is evident from the matrix elements

$$R_{mn}(\ell) = \int_{\ell T + \tau_m}^{(\ell+1)T + t_m} s_m^*(t - t_m)s_n(t + \ell T - \tau_n)\, dt \tag{5.77}$$

Since there is no inter-symbol interference, $\mathbf{R}(\ell) = 0$, $\forall |\ell| > 1$ and $\mathbf{R}(-1) = \mathbf{R}^\mathrm{H}$ (1). Due to the ordering of the user $\mathbf{R}^\mathrm{H}(1)$ is an upper triangular matrix with zero elements on the diagonal. The decorrelating detector front end consists of K filters matched to the normalized signature waveforms of the users. The output of this filter bank, sampled at the ℓ-*th* bit epoch is:

$$\mathbf{y}(\ell) = \int_{-\infty}^{+\infty} r(t)\mathbf{s}(t - \ell T)\, dt \tag{5.78}$$

The vector of sufficient statistics can also be represented as:

$$\begin{aligned} \mathbf{y}(\ell) = {}&\mathbf{R}(-1)\mathbf{E}\mathbf{\Phi}\ \ \mathbf{C}(\ell + 1)\mathbf{b}(\ell + 1) + \mathbf{R}(0)\mathbf{E}\mathbf{\Phi}\ \ \mathbf{C}(\ell)\mathbf{b}(\ell) \\ &+ \mathbf{R}(1)\mathbf{E}\mathbf{\Phi}\mathbf{C}(\ell - 1)\mathbf{b}(\ell - 1) + \mathbf{n}_y(\ell) \end{aligned} \tag{5.79}$$

The covariance matrix of the matched filter output noise vector sequence, $\{\mathbf{n}_y(\ell)\}$ is given by:

$$E\{\mathbf{n}_y^*(i)\mathbf{n}_y^\mathrm{T}(j)\} = \sigma^2 \mathbf{R}^*(i - j) \tag{5.80}$$

As in Equation (5.60) the decorrelator is a K-input K-output linear time-invariant filter with transfer function matrix:

$$\mathbf{G}(z) = [\mathbf{R}(-1)z + \mathbf{R}(0) + \mathbf{R}(1)z^{-1}]^{-1} \triangleq \mathbf{S}^{-1}(z) \tag{5.81}$$

The z-transform of the decorrelator output vector is:

$$\mathbf{P}(z) = \mathbf{E}\mathbf{\Phi}(\mathbf{C}\mathbf{b})(z) + N_p(z) \tag{5.82}$$

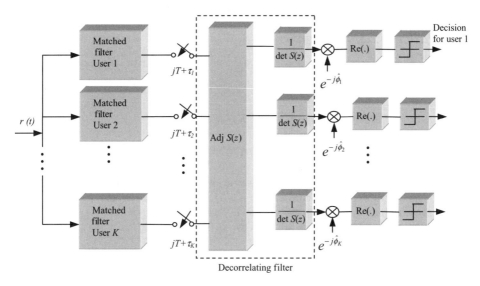

Figure 5.11 Coherent decorrelating multiuser detector.

where $N_p(z)$ is the z-transform of the output noise vector sequence having power spectral density

$$\sigma^2 \mathbf{S}^{-1}(z) = \sigma^2 \sum_{m=-\infty}^{\infty} \mathbf{D}(m) z^{-m} \qquad (5.83)$$

The receiver block diagram for coherent reception is shown in Figure 5.11. Performance results for the detector are shown in Figure 5.12. Significant improvement in the BER is evident.

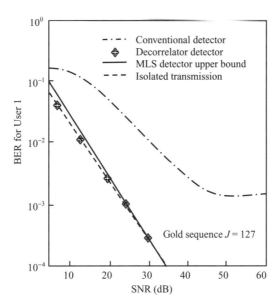

Figure 5.12 Bit error rate of user 1 for the two-user case with Rayleigh faded paths (same average path strength) and Gold sequences of period $J = 127$ [46].

5.2.8   Multiuser Detection in Frequency Selective
Rayleigh Fading Channels

By using analogy with Equation (5.64) the received signal in this case can be represented as:

$$r(t) = S(t, \mathbf{b}) + n(t)$$

$$S(t, \mathbf{b}) = \sum_{i=-M}^{M} \sum_{k=1}^{K} b_k(i) h_k(t - iT - \tau_k) \tag{5.84}$$

$$h_k(t) = c_k(t)^* u_k(t)$$

In Equation (5.84), $h_k(t)$ is the equivalent received symbol waveform of finite duration $[0, T_k]$ (convolution of equivalent low pass signature waveform $u_k(t)$ and the channel impulse response $c_k(t)$). We define the memory of this channel as v, the smallest integer such that $h_k(t) = 0$ for $t > (v + 1)T$, and all $k = 1 \cdots K$. The impulse response of the kth user channel is given by:

$$c_k(t) = \sum_{\ell=0}^{L-1} c_{k,l}(t) \delta(t - \tau_{k,\ell}) \tag{5.85}$$

When the signaling interval T is much smaller than the coherence time of the channel, the channel is characterized as slow fading, implying that the channel characteristics can be measured accurately. Since the channel is assumed to be Rayleigh fading, the coefficients $c_{k,\ell}(t)$ are modeled as independent zero mean complex-valued Gaussian random processes. We will use the following notation:

$$h_k(t) = \sum_{\ell=0}^{L-1} c_{k,l}(t) u_k(t - \tau_{k,l}) = \mathbf{c}_k^{\mathrm{T}}(t) \mathbf{u}_k(t) \tag{5.86}$$

For the single user vector of channel coefficients we use:

$$\mathbf{c}_k(t) = [c_{k,0}(t), c_{k,1}(t) \cdots c_{k,L-1}(t)]^{\mathrm{T}} \tag{5.87}$$

and for the signal vector of the delayed signature waveform:

$$\mathbf{u}_k(t) = [u_k(t - \tau_{k,0}), u_k(t - \tau_{k,1}) \cdots u_k(t - \tau_{k,L-1})]^{\mathrm{T}} \tag{5.88}$$

The equivalent low pass signature waveform is represented as

$$u_k(t) = \sqrt{E_k} s_k(t) e^{j\phi_k} \tag{5.89}$$

where E_k is the energy, $s_k(t)$ is the real-valued, unit-energy signature waveform with period T and ϕ_k is the carrier phase. In this case the received signal given by Equation (5.84) becomes:

$$r(t) = S(t, \mathbf{b}) + n(t) = \mathbf{b}^{\mathrm{T}} \mathbf{h}_t + n(t) \tag{5.90}$$

The equivalent data sequence is as in Equation (5.66):

$$\mathbf{b} = [b_1(-M) \cdots b_K(-M) \cdots b_1(M) \cdots b_K(M)]^{\mathrm{T}} \tag{5.91}$$

The equivalent waveform vector of NK elements is:

$$\mathbf{h}_t = [\mathbf{h}^{\mathrm{T}}(t + MT) \cdots \mathbf{h}^{\mathrm{T}}(t - MT)]^{\mathrm{T}} \tag{5.92}$$

with

$$\mathbf{h}(t) = [h_1(t - t_1) \cdots h_K(t - \tau_K)]^{\mathrm{T}} = \mathbf{C}^{\mathrm{T}}(t)\mathbf{u}(t) \tag{5.93}$$

where

$$\mathbf{C}(t) = \begin{bmatrix} \mathbf{c}_1(t) & 0 & 0 & \cdots \\ & \mathbf{c}_2(t) & 0 & \cdots \\ & & \ddots & \\ \cdots & 0 & 0 & \mathbf{c}_K(t) \end{bmatrix} \tag{5.94}$$

is a $KL \times K$ multichannel matrix. KL is the total number of fading paths for all K users and:

$$\mathbf{u}(t) = [\mathbf{u}_1(t - t_1) \cdots \mathbf{u}_K(t - \tau_K)]^T \tag{5.95}$$

is the equivalent signature vector of KL elements.

5.2.8.1 Multiuser Maximum Likelihood Sequence Detection

The log likelihood function in this case becomes:

$$L(\mathbf{b}) = 2\,\mathrm{Re}\{\mathbf{b}^{\mathrm{H}}\mathbf{y}\} - \mathbf{b}^{\mathrm{H}}\mathbf{H}\mathbf{b} \tag{5.96}$$

where superscript H denotes the conjugate transpose,

$$\mathbf{y} = \int_{-\infty}^{+\infty} r(t)\mathbf{h}_t^* \, \mathrm{d}t \tag{5.97}$$

is the output of the bank of matched filters sampled at the bit epoch of the users. Matrix $\mathbf{H}$ is an $N \times N$ block Toeplitz crosscorrelation waveform matrix with $K \times K$ block elements,

$$\mathbf{H}(i - j) = \int_{-\infty}^{+\infty} \mathbf{h}^*(t - iT)\mathbf{h}^{\mathrm{T}}(t - jT) \, \mathrm{d}t \tag{5.98}$$

5.2.8.2 Viterbi Algorithm

Since every waveform $h_k(t)$ is time limited to $[0, T_k]$, $T_k < (v + 1)T$, it follows that $\mathbf{H}(l) = 0, \forall |l| > v + 1$ and $\mathbf{H}(j) = \mathbf{H}^{\mathrm{H}}(j)$ for $j = 1 \cdots v + 1$. Due to the ordering of the users $\mathbf{H}^{\mathrm{H}}(v + 1)$ is an upper triangular matrix with zero elements on the diagonal. Provided that knowledge of a channel is available, the MLS detector may be implemented as a dynamic programming algorithm of the Viterbi type. The vector Viterbi algorithm is the modification of the one introduced for M-input M-output linear channels where the dimensionality of the state space is $2^{(v + 1)K}$. As in the case of the AWGN channel, a more efficient decomposition of the likelihood function results in an algorithm with a state space of dimension $2^{(v + 1)K - 1}$.

Frequency selective fading is described by the wide-sense stationary uncorrelated scattering model. The bandwidth of each signature waveform is much larger than the coherence bandwidth of the channel, $B_{\mathrm{w}} \gg (\Delta f)_{\mathrm{c}}$. The time varying frequency selective channel for each user can be represented as a tapped delay line with tap spacing $1/B_w$, so that Equation (5.86) becomes:

$$h_k(t) = \sum_{i=0}^{L-1} c_{k,i}(t)u_k\left(t - \frac{i}{B_w}\right)$$

$$= \mathbf{s}_k^T(t)\mathbf{E}_k\boldsymbol{\Phi}_k\mathbf{c}_k(t)$$

(5.99)

The signature waveform vector may be described as:

$$\mathbf{s}_k(t) = \left[s_k(t), s_k\left(t - \frac{i}{B_w}\right)\cdots s_k\left(t - \frac{L-i}{B_w}\right)\right]^T$$

(5.100)

and

$$\mathbf{E}_k = \sqrt{E_k}\mathbf{I}_L$$
$$\boldsymbol{\Phi}_k = e^{j\phi_k}\mathbf{I}_L$$

(5.101)

For a data symbol duration much longer than the multipath delay spread, $T \gg T_m$, any inter-symbol interference due to channel dispersion can be neglected. Based on the above discussion, the channel model is presented in Figure 5.13. If we use notation

$$\mathbf{b}(i) = [b_1(i)b_2(i)\cdots b_K(i)]^T, \quad i = -M\cdots M$$
$$\mathbf{s}(t) = [\mathbf{s}_1^T(t - \tau_1)\mathbf{s}_2^T(t - \tau_2)\cdots \mathbf{s}_K^T(t - \tau_K)]^T$$
$$\mathbf{E} = \text{diag}(\mathbf{E}_1, \mathbf{E}_2, \dots \mathbf{E}_K)$$
$$\boldsymbol{\Phi} = \text{diag}(\boldsymbol{\Phi}_1, \boldsymbol{\Phi}_2, \dots \boldsymbol{\Phi}_K)$$
$$\mathbf{h}^T(t) = [h_1(t - \tau_1)\cdots h_K(t - \tau_K)] = \mathbf{s}^T(t)\mathbf{E}\boldsymbol{\Phi}\mathbf{C}(t)$$

(5.102)

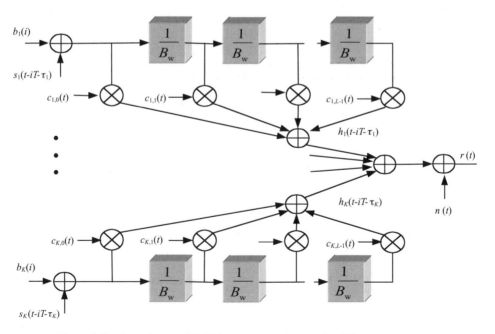

Figure 5.13   A synchronous CDMA frequency selective Rayleigh fading channel model.

Equation (5.84) becomes:

$$r(t) = \sum_{i=-M}^{M} \mathbf{h}^T(t - iT)\mathbf{b}(i) + n(t) = \sum_{i=-M}^{M} \mathbf{s}^T(t)\mathbf{E}\boldsymbol{\Phi}\mathbf{C}(t)\mathbf{b}(i) + n(t) \tag{5.103}$$

We define a $KL \times KL$ crosscorrelation matrix of normalized signature waveforms,

$$\mathbf{R}(l) = \int_{-\infty}^{+\infty} \mathbf{s}(t)\mathbf{s}^T(t + lT)\,dt \tag{5.104}$$

The asynchronous mode is evident from the structure of the $L \times L$ crosscorrelation matrix between the users m and n,

$$\mathbf{R}_{mn}(l) = \int_{lT+\tau_m}^{(l+1)T+\tau_m} \mathbf{s}_m(t - \tau_m)\,\mathbf{s}_n^T(t + lT - \tau_n)\,dt \tag{5.105}$$

Since there is no inter-symbol interference, $\mathbf{R}(l) = 0, \forall |l| > 1$ and $\mathbf{R}(-1) = \mathbf{R}^H(1)$. Due to the ordering of the users $\mathbf{R}^H(1)$ is an upper triangular matrix with zero elements on the diagonal.

The front end of the multiuser detector consists of KL filters matched to the normalized properly delayed signature waveforms of the users, as shown in Figure 5.14. The output of this filter bank sampled at the bit epochs is given by the vector

$$\mathbf{y}(l) = \int_{-\infty}^{+\infty} r(t)\mathbf{s}(t - lT)\,dt \tag{5.106}$$

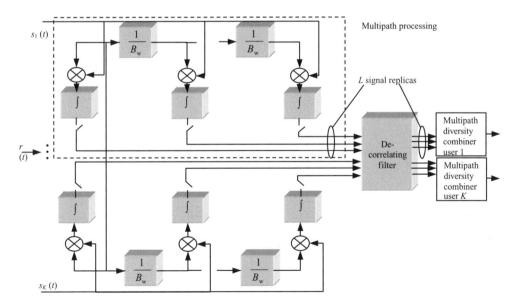

Figure 5.14   Multipath decorrelation.

The vector of sufficient statistics can also be expressed as:

$$
\begin{aligned}
\mathbf{y}(l) = {}& \mathbf{R}(-1)\mathbf{E}\boldsymbol{\Phi}\mathbf{C}(l+1)\mathbf{b}(l+1) + \mathbf{R}(0)\mathbf{E}\boldsymbol{\Phi}\mathbf{C}(l)\mathbf{b}(l) \\
& + \mathbf{R}(1)\mathbf{E}\boldsymbol{\Phi}\mathbf{C}(l-1)\mathbf{b}(l-1) + n(l)
\end{aligned}
\tag{5.107}
$$

The covariance matrix of the matched filter output noise vector is given by:

$$
E\{\mathbf{n}^*(i)\mathbf{n}^{\mathrm{T}}(j)\} = \sigma^2 \mathbf{R}(i-j)
$$

Taking the z-transform gives

$$
\mathbf{Y}(z) = \mathbf{S}(z)(\mathbf{E}\boldsymbol{\Phi}\mathbf{C}\mathbf{b})(z) + n(z)
\tag{5.108}
$$

where $(\mathbf{E}\boldsymbol{\Phi}\mathbf{C}\mathbf{b})(z)$ is the transform of sequence

$$
\left\{ \left[\sqrt{E_1}e^{j\phi_1}c_{1,0}(i)b_1(i) \cdots \sqrt{E_1}e^{j\phi_1}c_{1,L-1}(i)b_1(i) \cdots \sqrt{E_K}e^{j\phi_K}c_{K,L-1}(i)b_K(i) \right]^{\mathrm{T}} \right\}
\tag{5.109}
$$

$\mathbf{S}(z)$ is the equivalent transfer function of the CDMA multipath channel which depends only on the signature waveforms of the users. The multipath decorrelating (MD) filter is a KL-input KL-output linear time-invariant filter with transfer function matrix:

$$
\mathbf{G}(z) \overset{\Delta}{=} [\mathbf{S}(z)]^{-1} = \frac{\mathrm{adj}\mathbf{S}(z)}{\det \mathbf{S}(z)} = [\mathbf{R}(-1)z + \mathbf{R}(0) + \mathbf{R}(1)z^{-1}]^{-1}
\tag{5.110}
$$

The necessary and sufficient condition for the existence of a stable, but non-causal realization of the decorrelating filter is

$$
\det[\mathbf{R}(-1)e^{jw} + \mathbf{R}(0) + \mathbf{R}(1)e^{-jw}]^{-1} \neq 0 \; \forall w \in [0, 2\pi]
\tag{5.111}
$$

The z-transform of the decorrelating detector outputs is

$$
\mathbf{P}(z) = (\mathbf{E}\boldsymbol{\Phi}\mathbf{C}\mathbf{b})(z) + N_p(z)
\tag{5.112}
$$

$N_p(z)$ is the z-transform of a stationary, filtered Gaussian noise vector sequence. The z-transform of the noise covariance matrix sequence is equal to

$$
\sigma^2 [\mathbf{S}(z)]^{-1} = \sigma^2 \sum_{m=-\infty}^{\infty} \mathbf{D}(m)z^{-m}
\tag{5.113}
$$

The output of the decorrelating detector containing L signal replicas of user k may be expressed as:

$$
\mathbf{p}_k(l) = \mathbf{c}_k \sqrt{E_k}e^{j\phi_k}b_k(l) + \mathbf{n}_k(l)
\tag{5.114}
$$

The noise covariance matrix is given by:

$$
\sigma^2 [\mathbf{D}(0)]_{kk} = \sigma^2 \frac{1}{2\pi} \int_0^{2\pi} [\mathbf{S}(e^{-jw})]_{kk}^{-1} \, dw
\tag{5.115}
$$

5.2.8.3   Coherent Reception With Maximal Ratio Combining

Since the front end of the coherent multiuser detector contains the decorrelating filter, the noise components in the L branches of the kth user are correlated. The usual approach prior to combining is to introduce the whitening operation, where whitening filter $(\mathbf{T}^H)^{-1}$ is obtained by Cholesky decomposition $[\mathbf{D}(0)_{kk} = \mathbf{T}^T\mathbf{T}^*$. So, the output of the user of interest is given by

$$\begin{aligned}\mathbf{p}_{kw} &= \mathbf{f}\sqrt{E_k}e^{j\phi_k}b_k + \mathbf{n}_{kw} \\ &= p'_{kw}b_k + \mathbf{n}_{kw}\end{aligned} \tag{5.116}$$

where

$$\mathbf{f} = (\mathbf{T}^H)^{-1}\mathbf{c}_k \tag{5.117}$$

and $\mathbf{n}_{kw}$ is a zero mean Gaussian white noise vector with covariance matrix $\sigma^2 I_L$. The optimal combiner in this situation is the maximal ratio combiner (MRC). The receiver block diagram is shown in Figure 5.15. The output of the maximal ratio combiner can be represented as:

$$\hat{b}_k = \text{sgn}\left(\mathbf{p}_{kw} \cdot \hat{\mathbf{p}}'_{kw}*\right) \tag{5.118}$$

For illustration purposes, a CDMA cellular mobile radio system with 1.25 MHz bandwidth and 9600 bps data rate is used. The multipath intensity profile of the mobile radio channel is given by:

$$r(\tau) = \frac{P}{T_m}e^{-\frac{\tau}{T_m}} \tag{5.119}$$

where P is the total average received power and T_m is the multipath delay spread. Typical values of the multipath delay spread are $T_m = 0.5$ μs for the suburban environment and $T_m = 3$ μs for an urban environment. Therefore, we expect the multipath diversity reception with two branches in suburban areas and four to five branches in an urban setting. For the given parameters, inter-symbol interference is negligible, and the mobile radio channel can be described as a discrete multipath Rayleigh fading channel with mean square value of the path coefficients given by:

$$\overline{c_{k,l}^2} = \frac{1}{B_w}r\left(\frac{l}{B_w}\right) \quad l = 0\cdots L-1 \tag{5.120}$$

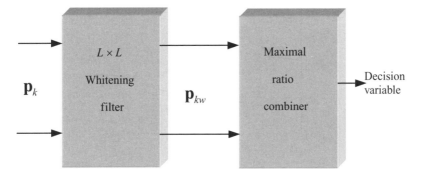

Figure 5.15   Maximal ratio combining after multipath decorrelation for coherent reception.

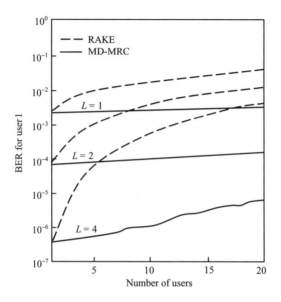

Figure 5.16 Error probability for a coherent RAKE and MD-MRC multiuser receiver for different multipath diversity order in a mobile radio channel using Gold signature sequences of length $J = 127$, average $i = 20$ dB [46].

The BER versus the number of users is shown in Figures 5.16 and 5.17. One can see that for a large product KL, multiuser detector performance starts to degrade due to noise enhancement caused by matrix inversion.

More details on multiuser detection can be found in [47–57]. The latest results in this field including the systems using multiple antennas, can be found in [58–80], and [81–124].

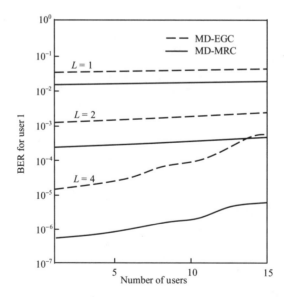

Figure 5.17 Multiuser receiver error probability for different multipath diversity orders in a mobile radio channel using Gold signature sequences of length $J = 127$, average SNR $= 20$ dB [46].

5.3 Minimum Mean Square Error (MMSE) Linear Multiuser Detection

If the amplitude of user k's signal in Equation (5.43) is A_k, then the vector of matched filter outputs $\mathbf{y}$ in Equation (5.46) can be represented as:

$$\mathbf{y} = \mathbf{RAb} + \mathbf{n} \tag{5.121}$$

where $\mathbf{A}$ is a diagonal matrix with elements A_k

$$\mathbf{A} = \text{diag}||A_k|| \tag{5.122}$$

If the multiuser detector transfer function is denoted as M then the minimum mean square error (MMSE) detector is defined as

$$\min_{M \in R^{KxK}} \mathbf{E}[||\mathbf{b} - M\mathbf{y}||^2] \tag{5.123}$$

One can show that the MMSE linear detector outputs the following decisions [84, 85, 88]:

$$\hat{b}_k = \text{sgn}\left(\frac{1}{A_k}([\mathbf{R} + \sigma^2 A^{-2}]^{-1}y)_k\right)$$
$$= \text{sgn}(([\mathbf{R} + \sigma^2 A^{-2}]^{-1}\mathbf{y})_k) \tag{5.124}$$

Therefore, the MMSE linear detector replaces the transformation $\mathbf{R}^{-1}$ of the decorrelating detector by

$$[\mathbf{R} + \sigma^2 \mathbf{A}^{-2}]^{-1} \tag{5.125}$$

where

$$\sigma^2 \mathbf{A}^{-2} = \text{diag}\left\{\frac{\sigma^2}{A_1^2}, \ldots, \frac{\sigma^2}{A_K^2}\right\} \tag{5.126}$$

As an illustration, for the two users case we have:

$$[\mathbf{R} + \sigma^2 \mathbf{A}^{-2}]^{-1} = \left[\left(1 + \frac{\sigma^2}{A_1^2}\right)\left(1 + \frac{\sigma^2}{A_2^2}\right) - \rho^2\right]^{-1} \begin{bmatrix} 1 + \frac{\sigma^2}{A_2^2} & -\rho \\ -\rho & 1 + \frac{\sigma^2}{A_1^2} \end{bmatrix} \tag{5.127}$$

In the asynchronous case, similarly to the solution in Section 5.3 the MMSE linear detector is a K-input, K-output, linear, time-invariant filter with transfer function

$$[\mathbf{R}^{\mathrm{T}}[1]z + \mathbf{R}[0] + \sigma^2 \mathbf{A}^{-2} + \mathbf{R}[1]z^{-1}]^{-1} \tag{5.128}$$

In Figure 5.18, the BER is presented versus the near–far ratio for different detectors. One can see that MMSE shows better performance than the decorrelator. In the figure, the signal to noise ratio of the desired user is equal to 10 dB.

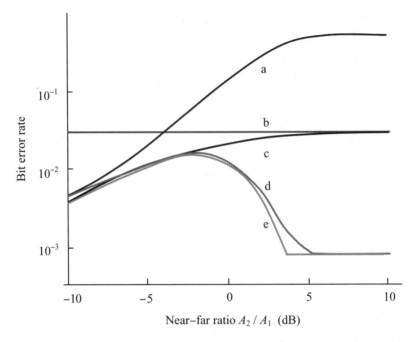

Figure 5.18   Bit error rate with two users and crosscorrelation $\rho = 0.8$: (a) single user matched filter; (b) decorrelator; (c) MMSE; (d) minimum (upper bound); (e) minimum (lower bound).

5.3.1   System Model in Multipath Fading Channels

In this section the channel impulse response and the received signal will be presented as

$$c_k(t) = \sum_{l=1}^{L_k} c_{k,l}^{(n)} \delta(t - \tau_{k,l}) \tag{5.129}$$

$$r(t) = \sum_{n=0}^{N_b-1} \sum_{k=1}^{K} \sum_{l=1}^{L} A_k b_k^{(n)} c_{k,l}^{(n)} s_k(t - nT - \tau_{k,l}) + n(t) \tag{5.130}$$

The received signal is time discretized, by antialias filtering and sampling $r(t)$ at the rate $1/T_s = S/T_c = SG/T$, where S is the number of samples per chip and $G = T/T_C$ is the processing gain. The received discrete time signal over a data block of N_b symbols is:

$$\mathbf{r} = \mathbf{SCAb} + \mathbf{n} \in C^{SGN_b} \tag{5.131}$$

where

$$\mathbf{r} = [\mathbf{r}^{T(0)}, \dots, \mathbf{r}^{T(N_b-1)}]^T \in C^{SGN_b} \tag{5.132}$$

is the input sample vector with

$$\mathbf{r}^{T(n)} = [r(T_s(nSG + 1)), \dots, r(T_s(n + 1)SG] \in C^{SGN_b} \tag{5.133}$$

$$S = [S^{(0)}, S^{(1)}, \ldots, S^{(N_b-1)}] \in \mathbf{R}^{SGN_b \times KLN_b}$$

$$= \begin{bmatrix} S^{(0)}(0) & 0 & \cdots & 0 \\ \vdots & S^{(1)}(0) & \ddots & \vdots \\ S^{(0)}(D) & \vdots & \ddots & 0 \\ 0 & S^{(1)}(D) & \ddots & S^{(N_b-1)}(0) \\ \vdots & \ddots & \ddots & \vdots \\ 0 & \cdots & 0 & S^{(N_b-1)}(D) \end{bmatrix} \tag{5.134}$$

is the sampled spreading sequence matrix and $D = \lceil (T + T_m)/T \rceil$. In a single path channel, $D = 1$ due to the asynchronicity of users. In multipath channels, $D \geq 2$ due to the multipath spread. The code matrix is defined with several components $(S^{(n)}(0), \ldots, S^{(n)}(D))$ for each symbol interval to simplify the presentation of the crosscorrelation matrix components. T_m is the maximum delay spread,

$$S^{(n)} = [s_{1,1}^{(n)}, \ldots, s_{1,L}^{(n)}, \ldots, s_{K,L}^{(n)}] \in R^{SGN_b \times KL} \tag{5.135}$$

where

$$s_{k,l}^{(n)} = \begin{cases} \mathbf{0}_{SGN_b \times 1}^{T} & n = 0, \tau_{k,l} = 0 \\ [[s_k(T_s(SG - \tau_{k,l} + 1)), \ldots, s_k(T_sSG)]^{T}, \mathbf{0}_{(SGN_b - \tau_{k,l}) \times 1}^{T}]^{T} & n = 0, \tau_{k,l} > 0 \\ [\mathbf{0}_{((n-1)SG + \tau_{k,l}) \times 1}^{T}, \mathbf{s}_k^{T}, \mathbf{0}_{(SG(N_b - n) - \tau_{k,l}) \times 1}^{T}]^{T} & 0 < n < N_b - 1 \\ [\mathbf{0}_{(SG(N_b - 1) + \tau_{k,l}) \times 1}^{T}, [s_k(T_s), \ldots, s_k(T_s(SG - \tau_{k,l}))]]^{T} & n = N_b - 1 \end{cases} \tag{5.136}$$

where $\tau_{k,l}$ is the time discretized delay in sample intervals and

$$\mathbf{s}_k = [s_k(T_s), \ldots, s_k(T_sSG)]^{T} \in R^{SG} \tag{5.137}$$

is the sampled signature sequence of the kth user. By analogy with Equation (5.94):

$$C = \mathrm{diag}[C^{(0)}, \ldots, C^{(N_b-1)}] \in C^{KLN_b \times KN_b} \tag{5.138}$$

is the channel coefficient matrix with

$$C^{(n)} = \mathrm{diag}[\mathbf{c}_1^{(n)}, \ldots, \mathbf{c}_K^{(n)}] \in C^{KL \times K} \tag{5.139}$$

and

$$\mathbf{c}_k^{(n)} = [c_{k,1}^{(n)}, \ldots, c_{k,L}^{(n)}]^{T} \in C^{L} \tag{5.140}$$

Equation (5.122) now becomes:

$$A = \mathrm{diag}[A^{(0)}, \ldots, A^{(N_b-1)}] \in R^{KN_b \times KN_b} \tag{5.141}$$

the matrix of total received average amplitudes with

$$A^{(n)} = \mathrm{diag}[A_1, \ldots, A_K] \in R^{K \times K} \tag{5.142}$$

The bit vector from Equation (5.91) becomes

$$\mathbf{b} = [\mathbf{b}^{T(0)}, \ldots, \mathbf{b}^{T(N_b-1)}]^{T} \in \aleph^{KN_b} \tag{5.143}$$

with the modulation symbol alphabet $\aleph$ (with BPSK $\aleph = \{-1, 1\}$) and

$$\mathbf{b}^{(n)} = [b_1^{(n)}, \ldots, b_K^{(n)}] \in \aleph^K \tag{5.144}$$

and $\mathbf{n} \in C^{SGN_b}$ is the channel noise vector. It is assumed that the data bits are independent identically distributed random variables independent from the channel coefficients and the noise process.

The crosscorrelation matrix from Equation (5.104) for the spreading sequences can be formed as:

$$\mathbf{R} = \mathbf{S}^T\mathbf{S} \in R^{KLN_b \times KLN_b} \tag{5.145}$$

$$= \begin{bmatrix}
\mathbf{R}^{(0,0)} & \cdots & \mathbf{R}^{(0,D)} & \mathbf{0}_{KL}\cdots & & \mathbf{0}_{KL} \\
\vdots & \ddots & \ddots & \ddots & & \vdots \\
\mathbf{R}^{(D,0)} & \ddots & \ddots & \ddots & & \mathbf{0}_{KL} \\
\mathbf{0}_{KL} & \ddots & \ddots & \ddots & & \mathbf{R}^{(N_b-D,N_b-1)} \\
\vdots & \ddots & \ddots & \ddots & & \vdots \\
\mathbf{0}_{KL} & \cdots & \mathbf{0}_{KL} & \cdots & & \mathbf{R}^{(N_b-1,N_b-1)}
\end{bmatrix}$$

where

$$\mathbf{R}^{(n,n-j)} = \sum_{i=0}^{D-j} \mathbf{S}^{T(n)}(i)\mathbf{S}^{(n-j)}(i+j), \quad j \in \{0, \ldots, D\} \tag{5.146}$$

and $\mathbf{R}^{(n-j,n)} = \mathbf{R}^{T(n,n-j)}$. The elements of the correlation matrix can be written as

$$\mathbf{R}^{(n,n')} = \begin{bmatrix}
\mathbf{R}_{1,1}^{(n,n')} & \cdots & \mathbf{R}_{1,K}^{(n,n')} \\
\vdots & \ddots & \vdots \\
\mathbf{R}_{K,1}^{(n,n')} & & \mathbf{R}_{K,K}^{(n,n')}
\end{bmatrix} \in R^{KL \times KL} \tag{5.147}$$

and

$$\mathbf{R}_{k,k'}^{(n,n')} = \begin{bmatrix}
R_{k1,k'1}^{(n,n')} & \cdots & R_{k1,k'L}^{(n,n')} \\
\vdots & \ddots & \vdots \\
R_{kL,k'1}^{(n,n')} & \cdots & R_{kL,k'L}^{(n,n')}
\end{bmatrix} \in R^{L \times L} \tag{5.148}$$

with

$$\mathbf{R}_{kl,k'l'}^{(n,n')} = \sum_{j=\tau_{k,l}}^{SG-1+\tau_{k,l}} s_k(T_s(j - \tau_{k,l}))s_{k'}(T_s(j - \tau_{k'l'} + (n' - n)SG)) = \mathbf{s}_{k,l}^{T(n)}\mathbf{s}_{k',l'}^{(n')} \tag{5.149}$$

which represents the correlation between users k and k', the lth and l'th paths, and between their nth and n'th symbol intervals.

5.3.2 MMSE Detector Structures

One of the conclusions in Section 5.2 was that noise enhancement in linear MUD causes system performance degradation for large products KL. In this section we consider a possibility for reducing the

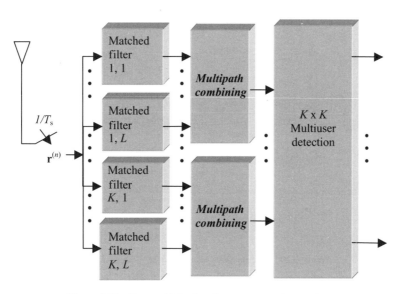

Figure 5.19 Postcombining interference suppression receiver.

size of the matrix to be inverted by using multipath combining prior to MUD. The structure is called a postcombining detector and the basic block diagram of the receiver is shown in Figure 5.19.

The starting point in the derivation of the receiver structure is the cost function $E\{|b - \hat{b}|^2\}$ where

$$\hat{b} = \mathbf{L}^{\mathrm{H}}_{[Post]}\mathbf{r} \tag{5.150}$$

The detector linear transform matrix is given as

$$\mathbf{L}_{[post]} = \mathbf{SC\Lambda}(\mathbf{\Lambda C}^{\mathrm{H}}\mathbf{RC\Lambda} + \sigma^2\mathbf{I})^{-1} \in C^{SGN_b \times KN_b} \tag{5.151}$$

This result is obtained by minimizing the cost function and derivation details may be found in any standard textbook on signal processing. Here $\mathbf{R} = \mathbf{S}^{\mathrm{T}}\mathbf{S}$ is the signature sequence crosscorrelation matrix defined by Equation (5.145). The output of the postcombining LMMSE receiver is

$$\mathbf{y}_{[post]} = (\mathbf{\Lambda C}^{\mathrm{H}}\mathbf{RC\Lambda} + \sigma^2\mathbf{I})^{-1}(\mathbf{SC\Lambda})^{\mathrm{H}}\mathbf{r} \in C^K \tag{5.152}$$

where $(\mathbf{SC\Lambda})^{\mathrm{H}}\mathbf{r}$ is the multipath (MR) combined matched filter bank output. For non-fading AWGN:

$$\mathbf{L}_{[post]} = \mathbf{S}(\mathbf{R} + \sigma^2(\mathbf{A}^{\mathrm{H}}\mathbf{A})^{-1})^{-1} \tag{5.153}$$

The postcombining LMMSE receiver in fading channels depends on the channel complex coefficients of all users and paths. If the channel is changing rapidly, the optimal LMMSE receiver changes continuously. The adaptive versions of LMMSE receivers have increasing convergence problems as the fading rate increases. The dependence on the fading channel state can be removed by applying a precombining interference suppression type of receiver. The receiver block diagram in this case is shown in Figure 5.20.

The transfer function of the detector is obtained by minimizing each element of the cost function

$$E\{|\mathbf{h} - \hat{\mathbf{h}}|^2\} \tag{5.154}$$

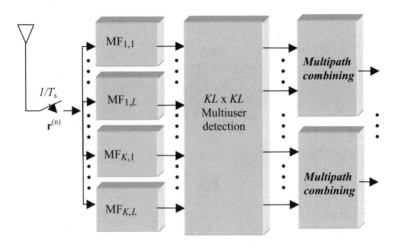

Figure 5.20 Precombining interference suppression receiver.

where

$$\mathbf{h} = \mathbf{CAb} \tag{5.155}$$

and

$$\hat{\mathbf{h}} = \mathbf{L}_{[\text{pre}]}^{\mathsf{T}} \mathbf{r} \tag{5.156}$$

is estimated.

The solution of this minimization is [125]:

$$\mathbf{L}_{[\text{pre}]} = \mathbf{S}(\mathbf{R} + \sigma^2 \mathbf{R}_h^{-1})^{-1} \in \mathbf{R}^{SGN_b \times KLN_b} \tag{5.157}$$

$$\mathbf{R}_h = \text{diag}[A_1^2 \mathbf{R}_{c_1}, \dots, A_K^2 \mathbf{R}_{c_k}] \in \mathbf{R}^{KLN_b \times KLN_b} \tag{5.158}$$

$$\mathbf{R}_{c_k} = \text{diag}[E[|c_{k,1}|^2], \dots, E[|c_{k,L}|^2]] \in \mathbf{R}^{L \times L} \tag{5.159}$$

$$\mathbf{y}_{[\text{pre}]} = (\mathbf{R} + \sigma^2 \mathbf{R}_h^{-1})^{-1} \mathbf{S}^{\mathsf{T}} \mathbf{r} \in C^{KL} \tag{5.160}$$

The two detectors are compared in Figure 5.21. The postcombining scheme performs better. The illustration of LMMSE–RAKE receiver performance in the near–far environment is shown in Figure 5.22 [126]. Considerable improvement compared to conventional RAKE is evident.

5.3.3   Spatial Processing

When combined with multiple receive antennas the receiver structures may have one of the forms shown in Figure 5.23 [125–129].

The channel impulse response for the kth user's ith sensor can now be written as:

$$c_{k,i}(t) = \sum_{l=1}^{L_k} c_{k,l}^{(n)} e^{j 2\pi \ \lambda^{-1} \langle e(\phi_{k,l}), \varepsilon_i \rangle} \delta(t - (\tau_{k,l,i})) \tag{5.161}$$

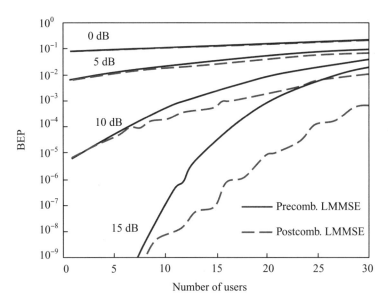

Figure 5.21 Bit error probabilities as a function of the number of users for the postcombining and precombining LMMSE detectors in an asynchronous two-path fixed channel with different SNRs and bit rate 16 kbit/s, Gold code of length 31, $td/T = 4.63 \times 10^{-3}$, maximum delay spread 10 chips [126].

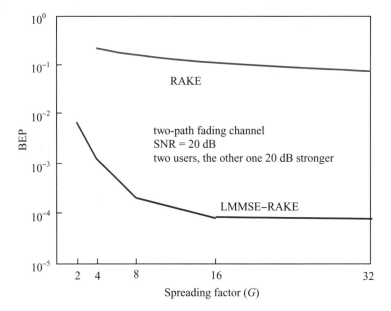

Figure 5.22 Bit error probabilities as a function of the near–far ratio for the conventional RAKE receiver and the precombining LMMSE (LMMSE–RAKE) receiver with different spreading factors (G) in a two-path Rayleigh fading channel with maximum delay spreads of 2 μs for $G = 4$, and 7 μs for other spreading factors. The average signal to noise ratio is 20 dB, the data modulation is BPSK, the number of users is 2, the other user has 20 dB higher power. Data rates vary from 128 kbit/s to 2.048 Mbit/s; no channel coding is assumed [126].

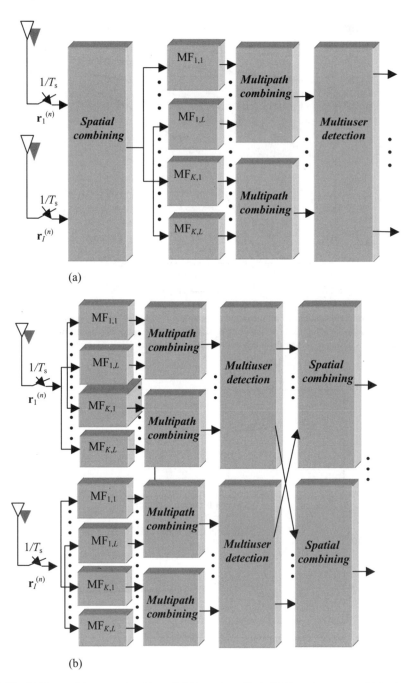

Figure 5.23   (a) The spatial–temporal multiuser (STM) receiver; (b) the TMS receive post combining interference suppression receiver with spatial signal processing (c) the SMT receiver; (d) the MST receiver. A precombining interference suppression receiver with spatial signal processing.

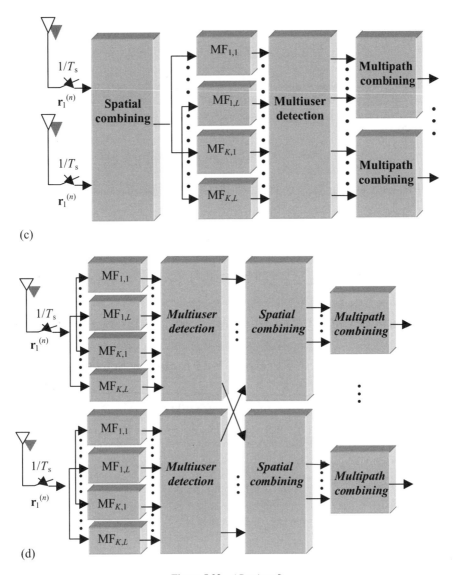

Figure 5.23 (*Continued*).

where L_k is the number of propagation paths (assumed to be the same for all users for simplicity; $L_k = L, \forall k$), $c_{k,l}^{(n)}$ is the complex attenuation factor of the kth user's lth path, $\tau_{k,l,i}$ is the propagation delay for the ith sensor, ε_i is the position vector of the ith sensor with respect to some arbitrarily chosen reference point, λ is the wavelength of the carrier, $\mathbf{e}(\phi_{k,l})$ is a unit vector pointing to direction $\phi_{k,l}$ (direction of arrival), and $\langle . , . \rangle$ indicates the inner product.

Assuming that the number of propagation paths is the same for all users, the channel impulse response can be written as:

$$c_{k,i}(t) = \sum_{l=1}^{L} c_{k,l}^{(n)} e^{j2\pi \ \lambda^{-1} \langle \mathbf{e}(\phi_{k,l}), \varepsilon_i \rangle} \delta(t - \tau_{k,l}) \tag{5.162}$$

The channel matrix for the ith sensor consists of two components

$$\mathbf{C}_i = \mathbf{C} \circ \boldsymbol{\Phi}_i \in C^{KLN_b \times KN_b} \tag{5.163}$$

where $\mathbf{C}$ is the channel matrix defined in Equation (5.139). $\circ$ is the Schur product defined as $\mathbf{Z} = \mathbf{X} \circ \mathbf{Y} \in C^{x \times y}$, i.e. all components of the matrix $\mathbf{X} \in C^{x \times y}$ are multiplied elementwise by the matrix $\mathbf{Y} \in C^{x \times y}$, $\boldsymbol{\Phi}_i = \mathrm{diag}(\tilde{\phi}_i) \otimes \mathbf{I}_{N_b}$ and $\tilde{\varphi}_i = \mathrm{diag}(\varphi_1, \dots, \varphi_K)$, $\varphi_k = [\phi_{k,1}, \dots, \phi_{k,L}]^{\mathrm{T}}$, is the matrix of the direction vectors

$$\varphi_i = [e^{j2\pi \; \lambda^{-1}\langle \mathbf{e}(\phi_{1,1}),\varepsilon_i \rangle}, \dots, e^{j2\pi \; \lambda^{-1}\langle \mathbf{e}(\phi_{K,L}),\varepsilon_i \rangle}]^{\mathrm{T}} \in C^{KL} \tag{5.164}$$

By using the previous notation one can show that the equivalent detector transform matrices are given as [125–128].

$$\mathbf{L}_{[\mathrm{STM}]} = \sum_{i=1}^{I} \mathbf{S}(\mathbf{C} \circ \boldsymbol{\Phi}_i) \cdot \left(\sum_{i=1}^{I} \mathbf{A}^{\mathrm{H}}(\boldsymbol{\Phi}_i^{\mathrm{H}} \circ \mathbf{C}^{\mathrm{H}})\mathbf{R}(\mathbf{C} \circ \boldsymbol{\Phi}_i)\mathbf{A} + \sigma^2 \mathbf{I} \right)^{-1}$$

$$\mathbf{L}_{[\mathrm{SMT}]} = \sum_{i=1}^{I} \mathbf{S}\,\boldsymbol{\Phi}_i \left(\sum_{i=1}^{I} \boldsymbol{\Phi}_i^{\mathrm{H}} \mathbf{R}\boldsymbol{\Phi}_i + \sigma^2 \mathbf{R}_h^{-1} \right)^{-1} \tag{5.165}$$

$$\mathbf{L}_{[\mathrm{MST}]i} = \mathbf{S}\left(\mathbf{R} + \sigma^2 \mathbf{R}_h^{-1} \right)^{-1}$$

$$\mathbf{L}_{[\mathrm{TMS}]} = \mathbf{SCA}\,(\mathbf{AC}^{\mathrm{H}}\mathbf{RCA} + \sigma^2 \mathbf{I})^{-1}$$

5.4   Single User LMMSE Receivers for Frequency Selective Fading Channels

5.4.1   Adaptive Precombining LMMSE Receivers

In this case the MSE criterion $E\{|\mathbf{h} - \widehat{\mathbf{h}}|^2\}$ requires that the reference signal $\mathbf{h} = \mathbf{CAb}$ is available in adaptive implementations. For adaptive single user receivers, the optimization criterion is presented for each path separately, i.e.

$$J_{k,l} = E\{|(\mathbf{h})_{k,l} - (\widehat{\mathbf{h}})_{k,l}|^2\} \tag{5.166}$$

The receiver block diagram is given in Figure 5.24 [89–94].

By using notation

$$\bar{\mathbf{r}}^{(n)} = [\mathbf{r}^{\mathrm{T}(n-D)}, \dots, r^{\mathrm{T}(n)}, \dots, r^{\mathrm{T}(n+D)}]^{\mathrm{T}} \in C^{MSG}$$

$$\mathbf{w}_{k,l}^{(n)} = [w_{k,l}^{(n)}(0), \dots, w_{k,l}^{(n)}(MSG - 1)]^{\mathrm{T}} \in C^{MSG} \tag{5.167}$$

$$y_{k,l}^{(n)} = \mathbf{w}_{k,l}^{\mathrm{H}(n)} \bar{\mathbf{r}}^{(n)}$$

the bit estimation is defined as

$$\hat{b}_k^{(n)} = \mathrm{sgn}\left(\sum_{l=1}^{L} \hat{c}_{k,l}^{(n)} y_{k,l}^{(n)} \right) \tag{5.168}$$

The filter coefficients $\mathbf{w}$ are derived using the MSE criterion ($E[|e_{k,l}^{(n)}|^2]$). This leads to the optimal filter coefficients $\mathbf{w}_{[\mathrm{MSE}]k,l} = \mathbf{R}_{\bar{r}}^{-1} \mathbf{R}_{\bar{r}d_{k,l}}$ where $\mathbf{R}_{\bar{r}d_{k,l}}$ is the crosscorrelation vector between the input vector $\bar{\mathbf{r}}$ and the desired response $d_{k,l}$, and $\mathbf{R}_{\bar{r}}$ is the input signal crosscorrelation matrix. Adaptive filtering can be implemented by using a number of algorithms.

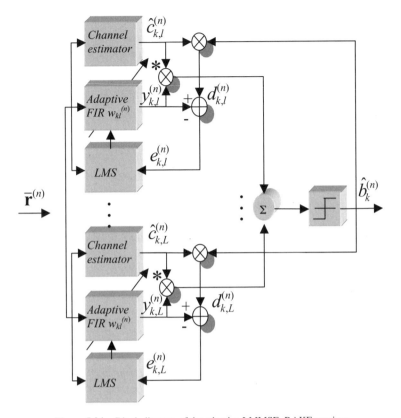

Figure 5.24   Block diagram of the adaptive LMMSE–RAKE receiver.

5.4.1.1   The Steepest Descent Algorithm

In this case we have

$$\mathbf{w}_{k,l}^{(n+1)} = \mathbf{w}_{k,l}^{(n)} - \mu \nabla_{k,l} \tag{5.169}$$

where ∇ is the gradient of

$$(J_{k,l} = E\{|c_{k,l}A_k b_k - \mathbf{w}_{k,l}^{\mathrm{H}}\mathbf{r}|^2\}) \tag{5.170}$$

This can be represented as

$$\nabla_{k,l} = \frac{\partial J_{k,l}}{\partial \operatorname{Re}\{\mathbf{w}_{k,l}\}} + \mathrm{j}\frac{\partial J_{k,l}}{\partial \operatorname{Im}\{\mathbf{w}_{k,l}\}} = 2\frac{\partial J_{k,l}}{\partial \mathbf{w}_{k,l}^*} \tag{5.171}$$

If the processing window $M = 1$ we have $\bar{\mathbf{r}}^{(n)} = \mathbf{r}^{(n)} \hat{=} \mathbf{r}$ and

$$\begin{aligned}\nabla_{k,l} &= -2E[\mathbf{r}(c_{k,l}A_k b_k)^*] + 2E[\mathbf{r}\mathbf{r}^{\mathrm{H}}]\mathbf{w}_{k,l} \\ &= -2\mathbf{R}_{rd_{k,l}} + 2\mathbf{R}_r\mathbf{w}_{k,l}\end{aligned} \tag{5.172}$$

where $d_{k,l} = c_{k,l}A_k b_k$. If we assume that $A_k = 1, \forall k$

$$\mathbf{w}_{k,l}^{(n+1)} = \mathbf{w}_{k,l}^{(n)} - 2\mu \left(\mathbf{R}_{rd_{k,l}} - \mathbf{R}_r \mathbf{w}_{k,l}^{(n)} \right) \tag{5.173}$$

As a stochastic approximation, Equation (5.172) can be represented as

$$\nabla_{k,l} \approx -2\mathbf{r}(c_{k,l}b_k)^* + 2\mathbf{r}\mathbf{r}^H \mathbf{w}_{k,l}^{(n)} = -2\mathbf{r}(c_{k,l}b_k)^* + 2\mathbf{r}y_{k,l}^*$$

From this equation, and assuming that $M > 1$, the LMS algorithm for updating the filter coefficients results in

$$\mathbf{w}_{k,l}^{(n+1)} = \mathbf{w}_{k,l}^{(n)} + 2\mu\bar{\mathbf{r}}^{(n)} \left(c_{k,l}^{(n)}b_k^{(n)} - y_{k,l}^{(n)} \right)^* \in C^{MSG} \tag{5.174}$$

We decompose Equation (5.174) into adaptive and fixed components as:

$$\mathbf{w}_{k,l}^{(n)} = \bar{\mathbf{s}}_{k,l} + \mathbf{x}_{k,l}^{(n)} \in C^{MSG} \tag{5.175}$$

where $\mathbf{x}_{k,l}^{(n)}$ is the adaptive filter component and

$$\bar{\mathbf{s}}_{k,l} = [0_{(DSG+\tau_{k,l})\times 1}^T, \mathbf{s}_k^T, \quad 0_{(DSG-\tau_{k,l})\times 1}^T]^T \tag{5.176}$$

is the fixed spreading sequence of the kth user with the delay $\tau_{k,l}$. In this case every branch from Figure 5.24 can be represented as shown in Figure 5.25.

In this case, Equation (5.47) gives:

$$\begin{aligned}
\mathbf{x}_{k,l}^{(n+1)} &= \mathbf{x}_{k,l}^{(n)} - 2\mu_{k,l}^{(n)} \left(c_{k,l}^{(n)}b_k^{(n)} - y_{k,l}^{(n)} \right)^* \bar{\mathbf{r}}^{(n)} = \mathbf{x}_{k,l}^{(n)} - 2\mu_{k,l}^{(n)}e_{k,l}^{*(n)}\bar{\mathbf{r}}^{(n)} \\
\mu_{k,l}^{(n)} &= \mu / \left(\bar{\mathbf{r}}^{H(n)}\bar{\mathbf{r}}^{(n)} \right); \quad 0 < \mu < 1 \\
e_{k,l}^{(n)} &= d_{k,l}^{(n)} - y_{k,l}^{(n)}
\end{aligned} \tag{5.177}$$

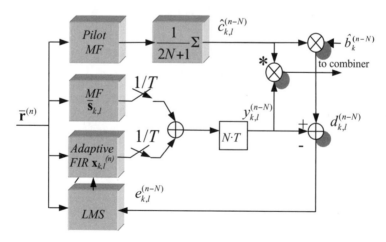

Figure 5.25 Block diagram of one receiver branch in the adaptive LMMSE–RAKE receiver.

The reference signal is

$$d_{k,l}^{(n)} = \hat{c}_{k,l}^{(n)} b_k^{(n)} \quad \text{or} \quad d_{k,l}^{(n)} = \hat{c}_{k,l}^{(n)} \hat{b}_k^{(n)} \tag{5.178}$$

and the channel estimator is using a pilot channel

$$\hat{c}_{k,l}^{(n)} = \frac{1}{2N+1} \sum_{i=-N}^{N} \bar{\mathbf{s}}_{p,l}^{\mathrm{T}} \bar{\mathbf{r}}^{(n-i)} \tag{5.179}$$

To illustrate the system operation the following example is used [126]: carrier frequency 2.0 GHz, symbol rate 16 kbit/s, 31 chip Gold code and rectangular chip waveform. A synchronous downlink with equal energy two-path ($L = 2$) Rayleigh fading channel with vehicle speeds of 40 km/h (which results in the maximum normalized Doppler shift of 4.36×10^{-3}) and maximum delay spread of ten chip intervals. The number of users examined was 1–30 including the unmodulated pilot channel. The average energy was the same for the pilot channel and user data channels. A simple moving average smoother of length eleven symbols was used in a conventional channel estimator. Perfect channel estimation and ideal truncated precombining LMMSE receivers were used in the analysis to obtain the bit error probability lower bounds. The receiver processing window is three symbols ($M = 3$) unless otherwise stated. The adaptive algorithm used in the simulations was normalized LMS with

$$\mu_{k,l}^{(n)} = \frac{1}{100 \cdot (2D+1)SG} \left(\bar{\mathbf{r}}_{k,l}^{\mathrm{H}(n)} \bar{\mathbf{r}}_{k,l}^{(n)} \right)^{-1} \tag{5.180}$$

The simulation results were produced by averaging over the BERs of randomly selected users with different delay spreads.

The simulation results are shown in Figure 5.26. In general, one can notice that the improvement gains are lower than in the case of multiuser detectors. In the presence of a strong near–far effect, the improvements should be more evident.

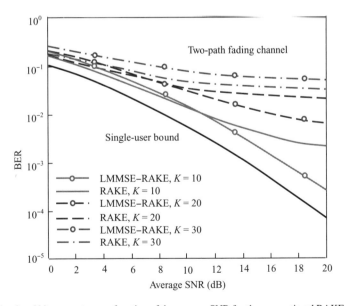

Figure 5.26   Simulated bit error rates as a function of the average SNR for the conventional RAKE and the adaptive LMMSE–RAKE in a two-path fading channel for vehicle speeds of 40 km/h with different numbers of users [126].

5.4.1.2   Blind Adaptive LMMSE–RAKE

In this case, in Equation (5.170) we use estimates of bits $\hat{b}_{k,l}$ instead of $b_{k,l}$ [95–98]:

$$\mathbf{x}_{k,l}^{(n+1)} = \mathbf{x}_{k,l}^{(n)} + 2\mu_{k,l}^{(n)}(c_{k,l}^{(n)}\hat{b}_{k,l}^{(n)} - y_{k,l}^{(n)})^* \bar{\mathbf{r}}^{(n)} \tag{5.181}$$

The MSE criterion now gives

$$\mathbf{w}_{[MSE]k,l} = \mathbf{R}_{\bar{r}}^{-1}\mathbf{R}_{\bar{r}d_{k,l}} = \mathbf{R}_{\bar{r}}^{-1}\bar{s}_{k,l}E\left\lfloor |c_{k,l}|^2 \right\rfloor \tag{5.182}$$

Similarly, the minimum output energy criteria defined as

$$\text{MOE}\left(E\left\lfloor |y_{k,l}|^2 \right\rfloor\right) \tag{5.183}$$

gives

$$\mathbf{w}_{[MOE]k,l} = \mathbf{R}_{\bar{r}}^{-1}\bar{s}_{k,l} / \left(\bar{\mathbf{s}}_{k,l}^T \mathbf{R}_{\bar{r}}^{-1}\bar{s}_{k,l}\right) \tag{5.184}$$

An implementation example can be seen in [130]. The stochastic approximation of the gradient of Equation (5.172) for the MOE criterion gives

$$\nabla_{k,l} = \bar{\mathbf{r}}^{(n)}\bar{\mathbf{r}}^{H(n)}\mathbf{w}_{k,l} \tag{5.185}$$

If we want to keep the useful signal autocorrelation unchanged, Equation (5.184) should be constrained to satisfy $\bar{\mathbf{s}}_{k,l}^T\mathbf{x}_{k,l}^{(n)} = 0$. The orthogonality condition is maintained at each step of the algorithm by projecting the gradient onto the linear subspace orthogonal to $\bar{s}_{k,l}^T$. In practice, this is accomplished by subtracting an estimate of the desired signal component from the received signal vector. An implementation can be seen in [131]. So, we have:

$$\mathbf{x}_{k,l}^{(n+1)} = \mathbf{x}_{k,l}^{(n)} - 2\mu_{k,l}^{(n)}\bar{\mathbf{r}}^{H(n)}\left(\bar{\mathbf{s}}_{k,l} + \mathbf{x}_{k,l}^{(n)}\right)\left(\bar{\mathbf{r}}^{(n)} - \mathbf{F}_{k,l}\left(\mathbf{F}_{k,l}^T\bar{\mathbf{r}}^{(n)}\right)\right) \tag{5.186}$$

where

$$\mathbf{F}_{k,l} = \begin{bmatrix} \mathbf{0}_{\tau_{k,l}\times 1}^T, & \mathbf{s}_k^T, & \mathbf{0}_{(2DSG-\tau_{k,l})\times 1}^T \\ \mathbf{0}_{(SG-\tau_{k,l})\times 1}^T, & \mathbf{s}_k^T, & \mathbf{0}_{((2D-1)SG-\tau_{k,l})\times 1}^T \\ \mathbf{0}_{(2DSG+\tau_{k,l})\times 1}^T, & [s_k(T_s),\ldots,s_k(T_s(SG-\tau_{k,l}))] \end{bmatrix}^T \in \mathbf{R}^{MSG\times M} \tag{5.187}$$

is a block diagonal matrix of sampled spreading sequence vectors. Effectively, M separate filters are adapted.

5.4.1.3   Griffiths's Algorithm

In this case instead of assuming that vector $\mathbf{R}_{\bar{r}d_{k,l}}$ is known, the instantaneous estimate for the covariance is used, i.e.

$$\mathbf{R}_{\bar{r}} \approx \bar{\mathbf{r}}^{(n)}\bar{\mathbf{r}}^{H(n)} \tag{5.188}$$

In this case the crosscorrelation is $\mathbf{R}_{\bar{r}d_{k,l}} = E[|c_{k,l}|^2]\bar{s}_{k,l}$, and Griffiths's algorithm results in:

$$\mathbf{x}_{k,l}^{(n+1)} = \mathbf{x}_{k,l}^{(n)} + 2\mu_{k,l}^{(n)}\left(E\left[|c_{k,l}|^2\right]\mathbf{F}_{k,l}\mathbf{1}_M - \bar{\mathbf{r}}_{k,l}^{*(n)}\left(\bar{\mathbf{s}}_{k,l} + \mathbf{x}_{k,l}^{(n)}\right)^H\bar{\mathbf{r}}^{(n)}\right) \tag{5.189}$$

In practice, the energy of multipath components ($E[|c_{k,l}|^2]$) is not known and must be estimated.

5.4.1.4 Constant Modulus Algorithm

In this case the optimization criterion is $E[(|y_{k,l}|^2 - \omega)^2]$ where ω is the so-called constant modulus (CM), set according to the received signal power, i.e $\omega = E[|c_{k,l}|^2]$ or $\omega^{(n)} = |c_{k,l}^{(n)}|^2$. By using the CM algorithm, it is possible to avoid the use of the data decisions in the reference signal in the adaptive LMMSE–RAKE receiver by taking the absolute value of the estimated channel coefficients ($|\hat{c}_{k,l}^{(n)}|$) in adapting the receiver. In the precombining LMMSE receiver framework, the cost function for the BPSK data modulation is

$$E[||\hat{\mathbf{h}}|^2 - |\mathbf{h}|^2|^2] \tag{5.190}$$

The stochastic approximation of the gradient for the CM criterion is

$$\nabla_{k,l}^{(n+1)} = \left(|y_{k,l}^{(n)}|^2 - |\hat{c}_{k,l}^{(n)}|^2\right) \bar{\mathbf{r}}^{(n)} \bar{\mathbf{r}}^{H(n)} \mathbf{w}_{k,l} \tag{5.191}$$

Hence, the constant modulus algorithm can be expressed as:

$$\mathbf{x}_{k,l}^{(n+1)} = \mathbf{x}_{k,l}^{(n)} - 2\mu_{k,l}^{(n)} y_{k,l}^{*(n)} \left(|y_{k,l}^{(n)}|^2 - |\hat{c}_{k,l}^{(n)}|^2\right) \bar{\mathbf{r}}^{(n)} \tag{5.192}$$

5.4.1.5 Constrained LMMSE–RAKE, Griffiths's and Constant Modulus Algorithms

The adaptive LMMSE–RAKE, the Griffiths's, and the constant modulus algorithm contain no constraints. By applying the orthogonality constraint $\bar{\mathbf{s}}_{k,l}^T \mathbf{x}_{k,l}^{(n)} = 0$ to each of these algorithms, an additional term $\bar{\mathbf{s}}_{k,l}^T \mathbf{x}_{k,l}^{(n)} \bar{\mathbf{s}}_{k,l}$ is subtracted from the new $\mathbf{x}_{k,l}^{(n+1)}$ update at every iteration. The constrained LMMSE–RAKE receiver becomes [28, 34]:

$$\mathbf{x}_{k,l}^{(n+1)} = \mathbf{x}_{k,l}^{(n)} + 2\mu_{k,l}^{(n)} \left(\hat{c}_{k,l}^{(n)} \hat{b}_k^{(n)} - y_{k,l}^{(n)}\right)^* \bar{\mathbf{r}}^{(n)} - \bar{\mathbf{s}}_{k,l}^T \mathbf{x}_{k,l}^{(n)} \bar{\mathbf{s}}_{k,l} \tag{5.193}$$

The Griffiths's and the constant modulus algorithms can also be defined in a similar way.

5.4.2 Blind Least Squares Receivers

All blind adaptive algorithms described in the previous section are based on the gradient of the cost function. In practical adaptive algorithms, the gradient is estimated, i.e. the expectation in the optimization criterion is not taken but is replaced in most cases by some stochastic approximation. In fact, the stochastic approximation used in LMS algorithms is accurate only for small step sizes μ. This results in rather slow convergence, which may be intolerable in practical applications.

Another drawback with the blind adaptive receivers presented above is the delay estimation. Those receiver structures as such support only conventional delay estimation based on matched filtering (MF). The MF-based delay estimation is sufficient for the downlink receivers in systems with an unmodulated pilot channel, since the zero mean MAI can be averaged out if the rate of fading is low enough. If CDMA systems do not have the pilot channel, it would be beneficial to use some near–far resistant delay estimators.

5.4.3 Least Squares (LS) Receiver

One possible solution to both the convergence and the synchronization problems is based on blind linear least squares (LS) receivers. The cost function in this case is:

$$J_{[LS]k,l} = \sum_{j=n-N+1}^{n} \left(c_{k,l}^{(j)} b_k^{(j)} - \mathbf{w}_{k,l}^{H(n)} \bar{\mathbf{r}}^{(j)}\right)^2 \tag{5.194}$$

where N is the observation window in symbol intervals. Filter weights are given as

$$\mathbf{w}_{k,l}^{(n)} = \hat{\mathbf{R}}_{\bar{r}}^{-1(n)} \bar{\mathbf{s}}_{k,l} \tag{5.195}$$

$\hat{\mathbf{R}}_{\bar{r}}^{-1(n)}$ denotes the estimated covariance matrix over a finite data block called the sample covariance matrix. This matrix can be expressed as

$$\hat{\mathbf{R}}_{\bar{r}}^{(n)} = \sum_{j=n-N+1}^{n} \bar{\mathbf{r}}^{(j)} \bar{\mathbf{r}}^{\mathrm{H}(j)} \tag{5.196}$$

Analogous to the MOE criterion, the LS criterion can be modified as:

$$J_{[\mathrm{LS'}]k,l} = \sum_{j=n-N+1}^{n} \left(\mathbf{w}_{k,l}^{\mathrm{H}(n)} \bar{\mathbf{r}}^{(j)} \right)^2, \quad \text{subject to } \mathbf{w}_{k,l}^{\mathrm{T}} \bar{\mathbf{s}}_{k,l} = 1 \tag{5.197}$$

which results in

$$\mathbf{w}_{k,l}^{(n)} = \frac{\hat{\mathbf{R}}_{\bar{r}}^{-1(n)} \bar{\mathbf{s}}_{k,l}}{\bar{s}_{k,l}^{\mathrm{T}} \hat{\mathbf{R}}_{\bar{r}}^{-1(n)} \bar{\mathbf{s}}_{k,l}} \tag{5.198}$$

The adaptation of the blind LS receiver means updating the inverse of the sample covariance. The blind adaptive LS receiver is significantly more complex than the stochastic gradient based blind adaptive receivers. Recursive methods, such as the recursive least squares (RLS) algorithm, for updating the inverse and iteratively finding the filter weights are known. Also, the methods based on eigendecomposition of the covariance matrix have been proposed to avoid explicit matrix inversion.

5.4.4 Method Based on the Matrix Inversion Lemma

The general relation

$$(\mathbf{A} + \mathbf{BCD})^{-1} = \mathbf{A}^{-1} - \mathbf{A}^{-1}\mathbf{B}(\mathbf{DA}^{-1}\mathbf{B} + \mathbf{C}^{-1})^{-1}\mathbf{DA}^{-1} \tag{5.199}$$

becomes

$$\hat{\mathbf{R}}_{\bar{r}}^{-1(n)} = \left(\hat{\mathbf{R}}_{\bar{r}}^{-1(n-1)} + \bar{\mathbf{r}}^{(n)} \bar{\mathbf{r}}^{\mathrm{H}(n)} \right)^{-1}$$
$$= R_{\bar{r}}^{-1(n-1)} - \frac{\mathbf{R}_{\bar{r}}^{-1(n-1)} \bar{\mathbf{r}}^{(n)} \bar{\mathbf{r}}^{\mathrm{H}(n)} \mathbf{R}_{\bar{r}}^{-1(n-1)}}{1 + \bar{\mathbf{r}}^{\mathrm{H}(n)} \mathbf{R}_{\bar{r}}^{-1(n-1)} \bar{\mathbf{r}}^{(n)}} \tag{5.200}$$

In time-variant channels, the old values of the inverses must be weighted by the so-called forgetting factor ($0 < \gamma < 1$), which results in:

$$\hat{\mathbf{R}}_{\bar{r}}^{-1(n)} = \frac{1}{\gamma} \left(\hat{\mathbf{R}}_{\bar{r}}^{-1(n-1)} - \frac{\hat{\mathbf{R}}_{\bar{r}}^{-1(n-1)} \bar{\mathbf{r}}^{(n)} \bar{\mathbf{r}}^{\mathrm{H}(n)} \hat{\mathbf{R}}_{\bar{r}}^{-1(n-1)}}{\gamma + \bar{\mathbf{r}}^{\mathrm{H}(n)} \hat{\mathbf{R}}_{\bar{r}}^{-1(n-1)} \bar{\mathbf{r}}^{(n)}} \right) \tag{5.201}$$

It is sufficient to initialize the algorithm as $\hat{\mathbf{R}}_{\bar{r}}^{-1(0)} = \mathbf{I}$. For illustration purposes a numerical example is shown in Figure 5.27 [126] and Table 5.3. System parameters are shown in the figure. In general, one can see that the blind algorithms are inferior when compared with LMMSE–RAKE using pilot symbols.

More information on the topic can be found in [99–106] and especially in the late publications that include MIMO channels [107–113].

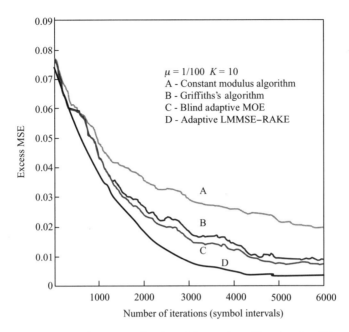

Figure 5.27   Excess mean squared error as a function of the number of iterations for different blind adaptive receivers in a two-path fading channel with vehicle speeds of 40 km/h, the number of active users $K = 10$, SNR $= 20$ dB, $\mu = 100^{-1}$ [126].

Table 5.3   The BERs of different blind adaptive receivers at an SNR of 20 dB in a two-path Rayleigh fading channel at vehicle speeds of 40 km/h. The acronyms used are: adaptive LMMSE–RAKE (LR), adaptive MOE (MOE), Griffiths's algorithm (GRA), constant modulus algorithm with average channel tap powers (CMA2), constrained adaptive LMMSE–RAKE (C-LR), constrained constant modulus algorithm (C-CMA), constrained Griffiths's algorithm (C-GRA), constrained constant modulus algorithm with average channel tap powers (C-CMA2) and conventional RAKE (RAKE) [126]

| Adaptive receiver | K = 30 | | K = 15 | | |
	$\mu = 100^{-1}$	$\mu = 10^{-1}$	$\mu = 100^{-1}$	$\mu = 10^{-1}$	$\mu = 2^{-1}$
LR	4.5×10^{-2}	3.9×10^{-1}	6.3×10^{-4}	7.2×10^{-4}	3.0×10^{-2}
MOE	2.8×10^{-2}	4.2×10^{-2}	6.0×10^{-4}	2.1×10^{-3}	9.1×10^{-2}
GRA	2.8×10^{-2}	4.7×10^{-2}	6.4×10^{-4}	3.3×10^{-3}	1.2×10^{-1}
CMA	3.9×10^{-2}	4.0×10^{-1}	1.2×10^{-3}	2.1×10^{-2}	5.0×10^{-1}
CMA2	3.3×10^{-2}	4.0×10^{-1}	1.8×10^{-3}	2.1×10^{-2}	5.0×10^{-1}
C-LR	3.2×10^{-2}	4.2×10^{-2}	6.3×10^{-4}	6.4×10^{-4}	1.9×10^{-3}
C-CMA	3.3×10^{-2}	5.0×10^{-1}	6.1×10^{-4}	3.8×10^{-1}	5.0×10^{-1}
C-GRA	2.8×10^{-2}	4.2×10^{-2}	6.1×10^{-4}	2.3×10^{-3}	9.7×10^{-2}
C-CMA2	2.9×10^{-2}	5.0×10^{-1}	7.7×10^{-4}	2.7×10^{-1}	5.0×10^{-1}
RAKE	3.1×10^{-2}	3.1×10^{-2}	7.1×10^{-3}	7.1×10^{-3}	7.1×10^{-3}

5.5   Signal Subspace-Based Channel Estimation for CDMA Systems

The practical CDMA systems use the pilot signals for channel estimation enabling even the use of a smoother for these purposes. While the solution is simple, it is very inefficient in the presence of Doppler or the near–far effect. The advanced channel estimation algorithms in CDMA systems are based either on Kalman-type [114, 120–122] estimators for channels with high dynamics, or a signal subspace-based approach in the presence of high levels of multiple access interference [116, 132–141]. The best performance is obtained if a joint detection of data and channel is used [117, 119]. Additional results including blind estimation are given in [123, 124, 142]. In this section we present a multiuser channel estimation problem through a signal subspace-based approach [116]. For these purposes the received signal for K users will be presented as:

$$r(t) = \sum_{k=1}^{K} r_k(t) + \eta_t, \quad -\infty < t < \infty \tag{5.202}$$

If the channel impulse response for user k is $h_k(t, \tau)$ we have

$$\begin{aligned} r_k(t) &= h_k(t, \tau)^* s_k(t) \\ &= \int_{-\infty}^{\infty} h_k(t, \alpha) s_k(\alpha) \, d\alpha \end{aligned} \tag{5.203}$$

If phase shift keying (PSK) is used to modulate the data, then the baseband complex envelope representation of the kth user's transmitted signal is given by

$$s_k(t) = \sqrt{2P_k} e^{j\phi_k} \sum_{i} e^{j(2\pi\,M)m_k^{(i)}} a_k(t - iT) \tag{5.204}$$

where P_k is the transmitted power, ϕ_k is the carrier phase relative to the local oscillator at the receiver, M is the size of the symbol alphabet, $m_k^{(i)} \in \{0, 1, \ldots, M - 1\}$ is the transmitted symbol, $a_k(t)$ is the spreading waveform, and T is the symbol duration. The spreading waveform is given by

$$4a_k(t) = \sum_{n=0}^{N-1} \prod_{Tc}(t - nT_c) a_k^{(n)} \tag{5.205}$$

where $\Pi_{Tc}(t)$ is a rectangular pulse, T_c is the chip duration ($T_c = T/N$), and $\{a_k^{(n)}\}$ for $n = 0, 1, \ldots, N - 1$ is a signature sequence (possibly complex-valued since the signature alphabet need not be binary). The chip matched filter can be implemented as an integrate-and-dump circuit, and the discrete time signal is given by

$$r[n] = \frac{1}{T_c} \int_{nT_c}^{(n+1)T_c} r(t) \, dt \tag{5.206}$$

Thus, the received signal can be converted into a sequence of WSS random vectors by buffering $r[n]$ into blocks of length N

$$\mathbf{y}_i = [r[iN]\, r[1 + iN] \cdots r[N - 1 + iN]]^{\mathrm{T}} \in \mathbb{C}^N \tag{5.207}$$

where the nth element of the ith observation vector is given by $y_{i,n} = r[n + iN]$. Although each observation vector corresponds to one symbol interval, this buffering was done without regard to the actual symbol intervals of the users. Since the system is asynchronous, each observation vector will contain at least the end of the previous symbol (left) and the beginning of the current symbol (right) for each user. The factors due to the power, phase and transmitted symbols of the kth user may be collected into a single complex constant $c_k^{(i)}$, e.g. some constant times $\sqrt{2P_k}e^{j[\phi_k+(2\pi/M)m_k^{(i)}]}$, and Equation (5.207) becomes:

$$\mathbf{y}_i = \sum_{k=1}^{K} [c_k^{(i-1)}\mathbf{u}_k^r + c_k^{(i)}\mathbf{u}_k^l] + \eta_i = \mathbf{A}\mathbf{c}_i + \eta_i \tag{5.208}$$

where $\eta_i = [\eta_{i,0}, \ldots, \eta_{i,N-1}]^T \in \mathbf{C}^N$ is a Gaussian random vector. Its elements are zero mean with variance $\sigma^2 = N_0/2T_c$ and are mutually independent. Vectors $\mathbf{u}_k^r$ and $\mathbf{u}_k^l$ are the right side of the kth user's code vector followed by zeros, and zeros followed by the left side of the kth user's code vector, respectively. In addition, we have defined $\mathbf{c}_i = [c_1^{(i-1)}c_1^{(i)}\cdots c_K^{(i-1)}c_K^{(i)}]^T \in \mathbf{C}^{2K}$ and the signal matrix. $\mathbf{A} = \lfloor \mathbf{u}_1^r\mathbf{u}_1^\ell \cdots \mathbf{u}_K^r\mathbf{u}_K^\ell \rfloor \in \mathbf{C}^{N\times 2K}$. We will start with the assumption that each user's signal goes through a single propagation path with an associated attenuation factor and propagation delay. We assume that these parameters vary slowly with time, so that for sufficiently short intervals the channel is approximately a linear time-invariant (LTI) system. The baseband channel impulse response can then be represented by a Dirac delta function as $h_k(t, \tau) = h_k(t) = \alpha_k\delta(t - \tau_k)$, $\forall \tau$, where α_k is a complex-valued attenuation weight and τ_k is the propagation delay. Since there is just a single path, we assume that α_k is incorporated into $c_k^{(i)}$ and concentrate solely on the delay.

Let us define $v \in \{0, \ldots, N-1\}$ and $\gamma \in [0, 1)$ such that $(\tau_k/T_c) \bmod N = v + \gamma$. If $\gamma = 0$, i.e. the received signal is precisely aligned with the chip matched filter and only one chip will contribute to each sample, the signal vectors become:

$$\begin{aligned}\mathbf{u}_k^r &= \mathbf{a}_k^r(v) \equiv [a_k^{(N-v)}\cdots a_k^{(N-1)}0\cdots 0]^T \\ \mathbf{u}_k^l &= \mathbf{a}_k^l(v) \equiv [0\cdots 0 \quad a_k^{(0)}\cdots a_k^{(N-v-1)}]^T\end{aligned} \tag{5.209}$$

Since the chip matched filter is just an integrator, the samples for a non-zero γ can be represented as:

$$\begin{aligned}\mathbf{u}_k^r &= (1-\gamma)\mathbf{a}_k^r(v) + \gamma\mathbf{a}_k^r(v+1) \\ \mathbf{u}_k^l &= (1-\gamma)\mathbf{a}_k^l(v) + \gamma\mathbf{a}_k^l(v+1)\end{aligned} \tag{5.210}$$

For the more general case of a multipath transmission channel with L distinct propagation paths, the impulse response becomes a series of delta functions

$$h_k(t, \tau) = h_k(t) = \sum_{p=1}^{L} \alpha_{k,p}\delta(t - \tau_{k,p}) \tag{5.211}$$

The signal vectors can be represented as

$$\begin{aligned}\mathbf{u}_k^r &= \sum_{p=1}^{L}\alpha_{k,p}[(1-\gamma_{k,p})\mathbf{a}_k^r(v_{k,p}) + \gamma_{k,p}\mathbf{a}_k^r(v_{k,p}+1)] \\ \mathbf{u}_k^l &= \sum_{p=1}^{L}\alpha_{k,p}[(1-\gamma_{k,p})\mathbf{a}_k^l(v_{k,p}) + \gamma_{k,p}\mathbf{a}_k^l(v_{k,p}+1)]\end{aligned} \tag{5.212}$$

If we introduce the following notation

$$\begin{aligned}\mathbf{U}_k^r &= \lfloor \mathbf{a}_k^r(0)\cdots \mathbf{a}_k^r(N-1) \rfloor \in \mathbf{C}^{N\times N} \\ \mathbf{U}_k^l &= [\mathbf{a}_k^l(0)\cdots \mathbf{a}_k^l(N-1)] \in \mathbf{C}^{N\times N}\end{aligned} \tag{5.213}$$

where the $\mathbf{a}_k^s$ are as defined in Equation (5.209), then the signal vectors may be expressed as a linear combination of the columns of these matrices

$$
\begin{aligned}
\mathbf{u}_k^r &= \mathbf{U}_k^r \mathbf{h}_k \\
\mathbf{u}_k^i &= \mathbf{U}_k^i \mathbf{h}_k
\end{aligned}
\tag{5.214}
$$

where $\mathbf{h}_k$ is the composite impulse response of the channel and the receiver front end, evaluated modulo the symbol period. Thus, the nth element of the impulse response is given by

$$
h_{k,n} = \sum_{j=0}^{\infty} \frac{1}{T_c} \int_{jT+nT_c}^{jT+(n+1)T_c} h_k(t)^* \prod_{T_c}(t)\,dt
\tag{5.215}
$$

For delay spread $T_m < T/2$, at most two terms in the summation will be non-zero.

5.5.1 Estimating the Signal Subspace

The correlation matrix of the observation vectors is given by:

$$
\mathbf{R} = E\left[\mathbf{y}_i \mathbf{y}_i^\dagger\right] = \mathbf{ACA}^\dagger + \sigma^2 \mathbf{I}
\tag{5.216}
$$

where $\mathbf{C} = E\left[\mathbf{c}_i \mathbf{c}_i^\dagger\right] \in C^{2K \times 2K}$ is diagonal. The correlation matrix can also be expressed in terms of its eigenvector decomposition:

$$
\mathbf{R} = \mathbf{VDV}^\dagger
\tag{5.217}
$$

where the columns of $\mathbf{V} \in C^{N \times N}$ are the eigenvectors of $\mathbf{R}$, and $\mathbf{D}$ is a diagonal matrix of the corresponding eigenvalues (λ_n). Details of eigenvector decomposition are given in Appendix 5.1. Furthermore,

$$
\lambda_n =
\begin{cases}
d_n + \sigma^2, & \text{if } n \le 2K \\
\sigma^2, & \text{otherwise}
\end{cases}
\tag{5.218}
$$

where d_n is the variance of the signal vectors along the nth eigenvector and we assume that $2K < N$. Since the $2K$ largest eigenvalues of $\mathbf{R}$ correspond to the signal subspace, $\mathbf{V}$ can be partitioned as $\mathbf{V} = [\mathbf{V}_S \; \mathbf{V}_N]$, where the columns of $\mathbf{V}_S = [\mathbf{v}_{S,1} \cdots \mathbf{v}_{S,2K}] \in C^{N \times 2K}$ form a basis for the signal subspace S_Y, and $\mathbf{V}_N = [\mathbf{v}_{N,1} \cdots \mathbf{v}_{N,N-2K}] \in C^{N \times N-2K}$ spans the noise subspace N_Y. Readers less familiar with eigenvalue decomposition are referred to Appendix 5.1. Since we would like to track slowly varying parameters, we form a moving average or a Bartlett estimate of the correlation matrix based on the J most recent observations:

$$
\hat{\mathbf{R}}_i = \frac{1}{J} \sum_{j=i-J+1}^{i} \mathbf{y}_j \mathbf{y}_j^\dagger
\tag{5.219}
$$

It is well known [143] that the maximum likelihood estimate of the eigenvalues and associated eigenvectors of $\mathbf{R}$ is just the eigenvector decomposition of $\hat{\mathbf{R}}_i$. Thus, we perform an eigenvalue decomposition of $\hat{\mathbf{R}}_i$ and select the eigenvectors corresponding to the $2K$ largest eigenvalues as a basis for $\hat{S}_Y$.

5.5.2   Channel Estimation

Consider the projection of a given user's signal vectors into the estimated noise subspace

$$
\left.
\begin{aligned}
\mathbf{e}_k^{\mathrm{r}} &= (\mathbf{u}_k^{\mathrm{r}\dagger}\hat{\mathbf{V}}_N)^{\mathrm{T}} \\
\mathbf{e}_k^{l} &= (\mathbf{u}_k^{l\dagger}\hat{\mathbf{V}}_N)^{\mathrm{T}}
\end{aligned}
\right\} \in \mathbf{C}^{N-2K}
\tag{5.220}
$$

If $\mathbf{u}_k^{\mathrm{r}}$ and $\mathbf{u}_k^{l}$ both lie in the signal subspace, then their sum $\mathbf{u}_k = \mathbf{u}_k^{\mathrm{r}} + \mathbf{u}_k^{l}$ must also be contained in $\mathbf{V}_S$. The projection of $\mathbf{u}_k$ into the estimated noise subspace

$$
\tilde{\mathbf{e}}_k = (\mathbf{u}_k^{\dagger}\hat{\mathbf{V}}_N)^{\mathrm{T}}
\tag{5.221}
$$

is a Gaussian random vector [115] and thus has probability density function

$$
p_{\tilde{\mathbf{e}}}(\tilde{\mathbf{e}}_k) = \frac{1}{\det[\pi\,\mathbf{K}]} \exp\{-\tilde{\mathbf{e}}_k^{\dagger}\mathbf{K}^{-1}\tilde{\mathbf{e}}_k\}
\tag{5.222}
$$

The covariance matrix $\mathbf{K}$ is a scalar multiple of the identity given by

$$
\mathbf{K} = \frac{1}{J}\mathbf{u}_k^{\dagger}\mathbf{Q}\mathbf{u}_k\mathbf{I}
\tag{5.223}
$$

and

$$
\mathbf{Q} = \sigma^2 \left[\sum_{k=1}^{2K} \frac{\lambda_k}{(\sigma^2 - \lambda_k)^2} \mathbf{v}_{S,k}\mathbf{v}_{S,k}^{\dagger} \right]
\tag{5.224}
$$

Therefore, within an additive constant, the log likelihood function of $\hat{\mathbf{e}}_k$ is:

$$
\begin{aligned}
\Lambda(\tilde{\mathbf{e}}_k) &= -(N - 2K)\ln(\mathbf{u}_k^{\dagger}\mathbf{Q}\mathbf{u}_k) - J\frac{\mathbf{e}_k^{\dagger}\mathbf{e}_k}{\mathbf{u}_k^{\dagger}\mathbf{Q}\mathbf{u}_k} \\
&= -(N - 2K)\ln(\mathbf{u}_k^{\dagger}\mathbf{Q}\mathbf{u}_k) - J\frac{\mathbf{u}_k^{\dagger}\mathbf{V}_N\mathbf{V}_N^{\dagger}\mathbf{u}_k}{\mathbf{u}_k^{\dagger}\mathbf{Q}\mathbf{u}_k}
\end{aligned}
\tag{5.225}
$$

The exact $\mathbf{V}_N$ and $\mathbf{Q}$ are unknown, but we may replace them with their estimates. The best estimates will minimize $\tilde{\mathbf{e}}_k$, which will result in the maximum of the likelihood function.

Unfortunately, maximizing this likelihood function is prohibitively complex for a general multipath channel, so we will consider only a single propagation path. In this case, the vector $\mathbf{u}_k$ is a function of only one unknown parameter: the delay τ_k. To form the timing estimate, we must solve

$$
\hat{\tau}_k = \arg\max_{\tau_k \in [0,T)} \Lambda(\mathbf{u}_k)
\tag{5.226}
$$

Ideally, we would like to differentiate the log likelihood function with respect to τ. However, the desired user's delay lies within an uncertainty region, $\tau_k \in [0, T]$, and $\mathbf{u}_k(\tau)$ is only piecewise continuous on this interval. To deal with these problems, we divide the uncertainty region into N cells of width T_c and consider a single cell, $c_v \equiv [vT_c, (v + 1)T_c)$. We again define $v \in \{0, \ldots, N - 1\}$ and $\gamma \in [0, 1)$ such that $(\tau/T_c) \bmod N = v + \gamma$, and for $\tau \in c_v$ the desired user's signal vector becomes

$$
\mathbf{u}_k\{\tau\} = (1 - \gamma)\mathbf{u}_k(v) + \gamma\mathbf{u}_k(v + 1)
\tag{5.227}
$$

and

$$\frac{d}{dt}\mathbf{u}_k(\tau) = \mathbf{u}_k(v+1) - \mathbf{u}_k(v)$$

$$= \text{a constant} \tag{5.228}$$

Thus, within a given cell, we can differentiate the log likelihood function and solve for the maximum in closed form. We then choose whichever of the N solutions yields the largest value for Equation (5.226). Details can be found in [116].

Under certain conditions, it may be possible to simplify this algorithm. Note that maximizing the log likelihood function (5.226) is equivalent to maximizing

$$\Lambda(\tilde{\mathbf{e}}_k) = -\frac{N-2K}{J}\ln(\mathbf{u}_k^\dagger \mathbf{Q}\mathbf{u}_k) - \frac{\mathbf{u}_k^\dagger \mathbf{V}_N \mathbf{V}_N^\dagger \mathbf{u}_k}{\mathbf{u}_k^\dagger \mathbf{Q}\mathbf{u}_k} \tag{5.229}$$

As $J \to \infty$, the leading term goes to zero; thus, for large observation windows, we can use the following approximation:

$$\Lambda(\tilde{\mathbf{e}}_k) \approx -\frac{\mathbf{u}_k^\dagger \mathbf{V}_N \mathbf{V}_N^\dagger \mathbf{u}_k}{\mathbf{u}_k^\dagger \mathbf{Q}\mathbf{u}_k} \tag{5.230}$$

This yields a much simpler expression for the stationary points [116]. The MUSIC (Multiple Signal Classification) algorithm is equivalent to Equation (5.29) when one only maximizes the numerator and ignores the denominator, i.e. one assumes $\mathbf{u}_k^\dagger \mathbf{Q}\mathbf{u}_k$ is equal to one in Equation (5.28) or Equation (5.29). This yields an even simpler approximation for the log likelihood function

$$\Lambda(\tilde{\mathbf{e}}_k) \approx -\mathbf{u}_k^\dagger \mathbf{V}_N \mathbf{V}_N^\dagger \mathbf{u}_k \tag{5.231}$$

which further simplifies the solution for the stationary points [116].

For illustration purposes, the simulation results for five users with length 31 Gold codes are presented in Figures 5.28–5.30.

A single desired user was acquired and tracked in the presence of strong multiple access interference (MAI). The power ratio between each of the four interfering users and the desired user is designated the MAI level.

We first compare the true log likelihood estimate, Equation (5.225), with the large observation window approximation, Equation (5.29), and the MUSIC algorithm, Equation (5.30). This is done for a window size of 200 symbols and with a varying SNR. Figure 5.28(a) shows the probability of acquisition for each method, where acquisition is defined as $|\tau_k - \hat{\tau}_k| < 1/2T_c$. Using the approximate log likelihood function results in almost no drop in performance. Furthermore, when the SNR is poor, both probabilistic approaches considerably outperform the MUSIC algorithm. In Figure 5.28(b), we compare the RMSE of the delay estimate once acquisition has occurred, i.e. after processing enough symbols to reach within half of one chip. The approximate log-likelihood function experiences a slight increase in error at low SNR, but again both probabilistic methods do better than MUSIC.

The same parameters as a function of the window size are shown in Figure 5.29. One can say that for $J > 100$ the performance curve settles down to steady state values. The RMSE versus MAI and SNR are shown in Figure 5.30. One can see that for an extremely wide ranging near–far effect the performance is good.

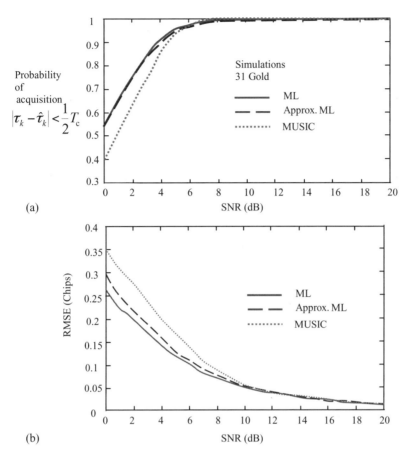

Figure 5.28 (a) Probability of acquisition for the maximum likelihood estimator, the approximate ML and the MUSIC algorithm [$K = 5$, $N = 31$, $J = 200$, MAI $= 20$ dB]; (b) Root mean squared error (RMSE) of the delay estimate in chips for the maximum likelihood (ML) estimator, the approximate ML and the MUSIC algorithm [$K = 5$, $N = 31$, $J = 200$, MAI $= 20$ dB].

5.6 Iterative Receivers for Layered Space–Time Coding

Space–time trellis codes have a potential drawback in that the maximum likelihood decoder complexity grows exponentially with the number of bits per symbol, thus limiting achievable data rates. In Chapter 4.10 we discussed a layered space–time (LST) architecture that can attain a tight lower bound on the MIMO channel capacity with significant reduction in complexity. In this section we further generalize the LST concept and discuss a variety of receiver schemes based on iterative decoding and interference cancellation algorithms as described in Sections 5.2–5.4.

5.6.1 LST Architectures

There is a number of various LST architectures, depending on the way the modulated symbols are assigned to transmit antennas. In uncoded LST structure, known as *vertical layered space–time* (VLST) or *vertical Bell Laboratories layered space–time* (VBLAST), the input information sequence, denoted by c, is first demultiplexed into n_T sub-streams and each of them is subsequently modulated by an M-level modulation scheme and transmitted from a transmit antenna. The signal processing chain related to an individual

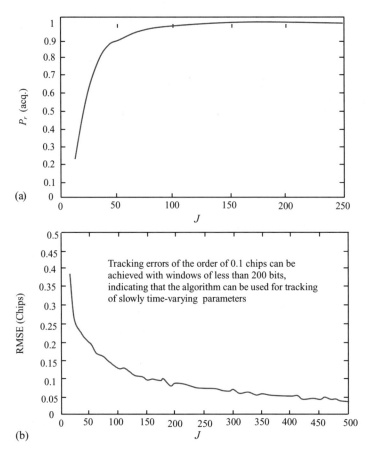

Figure 5.29 (a) Probability of acquisition and (b) root mean squared error (RMSE) of timing estimate in chips of the subspace-based maximum likelihood estimator for varying window size [$N = 31$, SNR $= 8$ dB, $K = 5$, MAI $= 20$ dB] [116] © 1996, IEEE.

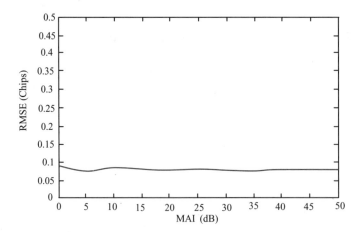

Figure 5.30 RMSE of the subspace-based maximum likelihood estimator for varying MAI level [$K = 5$, $N = 31$, $J = 200$, SNR $= 8$ dB].

sub-stream is referred to as a *layer*. The modulated symbols are arranged into a transmission matrix, denoted by $\mathbf{X}$, which consists of n_T rows and L columns, where L is the transmission block length. The tth column of the transmission matrix, denoted by x_t, consists of the modulated symbols $x_t^1, x_t^2, \ldots, x_t^{n_T}$, where $t = 1, 2, L$ At a given time t, the transmitter sends the tth column from the transmission matrix, one symbol from each antenna. That is, a transmission matrix entry x_t^i is transmitted from antenna i at time t. Vertical structuring refers to transmitting a sequence of matrix columns in the space–time domain This simple transmission process can be combined with conventional block or convolutional one-dimension codes, to improve the performance of the system. The term 'one-dimensional' refers to the space domain, while these codes can be multidimensional in the time domain. The block diagrams of various LST architectures with error control coding are shown in Figure 5.31.

In the *horizontal layered space–time* (HLST) architecture, shown in Figure 5.31(a), the information sequence is first encoded by a channel code and subsequently demultiplexed into n_T sub-streams. Each sub-stream is modulated, interleaved and assigned to a transmit antenna.

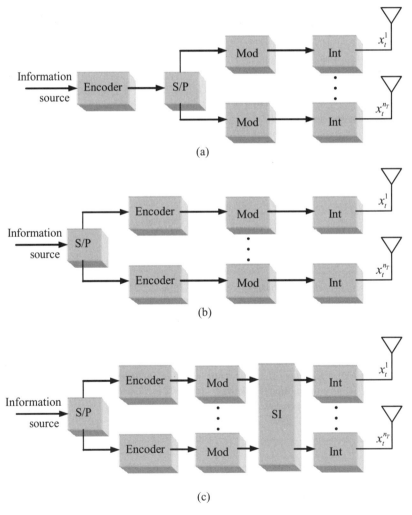

Figure 5.31   LST transmitter architectures with error control coding: (a) an HLST architecture with a single code; (b) an HLST architecture with separate codes in each layer; (c) DLST and TLST architectures.

If the modulator output symbols are denoted by x_t^i, where i represents the layer index and t is the time interval, the transmission matrix, formed from the modulator outputs, denoted by X, is given by

$$\mathbf{X} = \left[x_t^i \right] \tag{5.232}$$

In a system with three transmit antennas, the transmission matrix X is given by

$$\mathbf{X} = \begin{bmatrix} x_1^1 & x_2^1 & x_3^1 & x_4^1 & \cdots \\ x_1^2 & x_2^2 & x_3^2 & x_4^2 & \cdots \\ x_1^3 & x_2^3 & x_3^3 & x_4^3 & \cdots \end{bmatrix} \tag{5.233}$$

The sequence $x_1^1, x_2^1, x_3^1, x_4^1, \ldots$ is transmitted from antenna 1, the sequence $x_1^2, x_2^2, x_3^2, x_4^2, \ldots$ from antenna 2 and the sequence $x_1^3, x_2^3, x_3^3, x_4^3, \ldots$ transmitted from antenna 3.

An HLST architecture can also be implemented by splitting the information sequence into n_T substreams, as shown in Figure 5.31(b). After that, each sub-stream is encoded independently by a channel encoder, interleaved, modulated and then transmitted by a particular transmit antenna. We assume that channel encoders for various layers are identical although different coding in each sub-stream can be used.

A better performance is achieved by a *diagonal layered space–time* (DLST) architecture as discussed in Section 4.16, in which a modulated codeword for each encoder is distributed among the n_T antennas along the diagonal of the transmission array. For example, the DLST transmission matrix, for a system with three antennas, is formed from matrix $\mathbf{X}$ in (5.233), by delaying the ith row entries by $(i-1)$ time units, so that the first nonzero entries lie on a diagonal in $\mathbf{X}$. The entries below the diagonal are padded by zeros. Then the first diagonal is transmitted from the first antenna, the second diagonal from the second antenna, the third diagonal from the third antenna, and then the fourth diagonal from the first antenna, etc. Hence the codeword symbols of each encoder are transmitted over different antennas.

This can be represented by introducing a spatial interleaving SI after the modulators, as shown in Figure 5.31(c). The spatial interleaving operation on for the DLST scheme can be represented as:

$$\begin{bmatrix} x_1^1 & x_2^1 & x_3^1 & x_4^1 & x_5^1 & x_6^1 & \cdots \\ 0 & x_1^2 & x_2^2 & x_3^2 & x_4^2 & x_5^2 & \cdots \\ 0 & 0 & x_1^3 & x_2^3 & x_3^3 & x_4^3 & \cdots \end{bmatrix} \rightarrow \begin{bmatrix} x_1^1 & x_1^2 & x_1^3 & x_4^1 & x_4^2 & x_4^3 & \cdots \\ 0 & x_2^1 & x_2^2 & x_2^3 & x_5^1 & x_5^2 & \cdots \\ 0 & 0 & x_3^1 & x_3^2 & x_3^3 & x_6^1 & \cdots \end{bmatrix} \tag{5.234}$$

The rows of the matrix on the right-hand side of Equation (5.234) are obtained by concatenating the corresponding diagonals of the matrix on the left-hand side. The first row of this matrix is transmitted from the first antenna, the second row from the second antenna and the third row from the third antenna. The diagonal layering introduces space diversity and thus achieves a better performance than the horizontal one. It is important to note that there is a spectral efficiency loss in DLSV, since a portion of the transmission matrix on the left-hand side of Equation (5.234) is padded with zeros.

A *threaded layered space–time* (TLST) structure [144] is obtained from the HLSI by introducing a spatial interleaver SI prior to the time interleavers, as shown in Figure 5.31(c). In a system with $n_T = 3$, the operation of SI can be expressed as

$$\begin{bmatrix} x_1^1 & x_2^1 & x_3^1 & x_4^1 & \cdots \\ x_1^2 & x_2^2 & x_3^2 & x_4^2 & \cdots \\ x_1^3 & x_2^3 & x_3^3 & x_4^3 & \cdots \end{bmatrix} \rightarrow \begin{bmatrix} x_1^1 & x_2^3 & x_3^1 & x_4^1 & \ddots \\ x_1^2 & x_2^1 & x_3^3 & x_4^2 & \ddots \\ x_1^3 & x_2^3 & x_3^1 & x_4^3 & \ddots \end{bmatrix} \tag{5.235}$$

in which an element of the modulation matrix, shown on the left-hand side of Equation (5.235) denoted by x_t^i, represents the modulated symbol of layer i at time t. The matrix X' on the right-hand side of

Equation (5.235), is the TLST transmission matrix. The modulated symbols $x_1^1, x_2^3, x_3^2, x_4^1, \ldots$ generated by modulators in layers 1, 3, 2 and 1, respectively, are transmitted from antenna 1.

The spatial interleaver of the TLST can be represented by a cyclic-shift interleaver as follows. If we denote the left-hand side matrix in Equation (5.235) by $\mathbf{X}$, the first column of the transmission matrix $\mathbf{X}'$ is identical to the first column of the modulated matrix $\mathbf{X}$. The second column of $\mathbf{X}'$ is obtained by a cyclic shift of the second column of $\mathbf{X}$ by one position from the top to the bottom. The third column of $\mathbf{X}'$ is obtained by a cyclic shift of the third column of $\mathbf{X}$ by two positions, while the fourth column of $\mathbf{X}'$ is identical to the fourth column of $\mathbf{X}$, etc. In general, if we denote the entries of $\mathbf{X}'$ by $x_t^{i'}$, the mapping of x_t^i to $x_t^{i'}$ can be expressed as $x_t^{i'} = x_t^i, i' = [(i + t - 2) \mod n_T] + 1$.

The spectral efficiency HLST and TLST schemes is Rmn_T, where R is the code rate and m is the number of bits in a modulated symbol, while the spectral efficiency of the DLST is slightly reduced due to zero padding in the transmission matrix.

5.6.2   LST Receivers

In this section we discuss receiver structures for layered space–time architectures. For simplicity, horizontal layering with binary channel codes and BPSK modulation are assumed. Extension to nonbinary codes and to multilevel modulation schemes is straightforward.

The transmit diversity introduces spatial interference. The signals transmitted from various antennas propagate over independently scattered paths and interfere with each other upon reception at receiver. This interference can be represented by the following matrix operation:

$$\mathbf{r}_t = \mathbf{H}\mathbf{x}_t + \mathbf{n}_t \qquad (5.236)$$

where $\mathbf{r}_t$ is an n_R component column matrix of the received signals across the n_R receive antennas, $\mathbf{x}_t$ is the tth column in the transmission matrix $\mathbf{X}$ and $\mathbf{n}_t$ is an n_R-component column matrix of the AWGN noise signals from the receive antennas. The noise variance per receive antenna is σ^2. In a structure with spatial interleaving, vector $\mathbf{x}_t$ is the tth column of the matrix at the output of the spatial interleaver, denoted by $\mathbf{X}'$. In order to simplify the notation, we omit the subscripts in vectors $\mathbf{r}_t$, $\mathbf{x}_t$ and $\mathbf{n}_t$ and refer to them as $\mathbf{r}$, $\mathbf{x}$, and $\mathbf{n}$, respectively.

In accordance with the discussion presented so far in this chapter, an LST structure can be viewed as a synchronous CDMA in which the number of transmit antennas is equal to the number of users. Similarly, the interference between antennas is equivalent to multiple access interference (MAI) in CDMA systems, while the complex fading coefficients correspond to the spreading sequences. This analogy can be further extended to receiver strategies, so that multiuser receiver structures derived for CDMA in Sections 5.2–5.4 can be directly applied to LST systems. Under this scenario, the optimum receiver for an uncoded LST system is a maximum likelihood (ML) multiuser detector operating on a trellis. It computes *ML* statistics (see Equation (5.49)) as in the Viterbi algorithm. The complexity of this detection algorithm is exponential in the number of the transmit antennas.

For coded LST schemes, the optimum receiver performs joint detection and decoding on an overall trellis obtained by combining the trellises of the layered space–time code and the channel code. The complexity of the receiver is an exponential function of the product of the number of the transmit antennas and the code memory order. In this section we will examine a number of less complex receiver structures that have good performance/complexity trade-offs.

The original VLST receiver, described in Section 4.10, is based on a combination of interference suppression and cancellation. Conceptually, each transmitted sub-stream is considered to be the desired symbol and the remainder are treated as interferers. In [145] these interferers are suppressed by a zero forcing (ZF) approach which corresponds to decorrelating detector discussed in Section 5.2.2. This detection algorithm produces ZF-based decision statistics for a desired sub-stream from the received signal vector $\mathbf{r}$, which contains a residual interference from other transmitted sub-streams. Subsequently, a decision on the desired sub-stream is made from the decision statistics and its interference contribution

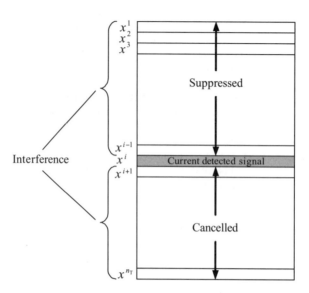

Figure 5.32   VLST detection based on combined interference suppression and successive interference cancellation (IS/SIC).

is regenerated and subtracted out from the received vector **r**. Thus **r** contains a lower level of interference and this will increase the probability of correct detection of other sub-streams.

This operation is illustrated in Figure 5.32 where, the first detected sub-stream is n_T. The detected symbol is subtracted from all other layers. These operations are repeated for the lower layers, until layer 1 is detected. Assuming that all symbols at previous layers have been detected correctly, the decision statistics for layer 1 will be free from interference. The soft decision statistics (see Section 2.3) from the detector at each layer are passed to a decision making device in a VBLAST system. In coded LST schemes, the decision statistics are passed to the channel decoder, which makes the hard decision on the transmitted symbol in this sub-stream. The hard symbol estimate is used to reconstruct the interference from this sub-stream, which then fed back to cancel its contribution while decoding the next sub-stream.

The ZF strategy is only possible if the number of receive antennas at least as large as the number of transmit antennas. Another drawback of this approach is that achievable diversity depends on a particular layer. If the ZF strategy is used in removing interference and if n_R receive antennas are available, it is possible to remove $n_i = n_R - d_o$ interferences with diversity order of d_o [146]. The diversity order can be expressed as $d_o = n_R - n_i$.

If the interference suppression starts at layer n_T, then at this layer $(n_T - 1)$ interferers need to be suppressed. Assuming that $n_R = n_T$, the diversity order in this layer 1. In the first layer, there are no interfernce to be suppressed, so the diversity order is $n_R = n_T$ As different layers have different diversity orders, diagonal layering is required to achieve equal performance of various encoded streams.

Apart from the original BLAST receivers we will consider minimum mean square error (MMSE) detectors, discussed in Section 5.3 and iterative receivers. The iterative receiver [147,148] based on the turbo processing principle introduced in Section 2.3.1, is the architecture with the best complexity/performance trade-off. Its complexity grows linearly with the number of transmit antennas and transmission rate.

5.6.3   QR Decomposition/SIC Detector

As discussed in Appendix 4.1, any $n_R \times n_T$ matrix **H**, where $n_R \geq n_T$, can be decomposed as $\mathbf{H} = \mathbf{U}_R \mathbf{R}$, where $\mathbf{U}_R$ is an $n_R \times n_T$ unitary matrix and **R** is an $n_T \times n_T$ upper triangular matrix, with entries

$(\mathbf{R}_{i,j})_t = 0$, for $i > j$, $i, j = 1, 2, \ldots, n_T$, represented as:

$$
\mathbf{R} =
\begin{bmatrix}
(R_{1,1})_t & (R_{1,2})_t & \cdots & (R_{1,n_T})_t \\
0 & (R_{2,2})_t & \cdots & (R_{2,n_T})_t \\
0 & 0 & \cdots & (R_{3,n_T})_t \\
\vdots & \vdots & \vdots & \vdots \\
0 & 0 & \cdots & (R_{n_T,n_T})_t
\end{bmatrix}
\tag{5.237}
$$

If we introduce an n_T-component column matrix y obtained by multiplying from the left the receive vector $\mathbf{r}$, given by Equation (5.237), by $\mathbf{U}_R^T$ then we have $\mathbf{y} = \mathbf{U}_R^T \mathbf{r}$ or $\mathbf{y} = \mathbf{U}_R^T \mathbf{H} \mathbf{x} + \mathbf{U}_R^T \mathbf{n}$. Substituting the **QR** decomposition of **H**, we get $\mathbf{y} = \mathbf{R}\mathbf{x} + \mathbf{n}'$, where $\mathbf{n}' = \mathbf{U}_R^T \mathbf{n}$ is an n_T-component column matrix of i.i.d AWGN noise signals. As **R** is upper-triangular, the ith component in **y** depends only on the ith and higher layer transmitted symbols at time t, as follows:

$$
y_t^i = (R_{i,i})_t \, x_t^i + n_t'^i + \sum_{j=i+t}^{n_T} (R_{i,j})_t \, x_t^j
\tag{5.238}
$$

Consider x_t^i as the current desired detected signal. Equation (5.238) shows that y_t^j contains a lower level of interference than in the received signal r_t, as the interference from x_t^l, for $1 < i$, are suppressed. The third term in Equation (5.238) represents contributions from other interferers $x_t^{i+1}, x_t^{i+2}, \ldots, x_t^{n_T}$, which can be cancelled by using the available decisions $\hat{x}_t^{i+1}, \hat{x}_t^{i+2}, \ldots, \hat{x}_t^{n_T}$, that have been detected. The decision statistics on x_t^i, denoted by y_t^i, can be rewritten as:

$$
y_t^i = \sum_{j=i}^{n_T} (R_{i,j})_t x_t^j + n_t'^i \quad i = 1, 2, \ldots, n_T
\tag{5.239}
$$

The estimate on the transmitted symbol x_t^i is given by

$$
x_{t^i}' = q \left(\frac{y_t^i - \displaystyle\sum_{j=i+1}^{n_T} (R_{i,j})_t \, \hat{x}_t^j}{(R_{i,i})_t} \right) \quad i = 1, 2, \ldots, n_T
\tag{5.240}
$$

where $q(x)$ denotes the hard decision on x. A QR factorization algorithms are discussed in Appendix 4.1.

For a system with three transmit antennas, the decision statistics for various layers can be expressed as

$$
y_t^1 = (R_{1,1})_t x_t^1 + (R_{1,2})_t x_t^2 + (R_{1,3})_t x_t^3 + n'^1
$$

$$
y_t^2 = (R_{2,2})_t x_t^2 + (R_{2,3})_t x_t^3 + n'^2
$$

$$
y_t^3 = (R_{3,3})_t x_t^3 + n'^3
$$

The estimate on the transmitted symbol x_t^3, denoted by $\hat{x}_t^3$, can be obtained from y_t^3 as

$$
\hat{x}_t^3 = q \left(\frac{y_t^3}{(R_{3,3})_t} \right)
$$

The contribution of $\hat{x}_t^3$ is cancelled from y_t^2 and the estimate on x_t^2 is obtained as

$$\hat{x}_i^2 = q \left(\frac{y_t^2 - (R_{2,3})_t \hat{x}_t^3}{(R_{2,2})_t} \right)$$

Finally, after cancelling out $\hat{x}_t^3$ and $\hat{x}_t^2$, we obtain for

$$\hat{x}_t^1 \; \hat{x}_t^1 = q \left(\frac{(y_t^1 - (R_{1,3})_t \hat{x}_t^3 - (R_{1,2})_t \hat{x}_t^2}{(R_{1,1})_t} \right)$$

The described algorithm applies to VBLAST. In coded LST schemes, the soft decision statistics on x_t^i, given by the arguments in the $q(\cdot)$ expressions are passed to the channel decoder, which estimates $\hat{x}_t^i$.

In the above example the decision statistic $y_t^{n_T}$ is computed first, then $y_t^{n_T - 1}$, and so on. The performance can be improved if the layer with the maximum SNR is detected first, followed by the one with the next largest SNR and so on [149].

5.6.4 MMSE/SIC Detector

In the MMSE detection algorithm introduced in Section 5.3, the expected value of the mean square error between the transmitted vector **x** and a linear combination of the received vector $\mathbf{W}^H \mathbf{r}$ *is* minimized. With this notation Equation (5.123) becomes:

$$\min E \left\{ \left(\mathbf{x} - \mathbf{W}^H \mathbf{r} \right)^2 \right\} \tag{5.241}$$

where **W** is an $n_R \times n_T$ matrix of linear combination coefficients given by:

$$\mathbf{W}^H = \left[\mathbf{H}^H \mathbf{H} + \sigma^2 \mathbf{I}_{n_T} \right]^{-1} \mathbf{H}^H \tag{5.242}$$

σ^2 is the noise variance and $\mathbf{I}_{n_T}$ is an $n_T \times n_T$ identity matrix. The decision statistics for the symbol sent from antenna i at time t is obtained as:

$$y_t^i = \mathbf{W}_i^H \mathbf{r} \tag{5.243}$$

where $\mathbf{W}_i^H$ is the ith row of $\mathbf{W}^H$ consisting of n_R components. The estimate of the symbol sent by antenna i, denoted by $\hat{x}_t^i$, is obtained by making a hard decision on y_t^i:

$$\hat{x}_i^t = q \left(y_i^t \right) \tag{5.244}$$

In an algorithm with interference suppression only, the detector calculates the hard decisions estimates by using Equations (5.243) and (5.244) for all transmit antennas. In a combined interference suppression and interference cancellation, the receiver starts from antenna n_T and computes its signal estimate by using Equations (5.243) and (5.244). The received signal **r** in this level is denoted by $\mathbf{r}^{n_T}$. For calculation of the next antenna signal ($n_T - 1$), the interference contribution of the hard estimate $\hat{x}_t^{n_T}$ is subtracted from the received signal $\mathbf{r}^{n_T}$ and this modified received signal, denoted by $\mathbf{r}^{n_T - 1}$, is used in computing the decision statistics for antenna ($n_T - 1$) in Equation (5.243) and its hard estimate from Equation (5.244). In the next level, corresponding to antenna ($n_T - 2$), the interference from $n_T - 1$ is subtracted from the received signal $\mathbf{r}^{n_T - 1}$ and this signal is used to calculate the decision statistics in Equation (5.243) for antenna ($n_T - 2$). This process continues for all other levels up to the first antenna. After detection of level i, the hard estimate $\hat{x}_t^i$ is subtracted from the received signal to remove its interference contribution, giving the received signal for level $i - 1$:

$$\mathbf{r}^{i-1} = \mathbf{r}^i - \hat{x}_t^i \mathbf{h}_i \tag{5.245}$$

where $\mathbf{h}_i$ is the ith column in the channel matrix $\mathbf{H}$, corresponding to the path attenuations from antenna i. The operation $\hat{x}_t^i \mathbf{h}_i$ in Equation (5.245) replicates the interference contribution caused by $\hat{x}_t^i$ in the received vector. $\mathbf{r}^{i-1}$ is the received vector free from interference coming from $\hat{x}_t^{n_T}, \hat{x}_t^{n_T-1}, \ldots, \hat{x}_t^i$. For estimation of the next antenna signal x_t^{i-1}, this signal $\mathbf{r}^{i-1}$ *is* used in Equation (5.243) instead of $\mathbf{r}$. Finally, a deflated version of the channel matrix is calculated, denoted by $\mathbf{H}_d^{i-1}$, by deleting a column from $\mathbf{H}_d^i$. The deflated matrix $\mathbf{H}_d^{i-1}$ at the $(n_T - i + 1)$th cancellation step is given by:

$$\mathbf{H}_d^{i-1} = \begin{bmatrix} h_{1,1} & h_{1,2} & \cdots & h_{1,i-1} \\ h_{2,1} & h_{2,2} & \cdots & h_{2,i-1} \\ \vdots & \vdots & \vdots & \vdots \\ h_{n_R,1} & h_{n_R,2} & \cdots & h_{n_R,i-1} \end{bmatrix} \tag{5.246}$$

This definition is needed as the interference associated with the current symbol has been removed. This deflated matrix $\mathbf{H}_d^{i-1}$ is used in Equation (5.242) for computing the MMSE coefficients and the signal estimate from antenna $i - 1$. Once the symbols from each antenna have been estimated, the receiver repeats the process on the vector $\mathbf{r}_{t+1}$ received at time $(t + 1)$. The algorithm can be summarized by the following pseudo code

---

MMSE/SIC Algorithm

Set $i = n_T$

and $\mathbf{r}^{n_T} = \mathbf{r}$

while $i \geq 1$

{

$\mathbf{W}^H = [\mathbf{H}^H \mathbf{H} + \sigma^2 \mathbf{I}_{n_T}]^{-1} \mathbf{H}^H$

$y_t^i = \mathbf{W}_i^H \mathbf{r}^i$

$\hat{x}_i^t = q(y_i^t)$

$\mathbf{r}^{i-1} = \mathbf{r}^i - \hat{x}_t^i \mathbf{h}_i$

Compute $\mathbf{H}_d^{i-1}$ by deleting column i from $\mathbf{H}_d^i$.

$\mathbf{H} = \mathbf{H}_d^{i-1}$

$i = i - 1$

}

The receiver can be implemented without the interference cancellation step of Equation (5.245), which will reduce system performance but some computational cost can be saved. Using cancellation requires that MMSE coefficients be recalculated at each iteration, as $\mathbf{H}$ is deflated. With no cancellation, the MMSE coefficients are only computed once, as n remains unchanged. The most computationally intensive operation detection algorithm is the computation of the MMSE coefficients. A direct calculation of the MMSE coefficients based on Equation (5.242), has a complexity polynomial in the number of

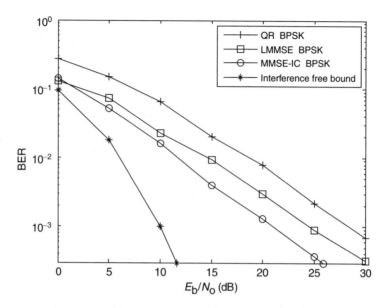

Figure 5.33 V-BLAST example, $n_T = 4$, $n_R = 4$, with QR decomposition, MMSE interference suppression and MMSE/SIC.

transmit antennas. However, on slow fading channels, it is possible to implement adaptive MMSE receivers with the complexity being linear in the number of transmit antennas.

The described algorithm is for uncoded LST systems. The same detector can be applied to coded systems. The receiver consists of the MMSE interference suppressor/canceller (MMSE/SIC) followed by the decoder. The decision statistics y_t^i, from Equation (5.243), is passed to the decoder which makes the decision on the symbol estimate $\hat{x}_t^i$.

The performance of a QR decomposition receiver (QR), the linear MMSE (LMMSE) detector and the performance of the last detected layer in MMSE/SIC are shown for a VBLAST structure with $n_T = 4$, $n_R = 4$ and BPSK modulation on a slow Rayleigh fading channel in Figure 5.33. The figure also shows the interference free (single layer) BER which is given by [150]

$$P_b = \left[\frac{1}{2} (1 - \mu) \right]^{n_R} \sum_{k=0}^{k=n_R-1} \left[\frac{1}{2} (1 + \mu) \right]^k \tag{5.247}$$

where $\mu = \sqrt{\dfrac{\frac{\gamma_b}{n_R}}{1 + \frac{\gamma_b}{n_R}}}$ and $\gamma_b = \dfrac{E_b}{N_o}$.

One of the disadvantages of the MMSE scheme with successive interference cancellation is that the first desired detected signal to be processed sees all the interference from the remaining $(n_T - 1)$ signals, whereas each antenna signal to be processed later sees less and less interference as the cancellation progresses. This problem can be alleviated either by ordering the layers to be processed in decreasing signal power or by assigning power to the transmitted signals according to the processing order. Another disadvantage of the successive scheme is that a delay of n_T computation stages is required to carry out the cancellation process.

The complexity of the LST receiver can be further reduced by replacing the MMSE interference suppressor by a matched filter, resulting in interference cancellation only.

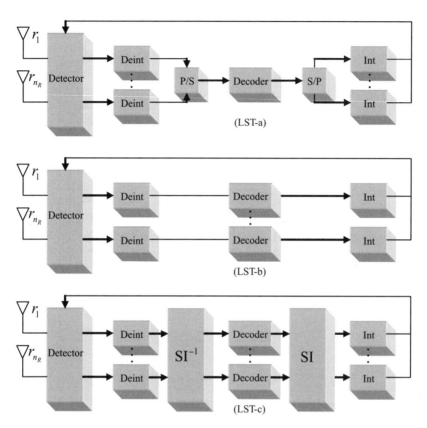

Figure 5.34 Block diagrams of iterative LSTC receivers; (LST-a) HLST with a single decoder; (LST-b) HLST with separate decoders; (LST-c) DLST and TLST receivers.

5.6.5   Iterative LST Receivers

The objective in the detection of space–time signals is to design a low-complexity detector, which can efficiently remove multilayer interference and approach the interference free bound. The iterative processing principle, as applied in turbo coding and discussed in Section 2.3.1, has been successfully extended to joint detection and decoding [151–158]. This receiver can be applied only in coded LST systems. Block diagrams of the iterative receivers for LST(a)–(c) architectures are shown in Figure 5.34. In all three receivers, the detector provides joint soft-decision estimates of the n_T transmitted symbol sequences. In LST(a) the detected sequence is decoded by a single decoder with soft inputs/soft outputs, while in LST (b) each of the detected sequences is decoded by a separate channel decoder with soft inputs/soft outputs. At each iteration, the decoder soft outputs are used to update the *a priori* probabilities of the transmitted signals. These updated probabilities are then used to calculate the symbol estimate in the detector. Note that each of the coded streams is independently interleaved to enable receiver convergence. In LST (c), apart from time interleaving/deinterleaving, there is space interleaving/deinterleaving across transmit antennas.

5.6.5.1   Iterative Receiver (IR)/PIC

A block diagram of the standard iterative receiver (IR) with a parallel interference canceller (PIC-STD) is shown in Figure 5.35 as IR/PIC structure. For simplicity we assume that an HLST architecture with

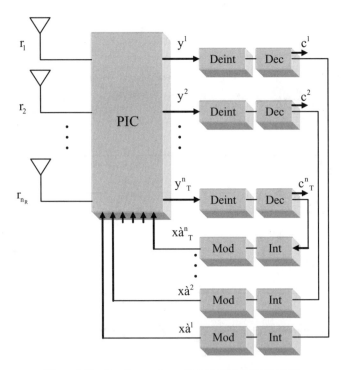

Figure 5.35   Iterative receiver with PIC-STD (IR/PIC-STD).

separate error control coding in each layer is used. In addition, the same convolutional codes with BPSK modulation are selected in each layer. In the first iteration, the PIC detectors are equivalent to a bank of matched filters. The detectors provide decision statistics of the n_T transmitted symbol sequences. The decision statistics in the first iteration, for antenna i and time t, denoted by $y_t^{i,1}$, is determined as $y_t^{i,1} = \mathbf{h}_i^H \mathbf{r}$, where $\mathbf{h}_i^H$ is the ith row of matrix $\mathbf{H}^H$. These decision statistics are passed to the respective decoders, which generate soft estimates on the transmitted symbols. In the second and later iterations, the decoder soft output is used to update the PIC detector decision statistics.

The decision statistics in the kth iteration at time t, for transmit antenna i denoted by $y_t^{i,k}$, is given by $y_t^{i,k} = \mathbf{h}_i^H(\mathbf{r} - \mathbf{H}\hat{\underline{\mathbf{X}}}_i^{k-1})$ where $\hat{\mathbf{X}}_i^{k-1}$ is an $n_T \times 1$ column matrix with the symbol estimates from the $(k-1)$th iteration a s elements, except for the ith element which is set to zero. It can be written as:

$$\hat{\mathbf{x}}_t^{k-1} = (\hat{x}_t^{1,k-1}, \ldots, \hat{x}_t^{i-1,k-1}, 0, \hat{x}_t^{i+1,k-1}, \ldots, \hat{x}_t^{n_T,k-1})^T$$

The detection outputs for layer i for a whole block of transmitted symbols form a vector, $\mathbf{y}^{i,k}$, which is interleaved and then passed to the ith decoder.

The decoder in the kth iteration calculates the log-likelihood ratios (LLR) for antenna i at time t, denoted by $\lambda_t^{i,k}$ and given by:

$$\lambda_t^{i,k} = \log \frac{P\left(x_t^{i,k} = 1 | \mathbf{y}^{i,k}\right)}{P\left(x_t^{i,k} = -1 | \mathbf{y}^{i,k}\right)}$$

where $P(x_t^{i,k} = j | \mathbf{y}^{i,k})$, $j = 1, -1$, are the symbol *a posteriori* probabilities (APP). The LLR can be calculated by the iterative MAP algorithm (Appendix 2.1). The symbol *a posteriori* probabilities

$P(x_t^{i,k} = j|\mathbf{y}^{i,k})$, $j = 1, -1$, can be expressed as:

$$P\left(x_t^{i,k} = 1|\mathbf{y}^{i,k}\right) = \frac{e^{\lambda_t^{i,k}}}{1 + e^{\lambda_t^{i,k}}}; \quad P\left(x_t^{i,k} = j|\mathbf{y}^{i,k}\right) = \frac{1}{1 + e^{\lambda_t^{i,k}}} \tag{5.248}$$

The estimates of the transmitted symbols are calculated by finding their mean:

$$\hat{x}_t^{i,k} = 1 \cdot P\left(x_t^{i,k} = 1|\mathbf{y}^{i,k}\right) + (-1) \cdot P\left(x_t^{i,k} = -1|\mathbf{y}^{i,k}\right)$$

By combining the above equations we express the symbol estimates as functions of the LLR:

$$\hat{x}_t^{i,k} = \frac{e^{\lambda_t^{i,k}} - 1}{e^{\lambda_t^{i,k}} + 1} \tag{5.249}$$

When the LLR is calculated on the basis of the *a posteriori* probabilities, it is obtained as

$$\lambda_t^{i,k} = \log \frac{\displaystyle\sum_{m,m'=0,x_t^i=1}^{m,m'=M_s-1} \alpha_j - 1\left(m'\right) p_t\left(x_t^i = 1\right) \exp\left(-\frac{\displaystyle\sum_{l=(j-1)n}^{jn}\left(y_l^{i,k} - x_l^i\right)^2}{2\left(\sigma^{i,k}\right)^2}\right) \beta_j(m)}{\displaystyle\sum_{m,m'=0,x_t^i=-1}^{m,m'=M_s-1} \alpha_j - 1\left(m'\right) p_t\left(x_t^i = -1\right) \exp\left(-\frac{\displaystyle\sum_{l=(j-1)n}^{jn}\left(y_l^{i,k} - x_l^i\right)^2}{2\left(\sigma^{i,k}\right)^2}\right) \beta_j(m)} \tag{5.250}$$

where $\lambda_t^{i,k}$ denotes the LLR ratio for the pth symbol within the jth codeword transmitted at time $t = (j - 1)n + p$, and n is the code symbol length, m' and m are the pair of states connected in the trellis, x_t^i is the tth BPSK modulated symbol in a code symbol connecting the states m' and m, $y_t^{i,k}$ is the detector output in iteration k for antenna i at time t, $(\sigma^{i,k})^2$ is the noise variance for layer i, and iteration k, M_s is the number of states in the trellis, and $\alpha(m')$ and $\beta(m)$ are the feed-forward and the feedback recursive variables, defined as for the LLR (Appendix 2.1).

In computing the LLR value in Equation (5.250) the decoder uses two inputs. The first input is the decision statistics, $y_t^{i,k}$, which depends on the transmitted signal x_t^i. The second input is the *a priori* probability on the transmitted signal x_t^i, computed as:

$$P_t\left(x_t^i = l\right) = \frac{1}{\sqrt{2\pi}\sigma} e^{-\frac{\left(y_t^{i,k} - l\mu_t^i\right)^2}{2\sigma^2}} \quad l = 1, -1 \tag{5.251}$$

where μ_t^i is the mean of the received amplitude after matched filtering, given by $\mu_t^i = \mathbf{h}_i^H \mathbf{h}_i$.

As $p_t(x_t^i = l)$ in Equation (5.252) depends also on x_t^i, the inputs to the decoder in iteration k, where $k > 1$, are correlated. This causes the decision statistics mean value, conditional on x_t^i, to be biased [141, 159]. The bias always has a sign opposite to x_t^i. Thus, the bias reduces the useful signal term and degrades the system performance. This bias is particularly significant for a large number of interferers. The bias effect can be eliminated by estimating the mean of the transmitted symbols based on the *a posteriori* extrinsic information ratio (EIR) instead of the LLR [147]. The extrinsic information represents the information on the coded bit of interest calculated from the *a priori* information on the other coded bits and the code constraints. The EIR does not include the metric for the symbol x_t^i that is being estimated.

That is:

$$\lambda_{t,e}^{i,k} = \log \frac{\displaystyle\sum_{m,m'=0,x_t^i=1}^{m,m'=M_s-1} \alpha_j - 1\left(m'\right) p_t\left(x_t^i = 1\right) \exp\left(-\frac{\displaystyle\sum_{l=(j-1)n,l\neq t}^{jn} \left(y_l^{i,k} - x_l^i\right)^2}{2\left(\sigma^{i,k}\right)^2}\right) \beta_j\left(m\right)}{\displaystyle\sum_{m,m'=0,x_t^i=-1}^{m,m'=M_s-1} \alpha_j - 1\left(m'\right) p_t\left(x_t^i = -1\right) \exp\left(-\frac{\displaystyle\sum_{l=(j-1)n,l\neq t}^{jn} \left(y_l^{i,k} - x_l^i\right)^2}{2\left(\sigma^{i,k}\right)^2}\right) \beta_j\left(m\right)} \qquad (5.252)$$

where $\lambda_{t,e}^{i,k}$ denotes the EIR for the pth symbol within the jth codeword transmitted at time $t = (j - 1)n + p$, $y_t^{i,k}$ is the detector output iteration k for antenna i, $\alpha(m')$ and $\beta(m')$ are defined as for the LLR (Appendix 2.1). However, excluding the contribution of the bit of interest reduces the extrinsic information SNR, which leads to a degraded system performance.

A decision statistics combining (DSC) method is effective in minimizing these effects. In the iterative parallel interference canceller with decision statistics combining (PIC-DSC) [147], shown in Figure 5.36, a DSC module is added to the PIC-DSC structure. The decision statistics of the PIC-DSC is generated as a weighted sum of the current PIC output and the DSC output from the previous operation. In each stage, except in the first one, the PIC output is passed to the DSC module. The DSC nodule performs recursive linear combining of the detector output in iteration k for layer i, denoted by $y^{i,k}$, with the DSC output

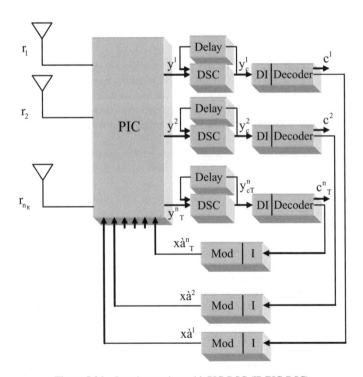

Figure 5.36   Iterative receiver with PIC-DSC (IR/PIC-DSC).

from the previous iteration for the same layer, denoted by $y_c^{i,k-1}$ The output of the decision statistics combiner, in iteration k and for layer i, denoted by $y_c^{i,k}$, is given by $y_c^{i,k} = p_1^{i,k} y_c^{i,k} + p_2^{i,k} y_c^{i,k-1}$ where $p_1^{i,k}$ and $p_2^{i,k}$ are the DSC weighting coefficients in stage k, respectively. They are estimated by maximizing the signal-to-noise plus interference ratio (SINR) at the output of DSC in iteration k under the assumption that $y^{i,k}$ and $y_c^{i,k-1}$ are Gaussian random variables with the conditional means $\mu^{i,k}$ and $\mu_c^{i,k-1}$, given that x^i is the transmitted symbol for antenna i, and variances $(\sigma^{i,k})^2$ and $(\sigma_c^{i,k-1})^2$, respectively.

Coefficients $p_1^{i,k}$ and $p_2^{i,k}$ can be normalized in the following way: $E\{y_c^{i,k}\} = p_1^{i,k}\mu^{i,k} + p_1^{i,k}\mu_c^{i,k-1} = 1$. The SINR at the output of the DSC for layer i and in iteration k is then given by:

$$SINR^{i,k} = \cfrac{1}{\left(p_1^{i,k}\right)^2 \left(\sigma^{i,k}\right)^2 + 2p_1^{i,k}\left(\cfrac{1 - p_1^{i,k}\mu^{i,k}}{\mu_c^{i,k-1}}\right)\rho_{k,k-1}^i + \left(\cfrac{1 - p_1^{i,k}\mu^{i,k}}{\mu_c^{i,k-1}}\right)^2 \left(\sigma_c^{i,k-1}\right)^2} \qquad (5.253)$$

where $\rho_{k,k-1}^i$ is the correlation coefficient for layer i, between the detector output in the kth and $(k-1)$th iterations defined as:

$$\rho_{k,k-1}^i = \frac{E\left\{\left(y^{i,k} - \mu^{i,k}x^i\right)\left(y_c^{i,k-i} - \mu_c^{i,k-1}x^i\right)|x^i\right\}}{\sigma^{i,k}\sigma_c^{i,k-1}} \qquad (5.254)$$

The optimal combining coefficient is given by

$$p_1^{i,k}{}_{opt} = \cfrac{\cfrac{\mu^{i,k}}{\left(\mu_c^{i,k-1}\right)^2}\left(\sigma_c^{i,k-1}\right)^2 - \cfrac{1}{\mu_c^{i,k-1}}\rho_{k,k-1}^i\sigma_c^{i,k-1}}{\left(\sigma^{i,k}\right)^2 - 2\cfrac{\mu^{i,k}}{\mu_c^{i,k-1}}\rho_{k,k-1}^i\sigma^{i,k}\sigma_c^{i,k-1} + \left(\cfrac{\mu^{i,k}}{\mu_c^{i,k-1}}\right)^2\left(\sigma_c^{i,k-1}\right)^2} \qquad (5.255)$$

In the derivation of the optimal coefficients we assume that $\mu^{k,i}$, $\mu_c^{k,i-1}$, $(\sigma^{k,i})^2$ and $(\sigma_c^{k,i-1})^2$ are the true conditional means and the true variances of the detector outputs.

The parameters required for the calculation of the optimal combining coefficients in Equation (5.255) are difficult to estimate, apart from the signal variances. However, in a system with a large number of interferers, which happens when the number of transmit antennas is large relative to the number of receive antennas, and for the APP based symbol estimates, the DSC inputs in the first few iterations are low correlated. Thus, it is possible to combine them, in a way similar to receiving diversity maximum ratio combining. Under these conditions, the weighting coefficient in this receiver can be obtained from Equation (5.255) by assuming that the correlation coefficient is zero and neglecting the reduction of the received signal conditional mean caused by interference. The DSC coefficients are then given by:

$$p_1^{i,k} = \frac{\left(\sigma_c^{i,k-1}\right)^2}{\left(\sigma_c^{i,k-1}\right)^2 + (\sigma^{i,k})^2} \qquad (5.256)$$

The DSC output, in the second and higher iterations, with coefficients from Equation (5.256) can be expressed as:

$$y_C^{i,k} = \frac{(\sigma_c^{i,k-1})^2 1}{(\sigma_c^{i,k-1})^2 + (\sigma^{i,k})^2}y^{i,k} + \frac{(\sigma^{i,k}1)^2}{(\sigma^{i,k})^2 + (\sigma_c^{i,k-1})^2}y_c^{i,k-1} \quad i > 1 \qquad (5.257)$$

The complexity of both PIC-STD and PIC-DSC is linear in the number of transmit antennas. We demonstrate the performance of an HLST scheme with separate $R = 1/2$, 4-state convolutional component encoders, the frame size of $L = 206$ symbols and BPSK modulation. In simulations decoding is performed by a MAP algorithm. The HLSTC with n_T transmit and nR receive antennas is denoted as an (n_T, n_R)

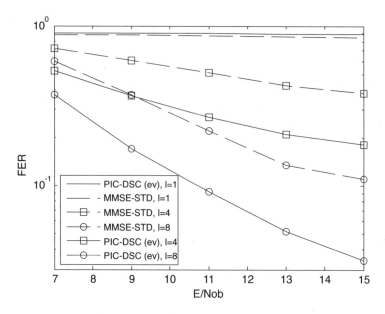

Figure 5.37   PER performance of $n_T = 6$, $n_R = 2$, $R = 1/2$, BPSK, a PIC-STD and PIC-DSC detection on a slow Rayleigh fading channel.

HI-STC. The channel is modelled as a frequency flat slow Rayleigh fading channel and the results are shown in the form of the frame error rate (FER) versus E_b/N_0 The SNR is related to E_b/N_0 as $SNR = \eta E_b/N_0$, where $\eta = Rmn_T$ is the spectral efficiency and m is the number of bits per modulation symbol. Figure 5.37 compares the performance of the PIC-STD with EIR and LLR based symbol estimates and the PIC-DSC for a (6.2) HLSTC. The spectral efficiency of the HI-STC is $\eta = 3$ bits/s/Hz. The results show that for the PIC-STD with LLR based symbol estimates the error floor is higher than for the other two schemes. With $l = 8$ iterations, the error floor for the PIC-STD(LLR) appears at FER of 0.1, while for the PIC-STD (EIR) the error floor is about 0.04. However, the PIC-DSC receiver has an error floor below 0.007.

5.6.5.2   Iterative F/B MMSE Receiver

We consider an iterative receiver with a multiuser detector consisting of a feed-forward (F) module which performs interference suppression followed by a feedback (B) module which performs parallel interference cancellation, as proposed in [144]. We refer to this receiver structure, presented in Figure 5.38, as an iterative F/B MMSE receiver. The decision statistics vector obtained at the output of the feedback module the kth iteration at time t, for layer i, denoted by $y_t^{i,k}$, is given by $y_t^{i,k} = (\mathbf{w}_f^{i,k})^H \mathbf{r} + w_b^{i,k}$, where $\mathbf{w}_f^{i,k}$ is an $n_R \times 1$ optimized feed-forward coefficients column matrix and $w_b^{i,k}$ is a single coefficient which represents cancellation term.

The coefficients $\mathbf{w}_f^{i,k}$ and $w_b^{i,k}$ are calculated by minimizing the mean square error between the transmitted symbol and its estimate, given by:

$$e = E\left\{ \left| (\mathbf{w}_f^{i,k})^H \mathbf{r} + w_b^{i,k} - x_t^i \right|^2 \right\} \qquad (5.258)$$

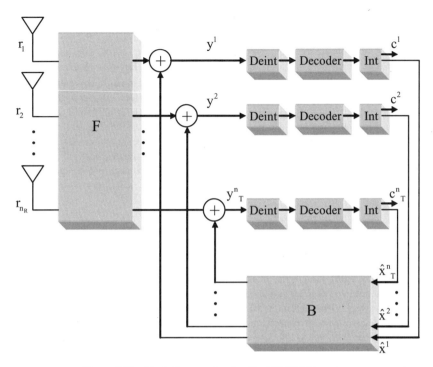

Figure 5.38   Block diagram of an iterative F/B MMSE receiver.

Let us denote by $\mathbf{x}^i$ a column matrix with $(n_T - 1)$ components, consisting of the transmitted symbols from all antennas except antenna i $(\mathbf{x}^i)^T = (x_t^1, x_t^2, \ldots, x_t^{i-1}, x_t^{i+1}, \ldots, x_t^{n_T})$.

Similarly, we define a vector $\hat{\mathbf{x}}^{i,k}$ of the symbol estimates from other antennas in the kth iteration $(\hat{\mathbf{x}}^{i,k})^T = (\hat{x}_t^{1,k}, \hat{x}_t^{2,k}, \ldots, \hat{x}_t^{i-1,k}, \hat{x}_t^{i+1,k}, \ldots, \hat{x}_t^{n_T,k})$. The decoder calculates the LLRs for the transmitted symbols at a particular time instant for each transmit antenna. These LLR values are used to calculate the transmitted symbol estimates $\hat{x}_t^{l,k}$, $l = 1, 2, \ldots, i - 1, i + 1, \ldots, n_T$ in $(\hat{\mathbf{x}}^k)^T$ as before.

Let us denote by h_i the ith of the channel matrix H, representing a column matrix with n_R complex channel gains for the ith transmit antenna, and by $\mathbf{H}^i$ an $n_R \times (n_T - 1)$ matrix composed of the complex channel gains for the other $(n_T - 1)$ transmit antennas. To simplify the notation, we define the following matrices:

$$\mathbf{A} = \mathbf{h}_i \mathbf{h}_i^H$$
$$\mathbf{B} = \mathbf{H}^i \left[\mathbf{I}_{n_T - 1} - \mathrm{diag}((\hat{\mathbf{x}}^{i,k},)^H \hat{\mathbf{x}}^{i,k}) + \hat{\mathbf{x}}^{i,k} (\hat{\mathbf{x}}^{i,k})^H \right] (\mathbf{H}^i)^H \tag{5.259}$$

where $\mathbf{I}_{n_T - 1}$ is an $(n_T - 1) \times (n_T - 1)$ identity matrix, and

$$\mathbf{D} = \mathbf{H}^i \hat{x}^{i,k} \qquad \mathbf{R}_n, = \sigma^2 \mathbf{I}_{n_R} \tag{5.260}$$

where σ^2 is the noise variance. The optimum feed-forward and feedback coefficients are given by

$$\left(\mathbf{w}_f^{i,k} \right)^H = \mathbf{h}_i^H (\mathbf{A} + \mathbf{B} + \mathbf{R}_n - \mathbf{D}\mathbf{D}^H)$$
$$w_b^{i,k} = - \left(\mathbf{w}_f^{i,k} \right)^H \mathbf{D} \tag{5.261}$$

The MMSE coefficients were derived assuming perfect interleaving and feedback symbol estimates based on the extrinsic information ratio (EIR) [158]. In the first iteration, since the *a priori* probabilities of the transmitted symbols are the same, the symbol estimates $\hat{x}^{i,1}$ are zeros. Thus in the first iteration the feed-forward coefficients $W_f^{i,1}$ are obtained in a similar way as in Equation (5.242) and the feedback coefficient $w_b^{i,1} = 0$. In the second and higher iterations, the symbol estimates, computed by the decoder as in Equation (5.249), are used to recalculate the new set of feed-forward and feedback coefficients as described above. In the case of hard decision decoding $(\hat{x}_t^{i,k})^2 = 1$ for all i and $k > 1$. The iterative MMSE receiver, which employs hard decision decoders, is equivalent to the receiver that performs linear MMSE suppression in the first iteration and parallel interference cancellation in the following iterations. This filter would be optimal in the MMSE sense if perfect symbol estimates were fed back.

The *a priori* probability on the transmitted signal x_t^i, used in the decoder, is computed as:

$$p_t\left(x_t^i = 1\right) = \frac{1}{\sqrt{2\pi}\sigma}e^{-\frac{(y_t^i - l)^2}{2\sigma^2}} \quad l = 1, -1 \tag{5.262}$$

It has been observed that the iterative MMSE receiver performs better if LLRs are used for symbol estimation instead of EIRs, though the MMSE filter coefficients were derived assuming EIR symbol estimation and uncorrelated decoder outputs. If LLRs are used there is a bias between symbol estimates. However, the bias effect is less relevant in the iterative MMSE receiver than in the iterative PIC receiver, since the MMSE detector performs interference suppression as well as cancellation. Thus the use of DSC in iterative MMSE receivers is less effective than for iterative PIC receivers.

Figure 5.39 compares the iterative MMSE and iterative PIC-DSC performance for an (8,2) HLSTC. The results show that the iterative PIC-DSC outperforms the iterative MMSE in terms of the achieved FER after two iterations. The error floor in the FBR performance of the MMSE-STD appears at $E_b/N_0 = 13$ dB and for FER $= 0.1$, while for the PIC-DSC receiver the error floor appears at $E_b/N_0 = 15$ dB and for FER $= 0.03$.

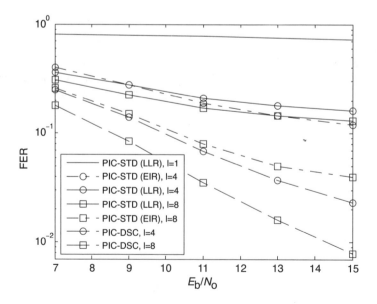

Figure 5.39   FER of a HLSTC with $n_T = 8$, $n_R = 2$, $R = 1/2$, iterative MMSE and iterative PIC-DSC receivers, BPSK modulation on a slow Rayleigh fading channel.

Appendix 5.1 Linear and Matrix Algebra

Definitions

Consider an $m \times n$ matrix $\mathbf{R}$ with elements r_{ij}, $i = 1, 2, \ldots, m$; $j = 1, 2, \ldots, n$. A shorthand notation for describing $\mathbf{R}$ is

$$[\mathbf{R}]_{ij} = \mathbf{r}_{ij}$$

The *transpose* of $\mathbf{R}$, which is denoted by $\mathbf{R}^T$, is defined as the $n \times m$ matrix with elements r_{ji} or

$$[\mathbf{R}^T]_{ij} = r_{ji}$$

A square matrix is one for which $m = n$. A square matrix is symmetric if $\mathbf{R}^T = \mathbf{R}$. The *rank* of a matrix is the number of linearly independent rows or columns, whichever is less. The *inverse* of a square $n \times n$ matrix is the square $n \times n$ matrix $\mathbf{R}^{-1}$ for which

$$\mathbf{R}^{-1}\mathbf{R} = \mathbf{R}\mathbf{R}^{-1} = \mathbf{I}$$

where $\mathbf{I}$ is the $n \times n$ identity matrix. The inverse will exist if and only if the rank of $\mathbf{R}$ is n. If the inverse does not exist, then $\mathbf{R}$ is singular. The *determinant* of a square $n \times n$ matrix is denoted by $\det(\mathbf{R})$. It is computed as

$$\det(\mathbf{R}) = \sum_{j=1}^{n} r_{ij} C_{ij}$$

where

$$C_{ij} = (-1)^{i+j} M_{ij}$$

M_{ij} is the determinant of the submatrix of $\mathbf{R}$ obtained by deleting the ith row and jth colunm and is termed the *minor* of r_{ij}. C_{ij} is the cofactor of r_{ij}. Note that any choice of i for $i = 1, 2, \ldots, n$ will yield the same value for $\det(\mathbf{R})$.

A *quadratic form* Q is defined as

$$Q = \sum_{i=1}^{n} \sum_{j=1}^{n} r_{ij} x_i x_j$$

In defining the quadratic form it is assumed that $r_{ji} = r_{ij}$. This entails no loss in generality since any quadratic function may be expressed in this manner. Q may also be expressed as

$$Q = \mathbf{x}^T \mathbf{R} \mathbf{x}$$

where $\mathbf{x} = [x_1 x_2 \ldots x_n]^T$ and $\mathbf{R}$ is a square $n \times n$ matrix with $r_{ji} = r_{ij}$ or $\mathbf{R}$ is a symmetric matrix.

A square $n \times n$ matrix $\mathbf{R}$ is *positive semidefinite* if $\mathbf{R}$ is symmetric and

$$\mathbf{x}^T \mathbf{R} \mathbf{x} \geq 0$$

for all $\mathbf{x} \neq \mathbf{0}$. If the quadratic form is strictly positive, then $\mathbf{R}$ is *positive definite*. When referring to a matrix as *positive definite* or positive semidefinite, it is always assumed that the matrix is symmetric.

The *trace* of a square $n \times n$ matrix is the sum of its diagonal elements or

$$\mathrm{tr}(\mathbf{R}) = \sum_{i=1}^{n} r_{ii}$$

A *partitioned* $m \times n$ matrix $\mathbf{R}$ is one that is expressed in terms of its submatrices. An example is the 2×2 partitioning

$$\mathbf{R} = \begin{bmatrix} \mathbf{R}_{11} & \mathbf{R}_{12} \\ \mathbf{R}_{21} & \mathbf{R}_{22} \end{bmatrix}$$

Each 'element' $\mathbf{R}_{ij}$ is a submatrix of $\mathbf{R}$. The dimensions of the partitions are given as

$$\begin{bmatrix} k \times l & k \times (n-l) \\ (m-k) \times l & (m-k) \times (n-l) \end{bmatrix}$$

Special Matrices

A *diagonal* matrix is a square $n \times n$ matrix with $r_{ij} = 0$ for $i \neq j$, in other words, all elements off the principal diagonal are zero. A diagonal matrix appears as:

$$\mathbf{R} = \begin{bmatrix} r_{11} & 0 & \cdots & 0 \\ 0 & r_{22} & \cdots & 0 \\ \vdots & \vdots & \ddots & \vdots \\ 0 & 0 & \cdots & r_{nn} \end{bmatrix}$$

A diagonal matrix will sometimes be denoted by $\mathrm{diag}(r_{11}, r_{22}, \ldots, r_{nn})$. The inverse of a diagonal matrix is found by simply inverting each element on the principal diagonal. A generalization of the diagonal matrix is the square $n \times n$ block diagonal matrix

$$\mathbf{R} = \begin{bmatrix} \mathbf{R}_{11} & \mathbf{0} & \cdots\cdots\cdots & \mathbf{0} \\ \mathbf{0} & \mathbf{R}_{22} & \cdots\cdots\cdots & \mathbf{0} \\ \vdots & & & \vdots \\ \mathbf{0} & \mathbf{0} & \cdots\cdots\cdots & \mathbf{R}_{kk} \end{bmatrix}$$

in which all submatrices $\mathbf{R}_{ii}$ are square and the other submatrices are identically zero. The dimensions of the submatrices need not be identical. For instance, if $k = 2$, $\mathbf{R}_{11}$ might have dimension 2×2 while $\mathbf{R}_{22}$ might be a scalar. If all $\mathbf{R}_{ii}$ are nonsingular, then the inverse is easily found as:

$$\mathbf{R}^{-1} = \begin{bmatrix} \mathbf{R}_{11}^{-1} & \mathbf{0} & \cdots\cdots\cdots & \mathbf{0} \\ \mathbf{0} & \mathbf{R}_{22}^{-1} & \cdots\cdots\cdots & \mathbf{0} \\ \vdots & & & \vdots \\ \mathbf{0} & \mathbf{0} & \cdots\cdots\cdots & \mathbf{R}_{kk}^{-1} \end{bmatrix}$$

Also, the determinant is:

$$\det(\mathbf{R}) = \prod_{i=1}^{n} \det(\mathbf{R}_{ii})$$

A square $n \times n$ matrix is *orthogonal* if

$$R^{-1} = R^{T}$$

For a matrix to be orthogonal the columns (and rows) must be orthonormal or, if

$$\mathbf{R} = [\mathbf{r}_1 \mathbf{r}_2 \ldots \mathbf{r}_n]$$

where $\mathbf{r}_i$ denotes the ith column, the conditions

$$\mathbf{r}_i^{T} \mathbf{r}_j = \begin{cases} 0 & \text{for} \quad i \neq j \\ 1 & \text{for} \quad i = j \end{cases}$$

must be satisfied.

An *idempotent* matrix is a square $n \times n$ matrix which satisfies

$$\mathbf{R}^2 = \mathbf{R}$$

This condition implies that $\mathbf{R}^l = \mathbf{R}$ for $l \geq 1$. An example is the projection matrix

$$\mathbf{R} = \mathbf{H}(\mathbf{H}^{T}\mathbf{H})^{-1}\mathbf{H}^{T}$$

where $\mathbf{H}$ is an $m \times n$ full rank matrix with $m > n$.

A square $n \times n$ *Toeplitz* matrix is defined as

$$[\mathbf{R}]_{ij} = r_{i-j}$$

or

$$\mathbf{R} = \begin{bmatrix} r_0 & r_{-1} & r_{-2} & \cdots & r_{-(n-1)} \\ r_1 & r_0 & r_1 & \cdots & r_{-(n-2)} \\ \vdots & \vdots & \vdots & \vdots & \vdots \\ r_{n-1} & r_{n-2} & r_{n-3} & \cdots & r_0 \end{bmatrix}$$

Each element along a northwest–southeast diagonal is the same. If, in addition, $r_{-k} = r_k$, then $\mathbf{R}$ is symmetric Toeplitz.

Matrix Manipulation and Formulas

Some useful formulas for the algebraic manipulation of matrices are summarized in this section. For $n \times n$ matrices $\mathbf{R}$ and $\mathbf{P}$ the following relationships are useful.

$(\mathbf{RP})^{T} = \mathbf{P}^{T}\mathbf{R}^{T}$

$(\mathbf{R}^{T})^{-1} = (\mathbf{R}^{-1})^{T}$

$(\mathbf{RP})^{-1} = \mathbf{P}^{-1}\mathbf{R}^{-1}$

$\det(\mathbf{R}^{T}) = \det(\mathbf{R})$

$\det(c\mathbf{R}) = c^n \det(\mathbf{R}) \quad (c \text{ a scalar})$

$\det(\mathbf{RP}) = \det(\mathbf{R})\det(\mathbf{P})$

$$\det(\mathbf{R}^{-1}) = \frac{1}{\det(\mathbf{R})}$$

$$\mathrm{tr}(\mathbf{RP}) = \mathrm{tr}(\mathbf{PR})$$

$$\mathrm{tr}(\mathbf{R}^{\mathsf{T}}\mathbf{P}) = \sum_{i=1}^{n}\sum_{j=1}^{n}[\mathbf{R}]_{ij}[\mathbf{P}]_{ij}$$

For vectors **x** and **y** we have

$$\mathbf{y}^{\mathsf{T}}\mathbf{x} = \mathrm{tr}(\mathbf{xy}^{\mathsf{T}})$$

It is frequently necessary to determine the inverse of a matrix analytically. To do so one can make use of the following formula. The inverse of a square $n \times n$ matrix is

$$\mathbf{R}^{-1} = \frac{\mathbf{C}^{\mathsf{T}}}{\det(\mathbf{R})}$$

where **C** is the square $n \times n$ matrix of cofactors **R**. The cofactor matrix is defined by

$$[\mathbf{C}]_{ij} = (-1)^{i+j}M_{ij}$$

where M_{ij} is the minor of r_{ij} obtained by deleting the ith row and jth column of **R**.

Another formula which is quite useful is the matrix inversion lemma

$$(\mathbf{R} + \mathbf{PCD})^{-1} = \mathbf{R}^{-1} - \mathbf{R}^{-1}\mathbf{P}(\mathbf{DR}^{-1}\mathbf{P} + \mathbf{C}^{-1})^{-1}\mathbf{DR}^{-1}$$

where it is assumed that **R** is $n \times n$, **P** is $n \times m$, **C** is $m \times m$, and **D** is $m \times n$ and that the indicated inverses exist. A special case known as *Woodbury's identity* results when **P** is an $n \times 1$ column vector **u**, **C** a scalar of unity, and **D** a $1 \times n$ row vector $\mathbf{u}^{\mathsf{T}}$. Then

$$(\mathbf{R} + \mathbf{uu}^{\mathsf{T}})^{-1} = \mathbf{R}^{-1} - \frac{\mathbf{R}^{-1}\mathbf{uu}^{\mathsf{T}}\mathbf{R}^{-1}}{1 + \mathbf{u}^{\mathsf{T}}\mathbf{R}^{-1}\mathbf{u}}$$

Partitioned matrices may be manipulated according to the usual rules of matrix algebra by considering each submatrix as an element. For multiplication of partitioned matrices, the submatrices which are multiplied together must be conformable. As an illustration, for 2×2 partitioned matrices

$$\mathbf{RP} = \begin{bmatrix} \mathbf{R}_{11} & \mathbf{R}_{12} \\ \mathbf{R}_{21} & \mathbf{R}_{22} \end{bmatrix} \begin{bmatrix} \mathbf{P}_{11} & \mathbf{P}_{12} \\ \mathbf{P}_{21} & \mathbf{P}_{22} \end{bmatrix} = \begin{bmatrix} \mathbf{R}_{11}\mathbf{P}_{11} + \mathbf{R}_{12}\mathbf{P}_{21} & \mathbf{R}_{11}\mathbf{P}_{12} + \mathbf{R}_{12}\mathbf{P}_{22} \\ \mathbf{R}_{21}\mathbf{P}_{11} + \mathbf{R}_{22}\mathbf{P}_{21} & \mathbf{R}_{21}\mathbf{P}_{12} + \mathbf{R}_{22}\mathbf{P}_{22} \end{bmatrix}$$

The transposition of a partitioned matrix is formed by transposing the submatrices of the matrix and applying T to each submatrix. For a 2×2 partitioned matrix

$$\begin{bmatrix} \mathbf{R}_{11} & \mathbf{R}_{12} \\ \mathbf{R}_{21} & \mathbf{R}_{22} \end{bmatrix}^{\mathsf{T}} = \begin{bmatrix} \mathbf{P}_{11}^{\mathsf{T}} & \mathbf{P}_{21}^{\mathsf{T}} \\ \mathbf{P}_{12}^{\mathsf{T}} & \mathbf{P}_{22}^{\mathsf{T}} \end{bmatrix}$$

The extension of these properties to arbitrary partitioning is straightforward. Determination of the inverses and determinants of partitioned matrices is facilitated by employing the following formulas. Let **R** be a square $n \times n$ matrix partitioned as

$$\mathbf{R} = \begin{bmatrix} \mathbf{R}_{11} & \mathbf{R}_{12} \\ \mathbf{R}_{21} & \mathbf{R}_{22} \end{bmatrix} = \begin{bmatrix} k \times k & k \times (n-k) \\ (n-k) \times k & (n-k) \times (n-k) \end{bmatrix}$$

Then,

$$\mathbf{R}^{-1} = \begin{bmatrix} \left(\mathbf{R}_{11} - \mathbf{R}_{12}\mathbf{R}_{22}^{-1}\mathbf{R}_{21}\right)^{-1} & -\left(\mathbf{R}_{11} - \mathbf{R}_{12}\mathbf{R}_{22}^{-1}\mathbf{R}_{21}\right)^{-1}\mathbf{R}_{12}\mathbf{R}_{22}^{-1} \\ -\left(\mathbf{R}_{22} - \mathbf{R}_{21}\mathbf{R}_{11}^{-1}\mathbf{R}_{12}\right)^{-1}\mathbf{R}_{21}\mathbf{R}_{11}^{-1} & \left(\mathbf{R}_{22} - \mathbf{R}_{21}\mathbf{R}_{11}^{-1}\mathbf{R}_{12}\right)^{-1} \end{bmatrix}$$

and

$$\begin{aligned} \det(\mathbf{R}) &= \det\left(\mathbf{R}_{22}\right)\det(\mathbf{R}_{11} - \mathbf{R}_{12}\mathbf{R}_{22}^{-1}\mathbf{R}_{21}) \\ &= \det(\mathbf{R}_{11})\det\left(\mathbf{R}_{22} - \mathbf{R}_{21}\mathbf{R}_{11}^{-1}\mathbf{R}_{12}\right) \end{aligned}$$

where the inverses of $\mathbf{R}_{11}$ and $\mathbf{R}_{22}$ are assumed to exist.

Theorems

Some important theorems are summarized in this section.

1. A square $n \times n$ matrix $\mathbf{R}$ is invertible (non-singular) if and only if its columns (or rows) are linearly independent or, equivalently, if its determinant is non-zero. In such a case, $\mathbf{R}$ is full rank. Otherwise, it is singular.
2. A square $n \times n$ matrix $\mathbf{R}$ is positive definite if and only if
   - it can be written as

$$\mathbf{R} = \mathbf{C}\mathbf{C}^{\mathrm{T}}$$

   where $\mathbf{C}$ is also $n \times n$ and is full rank and hence invertible; or
   - the principal minors are all positive. (The ith principal minor is the determinant of the submatrix formed by deleting all rows and columns with an index greater than i). If $\mathbf{R}$ can be written as in the previous equation, but $\mathbf{C}$ is not full rank or the principal minors are only non-negative, then $\mathbf{R}$ is positive semidefinite.
3. If $\mathbf{R}$ is positive definite, then the inverse exists and may be found from the previous equation as

$$\mathbf{R}^{-1} = (\mathbf{C}^{-1})^{T}(\mathbf{C}^{-1})$$

4. Let $\mathbf{R}$ be positive definite. If $\mathbf{P}$ is an $m \times n$ matrix of full rank with $m \leq n$, then $\mathbf{P}\mathbf{R}\mathbf{P}^{\mathrm{T}}$ is also positive definite.
5. If $\mathbf{R}$ is positive definite (positive semidefinite), then
   - the diagonal elements are positive (non-negative);
   - the determinant of $\mathbf{R}$, which is a principal minor, is positive

Eigendecompostion of Matrices

An *eigenvector* of a square $n \times n$ matrix $\mathbf{R}$ is an $n \times 1$ vector $\mathbf{v}$ satisfying

$$\mathbf{R}\mathbf{v} - \lambda\mathbf{v} \qquad\qquad (A5.1)$$

for some scalar λ, which may be complex. λ is the eigenvalue of $\mathbf{R}$ corresponding to the eigenvector v. It is assumed that the eigenvector is normalized to have unit length or $\mathbf{v}^{\mathrm{T}}\mathbf{v} = 1$. If $\mathbf{R}$ is symmetric, then one can always find n linearly independent eigenvectors, although they will not in general be unique. An example is the identity matrix for which any vector is an eigenvector with eigenvalue 1. If $\mathbf{R}$ is symmetric, then the eigenvectors corresponding to distinct eigenvalues are orthonormal or $\mathbf{v}_i^{\mathrm{T}}\mathbf{v}_j = \delta_{ij}$

and the eigenvalues are real. If, furthermore, the matrix is positive definite (positive semidefinite), then the eigenvalues are positive (non-negative). For a positive semidefinite matrix the rank is equal to the number of non-zero eigenvalues.

The defining previous relation can also be written as

$$\mathbf{R}[\mathbf{v}_1 \quad \mathbf{v}_2 \quad \cdots \quad \mathbf{v}_n] = [\lambda_1 \mathbf{v}_1 \quad \lambda_2 \mathbf{v}_2 \quad \cdots \quad \lambda_n \mathbf{v}_n]$$

or

$$\mathbf{R}\mathbf{V} = \mathbf{V}\Lambda$$

where

$$\mathbf{V} = [\mathbf{v}_1 \quad \mathbf{v}_2 \quad \cdots \quad \mathbf{v}_n]$$
$$\Lambda = \text{diag}(\lambda_1, \lambda_2, \ldots, \lambda_n)$$

If $\mathbf{R}$ is symmetric so that the eigenvectors corresponding to distinct eigenvalues are orthonormal and the remaining eigenvectors are chosen to yield an orthonormal eigenvector set, $\mathbf{V}$ is an orthonormal matrix. As such, its inverse is $\mathbf{V}^{\mathrm{T}}$, so that the previous equation becomes

$$\mathbf{R} = \mathbf{V}\Lambda\mathbf{V}^{\mathrm{T}}$$
$$= \sum_{i=1}^{n} \lambda_i \mathbf{v}_i \mathbf{v}_i^{\mathrm{T}}$$

Also, the inverse is easily determined as

$$\mathbf{R}^{-1} = \mathbf{V}^{\mathrm{T}-1}\Lambda^{-1}\mathbf{V}^{-1} = \mathbf{V}\Lambda^{-1}\mathbf{V}^{\mathrm{T}} = \sum_{i=1}^{n} \frac{1}{\lambda_i} \mathbf{v}_i \mathbf{v}_i^{\mathrm{T}}$$

A final useful relationship follows as

$$\det(\mathbf{R}) = \det(\mathbf{V})\det(\Lambda)\det(\mathbf{V}^{-1}) = \det(\Lambda) = \prod_{i=1}^{n} \lambda_i$$

Calculation of Eigenvalues and Eigenvectors

We can write Equation (A5.1) as

$$\mathbf{R}\mathbf{v} - \lambda\mathbf{v} = 0 \Rightarrow (\mathbf{R} - \lambda\mathbf{I})\mathbf{v} = 0$$

These are n linear algebraic equations in the n unknowns $v_1, v_2, v_3, \ldots, v_n$ (the components of $\mathbf{v}$). For these equations to have a solution $\mathbf{v} \neq \mathbf{0}$, the determinant of the coefficient matrix $\mathbf{R} - \lambda\mathbf{I}$ must be zero. As an example, if $n = 2$, we have

$$\begin{bmatrix} r_{11} - \lambda & r_{12} \\ r_{21} & r_{22} - \lambda \end{bmatrix} \begin{bmatrix} v_1 \\ v_2 \end{bmatrix} = \begin{bmatrix} 0 \\ 0 \end{bmatrix}$$

which can be written as

$$(r_{11} - \lambda)v_1 + r_{12}v_2 = 0$$
$$r_{21}v_1 + (r_{22} - \lambda)v_2 = 0$$

(A5.2)

Now, $\mathbf{R} - \lambda\mathbf{I}$ is singular if and only if its determinant $\det(\mathbf{R} - \lambda\mathbf{I})$ is zero. This can be written as

$$\det(R - \lambda\mathbf{I}) = \begin{vmatrix} r_{11} - \lambda r_{12} \\ r_{21} r_{22} - \lambda \end{vmatrix} = (r_{11} - \lambda)(r_{22} - \lambda) - r_{12}r_{21}$$

$$= \lambda^2 - (r_{11} + r_{22})\lambda + r_{11}r_{22} - r_{12}r_{21} = 0$$

This quadratic equation in λ is called the *characteristic equation* of $\mathbf{R}$. Its solutions are the eigenvalues λ_1 and λ_2 of $\mathbf{R}$. First determine these. Then use Equation (A5.2) with $\lambda = \lambda_1$ to determine the eigenvector $\mathbf{v}_1$ of $\mathbf{R}$ corresponding to λ_1. Finally, use Equation (A5.2) with $\lambda = \lambda_2$ to find an eigenvector $\mathbf{v}_2$ of $\mathbf{R}$ corresponding to λ_2. Note that if v is an eigenvector of $\mathbf{R}$ so is kv for any $k \neq 0$. As an example, suppose that

$$\mathbf{R} = \begin{bmatrix} -4.0 & 4.0 \\ -1.6 & 1.2 \end{bmatrix}$$

then we have

$$\det(\mathbf{R} - \lambda\mathbf{I}) = \begin{vmatrix} -4 - \lambda & 4 \\ -1.6 & 1.2 - \lambda \end{vmatrix} = \lambda^2 + 2.8\,\lambda + 1.6 = 0$$

It has the solutions $\lambda_1 = -2$ and $\lambda_2 = -0.8$. These are the eigenvalues of $\mathbf{R}$. Eigenvectors are obtained from Equation (A5.2). For $\lambda = \lambda_1 = -2$ we have, from Equation (A5.2):

$$(-4.0 + 2.0)v_1 + 4.0v_2 = 0$$

$$-1.6v_1 + (1.2 + 2.0)v_2 = 0$$

A solution of the equation is $v_1 = 2$, $v_2 = 1$. Hence, an eigenvector of $\mathbf{R}$ corresponding to $\lambda = \lambda_1 = -2$ is

$$\mathbf{v}_1 = \begin{bmatrix} 2 \\ 1 \end{bmatrix}$$

Similarly, for $\lambda = \lambda_2 = -0.8$

$$\mathbf{v}_2 = \begin{bmatrix} 1 \\ 0.8 \end{bmatrix}$$

So, for this example:

$$\mathbf{V} = [\mathbf{v}_1 \quad \mathbf{v}_2] = \begin{bmatrix} 2 & 1 \\ 1 & 0.8 \end{bmatrix}$$

References

[1] Glisic, S. (2003) *Adaptive WCDMA–Theory and Practice*, John Wiley & Sons, Chichester.
[2] Sarwate, S. V. and Pursley, M. B. (1980) Crosscorrelation Properties of Pseudorandom and Related Sequences, *Proceedings of the IEEE*, **68**, 593–613.
[3] Mezger, K. and Bouwens, R. J. (1972) An Ordered Table of Primitive Polynomials over GF(2) of Degrees 2 Through 19 for Use with Linear Maximal Sequence Generators, TM107, Cooley Electronics Laboratory, University of Michigan, Ann Arbor (AD 746876).
[4] Golomb, S. W. (1982) *Shift Register Sequences*, Aegean Park Press, Laguna Hills, CA.
[5] Lindholm, J. H. (1968) An Analysis of the Pseudo-randomness Properties of Subsequences of long *m*-sequences, *IEEE Transactions On Information Theory*, **14**(4), 569–576.

[6] Massey, J. L. (1969) Shift-Register Synthesis and BCH Decoding, *IEEE Transactions On Information Theory*, **15**(1), 122–127.

[7] Groth, E. J. (1971) Generation of Binary Sequences with Controllable Complexity, *IEEE Transactions On Information Theory*, **17**(3), 288–296.

[8] Antweiler, M. and Bömer, L. (1992) Complex sequence over GF(p^m) with a two-level autocorrelation function and large linear span, *IEEE Transactions On Information Theory*, **38**, 120–130.

[9] Games, R. (1986) The geometry of m-sequences: three-valued crosscorrelations and quadrics in finite projective geometry, *SIAM Journal of Algebraic Discrete Methods*, **7**, 43–52.

[10] Schoulz, R. and Welch, L. (1984) GMW sequences, *IEEE Transactions On Information Theory*, **IT-30**, 548–553.

[11] Simon, M., Omura, J., Scholtz, R. and Levitt, B. (1985) *Spread-Spectrum Communications*. Computer Science Press, New York.

[12] Welch, L. R. (1974) Lower bounds on the maximum correlation of signals, *IEEE Transactions On Information Theory*, **IT-20**, 397–399.

[13] Pursley, M. B. (1977) Performance evaluation for phase-coded spread-spectrum multiple-access communication–Part I: System analysis, *IEEE Transactions on Communications*, **COM-25**, 795–799.

[14] Pursley, M. B. and Roefs, H. F. A. (1979) Numerical evaluation of correlation parameters for optimal phases of binary shift-register sequences, *IEEE Transactions On Communications*, **COM-27**, 1597–1604.

[15] Pursley, M. B. and Sarwate, D. V. (1976) Bounds on aperiodic crosscorrelation for binary sequences, *Electronics Letters*, **12**, 304–305.

[16] Pursley, M. B. and Sarwate, D. V. (1977) Evaluation of correlations parameters for periodic sequences, *IEEE Transactions On Information Theory*, **IT-23**, 508–513.

[17] Pursley, M. B. and Sarwate, D. V. (1977) Performance evaluation for phase-coded spread-spectrum multiple access communication–Part II: Code sequence analysis, *IEEE Transactions On Communications*, **COM-25**, 800–803.

[18] Roefs, H. F. A. and Pursley, M. B. (1977) Correlation parameters of random binary sequences, *Electronics Letters*, **13**, 488–489.

[19] Sarwate, D. V. (1979) Bounds on crosscorrelation and autocorrelation of sequences, *IEEE Transactions On Information Theory*, **IT-25**, 720–724.

[20] Sarwate, D. V. and Pursley, M. B. (1977) New correlation identities for periodic sequences, *Electronics Letters*, **13**(2), 48–49.

[21] Scholtz, R. A. and Welch, L. R. (1978) Group characters: Sequences with good correlation properties, *IEEE Transactions On Information Theory*, **IT-24**, 537–545.

[22] Glisic, S. and Vucetic, B. (1997) *CDMA for Wireless Communication* Kluwer AP, Boston.

[23] Gold, R. (1966) Characteristic linear sequences and their coset functions, *SIAM Journal of Applied Mathematics*, **14**, 980–985.

[24] Gold, R. (1966) *Study of correlation properties of binary sequences*, AF Avionics Lab., Wright-Patterson AFB, OH, Technical Report AFAL-TR-66-234, (AD 488858).

[25] Gold, R. (1967) Optimal binary sequences for spread spectrum multiplexing, *IEEE Transactions On Information Theory*, **IT-13**, 619–621.

[26] Gold, R. (1968) Maximal recursive sequences with 3-valued recursive crosscorrelation functions, *IEEE Transactions On Information Theory*, **IT-14**, 154–156.

[27] Golay, M. (1961) Complementary Series, *IRE Transactions on Information Theory*, **IT 7**, 82–87.

[28] Deng, X. and Fan, P. (1999) New binary sequences with good aperiodic autocorrelations obtained by evolutionary algorithm, *IEEE Communications Letters*, **3**(10), 288–290.

[29] Deng, H. (1996) Synthesis of binary sequences with good autocorrelation and crosscorrelation properties by simulated annealing, *IEEE Transaction on Aerospace Electronics Systems*, **32**(1), 98–107.

[30] Hu, F., Fan, P. Z., Darnell, M. and Jin, F. (1997) Binary sequences with good aperiodic autocorrelation functions obtained by neural network search, *Electronics Letters*, **33**(8), 688–690.

[31] Mertens, S. (1996) Exhaustive search for low-autocorrelation binary sequences, *Journal of Physics A: Mathematical and General*, **29**(18), 473–481.

[32] Kocabas, S. E. and Atalar, A. (2003) Binary sequences with low aperiodic autocorrelation for synchronization purposes, *IEEE Communications Letters*, **7**(1), 36–38.

[33] Zhang Guohua and Zhou Quan (2002) Pseudonoise codes constructed by Legendre sequence, *Electronics Letters*, **38**(8), 376–377.

[34] Walther, U. and Ferrweis, G. P. (2001) PN-generators embedded in high performance signal processors, *The 2001 IEEE International Symposium on Circuits and Systems* (ISCAS 2001), 6–9 May 2001, **4**, 45–48.

[35] Leon, D., Balkir, S., Hoffman, M. W. and Perez, L. C. (2001) Robust chaotic PN sequence generation techniques, *The 2001 IEEE International Symposium on Circuits and Systems* (ISCAS 2001), 6–9 May 2001, **4**, 53–56.

[36] Fujisaki, H. (2001) Optimum binary spreading sequences of Markov chains, *Electronics Letters*, **37**(20), 1234–1235.

[37] Hongtao Zhang, Jichang Guo, Huiyun Wang, Runtao Ding and Wai-Kai Chen (2000) Oversampled chaotic map binary sequences: definition, performance and realization, *The 2000 IEEE Asia-Pacific Conference on Circuits and Systems* (IEEE APCCAS 2000), 4–6 December 2000, pp. 618–621.

[38] Giardina, C. and Rudrapatna, A. N. (2000) Quasi-Walsh PN sequences and their applications in robust CDMA communication systems, *2000 IEEE International Conference on Personal Wireless Communications*, 17–20 December 2000, pp. 24–27.

[39] Abraham, V. D. and Rao, C. D. V. P. (2000) Optimization of spreading code and estimation of channel capacity in multi-code CDMA downlink, *2000 IEEE International Conference on Personal Wireless Communications*, 17–20 December 2000, pp. 469–473.

[40] Leon, D., Balkir, S., Hoffman, M. and Perez, L.C. (2000) Fully programmable, scalable chaos-based PN sequence generation, *Electronics Letters*, **36**(16), 1371–1372.

[41] Yerdu, S. (1986) Optimum multiuser asymptotic efficiency, *IEEE Transactions On Communications*, **COM-34**, 890–897.

[42] Verdu, S. (1986) Minimum probability of error for asynchronous Gaussian multiple-access channels, *IEEE Transactions On Information Theory*, **IT-32**, 85–96.

[43] Varanasi, M. and Aazhang, B. (1990) Multistage detection in asynchronous code division multiple access communications, *IEEE Transactions On Communications*, **38**, 509–519.

[44] Varanasi, M. and Aazhang, B. (1991) Near optimum detection in synchronous code division multiple access systems, *IEEE Transactions On Communications*, **39**, 725–736.

[45] Varanasi, M. (1993) Noncoherent detection in a synchronous multiuser channels, *IEEE Transactions On Information Theory*, **37**(1), 157–176.

[46] Zvonar, Z. (1993) *Multiuser detection for Rayleigh fading channel*, Ph.D. thesis, Department of Electrical and Computer Engineering, Northeastern University, Boston, Massachusetts, USA.

[47] Xie, Z. *et al.* (1990) Multiuser signal detection using sequential decoding, *IEEE Transactions on Communications*, **38**, 578–583.

[48] Aazhang, B. *et al.* (1992) Neural networks for multiuser detection in code division multiple access communications, *IEEE Transactions On Communications*, **COM-40**, 1212–1222.

[49] Lupas, R. and Verdu, S. (1989) Linear multiuser detectors for synchronous code division multiple access channels, *IEEE Transactions On Information Theory*, **35**, 123–136.

[50] Lupas, R. and Verdu, S. (1990) Near–Far resistance of multiuser detectors in synchronous channels, *IEEE Transactions On Communications*, **38**(4), 496–508.

[51] Varanasi, M. and Aazhang, B. (1991) Optimally near–far resistant multiuser detection in differentially coherent synchronous channels, *IEEE Transactions On Information Theory*, **39**, 1006–1018.

[52] Xie, Z. *et al.* (1990) A family of suboptimum detectors for coherent multiuser communications, *IEEE ISAC*, **8**(4), 683–690.

[53] Xie, Z. *et al.* (1993) Joint signal detection and parameter estimation in multiuser communications, *IEEE Transactions On Communications*, **41**(7), 1208–1216.

[54] Zvonar, Z. *et al.* (1992) Optimum detection in synchronous multiple-access multipath Rayleigh fading channels, *Proceedings of the 26th Annual Conference on Information Sciences and Systems*, Princeton University, March 1992.

[55] Wijayasuriya, S. S. H., Norton, G. H. and McGeehan, J. P. (1992) Sliding Window Decorrelating Algorithm for DS-CDMA Receivers, *Electronics Letters*, **28**, 1596–1598.

[56] Wijayasuriya, S. S. H., McGeehan, J. P. and Norton, G. H. (1993) RAKE Decorrelating Receiver for DS-CDMA Mobile Radio Networks, *Electronics Letters*, **29**, 395–396.

[57] Tan, P. H. and Rasmussen, L. (2000) Linear interference cancellation in CDMA based on iterative techniques for linear equation systems, *IEEE Transactions On Communications*, **48**(12).

[58] Kapur, A. and Varanasi, M. K. (2003) Multiuser detection for overloaded CDMA systems, *IEEE Transactions On Information Theory*, **49**(7), 1728–1742.

[59] Buzzi, S., Lops, M. and Poor, H. V. (2003) Blind adaptive joint multiuser detection and equalization in dispersive differentially encoded CDMA channels, *IEEE Transactions on Signal Processing*, **51**(7), 1880–1893.

[60] Lim, H. S., Rao, M. V. C., Tan, A. W. C. and Chuah, H. T. (2003) Multiuser detection for DS-CDMA systems using evolutionary programming, *IEEE Communications Letters*, **7**(3), 101–103.

[61] Liping Sun and Guangrui Hu (2003) A new sign algorithm for interference suppression in DS-CDMA systems, *IEEE Communications Letters*, **7**(5), 233–235.

[62] Abe, T. and Matsumoto, T. (2003) Space–time turbo equalization in frequency-selective MIMO channels, *IEEE Transactions on Vehicular Technology*, **52**(3), 469–475.

[63] Yen, K. and Hanzo, L. (2003) Antenna-diversity-assisted genetic-algorithm-based multiuser detection schemes for synchronous CDMA systems, *IEEE Transactions On Communications*, **51**(3), 366–370.

[64] Jong-Hun Rhee, Moo-Yeon Woo and Dong-Ku Kim (2003) Multichannel joint detection of multicarrier 16-QAM DS/CDMA system for high-speed data transmission, *IEEE Transactions on Vehicular Technology*, **52**(1), 37–47.

[65] Yin, G. G., Krishnamurthy, V. and Ion, C. (2003) Iterate-averaging sign algorithms for adaptive filtering with applications to blind multiuser detection, *IEEE Transactions On Information Theory*, **49**(3), 657–671.

[66] de Lamare, R. C. and Sampaio-Neto, R. (2003) Adaptive MBER decision feedback multiuser receivers in frequency selective fading channels, *IEEE Communications Letters*, **7**(2), 73–75.

[67] Al-Bayati, A. K. S., Prakriya, S. and Prasad, S. (2003) Block modulus precoding for blind multiuser detection of DS-CDMA signals, *IEEE Transactions On Communications*, **51**(1), 52–56.

[68] Weihua Ye and Varshney, P. K. (2003) An equicorrelation-based multiuser communication scheme for DS-CDMA systems, *IEEE Transactions On Communications*, **51**(1), 43–47.

[69] Kuei-Chiang Lai and Shynk, J. J. (2003) Performance evaluation of a generalized linear SIC for DS/CDMA signals, *IEEE Transactions on Signal Processing*, **51**(6), 1604–1614.

[70] Kafle, P. L. and Sesay, A. B. (2003) Iterative semi-blind multiuser detection for coded mc-cdma uplink system, *IEEE Transactions On Communications*, **51**(7), 1034–1039.

[71] Zhiyu Xu and Cheng, R. S. (2003) A robust rank estimation algorithm of group-blind MMSE multiuser detectors for CDMA systems, *IEEE Transactions on Communications*, **51**(4), 547–552.

[72] Wei Zha and Blostein, S. D. (2003) Soft-decision multistage multiuser interference cancellation, *IEEE Transactions on Vehicular Technology*, **52**(2), 380–389.

[73] Ee-Lin Kuan and Hanzo, L. (2003) Burst-by-burst adaptive multiuser detection cdma: a framework for existing and future wireless standards, *Proceedings of the IEEE*, **91**(2), 278– 302.

[74] Brunel, L. and Boutros, J. J. (2003) Lattice decoding for joint detection in direct-sequence CDMA systems, *IEEE Transactions On Information Theory*, **49**(4), 1030–1037.

[75] Horlin, F. and Vandendorpe, L. (2003) CA-CDMA: channel-adapted CDMA for MAI/ISI-free burst transmission, *IEEE Transactions On Communications*, **51**(2), 275–283.

[76] Buzzi, S. and Lops, M. (2003) Performance analysis for the improved linear multiuser detectors in BPSK-modulated DS-CDMA systems, *IEEE Transactions On Communications*, **51**(1), 37–42.

[77] Huaiyu Dai and Poor, H. V. (2002) Iterative space–time processing for multiuser detection in multipath CDMA channels, *IEEE Transactions on Signal Processing*, **50**(9), 2116–2127.

[78] Hongbin Li and Jian Li (2002) Differential and coherent decorrelating multiuser receivers for space-time-coded CDMA systems, *Signal Processing*, **50**(10), 2529–2537.

[79] Reynolds, D., Xiaodong Wang and Poor, H. V. (2002) Blind adaptive space–time multiuser detection with multiple transmitter and receiver antennas, *IEEE Transactions on Signal Processing*, **50**(6), 1261–1276.

[80] Nafie, M. and Tewfik, A. H. (2002) A flexible receiver for CDMA multiuser communications, *IEEE Transactions on Signal Processing*, **50**(7), 1747–1758.

[81] Brown, T. and Kaveh, M. (1995) A decorrelating detector for use with antenna arrays, *International Journal of Wireless Information Networks*, **2**(4), 239–246.

[82] Jung, P. and Blanz, J. (1995) Joint detection with coherent receiver antenna diversity in CDMA mobile radio systems, *IEEE Transactions on Vehicular Technology*, **44**(1), 76–88.

[83] Jung, P., Blanz, J., Nasshan, M. and Baier, P. W. (1994) Simulation of the uplink of JD–CDMA mobile radio systems with coherent receiver antenna diversity, *Wireless Personal Communications*, **1**(2), 61–89.

[84] Madhow, U. and Honing, M. L. (1994) MMSE interference suppression for direct sequence spread-spectrum CDMA, *IEEE Transactions On Communications*, **42**(12), 3178–3188.

[85] Klein, A., Kaleh, G. K. and Baier, P. W. (1996) Zero forcing and minimum mean-square-error equalization for multiuser detection in code-division multiple access channels, *IEEE Transactions on Vehicular Technology*, **45**(2), 276–287.

[86] Berstein, X. and Haimovich, A. M. (1996) Space–time optimum combining for CDMA communications, *Wireless Personal Communications*, **3**(1–2), 73–89.

[87] Gray, S. D., Preisig, J. C. and Brady, D. (1997) Multiuser detection in a horizontal underwater acoustic channel using array observations, *IEEE Transactions on Signal Processing*, **45**(1), 148–160.

[88] Poor, H. V. and Verdú, S. (1997) Probability of error in MMSE multiuser detection, *IEEE Transactions On Information Theory*, **43**(3), 858–871.

[89] Rapajic, P. B. and Vucetic, B. S. (1995) Linear adaptive transmitter–receiver structures for asynchronous SCMA systems, *European Transactions on Telecommunications*, **6**(1), 21–27.

[90] Miller, S. L. (1995) An adaptive direct sequence code-division multiple-access receiver for multiuser interference rejection, *IEEE Transactions On Communications*, **43**, 1746–1755.

[91] Rapajic, P. B. and Vucetic, B. S. (1994) Adaptive receiver structures for asynchronous CDMA systems, *IEEE Journal on Selected Areas in Communications*, **12**(4), 685–697.

[92] Lee, K. B. (1996) Orthogonalization based adaptive interference suppression for direct sequence code-division multiple-access systems, *IEEE Transactions On Communications*, **44**(9), 1082–1085.

[93] Miller, S. L. (1996) Training analysis of adaptive interference suppression for direct sequence code-division multiple-access systems, *IEEE Transactions on Communications*, **44**(4), 488– 495.

[94] Honig, M. (1998) Adaptive linear interference suppression for packet DS-CDMA, *European Transactions on Telecommunications*, **9**(2), 173–181.

[95] Hong, M., Madhow, U. and Verdú, S. (1995) Blind adaptive multiuser detection, *IEEE Transactions On Information Theory*, **41**(3), 944–960.

[96] Park, S. C. and Dohery, J. F. (1997) Generalized projection algorithm for blind interference suppression in DS/CDMA communications, *IEEE Transactions on Circuits and Systems–Part II Analog and Digital Signal Processing*, **44**(6), 453–460.

[97] Schodorf, J. B. and Williams, D. B. (1997) A constrained optimisation approach to multiuser detection, *IEEE Transactions on Signal Processing*, **45**(1), 258–262.

[98] Wang, X. and Poor, H. V. (1998) Blind equalization and multiuser detection in dispersive CDMA channels, *IEEE Transactions On Communications*, **46**(1), 91–103.

[99] Iltis, R. A. (1998) Performance of constrained and unconstrained adaptive multiuser detectors for quasi-synchronous CDMA, *IEEE Transactions On Communications*, **46**(1), 135–143.

[100] Madhow, U. (1997) Blind adaptive interference suppression for the near–far resistant acquisition and demodulation of direct-sequence CDMA, *IEEE Transactions on Signal Processing*, **45**(1), 124–136.

[101] Mowbray, R. S., Pringle, R. D. and Grant, P. M. (1992) Increased CDMA system capacity through adaptive cochannel interference regeneration and cancellation, *IEE Proceedings I*, **139**, 515–524.

[102] Kohno, R., Imai, H., Hatori, M. and Pasupathy, S. (1990) Combination of an adaptive array antenna and a canceller of interference for direct sequence spread-spectrum multiple-access system, *IEEE Journal on Selected Areas in Communications*, **8**(4), 675–682.

[103] Nelson, L. B. and Poor, H. V. (1996) Iterative multiuser receivers for CDMA channels: An EM-based approach, *IEEE Transactions On Communications*, **44**(12), 1700–1710.

[104] Soong, A. C. K. and Krzymien, W. A. (1996) A novel CDMA multiuser interference cancellation receiver with reference symbol aided estimation of channel parameters, *IEEE Journal on Selected Areas in Communications*, **14**(8), 1536–1547.

[105] Wang, X. and Poor, H. V. (1998) Blind multiuser detection: A subspace approach, *IEEE Transactions On Information Theory*, **44**(2), 677–690.

[106] Juntti, M. and Glisic, S. (1997) Advanced CDMA for wireless communications, In Glisic, S. G. and Leppänen P. A. (eds) *Wireless Communications: TDMA Versus DCMA*, Kluwer, pp. 447–490.

[107] Zihua Guo and Ben Letaief, K. (2003) A low complexity reduced–rank MMSE receiver for DS/CDMA communications, *IEEE Transactions on Wireless Communications*, **2**(1), 59–68.

[108] Zhiyu Xu and Cheng, R. S. (2003) A robust rank estimation algorithm of group-blind MMSE multiuser detectors for CDMA systems, *IEEE Transactions On Communications*, **51**(4), 547–552.

[109] Mantravadi, A. and Veeravalli, V. V. (2002) MMSE detection in asynchronous CDMA systems: an equivalence result, *IEEE Transactions On Information Theory*, **48**(12), 3128–3137.

[110] Woodward, G., Ratasuk, R., Honig, M. L. and Rapajic, P. B. (2002) Minimum mean-squared error multiuser decision-feedback detectors for DS–CDMA, *IEEE Transactions on Communications*, **50**(12), 2104–2112.

[111] Ju Ho Lee and Hyung-Myung Kim (2002) Partial zero-forcing adaptive MMSE receiver for DS-CDMA uplink in multicell environments, *IEEE Transactions on Vehicular Technology*, **51**(5), 1066–1071.

[112] Schober, R., Gerstacker, W. H. and Lampe, A. (2002) Noncoherent MMSE interference suppression for DS-CDMA, *IEEE Transactions On Communications*, **50**(4), 577–587.

[113] Buzzi, S., Lops, M. and Tulino, A. M. (2002) A generalized minimum-mean-output-energy strategy for CDMA systems with improper MAI, *IEEE Transactions on Information Theory*, **48**(3), 761–767.

[114] Iltis, R. (1994) An EKF-based joint estimator for interference, multipath, and code delay in a DS spread-spectrum receiver, *IEEE Transactions On Communications* **42**, 1288–1299.

[115] Kay, S. (1993) *Fundamentals of Statistical Signal Processing–Estimation Theory*, Prentice Hall.

[116] Bensley, J. S. and Aazhang, B. (1996) Subspace-Based Channel Estimation for Code Division Multiple Access Communications and Systems, *IEEE Transactions on Communications*, **44**(8), 1009–1020.

[117] Xie, Z., Rushforth, C., Short, R. and Moon, T. (1993) Joint signal detection and parameter estimation in multiuser communications, *IEEE Transactions On Communications*, **41**, 1208–1216.

[118] Aazhang, B., Paris, B. and Orsak, G. (1992) Neural networks for multiuser detection in code-division multiple-access communications, *IEEE Transactions On Communications*, **40**, 1212–1222.

[119] Iltis, R. A. and Mailaender, L. (1994) – An adaptive multiuser detector with joint amplitude and delay estimation, *IEEE Journal on Selected Areas in Communications*, **12**(5), 774–785.

[120] Iltis, R. (1990) Joint estimation of PN code delay and multipath using the extended Kalman filter, *IEEE Transactions On Communications*, **38**, 1677–1685.

[121] Iltis, R. and Fuxjaeger, A. (1991) A digital DS spread-spectrum receiver with joint channel and Doppler shift estimation, *IEEE Transactions On Communications*, **39**, 1255–1267.

[122] Clark, A. P. and Harun, R. (1986) Assesment of Kalman-Filter Channel Estimators for an HF Radio Link, *IEE Proceedings*, **133** (pt.F), 513–521.

[123] Hatzinakos, D. and Nikias, C. L. (1989) Estimation of multipath channel response in frequency selective channels, *IEEE Journal on Selected Areas in Communications*, **SAC-7**, 12–19.

[124] Shalvi, O. and Weinstein, E. (1990) New criteria for blind deconvolation of nonminimum phase systems (channels), *IEEE Transactions On Information Theory*, **IT-36**, 312–321.

[125] Huang, H. C. (1996) *Combined multipath processing, array processing, and multiuser detection for DS-CDMA channels*, Ph.D. Thesis, Princeton University, Princeton, NJ, USA.

[126] Latva-aho, M. (1998) *Advanced receivers for wideband CDMA systems*, Ph.D. Thesis, University of Oulu, Finland.

[127] Miller, S. Y. (1989) *Detection and estimation in multiple-access channels*, Ph.D. Thesis, Princeton University, Princeton, NJ, USA.

[128] Miller, S. Y. and Schwartz, S. C. (1995) Integrated spatial-temporal detectors for asynchronous Gaussian multiple-access channels, *IEEE Transactions On Communications*, **43**, 396–411.

[129] Zvonar, Z. (1996) Combined multiuser detection and diversity reception for wireless CDMA systems, *IEEE Transactions on Vehicular Technology*, **45**(1), 205–211.

[130] Fanucci, L. *et al.* (2001) VLSI Implementation of CDMA Blind Adaptive Interference-Mitigating Detector, *IEEE JSAC*, **19**(2), 179–190.

[131] De Gaudenzi, R. *et al.* (1998) Design of a Low Complexity Adaptive Interference Mitigating Detector for DS/SS Receiver in CDMA Radio Networks, *IEEE Transactions On Communications*, **46**(1), 125–134.

[132] Amleh, K. and Hongbin Li (2003) An algebraic approach to blind carrier offset and code timing estimation for DS-CDMA systems, *IEEE Signal Processing Letters*, **10**(2), 32–34.

[133] Manikas, A. and Sethi, M. (2003) A space–time channel estimator and single-user receiver for code-reuse in DS-CDMA systems, *IEEE Transactions on Signal Processing*, **51**(1), 39–51.

[134] Bin Xu, Chenyang Yang and Shiyi Mao (2002) An improved blind adaptive multiuser detector in multipath CDMA channels based on subspace estimation, *IEEE 55th Vehicular Technology Conference* (VTC Spring 2002), 6–9 May 2002, **1**, 285–288.

[135] Fock, G., Schulz-Rittich, P., Schenke, A. and Meyr, H. (2002) Low complexity high resolution Subspace-based delay estimation for DS-CDMA, *IEEE International Conference on Communications* (ICC 2002), 28 April–2 May 2002, **1**, 31–35.

[136] Xiaojun Wu, Qinye Yin and Jianguo Zhang (2002) Subspace-based estimation method of uplink FIR channel in MC-CDMA system without cyclic prefix over frequency-selective fading channel, *IEEE International Symposium on Circuits and Systems* (ISCAS 2002), 26–29 May 2002, **1**, I-213–I-216.

[137] Yugang Ma, Li, K. H., Kot, A. C. and Ye, G. (2002) A blind code timing estimator and its implementation for DS-CDMA signals in unknown colored noise, *IEEE Transactions on Vehicular Technology*, **51**(6), 1600–1607.

[138] Zhengyuan Xu (2002) Asymptotic performance of subspace methods for synchronous multirate CDMA systems, *IEEE Transactions on Signal Processing*, **50**(8), 2015–2026.

[139] Lei Huang, Fu-Chun Zheng and Faulkner, M. (2002) Blind adaptive channel estimation for dual-rate DS/CDMA signals, *IEEE Communications Letters*, **6**(4), 129–131.

[140] Affes, S. N., Hansen, H. and Mermelstein, P. (2002) Interference subspace rejection: a framework for multiuser detection in wideband CDMA, *IEEE Journal on Selected Areas in Communications*, **20**(2), 287–302.

[141] Zhengyuan Xu (2002) Perturbation analysis for subspace decomposition with applications in Subspace-based algorithms, *IEEE Transactions on Signal Processing*, **50**(11), 2820–2830.

[142] Tugnait, J. (1994) Blind Estimation of Digital Communication Channel Impulse Response, *IEEE Transactions On Communications*, **42**(2/3/4), 1606–1616.

[143] Muirhead, R. (1982) *Aspects of Multivariate Statistical Theory*, Jon Wiley & Sons, New York.

[144] Gamal, H. El. and Hammons, A. R. (2001) The layered space–time architecture: a new perspective, *IEEE Transactions on Information Theory*, **47**, 2321–2334.

[145] Golden, G. D., Foschini, G. J., Valenzuela, R. A. and Wolniansky, P. W. (1999) Detection algorithm and initial laboratory results using the V-BLAST space–time communication architecture, *Electronics Letters*, **35**, 14–15.

[146] Winters, J. H., Salz, J. and Gitlin, R. D. (1994) The impact of antenna diversity on the capacity of wireless communication systems. *IEEE Transactions on Communications* **42**, 1740–1751.

[147] Marinkovic, S., Vucetic, B. and Ushirokawa, A. (2001) Space–time iterative and multistage receiver structures for CDMA mobile communication systems, *IEEE Journal on Selected Areas in Communications*, **19**, 1594–1604.

[148] Ariyavisitakul, S. L. (2000) Turbo space–time processing to improve wireless channel capacity, *IEEE Transactions on Communications*, **48**, 1347–1358.

[149] Foschini, G., Golden, G., Valenzuela, R. and Wolniansky, P. (2000) Simplified processing for high spectral efficiency wireless communication employing multi-element arrays, *IEEE Journal on Selected Areas in Communications*, **17**, 1841–1852.

[150] Proakis, J. G. (1995) *Digital Communications*, Series in Electrical and Computer Engineering, McGraw-Hill, New York.

[151] Alexander, P., Grant, A. J. and Reed, M. (1998) Iterative multiuser detection in codedivision multiple-access with error control coding, *European Transactions on Telecommunications*, special issue on *CDMA Techniques in Wireless Communications Systems*, **9**, 419–425.

[152] Moher, M. (1998) An iterative multiuser decode: for near-capacity communications, *IEEE Transactions on Communications*, **46**, 870–880.

[153] Reed, M. (1999) Iterative Receiver Techniques for Coded Multiple Access Communication Systems, PhD Thesis, The University of South Australia.

[154] Alexander, P., Reed, M., Asenstorfer, J. and Schlegel, C. (1998) Iterative multiuser detection for CDMA with FEC: near-single user performance, *IEEE Transactions on Communication*, **46**, 1693–1699.

[155] Alexander, P., Reed, M., Asenstorfer, J. and Schlegel, C. (1999) Iterative multiuser interference reduction: turbo CDMA, *IEEF Transactions on Communication*, **47**, 1008–1014.

[156] Wang, X. and Poor, H. V. (1999) Iterative (turbo) soft interference cancellation and decoding for coded CDMA, *IEEE Transactions on Communications*, **47**, 1046–1061.

[157] El Gamal, H. and Geraniotis, E. (2000) Iterative multiuser detection for coded CDMA signals in AWGN and fading channels, *IEEE Journal on Selected Areas in Communications*, **18**, 30–41

[158] El Gamal, H. and Hammons, R. (2001) A new approach to layered space–time and signal processing, *IEEE Transactions on Information Theory*, **47**, 2321–2334.

[159] Buehrel, R.M., Nicoloso, S.P. and Gollamudi, S. (1999) Linear versus nonlinear interference cancellation, *Journal of Communications and Networks*, **1**, 118–132.

6

Channel Estimation and Equalization

The basic concept of time division multiple access (TDMA) has been discussed in Chapter 1. Within this chapter we cover the basic enabling technologies for TDMA. Coding and modulation are covered in Chapters 2, 3 and 4 so that in this chapter we focus on the remaining topics, mainly TDMA-specific channel estimation and equalization. MIMO channel equalization will be discussed in Chapter 9 within the general problem of linear precoding.

6.1 Equalization in the Digital Data Transmission System

6.1.1 Zero-Forcing Equalizers

The basic problem of channel equalization is illustrated in Figure 6.1. Figure 6.1(a) presents the transmitted pulse $p(t)$, Figure 6.1(b) shows the pulse after propagation through the channel $p_c(t)$, while Figure 6.1(c) shows the difference between the received $p_c(t)$ and the equalized pulse $p_{eq}(t) \cong p(t)$.

Figure 6.2 illustrates the mutual impact (inter-symbol interference) of the non-equalized and equalized signals. Finally, Figure 6.3 illustrates a general structure and function of the equalizer. To summarize, the transmitted signal for full response signaling is created in such a way that the pulse goes through zero at the time instances $\pm kT$, $k \neq 0$ from the pulse maximum (Nyquist signaling) $p(\pm kT) = 0$, $k \neq 0$. That way the adjacent symbols sampled at those instances will not be affected. The degradation caused by the channel will result in the pulse $p_c(\pm kT) \neq 0$, $k \neq 0$ which produces inter-symbol interference. The equalizer is supposed to compensate for this degradation by regenerating a pulse $p_{eq}(\pm kT) \cong p(\pm kT) = 0$, $k \neq 0$.

This is represented by

$$p_{eq}(t) = \sum_{n=-N}^{N} C_n p_c(t - nT)$$

for $t = mT + \Delta t$

$$p_{eq}(mT + \Delta t) = \sum_{n=-N}^{N} C_n p_c((m-n)T + \Delta t)$$

$$= \begin{cases} 1 & m = 0 \\ 0 & m \neq 0 \end{cases} \quad m = 0, \pm 1, \pm 2, \ldots, \pm N$$

(6.1)

Advanced Wireless Communications & Internet: Future Evolving Technologies, Third Edition. Savo Glisic.
© 2011 John Wiley & Sons, Ltd. Published 2011 by John Wiley & Sons, Ltd.

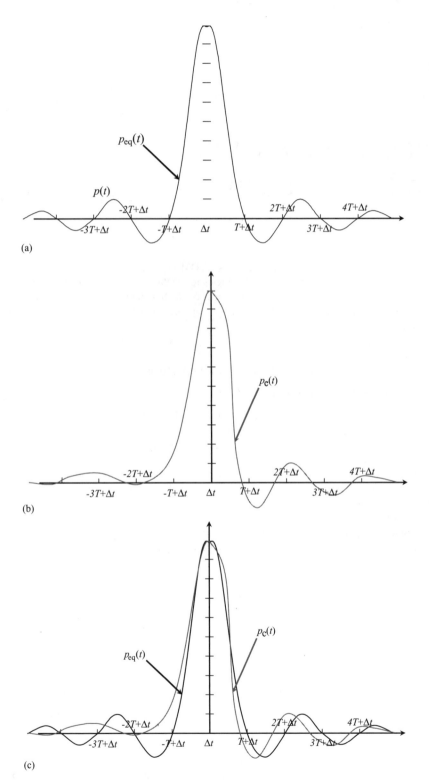

Figure 6.1   (a) Transmitted; (b) received; and (c) equalized pulses.

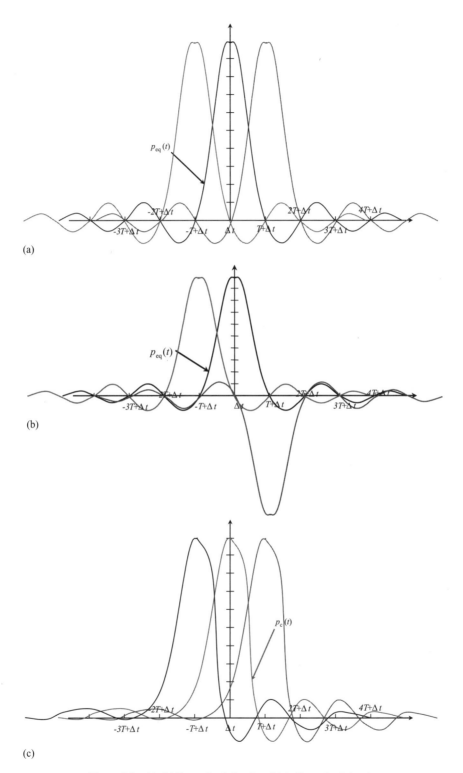

Figure 6.2 (a), (b) Transmitted signals and (c), (d) received signals.

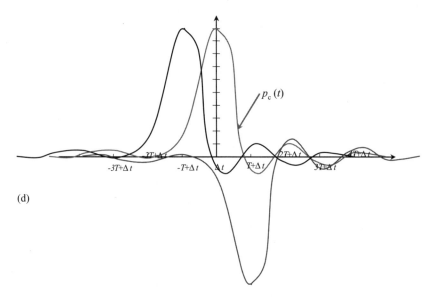

(d)

Figure 6.2 *(Continued).*

where $t = \Delta t$ is the sampling time for which $p_{eq}(t)$ is maximal. Equation (6.1) is implemented by the structure shown in Figure 6.3 with the basic building block shown in Figure 6.4. From time to time, throughout the rest of the book, we will explicitly emphasize the importance of this basic building block in order to be able to motivate some basic approaches in building up a common reconfigurable platform for different technologies.

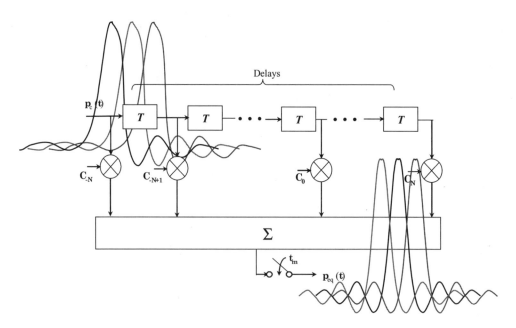

Figure 6.3 Transversal filter equalizer.

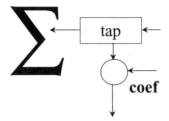

Figure 6.4 Transversal filter equalizer basic building block.

Equation (6.1) for $m = 0, \pm 1, \pm 2, \ldots, \pm N$ can be written in matrix form as $\mathbf{P}_{eq} = \mathbf{P}_c \, \mathbf{C}$, where $\mathbf{P}_{eq}$ and $\mathbf{C}$ are vectors or column matrices given by

$$
\mathbf{P}_{eq} = \begin{bmatrix} 0 \\ 0 \\ \cdot \\ \cdot \\ \cdot \\ 0 \\ 1 \\ 0 \\ 0 \\ \cdot \\ \cdot \\ \cdot \\ 0 \end{bmatrix} \begin{array}{l} \left. \vphantom{\begin{matrix}0\\0\\ \cdot \\ \cdot \\ \cdot \\0\end{matrix}} \right\} N \text{ zeros} \\[3em] \left. \vphantom{\begin{matrix}0\\0\\ \cdot \\ \cdot \\ \cdot \\0\end{matrix}} \right\} N \text{ zeros} \end{array} \qquad \mathbf{C} = \begin{bmatrix} C_{-N} \\ C_{-N+1} \\ \cdot \\ \cdot \\ \cdot \\ C_0 \\ C_1 \\ \cdot \\ \cdot \\ \cdot \\ C_N \end{bmatrix} \qquad (6.2)
$$

and $\mathbf{P}_c$ is the $(2N + 1) \times (2N + 1)$ matrix of channel responses of the form

$$
\mathbf{P}_c = \begin{bmatrix} p_c(0) & p_c(-1) & \cdots & p_c(-2N) \\ p_c(1) & p_c(0) & \cdots & p_c(-2N + 1) \\ p_c(2) & p_c(1) & \cdots & p_c(-2N + 2) \\ \cdot & \cdot & \cdot & \cdot \\ \cdot & \cdot & \cdot & \cdot \\ \cdot & \cdot & \cdot & \cdot \\ p_c(2N) & p_c(2N - 1) & \cdots & p_c(0) \end{bmatrix} \qquad (6.3)
$$

The solution for the equalizer coefficients is

$$
\mathbf{C} = \mathbf{P}_c^{-1} \mathbf{P}_{eq} \qquad (6.4)
$$

given $\mathbf{P}_c^{-1} \neq \mathbf{0}$. This way we force the $\mathbf{P}_{eq} \, (\pm \, kT) = \mathbf{0}$, hence the name zero-forcing equalizer. One should be aware that in a channel with noise $\mathbf{P}_c \Rightarrow \mathbf{P}_c + \mathbf{P}_n$, where $\mathbf{P}_n$ is the noise matrix given by (6.3) where channel samples are replaced with corresponding noise samples $p_c(\,) \Rightarrow p_n(\,)$. Equation (6.4) now results in

$$
\mathbf{P}_{eq} = \mathbf{P}_c \mathbf{C} + \mathbf{P}_n \mathbf{C} = \mathbf{P}_c \mathbf{C} + \mathbf{N}_{eq} \qquad (6.5)
$$
$$
\mathbf{C} = \mathbf{P}_c^{-1} \mathbf{P}_{eq} - \mathbf{P}_c^{-1} \mathbf{N}_{eq} \qquad (6.6)
$$

The second term represents the noise enhancement factor and is a limiting factor in achieving good performance.

6.1.1.1   Example

Suppose that the sample values for the channel response are

$$
\begin{array}{lll}
p_c(-5) = 0.01 & p_c(-4) = -0.02 & p_c(-3) = 0.05 \\
p_c(-2) = -0.1 & p_c(-1) = 0.2 & p_c(0) = 1 \\
p_c(1) = -0.1 & p_c(2) = 0.1 & p_c(3) = -0.05 \\
p_c(4) = 0.02 & p_c(5) = 0.005 &
\end{array}
$$

The channel response matrix is

$$
\mathbf{P}_c =
\begin{bmatrix}
1.0 & 0.2 & -0.1 & 0.05 & -0.02 \\
-0.1 & 1.0 & 0.2 & -0.1 & 0.05 \\
0.1 & -0.1 & 1.0 & 0.2 & -0.1 \\
-0.05 & 0.1 & -0.1 & 1.0 & 0.2 \\
0.02 & -0.05 & 0.1 & -0.1 & 1.0
\end{bmatrix}
$$

and its inverse

$$
\mathbf{P}_c^{-1} =
\begin{bmatrix}
0.996 & -0.170 & -0.117 & -0.083 & 0.056 \\
0.118 & 0.945 & 0.158 & 0.112 & -0.083 \\
-0.091 & 0.133 & 0.937 & -0.158 & 0.117 \\
0.028 & -0.095 & 0.133 & 0.945 & -0.170 \\
-0.002 & 0.028 & -0.091 & 0.118 & 0.966
\end{bmatrix}
$$

The coefficient vector is the center column of $\mathbf{P}_c^{-1}$. Therefore,

$$
\begin{array}{lll}
C_{-2} = 0.117 & C_{-1} = -0.158 & C_0 = 0.937 \\
C_1 = 0.133 & C_2 = -0.091 &
\end{array}
$$

The sample values of the equalized pulse response are given as

$$
p_{eq}(m) = \sum_{n=-2}^{2} C_n p_c(m-n)
$$

where $\Delta t = 0$. For this example we have,

$$
\begin{aligned}
p_{eq}(0) &= (0.117)(0.1) + (-0.158)(-0.1) + (0.937)(1) \\
&\quad + (0.133)(0.2) + (-0.091)(-0.1) \\
&= 1.0
\end{aligned}
$$

which checks with the desired value of unity. Similarly, it can be verified that $p_{eq}(-2) = p_{eq}(-1) = p_{eq}(1) = p_{eq}(2) = 0$.

Values of $p_{eq}(n)$ for $n < -2$ or $n < 2$ are not zero. For example,

$$p_{eq}(3) = (0.117)(0.005) + (-0.158)(0.02) + (0.937)(-0.05)$$
$$+ (0.133)(0.1) + (-0.091)(-0.1)$$
$$= -0.027$$

$$p_{eq}(-3) = (0.117)(0.2) + (-0.158)(-0.1) + (0.937)(0.05)$$
$$+ (0.133)(-0.02) + (-0.091)(0.01)$$
$$= 0.082$$

6.2   LMS Equalizer

Let us now, in the next iteration, revisit the same problem of the channel equalization by introducing more details. We will be dealing with a QPSK signal and equalizer, as shown in Figure 6.5.

6.2.1   Signal Model

The transmitted QPSK signal has the form

$$s_{tr}(t) = d_1(t) \cos \omega_0 t - d_2(t) \sin \omega_0 t \tag{6.7}$$

The received signal can be represented as

$$y(t) = s_{rec}(t) + n(t)$$
$$= s_{tr}(t) + \beta s_{tr}(t - \tau_m) + n_c(t)\cos(\omega_0 t + \alpha) - n_s(t)\sin(\omega_0 t + \alpha) \tag{6.7a}$$

We define the desired non-distorted complex signal at the receiver as

$$D(t) - (1 + \beta)[d_1(t) + jd_2(t)] \tag{6.8}$$

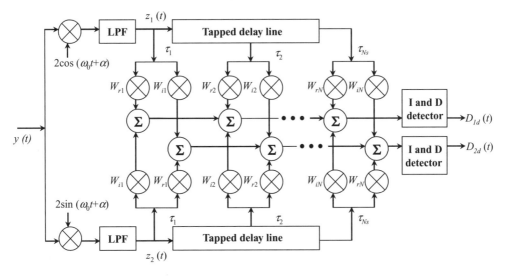

Figure 6.5   Transversal filter equalizer for QPSK.

The cost function for the MMSE equalizer is defined as

$$C = E\lfloor|d_e(t) - D(t)|^2\rfloor \tag{6.9}$$

If we define the complex vectors of equalizer coefficients **W** and LPF output signal samples **Z** (see Figure 6.5)

$$\mathbf{W} = \begin{bmatrix} W_1 \\ W_2 \\ \vdots \\ W_N \end{bmatrix} \qquad \mathbf{Z} = \begin{bmatrix} z(t) \\ z(t - \Delta) \\ \vdots \\ z(t - (N-1)\Delta) \end{bmatrix} \tag{6.10}$$

then the cost function becomes

$$C = E\lfloor|\mathbf{W}'\mathbf{Z} - D(t)|\rfloor^2 \tag{6.11}$$

By setting the gradient to zero we get

$$\mathbf{W}_{\text{opt}} = [E(\mathbf{Z}\mathbf{Z}')]^{-1} E(\mathbf{Z}\mathbf{D}^*(t))$$

A sample of performance results is shown in Figure 6.6.

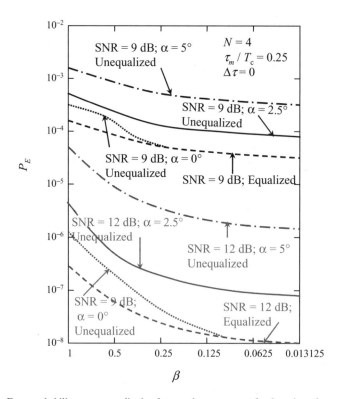

Figure 6.6   Error probability versus amplitude of a specular component for detection of equalized QPSK.

6.2.2   Adaptive Weight Adjustment

Setting the tap coefficients of the zero-forcing and LMS equalizers involves the solutions of a set of simultaneous equations. In the case of the zero-forcing equalizer, adjustment of the tap coefficients involves measuring the channel filter output at T second spaced sampling times in response to a test pulse, and solving for the tap gains. In the case of the LMS equalizer, these equations involve data, noise and multipath dependent parameters, which may be difficult to determine or may not be known at all. Known data assumes a kind of preamble (midamble) embedded into the data.

6.2.3   Automatic Systems

So-called *preset* algorithms, use a training sequence. In *adaptive* algorithms, adjustment of the coefficients is performed continuously during data transmission.

A zero-forcing equalizer can be solved iteratively for the coefficient vector $\mathbf{C}$. Let $\mathbf{C}$ at the kth iteration be $\mathbf{C}^{(k)}$, then the error in the solution is

$$\mathbf{E}^{(k)} = \mathbf{P}_\mathrm{c}\mathbf{C}^{(k)} - \mathbf{P}_\mathrm{eq} \tag{6.12}$$

where $\mathbf{E}^{(k)}$ is a vector with $2N + 1$ components, each of which represent the error in a component of $\mathbf{C}^{(k)}$. Each component of $\mathbf{C}^{(k)}$ can be adjusted in accordance with the error in it.

6.2.4   Iterative Algorithm

If A is a small positive constant, the adjustment algorithm for the jth component of $\mathbf{C}^{(k)}$ is

$$C_j^{(k+1)} = C_j^{(k)} - \mathrm{Asgn}(E_j^{(k)}) \tag{6.13}$$

The iteration process is continued until $C_j^{(k+1)}$ and $C_j^{(k)}$ differ by some suitable small increment. The algorithm converges under fairly broad restrictions.

6.2.5   The LMS Algorithm

The gradient of the mean square error with respect to the jth tap gain is twice the *correlation* between the equalizer output and the error between actual and desired outputs. Two problems arise in applying it to adaptive weight adjustment. First, it requires the expectation or average to be taken. Since this is not available, the unbiased but noisy estimate $z^*(t - j\Delta)\, e(t)$ can be used. Secondly, the undistorted output $D(t)$ is not available unless a known data sequence is transmitted.

An alternative to sending a known data sequence is to assume that the detected data is correct (which is true even in a fairly bad channel giving an error probability of only 10^{-2}) and using the detected data to reconstruct an estimate of the undistorted output $D(t)$. Equalizers using this method of data estimation are called *decision-directed*.

A suitable decision-directed algorithm for weight adjustment of the LMS equalizer is

$$\mathbf{W}(n + 1) = \mathbf{W}(n) - A\mathbf{Z}_T^* \left[d_e\left(t_n - \Delta\right) - D_d\left(t_n\right) \right] \tag{6.14}$$

More details on LMS-based equalizers can be found in [1–7].

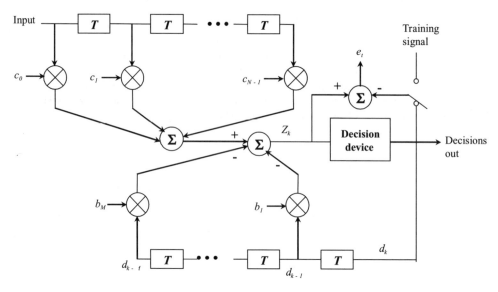

Figure 6.7   Decision feedback equalizer.

6.2.6   Decision Feedback Equalizer (DFE)

Non-linear equalizer structures may provide better performance under many circumstances. A simple non-linear equalizer is the decision feedback equalizer (DFE) which uses feedback of decisions on symbols already received to cancel the interference from symbols which have already been detected.

The basic idea is that, assuming past decisions are correct, the ISI contributed by these symbols can be canceled exactly by subtracting appropriately weighted past symbol values from the equalizer output. This is the purpose of the delay line with weights $b_1, b_2, \ldots, b_M$ shown in Figure 6.7. The forward transversal filter, with weights $c_0, c_1, \ldots, c_{N-1}$, then compensates for ISI over a smaller portion of the ISI-contaminated received signal. Both the feedback and feedforward coefficients can be adjusted simultaneously to minimize the mean square error.

Decision feedback equalizers are discussed in [8–41].

6.2.7   Blind Equalizers

What we need for designing a blind equalizer is to recover the transmitted message (a_t) from the received one (x_t) only, without any preamble for identification of the unknown channel.

The equivalent system models are given in Figure 6.8 and Figure 6.9. The cost function defined by Equation (6.11) is now modified in such a way that it does not require knowledge of data. Options are presented by Equations (6.15) and (6.16). The remaining details are the same as in the previous discussion.

$$J(W) = E\left(\frac{1}{2}c_t^2(W) - \alpha \, |c_t(W)|\right)$$

$$\alpha = \frac{Ea_t^2}{E|a_t|}$$

(6.15)

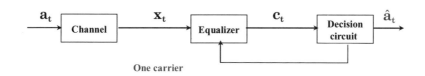

One carrier

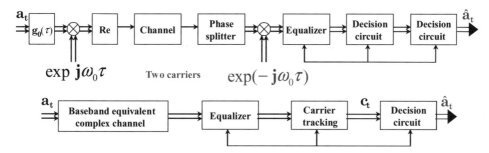

Two carriers

Baseband equivalent

Figure 6.8 System models.

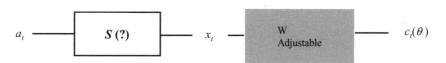

Figure 6.9 Equivalent model of the system from Figure 6.8.

For complex signals, parameters a_t, x_t, c_t, S and W in Figure 6.9 are complex. Re a_t and Im a_t are independent and the cost function is defined as

$$J(W) = E(\psi(\text{Re } c_t(W)) + \psi(\text{Im } c_t(W)))$$

$$\psi(x) = \frac{1}{2}x^2 - \alpha|x| \quad \text{(Sato function)}$$

$$\alpha = \left(\int x^2 v(\mathrm{d}x)\right)\left(\int |x|\, v(\mathrm{d}x)\right)^{-1}$$

(6.16)

Blind equalizers are discussed in [42–64].

6.3 Detection for a Statistically Known, Time Varying Channel

The system model is given in Figure 6.10. Parameters $h_T(\tau)$, $c(t, \tau)$ and $h_R(\tau)$ represent transmitter filter, channel and receive filter impulse responses respectively. The overall system pulse response is a convolution $f(\tau; t)$ of the three and is given by Equation (6.17).

$$f(\tau; t) = h_T(\tau) * c(\tau; t) * h_R(\tau)$$

$$r(kT_s) = r_k = \sum_n a_n f(kT_s - nT;\ kT_s) + w_k$$

(6.17)

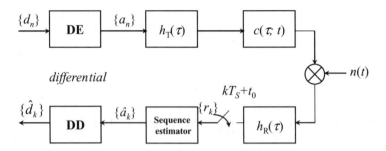

Figure 6.10   Block diagram of the communication system under consideration, in complex envelope form.

Equation (6.17) also represents the samples of the overall received signal at sampling points kT_s. The sampling interval is T_s and the symbol interval is T.

6.3.1   Signal Model

The convolution can also be presented as:

$$r_k = \sum_{n=0}^{L} a_{l-n} f_i(k) + w_k \qquad (6.18)$$

For the T-spaced sampling (TS) case, we have $l = k$, and $i = n$; for the fractional spaced sampling (FS) case, $l = [k/2]$, and $i = 2n + m$, with $m = (k + 1)$ mod 2, i.e. $m = 1$ for k even, and $m = 0$ for k odd. Tapped delay line system models for TS and FS are given in Figures 6.11 and 6.12.

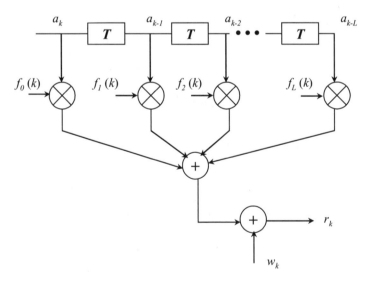

Figure 6.11   Tapped delay line model of the equivalent discrete time channel for the TS case. The blocks containing 'T' denote delays T_s.

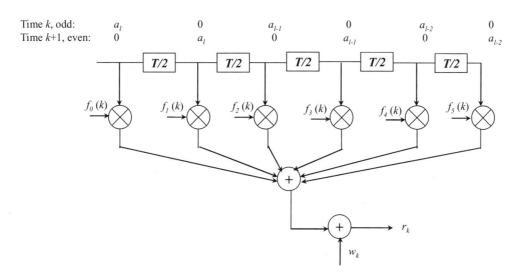

| Time k, odd: | a_l | 0 | a_{l-1} | 0 | a_{l-2} | 0 |
| Time $k+1$, even: | 0 | a_l | 0 | a_{l-1} | 0 | a_{l-2} |

Figure 6.12 Tapped delay line model of the equivalent discrete time channel for the FS case and a channel impulse response length of $3T$. The blocks containing '$T/2$' denote delays of $T/2$s, $l = [k/2]$, and $[x]$ denotes the smallest integer $\geq x$.

6.3.2   Channel Model

For the channel model we assume:

- The worst case channel in terms of T_M and f_D.
- For the worst case value of T_M we use 20 μs.
- With the IS-54 symbol duration of $T = 1/24\,000 \approx 41.7$ μs, this corresponds to roughly $T_M = T/2$, i.e. the echo delay is half a symbol duration.
- For the impulse response shape, we use a 'double-spike' spaced by T_M : $c(\tau; t) = c_0(t)\delta(\tau) + c_1(t)\delta(\tau - T_M)$ where $E[c_0\,(t)^2] = E[|c_1\,(t)|^2] = 0.5$ so that the average energy of the channel is normalized to one.
- The response $f(\tau; t)$ is then

$$f(\tau; t) = c_0(t)h(\tau)+c_1(t)h(t - T_M)$$

- $h(\tau)$ is the full raised cosine response, equal to the convolution of $h_T(\tau)$ and $h_R(\tau)$.
- We set t equal to nT_s and t equal to kT_s (assuming the sampling phase $t_0 = 0$).
- For the TS case, we have, with $T_s = T$,

$$f_0^{TS}(k) = c_0(k)h(0) + c_1(k)h(T/2)$$
$$f_n^{TS}(k) = c_1(k)h((2n - 1)T/2), \quad n = 1,\ 2,\dots$$

- The symmetry of $h(\tau)$ was used.
- For the TS case, we truncate the number of taps to $L + 1 = 3$. The normalized average tap energies are then $E[|f_0|^2] = 0.7717$, $E[|f_1|^2] = 0.2136$, and $E[|f_2|^2] = 0.0147$.

- In the FS case, with $t = nT/2$ and $k = kT/2$

$$f_0^{FS}(k) = c_0(k)h(0) + c_1(k)h(T/2)$$

$$f_1^{FS}(k) = c_1(k)h(T/2) + c_1(k)h(0)$$

$$f_{2n}^{FS}(k) = c_1(k)h((2n-1)T/2) \quad n = \pm 1, \pm 2, \ldots$$

$$f_{2n+1}^{FS}(k) = c_0(k)h((2n+1)T/2) \quad n = \pm 1, \pm 2, \ldots$$

- We retain the most significant taps, yielding $2(L+1) = 6$.
- The average (normalized) tap energies here are $E[|f_{-2}|^2] = E[|f_3|^2] = 0.00735$, $E[|f_{-1}|^2] = E[|f_2|^2] = 0.1068$, and $E[|f_0|^2] = E[|f_1|^2] = 0.38585$.
- In both the TS and FS cases, we choose the minimum number of taps such that the truncated response contained at least 98% of the energy of the untruncated response.
- The autocorrelation of the continuous time processes $c_0(t)$ and $c_1(t)$ is $r_c(\tau) = J_0(2\pi \tau f_D)$, where $J_0(x)$ is the zeroth order Bessel function of the first kind.
- For this model, for a vehicle speed of 30 m/s (67.1 mph) and a carrier frequency of 900 MHz, the maximum Doppler shift $f_D \approx 90$ Hz.
- For simulation purposes, the autocorrelation is approximated by something more easily synthesized, namely, the inverse Fourier transform of a Chebyshev Type I magnitude squared frequency response.
- For time separations $\tau \leq 50T$, the Chebyshev filter yields a very good approximation to the desired autocorrelation.

6.3.3   Statistical Description of the Received Sequence

The following assumptions are made

- The received sequence $\mathbf{r}_N = (r_1, r_2, \ldots, r_N)^T$.
- The transmitted sequence $\mathbf{a}_N = (a_1, a_2, \ldots, a_N)^T$.
- The pdf is complex Gaussian

$$p(\mathbf{r}_N|\mathbf{a}_N) = \frac{1}{\pi \det |\mathbf{C}(\mathbf{a}_N)|} \cdot \exp\left(-\mathbf{r}_N^H \mathbf{C}^{-1}(\mathbf{a}_N)\mathbf{r}_N\right) \tag{6.19}$$

- $\mathbf{C}(\mathbf{a}_N)$ denotes the covariance matrix of $\mathbf{r}_N$ given $\mathbf{a}_N$.
- The superscript H denotes Hermitian (conjugate transpose).
- The elements of $\mathbf{C}(\mathbf{a}_N)$, abbreviated C_a, are obtained by taking the expectation $E[\mathbf{r}_N \cdot \mathbf{r}_N^H|\mathbf{a}_N]$:

$$c_{ij} = \sum_{n=0}^{L} \sum_{m=0}^{L} a_{i-n} a_{j-m}^* E[f_n(i) f_m^*(j)] + \sigma_w^2 \delta_{ij} \tag{6.20}$$

The assumption that the channel is statistically known means that $E\lfloor f_n(i) f_m^*(j)\rfloor$ are known.

6.3.4   The ML Sequence (Block) Estimator for a Statistically Known Channel

The maximum likelihood sequence estimator (MLSE) $\hat{\mathbf{a}}$ is defined as

$$\hat{\mathbf{a}} = \arg \min_a \Lambda(\mathbf{a}, \mathbf{r}) \tag{6.21}$$

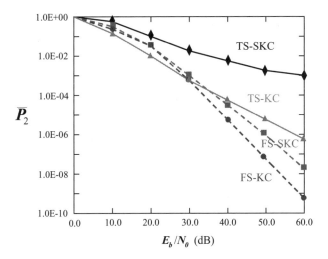

Figure 6.13 Plots of union upper bounds on the average sequence error probability versus E_b/N_0 for the TS and FS SKC detectors, and for the TS and FS known channel detector, for block length $N = 4$, using binary PSK modulation. Channel parameters are $f_D = 90$ Hz, $T_M = T/2$, and $t_0 = 0$.

where $\Lambda(\,)$ is the logarithm of Equation (6.19). The sequence metrics are

$$\Lambda(\mathbf{a}, \mathbf{r}) = \ln |\mathbf{C_a}| + \mathbf{r}^H \mathbf{C_a}^{-1} \mathbf{r} \tag{6.22}$$

Matrix inversion $\mathbf{C}^{-1}$ can be calculated by Cholesky decomposition (factorization). For a given vector of samples $\mathbf{r}$ and the known channel, the algorithm will search for $\mathbf{a}$ that minimizes Equation (6.21). Some results are given in Figures 6.13–6.17.

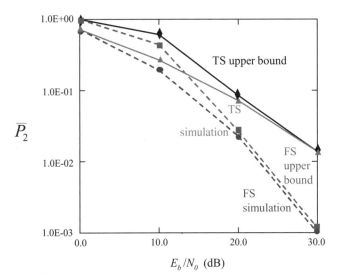

Figure 6.14 Plots of the union upper bounds and simulated average sequence error probability versus E_b/N_0 for the TS and FS SKC detectors, for block length $N = 4$, and binary PSK modulation. The channel parameters are $f_D = 90$ Hz, $T_M = T/2$, and $t_0 = 0$.

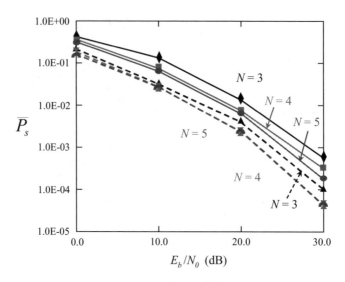

Figure 6.15 Plots of simulated average symbol error probability $\overline{P}_s$ versus E_b/N_0 for the FS SKC (solid lines) and FS KC (dashed lines) detectors, for block length $N = 3, 4, 5$ and binary PSK modulation. The channel parameters are $f_D = 90$ Hz, $T_M = T/2$, and $t_0 = 0$.

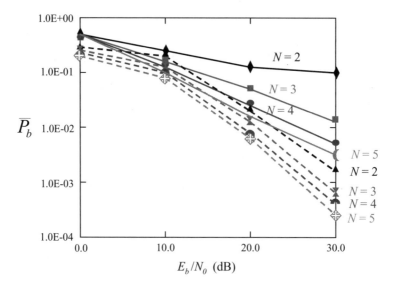

Figure 6.16 Plots of simulated average BEP $\overline{P}_b$ (after differential decoding) versus E_b/N_0 for the TS (solid lines) and FS (dashed lines) SKC detectors, for block length $N = 2 - 5$, and binary PSK modulation. The channel parameters are $f_D = 90$ Hz, $T_M = T/2$, and $t_0 = 0$.

6.4 LMS-Adaptive MLSE Equalization on Multipath Fading Channels

Within this section we are going to drop the assumption that the channel is statistically known and deal with the problem where both data and channel have to be estimated simultaneously.

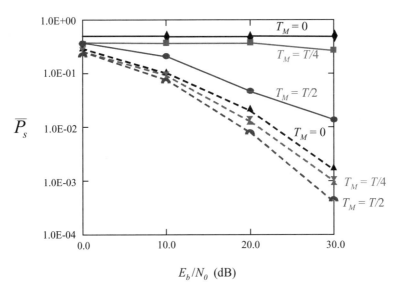

Figure 6.17 Plots of simulated average symbol error probability $\bar{P}_s$ versus E_b/N_0 for the TS and FS SKC detectors, for block length $N = 4$ and binary PSK modulation, showing the effect of mismatch between the estimated delay spread, $\hat{T}_M$, and the actual delay spread T_M. For all cases, $\hat{T}_M = T/2$, but the actual T_M is zero, $T/4$ and $T/2$. The other channel parameters are $f_D = 90$ Hz and $t_0 = 0$. Solid curves are TS results, dashed curves are FS results.

6.4.1 System and Channel Models

The system model is given in Figure 6.18.

The received signal can be represented as

$$r_k = \sum_{i=0}^{l} x_{k-i}\, g^*_{k,i} + \eta_k$$

$$(6.23)$$

$$r_k = \mathbf{g}^H_k \mathbf{x}_k + \eta_k$$

where

$$\mathbf{g}_k = [g_{k,0},\, g_{k,1},\, \ldots,\, g_{k,L}]^T,\, \mathbf{x}_k = [x_k,\, x_{k-1},\, \ldots,\, x_{k-L}]^T$$

$$\frac{1}{2}E\{g_{k,i}\, g^*_{k-l,j}\} = R_g(l,i,j)$$

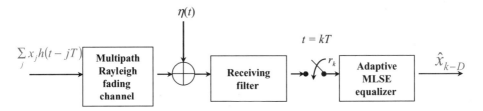

Figure 6.18 System model.

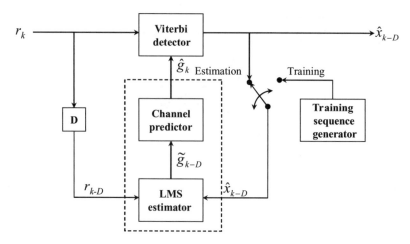

Figure 6.19   Adaptive MLSE equalizer.

Parameter $g_{k,i}$ represents the samples of the equivalent channel impulse response. The operation of the system is presented in Figure 6.19. For data estimation by the Viterbi algorithm, the system uses the channel estimates. The Viterbi algorithm delay is DT. This class of algorithms is known as *delayed decision-directed equalization.*

The metric used in the Viterbi detector is $|r_k - \sum_{i=0}^{L} x_{k-i}\hat{g}_{k,i}^*|^2$

6.4.2   Adaptive Channel Estimator and LMS Estimator Model

Assuming that the feedback decisions are correct, i.e. $\hat{x}_{k-D} = x_{k-D}$ at the input of the LMS estimator, and the order of the LMS estimator is $L + 1$ the following step is implemented: the LMS estimator updates $\tilde{g}_{k-D}$ each time by:

$$\tilde{\mathbf{g}}_{k-D} = \tilde{\mathbf{g}}_{k-D-1} + \mu \mathbf{x}_{k-D}\alpha_{k-D}^*$$

where

$$\alpha_{k-D} = r_{k-D} - \tilde{\mathbf{g}}_{k-D-1}^{H} x_{k-D} \tag{6.24}$$

6.4.3   The Channel Prediction Algorithm

The linear prediction generates

$$\hat{\mathbf{g}}_k = \sum_{j=0}^{q} \alpha_j \hat{\mathbf{g}}_{k-D-j} \tag{6.25}$$

where $\alpha_0, \ldots, \alpha_q$ are constants and q is a positive integer. Due to delay DT of the Viterbi algorithm, the straight line extrapolator is used

$$\hat{\mathbf{g}}_k = \hat{\mathbf{g}}_{k-D} + \frac{p}{q}(\hat{\mathbf{g}}_{k-D} - \hat{\mathbf{g}}_{k-D-q}) \tag{6.26}$$

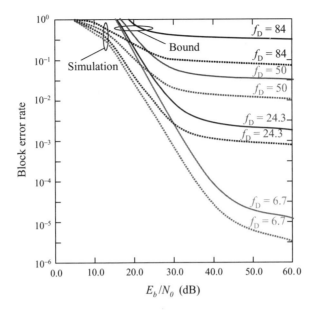

Figure 6.20 Analytical and simulation results for adaptive MLSE employing straight line extrapolation prediction on a two-tap Rayleigh fading channel with $f_D = 6.7, 24.3, 50$ and 84 Hz.

where p is the prediction step. Although there exist many other channel predictions from simulation results and comparisons, the straight line extrapolator gives fair performance and very simple implementation. If the autocorrelation function $R_g (l, i, j)$ is known by the receiver, the prediction coefficients α_j can be obtained from the Wiener solution. Some performance results are given in Figures 6.20–6.24.

MLSE equalizers are discussed in [65–75].

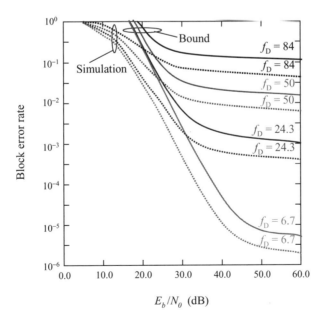

Figure 6.21 Analytical and simulation results for adaptive MLSE employing linear channel prediction on a two-tap Rayleigh fading channel with $f_D = 6.7, 24.3, 50$ and 84 Hz.

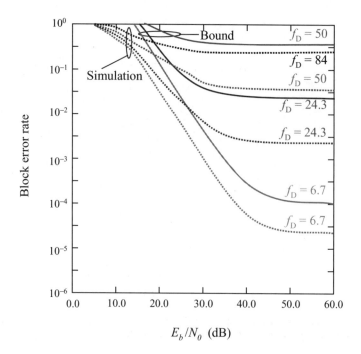

Figure 6.22   Analytical and simulation results for adaptive MLSE without channel prediction on a two-tap Rayleigh fading channel with $f_D = 6.7, 24.3, 50$ and 84 Hz. (Note that the analytical bound of the case of $f_D = 84$ Hz is greater than 1 and hence not shown).

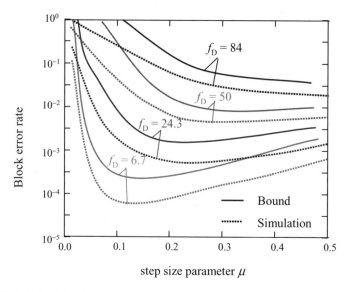

Figure 6.23   Analytical and simulation results obtained by varying the step size parameter μ at $E_b/N_0 = 35$ dB.

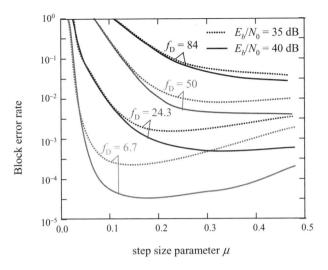

Figure 6.24   Analytical results obtained by varying the step size parameter μ at $E_b/N_0 = 35$ dB and 40 dB.

6.5   Adaptive Channel Identification and Data Demodulation

The estimation technique presented in Section 6.4 is rather inefficient when detection delay D or Doppler f_D increase. In such examples, a joint estimation of both data and channel would give better results. Unfortunately, pure joint estimation would be too complex. In this section we present an algorithm where the joint ML function is maximized by alternating the maximization process with respect to data (given the channel) and channel (given data from the previous iteration).

6.5.1   System Model

The received signal is given as

$$r_k = \sum_{l=0}^{L} h_l(k)a_{k-l} + n_k, k = 1, 2, \ldots, N$$

$$\mathbf{h}(k) = [h_0(k),\ h_1(k), \ldots, h_L(k)]^{\mathrm{T}}$$

(6.27)

The memory L of the channel is determined by the time spread $T_{\max}$ of the actual channel and the symbol period T_s:

$$L = \left[\frac{T_{\max}}{T_s} \right] + 1$$

6.5.2   Joint Channel and Data Estimation

The approach used in this section is based on [76]. The process is described by the following steps.

6.5.2.1   Block Sequence Estimation (BSE)

- The received sequence is fed into the metric computation unit in blocks of N_b symbols at a time.
- The data demodulation is also performed in blocks, rather than symbol by symbol as in the conventional VA.

- At time k the receiver processes a block of N_b data symbols $(r_k, \ldots, r_{k+N_b-1})$.
- After computing the path metrics at time $k + N_b - 1$, the survivor paths of each state are traced backwards in the trellis in order to detect a merge.
- If a merge occurs within the block, at the time $k + N_b - 1 - \delta(\delta < N_b)$, decisions are made on the data sequence $\{a_k, \ldots, a_{k+N_b-1-\delta}\}$.
- The initial point of the next block is set at time $k + N_b - 1 - \delta$ and the next N_b data symbols are processed.
- This means that the portion of the received symbols $(r_{k+N_b-\delta}, \ldots, r_{k+N_b-1})$ is processed twice.
- The state metrics at time $k + N_b - 1 - \delta$ are reset to a large number except for the metric of the state at which the merge occurred. This metric is set to zero. This way the chances of errors associated with illegal state transitions at the beginning of the new block are reduced. It is equivalent to restarting the VA from a known initial state.
- If a merge is not detected within the block $(\delta < N_b)$, the state with the minimum metric at the end of the block (at the time $k + N_b - 1$) is chosen and its survivor path is traced backwards in the trellis.
- A merge is assumed at time $k + N_b - \delta(\delta = \lceil N_b/2 \rceil)$ on the best survivor path.
- The state metrics at time $k + N_b - 1 - \delta$ are reset to a large number, except for the metric of the best survivor. This metric is set to zero.

6.5.2.2 Block Adaptive Channel Estimation

In the derivation of the adaptive schemes for the CIR estimation we will temporarily assume that the channel is unknown but static. The discrete time index k will be dropped from Equation (6.27) to simplify the notation. We will revisit the time varying CIR scenario in subsequent sections. In vector notation, Equation (6.27) can be written as:

$$\mathbf{r} = \mathbf{Ah} + \mathbf{n} \tag{6.28}$$

where

$$\mathbf{r} = [r_1, \ldots, r_N]^T, \mathbf{h} = [h_0, \ldots, h_L]^T, \mathbf{n} = [n_1, \ldots, n_N]^T, \quad \text{and}$$

$$\mathbf{A} = \begin{bmatrix} a_1 & a_0 & \cdots & a_{1-L} \\ a_2 & a_1 & \cdots & a_{2-L} \\ \vdots & \vdots & \vdots & \vdots \\ a_N & a_{N-1} & \cdots & a_{N-L} \end{bmatrix} \tag{6.29}$$

Let us use $\mathbf{a} = [a_1 - L, \ldots, a_N]^T$ to denote the transmitted data vector.

6.5.2.3 Maximum Likelihood (ML)

ML estimates of the channel $\mathbf{h}_{ML}$ and data $\mathbf{a}_{ML}$ are those that maximize the conditional probability density function (pdf) $p(\mathbf{r}|\mathbf{a}, \mathbf{h})$ or, equivalently, minimize $|\mathbf{r} - \mathbf{Ah}|^2$. With respect to $\mathbf{A}$ and $\mathbf{h}$

$$(\mathbf{a}_{ML}, \mathbf{h}_{ML}) = \underset{a \in A^{N+L-1}, \mathbf{h} \in R^L}{\arg \min} |\mathbf{r} - \mathbf{Ah}|^2 \tag{6.30}$$

The joint minimization over $\mathbf{a}$ and $\mathbf{h}$ cannot be computed in closed form. When either $\mathbf{a}$ or $\mathbf{h}$ is fixed, this minimization is a well known problem. When the channel $\mathbf{h}$ is fixed, the ML estimate of $\mathbf{a}$ is computed via the VA. When the data $\mathbf{a}$ is given, the minimization of $C(\mathbf{A}, \mathbf{h}) = |\mathbf{r} - \mathbf{Ah}|^2$ over $\mathbf{h}$ is a standard least squares problem which results in

$$\mathbf{h}_{ML} = (\mathbf{A}^H \mathbf{A})^{-1} \mathbf{A}^H \mathbf{r} \tag{6.31}$$

Thus, the joint minimization of $C(\mathbf{A}, \mathbf{h})$ over $\mathbf{A}$ and $\mathbf{h}$ can be viewed as an alternative minimization type of problem.

6.5.2.4 Iterative Procedure

The above algorithms can be summarized as:

1. Start with an initial estimate $\hat{\mathbf{h}}(0)$ of the channel.
2. Minimize $C(\hat{\mathbf{A}}, \hat{\mathbf{h}})$ with respect to $\hat{\mathbf{A}}$ via the VA to obtain $\hat{\mathbf{A}}(l)$.
3. Minimize $C(\hat{\mathbf{A}}, \hat{\mathbf{h}})$ w.r.t. $\hat{\mathbf{h}}(l)$ to obtain

$$\hat{\mathbf{h}}(l + 1) = [\hat{\mathbf{A}}^H(l)\hat{\mathbf{A}}(l)]^{-1}\hat{\mathbf{A}}^H(l)\mathbf{r} \tag{6.32}$$

4. Repeat Steps 2 and 3 until the algorithm converges.

Simulation studies have shown that this scheme can converge very fast to a global minimum (in approximately 5–6 iterations).

The above algorithm pays a heavy price as a tradeoff for its optimum performance. The receiver has to process a long received sequence (in the order of a few thousands of symbols) several times before it can provide the user with any reliable decisions about the transmitted data. This implies a significant decision delay. The algorithm is impractical for real-time implementation, especially when the channel is time varying. This is a strong motivation for constructing algorithms to perform the joint channel and data estimation recursively in time, which provides the user with reliable data estimates quickly enough to be able to adopt to channel variations too.

6.5.2.5 Iterative CIR Estimator

In this case we have

$$\hat{\mathbf{h}}(l + 1) = \hat{\mathbf{h}}(l) + \mu \left. \frac{\partial[\ln p(\mathbf{r}|\hat{\mathbf{A}}, \mathbf{h})]}{\partial \mathbf{h}} \right|_{\mathbf{h} = \hat{\mathbf{h}}(l)} \tag{6.33}$$

$$= \hat{\mathbf{h}}(l) + \mu \hat{\mathbf{A}}^H[\mathbf{r} - \hat{\mathbf{A}}\,\hat{\mathbf{h}}(l)]$$

If we can achieve perfect convergence, then

$$\hat{\mathbf{h}}(l + 1) = \hat{\mathbf{h}}(l) \Leftrightarrow \hat{\mathbf{A}}^H[\mathbf{r} - \hat{\mathbf{A}}\,\hat{\mathbf{h}}(l)] = 0 \Leftrightarrow \mathbf{h}(l)$$
$$= (\hat{\mathbf{A}}^H \hat{\mathbf{A}})^{-1} \hat{\mathbf{A}}^H \mathbf{r} \tag{6.34}$$

which is exactly the ML estimate of the channel. In Equation (6.32), $\hat{\mathbf{A}}$ and $\mathbf{r}$ extend over the whole data record. If, instead, we consider only a portion of data, i.e. a block of N_b symbols, then each time we update the CIR estimate we can perform the joint ML estimation of data and channel recursively in time. Notice that Equation (6.33) has the form of a block least mean square (BLMS) equation. This is a consequence of the quadratic exponent of the Gaussian probability distribution of the noise vector. Given the CIR estimate $\hat{\mathbf{h}}(l)$ at the lth recursion, the corresponding block of data estimates can be provided by the BSE, as was developed earlier. Starting from an initial guess, $\hat{\mathbf{h}}(0)$, for the CIR we construct a joint data and channel estimation scheme which operates on successive blocks of data in a time recursive manner.

6.5.2.6   BSE/BLMS Algorithm

In this case:

1. Start with an initial estimate $\hat{\mathbf{h}}(0)$ of the CIR. Initialize the block counter ($l \to 1$) and the symbol counter ($k \to 1$).
2. Receive the lth block of data $\mathbf{r}(l) = [r_k, \ldots, r_{k+N_b-1}]^T$, where N_b is the block length.
3. Apply the BSE on $\mathbf{r}(l)$, using $\hat{\mathbf{h}}(l-1)$ for metric computations, to obtain the lth block of data estimates:

$$\hat{\mathbf{a}}(l) = [\hat{a}_k, \ldots, \hat{a}_{k+N_b-1-\delta(l)}]^T$$

where $\delta(l)$ is the merging delay of the BSE for the lth block.

4. Using the data estimates $\hat{\mathbf{a}}(l)$ and the first $N_b - \delta(l)$ entries of the lth block of received symbols $\mathbf{r}_\delta(l) = [r_k, \ldots, r_{k+N_b-1-\delta(l)}]^T$, update the CIR estimate as follows:

$$\hat{\mathbf{h}}(l) = \hat{\mathbf{h}}(l-1) + \mu \hat{\mathbf{A}}^H(l)[\mathbf{r}_\delta(l) - \hat{\mathbf{A}}(l)\hat{\mathbf{h}}(l-1)] \tag{6.35}$$

The matrix $\hat{\mathbf{A}}(l)$ is given by:

$$\hat{\mathbf{A}}(l) = \begin{bmatrix} \hat{a}_k & \hat{a}_{k-1} & \cdots & \hat{a}_{k-L} \\ \hat{a}_{k+1} & \hat{a}_k & \cdots & \hat{a}_{k+1-L} \\ \vdots & \vdots & \vdots & \vdots \\ \hat{a}_{k+N_b-1-\delta(l)} & \hat{a}_{k+N_b-2-\delta(l)} & \cdots & \hat{a}_{k+N_b-1-\delta(l)-L} \end{bmatrix} \tag{6.36}$$

and μ is a step size parameter.

5. Reset the block and symbol counters to $l \to l+1$ and $k \to k + N_b - \delta(l)$ and go to Step 2 until all data have been processed. An issue is the choice of the step size parameter μ. A large value of μ would provide faster adaptation but could lead to divergence of the channel estimate. On the other hand, if μ is too small, we will lose a lot of data while trying to acquire the channel.

6.5.2.7   Recursive Least Squares (RLS) Channel Estimation

In this case the CIR estimation is based on the minimization of the weighted sum of the squared errors

$$J(k) = \sum_{l=0}^{k} \lambda^{k-l} e^2(l) = \sum_{l=0}^{k} \lambda^{k-l} |r_l - \hat{\mathbf{h}}^T(k) \cdot \hat{\mathbf{a}}(l)|^2 \tag{6.37}$$

where $\hat{\mathbf{h}}(k) = \lfloor \hat{h}_0(k), \ldots, \hat{h}_L(k) \rfloor^T$ is the CIR estimate at time k, $\hat{\mathbf{a}}(l) = [\hat{a}_l, \ldots, \hat{a}_{l-L}]^T$ is the vector of the estimated data at times $l, \ldots, l - L$ and r_l is the received signal at time l. The parameter λ, ($0 < \lambda \le 1$) gives more weight to recent errors. Thus, we allow for time varying channels. The RLS algorithm for the CIR estimation is described as [77–79]:

$$g(k) = \frac{\mathbf{P}(k)\hat{\mathbf{a}}^*(k)}{\lambda + \hat{\mathbf{a}}^T(k)\mathbf{P}(k)\hat{\mathbf{a}}^*(k)}$$

$$\mathbf{P}(k+1) = \frac{1}{\lambda}[\mathbf{P}(k) - g(k)\hat{\mathbf{a}}^T(k)\mathbf{P}(k)]$$

$$e(k) = r_k - \hat{\mathbf{h}}^T(k)\hat{\mathbf{a}}(k) \tag{6.38}$$

$$\hat{\mathbf{h}}(k+1) = \hat{\mathbf{h}}(k) + e(k)g(k)$$

$$k = 1, 2, \ldots$$

To initialize the algorithm we set $\mathbf{P}(0) = \varepsilon^{-1}\mathbf{I}$, where ε is a small positive constant.

6.5.2.8   BSE/RLS Algorithm

Incorporating the CIR update into the recursive joint channel and data estimation algorithm gives:

1. Start with an initial estimate $\hat{\mathbf{h}}(0)$ of the CIR. Initialize the block counter ($l \to 1$) and the symbol counter ($k \to 1$).
2. Receive the lth block of data $\mathbf{r}(l) = [r_k, \ldots, r_{k+N_b-1}]^{\mathrm{T}}$, where N_b is the block length.
3. Apply the BSE on $\mathbf{r}(l)$, using $\hat{\mathbf{h}}(l-1)$ for metric computations, to obtain the lth block of data estimates:

$$\hat{\mathbf{a}}(l) = [\hat{a}_k, \ldots, \hat{a}_{k+N_b-1-\delta(l)}]^{\mathrm{T}}$$

   where $\delta(l)$ is the merging delay of the BSE for the lth block.
4. Using the data estimates $\hat{\mathbf{a}}(l)$ and the first $N_b - \delta(l)$ entries of the lth block of received symbols $\mathbf{r}_\delta(l) = [r_k, \ldots, r_{k+N_b-1-\delta(l)}]^{\mathrm{T}}$, obtain the new CIR estimate $\hat{\mathbf{h}}(l)$ by executing the RLS recursion $N_b - \delta(l)$ times.
5. Reset the block and symbol counters to $l \to l+1$ and $k \to k + N_b - \delta(l)$ and go to Step 2 until all data have been processed.

A schematic diagram of adaptive BSE (BSE/BLMS or BSE/RLS) is shown in Figure 6.25.

The adaptive BSE derives its strength from the fact that, due to the variable decision delay, the data estimates used for CIR tracking are more likely to be taken from the ML path. The block processing of the data contributes to the averaging of the effects of symbol errors. Thus, the CIR estimator is fed with better data estimates, which improves its tracking capability compared to the conventional adaptive MLSE.

The possible data errors towards the end of each block, caused by computing the metrics for the whole block using the same CIR estimate, are alleviated by the fact that the last symbols of each block are processed again with the updated CIR estimate. If the fading rate gets higher, the block length N_b must be chosen carefully so that the CIR estimate is not outdated. A choice of $N_b \approx 5L$ is a good rule of thumb.

6.5.3   Data Estimation and Tracking for a Fading Channel

In a static channel environment, the BSE/BLMS and BSE/RLS as described so far, operate starting from an initial guess $\hat{\mathbf{h}}(0)$ of the CIR. For a fading channel, the task of CIR acquisition is much heavier,

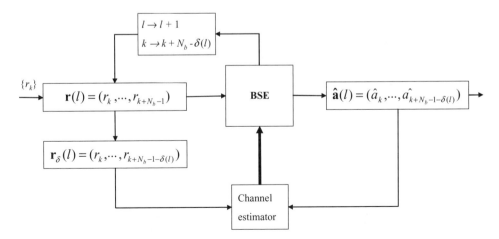

Figure 6.25   Schematic diagram for adaptive BSE [76] © 1998, IEEE.

especially when the fading rate becomes high. In such environments, the CIR acquisition is accomplished via a training sequence, which is periodically sent to the receiver as a portion of a fixed size data packet. This format is used in mobile communications. An example of such a data structure is the time division multiple access (TDMA) slot of the IS-54 North American Digital Cellular standard described briefly in Chapter 1.

6.5.3.1 Training

The BSE/BLMS, BSE/RLS can be applied in this signaling format with an appropriate adjustment in the CIR acquisition step. This is accomplished by feeding the above described adaptive algorithms (BLMS or LMS, RLS) with the training symbols at the header of the TDMA slot, producing the CIR estimate $\hat{\mathbf{h}}(0)$. As soon as the acquisition stage is over, the BSE/BLMS or BSE/RLS is activated for joint data estimation and channel tracking, as described in the previous subsections, for the duration of the entire TDMA slot. The performance of this scheme, obtained by simulation, is now presented.

6.5.3.2 Performance and Computational Complexity

In the evaluation of the system performance the same assumptions are used as in [76]:

- Static and fading channel environments.
- To cover both the static and time varying CIR, the SNR is defined as

$$\text{SNR} = 10 \log \left[\frac{\sigma_a^2}{\sigma^2} \sum_{i=0}^{L} E(|h_i|^2) \right]$$

  where σ_a^2 and σ^2 are the variances of the input data and additive noise, respectively.
- The above definition implies independence of the CIR variations from the transmitted data sequence, which is quite a reasonable assumption.

6.5.4 The Static Channel Environment

- CIR is assumed constant for the duration of the entire received data record.
- In the simulation there is no training sequence, i.e. the BSE operates in a blind fashion.
- The modulation format is binary phase shift keying (BPSK).
- No coding is assumed.
- The input alphabet is $\{-1, 1\}$, i.e. the transmitted data sequence $\{a_k\}$ consists of real numbers.
- Two particular channel examples are used whose impulse responses are given by the vectors $\mathbf{h}_1 = [0.407, 0.815, 0.407]$ and $\mathbf{h}_2 = [0.1897, 0.5097, 0.6847, 0.46, 0.1545]$, respectively.
- Both channels exhibit deep nulls in their magnitude response within the frequency band of interest.
- Channel 2 has non-linear phase characteristics.
- As a result, linear equalization methods exhibit very poor performance on these channels.
- The MLSE for the corresponding *known* channel is used in Figure 6.26 for comparison.

6.5.4.1 Simulations/BSE/BLMS

- The error rates are steady state after the channel acquisition has been completed.
- The acquisition (convergence) time is described independently.
- In both cases the block length is chosen to be $N_b = 8L$ and the step size parameter $\mu = 0.01$.
- The initial guesses for Channels 1 and 2 are $\hat{\mathbf{h}}(0) = [0, 1, 0]$ and $\hat{\mathbf{h}}(0) = [0, 0, 1, 0, 0]$, respectively.
- The probability of error is estimated over a data record of 100 000 symbols.

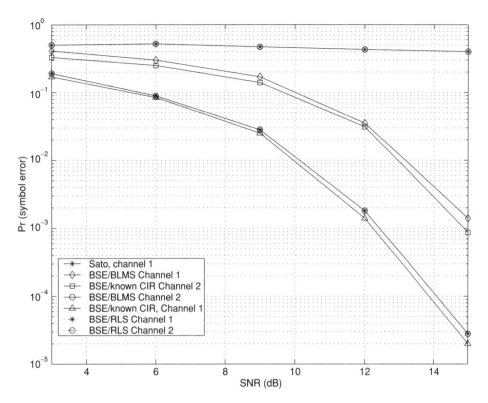

Figure 6.26 Probability of symbol error versus SNR using the BPSK waveform for the BSE/BLMS and the BSE/RLS algorithms over Channels 1 and 2.

- The performance is very close to that of the known CIR environment.
- For comparison purposes, a linear blind equalizer (Sato's algorithm) described by Equation (6.16) is used. As expected, it fails to converge at all for this type of channel.

6.5.4.2 The BSE/RLS

- Is simulated for Channels 1 and 2, starting with the same initial guesses for the channels as in BSE/BMLS.
- For the initialization of the BSE/RLS, the parameters $\lambda = 1$ and $\varepsilon = 0.001$ are used in P(0).
- The performance of the BSE/RLS in Figure 6.26 is also very close to that of the known channel.
- An improvement is observed for Channel 2 compared to the BSE/BLMS, especially at high SNR. This is no surprise, since acquisition capability (convergence speed) is better than the LMS at high SNR.
- The tradeoff for the improved performance of the BSE/RLS, however, is its increased computational complexity.

6.5.4.3 The Effect of the Block Length N_b

BSE/BLMS is simulated on Channel 1 for $N_b = 8L$, $5L$, $4L$ and $3L$ symbols and the results are shown in Figure 6.27.

- The performance degradation for smaller N_b must be attributed to the BSE algorithm.
- As the block length becomes smaller, the number of blocks in which no merge occurs increases.
- Bad data decisions are forced, reflecting upon the channel estimate too.

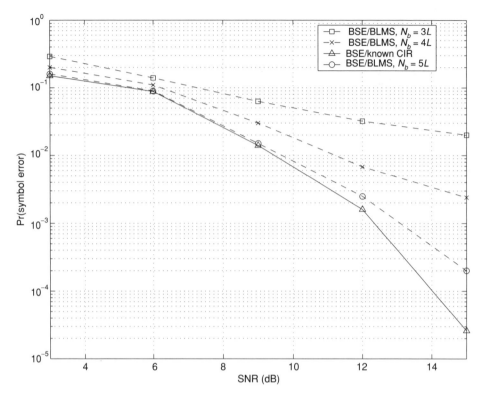

Figure 6.27 Probability of symbol error versus SNR using the BPSK waveform for the BSE/BLMS algorithm with various values of N_b over Channel 1.

6.5.4.4 The Convergence Properties

- 300 independent Monte Carlo runs of BSE/BLMS are performed and the average squared error $|\hat{\mathbf{h}} - \mathbf{h}|^2$ as the number of processed data varies between 100 and 1000 symbols is computed.
- Simulations are performed for block sizes $N_b = 5L$, $10L$ and $15L$ points, and for SNR $= 6$ and 12 dB.
- The results are shown in Figure 6.28.
- For Channel 1, independent of the block size, we approach the steady state mean squared error within the first few hundred symbol periods.
- For SNR $= 6$ dB the steady state is reached after processing about 700 symbols, while for SNR $= 12$ dB we need approximately 400 symbols. This allows us to specify the *acquisition time* for the BSE/BLMS at about 400 symbol periods.

6.5.5 The Time Varying Channel Environment

In this segment a TDMA with the following parameters is simulated.

- The carrier frequency is assumed to be 900 MHz and the bit rate 48.6 kb/s.
- For QPSK modulation, that translates to a baud rate of 24.3 ksymbols/s, or a symbol period $T_s = 41\,\mu s$.
- The mobile radio channel is assumed to be wideband with $L + 1$ taps.
- Each element of the CIR $\{h_i(k)\}_{i=0}^{L}$ is modeled as an independent low pass zero mean complex Gaussian random process with Rayleigh distributed amplitude and uniformly distributed phase in the interval $[-\pi, \pi]$.

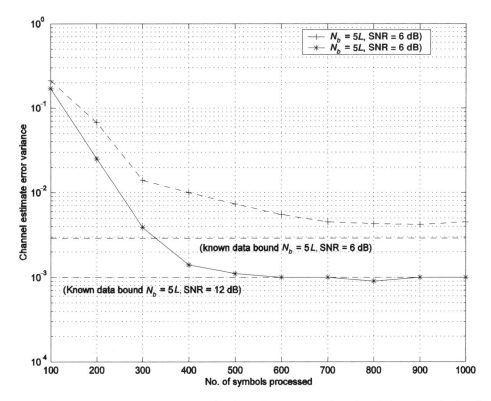

Figure 6.28 Channel estimate error variance for Channel 1 versus number of symbols processed using the BSE/BLMS algorithm at SNR equal to 6 and 12 dB.

- Each of the h_is is generated by passing a zero mean white complex Gaussian noise sequence through a digital second order Butterworth filter whose cutoff frequency f_d is determined by

$$f_d = T_s \left(f_c \frac{v}{v_c} \right)$$

where f_d is the normalized Doppler shift, f_c is the carrier frequency, and v and v_c are the vehicle speed and speed of light, respectively. This simulation procedure is widely used in the literature.
- For these system specifications the performance of all algorithms is investigated for vehicle speeds ranging from $v = 60$ mph (100 km/h) up to 150 mph (248 km/h).
- The corresponding normalized Doppler shift ranges from $f_d = 0.0034$ to 0.0085.
- A three-tap CIR ($L = 2$) is used. For the specifications of the IS-54 standard ($T_s = 41$ μs), this channel model covers a maximum time spread of $T_{max} = 123$ μs.
- The CIR elements $h_0(k)$, $h_1(k)$, $h_2(k)$ are constructed so as to have equal power $\sigma_h^2 = 1/3$ and to be independent of each other.
- This corresponds to a fading scenario where no *line of sight* is present (Rayleigh fading).

Figure 6.29 compares the performance of all algorithms tested: the BSE; the conventional adaptive MLSE; a variable D (delay in the CIR estimation loop) version of the conventional adaptive MLSE; and the Per Survivor Processing (PSP) known CIR environment. PSP estimates the channel only for the surviving trajectories in VA, and for those trajectories data is known. This corresponds to an ideal receiver capable of estimating the CIR perfectly, with zero delay, which is unrealistic when the channel is varying rapidly.

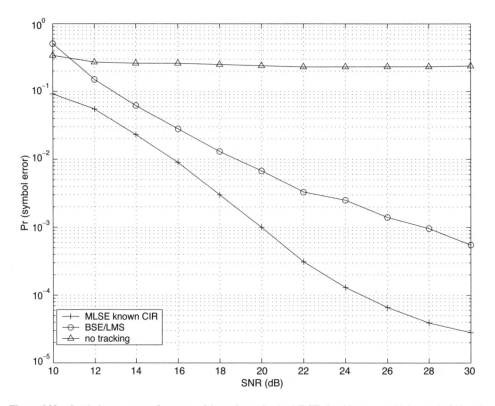

Figure 6.29   Symbol error rate performance of the various adaptive MLSE algorithms at a vehicle speed of 60 mph.

The adaptive D version of the conventional MLSE is derived by detecting a merge within N_b epochs from the current symbol and setting D equal to the merging delay. If no merge is detected, D is set to $N_b - 1$. The symbol error rate is computed as the fraction of data symbols that are in error over a time interval of 5000 consecutive TDMA slots, each packet of which has the structure shown in Chapter 1. For convenience, only the initial training symbols and the information symbols inside the TDMA slot were considered.

For LMS channel acquisition and tracking, the optimum value of the step size parameter was found to be $\mu = 0.06$ for $f_d = 0.0034$ and $\mu = 0.12$ for $f_d = 0.0085$, at SNR = 20 dB. The optimum value of the weight parameter was found to be $\lambda = 0.73$ for $f_d = 0.0085$ at SNR = 20 dB. For all of the simulated algorithms presented, the initial CIR guess at the beginning of the first TDMA slot was set at $\hat{\mathbf{h}}(0) = [0, \ldots, 0]$. The last CIR estimate computed at the end of each subsequent slot is utilized to initialize the LMS or RLS at the beginning of the next slot.

6.5.5.1   Performance

In Figure 6.29, the performance of all the adaptive MLSE algorithms is presented for $f_d = 0.0034$ (82.6 Hz), which corresponds to a vehicle speed of 60 mph. The following parameters were used.

- An LMS update was used for the CIR acquisition and tracking. The step size parameter was set at $\mu = 0.06$ for all algorithms.
- For the conventional adaptive MLSE, the optimum performance was achieved for a delay in the CIR estimation loop equal to $D = 4$.

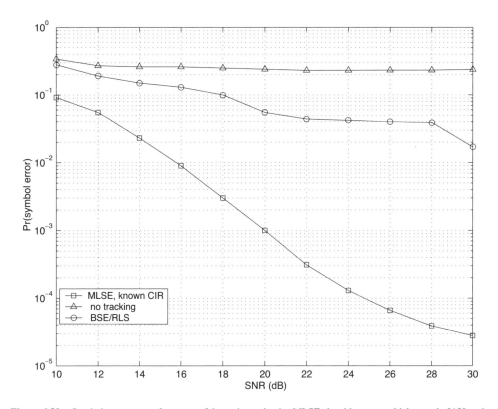

Figure 6.30 Symbol error rate performance of the various adaptive MLSE algorithms at a vehicle speed of 150 mph. The results from the use of either the LMS or the RLS channel estimates are shown.

- The decision delay was set at $\delta = 5L$ for both the PSP, the conventional MLSE and the adaptive D MLSE.
- For the BSE/LMS a block length $N_b = 11$ was found to yield the best performance.
- Both the PSP and the BSE/LMS exhibit a clear advantage over the conventional adaptive MLSE at SNR above 20 dB.
- The adaptive D MLSE also exhibits a performance advantage over the conventional MLSE and gets closer to the BSE at high SNR.
- For comparison purposes a receiver which performs no tracking of the CIR after the initial acquisition mode was also simulated. It reaches a floor of about 2×10^{-1} at SNR = 20 dB. This clearly enforces the necessity of the use of CIR tracking algorithms even at this fading rate.

Figure 6.30 presents the performance of all algorithms for $f_d = 0.0085$ (206.5 Hz), and a speed of 150 mph. Both the LMS and RLS were used for CIR acquisition and tracking. For the LMS, the step size parameter was set at $\mu = 0.12$ and for the RLS, the weight parameter was set at $\lambda = 0.73$, for all algorithms. The optimum parameters were again found to be $D = 4$ and $N_b = 11$. The PSP and the BSE exhibit a clear advantage over the conventional and adaptive D MLSE at high SNR.

A much poorer performance was exhibited by all algorithms compared to the smaller vehicle speed in Figure 6.29. The performance degradation must be attributed to the higher fading rate making it very difficult to track the channel variations. The RLS exhibited slightly better performance at high SNR, which was anticipated. Absence of CIR tracking yields a floor of 0.5 probability of symbol error at SNR = 10 dB.

Figure 6.31 presents the error rates for the information symbols inside the TDMA slot for $f_d = 0.0085$ at SNR = 20 dB with LMS update. In the figure, only every other information symbol is plotted. The error

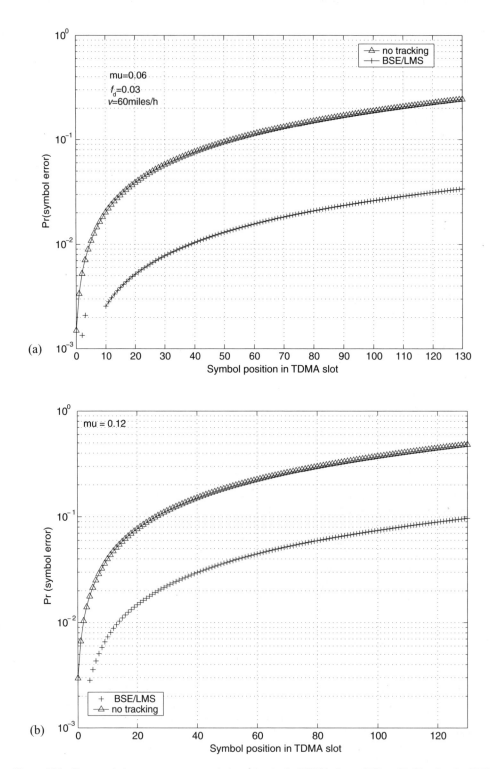

Figure 6.31 Data symbol error rate versus symbol position in the TDMA slot at SNR = 20 dB using the LMS channel acquisition and tracking algorithms (a) $v = 60$ mph; (b) $v = 150$ mph.

rates tend to increase toward the end of the slot, even with CIR tracking algorithms. If no CIR tracking is employed, the error rates become unacceptable after the first 10–20 symbols within the TDMA slot. For BSE/LMS at 60 mph and error rate 10^{-3} there is a 10 dB worse performance from the lower bound of a known channel. PSP has manifested a similar performance. At 150 mph the error rate of all algorithms levels off at residual error rates larger than 10^{-2}, even at 30 dB. All algorithms fail at a velocity of 150 mph.

6.5.5.2 Computational Complexity and Reconfiguration Efficiency

Computational complexity is measured by the required number of multiplications, N_{mul}, and additions, N_{add}, per data sample. The complexity is computed for a CIR of length $L + 1$ and an alphabet size M. The exact number of required operations per data symbol for the BSE cannot be expressed in closed form because of the variable merging delay inside each block. If $\bar{\delta}$ denotes the average merging delay over all blocks processed, then the required operations for metric computations per data symbol must be multiplied by a factor

$$f_{ex} = \frac{1}{[1 - \bar{\delta}/N_b]}$$

to account for processing the last part of each block twice. Parameter $\bar{\delta}$ is obtained from simulations.

The three-tap fading channel $(L = 2)$ of the previous section is used and 5000 frames of PSK data $(M = 4)$ are processed, amounting to 82 953 processed blocks, each of length $N_b = 11$ symbols. The fading rate is set at $f_d = 0.0034$ and the SNR is 30 dB. $\bar{\delta} = 2.9453$ symbols is found, which gives $f_{ex} = 1.365662$.

Table 6.1 presents the number of complex operations required per data symbol for metric calculations in the adaptive MLSE algorithm. Table 6.2 presents the required number of complex operations per symbol for this specific example, with both LMS and RLS updates for all channel estimators. For the Sato algorithm, discussed in Section 6.2, a tapped delay line equalizer with $L_{sa} = 16$ taps is assumed. From Tables 6.1 and 6.2, the conventional adaptive MLSE requires the smallest number of operations of all MLSE-based techniques, but significantly more operations than linear equalizers such as Sato's algorithm. The BSE requires an increased number of operations for the metric update compared to the conventional adaptive MLSE due to reprocessing the last part of each data block. A hardware implementation of the BSE would be more difficult due to the necessary checks for the detection of a merge inside each block. For the CIR estimation part, the PSP requires considerably more operations than all other techniques due to the multiple CIR estimates that it preserves (one for each state). This overhead becomes even larger if the RLS estimation algorithm is used for CIR tracking.

For the calculation of reconfiguration efficiency, relations defined in Chapters 2 and 3 are still valid. For practical applications the relative complexity D_r can be calculated by using data directly from Tables 6.1 and 6.2.

More details on the topic can be found in [80–93].

Table 6.1 Complex operations per data symbol for adaptive MLSE algorithms (metric calculations)

		Conventional adaptive MLSE	BSE	PSP	Sato
Metric	N_{mul}	$M^{L+1}(L+1)$	$M^{L+1}(L+1)f_{ex}$	$M^{L+1}(L+1)$	0
Update	N_{add}	$M^{L+1}(L+1)$	$M^{L+1}(L+1)f_{ex}$	$M^{L+1}(L+1)$	0
CIR update	N_{mul}	$2(L+1)$	$2(L+1)$	$M^{L+1}(L+1)$	$2(L_{sa}+1)$
(LMS)	N_{add}	$2(L+1)$	$2(L+1)$	$M^{L+1}(L+1)$	$2(L_{sa}+1)$
CIR update	N_{mul}	$4(L^2+L)$	$4(L^2+L)$	$M^L 4(L^2+L)$	
(RLS)	N_{add}	$3(L^2+L)$	$3(L^2+L)$	$M^L 3(L^2+L)$	

Table 6.2 Complex operations per data symbol for $M = 4$ and $L = 2$ (channel estimations)

		Conventional adaptive MLSE	BSE	PSP	Sato
Metric	N_{mul}	192	262	192	0
Update	N_{add}	192	262	192	0
CIR update	N_{mul}	6	6	96	34
(LMS)	N_{add}	6	6	96	34
CIR update	N_{mul}	24	24	384	
(RLS)	N_{add}	24	24	384	

6.6 Turbo Equalization

In this section we discuss how the problem from Section 6.5 can be solved by using the turbo principle discussed in Chapter 2.

6.6.1 Signal Format

The transmitted signal $s(t)$ is provided by the output of a filter whose impulse response is $h_e(t)$. The signal emitted can be expressed in the form:

$$s(t) = A \sum_k c_k h_e (t - kT) \exp j(2\pi f_0 t + \varphi_0) \tag{6.39}$$

In a multipath channel, the received signal $y(t)$, can be written as follows:

$$y(t) = \sum_{m=0}^{M-1} A_m(t) \sum_k c_k h_e(t - \tau_m - kT) \exp j(2\pi f_0 t + \varphi_0) + w(t) \tag{6.40}$$

where $A_m(t)$ are complex-valued independent multiplicative noise processes. The receiver matched filter output

$$R_n \triangleq R(nT) = \sum_{m=0}^{M-1} A_m(n) \sum_k c_{n-k} h_s(kT - \tau_m) + w_n \tag{6.41}$$

where $A_m(n)$ equals $A_m(nT)$ by definition, w_n denotes the response of the receiving matched filter to the noise $w(t)$, sampled at time $nT \cdot h_s(t)$ is defined by $h_s(t) = h_e(t) \times h_e^*(-t)$ and satisfies the Nyquist criterion.

$$\Gamma_k(n) = \sum_{m=0}^{M-1} A_m(n) h_s(kT - \tau_m) \tag{6.42}$$

Let us suppose that the ISI is limited to $(L_1 + L_2)$ symbols. Equation (6.41) may be written in the form:

$$R_n = \sum_{k=-L_2}^{L_1} \Gamma_k(n) c_{n-k} + w_n \tag{6.43}$$

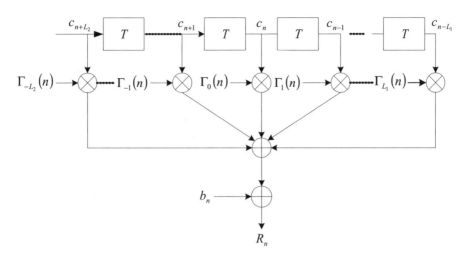

Figure 6.32 Equivalent discrete time model of a channel with inter-symbol interference.

6.6.2 Equivalent Discrete Time Channel Model

Quantities $\Gamma_k(n)$ are expressed as a linear combination of the multiplicative noises $A_m(n)$. Therefore, they are Gaussian in the case of a Rayleigh-type channel and constant in the case of a Gauss-type channel. Consequently, the set of modules made up of the modulator, the transmission channel and the demodulator can be represented by an equivalent discrete time channel as shown in Figure 6.32.

6.6.3 Equivalent System State Representations

By new indexing in Equation (6.43) we have:

$$R_n = \sum_{k=0}^{L_1+L_2} \Gamma_{k-L_2}(n)c_{n+L_2-k} + b_n \tag{6.44}$$

If we denote $S_n = (c_{n+L_2}, \ldots, c_{n-L_1+1})$ the state of the equivalent discrete time channel at time nT sample R_n depends on the channel state S_{n-1} and on the symbol c_{n+L_2}. Therefore, the equivalent discrete time channel can be modeled as a Markov chain and its behavior can be represented by the trellis diagram shown in Figure 6.33.

6.6.4 Turbo Equalization

In order to use a soft-input channel decoder, the symbol detector has to provide information about the reliability of the symbols estimated. This information may be obtained by using a soft-output Viterbi algorithm (SOVA) (the name often used for the algorithm described in Appendix 2.1), that associates an estimation of the logarithm of its likelihood ratio (LLR), $\Lambda_1(c_n)$, to each symbol c_n detected:

$$\Lambda_1(c_n) = \log \frac{\Pr\{c_n = +1|\mathbf{R}\}}{\Pr\{c_n = -1|\mathbf{R}\}} \tag{6.45}$$

where $\mathbf{R}$ denotes the vector of samples that constitutes the observation. After deinterleaving, the SOVA decoder provides a new LLR value of c_k, $\Lambda_2(c_k)$, that may be derived by analogy with the calculations

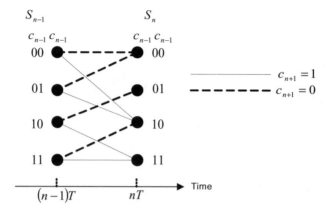

Figure 6.33   Trellis diagram for $L_1 = L_2 = 1$.

used in Chapter 2 and expressed in the form:

$$\Lambda_2(c_k) = \Lambda_1(c_k) + z_k \tag{6.46}$$

where z_k is the extrinsic information associated with symbol c_k and provided by the channel decoder (Figure 6.34).

The extrinsic information z_k is another estimation of the LLR of symbol c_k conditioned on the decoding step:

$$z_k = \log \frac{\Pr\{c_k = +1 | \text{decoding}\}}{\Pr\{c_k = -1 | \text{decoding}\}} \tag{6.47}$$

Hence, z_k may be used through a feedback loop by the symbol detector after interleaving. This is the basis of the turbo equalization principle.

6.6.5   Viterbi Algorithm

To evaluate the LLR of symbol c_{n-L_1}, the Viterbi algorithm used in the detector has to calculate a metric

$$\lambda_n^i = |R_n - r_n^i|^2 - 2\sigma_w^2 \log \Pr\{c_{n-L_1} = i\} \quad i = \pm 1 \tag{6.48}$$

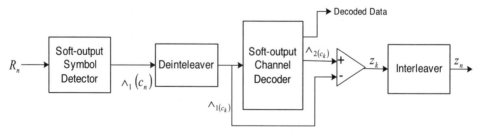

Figure 6.34   Extrinsic information deriving scheme.

where:

$$r_n^i = \sum_{k=0}^{L_1+L_2-1} \hat{\Gamma}_{k-L_2}(n) \cdot c_{n+L_2-k} + \hat{\Gamma}_{L_1}(n) \cdot i \quad i = \pm 1 \tag{6.49}$$

$\hat{\Gamma}_{k-L_2}(n)$, $0 \le k \le L_1 + L_2$ represents an estimation of quantity $\Gamma_{k-L_2}(n)$ and σ_w^2 denotes the variance of noise w_n, that is $\sigma_w^2 = E[|w_n|^2]$.

The *a priori* probabilities $\Pr\{c_{n-L_1} = i\}$ used in Equation (6.48) may be estimated from the extrinsic information z_{n-L_1}, if we assume that

$$z_{n-L_1} = \log \frac{\Pr\{c_{n-L_1} = +1\}}{\Pr\{c_{n-L_1} = -1\}} \tag{6.50}$$

From Equation (6.50)

$$\Pr\{c_{n-L_1} = +1\} \approx \frac{\exp z_{n-L_1}}{1 + \exp z_{n-L_1}} \tag{6.51}$$

$$\Pr\{c_{n-L_1} = -1\} \approx \frac{1}{1 + \exp z_{n-L_1}}$$

Using Equations (6.51) and (6.48), metrics λ_n^i are equal to:

$$\begin{aligned} \lambda_n^{+1} &= |R_n - r_n^{+1}|^2 - \gamma z_{n-L_1} \\ \lambda_n^{-1} &= |R_n - r_n^{-1}|^2 \end{aligned} \tag{6.52}$$

Note that the common term $\log(1 + \exp z_{n-L_1})$ has been suppressed in Equation (6.52). Coefficient γ is a weight introduced to take into account variance σ_w^2 and the fact that the extrinsic information is only an *estimation* of the *a priori* probability. Its value depends on the signal to noise ratio, that is to say the reliability of the extrinsic information.

6.6.6 Iterative Implementation of Turbo Equalization

The different processing stages in the turbo equalizer present a non-zero internal delay, so turbo equalizing can only be implemented in an iterative way. At each iteration q, a new value of extrinsic information is calculated and used by the symbol detector at the next iteration. Therefore, the turbo equalizer can be implemented in a modular pipelined structure, where each module is associated with one iteration. Then, performance in bit error rate (BER) terms is a function of the number of chained modules.

When extrinsic information is used by the symbol detector, it can be proved that, at iteration q, the LLR of symbol c_n, $\Lambda_1^q(c_n)$, may be expressed as:

$$\Lambda_1^q(c_n) = \hat{\Lambda}_1^q(c_n) + \gamma^q z_n^{q-1} \tag{6.53}$$

where $\hat{\Lambda}_1^q$ is a term depending on the samples of observation $\mathbf{R}$ and on z_n^{q-1}, $k \ne n$, and z_n^{q-1} denotes the extrinsic information of symbol c_n determined at iteration $q - 1$. If we apply the same approach as in turbo decoding described in Chapter 2, the quantity $\gamma^q z_n^{q-1}$ provided by the channel decoder at the previous iteration has to be subtracted from $\Lambda_1^q(c_n)$ (see Equation (6.52)), as illustrated in Figure 6.35. Hence, after deinterleaving, the channel decoder input is in fact equal to:

$$\tilde{\Lambda}_1^q(c_n) = \Lambda_1^q(c_n)\big|_{z_k^{q-1}=0} \tag{6.54}$$

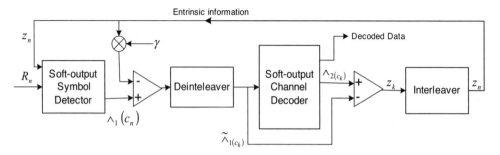

Figure 6.35 Principle of turbo equalization (under the zero internal delay assumption).

At the channel decoder output, the extrinsic information z_k^q may also be written as follows, using Equation (6.46):

$$z_k^q = \Lambda_2^q(c_n)\big|_{\tilde{\Lambda}_1^q(c_n)=0} \tag{6.55}$$

6.6.7 Performance

Performance of this device has been evaluated for a rate $R = 1/2$ recursive systematic encoder with constraint length $K = 5$ and generators $G_l = 23$, $G_2 = 35$. Bits were interleaved in a non-uniform matrix whose dimensions are 64 by 64. The modulation used was a BPSK modulation, with a Nyquist filter whose transfer function $H_s(f)$ was a raised cosine with a rolloff $\alpha = 1$, on both Gaussian and Rayleigh channels.

For both channels, $M = 5$ independent paths were considered, each with a mean power $P_m = E[|A_m(n)|^2]$, so that the total mean power was normalized: $\sum_{m=0}^{M-1} P_m = 1$. The delays τ_m were chosen as multiples of $T (\tau_m = mT$, $\Gamma_k(n) = A_k(n)$ since $h_s[(k-m)T] = \delta_{k-m,0})$. The coefficients for the Gaussian channel were chosen equal to:

$$\Gamma_0(n) = \sqrt{0.45}, \quad \Gamma_1(n) = \sqrt{0.25}, \quad \Gamma_2(n) = \sqrt{0.15},$$
$$\Gamma_3(n) = \sqrt{0.10}, \quad \Gamma_4(n) = \sqrt{0.05}$$

For the Rayleigh channel, the five paths had equal mean power ($P_i = 1/M$, $\forall i \in [1, M]\}$. A parameter BT, which is the product of the Doppler bandwidth and the symbol duration, fixes the variation velocity of the channel: the smaller BT is, the more slowly the channel parameters vary during a time interval symbol.

The discrete time equivalent channel was modeled by a 16-state trellis, and the symbol detector was working on the SOVA algorithm. The channel coefficients $\Gamma_k(n)$ were supposed perfectly known.

After deinterleaving, the soft estimations provided by the SOVA detector were used by the decoder, which also worked on a 16-state trellis and the SOVA algorithm. The extrinsic information extracted from the decoder was used by the symbol detector according to the principle depicted in Figure 6.35.

The BER was computed as a function of signal to noise ratio E_b/N_0, where E_b is the mean energy received per information bit d_k and N_0 is the noise power bilateral spectral density. The signal to noise ratio E_b/N_0 may be expressed as:

$$\frac{E_b}{N_0} = \frac{\sum_{m=0}^{M-1} P_m}{\sigma_b^2} \tag{6.56}$$

The results are shown in Figures 6.36–6.38 for different numbers of iterations n. One can see that even for three iterations the performance curves approach the BER curve with no ISI.

More details on turbo equalization can be found in [94–103].

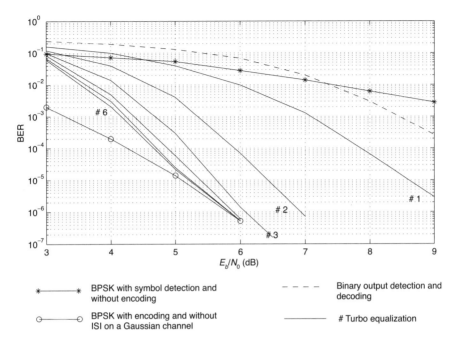

Figure 6.36 Performance of turbo equalization over a Gaussian channel (convolutional encoding with $K = 5$).

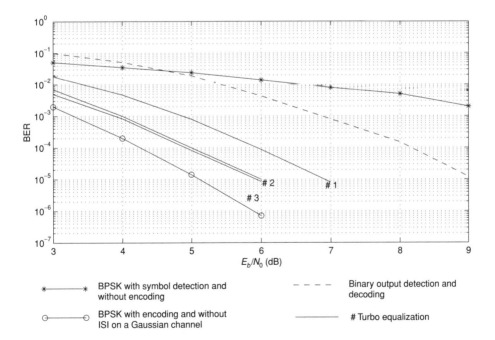

Figure 6.37 Performance of turbo equalization over a Rayleigh channel with $BT = 0.1$ (convolutional encoding with $K = 5$).

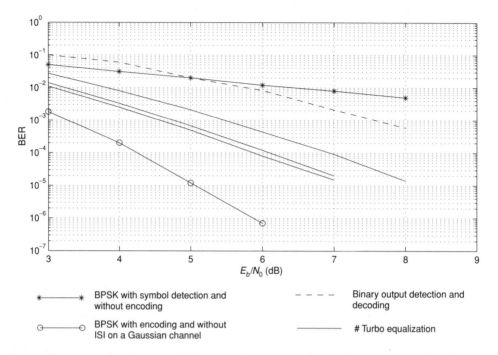

Figure 6.38   Performance of turbo equalization over a Rayleigh channel with $BT = 0.001$ (convolutional encoding with $K = 5$).

6.7   Kalman Filter Based Joint Channel Estimation and Data Detection Over Fading Channels

In this section we consider the problem of joint channel and data estimation in a fading channel based on using a Kalman-type estimator. The general system block diagram is shown in Figure 6.39. More details can be found in [104–107]. The presentation in this section is based on [104]. The channel model generator is shown in Figure 6.40.

6.7.1   Channel Model

If the filter $P(z)$ in Figure 6.40 is modeled as

$$P(z) = \frac{D}{1 - Az^{-1} - Bz^{-2} - Cz^{-3}} \tag{6.57}$$

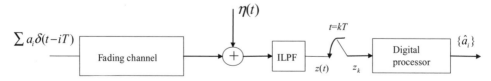

Figure 6.39   The signal model for the baseband communication system.

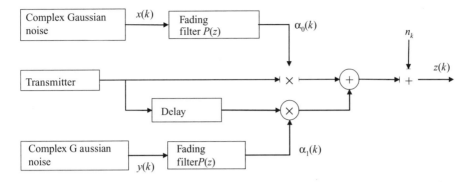

Figure 6.40   The fading channel model.

then at sampling time k, the CIR, $\mathbf{h}_k$, a complex Gaussian vector, is

$$\mathbf{h}_k = (h_{k,0}, h_{k,1}, \ldots, h_{k,\beta})^{\mathrm{T}} \tag{6.58}$$

truncated to a finite length of $(\beta + 1)$. By considering the third-order approximation of Equation (6.57), an autoregressive (AR) representation for the CIR can be introduced as

$$\mathbf{h}_k = A\mathbf{I}\mathbf{h}_{k-1} + B\mathbf{I}\mathbf{h}_{k-2} + C\mathbf{I}\mathbf{h}_{k-3} + D\mathbf{I}\mathbf{w}_k \tag{6.59}$$

where $\mathbf{I}$ is the identity matrix, $\mathbf{w}_k$ is a zero mean white complex circularly symmetric Gaussian process with the covariance matrix defined as $E(\mathbf{w}_k\mathbf{w}_l^{\mathrm{T}}) = \mathbf{Q}\delta_{kl}$ and $\mathbf{w}_l^{\mathrm{T}}$ is the conjugate transpose of $\mathbf{w}_l$. CIR at time k depends on its three consecutive previous values.

The state of such a system is a vector composed of three consecutive channel impulse responses as

$$\mathbf{x}_k = (\mathbf{h}_k^{\mathrm{T}}, \mathbf{h}_{k-1}^{\mathrm{T}}, \mathbf{h}_{k-2}^{\mathrm{T}})^{\mathrm{T}} \tag{6.60}$$

Using Equations (6.60) and (6.59) gives

$$\mathbf{x}_{k+1} = \begin{bmatrix} A\mathbf{I} & B\mathbf{I} & C\mathbf{I} \\ \mathbf{I} & 0 & 0 \\ 0 & \mathbf{I} & 0 \end{bmatrix} \mathbf{x}_k + \begin{bmatrix} D\mathbf{I} \\ 0 \\ 0 \end{bmatrix} \mathbf{w}_k \tag{6.61}$$

$$\mathbf{x}_{k+1} = \mathbf{F}\mathbf{x}_k + \mathbf{G}\mathbf{w}_k \tag{6.62}$$

where $\mathbf{F}$ and $\mathbf{G}$ are $3(\beta + 1) \times 3(\beta + 1)$ and $3(\beta + 1) \times (\beta + 1)$ matrices given in Equation (6.61). $\mathbf{F}$ is called the state transition matrix and $\mathbf{G}$ is the process noise coupling matrix.

6.7.2   The Received Signal

If we introduce a $1 \times (\beta + 1)$ vector $\mathbf{H}_k$

$$\mathbf{H}_k = (a_k, a_{k-1}, a_{k-2}, \ldots, a_{k-\beta}, 0, \ldots, 0) \tag{6.63}$$

where a_k is the transmitted data sequence, then the received signal z_k becomes

$$\mathbf{z}_k = \mathbf{H}_k\mathbf{x}_k + \mathbf{n}_k \tag{6.64}$$

The Kalman filter is optimum for minimizing the mean square estimation error [108], in the above linear time varying system. The algorithm is complex and, in practice, suboptimal methods are more advantageous due to their implementation simplicity.

6.7.3   Channel Estimation Alternatives

In this segment the same assumptions are used as in [104]:

- To avoid the decision delay, the PSP method [109] is used. In this method, there is a channel estimate for every possible sequence.
- The estimated channel impulse response will be used to compute the branch metrics in the trellis diagram of the Viterbi algorithm.
- The number of required estimators is limited to the number of survivor branches (or the number of states) in the Viterbi algorithm trellis diagram.
- In PSP each surviving path keeps and updates its own channel estimate. This method eliminates the problem of decision delay, and in order to employ the best available information for data detection the data sequence of the shortest path is used for channel estimation along the same path.
- The data communication system is based on the IS-136 standard presented in Chapter 1.
- The modulation is QPSK with four possible symbols ($\pm 1 \pm j$) and a symbol rate of 25 ksymbols/s.
- The differentially encoded data sequence is arranged into 162 symbol frames.
- The first 14 symbols of each frame make up a training preamble sequence to help the adaptation of the channel estimator.
- For the shaping filter at the transmitter, a finite impulse response (FIR) filter approximates a raised cosine frequency response with an excess bandwidth of 25% (slightly different from the 35% selected in IS-136).
- A two-ray fading channel model as described in Figure 6.40, where one ray has a fixed delay equal to one symbol period.
- The multiplicative coefficients of α_0 and α_1 are produced at the output of two fading filters, where the inputs are two independent zero mean complex Gaussian processes with equal variances.
- The length of the discrete impulse response of the shaping filter is set equal to the symbol interval so that the ISI at the receiver is only due to the multipath nature of the channel.
- The total length of the CIR is two symbol intervals, i.e. $\beta + 1 = 2$ if there is one sample per symbol interval.
- Therefore, there is ISI between two neighboring symbols and there are four possible states in the trellis diagram.
- The LMS algorithm, RLS algorithm or the Kalman filter can be used to estimate the channel impulse response.

6.7.4   Implementing the Estimator

6.7.4.1   The Kalman Filter Algorithm [108]

The measurement update equations are:

$$
\begin{aligned}
\hat{\mathbf{x}}_{k|k} &= \hat{\mathbf{x}}_k + \mathbf{K}_k(z_k - \mathrm{H}_k\hat{\mathbf{x}}_k) \\
\mathbf{K}_k &= \mathbf{P}_k\mathbf{H}_k^{\mathrm{T}}\mathbf{R}_k^{-1} \\
\mathbf{R}_k &= \mathbf{H}_k\mathbf{P}_k\mathbf{H}_k^{\mathrm{T}} + \mathrm{N}_0 \\
\mathbf{P}_{k|k} &= \mathbf{P}_k - \mathbf{K}_k\mathbf{H}_k\mathbf{P}_k
\end{aligned}
\tag{6.65}
$$

The time update equations are:

$$\hat{\mathbf{x}}_{k+1} = \mathbf{F}\hat{\mathbf{x}}_{k|k}$$
$$\mathbf{P}_{k+1} = \mathbf{F}\mathbf{P}_{k|k}\mathbf{F}^T + \mathbf{G}\mathbf{Q}\mathbf{G}^\mathrm{T} \tag{6.66}$$

6.7.4.2   The RLS Algorithm [110]

$$\hat{\mathbf{x}}_{k+1} = \hat{\mathbf{x}}_k + \mathbf{K}_k(z_k - \mathbf{H}_k\hat{\mathbf{x}}_k)$$
$$\mathbf{K}_k = \mathbf{P}_k\mathbf{H}_k^\mathrm{T}\mathbf{R}_k^{-1}$$
$$R_k = \mathbf{H}_k\mathbf{P}_k\mathbf{H}_k^\mathrm{T} + \lambda \tag{6.67}$$
$$\mathbf{P}_{k+1} = \lambda^{-1}(\mathbf{P}_k - \mathbf{K}_k\mathbf{H}_k\mathbf{P}_k)$$

In the above equations the measurement updated estimate $\hat{\mathbf{x}}_{k|k}$ is the linear least squares estimate of x_k given observations $\{z_0, z_1, \ldots, z_k\}$. $\hat{\mathbf{x}}_{k+1}$ is the time updated estimate of $\mathbf{x}_k$ given observations $\{z_0, z_1, \ldots, z_k\}$. The corresponding error covariance matrices of these estimations are

$$\mathbf{P}_{k|k} = E\lfloor(\mathbf{x}_k - \hat{\mathbf{x}}_{k|k})(\mathbf{x}_k - \hat{\mathbf{x}}_{k|k})^\mathrm{T}\rfloor$$
$$\mathbf{P}_k = E\lfloor(\mathbf{x}_k - \hat{\mathbf{x}}_k)(\mathbf{x}_k - \hat{\mathbf{x}}_k)^\mathrm{T}\rfloor \tag{6.68}$$

In the RLS algorithm, λ is a parameter called the forgetting factor.

6.7.5   The Kalman Filter

The estimator is based on two parts: *measurement update equations* and *time update equations*. The RLS algorithm is essentially identical to the measurement update equations of the Kalman filter. The Kalman filter can be used for channel estimation when some *a priori* information about the channel is available at the receiver (i.e. the **F** and **G** matrices). The RLS algorithm, which is a suboptimal method, does not require this *a priori* information and its computational complexity is less than the Kalman filter.

6.7.6   Implementation Issues

At the same precision, mathematically equivalent implementations can have different numerical stabilities, and some methods of implementation are more robust against roundoff errors.

In the Kalman filter and the RLS algorithm the estimation depends on the correct computation of the error covariance matrix. In an ill-conditioned problem, the solution will not be equal to the covariance matrix of the actual estimation uncertainty. Factors contributing to this problem are: large matrix dimensions, a growing number of arithmetic operations and poor machine precision. Solutions to these problems are factorization methods and square root filtering [110, 111].

6.7.6.1   Square Root Filtering

Some implementations are more robust against roundoff errors and ill-conditioned problems. The so-called square root filter implementations have generally better error propagation bounds than the conventional Kalman filter equations [112]. In the square root forms of the Kalman filter, matrices are factorized and *triangular square roots* are propagated in the recursive algorithm to preserve the symmetry of the covariance (information) matrices in the presence of roundoff errors.

Different techniques are used for changing the dependent variable of the recursive estimation algorithm to factors of the covariance matrix. A Cholesky factor of a symmetric non-negative definite matrix $\mathbf{M}$ is a matrix $\mathbf{C}$ such that $\mathbf{CC}^T = \mathbf{M}$. Cholesky decomposition algorithms solve for a diagonal factor and either a lower triangular factor $\mathbf{L}$ or an upper triangular factor $\mathbf{U}$ such that $\mathbf{M} = \mathbf{UD}_U\mathbf{U}^T = \mathbf{LD}_L\,\mathbf{L}^T$, where $\mathbf{D}_L$ and $\mathbf{D}_U$ are diagonal factors with non-negative diagonal elements.

The square root methods propagate the L-D or U-D factors of the covariance matrix rather than the covariance matrix. The propagation of square root matrices implicitly preserves the Hermitian symmetry and non-negative definiteness of the computed covariance matrix.

The condition number $\kappa(\mathbf{P}) = [\text{eigenvalue}_{max}(\mathbf{P})\ \text{eigenvalue}_{min}(\mathbf{P})]$ of the covariance matrix $\mathbf{P}$ can be written as

$$\kappa(\mathbf{P}) = \kappa(\mathbf{LDL}^T) = \kappa(\mathbf{BB}^T) = [\kappa(\mathbf{B})]^2$$

where $\mathbf{B} = \mathbf{LD}^{1/2}$. The condition number of $\mathbf{B}$ used in the square root method is much smaller than the condition number of $\mathbf{P}$ and this leads to improved numerical robustness of the algorithm.

In the square root method, the dynamic range of the numbers entering into computations will be reduced. Loosely speaking, we can say that the computations which involve numbers ranging between 2^{-N} to 2^{+N} will be reduced to ranges between $2^{-N/2}$ to $2^{+N/2}$, which would halve the length of required mantissa used in signal processing. All of these will directly affect the accuracy of computer computations.

Performance results are shown in Figures 6.41 and 6.42. Different Kalman algorithms will demonstrate superior performance only if the precision in the calculation of Equation (6.65) is high enough. From Figures 6.41 and 6.42, this means if the quantization is precise enough, requiring a mantissa length of 20 bits or higher.

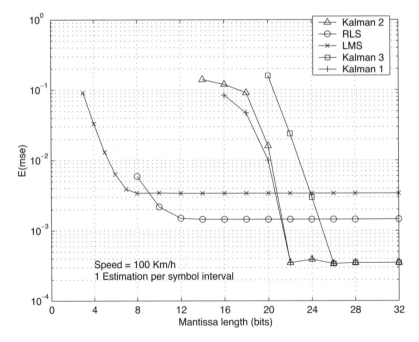

Figure 6.41   The effects of changing the word length on estimation error $E(\text{mse})$ ($E_b/N_0 = 15$ dB). Kalman 1 represents the WGS method, Kalman 2 is for the correlation method and Kalman 3 is the direct method, the PSP method is employed for detection.

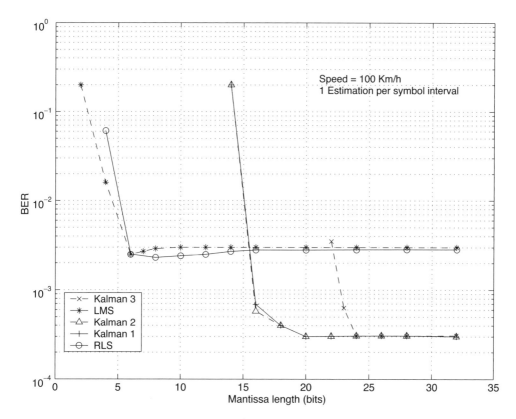

Figure 6.42 The effects of changing the word length on BER ($E_b/N_0 = 15$ dB). Kalman 1 represents the WGS method, Kalman 2 is for the correlation method and Kalman 3 is the direct method, the PSP method is employed for detection.

6.8 Equalization Using Higher Order Signal Statistics

In order to speed up the algorithms, equalization based on higher order signal statistics may be used [113–119]. In this section we discuss a group of such algorithms derived from the cost function defined as average entropy of the set of the signal samples [113]. Maximization of the average entropy is obtained by minimizing the mutual information of the set of signal samples which result in minimum ISI. The block diagram of the equalizer based on joint entropy minimization (JEM) principles is shown in Figure 6.43.

6.8.1 Problem Statement

For input data symbols $s(n)$, discrete time channel output $x(n)$ and the channel response $c(n)$, the received baseband symbol can be represented as

$$x(n) = \sum_{k=0}^{N} c(k)s(n-k) \tag{6.69}$$

We will assume that input symbols are independently identically distributed (i.i.d.) and zero mean, which is true in a digital communication system, and that $c(0) = 1$.

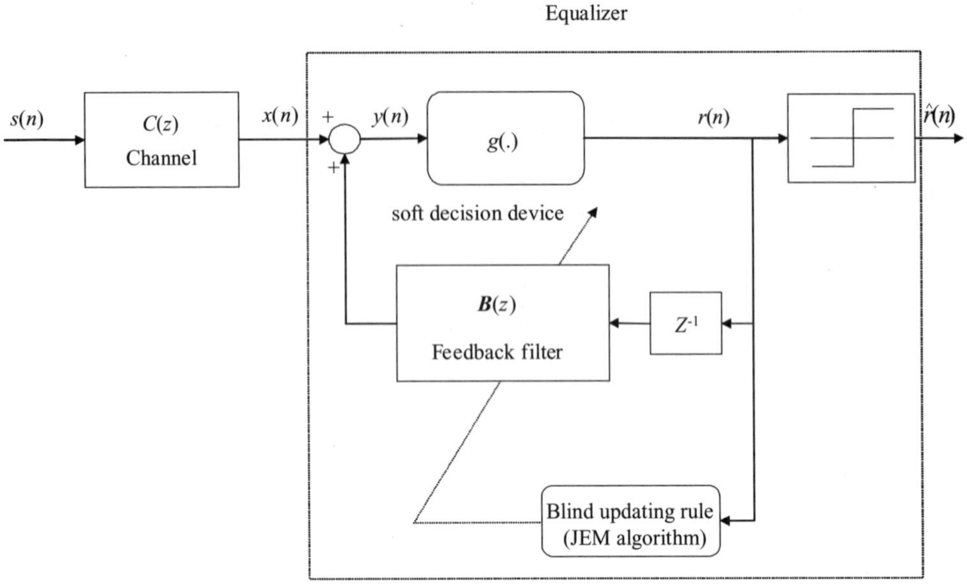

Equalizer

Figure 6.43   Block diagram of channel and JEM-DFE structure.

Signal $y(n)$ at the output of the equalizer (see Figure 6.43) can be presented as

$$y(n) = x(n) + \sum_{k=1}^{N} b(k)r(n-k)$$

$$= s(n) + \sum_{k=1}^{N} c(k)s(n-k) + \sum_{k=1}^{N} b(k)r(n-k) \qquad (6.70)$$

$$= s(n) + \sum_{k=1}^{N} c(k)s(n-k) + \sum_{k=1}^{N} b(k)g(y(n-k))$$

Non-linearity $g(\cdot)$ is a strictly monotone (increasing or decreasing) differentiable function. A more general DFE structure usually consists of a feedforward finite impulse response (FIR) filter followed by the feedback FIR filter given in Figure 6.43. As discussed earlier in the chapter, the main purpose of the feedforward filter is to eliminate the precursor ISI and the feedback filter cancels the postcursor ISI. For simplicity, only the feedback part is considered. This is not a hurdle since it is possible to separate the adaptation of the feedback part from the feedforward part.

6.8.2   Signal Model

By introducing auxiliary inputs and outputs we have

$$\mathbf{y}(n) = \mathbf{B} \cdot \mathbf{x}(n) \qquad (6.71)$$

$$\mathbf{x}(n) = [x(n) \quad r(n-1) \quad \dots \quad r(n-N)]^{\mathrm{T}}$$

$$\mathbf{y}(n) = [y(n) \quad r(n-1) \quad \dots \quad r(n-N)]^{\mathrm{T}} \qquad (6.72)$$

and

$$
\mathbf{B} = \begin{bmatrix} 1 & b(1) & \dots & b(N) \\ 0 & 1 & \ddots & 0 \\ \vdots & \ddots & \ddots & \ddots \\ 0 & \ddots & \dots & 1 \end{bmatrix}
\tag{6.73}
$$

In the presence of ISI (i.e. dependence), the entropy of the received symbols is smaller than when they are independent. Thus, maximizing the joint entropy of the equalizer outputs will result in removing the ISI.

6.8.3 Derivation of Algorithms for DFE

So, the algorithm is based on maximizing the joint entropy of the equalizer output $\mathbf{r}(n)$. The joint entropy of $\mathbf{r}(n) = g(\mathbf{y}(n))$ denoted by $\mathbf{H}[r_1(n), \dots, r_{N+1}(n)]$ is defined as [78, 120 (Chapter 6)]:

$$
\begin{aligned}
\mathbf{H}[r_1(n), \dots, r_{N+1}(n)] &\underline{\underline{\triangle}} - E\{\ln f_{\mathbf{r}}(\mathbf{r}(n))\} \\
&= E\{\ln |J|\} - E\{\ln f_{\mathbf{y}}(\mathbf{y}(n))\} \\
&= E\{\ln |J|\} - E\{\ln\{|B|^{-1} f_{\mathbf{x}}(\mathbf{x}(n))\}\} \\
&= E\{\ln |J|\} - E\{\ln f_{\mathbf{x}}(\mathbf{x}(n))\}
\end{aligned}
\tag{6.74}
$$

where $|J|$ is the absolute value of the Jacobian of the transformation

$$
J = \det \begin{bmatrix} \dfrac{\partial r_1(n)}{\partial y_1(n)} & \dots & \dfrac{\partial r_1(n)}{\partial y_{N+1}(n)} \\ \ddots & \dots & \ddots \\ \dfrac{\partial r_{N+1}(n)}{\partial y_1(n)} & \dots & \dfrac{\partial r_{N+1}(n)}{\partial y_{N+1}(n)} \end{bmatrix}
\tag{6.75}
$$

$$
J = \det \begin{bmatrix} \dfrac{\partial r(n)}{\partial y(n)} & \dots & \dfrac{\partial r(n)}{\partial r(n-N)} \\ \ddots & \dots & \ddots \\ \dfrac{\partial r(n-N)}{\partial y(n)} & \dots & \dfrac{\partial r(n-N)}{\partial r(n-N)} \end{bmatrix}
$$

$$
= \begin{bmatrix} \dfrac{\partial r(n)}{\partial y(n)} & \dots & 0 \\ \ddots & \dots & \ddots \\ \dfrac{\partial r(n-N)}{\partial y(n)} & \dots & 1 \end{bmatrix} = \dfrac{\partial r(n)}{\partial y(n)}
\tag{6.76}
$$

The quantity $r_i(n)$ $[x_i(n)]$ is the ith component of the vector $\mathbf{r}(n)$ $[\mathbf{x}(n)]$ and $f_r(\mathbf{r}(n))$ $[f_x(\mathbf{x}(n))]$ is the joint density function of the input vector $\mathbf{r}(n)$ $[\mathbf{x}(n)]$.

Assume that the previous decisions are correct, i.e.

$$
\text{Assumption: } r(n-k) = s(n-k), \ k = 1, \dots, N
\tag{6.77}
$$

Under Assumption (6.77), $r(n-i)$ are independent of $r(n-j)$, $i > j$ and thus Equation (6.75) follows from Equation (6.75). The joint entropy can also be expressed as [120, Chapter 15]:

$$\mathbf{H}[r_1(n), \ldots, r_{N+1}(n)] = \sum_{i=1}^{N+1} \mathbf{H}[r_i(n)] - \mathbf{I}[r_1(n), \ldots, r_{N+1}(n)] \tag{6.78}$$

Maximizing the joint entropy of $\mathbf{r}(n)$ is the same as maximizing the first term in Equation (6.78) while minimizing the mutual information $\mathbf{I}[r_1(n), \ldots, r_{N+1}(n)]$. Since the previous decisions that the outputs of the non-linear function $g(\cdot)$ are the same as the transmitted symbols $s(n-k)$, $k = 1, \ldots, N$, $E\{\ln f_x(\mathbf{x}(n))\}$ in Equation (6.74) can be considered to be a constant with respect to the feedback filter coefficients $b(k)$, $k = 1, \ldots, N$. Then, maximizing $E\{\ln |J|\}$ is equivalent to maximizing $\mathbf{H}[r_1(n), \ldots, r_{N+1}(n)]$. In doing so, the statistical dependence between the current output of the non-linear function $r(n)$ and the previous outputs $(r(n-k), k = 1, \ldots, N)$ can be reduced, which leads to ISI suppression.

6.8.4 The Equalizer Coefficients

The joint entropy $\mathbf{H}[r_1(n), \ldots, r_{N+1}(n)]$ is a non-linear function of the unknown equalizer coefficients and does not lead easily to a closed form solution. A gradient descent algorithm is used for maximizing Equation (6.74). The equalizer coefficients are then updated iteratively

$$b^{n+1}(k) = b^n(k) + \mu E \left\{ \frac{\partial \ln |J|}{\partial b^n(k)} \right\} \tag{6.79}$$

where μ is the positive step size. This equation can be further simplified based on the choice of non-linearity used in the decision device.

6.8.5 Stochastic Gradient DFE Adaptive Algorithms

Despite its superiority over the linear equalizer and its popularity, a major drawback of the DFE is that it suffers from error propagation. Any previous decision errors at the output of the decision device will produce the symbol estimate $y(n)$, whose ISI is not completely eliminated by the feedback filter. This error in turn will affect future decisions. An intuitive choice is one where we replace the hard limiter with a softer function. It is hoped that the errors in $r(n)$ can be reduced by using soft decisions.

The update equation given in (6.79) depends on the mapping used in the soft decision device. Various non-linear functions may be used for the soft decision device (function $g(\cdot)$ shown in Figure 6.43). We use two different functions for $g(\cdot)$ to derive new blind algorithms for DFE.

6.8.5.1 Equivalence of JEM and ISIC

First choice for the non-linear function is

$$g(n) = \alpha \cdot \tanh[\beta \cdot n] \tag{6.80}$$

resulting in the following

$$\frac{\partial \ln |J|}{\partial b(k)} = \frac{\partial \ln \left| \dfrac{\partial r(n)}{\partial y(n)} \right|}{\partial b(k)}$$

$$= -2\beta \tanh[\beta \cdot y(n)] r(n-k)$$

$$= -\frac{2\beta}{\alpha} r(n) r(n-k) \tag{6.81}$$

The gradient is a function of the non-linearity used in the decision device that provides a tool via which the algorithm characteristics can be varied. A new adaptive blind algorithm for DFE based on

JEM can be obtained as follows:

$$\text{JEM-1:} \quad b^{n+1}(k) = b^n(k) - \mu r(n)r(n-k) \tag{6.82}$$

Taylor series (TS) expansion of $r(n)$ leads to

$$r(n) = \alpha \cdot \tanh[\beta \cdot y(n)]$$
$$= \alpha\beta y(n) - \frac{\alpha\beta^3}{3}y^3(n) + \frac{2\alpha\beta^5}{15}y^5(n) + O(y^6(n)) \tag{6.83}$$

For simplicity only the first few terms of the expansion are considered. Further variations of JEM-1 can be obtained by substituting Equation (6.83) into Equation (6.82). By using only the first term in Equation (6.83) with $\alpha = 3$ and $\beta = 1/3$, a simpler update equation for the feedback filter coefficients is

$$\text{JEM-2:} \quad b^{n+1}(k) = b^n(k) - \mu y(n)r(n-k) \tag{6.84}$$

JEM-2 is exactly the same as the ISIC algorithm [121] which uses soft decision feedback. Here, the same update equation is arrived at via the motivation to maximize the entropy of the observations at the output of a soft decision device.

6.8.5.2 Equivalence of JEM and Decorrelation Criterion

Another possible variation of Equation (6.82) is to use the TS approximation for both $r(n)$ and $r(n-k)$.

$$r(n)r(n-k) = \alpha^2\beta^2 y(n)y(n-k) - \frac{\alpha^2\beta^4}{3}y^3(n)y(n-k) + y^3(n-k)y(n)\} + \cdots \tag{6.85}$$

By using only the first term in Equation (6.85) with $\alpha = 3$ and $\beta = 1/3$, we have

$$\text{JEM-3:} \quad b^{n+1}(k) = b^n(k) - \mu y(n)y(n-k) \tag{6.86}$$

JEM-3 is exactly the same as that for the adaptive blind algorithm based on the *decorrelation criterion* (DECA) [122].

6.8.5.3 JEM and CMA-DFE

Another function that has transfer characteristics similar to the hyperbolic tangent is the function with a cubic non-linearity

$$g(y(n)) = \alpha \cdot y(n) + \beta \cdot y^3(n) \tag{6.87}$$

With the above mapping we have

$$\frac{\partial \ln |J|}{\partial b(k)} = \frac{6\beta y(n)}{\alpha + 3\beta y^2(n)}r(n-k) \tag{6.88}$$

$$\frac{\partial \ln |J|}{\partial b(k)} = \left[\frac{6\beta}{\alpha}y(n)\left\{1 - \frac{3\beta}{\alpha}y^2(n) + \frac{9\beta^2}{\alpha}y^4\right\} + O(y^7(n))\right]r(n-k) \tag{6.89}$$

Dropping terms of order greater than three gives:

$$\text{JEM-4:} \quad b^{n+1}(k) = b^n(k) + \mu y(n)\left\{1 - \frac{3\beta}{\alpha}y^2(n)\right\} \cdot r(n-k) \tag{6.90}$$

With $\alpha = 3$ and $\beta = 1$, Equation (6.90) coincides with the CMA-DFE algorithm, provided the soft decisions given by Equation (6.87) are used in the CM algorithm for a DFE. The original CMA-DFE uses hard decisions [123, 124].

6.8.6   Convergence Analysis

6.8.6.1   Alternative JEM-DFE Scheme and Bussgang-Type Algorithm

Since soft decisions can possibly smooth the error surface and allow the algorithm to escape from local minima, the authors in [8] suggest the use of soft decisions at the start of equalization. These can then be replaced with a hard decision device after a few iterations. It is difficult to decide when one can switch from soft to hard decisions. A simple technique based on a modified JEM-DFE scheme is suggested to overcome this difficulty (Figure 6.44).

For the update equations (for example, (6.82) and (6.84)), the soft decision device outputs, $r(n)$ and $r(n - k)$, are used. On the other hand, the hard decisions $\hat{r}(n - k)$ in Figure 6.43 are used as inputs to the feedback filter instead of $r(n - k)$; see Figure 6.43. In this way, soft and hard decision modes can be combined. A structure similar to JEM-DFE (see Figure 6.44) was used in [124] and will be used in what follows.

6.8.6.2   JEM-DFE Structure in Figure 6.44 and Classical Bussgang-Type DFE

DFE can be as a linear considered equalizer [124]. The usual DFE structure consists of the feedforward filter followed by the feedback filter. The symbol estimate $y(n)$ can be expressed as:

$$
\begin{aligned}
y(n) &= \sum_{k=-N_f}^{0} f(k)x(n - k) + \sum_{k=1}^{N_b} b(k)\hat{r}(n - k) \\
&= \sum_{k=-N_f}^{N_b} w(k)\hat{x}(n - k)
\end{aligned}
\tag{6.91}
$$

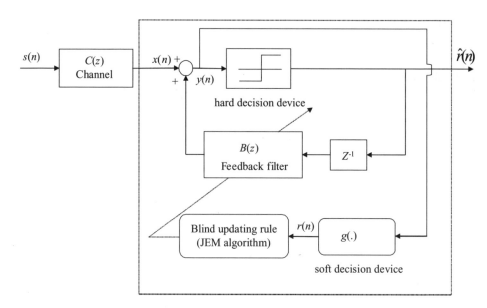

Figure 6.44   Block diagram of channel and alternative JEM-DFE structure.

The input to the equalizer is

$$\tilde{x}(n-k) = \begin{cases} x(n-k), & \text{when } -N_f \le k \le 0 \\ \hat{r}(n-k), & \text{when } 1 \le k \le N_b \end{cases} \tag{6.92}$$

The coefficients of the equalizer are

$$\omega(n-k) = \begin{cases} f(n-k), & \text{when } -N_f \le k \le 0 \\ b(n-k), & \text{when } 1 \le k \le N_b \end{cases} \tag{6.93}$$

and N_f and N_b are the feedforward and feedback filter lengths, respectively.

6.8.6.3 Bussgang-Type Algorithm [125 (Chapter 2)]

The algorithm is defined by

$$w^{n+1}(k) = w^n(k) + \mu\{\hat{g}(y(n)) - y(n)\}\tilde{x}(n-k) \tag{6.94}$$

where $\hat{g}(y(n))$ is some non-linear function of $y(n)$. Provided that N_f and N_b in Equation (6.94) are large enough, the symbol estimate $\{y(n)\}$ is *Bussgang* (see [126, Chapter 18], [127]). A stochastic process $\{y(n)\}$ is said to be a *Bussgang process* if it satisfies the condition $E\{y(n)y(n+k)\} = E\{y(n) g(y(n+k))\}$ where the function $g(\cdot)$ is a zero memory non-linearity.

Since the feedforward and feedback parts can be updated separately, consider only the feedback part of Equation (6.94). Then, the Bussgang-type DFE algorithm for the feedback part is

$$b^{n+1}(k) = b^n(k) + \mu\{\tilde{g}(y(n)) - y(n)\}r(n-k) \tag{6.95}$$

Comparing Equation (6.95) to Equations (6.82) and (6.90), $-r(n)$ in Equation (6.82), $-y(n)$ in Equation (6.84) and $y(n)\{1 - (3\beta/\alpha) y^2(n)\}$ in Equation (6.90) may be viewed to be equivalent to the error term $\{\hat{g}(y(n)) - y(n)\}$ in Equation (6.95). Thus, JEM-1, JEM-2 and JEM-4 can be considered to be special cases of the Bussgang-type DFE. By comparing Equation (6.95) to Equation (6.86), however, we note that JEM-3 cannot be categorized as a Bussgang-type algorithm.

6.8.6.4 Convergence of JEM Algorithms

JEM-1 and JEM-2 are guaranteed to converge for super-Gaussian inputs but not for sub-Gaussian. The input $\{s(n)\}$ is sub-Gaussian (super-Gaussian) if the kurtosis $\gamma_s = E\{s(n)^4\} - 3[E\{s(n)^2\}]^2$ is less than zero (greater than zero). For example, for a BPSK signal, $\gamma_s = -2$ and hence it is sub-Gaussian.

Extensive simulations show, however, that these algorithms converge to the right solution in almost all the cases. Convergence of JEM-3 is not considered here since it cannot be categorized as a Bussgang-type algorithm. JEM-4 is guaranteed to converge. When $\alpha = \infty$ and $\beta = 1/\alpha = 0$, JEM-2 coincides with JEM-3.

6.8.7 Kurtosis-Based Algorithm

The JEM approach cannot guarantee that the maximization of the joint entropy leads to minimization of the mutual information. To compute the mutual information, the probability density function of $r(n)$, the output of the decision device, is required. The approximation of the mutual information (or contrast function) was derived by Comon (see [128]). Maximization of the contrast function, or minimization of the mutual information of the equalizer outputs, is equivalent to minimization of the sum of the autocumulants squared. This result was obtained by using the Edgeworth expansion to separate the constant mixtures of the independent input signals [128].

6.8.7.1   Modified Signal Model

Under Assumption (6.77), Equation (6.72) can be rewritten as:

$$
\begin{aligned}
\mathbf{x}(n) &= [x(n) \quad s(n-1) \quad \cdots \quad s(n-N)]^{\mathrm{T}} \\
\mathbf{y}(n) &= [y(n) \quad s(n-1) \quad \cdots \quad s(n-N)]^{\mathrm{T}}
\end{aligned}
\tag{6.96}
$$

The new equivalent channel model is

$$
\mathbf{x}(n) = \mathbf{C} \cdot \mathbf{s}(n)
\tag{6.97}
$$

where

$$
\mathbf{s}(n) = [s(n) \quad s(n-1) \quad \cdots \quad s(n-N)]^{\mathrm{T}}
\tag{6.98}
$$

and

$$
\mathbf{C} =
\begin{bmatrix}
1 & c(1) & \cdots & c(N) \\
0 & 1 & \ddots & 0 \\
\vdots & \ddots & \ddots & \ddots \\
0 & \ddots & \cdots & 1
\end{bmatrix}
\tag{6.99}
$$

The matrix $\mathbf{C}$ is full rank so there exists a unique solution to Equation (6.97). Here, the mixture matrix $\mathbf{C}$ and the source vector $\mathbf{s}(n)$ are partially unknown. Equalization is attained by estimating the matrix $\mathbf{B}$ in Equation (6.71). It can be easily shown that $\mathbf{B}$ is the inverse matrix of $\mathbf{C}$ when $c(n-k) = -b(n-k)$, $k = 1, \ldots, N$. To find the solution $\mathbf{B}$, several contrast functions can be used (see [128] and [129]).

6.8.7.2   Contrast Function Proposed in [129]

In this case we have

$$
J = \sum_{k=1}^{N+1} C_{4y_k}(0,0,0), = C_{4y}(0,0,0) + \text{Constant}
\tag{6.100}
$$

where $C_{4y_k}(0,0,0)$ is the kurtosis (all zero lag of the fourth-order cumulant) of $y(n)$. A criterion similar to one defined in Equation (6.100) for a single output FIR channel was originally proposed for multiinput multioutput (MIMO) channel equalization with MIMO-DFEs in [130]. The cost function for the present simplified case with only a single channel is

$$
J_{AC} = C_{4y}(0,0,0)
\tag{6.101}
$$

The update equation for the feedback filter coefficients based on the criterion in Equation (6.101) is:

$$b^{n+1}(k) = b^n(k) + \mu\,\text{sgn}\{\gamma_s\}$$
$$\cdot\,[\hat{E}\{y(n)^3 r(n-k)\} - 3\hat{E}\{y(n)^2\}$$
$$\cdot\,\hat{E}\{y(n)r(n-k)\}]$$

(6.102)

where $\hat{E}\{\cdot\}$ is the estimate of $E\{\cdot\}$. Note that the contrast-based cost function of Equation (6.100) is similar to the autocumulant (AC) criterion of Equation (6.101). Since the preprocessing stage (prewhitening of the observations) is missing in the AC approach, however, the proposed criterion can only be thought of as approximately minimizing the mutual information. Simulation examples indicate that the equalizer based on the AC criterion converges much faster than most existing algorithms.

6.8.7.3 Performance Examples [113]

Example 1
An FIR non-minimum phase channel with impulse response given by

$$c(n) = \delta[n] + 2\delta[n-1]$$

is used. The channel input $s(n)$ is a sequence of 2000 independent BPSK symbols, uniformly distributed over $\{1, -1\}$. The received symbols $x(n)$ consist of the distorted symbols $s(n)$ and additive noise at 20 dB. This data is used to update the one-tap feedback filter $(b(1))$, which is initialized with a zero. The results are shown in Figures 6.45 and 6.46.

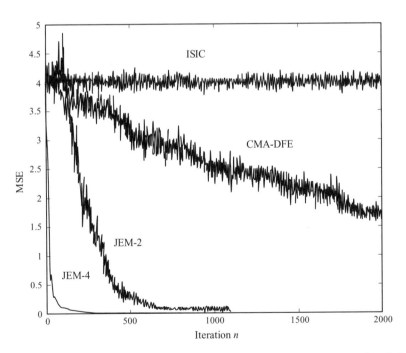

Figure 6.45   MSE comparison of ISIC, JEM-2, CMA-DFE and JEM-4: $(g(\cdot) = 3 \cdot \tanh[1/3y(n)])$ for JEM-2, $g(y(n)) = 3 \cdot y(n) + y^3(n)$ for JEM-4, $\mu_{ISIC} = \mu_{JEM-2} = \mu_{CMA-DFE} = 0.01$, $\mu_{JEM-4} = 0.001$ averaged over 100 Monte Carlo runs. SNR $= 20$ dB [113] © 1998, IEEE.

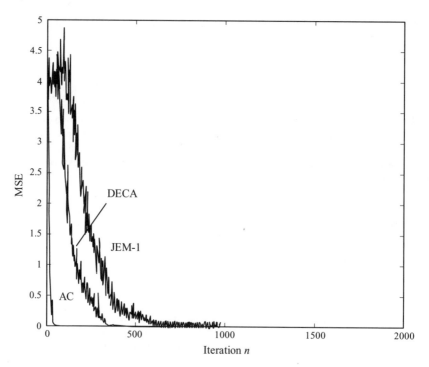

Figure 6.46 MSE comparison of AC, decorrelation algorithm (DECA) and JEM-1: $g(\cdot) = 3 \cdot \tanh[1/3 y(n)]$, $\mu_{AC} = \mu_{DECA} = \mu_{JEM-1} = 0.01$, averaged over 100 Monte Carlo runs. SNR $= 20$ dB [113] © 1998, IEEE.

Example 2

The performance of the adaptive blind algorithms was investigated with a multipath fading channel with impulse response

$$c(t) = \sum_{i=1}^{L_d} \varepsilon_i p(t - \tau_i)$$

where $p(t)$ is the pulse shape (raised cosine pulse). The delay $\{\tau_i\}$ is statistically independent and uniformly distributed in $[0, 3T]$ where T is the symbol interval. The attenuations $\{\varepsilon_i\}$ are independent zero mean Gaussian variables and the number of paths is six.

The equivalent discrete time baseband channel has a length of four symbol intervals ($c(k)$, $k = 0, \ldots, 3$). The channel coefficients $c(k)$ were normalized by $c(0)$. An array of three hundred sets of channels drawn from channel parameter distributions was generated. 10 000 independent uniformly distributed input symbol sequences $\{s(n)\}$ over $\{1, -1\}$ were used for each channel. For the sake of fair comparison, the step size was chosen to be the same for all the algorithms.

6.8.8   Performance Results

Table 6.3 shows the results. Non-JEM-type algorithms yield about 5 dB–22 dB improvement in SIR. JEM-type algorithms exhibit an 11 dB–59 dB improvement. The JEM-4 algorithm is sensitive to step size. The third column of Table 6.3 shows the number of successful convergences. The JEM-4 algorithm became unstable with 115 simulated channels among 300. When the robustness and performance are considered simultaneously, JEM-1, JEM-2 or DECA are winners among the algorithms.

Table 6.3 SIR Comparison of (blind) DFE algorithms: $g(\cdot) = 3\tanh[1/3y(n)]$ for JEM-1 and JEM-2, $g(\cdot) = 3y(n) + y^3(n)$ for JEM-4, and $\mu = 1e^{-4}$ for all algorithms, averaged over 185 simulated FIR channels: noise-free case [113] © 1998, IEEE

	SIR (dB)	Number of successes among a set of 300 channels
Unequalized	0.80	300
ISIC	4.87	300
JEM-2	5.70	300
CMA-DFE	10.88	295
JEM-4	58.92	185
DECA	6.16	300
AC	22.14	292
JEM-1	5.07	300

References

[1] Wen-Rong Wu and Yih-Ming Tsuie (2002) An LMS-based decision feedback equalizer for IS-136 receivers, *IEEE Transactions on Vehicular Technology*, **51**(1), 130–143.

[2] Zerguine, A., Shafi, A. and Bettayeb, M. (2001) Multilayer perceptron-based DFE with lattice structure, *IEEE Transactions on Neural Networks*, **12**(3), 532–545.

[3] Ki Yong Lee (1996) Complex fuzzy adaptive filter with LMS algorithm, *IEEE Transactions on Signal Processing*, **44**(2), 424–427.

[4] Reuter, M. and Zeidler, J. R. (1999) Nonlinear effects in LMS adaptive equalizers, *IEEE Transactions on Signal Processing*, **47**(6), 1570–1579.

[5] Xiaohua Li and Fan, H. (2000) Linear prediction methods for blind fractionally spaced equalization, *IEEE Transactions on Signal Processing*, **48**(6), 1667–1675.

[6] Gi Hun Lee, Jinho Choi, Rae-Hong Park, Iickho Song, Jae Hyuk Park and Byung-Uk Lee (1994) Modification of the reference signal for fast convergence in LMS-based adaptive equalizers, *IEEE Transactions on Consumer Electronics*, **40**(3), 645–654.

[7] Mao-Ching Chiu and Chi-chao Chao (1996) Analysis of LMS-adaptive MLSE equalization on multipath fading channels, *IEEE Transactions On Communications*, **44**(12), 1684–1692.

[8] Raghunath, K. J. and Parhi, K. K. (1993). Parallel adaptive decision feedback equalizers, *IEEE Transactions on Signal Processing*, **41**(5), 1956–1961.

[9] Shanbhag, N. R. and Parhi, K. K. (1995) Pipelined adaptive DFE architectures using relaxed look-ahead, *IEEE Transactions on Signal Processing*, **43**(6), 1368–1385.

[10] Al-Dhahir, N. and Cioffi, J. M. (1996) Efficient computation of the delay-optimized finite length MMSE-DFE, *IEEE Transactions on Signal Processing*, **44**(5), 1288–1292.

[11] Al-Dhahir, N. and Cioffi, J. M. (1997) Mismatched finite-complexity MMSE decision feedback equalizers, *IEEE Transactions on Signal Processing*, **45**(4), 935–944.

[12] Chen, S., Mulgrew, B. and Hanzo, L. (2000) Asymptotic Bayesian decision feedback equalizer using a set of hyperplanes, *IEEE Transactions on Signal Processing*, **48**(12), 3493–3500.

[13] Afkhamie, K. H., Zhi-Quan Luo and Kon Max Wong (2001) Interior point least squares estimation: transient convergence analysis and application to MMSE decision-feedback equalization, *IEEE Transactions on Signal Processing*, **49**(7), 1543–1555.

[14] Liavas, A. P. (2002) On the robustness of the finite-length MMSE-DFE with respect to channel and second-order statistics estimation errors, *IEEE Transactions on Signal Processing*, **50**(11), 2866–2874.

[15] Berberidis, K. and Karaivazoglou, P. (2002) An efficient block adaptive decision feedback equalizer implemented in the frequency domain, *IEEE Transactions on Signal Processing*, **50**(9), 2273–2285.

[16] Sheng Chen (2002) Importance of sampling simulation for evaluating lower-bound symbol error rate of the Bayesian DFE with multilevel signaling schemes, *IEEE Transactions on Signal Processing*, **50**(5), 1229–1236.

[17] Hunsoo Choo, Muhammad, K. and Roy, K. (2003) Two's complement computation sharing multiplier and its applications to high performance DFE, *IEEE Transactions on Signal Processing*, **51**(2), 458–469.

[18] Monsen, P. (1977) Theoretical and Measured Performance of a DFE Modem on a Fading Multipath Channel, *IEEE Transactions On Communications*, **25**(10), 1144–1153.

[19] Cantoni, A. and Butler, P. (1976) Stability of Decision Feedback Inverses, *IEEE Transactions On Communications*, **24**(9), 970–977.

[20] Ehrman, L. and Monsen, P. (1977) Troposcatter Test Results for a High-Speed Decision-Feedback Equalizer Modem, *IEEE Transactions On Communications*, **25**(12), 1499–1504.

[21] Taylor, D. and Shafi, M. (1984) Decision Feedback Equalization for Multipath Induced Interference in Digital Microwave LOS Links, *IEEE Transactions on Communications*, **32**(3), 267–279.

[22] Falconer, D., Sheikh, A., Eleftheriou, E. and Tobis, M. (1985) Comparison of DFE and MLSE Receiver Performance on HF Channels, *IEEE Transactions On Communications*, **33**(5), 484–486.

[23] Fuyun Ling and Proakis, J. (1985) Adaptive Lattice Decision-Feedback Equalizers–Their Performance and Application to Time-Variant Multipath Channels, *IEEE Transactions On Communications*, **33**(4), 348–356.

[24] Leclert, A. and Vandamme, P. (1985) Decision Feedback Equalization of Dispersive Radio Channels, *IEEE Transactions On Communications*, **33**(7), 676–684.

[25] Shafi, M. and Moore, D. (1986) Further Results on Adaptive Equalizer Improvements for 16 QAM and 64 QAM Digital Radio, *IEEE Transactions On Communications*, **34**(1), 59–66.

[26] Kennedy, R. and Anderson, B. (1987) Recovery Times of Decision Feedback Equalizers on Noiseless Channels, *IEEE Transactions On Communications*, **35**(10), 1012–1021.

[27] Kennedy, R., Anderson, B. and Bitmead, R. (1987) Tight Bounds on the Error Probabilities of Decision Feedback Equalizers, *IEEE Transactions On Communications*, **35**(10), 1022–1028.

[28] Kennedy, R. and Anderson, B. (1987) Error Recovery of Decision Feedback Equalizers on Exponential Impulse Response Channels, *IEEE Transactions On Communications*, **35**(8), 846–848.

[29] Kennedy, R. A., Anderson, B. D. O. and Bitmead, R. R. (1989) Channels leading to rapid error recovery for decision feedback Equalizers, *IEEE Transactions On Communications*, **37**(11), 1126–1135.

[30] Zhou, K., Proakis, J. G. and Ling, F. (1990) Decision-feedback equalization of time-dispersive channels with coded modulation, *IEEE Transactions On Communications*, **38**(1), 18–24.

[31] Altekar, S. A. and Beaulieu, N. C. (1991) On the tightness of two error bounds for decision feedback Equalizers, *IEEE Transactions On Information Theory*, **37**(3), 638–639.

[32] Lin, D. W. (1991) Minimum mean-squared error decision-feedback equalization for digital subscriber line transmission with possibly correlated line codes, *IEEE Transactions On Communications*, **39**(8), 1197–1206.

[33] Williamson, D., Kennedy, R. A. and Pulford, G. W. (1992) Block decision feedback equalization, *IEEE Transactions On Communications*, **40**(2), 255–264.

[34] Pahlavan, K., Howard, S. J. and Sexton, T. A. (1993) Decision feedback equalization of the indoor radio channel, *IEEE Transactions On Communications*, **41**(1), 164–170.

[35] Jaekyun Moon and Sian She (1994) Constrained-complexity equalizer design for fixed delay tree search with decision feedback, *IEEE Transactions on Magnetics*, **30**(5), 2762–2768.

[36] Stojanovic, M., Proakis, J. G. and Catipovic, J. A. (1995) Analysis of the impact of channel estimation errors on the performance of a decision-feedback equalizer in fading multipath channels, *IEEE Transactions On Communications*, **43**(2,3,4), 877–886.

[37] McEwen, P. A. and Kenney, J. G. (1995) Allpass forward equalizer for decision feedback equalization, *IEEE Transactions on Magnetics*, **31**(6), 3045–3047.

[38] Russell, M. and Jan W. M. Bergmans (1995) A technique to reduce error propagation in *M*-ary decision feedback equalization, *IEEE Transactions On Communications*, **43**(12), 2878.

[39] Cioffi, J. M., Dudevoir, G. P., Vedat Eyuboglu, M. and Forney, G. D., Jr. (1995) MMSE decision-feedback equalizers and coding. I. Equalization results, *IEEE Transactions On Communications*, **43**(10), 2582–2594.

[40] Sheng Chen, McLaughlin, S., Mulgrew, B. and Grant, P. M. (1995) Adaptive Bayesian decision feedback equalizer for dispersive mobile radio Channels, *IEEE Transactions On Communications*, **43**(5), 1937–1946.

[41] Mathew, G., Farhang-Boroujeny, B. and Wood, R. W. (1997) Design of multilevel decision feedback Equalizers, *IEEE Transactions on Magnetics*, **33**(6), 4528–4542.

[42] Porat, B. and Friedlander, B. (1991) Blind equalization of digital communication channels using high-order moments, *IEEE Transactions on Signal Processing*, **39**(2), 522–526.

[43] Vembu, S., Verdu, S., Kennedy, R. A. and Sethares, W. (1994) Convex cost functions in blind equalization, *IEEE Transactions on Signal Processing*, **42**(8), 1952–1960.

[44] Ye Li and Zhi Ding (1995) Convergence analysis of finite length blind adaptive equalizers, *IEEE Transactions on Signal Processing*, **43**(9), 2120–2129.

[45] Ye Li and Liu, K. J. R. (1996) Static and dynamic convergence behavior of adaptive blind Equalizers, *IEEE Transactions on Signal Processing*, **44**(11), 2736–2745.

[46] Zhi Ding (1997) On convergence analysis of fractionally spaced adaptive blind equalizers, *IEEE Transactions on Signal Processing*, **45**(3), 650–657.

[47] Shtrom, V. and Fan, H. (1998) New class of zero-forcing cost functions in blind equalization, *IEEE Transactions on Signal Processing*, **46**(10), 2674–2683.

[48] Giannakis, G. B. and Tepedelenlioglu, C. (1999) Direct blind equalizers of multiple FIR channels: a deterministic approach, *IEEE Transactions on Signal Processing*, **47**(1), 62–74.

[49] Tugnait, J. K. and Bin Huang (1999) Second-order statistics-based blind equalization of IIR single-input multiple-output channels with common zeros, *IEEE Transactions on Signal Processing*, **47**(1), 147–157.

[50] Mannerkoski, J. and Koivunen, V. (2000) Autocorrelation properties of channel encoded sequences–applicability to blind equalization, *IEEE Transactions on Signal Processing*, **48**(12), 3501–3507.

[51] Junyu Mai and Sayed, A. H. (2000) A feedback approach to the steady-state performance of fractionally spaced blind adaptive equalizers, *IEEE Transactions on Signal Processing*, **48**(1), 80–91.

[52] Borah, D. K., Kennedy, R. A., Zhi Ding and Fijalkow, I. (2001) Sampling and prefiltering effects on blind equalizer design, *IEEE Transactions on Signal Processing*, **49**(1), 209–218.

[53] Lopez-Valcarce, R. and Dasgupta, S. (2001) Blind channel equalization with colored sources based on second-order statistics: a linear prediction approach, *IEEE Transactions on Signal Processing*, **49**(9), 2050–2059.

[54] Upez-Valcarce, R. and Dasgupta, S. (2001) Blind equalization of nonlinear channels from second-order statistics, *IEEE Transactions on Signal Processing*, **49**(12), 3084–3097.

[55] Luo, Z.-Q. T., Mei Meng, Wong, K. M. and Jian-Kang Zhang (2002) A fractionally spaced blind equalizer based on linear programming, *IEEE Transactions on Signal Processing*, **50**(7), 1650–1660.

[56] Prakriya, S. (2002) Eigenanalysis-based blind methods for identification, equalization, and inversion of linear time-invariant channels, *IEEE Transactions on Signal Processing*, **50**(7), 1525–1532.

[57] Benveniste, A. and Goursat, M. (1984) Blind Equalizers, *IEEE Transactions On Communications*, **32**(8), 871–883.

[58] Mathis, H. and Douglas, S. C. (2003) Bussgang blind deconvolution for impulsive signals *IEEE Transactions on Signal Processing*, **51**(7), 1905–1915.

[59] Karaoguz, J. and Ardalan, S. H. (1991) Use of blind equalization for teletext broadcast systems, *IEEE Transactions on Broadcasting*, **37**(2), 44–54.

[60] Ding, Z., Kennedy, R. A., Anderson, B. D. O. and Johnson, C. R., Jr. (1993) Local convergence of the Sato blind equalizer and generalizations under constraints, *IEEE Transactions On Information Theory*, **39**(1), 129–144.

[61] Verdu, S., Anderson, B. D. O. and Kennedy, R. A. (1993) Blind equalization without gain identification, *IEEE Transactions On Information Theory*, **39**(1), 292–297.

[62] Tugnait, J. K. (1994) Blind estimation of digital communication channel impulse response, *IEEE Transactions On Communications*, **42**(2,3,4), 1606–1616.

[63] Tugnait, J. K. (1996) Blind equalization and estimation of FIR communications channels using fractional sampling, *IEEE Transactions On Communications*, **44**(3), 324–336.

[64] Dogancay, K. and Kennedy, R. A. (1999) Least squares approach to blind channel equalization, *IEEE Transactions On Communications*, **47**(11), 1678–1687.

[65] Won Lee and Hill, F. (1977) A Maximum-Likelihood Sequence Estimator with Decision-Feedback Equalization, *IEEE Transactions On Communications*, **25**(9), 971–979.

[66] Falconer, D., Sheikh, A., Eleftheriou, E. and Tobis, M. (1985) Comparison of DFE and MLSE Receiver Performance on HF Channels, *IEEE Transactions On Communications*, **33**(5), 484–486.

[67] Sheen, W.-H. and Stuber, G. L. (1991) MLSE equalization and decoding for multipath-fading channels, *IEEE Transactions On Communications*, **39**(10), 1455–1464.

[68] Mao-Ching Chiu and Chi-chao Chao (1996) Analysis of LMS-adaptive MLSE equalization on multipath fading channels, *IEEE Transactions On Communications*, **44**(12), 1684–1692.

[69] Yonghai Gu and Tho Le-Ngoc (1996) Adaptive combined DFE/MLSE techniques for ISI channels, *IEEE Transactions On Communications*, **44**(7), 847–857.

[70] Hamied, K. A. and Stuber, G. L. (1996) An adaptive truncated MLSE receiver for Japanese personal digital cellular, *IEEE Transactions on Vehicular Technology*, **45**(1), 41–50.

[71] Jiunn-Tsair Chen, Paulraj, A. and Reddy, U. (1999) Multichannel maximum-likelihood sequence estimation (MLSE) equalizer for GSM using a parametric channel model, *IEEE Transactions On Communications*, **47**(1), 53–63.

[72] Jiunn-Tsair Chen and Yeong-Cheng Wang (2001) Adaptive MLSE equalizers with parametric tracking for multipath fast-fading channels, *IEEE Transactions On Communications*, **49**(4), 655–663.

[73] Gerstacker, W. (2002) Equalization concepts for EDGE, *IEEE Transactions on Communications* **1**(1).

[74] Dual-Hallen, A. (1989) Delayed Decision-Feedback Sequence Estimation, *IEEE Transactions On Communications* **37**(5).

[75] Eyuboglu, M. V. (1988) Reduced-State Sequence Estimation with Set Partitioning and Decision Feedback, *IEEE Transactions On Communications*, **36**(1).

[76] Vaidis, T. and Weber, C. L. (1998) Block adaptive techniques for channel identification and data demodulation over band-limited channels, *IEEE Transactions On Communications*, **46**(2), 232–243.

[77] Sayed, A. H. and Kailath, T. (1994) A state-space approach to adaptive RLS filtering, *IEEE Signal Processing Magazine*, July.

[78] Kennedy, R. A. (1989) Design and optimization of nonlinear mapping in decision feedback equalization, in *Proceedings of the 35th Conference on Decision and Control*, Kobe, Japan, December 1996, pp. 1888–1889.

[79] Qureshi, S. U. H. (1985) Adaptive equalization, *Proceedings of IEEE*, **53**, 1349–1387.

[80] Qureshi, S. (1985) Adaptive equalization, *Proceedings of IEEE*, **73**, 1349–1387.

[81] Proakis, J. G. (1991) Adaptive equalization for TDMA digital mobile radio, *IEEE Transactions on Vehicular Technology*, **40**, 333–341.

[82] Magee, F. R. Jr. and Proakis, J. G. (1973) Adaptive maximum likelihood estimation for digital signaling in the presence of intersymbol interference, *IEEE Transactions On Information Theory*, **IT–19**, 120–124.

[83] Ungerboeck, G. (1974) Adaptive maximum likelihood receivers for carrier modulated data transmission systems, *IEEE Transactions On Communications*, **COM–22**, 624–636.

[84] Qureshi, S. and Newhall, E. E. (1973) An adaptive receiver for data transmission over time dispersive channels, *IEEE Transactions On Information Theory*, **IT–19**, 448–457.

[85] Qureshi, U. H. (1973) *An Adaptive Decision–Feedback Receiver Using Maximum Likelihood Sequence Estimation*. ICC, Seattle, WA.

[86] Raheli, R., Polydoros, A. and Tzou, C. K. (1995) Per survivor processing: A general approach to MLSE in uncertain environments, *IEEE Transactions On Communications*, **43**.

[87] Seshadri, N. (1994) Joint data and channel estimation using blind trellis search techniques, *IEEE Transactions On Communications*, **42**, 1000–1011.

[88] Forney, G. D. Jr. (1972) Maximum likelihood sequence estimation of digital sequences in the presence of intersymbol interference, *IEEE Transactions On Information Theory*, **IT–18**, 363–378.

[89] D'Avella, R., Moreno, L. and Sant'Agostino, M. (1989) An adaptive MLSE receiver for TDMA digital mobile radio, *IEEE Journal on Selected Areas in Communications*, **7**, 122–129.

[90] Chevillat, P. R. and Eleftheriou, E. (1989) Decoding of trellis–encoded signals in the presence of intersymbol interference and noise, *IEEE Transactions On Communications*, **37**, 669– 676.

[91] Heller, J. A. and Jacobs, I. M. (1971) Viterbi decoding for satellite and space communication, *IEEE Transactions On Communications*, **COM–19**, 835–848.

[92] Gosh, M. and Weber, C. L. (1992) Maximum likelihood blind equalization, *Optical Engineering* **31**(6), 1224–1229.

[93] Sato, Y. (1975) A method of self–recovering equalization for multilevel amplitude modulation systems, *IEEE Transactions On Communications*, **COM–23**, 679–682.

[94] Bee Leong Yeap, Choong Hin Wong and Hanzo, L. (2003) Reduced complexity in-phase/quadrature-phase, *M*-QAM turbo equalization using iterative channel estimation, *IEEE Transactions on Wireless Communications*, **2**(1), 2–10.

[95] Nelson, J., Singer, A. and Koetter, R. (2003) Linear turbo equalization for parallel ISI Channels, *IEEE Transactions On Communications*, **51**(6), 860–864.

[96] Xiaodong Wang and Rong Chen (2001) Blind turbo equalization in Gaussian and impulsive noise, *IEEE Transactions on Vehicular Technology*, **50**(4), 1092–1105.

[97] Yee, M. S., Liew, T. H. and Hanzo, L. (2001) Burst-by-burst adaptive turbo-coded radial basis function-assisted decision feedback equalization, *IEEE Transactions On Communications*, **49**(11), 1935–1945.

[98] Bee Leong Yeap, Tong Hooi Liew, Hamorsky, J. and Hanzo, L. (2002) Comparative study of turbo equalization schemes using convolutional turbo, and block-turbo codes, *IEEE Transactions on Wireless Communications*, **1**(2), 266–273.

[99] Mong-Suan Yee, Yeap, B. L. and Hanzo, L. (2003) Radial basis function-assisted turbo equalization, *IEEE Transactions On Communications*, **51**(4), 664–675.

[100] Okada, T. and Iwanami, Y. (2002) Turbo equalizer detection for GFSK digital FM signals, *IEEE International Conference on Communications* (ICC 2002), 28 April–2 May 2002, **5**, 2952–2956.

[101] Dejonghe, A. and Vandendorpe, L. (2002) Turbo-equalization for multilevel modulation: an efficient low-complexity scheme, *IEEE International Conference on Communications* (ICC 2002), 28 April–2 May 2002, **3**, 1863–1867.

[102] Tuchler, M., Otnes, R. and Schmidbauer, A. (2002) Performance of soft iterative channel estimation in turbo equalization, *IEEE International Conference on Communications* (ICC 2002), 28 April–2 May 2002, **3**, 1858–1862.

[103] Laot, C., Glavieux, A. and Labat, J. (2001) Turbo equalization: adaptive equalization and channel decoding jointly optimized, *IEEE Journal on Selected Areas in Communications*, **19**(9), 1744–1752.

[104] Omidi, M. J., Gulak, P. G. and Pasupathy, S. (1998) Parallel structures for joint channel estimation and data detection over fading channels, *IEEE Journal on Selected Areas in Communications*, **16**(9), 1616–1629.

[105] Rong Chen, Xiaodong Wang and Liu, J. S. (2000) Adaptive joint detection and decoding in flat-fading channels via mixture Kalman filtering, *IEEE Transactions On Information Theory*, **46**(6), 2079–2094.

[106] Giridhar, K., Shynk, J. J., Iltis, R. A. and Mathur, A. (1996) Adaptive MAPSD algorithms for symbol and timing recovery of mobile radio TDMA signals, *IEEE Transactions on Communications*, **44**(8), 976–987.

[107] Rollins, M. E. and Simmons, S. J. (1997) Simplified per-survivor Kalman processing in fast frequency-selective fading channels, *IEEE Transactions On Communications*, **45**, 514–553.

[108] Anderson, B. D. O. and Moore, J. B. (1979) *Optimal Filtering*. Prentice Hall, Englewood Cliffs, NJ.

[109] Raheli, R., Polydoros, A. and Tzou, C. (1995) Per-survivor processing: A general approach to MLSE in uncertain environments, *IEEE Transactions On Communications*, **43**, 354–364.

[110] Rao, P. and Bayoumi, M. A. (1991) An algorithm specific VLSI parallel architecture for Kalman filter, in *VLSI Signal Processing IV. IEEE*, Piscataway, NJ, pp. 264–273.

[111] Bayoumi, M., Rao, P. and Alhalabi, B. (1992) VLSI parallel architecture for Kalman filter – an algorithm specific approach, *Journal of VLSI Signal Processing*, **4**(2,3), 147–163.

[112] Grewal, M. S. and Andrews, A. P. (1993) *Kalman Filtering, Theory and Practice*. Prentice Hall, Englewood Cliffs, NJ.

[113] Young-Hoon Kim and Shamsunder, S. (1998) Adaptive algorithms for channel equalization with soft decision feedback, *IEEE Journal on Selected Areas in Communications*, **16**(9), 1660–1669.

[114] Jie Zhu, Xi-Ren Cao and Ruey-Wen Liu (1999) A blind fractionally spaced equalizer using higher order statistics, *IEEE Transactions on Circuits and Systems II: Analog and Digital Signal Processing*, **46**(6), 755–764.

[115] Chong-Yung Chi and Chii-Horng Chen (2001) Cumulant-based inverse filter criteria for MIMO blind deconvolution: properties, algorithms, and application to DS/CDMA systems in multipath, *IEEE Transactions on Signal Processing*, **49**(7), 1282–1299.

[116] Behbahani, A. R. S. and Asjadi, H. (1996) Blind equalization based on third-order cumulant for 4-level and 8-level PAM, *Record of the 5th IEEE International Conference on Universal Personal Communications*, 29 September–2 October 1996, **1**, 136–140.

[117] Hatzinakos, D. and Nikias, C. L. (1991) Blind equalization using a tricepstrum-based algorithm, *IEEE Transactions On Communications*, **39**(5), 669–682.

[118] Tugnait, J. K. (1995) Blind equalization and estimation of digital communication FIR channels using cumulant matching, *IEEE Transactions On Communications*, **43**(2,3,4), 1240–1245.

[119] Chih-Chun Feng and Chong-Yung Chi (1999) Performance of cumulant based inverse filters for blind deconvolution, *IEEE Transactions on Signal Processing*, **47**(7), 1922–1935.

[120] Papoulis, A. (1991) *Probability, Random Variables, and Stochastic Processes*, 3rd edition, McGraw Hill, New York.

[121] Marcos, S., Cherif, S. and Jaidane, M. (1995) Blind cancellation of intersymbol interference in decision feedback equalizers, in *Proceedings of ICASSP*, pp. 1073–1076.

[122] Kamel, R. E. and Bar-Ness, Y. (1993) Blind decision feedback equalization using the decorrelation criterion, in *Proceedings of GLOBECOM*, pp. 87–91.

[123] Kennedy, R. A. (1993) Blind adaptation of decision feedback equalizers: Gross convergence properties, *International Journal of Adaptive Control Signal Processes*, **7**, 497–523.

[124] Papadias, C. B. and Paulraj, A. (1995) Decision–feedback equalization and identification of linear channels using blind algorithms of the Bussgang type, in *Proceedings of Asilomar Coni Signals, Systems, Computers*, Pacific Grove, CA, 1995, pp. 335–340.

[125] Haykin, S. (Ed.) *Blind Deconvolution*, Prentice Hall, Englewood Cliffs, NJ.

[126] Haykin, S. (1996) *Adaptive Filter Theory*, 3rd edition, Prentice Hall, Englewood Cliffs, NJ.

[127] Benveniste, A., Goursat, M. and Ruget, G. (1980) Robust identification of a nonminimum phase system: Blind adjustment of a linear equalizer in data Communications, *IEEE Transactions on Automatic Control*, **AC–25**, 385–399.

[128] Comon, P. (1994) Independent component analysis, a new concept?, *Signal Processing*, **36**, 287–314.

[129] Moreau, E. and Macchi, O. (1993) New self-adaptive algorithms for source separation based on contrast functions, in *Proceedings of the IEEE Signal Processing Workshop on Higher-Order Statistics*, June 1993, pp. 215–219.

[130] Kim, Y.-H. and Shamsunder, S. (1998) Multichannel algorithms for simultaneous equalization and interference suppression, in *Wireless Personal Communications*, Kluwer, Norwell, MA, pp. 219–237.

7

Orthogonal Frequency Division Multiplexing – OFDM and Multicarrier CDMA

The basic principles of the orthogonal frequency division multiplexing (OFDM) concept are presented in Chapter 1. Within this chapter we introduce further details with the main focus on synchronization, channel estimation, space–time frequency coding, pick to average power ratio (PAPR) and some comparisons of efficiency in dealing with multipath propagation impacts by using equalization and TDMA or OFDM signal formats.

7.1 Timing and Frequency Offset in OFDM

As already indicated in Chapter 1, the transmitted OFDM signal can be represented as

$$x(t) = \sum_{k=-k_1}^{N+k_2+1} \sum_{n=0}^{N-1} D_k e^{j2\pi \frac{nk}{N}} w\left(t - \frac{k}{f_s}\right) \tag{7.1}$$

for $-k_1/f_s < t < (N + k_2)/f$, where k_1 and k_2 are pre- and postfix lengths and $w(t)$ is the time domain window function. The received signal $r(t)$ is filtered and sampled at the rate, or multiples of, $1/T$.

The sampled signal at the output of the receiver FFT with ideal channel can be represented as a convolution:

$$y_n = \left[\sum_{k=-\infty}^{\infty} X_c\left(f + \frac{Nk}{T}\right)\right] \otimes TW(fT)|_{f=\frac{n}{T}} \tag{7.2}$$

where $X_c(f)$ is the Fourier transform of the periodically repeated analog equivalent of the signal generated by the transmitter's IFFT.

$X_c(f)$ is a line spectrum at k/T and $W(f)$ is the Fourier transform of window function $w(t)$.

Assuming that the sampling time has a relative phase offset of τ and that the offset does not change during one OFDM symbol, the sampled received signal for a non-dispersive channel can be

---

Advanced Wireless Communications & Internet: Future Evolving Technologies, Third Edition. Savo Glisic.
© 2011 John Wiley & Sons, Ltd. Published 2011 by John Wiley & Sons, Ltd.

simplified to:

$$y_k = \sum_{n=0}^{N-1} D_n e^{j\varphi} e^{j2\pi \frac{n}{N} f_c t} \Big|_{t=\frac{k+\tau}{f_s}}$$

(7.3)

where φ represents envelope delay distortion. After the Fourier transform at the receiver we have

$$\tilde{D}_m = \frac{1}{N} \sum_{k=0}^{N-1} \sum_{n=0}^{N-1} D_n e^{j2\pi \frac{n}{N}k} e^{j(\varphi+2\pi \frac{n}{N}\tau)} e^{-j2\pi \frac{m}{N}k}$$

$$= \sum_{n=0}^{N-1} D_n e^{j(\varphi+2\pi \frac{n}{N}\tau)} \sum_{k=0}^{N-1} e^{-j2\pi \frac{k}{N}(n-m)} = \begin{cases} 0, \ n \neq m \\ D_m e^{j(\varphi+2\pi \, n\tau/N)}, \ n = m \end{cases}$$

(7.4)

The timing phase (error), or envelope delay distortion, does not violate the orthogonality of the subcarriers and the effect of the timing offset is a phase rotation which linearly changes with subcarriers' orders. On the other hand, envelope delay results in the same amount of rotation for all subcarriers.

It is straightforward to show that in a more general case, with pulse shaping filter with rolloff factor α and dispersive channel with impulse response $h(t)$, the detected data is attenuated and phase rotated such that $\tilde{D}_m = \gamma_m(\tau) D_m$, where:

$$\gamma_m(\tau) = \begin{cases} H\left(\dfrac{m}{NT}\right) e^{j2\pi \frac{m}{NT}\tau}, \ 0 \leq \dfrac{m}{N} \leq \dfrac{1-\alpha}{2} \\[2mm] H\left(\dfrac{m}{NT}\right) e^{j2\pi \frac{m}{NT}\tau} + H\left(\dfrac{m-N}{NT}\right) e^{j2\pi \frac{m-N}{NT}\tau}, \ \dfrac{1-\alpha}{2} \leq \dfrac{m}{N} \leq \dfrac{1+\alpha}{2} \\[2mm] H\left(\dfrac{m-N}{NT}\right) e^{j2\pi \frac{m-N}{NT}\tau}, \ \dfrac{1+\alpha}{2} \leq \dfrac{m}{N} \leq 1 \end{cases}$$

(7.5)

where $H(m/NT)$ is the Fourier transform of $h(t)$ at frequency m/NT. The estimation of τ and φ will be discussed later in this section.

We will now see that frequency offset amounts to inter-channel interference which is similar to inter-symbol interference of a single carrier signal due to a timing jitter. In a non-dispersive channel with rectangular pulse shaping, the interference caused by frequency offset could be too constraining. The sampled signal is:

$$y_k = \sum_{n=0}^{N-1} D_k e^{j2\pi t\left(\frac{n}{N} f_s + \delta f\right)} \Big|_{t=\frac{k+\tau}{f_s}} = \sum_{n=0}^{N-1} D_k e^{j2\pi \left(\frac{n}{N} + \frac{\delta f}{f_s}\right)k}$$

(7.6)

After DFT we have

$$\tilde{D} = D_m \left(\frac{e^{j2\pi \Delta f} - 1}{e^{j2\pi \frac{\Delta f}{N}} - 1}\right) + \sum_{n=0}^{N-1} D_n \sum_{k=0}^{N-1} e^{j2\pi \frac{k}{N}(n-m+\Delta f)} + N_m$$

(7.7)

where $\Delta f = n\delta/f_s$. So, due to the frequency offset we have attenuation of the desired signal (the first term of Equation (7.7)), and an interference between different symbols of several subcarriers (the second term of Equation (7.7)).

In order to avoid frequency offset, the window function

$$w_n = w(t)|_{t=nT}$$

(7.8)

must be such that zero crossings of its Fourier transform are at multiples of symbol frequency

$$W_m = W(\omega)|_{\omega=2\pi \, mf_s} = \delta_m \tag{7.9}$$

A generalized *sinc* function of the form

$$\frac{\sin \omega n}{\omega n} \times g(n) \tag{7.10}$$

for any differentiable function, $g(t)$ satisfies the condition.

Another desired property of a sinc function is its low rate of change in the vicinity of in-frequency sampling points. A typical example is the raised cosine function in time:

$$w_{\text{rc}} = \begin{cases} T, & 0 \le |t| \le \dfrac{1-\beta}{2T} \\[3mm] \dfrac{\sin \pi \dfrac{t}{T}}{\pi \dfrac{t}{T}} \times \dfrac{\cos \beta\pi \dfrac{t}{T}}{1 - 4\beta^2 \dfrac{t^2}{T^2}}, & \dfrac{1-\beta}{2T} \le |t| \le \dfrac{1+\beta}{2T} \\[3mm] 0, & \text{elsewhere} \end{cases} \tag{7.11}$$

where β is the rolloff factor for time domain pulse shaping. A higher rolloff factor requires a longer cyclic extension and a guard interval, which consumes higher bandwidth. The results for adjacent channel interference (ACI) caused by frequency offset in systems with different windowing functions are shown in Figure 7.1.

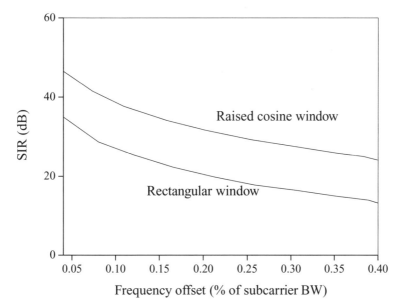

Figure 7.1   ACI caused by frequency offset.

7.1.1 Robust Frequency and Timing Synchronization for OFDM

7.1.1.1 Symbol Timing Estimation Algorithm

The symbol timing recovery relies on searching for a training symbol with two identical halves in the time. Consider the first training symbol where the first half is identical to the second half (in time order), except for a phase shift caused by the carrier frequency offset. If the conjugate of a sample from the first half is multiplied by the corresponding sample from the second half ($T/2$ seconds later), the effect of the channel should cancel, and the result will have a phase of approximately $\phi = \pi\, T\Delta f$. At the start of the frame, the products of each of these pairs of samples will have approximately the same phase, so the magnitude of the sum will be a large value. Let us use L complex samples in one half of the first training symbol (excluding the cyclic prefix), and let the sum of the pairs of products be

$$P(d) = \sum_{m=0}^{L-1} (r_{d+m}^* r_{d+m+L}) \tag{7.12}$$

This can be implemented with the iterative formula

$$P(d+1) = P(d) + (r_{d+L}^* r_{d+2L}) - (r_d^* r_{d+L}) \tag{7.13}$$

where d is a time index corresponding to the first sample in a window of $2L$ samples. This window slides along in time as the receiver searches for the first training symbol. The received energy for the second half symbol is defined by

$$R(d) = \sum_{m=0}^{L-1} |r_{d+m+L}|^2 \tag{7.14}$$

This can also be calculated iteratively. $R(d)$ may be used as part of an automatic gain control (AGC) loop. A timing metric can be defined as

$$M(d) = \frac{|P(d)|^2}{(R(d))^2} \tag{7.15}$$

Equation (7.15) is shown in Figure 7.2 and Figure 7.3.

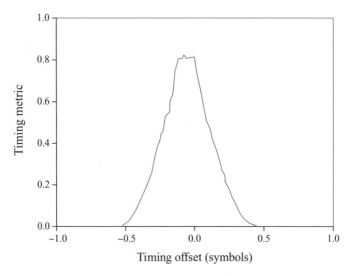

Figure 7.2 The timing metric for the AWGN channel (SNR $= 10$).

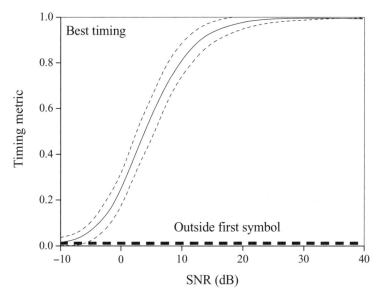

Figure 7.3   Expected value of timing metric with $L = 512$. Dashed lines indicate three standard deviations.

For the results in these figures, OFDM symbols are generated with 1000 frequencies, –500 to 499, and slightly oversampled at a rate of 1024 samples for the useful part of each symbol. In an actual hardware implementation, the ratio of the sampling rate to the number of frequencies would be higher to ease filtering requirements. The guard interval is set to about 10% of the useful part, which is 102 samples.

7.1.1.2   Carrier Frequency Offset Estimation Algorithm

The main difference between the two halves of the first training symbol will be a phase difference of

$$\phi = \pi \, T \Delta f \tag{7.16}$$

which can be estimated by

$$\hat{\phi} = \text{angle}(P(d)) \tag{7.17}$$

The second training symbol contains a PN sequence on the odd frequencies to measure these subchannels, and another PN sequence on the even frequencies to help determine frequency offset. If $|\hat{\phi}|$ can be guaranteed to be less than π, then the frequency offset estimate is

$$\Delta \hat{f} = \hat{\phi}/(\pi \, T) \tag{7.18}$$

and the even PN frequencies on the second training symbol would not be needed. Otherwise, the actual frequency offset would be

$$\frac{\phi}{\pi \, T} + \frac{2z}{T} \tag{7.19}$$

where z is an integer. By partially correcting the frequency offset, adjacent carrier interference (ACI) can be avoided, and then the remaining offset of $2z/T$ can be found. After the two training symbols are frequency corrected by $\hat{\phi}/(\pi \, T)$ (by multiplying the samples by $\exp(-j2t\hat{\phi}/T)$, let their FFTs be $x_{1,k}$

and $x_{2,k}$, and let the differentially-modulated PN sequence on the even frequencies of the second training symbol be v_k. The PN sequence v_k will appear at the output except it will be shifted by $2z$ positions because of the uncompensated frequency shift of $2z/T$. Note that because there is a guard interval and there is still a frequency offset, even if there were no differential modulation between training symbols 1 and 2 ($v_k = 1$) there would still be a phase shift between $x_{1,k}$ and $x_{2,k}$ of $2\pi\ (T + T_g)2z/T$. Since at this point the integer z is unknown, this additional phase shift is unknown. Since the phase shift is the same for each pair of frequencies, a metric similar to the previous one can be used.

Let X be the set of indices for the even frequency components, $X = \{-W, -W + 2, \ldots, -4, -2, 2, 4, \ldots, W - 2, W\}$. The number of even positions shifted can be calculated by finding $\hat{g}$ to maximize

$$B(g) = \frac{\left|\Sigma_{k \in X} x^*_{1,k+2g} v^*_k x_{2,k+2g}\right|^2}{2\left(\Sigma_{k \in X} |x_{2,k}|^2\right)^2} \tag{7.20}$$

with integer g spanning the range of possible frequency offsets and W being the number of even frequencies with the PN sequence.

Then the frequency offset estimate would be:

$$\Delta\hat{f} = [\hat{\phi}/(\pi\ T)] + (2\hat{g}/T) \tag{7.21}$$

7.1.1.3   Variance of Carrier Frequency Offset Estimator

From Equations (7.18) and (7.20) we have:

$$\text{var}[\hat{\phi}/\pi] = \frac{1}{\pi^2 \cdot L \cdot \text{SNR}} \tag{7.22}$$

One should keep in mind the Cramer–Rao bound is

$$\text{var}[\hat{\phi}/\pi] \geq \frac{1}{\pi^2 \cdot L \cdot \text{SNR}} \tag{7.23}$$

At the correct frequency offset, all the signal products

$$s^*_{1,h+g_{\text{correct}}} v^*_h s_{2,h+g_{\text{correct}}}$$

have the same phase and

$$\mu_B = E\ [B(g_{\text{correct}})] = \frac{\sigma^4_s}{(\sigma^2_s + \sigma^2_n)^2}$$

$$\text{var}[B(g_{\text{correct}})] = \frac{\sigma^4_s[(2 + 2\mu_B)\sigma^2_s\sigma^2_n + (1 + 4\mu_B)\sigma^4_n]}{W(\sigma^2_s + \sigma^2_n)^4} \tag{7.24}$$

At an incorrect frequency offset the signal products no longer add phase, and $B(g_{\text{incorrect}})$ has a chi-squared distribution with two degrees of freedom with

$$E\ [B(g_{\text{correct}})] = \frac{1}{2W}\left(1 + \frac{\sigma^2_s}{\sigma^2_s + \sigma^2_n}\right) < \frac{1}{W}$$

$$\text{var}[B(g_{\text{correct}})] = \frac{1}{4W^2}\left(1 + \frac{3\sigma^4_s + 2\sigma^2_s\sigma^2_n}{(\sigma^2_s + \sigma^2_n)^2}\right) < \frac{1}{W^2} \tag{7.25}$$

7.2 Fading Channel Estimation for OFDM Systems

7.2.1 Statistics of Mobile Radio Channels

The channel impulse response will be represented as

$$h(t, \tau) = \sum_k \gamma_k(t)\delta(\tau - \tau_k) \tag{7.26}$$

Assume that $\gamma_k(t)$ has the same normalized time correlation function $r_t(\Delta t)$ for all k, and power spectrum $p_t(\Omega)$.

$$r_{\gamma k}(\Delta t) \triangleq E\{\gamma_k(t + \Delta t)\gamma_k^*(t)\} = \sigma_k^2 r_t(\Delta t) \tag{7.27}$$

where σ_k^2 is the average power of the kth path.

The frequency response of the time-varying radio channel at time t is

$$H(t, f) \triangleq \int_{-\infty}^{\infty} h(t, \tau)e^{-j2\pi f\tau}d\tau = \sum_k \gamma_k(t)e^{-j2\pi f\tau_k} \tag{7.28}$$

The time–frequency correlation is defined as

$$\begin{aligned}
r_H(\Delta t, \Delta f) &\triangleq E\{H(t + \Delta t, f + \Delta f)H^*(t, f)\} \\
&= \sum_k r_{\gamma k}(\Delta t)e^{-j2\pi \Delta f\tau_k} \\
&= r_t(\Delta t)\left(\sum_k \sigma_k^2 e^{-j2\pi \Delta f\tau_k}\right) \\
&= \sigma_H^2 r_t(\Delta t)r_f(\Delta f)
\end{aligned} \tag{7.29}$$

where $\sigma_H^2 \triangleq \sum_k \sigma_k^2$ is the total average power of the channel impulse response. The frequency correlation is defined as

$$r_f(\Delta f) = \sum_k \frac{\sigma_k^2}{\sigma_H^2}e^{-j2\pi \Delta f\tau_k} \tag{7.30}$$

where $r_t(0) = r_f(0) = 1$. Without loss of generality, we also assume that $\sigma_H^2 = 1$, so that it can be omitted. For block length T_f and tone spacing (subchannel spacing) Δf, the correlation function for different blocks and tones is

$$r_H[n, k] = r_t[n]r_f[k] \tag{7.31}$$

where

$$r_t[n] \triangleq r_t(nT_f) \quad r_f[k] \triangleq r_f(k\Delta f)$$

From Jakes's model of the channel we have:

$$r_t[n] = J_0(n\omega_d) \triangleq r_J[n]$$

$$p_J(\omega) = \begin{cases} \dfrac{2}{\omega_d}\dfrac{1}{\sqrt{1 - (\omega/\omega_d)^2}}, & \text{if } |\omega| < \omega_d \\ 0, & \text{otherwise} \end{cases} \tag{7.32}$$

As an example, for carrier frequency $f_c = 2\,\text{GHz}$, $f_d = 184\,\text{Hz}$ when the user is moving at 60 mph.

7.2.2   Diversity Receiver

In the case of space (antenna) diversity, the signal from the mth antenna at the kth tone and the nth block can be represented as

$$x_m[n, k] = H_m[n, k]a[n, k] + \omega_m[n, k] \tag{7.33}$$

where $\omega_m[n, k]$ is additive Gaussian noise from the mth antenna at the kth tone and the nth block, with zero mean and variance ρ. Let us assume that $\omega_m[n, k]$ is independent for different ns, ks, or ms. $H_m[n, k]$, the frequency response at the kth tone and the nth block corresponding to the mth antenna, is assumed independent for different ms, but with the same statistics. $a[n, k]$ is the signal modulating the kth tone during the nth block and is assumed to have unit variance and be independent for different ks and ns.

With knowledge of the channel parameters, $a[n, k]$ can be estimated as $y[n, k]$ by an MMSE combiner

$$y[n, k] = \frac{\sum_{m=1}^{p} H_m^*[n, k]x_m[n, k]}{\sum_{m=1}^{p} |H_m[n, k]|^2} \tag{7.34}$$

The transceiver (transmitter/receiver) block diagram is given in Figure 7.4.

7.2.3   MMSE Channel Estimation

If the reference $a[n, k]$ is ideal (pilot symbols), a temporal estimation of $H[n, k]$ is

$$\tilde{H}[n, k] = x[n, k]a^*[n, k] \doteq H[n, k] + \omega[n, k]a^*[n, k] \tag{7.35}$$

where * denotes the complex conjugate. $\tilde{H}[n, k]$s for different ns and ks are correlated; therefore, an MMSE channel estimator can be constructed as follows:

$$\hat{H}[n, k] = \sum_{l} \sum_{m=-\infty}^{0} c[m, l, k]\tilde{H}[n - m, k - l] \tag{7.36}$$

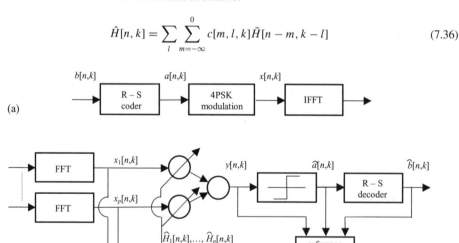

Figure 7.4   (a) Transmitter; (b) diversity receiver.

where $c[m, l, k]$s are selected to minimize

$$\mathrm{MSE}(\{c[m, l, k]\}) = E|\tilde{H}[n, k,] - H[n, k]|^2 \tag{7.37}$$

If K is the number of tones in each OFDM block, then we will be using the following notation

$$c[m, k] \stackrel{\Delta}{=} \begin{pmatrix} c[m, k-1, k] \\ \vdots \\ c[m, 0, k] \\ \vdots \\ c[m, -K+k, k] \end{pmatrix} \qquad \begin{aligned} c(\omega; k) &\stackrel{\Delta}{=} \sum_{n=-\infty}^{0} c[n, k]e^{-jn\omega} \\ \mathbf{C}(\omega) &\stackrel{\Delta}{=} (c(\omega; 1), c(\omega; 2), \cdots, c(\omega; K)) \end{aligned} \tag{7.38}$$

Starting from the fact that the projection of the estimation error on $H[n, k]$ is zero, $E\{(\hat{H}[n, k] - H[n, k])\tilde{H}^*[n - m, k - l]\} = 0$ (for orthogonality principles see Chapter 5), the estimation coefficients are given by [1]:

$$\mathbf{C}(\omega) = \mathbf{U}^H \Phi(\omega) \mathbf{U} \tag{7.39}$$

where $\Phi(\omega)$ is a diagonal matrix with the lth element

$$\Phi_l(\omega) = 1 - \frac{1}{M_l(\omega)\gamma_l[0]} \tag{7.40}$$

and $M_l(\omega)$ is a stable one-sided FT

$$M_l(\omega) = \sum_{n=0}^{\infty} \gamma_l[n]e^{-jn\omega} \tag{7.41}$$

$$M_l(\omega)M_l(-\omega) = \frac{d_l}{\rho}p_t(\omega) + 1 \tag{7.42}$$

The DC component $\gamma_l[0]$ in $M_l(\omega)$ can be found by

$$\gamma_l^2[0] = \exp\left\{\frac{1}{2\pi}\int_{-\pi}^{\pi}\ln\left[\frac{d_l}{\rho}p_t(\omega) + 1\right]d\omega\right\} \tag{7.43}$$

The d_ls and u_ls are the corresponding eigenvalues and eigenvectors of the frequency domain correlation matrix $\mathbf{R}_f$,

$$\mathbf{R}_f \stackrel{\Delta}{=} \begin{bmatrix} r_f[0] & r_f[1] & \cdots & r_f[K-1] \\ r_f[-1] & r_f[0] & \cdots & r_f[K-2] \\ \vdots & \vdots & \ddots & \vdots \\ r_f[-K+1] & r_f[-K+2] & \cdots & r_f[0] \end{bmatrix} \tag{7.44}$$

$\mathbf{U} = (u_1, \ldots, u_k)$ is a unitary matrix.

$$\mathbf{R}_f = \overline{\mathbf{U}}^H \mathbf{D} \overline{\mathbf{U}} \quad \text{or} \quad \overline{\mathbf{U}}\mathbf{R}_f\overline{\mathbf{U}}^H = \mathbf{D} \tag{7.45}$$

and $\mathbf{D} = \mathrm{diag}\{d_1, \ldots, d_K\}$ and $\Sigma_k d_k = K$. The processing is illustrated in Figure 7.5.

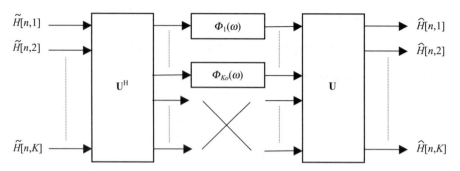

Figure 7.5   Channel estimator for OFDM systems.

The unitary linear inverse transform $\mathbf{U}^H$ and transform $\mathbf{U}$ in the figure perform the eigendecomposition of the frequency domain correlation. The estimator turns off the zero or small d_ls to reduce the estimation noise. For those large d_ls, linear filters are used to take advantage of the time domain correlation.

One can show [1] that for *Jakes's model*:

$$\overline{\mathrm{MMSE}}_J(\omega_d) = \frac{\rho}{K} \sum_{l-1}^{K} \left(1 - \left(\frac{\alpha_l}{2}\right)^{-(\omega_d)\pi} \exp\left\{-\frac{\omega_d b(\alpha_l)}{\pi}\right\}\right) \tag{7.46}$$

$$\alpha_l \stackrel{\Delta}{=} \frac{2d_l}{\omega_d \rho}$$

$$b(\alpha_l) \stackrel{\Delta}{=} \begin{cases} \dfrac{\pi}{2}\alpha_l + \sqrt{1 - \alpha_l^2}\, \ln\dfrac{1 + \sqrt{1 - \alpha_l^2}}{\alpha_l}, & \text{if } \alpha_l < 1 \\[2ex] \dfrac{\pi}{2}\alpha_l - \sqrt{\alpha_l^2 - 1}\left(\dfrac{\pi}{2} - \arcsin\dfrac{1}{\alpha_l}\right), & \text{if } \alpha_l \geq 1 \end{cases}$$

7.2.4   FIR Channel Estimator

For a reader less familiar with eigenvalue decomposition, discussed in Chapter 5, we provide a simplified interpretation of the processing by using an approximation. Instead of proper eigenvectors we use DFT matrix $\mathbf{W}$ defined

$$\mathbf{W} \stackrel{\Delta}{=} \frac{1}{\sqrt{K}} \begin{pmatrix} 1 & 1 & \cdots & 1 \\ 1 & e^{j(2\pi/K)} & \cdots & e^{j(2\pi(K-1)/K)} \\ \vdots & \vdots & \ddots & \vdots \\ 1 & e^{j(2\pi(K-1)/K)} & \cdots & e^{j(2\pi(K-1)(K-1)/K)} \end{pmatrix} \tag{7.47}$$

and with notation

$$\bar{\mathbf{R}}_t = \begin{pmatrix} \bar{r}_t[0] & \bar{r}_t[1] & \cdots & \bar{r}_t[L-1] \\ \bar{r}_t[-1] & \bar{r}_t[0] & \cdots & \bar{r}_t[L-2] \\ \vdots & \vdots & \ddots & \vdots \\ \bar{r}_t[-L+1] & \bar{r}_t[-L+2] & \cdots & \bar{r}_t[0] \end{pmatrix} \tag{7.48}$$

$$\bar{\mathbf{r}}_t = (\bar{r}_t[0], \ \bar{r}_t[1], \cdots, \bar{r}_t[L-1])^\mathrm{T}$$

$$\bar{r}_t[n] = \frac{\sin(n\omega_d)}{n\omega_d}$$

we have, for the coefficient matrix of the designed FIR channel estimator,

$$\bar{\mathbf{C}}(\omega) = \mathbf{W}^H \overline{\mathbf{\Phi}}(\omega) \mathbf{W} \qquad (7.49)$$

with

$$\overline{\mathbf{\Phi}}(\omega) = \text{diag}\{\underbrace{c(\omega), \ldots, c(\omega)}_{K_0 \text{ elements}}, 0, \ldots 0\}$$

If the maximum delay spread is t_d, then for all $l \leq K_0$ ($K_0 = K t_d / T_s$), $d_l \approx 0$, where Ts is the symbol interval. In Equation (7.49), $c(\omega)$ is the FT of c_n given by

$$(c_0, c_1, c_2, \ldots, c_{L-1})^T = \left(\overline{\mathbf{R}}_t + \frac{K_o \rho}{K} \mathbf{I} \right)^{-1} \bar{r}_t$$

and L is the length of the FIR estimator. The estimation error

$$\overline{\text{MSE}} = \frac{K_o c_0 \rho}{K} \qquad (7.50)$$

For the robust FIR channel estimator, the $\mathbf{U}$ in Figure 7.5 is the DFT matrix $\mathbf{W}$ and the $\Phi_k(\omega)$s for $k = 1, \ldots, K$ are $c(\omega)$.

7.2.5   System Performance

The set of assumptions used to generate the performance curves is the same as in [1]: a two-way Rayleigh fading with delay from 0 to 40 μs and Doppler frequency from 10 to 200 Hz; the channels corresponding to different receivers have the same statistics; two antennas are used for receiver diversity. For the OFDM signal, it is assumed that the entire channel bandwidth, 800 kHz, is divided into 128 subchannels. The four subchannels on each end are used as guard tones and the rest (120 tones) are used to transmit data. To make the tones orthogonal to each other, the symbol duration is 160 μs. An additional 40μs guard interval is used to provide protection from ISI, the length $T_f = 200$ ms and a subchannel symbol rate $r_b = 5$ kBd.

To compare the performance of the OFDM system with and without the channel estimation, PSK modulation with coherent demodulation and differential PSK (DPSK) modulation with differential demodulation are used, respectively. The (40, 20) RS code, with each code symbol consisting of three quadrature PSK/differential quadrature PSK (QPSK/DQPSK) symbols grouped in frequency, is used in the system. Each OFDM block forms an RS code word. The RS decoder erases ten symbols, based on signal strength, and corrects five additional random errors. Hence, the simulated system can transmit data at 1.2 Mb/s before decoding or 600 kb/s after decoding, over an 800 kHz channel.

7.2.6   Reference Generation

Four different ways to generate the reference signal are used:

1. *Undecoded/decoded dual mode reference*: If the RS decoder can successfully correct all errors in an OFDM block, the reference for the block can be generated by the decoded data; hence $\bar{a}[n, k] = a[n, k]$. *Otherwise*, $\bar{a}[n, k] = \hat{a}[n, k]$.
2. *Undecoded reference*: $\bar{a}[n, k] = \hat{a}[n, k]$, no matter whether the RS decoder can successfully correct all errors in a block or not.

3. *Decoded/CMA dual mode reference*: The constant modulus algorithm (CMA) is used to generate a reference for the OFDM channel estimator. If the RS decoder can successfully correct all
   errors in a block, the reference for the block can be generated from the decoded data; hence
   $\tilde{a}[n, k] = a[n, k]$. Otherwise, the reference can use the projection of $y[n, k]$ on the unit circle,
   i.e. $\tilde{a}[n,k] = y[n,k]/\text{mod}(y[n,k])$.
4. *Error removal reference*: If the RS decoder can successfully correct all errors in a block, the reference
   for the block can be generated by the decoded data. Otherwise, the $\tilde{H}[n - 1, k]$s are used instead of
   the $\tilde{H}[n, k]$s *for* $k = 1, \ldots, K$ respectively.

The results are shown in Figure 7.6.

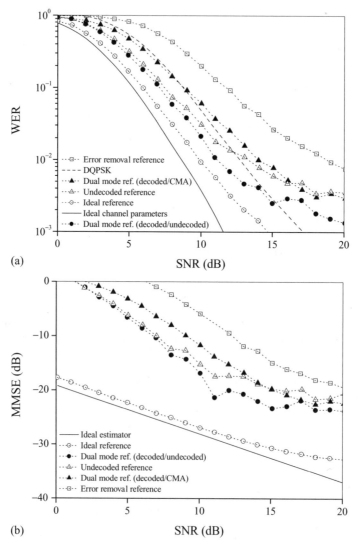

Figure 7.6 (a) WER and (b) MMSE of a robust estimator with different references versus SNR for the system with
1% training blocks when the 50-tap channel estimator matches the channel with $f_d = 200$ Hz and $t_d = 5$ μs.

In general a careful study of the results based on the previous discussion suggests the following conclusions:

- If a channel estimator is designed to match the channel with 40 Hz maximum Doppler frequency and 20 μs maximum delay spread, then for all channels with $f_d \leq 40$ Hz and $t_d \leq 20$ μs, the system performance is not worse than the channel with $f_d = 40$ Hz and $t_d = 20$ μs. For channels with $f_d > 40$ Hz or $t_d > 20$ μs, such as $f_d = 80$ Hz and $t_d = 20$ μs or $f_d = 40$ Hz and $t_d = 40$ μs, the system performance degrades dramatically.
- If the estimator is designed to match a Doppler frequency or delay spread larger than the actual ones, the system performance degrades only slightly compared with estimation that exactly matches the channel Doppler frequency and delay spread.

More details on channel estimation can be found in [2–11].

7.3 64 DAPSK and 64 QAM Modulated OFDM Signals

In this section we discuss in more detail specific, high constellation, modulation schemes for OFDM signals. The transmission system and two options for the signal constellation are shown in Figures 7.7 and 7.8.

In this section we consider an OFDM system with $N = 1024$ subcarriers and parameters the same as in [12]. Each OFDM symbol transfers 6144 bits in total. The convolutional code has the memory length $m = 6$. The bit interleaver is a block interleaver with 83 rows and 74 columns, which means that only two bits of an OFDM symbol are not involved in the interleaving process. The performance of convolutional codes with code rates 1/2, 2/3 and 3/4 has been analyzed.

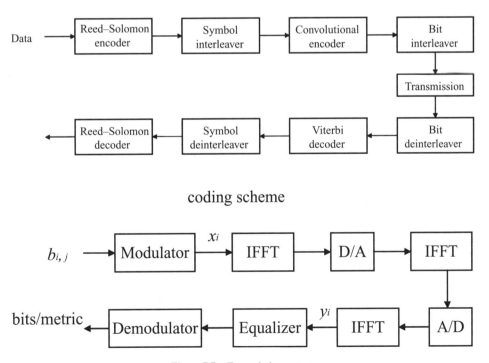

coding scheme

Figure 7.7 Transmission system.

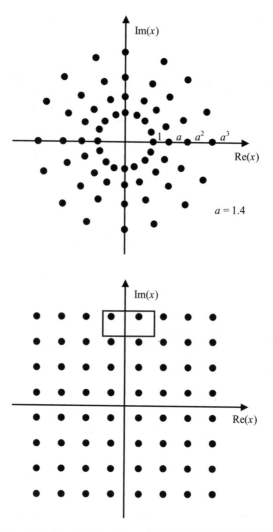

Figure 7.8 Constellation diagrams of 64 DAPSK and 64 QAM.

A (204, 188) RS code with $p = 8$ b/symbol has been chosen for the DTVB application. The objective of the concatenated code is to fulfil the requirement of a residual bit error rate (BER) of $10E^{-11}$ at the output of the RS decoder. For this, the BER at the output of the Viterbi decoder is required to be lower than $2 \times 10E^{-4}$ [13]. The above parameters are used in the system for digital terrestrial video broadcasting [14–17]. Such a system can transmit 34 Mbits/s over an 8 MHz radio channel.

The phase modulation is independent of the amplitude and identical to the well known 16 DPSK.

The input bits $b_{i,0}$, $b_{i,1}$, $b_{i,2}$, $b_{i,3}$ are used for this differential phase modulation in the 64 DAPSK scheme. The amplitude states $|x_i|$ are chosen from the constellation diagram depending on the previous amplitude state $|\tilde{x}_i|$ and the two information bits $b_{i,5}$ and $b_{i,4}$ according to Table 7.1. This means, e.g. for a subcarrier i if the amplitude in the previous OFDM symbol was $|\tilde{x}_i|$ and the input information bits $b_{i,5}$ and $b_{i,4}$ were both zero, then the amplitude in the current OFDM symbol would be $|x_i| = |\tilde{x}_i|$ again. Table 7.1 shows the amplitude state $|x_i|$ for the subcarrier in the current OFDM symbol.

Table 7.1 Differential amplitude modulation for 64 DAPSK. Choice of the current amplitude state $|x_i|$ depending on the previous state $|\tilde{x}_i|$ and the amplitude bits

	Amplitude bits $b_{i,4}\ b_{i,5}$			
$|\tilde{x}_i|$	00	01	11	10
1	1	a	a^2	a^3
a	a	a^2	a^3	1
a^2	a^2	a^3	1	a
a^3	a^3	1	a	a^2

In each OFDM receiver after block synchronization, analog-to-digital (A/D) conversion, and removing of the guard interval, a fast Fourier transform (FFT) will produce the complex output states y_i for each subcarrier i. If the coherent 64 QAM is used, after the FFT a channel equalization must be performed.

This means that the channel transfer factor α_i is assumed to be known exactly for each subcarrier i. With this information, the transmitted state x_i of each subcarrier in the 64 QAM constellation diagram is evaluated by simple quotient

$$x_i \approx y_i / \alpha_i$$

in order to get the coded bit sequence for hard decision decoding or the metric increments for soft decision decoding.

For the non-coherent 64 DAPSK demodulation, first the quotient

$$r_i = y_i / \tilde{y}_i \approx x_i / \tilde{x}_i$$

of the currently received state y_i and the preceding state $\tilde{y}_i$ of the same subcarrier i in the receiver (FFT output) are calculated. The resulting complex quotient r_i is evaluated in order to get the phase and amplitude bits $b_{i,0}, b_{i,1}, \ldots, b_{i,5}$. This quotient r_i is nearly independent of channel transfer factor α_i if the radio channel does not change the transmission behavior too quickly. Therefore, pilot symbols, channel estimation and equalization are not needed in the 64 DAPSK receiver, which reduces the computation complexity.

The four bits $b_{i,0}, b_{i,1}, b_{i,2}, b_{i,3}$ are determined depending on the phase difference between y_i and $\tilde{y}_i$ only. This phase demodulation process is the same as the demodulation of 16 DPSK.

For the amplitude demodulation, Table 7.2 shows how the amplitude bits $b_{i,4}$ and $b_{i,5}$ are obtained. For the evaluation of $|r_i|$, simple amplitude thresholds are used.

Performance results are given in Figure 7.9 and Tables 7.3–7.5. On average, 4 dB SNR degradation must be accepted in order to have much simpler-to-implement DAPSK. More details on coherent APSK and DAPSK can be found in [12–22].

Table 7.2 Evaluation of the amplitude information bits $b_{i,5}$ and $b_{i,4}$ in the 64 DAPSK demodulation

	$|r_i| = |y_i / \tilde{y}_i|$						
	a^{-3}	a^{-2}	a^{-1}	1	a^1	a^2	a^3
$b_{i,4}\ b_{i,5}$	01	11	10	00	01	11	10
thresholds		$a^{-2.5}$	$a^{-1.5}$	$a^{-0.5}$	$a^{-0.5}$	$a^{1.5}$	$a^{-2.5}$
	$-a^{-2.5}$	$a^{-1.5}$	$a^{-0.5}$	$a^{0.5}$	$a^{1.5}$	$a^{2.5}$	$a^{3.5}$

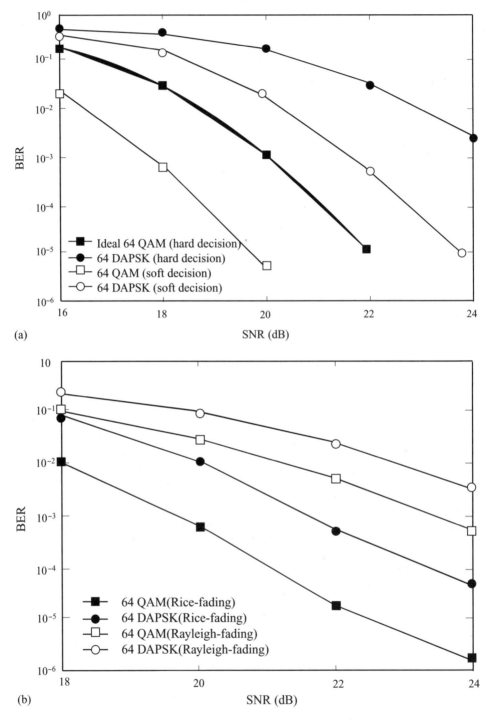

Figure 7.9   (a) Performance of ideal 64 QAM and 64 DAPSK (ring ratio $a = 1.4$) with a convolutional code $m = 6$, $R = 3/4$ for hard and soft decision decoding in an AWGN channel; (b) performance of ideal 64 QAM (convolutional code $R = 3/4$) and 64 DAPSK ($R = 2/3$) over the Rayleigh and Rice-fading channel (fixed user data rate).

Table 7.3   SNR [dB] required for BER $= 2 \times 10^{-4}$

Code rate R	Metric	Ideal 64 QAM			64 DAPSK		
		AWGN channel	Rice channel	Rayleigh channel	AWGN channel	Rice channel	Rayleigh channel
1/2	hard decision	17.3	19.6	24.1	21.4	23.7	28.6
2/3	hard decision	19.5	22.4	28.6	24.0	26.5	33.2
3/4	hard decision	20.9	23.9	32.8	25.7	28.2	37.4
1/2	soft decision	13.6	15.6	18.6	17.7	19.5	22.3
2/3	soft decision	16.8	19.1	22.4	20.6	22.8	26.5
3/4	soft decision	18.5	20.8	24.9	22.5	24.8	29.3

Table 7.4   Difference of required SNR [dB] for 64 QAM (ideal channel estimation) and 64 DAPSK modulation at BER $= 2 \times 10^{-4}$, fixed code rates

Code rate R	Hard decision			Soft decision		
	AWGN channel	Rice channel	Rayleigh channel	AWGN channel	Hire channel	Rayleigh channel
1/2	4.1	4.1	4.5	4.1	3.9	3.7
2/3	4.5	4.1	4.6	3.8	3.7	4.1
3/4	4.8	4.3	4.6	4.0	4.0	4.4

Table 7.5   Difference of required SNR [dB] for 64 QAM (ideal channel estimation) with $R = 3/4$ and 64 DAPSK modulation with $R = 2/3$ at BER $= 2 \times 10^{-4}$, fixed user data rate

Code rate R	Hard decision			Soft decision		
	AWGN channel	Rice channel	Rayleigh channel	AWGN channel	Rice channel	Rayleigh channel
QAM: 3/4 DAPSK: 2/3	3.1	2.6	0.4	2.1	2.0	1.6

7.4   Space–Time Coding with OFDM Signals

In this section we discuss space–time coding with an OFDM signal. The system block diagram is shown in Figure 7.10.

In the system, Alamouti's $\mathbf{G}_2$ space–time block code is used

$$\mathbf{G}_2 = \begin{pmatrix} x_1 & x_2 \\ -\bar{x}_2 & \bar{x}_1 \end{pmatrix}$$

For comparison the system parameters used in this section are the same as in [23] and are specified in Tables 7.6–7.8.

7.4.1   Signal and Channel Parameters

A two-ray channel impulse response having equal amplitudes and differential delay of 5 μs is used. The average signal power received from each transmitter antenna is the same. All multipath components

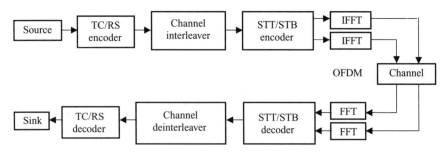

Figure 7.10 Turbo convolutional (TC) or Reed–Solomon (RS) code used with space–time trellis (STT) or space–time block (STB) encoder.

Table 7.6 Modulation parameters

Modulation scheme	Bits per symbol BPS	Decoding algorithm	No. of states	No. of transmitters	No. of termination symbols
4PSK	2	VA	4	2	1
			8	2	2
			16	2	2
			32	2	3
8PSK	3	VA	8	2	1
			16	2	2
			32	2	2

Table 7.7 The parameters associated with the turbo convolutional (n, k, K) TC(2, 1, 3) code

Code	Code rate R	Modulation mode	BPS	Random turbo interleaver depth	Random separation interleaver depth
				128 carriers	
		16 QAM	2	256	512
		64 QAM	3	384	768
TC(2, 1, 3)	0.50			512 carriers	
		QPSK	1	512	1024
		16 QAM	2	1024	2048

Table 7.8 The coding parameters of the Reed–Solomon codes

Code	Galois field	Rate	Correctable symbol errors
RS(105, 51)	2^{10}	0.49	27
RS(153, 102)	2^{10}	0.67	25

undergo independent Rayleigh fading; Jakes's model. The receiver has a perfect knowledge of the CIR. A 128 subcarrier (160 μs) OFDM signal with a cyclic extension of 32 samples (40 μs) is used. The results are presented in Figure 7.11.

For increased delay spread and Doppler, the variation of the frequency domain fading envelope will eventually destroy the orthogonality of the space–time block code $\mathbf{G}_2$ (see Figure 7.13). For this reason the two transmission instants of the space–time block code $\mathbf{G}_2$ will have to be allocated to the same OFDM symbol. In the previous example they were allocated to the adjacent subcarriers. The transmission

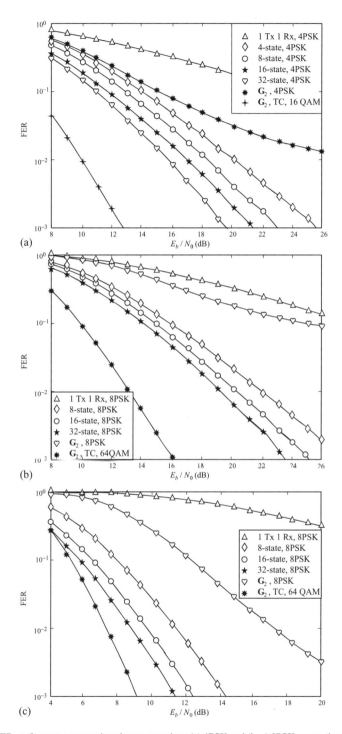

Figure 7.11   FER performance comparison between various (a) 4PSK and (b, c) 8PSK space–time trellis codes and the space–time block code G_2 concatenated with the TC(2, 1, 3) code using (a, b) one or (c) two receivers and the 128 subcarrier OFDM modem over a channel having a CIR characterized by two equal-power rays separated by a delay spread of 5 μs. The maximum Doppler frequency was 200 Hz. The effective throughput was (a) 2 BPS or (b, c) 3 BPS and the coding parameters are shown in Tables 7.6–7.7.

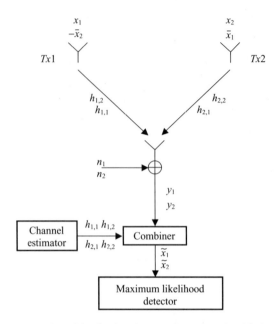

Figure 7.12 Baseband representation of the simple twin-transmitter space–time block code G_2 using one receiver over varying fading conditions.

system for the time-varying channel is now modeled in Figure 7.12. The received signals are given by:

$$
\begin{aligned}
\tilde{x}_1 &= \bar{h}_{1,1} y_1 + h_{2,2} \bar{y}_2 \\
&= \bar{h}_{1,1} h_{1,1} x_1 + \bar{h}_{1,1} h_{2,1} x_2 + \bar{h}_{1,1} n_1 - h_{2,2} \bar{h}_{1,2} x_2 + h_{2,2} \bar{h}_{2,2} x_1 + h_{2,2} \bar{n}_2 \\
&= (|h_{1,1}|^2 + |h_{2,2}|^2) x_1 + (\bar{h}_{1,1} h_{2,1} - h_{2,2} \bar{h}_{1,2}) x_2 + \bar{h}_{1,1} n_1 + h_{2,2} \bar{n}_2 \\
\tilde{x}_2 &= \bar{h}_{2,1} y_1 + h_{1,2} \bar{y}_2 \\
&= \bar{h}_{2,1} h_{1,1} x_1 + \bar{h}_{2,1} h_{2,1} x_2 + \bar{h}_{2,1} n_1 - h_{1,2} \bar{h}_{1,2} 1 x_2 - h_{1,2} \bar{h}_{2,2} x_1 - h_{1,2} \bar{n}_2 \\
&= (|h_{2,1}|^2 + |h_{1,2}|^2) x_2 + (\bar{h}_{2,1} h_{1,1} - h_{1,2} \bar{h}_{2,2}) x_1 + \bar{h}_{2,1} n_1 - h_{1,2} \bar{n}_2
\end{aligned}
\tag{7.51}
$$

The signal to interference ratio (SIR) for signal x_1 is:

$$
\text{SIR} = \frac{|h_{1,1}|^2 + |h_{2,2}|^2}{\bar{h}_{1,1} h_{2,1} - h_{2,2} \bar{h}_{1,2}}
\tag{7.52}
$$

and for signal x_2 is:

$$
\text{SIR} = \frac{|h_{2,1}|^2 + |h_{1,2}|^2}{\bar{h}_{2,1} h_{1,1} - h_{1,2} \bar{h}_{2,2}}
\tag{7.53}
$$

The two transmission instants are no longer assumed to be associated with the same complex transfer function values. Performance curves are given in Figures 7.13 and 7.14.

The fading amplitude variation versus time is slower than that versus the subcarrier index within the OFDM symbols. This implies that the SIR attained would be higher, if we were to allocate the two transmission instants of the space–time block code G_2 to the same subcarrier of consecutive OFDM symbols. This increase in SIR is achieved by doubling the delay of the system, since in this scenario two consecutive OFDM symbols have to be decoded.

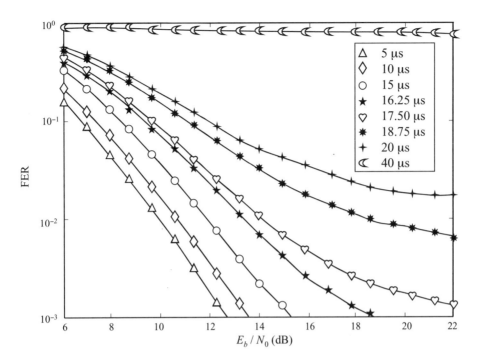

Figure 7.13 FER performance of the space–time block code G_2 concatenated with the TC(2, 1, 3) code using one receiver, the 128 subcarrier OFDM modem and 16 QAM. The CIR exhibits two equal-power rays separated by various delay spreads and a maximum Doppler frequency of 200 Hz. The coding parameters are shown in Tables 7.6–7.9.

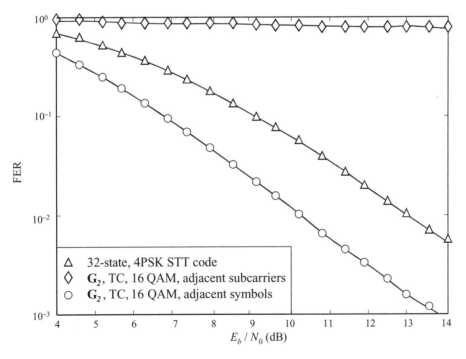

Figure 7.14 FER performance comparison between adjacent subcarriers and adjacent OFDM symbols allocation for the space–time block code G_2 concatenated with the TC(2, 1, 3) code using one receiver, the 128 subcarrier OFDM modem and 16 QAM over a channel having a CIR characterized by two equal-power rays separated by a delay spread of 40 μs. The maximum Doppler frequency was 100 Hz. The coding parameters are shown in Tables 7.6–7.9.

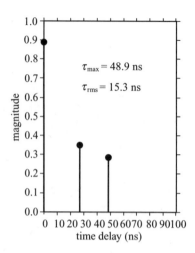

Figure 7.15   Short WATM channel impulse response.

7.4.2   The Wireless Asynchronous Transfer Mode System

The CIR used in this case has three paths and is referred to as the shortened WATM CIR, as shown in Figure 7.15. A 512 subcarrier (2.2756 μs) OFDM time domain signal with a cyclic extension of 64 samples (0.2844 μs) is used. The sampling rate is 225 Msamples/s and the carrier frequency is 60 GHz. The results are shown in Figure 7.16.

7.4.3   Space–Time Coded Adaptive Modulation for OFDM

All subcarriers in an adaptive OFDM (AOFDM) symbol are split into blocks of adjacent subcarriers, referred to as subbands. The same modulation scheme is employed for all subcarriers of the same subband. This substantially simplifies the task of signaling the modulation modes, since there are typically four modes and, for example, 32 subbands, requiring a total of 64 AOFDM mode signaling bits. The system is referred to as subband adaptive OFDM.

The system is presented in Figure 7.17. The choice of the modulation scheme to be used by the transmitter for its next OFDM symbol is determined by the channel quality estimate of the receiver, based on the current OFDM symbol. The following assumptions are used: the average signal power received from each transmitter antenna is the same; all multipath components undergo independent Rayleigh fading; the receiver has a perfect knowledge of the CIR and perfect signaling of the AOFDM modulation modes is available.

Optimized switching levels for adaptive modulation over a Rayleigh fading channel, shown in the instantaneous channel SNR (dB) are given in Table 7.9. The performance results are given in Figure 7.18.

7.4.4   Turbo and Space–Time Coded Adaptive OFDM

In this case the system parameters are as defined in Table 7.10. Performance results are shown in Figure 7.19.

For details on space–time coding with OFDMA signals and same set of parameters see [23–31].

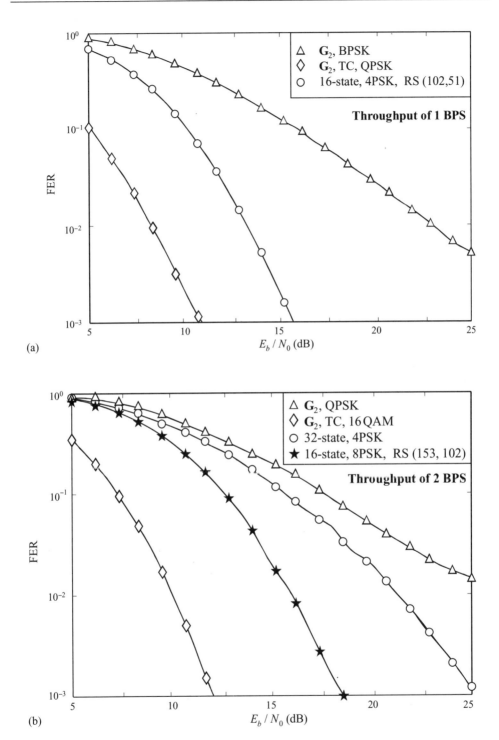

Figure 7.16   FER performance comparison between the TC (2, 1, 3) coded space–time block code G_2 and the RS(102, 51) GF(2^{10}) coded 16-state 4PSK space–time trellis code using one 512 subcarrier OFDM receiver over the shortened WATM channel at an effective throughput of (a) 1 BPS and (b) 2BPS. The coding parameters are shown in Tables 7.6–7.9.

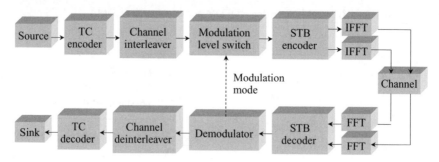

Figure 7.17 System overview of the turbo coded and space–time coded adaptive OFDM.

Table 7.9 Switching SNR

System	NoTx	BPSK	QPSK	16 QAM	64 QAM
Speech	$-\infty$	3.31	6.48	11.61	17.64
Data	$-\infty$	7.98	10.42	16.76	26.33

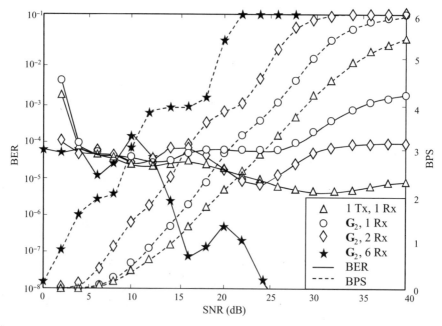

Figure 7.18 BER and BPS performance of 16 subband AOFDM employing the space–time block code G_2 using multiple receivers for a target BER of 10^{-4} over the shortened WATM channel shown in Figure 7.15.

Table 7.10 Coding rates and switching levels (dB) for TC (2, 1, 3) and space–time coded adaptive OFDM over the shortened WATM channel of Figure 7.15 for a target BER of 10^{-4}

	NoTx	BPSK	QPSK	16 QAM	64 QAM
		Half rate TC (2, 1, 3)			
Rate	–	0.50	0.50	0.50	0.50
Thresholds (dB)	$-\infty$	-4.0	-1.3	5.4	9.8
		Variable rate TC (2, 1, 3)			
Rate	–	0.50	0.67	0.75	0.90
Thresholds (dB)	$-\infty$	-4.0	2.0	9.70	21.50

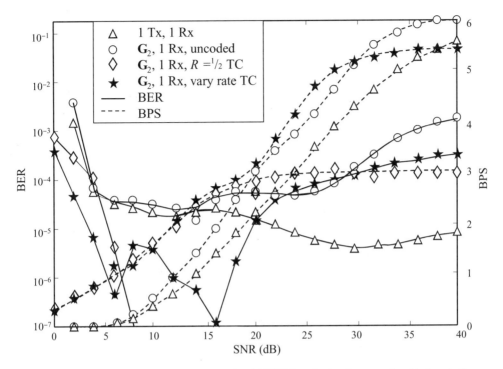

Figure 7.19 BER and BPS performance of 16 subband AOFDM employing the space–time block code G_2 concatenated with both half rate and variable rate TC(2, 1, 3) at a target BER of 10^{-4} over the shortened WATM channel.

7.5 Layered Space–Time Coding for MIMO OFDM

In this section we extend the discussion on space–time coded OFDM to the case with $n_t = n_r = 4$ assuming the Jakes fading model and a layered architecture. The channel modeling is discussed in Chapter 14, but in this section we assume the channel estimation procedures as in [29] and [32] and the TU channel model considered in [32]. The OFDM signals assume a channel bandwidth of 1.25 MHz, which is divided into 256 subchannels. Two subchannels at each end of the band are used as guard tones, with the other 252 tones used to transmit data. The symbol duration is taken to be 204.8 μs so that the tones are orthogonal. A 20.2 μs guard interval is used to provide protection from inter-symbol interference, making the block duration $T_f = 225$ μs. The subchannel symbol rate is $r_b = 4.44$ kbaud. Systems with the same signal parameters are discussed, for example, in [32, 33].

7.5.1 System Model (Two Times Two Transmit Antennas)

First we consider the $n_g = 2$ (groups of antennas) MIMO-OFDM implementation illustrated in Figure 7.20. In this case, two antenna space–time codes are employed that use 16 states and QPSK modulation. Data is grouped into blocks of 500 information bits, called words. Each word is coded into 252 symbols to form an OFDM block. Since this system uses $n_g = 2$, it can transmit two of these data blocks (1000 bits total) in parallel. Each time slot consists of ten OFDM blocks with the first block used for training and the following nine blocks used for data transmission. This leads to a system capable of transmitting 4 Mbit/s using 1.25 MHz of bandwidth, so the transmission efficiency is 3.2 bit/s/Hz.

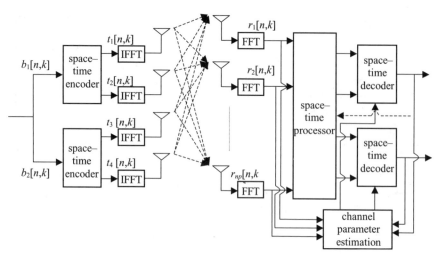

Figure 7.20   MIMO–OFDM using $n_g = 2$ individual space–time encoders, each using $n_t / n_g - 2$ transmit antennas.

7.5.2   Interference Cancellation

In general, all the interference cancellation schemes discussed in Chapter 5, modified for OFDM signals, can be used. One option has already been discussed in Chapter 4. An initial study of the system outlined in this section was provided in [34]. In [34] several interference cancellation approaches were described and performance was evaluated. Here, we focus on the successive interference cancellation approach based on signal quality (see Chapter 4). Basically, the strongest signal is detected first, subtracted from the input signal and then the procedure is repeated. We also assume that interleaving is employed. For the two-antenna, 16-state code given in Chapter 4 [3 (Figure 5)], the word error rate (WER) is given in Figure 7.21 for the case where the channel has a TU delay profile and for Doppler frequencies of 5, 40, 100, and 200 Hz. The other two curves in Figure 7.21 illustrate the performance improvement that can be obtained using the improved convolutional space–time codes given in Section 4.6. One of these codes was designed to be optimal for the quasi-static fading model from Section 4.2 (CC1). The other code was designed to be optimal for the rapid fading model in Section 4.2 (CC2). The improved codes from Section 4.6 are optimal codes based on the criterion given in Section 4.2. All these details are available in Chapter 4. In summary:

$$\begin{bmatrix} D + 2D^2 & 1 + 2D^2 \\ 2D & 2 \end{bmatrix} \text{ Figure 4.7 (TC)}$$

$$\begin{bmatrix} D + D^2 & 2 + D \\ 2 + D & 2 + 2D + 2D^2 \end{bmatrix} \text{ quasi-static code Section 4.6 (CC1)}$$

$$\begin{bmatrix} 2D^2 & 2 + D + 2D^2 \\ 2 + D & 2D + 2D^2 \end{bmatrix} \text{ rapid fading code Section 4.6 (CC2)}$$

7.5.3   Four Transmit Antennas

Next we investigate the approach that uses four-antenna space–time codes. We consider 16-state and 256-state codes, designed using an *ad hoc* approach. The codes are presented in Table 7.11 and the performance results are given in Figure 7.22.

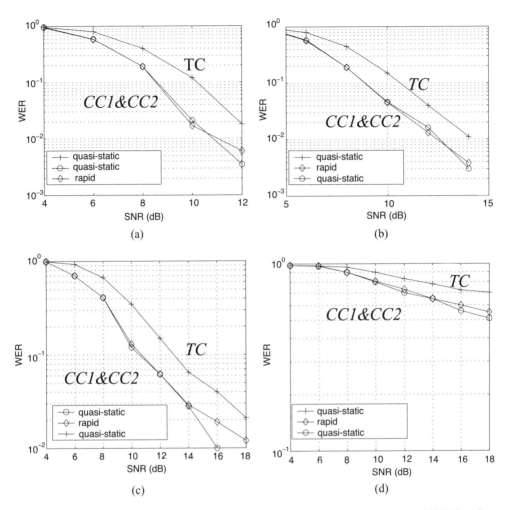

Figure 7.21   Performance curves. (a) 5 Hz Doppler; (b) 40 Hz Doppler; (c) 100 Hz Doppler; (d) 200 Hz Doppler.

Table 7.11   Generator matrices for the four transmit antenna codes used in figure 7.22. These codes are in GF(4) with elements denoted by $\{0, 1, a, 1 + a\}$ for the 256-state code

$$
\begin{bmatrix}
(1+a)+D & a+(1+a)D & a+D & 1+(1+a)D \\
a+(1+a)D & a+D & 1+(1+a)D & 1+(1+a)D \\
a+D & 1+(1+a)D & 1+(1+a)D & (1+a)+aD \\
1+(1+a)D & 1+(1+a)D & (1+a)+aD & 1+aD
\end{bmatrix}
$$

$$
\begin{bmatrix}
(1+a)+(1+a)D+aD^2 & (1+a)D+aD^2 & 1+D^2 & 1+D^2 \\
(1+a)D+aD^2 & 1+D^2 & 1+(1+a)D+(1+a)D^2 & (1+a)+aD+(1+a)D^2 \\
1+D^2 & 1+(1+a)D+(1+a)D^2 & (1+a)+aD+(1+a)D^2 & (1+a)+aD \\
1+(1+a)D+(1+a)D^2 & (1+a)+aD+(1+a)D^2 & (1+a)+aD & D+D^2
\end{bmatrix}
$$

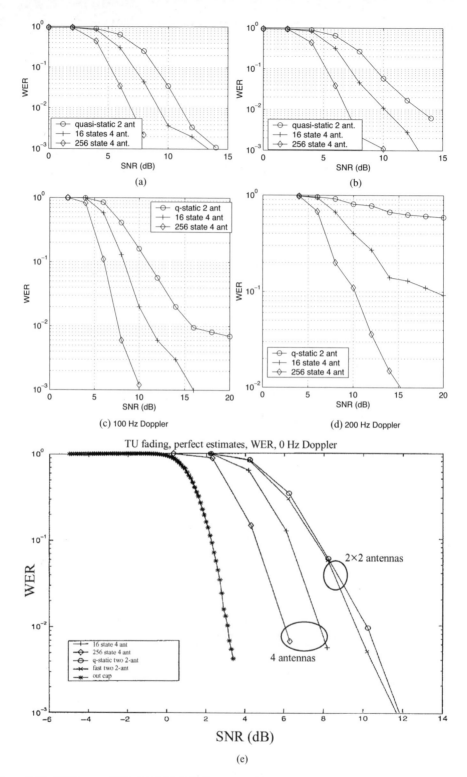

Figure 7.22 WER versus SNR of MIMO–OFDM systems with $n_t = n_r = 4$, TU channel with different Doppler frequencies. Here we compare the best code from the last figure with codes designed for four transmit antenna cases. See Table 7.11 and Figure 7.21 for details on the codes. (a) 5 Hz Doppler; (b) 40 Hz Doppler; (c) 100 Hz Doppler; (d) 200 Hz Doppler; (e) comparisons of WER for best MIMO–OFDM systems with perfect estimates and no Doppler.

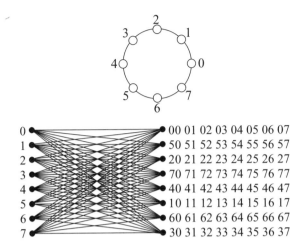

Figure 7.23   Eight-state 8PSK STTC with two transmit antennas and a spectral efficiency of 3 bits/s/Hz.

7.6   Space–Time Coded TDMA/OFDM Reconfiguration Efficiency

In this section we compare two options for combating the fading in digital wireless communications, mainly the TDMA-based concept with equalization versus the OFDM concept. The structure of the code and the equivalent model of the encoder are shown in Figures 7.23 and 7.24 respectively. This structure is considered for the EDGE system. The presentation in this section is based mainly on [35–37].

7.6.1   Frequency Selective Channel Model

The structure of this space–time trellis code (STTC) can be exploited to reduce the complexity of joint equalization and decoding in a frequency selective channel. This is achieved by embedding the space–time encoder in Figure 7.24 in the two channels $\mathbf{h}_1$ and $\mathbf{h}_2$, resulting in an equivalent single-input single-output (SISO) data-dependent channel impulse response (CIR) with memory $(v - 1)$, whose delay

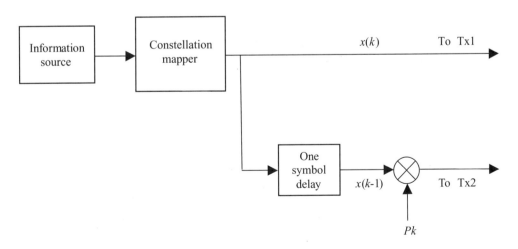

Figure 7.24   STTC with two transmit antennas.

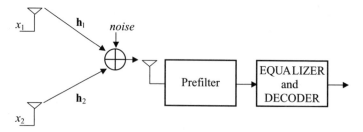

Figure 7.25   Receiver structure for STTC joint equalization/decoding with two transmit and one receive antennas.

D-transform is given by

$$h_{\text{eqv}}^{\text{STTC}}(k, D) = \mathbf{h}_1(0) + \sum_{m=1}^{v} (\mathbf{h}_1(m) + p_k\mathbf{h}_2(m - 1))D^m + p_k\mathbf{h}_2(v))D^{v+1} \tag{7.54}$$
$$= h_1(D) + p_k Dh_2(D)$$

where D is the delay operator and $p_k = \pm 1$ is data dependent. Therefore, trellis-based joint space–time equalization and decoding with 8^{v+1} states can be performed on this equivalent channel. The traditional trellis equalization would require 8^{2v} states, and STTC decoding requires eight states. The receiver block diagram is shown in Figure 7.25.

7.6.2   Front End Prefilter

The objective of the prefilter is to shorten and shape the effective CIR seen by the equalizer to reduce its complexity (since the number of equalizer trellis states is exponential in the CIR memory).

7.6.3   Time-Invariant Channel

First, we describe the prefilter design problem for a time-invariant channel with memory v and then extend it to the data-dependent time-varying channel case. Assume that the FIR prefilter has N_f taps and denote its impulse response by the vector $\mathbf{w}$. Then, the impulse response of the effective channel at the output of the prefilter is given by $\mathbf{h}_{\text{eff}} = \mathbf{Hw}$, where $\mathbf{H}$ is the $(N_f + v) \times N_f$ Toeplitz convolution matrix. Let the vector $\mathbf{h}_{\text{win}}$ contain the $(N_b + 1)$ taps (where $N_b < v$) of $\mathbf{h}_{\text{eff}}$ to retain after shortening (whose energy is to be maximized), and let $\mathbf{h}_{\text{wall}}$ contain the remaining taps (whose energy is to be minimized). Then

$$\mathbf{h}_{\text{win}} = \mathbf{J}_{\text{win}}\mathbf{h}_{\text{eff}} = \underbrace{\mathbf{J}_{\text{win}}\mathbf{H}}_{\mathbf{H}_{\text{win}}}\mathbf{w} = \mathbf{H}_{\text{win}}\mathbf{w} \tag{7.55}$$

where the $(N_b + 1) \times N_f + v)$-dimensional matrix $\mathbf{J}_{\text{win}}$ is constructed using columns of the identity matrix corresponding to tap positions of $\mathbf{h}_{\text{win}}$ within $\mathbf{h}_{\text{eff}}$. And

$$\mathbf{h}_{\text{wall}} = \mathbf{J}_{\text{wall}}\mathbf{h}_{\text{eff}} = \underbrace{\mathbf{J}_{\text{wall}}\mathbf{H}}_{\mathbf{H}_{\text{wall}}}\mathbf{w} = \mathbf{H}_{\text{wall}}\mathbf{w} \tag{7.56}$$

where the $(N_f + v - N_b - 1) \times (N_f + v)$-dimensional matrix $\mathbf{J}_{\text{wall}}$ is constructed from the columns of the identity matrix corresponding to tap positions of $\mathbf{h}_{\text{wall}}$ within $\mathbf{h}_{\text{eff}}$.

The prefilter design criterion maximizes the shortening signal to noise ratio (SSNR), the desired signal energy of the shortened channel contained in $\mathbf{h}_{\text{win}}$ divided by the residual ISI energy in $\mathbf{h}_{\text{wall}}$ plus the noise energy at the prefilter output.

7.6.4 Optimization Problem

The problem reduces to the generalized eigenvector problem (for specific details see [36] and the references therein):

$$\max_{\mathbf{w}} \mathbf{w}^*\mathbf{B}\mathbf{w} \quad \text{subject to } \mathbf{w}^*\mathbf{A}\mathbf{w} = 1 \tag{7.57}$$

where $(\cdot)^*$ denotes the complex conjugate transpose operation, $\mathbf{B} = \mathbf{H}_{\text{win}}^*\mathbf{H}_{\text{win}}$, $\mathbf{A} = \mathbf{H}_{\text{wall}}^*\mathbf{H}_{\text{wall}} + \mathbf{R}_{zz}$, and $\mathbf{H}_{zz}$ is the noise autocorrelation matrix at the prefilter input. The solution has the form

$$\mathbf{w}_{\text{opt}} = (\mathbf{L}_A^*)^{-1}\mathbf{u}_{\text{max}} \tag{7.58}$$

here, $\mathbf{A} = \mathbf{L}_A \, \mathbf{L}_A^*$ is the Cholesky factorization of the matrix $\mathbf{A}$, and $\mathbf{u}_{\text{max}}$ is the unitnorm eigenvector of matrix $(\mathbf{L}_A)^{-1}\mathbf{B}(\mathbf{L}_A)^{-1}$ that corresponds to its largest eigenvalue λ_{max}. The resulting optimal SSNR is given by

$$\text{SSNR}_{\text{opt}} = \frac{\mathbf{w}_{\text{opt}}^*\mathbf{B}\mathbf{w}_{\text{opt}}}{\mathbf{w}_{\text{opt}}^*\mathbf{A}\mathbf{w}_{\text{opt}}} = \lambda_{\text{max}} \tag{7.59}$$

Equation (7.59) provides the optimal prefilter for a time-invariant channel.

7.6.5 Average Channel

For the eight-state 8PSK STTC with two transmit antennas, we can design the prefilter for the *average* of the equivalent channel given in Equation (7.54). It can be shown [36] that the matrices $\mathbf{A}$ and $\mathbf{B}$ in this case are modified to

$$\begin{aligned}
\mathbf{B} &= (\mathbf{H}_{\text{win}}^1)^*\mathbf{H}_{\text{win}}^1 + (\mathbf{H}_{\text{win}}^2)^*\mathbf{H}_{\text{win}}^2 \\
\mathbf{A} &= (\mathbf{H}_{\text{wall}}^1)^*\mathbf{H}_{\text{wall}}^1 + (\mathbf{H}_{\text{wall}}^2)^*\mathbf{H}_{\text{wall}}^2 + \mathbf{R}_{zz}
\end{aligned} \tag{7.60}$$

where $\mathbf{H}_{\text{win}}^i$ and $\mathbf{H}_{\text{wall}}^i (i = 1, 2)$ are matrices corresponding to the constant channels $\mathbf{h}_1$ and $\mathbf{h}_2$ between the two transmit and the single receive antennas. The main attractive feature of the prefilter is that it is a single time-invariant (over a transmission block) FIR filter that shortens both channels $\mathbf{h}_1$ and $\mathbf{h}_2$ simultaneously without excessive noise enhancement.

7.6.6 Prefiltered M-BCJR Equalizer

The algorithm as proposed in [38], is a reduced complexity version of the Bahl–Cocke–Jelinek–Raviv (BCJR) forward–backward algorithm presented in Appendix 2.1, also elaborated in [39], where at each trellis step, only the M active states associated with the highest metrics are retained. An improved version of the M-BCJR algorithm was proposed in [40] based on a log domain implementation of the BCJR algorithm and operates as follows [37].

The forward and backward recursions independently select trees of active nodes without restricting one to be a subtree of the other. To form soft decisions at any time instant, we use all edges with at least one active node.

Let L be the number of trellis steps; $Y_1 Y_2 \ldots Y_L$ the received outputs; s_t the trellis state at time t; S the number of trellis states and u_t the input at time t. The quantity calculated by the algorithm is not $Pr(u_t = u | Y_1 \ldots Y_L)$ as in BCJR, but an approximation, as detailed in what follows.

Using the channel observations and the channel description, calculate for each trellis step t the quantities

$$\gamma_t(i, j) = Pr(s_t = j; \ Y_t \ | \ s_{t-1} = i) \tag{7.61}$$

The Forward Recursion (see Appendix 2.1)
1. For $t = 0$ (initialization) $\alpha_0(0) = 1$, $\alpha_0(i) = 0$ for $i = 1 \ldots S$.
2. For $t = 1, \ldots, L - 1$
   - $\alpha_t(i) = \sum_j \alpha_{t-1}(j)\gamma_t(i, j)$
   - Let A_t denote the M largest αs at time t. Any sorting algorithm can be used to construct A_t. Set $\alpha_t(i) = 0 \ if \ \alpha_t(i) \notin A_t$.

The Backward Recursion
1. For $t = L$ (initialization) $\beta_L(0) = 1$, $\beta_L(i) = 0$ for $i = 1 \ldots S$.
2. For $t = L - 1, \ldots, 1$
   - $\beta_t(i) = \sum_j \beta_{t+1}(j)\gamma_{t+1}(i, j)$
   - Let B_t denote the M largest βs at time t. Any sorting algorithm can be used to construct B_t. Set $\beta_t(i) = 0 \ if \ \beta_t(i) \notin B_t$.

The probabilities of error E_α and E_β (in the sense that the correct state is not included in the M selected states) can be calculated as follows:

$$E_\alpha = Q\left(\sqrt{\frac{|\mathbf{h}(0)|^2 d_{\min}^2}{2\sigma_z^2}}\right); \quad E_\beta = Q\sqrt{\frac{|\mathbf{h}(v)|^2 d_{\min}^2}{2\sigma_z^2}} \tag{7.62}$$

where $Q(\cdot)$ is the standard Q function, $d_{\min}$ the minimum Euclidean distance between any two constellation points and σ_z^2 the noise variance.

7.6.7   Decision

To make a decision at step $0 < t < L$ on the input $u_t = u$, do the following:

1. Set $P(u_t = u) = 0$.
2. For all edges (i, j) that have input u
   - if $\beta_t(j) \neq 0$ and $\alpha_{t-1}(i) \neq 0$, then $P(u_t = u) + = \alpha_{t-1}(i)\gamma(i, j)\beta_t(j)$;
   - if $\beta_t(j) \neq 0$ and $\alpha_{t-1}(i) = 0$, then $P(u_t = u) + = E_\alpha\gamma(i, j)\beta_t(j)$;
   - if $\beta_t(j) = 0$ and $\alpha_{t-1}(i) \neq 0$, then $P(u_t = u) + = \alpha_{t-1}(i)\gamma(i, j)E_\beta$.

The performance of the M-BCJR equalizer/decoder is further improved, especially for small M, by using the prefilter of the previous subsection to concentrate the channel energy in a smaller number of taps. In fact, two different prefilters should be used for the forward and backward recursions since the forward recursion favors a close to minimum phase channel, whereas the backward recursion favors a close to maximum phase channel [40]. The value of M and the number of prefilter taps can be jointly optimized to achieve the best performance complexity tradeoffs.

7.6.8   Prefiltered MLSE/DDFSE Equalizer Complexity

To evaluate reconfiguration efficiency we need to estimate the complexity of the algorithm. For a size 2^b signal constellation, n_i transmit antennas and MIMO channel memory of v, the MIMO MLSE equalizer has $2^{bn_i v}$ states in general. The number of equalizer states can be reduced by using the STTC trellis structure as shown in [41] or by a MIMO channel-shortening prefilter [42]. However, this complexity is still too high for large signal constellations, even for two transmit antennas and short-to-moderate MIMO channel memory. For example, for an 8PSK constellation and the EDGE TU channel (where V = 3), the number of full MLSE equalizer states is equal to $8^6 = 262\,144$ states.

7.6.9   Delayed Decision Feedback Sequence Estimation (DDFSE)

In order to reduce complexity, DDFSE, as discussed in Chapter 6, was introduced in [43]. This is a hybrid scheme between MLSE and decision feedback equalization (DFE) for channels with long memory. Basically, the CIR is divided into a leading part and a tail. Then, an MLSE equalizer is constructed based on the leading part, and the interfering effect of the CIR tail is canceled by feedback using previous (hard) decisions (assumed correct).

At time k, the branch metric $\xi(k)$ of the DDFSE equalizer/decoder is given by

$$\xi(k) = \left| y(k) - \sum_{i=o}^{n} \mathbf{h}_{\text{eqv}}^{\text{STTC}}(i) x(k-i) - \sum_{i=n+1}^{v+1} \mathbf{h}_{\text{eqv}}^{\text{STTC}}(i) \hat{x}(k-i) \right|^2 \tag{7.63}$$

where

    $y(k)$ is the kth received symbol;

    n is the design parameter ($0 \leq n \leq v$) that determines the number of DDFSE trellis states;

    $\mathbf{h}_{\text{eqv}}^{\text{STTC}}$ is the impulse response vector of the equivalent channel;

    $x(k)$ are all possible input symbols according to different transitions along the path history;

    $\hat{x}(k)$ are the previous hard symbol decisions along the path history.

7.6.10   Equalization Schemes for STBC

The focus is on the case of two transmit antennas where full-rate Alamouti-type space–time block codes can be constructed for any signal constellation.

7.6.10.1   Time-Reversal Space–Time Block Coding (TR-STBC)

TR-STBC was introduced in [44] as an extension of the Alamouti STBC scheme to frequency selective channels by imposing the Alamouti orthogonal structure at a *block, not symbol,* level, as in the flat fading channel case. More specifically, the transmitted blocks from antennas one and two at time $(k+1)$ (which were denoted by $\mathbf{x}_1^{(k+1)}$ and $\mathbf{x}_2^{(k+1)}$, respectively) are generated by the encoding rule (for $k = 0, 2, 4, \ldots$)

$$\mathbf{x}_1^{(k+1)} = -\mathbf{J}\bar{\mathbf{x}}_2^{(k)}; \quad \mathbf{x}_2^{(k+1)} = \mathbf{J}\bar{\mathbf{x}}_1^{(k)} \tag{7.64}$$

where $\mathbf{J}$ is the time reversal matrix that consists of ones on the main antidiagonal and zeros everywhere else. To eliminate inter-block interference (IBI) between adjacent blocks due to channel memory, length-v all-zero guard sequences are inserted between information blocks.

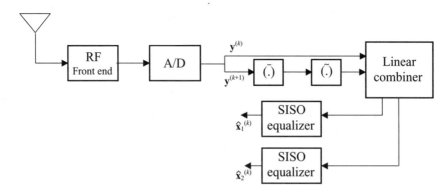

Figure 7.26 TR-STBC receiver block diagram. The operations $(\bar{\cdot})$ and $(\tilde{\cdot})$ denote complex conjugation and time reversal, respectively [37].

The TR-STBC receiver in Figure 7.26 employs linear combining techniques (a spatio-temporal matched filter) to eliminate the mutual interference effects between the two transmit antennas *while still achieving the maximum diversity gain* of $\mathbf{h}_1^2 + \mathbf{h}_2^2$ (where · denotes the norm of a vector).

TR-STBC uses a combination of complex conjugation, time reversal, and matched filtering operations, as described in detail in [29], to convert the two-input single-output channel to two single-input single-output (SISO) channels, each with an equivalent impulse response given by

$$h_{eqv}^{TR-STBC}(D) = h_1(D)\bar{h}_1(\bar{D}^{-1}) + h_2(D)\bar{h}_2(\bar{D}^{-1}) \tag{7.65}$$

to which standard SISO equalization schemes can be applied. In Equation (7.65), $h_i(D)$ is the D-transform of $\mathbf{h}_i(k)$, and $\bar{h}_i(\bar{D}^{-1})$ is the D-transform of $\mathbf{h}_i(-k)$ for $i = 1, 2$. A whitened matched filter (WMF) front end can be used to convert $h_{eqv}^{TR-STBC}(D)$ to its minimum phase equivalent followed by trellis or feedback equalization, as will be further discussed later in the section. TR-STBC assumes that the two channels $h_1(D)$ and $h_2(D)$ are fixed over two consecutive transmission blocks and perfectly known at the receiver. In the next two subsections, we describe two alternative STBC joint equalization/decoding schemes that use frequency domain processing.

7.6.10.2   OFDM-STBC

In this case at the receiver end in Figure 7.27, received blocks are processed in pairs where their FFTs are computed and linearly combined. Finally, gain and phase adjustment is performed using minimum mean square error frequency domain equalization (MMSE-FDE) with a single complex tap for each subchannel, followed by a decision device. While the use of the Alamouti STBC modifies the channel frequency gain at subchannel i from $|H(i)|^2$ to $|H_1(i)|^2 + |H_2(i)|^2$, which provides increased immunity against fading, decision errors occurring at any subchannel result in an irreducible error floor.

OFDM has two main drawbacks, namely, a high peak to average ratio (PAR), which results in larger backoff with non-linear amplifiers, and high sensitivity to frequency errors and phase noise [45]. An alternative equalization scheme that overcomes these two drawbacks of OFDM while retaining its reduced implementation complexity advantage (due to the use of FFT) is the single-carrier (SC) FDE [46], which has been extended to receive diversity systems in [47] and to Alamouti-type STBC transmit diversity systems in [48], and is described next.

These schemes can be extended to more than two transmit antennas using the theory of *orthogonal designs* presented in Chapter 4.

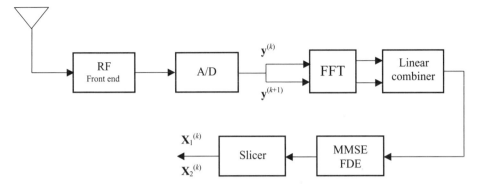

Figure 7.27 OFDM-STBC receiver block diagram.

7.6.11 Single-Carrier Frequency Domain Equalized Space–Time Block Coding SC FDE STBC

The SC FDE, shown in Figure 7.28, is distinct from OFDM in that the IFFT block is moved to the receiver end and placed before the decision device. As noted in [46], this causes the effects of deep nulls in the channel frequency response to be spread out, by the IFFT operation, over all symbols, thus reducing their effect and improving performance.

7.6.11.1 Encoder

Denote the nth symbol of the kth transmitted block (of length N) from antenna i by $\mathbf{x}_i^{(k)}(n)$. Then, the FDE-STBC encoding rule is given by [48]:

$$
\begin{aligned}
\mathbf{x}_1^{(k+1)}(n) &= -\bar{\mathbf{x}}_2^{(k)}((-n)_N) \\
\mathbf{x}_2^{(k+1)}(n) &= -\bar{\mathbf{x}}_1^{(k)}((-n)_N)
\end{aligned}
\tag{7.66}
$$
$$
\text{for } n = 0,\ 1, \ldots, N-1 \text{ and } k = 0,\ 2,\ 4, \ldots
$$

where $(\cdot)_N$ denotes the modulo-N operation that distinguishes this encoding scheme from TR-STBC, Equation (7.64). In addition, a cyclic prefix (CP) is added to each transmitted block to eliminate IBI and make the two channel matrices circulant. Taking the discrete Fourier 66 transform (DFT) of

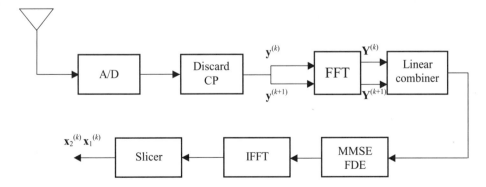

Figure 7.28 FDE-STBC receiver block diagram.

Equation (7.66), we see

$$\mathbf{X}_1^{(k+1)}(m) = -\bar{\mathbf{X}}_2^{(k)}; \quad \mathbf{X}_2^{(k+1)}(m) = -\bar{\mathbf{X}}_1^{(k)}$$
$$\text{for } m = 0, 1, \ldots, N - 1 \text{ and } k = 0, 2, 4, \ldots \tag{7.67}$$

which reveals that this is also a *block-level* implementation of the symbol-level Alamouti encoding rule.

7.6.11.2   Receiver

After analog-to-digital (A/D) conversion, the CP part of each received block is discarded. Mathematically, we can express the input–output relationship over the *jth* received block as

$$\mathbf{y}^{(j)} = \mathbf{H}_1^{(j)}\mathbf{x}_1^{(j)} + \mathbf{H}_2^{(j)}\mathbf{x}_2^{(j)} + \mathbf{z}^{(j)} \tag{7.68}$$

where $\mathbf{H}_1^{(j)}$ and $\mathbf{H}_2^{(j)}$ are $N \times N$ circulant matrices whose first columns are equal to $\mathbf{h}_1^{(j)}$ and $\mathbf{h}_2^{(j)}$, respectively, appended by $(N - v - 1)$ zeros and $\mathbf{z}^{(j)}$ is the noise vector. Since $\mathbf{H}_1^{(j)}$ and $\mathbf{H}_2^{(j)}$ are circulant matrices, they admit the eigendecompositions

$$\mathbf{H}_1^{(j)} = \mathbf{Q}^*\boldsymbol{\Lambda}_1^{(j)}\mathbf{Q}; \quad \mathbf{H}_2^{(j)} = \mathbf{Q}^*\boldsymbol{\Lambda}_2^{(j)}\mathbf{Q} \tag{7.69}$$

where $\mathbf{Q}$ is the orthonormal FFT matrix and $\boldsymbol{\Lambda}_1^{(j)}$ (respectively $\boldsymbol{\Lambda}_2^{(j)}$) is a diagonal matrix whose (n, n) entry is equal to the nth FFT coefficient of $\mathbf{h}_1^{(j)}$(resp. $\mathbf{h}_2^{(j)}$). Therefore, applying the FFT to $\mathbf{y}^{(j)}$, we get (for $j = k, k + 1$)

$$\mathbf{Y}^{(j)} = \mathbf{Q}\mathbf{y}^{(j)} = \boldsymbol{\Lambda}_1^{(j)}\mathbf{X}_1^{(j)} + \boldsymbol{\Lambda}_2^{(j)}\mathbf{X}_2^{(j)} + \mathbf{Z}^{(j)} \tag{7.70}$$

7.6.11.3   Processing

The length-N blocks at the FFT output are then processed in pairs, resulting in the two blocks (we drop the time index from the channel matrices since they are assumed fixed over the two blocks under consideration):

$$\underbrace{\begin{bmatrix} \mathbf{Y}^{(k)} \\ \bar{\mathbf{Y}}^{(k+1)} \end{bmatrix}}_{Y} = \underbrace{\begin{bmatrix} \boldsymbol{\Lambda}_1 & \boldsymbol{\Lambda}_2 \\ \bar{\boldsymbol{\Lambda}}_2 & -\bar{\boldsymbol{\Lambda}}_2 \end{bmatrix}}_{\Lambda} \underbrace{\begin{bmatrix} \mathbf{X}_1^{(k)} \\ \mathbf{X}_2^{(k)} \end{bmatrix}}_{X} + \underbrace{\begin{bmatrix} \mathbf{Z}^{(k)} \\ \bar{\mathbf{Z}}^{(k+1)} \end{bmatrix}}_{Z} \tag{7.71}$$

where $\mathbf{X}_1^{(k)}$ and $\mathbf{X}_2^{(k)}$ are the FFTs of the information blocks $\mathbf{x}_1^{(k)}$ and $\mathbf{x}_2^{(k)}$, respectively, and $\mathbf{Z}$ is the noise vector. To eliminate *inter-antenna interference*, the linear combiner $\boldsymbol{\Lambda}_*$ is applied to $\mathbf{Y}$. Due to the orthogonal Alamouti-like structure of $\boldsymbol{\Lambda}$, a second-order diversity gain is achieved. By *Alamouti-like* we mean any 2×2 complex orthogonal matrix of the form

$$\begin{bmatrix} c_1 & c_2 \\ \pm\bar{c}_2 & \mp\bar{c}_1 \end{bmatrix}$$

Then, the two decoupled blocks at the output of the linear combiner are equalized separately, using the MMSE-FDE [46], which consists of N complex taps per block that mitigate *inter-symbol interference*. Finally, the MMSE-FDE output is transformed back to the time domain using the inverse FFT where decisions are made. Note that the SC MMSE-FDE is equivalent to block MMSE linear equalization [49]; hence, its performance can be improved at the expense of increased complexity by adding a feedback section as discussed in [50].

7.6.11.4   Channel Estimator

Formulate the channel estimation problem for the two-transmit one-receive scenario as in Chapter 4. The analysis can be easily generalized to multiple transmit/receive antennas. Transmit two training sequences s_1 and s_2 from the first and second antennas simultaneously in synchronized data blocks, where each block consists of N information symbols and N_t training symbols. For two transmit antennas, the receiver uses the $2N_t$ known training symbols to estimate the $2(v + 1)$ unknown channel coefficients. The observed training sequence output, which does not have interference from information or preamble symbols, can be expressed as

$$y = \begin{bmatrix} S_1 & S_2 \end{bmatrix} \begin{bmatrix} h_1 \\ h_2 \end{bmatrix} + z = Sh + z \qquad (7.72)$$

where the column vectors y and z are of size $(N_t - v)$, S_1, and S_2, are Toeplitz matrices of size $(N_t - v) \times (v + 1)$ that contain training symbols. The MMSE channel estimate assuming that S has full column rank, is given by [51]:

$$\hat{h} = \begin{bmatrix} \hat{h}_1 \\ \hat{h}_2 \end{bmatrix} = (S^*S)^{-1}S^*y \qquad (7.73)$$

where $(\cdot)^{-1}$ denotes the inverse. The estimation error (mean square error) is given as

$$\text{MSE} = E\lfloor (h - \hat{h})^*(h - \hat{h}) \rfloor = \sigma_z^2 \text{tr}((S^*S)^{-1}) \qquad (7.74)$$

where we assume that white noise with variance σ_z^2 and tr $(\cdot)$ denotes the trace of a matrix. The channel estimation MMSE is equal to

$$\text{MMSE} = \frac{\sigma_z^2 (v + 1)}{(N_t - v)} \qquad (7.75)$$

which is achieved if and only if

$$S^*S = \begin{bmatrix} S_1^*S_1 & S_2^*S_1 \\ S_1^*S_2 & S_2^*S_2 \end{bmatrix} = (N_t - v)I_{v+1} \qquad (7.76)$$

where I_{v+1} is the identity matrix of size $v + 1$. Two optimal training sequences that satisfy Equation (7.76) have an impulse-like autocorrelation sequence and zero cross-correlation. In this case, computing the channel estimates using Equation (7.73) reduces to a simple crosscorrelation (matrix–vector product).

7.6.11.5   Training Sequences

For implementation purposes (to avoid non-linear amplifier distortion), it is desirable to use training sequences with constant amplitude. Optimal constant amplitude training sequences can be constructed from a Pth root-of-unity alphabet $A_p = \{e^{i2\pi k/P} | k = 0, 1, \ldots, P - 1\}$ (where $i = \sqrt{-1}$)without constraining the alphabet size P. Such sequences are the perfect roots-of-unity sequences (PRUS), which are also known as polyphase sequences. Chu [52] showed that for any training sequence length N_t, there exists a PRUS with alphabet size $P = 2N_t$. In [53] and [54] the interested reader can find details on how to design PSK-type training sequences for dual-antenna transmissions with negligible performance loss from PRUS.

7.6.11.6   Performance Results

For the performance results generation signal and channel parameters as in [37] are used. An 8PSK modulation with two transmit and one receive antennas on the TU EDGE channel is used. The overall CIR length is effectively four symbol periods, i.e. $v = 3$.

In EDGE, fading can be safely assumed to be quasi-static, i.e. the CIR can be assumed constant for the duration of a transmission block. This is due to the fact that the coherence time of the channel at around 1 GHz carrier frequency is much larger than the block duration of 577 µs, even for highway speeds. This eliminates the need for channel tracking at the receiver. In addition, assuming ideal frequency hopping, the fading process is independent from block to block. The noise samples are generated as independent samples of a zero mean complex Gaussian random variable with a variance of 1/SNR per complex dimension. The reason for doubling the noise variance (compared with the single transmit antenna case) is that with two-antenna transmissions, we assume that the total transmitted power is the same as in the single antenna case and is divided equally between the two antennas. The average energy of the symbols transmitted from each antenna is normalized to one so that the signal to noise ratio is SNR. The results are shown in Figure 7.29.

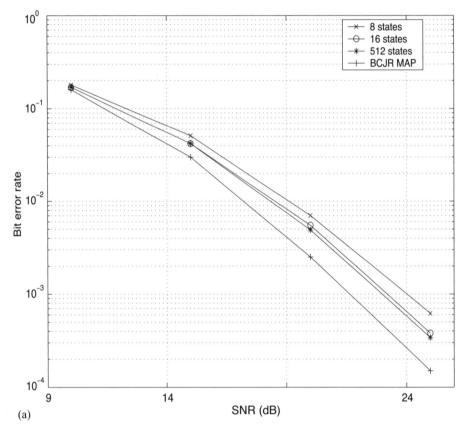

(a)

Figure 7.29   (a) BER performance of two-transmit one-receive eight-state 8PSK STTC with prefiltered M-BCJR equalizer as a function of M (the number of active states). BER of a 4096-state full BCJR-MAP equalizer is shown as a benchmark; (b) BER performance of two-transmit one-receive eight-state 8PSK STTC with prefiltered 64-state DDFSE, prefiltered 16-state M-BCJR, and full BCJR-MAP equalizers; (c) BER performance of two-transmit one-receive eight-state 8PSK STTC with prefiltered 16-state M-BCJR with perfect and estimated CSI. Full BCJR-MAP equalizer performance shown as BER lower bound; (d) BER performance of two-transmit one-receive OFDM-STBC, FDE-STBC, and TR-STBC. For OFDM and FDE-STBC, a size 64 FFT is assumed. For TR-STBC, an ideal whitened matched filter front end and a three-tap feedback filter are assumed; (e) effect of channel estimation on performance of SC FDE-STBC; (f) BER performance of two-transmit one-receive TR-STBC with 512-state full BCJR-MAP, eight-state M-BCJR, and SISO MMSE-DFE with $N_b = 3$ feedback taps. An ideal whitened matched filter front end is assumed.

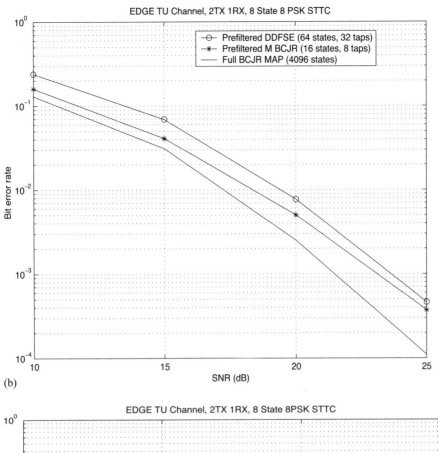

(b)

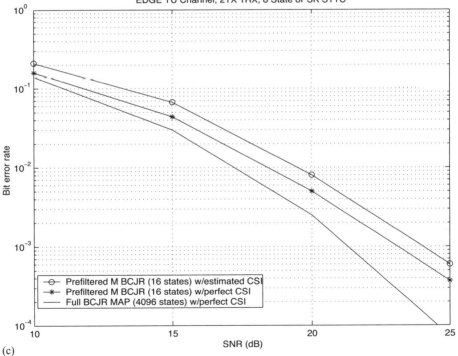

(c)

Figure 7.29 (*Continued*).

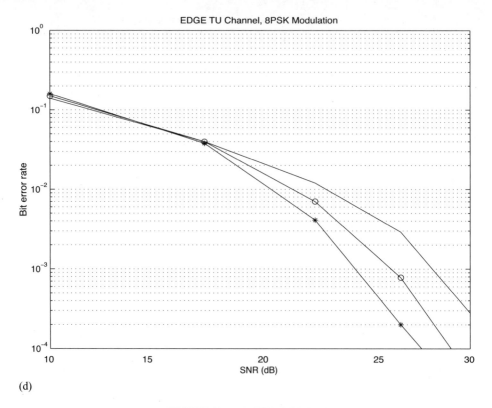

(d)

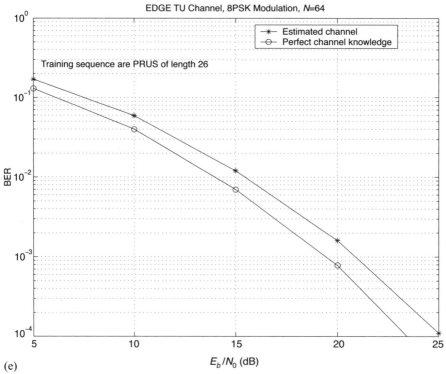

(e)

Figure 7.29 *(Continued).*

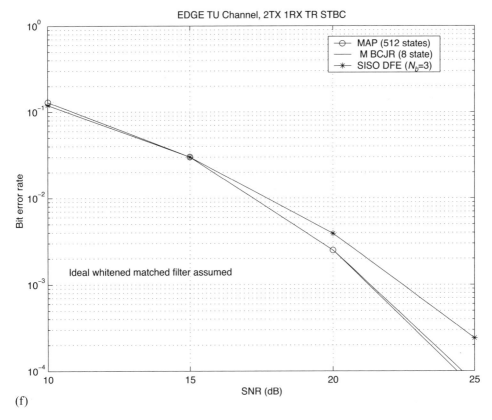

(f)

Figure 7.29 (*Continued*).

For reconfiguration efficiency, relations from Chapter 4 are applicable where values for D and g_{12} can be derived from results presented in Tables 7.12 and 7.13. For two configurations, g_{12} is obtained as a difference of the corresponding entries in column 2 and D_r as a ratio of the corresponding complexity numbers from column 3.

In summary, for STTCs, the prefiltered M-BCJR equalizer/decoder outperforms the prefiltered DDFSE equalizer/decoder and has lower implementation complexity. For space–time block codes, TR-STBC achieves the best performance among the three investigated schemes. OFDM-STBC has the highest PAR and is the most sensitive to frequency errors but is also the most flexible among the three schemes in its support of multirate and multiQOS requirements. The three STBC schemes suffer the same amount of overhead (in the form of an all zero guard sequence for TR-STBC and a cyclic prefix guard sequence for FDE-STBC and OFDM-STBC).

Table 7.12 Performance and complexity comparison summary between the equalization schemes for the eight-state 8PSK STTC over the TU EDGE channel

Equalization scheme	SNR (dB) at BER $= 10^{-3}$	Receiver complexity (per block)
Full BCJR-MAP	21.3	4096 states (each direction)
Prefiltered M-BCJR	23.1	16 states (each direction) 8-tap prefilter (each direction)
Prefiltered DDFSE	23.6	64 states and 32-tap prefilter

Table 7.13 Performance and complexity comparison summary between the Alamouti-type STBC equalization schemes over the TU EDGE channel, assuming 8PSK modulation and block size of 64

Equalization scheme	SNR (dB) at BER $= 10^{-3}$	Receiver complexity (per block)
Full BCJR-MAP	21.3	512 states (each direction) and 20-tap WMF
TR-STBC	22.2	20-tap WMF and 3 feedback taps
FDE-STBC	24.2	Size 64 FFT/IFFT and 64-tap FDE
OFDM-STBC	26.5	Size 64 FFT and 64-tap FDE

7.7 Multicarrier CDMA System

In Chapter 1 we discussed a variety of different structures for MC CDMA at the introductory level. In a number of the following sections we will provide more in-depth discussion on the performance of these systems. In the system shown in Figure 7.30 [55, 56] the MC-CDMA BPSK signal transmitted by the kth user is:

$$s_k(t) = \text{Re}\left\{ \sum_{n=-\infty}^{+\infty} u_k[n]h(t - nT_c - \tau_k) \sum_{l=0}^{L-1} e^{j(\omega_l t + \psi_{k,l})} \right\} \tag{7.77}$$

where

$$u_k[n] = AC_{pk}[n] + BC_{dk}[n]d_k[n] \tag{7.78}$$

A and B are the signal amplitudes of the pilot and the data channel respectively, $d_k[n]$ is binary data, $h(i)$ is the impulse response of the chip wave-shaping filter, ω_l and $\psi_{k,l}$ are the carrier frequency and carrier phase of the lth subcarrier, respectively, and L is the number of subcarriers.

The signal is transmitted through a fading channel. The bandwidth of the subcarriers in this section are selected such that each subcarrier experiences independent, slowly varying, flat Rayleigh fading. Assuming perfect average power control, the received signal is given by

$$r(t) = \text{Re}\left\{ \sum_{k=0}^{K-1} \sum_{n=-\infty}^{+\infty} u_k[n]h(t - nT_c - \tau_k) \sum_{l=0}^{L-1} \alpha_{k,l}e^{j(\omega_l t + \theta_{k,l})} \right\} + n_w(t) \tag{7.79}$$

where K is the total number of users, τ_k is the relative time delay of user k, $\alpha_{k,1}$ and $\theta_{k,1}$ are the fading amplitude and phase, respectively, of the lth path for the kth user, and $n_w(t)$ is zero mean white Gaussian noise with two-sided spectral density $\eta_0/2$.

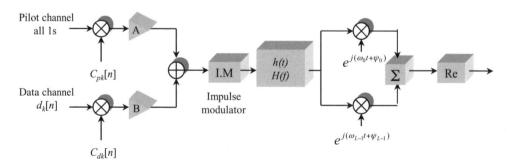

Figure 7.30 The complex transmitter block diagram [55].

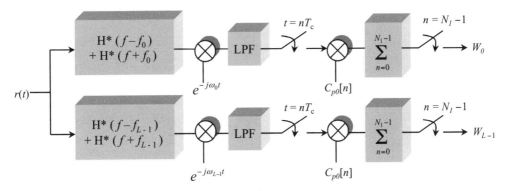

Figure 7.31   The complex channel estimator block diagram [55].

The channel estimator based on a pilot signal is shown in Figure 7.31. Assuming

$$X(f) \equiv |H(f)|^2 = \begin{cases} \dfrac{1}{W}, & -\dfrac{W}{2} < f < \dfrac{W}{2} \\ 0, & \text{otherwise} \end{cases} \tag{7.80}$$

$$W = \frac{1}{T_c}$$

$$F^{-1}|H(f)|^2 \equiv x(t)$$

and

$$\int_{-\infty}^{+\infty} |H(f)|^2 \mathrm{d}f \equiv 1$$

the lth complex channel estimate is given by

$$W_l = \alpha_{0,l} e^{j\theta_{0,l}} A N_l + I_{p,l} + N_{p,l} \tag{7.81}$$

We have used the fact that the spreading sequences $C_{pk}[n]$ and $C_{dk}[n]$ are orthogonal in an estimation interval.

7.7.1   Data Demodulation

Each subcarrier of the received signal is chip-matched filtered, demodulated, despread by the corresponding data spreading sequence, and then integrated over the bit interval of N chips (see Figure 7.32). The lth demodulator output before combining is

$$Y_l = \alpha_{0,l} e^{j\theta_{0,l}} B d[0] N + I_{p,l} + N_{d,l} \tag{7.82}$$

where the interference term is

$$I_{d,l} = \sum_{n=0}^{N-1} C_{d0}[n] R_n \tag{7.83}$$

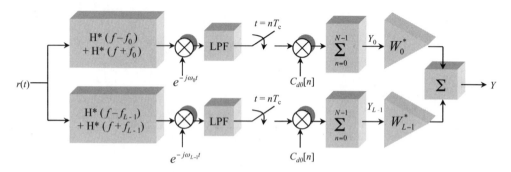

Figure 7.32   The complex data demodulation block diagram.

and the noise term is

$$N_{d,l} = \sum_{n=0}^{N-1} [n_w(t) * h^*(t)]_{t=nT_c} \cdot C_{d0}[n] \tag{7.84}$$

The combined signal is given by

$$Y = \sum_{l=0}^{L-1} W_l^* Y_l \tag{7.85}$$

and the final decision statistic is

$$Z = \text{Re}\{Y\} = \sum_{l=0}^{L-1} \left[\frac{1}{2} W_l^* Y_l + \frac{1}{2} W_l Y_l^* \right] \tag{7.86}$$

Bit error rate analysis for such a system can be found in [55].

7.7.2   Performance Examples

In this example the signal parameters are the same as in [55]. The bandwidth of each subcarrier is fixed to be the coherence bandwidth of the channel. The total bandwidth is proportional to the number of subcarriers. To make the comparison between different bandwidths, the total transmit power is kept constant, that is, decreasing the transmit power per subcarrier as the number of subcarriers increases. Traditionally, assuming perfect channel estimation, the probability of error improves monotonically with the number of subcarriers [56]. However, when there is estimation error, the situation is different.

The probability of error is plotted against the number of subcarriers L in Figure 7.33(a) with ten total users, the estimation interval N_i equaling 64 chips, and E_b/η_0 of 4, 7, 10 and 13 dB. A processing gain of 64 is used, and there is equal power in the pilot and the data channel. As the number of subcarriers increases, the bit error rate first improves and then degrades. The increasing L helps performance by introducing diversity gain. At the same time, as L goes up, the transmit energy-per-band goes down; this causes more estimation error, and in turn, results in performance degradation. Thus, an optimal value of L exists. When we increase the E_b/η_0, the optimal L becomes larger, because the higher signal to noise ratio reduces the degradation due to the estimation error. Some additional results are shown in Figures 7.33 (b) and (c).

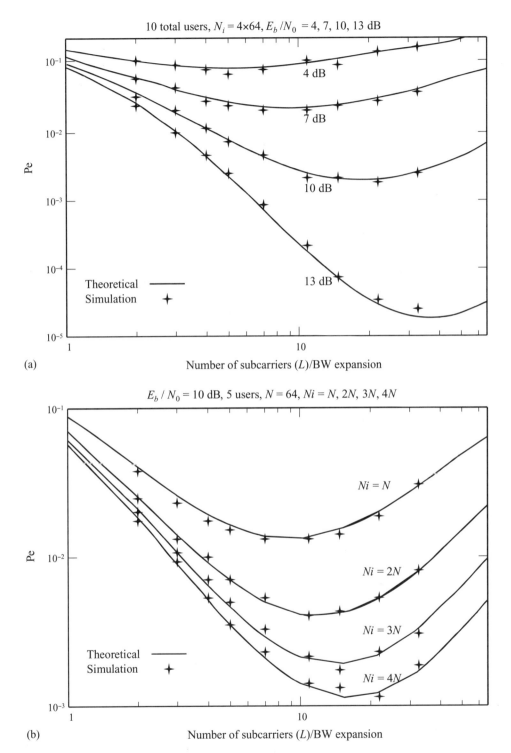

Figure 7.33   (a) Probability of error versus number of subcarriers for varying E_b/η_0; (b) probability of error versus number of subcarriers for different estimation intervals; (c) probability of error versus number of subcarriers for different numbers of users.

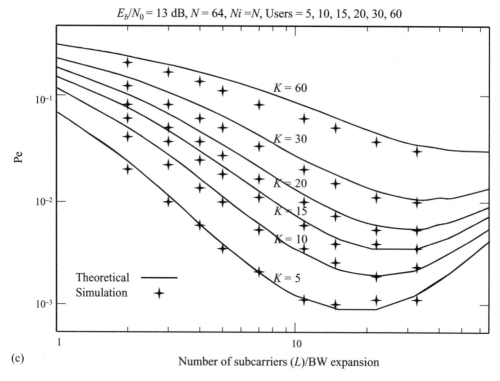

$E_b/N_0 = 13$ dB, $N = 64$, $Ni = N$, Users = 5, 10, 15, 20, 30, 60

(c)

Number of subcarriers (L)/BW expansion

Figure 7.33 (*Continued*).

7.8   Multicarrier DS-CDMA Broadcast Systems

As pointed out in Chapter 1, multicarrier direct sequence code division multiple access (DS-CDMA) systems can be classified into two categories: those with overlapping bandwidths [57, 58] and those with disjoint bandwidths [56]. In this section, the non-overlapping bandwidth system is considered, employed in the forward link (base-to-mobile link) of a cellular system, wherein all user signals are synchronous. Using this type of multicarrier DS-CDMA to generate a wideband CDMA waveform, in particular by choosing the bandwidth of a subcarrier equal to that of a narrowband CDMA waveform, we can achieve some degree of compatibility between wideband (e.g. UMTS/WCDMA) and narrowband (e.g. UMTS/cdma2000) CDMA systems. The spectra of single-carrier and multicarrier CDMA are shown in Figures 7.34(a) and (b), respectively.

A base station transmitter using a multicarrier DS-CDMA system is shown in Figure 7.35. The transmitted signal is given by

$$s(t) = Re \left[\sum_{m=1}^{M} S_m(t) \sqrt{2} e^{j(\omega_m t + \theta_m)} \right] \tag{7.87}$$

$S_m(t)$ is given by

$$S_m(t) = \sum_{k=1}^{K} \sqrt{E_c} \sum_{n=-\infty}^{+\infty} d_v^{(k)} c_n^{(k)} e^{j\phi_{k,m}} h(t - nT_c) \tag{7.88}$$

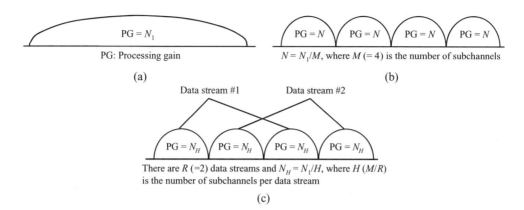

PG: Processing gain

(a)

$N = N_1/M$, where M (= 4) is the number of subchannels

(b)

Data stream #1   Data stream #2

There are R (=2) data streams and $N_H = N_1/H$, where H (M/R) is the number of subchannels per data stream

(c)

Figure 7.34   Spectra of (a) single-carrier CDMA; (b) multicarrier CDMA; and (c) hybrid multicarrier CDMA/FDM.

where $d_v^{(k)}$ and $c_n^{(k)}$ are the data and the spreading sequences of the kth user, respectively, $v = \lfloor n/N \rfloor$ where $\lfloor x \rfloor$ is the largest integer less than or equal to x and N is the number of chips of the spreading sequence per data bit), $h(t)$ is the impulse response of the chip wave-shaping filter, $\phi_{k,m}$ is a carrier phase randomly chosen by a base station for the mth subchannel of the kth user, and $1/T_c$ is the chip rate. Both θ_m and $\phi_{k,m}$ are uniformly distributed over $[0, 2\pi)$. The energy per data bit is defined as $E_b \equiv E_c NM$. Note that the need for $\phi_{k,m}$ to be random for different users and different subchannels is to obtain a proper processing gain in the multicarrier system. The reason for this is as follows. A multicarrier system is known to provide an effective processing gain of MN [59] for an asynchronous communication link, where the carrier phase difference between different users' signals can be assumed to be random, and the phases of any given user are uncorrelated in different subchannels. Therefore, coherent combining of the despread signals from M correlators increases the processing gain by M times that obtained by despreading the signals in each subchannel. However, for the forward link, without the $\phi_{k,m}$, the interference components of the M correlator outputs are also combined coherently, and consequently, there is no increased processing gain achieved by the coherent combining of the M correlator outputs, only a diversity gain. With the random $\phi_{k,m}$, it is possible for the receivers to demodulate the data coherently by using a common pilot signal from a base station if each receiver is provided the values of the $\phi_{k,m}$ by the base station when it establishes a communication link.

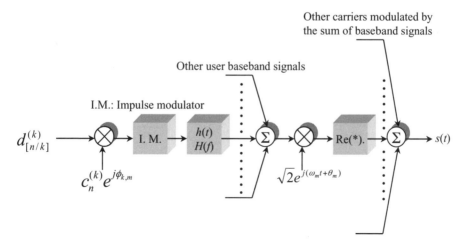

Figure 7.35   Block diagram of a multicarrier CDMA base station transmitter.

It is assumed that $c_n^{(k)} = a_n b_n^{(k)}$, where a_n is a random sequence commonly used by all users and $b_n^{(k)}$ is a member of either an orthogonal or a quasi-orthogonal code set assigned to the kth user. We also have

$$E\{c_n^{(k)} c_{n+i}^{(k')}\} = 0 \quad \text{for } i \neq 0 \tag{7.89}$$

for all k and k'. It is also assumed that the $b_n^{(k)}$ satisfy the following relationship for all k:

$$\frac{1}{K-1} \sum_{\substack{k'=1 \\ k' \neq k}}^{K} \left(\sum_{n=0}^{N-1} b_n^{(k)} b_n^{(k')} \right)^2 = (1-q)N \tag{7.90}$$

where q is a measure of the orthogonality of the set of quasi-orthogonal codes. The quasi-orthogonal codes with these characteristics are standardized [60].

Further, we assume that the channel is a slowly varying frequency selective Rayleigh fading channel and is modeled as a finite length tapped delay line. The complex low-pass equivalent response for the mth subchannel is given by

$$c_m(t) = \sum_{i=0}^{L-1} \alpha_{m,i} e^{j\beta_{m,i}} \delta(t - iT_c) \tag{7.91}$$

where L is the number of resolvable paths, $\alpha_{m,i}$ are independent but not necessarily identically distributed Rayleigh random variables, and the $\beta_{m,i}$ are independently, identically distributed (i.i.d.), uniform random variables over $[0, 2\pi)$. A unit energy constraint is assumed, i.e. $\sum_{i=0}^{L-1} E\{\alpha_{m,i}^2\} = 1$. Then, for a constant multipath intensity profile (MIP), the second moment of each path of a subchannel is given by $E\{\alpha_{m,i}^2\} = 1/L$. For an exponential MIP, the second moments are assumed to be related to the second moment of the initial path strength by $E\{\alpha_{m,i}^2\} = E\{\alpha_{m,0}^2\} \exp(-ri)$, where r is the MIP decay factor. For comparison of the performance of single-carrier and multicarrier systems, we use the facts that $r = Mr_1$ and $L = L_1/M$, where r_1 and L_1 are the MIP decay factor and the number of resolvable paths, respectively, for the single-carrier system [61].

Since all signals that arrive from the base station at a given mobile unit propagate over the same path, they all fade in unison. The received signal at the desired mobile unit is then given by

$$r(t) = \text{Re}[R_c(t)] + n(t) \tag{7.92}$$

where $R_c(t)$ is the complex representation of the received signal given by:

$$R_c(t) = \sum_{m=1}^{M} \sum_{i=0}^{L-1} \alpha_{m,i} S_m(t - iT_c) \sqrt{2} e^{j(\omega_m t + \theta'_{m,i})} \tag{7.93}$$

$n(t)$ is additive white Gaussian noise (AWGN) with a two-sided power spectral density of $\eta_0/2$, and $\theta'_{m,i} = \theta_m + \beta_{m,i}$. The corresponding expression for the single-carrier system is obtained by setting $M = 1$ and L equal to L_1.

The receiver of the desired user ($k = 1$) is shown in Figure 7.36, where a RAKE receiver in each subchannel, and perfect phase recovery of each carrier from the pilot signal detector is assumed [59]. The chip wave-shaping filter given in [59] is assumed, where $X(f) \equiv |H(f)|^2$ is a raised cosine filter. The DS waveforms do not overlap and therefore, adjacent channel interference may be ignored.

Each data stream modulates H ($=M/R$) disjoint carriers, and the number of chips of the spreading sequence per data bit is N_1/H. To make a fair comparison of the performance, given a fixed information rate and total bandwidth, the relationship $HRL = ML = L_1$ must be satisfied. Bit error rate analysis for such a system can be found in [62] and some results are shown in Figure 7.37.

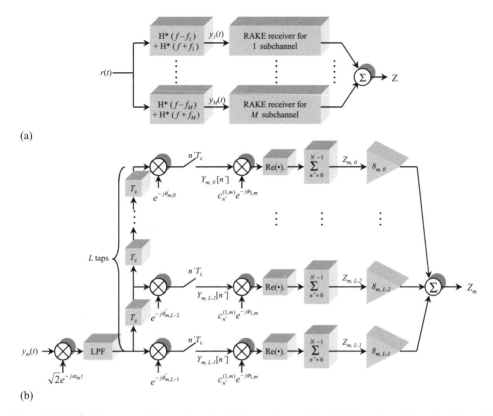

Figure 7.36 Block diagrams of (a) a multicarrier CDMA receiver and (b) a RAKE receiver of the mth subchannel for the first user.

7.9 Frame By Frame Adaptive Rate Coded Multicarrier DS-CDMA System

In this section we discuss an adaptive rate convolutionally coded multicarrier direct sequence code division multiple access (DS-CDMA) system. In order to accommodate a number of coding rates easily and make the encoder and decoder structure simple, the rate compatible punctured convolutional (RCPC) code discussed in Chapter 2 is used. We choose the coding rate that has the highest data throughput in the signal to interference and noise ratio (SINR) sense. To achieve maximum data throughput, a rate adaptive system is used, based on the channel state information (the signal to interference to noise ratio, SINR, estimate). The SINR estimate is obtained by the soft decision Viterbi decoding metric. It will be demonstrated that the rate adaptive convolutionally coded multicarrier DS-CDMA system can enhance spectral efficiency and provide frequency diversity.

7.9.1 Transmitter

The power spectral density of a convolutionally coded orthogonal multicarrier (CC-OM) signal and the transmitter for the rate adaptive CC-OM DS-CDMA system considered in this section are shown in Figure 7.38 and Figure 7.39 respectively. For user k, the (information) bits $\{b_k^i\}$, each with duration T_b, are encoded by the RCPC encoder of rate r. The relationship between T_b and the duration T_s of a coded binary symbol can be written as

$$T_s = rMT_b \tag{7.94}$$

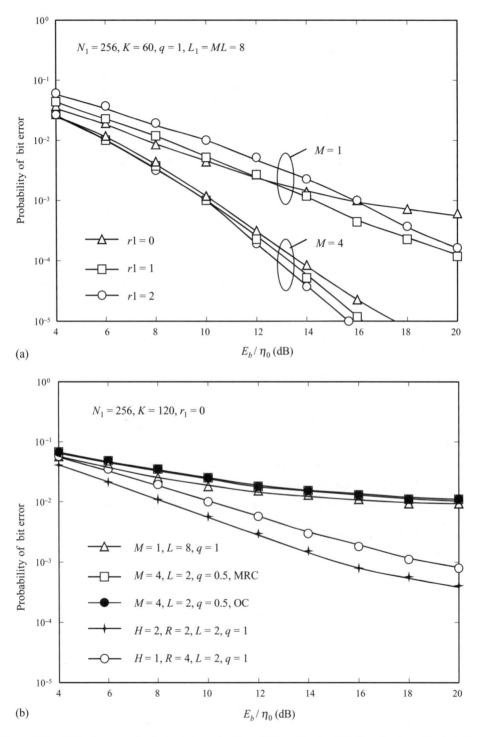

Figure 7.37   (a) Performance of a single-carrier and a Figure 7.37 multicarrier CDMA system in multipath fading channel when both systems employ an orthogonal code set; (b) performance comparison of a single-carrier CDMA, a multicarrier CDMA and a hybrid multicarrier CDMA/FDM system for $K = 120$ in a multipath fading channel with a constant MIP; (c) probability of bit error versus K for $E_b/\eta_0 = 15$ (decibels).

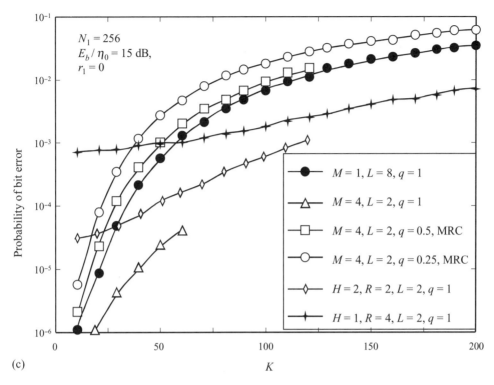

(c)

Figure 7.37   (*Continued*).

where M is the number of subchannels. The M coded binary symbols are allocated to M subchannels to get frequency diversity. They are interleaved to get time diversity as well as frequency diversity, and are spread by each user's pseudonoise (PN) signature waveform $c_k(t)$ with chip duration $T_c = T_s/N$, where N is the processing gain of the DS narrowband waveforms modulated by subcarriers. For CC-OM systems, we have $M = (2B_T/B_S) - 1$, where B_T and B_S are the total and subchannel bandwidths, respectively. Since we fix the subchannel bandwidth B_S (or equivalently, the symbol duration T_s) in this section, T_b varies according to Equation (7.94) when the code rate r changes.

The transmitted signal $s_k(t)$ of user k can be written as

$$s_k(t) = \sqrt{2P} \sum_{j=-\infty}^{\infty} \sum_{m=1}^{M} x_{k,m}^j c_k(t - jT_s) \cos(\omega_m t + \varphi_{k,m}) \qquad (7.95)$$

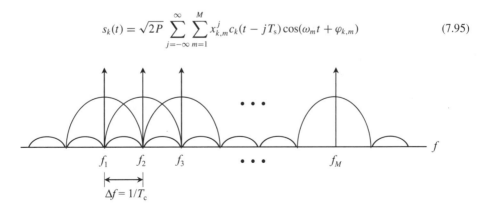

Figure 7.38   A typical CC-OM power spectral density.

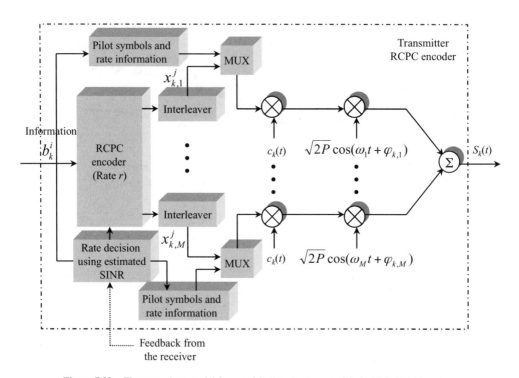

Figure 7.39   The transmitter model for user k in the adaptive rate CC-OM DS-CDMA system.

The channel is assumed to be frequency selective Rayleigh fading and not to vary during one symbol duration. However, the subchannels are assumed to be non-selective by choosing the number of subcarriers appropriately as [56]:

$$MT_c \geq T \tag{7.96}$$

where T is the maximum delay spread of the channel. Then the complex low-pass impulse response of the subchannels of user k can be modeled as

$$h_{k,m}(t) = \alpha_{k,m} e^{j\beta_{k,m}} \delta(t) \tag{7.97}$$

where $\alpha_{k,m}$ is the fading amplitude and $\beta_{k,m}$ is the random phase of the mth subchannel, $m = 1, 2, \ldots$, M. The phases $\{\beta_{k,m}\}$ are i.i.d. uniform random variables on $[0, 2\pi)$. In general, the fading amplitudes $\{\alpha_{k,m}\}$ are correlated, but we can assume that they are i.i.d. Rayleigh random variables once the coded symbols are properly interleaved in the time domain.

Frame by frame transmission is assumed. Such a frame by frame transmission is typical of many cellular systems. The frame discussed in this section is shown in Figure 7.40. Each frame of duration T_f consists of a header of duration T_p and data symbols of duration T_d. We have $T_d = N_s T_s$ and $T_p = N_p T_s$, where N_s is the number of data symbols and N_p is normally 6–10. The header contains pilot symbols and information on the rate and channel state. The function of the MUX in Figure 7.39 is to combine the header and data symbols to make frames as shown in Figure 7.40.

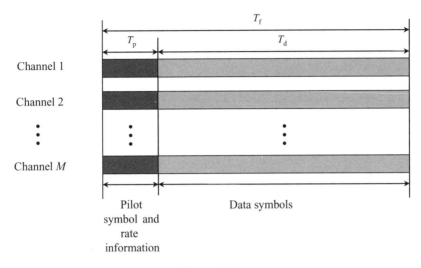

Figure 7.40   The frame structure.

7.9.2   *Receiver*

The receiver for the adaptive rate CC-OM DS-CDMA system in this section is shown in Figure 7.41. Let us assume that there are K users *in a cell* and power control is employed. Then the received signal *at the base station* can be written as

$$r(t) = \sqrt{2P} \sum_{j=-\infty}^{\infty} \sum_{k=1}^{K} \sum_{m=1}^{M} \alpha_{k,m} x_{k,m}^{j} c_k(t - \tau_k - jT_s) \cos(\omega_m t + \phi_{k,m}) + n(t) \qquad (7.98)$$

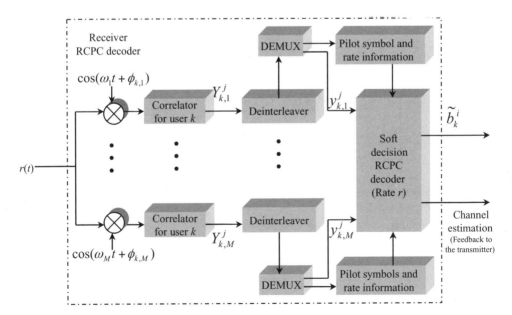

Figure 7.41   The receiver model for user k in the adaptive rate CC-OM DS-CDMA system.

7.9.3   Rate-Compatible Punctured Convolutional (RCPC) Codes

RCPC codes are discussed in Chapter 2. For this section, some additional details are specified. Let the code rate and constraint length of the parent code be $R = 1/n$ and L_c, respectively. The parent code is completely specified by the n generator polynomials $G^j(D) = g_0^j + g_1^j D + \cdots + g_{L_c-1}^j D^{L_c-1}$, $j = 1, 2, \ldots, n$, where $g_i^j \in \{0, 1\}$. The puncturing is done according to the rate compatibility criterion, which requires that lower rate codes use the same coded bits as the higher rate codes plus one or more additional bit(s). The bits to be punctured are described by an $n \times p$ puncturing matrix **P** consisting of zeros and ones, where p is called the puncturing period. At time instant t, the output from each generator $G^j(D)$ is transmitted if $P(j, t \bmod p) = 1$ and punctured otherwise. Here, $\mathbf{P}(a, b)$ denotes the element on row a and column b in the matrix **P**. The number p of columns determines the number of code rates and the rate resolution that can be obtained. Generally, from a parent code of rate $1/n$, we can obtain a family of $(n - l)p$ different codes with rates

$$r = \frac{p}{np}, \frac{p}{np-1}, \ldots, \frac{p}{p+1} \tag{7.99}$$

The code rate of RCPC codes can be changed during even one information bit transmission and, thus, unequal error protection can be obtained [8]. In this section, however, the code rate of RCPC codes is changed frame by frame, not bit by bit, because we assumed frame by frame transmission. An example of the RCPC encoder is shown in Figure 7.42.

7.9.4   Rate Adaptation

A threshold-based adaptation scheme is used which adaptively changes the coding rate depending upon the SINR estimated. Let $\theta_0 = -\infty$, $\theta_1, \theta_2, \ldots, \theta_Q = \infty$ be the SINR threshold values, which are chosen such that between θ_{j-1} and θ_j the channel coding rate r_j has the highest throughput. Here, Q is the number of possible code rates. Then, as discussed in Chapter 3, the transmitter mode (rate) adaptation

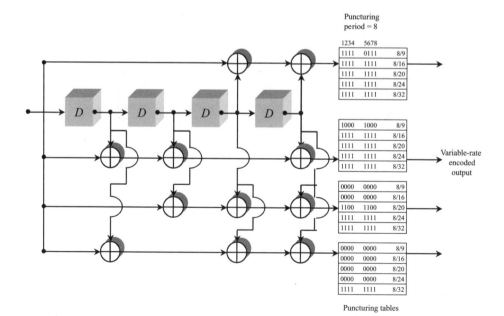

Figure 7.42   A five-rate RCPC encoder from a rate 1/4 parent code [63] (*D*: delay of duration T_b).

scheme can be defined as follows:

$$\text{Choose } r_j \text{ if } \theta_{j-1} \leq \text{SINR} < \theta_j, \quad j = 1, \ldots, Q \tag{7.100}$$

In this method, bit by bit adaptation is not assumed due to the feedback delay. Instead, we choose the adaptation interval T_a in such a way that T_a is long enough to allow the transmission of at least one frame and short enough to react quickly to the possible change of the SINR. The transmitter can then adapt its data rate every $[T_a/T_f]T_f$. This allows efficient error recovery through ARQ mechanisms even with dynamic rate adaptation. The rate at which the transmitter reacts to the changes in the SINR depends on the SINR estimate and feedback delay in the system [64].

7.9.4.1 Example

RCPC codes with rate 1/4 convolutional codes of constraint length $L_c = 5$ and 9 are used as the parent codes [64, 65], as the error control capability varies when the constraint length changes, we use two values of L_c. The number, M, of subcarriers is four and nine. The processing gain N is 192 and 96, for $M = 4$ and $M = 9$, respectively, when the total bandwidth

$$B_T = N(M + 1)/T_s \tag{7.101}$$

is fixed. We assume each frame contains 144 symbols with $T_f = 10$ ms and $T_p = 0$ (that is, we assume perfect feedback to simplify the simulations).

The adaptive rate CC-OM DS-CDMA system was implemented based on the above discussion in [66]. When $L_c = 5$, fixing the coding rate to 1/2 allows us to get the highest throughput and the result is Figure 7.43(a). When $L_c = 9$, using the thresholds $\theta_1 = 2.5$ dB and $\theta_2 = 5.5$ dB, we get Figure 7.43(b). In these figures, the throughput of the conventional system (fixed rate with 1/M) is also shown. It is clear that we can get much higher throughput with the adaptive system.

7.10   Intermodulation Interference Suppression in Multicarrier CDMA Systems

In this section, a coded multicarrier direct sequence code division multiple access (DS-CDMA) system is presented that, by the use of a minimum mean squared error receiver, achieves frequency diversity (instead of path diversity as in a conventional single carrier (SC) RAKE DS-CDMA). It also has the ability to suppress the intermodulation distortion and partially compensate for the signal distortion introduced by a non-linear amplifier at the transmitter. A frequency selective Rayleigh fading channel is decomposed into M frequency non-selective channels, based on the channel coherence bandwidth. A rate 1/M convolutional code, after being interleaved, is used to modulate M different DS-CDMA waveforms. The system is shown to effectively combat intermodulation distortion in the presence of multiple access interference.

7.10.1   Transmitter

In this section, the transmitter shown in Figure 7.44 is considered. The input signal to the power amplifier for the kth user $s_k(t)$ is given by

$$s_k(t) = A_k \sum_{i=-\infty}^{\infty} \sum_{m=1}^{M} d_{k,m}^{(i)} p_{k,m}(t - iY - \tau_k) \cos(\omega_m t + \theta_{k,m}) \tag{7.102}$$

and $p_k(t)$ is a spreading (or signature) waveform given by

$$p_{k,m}(t) = \sum_{n=0}^{N-1} c_{k,m}^{(n)} h(t - nMT_c) \tag{7.103}$$

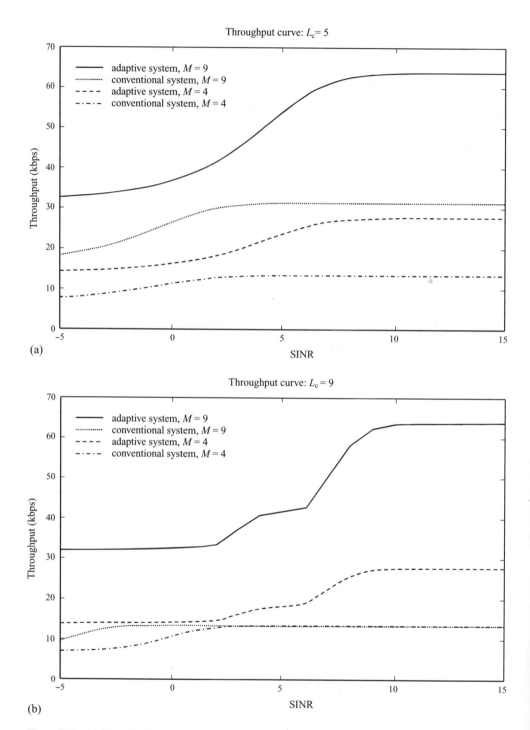

Figure 7.43 (a) The adaptive throughput curves and the conventional throughput curves when $L_c = 5$; (b) the adaptive throughput curves and the conventional throughput curves when $L_c = 9$.

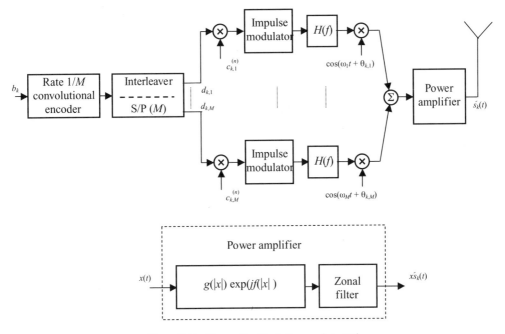

Figure 7.44   Transmitter block diagram for user k.

In Equation (7.103), $c_{k,m}^{(n)} \in \{\pm 1\}$ is the nth chip of the spreading sequence, N is the processing gain, which is taken to be equal to the period of the spreading sequence, $h(t)$ is the impulse response of the chip wave-shaping filter, $1/MT_c$ is the chip rate of the band-limited MC DS-CDMA system, and $1/T_c$ is the chip rate of a band-limited single carrier (SC) DS-CDMA system that occupies the same spread bandwidth as does the MC system. That is, $T = NMT_c$.

7.10.2   Non-Linear Power Amplifier Model

For an input signal formed by the sum of M subcarrier signals, i.e.

$$x(t) = Re \left\{ \sum_{m=1}^{M} a_m(t) \exp[j\omega_0 t + j\psi_m(t)] \right\}$$ (7.104)

$$= Re \{A(t) \exp[j\omega_0 t + j\Psi(t)]\}$$

the output signal of a non-linear power amplifier can be represented by

$$\hat{x}(t) = Re \{g[A(t)] \exp[j\omega_0 t + j\Psi(t) + jf[A(t)]]\}$$ (7.105)

where $g(A)$ and $f(A)$ represent the AM/AM (see Figure 7.45) and AM/PM conversion characteristics of the non-linear power amplifier.

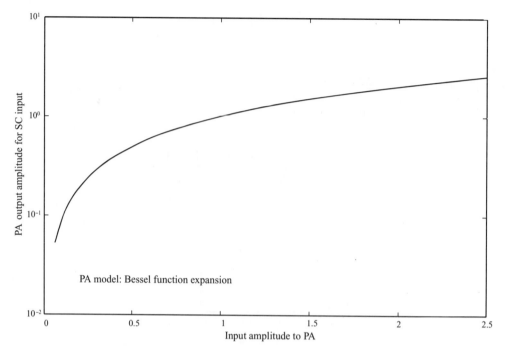

Figure 7.45   Power amplifier transfer function for single carrier.

7.10.3   MMSE Receiver

As shown in Figure 7.46, M MMSE filters are combined with a soft-decision Viterbi decoder so that the soft outputs from the M MMSE filters are parallel-to-serial converted, deinterleaved, and decoded. The received DS-CDMA signal after the LPF for each subcarrier is despread (either partially or fully) over consecutive F chips, which is characterized by a parameter $N_t = \lceil N/F \rceil$. The decision symbols needed by the M MMSE filters can be obtained via an interleaver with a serial-to-parallel converter and an encoder which is identical to that used in the transmitter. The tap weight vector of the mth MMSE filter ω_m is chosen so as to minimize the conditional mean square error, conditioned on all parameters of the desired user (user 1) and certain parameters of the MAI and IM, i.e. with $q = \{\alpha_1, \breve{\theta}_1, \rho^{\pm}_{k,I(m,h)}\}$,

$$\text{MSE} = E\{(\omega'_m z_m - d_{1,m})^2 \mid \mathbf{q}\} \tag{7.106}$$

For the notation see Figure 7.46. Note that we omit the superscript i, which denotes the estimated bit, for notational simplicity. From Chapter 5, the optimum tap weight vector for Equation (7.106) is given by:

$$(\omega_m)_{\text{opt}} = \mathbf{R}_m^{-1} \mathbf{a}_m \tag{7.107}$$

where

$$\mathbf{R}_m = E\{z_m z'_m \mid \mathbf{q}\} \tag{7.108}$$

and

$$\mathbf{a}_m = E\{d_{1,m} z_m \mid \mathbf{q}\} \tag{7.109}$$

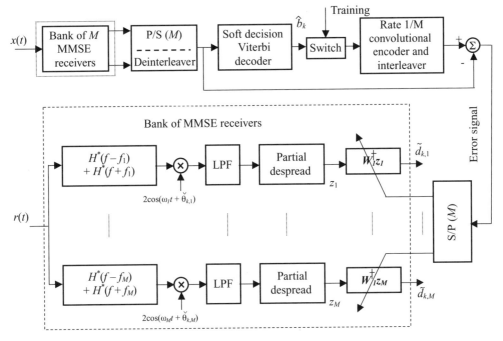

Figure 7.46   Receiver block diagram for user k.

7.10.3.1   Example

For the numerical example, we use the same parameters as in [67], the PA model given in the previous section and $M = 4$, N $= 32$, $R = 1/4$ and constraint length $K = 3$ coded MC-CDMA system and an SC coded RAKE system with the same coding scheme and the same bandwidth as the coded MC-CDMA system. Note that there are not any IM terms in the SC system, but the non-linear distortion introduced by the PA is taken into account. The same PA output power is assumed for both the SC input signal and the MC input signal. The system performance is obtained by simulation. In Figures 7.47–7.48, the maximal number of resolvable paths for the SC is denoted L_p, and L_t is the actual number of RAKE taps.

7.11   Successive Interference Cancellation in Multicarrier DS-CDMA Systems

This section presents a successive interference cancellation (SIC) scheme for a multicarrier (MC) asynchronous DS-CDMA system, wherein the output of a convolutional encoder modulates band-limited spreading waveforms at different subcarrier frequencies. In every subband, the SIC receiver successively detects the interferers signals and subtracts them from that of the user of interest. The SIC receiver employs maximal ratio combining (SIC-MRC) for detection of the desired user, and feeds a soft decision Viterbi decoder.

7.11.1   System and Channel Model

A K-user asynchronous communication system is presented. We assume knowledge of the time delays and the spreading sequences of all the users, but no knowledge of the channel gains of the interferers. The user data symbols are input to a rate $1/M$ convolutional encoder. The output code symbols are

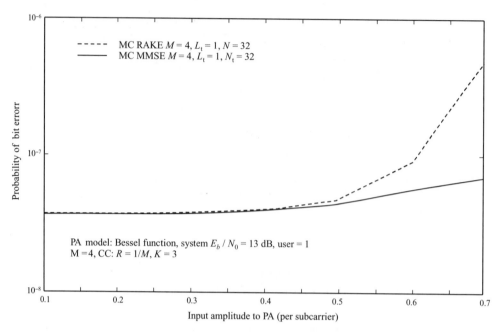

Figure 7.47 Probability of bit error in the presence of IMD (different PN (DPN) code for each subcarrier).

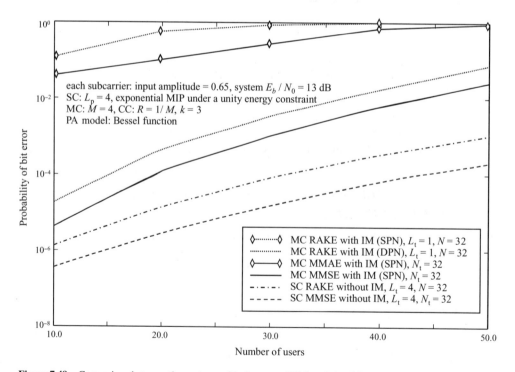

Figure 7.48 Comparison between the systems with the same (SPN) and the different (DPN) PN codes for each subcarrier.

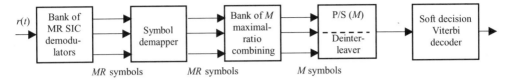

Figure 7.49 Receiver block diagram for the desired user 1.

interleaved and serial-to-parallel (S/P) converted such that M parallel code symbols may be transmitted simultaneously. Then, each of the M code symbols is replicated by a rate $1/R$ repetition code, and transmitted over MR subcarriers.

For example, if $M = 4$ and $R = 2$, then the $M = 4$ code symbols are mapped to a total of eight subcarriers in the following way: the first code symbol is transmitted on the first and the fifth subcarriers, the second code symbol is transmitted on the second and the sixth subcarriers, the third code symbol is transmitted on the third and the seventh subcarriers, and the fourth code symbol is transmitted on the fourth and the eighth subcarriers. Therefore, the minimum subcarrier distance for the same code symbol is maximized. The mapped code symbols are then multiplied by the spreading sequence assigned to the given user. The transmitted signal for the kth user is:

$$S_k(t) = \sqrt{2E_{ck}} \left\{ \sum_{j=-\infty}^{+\infty} a_k(t - jT - \tau_k) \times \sum_{m=1}^{MR} b^j_{k,[m]_M} \cos(\omega_m t + \theta_{k,m}) \right\} \qquad (7.110)$$

where

$$a_k(t) = \sum_{n=0}^{N-1} c_k^{(n)} h(t - nT_c)$$

We assume that the channel in each subband is a slow-varying frequency non-selective Rayleigh channel with transfer function

$$\zeta_{k,m} = \alpha_{k,m} \exp(j\beta_{k,m})$$

The received signal is given by

$$r(t) = \sum_{k=1}^{K} \sqrt{2E_{ck}} \sum_{j=-\infty}^{+\infty} a_k(t - jT - \tau_k) \times \sum_{m=1}^{MR} b^j_{k,[m]_M} \alpha_{k,m} \cos(\omega_m t + \theta'_{k,m}) + n_w(t) \qquad (7.111)$$

The receiver structure for the desired user, user 1, is shown in Figures 7.49–7.50. Interferers $2, 3, \ldots, K$ are renumbered as $1_m, 2_m, \ldots, (K-1)_m$, which defines the cancellation order such that in the m-th subband, interferer 1_m is the strongest, 2_m is the second strongest, and so on. The cancellation order in each subband is different, i.e. for every subband there is a distinct successive interference cancellation order.

7.11.1.1 Example

In the following numerical example the signal parameters are the same as in [68]. Gold sequences are used with $N = 31$ when convolutional coding is employed, and the constraint length of the convolutional code is three. Ideal power control is assumed, i.e. $E_{ck} \triangleq E_c$ for every user k. Also, $X(f)$ is a raised cosine function with a rolloff factor $\alpha = 0.5$, and all of the comparisons are based on the same transmitted data rate. The

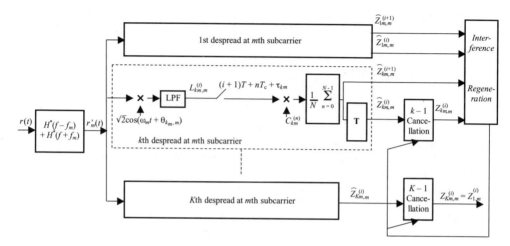

Figure 7.50 SIC demodulator at mth subcarrier for the desired user 1.

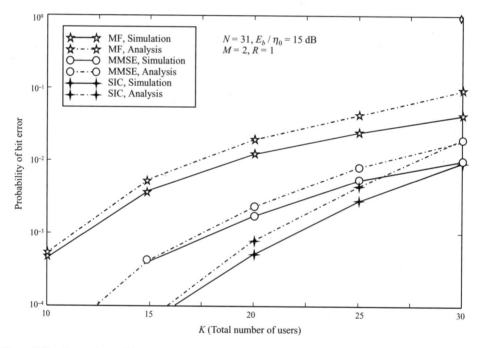

Figure 7.51 Comparisons of the analytical bounds with simulation results for the MF and SIC receivers in convolutionally coded MC CDMA, where perfect CSI is assumed.

analytical results presented in [68] are averaged over 1000 realizations, and the cancellation order in the simulation of SIC on the mth subband is based upon the decreasing order of $|Z_{k_m,m}(i)|$, i.e. $|\hat{Z}_{1_m,m}(i)| \geq \cdots \geq |\hat{Z}_{(K-1)_m,m}(i)|$. To simulate the time-correlated Rayleigh fading channel, the Jakes model (see Chapter 14) is used with a bit rate of 20 000 bits/s and a maximum Doppler frequency of 100 Hz, while a block interleaver/deinterleaver is employed to separate adjacent data bits into a block size of 30×30, which results in a delay of 45 ms. Some results are given in Figure 7.51.

7.12   MMSE Detection of Multicarrier CDMA

The results from the previous section have already demonstrated that an MMSE detector performs better than a receiver with SIC. For this reason, in this section the minimum mean squared error (MMSE) detection of multicarrier code division multiple access (CDMA) signals is presented in more detail. The performance of two different design strategies for MMSE detection is compared. In one case, the MMSE filters are designed separately for each carrier, while in the other case the optimization of the filters is done jointly. Naturally, the joint optimization produces a better receiver, but the difference in performance is shown to be substantial. The multicarrier CDMA performance is then compared to that of a single-carrier CDMA system on a frequency selective fading channel. A mechanism to track the channel fading parameters for all the users' signals is presented which enables joint optimization of the receiver filters in a time-varying channel.

Simulation results show that the performance of this receiver is close to the ideal theoretical results for moderate vehicle speeds. Performance begins to degrade when the normalized Doppler rate is higher than about 1%.

The received signal on the system's reverse link on the mth carrier is given by

$$r_m(t) = \mathrm{Re}\left\{\sum_{i=-\infty}^{+\infty}\sum_{k=1}^{K}\sqrt{\frac{2P_k}{M}}\gamma_{k,m}(i)d_k(i) \times c_{k,m}(t - iT_b - \tau_k)\exp(j\omega_m t)\right\} + n_m(t) \qquad (7.112)$$

The received signal is processed with a chip-matched filter, which consists of an integrator with duration MT_c. The samples are stored for one bit interval, giving a column vector of length N/M:

$$r_m(i) = \sum_{k=1}^{K}\sqrt{\frac{P_k}{P_1}}\gamma_{k,m}(i)[d_k(i)\mathbf{f}_{k,m} + d_k(i-1)\mathbf{g}_{k,m}] + \mathbf{n}_m(i) \qquad (7.113)$$

where $\mathbf{f}_{k,m}$ and $\mathbf{g}_{k,m}$ depend on the left- and right-cyclic shifts of $\mathbf{c}_{k,m}$, the spreading code of the kth user on the mth carrier.

A block diagram of a general linear receiver is shown in Figure 7.52. Each of the M received vectors is processed with a receiver filter $\mathbf{w}_m(i)$ to form a statistic $Z_m(i) = \mathbf{w}_m^{\mathrm{H}}(i)\mathbf{r}_m(i)$, for $m = 1, 2, \ldots, M$. Note the time dependence of the filters in the time-varying fading channel. The individual statistics are summed to form an overall decision statistic $Z(i) = \sum_{m=1}^{M} Z_m(i)$. Equivalently, we can define an overall receiver filter as

$$\mathbf{w}(i) = \left[\mathbf{w}_1^{\mathrm{T}}(i), \mathbf{w}_2^{\mathrm{T}}(i), \ldots, \mathbf{w}_M^{\mathrm{T}}(i)\right]^{\mathrm{T}} \qquad (7.114)$$

and an overall received vector as

$$\mathbf{r}(i) = \left[\mathbf{r}_1^{\mathrm{T}}(i), \mathbf{r}_2^{\mathrm{T}}(i), \ldots, \mathbf{r}_M^{\mathrm{T}}(i)\right]^{\mathrm{T}} \qquad (7.115)$$

which gives $Z(i) = \mathbf{w}^{\mathrm{H}}(i)\mathbf{r}(i)$.

We next consider two different design strategies for performing MMSE detection. The best performance is obtained when the filters $\mathbf{w}_1(i), \mathbf{w}_2(i), \cdots, \mathbf{w}_M(i)$ are designed jointly so as to minimize the composite mean squared error

$$J = E[|d_1(i) - \mathbf{w}^{\mathrm{H}}(i)\,\mathbf{r}\,(i)|^2] \qquad (7.116)$$

This gives the well-known Wiener solution $\mathbf{w}(i) = \mathbf{R}^{-1}(i)\,\mathbf{p}\,(i)$, with

$$\mathbf{R}(i) = E\left\lfloor\mathbf{r}(i)\mathbf{r}^{\mathrm{H}}(i)\right\rfloor \quad \text{and} \quad \mathbf{p}(i) = E\left\lfloor d_1^*(i)\,\mathbf{r}\,(i)\right\rfloor \qquad (7.117)$$

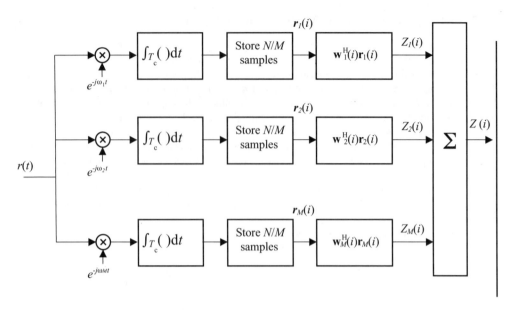

Figure 7.52   General receiver for multicarrier CDMA.

representing the correlation matrix and steering vector, respectively. These can be further decomposed as

$$\mathbf{R}(i) = \begin{bmatrix} \mathbf{R}_{1,1}(i) & \mathbf{R}_{1,2}(i) & \cdots & \mathbf{R}_{1,M}(i) \\ \mathbf{R}_{2,1}(i) & \mathbf{R}_{2,2}(i) & \cdots & \mathbf{R}_{2,M}(i) \\ \vdots & \vdots & \ddots & \vdots \\ \mathbf{R}_{M,1}(i) & \mathbf{R}_{M,2}(i) & \cdots & \mathbf{R}_{M,M}(i) \end{bmatrix} \tag{7.118}$$

where the individual submatrices are defined as

$$\mathbf{R}_{m,n}(i) = E\left[\mathbf{r}_m(i)\mathbf{r}_n^{H}(i)\right] \tag{7.119}$$

and

$$\mathbf{p}(i) = \left[\mathbf{p}_1^{T}(i), \mathbf{p}_2^{T}(i), \ldots, \mathbf{p}_M^{T}(i)\right]^{T} \tag{7.120}$$

with

$$\mathbf{p}_m(i) = E\left[d_1^{*}(i)\mathbf{r}_m(i)\right] \tag{7.121}$$

An alternative suboptimal approach is to design the M filters separately by choosing each of the filters $\mathbf{w}_1(i), \mathbf{w}_2(i), \ldots, \mathbf{w}_M(i)$ to minimize the individual mean squared error quantities $J = E[|d_1(i) - \mathbf{w}_m^{H}(i)\mathbf{r}_m(i)|^2]$, which leads to $\mathbf{w}_m(i) = \mathbf{R}_{m,m}^{-1}(i)\mathbf{p}_m(i)$. Together with the previous notation, an overall

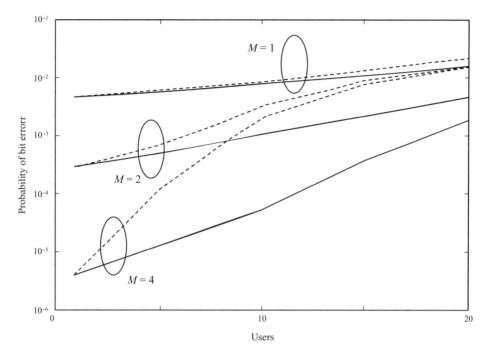

Figure 7.53 Probability of error versus number of users for the multicarrier CDMA system, with $E_b/N_0 = 17$ dB, composite processing gain of 32 chips/bit and M carriers. The solid lines are the Wiener solutions when all users' fading processes are tracked, the dashed lines are the Wiener solutions when only the desired user's fading processes are tracked.

filter using this design strategy can then be written as $\mathbf{w}(i) = \mathbf{R}_B^{-1}(i)\mathbf{p}(i)$ where

$$\mathbf{R}_B(i) = \begin{bmatrix} \mathbf{R}_{1,1}(i) & 0 & \cdots & 0 \\ 0 & \mathbf{R}_{2,2}(i) & \cdots & 0 \\ \vdots & \vdots & \ddots & \vdots \\ 0 & 0 & \cdots & \mathbf{R}_{M,M}(i) \end{bmatrix} \tag{7.122}$$

An example of performance results is given in Figure 7.53. One can see that the joint detector demonstrates better performance.

Before proceeding to present a tracking algorithm, it is worthwhile to compare the performance of a multicarrier CDMA system to that of a single-carrier system which realizes diversity inherently by operating on a frequency selective fading channel. In order to get a fair comparison to the multicarrier case, the system will be assumed to employ spreading waveforms of length N chips/bit, where N in the multicarrier case is the composite processing gain. The received signal will consist of M resolvable components, delayed with respect to one another by a sufficient number of chips. Results for this case are shown for comparison to the multicarrier case in Figure 7.54.

7.12.1   Tracking the Fading Processes

The presentation in this section is based on [69–71]. The received vector on the mth carrier, from Equation (7.113), may be written as

$$\mathbf{r}_m(i) = (\mathbf{F}_m\mathbf{D}(i) + \mathbf{G}_m\mathbf{D}(i-1))\mathbf{P}\mathbf{\Gamma}_m(i) + \mathbf{n}_m(i) \tag{7.123}$$

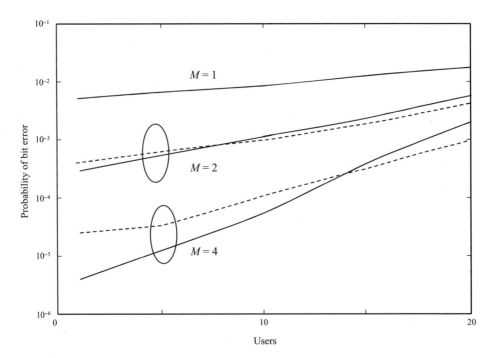

Figure 7.54 Probability of error versus number of users for the multicarrier CDMA system (solid curves) and the single-carrier system on a frequency selective fading channel (dashed curves). For each $E_b/N_0 = 17$ dB, the composite processing gain is 32 chips/bit, and M is the number of carriers used (for multicarrier) or the number of resolvable paths present (for frequency selective). The Wiener solution is formed, with all users' fading processes tracked perfectly.

where the matrices in this expression are defined as

$$\mathbf{F}_m = \left[\mathbf{f}_{1,m}, \mathbf{f}_{2,m}, \ldots, \mathbf{f}_{K,m} \right]$$

$$\mathbf{G}_m = [\mathbf{g}_{1,m}, \mathbf{g}_{2,m}, \ldots, \mathbf{g}_{K,m}]$$

$$\mathbf{D}(i) = \begin{bmatrix} d_1(i) & & & \\ & d_2(i) & & \\ & & \ddots & \\ & & & d_K(i) \end{bmatrix}$$

$$\mathbf{P}(i) = \begin{bmatrix} 1 & & & \\ & \sqrt{P_2/P_1} & & \\ & & \ddots & \\ & & & \sqrt{P_K/P_1} \end{bmatrix} \tag{7.124}$$

and the column vector of fading coefficients on the mth carrier, which is to be tracked, is

$$\mathbf{\Gamma}_m (i) = [\gamma_{1,m} (i), \gamma_{2,m} (i), \ldots, \gamma_{K,m}(i)]^{\mathrm{T}} \tag{7.125}$$

Also, in Equation (7.123), recall that $\mathbf{n}_m(i)$ is a column vector of independent, complex Gaussian noise samples, with the real and imaginary parts independent from each other and each with variance

$\sigma^2 = N/(2E_b/N_0)$. Then, Equation (7.123) can be rewritten as

$$\mathbf{r}_m(i) = \mathbf{\Lambda}_m(i)\mathbf{\Gamma}_m(i) + \mathbf{n}_m(i) \tag{7.126}$$

with

$$\mathbf{\Lambda}_m(i) = (\mathbf{F}_m\mathbf{D}(i) + \mathbf{G}_m\mathbf{D}(i-1))\mathbf{P} \tag{7.127}$$

For estimation purposes, we next assume that the fading processes are essentially constant over a window of L bit intervals, which gives

$$\mathbf{r}_m^{(L)}(i) = \mathbf{\Lambda}_m^{(L)}(i)\mathbf{\Gamma}_m(i) + \mathbf{n}_m^L(i) \tag{7.128}$$

where matrices from L bit intervals have been concatenated to form

$$\mathbf{r}_m^{(L)}(i) = \left[\mathbf{r}_m^T(i-(L-1)), \ldots, \mathbf{r}_m^T(i)\right]$$
$$\mathbf{\Lambda}_m^{(L)}(i) = \left[\mathbf{\Lambda}_m^T(i-(L-1)), \ldots, \mathbf{\Lambda}_m^T(i)\right] \tag{7.129}$$
$$\mathbf{n}_m^{(L)}(i) = \left[\mathbf{n}_m^T(i-(L-1)), \ldots, \mathbf{n}_m^T(i)\right]$$

We now assume that the receiver has knowledge of the data bits of all of the users. This would be reasonable either when the receiver is in training mode, or when decision feedback is used. In this case, the maximum likelihood estimate of the vector $\mathbf{\Gamma}_m(i)$ minimizes the quadratic cost function

$$C(\mathbf{\Gamma}_m(i)) = \left\|\mathbf{r}_m^{(L)}(i) - \mathbf{\Lambda}_m^{(L)}\mathbf{\Gamma}_m(i)\right\|^2$$

giving the least squares solution

$$\begin{aligned}\mathbf{\Gamma}_{m,ML}(i) &= \left(\left[\mathbf{\Lambda}_m^{(L)}(i)\right]^T \left[\mathbf{\Lambda}_m^{(L)}(i)\right]\right)^{-1} \times \left[\mathbf{\Lambda}_m^{(L)}(i)\right]^T \left[\mathbf{r}_m^{(L)}(i)\right] \\ &= \mathbf{Q}_m^{-1}(i)\mathbf{S}_m(i)\end{aligned} \tag{7.130}$$

where

$$\begin{aligned}\mathbf{Q}_m(i) &= \left[\mathbf{\Lambda}_m^{(L)}(i)\right]^T \left[\mathbf{\Lambda}_m^{(L)}(i)\right] \\ &= \sum_{j=0}^{L-1} \mathbf{\Lambda}_m^T(i-j)\mathbf{\Lambda}_m(i-j) \\ \mathbf{S}_m(i) &= \left[\mathbf{\Lambda}_m^{(L)}(i)\right]^T \left[\mathbf{r}_m^{(L)}(i)\right] \\ &= \sum_{j=0}^{L-1} \mathbf{\Lambda}_m^T(i-j)\mathbf{r}_m(i-j)\end{aligned} \tag{7.131}$$

This form of the matrices suggests recursive estimates using exponentially weighted windows

$$\begin{aligned}\mathbf{Q}_m(i) &= \lambda\mathbf{Q}_m(i-1) + \mathbf{\Lambda}_m^T(i)\mathbf{\Lambda}_m(i) \\ \mathbf{S}_m(i) &= \lambda\mathbf{S}_m(i-1) + \mathbf{\Lambda}_m^T(i)\mathbf{r}_m(i)\end{aligned} \tag{7.132}$$

where $0 < \lambda < 1$ is the forgetting factor. Once the fading has been estimated according to this procedure, an estimate of the true Wiener solution of the tap weights may be formed.

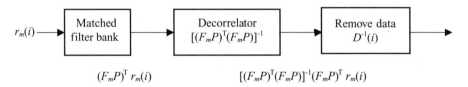

$$(F_mP)^{\mathrm{T}} r_m(i) \qquad\qquad [(F_mP)^{\mathrm{T}}(F_mP)]^{-1}(F_mP)^{\mathrm{T}} r_m(i)$$

Figure 7.55   Block diagram of the estimator for fading processes using a single bit observation window.

To obtain some insight into the operation of this channel estimator, consider a single bit estimator, that is, let $\lambda = 0$. We have

$$
\begin{aligned}
\boldsymbol{\Gamma}_{m,ML}(i) &= \left[\boldsymbol{\Lambda}_m^{(L)}(i)\boldsymbol{\Lambda}_m^{(L)}(i)\right]^{-1}\left[\boldsymbol{\Lambda}_m^{(L)}(i)\mathbf{r}_m^{(L)}(i)\right] \\
&= \boldsymbol{\Gamma}_m(i) + \left[\boldsymbol{\Lambda}_m^{(L)}(i)\boldsymbol{\Lambda}_m^{(L)}(i)\right]^{-1}\left[\boldsymbol{\Lambda}_m^{(L)}(i)\mathbf{n}_m^{(L)}(i)\right]
\end{aligned}
\tag{7.133}
$$

Thus, the estimate of $\boldsymbol{\Gamma}_m(i)$ is equal to the true value plus a term due only to the thermal noise, and independent of the multiaccess interference.

Furthermore, if the system were synchronous, then the matrix $\mathbf{G}_m$ would be a zero matrix, and the estimate of $\boldsymbol{\Gamma}_m(i)$ could be written as

$$
\boldsymbol{\Gamma}_{m,ML}(i) = \mathbf{D}^{-1}(i)[(\mathbf{F}_m\mathbf{P})^{\mathrm{T}}(\mathbf{F}_m\mathbf{P})]^{-1}[(\mathbf{F}_m\mathbf{P})^{\mathrm{T}}\mathbf{r}_m^{(L)}(i)]
\tag{7.134}
$$

The estimator could then be visualized as in Figure 7.55. The received signal is processed first with a matched filter bank. The MAI is then removed with a decorrelator. Finally, the data are removed, leaving an unbiased estimate of $\boldsymbol{\Gamma}_m(i)$.

One should keep in mind that the matrix $(\mathbf{F}_m\mathbf{P})^{\mathrm{T}}(\mathbf{F}_m\mathbf{P})$ will be invertible only if the columns of $\mathbf{F}_m$ are linearly independent. This condition will be violated as the number of users surpasses N/M and the decorrelator will not exist. However, as the observation window was increased, which would obviously be done in order to get good estimates of the fading processes, the existence of the decorrelator would be almost certain. A similar interpretation of the estimator would still apply, that is, matched filter/decorrelator/data removal.

The performance of this algorithm will be illustrated first for a multicarrier CDMA system with the same parameters as in [69]. A user with a data rate of 10 kHz, a carrier frequency of 900 MHz, a processing gain of 32 chips/bit, and an E_b/N_0 of 17 dB was simulated. With 15 users present, results for the average probability of bit error are shown in Figure 7.56(a) as a function of the vehicle speed, which was varied from 20 mph up to 200 mph. Single-carrier, two-carrier and four-carrier systems were considered. Results for the ideal Wiener solution are shown for comparison. It is seen that the performance is very close to ideal for vehicle speeds below about 80 mph, or a normalized Doppler frequency under 1%.

In Figure 7.56(b), the identical system was simulated, this time with a fixed vehicle speed of 80 mph, and with the number of users varying between 1 and 30. Again, the results are seen to be close to ideal at this speed.

As mentioned previously, it is straightforward to extend this tracking algorithm to the frequency selective case. Without going into such details, in Figure 7.56(c), an identical system to that used to generate the results shown in Figure 7.56(a) for the multicarrier case was applied to a single-carrier system operating on a frequency selective channel with M resolvable paths. The average probability of bit error is shown as a function of the vehicle speed for 15 users. Again, the tracking algorithm gives results which are very close to ideal for vehicle speeds below about 80 mph, or a normalized Doppler frequency under 1%, In Figure 7.56(d), the vehicle speed was fixed at 80 mph, and the number of users was varied between 1 and 30, the results are again close to ideal at this speed.

Additional discussions on MMSE detectors for MC CDMA systems can be found in [69–81].

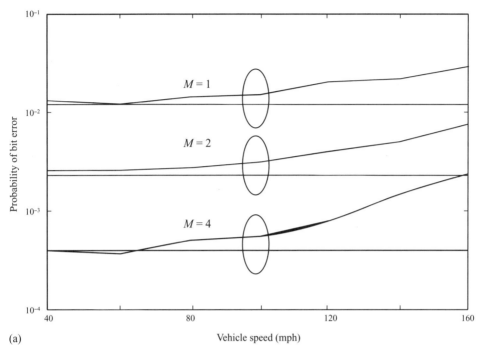

(a)   Vehicle speed (mph)

Figure 7.56   (a) Probability of error versus vehicle speed for multicarrier CDMA system with $E_b/N_0 = 17$ dB, composite processing gain of 32 chips/bit, 15 asynchronous users, bit rate of 10 000 bits/s, and carrier frequency of 900 MHz. M is the number of carriers used. Curves show the performance of the tracking algorithm, and solid straight lines show the Wiener solutions for 15 users. (b) Probability of error versus number of users for multicarrier CDMA system with $E_b/N_0 = 17$ dB, composite processing gain of 32 chips/bit, vehicle speed of 80 mph, bit rate of 10 000 bits/s, and carrier frequency of 900 MHz. M is the number of carriers used. Dashed lines show the performance of the tracking algorithm, and solid lines show the Wiener solutions. (c) Probability of error versus vehicle speed for single-carrier CDMA system with M resolvable paths. The system has 15 users, $E_b/N_0 = 17$ dB, composite processing gain of 32 chips/bit, bit rate of 10 000 bits/s, and carrier frequency of 900 MHz. Curves show the performance of the tracking algorithm, and solid straight lines show the Wiener solutions for 15 users. (d) Probability of error versus number of users for single-carrier system with M resolvable paths. System has $E_b/N_0 = 17$ dB, composite processing gain of 32 chips/bit, vehicle speed of 80 mph, bit rate of 10 000 bits/s, and carrier frequency of 900 MHz. Dashed lines show the performance of the tracking algorithm, and solid lines show the Wiener solutions.

7.13   Approximation of Optimum Multiuser Receiver for Space–Time Coded Multicarrier CDMA Systems

In this section we extend the discussion on multiuser detection for MC CDMA to the case when space–time coding is included. The presentation is based on [82–84]. The structure of the MC CDMA modulator of the kth user is illustrated in Figure 7.57. BPSK symbols of the kth user are first serial-to-parallel converted, by grouping every P symbols into a vector

$$\underline{s}^k \overset{\Delta}{=} [s^k[1], \ldots, s^k[P]]^{\mathrm{T}} \tag{7.135}$$

In the next step, symbol $s^k[p]$ in $\underline{s}^k$ is spread by a spreading sequence $.c^k[p], p = 1, \ldots, p$ The sequences corresponding to all P binary phase shift keying (BPSK) symbols are represented by an N-vector as

$$\underline{c}^k \overset{\Delta}{=} [c^k[1], \ldots, c^k[P]]^{\mathrm{T}} \tag{7.136}$$

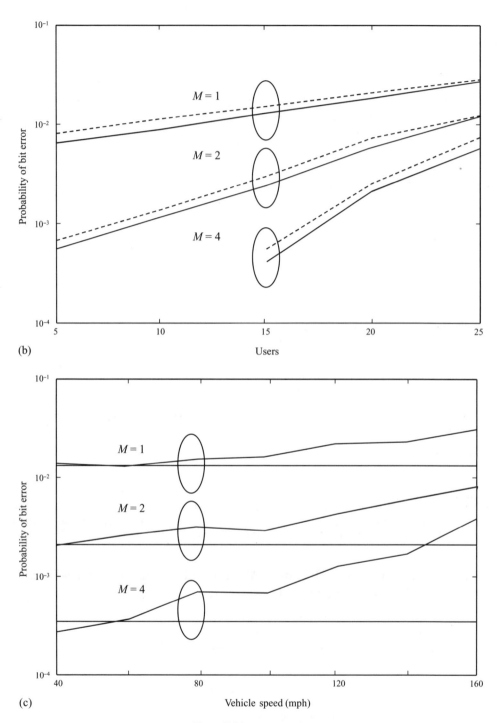

Figure 7.56   (*Continued*).

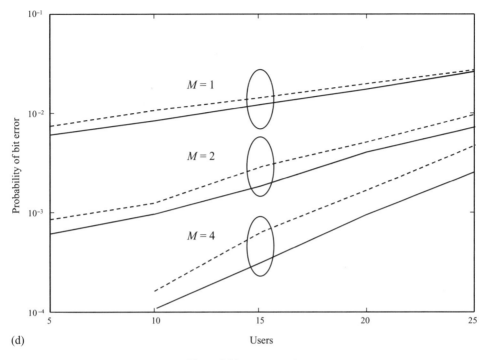

Figure 7.56 (*Continued*).

and the spread signals are represented by a componentwise product of the above vectors

$$\underline{s}^k \underline{c}^k \triangleq [s^k[1]c^k[1]^{\mathrm{T}}, \ldots, s^k[P]c^k[P]^{\mathrm{T}}]^{\mathrm{T}} \tag{7.137}$$

In order to avoid the strong correlation among the G subcarriers occupied by a particular symbol, all N spread chips in Equation (7.137) are interleaved before they are transmitted from the N subcarriers. The interleaving function is denoted by **T** in Figure 7.57; for simplicity, the interleaver is assumed to be the same for all K users.

The Alamouti space–time block codes (STBC), discussed in Chapter 4, are employed to further increase the system capacity. In Chapter 4, the STBC was used to transmit scalar symbols. In this section, we extend it to the vector form and apply it in MC CDMA systems. In vector form, the simplest (2×2) STBC discussed in Chapter 4, as illustrated in Figure 7.58, takes two time slots to transmit two

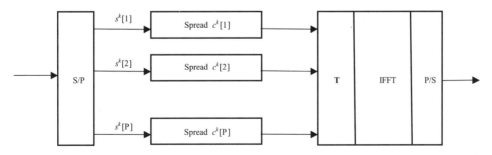

Figure 7.57 MC CDMA modulator structure of the kth user, with interleaver T.

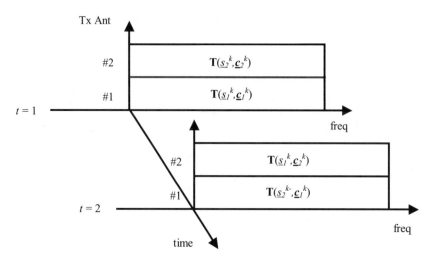

Figure 7.58   Transmitted signal structure of a particular STBC slot of the kth user in an STBC MC CDMA system.

symbol vectors $\underline{s}_1^k$, $\underline{s}_2^k$ [cf. Equation (7.135)]. At the first time slot, signals $\mathbf{T}(\underline{s}_1^k \underline{c}_1^k)$ are transmitted from the *first* antenna on N subcarriers and $\mathbf{T}(\underline{s}_2^k \underline{c}_2^k)$ is transmitted from the *second* antenna. At the second time slot, $\mathbf{T}(-\underline{s}_2^k \underline{c}_1^k)$ is transmitted from the *first* antenna and $\mathbf{T}(\underline{s}_1^k \underline{c}_2^k)$ is transmitted from the *second* antenna. From Figure 7.58, we can see that different spreading sequences $\underline{c}_1^k$ and $\underline{c}_2^k$ are assigned to different transmitter antennas; this structure is shown to be an efficient way to resolve the so-called 'antenna-ambiguity' in blind algorithms, as will be discussed. The structure of the transmitter for the kth user is given in Figure 7.59.

7.13.1   Frequency Selective Fading Channels

The system with one receiver antenna is considered and the extension to multiple receiver antennas is straightforward. The time domain channel impulse response of the kth user between the jth transmitter antenna and the receiver antenna will be modeled as

$$h_j^k(\tau) = \sum_{l=0}^{L-1} \alpha_j^k(l) \delta\left(\tau - \frac{l}{\Delta_f}\right) \tag{7.138}$$

where $\delta(\cdot)$ is the Kronecker delta function; $L \triangleq \lceil \tau_m \Delta_f + 1 \rceil$, with τ_m being the maximum multipath spread of all K users (note that we have assumed the synchronous transmission of all k users) and Δ_f being the whole bandwidth of multicarrier systems; $\alpha_j^k(l)$ is the complex amplitude of the lth tap associated with the jth transmitter antenna of the kth user, whose relative delay is l/Δ_f.

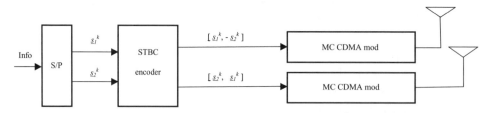

Figure 7.59   Transmitter structure of the kth user in an STBC MC CDMA system.

For MC CDMA systems with proper cyclic extensions and sample timing, with tolerable leakage, the channel frequency response of the kth user at its jth transmitter antenna and at the nth subcarrier can be expressed as

$$H_j^k[n] \overset{\Delta}{=} H_j^k(n\Delta_f)$$

$$= \sum_{l=0}^{L-1} \alpha_j^k(l) \exp\left(-\frac{j2\pi\,nl}{N}\right) \tag{7.139}$$

$$= \mathbf{w}_\mathrm{f}^\mathrm{H}(n)\mathbf{h}_j^k$$

where

$$\mathbf{h}_j^k \overset{\Delta}{=} [\alpha_j^k(0), \alpha_j^k(1), \dots, \alpha_j^k(L-1)]^\mathrm{T} \tag{7.140}$$

contains the time response of all L taps; and

$$\mathbf{w}_\mathrm{f}(n) \overset{\Delta}{=} \left[1, \exp\left(-\frac{j2\pi\,n}{N}\right), \dots, \exp\left(-\frac{j2\pi\,n(L-1)}{N}\right)\right]^\mathrm{H} \tag{7.141}$$

contains the corresponding discrete Fourier transform (DFT) coefficients.

7.13.2 Receiver Signal Model of STBC MC CDMA Systems

The transmitted signals of all K users propagate through their respective frequency selective fading channels and finally reach the receiver antenna. We assume that the fading processes associated with different transmitter–receiver antenna pairs are uncorrelated. At the receiver, after matched filtering, the discrete Fourier transform (DFT) is applied to the received chip rate sampled discrete time signals. Consider all the DFT-ed signals in one signal frame, which spans M STBC slots, or equivalently, $2M$ time slots [recall that each STBC slot consists of two neighboring time slots (see Figure 7.58).] Using Equation (7.139) and assuming that the fading processes are time invariant within one signal frame, the received signal model can be represented as

$$\mathbf{y}_i = \sum_{k=1}^{K} \underbrace{\begin{bmatrix} \mathbf{S}_{i,1}^k \mathbf{C}_1^k & \mathbf{S}_{i,2}^k \mathbf{C}_2^k \\ -\mathbf{S}_{i,2}^k \mathbf{C}_1^k & \mathbf{S}_{i,1}^k \mathbf{C}_2^k \end{bmatrix}}_{\mathbf{X}_i^k} \underbrace{\begin{bmatrix} \mathbf{TW}_\mathrm{f} & 0 \\ 0 & \mathbf{TW}_\mathrm{f} \end{bmatrix}}_{\mathbf{W}} \times \underbrace{\begin{bmatrix} \mathbf{h}_1^k \\ \mathbf{h}_2^k \end{bmatrix}}_{\mathbf{h}^k} + \mathbf{v}_i$$

$$= \sum_{k=1}^{K} \mathbf{X}_i^k \mathbf{W} \mathbf{h}^k + \mathbf{v}_i, \quad i = 0, \dots, M-1 \tag{7.142}$$

with

$$\mathbf{C}_j^k \overset{\Delta}{=} \mathrm{diag}\{\underline{\mathbf{c}}_j^k\}_{N\times N}$$

$$\underline{\mathbf{c}}_j^k \overset{\Delta}{=} [\mathbf{c}_j^k[1], \dots, \mathbf{c}_j^k[P]^\mathrm{T}]^\mathrm{T}, \quad j = 1, 2$$

$$\mathbf{S}_{i,j}^k \overset{\Delta}{=} \mathrm{diag}\{\underline{\mathbf{s}}_{i,j}^k \otimes \mathbf{I}_\mathrm{G}\}$$

$$= \mathrm{diag}\{\mathbf{s}_{i,j}^k[1]\mathbf{I}_\mathrm{G}, \dots, \mathbf{s}_{i,j}^k[P]\mathbf{I}_\mathrm{G}\}_{N\times N}, \quad j = 1, 2$$

$$\mathbf{W}_\mathrm{f} \overset{\Delta}{=} [\mathbf{w}_\mathrm{f}(0), \mathbf{w}_\mathrm{f}(1), \dots, \mathbf{w}_\mathrm{f}(N-1)]_{N\times L}^\mathrm{H}$$

where $\otimes$ is the Kronecker matrix product; $\underline{s}_{i,j}^k$, $j = 1, 2$ are the symbol vectors input to the STBC encoder at the ith STBC slot; $\underline{c}_j^k$ is the spreading sequence assigned to the jth transmitter antenna of the kth user; $\mathbf{T}$ is an $(N \times N)$ permutation matrix, which acts as an interleaver mapping the N chips to their assigned subcarriers; $\mathbf{y}_i$ is the received signal during the ith STBC slot; $\mathbf{v}_i$ is circularly symmetric complex Gaussian ambient noise, with covariance matrix $\sigma^2 \mathbf{I}_{2N}$.

Note that Equation (7.142) can be used to describe both the slow-fading (when M is large) and the fast-fading (when M is small, e.g. $M = 1$) cases. In this section, the fading channels are assumed to be static during two neighboring time slots (i.e. one STBC slot), this is the only limitation of the signal model in Equation (7.142). Due to the orthogonality property of the STBC, i.e. $\mathbf{X}_i^{k^{\mathrm{H}}} \mathbf{X}_i^k = 2\mathbf{I}_{2N}$, the orthogonality property of the OFDM multicarrier modulation, i.e. $\mathbf{W}_{\mathrm{f}}^{\mathrm{H}} \mathbf{W}_{\mathrm{f}} = N \cdot \mathbf{I}_L$, the fact that the permutation matrix satisfies $\mathbf{T}^{\mathrm{T}} \mathbf{T} = \mathbf{I}_N$ and the definitions given in Equation (7.142), we have

$$\mathbf{W}^{\mathrm{H}} \mathbf{X}_i^{k^{\mathrm{H}}} \mathbf{X}_i^k \mathbf{W} = \begin{bmatrix} \mathbf{W}_{\mathrm{f}}^{\mathrm{H}} \mathbf{T}^{\mathrm{T}} & 0 \\ 0 & \mathbf{W}_{\mathrm{f}}^{\mathrm{H}} \mathbf{T}^{\mathrm{T}} \end{bmatrix} \begin{bmatrix} 2\mathbf{I}_N & 0 \\ 0 & 2\mathbf{I}_N \end{bmatrix} \times \begin{bmatrix} \mathbf{T} \mathbf{W}_{\mathrm{f}} & 0 \\ 0 & \mathbf{T} \mathbf{W}_{\mathrm{f}} \end{bmatrix} \qquad (7.143)$$

$$= 2N \cdot \mathbf{I}_{2L}$$

As we know from orthogonal design, discussed in Chapter 4, the structure of Equation (7.143) can be exploited to reduce the computational complexity of the optimal receiver for the STBC MC CDMA system. Note that the interleaver function $\mathbf{T}$ is the same for all K users; hence, the signal model Equation (7.142) can be written in an alternative form, which decouples the signal components corresponding to different symbols

$$\mathbf{y}_i[p] = \sum_{k=1}^{K} \mathbf{X}_i^k[p] \mathbf{W}[p] \mathbf{h}^k + \mathbf{v}_i[p], \quad i = 0, \ldots, M - 1 \qquad (7.144)$$

$$p = 1, \ldots, P$$

where $\mathbf{X}_i^k[p]$ is a $(2G \times 2G)$ submatrix decimated from $\mathbf{X}_i^k$, which contains all the rows and columns related to the symbol $s_{i,1}^k[p]$ and $s_{i,2}^k[p]$; $\mathbf{y}_i[p]$, $\mathbf{W}[p]$ and $\mathbf{v}_i[p]$ are then the corresponding decimations from $\mathbf{y}_i$, $\mathbf{W}$, and $\mathbf{v}_i$. Equation (7.144) will be used later in deriving the conditional posterior distributions of the unknown symbols [see Equation (7.158)].

7.13.3   Blind Approach

Since the channel state information is unknown to the receiver, there are two types of ambiguity inherent in the design of blind receivers for the STBC MC CDMA system: phase ambiguity and antenna ambiguity.

To resolve the phase ambiguity, differential encoding is employed before the STBC encoding. For each signal frame, a block of BPSK bits $\mathbf{b}^k = \{b^k[1], b^k[2], \ldots, b^k[2MP - 1]\}$ is input to the differential encoder, and the output $\mathbf{d}^k = \{d^k[1], \ldots, d^k[2MP]\}$ is given by

$$\begin{cases} d^k[1] = 1 \\ d^k[n] = d^k[n-1]b^k[n-1], \quad n = 2, \ldots, 2MP \end{cases} \qquad (7.145)$$

These differentially encoded bits $\{d^k[n]\}_n$ are the same set of bits $\{s_{i,j}^k[p]\}_{i,j,p}$ [defined by Equation (7.142)] input to the STBC encoder, where they are related as $s_{i,j}^k[p] = d^k[n]|_{n=(2i+j)P+p}$. Henceforth, the index n of $d^k[n]$ is understood as an implicit function of the index (i,j,p) of $s_{i,j}^k[p]$.

One possible approach to resolving the antenna ambiguity is to employ the differential space–time modulation as discusssed in Chapter 4; however, in that case, the signal constellation will be changed. Fortunately, in the STBC MC CDMA system, the antenna ambiguity can be resolved by using different spreading sequences on different transmitter antennas. Note that the usage of orthogonal sequences will result in a maximum 50% system loading (defined as K/G) in the (2×2) STBC MC CDMA system. The problem with this is that the channel frequency selectivity destroys the orthogonality of the sequences. For this reason, in this section, random sequences are assumed and a sufficient number of these sequences, such that they impose no constraint on the system loading. When the system employs the outer channel code, it is possible to use the same spreading sequence at different antennas of the same user. In this case, the antenna ambiguity can be resolved by exploiting the coding structure. For example, for the system considered here, due to the antenna ambiguity, we have two possible code bit sequences at the output of the multiuser detector. We can send both of them to the channel decoder and count the number of bit corrections. The correct code bit sequence will have only a few bit errors, whereas the incorrect one will have many errors.

7.13.4   Bayesian Optimal Blind Receiver

From here on the following matrix notation will be used

$$\mathbf{Y} \overset{\Delta}{=} \{\mathbf{y}_i\}_{i=0}^{M-1}; \mathbf{H} \overset{\Delta}{=} \{\mathbf{h}^k\}_{k=1}^{K}; \mathbf{D} \overset{\Delta}{=} \{\mathbf{d}^k\}_{k=1}^{K}; \mathbf{B} \overset{\Delta}{=} \{\mathbf{b}^k\}_{k=1}^{K}$$

The optimal blind receiver estimates the *a posteriori* probabilities (APP) of the multiuser data bits

$$P[b^k[n] = +1 | \mathbf{Y}], \quad n = 1, 2, \ldots, MP - 1, \quad k = 1, \ldots, K \tag{7.146}$$

based on the received signals $\mathbf{Y}$, the signal structure Equation (7.142), the spreading sequences of all users, and the prior information of $\mathbf{B}$, without knowing the channel response $\mathbf{H}$ and the noise variance σ^2.

So, the unknown parameters have to be averaged out from the *a posteriori* function. The Bayesian solution to Equation (7.146) is given by

$$P[b^k[n] = +1 | \mathbf{Y}]$$

$$= \sum_{\mathbf{B}:b^k[n]=+1} \int p[\mathbf{Y}|\mathbf{H}, \sigma^2, \mathbf{B}] p[\mathbf{H}] p[\sigma^2] p[\mathbf{B}] \, \mathrm{d}\mathbf{H} \, \mathrm{d}\sigma^2 \tag{7.147}$$

where $P[\mathbf{Y}|\mathbf{H}, \sigma^2, \mathbf{B}]$ is a Gaussian density function [see Equation (7.142)]; $p[\mathbf{H}]$, $p[\sigma^2]$ and $p[\mathbf{B}]$ are prior distributions of the independent and unknown quantities $\mathbf{H}$, σ^2 and $\mathbf{B}$ respectively. Clearly the computation in Equation (7.147) involves a very high-dimensional integral which is certainly unfeasible for any practical implementations. Thus, the Gibbs sampler, a Monte Carlo method, is used to calculate the *a posteriori* probabilities of the unknown symbols.

7.13.5   Blind Bayesian Monte Carlo Multiuser Receiver Approximation

In this section, we consider the problem of computing the *a posteriori* bit probabilities in Equation (7.146). The problem is solved under a Bayesian framework, by treating the unknown quantities as realizations of random variables with some prior distributions. The Gibbs sampler [85–89] is then employed to compute the Bayesian estimates.

7.13.6   Gibbs Sampler

The Gibbs sampler [85–89] is a Markov chain Monte Carlo (MCMC) procedure for numerical Bayesian computation. Let $\theta = [\theta_1, \ldots, \theta_d]^\mathrm{T}$ be a vector of unknown parameters. Let $\mathbf{Y}$ be the observed data. To generate random samples from the joint posterior distribution $p[\theta|\mathbf{Y}]$, given the samples at the $(j-l)$ th iteration, $\theta^{(j-1)} = [\theta_1^{(j-1)}, \ldots, \theta_d^{(j-1)}]^\mathrm{T}$ at the jth iteration, the Gibbs algorithm iterates as follows to obtain samples $\theta^{(j)} = [\theta_1^{(j)}, \ldots, \theta_d^{(j)}]^\mathrm{T}$.

For $i = 1, \ldots, d$, draw $\theta_i^{(j)}$ from the conditional distribution

$$p\left[\theta_i | \theta_1^{(j)}, \ldots, \theta_{i-1}^{(j)}, \theta_{i+1}^{(j)}, \ldots, \theta_d^{(j)}, \mathbf{Y}\right]$$

It is known that under regularity conditions [86–89]:

1. The distribution of $\theta_i^{(j)}$ converges geometrically to $p[\theta|\mathbf{Y}]$, as $j \to \infty$;
2. $(1/J) \sum_{j=1}^{J} f(\theta^{(j)}) \xrightarrow{a.s.} \int f(\theta) p[\theta|\mathbf{Y}] \, d\theta$, as $J \to \infty$, for any integrable function f.

Hence, the marginal *a posteriori* distribution of any parameter θ_i can be computed easily from the samples drawn by the Gibbs sampler.

7.13.7　Prior Distributions

For simplicity, we choose the sampling space, the set of unknown parameters sampled by the Gibbs sampler, to be $\{\mathbf{D}, \mathbf{H}, \sigma^2\}$, which are assumed to be independent of each other. Next, we specify their prior distributions $p[\mathbf{H}]$, $p[\sigma^2]$ and $p[\mathbf{D}]$.

1. For the unknown channel $\mathbf{h}^k$, a complex Gaussian prior distribution is assumed

$$p[\mathbf{h}^k] \sim N_c(\mathbf{h}_{k,0}, \; \boldsymbol{\Sigma}_{k0}) \tag{7.148}$$

　　Note that large value of Σ_{k0} corresponds to less informative prior distributions.
2. For the noise variance σ^2, an inverse chi squared prior distribution is assumed

$$p[\sigma^2] \sim \chi^{-2}(2\nu_0, \lambda_0) \tag{7.149}$$

　　a small value of $2\nu_0$ corresponds to the less informative prior distributions.
3. The data bit sequence $\mathbf{d}^k$ is a Markov chain, encoded from $\mathbf{b}^k$. Its prior distribution can be expressed as

$$\begin{aligned}
p[\mathbf{d}^k] &= p(d^k[1]) p(d^k[2]|d^k[1]) \cdots p(d^k[2PM]|d^k[2PM-1]) \\
&= p(d^k[1]) p(b^k[1] = d^k[2]d^k[1]) \\
&\quad \cdots p(b^k[2PM-1] = d^k[2PM]d^k[PM-1]) \\
&= \frac{1}{2} \prod_{n=2}^{PM} \frac{\exp(\rho^k[n-1]d^k[n-1]d^k[n])}{1 + \exp(\rho^k[n-1]d^k[n-1]d^k[n])}
\end{aligned} \tag{7.150}$$

where $\rho^k[n]$ denotes the *a priori* log likelihood ratio (LLR) of $b^k[n]$,

$$\rho^k[n] \triangleq \log \frac{P(b^k[n] = +1)}{P(b^k[n] = -1)} \tag{7.151}$$

Note that in Equation (7.150), we set $p(d^k[1]) = 1/2$ to account for the phase ambiguity in $d^k[1]$.

7.13.8　Conditional Posterior Distributions

The following conditional posterior distributions are required by the Bayesian multiuser detector. The derivations can be found in Appendix 7.1 [83].

1. The conditional distribution of the kth user's channel response $\mathbf{h}^k$ given σ^2, $\mathbf{D}$, $\mathbf{H}^k$, and $\mathbf{Y}$ is

$$p[\mathbf{h}^k \,|\, \mathbf{D}, \sigma^2, \mathbf{H}^k, \mathbf{Y}] \sim N_c(\mathbf{h}_{k*}, \mathbf{\Sigma}_{k*}) \tag{7.152}$$

where $\mathbf{H}^k \triangleq \mathbf{H}\backslash\mathbf{h}^k$ with

$$\mathbf{\Sigma}_{k*}^{-1} \triangleq \mathbf{\Sigma}_{k0}^{-1} + \frac{1}{\sigma^2} \sum_{i=0}^{M-1} \mathbf{W}^H \mathbf{X}_i^{k^H} \mathbf{X}_i^k \mathbf{W}$$

$$= \mathbf{\Sigma}_{k0}^{-1} + \frac{2MN}{\sigma^2} \mathbf{I}_{2L} \tag{7.153}$$

and

$$\mathbf{h}_{k*} \triangleq \mathbf{\Sigma}_{k*} \left[\mathbf{\Sigma}_{k0}^{-1} \mathbf{h}_{k0} + \frac{1}{\sigma^2} \sum_{i=0}^{M-1} \mathbf{W}^{k^H} \mathbf{X}_i^k \left(\mathbf{y}_i - \sum_{l \neq k} \mathbf{X}_i^l \mathbf{W}^l \mathbf{h}^l \right) \right] \tag{7.154}$$

Equation (7.153) follows from the orthogonality property in Equation (7.143).

2. The conditional distribution of the noise variance σ^2 given $\mathbf{H}$, $\mathbf{D}$, and $\mathbf{Y}$ is given by

$$p[\sigma^2 | \mathbf{H}, \mathbf{D}, \mathbf{Y}] \sim \chi^{-2} \left(2[v_0 + MN], \frac{v_0 \lambda_0 + s^2}{v_0 + 2MN} \right) \tag{7.155}$$

with

$$s^2 \triangleq \sum_{i=0}^{M-1} \left\| \mathbf{y}_i - \sum_{k=1}^{K} \mathbf{X}_i^k \mathbf{W}^k \mathbf{h}^k \right\|^2 \tag{7.156}$$

3. The conditional distribution of the data bit $d^k[n]$, given $\mathbf{H}$, σ^2, $\mathbf{D}_n^k$ and $\mathbf{Y}$ can be obtained from

$$\frac{P[d^k[n] = +1 | \mathbf{H}, \sigma^2, \mathbf{D}_n^k, \mathbf{Y}]}{P[d^k[n] = -1 | \mathbf{H}, \sigma^2, \mathbf{D}_n^k, \mathbf{Y}]}$$

$$= \exp \left[d^k[n+1]\rho^k[n] + d^k[n-1]\rho^k[n-1] - \frac{\Delta s^2}{\sigma^2} \right] \tag{7.157}$$

where $\mathbf{D}_n^k \triangleq \mathbf{D}\backslash d^k[n]$ and

$$\Delta s^2 \triangleq \left\| \mathbf{y}_i[p] - \sum_l \mathbf{X}_i^l \mathbf{W}[p]\mathbf{h}^l \right\|^2_{s_{i,j}^k[p]=+1}$$

$$- \left\| \mathbf{y}_i[p] - \sum_l \mathbf{X}_i^l \mathbf{W}[p]\mathbf{h}^l \right\|^2_{s_{i,j}^k[p]=-1} \tag{7.158}$$

On the right-hand side of Equation (7.157), the first two items correspond to the prior information of the differentially encoded bits $\mathbf{D}$, through which the prior information of data bits $\mathbf{B}$ is incorporated.

7.13.9   Gibbs Multiuser Detection

Given the initial values of the unknown quantities $\{\mathbf{H}^{(0)}, \sigma^{2(0)}, \mathbf{D}^{(0)}\}$ drawn from their prior distributions (Equations 7.148–7.150), at the jth iteration, the Gibbs multiuser detector operates as follows.

1. For $k = 1, \ldots, K$, draw $\mathbf{h}^{k(j)}$ from $p[\mathbf{h}^k|\mathbf{H}^{k(j-1)}, \sigma^{2(j-1)}, \mathbf{D}^{(j-1)}, \mathbf{Y}]$, given by Equation (7.152), with $\mathbf{H}^{k(j-1)} \triangleq \{\mathbf{h}^{1(j)}, \ldots, \mathbf{h}^{(k-1)(j)}, \mathbf{h}^{(k+1)(j-1)}, \ldots, \mathbf{h}^{K(j-1)}\}]$.
2. Draw $\sigma^{2(j)}$ from $p[\sigma^2|\mathbf{H}^{(j)}, \mathbf{D}^{(j-1)}, \mathbf{Y}]$ given by Equation (7.155).
3. For $n = 1, \ldots, 2PM$, and for $k = 1, \ldots, K$, draw $d^{k(j)}[n]$ from $P[d^k[n]|\mathbf{H}^{(j)}, \sigma^{2(j)}, \mathbf{D}_n^{k(j-1)}, \mathbf{Y}]$ given by Equation (7.157), where $\mathbf{D}_n^{k(j-1)} \triangleq \{d^{1(j)}[1], \ldots, d^{1(j)}[2PM], \ldots, d^{k(j)}[n-1], d^{k(j-1)}[n+1], \ldots, d^{K(j-1)}[2PM]\}$.

To ensure convergence, the Gibbs iteration is usually carried out for $(J_0 + J)$ iterations. The initial J_0 iterations represent the burn-in period and only the samples from the last J iterations are used to calculate the Bayesian interference. In particular, the posterior distribution of the multiuser data bits $b^k[n]$ can be obtained by

$$P[b^k[n] = +1|\mathbf{Y}] \cong \frac{1}{J} \sum_{j=J_0+1}^{J_0+J} \delta_{kn}^{(j)}, \quad k = 1, \ldots, K; \quad n = 1, \ldots, 2PM - 1 \tag{7.159}$$

where $\delta_{kn}^{(j)}$ is an indicator such that

$$\delta_{kn}^{(j)} = \begin{cases} 1, & \text{if } d^{k(j)}[n]d^{k(j)}[n-1] = +1 \\ 0, & \text{if } d^{k(j)}[n]d^{k(j)}[n-1] = -1 \end{cases} \tag{7.160}$$

7.13.10   Sampling Space of Data

As we are interested in computing the posterior probabilities of the multiuser data bits, direct sampling can be done on the data bits $\mathbf{B}$. Note that the conditional posterior distribution of $b^k[n]$, given $\mathbf{H}$, σ^2, $\mathbf{B}_n^k$ and $\mathbf{Y}$, involves a large number of received signals, i.e. $\{\mathbf{y}_i, \ldots, \mathbf{y}_{M-1}\}$. This long memory in the receiver signal processing will increase the computational complexity and decrease the convergence speed of the Gibbs procedure. To avoid these disadvantages, the Gibbs procedure described in this section samples the differentially encoded bits $\mathbf{D}$, and outputs a sampling sequence $\{d^{k(j)}[n]\}$. It is shown in Equation (7.159) that the marginal posterior probability of $b^k[n]$ can be computed easily from the output samples $\{d^{k(j)}[n]\}$.

7.13.11   The Orthogonality Property

The dominant computations involved in the Gibbs sampler are Equations (7.153–7.154) By exploiting the orthogonality property, Equation (7.143), in STBC MC CDMA systems, the matrix Σ_{k*}^{-1} in Equation (7.153) is simply a constant matrix. In addition to this, with no matrix inversion involved in computing $\mathbf{h}_{k*}$, the numerical stability is also improved.

7.13.12   Blind Turbo Multiuser Receiver

In this section, we consider employing iterative multiuser detection and decoding to improve the performance of the Bayesian multiuser receiver in a coded STBC MC CDMA system. Because it utilizes the *a priori* bit probabilities, and it produces the *a posteriori* bit probabilities, the Bayesian multiuser detector is well suited for iterative processing, which allows the multiuser detector to refine its processing based on the information from the decoding stage and vice versa. The kth user's transmitter structure is shown in Figure 7.60, with block of information bits $\{a^k[l]\}$ encoded using some channel code (e.g. block code, convolutional code or turbo code). A code bit interleaver is used to reduce the influence of error bursts at the input of the channel decoder. The interleaved code bits are then mapped to BPSK

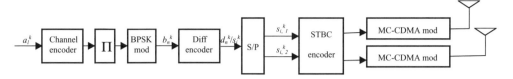

Figure 7.60 Transmitter structure of the kth user in an STBC MC CDMA system employing outer channel code, where Π represents an interleaver.

symbols $\{b^k[n]\}$. These BPSK symbols are differentially encoded to yield the symbol stream $\{d^k[n]\}$, which is then serial-to-parallel converted and reorganized in to a vector form as $\{\underline{\mathbf{s}}_{i,j}^k\}$ to feed into the STBC encoder followed by the MC CDMA modulator, and finally transmitted from two antennas.

An iterative (turbo) receiver structure is shown in Figure 7.61. It consists of two stages: the Bayesian multiuser detector developed in the previous sections, followed by a soft-input soft-output channel decoder. The two stages are separated by a deinterleaver and an interleaver. Assume that $\{b^k[n]\}$ is mapped into $\{b^k[\pi(n)]\}$ after deinterleaving.

In the first stage, the blind Bayesian multiuser detector incorporates the *a priori* information $\{\lambda_2^p(b^k[n])\}$, which is computed by the channel decoder in the previous iteration. At the first iteration, it is assumed that all code bits are equally likely. At the output of the blind Bayesian multiuser detector, the *a posteriori* LLR is given by

$$\Lambda_1(b^k[n]) \triangleq \log \frac{P[b^k[n] = +1|\mathbf{Y}]}{P[b^k[n] = -1|\mathbf{Y}]} \tag{7.161}$$

According to the 'turbo principle' described in Chapter 2, the *a priori* information $\{\lambda_1^p(b^k[n])\}$ should be subtracted from the *a posteriori* LLR $\Lambda_1(b^k[n])$ to obtain the extrinsic information to deliver to the channel decoders. However, the posterior distribution delivered by the blind Bayesian multiuser detector is a quantized value instead of the true value, due to the finite number of samples. Therefore, to ensure numerical stability, the posterior LLR $\Lambda_1(b^k[n])$ is regarded as the approximated extrinsic information, deinterleaved and fed back to the channel decoder of the kth user.

The soft-input soft-output channel decoder, using the MAP decoding algorithm (Appendix 7.1), computes the *a posteriori* LLR of each code bit of the kth user

$$\Lambda_2(b^k[\pi(n)]) \triangleq \log \frac{P[b^k[n] = +1|\{\Lambda_1^p(b^k[\pi(i)])\}_{i-1}^{PM-1}]}{P[b^k[n] = -1|\{\Lambda_1^p(b^k[\pi(i)])\}_{i-1}^{PM-1}]} \tag{7.162}$$

$$= \lambda_2(b^k[\pi(n)]) + \Lambda_1^p(b^k[\pi(n)])$$

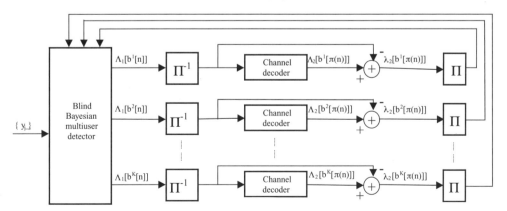

Figure 7.61 Turbo Bayesian multiuser receiver structure, where Π denotes an interleaver and Π^{-1} denotes the corresponding deinterleaver.

It is seen from Equation (7.162) that the output of the MAP decoder is the sum of the prior information $\Lambda_1^p(b^k[\pi(n)])$, and the *extrinsic* information $\lambda_2^p(b^k[\pi(n)])$ delivered by the channel decoder. After interleaving, the extrinsic information delivered by the channel decoder $\{\lambda_2^p(b^k[n])\}_{n=1}^{PM-1}$ is then fed back to the blind Bayesian multiuser detector as the refined prior information $\rho^k[n]$ [see Equation (7.151)] for the next iteration.

7.13.13 Decoder-Assisted Convergence Assessment

Although it is desirable to have the Gibbs sampler reach convergence within the burn-in period (J_0 iterations), this may not always be the case. Hence, we need some mechanism to detect the convergence. In the coded system considered here, the blind multiuser detector is followed by a bank of channel decoders, we can assess convergence by monitoring the number of bit corrections made by the channel decoders [90]. The number of corrections is determined by comparing the signs of the code bit LLR at the input and output of the MAP channel decoder. If this number exceeds some predetermined threshold, then we decide convergence is not achieved, in which case, the Gibbs multiuser detector will be applied again to the same data block.

7.13.14 Performance Example

In this section, we present some computer simulation results to illustrate the performance of the Bayesian multiuser receivers in the STBC MC CDMA system, where there are two transmitter antennas, one receiver antenna, N subcarriers and K users. The signal parameters are the same as in [83]. All spreading sequences used in simulations are randomly and independently generated for each transmitter antenna of each user. The frequency selective fading channels are assumed to be uncorrelated and have the same statistics for different transmitter–receiver antenna pairs. For simplicity, all L taps of a particular fading channel are assumed to be of equal power and normalized such that $\sum_{j=1}^{2} ||\mathbf{h}_j||^2 = 1$, and have delays $\tau_i = (l/\Delta_f), l = 0, 1, \ldots, L - 1$; and all K users in the system are assumed to have equal transmission power. Such a system setup is also the worst case scenario from the interference mitigation point of view (Chapter 5). For STBC MC CDMA systems employing outer channel codes, a four-state, rate-1/2 convolutional code with generator (5,7) in octal notation is chosen for all users. For each block of received signals, $J_0 + J = 100$ samples are drawn by the Gibbs sampler, with the first $J_0 = 50$ samples discarded. As discussed in the previous section, at the end of the 100 Gibbs iterations, the convergence of the Gibbs sampler is tested. In very few cases, when the Gibbs sampler is not convergent, it is restarted for another round of 100 Gibbs iterations.

The performance is demonstrated in two forms: one (Figure 7.62) is in terms of the bit error rate (BER) and OFDM word error rate (WER) versus the number of users at a particular signal to noise ratio (SNR), where SNR $= (1/\sigma^2)$ [cf. Equation (7.142)]; the other (Figure 7.63) is in terms of the BER/WBR versus SNR for the system with a particular number of users. Figure 7.62. demonstrates the expected effects of a multiuser detector where BER does not change significantly with the increase in the number of users. These curves are rather close to the single user (SU) case which is also expected.

7.14 Parallel Interference Cancellation in OFDM Systems in Time-Varying Multipath Fading Channels

In order to increase the system capacity the 4G systems will try to bring as much as possible of WLAN technology into cellular wireless networks with high mobility. Time varying multipath channel will generate inter-channel interference (ICI) and how to deal with this problem in a systematic way will be the topic of the next few sections.

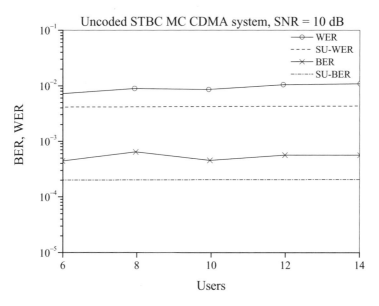

Figure 7.62 BER and OFDM WER of an STBC MC CDMA system in two-tap frequency selective fading channels, where $N = 16$, $G = 16$, $L = 2$, SNR $= 10$ dB.

In this section we further specify the notation introduced in Section 7.1. The sequence $S_n(m)$ (the frequency-domain symbol) is fed to an IDFT, producing the OFDM signal $s_n(k)$ (the time-domain symbol) with:

$$s_n(k) = IDFT\{S_n(m)\} = \frac{1}{N}\sum_{m=0}^{N-1} S_n(m)e^{\frac{j2\pi km}{N}}, \qquad k = 0, \ldots, N-1 \qquad (7.163)$$

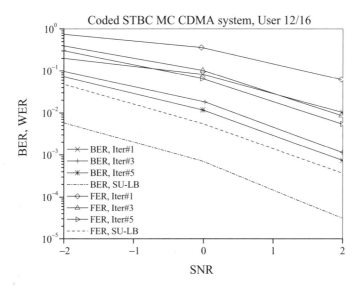

Figure 7.63 BER and OFDM WER of an STBC MC CDMA system employing outer convolutional channel code in five-tap frequency selective fading channels, where $N = 256$, $G = 16$, $L = 5$, $K = 12$.

The IDFT output sequence with the addition of the guard interval is:

$$s_n^g(k) = s_n(k + N - G)_N, \quad 0 \le k \le N + G - 1 \tag{7.164}$$

where G is the length of the guard interval and $(k)_N$ denotes the residue of k modulo N. As this waveform is transmitted over the multipath channel, the received sampling data $y_n^g(k)$ at the kth interval of the nth OFDM symbol can be expressed as:

$$y_n^g(k) = s_n^g(k) * h_n^g(k, k) + v_n^g(k) = = \sum_{l=0}^{k} s_n^g(k - l)h_n^g(k, l)$$

$$+ \sum_{l=k+1}^{L} s_{n-1}^g(k - l + N + G)h_n^g(k, l) + v_n^g(k) \tag{7.165}$$

where $*$ denotes the convolution operation, $L = [T_m/T_s]$ represents the maximum delay spread, $v_n^g(l)$ the ambient channel noise, and $h_n^g(k, l)$ the equivalent discrete channel at position l and instant k.

After removal of the guard interval, we get:

$$y_n(k) = y_n^g(k + G) = \sum_{l=0}^{L} s_n(k - l)_N h_n(k, l) + v_n(k); \quad 0 \le k \le N - 1 \tag{7.166}$$

where $h_n(k, l) = h_n^g(k + G, l)$ and $v_n(k) = v_n^g(k + G)$. The DFT, at the receiver gives:

$$Y_n(m) = DFT\{y_n(k)\} = \sum_{k=0}^{N-1} y_n(k)e^{-\frac{j2\pi km}{N}} = \frac{1}{N} \sum_{l=0}^{L} \sum_{d=0}^{N-1} S_n(d)H_n^l(m - d)_N e^{-\frac{j2\pi km}{N}}$$

$$+ V_n(m) = \alpha_n(m, m)S_n(m) + \sum_{d=0,d\neq m}^{N-1} \alpha_n(m, d)S_n(d) + V_n(m), \tag{7.167}$$

for $m = 0, \ldots, N - 1$

where $H_n^l(m - d)_N$ is the frequency response of a time-varying muhipath channel with:

$$H_n^l(m - d)_N = \sum_{k=0}^{N-1} h_n(k, l)e^{-\frac{j2\pi(m-d)k}{N}}$$

$$\alpha_n(m, d) = \frac{1}{N} \sum_{l=0}^{L} H_n^l(m - d)_N e^{-\frac{j2\pi ld}{N}} \tag{7.168}$$

$$V_n(m) = DFT\{v_n(l)\}.$$

Numerical analysis of these coefficients shows that the frequency responses of a time-varying channel, $E\left|H_n^0(0)\right| \cdots E\left|H_n^0(N - 1)\right|$, and the complex weighting coefficients, $E\left|\alpha_n(0, 0)\right| \cdots E\left|\alpha_n(0, N - 1)\right|$, for different Doppler spread cases, are in the lower frequency bands, and the weighting coefficients decay smoothly. That is, the ICI comes from only a few neighboring subcarriers. In the conventional receivers the ICI is modeled as a white Gaussian process and (7.167) can be written as:

$$Y_n(m) = \alpha_n(m, m)S_n(m) + J_n(m)$$

$$J_n(m) = \sum_{d=0,d\neq m}^{N-1} \alpha_n(m, d)S_n(d) + V_n(m) \tag{7.169}$$

A conventional equalization method uses a one-tap equalizer to equalize the distorted samples and, the symbol samples are equalized by $w_n(m)$ as:

$$z_n(m) = w_n(m)Y_n(m), \qquad m = 1, \ldots, N-1 \qquad (7.170)$$

and, the decision $dec(*)$ output is $\hat{S}_n(m) = dec(z_n(m))$.

The zero-forcing (ZF) and the MMSE solutions can be obtained as:

$$w_n(m) = \frac{1}{\alpha_n(m, n)} \qquad (7.171a)$$

and

$$w_n(m) = \frac{\alpha_n^*(m, m)}{|\alpha_n(m, m)|^2 + \dfrac{\sigma_J^2}{E_S}} \qquad (7.171b)$$

respectively, where σ_J^2 is the variance of noise $J_n(m)$ and E_S is the variance of the transmitted data symbols. Note that the MMSE solution reduces to the ZF solution for $\sigma_J^2 = 0$. In the case of a rapidly time-varying channel, the adaptation of the equalizer's tap coefficients has to be carried out via a channel estimator. Pilot-based correction is an efficient method for this, and several types of pilot arrangement for time-frequency domains have been studied. If the equalizer considers the ICI as an additive Gaussian random process, the performance of equalizer degrades significantly for larger channel variation. As will be demonstrated latter, a parallel interference cancellation (PIC) equalizer consisting of a set of prefilters in the first stage and a set of ICI cancellation filters in the second stage, will perform better. In order to subtract the ICI term from the symbol $Yn(m)$, we need to know all the symbols $Sn(j)$ ($j \neq m$) of the ICI term. However, we cannot get prior knowledge of the symbols of the ICI term. In order to obtain the accurate symbols of the ICI term, the received signal vector $Yn(m)$ is fed to the prefilter to obtain an initial decision in the first stage. Then, the initial decision is fed to the ICI cancellation filter to cancel the ICI term in the second stage, as shown in Figure 7.64.

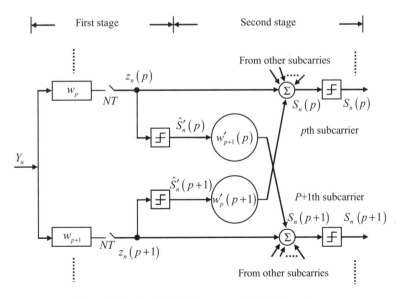

Figure 7.64   Parallel interference canceller (PIC) equalizer.

For the pth subcarrier, in the first stage of PIC equalization the sampled signal vector, $(Y_n(0), \ldots, Y_n(N-1)$, is fed to the prefilter $w_p(m)$. The prefilter output $z_n(p)$ at the nth symbol interval is:

$$z_n(p) = \sum_{m=0}^{N-1} Y_n(m)w_p(m) = \sum_{m=0}^{N-1} \alpha_n(m, p)S_n(p)w_p(m)$$

$$+ \sum_{m=0}^{N-1}\sum_{d=0,d\neq p}^{N-1} \alpha_n(m, d)S_n(d)w_p(m) + \sum_{m=0}^{N-1} V_n(m)w_p(m), \qquad 0 \le p \le N-1 \tag{7.172}$$

The first term is the desired signal, the second the ICI, and the third the noise. If the prefilters compensate for the multiplicative distortion completely, the output signal $z_n(p)$ of the pth subcarrier contains only the desired signal, ICI, and noise. Then, an initial decision is made as $\hat{S}'_n(p)$, given by:

$$\hat{S}'_n(p) = dec(z_n(p)), \quad p = 0, \ldots, N-1 \tag{7.173}$$

where $dec()$ denotes the decision function.

In the second stage, the ICI cancellation filters utilize the initial decisions $\hat{S}'_n(p); p = 1, \ldots, K$, to cancel the ICI. For the pth desired subcarrier, the initial decisions $\hat{S}'_n(d)$, for $d = 1, \ldots, p-1$, $p+1, \ldots, N$, are passed through the ICI cancellation filters $w'_p(d)$, where the subscript p denotes the pth desired subcarrier and $d = 1, \ldots, N, d \neq p$ denotes the input of the pth ICI cancellation filter from the dth subcarrier's initial decision. The final decision is made in the second stage of PICs as:

$$\tilde{S}_n(p) = z_n(p) + \sum_{d=0,d\neq p}^{N-1} \hat{S}'_n(d)w'_p(d) = \sum_{m=0}^{N-1} \alpha_n(m, p)S_n(p)w_p(m)$$

$$+ \sum_{d=0,d\neq p}^{N-1}\left[\sum_{m=0}^{N-1} \alpha_n(m, d)S_n(d)w_p(m) + \hat{S}'_n(d)w'_p(d)\right] + \sum_{m=0}^{N-1} V_n(m)w_p(m)$$

and

$$\hat{S}_n(p) = dec\left(\{\tilde{S}_n(p)\}\right) \tag{7.174}$$

The PIC equalizer is more complex to implement than is a one-tap equalizer. Since most of the energy of the time-varying channel is concentrated in the neighborhood of the dc component in the frequency domain, and the ICI term mainly comes from only a few neighboring subcarriers, the ICI terms that do not significantly affect $Yn(m)$ in (7.167) can be ignored as:

$$\alpha_n(m, d) \approx 0, \quad \text{for} \quad |m - d| > q \tag{7.175}$$

Therefore, we can simplify the PIC equalizer by using $2q + 1$-taps' prefilter and $2q$-taps' ICI cancellation filter in each subcarrier. The outputs in the first stage and the second stage become:

$$z_n(p) = \sum_{m=p-q}^{p+q} Y_n(m)w_p(m) = \sum_{m=p-q}^{p+q} \alpha_n(m, p)S_n(p)w_p(m)$$

$$+ \sum_{m=p-q}^{p+q}\sum_{d=0,d\neq p}^{N-1} \alpha_n(m, d)S_n(d)w_p(m) + \sum_{m=p-q}^{p+q} V_n(m)w_p(m) \tag{7.176}$$

$$\tilde{S}_n(p) = z_n(p) + \sum_{d=p-q,d\neq p}^{p+q} \hat{S}'_n(d)w'_p(d) =$$

$$= \sum_{m=p-q}^{p+q} \alpha_n(m, p) S_n(p) w_p(m) + \sum_{d=p-q, d\neq p}^{p+q} \left[\sum_{m=p-q}^{p+q} \alpha_n(m, d) S_n(d) w_p \right.$$

$$\left. (m) + \hat{S}'_n(d) w'_p(m) \right] + \sum_{m=p-q}^{p+q} \sum_{d=0, d\neq p-q \ldots p+q}^{N-1} \alpha_n(m, d) S_n(d) w_p(m)$$

$$+ \sum_{m=p-q}^{p+q} V_n(m) w_p(m) \approx \sum_{m=p-q}^{p+q} \alpha_n(m, d) S_n(d) w_p(m)$$

$$+ \sum_{d=p-q, d\neq p}^{p+q} \left[\sum_{m=p-q}^{p+q} \alpha_n(m, d) S_n(d) w_p(m) + \hat{S}'_n(d) w'_p(m) \right]$$

$$+ \sum_{m=p-q}^{p+q} V_n(m) w_p(m) \tag{7.177}$$

In order to improve convergence, a cost function developed from the error signal is used. If error signals are defined as

$$e'_{p,n} = S_n(p) - z_n(p) = S_n(p) - \mathbf{y}^T_{p,n} \mathbf{W}_p$$

$$e_{p,n} = S_n(p) - \tilde{S}_n(p) = S_n(p) - (\mathbf{y}^T_{p,n} \mathbf{W}_p + \hat{\mathbf{S}}'^T_{p,n} \mathbf{W}'_p)$$

where

$$\mathbf{y}_{p,n} = [Y_n(p-q), \ldots, Y_n(p+q)]^T$$

$$\hat{\mathbf{S}}'_{p,n} = [\hat{S}'_n(p-q), \ldots, \hat{S}'_n(p-1), \hat{S}'_n(p+1), \ldots, \hat{S}'_n(p+q)]^T \tag{7.178}$$

$$\mathbf{W}_p = [w(p-q), \ldots, w(p+q)]^T$$

$$\mathbf{W}'_p = [w'(p-q), \ldots, w'(p-1), w'(p+1), \ldots, w'(p+q)]^T$$

then, under the assumption of perfect decision, the MSE cost function has the following form

$$U_{p,n} = \gamma E\left\{ |e'_{p,n}|^2 \right\} + (1-\gamma) E\left\{ |e'_{p,n}|^2 \right\}$$

$$= \gamma [E_S \mathbf{I} - \mathbf{R}_{sy} \mathbf{W}_p - \mathbf{W}^H_p \mathbf{R}^H_{sy} + \mathbf{W}^H_p \mathbf{R}_{yy} \mathbf{W}_p]$$

$$+ (1+\gamma)[E_s \mathbf{I} - \mathbf{R}_{sy} \mathbf{W}_p - \mathbf{W}^H_p \mathbf{R}^H_{sy} + \mathbf{W}^H_p \mathbf{R}_{yy} \mathbf{W}_p \tag{7.179}$$

$$+ \mathbf{W}^H_p \mathbf{R}_{ys} \mathbf{W}'_p + \mathbf{W}'^H_p \mathbf{R}_{ys} \mathbf{W}_p + E_s \mathbf{W}'^H_p \mathbf{W}'_p]$$

with $0 \leq \gamma \leq 1$ for $p = 0, \ldots, N-1$, where E{} denotes the expectation operation, E_S is the variance of transmitted data symbols, $\mathbf{R}_{yy} = E[\mathbf{y}^*_{p,n} \mathbf{y}^T_{p,n}] = E_s \sum_{d=0}^{N-1} \mathbf{R}(d) + \sigma^2_V \mathbf{I}$ and:

$$\mathbf{R}(d) = \begin{bmatrix} |\alpha_n(p-q, d)|^2 & \cdots & \alpha^*_n(p-q, d)\alpha_n(p+q, d) \\ & \cdot & \\ \alpha^*_n(p, d)\alpha_n(p-q, d) & \cdots & \alpha^*_n(p, d)\alpha_n(p+q, d) \\ & \cdot & \\ \alpha^*_n(p+q, d)\alpha_n(p-q, d) & \cdots & |\alpha_n(p+q, d)|^2 \end{bmatrix}$$

$$\mathbf{R}_{ys} = E\left[\mathbf{y}^*_{p,n}\mathbf{S}^T_{p,n}\right]$$

$$= E_S \begin{bmatrix} \alpha^*_n(p-q,p-q) & \cdots & \alpha^*_n(p-q,p-1) & \alpha^*_n(p-q,p+1) & \cdots & \alpha^*_n(p-q,p+q) \\ & \ddots & & & & \\ \alpha^*_n(p,p-q) & \cdot & \alpha^*_n(p,p-1) & \alpha^*_n(p,p+1) & \cdot & \alpha^*_n(p,p+q) \\ & & & & & \\ \alpha^*_n(p+q,p-q) & \cdots & \alpha^*_n(p+q,p-1) & \alpha^*_n(p,p+1) & \cdots & \alpha^*_n(p+q,p+q) \end{bmatrix}$$

$$\mathbf{R}_{sy} = E\left[\mathbf{S}^H_n(p)\mathbf{y}^T_{p,n}\right] = E_S\left[\alpha_n(p-q,p)\ldots\alpha_n(p,p)\ldots\alpha_n(p+q,p)\right] \tag{7.180}$$

MMSE solution is derived from $\partial U_{p,n}/\partial\mathbf{W}_p = 0$ and $\partial U_{p,n}/\partial\mathbf{W}'_p = 0$ resulting in

$$\frac{\partial U_{p,n}}{\partial\mathbf{W}_p} = \gamma\left(-2\mathbf{R}^H_{sy} + 2\mathbf{R}_{yy}\mathbf{W}_p\right) + (1+\gamma)\left(-2\mathbf{R}^H_{sy} + 2\mathbf{R}_{yy}\mathbf{W}_p + 2\mathbf{R}_{ys}\mathbf{W}'_p\right) = 0$$

$$\frac{\partial U_{p,n}}{\partial\mathbf{W}'_p} = (1-\gamma)\left(2\mathbf{R}^H_{ys}\mathbf{W}_p + 2E_S\mathbf{W}'_p\right) = 0 \tag{7.181}$$

So, the MMSE optimum tap coefficients estimates are

$$\mathbf{W}_{Popt} = \left[\mathbf{R}_{yy} + (\gamma - 1)\frac{\mathbf{R}_{ys}\mathbf{R}^H_{ys}}{E_S}\right]\mathbf{R}^H_{sy} \quad ; \quad W'_{Popt} = -\frac{R^H_{YS}}{E_S}W_{Popt} \tag{7.182}$$

In order to compute W_{Popt} and W'_{Popt} accurately, we need to know the impulse response of the channel.

Performance curves are shown for the following simulation scenario. The data modulation scheme is 16-QAM and the total number of subcarriers is $N = 64$. A frequency-selective Rayleigh channel is generated by using Jake's model for two-path with a multipath spread of 2 μs. The carrier frequency is 5 GHz, the symbol rate of 16-QAM is 0.47 Msymbols/s (bit rate is 1.88 Mb/s). The OFDM block is composed of 68 samples, among them one for cyclic prefix and three for pilot signals ($N_p = 1$). The relative velocity between the transmitter and receiver is up to 216 km/h, resulting in $f_{dmax}NT_s$ equal to 0.032 for 54 km/h, 0.064 for 108 km/h, and 0.128 for 216 km/h.

Figure 7.65(a) shows the ratio of MSE_1 over MSE_2 versus γ, where MSE_1, is at the output of the prefilters, and MSE_2 is at the output of the ICI cancellation filters. MSE_1 represents the performance of the initial decision of the equalizer and MSE_2 represents the performance of the overall equalizer.

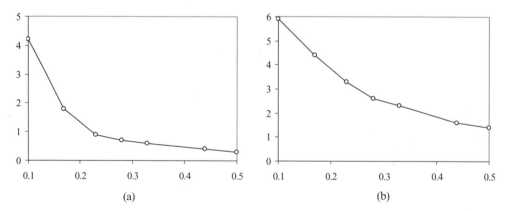

Figure 7.65   Simulation results of SER and $\text{MSE}_1/\text{MSE}_2$ versus γ for $f_{d\max}NT_S = 0.032$ and $q = 1$.

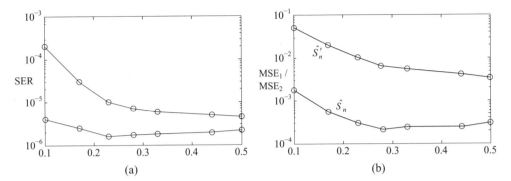

Figure 7.66   Simulation results of SER and MSE_1/MSE_2 versus γ for $f_{d_{max}}NT_S = 0.128$ and $q = 2$ performance.

Figure 7.65(b) presents the symbol error rate for the two cases. Figure 7.66 shows the same parameters for higher Doppler.

7.15   Zero Forcing OFDM Equalizer in Time-Varying Multipath Fading Channels

Samples of an OFDM signal, implemented by an inverse fast Fourier transform (IFFT), can be expressed as follows:

$$x_n = \sum_{m=0}^{N-1} X_m e^{j2\pi nm/N} \qquad 0 \leq n \leq N \tag{7.183}$$

where x_n represents the nth sample of the output of the IFFT. For the multipath fading channel consisting of L discrete paths, the received signal can be written as:

$$y_n = \sum_{l=0}^{L-1} h_{n,l} x_{n-l} + w_n = h_{n,0} x_n + h_{n,1} x_{n-1} + \cdots + h_{n,L-1} x_{n-L+1} + w_n$$

which after FFT becomes:

$$Y_m = \sum_{k=0}^{N-1}\sum_{l=0}^{L-1} X_k H_l^{(m-k)} e^{-j2\pi lk/N} + W_m =$$

$$= \left[\sum_{l=0}^{L-1} H_l^0 e^{-j2\pi lm/N}\right] X_m + \sum_{k\neq m}^{N-1}\sum_{l=0}^{L-1} X_k H_l^{(m-k)} e^{-j2\pi lk/N} + W_m = \tag{7.184}$$

$$= \alpha_m X_m + \beta_m + W_m, \ 0 \leq m \leq N-1$$

where W_m denotes the FFT of w_n, and $H_l^{(m-k)}$ represents the FFT of a time-variant multipath channel $h_{n,l}$ given as :

$$H_l^{(m-k)} = \frac{1}{N} \sum_{n=0}^{N-1} h_{n,l} e^{-j2\pi n(m-k)/N} \tag{7.185}$$

Here, α_m and β_m represent the multiplicative distortion at the desired subchannel and the ICI, respectively. If the channel is assumed to be time invariant during a block period, $H_l^{(m-k)}$ in Equation (7.185) vanishes, implying that there exists no ICI for time-invariant channels. In this case, Y_m in Equation (7.184) contains only the multiplicative distortion, which can be easily compensated for by a one-tap frequency-domain equalizer.

In the general case where the multipath channel cannot be regarded as time invariant during a block period, Equation (7.184) can be expressed in vector form as:

$$\mathbf{Y} = \mathbf{HX} + \mathbf{W} \tag{7.186}$$

where $\mathbf{Y} = [Y_0, \ldots, Y_{N-1}]^T$, $\mathbf{X} = [X_0, \ldots, X_{N-1}]^T$, $\mathbf{W} = [W_0, \ldots, W_{N-1}]^T$, and:

$$H = \begin{bmatrix} a_{0,0} & a_{0,1} & \cdots & a_{0,N-1} \\ a_{1,0} & a_{1,1} & \cdots & a_{1,N-1} \\ \cdot & \cdot & \cdots & \cdot \\ a_{N-1,0} & a_{N-1,1} & \cdots & a_{N-1,N-1} \end{bmatrix} \tag{7.187}$$

Here, $a_{m,k}$ in Equation (16.187) is defined as

$$a_{m,k} = H_0^{(m-k)} + H_1^{(m-k)} e^{-j\pi k/N} + \cdots + H_{L-1}^{(m-k)} e^{-j\pi k(L-1)/N} m \qquad 0 \le (m,k) \le N-1 \tag{7.188}$$

In order to solve for $\mathbf{X}$ in (7.186), we need to estimate the channel matrix $\mathbf{H}$ and calculate its matrix inverse. Since $\mathbf{H}$ can have a large size, it is difficult to process in real time. For that reason, similar approximations like those indicated in the previous section in (7.175) will be used. Here we provide additional arguments.

If the multipath fading channel is slowly time varying (e.g., $\Delta f_D = T_b \bullet f_D < 0.1$), the time variations of the CIR, $h_{n,j}$, for all L paths, can be approximated by straight lines with low slopes during a block period T_b. For the channel with $\Delta f_D > 0.1$, the assumption that the CIR varies in a linear fashion during a block period no longer holds and gives rise to an error floor. When the multipath fading channel is slowly time varying, the matrix equation in (7.186) can be greatly simplified. Since most energy of the straight line with a low slope is concentrated in the neighborhood of the dc component in the frequency domain, the ICI terms which do not significantly affect Y_m in (7.187) can be ignored, i.e.:

$$a_{m,k} = 0 \quad \text{for} \quad |m-k| > q/2 \tag{7.189}$$

where q denotes the number of dominant ICI terms. Figure 7.67(a) shows the time variation of the CIR in a block and Figure 7.67(b) the corresponding magnitude response (absolute value of the Fourier transform) for three different Doppler frequencies. From this figure, one can see that the time variation of the CIR can be approximated as a straight line, and most of the energy is concentrated in the neighborhood of the dc component. By using the approximation in Equation (7.189), we have:

$$\mathbf{H}' = \begin{bmatrix} a_{0,0} & a_{0,1} & \cdots & a_{0,\frac{q}{2}} & 0 & \cdot & & \cdot & & \cdot & 0 & & \cdot \\ a_{1,0} & a_{1,1} & \cdot & & \cdot & \cdot & & \cdot & & \cdot & & \cdot \\ \cdot & \cdot & \cdot & & \cdot & \cdot & & \cdot & & \cdot & & \cdot \\ a_{\frac{q}{2},0} & \cdot & \cdot & & \cdot & \cdot & & \cdot & & \cdot & & \cdot \\ 0 & \cdot & \cdot & & \cdot & \cdot & & \cdot & & \cdot & 0 & & \cdot \\ \cdot & & \cdot & & \cdot & \cdot & & \cdot & & a_{N-1-\frac{q}{2},N-1} & & \cdot \\ \cdot & & \cdot & & \cdot & & \cdot & & \cdot & & \cdot \\ \cdot & \cdot & \cdot & & \cdot & & \cdot & a_{N-2,N-2} & a_{N-2,N-1} & & \cdot \\ 0 & \cdot & \cdot & \cdot & 0 & a_{N-1,N-1-\frac{q}{2}} & \cdot & a_{N-1,N-2} & a_{N-1,N-1} & & \cdot \end{bmatrix} \tag{7.190}$$

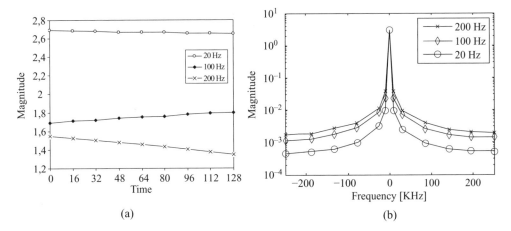

Figure 7.67 The characteristic of a CIR in a slowly time-varying environment. (a) Time variation of the CIR for different Doppler frequencies within a block period, and (b) corresponding magnitude responses for different Doppler frequencies.

The matrix in Equation (7.190) has nonzero element around diagonal $h'_{ij} \neq 0$, $i - q/2 < j < i + q/2$. Since the matrix becomes a sparse matrix for $q \ll N$, it is not efficient to calculate the matrix inverse to estimate the transmitted sequence. By transforming the matrix $\mathbf{H}'$ of order $N \times N$ to a block-diagonal matrix $\mathbf{H}$ of order $(N - q)(q + 1) \times (N - q)(q + 1)$ as shown in Figure 7.68, we obtain:

$$\mathbf{H} = \begin{bmatrix} A_0 & \cdot & \cdot & 0 \\ \cdot & A_1 & \cdot & \cdot \\ \cdot & \cdot & \ddots & \cdot \\ 0 & \cdot & \cdot & A_{N-1-q} \end{bmatrix} \tag{7.191}$$

where $\mathbf{A}_n$ is

$$\mathbf{A}_n = \begin{bmatrix} a_{n,n} & a_{n,n+1} & \cdots & a_{n,n+\frac{q}{2}} & 0 & \cdots & & 0 \\ a_{n+1,n} & a_{n+1,n+1} & & & & & & \\ \cdots & & & & & & & \\ a_{n+\frac{q}{2},n} & & \cdot & & \cdot & & & \\ 0 & & & \cdot & & \cdots & & 0 \\ & & & & \cdot & & \cdots & a_{n+\frac{q}{2},n+\frac{q}{2}} \\ \cdots & & & & & \cdots & & \cdots \\ & & & & & \cdot & a_{n+q-1,n+q-1} & a_{n+q-1,n+q} \\ 0 & & \cdots & & a_{n+q,n+\frac{q}{2}} & \cdots & a_{n+q,n+q-1} & a_{n+q,n+q} \end{bmatrix} \tag{7.192}$$

and input–output relationship of the multipath channel becomes:

$$\mathbf{Y} = \mathbf{HX} + \mathbf{W} \tag{7.193}$$

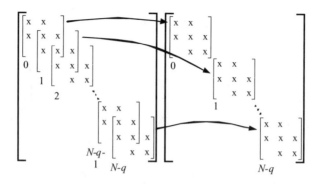

Figure 7.68   Transformation of matrix H' to H.

where

$$\mathbf{X} = [X_0 \, X_1 \quad \cdots \quad X_{N-1-q}]^T, \qquad \mathbf{X}_n = [X_n \, X_{n+1} \quad \cdots \quad X_{n+q}]^T$$

$$\mathbf{Y} = [Y_0 \, Y_1 \quad \cdots \quad Y_{N-1-q}]^T, \qquad \mathbf{Y}_n = [Y_n \, Y_{n+1} \quad \cdots \quad Y_{n+q}]^T$$

$$\mathbf{W} = [W_0 \, W_1 \quad \cdots \quad W_{N-1-q}]^T \quad \text{and} \quad \mathbf{W}_n = [W_n \, W_{n+1} \quad \cdots \quad W_{n+q}]^T$$

Multiplying (7.193) by the inverse of **H**, using is estimated value of **H**, gives:

$$\tilde{\mathbf{X}} = \tilde{\mathbf{H}}^{-1}\mathbf{Y} \rightarrow \tilde{\mathbf{X}}_n = \tilde{\mathbf{A}}_n^{-1}\mathbf{Y}_n, \quad 0 \le n \le N - 1 - q \tag{7.194}$$

where

$$\tilde{\mathbf{H}}^{-1} = \begin{bmatrix} \tilde{\mathbf{A}}_0^{-1} & & & 0 \\ & \tilde{\mathbf{A}}_1^{-1} & & \\ & & \cdots & \\ 0 & & & \tilde{\mathbf{A}}_{N-1-q}^{-1} \end{bmatrix}$$

Finally, the transmitted symbols $\tilde{X}_{\frac{q}{2}+1}, \ldots, \tilde{X}_{N-2-\frac{q}{2}}$ are estimated by selecting the elements in the middle of $\tilde{X}_{n,1} \le n \le N - 2 - q(N \gg q)$. The remaining symbols $\tilde{X}_0, \ldots, \tilde{X}_{\frac{q}{2}}(\tilde{X}_{N-1+\frac{q}{2}}, \ldots, \tilde{X}_{N-1})$ are estimated by taking the first (last) $q/2$ elements of $\tilde{X}_0(\tilde{X}_{N-1-q})$.

The size of the matrix inverse is reduced to $(q + 1) \times (q + 1)$, implying that the transmitted sequence can be obtained with a moderate amount of computational complexity for a small value of q. Thus, the linear approximation assumption discussed in this paper has transformed a large-size (N) matrix inverse problem into ($N - q$) small-size (q) matrix inverse problems. The solution ($g = 2$) for the transmitted sequence in a multipath fading channel is listed in Table 7.14. The required multiplications and additions for $q = 2$ are $6N + 2$ and $3N$ respectively.

In order to construct the matrix equation in Equation (7.193), it is necessary to estimate the channel matrix H. To demonstrate the effectiveness of the above equalizer for time-variant multipath channels, the simulations were performed for the scenario similar to the one used to generate Figures 7.65 and 7.66. A two-path fading channel with a multipath spread of 2 μs, bandwidth 500 kHz, carrier at 1 GHz, 64 subbands, the size of FFT became 64, OFDM block 68 samples, one for a cyclic prefix and three for pilot signals ($N_p = 1$), modulation 16-QAM, the data rate 1.88 Mb/s (64 subcarriers × 4 bits per

Table 7.14 Frequency-domain equalization ($q = 3$) for an OFDM system in a multipath fading channel

$\tilde{X}$

$\tilde{X}_0 = (b_{0,0}Y_0 + b_{0,1}Y_1 + b_{0,2}Y_2)/\Delta_0$

.

$\tilde{X}_{n+1} = (b_{n+1,n}Y_n + b_{n+1,n+1}Y_{n+1} + b_{n+1,n+2}Y_{n+2})/\Delta_n$

.

$\tilde{X}_{N-1} = (b_{N-1,N-3}Y_{N-3} + b_{N-1,N-2}Y_{N-2} + b_{N-1,N-1}Y_{N-1})/\Delta_{N-2}$

$\Delta_n = \tilde{a}_{n,n}\tilde{a}_{n+1,n+1}\tilde{a}_{n+2,n+2} - \tilde{a}_{n,n}\tilde{a}_{n+2,n+1}\tilde{a}_{n+1,n+2} - \tilde{a}_{n+2,n+2}\tilde{a}_{n+1,n}\tilde{a}_{n,n+1}$

b	$\tilde{a}$
$b_{0,0}$	$\tilde{a}_{1,1}\tilde{a}_{2,2} - \tilde{a}_{1,2}\tilde{a}_{2,1}$
$b_{0,1}$	$-\tilde{a}_{0,1}\tilde{a}_{2,2}$
$b_{0,2}$	$\tilde{a}_{0,1}\tilde{a}_{1,2}$
.	.
.	.
$b_{n+1,n}$	$-\tilde{a}_{n+1,n}\tilde{a}_{n-2,n+2}$
$b_{n,n}$	$-\tilde{a}_{n,n}\tilde{a}_{n+2,n+2}$
$b_{n+1,n+2}$	$-\tilde{a}_{n,n}\tilde{a}_{n-1,n+2}$
.	.
$b_{N-1,N-3}$	$\tilde{a}_{N-2,N-3}\tilde{a}_{N-1,N-2}$
$b_{N-1,N-2}$	$-\tilde{a}_{N-2,N-3}\tilde{a}_{N-1,N-2}$
$b_{N-1,N-1}$	$\tilde{a}_{N-2,N-3}\tilde{a}_{N-2,N-2} - \tilde{a}_{N-3,N-2}\tilde{a}_{N-2,N-2}$

symbol/136 μs), Doppler up to 200 Hz → $\Delta f_D = 0.272\%$ for 20 Hz, 1.36% for 100 Hz, and 2.72% for 200 Hz. Perfect carrier and symbol synchronizations were assumed.

The results for conventional frequency domain equalizer, with one tap in an OFDM system which compensates for the frequency-selectivity of a multipath fading channel, assuming that the channel is stationary over the period of an FFT block, are denoted in Figure 7.69 by C.

The results for zero forcing equalizer, described in this section are denoted in Figure 7.69 by P.

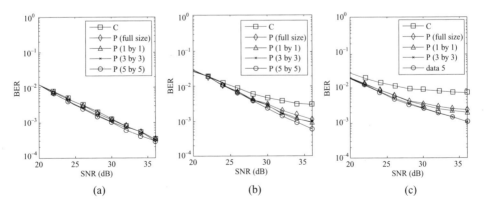

Figure 7.69 BER for the two types of equalizers in time-variant channels with three different Doppler frequencies: (a) $f_D = 20$ Hz, (b) $f_D = 100$ Hz. (c) $f_D = 200$ Hz.

7.16   Channel Estimation for OFDM Systems Using Multiple Receive Antennas

Figure 7.70 represents a block diagram of a space division multiple access (SDMA) *uplink* scenario, as observed on an OFDM subcarrier basis. Each of the L users is equipped with a single transmit antenna and the BS's receiver has P-element antenna. For simplicity, we have omitted the subcarrier index k (K subcarriers, $k = 0, \ldots, K - 1$). A decision-directed channel estimator (DDCE) aided OFDM receiver is shown in Figure 7.71 and parallel interference canceller (PIC)-assisted channel transfer function estimator in Figure 7.72.

The complex output signal $x_p[n,k]$ of the pth receiver-antenna element in the kth subcarrier of the nth OFDM symbol is given by:

$$x_p[n, k] = \sum_{i=1}^{L} H_p^{(i)}[n, k] s^{(i)}[n, k] + n_p[n, k] \tag{7.195}$$

In vector notation, we have:

$$\mathbf{x}_p[n] = \sum_{i=1}^{L} \mathbf{S}^{(i)}[n] \mathbf{H}_p^{(i)}[n] + \mathbf{n}_p[n] \tag{7.196}$$

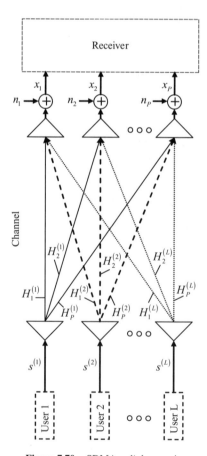

Figure 7.70   SDMA *uplink* scenario.

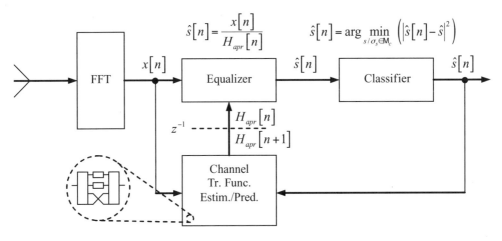

Figure 7.71   DDCE-aided OFDM receiver.

From Equation (7.196) an *a posteriori* (apt) estimate of the channel is:

$$\tilde{\mathbf{H}}^{(i)}_{apt}[n] = \breve{\mathbf{S}}^{(j)-1}[n] \left(\mathbf{x}[n] - \sum_{\substack{i=1 \\ i \neq j}}^{L} \breve{\mathbf{S}}^{(i)}[n]\widehat{\mathbf{H}}^{(i)}_{apr}[n] \right) \tag{7.197}$$

and its prediction (*a priori* channel transfer-factor estimation) as:

$$\widehat{\mathbf{H}}^{(i)}_{apr}[n] = f\left(\tilde{\mathbf{H}}^{(i)}_{apt}[n-1], \dots \tilde{\mathbf{H}}^{(i)}_{apt}[n - N^{[t]}_{tap}] \right) \tag{7.198}$$

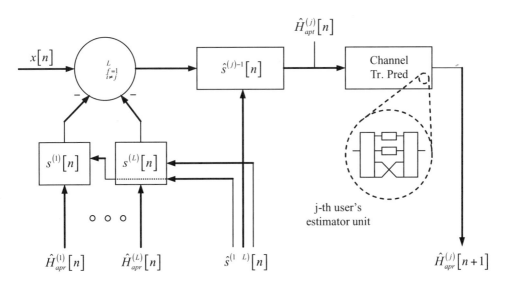

Figure 7.72   PIC-assisted channel transfer-function estimation.

with optimum predictor coefficients obtained as:

$$\tilde{c}_{\text{pre}}^{(j)}|\text{opt} = \left[\mathbf{R}^{[t](j)} + \frac{K_0}{\text{Trace}\left(\gamma^{[f](j)}\mathbf{I}_{K_0}^{(j)}\right)}\frac{\alpha_j}{\sigma_j^2} \times \left(\sum \sigma_j^2 \overline{MSE}_{\text{apr}}^i + \sigma_n^2\right)\mathbf{I}\right]^{-1} \cdot \mathbf{r}^{[t](j)} \qquad (7.199)$$

$$\mathbf{R}_{\text{apt}}^{[f](i)} = E\left\{\tilde{\mathbf{H}}_{\text{apt}}^{(i)}\tilde{\mathbf{H}}_{\text{apt}}^{(i)}\right\},$$

$$\overline{MSE}_{\text{apr}}^{(j)}[n] = \frac{1}{K}\text{Trace}\left(R_{\Delta\widehat{\mathbf{H}}_{apr}^{(j)}}[n]\right) \qquad (7.200)$$

$$\Delta\widehat{\mathbf{H}}_{\text{apr}}^{j}[n] = \mathbf{H}^j[n] - \widehat{\mathbf{H}}_{\text{apr}}^{j}[n]$$

where $\mathbf{R}_{\Delta\widehat{\mathbf{H}}_{apr}^{(j)}}[n] \in \mathbb{C}^{K \times K}$ denotes the autocorrelation matrix of the vector $\Delta\widehat{\mathbf{H}}^{(j)}[n]$ of *a priori* channel transfer-factor estimation errors. Matrix $\mathbf{R}^{[f](i)} = E\left\{\mathbf{H}^{(i)}\mathbf{H}^{(i)}\right\}$ and $\mathbf{R}^{[f](i)} = \mathbf{U}^{[f](i)}\mathbf{\Lambda}^{[f](i)}\mathbf{U}^{[f](i)\text{H}} \cdot \gamma^{[f](j)}$ is the decomposition of the *j*th user's channel's space-frequency correlation matrix $\mathbf{R}^{[f](j)}$ with respect to the unitary transform matrix $\tilde{\mathbf{U}}^{[f](j)}$, which is formulated as $\gamma^{[f](j)} = \tilde{\mathbf{U}}^{[f](j)}\mathbf{R}^{[f](j)}\tilde{\mathbf{U}}^{[f](j)}$, and $\mathbf{I}_{K_0}^{(j)}$ is a sparse identity matrix having unity entries only at those K_0 number of positions, which are associated with a significant value of $\gamma^{[f](j)}$. Hence, we note that the evaluation of Trace($\gamma^{[f](j)}\mathbf{I}_{K_0}^{(j)}$) requires the knowledge of $\mathbf{R}^{[f](j)}$ which is not directly available in practice.

For the simulation results shown in Figure 7.73, four simultaneous equal-power OFDM users with one transmit antenna and the same modulation scheme and a BS with four receiving antennas were assumed. The channels between the different transmit antennas and each receiver antenna were assumed to have the same Doppler power spectrum. For the channel's space–time correlation function the Jakes model is used, having an OFDM-symbol-normalized Doppler frequency of $F_D = 0.007$, which corresponds to a vehicular speed of 50 km/h, or equivalently, 31.25 miles/h in the context of the indoor wireless

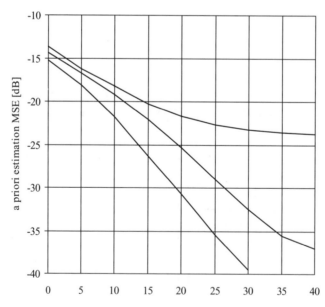

Figure 7.73 *A priori* channel-estimation MSE versus SNR performance PIC-assisted DDCE, using the optimum recursive predictor coefficients evaluated with the aid of the iterative approach. Fr.-Inv. Fad. SWATM from Section 7.4. L Rec.-Antennas, L Users, MPSK.

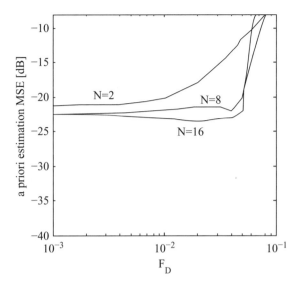

Figure 7.74   A priori channel-estimation MSE versus OFDM-symbol-normalized Doppler-frequency performance exhibited by the PIC-assisted DDCB, using optimum recursive predictor coefficients. The predictor coefficients were optimized for $\tilde{F}_D = 0.05$, using the iterative approach.

asynchronous transfer mode (WATM) system's parameters (see Section 7.4). Furthermore, we considered 'frame-invariant' fading, where the fading envelope of each CIR-related tap has been kept constant during each OFDM symbol's transmission period.

The sparse identity matrix $\mathbf{I}_{K_0}^{(j)}$ could be designed for retaining the first K_0 CIR-related coefficients of $\gamma^{[f](j)}$–rather than the K_0 largest one–or alternatively, for retaining the first K_0^{I} and the last K_0^{II} CIR-related coefficients of $\gamma^{[f](j)}$, where $K_0 = K_0^{I} + K_0^{II}$.

The simulation results with the same parameters versus Doppler frequency are shown in Figure 7.74.

7.17   Turbo Processing for an OFDM-Based MIMO System

In this section we will discuss performance of a receiver using constrained list sphere decoder (CLSD)-based soft-detector and the space-time bit-interleaved coded modulation (STBICM) approach.

Let $\mathbf{H} = [h_{n,m}] \in \mathbf{C}^{N \times M}$ denote the MIMO channel matrix for the kth subcarrier, where $h_{n,m}$ denotes the channel gain from the mth Tx antenna to the nth Rx antenna for subcarrier k. For simplicity index k is dropped. A received data vector $\mathbf{y}$ for this subcarrier, can be written as:

$$\mathbf{y} = \mathbf{Hx} + \mathbf{w} \tag{7.201}$$

where $\mathbf{x} = [x_1 x_2]^T$ is the 2×1 symbol vector transmitted on this subcarrier and $\mathbf{w} \sim N(0, \sigma^2 \mathbf{I}_N)$ is the additive white circularly symmetric complex Gaussian noise with variance σ^2.

We consider the general case of MIMO soft-detection where $\mathbf{H}$ is:

$N \times M$. Let $\beta = BM$ and $\mathbf{b} = [b_1 b_2 \ldots b_\beta]T$, with $b_i \in \{-1, +1\}$, $i = 1, 2, \ldots, \beta$, be the bit vector that maps to $\mathbf{x}$. Then $\mathbf{x}$ can be expressed as $\mathbf{x(b)}$ to stress its dependence on $\mathbf{b}$.

The bit metric for the ith bit, $i = 1, 2, \ldots, \beta$, is defined as:

$$l_D(i) = \ln \frac{P(b_i = +1 \, |\mathbf{y}, \mathbf{H})}{P(b_i = -1 \, |\mathbf{y}, \mathbf{H})} \tag{7.202}$$

which, by using the Bayes' theorem, can be written as

$$l_D(i) = \ln \frac{P(b_i = +1)}{P(b_i = -1)} + \ln \frac{P(b_i = +1, \mathbf{H})}{P(b_i = -1, \mathbf{H})} \overset{\triangle}{=} l_A(i) + l_E(i) \qquad (7.203)$$

Here, $l_A(i)$ and $l_E(i)$ are referred to as the *a priori* and extrinsic information, respectively. The *a priori* information can be obtained from the BCJR algorithm presented in Appendix 2.1, a soft-in/soft-out decoder, and the extrinsic information can be calculated by the STBICM-based soft-detector:

$$l_E(i) \approx \frac{1}{2} \max_{\mathbf{b} \in B_{i,+1}} \left\{ -\frac{1}{\sigma^2} \|y - \mathbf{Hx(b)}\|^2 + \mathbf{b}^T \mathbf{I}_A - l_A(i) \right\} -$$
$$-\frac{1}{2} \max_{\mathbf{b} \in B_{i,-1}} \left\{ -\frac{1}{\sigma^2} \|y - \mathbf{Hx(b)}\|^2 + \mathbf{b}^T \mathbf{I}_A + l_A(i) \right\} \qquad (7.204)$$

where $B_{i,+1}$ and $B_{i,-1}$ are the set of $2^{\beta-1}$ bit vectors with b_i being $+1$ and -1, respectively, and $\mathbf{I}_A = [l_A(1)l_A(2)\dots l_A(\beta)]^T$.

In the iterative decoding process, as shown in Figure 7.75, the *a priori* information $l_A(i)$ for the MIMO detector i.e., the inner decoder, is obtained from the soft output of the BCJR decoder, i.e., the outer decoder. Specifically, it is from the incremental soft information (denoted as L_E in the figure) due to the BCJR decoder. Similarly, the input for the BCJR decoder (denoted as L_A in the figure) is the deinterleaved extrinsic information l_E, which is the incremental soft information from the MIMO detector. The iterative decoding starts from the MIMO detection with the *a priori* information being 0.

The optimal but extremely inefficient STBICM-based soft detector of Equation (7.204) can be simplified by using list sphere decoder (LSD). The LSD-based soft-detector keeps the STBICM framework while improving its efficiency by searching in much smaller subsets $\tilde{B}_{i,+1} \subset B_{i,+1}$ and $\tilde{B}_{i,-1} \subset B_{i,-1}$ with $|\tilde{B}_{i,+1}| \ll 2^{\beta-1}$ and $|\tilde{B}_{i,11}| \ll 2^{\beta-1}$.

The LSD-based soft detector is implemented in the following two steps.

(i) Obtain a set $\tilde{B}$ of vectors $\mathbf{b}$, referred to as the candidate pool, which satisfies

$$\begin{cases} \|\mathbf{y} - \mathbf{Hx(b)}\|^2 \leq d_l & \forall \mathbf{b} \in \tilde{B} \\ \|\mathbf{y} - \mathbf{Hx(b)}\|^2 > d_l & \forall \mathbf{b} \notin \tilde{B} \end{cases} \qquad (7.205)$$

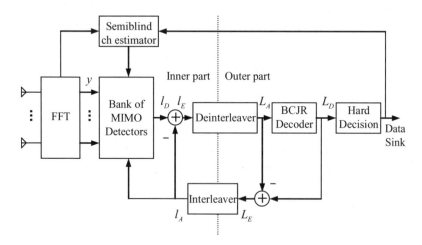

Figure 7.75   Turbo processing for the OFDM-based MIMO system.

by using the LSD algorithm with a fixed sphere radius d_l determined by the antenna numbers and noise variance.

(ii) Calculate $\tilde{B}_{i,+1} = B_{i,+1} \cap \tilde{B}$ and $\tilde{B}_{i,-1} = B_{i,-1} \cap \tilde{B}$ for each $i = 1, 2, \ldots, \beta$ and obtain the bit metric using Equation (7.204) with $B_{i,+1}$ and $B_{i,-1}$ being replaced by $\tilde{B}_{i,+1}$ and $\tilde{B}_{i,-1}$, respectively.

The LSD-based soft detector is orders of magnitude more efficient than the STBICM-based soft detector. However, this efficiency is obtained at the cost of some performance degradation due to the limited candidate pool used by LSD. Due to the limited candidate pool, the *a priori* information from the outer decoder can be quite poor. When the poor *a priori* information is used blindly without being censored, the performance of the iterative processing can degrade with the increase in iteration number.

The detection/decoding can be improved by constraining the *a priori* information from the outer decoder. We follow the simple rules below to constrain the *a priori* information:

(R1) The larger the sphere radius, the larger the maximum allowable *a priori* information, denoted l_{max}.

(R2) The larger the value of β, the smaller the maximum allowable *a priori* information l_{max}.

(R3) The higher the SNR, the larger the maximum allowable *a priori* information l_{max}.

R1 is chosen because for a given SNR and a given number of candidates in a sphere, the larger the radius, the better the channel. We give more freedom or larger maximum allowable *a priori* information l_{max} to a better channel. R2 is chosen because the more complicated the constellation and the larger the number of transmitted symbols, the larger the $\mathbf{b}^T \mathbf{l}_A$. We put more constraint on the maximum allowable value of $\mathbf{l}_A$ to prevent the value of $\mathbf{b}^T \mathbf{l}_A$ from growing too large. Finally, R3 is chosen because the higher the SNR or the smaller the noise variance σ^2, the more trust we can put on the larger *a priori* information from the outer decoder.

Based on the aforementioned rules, the value of the *a priori* information for our MIMO system is constrained as:

$$l_{max} = \frac{d}{\min\{1, k_C \sigma^2\}} \qquad (7.206)$$

where d is the radius of the sphere and k_C is a constant related to β. For example, for 64-QAM, we choose $k_C = 4 \times 42$ for above MIMO system, where 42 is the average power of the 64-QAM constellation. We choose d so that the sphere contains five candidates.

Once we have determined the maximum allowable *a priori* information l_{max}, we can constrain the *a priori* information from the outer decoder

$$\mathbf{l}_A^{(C)} = \begin{cases} \mathbf{l}_A, & \max\{\mathbf{l}_A\} \le l_{max} \\ \dfrac{l_{max}}{\max\{l_A\}} \mathbf{l}_A, & \max\{\mathbf{l}_A\} > l_{max} \end{cases} \qquad (7.207)$$

This constraining scheme in Equation (7.207) is by no means the best; however, it is very simple and can lead to excellent results in WLAN applications. We also clip the soft outputs of the CLSD-based soft detector (by ± 50) in addition to constraining the *a priori* information.

Some performance illustrations are given in Figure 7.76.

7.18 PAPR Reduction of OFDM Signals

It has been already pointed out in Section 7.10 that nonlinear effects can cause clipping of OFDM signal due to a large peak-to-power ratio (PAPR). In fading channels, this problem becomes even more serious.

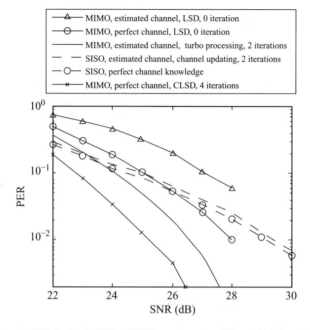

Figure 7.76   PER versus SNR for the MIMO and SISO systems at the 108- and 54-Mbps data rates, respectively.

A possible technique for PAPR reduction in orthogonal frequency division multiplexing (OFDM) systems is utilization of signal scrambling. Golay sequences (with dual capabilities of error correction and peak reduction) [91] and partial transmit sequences (PTS) [92, 93] have been suggested for these purposes. Computation of optimal PTS weight factors via exhaustive search requires exponential complexity in the number of subblocks; consequently, many suboptimal strategies have been developed.

Suboptimal PTS strategies include the following. The iterative flipping algorithm (FA) [94] has complexity linearly proportional to the number of subblocks, and each phase factor is individually optimized regardless of the optimal value of other phases. A neighborhood search is proposed in [95] using gradient descent search. In [96] dual layered phase sequencing is used to reduce complexity, at the price of PAPR performance degradation. In [93] a suboptimal strategy is developed by modifying the problem into an equivalent problem of minimizing the sum of phase-rotated vectors. An initial set of phase vectors is computed by reducing the peak amplitude of each sample and the best phase vector of the set is chosen as the final solution. Finally, [97] gives an orthogonal projection-based approach for computing PTS phase factors.

In this section we present an efficient algorithm for computing the optimal PTS weights that has lower complexity than does exhaustive search.

For the system model, let $\mathbf{X} = [X_1, \ldots, X_N]^T$ be a block of N symbols being transmitted, where each symbol is modulated to one of the carrier frequencies $\{f_n, n = 1, \ldots, N\}$ where $f_n = n\Delta f$, $\Delta f = 1/NT$ and T is the signal period. The complex envelope of the transmitted signal is:

$$x(t) = \frac{1}{\sqrt{N}} \sum_{n=1}^{N} X_n e^{j2\pi f_n t} \quad 0 \leq t < NT \tag{7.208}$$

where $j = \sqrt{-1}$. The PAPR of the OFDM signal $x(t)$ is defined as:

$$\text{PAPR} = \frac{\max |x(t)|^2}{E\left[|x(t)|^2\right]} \tag{7.209}$$

$E[|x(t)|^2] = 1$ for unitary signal constellations. The signal in Equation (7.208) can be oversampled by generating LN samples, where L > 1 is the oversampling factor. These samples can be computed by using appropriate zero-padding and using an inverse fast Fourier transform (IFFT).

In the PTS approach, **X** is divided into M disjoint sub-blocks $\mathbf{X}_m (1 \leq m < M)$ of length U where $N = MU$ for some integers M and U. For $m = 1, \ldots, M$, let $[R_{1,m}, \ldots, R_{LN,m}]^T$ be the zero-padded IFFT of $\mathbf{X}_m$ (all zeros except for the block m). PTS combines phase-rotated versions of these sub-block IFFTs in order to minimize the PAPR. The signal samples at the PTS output can be written as:

$$\mathbf{x}' = \underbrace{\begin{bmatrix} R_{1,1} & R_{1,2} & \cdots & R_{1,M} \\ R_{2,1} & R_{2,2} & & \cdot \\ \cdots & & & \cdot \\ R_{LN,1} & \cdot & \cdot & R_{LN,M} \end{bmatrix}}_{\mathbf{R}} \cdot \underbrace{\begin{bmatrix} b_1 \\ b_2 \\ \cdots \\ b_M \end{bmatrix}}_{\mathbf{b}} \qquad (7.210)$$

where $\mathbf{x}' = [x_1', \ldots, x_{LN}']$ is the block of optimized signal samples. The optimization problem is to find optimum phases b_i according to:

$$\{b_1^*, \ldots, b_M^*\} = \underbrace{\arg\min}_{\{b_1, \ldots, b_M\}} \left(\underbrace{\max}_{1 \leq k < LN} \left| \sum_{m=1}^{M} b_m R_{k,m} \right| \right) \qquad (7.211)$$

where $b_i \in P = \left\{ e^{\frac{j2\pi k}{q}}, k = 0, \ldots, q-1 \right\}$. The last phase factor can be fixed ($b_M = 1$) without loss of generality. Therefore, q^{M-1} distinct possible vectors **b** should be tested to solve Equation (7.211). Accordingly, in an exhaustive search approach, the computational complexity increases exponentially with the number of sub-blocks.

The optimization algorithm that solves Equation (7.211) with lower complexity, is motivated by the shortest vector problem (SVP) in a lattice [100]. An M-dimensional lattice is the set of vectors (lattice points) $\{\mathbf{Ab} | b_i \in \mathbb{Z} \}$, where $\mathbf{b} = (b_1, \ldots, b_M)$ and the columns of matrix $\mathbf{A} \in \mathbb{R}^{N \times M}$ are called the *basis* for the lattice. The SVP requires finding the shortest non-zero vector in the lattice, where the length can be measured in any l_p ($p \geq 1$) norm. The l_p norm of a vector $\mathbf{x} = (x_1, x_2, \ldots, x_N)$ is defined as $\|\mathbf{x}\|_p = \left(\sum |x_i|^p \right)^{1/p}$ and $\|\mathbf{x}\|_\infty = \max_i |x_i|$. Fincke and Phost [98] have developed an efficient algorithm for SVP in l_2 (i.e., Euclidean distance) by enumerating all the lattice points inside a sphere centered at the origin. This is one example of sphere decoding that has wide application in communication problems (see [99] for a detailed survey). The signal vectors (7.210) can be interpreted as lattice points generated by **R**. However, (7.211) is equivalent to the SVP in l_∞ norm. As such, the original Fincke–Phost sphere decoder (FPSD) cannot be directly applied to problem at hand. Nevertheless, the basic premise of FPSD – to generate *only* lattice points x for which $\|x\|_2 \leq \mu$ – can be adapted; consequently, only lattice points for which $\|\mathbf{x}\|_\infty \leq \mu$ are generated, and this is equivalent to $|x_k| \leq \mu \forall k$. We refer to this as a sphere decoder based algorithm (SDBA) (see Appendix 4.2 or [100]).

Let $\mathbf{x}' = [x_1', \cdots, x_{LN}']$ be defined as in Equation (7.210) and let R_k represent the kth row of the matrix **R**. Then each element of $\mathbf{x}'$ can be expressed as $x_k' = R_k \cdot \mathbf{b}$. To find the PAPR of the OFDM signal, the amplitude of x_k' is computed according to

$$|x_k'|^2 = x_k'^H \cdot x_k' = \mathbf{b}^H \cdot \mathbf{R}_k^H \cdot \mathbf{R}_k \cdot \mathbf{b}$$
$$= \mathbf{b}^H \cdot \underbrace{\left[\mathbf{R}_k^H \cdot \mathbf{R}_k + \alpha^2 \mathbf{I} \right]}_{\mathbf{A_k}} \cdot \mathbf{b} - \alpha^2 \mathbf{b}^H \cdot \mathbf{b} = \mathbf{b}^H \cdot \mathbf{A}_k \cdot \mathbf{b} - \alpha^2 M \qquad (7.212)$$

where α is an arbitrary non-zero real number and $(\cdot)^H$ denotes conjugate transpose. The resulting $M \times M$ matrix $\mathbf{A}_k$ is positive-definite due to the addition of $\alpha^2 \mathbf{I}$, and therefore can be Cholesky factorized as

$\mathbf{A}_k = \mathbf{Q}_k^H \cdot \mathbf{Q}_k$ where $\mathbf{A}_k$ is an upper-triangular matrix. Substituting $\mathbf{A}_k$ into Equation (7.212) gives:

$$\left|x_k'\right|^2 = \mathbf{b}^H \cdot \mathbf{Q}_k^H \cdot \mathbf{Q}_k \cdot \mathbf{b} - \alpha^2 M = \|\mathbf{Q}_k \cdot \mathbf{b}\|^2 - \alpha^2 M \tag{7.213}$$

where the signal sample is now a function of the phase vector $\mathbf{b}$.

If we wish to limit the PAPR Equation (7.209) to $\mu^2 E\left[|x(t)|^2\right]$ for some positive number μ the candidate phase vectors can be generated from Equation (7.213) subject to the following constraint:

$$\left\| \begin{bmatrix} Q_{1,1}^k & \cdots & sQ_{1,M}^k \\ 0 & Q_{2,2}^k & \vdots \\ \vdots & \ddots & \vdots \\ 0 & 0 & Q_{M,M}^k \end{bmatrix} \cdot \begin{bmatrix} b_1 \\ b_2 \\ \vdots \\ b_M \end{bmatrix} \right\|^2 < \mu^2 + \alpha^2 M \tag{7.214}$$

for $1 \le k \le LN$. Sphere decoding only searches among those candidates that lie inside the sphere of radius $\mu^2 + \alpha^2 M$ and, therefore, reduces the complexity of the search. If we rewrite Equation (7.214) as

$$\sum_{v=1}^{M} \left| \sum_{u=0}^{M} Q_{v,u}^k b_u \right|^2 < \mu^2 + \alpha^2 M, \; 1 \le k \le LN \tag{7.215}$$

then in order to satisfy Equation (7.215), the following set of inequalities must be satisfied for $1 \le k \le LN$:

$$\left|Q_{M,M}^k b_M\right|^2 < \mu^2 + \alpha^2 M,$$

$$\sum_{v=M-1}^{M} \left| \sum_{u=v}^{M} Q_{v,u}^k b_u \right|^2 < \mu^2 + \alpha^2 M$$

$$\sum_{v=M-2}^{M} \left| \sum_{u=v}^{M} Q_{v,u}^k b_u \right|^2 < \mu^2 + \alpha^2 M \tag{7.216}$$

$$\vdots$$

$$\sum_{v=1}^{M} \left| \sum_{u=v}^{M} Q_{v,u}^k b_u \right|^2 < \mu^2 + \alpha^2 M$$

Note that the first equation contains b_M only, the second b_{M-1} and b_M only, and so on. We fix $b_M = 1$ without loss of generality. However, the first term in Equation (7.216) constrains the parameter μ (which specifies the achievable PAPR reduction). We use the second line of Equation (7.216) and b_M to generate candidates for b_{M-1}. These candidates and (7.218) again are used to generate candidates for b_{M-2}. This process is repeated until the candidates for the whole phase vector $\mathbf{b}$ are generated. The resulting number of candidates is substantially smaller than q^{M-1}. Therefore, the search space is reduced, compared with exhaustively searching all q^{M-1} phase vectors, which reduces complexity.

For simulation an OFDM signals with 512 8-PSK subcarriers, $L = 4$ and $\alpha = \sqrt{1/M}$ is used. Figure 7.77 compares the complementary cumulative density functions (CCDF's) of the PAPR where the number of sub-blocks in PTS is $M = 8$ and 4. Also, the PTS phase factors are chosen from $P = \{+1, -1\}$. Note that both the SDBA algorithm presented in this section and exhaustive search perform identically, verifying that the SDBA algorithm is optimal, resulting in approximately 1-dB additional reduction compared with the FA.

Instead of scrambling, care can be taken over low peak-to-average power ratios in the stage of channel coding [101–120].

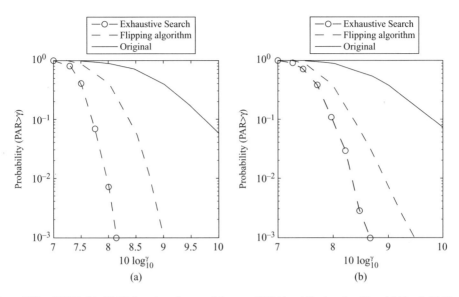

Figure 7.77   CCDF of the PAPR for exhaustive search (same as SDBA) and flipping algorithm, (a) $M = 8$, (b) $M = 4$.

Appendix 7.1

Derivation of Equation (7.152)

As explained earlier, $d^k[n]$ is also denoted as $s_{i,j}^k[p]$, and among all the received signals, only $y_i[p]$ is related to $s_{i,j}^k[p]$. Hence, Δs^2 can be further simplified to

$$\Delta s^2 \triangleq \left\| y_i[p] + \sum_l X_i^l[p]W[p]h^l \right\|^2_{s_{i,j}^k[p]=+1}$$

$$- \left\| y_i[p] + \sum_l X_i^l[p]W[p]h^l \right\|^2_{s_{i,j}^k[p]=-1}$$

$$p[\mathbf{h}^k | \mathbf{H}^k, \sigma^2, \mathbf{D}, \mathbf{Y}] \propto p[\mathbf{Y}|\mathbf{H}, \sigma^2, \mathbf{D}]p[\mathbf{h}^k]$$

$$\propto \exp\left\{ -\frac{1}{\sigma^2} \sum_{i=0}^{M-1} \left\| y_i + \sum_{l=1}^{K} X_i^l \mathbf{W} h^l \right\|^2 \right\} \exp\left\{ -(\mathbf{h}^k - \mathbf{h}_{k0})^H \Sigma_{k0}^{-1}(\mathbf{h}^k - \mathbf{h}_{k0}) \right\}$$

$$\propto \exp\left\{ -\mathbf{h}^{k^H} \underbrace{\left(\Sigma_{k0}^{-1} + \frac{1}{\sigma^2} \sum_{i=0}^{M-1} \mathbf{W}^H \mathbf{X}_i^{k^H} \mathbf{X}_i^k \mathbf{W} \right)}_{\Sigma_{k*}^{-1}} \mathbf{h}^k \right.$$

$$\left. + 2R\left\{ \mathbf{h}^{k^H} \underbrace{\left[\Sigma_{k0}^{-1}\mathbf{h}_{k0} + \frac{1}{\sigma^2} \sum_{i=0}^{M-1} \mathbf{W}^H \mathbf{X}_i^{k^H}\left(y_i - \sum_{l\neq k} \mathbf{X}_i^k \mathbf{W} h^l \right) \right]}_{\Sigma_{k*}^{-1}h_{k*}} \right\} \right\}$$

$$\propto \exp\left\{ -(\mathbf{h}^k - \mathbf{h}_{k*})^H \Sigma_{k*}^{-1}(\mathbf{h}^k - \mathbf{h}_{k*}) \right\} \sim N_c(H_{k*}, \Sigma_{k*})$$

(A.7.1)

Derivation of Equation (7.155)

$$p[\sigma^2|\mathbf{H}, \mathbf{D}, \mathbf{Y}] \propto p[\mathbf{Y}|\mathbf{H}, \sigma^2|\mathbf{D}]p[\sigma^2]$$

$$\propto \left(\frac{1}{\sigma^2}\right)^{2MN} \exp\left(-\frac{1}{\sigma^2}\underbrace{\sum_{i=0}^{M-1}\left\|\mathbf{y}_i + \sum_{k=1}^{K}\mathbf{X}_i^k\mathbf{W}\mathbf{h}^k\right\|^2}_{s^2}\right) \times \left(\frac{1}{\sigma^2}\right)^{v_0+1}\exp\left(-\frac{v_0\lambda_0}{\sigma^2}\right)$$

(A.7.2)

$$= \left(\frac{1}{\sigma^2}\right)^{v_0+2MN+1}\exp\left(-\frac{v_0\lambda_0 + s^2}{\sigma^2}\right)$$

$$\sim \chi^{-2}\left(2[v_0 + 2MN], \frac{v_0\lambda_0 + s^2}{v_0 + 2MN}\right)$$

Derivation of Equation (7.157)

$$p[d^k[n] = +1|\mathbf{H}, \sigma^2, \mathbf{D}_n^k, \mathbf{Y}] = \frac{p[d^k[n] = +1|\mathbf{H}, \sigma^2, \mathbf{D}_n^k|\mathbf{Y}]}{p[\mathbf{H}, \sigma^2, \mathbf{D}_n^k|\mathbf{Y}]}$$

$$= p[\mathbf{Y}|d^k[n] = +1, \mathbf{D}_n^k, \mathbf{H}, \sigma^2]\frac{P[d^k[n] = +1|\mathbf{D}_n^k]P[\mathbf{H}]P[\sigma^2]}{p[\mathbf{H}, \sigma^2, \mathbf{D}_n^k|\mathbf{Y}]}$$

$$\Rightarrow \frac{p[d^k[n] = +1|\mathbf{H}, \sigma^2, \mathbf{D}_n^k, \mathbf{Y}]}{p[d^k[n] = -1|\mathbf{H}, \sigma^2, \mathbf{D}_n^k, \mathbf{Y}]} = \frac{P[d^k[n] = +1, \mathbf{D}_n^k]p[\mathbf{Y}|d^k[n] = +1, \mathbf{D}_n^k, \mathbf{H}, \sigma^2]}{P[d^k[n] = -1, \mathbf{D}_n^k]p[\mathbf{Y}|d^k[n] = -1, \mathbf{D}_n^k, \mathbf{H}, \sigma^2]}$$

$$= \frac{P[d^k[n + 1]|d^k[n] = +1]}{P[d^k[n + 1]|d^k[n] = -1]} \times \frac{P[d^k[n] = +1|d^k[n - 1]]}{P[d^k[n] = -1|d^k[n - 1]]}$$

(A.7.3)

$$\times \exp\left\{-\frac{1}{\sigma^2}\underbrace{\sum_{t=0}^{M-1}\left[\left\|\mathbf{y}_t + \sum_l\mathbf{X}_t^l\mathbf{W}\mathbf{h}^l\right\|^2_{d^k[n]=+1} - \left\|\mathbf{y}_t + \sum_l\mathbf{X}_t^l(d^l)\mathbf{W}\mathbf{h}^l\right\|^2_{d^k[n]=-1}\right]}_{\Delta s^2}\right\}$$

$$= \exp\left[\rho^k[n]d^k[n + 1] + \rho^k[n - 1]d^k[n - 1] - \frac{\Delta s^2}{\sigma^2}\right].$$

References

[1] Li, Y. G., Cimini, L. J. Jr, and Sollenberger, N. R. (1998) Robust channel estimation for OFDM systems with rapid dispersive fading channels, *IEEE Transactions On Communications*, **46**, 902–915.
[2] Jia-Chin Lin (2003) Maximum-likelihood frame timing instant and frequency offset estimation for OFDM communication over a fast Rayleigh-fading channel, *IEEE Transactions on Vehicular Technology*, **52**(4), 1049–1062.
[3] Chengyang Li and Roy, S. (2003) Subspace-based blind channel estimation for OFDM by exploiting virtual carriers, *IEEE Transactions on Wireless Communications*, **2**(1), 141–150.
[4] Seog Geun Kang, Yong Min Ha and Eon Kyeong Joo (2003) A comparative investigation on channel estimation algorithms for OFDM in mobile Communications, *IEEE Transactions on Broadcasting*, **49**(2), 142–149.
[5] Zheng Yuanjin (2003) A novel channel estimation and tracking method for wireless OFDM systems based on pilots and Kalman filtering, *IEEE Transactions on Consumer Electronics*, **49**(2), 275–283.
[6] Xiaobo Zhou and Xiaodong Wang (2003) Channel estimation for OFDM systems using adaptive radial basis function networks, *IEEE Transactions on Vehicular–chnology*, **52**(1), 48–59.

[7] Deneire, L., Vandenameele, P., van der Perre, L., Gyselinckx, B. and Engels, M. (2003) A low-complexity ML channel estimator for OFDM, *IEEE Transactions On Communications*, **51**(2), 135–140.

[8] Muquet, B., de Courville, M. and Duhamel, P. (2002) Subspace-based blind and semi-blind channel estimation for OFDM systems, *IEEE Transactions on Signal Processing*, **50**(7), 1699–1712.

[9] Luise, M., Marselli, M. and Reggiannini, R. (2002) Low-complexity blind carrier frequency recovery for OFDM signals over frequency-selective radio channels, *IEEE Transactions On Communications*, **50**(7), 1182–1188.

[10] Coleri, S., Ergen, M., Puri, A. and Bahai, A. (2002) Channel estimation techniques based on pilot arrangement in OFDM systems, *IEEE Transactions on Broadcasting*, **48**(3), 223–229.

[11] Landstrom, D., Wilson, S. K., van de Beek, J.-J., Odling, P. and Borjesson, P. O. (2002) Symbol time offset estimation in coherent OFDM systems, *IEEE Transactions On Communications*, **50**(4), 545–549.

[12] May, T., Rohling, H. and Engels, V. (1998) Performance analysis of Viterbi decoding for 64-DAPSK and 64-QAM modulated OFDM signals, *IEEE Transactions On Communications*, **46**(2), 182–190.

[13] Shäfer, R. (1995) Terrestrial transmission of DTVB signals–The European specification. in *Proceedings of the International Broadcasting Convention*, Amsterdam, The Netherlands, 1995, pp. 79–84.

[14] Bagels, V. and Rohling, H. (1995) Multilevel differential modulation techniques (64-DAPSK) for multicarrier transmission systems, *European Transactions on Telecommunications*, **6**, 633–640.

[15] Rohling, H. and Engels, V. (1995) Differential amplitude phase shift keying (DAPSK)–A new modulation method for DTVB, in *Proceedings of the International Broadcasting Convention*, Amsterdam, The Netherlands, 1995, pp. 102–108.

[16] Monnier, R., Rault, J. B. and de Couasnon, T. (1992) Digital television broadcasting with high spectral efficiency, in *Proceedings of the International Broadcasting Convention*, Amsterdam, The Netherlands, 1992, pp. 380–384.

[17] Marti, B., Bernard, P., Lodge, N. and Shäfer, R. (1993) European activies on digital television broadcasting–From company to cooperative projects, *EBU Technical Review*, pp. 22–29.

[18] Chow, Y. C., Nix, A. R. and McGeehan, J. P. (1992) Analysis of 16-APSK modulation in AWGN and Rayleigh fading channel, *Electronics Letters*, **28**, 1608–1610.

[19] Giallorenci, T. R. and Wilson, S. G. (1995) Noncoherent demodulation techniques for trellis-coded M-DPSK signals, *IEEE Transactions On Communications*, **43**, 2370–2380.

[20] Simon, M. K. and Divsalar, D. (1988) The performance of trellis coded multilevel DPSK on a fading mobile satellite channel, *IEEE Transactions on Vehicular Technology*, **37**, 78–91.

[21] Divsalar, D. and Simon, M. K. (1990) Multiple-symbol differential detection of MPSK, *IEEE Transactions On Communications*, **38**, 300–308.

[22] Divsalar, D., Simon, M. K. and Shahshahani, M. (1990) The performance of trellis coded MDPSK with multiple symbol detection, *IEEE Transactions On Communications*, **38**, 1391–1403.

[23] Hanzo, L. *et al.* (2002) *Adaptive Wireless Transceivers*. John Wiley & Sons, Ltd, Chichester.

[24] Zhiqiang Liu and Giannakis, G. B. (2003) Block differentially encoded OFDM with maximum multipath diversity, *IEEE Transactions on Wireless Communications*, **2**(3), 420–423.

[25] Zhiqiang Liu, Yan Xin and Giannakis, G. B. (2002) Space–time-frequency coded OFDM over frequency-selective fading channels, *IEEE Transactions on Signal Processing*, **50**(10), 2465–2476.

[26] Stamoulis, A., Zhiqiang, L. and Giannakis, G. B. (2002) Space–time block-coded OFDMA with linear precoding for multirate services, *IEEE Transactions on Signal Processing*, **50**(1), 119–129.

[27] Kuo-Hui Li and Ingram, M. A. (2002) Space–time block-coded OFDM systems with RF beamformers for high-speed indoor wireless communications, *IEEE Transactions On Communications*, **50**(12), 1899–1901.

[28] Lu, B., Xiaodong Wang and Ye Li (2002) Iterative receivers for space–time block-coded OFDM systems in dispersive fading channels, *IEEE Transactions on Wireless Communications*, **1**(2), 213–225.

[29] Ye Li (2002) Simplified channel estimation for OFDM systems with multiple transmit antennas, *IEEE Transactions on Wireless Communications*, **1**(1), 67–75.

[30] Lu, B., Xiaodong Wang and Narayanan, K. R. (2002) LDPC-based space–time coded OFDM systems over correlated fading channels: Performance analysis and receiver design, *IEEE Transactions On Communications*, **50**(1), 74–88.

[31] Blum, R. S., Ye Geoffrey Li, Winters, J. H. and Qing Yan (2001) Improved space–time coding for MIMO-OFDM wireless communications, *IEEE Transactions on Communications*, **49**(11), 1873–1878.

[32] Li, Y. G., Seshadri, N. and Ariyavisitakul, S. (1999) Channel estimation for OFDM systems with transmitter diversity in mobile wireless channels, *IEEE Journal on Selected Areas in Communications*, **17**, 461–471.

[33] Blum, R. S., Ye Geoffrey Li, Winters, J. H. and Qing Yan (2001) Improved space–time coding for MIMO-OFDM wireless Communications, *IEEE Transactions On Communications*, **49**(11), 1873–1878.

[34] Li, Y. G., Winters, J. H. and Sollenberger, N. R. (2001) Signal detection for MIMO-OFDM wireless Communications, *IEEE International Conference on Communications*, June 2001.

[35] Al-Dhahir, N., Uysal, M. and Georghiades, C. N. (2001) Three space–time block-coding schemes for frequency-selective fading channels with application to EDGE, *IEEE 54th Vehicular Technology Conference*, 7–11 October 2001, **3**, 1834–1838.

[36] Younis, W. and Al-Dhahir, N. (2002) Joint prefiltering and MLSE equalization of space–time-coded transmissions over frequency-selective channels, *IEEE Transactions on Vehicular Technology*, **51**(1), 144–154.

[37] Al-Dhahir, N. (2002) Overview and comparison of equalization schemes for space–time-coded signals with application to EDGE, *IEEE Transactions on Signal Processing*, **50**(10), 2477–2488.

[38] Franz, V. and Anderson, J. (1998) Concatenated decoding with a reduced-search BCJR algorithm, *IEEE Journal on Selected Areas in Communications*, **16**, 186–195.

[39] Bahl, L., Cocke, J., Jelinek, F. and Raviv, J. (1974) Optimal decoding of linear codes for minimizing symbol error rate, *IEEE Transactions On Information Theory*, **IT-20**, 284–287.

[40] Fragouli, C., Al-Dhahir, N., Diggavi, S. and Turin, W. (2002) Prefiltered space–time M-BCJR equalizer for frequency-selective channels, *IEEE Transactions on Communications*, **50**, 742–753.

[41] Naguib, A. and Seshadri, N. (2000) MLSE and equalization of space–time coded signals, in *Proceedings of Vehicular Technology Conference*, May 2000, pp. 1688–1693.

[42] Al-Dhahir, N. (2001) FIR channel-shortening equalizers for MIMO ISI channels, *IEEE Transactions On Communications*, **50**, 213–218.

[43] Duel-Halien, A. and Heegard, C. (1989) Delayed decision-feedback sequence estimation, *IEEE Transactions On Communications*, **36**, 428–436.

[44] Lindskog, E. and Paulraj, A. (2000) A transmit diversity scheme for delay spread channels, in *Proceedings of the International Conference on Communications*, June 2000, pp. 307–311.

[45] Pollet, T., Van Bladel, M. and Moeneclaey, M. (1995) BER sensitivity of OFDM systems to carrier frequency offset and wiener phase noise, *IEEE Transactions On Communications*, **44**, 191–193.

[46] Sari, H., Karam, G. and Jeanclaude, I. (1995) Transmission techniques for digital terrestrial TV broadcasting, *IEEE Communications Magazine*, **33**, 100–109.

[47] Clark, M. V. (1998) Adaptive frequency-domain equalization and diversity combining for broadband wireless Communications, *IEEE Journal on Selected Areas in Communications*, **16**, 1385–1395.

[48] Al-Dhahir, N. (2001) Single-carrier frequency-domain equalization for space–time block-coded transmissions over frequency-selective fading channels, *IEEE Communications Letters*, **5**, 304–306.

[49] Kaleh, G. (1995) Channel equalization for block transmission systems, *IEEE Journal on Selected Areas in Communications*, **13**, 110–121.

[50] SC FDE PHY Layer Sys. Proposal for Sub 11 GHz BWA (Online). Available: http://www.ieee802.org/16/tg3/contrib/802 163p-0131r2.pdf.

[51] Crazier, S., Falconer, D. and Mahmoud, S. (1991) Least sum of squared errors (LSSE) channel estimation, *Proceedings of the Institute of Electronic Engineers F*, pp. 371–378.

[52] Chu, D. (1972) Polyphase codes with good periodic correlation properties, *IEEE Transactions On Information Theory*, **IT-18**, 531–532.

[53] Fragouli, C., Al-Dhahir, N. and Turin, W. (2002) Reduced-complexity training schemes for multiple-antenna broadband transmissions, in *Proceedings of WCNC*, **1**, 78–83.

[54] Fragouli, C., Al-Dhahir, N. and Turin, W. (2002) Finite-alphabet constant-amplitude training sequences for multiple-antenna broadband transmissions, in *Proceedings of the International Control Conference*, **1**, 6–10.

[55] Chong, L. L. and Milstein, L. B. (2000) Error rate of a multicarrier CDMA system with imperfect channel estimates, *IEEE International Conference on Communications* (ICC 2000), 18–22 June 2000, **2**, 934–938.

[56] Kondo, S. and Milstein, L. B. (1996) On the performance of multicarrier DS CDMA systems, *IEEE Transactions On Communications*, **44**, 238–246.

[57] Fazel, K. and Fettweis, G. P. (Eds.) (1997) *Multi-Carrier Spread Spectrum*. Kluwer, Boston, MA.

[58] Sourour, E. and Nakagawa, M. (1996) Performance of orthogonal multicarrier CDMA in a multipath fading channel, *IEEE Transactions On Communications*, **44**, 356–367.

[59] Kondo, S. and Milstein, L. B. (1993) On the use of multicarrier direct sequence spread spectrum systems, in *Proceedings of IEEE MILCOM*, Boston, MA, October 1993, pp. 52–56.

[60] TIA/TR45.5. (1998, July). *The cdma2000 ITU-R RTT candidate submission (0.18)* (Online). Available: http://www.itu.Int/imt/2-radio-dev/proposals/index.html.

[61] Eng, T. and Milstein, L. B. (1994) Comparison of hybrid FDMA/CDMA systems in frequency selective Rayleigh fading, *IEEE Journal on Selected Areas in Communications*, **12**, 938–951.

[62] Dongwook Lee and Milstein, L. B. (1999) Comparison of multicarrier DS-CDMA broadcast systems in a multipath fading channel, *IEEE Transactions On Communications*, **47**(12), 1897–1904.

[63] Hagenauer, J. (1988) Rate-compatible punctured convolutional codes (RCPC Codes) and their applications, *IEEE Transactions On Communications*, **36**, 389–400.

[64] Balachandran, K., Kadaba, S. R. and Nanda, S. (1999) Channel quality estimation and rate adaptation for cellular mobile radio, *IEEE Journal on Selected Areas in Communications*, **17**, 1244–1256.

[65] Frenger, P., Orten, P., Ottosson, T. and Svensson, A. (1999) Rate-compatible convolutional codes for multirate DS-CDMA systems, *IEEE Transactions On Communications*, **47**, 828–836.

[66] Jumi Lee, Iickho Song, So Ryoung Park, and Seokho Yoon (2001) Analysis of an adaptive rate convolutionally coded multicarrier DS/CDMA system, *IEEE Transactions on Vehicular Technology*, **50**(4), 1014–1023.

[67] Weiping Xu and Milstein, L. B. (2001) On the use of Interference suppression to reduce Intermodulation distortion in multicarrier CDMA systems, *IEEE Transactions On Communications*, **49**(1), 130–141.

[68] Lin Fang and Milstein, L. B. (2001) Performance of successive Interference cancellation in convolutionally coded multicarrier DS/CDMA systems, *IEEE Transactions On Communications*, **49**(12), 2062–2067.

[69] Miller, S. L. and Rainbolt, B. J. (2000) MMSE detection of multicarrier CDMA, *IEEE Journal on Selected Areas in Communications*, **18**(11), 2356–2362.

[70] Kalofonos, D. N., Stojanovic, M. and Proakis, J. G. (1998) On the performance of adaptive MMSE detectors for a MC-CDMA system in fast fading Rayleigh channels, *The Ninth IEEE International Symposium on Personal, Indoor and Mobile Radio Communications*, 8–11 September 1998, **3**, 1309–1313.

[71] Kalofonos, D. N., Stojanovic, M. and Proakis, J. G. (2003) Performance of adaptive MC-CDMA detectors in rapidly fading Rayleigh channels, *IEEE Transactions on Wireless Communications*, **2**(2), 229–239.

[72] Jinghong Ma and Tugnait, J. K. (2002) Blind detection of multirate asynchronous CDMA signals in multipath channels, *IEEE Transactions on Signal Processing*, **50**(9), 2258–2272.

[73] Kunjie Wang, Pingping Zong and Bar-Ness, Y. (2001) A reduced complexity partial sampling MMSE receiver for asynchronous MC-CDMA systems, *IEEE Global Telecommunications Conference*, November 2001, **2**, 728–732.

[74] Pingping Zong, Kunjie Wang and Bar-Ness, Y. (2001) Partial sampling MMSE Interference suppression in asynchronous multicarrier CDMA system, *IEEE Journal on Selected Areas in Communications*, **19**(8), 1605–1613.

[75] Weiping Xu and Milstein, L. B. (1998) MMSE Interference suppression for multicarrier DS-CDMA in frequency selective channels, *IEEE Global Telecommunications Conference*, 1998, 8–12 November 1998, **1**, 259–264.

[76] Petre, F., Vandenameele, P., Bourdoux, A., Gyselinckx, B., Engels, M., Moonen, M. and DeMan, H. (2000) Combined MMSE/pcPIC multiuser detection for MC-CDMA, *IEEE 51st Vehicular Technology Conference Proceedings*, Tokyo, 15–18 May 2000, **2**, 770–774.

[77] Petre, F., Engels, M., Moonen, M., Gyselinckx, B. and De Man, H. (2001) Adaptive MMSE/pcPIC-MMSE multiuser detector for MC-CDMA satellite system, *IEEE International Conference on Communications*, 11–14 June 2001, **9**, 2640–2644.

[78] Hyung-Yun Kong and Chang-Hee Lee (1999) Design of MC-CDMA system based on non-linear MMSE, *Proceedings of the IEEE Region 10 Conference*, 15–17 September 1999, **1**, 53–56.

[79] Pingping Zong, Kunjie Wang and Bar-Ness, Y. (2001) A novel partial sampling MMSE receiver for uplink multicarrier CDMA, *IEEE VTS 53rd Vehicular Technology Conference*, 6–9 May 2001, **1**, 741–745.

[80] Helard, J.-F., Baudais, J.-Y. and Citerne, J. (2000) Linear MMSE detection technique for MC-CDMA, *Electronics Letters*, **36**(7), 665–666.

[81] Namgoong, J., Wong, T. F. and Lehnert, J. S. (1999) Subspace MMSE receiver for multicarrier CDMA, *IEEE Wireless Communications and Networking Conference* (WCNC 1999), 21–24, September 1999, **1**, 90–94.

[82] Zigang Yang, Lu, B. and Xiaodong Wang (2001) Blind Bayesian multiuser receiver for space–time coded MC-CDMA system over frequency-selective fading channel, *IEEE Global Telecommunications Conference*, 25–29 November 2001, **2**, 781–785.

[83] Zigang Yang, Ben Lu and Xiaodong Wang (2001) Bayesian Monte Carlo multiuser receiver for space–time coded multicarrier CDMA systems, *IEEE Journal on Selected Areas in Communications*, **19**(8), 1625–1637.

[84] Lie-Liang Yang and Hanzo, L. (2002) Broadband MC DS-CDMA using space–time and frequency-domain spreading, *IEEE 56th Vehicular Technology Conference*, 24–28 September 2002, **3**, 1632–1636.

[85] Gelfand, A. and Smith, A. (1990) Sampling-based approaches to calculating marginal densities, *Journal of the American Statistical Association*, **85**, 398–409.

[86] Geman, S. and Geman, D. (1984) Stochastic relaxation, Gibbs distribution, and the Bayesian restoration of images, *IEEE Transactions on Pattern Analysis and Machine Intelligence*, **PAMI-6**, 721–741.

[87] Chan, K. (1993) Asymptotic behavior of the Gibbs sampler, *Journal of the American Statistical Association*, **88**, 320–326.

[88] Liu, J., Wong, W. and Hong, A. (1995) Covariance structure and convergence rate of the Gibbs sampler with various scans, *Journal of the Royal Statistical Society Series B*, **57**, 157–169.

[89] Robert, C. and Casella, G. (1999) *Monte Carlo Statistical Methods*. Springer-Verlag, New York.

[90] Wang, X. and Chen, R. (2000) Adaptive Bayesian multiuser detection for synchronous CDMA with Gaussian and impulsive noise, *IEEE Transactions on Signal Processing*, **48**, 2013–2028.

[91] Davis, J. A. and Jedwab, J. (1999) Peak-to-average power control in OFDM, Golay complementary sequences, and Reed-Muller codes, *IEEE Transactions on Information Theory*, **45**, 2397–2417.

[92] Muller, S. H. and Huber, J. B. (1997) OFDM with reduced peak-to-average power ratio by optimum combination of partial transmit sequences, *Electronics Letters*, **33**, 368–369.

[93] Tellambura, C. (2001) Improved phase factor computation for the PAR reduction of an OFDM signal using PTS, *IEEE Communications Letters*, **5**, 135–137.

[94] Cimini, L. J. and Sollenberger, N. R. (2000) Peak-to-average power ratio reduction of an OFDM signal using partial transmit sequences, *IEEE Communications Letters*, **4**, 86–88.

[95] Han, S. H. and Lee, J. H. (2004) PAPR reduction of OFDM signals using a reduced complexity PTS technique, *IEEE Signal Processing Letters*, **11**, 887–890.

[96] Ho, W. S., Madhukumar, A. S. and Chin, F. (2003) Peak-to-average power reduction using partial transmit sequences: a suboptimal approach based on dual layered phase sequencing, *IEEE Transactions in Vehicular Technology*, 1268–1272.

[97] Chen, H. and Pottie, G. J. (2002) An orthogonal projection-based approach for PAR reduction in OFDM, *IEEE Communications Letters*, **6**, 169–171.

[98] Fincke, U. and Phost, M. (1985) Improved methods for calculating vectors of short length in a lattice, inlcuding a complexity analysis, *Mathematical Computation*, **44**, 463–471.

[99] Mow, W. H. (2003) Universal lattice decoding: principle and recent advances, *Wireless Communication and Mobile Computing*, **3**, 553–569.

[100] Alavi, A., Tellambura, C. and Fair, I. (2005) PAPR reduction of OFDM signals using partial transmit sequence: an optimal approach using sphere decoding, *IEEE Communications Letters*, **9**, 982–984

[101] Davis, J. A. and Jedwab, J. (1997) Peak-to-mean power control and error correction for OFDM transmission using Golay sequences and Reed–Muller codes, *Electronics Letters*, **33**, 267–268.

[102] Davis, J. A. and Jedwab, J. (1999) Peak-to-mean power control in OFDM, Golay complementary sequences and Reed–Muller codes, *IEEE Transaction Information Theory*, **45**, 2397–2417.

[103] Friese, M. (1996) Multicarrier modulation with low peak-to-mean average power ratio, *Electronics Letters*, **32**, 713–714.

[104] Ho, T. F. and Wei, V. K. (1995) Synthesis of low-crest waveforms for multicarrier CDMA systems, in *Proceedings of IEEE GLOBECOMM*, pp. 131–135.

[105] Jedwab, J. (1997) Comment: M-sequences for OFDM peak-to-average power ratio reduction and error correction, *Electronics Letters*, **33**, 1293–1294.

[106] Jones, A. E., Wilkinson, T. A. and Barton, S. K. (1994) Block coding scheme for reduction of peak to mean envelope power ratio of multicarrier transmission schemes, *Electronics Letters*, **30**, 2098–2099.

[107] Jones, A. E. and Wilkinson, T. A. (1996) Combined coding error control and increased robustness to system nonlinearities in OFDM, in *Proceedings IEEE 46th Vehicular Technology Conference*, Atlanta, GA, pp. 904–908.

[108] Kumar, P. V., Helleseth, T. and Calderbank, A. R. (1995) An upper bound for Weil exponential sums over Galois rings and applications, *IEEE Transactions on Information Theory*, **41**, 456–468.

[109] Li, X. and Ritcey, J. A. (1997) M-sequences for OFDM peak-to-average power ratio reduction and error correction, *Electronics Letters*, **33**, 554–555.

[110] Müller, S. H., Bäuml, R. W., Fischer, R. F. H. and Huber, J. B. (1997) OFDM with reduced peak-to-average power ratio by multiple signal representation, *Annals Télécommunication*, **52**, 58–67.

[111] van Nee, R. D. J. (1996) OFDM codes for peak-to-average power reduction and error correction, in *Proceedings IEEE GLOBECOM*, London, pp. 740–744.

[112] Ochiai, H. and Imai, H. (1997) Block coding scheme based on complementary sequences for multicarrier signals, *IEICE Transactions Fundamentals*, 2136–2143.

[113] Paterson, K. G. (2000) Generalized Reed–Muller codes and power control in OFDM modulation, *IEEE Transactions on Information Theory*, **46**, 104–120.

[114] Shepherd, S., Orriss, J. and Barton, S. (1998) Asymptotic limits in peak envelope power reduction by redundant coding in orthogonal frequency-division multiplex modulation, *IEEE Transactions on Communication*, **46**, 5–10.

[115] Tarokh, V. and Jafarkhani, H. (2000) On reducing the peak to average power ratio in multicarrier communications, *IEEE Transactions on Communication*, **48**, 37–44.

[116] Wilkinson, T. A. and Jones, A. E. (1995) Minimization of the peak to mean envelope power ratio of multicarrier transmission schemes by block coding, in *Proceedings of IEEE 45th Vehicular Technology Conference*, July, pp. 825–829.

[117] Wulich D. (1996) Reduction of peak-to-mean ratio of multicarrier modulation using cyclic coding, *Electronics Letters*, **32**, 432–433.

[118] Paterson, K. G. and Tarokh, V. (2000) On the existence and construction of good codes with low peak-to-average power ratios, *IEEE Transactions on Information Theory*, **46**, 1974–1987.

[119] MacWilliams, F. J. and Sloane, N. J. A. (1986) *The Theory of Error-Correcting Codes*, 2nd ed., North Holland, Amsterdam.

[120] Moreno, O. and Moreno, C. J. (1994) The MacWilliams–Sloane conjecture on the tightness of the Carlitz–Uchiyama bound and the weights of duals of BCH codes, *IEEE Transactions on Information Theory*, **40**, 1894–1907.

8

Ultra Wide Band Radio

In this chapter we discuss technology which is based on the spread spectrum concept, such as CDMA, as described in Chapter 5. The difference is that the pulse (called chip in Chapter 5) period used in this field is below 1 ns, resulting in a bandwidth of over 1 GHz, hence the name *Ultra Wide Band (UWB) Radio*. The second important characteristic is that the signal can be transmitted with no carrier. This is why very often the system is also referred to as *Impulse Radio* (IR). The above characteristics of the signal will require the modification of the signal format and detection concepts. In addition to these issues the chapter will also cover the basic characteristics of the UWB channel.

8.1 UWB Multiple Access in a Gaussian Channel

A typical time-hopping format used in this case can be represented as [1–31]:

$$s_{\mathrm{tr}}^{(k)}(t^{(k)}) = \sum_{j=-\infty}^{\infty} \omega_{\mathrm{tr}}(t^{(k)} - jT_{\mathrm{f}} - c_j^{(k)}T_{\mathrm{c}} - \delta d_{[j/N_{\mathrm{s}}]}^{(k)}) \tag{8.1}$$

where $t^{(k)}$ is the kth transmitter's clock time and T_{f} is the *pulse repetition time*. The transmitted pulse waveform ω_{tr} is referred to as a *monocycle*. To eliminate collisions due to multiple access, each user (indexed by k) is assigned a distinctive time shift pattern $\{c_j^{(k)}\}$ called a *time-hopping sequence*. This provides an additional time shift of $c_j^{(k)}T_{\mathrm{c}}$ seconds to the jth monocycle in the pulse train, where T_{c} is the duration of addressable time delay bins. For a fixed T_{f}, the *symbol rate* R_{s} determines the number N_{s} of monocycles that are modulated by a given binary symbol as $R_{\mathrm{s}} = (1/N_{\mathrm{s}}T_{\mathrm{f}})s^{-1}$. The modulation index δ is chosen to optimize performance. For performance prediction purposes, most of the time the data sequence $\{d_j^{(k)}\}_{j=-\infty}^{\infty}$ is modeled as a wide-sense stationary random process composed of equally likely symbols. For data, a pulse position modulation is used.

8.1.1 The Multiple Access Channel

When K users are active in the multiple access system, the composite received signal at the output of the receiver's antenna is modeled as

$$r(t) = \sum_{k=1}^{K} A_k s_{\mathrm{rec}}^{(k)}(t - t_k) + n(t) \tag{8.2}$$

Advanced Wireless Communications & Internet: Future Evolving Technologies, Third Edition. Savo Glisic.
© 2011 John Wiley & Sons, Ltd. Published 2011 by John Wiley & Sons, Ltd.

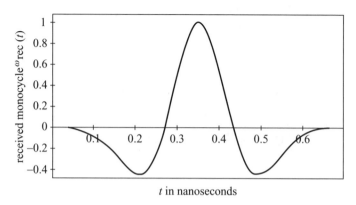

t in nanoseconds

Figure 8.1   A typical ideal received monocycle $\omega_{\text{rec}}(t)$ at the output of the antenna subsystem as a function of time in nanoseconds.

The antenna/propagation system modifies the shape of the transmitted monocycle $\omega_{\text{tr}}(t)$ to $\omega_{\text{rec}}(t)$ at its output. An idealized received monocycle shape $\omega_{\text{rec}}(t)$ for a free-space channel model with no fading is shown in Figure 8.1.

8.1.2   Receiver

The optimum receiver for a single bit of a binary modulated impulse radio signal in additive white Gaussian noise (AWGN) is a correlation receiver defined as

$$decide\ d_0^{(1)} = 0 \quad if$$

$$\overbrace{\sum_{j=0}^{N_s-1} \int_{\tau_1+jT_f}^{\tau_1+(j+1)T_f}}^{pulse\ correlator\ output\ \triangleq\ \alpha_j(u)} r(u,t)\ \upsilon\ (t-\tau_1-jT_f-c_j^{(1)}T_c)\ dt > 0 \tag{8.3}$$

$$\underbrace{\phantom{\sum_{j=0}^{N_s-1} \int_{\tau_1+jT_f}^{\tau_1+(j+1)T_f} r(u,t)\ \upsilon\ (t-\tau_1-jT_f-c_j^{(1)}T_c)\ dt}}_{test\ statistic\ \triangleq\ \alpha(u)}$$

where $\upsilon(t) \triangleq \omega_{\text{rec}}(t) - \omega_{\text{rec}}(t-\delta)$.

The *optimal detection* in a multiuser environment, with knowledge of all time-hopping sequences, leads to complex parallel receiver designs [2]. However, if the number of users is large and no such multiuser detector is feasible, then it is reasonable to approximate the combined effect of the other users dehopped interfering signals as a Gaussian random process [2]. Hence, the single-link reception algorithm (8.3) as shown in Figure 8.2 can be used for practical implementations. The test statistic in Algorithm (8.3) consists of summing the N_s correlations α_j of the correlators template signal $\upsilon(t)$ at various time shifts with the received signal $r(t)$.

For the monocycle waveform of Figure 8.1, the optimum choice of δ is 0.156 ns. By choosing $\delta = 0.156$ ns and $T_f = 100$ ns, we achieve the results shown in Figure 8.3. For the evaluation of the system efficiency, formulas from Chapter 4 are applicable with

$$D_r = K, \quad \left(\frac{k_{02}}{k_{01}}\right) = K \text{ and } -g_{12} = \text{Additional Required Power (ARP)}$$

Both $K = $ total number of users and ARP are available in Figure 8.3.

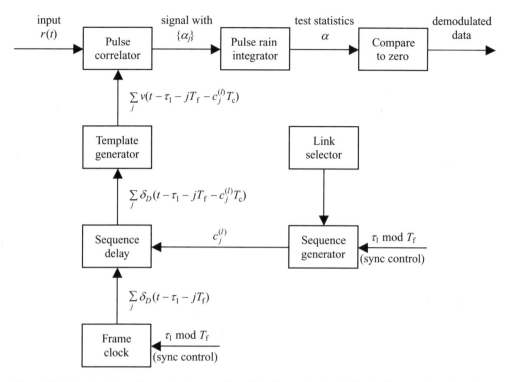

Figure 8.2   Receiver block diagram for the reception of the first user's signal. Clock pulses are denoted by Dirac delta functions $\delta_D(\cdot)$ [31] © 2001, IEEE.

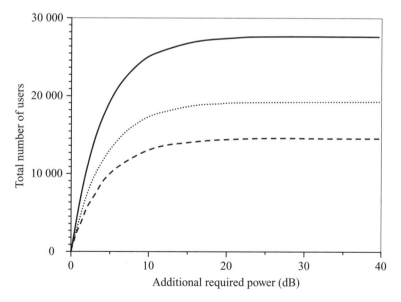

Figure 8.3   Total number of users versus additional required power (decibels) for the impulse radio example. Ideal power control is assumed at the receiver. Three different BER performance levels with the data rate set at 19.2 kb/s are considered.

8.2   The UWB Channel

8.2.1   Energy Capture

In Section 8.1 a Gaussian channel was assumed. For a UWB signal, a high resolution of multipath channel is expected. In this section we discuss some characteristics of a UWB signal in such a channel.

8.2.2   The Received Signal Model

In general, the received signal can be presented as

$$r(u, t) = s(u_s, t) + n(u_n, t) \qquad (8.4)$$

where u characterizes the set of parameters defining the environment (position of the receiver in the room). The RAKE correlator structure is modeled as

$$\sum_{i=1}^{L_p} c_i \omega(t - \tau_i) \qquad (8.5)$$

For experimental purposes, the pulse in Figure 8.1 that can be represented as $\omega_{rec}(t + 1.0) = \lfloor 1 - 4\pi(t/\tau_m)^2 \rfloor \exp\lfloor -2\pi(t/\tau_m)^2 \rfloor$ with $t_m = 0.78125$ is used. The *ML estimates* of the amplitude vector $\hat{c}(\tilde{u})$ and delay vector $\hat{\tau}(\tilde{u})$ based on a specific observation $r(\tilde{u}, t)$ are the values c and τ which minimize the following mean squared error:

$$E(\tilde{u}, L_p) = \int_0^T \left| r(\tilde{u}, t) - \sum_{i=1}^{L_p} c_i \omega(t - \tau_i) \right|^2 dt \qquad (8.6)$$

The minimum value of the above mean squared error is denoted by $E_{min}(u, L_p)$. The *energy capture*, a function of L_p for each observation $r(\tilde{u}, t)$, is defined mathematically as

$$EC(\tilde{u}, L_p) = 1 - \underbrace{\frac{E_{min}(\tilde{u}, L_p)}{E_{tot}(\tilde{u})}}_{\triangleq \text{ normalized MSE}} \qquad (8.7)$$

8.2.3   The UWB Signal Propagation Experiment 1

A UWB signal propagation experiment performed in a typical modern office building [4] is described. The bandwidth of the signal used in this experiment is in excess of 1 GHz, resulting in a differential path delay resolution of less than a nanosecond.

The transmitter is kept stationary in the central location of the building. Multipath profiles are measured using a digital sampling oscilloscope on one floor at various locations in 14 different rooms and hallways. In each office, multipath measurements are made at 49 different locations. They are arranged spatially in a level 7 × 7 square grid with six inch (15 cm) spacing, covering 3′ × 3′ (90 × 90 cm).

Measurements from three different offices are used in the following discussions as typical examples of propagation environments. In these offices, the receiving antennas are located 6, 10 and 17 m away from the transmitter, representing typical UWB signal transmissions characterized as a 'high signal to noise ratio (SNR)' environment, 'low SNR' environment, and 'extremely low SNR' environment, respectively. The transmitter and receiving antenna are located in different rooms in these examples. Detailed results of this UWB signal propagation experiment can be found in [4, 13]. The results are shown in Figure 8.4 in the form of upper and lower bound curves.

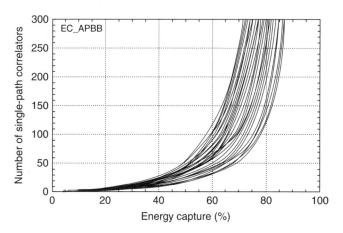

Figure 8.4 The required diversity level, L_p, in a UWB RAKE receiver as a function of percentage energy capture for each of the 49 received waveforms in an office representing a 'high SNR' environment (courtesy Moe Win, Massachusetts Institute of Technology) [4] © 1997, IEEE.

8.2.4 UWB Propagation Experiment 2

The propagation experiment described in this section uses two vertically polarized diamond-dipole antennas [6], each 1.65 m above the floor and 1.05 m below the ceiling in an office/laboratory environment [7]. The equivalent received pulse at 1 m in free space can be estimated as the 'direct path' signal in an experiment in which there is no multipath signal, as shown in Figure 8.5.

A collection of results of recovered signal locations (delay and azimuth) is shown in Figure 8.6.

8.2.5 Clustering Models for the Indoor Multipath Propagation Channel

A number of models for the indoor multipath propagation channel [8–12] have reported a clustering of multipath components, in both time and angle. In the model presented in [11], the received signal

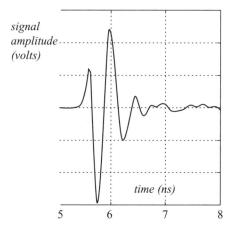

Figure 8.5 Transmitted pulse shape captured at 1 m separation from the transmit antenna.

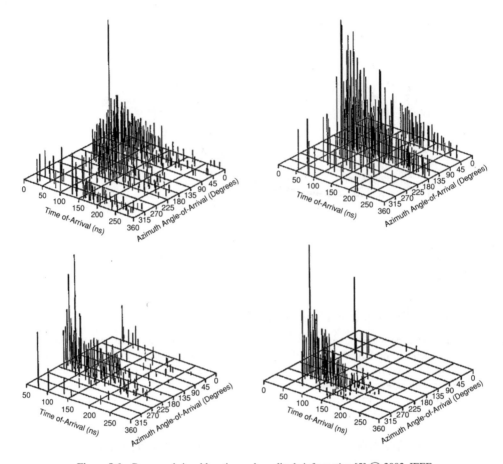

Figure 8.6   Recovered signal location and amplitude information [5] © 2002, IEEE.

amplitude β_{kl} is a Rayleigh distributed random variable with a mean square value that obeys a double exponential decay law, according to

$$\overline{\beta_{kl}^2} = \overline{\beta^2(0,0)}\, e^{-T_l/\Gamma}\, e^{-\tau_{kl}/\gamma} \tag{8.8}$$

where $\overline{\beta^2(0,0)}$ describes the average power of the first arrival of the first cluster, T_l represents the arrival time of the lth cluster, and τ_{kl} is the arrival time of the kth arrival within the lth cluster, relative to T_1. The parameters Γ and γ determine the inter-cluster signal level rate of decay and the intra-cluster rate of decay, respectively. The parameter Γ is generally determined by the architecture of the building, while γ is determined by objects close to the receiving antenna, such as furniture. The results presented in [11] make the assumption that the channel impulse response as a function of time and azimuth angle is a separable function, or

$$h(t, \theta) = h(t)\, h(\theta) \tag{8.9}$$

from which independent descriptions of the multipath time-of-arrival and angle-of-arrival are developed. This is justified by observing that the angular deviation of the signal arrivals within a cluster from the cluster mean does not increase as a function of time.

The cluster decay rate Γ and the ray decay rate γ can be interpreted for the environment in which the measurements were made. For the results, presented later in this section, at least one wall separates the transmitter and the receiver. Each cluster can be viewed as a path that exists between the transmitter and the receiver, along which signals propagate. This cluster path is generally a function of the architecture of the building itself. The component arrivals within a cluster vary because of secondary effects, e.g. reflections from the furniture or other objects. The primary source of degradation in the propagation through the features of the building is captured in the decay exponent Γ. Relative effects between paths in the same cluster do not always involve the penetration of additional obstructions or additional reflections, and therefore tend to contribute less to the decay of the component signals. Results for $p(\theta)$ generated from the data in Figure 8.6 are shown in Figure 8.7.

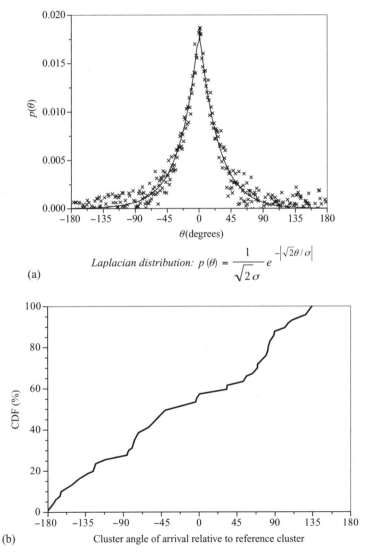

Laplacian distribution: $p(\theta) = \dfrac{1}{\sqrt{2}\sigma} e^{-\left|\sqrt{2}\theta/\sigma\right|}$

(a)

(b)

Figure 8.7 Ray arrival angles at $1°$ of resolution and a best fit Laplacian density with $\sigma = 38°$; (b) distribution of the cluster azimuth angle of arrival, relative to the reference cluster [5] © 2002, IEEE.

Table 8.1   Channel parameters

Parameter	UWB [5]	Spencer et al. [11]	Spencer et al. [11]	Saleh–Valenzuela [10]
Γ	27.9 ns	33.6 ns	78.0 ns	60 ns
γ	84.1 ns	28.6 ns	82.2 ns	20 ns
$1/\Lambda$	45.5 ns	16.8 ns	17.3 ns	300 ns
$1/\lambda$	2.3 ns	5.1 ns	6.6 ns	5 ns
σ	37°	25.5°	21.5°	...

Interarrival times are hypothesized [11] to follow exponential rate laws, given by

$$p\left(T_l \,|\, T_{l-1}\right) = \Lambda e^{-\Lambda(T_l - T_{l-1})}$$
$$p(T_{kl} | T_{k-1,l}) = \lambda e^{-\lambda(T_l - T_{l-1})}$$

where Λ is the cluster arrival rate and λ is the ray arrival rate. Channel parameters are summarized in Table 8.1.

8.2.6   Path Loss Modeling

In this segment we are interested in a transceiver operating at approximately 2 GHz center frequency with a bandwidth in excess of 1.5 GHz, which translates to sub-nanosecond time resolution in the CIRs.

8.2.6.1   Measurement Procedure

The measurement campaign is described in [14] and is conducted in a single-floor, hard-partition office building (fully furnished). The walls are constructed of drywall with vertical metal studs; there is a suspended ceiling ten feet (three metres) in height with carpeted concrete floor. Measurements are conducted with a stationary receiver and mobile transmitter, both transmit and receive antennas are five feet (1.5 metres) above the floor. For each measurement, a 300 ns time domain scan is recorded and the LOS distance from transmitter to receiver is recorded. A total of 906 profiles are included in the dataset with seven different receiver locations recorded over the course of several days. Except for a reference measurement made for each receiver location, all successive measurements are NLOS links, chosen randomly throughout the office layout, that penetrate anywhere from one to five walls. The remainder of the datapoints are taken in a similar fashion.

8.2.6.2   Path Loss Modeling

The average path loss for an arbitrary T-R separation is expressed using the power law as a function of distance. The *indoor* environment measurements show that at any given d, shadowing leads to signals with a path loss that is lognormally distributed about the mean [15, 16]. That is:

$$\text{PL}\,(d) = \text{PL}_0\,(d_0) + 10N \log\left(\frac{d}{d_0}\right) + X_\sigma \tag{8.10}$$

where N is the path loss exponent, X_σ is a zero mean lognormally distributed random variable with standard deviation σ (dB) and PL_0 is the free space path loss at reference distance, d_0. Some results are shown in Figure 8.8.

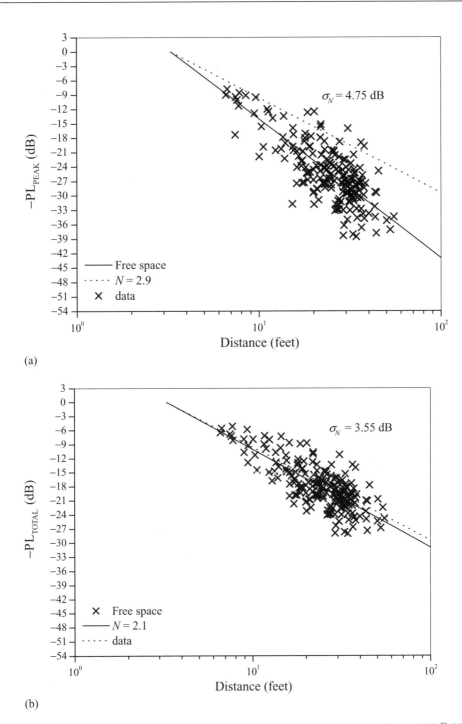

(a)

(b)

Figure 8.8 (a) Peak PL vs. distance; (b) total PL vs. distance; (c) peak PL + RAKE gain vs. distance [14] © 2002, IEEE.

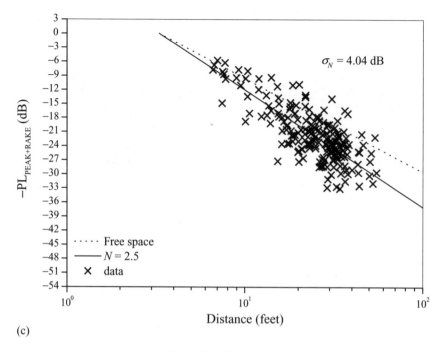

Figure 8.8 *(Continued)*.

Assuming a simple RAKE with four correlators where each component is weighted equally, we can calculate the path loss versus distance using the peak CIR power plus RAKE gain, $PL_{PEAK+RAKE}$, for each CIR, as shown in Figure 8.8(c). The exponent N obtained from performing a least squares fit is 2.5, with a standard deviation of 4.04 dB. The results for delays are shown in Figure 8.9.

8.2.6.3 In-Home Channel

For the in-home channel, Equation (8.10) can also be used to model path loss. Some results are shown in Table 8.2 [17–20].

Table 8.3 presents the results for delay spread in the in-home channel [17].

8.3 UWB System with *M*-ary Modulation

8.3.1 *Performance in a Gaussian Channel*

We will first assume that the transmitted pulse and the received signal are $p_{TX}(t) \triangleq \int_{-\infty}^{t} p(\xi)d\xi$ and $p(t) + n(t)$ respectively (we ignore effects of propagation). The effect of the antenna system in the transmitted pulse is modeled as a differentiation operation. The noise $n(t)$ is AWGN with two-sided power density $N_0/2$. The UWB pulse $p(t)$ has duration T_p and energy $E_p = \int_{-\infty}^{\infty} [p(t)]^2 dt$. The normalized signal correlation function of $p(t)$ is

$$\gamma_p(\tau) \triangleq \frac{1}{E_p} \int_{-\infty}^{\infty} p(\tau) p(t - \tau) dt > -1 \quad \forall \tau \tag{8.11}$$

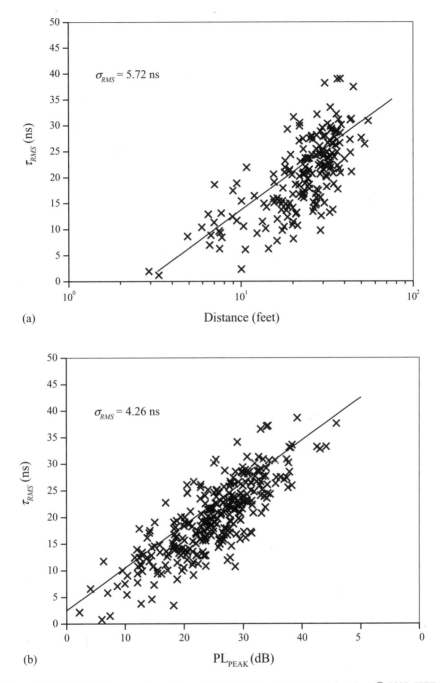

Table 8.2   Statistical values of the path loss parameters

	LOS		NLOS	
	Mean	Standard deviation	Mean	Standard deviation
PL_0 (dB)	47		51	
N	1.7	0.3	3.5	0.97
σ (dB)	1.6	0.5	2.7	0.98

Table 8.3   Percentage of power contained in profile, number of paths, mean excess delay and RMS delay spread for 5, 10, 15, 20 and 30 dB threshold level

	50% NLOS				90% NLOS			
Threshold	% Power	L	τ_m (ns)	τ_{RMS} (ns)	% Power	L	τ_m (ns)	τ_{RMS} (ns)
5 dB	46.8	7	1.95	1.52	46.9	8	2.2	1.65
10 dB	89.2	27	7.1	5.77	86.5	31	8.1	6.7
15 dB	97.3	39	8.6	7.48	96	48	10.3	9.3
20 dB	99.4	48	9.87	8.14	99.5	69	12.2	11
30 dB	99.97	60	10.83	8.43	99.96	82	12.4	11.5

Parameter $\gamma_{min} \overset{\Delta}{=} \gamma_p(\tau_{min})$ is defined as the minimum value of $\gamma_p(\tau)$, $\tau \in (0, T_p]$. The transmitted signal is a PPM signal, and each is composed of N_s time-shifted pulses

$$\Psi_{TX}^{(j)}(t) = \sum_{k=0}^{N_s-1} p_{TX}\left(t - kT_f - \alpha_j^k \tau_{min}\right) \tag{8.12}$$

$j = 1, 2, \ldots, M$. In the absence of noise, the received signals are composed of N_s time-shifted UWB pulses

$$\Psi_j(t) = \sum_{k=0}^{N_s-1} p\left(t - kT_f - \alpha_j^k \tau_{min}\right) \tag{8.13}$$

Each $\Psi_j(t)$ represents the jth signal in an ensemble of M signals, each signal identified by the sequence of time shifts $a_j^k \tau_{min} \in \{0, \tau_{min}\}$ (this choice of time shifts allows us to produce M-ary PPM signals, which are equally correlated). The a_j^k is a 0, 1 pattern representing the jth cyclic shift of an m-sequence of length N_s. Since there are at most N_s cyclic shifts in an m-sequence, we require that $2 \leq M < N_s$.

The pulse duration satisfies $T_p + \tau_{min} < T_f$, where T_f is the time shift value corresponding to the frame period. Each signal $\Psi_j(t)$ has duration $T_s \overset{\Delta}{=} N_s T_f$ and energy $E_\Psi = N_s E_p$. The signals in Equation (8.133) have normalized correlation values

$$\beta_{ij} \overset{\Delta}{=} \frac{\int_{-\infty}^{\infty} \Psi_i(t)\Psi_j(t)\,dt}{E_\Psi} = \beta = \frac{1 + \gamma_{min}}{2} \tag{8.14}$$

for all $i \neq j$, i.e. they are equally correlated.

The optimum receiver is a bank of filters matched to the M signals $\Psi_j(t)$, $j = 1, 2, \ldots, M$. The receiver is assumed to be perfectly synchronized with the transmitter.

The union bound on the bit error probability using these equally correlated signals can be written [22]:

$$\text{UBPb} = \frac{1}{M} \sum_{\substack{i=1 \\ i \neq j}}^{M} \sum_{j=1}^{M} Q\left(\sqrt{\frac{E_\Psi}{N_0}(1-\beta)} \right) = \frac{M}{2} \int_{\sqrt{\log_2(M)\text{SNRb}}}^{\infty} \frac{\exp(-\xi^2/2)}{\sqrt{2\pi}} d\xi \qquad (8.15)$$

where

$$\text{SNRb} = \frac{1}{\log_2(M)} \frac{E_\Psi}{N_0}(1-\beta) \qquad (8.16)$$

is the received bit SNR and $Q(\xi)$ is the Gaussian tail function.

As an example for $p(t)$, we consider a UWB pulse that can be modeled by properly scaling the second derivative of a Gaussian function $\exp(-2\pi[t/t_n]^2)$. In this case, we have

$$p_{\text{TX}}(t) = t \exp\left(-2\pi \left[\frac{t}{t_n}\right]^2 \right) \qquad (8.17)$$

$$p(t) = \left[1 - 4\pi \left[\frac{t}{t_n}\right]^2 \right] \exp\left(-2\pi \left[\frac{t}{t_n}\right]^2 \right) \qquad (8.18)$$

where the value $\tau_n = 0.7531$ ns was used to fit the model $p(t)$ to a measured waveform $p_m(t)$ from a particular experimental radio link [23]. This resulted in $T_p \approx 2.0$ ns.

The normalized signal correlation function corresponding to $p(t)$ is calculated using Equation (8.11) to give:

$$\gamma_p(t) = \left[1 - 4\pi \left[\frac{t}{t_n}\right]^2 + \frac{4\pi^2}{3} \left[\frac{t}{t_n}\right]^4 \right] \exp\left(-\pi \left[\frac{t}{t_n}\right]^2 \right) \qquad (8.19)$$

For this $\gamma_p(t)$, we have $\tau_{\min} = 0.4073$ ns and $\gamma_{\min} = -0.6183$, so $\beta = 0.1909$ in Equation (8.14). Both $p(t - T_p/2)$ and $\gamma_p(\tau)$ are depicted in Figure 8.10. Figure 8.11 shows the spectrum of the impulse $p(t)$. The 3 dB bandwidth of the pulse is close to 1 GHz. The center frequency is around 1.1 GHz.

The specific values of N_s and T_f do not affect SNRb, as long as $M < N_s$ and $T_p + \tau_{\min} < T_f$. Hence, we set arbitrarily $T_f = 500$ ns and $N_s > 1000$ [21]. The BER in AWGN can now be calculated using UBPb from Equation (8.15). Results for different values of M are shown in Figure 8.12. Values as large as $M = 128$ are easily obtained with the PPM signal design in Equation (8.13), allowing us to exploit the benefits of M-ary modulation without an excessive increase in the complexity of the receiver [24].

8.3.2 Performance in a Dense Multipath Channel

In this section we discuss performance of M-ary UWB signals in a dense multipath channel with AWGN. The channel can be, for example, an indoor radio channel as discussed in Section 8.2. In the analysis, we will assume that the transmitter is placed at a certain fixed location, and the receiver is placed at a variable location denoted u_0. The transmitted pulse is the same pulse $p_{\text{TX}}(t)$ as in the AWGN case, and the received UWB signal is $\sqrt{E_a}\tilde{p}(u_0, t) + n(t)$. The pulse $\sqrt{E_a}\tilde{p}(u_0, t)$ is a multipath spread version of $p(t)$ received at position u_0 with average duration $T_a \gg T_p$. The pulse has 'random' energy $\tilde{E}(u_0) \triangleq E_a\tilde{\alpha}^2(u_0)$, where E_a is the average energy and

$$\tilde{\alpha}(u_0) \triangleq \int_{-\infty}^{\infty} [\tilde{p}(u_0, t)]^2 dt \qquad (8.20)$$

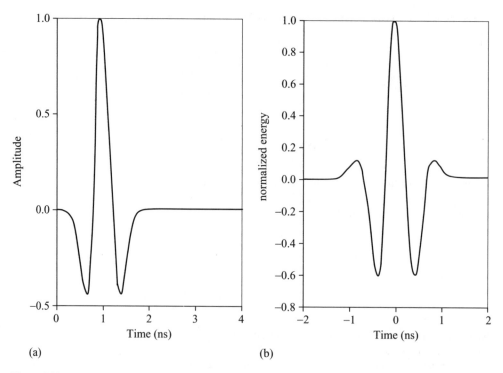

Figure 8.10   (a) The pulse $p\,(t - T_p/2)$ as a function of time $0 \leq t \leq 4$ ns; (b) the signal autocorrelation $\gamma_p\,(\tau)$ as a function of time shift $-2 \leq \tau \leq 2$ ns [21] © 2001, IEEE.

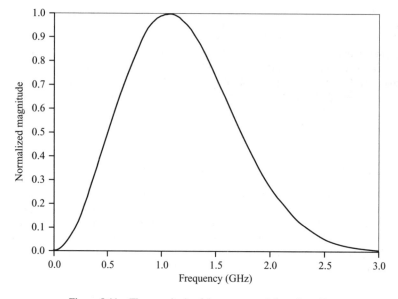

Figure 8.11   The magnitude of the spectrum of the pulse $p(t)$.

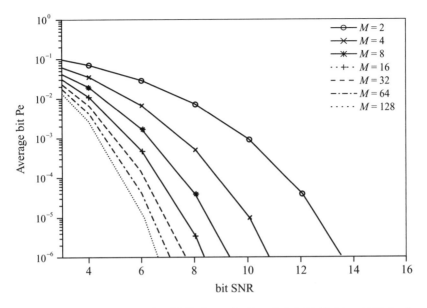

Figure 8.12   The UBPb in Equation (8.15). Curves for M are 2, 4, 8, 16, 32, 64, and 128 signals.

is the normalized energy. The pulse has normalized signal correlation

$$\tilde{\gamma}(u_0, \tau) \triangleq \frac{\int_{-\infty}^{\infty} \tilde{p}(u_0, t)\tilde{p}(u_0, t - \tau)\,dt}{\int_{-\infty}^{\infty} [\tilde{p}(u_0, t)]^2\,dt} \tag{8.21}$$

The transmitted signals are the same $\Psi_{\mathrm{TX}}^{(j)}(t)$ as given in Equation (8.12). In the absence of noise, the received signals are composed of N_s time-shifted UWB pulses

$$\widetilde{\Psi}_j(u_0, t) = \sum_{k=0}^{N_s - 1} \sqrt{E_a}\,\tilde{p}(u_0, t - kT_f - \alpha_j^k \tau_{\min}) \tag{8.22}$$

for $j = 1, 2, \ldots, M$. The UWB PPM signal $\widetilde{\Psi}_j(u_0, t)$ is a multipath spread version of $\Psi_j(t)$ received at position u_0. Assume that $\widetilde{\Psi}_j(u_0, t)$ has fixed duration $T_s \approx N_s T_f$, provided that $T_a + \tau_{\min} < T_f$. The signals in Equation (8.22) have 'random' energy

$$\tilde{E}_\Psi(u_0) = \int_{-\infty}^{\infty} \left[\widetilde{\Psi}_j(u_0, \xi)\right]^2 d\xi = \bar{E}_\Psi \tilde{\alpha} u_0 \tag{8.23}$$

for $j = 1, 2, \ldots, M$, where $\bar{E}_\Psi = N_s E_a$ is the average energy. The signals in Equation (8.22) have normalized correlation values

$$\tilde{\beta}_{ij}(u_0) \triangleq \frac{\int_{-\infty}^{\infty} \widetilde{\Psi}_i(u_0, \xi)\,\widetilde{\Psi}_j(u_0, \xi)\,d\xi}{\tilde{E}_\Psi(u_0)} = \tilde{\beta}(u_0) = \frac{1 + \tilde{\gamma}(u_0, \tau_{\min})}{2} \tag{8.24}$$

for all $i \neq j$, i.e. they are equally correlated. The multipath effects change with the particular position u_0, and therefore the M-ary set of received signals $\{\widetilde{\Psi}_j(u_0, t)\}_{j=1}^{j=M}$ also changes with the particular position u_0.

8.3.3   Receiver and BER Performance

Conditioned on a particular physical location u_0, the optimum receiver (matched filter) is a kind of perfect Rake receiver that is able to construct a reference signal $\widetilde{\Psi}_j(u_0, T_s - t)$ that is perfectly matched to the signal received $\widetilde{\Psi}_j(u_0, t)$ over the multipath conditions at that location u_0. We will assume that the receiver is perfectly synchronized with the transmitter. Performance analysis for the perfect Rake receiver can be calculated using standard techniques, conditioned on a particular physical location u_0, the union bound on the bit error probability using these equally correlated signals can be written

$$\text{UBPb}(u_0) = \frac{M}{2} \int_{\sqrt{\log_2(M)\text{SNRb}(u_0)}}^{\infty} \frac{\exp(-\xi^2/2)}{\sqrt{2\pi}} d\xi \tag{8.25}$$

where

$$\begin{aligned}
\text{SNRb}(u_0) &= \frac{1}{\log_2(M)} \frac{\bar{E}_\Psi(u_0)}{N_0}(1 - \tilde{\beta}(u_0)) \\
&= \frac{1}{\log_2(M)} \frac{\bar{E}_\Psi \tilde{\alpha}^2(u_0)}{N_0}(1 - \tilde{\beta}(u_0))
\end{aligned} \tag{8.26}$$

is the received bit SNR [25].

8.3.4   Time Variations

The $\tilde{\beta}(u_0)$ value accounts for changes in the correlation properties of the received signals. These changes in $\tilde{\beta}(u_0)$ translate into changes in the Euclidean distance between signals. Therefore, the $(1 - \tilde{\beta}(u_0))$ value accounts for energy variations at the output of the perfect matched filter due to distortions in the shape of the signal correlation function caused by multipath. The $\tilde{\alpha}^2(u_0)$ value accounts for variations in the received signal energy due to fading caused by multipath. The average performance can be obtained by taking the expected value $\mathbf{E}_u(\cdot)$ over all values of u_0

$$\overline{\text{UBPb}}\left(\frac{\bar{E}_\Psi}{N_0}\right) = \mathbf{E}_u\{\text{UBPb}(u)\} \tag{8.27}$$

where

$$\left(\frac{\bar{E}_\Psi}{N_0}\right) \triangleq \mathbf{E}_u\{\text{SNRb}(u)\} \tag{8.28}$$

is the average received bit SNR. This BER analysis provides a theoretical matched filter bound for the best performance attainable when the multipath channel is perfectly estimated.

Instead of using a channel model, the calculations based on the *received* waveforms can be used. This is possible since the expected value in Equation (8.27) is taken with respect to the quantities $\tilde{\alpha}^2(u_0)$ and $\tilde{\beta}(u_0)$. Histograms of these quantities can be calculated for a particular indoor channel environment and a first approximation can be obtained using the sample mean value

$$\overline{\text{UBPb}}\left(\frac{\bar{E}_\Psi}{N_0}\right) \approx \frac{1}{u_*} \sum_{u_0=1}^{u_*} \text{UBPb}(u_0) \tag{8.27a}$$

This calculation represents a rough approximation to the performance of UWB signals in the presence of dense multipath in a particular indoor radio environment. The histograms for $\tilde{\alpha}^2(u_0)$ and $\tilde{\beta}(u_0)$ can be derived from their definitions in Equations (8.20) and (8.24) using the ensemble of pulse responses

$$\{\tilde{p}(u_0, t)\}, \quad u_0 = 1, 2, \ldots, u_*$$

taken in a measurement experiment in the multipath channel of interest.

8.3.5 *Performance Example*

The channel responses $\tilde{p}(u_0, t)$ were measured in eight different rooms and hallways in a typical office building described in Section 8.2 (see also [21]). In every room and hallway, 49 different locations are arranged spatially in a 7×7 square grid with six inch spacing. At every location u_0, the $T_a = 300$ ns-long pulses $\tilde{p}(u_0, t)$ are recorded, keeping the transmitter, the receiver and the environment stationary.

The UWB transmitter is placed at a fixed location inside the building. It consists of a step recovery diode-based pulser connected to a UWB omnidirectional antenna. The pulser produces a train of UWB 'Gaussian monocycles' $p_{TX}(t)$. The train of $p_{TX}(t)$ is transmitted as an excitation signal to the propagation channel. The train has a repetition rate of 500 ns with a tightly controlled average monocycle-to-monocycle interval. The clock driving the pulser has resolution in the order of picoseconds.

The $\tilde{p}(u_0, t)$ represents the convolution of $p_{TX}(t)$ with the channel impulse response at location u_0. The 500 ns repetition rate is long enough to make sure that pulse responses $\tilde{p}(u_0, t)$ corresponding to adjacent impulses $p_{TX}(t)$ do not overlap. The receiver consists of a UWB antenna and a low-noise amplifier. The output of this amplifier is captured using a high-speed digital sampling scope. The scope takes samples in windows of 50 ns at a sampling rate of 20.5 GHz. Noise in the measured $\tilde{p}(u_0, t)$ is reduced by averaging 32 consecutive received pulses measured at exactly the same location u_0. These samples are sent to a data storage and processing unit.

A total of $u_* = 392$ channel pulse responses $\tilde{p}(u_0, t)$ were measured. An equal number of normalized energy values $\tilde{\alpha}^2(u_0)$ and normalized correlation functions $\tilde{\gamma}(u_0, \tau)$ were calculated using Equations (8.20) and (8.21), respectively. Figure 8.13 shows histograms of $\tilde{\alpha}^2(u_0)$, $\tilde{\beta}(u_0)$ and the product $\tilde{\alpha}^2(u_0) \times (1 - \tilde{\beta}^2(u_0))$. The measured $\tilde{p}(u_0, t)$ have $T_a \approx 300$ ns. The rest of the parameter values are the same as those used in the AWGN case, i.e. $\tau_{\min} = 0.4073$ ns, $T_f = 500$ ns and $N_s > 1000$. With these values, the conditions $M < N_s$ and $T_a + \tau_{\min} < T_f$ will be satisfied.

For the results from Figure 8.13, the average BER curves, Equation (8.27), are shown in Figure 8.14.

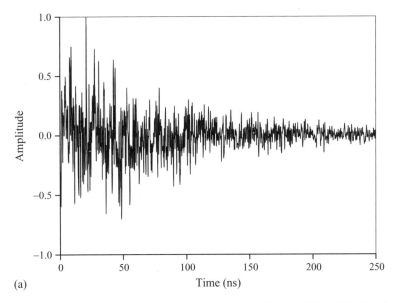

(a)

Figure 8.13   (a) Normalized correlation function $\tilde{\gamma}(u_0, t)$ of the pulse $\tilde{p}(u_0, t)$ in Figure 8.5; (b) a closer view of the correlation in (a). The spreading caused by multipath is notorious. The long tails in the correlation function are the effect of the pulse spreading [21]. The histogram of the normalized values of (c) the received energy $\tilde{\alpha}^2(u_0)$; (d) the correlation value $\tilde{\beta}(u_0)$; and (e) the product $\tilde{\alpha}^2(u_0) \times (1 - \tilde{\beta}^2(u_0))$. The ordinate represents appearance frequency, and the abscissa represents the value of the parameter. The size of the sample is $u_0 = 392$ [21] © 2001, IEEE.

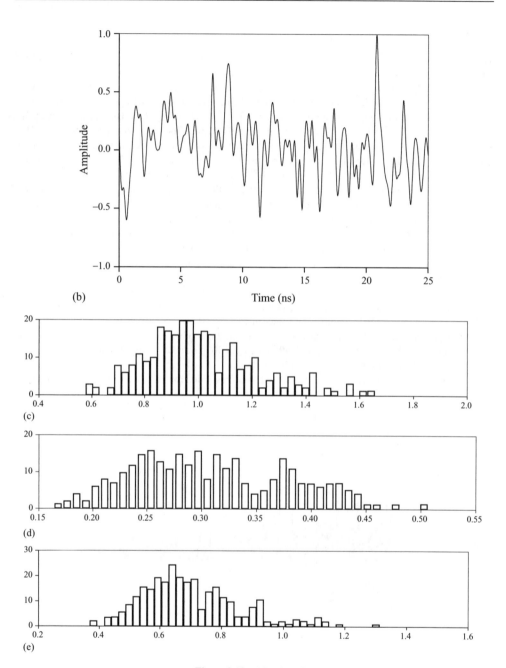

Figure 8.13   (*Continued*).

8.4   *M*-ary PPM UWB Multiple Access

The signal format introduced in Section 8.3 is now used for the multiple access system [26]. The TH PPM signal conveying information exclusively in the time shifts is now represented as

$$x^{(v)}(t) = \sum_{k=0}^{\infty} \omega \left(t - kT_{\mathrm{f}} - c_k^{(v)} T_{\mathrm{c}} - \delta_{d_{[k/N_{\mathrm{s}}]}^{(v)}}^k \right) \qquad (8.29)$$

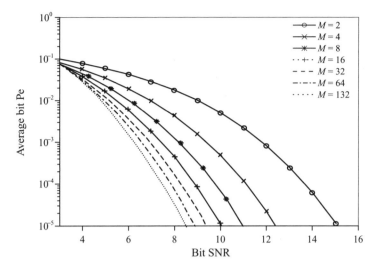

Figure 8.14 $\overline{\text{UBPb}}(\bar{E}_\Psi/N_0)$ in Equation (8.27a); $M = 2, 4, 8, 16, 32, 64$ and 132 signals.

The superscript (v), $1 \le v \le N_u$, indicates user-dependent quantities, where N_u is the number of simultaneous active users. The index k is the number of time hops that the signal $x^{(v)}(t)$ has experienced, and also the number of pulses that have been transmitted. T_f is the frame (pulse repetition) time and equals the average time between pulse transmissions. The notation $[q]$ stands for the integer part of q. The $\{c_k^{(v)}\}$ is the pseudorandom time-hopping sequence assigned to user v. It is periodic with period N_p (i.e. $c_{k+lN_p}^{(v)} = c_k^{(v)}$, for all k, l integers) and each sequence element is an integer in the range $0 \le c_k^{(v)} \le N_h$. For a given time shift parameter T_c, the time-hopping code provides an additional time shift to the pulse in every frame, each time shift being a discrete time value $c_k^{(v)} T_c$, with $0 \le c_k^{(v)} T_c \le N_h T_c$. The time shift corresponding to the data modulation is

$$\delta^k_{d^{(v)}_{[k/N_s]}} \in \{\tau_1 = 0 < \tau_2 < \cdots < \tau_\eta\} \tag{8.30}$$

with $\eta \ge 2$ an integer. The data sequence $\{d_m^{(v)}\}$ of user v is an M-ary symbol stream, $1 \le d_m^{(v)} \le M$, that conveys information in some form. The system under study uses fast time-hopping, which means that there are $N_s > 1$ pulses transmitted per symbol. The data symbol changes only every N_s hops. Assuming that a new data symbol begins with pulse index $k = 0$, the data symbol index is $[k/N_s]$.

We will use the following notation

$$H_m^{(v)}(t) \triangleq \sum_{k=mN_s}^{(m+1)N_s-1} T_c c_k^{(v)} p(t - kT_f)$$

$$p(t) = \begin{cases} 1, & \text{if } 0 \le t \le T_f \\ 0, & \text{otherwise} \end{cases} \tag{8.31}$$

and

$$S_i(t) \triangleq \sum_{k=0}^{N_s-1} \omega(t - kT_f - \delta_i^k) \tag{8.32}$$

for $i = 1, 2, \ldots, M$, then Equation (8.29) can be written

$$x^{(v)}(t) = \sum_{m=0}^{\infty} S_{d_m^{(v)}} \left(t - mN_s T_f - H_m^{(v)}(t)\right) \triangleq \sum_{m=0}^{\infty} X^{(v)}_{m,d_m^{(v)}}(t) \tag{8.29a}$$

The signal $S_i(t)$ in Equation (8.32) is the received signal corresponding to the transmitted signal $\int_{-\infty}^{t} S_i(\xi)d\xi = \sum_{k=0}^{N_s-1} \omega_{TX}(t - kT_f - \delta_i^k)$. With this notation, the signal correlation function is defined as

$$
\begin{aligned}
R_{ij} &\triangleq \int_{-\infty}^{\infty} X_{m,i}^{(v)}(\xi) \, X_{m,i}^{(v)}(\xi) \, d\xi \\
&= \int_{-\infty}^{\infty} S_i(\xi) \, S_j \, d\xi \; = E_\omega \sum_{k=0}^{N_s-1} \gamma_\omega \left(\delta_i^k - \delta_j^k\right)
\end{aligned}
\tag{8.33}
$$

since the pulses are non-overlapping. The energy in the ith signal $X_{m,i}^{(v)}$ is $E_S = R_{ii} = N_s E_\omega$, and the normalized correlation value is

$$
\alpha_{i,j} \triangleq \frac{R_{ij}}{E_S} = \frac{1}{N_s} \sum_{k=0}^{N_s-1} \gamma_\omega \left(\delta_i^k - \delta_j^k\right) \geq \gamma_{\min}
\tag{8.34}
$$

8.4.1   M-ary PPM Signal Sets

The PPM signal $S_i(t)$ in Equation (8.32) represents the ith signal in an ensemble of M information signals, each signal completely identified by the pulse shape $\omega(t)$ and the sequence of time shifts $\{\delta_i^k\}, k = 0, 1, 2, \ldots, N_s - 1$. The most interesting M-ary PPM sets from the practical point of view are orthogonal (OR), equally correlated (EC), and N-orthogonal (NO) signal sets.

In general, these signal designs have the property that the structure of the M-ary autocorrelation matrix is preserved for different $\omega(t)$. This is important because $\omega(t)$ is, in general, a non-standard pulse, and these signal designs reduce the dependence of the MA performance on the shape of $\omega(t)$. The time shift patterns defining each M-ary PPM signal set and their respective correlation properties are studied in detail in [27–29] and are summarized in Table 8.4. In the EC case, the a_i^k is a 0, 1 pattern representing the ith cyclic shift, $i = 1, 2, \ldots, M$, of an m-sequence [31] of length $N_s = 2^m - 1, m \geq 1$, and $N_s \geq M$.

The M-ary correlation receiver, shown in Figure 8.15, consists of M filters matched to the signals $\{X_{m,j}^{(1)}(t - \tau^{(1)})\}, j = 1, 2, \ldots M, t \in T_m$, followed by samplers and a decision circuit that selects the maximum among the decision variables

$$
\int_{t \in T_m} r(t) \, X_{m,j}^{(1)}(t - \tau^{(1)})dt, \quad j = 1, 2, \ldots M
\tag{8.35}
$$

8.4.2   Performance Results

In this subsection, we illustrate the theoretical MA performance of this system for a specific $\omega(t)$ under perfect power control (i.e. $A_{(v)} = A^{(1)}$ for $v = 2, 3, \ldots, N_u$). The system parameters are the same as in [26]. The $\omega(t)$ considered here is the second derivative of a Gaussian function

$$
\omega(t) = \left[1 - 4\pi \left[\frac{t}{t_n}\right]^2\right] \exp\left(-2\pi \left[\frac{t}{t_n}\right]^2\right)
\tag{8.36}
$$

where the value t_n is used to fit the model $\omega(t)$ to a measured waveform from a particular experimental radio link. The normalized signal correlation function corresponding to $\omega(t)$ in Equation (8.36) is

$$
\gamma_\omega(\tau) = \left[1 - 4\pi \left[\frac{t}{t_n}\right]^2 + \frac{4\pi^2}{3}\left[\frac{t}{t_n}\right]^4\right] \exp\left(-\pi \left[\frac{t}{t_n}\right]^2\right)
\tag{8.37}
$$

In this case, T_ω and $\tau_{\min}$ depend on t_n, and $\gamma_{\min} = -0.6183$ for any t_n. Using $t_n = 0.4472$ ns, we get $T_\omega \approx 1.2$ ns and $\tau_{\min} = 0.2149$ ns. Figure 8.16 depicts $\omega(t - T_\omega/2)$, $\gamma_\omega(\tau)$ and the spectrum of $\omega(t)$. The 3 dB bandwidth of $\omega(t)$ is in excess of 1 GHz.

Given $\omega(t)$, the signal design is complete when we specify N_s, T_f and δ_j^k in Table 8.4. For OR signals $T_{OR} = 2T_\omega$, for EC signals $\tau_1 = 0$ and $\tau_2 = \tau_{\min}$, for NO1 signals $\tau_1 = 0$, $\tau_2 = \tau_{\min}$ and $T_{NO1} \triangleq t_{\min} + 2T_\omega$, and for NO2 signals $\tau_1 = 0$, $\tau_2 = \tau_{\min}$ and $T_{NO2} \triangleq t_{\min} + 2T_\omega$.

Table 8.4 Time-shift patterns and normalized correlation values of the M-ary PPM signals under study. Orthogonal (OR), equally correlated (EC), N-orthogonal design 1 (NO1) and N-orthogonal design 2 (NO2) [26] © 2001, IEEE

Type of signal	Time shift pattern $\{\delta_i^k\}$ $i = 1, 2, \ldots, M$ $k = 0, 1, 2, \ldots, N_s - 1$	Normalized correlation coeficients
OR	$\delta_i^k = [(k+i-1) \bmod M]\, T_{OR}$ $T_{OR} = 2T_\omega$	$\alpha_{ij}^{(OR)} = \begin{cases} 1, & i = j \\ 0, & i \neq j \end{cases}$
EC	$\delta_i^k = a_i^k \tau_2$ $\tau_2 \in (0, T_\omega]$ $a_i^k \in \{0, 1\}$	$\alpha_{ij}^{(EC)} = \begin{cases} 1, & i = j \\ \lambda, & i \neq j \end{cases}$ $\;\;\lvert \lambda \rvert < 1$
NO1	$\delta_i^k = \tau_1 + \left[(k + 2\bar{I}) \bmod L\right] T_{NO1}$ $L \triangleq \left[\dfrac{M}{2}\right]$ $I = i - \left[\dfrac{i-1}{2}\right] 2$ $\bar{I} \triangleq \left[\dfrac{i-1}{2}\right]$ $T_{NO1} \triangleq \tau_2 + T_{OR}$ $0 = \tau_1 < \tau_2 < T_\omega$	$\alpha_{ij}^{(NO1)} = \begin{cases} 1, & i = j \\ 0, & \left[\dfrac{i-1}{2}\right] \neq \left[\dfrac{j-1}{2}\right] \\ \beta_{ij}, & \left[\dfrac{i-1}{2}\right] = \left[\dfrac{j-1}{2}\right] \end{cases}$ $\beta_{ij} = \gamma_\omega(\tau_J - \tau_I)$ $J = j - \left[\dfrac{j-1}{2}\right] 2$
NO2	$\delta_i^k = a_i^k t_2 + \left[(k + 2\bar{I}) \bmod L\right] T_{NO2}$ $T_{NO2} \triangleq t_2 + T_{OR}$ $0 = t_1 < t_2 < T_\omega$	$\alpha_{ij}^{(NO2)} = \begin{cases} 1, & i = j \\ 0, & \left[\dfrac{i-1}{2}\right] \neq \left[\dfrac{j-1}{2}\right] \\ \lambda, & \left[\dfrac{i-1}{2}\right] = \left[\dfrac{j-1}{2}\right] \end{cases}$

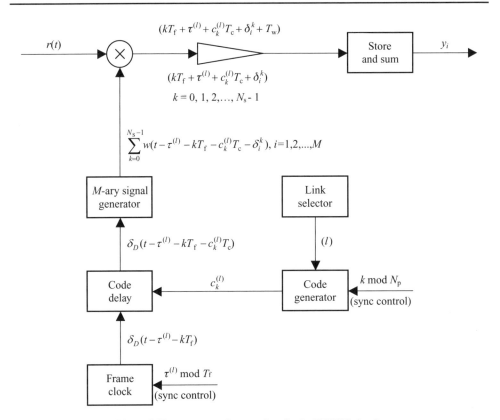

Figure 8.15 M-ary correlator receiver for the TH PPM signals.

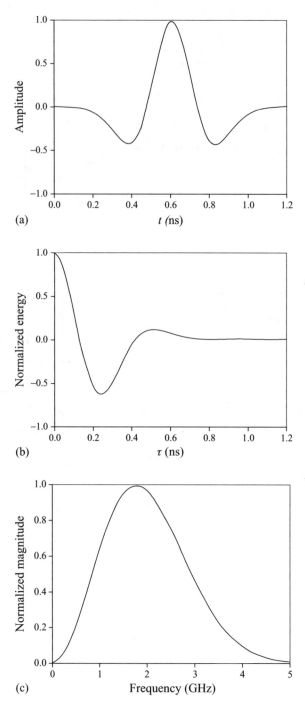

Figure 8.16   (a) Pulse $\omega(t - T_\omega/2)$ as a function of time t; (b) signal autocorrelation function $\gamma_\omega(\tau)$ as a function of time shift τ; (c) magnitude of the spectrum of the pulse $\omega(t)$ [26] © 2001, IEEE.

To choose a value for T_f, we need $0 < N_h T_c < (T_f/2) - 2(T_\omega + \tau_\eta)$. Also notice that $\tau_\eta = 16 T_{OR}$ for $M = 16$ (the maximum value of M, in this example). By choosing $T_f = 100$ ns, we have that $0 < N_h T_c < 16$ ns.

To choose a value for N_s, notice that the M-ary PPM signal designs considered in [27] require that $N_s = 1/(R_s T_f) = \log_2 (M)/(R_b T_f) = \log_2 (M) N_s^{(2)}$. Hence, for a fixed T_f, the value of N_s is determined by R_b. However, in particular, the EC PPM signal design additionally requires $N_s \geq M$. Combining these two requirements on N_s we have that, in the EC PPM case, both R_b and N_s satisfy the relation $(\log_2 (N_s)/N_s) \geq R_b T_f$. In this example, we use $R_b = 100$ kb/s, $N_s^{(2)} = 100$, $2 \leq M \leq 16$, and $N_s = \log_2 (M) 100$; hence, $(\log_2 (N_s)/N_s) \geq R_b T_f$ holds, and both relations $N_s \geq M$ and $N_s = \log_2 (M)/R_b T_f$ are satisfied.

For a single-link communications bit, $E_b/N_0 = 14.30$ dB, $\text{SNRb}_{OR}(1) = 14.30$ dB [27], and $\text{SNRb}_{TSK}(1) = 16.40$ dB. The BER results are given in Figure 8.17. If we define $\text{SNR}(1)$ as the required signal to noise ratio for a certain BER with only one user in the network and $\text{SNR}(N_u)$ as the required initial signal to noise ratio with one user which, after adding $N_u - 1$ additional users, will still give $\text{SNR}(1)$, then we define the degradation factor DF as $\text{DF} = \text{SNR}(N_u)/\text{SNR}(1)$. In the next example we demonstrate some results for $N_u(\text{DF})$ when using the pulse $\omega(t)$ in Equation (8.36) with $t_n = 0.4472$ ns, $T_\omega = 1.2$ ns, $\tau_{min} = 0.2419$ ns, $\tau_1 = 0$, $\tau_2 = \tau_{min}$, and $T_f = 100$ ns. Figure 8.18 shows $N_u(\text{DF})$ for different values of DF using $R_b = 100$ kb/s, $N_s^{(2)} = 100$, $2 \leq M \leq 256$, and $N_s = \log_2 (M) 100$. Notice that $(\log_2(N_s)/N_s) \geq R_b T_f$ still holds for $M = 256$.

From $N_u(\text{DF})$ we can also find $R_b(\text{DF})$ for a particular value of N_u. Figure 8.19 shows $R_b(\text{DF})$ for different values of DF using $N_u = 1000$ active users.

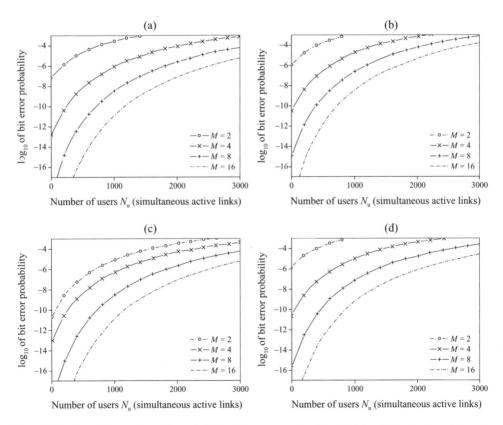

Figure 8.17 Base 10 logarithm of the probability of bit error as a function of N_u for different values of M, using $R_b = 100$ kb/s. (a) OR PPM signals, $\text{SNRb}_{OR}(1) = 14.30$ dB; (b) EC PPM signals, $\text{SNRb}_{EC}(1) = 13.39$ dB; (c) NO PPM signals, design 1, $\text{SNRb}_{OR}(1) = 14.30$ dB and $\text{SNRb}_{TSK}(1) = 16.40$ dB; (d) NO PPM signals, design 2, $\text{SNRb}_{OR}(1) = 14:30$ dB.

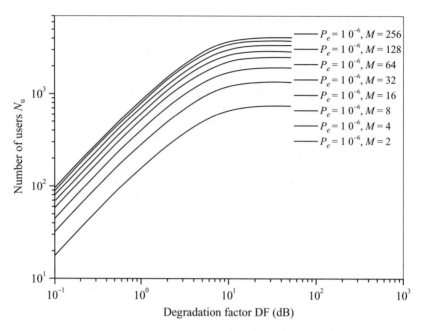

Figure 8.18   Number of simultaneous active links (users) N_u(DF) for EC PPM signals for $2 \leq M \leq 256$ with $P_e(1) =$ $\mathrm{UBP}_b^{(EC)}(1) \approx 10^{-6}$ and $R_b = 100$ kb/s.

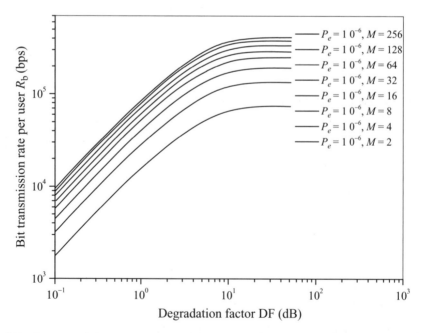

Figure 8.19   Data transmission rate per user R_b(DF) for EC PPM signals for $2 \leq M \leq 256$ with $P_e(1) =$ $\mathrm{UBP}_b^{(EC)}(1) \approx 10^{-6}$ and $N_u = 1000$ active users.

Table 8.5 Values of C_{sup} in (Gb/s) calculated using three different pulse widths. Also included are the values of N_{sup} for $R_{\text{b}} = 100$ kb/s [26]

Set I of parameters	Set II of parameters	Set III of parameters
$t_n = 0.2877$ ns	$t_n = 0.4472$ ns	$t_n = 0.7531$ ns
$T_\omega = 0.75$ ns	$T_\omega = 1.2$ ns	$T_\omega = 2.0$ ns
$T_{\min} = 0.1556$ ns	$T_{\min} = 0.2419$ ns	$T_{\min} = 0.4073$ ns
$C_{\text{sup}}^{(\text{I})} = 3.6394$ (Gbps)	$C_{\text{sup}}^{(\text{II})} = 2.3412$ (Gbps)	$C_{\text{sup}}^{(\text{III})} = 1.3903$ (Gbps)
$N_{\text{sup}}^{(\text{I})} = 36394$ (users)	$N_{\text{sup}}^{(\text{II})} = 23412$ (users)	$N_{\text{sup}}^{(\text{III})} = 13903$ (users)

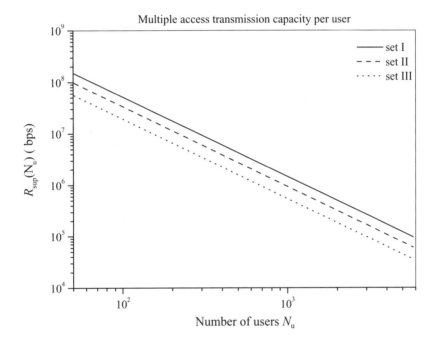

Figure 8.20 Upper bound on the bit transmission rate per user R_{sup} (N_{u}) in b/s, calculated using the sets I, II and III of parameters in Table 8.5.

The values of the upper bound on maximum capacity C_{sup} are given in Table 8.5. The upper bound on maximum data rate is shown in Figure 8.20.

8.5 Coded UWB Schemes

We can consider the simple time-hopping spread spectrum from Section 8.4 as a coded system in which a simple repetition block code with rate $1/N_s$ is used. As outlined in Chapter 2 the efficiency of the repetition code is very low. Thus, by applying a near optimal code instead of the above simple repetition code, we should expect the system performance to improve significantly. In [32, 33], a class of low-rate superorthogonal convolutional codes that have near optimal performance is introduced. In a superorthogonal code with constraint length K, the rate is equal to $1/2^{K-2}$. Since in the TH CDMA (UWB) system, N_s pulses are sent for each data bit, we must set $2^{K-2} = N_s$ or $K = 2 + \log_2 N_s$. The location of each pulse in each frame is determined by the user-dedicated pseudorandom sequence, along with the code symbol corresponding to that frame.

Decoding is performed using the Viterbi algorithm. The state diagram of this decoder consists of 2^{K-1} states. Two branches, corresponding to bit zero and bit one, exit from each state in the trellis diagram. To update the state metrics, it is first necessary to calculate the branch metrics, using the received signal $r(t)$. For this purpose, in each frame j, the quantity:

$$\alpha_j \triangleq \int_{\tau_1+jT_f}^{\tau_1+(j+1)T_f} r(t)v\left(t - \tau_1 - jT_f - c_j^{(1)}T_c\right) dt$$

(the pulse correlator output) is obtained. Because of the special form of the Hadamard–Walsh sequence that is used in the structure of superorthogonal codes, the branch metrics can be simply evaluated based on the outputs of pulse correlators $\alpha_j s$ [33]. This is also elaborated in Chapter 4 for space–time codes from orthogonal designs. The processing complexity of this decoder grows only linearly with K (or logarithmically with N_s); the required memory, however, grows exponentially with K (or equally linearly with N_s) [33]. Since in time-hopping, spread spectrum applications, the value of K is relatively low (the typical value is in the range 3–12), the system can be considered to be practical.

8.5.1   Performance

The path generating function of the code for a superorthogonal code is computed as [33, 34]:

$$T_{SO}(\gamma, \beta) = \frac{\beta W^{K+2}(1 - W)}{1 - W[1 + \beta(1 + W^{K-3} - 2W^{K-2})]} \tag{8.38}$$

in which $W = \gamma^{2^{K-3}}$ and K is the constraint length of the code. Expanding the above expression, we get a polynomial in γ and β. The coefficient and the powers of γ and β in each term of the polynomial indicate the number of paths and output–input path weights respectively. The free distance of this code is obtained from the first term of the expansion as $d_f = 2^{K-3} (K + 2) = N_s (\log_2 N_s + 4)/2$.

8.5.2   The Uncoded System as a Coded System with Repetition

In this case, the distance will be N_s. Comparing the free distances of these two schemes, it is clear that the coded scheme outperforms the uncoded scheme significantly.

An upper bound on the probability of error per bit for a memoryless channel is obtained using the union bound as follows:

$$P_b < \left.\frac{dT_{SO}(W, \beta)}{d\beta}\right|_{\beta=1} = \frac{W^{K+2}}{(1-2W)^2}\left(\frac{1-W}{1-W^{K-2}}\right)^2 \tag{8.39}$$

where $W = Z^{2^{K-3}}$. The parameter Z is calculated from the Bhattacharyya bound as

$$Z = \int_{-\infty}^{\infty} \sqrt{p_0(y)\, p_1(y)}\, dy \tag{8.40}$$

where $p_0(y)$ and $p_1(y)$ are pdfs of the pulse correlator output conditioned on the input symbol being zero and one, respectively.

A lower bound on the probability of error per bit is obtained by considering only the first term of the path generating function, Equation (8.38), The result is

$$P_b \geq P_{d_f} \tag{8.41}$$

where P_{d_f} is the probability of pairwise error in favor of an incorrect path that differs in d_f symbols from the correct path over the unmerged span in the trellis diagram.

For $\omega_{rec}(t + T_\omega/2) = \lfloor 1 - 4\pi(t/\tau_m)\rfloor \exp\left(-2\pi\,(t/\tau_m)^2\right)$, $\tau_m = 0.2877$ and δ and T_f set to 0.156 and 100 ns respectively, the performance curves are given in Figures 8.21 and 8.22 for different R_s.

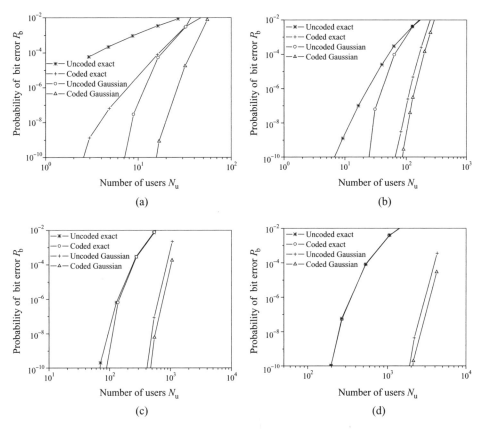

Figure 8.21 Probability of bit error as a function of number of users for synchronous uncoded and coded (upper bound) schemes in exact and Gaussian cases (a) at $R_s = 5$ Mb/s ($N_s = 2$); (b) at $R_s = 1.25$ Mb/s ($N_s = 8$); (c) at $R_s = 325$ kb/s ($N_s = 32$); (d) at $R_s = 78.1$ Mb/s ($N_s = 128$).

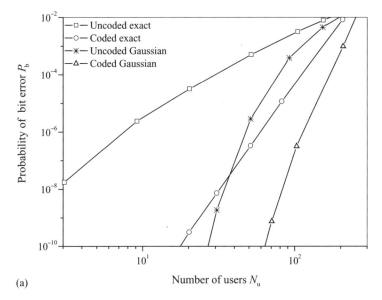

Figure 8.22 Probability of bit error as a function of number of users for asynchronous uncoded and coded (upper bound) schemes in exact and Gaussian cases (a) at $R_s = 5$ Mb/s ($N_s = 2$); (b) at $R_s = 2.5$ Mb/s ($N_s = 4$); (c) at $R_s = 1.25$ Mb/s ($N_s = 8$).

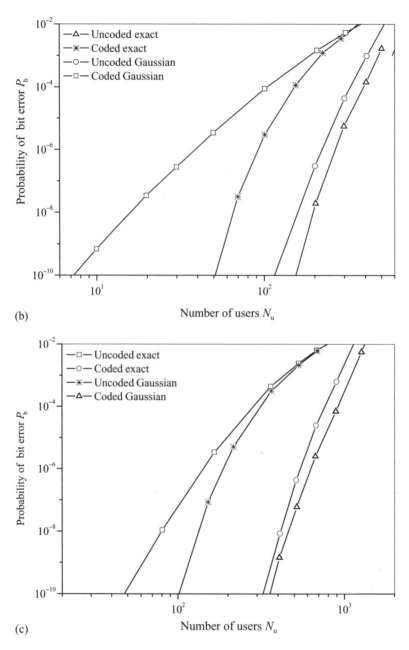

(b)

(c)

Figure 8.22 (*Continued*).

8.6 Multiuser Detection in UWB Radio

In this section we consider a system based on using orthogonal codes in a synchronous or quasi-synchronous context, coupled with TDD for transmissions through frequency selective channels. This is achieved by assigning to each user different orthogonal time-hopping sequences and designating two time slots for transmission: one for the uplink and one for the downlink. Binary PPM modulation from

Section 8.5 is used with eight users ($N_c' = 8$) and two symbols per burst ($K = 2$). As per [35], frame duration $T_f = 100$ ns and there is a maximum delay spread equal to 100 ns.

Each symbol is repeated N_f frames. A time guard of duration T_c is used, hence, $N_g = 1$. The chip duration $T_c = T_f/(N_c' + N_g) = 11.11$ ns. The sampling rate is equal to t_c which leads to a channel of length $L_1 = 9$. The channel model from Section 8.2 is used. This model is based on clusters of rays. The received signal is composed of attenuated and delayed versions of the transmitted signal arriving in clusters. The times of arrival are modeled as a Poisson process and the amplitudes as Gaussian. For the simulations, the receivers will be assumed to be synchronized on the strongest path. In the PPM-IRMA system, the received signal is the second derivative of the Gaussian function $\sqrt{\tau^3/3}(2/\pi)^{1/4}\exp(-t^2/\tau^2)$ (normalized to have $r_\omega(0) = 1$); hence, we have $r_\omega(t) = \exp(-t^2/(2\tau^2))\lfloor 1 - 2(t/\tau)^2 + (t/\tau)^4/3\rfloor$ where $r_\omega(t)$ is the correlation function of $\omega(t)$ and the parameter $\tau = 0.1225$ ns is adjusted to yield a pulse width equal to 0.7 ns.

For multiuser detection schemes (MF, zero forcing (ZF)-decorrelator and MMSE), based on the same logic as in Chapter 5, the results are shown in Figures 8.23 and 8.24. Details of the derivation of the detector transfer function are given in [36].

8.7 UWB with Space–Time Processing

8.7.1 Signal Model

In this segment, the Generalized Gaussian Pulse (GGP) $\Omega(t)/E_0$ shown in Figure 8.25 will be used.

The self-steering array beamforming system for UWB impulse waveforms is shown in Figure 8.26.

The response of the ith sensor to the received wavefront can be expressed in terms of the voltage signal as

$$v_i(t, \phi) = \Omega(t - \tau_i(\phi))$$
$$= [E_0/(1 - \alpha)] (\exp\{-4\pi[(t - \tau_i(\phi))/\Delta T]^2\}$$
$$- \alpha \exp\{-4\pi[\alpha(t - \tau_i(\phi))/\Delta T]^2\})$$

(8.42)

The relative time delay $\tau_i(\phi)/\Delta T$ is a function of the angle of incidence ϕ and the distance d between adjacent sensors:

$$\tau_i(\phi)/\Delta T = (id/c\Delta T)\sin\phi = (i/2m)\rho\sin\phi$$

(8.43)

$$\rho = 2md/c\Delta T = L/c\Delta T = L\Delta f/c$$

(8.44)

In order to achieve electronic beamsteering for enhancing the quality of signal reception from a desired look direction, e.g. $\phi = \phi_0$ the variable delay circuit VDC_i applies to the incoming signal $\Upsilon_i(t)$ a time delay $\tilde{\tau}_i = (id/c)\sin\phi_0$. In analogy to Equation (8.44), the relative time delay $\tilde{\tau}_i/\Delta T$ can also be expressed in terms of ρ:

$$\tilde{\tau}_i/\Delta T = (id/c\Delta T)\sin\phi_0 = (i/2m)\rho\sin\phi_0$$

(8.45)

Finally, the delayed signals $\Upsilon(t + \tilde{\tau}_i - \tau_i(\phi))$, $i = 0, \pm 1, \pm 2, \ldots, \pm m$ from the VDC_is are summed by SUM1, SUM2 and SUM3 to produce the beamformer's response function

$$\tilde{\Upsilon}(t, \phi) = \sum_{i=-m}^{m} \int \Upsilon(t + \tilde{\tau}_l - \tau_i(\phi))$$

(8.46)

Some results are shown in Figures 8.27 to 8.31.

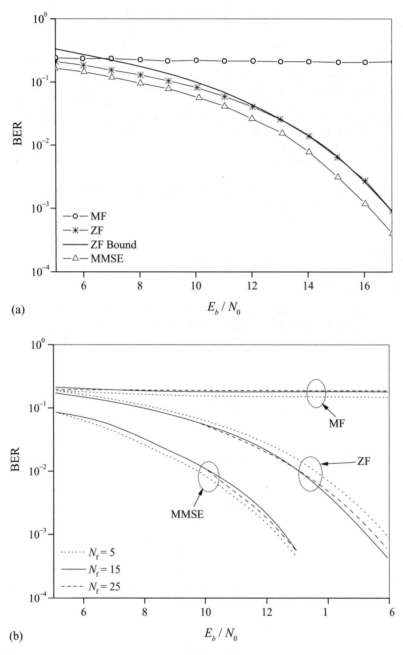

Figure 8.23 Comparing the different receivers (MF, ZF, MMSE) for the binary PPM-IRMA scheme with $K = 2$, $T_g = 1$ and eight users. (a) For $N_f = 1$; (b) for $N_f = 5$, $N_f = 15$ and $N_f = 25$.

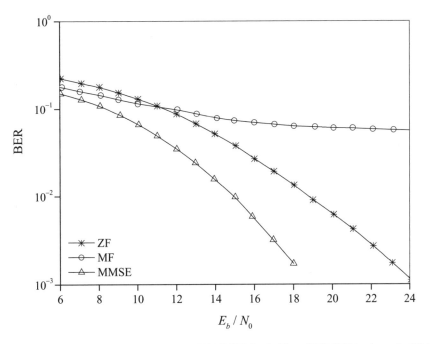

Figure 8.24   Comparing the different receivers (MF, ZF, MMSE) for the binary PPM-IRMA scheme for 100 Monte Carlo channel trials with the same parameters as in Figure 8.23(a).

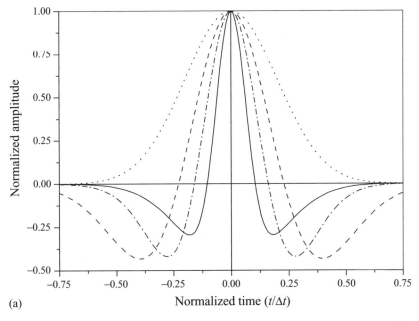

(a)

Figure 8.25   (a) Normalized time variation of the generalized Gaussian pulse $\Omega(t)/E_0$; (b) autocorrelation function $\Upsilon(t)/\Upsilon(0)$; (c) energy density spectrum $\Psi(f) = |\Lambda(f)|^2$ for values of the scaling parameter $\alpha = 0$ (dotted line), $\alpha = 0.75$ (dashed line), $\alpha = 1.5$ (dashed-dotted line) and $\alpha = 3$ (solid line).

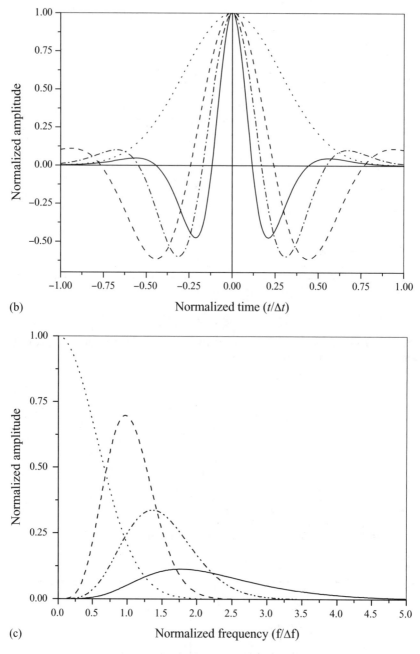

(b)

Normalized time $(t/\Delta t)$

(c)

Normalized frequency $(f/\Delta f)$

Figure 8.25   (*Continued*).

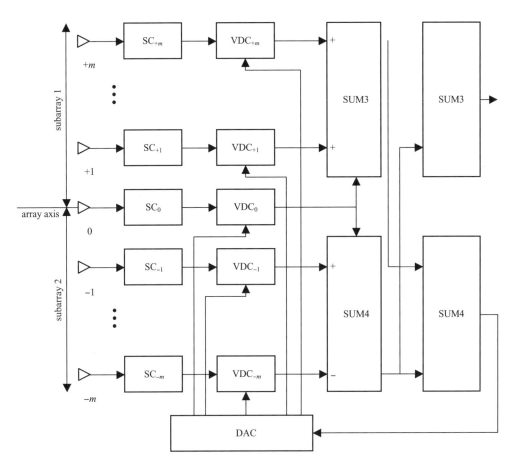

Figure 8.26 A self-steering array beamforming system for UWB impulse waveforms. The beamformer includes $M = 2m + 1$ sensors, sliding correlator (SC$_i$), variable delay circuit (VDC$_i$), delay adjustment computer (DAC) and summer circuits (SUM) [37] © 2002, IEEE.

8.7.2   The Monopulse Tracking System

The monopulse signal delivered to DAC by SUM4 is:

$$\Upsilon_\Delta(t, \phi) = \sum_{i=0}^{+m} \Upsilon(t + \tilde{\tau}_i - \tau_i(\phi)) - \sum_{i=-m}^{+0} \Upsilon(t + \tilde{\tau}_i - \tau_i(\phi)) \tag{8.47}$$

Angle-of-arrival estimation based on slope processing is illustrated in Figure 8.32. Basically, for different signs of ϕ, the control signal $\Upsilon_\Delta(t, \phi)$ has different slopes [see Figure 8.32(a)]. For larger ϕ, the control signal $\Upsilon_\Delta(t, \phi)$ is smaller. Based on this, the control mechanism is defined so that the delays of the VDC$_i$s are adjusted by DAC according to the relationship

$$\begin{aligned} \tilde{\tau}_i > \tilde{\tau}_{i+1}, \quad &\text{for} \quad -\pi/2 \leq \phi < \phi_0 \\ \tilde{\tau}_i < \tilde{\tau}_{i+1}, \quad &\text{for} \quad \phi_0 < \phi \leq \pi/2 \end{aligned} \tag{8.48}$$

More details can be found in [38].

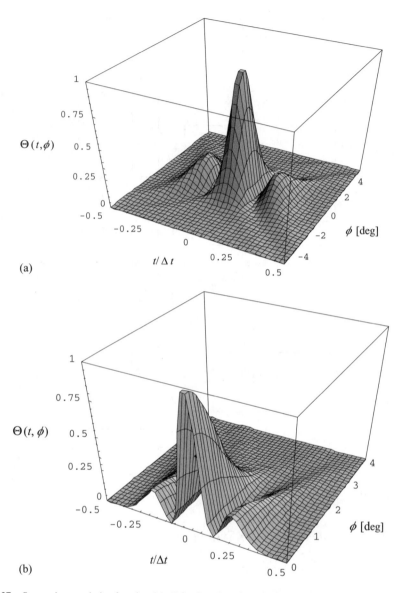

(a)

(b)

Figure 8.27   Space–time resolution function $\Theta(t,\phi)$ for the value of spacial frequency bandwidth $\rho = 10$. (a) Scaling parameter $\alpha = 3$ and the angular range $-5^0 \leq \phi \leq +5^0$; (b) $0^0 \leq \phi \leq +3^0$.

8.8   Beamforming for UWB Radio

8.8.1   Circular Array

The received signal is modeled as a Gaussian pulse [38–40]

$$\Omega(t) = (E/\Delta T) \exp\lfloor -\pi (t/\Delta T)^2 \rfloor \tag{8.49}$$

For the geometry shown in Figure 8.33 we have:

$$R_n = (r^2 + a^2 - 2ar \cos \Psi_n)^{1/2}$$

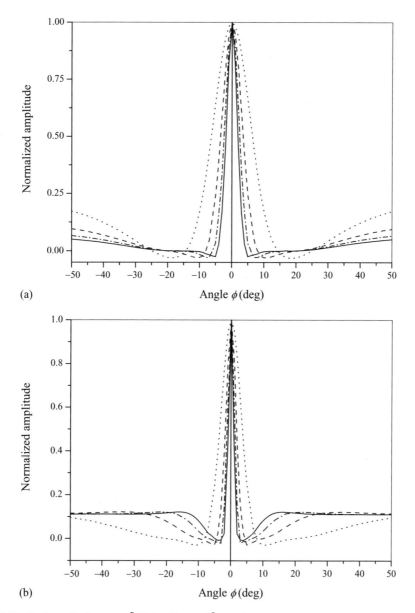

Figure 8.28 Peak amplitude pattern $\tilde{A}(\phi)$ for: (a) $\phi_0 = 0^0$, $\alpha = 3$, $d = c\Delta\ T/2$, and $M = 5$ (dotted line), 9 (dashed line), 13 (dashed-dotted line) and 17 (solid line); and (b) $M = 9$ and $d = c\Delta\ T/2$ (dotted line), $c\Delta\ T$ (dashed line), $3c\Delta\ T/2$ (dashed-dotted line) and $2c\Delta\ T$ (solid line).

which, for $r \gg a$, can be approximated as

$$R_n = r - a\cos\psi_n = r - a\sin\theta\cos(\phi - \phi_n)$$

with $\phi_n = 2\pi\left(\frac{n}{N}\right)$, $n = 1, 2, \ldots, N$.

We assume that the signal source is located in the far field at the point $P(r, \theta, \phi)$; $r \gg a$. So at the nth array element we have:

$$V_n(t) = \Omega(t + \tau_n) = (E/\Delta T)\ \exp\left\lfloor -\pi\left[(t + \tau_n)/\Delta T\right]^2\right\rfloor \tag{8.50}$$

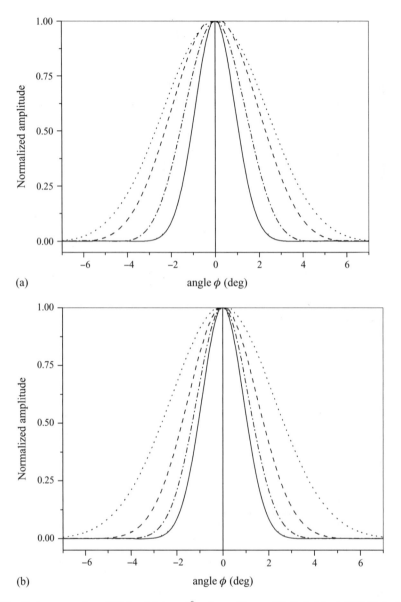

(a)

(b)

Figure 8.29   Peak power pattern $P(\phi)$ for: (a) $\phi_0 = 0^0$ and $\rho = 10$, $\alpha = 0.5$ (dotted line), 0.75 (dashed line), 1.5 (dashed-dotted line) and 3 (solid line); and (b) $\alpha = 3$ and $\rho = 4$ (dotted line), 6 (dashed line), 8 (dashed-dotted line) and 10 (solid line).

with $\tau_n = (a/c) \sin \theta \cos(\phi - \phi_n)$. To steer the peak of the main beam of the array in the (θ_0, ϕ_0) direction, a delay

$$\alpha_n = -\tau_n (\theta_0, \phi_0) = - (a/c) \sin \theta_0 \cos (\phi_0 - \phi_n) \tag{8.51}$$

must be applied to the voltage signal $V_n(t)$. So, the overall received signal is

$$V_T (t, \theta, \phi) = \sum_{n=1}^{N} \Omega (t + \tau_n + \alpha_n) \tag{8.52}$$

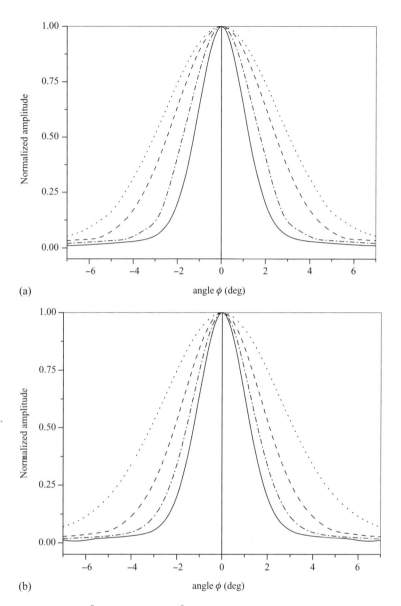

Figure 8.30 Energy pattern $\tilde{W}(\phi)$ for: (a) $\phi_0 = 0^0$ and $\rho = 10$, $\alpha = 0.5$ (dotted line), 0.75 (dashed line), 1.5 (dashed-dotted line) and 3 (solid line); and (b) $\alpha = 3$ and $\rho = 4$ (dotted line), 6 (dashed line), 8 (dashed-dotted line) and 10 (solid line).

We will use notation

$$\rho_c = (a/c\Delta T) = 2\Delta f a/c, \quad \Delta f = 1/2\Delta T \qquad (8.53)$$
$$\eta_s = \sin\theta\sin\phi - \sin\theta_0\sin\phi_0$$
$$\eta_c = \sin\theta\cos\phi - \sin\theta_0\cos\phi_0$$
$$\eta_0 = [(\eta_c)^2 + (\eta_s)^2]^{1/2}$$
$$\cos\xi = \frac{\eta_c}{\eta_0}$$
$$\xi = \arctan\frac{\eta_s}{\eta_c}$$

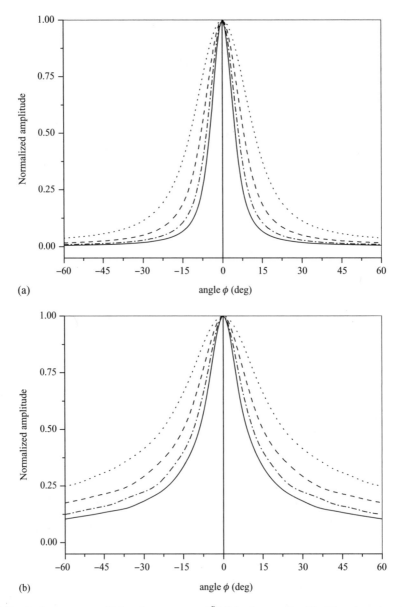

(a)

(b)

Figure 8.31 Peak power pattern $P(\phi)$ and energy pattern $\tilde{W}(\phi)$ for the case of an ideal Gaussian pulse with $\alpha = 0$. The plots are calculated for $\rho = 4$ (dotted line), 6 (dashed line), 8 (dashed-dotted line) and 10 (solid line).

Equations (8.52) and (8.54) together result in:

$$(\alpha_n + \tau_n)/\Delta T = \rho_c \eta_0 \cos{(\xi - \phi_n)} \qquad (8.54)$$

and we have

$$V_T(t, \theta, \phi) = (E/\Delta T) \sum_{n=1}^{N} \exp{\left\{-\pi \left[(t/\Delta T) + \rho_c \eta_0 \cos{(\xi - \phi_n)}\right]^2\right\}} \qquad (8.55)$$

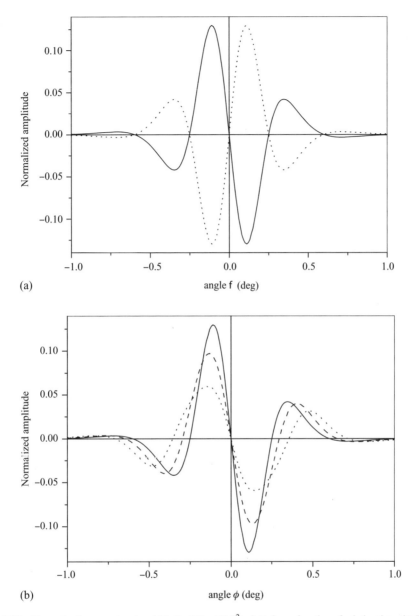

(a)

(b)

Figure 8.32 Normalized monopulse signal $\Upsilon_\Delta(t, \phi)/(m\,\Delta T\,E_0^2)$ plotted as a function of relative time $t/\Delta T$ for (a) $\alpha = 3$, $\rho = 10$, and the values of the angle of incidence $\phi = +2^0$ (solid line), and $\phi = -2^0$ (dotted line) and repeated for (b) $\phi = 2^0$ (solid line), $\phi = 3^0$ (dashed line), $\phi = 4^0$ (dotted line).

For a main beam in the vertical direction along the z-axis $\theta_0 = 0$ which yields $\alpha_n = \alpha_{nv} = 0$, $\eta_0 = \eta_{0v} = \sin\theta$, $\xi = \xi_v = \phi$ and we have:

$$V_v(t, \theta, \phi) = (E/\Delta T) \sum_{n=1}^{N} \exp\left\{-\pi\left[(t/\Delta T) + \rho_c \sin\theta \cos(\phi - \phi_n)\right]^2\right\} \qquad (8.56)$$

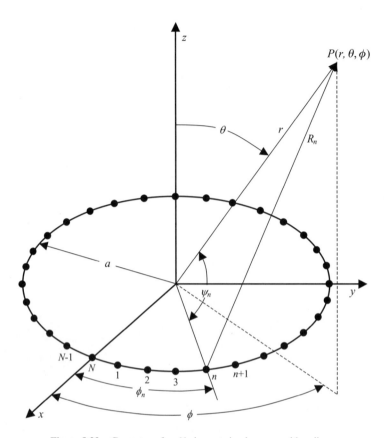

Figure 8.33 Geometry of an N-element circular array with radius a.

A main beam can be formed in the principal horizontal plane ($z = 0$) by setting $\theta_0 = \pi/2$ which gives

$$\alpha_n = \alpha_{nh} = -(a/c)\cos(\phi_0 - \phi_n)$$

$$\eta_0 = \eta_{0h} = 2\sin\left(\frac{\phi_0 - \phi_n}{2}\right)$$

$$\xi = \xi_h = \frac{\pi + \phi_0 + \phi_n}{2}$$

(8.57)

and we have

$$V_h(t, \phi) = (E/\Delta T) \times \sum_{n=1}^{N} \exp\left\{-\pi\left[(t/\Delta T) + 2\rho_c \sin\left(\frac{\phi - \phi_0}{2}\right) \times \cos\left(\frac{\pi + \phi_0 + \phi - 2\phi_n}{2}\right)\right]^2\right\}$$

(8.58)

A graph of $V_v(t, \theta, \phi)/N(E/\Delta T)$ is shown in Figure 8.34.

By using Figure 8.34, we define the peak amplitude pattern

$$A(\theta, \phi) = V_T(0, \theta, \phi)/N(E/\Delta T) = \frac{1}{N}\sum_{n=1}^{N} \exp\left\{-\pi\left[\rho_c\eta_0\cos(\xi - \phi_n)\right]^2\right\}$$

(8.59)

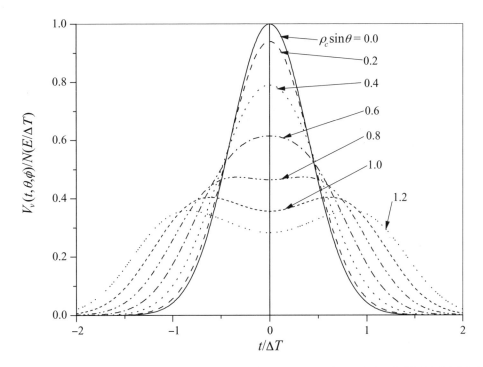

Figure 8.34   The time variation of the normalized voltage signal $V_V(t, \theta, \phi)$ for $\rho_c \sin \theta = 0, 0.2, 0.4, 0.6, 0.8, 1$, and 1.2.

Similarly, we have, for the vertical and the horizontal peak amplitude patterns, the following expressions

$$A_v(\theta, \phi) = V_v(0, \theta, \phi)/N(E/\Delta T) = \frac{1}{N} \sum_{n=1}^{N} \exp\left\{-\pi \left[\rho_c \sin \theta \cos (\phi - \phi_n)\right]^2\right\}$$

$$A_h(\phi) = \frac{1}{N} \sum_{n=1}^{N} \exp\left\{-\pi \left[2\rho_c \sin\left(\frac{\phi - \phi_0}{2}\right) \cos\left(\frac{\pi + \phi_0 + \phi - 2\phi_n}{2}\right)\right]^2\right\} \qquad (8.60)$$

In a similar way, one can show that for sinusoidal waves with the time variation $\exp(j\omega t)$, the *Array factor*, AF, of an N-element circular array is given by [41, 42]:

$$\text{AF}(\theta, \phi) = \sum_{n=1}^{N} I_n \exp\left[j\rho_s \eta_0 \cos(\xi - \phi_n)\right] \qquad (8.61)$$

where

$$\rho_s = 2\pi a/\lambda, \quad \lambda = \text{wavelength} \qquad (8.62)$$

I_n is the amplitude excitation of the array element n. For constant $I_n = I$ we have:

$$\text{AF}(\theta, \phi) = NI \sum_{n=1}^{N} J_{mN}(\rho_s \eta_0) \exp\left[jmN\left(\frac{\pi}{2} - \xi\right)\right] \qquad (8.63)$$

where $J_x()$ is the Bessel function of the first kind. The vertical and the horizontal amplitude patterns are now given as:

$$AF_v(\theta, \phi) = NI \sum_{n=1}^{N} J_{mN} \left(\rho_s \sin\theta\right) \exp\left[jmN\left(\frac{\pi}{2} - \phi\right)\right] \tag{8.64}$$

$$AF_h(\phi) = NI \sum_{n=1}^{N} J_{mN} \left(2\rho_s \sin\frac{\phi}{2}\right) \exp\left[-\frac{jmN\phi}{2}\right]$$

For periodic sinusoidal signals, the square of the amplitude pattern, $AF(\theta, \phi)$, represents the average power pattern $P_{av}(\theta, \phi)$. We also define the vertical and horizontal power patterns as

$$P_{av,v}(\theta, \phi) = [AF_v(\theta, \phi)]^2$$
$$P_{av,h}(\theta, \phi) = [AF_h(\theta, \phi)]^2 \tag{8.65}$$

The peak power pattern is defined as

$$P(\theta, \phi) = [A(\theta, \phi)]^2 = \frac{1}{N^2} \left[\sum_{n=1}^{N} \exp\{-\pi [\rho_c \eta_0 \cos(\xi - \phi_n)]^2\}\right]^2 \tag{8.66}$$

We also define the vertical and horizontal peak power patterns as

$$P_v(\theta, \phi) = [A_v(\theta, \phi)]^2$$
$$P_h(\phi) = [A_h(\phi)]^2 \tag{8.67}$$

The average power or energy pattern is defined as

$$W(\theta, \phi) = \frac{\int\limits_{-\infty}^{\infty} [V_T(t, \theta, \phi)]^2 \, dt}{\int\limits_{-\infty}^{\infty} [V_T(t, \theta_0, \phi_0)]^2 \, dt} = \frac{\int\limits_{-\infty}^{\infty} \left[\sum_{n=1}^{N} \exp\{-\pi [(t/\Delta T) + \rho_c \eta_0 \cos(\xi - \phi_n)]^2\}\right]^2 dt}{N^2 \int\limits_{-\infty}^{\infty} \exp\{-2\pi [(t/\Delta T)]^2\} dt} \tag{8.68}$$

As before, for the vertical and horizontal components, we have

$$W_v(\theta, \phi) = \frac{\int\limits_{-\infty}^{\infty} \left[\sum_{n=1}^{N} \exp\left\{-\pi [(t/\Delta T) + \rho_c \sin\theta \cos(\phi - \phi_n)]^2\right\}\right]^2 dt}{N^2 \int\limits_{-\infty}^{\infty} \exp\left\{-2\pi [(t/\Delta T)]^2\right\} dt} \tag{8.69}$$

$$W_h(\theta, \phi) = \frac{\int\limits_{-\infty}^{\infty} \left[\sum_{n=1}^{N} \exp\left\{-\pi \left[(t/\Delta T) + 2\rho_c \sin\left(\frac{\phi - \phi_0}{2}\right) \cos\left(\frac{\phi + \phi_0 + \pi - 2\phi_n}{2}\right)\right]^2\right\}\right]^2 dt}{N^2 \int\limits_{-\infty}^{\infty} \exp\{-2\pi [(t/\Delta T)]^2\} dt}$$

$$\tag{8.70}$$

Some numerical results are shown in Figures 8.35–8.44.

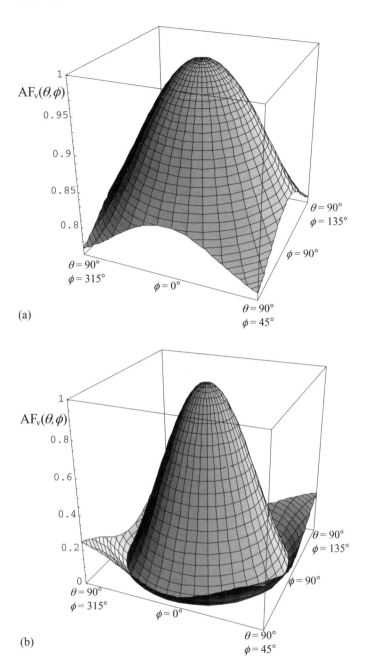

Figure 8.35 Normalized amplitude pattern $AF_v(\theta, \phi)$ given in Equation (8.65) for an infinitely extended periodic sinusoidal wave received by the circular array in Figure 8.33 with $N = 16$ elements and (a) $\rho_s = 1$; (b) $\rho_s = 3$; (c) $\rho_s = 6$; (d) $\rho_s = 12$.

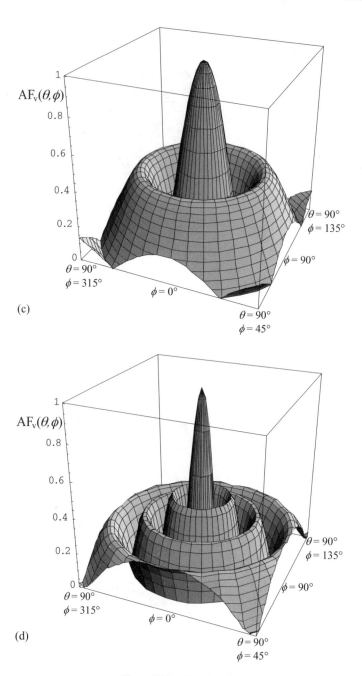

Figure 8.35 (*Continued*).

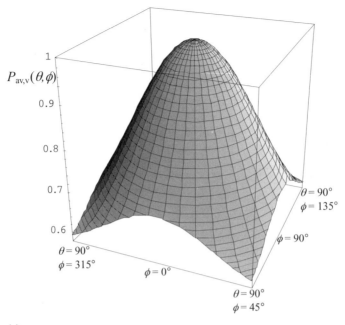

(a)

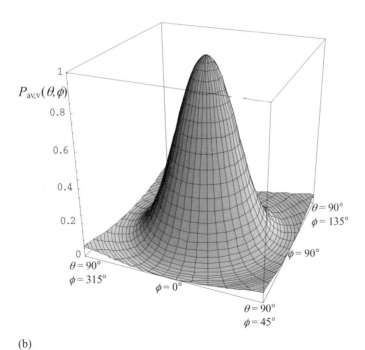

(b)

Figure 8.36 Normalized average power pattern $P_{av,v}(\theta, \phi)$ given in Equation (8.66) for an infinitely extended periodic sinusoidal wave received by the circular array in Figure 8.33 with $N = 16$ elements and (a) $\rho_s = 1$; (b) $\rho_s = 3$; (c) $\rho_s = 6$; (d) $\rho_s = 12$.

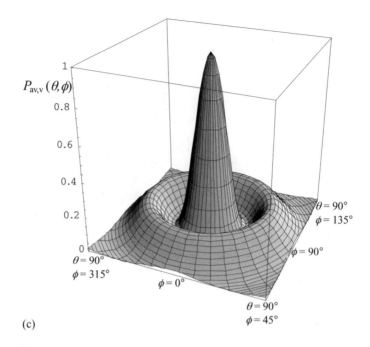

(c)

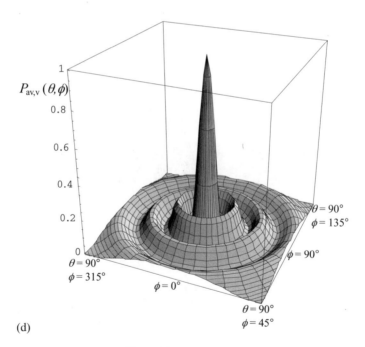

(d)

Figure 8.36   (*Continued*).

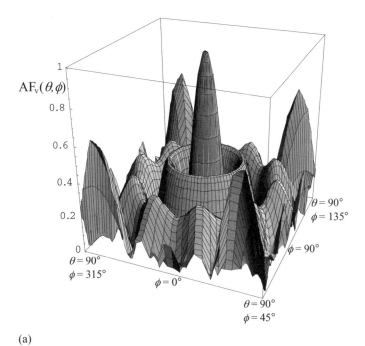

(a)

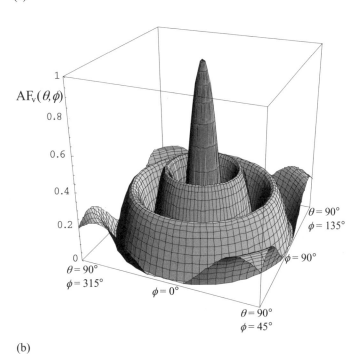

(b)

Figure 8.37 Normalized amplitude pattern $AF_v(\theta, \phi)$ given in Equation (8.65) for an infinitely extended periodic sinusoidal wave received by the circular array in Figure 8.33 with $\rho_s = 12$ and (a) $N = 10$ elements; (b) $N = 32$ elements.

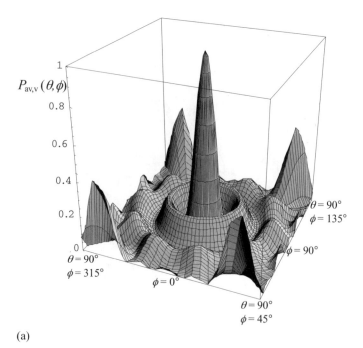

(a)

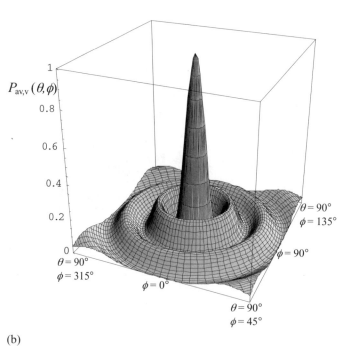

(b)

Figure 8.38 Normalized average power pattern $P_{av,v}(\theta, \phi)$ given in Equation (8.66) for an infinitely, extended periodic sinusoidal wave received by the circular array in Figure 8.33 with $\rho_s = 12$ and (a) $N = 10$ elements; (b) $N = 32$ elements.

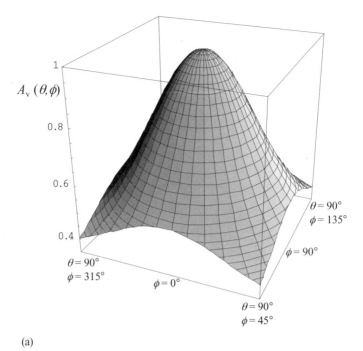

(a)

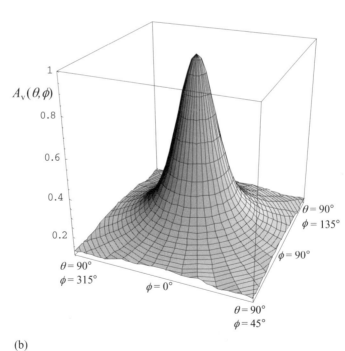

(b)

Figure 8.39 Peak amplitude pattern $A_v(\theta, \phi)$ given in Equation (8.61) for non-sinusoidal Gaussian pulses received by the circular array in Figure 8.33 with $N = 16$ elements and (a) $\rho_s = 1$; (b) $\rho_s = 3$; (c) $\rho_s = 6$; (d) $\rho_s = 12$.

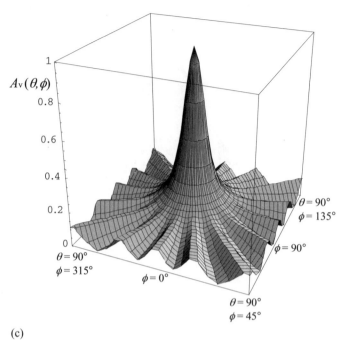

(c)

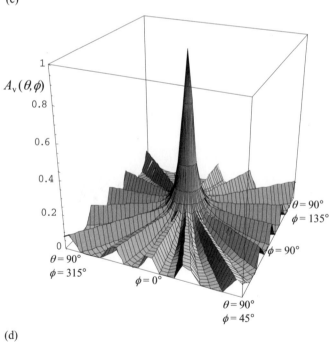

(d)

Figure 8.39   (*Continued*).

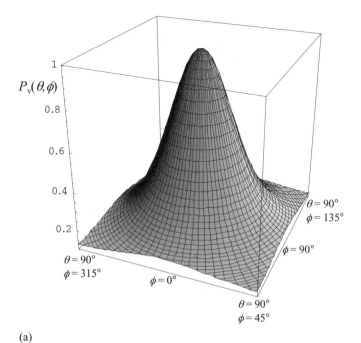

(a)

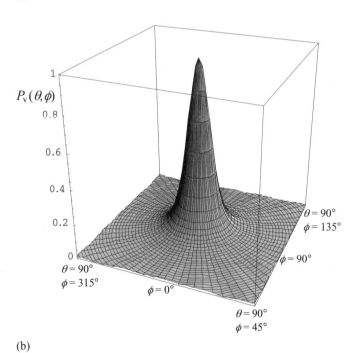

(b)

Figure 8.40   Peak power pattern $P_v(\theta, \phi)$ given in Equation (8.68) for non-sinusoidal Gaussian pulses received by the circular array in Figure 8.33 with $N = 16$ elements and (a) $\rho_s = 1$; (b) $\rho_s = 3$; (c) $\rho_s = 6$; (d) $\rho_s = 12$.

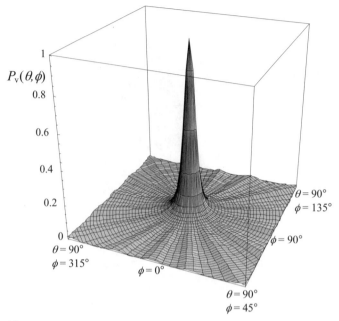

(c)

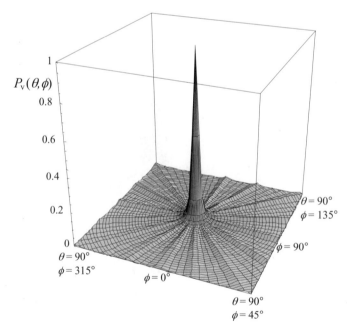

(d)

Figure 8.40   (*Continued*).

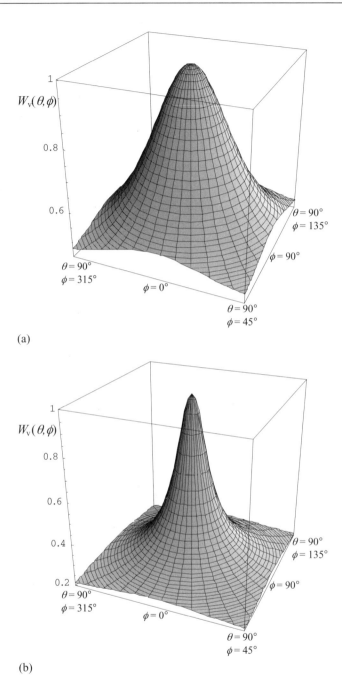

(a)

(b)

Figure 8.41 Energy pattern $W_v(\theta, \phi)$ given in Equation (8.70) for non-sinusoidal Gaussian pulses received by the circular array in Figure 8.33 with $N = 16$ elements and (a) $\rho_s = 1$; (b) $\rho_s = 3$; (c) $\rho_s = 6$; (d) $\rho_s = 12$.

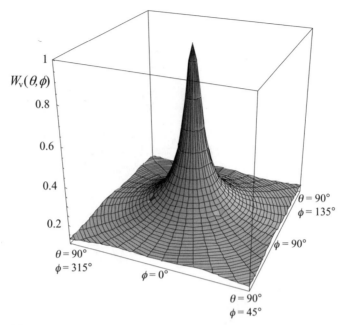

(c)

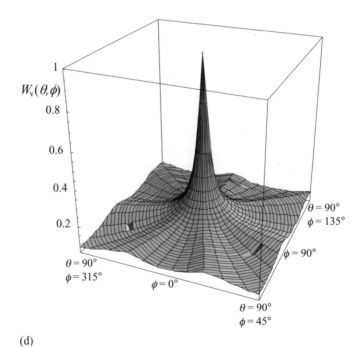

(d)

Figure 8.41   (*Continued*).

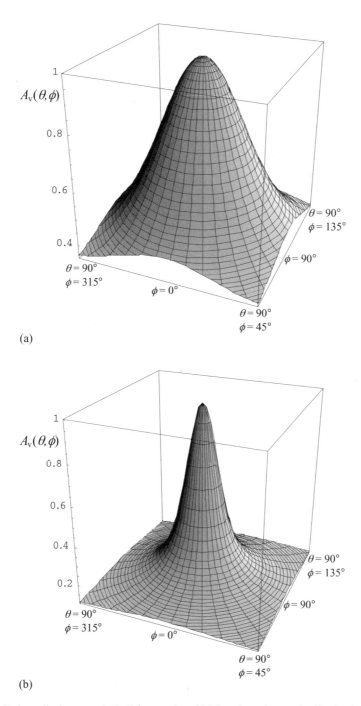

Figure 8.42 Peak amplitude pattern $A_v(\theta, \phi)$ for non-sinusoidal Gaussian pulses received by the circular array with a large number ($N \to \infty$) of elements and (a) $\rho_s = 1$; (b) $\rho_s = 3$; (c) $\rho_s = 6$; (d) $\rho_s = 10$.

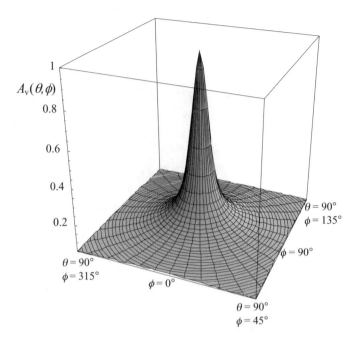

(c)

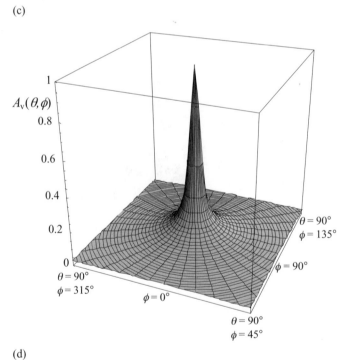

(d)

Figure 8.42   *(Continued)*.

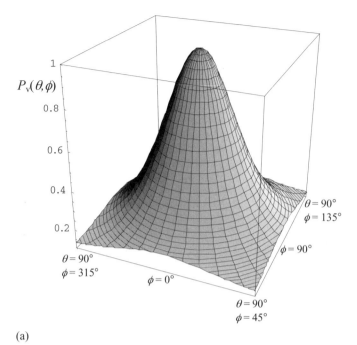

(a)

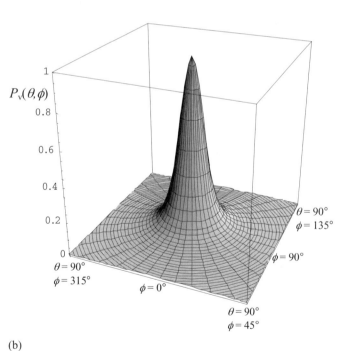

(b)

Figure 8.43 Peak power pattern $P_v(\theta, \phi)$ for non-sinusoidal Gaussian pulses received by the circular array with a large number ($N \rightarrow \infty$) of elements and (a) $\rho_s = 1$; (b) $\rho_s = 3$; (c) $\rho_s = 6$; (d) $\rho_s = 10$.

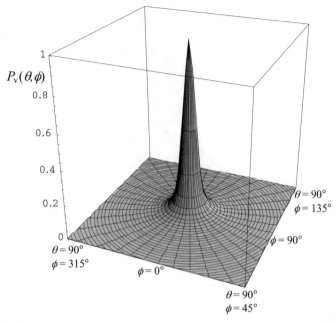

(c)

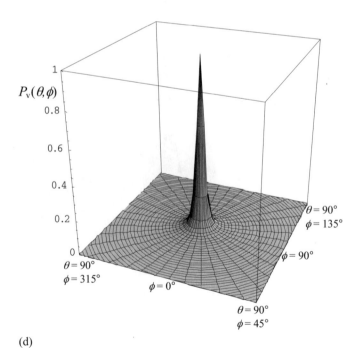

(d)

Figure 8.43   (*Continued*).

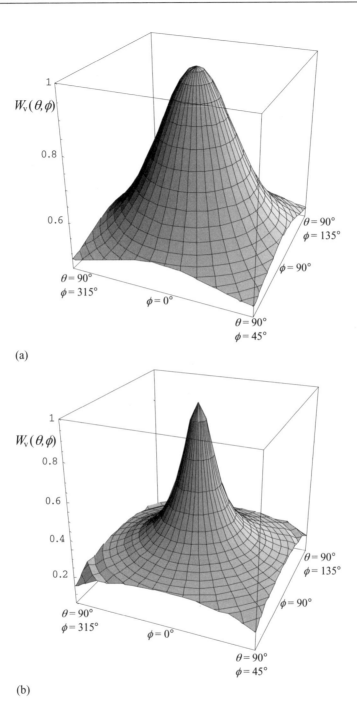

(a)

(b)

Figure 8.44 Energy pattern $W_v(\theta, \phi)$ for non-sinusoidal Gaussian pulses received by the circular array with a large number ($N \rightarrow \infty$) of elements and (a) $\rho_s = 1$; (b) $\rho_s = 3$; (c) $\rho_s = 6$; (d) $\rho_s = 10$.

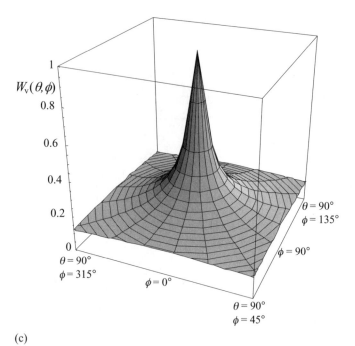

(c)

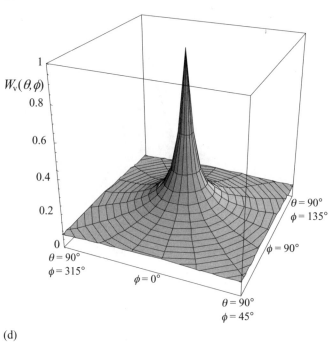

(d)

Figure 8.44   (*Continued*).

A slope pattern $S(\theta, \phi)$, a *vertical slope pattern* $S_v(\theta, \phi)$ and a *horizontal slope pattern* $S_h(\phi)$ can be derived by using a linear regression algorithm to calculate the slope of the ramp that best fits the rising section of the Gaussian pulse $V_T(t, \theta, \phi)$, $V_v(t, \theta, \phi)$ and $V_h(t, \phi)$. A plot of the ramp slope versus angle results in a slope pattern.

More details on practical aspects of UWB antenna and receiver design can be found in [43–88].

References

[1] Win, M. Z. and Scholtz, R. A. (1998) Impulse radio: How it works, *IEEE Communications Letters*, **2**, 36–38.

[2] Scholtz, R. A. (1993) Multiple access with time-hopping impulse modulation, in *Proceedings of MILCOM*, October 1993, pp. 447–450.

[3] Win, M. Z. and Scholtz, R. A. (1998) On the energy capture of ultrawide bandwidth signals in dense multipath environments, *IEEE Communications Letters*, **2**(9), 245–247.

[4] Win, M. Z., Scholtz, R. A. and Barnes, M. A. (1997) Ultra-wide bandwidth signal propagation for indoor wireless communications, in *Proceedings of the IEEE International Conference on Communications*, Montreal, Canada, June 1997, pp. 56–60.

[5] Cramer, R. J.-M., Scholtz, R. A. and Win, M. Z. (2002) Evaluation of an ultra-wide-band propagation channel, *IEEE Transactions on Antennas and Propagation*, **50**(5), 561–570.

[6] Schantz, H. G. and Fullerton, L. (2001) The diamond dipole: A Gaussian impulse antenna, *Proceedings of the IEEE AP-S International Symposium*, July 8–13, 2001.

[7] Scholtz, R. A. and Win, M. Z. (1997) Impulse radio, in *Wireless Communications: TDMA versus CDMA*, S. G. Glisic and P. A. Leppänen (Eds.) Kluwer, Norwell, MA.

[8] Hashemi, H. (1993) The indoor radio propagation channel, *Proceedings of the IEEE*, **81**, 943–968.

[9] Liberti, J. C. and Rappaport, T. S. (1999) *Smart Antennas for Wireless Communications: IS-95 and Third Generation CDMA Applications.* Prentice Hall, Englewood Cliffs, NJ.

[10] Saleh, A. A. M. and Valenzuela, R. A. (1987) A statistical model for indoor multipath propagation, *IEEE Journal on Selected Areas in Communications*, **5**, 128–137.

[11] Spencer, Q., Rice, M., Jeffs, B. and Jensen, M. (1997) A statistical model for the angle-of-arrival in indoor multipath propagation, in *Proceedings of the IEEE Vehicular Technology Conference*, 1415–1419.

[12] Spencer, Q., Jeffs, B., Jensen, M. and Swindlehurst, A. (2000) Modeling the statistical time and angle of arrival characteristics of an indoor multipath channel, *IEEE Journal on Selected Areas in Communications*, **18**, 347–360.

[13] Vaughan, R. G. and Scott, N. L. (1999) Super-resolution of pulsed multipath channels for delay spread characterization, *IEEE Transactions On Communications*, **47**, 343–347.

[14] Yano, S. M. (2002) Investigating the ultra-wideband indoor wireless channel, *IEEE 55th Vehicular Technology Conference* (VTC Spring 2002) 6–9 May 2002, **3**, 1200–1204.

[15] Rappaport, T. S. (1966) *Wireless Communications: Principles and Practices*, Prentice Hall.

[16] Cassioli, D. *et al.* (2001) A statistical model for the UWB indoor channel, *IEEE VTC2001*, **2**, 1159–1163.

[17] Ghassemzadeh, S. S., Jana, R., Rice, C. W., Turin, W. and Tarokh, V. (2002) A statistical path loss model for in-home UWB channels, *IEEE Conference on Ultra Wideband Systems and Technologies*, 21–23 May 2002, pp. 59–64.

[18] Turin, W., Jana, R., Ghassemzadeh, S. S., Rice, C. W. and Tarokh, T. (2002) Autoregressive modelling of an indoor UWB channel, *IEEE Conference on Ultra Wideband Systems and Technologies*, 21–23 May 2002, pp. 71–74.

[19] Ghassemzadeh, S. S. and Tarokh, V. (2003) UWB path loss characterization in residential environments, *IEEE Radio Frequency Integrated Circuits (RFIC) Symposium*, 8–10 June 2003, pp. 501–504.

[20] Ghassemzadeh, S. S. and Tarokh, V. (2003) UWB path loss characterization in residential environments, *IEEE MTT-S International Microwave Symposium Digest*, 8–13 June 2003, **1**, 365–368.

[21] Ramirez-Mireles, F. (2001) On the performance of ultra-wide-band signals in Gaussian noise and dense multipath, *IEEE Transactions on Vehicular Technology*, **50**(1), 244–249.

[22] Gagliardi, R. M. (1988) *Introduction to Telecommunications Engineering.* John Wiley & Sons, Inc., New York, pp. 357–437.

[23] Win, M. Z., Ramírez-Mireles, F., Scholtz, R. A. and Barnes, M. A. (1997) Ultra-wide bandwidth (UWB) signal propagation for outdoor wireless communications, in *Proceedings of the IEEE VTC Conference*, May 1997, pp. 251–255.

[24] Ramírez-Mireles, F. and Scholtz, R. A. (1998) Time-shift-keyed equicorrelated signal sets for impulse radio *M*-ary modulation, in *Proceedings of the IEEE Wireless Conference*, pp. 404–408.

[25] Ramírez-Mireles, F. and Scholtz, R. A. (1997) Performance of equicorrelated ultra-wideband pulse-position-modulated signals in the indoor wireless impulse radio channel, in *Proceedings of the IEEE PACRIM Conference*, pp. 640–644.

[26] Ramírez-Mireles, F. (2001) Performance of ultrawideband SSMA using time hopping and *M*-ary PPM, *IEEE Journal on Selected Areas in Communications*, **19**(6), 1186–1196.

[27] Ramírez-Mireles, F. (1998) *Multiple-access with ultra-wideband impulse radio modulation using spread spectrum time-hopping and block waveform pulse-position-modulated signals*, Ph.D. dissertation, Communication Sciences Institute, Electrical Engineering Department, University of Southern California.

[28] Georghiades, C. N. (1988) On PPM sequences with good autocorrelation properties, *IEEE Transactions On Information Theory*, **34**, 571–576.

[29] Gagliardi, R., Robbins, J. and Taylor, H. (1987) Acquisition sequences in PPM communications, *IEEE Transactions On Information Theory*, **33**, 738–744.

[30] Golomb, S. W. (1991) Construction of signals with favorable correlation properties, in *Surveys in Combinatorics*, Cambridge University Press, Cambridge.

[31] Forouzan, A. R., Nasiri-Kenari, M. and Salehi, J. A. (2002) Performance analysis of time-hopping spread-spectrum multiple-access systems: uncoded and coded schemes, *IEEE Transactions on Wireless Communications*, **1**, 671–681.

[32] Viterbi, A. J. (1990) Very low-rate convolutional codes for maximum theoretical performance of spread-spectrum multiple-access channels, *IEEE Journal on Selected Areas in Communications*, **8**, 641–649.

[33] Viterbi, A. (1995) *CDMA: Principles of Spread-Spectrum Communication*. Addison Wesley, Reading, MA.

[34] Shaft, P. D. (1977) Low-rate convolutional code application in spread-spectrum communications, *IEEE Transactions On Communications*, **COM-25**, 815–821.

[35] Win, M. Z., Qiu, X., Scholtz, R. A. and Li, V. O. K. (1999) ATM-based TH-SSMA network for multimedia PCS, *IEEE Journal on Selected Areas in Communications*, **17**, 824–836.

[36] Le Martret, C. J. and Giannakis, G. B. (2002) All-digital impulse radio with multiuser detection for wireless cellular systems, *IEEE Transactions On Communications*, **50**, 1440–1450.

[37] Hussain, M. G. M. (2002) Principles of space–time array processing for ultrawide-band impulse radar and radio communications, *IEEE Transactions on Vehicular Technology*, **51**, 393–403.

[38] Hussain, M. G. M. (1988) Performance analysis and advancement of self-steering arrays for nonsinusoidal waves–I, II, *IEEE Transactions on Electromagnetic Compatibility*, **30**, 161–174.

[39] Hussain, M. G. M. (1988) A self-steering array for nonsinusoidal waves based on array impulse response measurement, *IEEE Transactions on Electromagnetic Compatibility*, **30**, 154–160.

[40] Hussain, M. G. M., Al-Halabi, M. M. M. and Omar, A. A. (1989) Antenna patterns of nonsinusoidal waves with the time variation of a Gaussian pulse–Part III, *IEEE Transactions on Electromagnetic Compatibility*, **31**, 34–47.

[41] Balanis, C. A. (1982) Antenna Theory Analysis and Design, Harper and Row, New York, pp. 274–279.

[42] Ma, M. T. (1974) Theory and Application of Antenna Arrays, John Wiley & Sons New York, pp. 191–202.

[43] Hussain, M. G. M. (1988) Antenna patterns for nonsinusoidal waves with the time variation of a Gaussian pulse–Part I and II, *IEEE Transactions on Electromagnetic Compatibility*, **30**, 504–522.

[44] Rajeswaran, A., Somayazulu, V. S. and Foerster, J. R. (2003) RAKE performance for a pulse based UWB system in a realistic UWB indoor channel, *IEEE International Conference on Communications*, **4**, 2879–2883.

[45] Canadeo, C. M., Temple, M. A., Baldwin, R. O. and Raines, R. A. (2003) UWB multiple access performance in synchronous and asynchronous networks, *Electronics Letters*, **39**(11), 880–882.

[46] Miller, L. E. (2003) Autocorrelation functions for Hermite-polynomial ultra-wideband pulses, *Electronics Letters*, **39**(11), 870–871.

[47] Porcino, D. and Hirt, W. (2003) Ultra-wideband radio technology: potential and challenges ahead, *IEEE Communications Magazine*, **41**(7), 66–74.

[48] Aiello, G. R. and Rogerson, G. D. (2003) Ultra-wideband wireless systems, *IEEE Microwave Magazine*, **4**(2), 36–47.

[49] Hyun-Jin Park, Mi-Jeong Kim, Yoon-Jae So, Young-Hwan You and Hyoung-Kyu Song (2003) UWB communication system for home entertainment network, *IEEE Transactions on Consumer Electronics*, **49**(2), 302–311.

[50] Durisi, G. and Benedetto, S. (2003) Performance evaluation of TH-PPM UWB systems in the presence of multiuser interference, *IEEE Communications Letters*, **7**(5), 224–226.

[51] Parr, B., ByungLok Cho, Wallace, K. and Zhi Ding (2003) A novel ultra-wideband pulse design algorithm, *IEEE Communications Letters*, **7**(5), 219–221.

[52] Won Namgoong (2003) A channelized digital ultrawideband receiver, *IEEE Transactions on Wireless Communications*, **2**(3), 502–510.

[53] Nakagawa, M., Honggang Zhang and Sato, H. (2003) Ubiquitous homelinks based on IEEE 1394 and ultra wideband solutions, *IEEE Communications Magazine*, **41**(4), 74–82.

[54] Saberinia, E. and Tewfik, A. H. (2003) Single and multi-carrier UWB communications, *Seventh International Symposium on Signal Processing and Its Applications*, July 1–4, 2003, **2**, 343–346.

[55] Saberinia, E. and Tewfik, A. H. (2003) Receiver structures for multi-carrier UWB systems, *Seventh International Symposium on Signal Processing and Its Applications*, July 1–4, 2003, **1**, 313–316.

[56] Pidre Mosquera, J. M. and Isasa, M. V. (2003) Planar resistively loaded UWB dipoles analysis and comparison, *IEEE Society International Conference on Antennas and Propagation*, June 22–27, 2003, **3**, 636–639.

[57] Taniguchi, T. and Kobayashi, T. (2003) An omnidirectional and low-vswr antenna for the FCC-approved UWB frequency band, *IEEE Society International Conference on Antennas and Propagation*, June 22–27, 2003, **3**, 460–463.

[58] Schantz, H. G. (2003) UWB magnetic antennas, *IEEE Society International Conference on Antennas and Propagation*, June 22–27, 2003, **3**, 604–607.

[59] Ogawa, T., Tomiki, A. and Kobayashi, T. (2003) Development of two kinds of UWB sources for propagation, EMC and other experimental studies: impulse radio and direct-sequence spread spectrum, *IEEE Society International Conference on Antennas and Propagation*, June 22–27, 2003, **3**, 273–276.

[60] Xianming Qing, Wah Chia, M. Y. and Xuanhui Wu (2003) Wide-slot antenna for uwb applications, *IEEE Society International Conference on Antennas and Propagation*, June 22–27, 2003, **1**, 834–837.

[61] Kerkhoff, A. and Hao Ling (2003) Design of a planar monopole antenna for use with ultra-wideband (uwb) having a band-notched characteristic, *IEEE Society International Conference on Antennas and Propagation*, June 22–27, 2003, **1**, 830–833.

[62] Zwierzchowski, S. and Jazayeri, P. (2003) A systems and network analysis approach to antenna design for uwb communications, *IEEE Society International Conference on Antennas and Propagation*, June 22–27, 2003, **1**, 826–829

[63] Zhi Ning Chen, Xuan Hui Wu, Ning Yang and Chia, M. Y. W (2003) Design considerations for antennas in uwb wireless communication systems, *IEEE Society International Conference on Antennas and Propagation*, June 22–27, 2003, **1**, 822–825.

[64] Kwan-ho Lee, Chi-Chih Chen, Teixeira, F. L. and Lee, R. (2003) Numerical study of a uwb dual-polarized feed design for enhanced tapered chambers, *IEEE Society International Conference on Antennas and Propagation*, June 22–27, 2003, **1**, 265–268.

[65] Ghassemzadeh, S. S. and Tarokh, V. (2003) UWB path loss characterization in residential environments, *IEEE Radio Frequency Integrated Circuits (RFIC) Symposium*, June 8–10, 2003, pp. 501–504.

[66] Aiello, G. R. (2003) Challenges for ultra-wideband (UWB) CMOS integration, *IEEE Radio Frequency Integrated Circuits (RFIC) Symposium*, June 8–10, 2003, pp. 497–500.

[67] Woo Cheol Chung and Dong Sam Ha (2003) On the performance of bi-phase modulated uwb signals in a multipath channel, *IEEE Vehicular Technology Conference*, April 22–25, 2003, **3**, 1654–1658.

[68] Piazzo, L. and Romme, J. (2003) Spectrum control by means of the th code in uwb systems, *IEEE Vehicular Technology Conference*, April 22–25, 2003, **3**, 1649–1653.

[69] Guangrong Yue, Lijia Ge and Shaoqian Li (2003) Performance of uwb time-hopping spread-spectrum impulse radio in multipath environments, *IEEE Vehicular Technology Conference*, April 22–25, 2003, **3**, 1644–1648.

[70] Terri, M., Hong, A., Guibe, G. and Legrand, F. (2003) Major characteristics of UWB indoor transmission for simulation, *IEEE Vehicular Technology Conference*, April 22–25, 2003 **1**, 19–23.

[71] Yongfu Huang, Xiangning Fan, Jiang Wang and Guangguo Bi (2003) Analysis of the energy dynamic of UWB signal in multi-path environments, *IEEE Vehicular Technology Conference*, April 22–25, 2003, **1**, 15–18.

[72] Ray-Rong Lao, Jeon-Hwan Tarng and Chiuder Hsiao (2003) Transmission coefficients measurement of building materials for UWB systems in 3–10 GHz, *IEEE Vehicular Technology Conference*, April 22–25, 2003, **1**, 11–14.

[73] Alvarez, A., Valera, G., Lobeira, M., Torres, R. and Garcia, J. L. (2003) New channel impulse response model for UWB indoor system simulations, *IEEE Vehicular Technology Conference*, April 22–25, 2003, **1**, 1–5.

[74] Nakache, Y.-P. and Molisch, A. F. (2003) Spectral shape of UWB signals influence of modulation format, multiple access scheme and pulse shape, *IEEE Vehicular Technology Conference*, April 22–25, 2003, **4**, 2510–2514.

[75] Saberinia, E. and Tewfik, A. H. (2003) Generating UWB-OFDM signal using sigma-delta modulator, *IEEE Vehicular Technology Conference*, April 22–25, 2003, **2**, 1425–1429.

[76] Durisi, G. and Benedetto, S. (2003) Performance evaluation and comparison of different modulation schemes for UWB multi access systems, *IEEE International Conference on Communications*, 11–15 May 2003, **3**, 2187–2191.

[77] Nassar, C. R., Fang Zhu and Zhiqiang Wu (2003) Direct sequence spreading UWB systems: frequency domain processing for enhanced performance and throughput, *IEEE International Conference on Communications*, 11–15 May 2003, **3**, 2180–2186.

[78] Baccarelli, E. and Biagi, M. (2003) An adaptive codec for multi-user interference mitigation for UWB-based WLANs, *IEEE International Conference on Communications*, 11–15 May 2003, **3**, 2020–2024.

[79] Cassioli, D., Win, M. Z., Vatalaro, F. and Molisch, A. F. (2003) Effects of spreading bandwidth on the performance of UWB RAKE receivers, *IEEE International Conference on Communications*, 11–15 May 2003, **5**, 3545–3549.

[80] Kusuma, J., Maravic, I. and Vetterli, M. (2003) Sampling with finite rate of innovation: Channel and timing estimation for UWB and GPS, *IEEE International Conference on Communications*, 11–15 May 2003, **5**, 3540–3544.

[81] Yamamoto, N. and Ohtsuki, T. (2003) Adaptive internally turbo-coded ultra wideband-impulse radio (AITC-UWB-IR) system, *IEEE International Conference on Communications*, 11–15 May 2003, **5**, 3535–3539.

[82] Weisenhorn, M. and Hirt, W. (2003) Performance of binary antipodal signaling over the indoor UWB MIMO channel, *IEEE International Conference on Communications*, 11–15 May 2003, **4**, 2872–2878.

[83] Saberinia, E. and Tewfik, A. H. (2003) N-tone sigma-delta uwb-ofdm transmitter and receiver, *IEEE International Conference on Acoustics, Speech, and Signal Processing* (ICASSP'03), April 6–10, 2003, **4**, IV-129–IV-132.

[84] Huaning Niu, Ritcey, J. A. and Hai Liu (2003) Performance of uwb RAKE receivers with imperfect tap weights, *IEEE International Conference on Acoustics, Speech, and Signal Processing* (ICASSP'03), April 6–10, 2003, **4**, IV-125–IV-128.

[85] Mo, S. S., Gelman, A. D. and Gopal, J. (2003) Frame synchronization in UWB using multiple SYNC words to eliminate line frequencies, *IEEE Wireless Communications and Networking Conference*, 16–20 March 2003, **2**, 773–778.

[86] Canadeo, C. M., Temple, M. A., Baldwin, R. O. and Raines, R. A. (2003) Code selection for enhancing UWB multiple access communication performance using TH-PPM and DS-BPSK modulations, *IEEE Wireless Communications and Networking Conference*, 16–20 March 2003, **2**, 678–682.

[87] Boubaker, N. and Letaief, K. B. (2003) A low complexity MMSE-RAKE receiver in a realistic UWB channel and in the presence of NBI, *IEEE Wireless Communications and Networking Conference*, 16–20 March 2003, **1**, 233–237.

[88] Siwiak, K., Bertoni, H. and Yano, S. M. (2003) Relation between multipath and wave propagation attenuation, *Electronics Letters*, **39**(1), 142–143.

9

Linear Precoding for MIMO Channels

9.1 Space–Time Precoders and Equalizers for MIMO Channels

In this section the problem of equalization, which was initiated in Chapter 6 for single-input–single-output (SISO) systems, is extended to multiple-input–multiple-output (MIMO) channels.

9.1.1 ISI Modeling in MIMO Channels

Let us assume a MIMO channel with p transmitters and q receivers. Let $s_j(k)$ denote the sequence from transmitter $j(j = 1, \ldots, p)$, and let $h_{ij}(k)$ be the channel response from input j to output i $(i = 1, \ldots, q)$. Consequently, the output sequence at the receiver i is:

$$x_i(k) = \sum_{j=1}^{p} \sum_{l=0}^{d} h_{ij}(l) s_j(k-l) \tag{9.1}$$

where d denotes the maximal ISI degree. The above convolution can be equivalently expressed in a z-transform notation:

$$\mathbf{H}(D)\,\mathbf{s}(D) = \mathbf{x}(D) \tag{9.2}$$

where $s(D) = [s_1(D)\,s_2(D)\,s_p(D)^T]$, $\mathbf{x}(D) = [x_1(D) \ldots x_q(D)]^T$, and $\mathbf{H}(D) = \{h_{ij}(D)\}$ are the z-transform vectors (matrix) of the corresponding sequences or impulse responses where we use a delay operator $D = z^{-1}$. A p-input–q-output MIMO system can then be represented by the $q \times p$ matrix $\mathbf{H}(D)$, called the *transfer function* of the MIMO system.

A transfer function $\mathbf{H}(D)$ is perfectly recoverable (PR) if and only if there exists a polynomial matrix $\mathbf{G}(D)$ such that:

$$\mathbf{G}(D)\,\mathbf{H}(D) = \mathrm{Diag}\left[D^{k_j}\right] \tag{9.3}$$

where k_j denotes the necessary delays incurred on the recovered jth source signal. The equality in Equation (9.3) is referred to as a (generalized) Bezout identity and $\mathbf{G}(D)$ is referred to as a Bezout inverse of $\mathbf{H}(D)$, which can perfectly recover the original signals in the absence of noise. Given a MIMO system in Equation (9.1), the jth source signal is said to be perfectly recoverable (PR) with a ρ_j-tap

Advanced Wireless Communications & Internet: Future Evolving Technologies, Third Edition. Savo Glisic.
© 2011 John Wiley & Sons, Ltd. Published 2011 by John Wiley & Sons, Ltd.

equalizer (i.e., degree $= \rho_j - 1$) if there exists a (row) polynomial vector $\mathbf{g}$ (D) with degree lower than ρ_j such that:

$$\mathbf{g}(D)\,\mathbf{x}(D) = s_j(D)\,D^{k_j} \tag{9.4}$$

A polynomial matrix $\mathbf{C}$ (D) is said to be a (right) common factor of the columns in $\mathbf{H}$ (D) if $\mathbf{H}$ $(D) = \mathbf{H}'(D)\,\mathbf{C}(D)$ and $\mathbf{H}'(D)$ is itself a polynomial matrix. Moreover, a polynomial matrix is unimodular if and only if its determinant is a constant. A polynomial matrix is said to be (right) coprime if and only if there exists no nonunimodular right common factor. The right coprimeness of $\mathbf{H}$ (D) requires it to have more rows than columns, i.e., $q > p$, except when it is a unimodular matrix. In the sequel we will use some results from matrix theory [1]:

(R1) A MIMO system with transfer function $\mathbf{H}$ (D) is perfectly recoverable (PR) if and only if $\mathbf{H}$ (D) is (right) coprime, except for a common factor with determinant D^k. (This property will be called delay-permissive coprimeness.)

(R2) A MIMO system with transfer function $\mathbf{H}$ (D) is perfectly recoverable if and only if $\mathbf{H}$ (λ) has full column rank for any (complex value), $\lambda \neq 0$.

For any p-input–q-output MIMO transfer function $\mathbf{H}$ (D), there exists a unimodular polynomial matrix $\mathbf{U}$ (D) [2] such that $\mathbf{H}$ $(D)\,\mathbf{U}$ $(D) = \hat{\mathbf{H}}$ (D) where $\hat{\mathbf{H}}$ (D) is a minimal basis (MB). The $q \times p$ polynomial matrix $\hat{\mathbf{H}}$ (D) is called a *minimal basis* if and only if $\hat{\mathbf{H}}$ (D) is column reduced, i.e. its (column-wise) highest-degree coefficient matrix has full column rank. In the sequel, we will denote the kth column degree of the MB $\hat{\mathbf{H}}$ (D) by μ_k.

The *McMillan degree* (which is denoted as η) of a $q \times p$ $(p < q)$ polynomial matrix $\mathbf{H}$ (D) is defined as the highest degree of the determinants of all the $p \times p$ minors in $\mathbf{H}$ (D). It can be shown [2] that $\eta = \sum_{k=1}^{p} \mu_k$.

Given a $q \times p$ $(p < q)$ polynomial matrix $\mathbf{H}$ (D) with full column rank (on the polynomial ring), a $(q - p) \times q$ polynomial matrix $\mathbf{N}$ (D) is called a *null-space minimal basis (NMB)* of $\mathbf{H}$ (D) if and only if:

1. $\mathbf{N}$ (D) is in the null space of $\mathbf{H}$ (D), i.e., $\mathbf{N}$ $(D)\,\mathbf{H}$ $(D) = \mathbf{0}$;
2. $\mathbf{N}$ (D) is left-coprime;
3. $\mathbf{N}$ (D) is row reduced, i.e., its (row-wise) highest-degree coefficient matrix has full row rank.

We denote the row degrees of $\mathbf{N}$ (D) by $\{v_i\}_{i=1}^{q-p}$. The degree of the minimal basis of the null space of $\mathbf{H}$ (D) is defined as:

$$v = \max_i \{v_i\}$$

Any polynomial vector in the null subspace of $\mathbf{H}$ (D) can be expressed as a (polynomial-wise) combination of the row vectors in $\mathbf{N}$ (D). For more properties of minimal basis, see [2, Chapter 6] and [3].

The MIMO equivalent matrix Γ^ρ $[\mathbf{H}]$ is a block Toeplitz matrix (with ρ block-rows, i.e. $\rho \times q$ rows):

$$\Gamma^\rho [\mathbf{H}] = \begin{bmatrix} \mathbf{H}_d & \mathbf{H}_{d-1} & \cdots & \mathbf{H}_0 & 0 & \cdots & 0 \\ 0 & \mathbf{H}_d & \mathbf{H}_{d-1} & \cdots & \mathbf{H}_0 & \cdots & 0 \\ \vdots & \ddots & \ddots & \ddots & \ddots & \ddots & \vdots \\ 0 & \cdots & 0 & \mathbf{H}_d & \mathbf{H}_{d-1} & \cdots & \mathbf{H}_0 \end{bmatrix} \tag{9.5}$$

where $\mathbf{H}_i$ is the ith degree coefficient matrix of $\mathbf{H}$ (D), i.e. $\mathbf{H}$ $(D) = \mathbf{H}_0 + D\mathbf{H}_1 + \cdots + D^d\,\mathbf{H}_d$. The *equivalent* matrix can provide an effective tool for testing the coprimeness and, hence, the recoverability (PR) of the MIMO system.

Reference [4] defines equivalent matrix test for MIMO recoverability as follows:

a) A MIMO system $\mathbf{H}(D)$ is coprime if and only if there exists an integer ρ such that

$$\text{rank } \{\mathbf{\Gamma}^{\rho}[\mathbf{H}]\} = \eta + p \times \rho \qquad (9.6)$$

where η is the McMillan degree of $\mathbf{H}(D)$. (The smallest integer ρ to meet the above equality is equal to the NMB degree of $\mathbf{H}(D)$, i.e., v).

b) A MIMO system $\mathbf{H}(D)$ is PR if and only if there exists an integer ρ such that

$$\text{rank } \{\mathbf{\Gamma}^{\rho}[\mathbf{H}]\} = \eta' + p \times \rho \qquad (9.7)$$

where η' is the reduced McMillan degree of $\mathbf{H}(D)$, which can be obtained by simply subtracting the degree contributed by those pure delay factors from the full McMillan degree. More precisely, $\eta' = \eta -$ degree associated with pure delay factors.

Again, the smallest integer to meet the above equality is equal to the *NMB* degree v [4].

9.1.2 MIMO System Precoding and Equalization

The above system theory serves as a theoretical foundation for flexible transceiver design of MIMO systems. The analysis can be logically divided into the following categories:

a) $p < q$, and the channel is known to receiver;
b) $p > q$, and the channel is fed back to the transmitter;
c) $p \geq q$, and the channel is known to the receiver;
d) when $p < q$, the channel is known to the transmitter and the receiver.

Case (d) is usually handled without using transmitter channel knowledge, and hence, it can be treated as (a). As outlined in the flowchart in Figure 9.1, case (c) will be treated in Section 9.1.3, while cases (a) and (b) are treated in Sections 9.1.2.1 and 9.1.2.2 respectively.

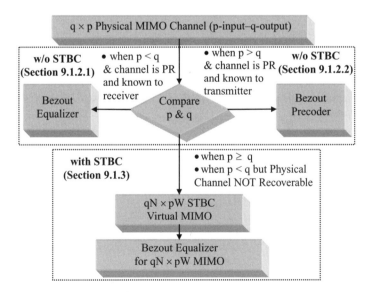

Figure 9.1 Design strategy is outlined in the flow chart. The pure Bezout solutions are outlined in the upper dashed box. The joint Bezout-STBC solutions will be treated in the next section as shown in the lower dashed box.

According to (R1) and (R2) of the previous section, when $p < q$ and the channel is known to receiver, there exists a PR Bezout equalizer if and only if the $q \times p$ transfer function $\mathbf{H}(D)$ is delay-permissive right coprime. More exactly, $\mathbf{G}(D)$ is a PR Bezout equalizer if and only if $\mathbf{G}(D)$ satisfies

$$\mathbf{G}(D)\,\mathbf{H}(D) = \mathrm{Diag}\left[D^{k_j}\right]$$

Similarly, when $p > q$ and the channel is known to the transmitter, there exists a PR Bezout precoder if and only if the $q \times p$ transfer function $\mathbf{H}(D)$ is delay-permissive left coprime. More exactly, $\mathbf{F}(D)$ is a PR Bezout precoder if and only if $\mathbf{F}(D)$ satisfies

$$\mathbf{H}(D)\,\mathbf{F}(D) = \mathrm{Diag}\left[D^{k_j}\right]$$

With such a Bezout precoder, the symbols received by the receiver will not only be ISI-free but also ICI-free (containing only the desired stream's information).

9.1.2.1 MIMO Bezout Equalizer

To analyze the system performance under noisy conditions, we need to investigate the SNR in Bezout inverse system.

The post-processing SNR is defined as the SNR at the output of the equalizer. Assuming BPSK constellation and that the bit is decided by zero thresholding, the BER can be determined as BER = $Q\left(\sqrt{\mathrm{SNR}}\right)$.

The Bezout equalization (*when $p < q$ and channel is known to receiver*) is illustrated in Figure 9.2. For a Bezout equalizer system $\mathbf{G}(D)\,\mathbf{H}(D) = \mathrm{Diag}\,[D^{kj}]$, let us denote one row of $\mathbf{G}(D)$ as $\mathbf{g}(D) = \mathbf{g}_0 + D\mathbf{g}_1 + \cdots + D^{\rho-1}\,\mathbf{g}_{\rho-1}$, and let $\vec{\mathbf{g}}$ denote the expanded row vector:

$$\vec{\mathbf{g}} \equiv \left[\mathbf{g}_{\rho-1} \cdots \mathbf{g}_1 \quad \mathbf{g}_0\right] \tag{9.8}$$

Obviously, the equalizer system can be designed separately for each individual stream. For the design of an individual equalizer $\mathbf{g}(D)$, note that the i.i.d. additive-white-Gaussian noise (AWGN) will be filtered by $\mathbf{g}(D)$ before its effect appears at the *output* of the equalizer. This leads to the post-processing noise power: $\sigma_n^2 \left\|\vec{\mathbf{g}}\right\|^2 = N_0/2 \left\|\vec{\mathbf{g}}\right\|^2$, where N_0 is the noise spectral density. Therefore, the SNR pertaining to a Bezout equalizer can be derived as:

$$\text{equalizer SNR} = \frac{2E_b}{N_0\|\vec{\mathbf{g}}\|^2} \tag{9.9}$$

where E_b is the transmit energy per bit. This immediately suggests a design criterion of minimizing the 2-norm of the equalizer.

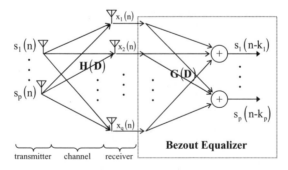

Figure 9.2   Bezout equalizer G (D) H (D) = Diag $[D^{kj}]$ with $p < q$.

From Figure 9.2, we note that the problem of designing an optimal Bezout equalizer for all the source inputs can be decoupled into the task of separately designing many individual equalizers, one for each input. The strategy of designing an optimal (individual) Bezout equalizer (for the jth input) is to find $\mathbf{g}$ (D) with minimal two-norm $\|\vec{g}\|$ such that:

$$\mathbf{g}(D)\mathbf{H}(D) = \left[0 \cdots D^{k_j} \cdots 0\right], j = 1, \ldots, p$$

Equivalently, in a resultant matrix notation:

$$\vec{g}\mathbf{\Gamma}^\rho[\mathbf{H}] = \mathbf{e}_i \tag{9.10}$$

where $\mathbf{e}_i$ is a row vector with all elements being zero except an entry 1 at $j + p(d + \rho - 1 - k_j)$, $k_j = 0$, $\ldots, d + \rho - 1$. The derivation of Bezout inverse depends not only on the equalizer order but also on the system delay k_j. The solution of an individual Bezout inverse is given in Equation (9.10). Take singular value decomposition (SVD) on $\mathbf{\Gamma}^\rho[\mathbf{H}] : \mathbf{\Gamma}^\rho[\mathbf{H}] = \mathbf{U\Sigma V}$, where $\mathbf{\Sigma}$ is a square diagonal matrix of positive singular values. Then, a Bezout inverse for Equation (9.10) exists if (and only if) there is a solution $\mathbf{b}$ for $\mathbf{bV} = \mathbf{e_i}$, and thereafter, the Bezout solution $\vec{g} = \mathbf{b\Sigma}^{-1}\mathbf{U}^H = \mathbf{e}_i\mathbf{V}^H\mathbf{\Sigma}^{-1}\mathbf{U}^H$ yields a minimum-norm solution with the optimal 2-norm

$$\|\vec{g}\|^2 = \mathbf{e}_i\mathbf{V}^H\mathbf{\Sigma}^{-2}\mathbf{Ve}_i^H \tag{9.11}$$

Note that the system delay k_j provides an extra (and important) degree of freedom for the Bezout inverse. Selection of k_j is just the same as selecting i since $i = j + p(d + \rho - 1 - k_j)$. The best integer i minimizing the 2-norm in Equation (9.11) is:

$$I^* = \arg\min_i\left\{\left(\mathbf{V}^H\mathbf{\Sigma}^{-2}\right)_{ii} \middle| i = j\,(\mathrm{mod}\,p), \quad \mathbf{e}_i \in \mathrm{RowSpan}\,\{\mathbf{V}\}\right\} \tag{9.12}$$

9.1.2.2 Bezout MIMO Precoder

When $p > q$ and the channel is known to transmitter a Bezout precoder system exists as illustrated in Figure 9.3. Mathematically

$$\mathbf{H}(D)\mathbf{F}(D) = \mathrm{Diag}\left[D^{k_j}\right] \tag{9.13}$$

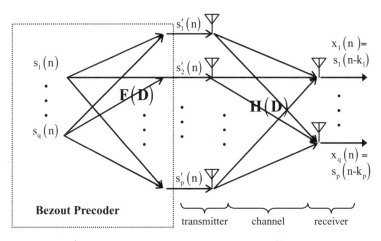

Figure 9.3 Bezout precoder: $\mathbf{H}(D)\mathbf{F}(D) = \mathrm{Diag}\,[D^{kj}]$ with $p < q$.

Let us denote one column of $\mathbf{F}(D)$ as $\mathbf{f}(D) = \mathbf{f}_0 + D\mathbf{f}_1 + \cdots + D^{\rho-1}\,\mathbf{f}_{\rho-1}$, and its expanded column vector $\vec{\mathbf{f}} \equiv [\mathbf{f}_{\rho-1}^T \quad \cdots \quad \mathbf{f}_1^T \quad \mathbf{f}_0^T]^T$. Let σ_s^2 be the input variance for one stream. Each precoding column vector $\mathbf{f}(D)$ amplifies the transmitter's power to $\sigma_s^2\|\vec{\mathbf{f}}\|^2$. In order to have an ISI-free communication, we must have $\mathbf{H}(D)\mathbf{f}(D) = [0 \quad \cdots \quad D^{k_j} \quad \cdots \quad 0]^T$, which produces a signal power of σ_s^2 at each receiver. Moreover, we must set $\sigma_s^2 = E_b/\|\vec{\mathbf{f}}\|^2$ in order to normalize the transmit energy per bit to E_b. In this case, the SNR for the Bezout precoder becomes

$$\text{precoder SNR} = \frac{2E_b}{N_0\|\vec{\mathbf{f}}\|^2} \tag{9.14}$$

Note that the noise power at each receiver is $N_0/2$.

Equation (9.14) establishes a useful duality between the optimal designs of Bezout equalizer and precoder systems. The optimal precoder design for a MIMO system $\mathbf{H}(D)$ is equivalent to the optimal equalizer design for a MIMO with transfer function $\mathbf{H}(D)^H$. Similarly to the previous equalizer design, design of an optimal Bezout precoder can be decoupled into the task of separately designing many individual precoders: one for each output.

An optimal (individual) Bezout precoder can be obtained by finding $f(D)$, with minimal two-norm such that $\mathbf{H}(D)\mathbf{f}(D) = [0 \cdots D^{kj} \cdots 0]^H$. Equivalently, in a resultant matrix notation

$$\mathbf{\Gamma}^\rho \left[\mathbf{H}^H\right]^H \vec{\mathbf{f}} = \mathbf{e}_i \tag{9.15}$$

where $\mathbf{e}_i$ is now a column vector with all zeros, except an entry 1 at $i = j + q(d + \rho - 1 - k_j)$, $k_j = 0, \ldots,$ $d + \rho - 1$. Take SVD on $\mathbf{\Gamma}^\rho\,[\mathbf{H}^H]^H$ as

$$\mathbf{\Gamma}^\rho[\mathbf{H}^H]^H = \mathbf{U}\mathbf{\Sigma}\mathbf{V} \tag{9.16}$$

For a given index i, a Bezout precoder $\vec{\mathbf{f}}$ for Equation (9.15) exists iff a solution $\mathbf{b}$ for $\mathbf{U}\mathbf{b} = \mathbf{e}_i$ exists. Similarly, the optimal Bezout precoder is $\vec{\mathbf{f}} = \mathbf{V}^H\mathbf{\Sigma}^{-1}\mathbf{b} = \mathbf{V}^H\mathbf{\Sigma}^{-1}\mathbf{U}^H\mathbf{e}_i$ with 2-norm:

$$\left\|\vec{\mathbf{f}}\right\|^2 = \left(\mathbf{U}\mathbf{\Sigma}^{-2}\mathbf{U}^H\right)_{ii} \tag{9.17}$$

The optimal integer i^* corresponding to the optimal delay k_j is:

$$i* = \arg\min_i \left\{ (\mathbf{U}\mathbf{\Sigma}^{-2}\mathbf{U}^H)_{ii} \,\middle|\, \hat{i} = j \,(\text{mod } q)\, \mathbf{e}_i \in \text{Column Span}\,\{\mathbf{U}\} \right\} \tag{9.18}$$

9.1.3 Precoder and Equalizer Design for STBC Systems

In many down-link communication scenarios, we have fewer receivers than transmitters (i.e., $p \geq q$). In this case it may be effective to adopt a design combining the Bezout and STBC techniques.

STBC, introduced in Chapter 4, can be regarded as a combination of channel coding techniques and equalization techniques. The performance can be enhanced by incorporating some structured redundancy at the transmitter side, e.g. with zero-padding or cyclic-prefixing. The basic STBC system used in this section is illustrated in Figure 9.4. Given a $q \times p$ transfer function of a (p-input–q-output) physical channel model, the STBC system can generate an expanded virtual $q' \times p'$ transfer function with $p' = pW$ and $q' = qN$. Namely, it results in a $qN \times pW$ *virtual transfer function*. Here, N denotes the block size, and W is the number of symbols sent per transmitter per block. For convenience, *this will be referred to as a N,W-STBC MIMO system*.

If the original channel is PR, there is no need to incorporate redundancy; then, we can set $W = N$. Such a (N,N)-STBC system is obtained by adding interleaving and deinterleaving operators to the transmission channel inputs and outputs, respectively.

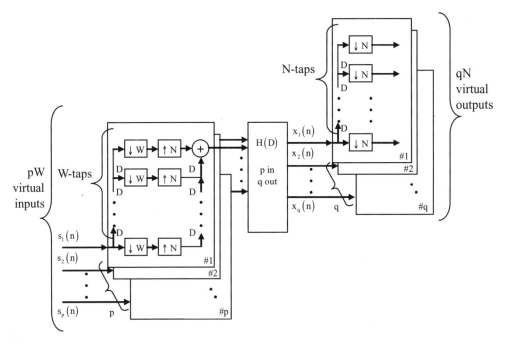

Figure 9.4   Basic (N,W)-STBC system. It creates a $Nq \times pW$ virtual transfer function from a p-input–q-output physical MIMO channel.

The virtual transfer function corresponding to a physical SISO channel has been well studied [5, 8, 9]. Given a transfer function $\mathbf{H}(D)$ of a p-input–q-output MIMO physical channel $\mathbf{H}(D) = \sum_{n=0}^{d} \mathbf{H}_n D^n$ and $\mathbf{H}_{j/N}(D)$ is its jth forward polyphase component out of N:

$$\mathbf{H}_{j/N}(D) = \sum_{n=0}^{[d/N]} \mathbf{H}_{N_{n+j}} D^n \tag{9.19}$$

where $0 \leq j \leq N - 1$. Following [5], we can obtain a $qN \times pN$ *virtual transfer function* denoted as $\bar{\mathbf{H}}(D)$:

$$\bar{\mathbf{H}}(D) = \begin{bmatrix} \mathbf{H}_{0/N}(D) & D\mathbf{H}_{N-1/N}(D) & \cdots & D\mathbf{H}_{1/N}(D) \\ \mathbf{H}_{1/N}(D) & \mathbf{H}_{0/N}(D) & \cdots & D\mathbf{H}_{2/N}(D) \\ \cdot & \cdot & \cdot & \cdot \\ \cdot & \cdot & \cdot & \cdot \\ \mathbf{H}_{N-2/N}(D) & \mathbf{H}_{N-3/N}(D) & \cdots & D\mathbf{H}_{N-1/N}(D) \\ \mathbf{H}_{N-1/N}(D) & \mathbf{H}_{N-2/N}(D) & \cdots & \mathbf{H}_{0/N}(D) \end{bmatrix} \tag{9.20}$$

If $\mathbf{H}(D)$ is a $q \times p$ (assuming $p \leq q$) transfer function of a physical MIMO channel and $\bar{\mathbf{H}}(D)$ the virtual transfer function of the (N,N)-STBC system, which is given in Equation (9.20), then there is a one-to-one correspondence between the common zeros of $\mathbf{H}(D)$ (denoted by $\{\lambda_i\}$) and that of $\bar{\mathbf{H}}(D)$ (denoted by $\{\eta_i\}$). More exactly [4]:

$$\eta_i = \lambda_i^N \tag{9.21}$$

and the (N,N)-STBC preserves the PR property of the original transfer function [6].

Problems arise for (N,N)-STBC systems when the original $\bar{\mathbf{H}}(D)$ is not PR or when it is PR but not robustly recoverable. An effective solution is via a redundant *precoding matrix* $\mathbf{F}$:

$$\tilde{\mathbf{H}}(D) = \bar{\mathbf{H}}(D)\mathbf{F} \tag{9.22}$$

The simplest (N, W)-STBC precoding is represented by a *precoding matrix*:

$$\mathbf{F} = \left[\mathbf{I}_{pW \times pW} \mid \mathbf{0}_{pW \times p(N-W)}\right]^T \tag{9.23}$$

This will lead to a $qN \times pW$ *virtual transfer function*, which is denoted as $\tilde{\mathbf{H}}(D)$

$$\tilde{\mathbf{H}}(D) = \begin{bmatrix} \mathbf{H}_{0/N}(D) & D\mathbf{H}_{N-1/N}(D) & \cdots & D\mathbf{H}_{N-W+1/N}(D) \\ \mathbf{H}_{1/N}(D) & \mathbf{H}_{0/N}(D) & \cdots & D\mathbf{H}_{N-W+2/N}(D) \\ \cdot & \cdot & \cdot & \\ \cdot & \cdot & \cdot & D\mathbf{H}_{N-1/N}(D) \\ & & & \mathbf{H}_{0/N}(D) \\ \cdot & \cdot & \cdot & \\ \mathbf{H}_{N-2/N}(D) & \mathbf{H}_{N-3/N}(D) & \cdots & \mathbf{H}_{N-W-1/N}(D) \\ \mathbf{H}_{N-1/N}(D) & \mathbf{H}_{N-2/N}(D) & \cdots & \mathbf{H}_{N-W-1/N}(D) \end{bmatrix} \tag{9.24}$$

Since W is the number of symbols sent per transmitter per block, the number of symbols being transmitted during one block period is pW; thus, the data transmission rate is pW/N. By using fewer columns than rows in the *precoding matrix* $\mathbf{F}$, some redundancy can be incorporated for the purpose of enhancing the robust recoverability of the new transfer function $\tilde{\mathbf{H}}(D)$.

If $\mathbf{H}(D)$ is PR, then $\tilde{\mathbf{H}}(D)$ in Equation (9.24) is PR if and only if the precoding matrix $\mathbf{F}$ has full column rank. In particular, for the (N,N)-STBC system, the $qN \times pW$ virtual transfer function $\tilde{\mathbf{H}}(D)$ is PR, provided that $\mathbf{H}(D)$ is PR [4].

9.2   Linear Precoding Based on Convex Optimization Theory

In this section, we discuss the design of a centralized precoder given fixed linear MIMO transmitter, channel, and receiver as shown in Figure 9.5. We define a precoder as a linear transformation on the transmitted symbols. If the precoded symbols are sent as is to the channel, then the precoder is the transmitter itself. However, in general, the precoded symbols may be transformed again before the channel. We refer to this transformation as the transmitter, and we assume that it is a fixed design

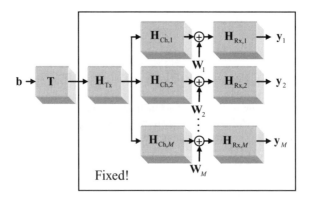

Figure 9.5   Block diagram of a precoder for a fixed MIMO receiver.

parameter. The output of the transmitter is then sent over a fixed MIMO channel (or channels) and is received using a fixed linear receiver (or receivers).

A precoder that applied a linear transformation on the transmitted symbols prior to the spreading was derived in [11]. This precoder inverted the channel at the transmitter side and is usually referred to as the transmit zero-forcing (ZF) precoder. The main drawback of the transmit ZF precoder is noise enhancement. Better solutions are transmit MF precoders and transmit rakes, which perform better in low SNR and transmit minimum mean-squared error (MMSE) precoders that compensate for the performance in the different SNR regions [12–16].

Variants of the previous precoders are discussed in [17–22]. Linear precoders based on an approximate maximum-likelihood approach and maximum asymptotic multiuser efficiency with different power constraints are derived in [23]. A linear precoding technique based on a decomposition approach is proposed in [24], and a linear precoder design for nonlinear maximum-likelihood (ML) receivers is discussed in [25]. Among the nonlinear precoders are the Tomlinson–Harashima precoder (THP) [26] and the 'Dirty Paper' precoder [27]. The nonlinear precoder that optimizes the transmitted symbols vector itself is derived in [28].

The problem of precoder design is highly related to other problems in the literature. In this section, we consider the design of linear precoders for fixed linear receivers. A related problem is the problem of jointly optimizing the precoder/transmitter and the receiver, which has been treated, e.g., in [29–37]. Another related problem is the joint design of rank-one transmit beamforming design and optimal power control [38–40].

In this section we discuss precoder design based on the framework of convex optimization theory [41], which allows for efficient numerical solutions using standard optimization packages [42]. It was shown in [10] that the power optimization problem can be formulated as a second-order cone program (SOCP) [43] or a semidefinite program (SDP) [44], known as a linear matrix inequalities (LMI) program. The SINR optimization can also be formulated as a standard conic program known as the generalized eigenvalue problem (GEVP) [45].

9.2.1 Generalized MIMO Systems

We assume that at each time instant, a block of symbols is modulated and transmitted over the channels. The possibly distorted output is then processed at the receivers in a linear fashion, as depicted in Figure 9.5. Denoting by y_i the length L output of the mth receiver, for $m = 1, \ldots, M$, we have

$$
\underbrace{\begin{bmatrix} \mathbf{y}_1 \\ \vdots \\ \mathbf{y}_M \end{bmatrix}}_{y} = \underbrace{\begin{bmatrix} \mathbf{H}_{Rx,1}\mathbf{H}_{Ch,1} \\ \vdots \\ \mathbf{H}_{Rx,M}\mathbf{H}_{Ch,M} \end{bmatrix}}_{H_{RxCh}} \mathbf{H}_{Tx}\mathbf{Tb} + \underbrace{\begin{bmatrix} \mathbf{H}_{Rx,1}\mathbf{w}_1 \\ \vdots \\ \mathbf{H}_{Rx,M}\mathbf{w}_M \end{bmatrix}}_{w}
\tag{9.25}
$$

where the matrices $\mathbf{H}_{Rx,m}$ and $\mathbf{H}_{Ch,m}$ denote the receiver and channel associated with the mth user, the matrix $\mathbf{H}_{Tx}$ is the centralized transmitter, $\mathbf{b}$ is the length $K = M \cdot L$ vector of independent, and unit variance transmitted symbols, and w_i are the noise vectors. The noise vectors may be correlated, and the channels are completely arbitrary. The only restriction is that the transmitter is centralized and has access to all of the K transmit components.

We assume a single stream per MIMO dimension, i.e., the length of $\mathbf{b}$ is equal to the length of $\mathbf{y}$. However, these streams can be dedicated to a single user or to multiple users. A single user, point-to-point communication system using L multiple receive and L transmit antennas is a special case of Equation (9.25) with $M = 1$. The downlink channel of a CDMA system with L users is a special case of Equation (9.25) with $L = 1$, where $\mathbf{H}_{Tx}$ is a signature matrix whose columns are the signatures of each of the users, $\mathbf{H}_{Ch,i} = \mathbf{I}$ and $\mathbf{H}_{Rx,i}$ are row vectors representing the linear receive filters of each of the users. A multiuser system with *transmit beamforming* in which L transmit antennas signal to L users each using a single receive antenna is a special case of Equation (9.25) with $L = 1$. Here, $\mathbf{H}_{Tx}$ is a beamforming matrix whose columns are the antenna weights of each of the L users, $\mathbf{H}_{Ch,i}$ are row vectors that represent the paths from the transmit antennas to the ith receive antenna, and $\mathbf{H}_{Rx,i}$ are arbitrary scalars.

In the sequel, we will assume that the transmitter $\mathbf{H}_{Tx}$, the channels $\mathbf{H}_{Ch,i}$ and the receivers $\mathbf{H}_{Rx,i}$ are fixed and cannot be altered due to budget restrictions, standardization, or physical problems. Given this fixed structure, we will try to improve the performance by introducing a linear precoder. The precoder, denoted by $\mathbf{T}$, linearly transforms the original symbol vector prior to the transmission.

As suggested by Equation (9.25), for ease of representation, we will use the following notation:

$$\mathbf{y} = \mathbf{HTb} + \mathbf{w} \qquad (9.26)$$

with $\mathbf{H} = \mathbf{H}_{RxCh}\,\mathbf{H}_{Tx}$ and the rest of the variables are defined in Equation (9.25).

The system performance (or quality of service (QoS)) is related to the output SINRs, and in particular to the worst SINR. The output SINR of the ith subchannel is defined as

$$\text{SINR}_i = \frac{\left|[\mathbf{HT}]_{i,i}\right|^2}{\sum_{j \neq i} \left|[\mathbf{HT}]_{i,j}\right|^2 + \sigma_i^2} \qquad (9.27)$$

for $i = 1, \ldots, K$, where $\sigma_i^2 = E\left\{|w_i|^2\right\} > 0$. Other criteria deal with the use of system resources, e.g., peak-to-average ratio, or maximal transmitted power. The most common resource measure is average transmitted power, which is defined as:

$$P = E\left\{\|\mathbf{H}_{Tx}\mathbf{Tb}\|^2\right\} = \text{Tr}\left\{\mathbf{T}^H \mathbf{H}_{Tx}^H \mathbf{H}_{Tx}\mathbf{T}\right\} \qquad (9.28)$$

The SINR metric and average power metric conflict, and one cannot maximize the SINRs while also minimizing the power, and vice versa. Depending on the application, the designer must decide which criterion is stricter. We therefore consider one of the following two complementary strategies. The first optimization strategy seeks to minimize the average transmitted power subject to QoS constraints. Given the required QoS, the system tries to satisfy it with minimum transmitted power [34, 39] (see also Chapter 16) as follows:

$$P(\gamma_o) = \begin{cases} \min_{\mathbf{T}} & \text{Tr}\left\{\mathbf{T}^H \mathbf{H}_{Tx}^H \mathbf{H}_{Tx}\mathbf{T}\right\} \\[2mm] \text{s.t.} & \dfrac{\left|[\mathbf{HT}]_{i,i}\right|^2}{\sum_{j \neq i} |[\mathbf{HT}]_{i,i}|^2 + \sigma_i^2} \geq \gamma_o \end{cases} \qquad (9.29)$$

where $\gamma_o > 0$ is the given worst SINR constraint.

The second strategy is maximizing the minimal SINR subject to a power constraint [33, 46] (see also Chapter 16). In this case, the problem can be formulated as:

$$S(P_o) = \begin{cases} \max_{\mathbf{T}} & \min_i \dfrac{|[\mathbf{HT}]_{i,i}|^2}{\sum_{j \neq i} |[\mathbf{HT}]_{i,i}|^2 + \sigma_i^2} \\[2mm] \text{s.t.} & \text{Tr}\left\{\mathbf{T}^H \mathbf{H}_{Tx}^H \mathbf{H}_{Tx}\mathbf{T}\right\} \leq P_o \end{cases} \qquad (9.30)$$

where $P_o > 0$ is the given power constraint.

9.2.2   Convex Optimization

In the sequel the following notation is used. Boldface capital letters denote matrices, boldface lower-case letters denote column vectors, and standard lower-case letters denote scalars. The superscripts $(\cdot)^T$,

$(\cdot)^*$, $(\cdot)^H$, $(\cdot)^{-1}$, and $(\cdot)^\dagger$ denote the transpose, the complex conjugate, the Hermitian, the matrix inverse operators, and the Moore–Penrose pseudoinverse, respectively. $[\mathbf{X}]_{i,j}$ denotes the (ith, jth) element of the matrix $\mathbf{X}$. By $\{x_i\}$, we denote a diagonal matrix with x_i being the (ith, ith) element; by vec($\mathbf{X}$), we denote stacking the elements of $\mathbf{X}$ in one long column vector; by e_i, we denote a zeros vector with a one at the ith element; by 1, we denote an all ones vector; and by $\mathbf{I}$, we denote the identity matrix of appropriate size. $\mathrm{Tr}\{\cdot\}$, $\Re\{\cdot\}$, $|\cdot|$, $\|\cdot\|$, and $\|\cdot\|_\infty$ denote the trace operator, the real part, the absolute value, the standard Euclidean norm, and the induced row sum matrix norm, respectively. Finally, $\mathbf{X} \succ 0$ denotes that the matrix $\mathbf{X}$ is a Hermitian positive semidefinite matrix, and $\mathrm{N}\{\cdot\}$ denotes the Null space operator.

The most widely used field in optimization is convex optimization. A convex program is a program with a convex objective function and convex constraints. It is well known that in such programs a local minimum is also a global minimum. The most common convex program is the linear program (LP) [41], i.e. an optimization with a linear objection function and linear (affine) constraints. Recently, conic programs, i.e. LPs with generalized inequalities are also used. The two standard conic programs are SOCP and SDP optimization. The standard form of an SOCP is [43]:

$$\text{SOCP}: \begin{cases} \min_x & \Re\left\{\mathbf{f}^H\mathbf{x}\right\} \\ \text{s.t.} & \begin{bmatrix} \mathbf{c}_i^H\mathbf{x} + d_i \\ \mathbf{A}_i^H\mathbf{x} + \mathbf{b}_i \end{bmatrix} \succ_K 0, \quad i = 1,\ldots,N \end{cases} \tag{9.31}$$

where the optimization variable is the vector $\mathbf{x}$ of length n and $\mathbf{f}$, $\mathbf{A}_i$, $\mathbf{b}_i$, $\mathbf{c}_i$, and d_i for $i = 1,\ldots,N$ are the data parameters of appropriate sizes. The notation $\succ_K$ denotes the following generalized inequality:

$$\begin{bmatrix} z \\ \mathbf{z} \end{bmatrix} \succ_K 0 \Leftrightarrow \|z\| \le z \tag{9.31a}$$

The standard form of an SDP is [44]:

$$\text{SDP}: \begin{cases} \min_x & \Re\left\{\mathbf{f}^H\mathbf{x}\right\} \\ \text{s.t.} & \mathbf{A}(\mathbf{x}) \succ 0 \end{cases} \tag{9.32}$$

where $\mathbf{A}(\mathbf{x}) = \mathbf{A}_0 + \sum_{i=1}^n x_i \mathbf{A}_i$ is a Hermitian matrix that depends affinely on $\mathbf{x}$. The data parameters are the Hermitian matrices $\mathbf{A}_i$ for $i = 0,\ldots,n$. The notation $\succ$ denotes the positive semidefinite generalized inequality. A simple case of an SDP is an SOCP. For example, each of SOC constraints in Equation (9.31) can be written as an LMI [21], as follows:

$$\begin{bmatrix} \mathbf{c}_i^H\mathbf{x} + d_i & \mathbf{x}^H\mathbf{A}_i + \mathbf{b}_i^H \\ \mathbf{A}_i^H\mathbf{x} + \mathbf{b}_i & \left(\mathbf{c}_i^H\mathbf{x} + d_i\right)\mathbf{I} \end{bmatrix} \succ 0\% \tag{9.31c}$$

A common optimization package designed to solve SOCP and SDP is *SEDUMI* [42].

Among nonconvex problems is the GEVP [45], which can still be efficiently solved. Its standard form is:

$$\text{GEVP}: \begin{cases} \min_{\beta,x} & \beta \\ \text{s.t.} & \beta\mathbf{B}(\mathbf{x}) - \mathbf{A}(\mathbf{x}) \succ 0 \\ & \mathbf{B}(\mathbf{x}) \succ 0 \\ & \mathbf{C}(\mathbf{x}) \succ 0 \end{cases} \tag{9.33}$$

where β is a real-valued optimization variable and $\mathbf{A}(\mathbf{x}) = \mathbf{A}_0 + \Sigma_{i=1}^n x_i\mathbf{A}_i$, $\mathbf{B}(\mathbf{x}) = \mathbf{B}_0 + \Sigma_{i=1}^n x_i\mathbf{B}_i$, and $\mathbf{C}(\mathbf{x}) = \mathbf{C}_0 + \Sigma_{i=1}^n x_i\mathbf{C}_i$ are Hermitian matrices that depend affinely on $\mathbf{x}$. The data parameters are the Hermitian matrices $\mathbf{A}_i$, $\mathbf{B}_i$ and $\mathbf{C}_i$ for $i = 0,\ldots,n$. The name of the GEVP arises from its resemblance to the well-known problem of minimizing the maximal generalized eigenvalue of the pencil $[\mathbf{A}, \mathbf{B}]$, i.e.

minimizing the largest β such that $\mathbf{Av} = \beta \mathbf{Bv}$. It is easy to show that this problem can be expressed as

$$\begin{cases} \min_\beta & \beta \\ \text{s.t.} & \beta \mathbf{B} - \mathbf{A} \succ 0 \end{cases} \tag{9.34}$$

which is a simple SDP. The GEVP generalizes this program to the case where $\mathbf{A}$ and $\mathbf{B}$ also depend on the optimization variables.

9.2.3   Precoding for Power Optimization

The first important property of any optimization problem is its feasibility (admissibility), i.e., whether a solution exists. In other words, we need to verify whether for a given γ_o there exists a $\mathbf{T}$ such that the constraint in Equation (9.29) is satisfied:

$$\min_i \frac{\left| [\mathbf{HT}]_{i,i} \right|^2}{\sum_{j \neq i} \left| [\mathbf{HT}]_{i,j} \right|^2 + \sigma_i^2} \geq \gamma_o \tag{9.29a}$$

Since the noise variances are positive, the SINRs are strictly lower than the signal-to-interference ratios (SIRs), i.e.,

$$\frac{\left| [\mathbf{HT}]_{i,i} \right|^2}{\sum_{j \neq i} \left| [\mathbf{HT}]_{i,j} \right|^2 + \sigma_i^2} < \frac{\left| [\mathbf{HT}]_{i,i} \right|^2}{\sum_{j \neq i} \left| [\mathbf{HT}]_{i,j} \right|^2} \tag{9.35}$$

for $i = 1, \ldots, K$. By scaling $\mathbf{T}$ to $a\mathbf{T}$ for large enough $a > 0$, the difference between the SIRs and the SINRs can be made insignificant.

Therefore, for the sake of examining the feasibility, the interesting metrics are the SIRs.

It was shown in [10] that there exists a $\mathbf{T}$ such that:

$$\min_i \frac{\left| [\mathbf{HT}]_{i,i} \right|^2}{\sum_{j \neq i} \left| [\mathbf{HT}]_{i,j} \right|^2} \geq \gamma_o \tag{9.36}$$

only if

$$\gamma_o \leq \frac{1}{\frac{K}{\text{rank}(\mathbf{H}) - 1}} \tag{9.37}$$

If the effective channel $\mathbf{H}$ is full rank, then the condition results in $\gamma_o \leq \infty$, i.e. any SIR is feasible. This is easily verified as the condition in Equation (9.29) can be satisfied by choosing $\mathbf{T} = a\mathbf{H}^{-1}$ for large enough $a > 0$. This choice of precoder inverts the channel and eliminates all interference. When the effective channel is rank deficient, the interference cannot be eliminated, and there is an upper bound on the maximal SIRs.

Experimenting with arbitrary channels shows that in almost all practical channels the bound can be achieved even for a fixed suboptimal receiver. For example, consider a rank $K - 1$ channel $\mathbf{H}$ with the normalized null vector $\mathbf{u} \in N\{\mathbf{H}^H\}$. Except for the case in which $u_i = 0$ for some $i = 1, \ldots, K$, the bound can always be attained by choosing

$$\mathbf{T} = \mathbf{H}^\dagger \text{diag}\left\{ 1/u_i^* \right\} \mathbf{Q} \tag{9.38}$$

where $\mathbf{Q}$ is a matrix with unit diagonal elements and $[\mathbf{Q}]_{i,j} = -1/(\{K-1\})$ for the nondiagonal $i \neq j$ elements. This can be shown by considering the following chain:

$$\mathbf{HT} = \mathbf{HH}^{\dagger}\mathrm{diag}\left\{1/u_i^*\right\}\mathbf{Q} = \mathrm{diag}\left\{1/u_i^*\right\}\mathbf{Q} \tag{9.39}$$

where we have used $\mathbf{HH}^{\dagger} = \mathbf{I} - \mathbf{uu}^H$ and the fact that $\mathbf{1} \in \mathrm{N}\{\mathbf{Q}\}$. Substituting the above $\mathbf{HT}$ into the SIRs yields the maximal SIRs in rank $K-1$ channels, as follows:

$$\frac{\left|[\mathbf{HT}]_{i,i}\right|^2}{\sum_{j \neq i}\left|[\mathbf{HT}]_{i,j}\right|^2} = \frac{1}{\frac{K}{K-1}-1}, \qquad i = 1,\ldots,K \tag{9.40}$$

9.2.3.1   Conic Optimization Solution

We now show that the P problem of Equation (9.29) can be represented as a standard conic optimization program. Thus, using off-the shelf optimization packages, we can numerically verify its feasibility and find its optimal solution. In order to use the standard forms of the conic programs, we must cast our problem constraints using the standard notations described in Section 9.2.2.

Using a real-valued slack variable P_o, the program can be rewritten as:

$$P(\gamma_o): \begin{cases} \min_{\mathbf{T},P_o} & P_o \\ \mathrm{s.t.} & \dfrac{\left|[\mathbf{HT}]_{i,i}\right|^2}{\sum_{j \neq i}\left|[\mathbf{HT}]_{i,j}\right|^2 + \sigma_i^2} \geq \gamma_o, \quad i = 1,\ldots,K \\ & \mathrm{Tr}\left\{\mathbf{T}^H\mathbf{H}_{Tx}^H\mathbf{H}_{Tx}\mathbf{T}\right\} \leq P_o \end{cases} \tag{9.41}$$

The argument $\mathbf{T}$ of the P program is defined up to a diagonal phase scaling on the right, i.e. if $\mathbf{T}$ is optimal, then $\mathbf{TT}\mathrm{diag}\left\{e^{j\phi_i}\right\}$, where ϕ_i for $i = 1,\ldots,K$ are arbitrary phases, is also optimal. This is easy to verify, as the phases do not change the objective nor the constraints. Therefore, we can restrict ourselves to precoders in which $[\mathbf{HT}]_{i,i} \geq 0$ for $i = 1,\ldots,K$, i.e. each has a nonnegative real part and a zero imaginary part. Taking this into account, we now recast the SINR constraints in standard form. Rearranging the constraints and using matrix notations, the constraints yield

$$\left(1 + \frac{1}{\gamma_o}\right)\left|[\mathbf{HT}]_{i,i}\right|^2 \geq \left\|\frac{\mathbf{T}^H\mathbf{H}^H\mathbf{e}_i}{\sigma_i}\right\|^2, \qquad i = 1,\ldots,K \tag{9.42}$$

Since $[\mathbf{HT}]_{i,i} \geq 0$ for $i = 1,\ldots,K$, we can take the square root of $\left|[\mathbf{HT}]_{i,i}\right|^2$, resulting in

$$\sqrt{1 + \frac{1}{\gamma_o}}\,[\mathbf{HT}]_{i,i} \geq \left\|\frac{\mathbf{T}^H\mathbf{H}^H\mathbf{e}_i}{\sigma_i}\right\|^2, \qquad i = 1,\ldots,K \tag{9.43}$$

which can be written as the SOCs

$$\begin{bmatrix} \sqrt{1 + \dfrac{1}{\gamma_o}}\,[\mathbf{HT}]_{i,i} \\ \mathbf{T}^H\mathbf{H}^H\mathbf{e}_i \\ \sigma_i \end{bmatrix} \succ_K 0, \qquad i = 1,\ldots,K \tag{9.44}$$

Similarly, the power constraint in Equation (9.41) can be reformulated using the vec(·) operator as $||vec\,(\mathbf{H}_{Tx}\mathbf{T})|| \leq \sqrt{P_o}$, which is equivalent to the SOC:

$$\begin{bmatrix} \sqrt{P_o} \\ vec\,(\mathbf{H}_{Tx}\mathbf{T}) \end{bmatrix} \succ_K 0 \tag{9.45}$$

Using $p = \sqrt{P_o}$, the program in Equation (9.27) can now be cast in the standard SOCP form [43] as follows:

$$P(\gamma_o): \begin{cases} \min_{T,p} \quad p \\ \text{s.t.} \quad \begin{bmatrix} \sqrt{1 + \dfrac{1}{\gamma_o}}[\mathbf{HT}]_{i,i} \\ \mathbf{T}^H\mathbf{H}^H\mathbf{e}_i \\ \sigma_i \end{bmatrix} \succ_K 0, \qquad i = 1,\ldots,K \\ \begin{bmatrix} p \\ vec\,(\mathbf{H}_{Tx}\mathbf{T}) \end{bmatrix} \succ_K 0 \end{cases} \tag{9.46}$$

and it can be efficiently solved using any standard SOCP package [42]. Such a solver can also numerically determine the feasibility of the optimization problem. A similar approach was taken in [40] in the context of transmit beamforming.

As explained in Section 9.2.2, each SOC constraint can be replaced with an SDP constraint using Equation (9.32). Thus, the problem can also be expressed as a standard SDP, as follows:

$$P(\gamma_o): \begin{cases} \min_{T,p} \quad p \\ \text{s.t.} \quad \mathbf{A}_i(\mathbf{T}) \succ 0, \quad i = 1,\ldots,K \\ \quad \mathbf{C}(\mathbf{T}) \succ 0 \end{cases} \tag{9.47}$$

where

$$\mathbf{A}_i(\mathbf{T}) = \begin{bmatrix} \sqrt{1 + \dfrac{1}{\gamma_o}}[\mathbf{HT}]_{i,i} & \begin{bmatrix} \mathbf{e}_i^H \mathbf{HT} & \sigma_i \end{bmatrix} \\ \begin{bmatrix} \mathbf{T}^H\mathbf{H}^H\mathbf{e}_i \\ \sigma_i \end{bmatrix} & \sqrt{1 + \dfrac{1}{\gamma_o}}[\mathbf{HT}]_{i,i}\,\mathbf{I} \end{bmatrix} \tag{9.48}$$

for $i = 1,\ldots,K$, and

$$\mathbf{C}(\mathbf{T}) = \begin{bmatrix} p & vec^H(\mathbf{H}_{Tx}\mathbf{T}) \\ vec(\mathbf{H}_{Tx}\mathbf{T}) & p\mathbf{I} \end{bmatrix} \tag{9.49}$$

Solving SOCPs via SDP is not very efficient. Interior point methods that solve SOCP directly have a much better worst case complexity than do their SDP counterparts [43].

Define now the dual variables $\lambda i > 0$ for $i = 1 \ldots K$ and denote $\Lambda = \text{diag}\{\lambda_i\}$ and $\mathbf{G}(\lambda_i) = \mathbf{H}^H \Lambda \mathbf{H} + \mathbf{H}_{Tx}^H \mathbf{H}_{Tx}$. If there exist $\lambda_i > 0$ such that

$$\gamma_o = \frac{1}{\left[\Lambda^{\frac{1}{2}}\mathbf{HG}^\dagger(\lambda_i)\mathbf{H}^H\Lambda^{\frac{1}{2}}\right]_{i,i}^{-1} - 1}, \qquad i = 1,\ldots,K \tag{9.50}$$

holds, then the program is strictly feasible [10]. Also, if the condition in Equation (9.50) holds, then the optimal $\mathbf{T}$ is of the form

$$\mathbf{T} = \mathbf{G}^{\dagger}(\lambda_i)\mathbf{H}^H \mathbf{\Lambda}^{\frac{1}{2}} \operatorname{diag}\{\delta_i\} \tag{9.51}$$

where δ_i are the positive weights that allocate the power between the users, as follows:

$$\delta_i = \sqrt{\sum_j \left[\left(\frac{\gamma_o}{1+\gamma_o}\mathbf{I} - \mathbf{F}\right)^{-1}\right]_{i,j} \lambda_j \sigma_j^2} \tag{9.52}$$

$$[\mathbf{F}]_{i,j} = \left[\mathbf{\Lambda}^{\frac{1}{2}}\mathbf{H}\mathbf{G}^{\dagger}(\lambda_i)\mathbf{H}^H\mathbf{\Lambda}^{\frac{1}{2}}\right]_{i,j}^2 \tag{9.53}$$

for $i, j = 1, \ldots, K$. This structure of T is unique within the range of $\mathbf{H}_{Tx}^H$. At this optimal solution, all the constraints are active, i.e. there are equal SINRs for all the subchannels. The optimal objective value is

$$P_o = \sum_i \lambda_i \sigma_i^2 \tag{9.54}$$

Equations (9.50)–(9.54) provide a simple strategy for designing the precoder.

Given a feasible γ_o, all one has to do is find $\lambda_i > 0$ which satisfy Equation (9.50). Once these are found, $\mathbf{T}$ can be derived through Equations (9.51)–(9.54). In some special cases, these variables can be derived in closed form. Otherwise, [10] provides two alternative methods for finding these variables. The structure of (9.50) motivates a fixed-point iteration for finding λ_i. By rearranging Equation (9.50), we arrive at the following simple iteration:

$$\lambda_i^{(n+1)} = \frac{\gamma_o}{1+\gamma_o} \frac{1}{\left[\mathbf{H}\mathbf{G}^{\dagger}\left(\lambda_i^{(n)}\right)\mathbf{H}^H\right]_{i,i}}, \quad i = 1, \ldots, K \tag{9.55}$$

Clearly, the optimal λ_i satisfy this fixed point. If $P(\lambda_o)$ is feasible, then the above iteration will converge from any $\gamma_i^{(0)}$ to a set $\gamma_i^{(n)} > 0$ that satisfies Equation (9.50) [10].

9.2.4   Precoder for SINR Optimization

The power optimization problem of Equation (9.29) and the SINR optimization problem of Equation (9.30) are inverse problems [10]:

$$\gamma_o = S(P(\gamma_o))$$
$$P_o = P(S(P_o)) \tag{9.56}$$

In addition, the optimal objective value of each program is continuous and strictly monotonic increasing in its input argument

$$\gamma_o > \tilde{\gamma}_o \Rightarrow P(\gamma_o) > P(\tilde{\gamma}_o)$$
$$P_o > \tilde{P}_o \Rightarrow S(P_o) > S(\tilde{P}_o) \tag{9.57}$$

Using Equations (9.56)–(9.57), we can solve $S(P_o)$ for a given P_o by iteratively solving $P(\gamma_o)$ for different γ_os. Due to the inversion property, if $P(\gamma)$, then its solution will be optimal also for $S(P_o)$. The strict monotonicity and continuity guarantees that a simple one-dimensional bisection search will efficiently find the required γ_o. This procedure is summarized in the following algorithm (see also [47, 10]).

$S(P_0)$

1. $\gamma_{\max} \leftarrow$ MaxSINR
2. $\gamma_{\min} \leftarrow$ MinSINR
3. **repeat**
4. $\gamma_o \leftarrow (\gamma_{\min} + \gamma_{\max})/2$
5. $\hat{P}_o \leftarrow P(\gamma_o)$
6. **if** $\hat{P}_o \leq P_o$
7. **then** $\gamma_{\min} \leftarrow \gamma_o$
8. **else** $\gamma_{-\max} \leftarrow \gamma_o$
9. **until** $\hat{P}_o = P_o$
10. **return** γ_o

where MinSINR and MaxSINR define a range of relevant SINRs for a specific application, and where we have used the convention that $\infty = P(\gamma_o)$ if it is infeasible. Theoretically, this means that the SINR optimization problem can be solved through the previous results concerning the power optimization.

The SINR optimization can be also cast as a standard GEVP program. Using a real-valued slack variable γ_o, the problem can be rewritten as:

$$S(P_o): \begin{cases} \max_{T,\gamma_o} & \gamma_o \\ \text{s.t.} & \dfrac{\left|[HT]_{i,i}\right|^2}{\sum_{j \neq i} \left|[HT]_{i,j}\right|^2 + \sigma_i^2} = \gamma_o, \quad i = 1,\ldots,K \\ & \operatorname{Tr}\left\{T^H H_{Tx}^H H_{Tx} T\right\} \leq P_o \end{cases} \qquad (9.58)$$

Although Equation (9.58) seems similar to Equation (9.41), it turns out to be more complicated. This is because the SINR matrix inequalities in Equation (9.48) are linear in $\beta = \sqrt{1 + 1/\gamma_o}$ or in T, but not in both simultaneously. Thus, when β is an optimization variable and not a parameter, these constraints are no longer LMIs. In fact, the sets which they define are not convex. Even so, we can still express them using generalized matrix inequalities as in Equations (9.47) and (9.48). If we rewrite the $A_i(T)$'s in Equation (9.48) and separate out the terms which are linear, we have:

$$A_i(T) = \beta A_i^1(T) - A_i^2(T) \qquad (9.59)$$

where $A_i^1(T)$ and $A_i^2(T)$ are matrices that depend affinely on T, as follows:

$$A_i^1(T) = \begin{bmatrix} [HT]_{i,i} & 0 \\ 0 & [HT]_{i,i}\, I \end{bmatrix}$$

$$A_i^2(T) = \begin{bmatrix} 0 & -\left[e_i^H HT \quad \sigma_i\right] \\ -\begin{bmatrix} T^H H^H e_i \\ \sigma_i \end{bmatrix} & 0 \end{bmatrix} \qquad (9.60)$$

Using Equation (9.59), we can express S in the standard GEVP form

$$S(P_o): \begin{cases} \min_{T,\beta} & \beta \\ \text{s.t.} & \beta A_i^1(T) \succ A_i^2(T), \quad i = 1,\ldots,K \\ & A_i^1(T) \succ 0, \qquad\qquad i = 1,\ldots,K \\ & C(T) \succ 0 \end{cases} \qquad (9.61)$$

which can be solved using appropriate software [45].

The SINR optimization problem can also be solved using the conditions in (9.50) by *fixed-point iteration for finding* λ_i. As explained in Equations (9.56)–(9.57), S and P are inverse problems. Thus, the optimal solution of the SINR optimization is also optimal for an inverse power optimization problem, and therefore must satisfy its optimality conditions as well. Thus, to optimize the SINRs, we need to find $\lambda_i > 0$ that satisfy Equations (9.50) and (9.54). Unfortunately, in this case, γ_o is an optimization variable and not a parameter and has to be found as well. This can be overcome by adjusting the fixed-point iteration in Equation (9.55), as follows:

$$\tilde{\lambda}_i = \frac{1}{\left[\mathbf{H}\mathbf{G}^\dagger \left(\lambda_i^{(n)} \right) \mathbf{H}^H \right]_{i,i}}, \qquad i = 1, \ldots, K \tag{9.62}$$

and then normalizing the result so that it will satisfy Equation (9.54):

$$\lambda_i^{(n+1)} = \frac{P_o \tilde{\lambda}_i}{\sum_j \sigma_j^2 \tilde{\lambda}_j}, \qquad i = 1, \ldots, K \tag{9.63}$$

If this iteration converges to a fixed point $\lambda_i^{(n)} > 0$, then it will satisfy Equations (9.50) and (9.54). Numerous numerical simulations with arbitrary initial points and parameters show a rapid convergence rate.

9.2.5   Performance Example

Consider a multiuser precoded downlink system. At each symbol's period, the base station transmits using an $N \times K$ nonorthogonal signatures matrix $\mathbf{H}_{Tx} = \mathbf{S}$. The maximal average transmitted power is $P_o = K$, and the cross correlations between the signatures are denoted by $\rho_{i,j} = [\mathbf{S}^H \mathbf{S}]_{i,j}$ with $\rho_{i,i} = 1$ for all i. For simplicity, we assume ideal channels, i.e. $\mathbf{H}_{Ch,i} = \mathbf{I}$, and equal noise variances, i.e., $\sigma_i^2 = \sigma^2$. Denoting by $\mathbf{y}$ the output vector of the multiple user receiver, we have

$$\mathbf{y} = \mathbf{H}_{Rx}\mathbf{S}\mathbf{T}\mathbf{b} + \mathbf{H}_{Rx}\mathbf{w} \tag{9.64}$$

where from Chapter 5, $\mathbf{H}_{Rx}$ is one of the standard filters, as follows:

- MF receiver: $\mathbf{H}_{Rx} = \mathbf{S}^H$;
- ZF receiver: $\mathbf{H}_{Rx} = (\mathbf{S}^H\mathbf{S})^{-1}\,\mathbf{S}^H$;
- MMSE receiver: $\mathbf{H}_{Rx} = \mathbf{S}^H \left(\mathbf{S}\mathbf{S}^H + \sigma^2\mathbf{I}^{-1}\right.$.

An interesting result of the precoder is its performance in an equal power and equal cross correlations multiuser system, i.e., $\rho_{i,j} = \rho$ for all $j \neq i$ and $\sigma_i^2 = \sigma^2$.

When the matrices $\mathbf{H}$ and $\mathbf{H}_{Tx}$ have equal diagonal elements and equal off-diagonal elements, and the variances $\sigma_i^2 = \sigma^2$ are equal, due to the symmetry, it is clear that choosing $\lambda_i = P_o/(K\sigma^2)$ will satisfy the conditions in Equation (9.59). Therefore, the solution for the SINR optimization problem is:

$$\gamma_o = \frac{1}{\left[\mathbf{H} \left(\mathbf{H}^H\mathbf{H} + \frac{K\sigma^2}{P_o}\mathbf{H}_{Tx}^H\mathbf{H}_{Tx} \right)^{-1} \mathbf{H}^H \right]_{i,i}} - 1 \tag{9.65}$$

$$\mathbf{T} = c \left[\mathbf{H}^H\mathbf{H} + \frac{K\sigma^2}{P_o}\mathbf{H}_{Tx}^H\mathbf{H}_{Tx} \right]^\dagger \mathbf{H}^H \tag{9.66}$$

where c is a constant that scales the matrix to satisfy the power constraint.

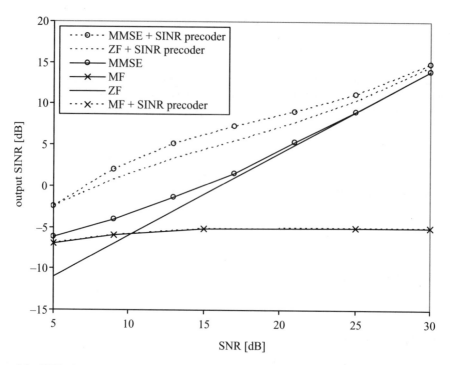

Figure 9.6   SINR of a symmetric system with equal cross correlations. Due to the symmetry, all users have equal output SINRs © 1989, IEEE.

In Figure 9.6, we plot the output SINRs given by Equation (9.65) for the three linear receivers. For comparison, we also plot the output SINRs that result from similar systems without a precoder [48]. As expected, using the precoder always improves the output SINR.

9.3   Convex Optimization-Theory-Based Beamforming

In this section we discuss the joint design of transmit and receive beamforming for multicarrier multiple-input–multiple-output (MIMO) channels under a variety of design criteria. As already indicated in Section 9.2, this linear processing is commonly termed linear precoding at the transmitter and equalization at the receiver. Instead of considering each design criterion in a separate way, we generalize the existing results by developing a unified framework based on considering two families of objective functions that embrace most reasonable criteria to design a communication system: Schur-concave and Schur-convex functions. Once the optimal structure of the transmit–receive processing is known, as in Section 9.2, the design problem simplifies and can be formulated within the framework of convex optimization theory, in which a number of interesting design criteria can be easily accommodated and efficiently solved, even though closed-form expressions may not exist.

A general convex optimization problem (convex program), introduced in Section 9.2, will be further specified in this section as follows [65]:

$$
\min_{\mathbf{x}} f_0(\mathbf{x})
$$
$$
\text{s.t.} \quad f_i(\mathbf{x}) \leq 0, \quad 1 \leq i \leq m
$$
$$
h_i(\mathbf{x}) = 0, \quad 1 \leq i \leq p
$$

where $\mathbf{x} \in R^n$ is the optimization variable, $f_0(\mathbf{x}), \ldots, f_m(\mathbf{x})$ are convex functions, and $h_1(\mathbf{x}), \ldots, h_p$ $(\mathbf{x})$ are linear functions (affine functions). The function f_0 is the objective function or cost function. The inequalities $f_i(\mathbf{x}) \leq 0$ are called inequality constraints, and the equations $h_i(\mathbf{x}) = 0$ are called equality constraints. When the functions f_i and h_i are linear (affine), the problem is called a linear program (LP) and is much simpler to solve.

In general, as already indicated in Section 9.2, some manipulations are required to convert the problem into a convex one (unfortunately, this is not always possible). The interest of expressing a problem in convex form is that although an analytical solution may not exist and the problem may be difficult to solve (it may have hundreds of variables and a nonlinear nondifferentiable objective function), it can still be solved (numerically) very efficiently [64]. Another interesting feature of expressing a problem in convex form is that additional constraints can be straightforwardly added, as long as they are convex.

In some cases, convex optimization problems can be analytically solved using the Karush–Kuhn–Tucker (KKT) optimality conditions, and closed-form expressions can be obtained. In general, however, one must resort to iterative methods [64, 67]. *Interior-point methods* can be used to iteratively solve convex problems [72]. In addition, the difference between the objective value at each iteration and the optimum value can be upper bounded using duality theory [64, 67]. This allows the utilization of nonheuristic stopping criteria such as stopping when some prespecified resolution has been reached. Another interesting family of iterative methods is *cutting-plane methods* [64].

In the sequel we will also use some results from majorization theory. For these purposes we need the following definitions (D) [63]:

(D1) For any $\mathbf{x} \in R^n$, let $x_{[1]} \geq \cdots \geq x_{[n]}$ denote the components of $\mathbf{x}$ in decreasing order (also termed order statistics of $\mathbf{x}$).

(D2) Let, $\mathbf{x}, \mathbf{y} \in R^n$. Vector $\mathbf{x}$ is majorized by vector $\mathbf{y}$ (or $\mathbf{y}$ majorizes $\mathbf{x}$) if

$$\sum_{i=1}^{k} x_{[i]} \leq \sum_{i=1}^{k} y_{[i]}, \quad i \leq k \leq n-1, \sum_{i=1}^{n} x_{[i]} = \sum_{i=1}^{n} y_{[i]}$$

and represent it by $\mathbf{x} \prec \mathbf{y}$.

(D3) A real-valued function ϕ defined on a set $A \subseteq R^n$ is said to be Schur-convex on A if $\mathbf{x} \prec \mathbf{y}$ on A $\Rightarrow \phi(\mathbf{x}) \leq \phi(\mathbf{y})$ Similarly, ϕ is said to be Schur-concave on A if $\mathbf{x} \prec \mathbf{y}$ on A $\Rightarrow \phi(\mathbf{x}) \geq \phi(\mathbf{y})$. As a consequence, if ϕ is Schur-convex on A, then $-\phi$ is Schur-concave on A and vice versa. It is important to remark that the sets of Schur-concave and Schur-convex functions do not form a partition of the set of all functions. In fact, neither are the two sets disjoint (the intersection is not empty), nor do they cover the entire set of all functions. By using these definitions we additionally introduce the following results [63]:

(R1) Let $\mathbf{R}$ be an $n \times n$ Hermitian matrix with diagonal elements denoted by the vector $\mathbf{d}$ and eigenvalues denoted by the vector λ. Then $\mathbf{d} \prec \lambda$

(R2) Let $\mathbf{x} \in R^n$ and $1 \in R^n$ denote the constant vector with $1_i \triangleq \sum_{j=1}^{n} x_j / n$. Then $1 \prec \mathbf{x}$

(R3) For any $\mathbf{x} \in R^n$, there exists a real symmetric (and therefore Hermitian) matrix with equal diagonal elements and eigenvalues given by $\mathbf{x}$.

9.3.1 *Multicarrier MIMO Signal Model*

We analyse a communication system with n_T transmit and n_R receive antennas. To cope with the frequency-selectivity of the channel, we use a multicarrier signal so that we have

$$\mathbf{y}_k = \mathbf{H}_k \mathbf{s}_k + \mathbf{n}_k, \quad 1 \leq k \leq N \tag{9.67}$$

where k denotes the carrier index, N is the number of carriers, $\mathbf{s}_k \in R^{nT \times 1}$ is the transmitted vector, $\mathbf{H}_k \in R^{nR \times nT}$ is the channel matrix, $\mathbf{y}_k \in R^{nR \times 1}$ is the received signal vector, and $\mathbf{n}_k \in R^{nR \times 1}$ is $n_k \sim CN(0, \mathbf{R}_{nk})$ a zero-mean circularly symmetric complex Gaussian noise vector with arbitrary covariance matrix $\mathbf{R}_{nk}$. The channel is assumed fixed during the transmission of a block and known at both sides of the communication link as well as the noise covariance matrix.

At each carrier k, the matrix channel has $K_k \leq \min(n_T, n_R)'$ channel eigenmodes or spatial subchannels (i.e., nonvanishing singular values of the channel matrix) [50] (see also Section 4.14). We can use them as a means of spatial multiplexing [51] to transmit simultaneously L_k symbols by having L_k *established substreams*. Notice that established substreams and spatial subchannels (or channel eigenmodes) are different concepts that may or may not coincide, depending on whether the channel is diagonalized or not. This will be further elaborated in the next section. In a practical system, we will typically have to have $L_k \leq K_k$ an acceptable performance. In the sequel boldface captital letters denote matrices, boldface lower-case letters denote column vectors, and italics denote scalars. $[\mathbf{X}]_{i,j}$ (also $[\mathbf{X}]_{ij}$) and $[\mathbf{X}]_{:j}$ denote the $(i$th, jth) element and jth column of matrix $\mathbf{X}$, respectively. By $\mathbf{A} \geq \mathbf{B}$, we mean that $\mathbf{A} - \mathbf{B}$ is positive semidefinite. The trace, determinant, and Frobenius norm of a matrix are denoted by $Tr(\cdot)$, $|\cdot|$, and $\|\cdot\|_F$, respectively. By $\text{diag}(\{\mathbf{X}_k\})$, we denote a block-diagonal matrix with diagonal blocks given by the set $\{\mathbf{X}_k\}$. The gradient of a function with respect to $\mathbf{x}$ is written as $\nabla_x f(\mathbf{x})$. We define $(x)^+ \triangleq \max(0, x)$. The transmitted vector at the kth carrier after linear precoding can be now represented as (see Figure 9.7(a)).

$$s_k = \mathbf{B}_k \mathbf{x}_k = \sum_{i=1}^{L_k} \mathbf{b}_{k,i} x_{k,i} \qquad (9.68)$$

where $\mathbf{x}_k \in R^{Lk \times 1}$ represents the L_k transmitted symbols. We assume zero-mean unit-energy uncorrelated (white) symbols, i.e., $E[\mathbf{x}_k \mathbf{x}_k^H] = \mathbf{I}L_k)$. $\mathbf{B}_k \in R^{nT \times Lk}$ is the precoding matrix, $\mathbf{b}_{k,i} \triangleq [\mathbf{B}_k]_{:,i}$, and $x_{k,i} \triangleq [\mathbf{x}_k]_i$. Each column of $\mathbf{B}_k$ can be considered as a different beamvector corresponding to each transmitted

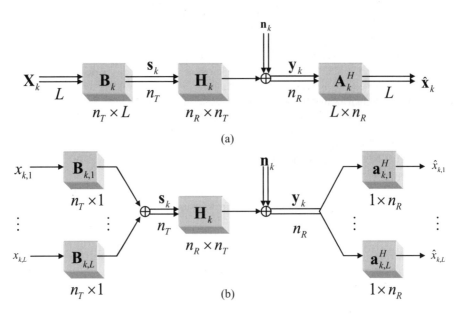

Figure 9.7 Matrix processing and multiple beamforming interpretations of the communication system. (We assume for the clarity of the figure that $L_k = L \forall k$.) (a) Matrix processing interpretation at carrier k. (b) Multiple beamforming interpretation at carrier k.

symbol, resulting into multiple beamforming architecture shown in Figure 9.7(b). Note that if only one symbol is transmitted per carrier ($L_k = 1 \forall k$), then Equation (9.68) reduces to a classical beamforming structure with a single beamvector: $\mathbf{s}_k = \mathbf{b}_k \mathbf{x}_k$. The transmitter is constrained in its average total transmit power:

$$\sum_{k=1}^{N} \mathrm{E}\left[\|\mathbf{B}_k \mathbf{x}_k\|^2\right] = \sum_{k=1}^{N} \|\mathbf{B}_k\|_F^2 \leq P_T \tag{9.69}$$

where P_T is the power in units of energy per block-transmission (or, equivalently, per OFDM symbol).

The received vector at the kth carrier after the equalizer is

$$\hat{\mathbf{x}}_k = \mathbf{A}_k^H \mathbf{y}_k \tag{9.70}$$

where $\mathbf{A}_k^H \in R^{L_k \times n_R}$ is the equalizer matrix, and $\hat{\mathbf{x}}_k \in R^{L_k \times 1}$ is the estimation of $\mathbf{x}_k$. Again, each column of $\mathbf{A}_k$ can be interpreted as a beamvector adapted to each spatial channel substream at carrier k, i.e., $\hat{x}_{k,i} = \mathbf{a}_{k,i}^H \mathbf{y}_k$ [see Figure 9.7(b)].

In the sequel, only independent processing at each carrier has been considered, called the *carrier-noncooperative approach* [49] [see Figure 9.8(a)]. This scheme, however, can be further generalized by allowing cooperation among carriers, called *carrier-cooperative approach* [49] [see Figure 9.8(b)]. The signal model is obtained by stacking the vectors corresponding to all carriers (e.g., $\mathbf{x}^T = [\mathbf{x}_1^T, \ldots, \mathbf{x}_N^T]$), by considering global transmit and receive matrices $\mathbf{B} \in R^{(n_T N) \times L_T}$ (the transmit power constraint reduces to $\|\mathbf{B}\|_F^2 \leq P_T$) and $\mathbf{A}^H \in R^{L_T \times (n_R N)}$, where $L_T = \sum_{k=1}^{N} L_k$ is the total number of transmitted symbols, and by defining the global channel as $\mathbf{H} = \mathrm{diag}(\{\mathbf{H}_k\}) \in R^{(n_R N) \times (n_T N)}$. This model can cope with intermodulation terms, unlike the noncooperative model, that implicitly assumes orthogonal carriers.

9.3.2 Channel Diagonalization

For some specific design criteria, the system optimization problem is greatly simplified because the channel turns out to be diagonalized by the precoder-equalizer processing, which allows a *scalarization* of the problem (all matrix equations are substituted with scalar ones). Examples are the minimization of the (weighted) sum of the MSEs of all channel spatial substreams [52–54], the minimization of the determinant of the MSE matrix [55], and the maximization of the mutual information [50, 56, 57].

In [49], these results are generalized by developing a unified framework. Instead of analyzing each design criterion in a separate way, the design is based on the minimization of some arbitrary objective function of the MSEs of all channel substreams $f_0(\{\mathrm{MSE}_{k,i}\})$, where $\mathrm{MSE}_{k,i}$ is the MSE of the ith spatial substream at the kth carrier (objective functions of the SINRs and of the BERs are also readily incorporated).

9.3.2.1 Optimum Equalizer

To design the system, we first easily derive the optimum equalizer $\mathbf{A}_k$'s, assuming the precoders $\mathbf{B}_k$'s fixed, and then deal with, the derivation of the optimal precoders $\mathbf{B}_k$'s. This two-step derivation has been independently used in [49] and [58]. The MSE matrix at the kth carrier is defined as the covariance matrix of the error vector (given by $e_k \triangleq \hat{\mathbf{x}}_k - \mathbf{x}_k$):

$$\mathbf{E}_k(\mathbf{B}_k, \mathbf{A}_k) \triangleq \mathrm{E}\left[(\hat{\mathbf{x}}_k - \mathbf{x}_k)(\hat{\mathbf{x}}_k - \mathbf{x}_k)^H\right]$$
$$= \mathbf{A}_k^H \mathbf{R}_{y_k} \mathbf{A}_k + \mathbf{I} - \mathbf{A}_k^H \mathbf{H}_k \mathbf{B}_k - \mathbf{B}_k^H \mathbf{H}_k^H \mathbf{A}_k \tag{9.71}$$

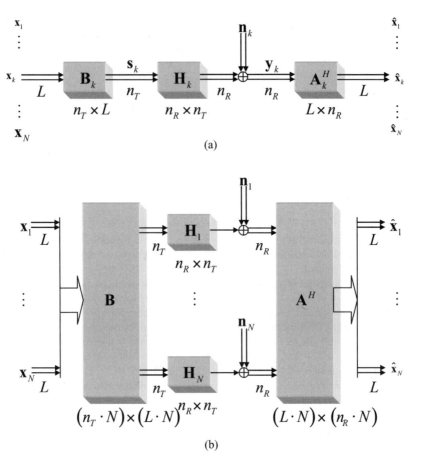

Figure 9.8   Carrier-cooperative versus carrier-noncooperative approaches. (We assume for clarity of the figure that $L_k = L \forall k$.) (a) Carrier-noncooperative approach. (b) Carrier-cooperative approach.

where $\mathbf{R}_{y_k} \triangleq E\left[\mathbf{y}_k\mathbf{y}_k^H\right] = \mathbf{H}_k\mathbf{B}_k\mathbf{B}_k^H\mathbf{H}_k^H + \mathbf{R}_{n_k}$. The MSE of the ($k$th), ($i$th) substream is the ith diagonal element of $\mathbf{E}_k$, i.e.,

$$\text{MSE}_{k,i}\left(\mathbf{B}_k, \mathbf{a}_{k,i}\right) = [\mathbf{E}_k]_{ii} = \mathbf{a}_{k,i}^H\mathbf{R}_{y_k}\mathbf{a}_{k,i} + 1 - \mathbf{a}_{k,i}^H\mathbf{H}_k\mathbf{b}_{k,i} - \mathbf{b}_{k,i}^H\mathbf{H}_k^H\mathbf{a}_{k,i} \qquad (9.72)$$

where $\mathbf{a}_{k,i}$ (or $\mathbf{b}_{k,i}$) is the ith column of $\mathbf{A}_k$ (or $\mathbf{B}_k$). Expression (9.72) is nonconvex in $(\mathbf{B}_k, \mathbf{a}_{k,i})$ whereas, for a given $\mathbf{B}_k$, $\text{MSE}_{k,i}$ is convex in $\mathbf{a}_{k,i}$ and independent of the other columns of $\mathbf{A}_k$ and of the other carriers. This means that each $\mathbf{a}_{k,i}$ can be independently optimized. To obtain the optimal equalizer $\mathbf{A}_k^{\text{opt}}$ in a more direct way, it suffices to find $\mathbf{A}_k$ such that the diagonal elements of $\mathbf{E}_k$ are minimized. This can be done regardless of the specific choice of the objective function f_0 since we know it is increasing in each argument. Alternatively, we can obtain $\mathbf{A}_k^{\text{opt}}$ so that $\mathbf{E}_k\left(\mathbf{B}_k, \mathbf{A}_k^{\text{opt}}\right) \preceq \mathbf{E}_k(\mathbf{B}_k, \mathbf{A}_k)$, which in particular implies that the diagonal elements are minimized (in fact, both criteria are equivalent as shown in [59]). In other words, we want to solve min $\mathbf{c}^H\mathbf{E}_k(\mathbf{B}_k, \mathbf{A}_k)\mathbf{c}$, $\forall\mathbf{c}$. Setting the gradient of $\mathbf{c}^H\mathbf{E}_k\mathbf{c} = \text{Tr}(\mathbf{E}_k\mathbf{cc}^H)$ to zero $\nabla_{\mathbf{A}_k^*}\text{Tr}(\mathbf{E}_k\mathbf{cc}^H) = \mathbf{R}_{y_k}\mathbf{A}_k\mathbf{cc}^H - \mathbf{H}_k\mathbf{B}_k\mathbf{cc}^H = 0$, $\forall\mathbf{c}$ and particularizing for all the vectors of the canonical base, it follows that:

$$\mathbf{A}_k^{\text{opt}} = \left(\mathbf{H}_k\mathbf{B}_k\mathbf{B}_k^H\mathbf{H}_k^H + \mathbf{R}_{n_k}\right)^{-1}\mathbf{H}_k\mathbf{B}_k \qquad (9.73)$$

Expression (9.73) is the linear minimum MSE (LMMSE) receiver or *Wiener filter* [see Section 5]. Using the optimal equalizer $\mathbf{A}_k^{\text{opt}}$, we obtain the following concentrated error matrix:

$$\mathbf{E}_k\left(\mathbf{B}_k\right) \overset{\Delta}{=} \mathbf{E}_k\left(\mathbf{B}_k, \mathbf{A}_k^{\text{opt}}\right) = \mathbf{I} - \mathbf{B}_k^H \mathbf{H}_k^H \left(\mathbf{H}_k \mathbf{B}_k \mathbf{B}_k^H \mathbf{H}_k^H + \mathbf{R}_{n_k}\right)^{-1} \mathbf{H}_k \mathbf{B}_k$$

$$= \left(\mathbf{I} + \mathbf{B}_k^H \mathbf{R}_{H_k} \mathbf{B}_k\right)^{-1} \qquad (9.74)$$

where we have used the matrix inversion lemma, $(\mathbf{A} + \mathbf{BCD})^{-1} = \mathbf{A}^{-1} - \mathbf{A}^{-1} \mathbf{B}(\mathbf{DA}^{-1}\mathbf{B} + \mathbf{C}^{-1})^{-1}\mathbf{DA}^{-1}$ and we have defined $\mathbf{R}_{Hk} \overset{\Delta}{=} \mathbf{H}_k^H \mathbf{R}_{n_k}^{-1} \mathbf{H}_k$ (note that the eigenvectors and eigenvalues of $\mathbf{R}_{H_k}$ are the right singular vectors and the squared singular values, respectively, of the whitened channel $\mathbf{R}_{n_k}^{-1/2}\mathbf{H}_k$). However, many objective functions are naturally expressed as functions of the SINR of each substream. The SINR at the kth carrier and the ith spatial substream is:

$$\text{SINR}_{k,i} \overset{\Delta}{=} \frac{\left|\mathbf{a}_{k,i}^H \mathbf{H}_k \mathbf{b}_{k,i}\right|^2}{\mathbf{a}_{k,i}^H \mathbf{R}_{k,i} \mathbf{a}_{k,i}} \leq \mathbf{b}_{k,i}^H \mathbf{H}_k^H \mathbf{R}_{k,i}^{-1} \mathbf{H}_k \mathbf{b}_{k,i} \qquad (9.75)$$

where $\mathbf{R}_{k,i} \overset{\Delta}{=} \mathbf{H}_k \mathbf{B}_k \mathbf{B}_k^H \mathbf{H}_k^H + \mathbf{R}_{nk} - \mathbf{H}_k \mathbf{b}_{k,i} \mathbf{b}_{k,i}^H \mathbf{H}_k^H$ is the interference-plus-noise covariance matrix seen by the (kth, ith) substream, the inequality comes from Cauchy–Schwarz's inequality [60] [with vectors $(\mathbf{R}_{k,i}^{-1/2}\mathbf{H}_k\mathbf{b}_{k,i})$ and $(\mathbf{R}_{k,i}^{1/2}\mathbf{a}_{k,i})$], and the upper bound is achieved by $\mathbf{a}_{k,i} \propto \mathbf{R}_{k,i}^{-1}\mathbf{H}_k\mathbf{b}_{k,i} \propto \mathbf{R}_{y_k}^{-1}\mathbf{H}_k\mathbf{b}_{k,i}$, i.e. the Wiener filter again. Noting that the MSE can be expressed as:

$$\text{MSE}_{k,i} = \left[\left(\mathbf{I} + \mathbf{B}_k^H \mathbf{H}_k^H \mathbf{R}_{n_k}^{-1} \mathbf{H}_k \mathbf{B}_k\right)^{-1}\right]_{ii} = \frac{1}{1 + \mathbf{b}_{k,i}^H \mathbf{H}_k^H \mathbf{R}_{k,i}^{-1} \mathbf{H}_k \mathbf{b}_{k,i}} \qquad (9.76)$$

the SINR can be easily related to the MSE as:

$$\text{SINR}_{k,i} = \frac{1}{\text{MSE}_{k,i}} - 1 \qquad (9.77)$$

So, maximizing the SINR is equivalent to minimizing the MSE. The performance of a digital communications system is given by the bit error rate (BER). Under the Gaussian assumption, the symbol error probability P_e will be analytically expressed as in the previous section as a function of the SINR P_e (SINR) $= \alpha Q\left(\sqrt{\beta \text{SINR}}\right)$ where α and β are constants that depend on the signal constellation, and Q is the Q-function defined as $Q(x) = (1/\sqrt{2\pi}) \int_x^\infty e^{-\lambda^2/2} d\lambda$. It is sometimes convenient to use the Chernoff upper bound of the tail of the Gaussian distribution function $Q(x) \leq (1/2)e^{-x^2/2}$ [61] to approximate the symbol error probability as $Pe \approx (1/2)\alpha e^{-\beta/2\text{SINR}}$ (which becomes a good approximation for high values of the SINR). The BER can be approximately obtained from the symbol error probability (assuming that a Gray encoding is used to map the bits into the constellation points) as BER $\approx P_e/k$ where $k = \log_2 M$ is the number of bits per symbol, and M is the constellation size. It can be seen from Figure 9.9 that BER is a convex function.

9.3.2.2 Optimum Precoder

To obtain the set of precoder matrices $\{\mathbf{B}_k\}$, we now consider the minimization of an arbitrary objective function of the diagonal elements of Equation (9.74). We start with optimization problem defined by:

$$\begin{aligned} &\min_{\mathbf{B}} f_0(\mathbf{d}(\mathbf{E}(\mathbf{B}))) \\ &\text{s.t.} \quad \text{Tr}(\mathbf{BB}^H) \leq P_T \end{aligned} \qquad (9.78)$$

where matrix $\mathbf{B} \in R^{n_T \times L}$ is the optimization variable, $\mathbf{d}(\mathbf{E}(\mathbf{B}))$ is the vector of diagonal elements of the MSE matrix $\mathbf{E}(\mathbf{B}) = (\mathbf{I} + \mathbf{B}^H \mathbf{R}_H \mathbf{B})$ [the diagonal elements of $\mathbf{E}(\mathbf{B})$ are assumed in decreasing order

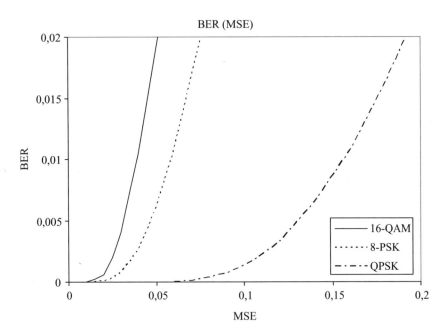

Figure 9.9 Convexity of the BER as a function of the MSE for the range of BER $\leq 2 \times 10^{-2}$ © 2000, IEEE.

w.l.o.g.], $\mathbf{R}_H \in R^{n_T \times n_T}$ is a positive semidefinite Hermitian matrix, and $f_0 : R^L \to R$ is an arbitrary objective function (increasing in each variable). It then follows that there is an optimal solution $\mathbf{B}$ of at most rank $\breve{L} \triangleq \min(L, \text{rank}(\mathbf{R}_H))$ rank with the following structure [49]:

- If f_0 is Schur-concave, then

$$\mathbf{B} = \mathbf{U}_{H,1} \sum_{B,1} \tag{9.79}$$

where $\mathbf{U}_{H,1} \in R^{n_T \times \breve{L}}$ has as columns the eigenvectors of $\mathbf{R}_H$ corresponding to the $\breve{L}$ largest eigenvalues in increasing order, and $\Sigma_{B,1} = [0 \text{ diag}(\{\sigma_{B,i}\})] \in R^{\breve{L} \times L}$ has zero elements, except along the rightmost main diagonal (which can be assumed real).
- If f_0 is Schur-convex, then

$$\mathbf{B} = \mathbf{U}_{H,1} \sum_{B,1} \mathbf{V}_B^H \tag{9.80}$$

where $\mathbf{U}_{H,1}$ and $\Sigma_{B,1}$ are defined as before, and $\mathbf{V}_B \in R^{L \times L}$ is a unitary matrix such that $(\mathbf{I} + \mathbf{B}^H \mathbf{R}_H \mathbf{B})^{-1}$ has identical diagonal elements. This rotation can be computed using the algorithm given in [62] or with any rotation matrix $\mathbf{Q}$ that satisfies $|\mathbf{Q}_{ik}| = |\mathbf{Q}_{il}|$, $\forall i, k, l$, such as the discrete Fourier transform (DFT) matrix or the Hadamard matrix (when the dimensions are appropriate such as a power of two [61]).

For the simple case in which only one symbol per carrier is transmitted at each transmission, i.e. a single spatial eigenmode $L = 1$ is utilized, Equations (9.78)–(9.80) simplify, and the diagonal structure simply means that the spatial subchannel (eigenmode) with highest gain is used.

For Schur-concave objective functions, the global communication process including pre- and post-processing $\mathbf{A}^H \mathbf{HB}$ is fully diagonalized [see Figure 9.10(b)] as well as the MSE matrix $\mathbf{E}$. Among

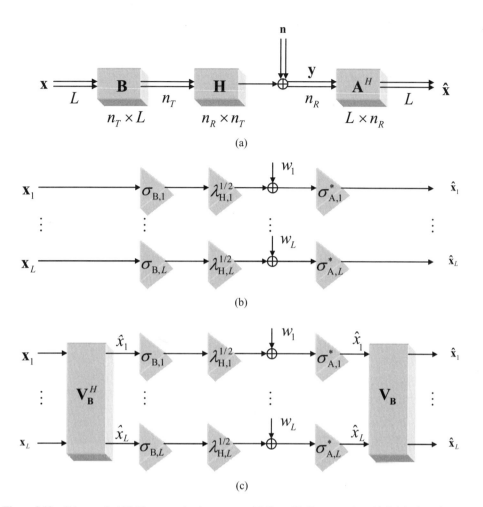

Figure 9.10 Scheme of a MIMO communication system with linear Tx-Rx processing. (a) Original matrix system. (b) Fully diagonalized system. (c) Diagonalized (up to a rotation) system © 2000, IEEE.

the L established substreams, only $\breve{L}$ are associated with nonzero channel eigenvalues, whereas the remainder $L_0 = L - \breve{L}$ are associated with zero eigenvalues. Having in mind that $\mathbf{A} = (\mathbf{H}\mathbf{B}\mathbf{B}^H \mathbf{H}^H + \mathbf{R}_n)^{-1} \mathbf{H}\mathbf{B} = \mathbf{R}_n^{-1} \mathbf{H}\mathbf{B}(\mathbf{I} + \mathbf{B}^H \mathbf{H}^H \mathbf{R}_n^{-1} \mathbf{H}\mathbf{B}^{-1}$ the global communication process is $\hat{\mathbf{x}} = (\mathbf{I} + \sum_{B,1}^{H} \mathbf{D}_{H,1})^{-1} \sum_{B,1}^{H} \mathbf{D}_{H,1}^{1/2}(\mathbf{D}_{H,1}^{1/2} \sum_{B,1} \mathbf{x} + \mathbf{w})$ or,

$$\hat{x}_i = \begin{cases} 0, & 1 \leq i \leq L_0 \\ \dfrac{\sigma_{B,(i-L_0)}^2 \lambda_{H,(i-L_0)}}{1 + \sigma_{B,(i-L_0)}^2 \lambda_{H,(i-L_0)}} x_i \\ \quad + \dfrac{\sigma_{B,(i-L_0)} \lambda_{H,(i-L_0)}^{1/2}}{1 + \sigma_{B,(i-L_0)}^2 \lambda_{H,(i-L_0)}} w_i, & L_0 < i \leq L \end{cases} \tag{9.81}$$

where $\mathbf{D}_{H,1} = \mathrm{diag}(\{\lambda_{H,i}\}_{k=1}^{\breve{L}})$, the $\lambda_{H,i}$s are the $\breve{L}$ largest eigenvalues of $\mathbf{R}_H$ in increasing order, the $\sigma_{B,i}^2$s represent the allocated power, and w is a normalized equivalent white noise.

The MSE matrix is $\mathbf{E} = (\mathbf{I} + \sum_{B,1}^{H} \mathbf{D}_{H,1} \sum_{B,1})^{-1}$, and the corresponding MSEs are given by:

$$
\mathrm{MSE}_i = \begin{cases} 1, & 1 \le i \le L_0 \\ \dfrac{1}{1 + \sigma_{B,(i-L_0)}2\lambda_{H,(i-L_0)}}, & L_0 < i \le L \end{cases} \tag{9.82}
$$

Similarly, using Equation (9.77) we have

$$
\mathrm{SINR}_i = \begin{cases} 0, & 1 \le i \le L_0 \\ \sigma_{B,(i-L_0)}2\lambda_{H,(i-L_0)}, & L_0 < i \le L \end{cases} \tag{9.83}
$$

For Schur-convex objective functions, the global communication process including pre- and post-processing $\mathbf{A}^H \mathbf{HB}$ is diagonalized only up to a very specific rotation of the data symbols [see Figure 9.10(c)], and the MSE matrix $\mathbf{E}$ is nondiagonal with equal diagonal elements (equal MSEs). In particular, assuming a pre-rotation of the data symbols at the transmitter $\hat{\mathbf{x}} = \mathbf{V}_B^H \mathbf{x}$ and a post-rotation of the estimates at the receiver $\tilde{\hat{\mathbf{x}}} = \mathbf{V}_B^H \hat{\mathbf{x}}$, the same diagonalizing results of Schur-concave functions apply [see Figure 9.10(c)]. Since the diagonal elements of the MSE matrix $\mathbf{E} = (\mathbf{I} + \mathbf{B}^H \mathbf{R}_H \mathbf{B})^{-1}$ are equal whenever the appropriate rotation is included, the MSEs are identical and given by:

$$
\mathrm{MSE}_i = \frac{1}{L}\mathrm{Tr}\,(\mathbf{E}) = \frac{1}{L}\left(L_0 + \sum_{j=1}^{\check{L}} \frac{1}{1 + \sigma_{B,j}^2 \lambda_{H,j}}\right), \qquad 1 \le i \le L \tag{9.84}
$$

Similarly by using Equation (9.77) we have:

$$
\mathrm{SINR}_i = \frac{L}{L_0 + \sum_{j=1}^{L} \dfrac{1}{1 + \sigma_{B,j}^2 \lambda H, j}} - 1, \qquad 1 \le i \le L \tag{9.85}
$$

The result defined by Equations (9.78)–(9.80) is easily extended to the multicarrier case as follows. For any carrier k, consider the matrices corresponding to the rest of the carriers $\{\mathbf{B}_l\}_{l \ne k}$ fixed, and Equations (9.78)–(9.80) can be directly invoked to show the optimal structure for $\mathbf{B}_k$.

9.3.3 Convex Optimization-Based Beamforming

In the sequel we use notation $z_{k,i} \triangleq \sigma_{B_k,i}^2$ and $\lambda_{k,i} \triangleq \lambda_{H_k,i}$. Note also that for Schur-concave functions with $L_k > \mathrm{rank}(\mathbf{R}_{Hk})$, the $L_k - \check{L}_k$ substreams associated with zero eigenvalues are simply ignored in the optimization process.

9.3.3.1 Minimization of the ARITH-MSE

For the minimization of the (weighted) arithmetic mean of the MSEs (ARITH-MSE) the objective function is

$$
f_0(\{\mathrm{MSE}_{k,i}\}) = \sum_{k,i} (w_{k,i}\mathrm{MSE}_{k,i}) \tag{9.86}
$$

The function $f_0(\{x_i\}) = \Sigma_i(w_i x_i)$ (assuming $x_i \ge x_{i+1}$) is minimized when the weights are in increasing order $w_i \le w_{i+1}$, and it is then a Schur-concave function [49]. So, the objective function Equation (9.86) is Schur-concave on each carrier k, and by Equations (9.78)–(9.80), the diagonal structure is optimal, and the MSEs are given by Equation (9.82). The problem in convex form (the objective is convex and the constraints linear) is:

$$\min_{\{z_{k,i}\}} \sum_{k,i} w_{k,i} \frac{1}{1 + \lambda_{k,i} \, z_{k,i}}$$

$$\text{s.t.} \sum_{k,i} z_{k,i} \leq P_T \tag{9.87}$$

$$z_{k,i} = 0, \qquad 1 \leq k \leq N, \ 1 \leq i \leq \breve{L}_k$$

This particular problem can be solved very efficiently because the solution has a water-filling interpretation (from the KKT optimality conditions):

$$z_{k,i} = \left(\mu^{-1/2} w_{k,i}^{1/2} \lambda_{k,i}^{-1/2} - \lambda_{k,i}^{-1} \right)^+ \tag{9.88}$$

where $\mu^{-1/2}$ is the *water-level* chosen to satisfy the power constraint with equality.

9.3.3.2 Minimization of the GEOM-MSE

The objective function corresponding to the minimization of the weighted geometric mean of the MSEs (GEOM-MSE) is

$$f_0(\{\text{MSE}_{k,i}\}) = \prod_{k,i} \left(\text{MSE}_{k,i} \right)^{w_{k,i}} \tag{9.89}$$

The function $f_0(\{x_i\}) = \prod_i x_i^{w_i}$ (assuming $x_i \geq x_{i+1} > 0$) is minimized when the weights are in increasing order $w_i \leq w_{i+1}$, and it is then a Schur-concave function [49]. So, the objective function, Equation (9.90), is Schur-concave on each carrier k and by Equations (9.78)–(9.80), the diagonal structure is optimal, and the MSEs are given by Equation (9.82). The problem in convex form (since the objective is log-convex, it is also convex [64]) is:

$$\min_{\{z_{k,i}\}} \prod_{k,i} \left(\frac{1}{1 + \lambda_{k,i} z_{k,i}} \right)^{w_{k,i}}$$

$$\text{s.t.} \sum_{k,i} z_{k,i} \leq P_T \tag{9.90}$$

$$z_{k,i} \geq 0, \qquad 1 \leq k \leq N, \ 1 \leq i \leq \breve{L}_k$$

This problem also has a water-filling solution (from the KKT optimality conditions):

$$z_{k,i} = \left(\mu^{-1} w_{k,i} - \lambda_{k,i}^{-1} \right)^+ \tag{9.91}$$

where μ^{-1} is the water-level chosen to satisfy the power constraint with equality. Note that for $w_{k,i} = 1$, Equation (9.91) becomes the classical capacity-achieving water-filling solution [50, 56] [see also Equation (4.201)].

9.3.3.3 Minimization of |E|

The minimization of the determinant of the MSE matrix was considered in [55]. This particular criterion is easily accommodated in the above framework as a Schur-concave function of the diagonal elements of the MSE matrix $\mathbf{E}$. For the carrier-noncooperative case, simply consider the global MSE matrix defined as $\mathbf{E} = \text{diag}(\mathbf{E}_1, \ldots, \mathbf{E}_N)$. Using the fact that $\mathbf{X} \geq \mathbf{Y} \Rightarrow |\mathbf{X}| \geq |\mathbf{Y}|$, it follows that $|\mathbf{E}|$ is minimized for the choice of the equalizer matrix given by Equation (9.73). From Equation (9.74), it is clear that $|\mathbf{E}|$ does not change if the precoder matrices $\mathbf{B}_k$s are post-multiplied by a unitary matrix (a rotation).

Therefore, we can always choose a rotation matrix so that $\mathbf{E}$ is diagonal without loss of optimality (see also [15]), and then

$$|\mathbf{E}| = \prod_j \lambda_j(\mathbf{E}) = \prod_j [\mathbf{E}]_{jj} \qquad (9.92)$$

Therefore, the minimization of $|\mathbf{E}|$ is equivalent to the minimization of the (unweighted) product of the MSEs as in Section 9.3.3.2.

9.3.3.4 Maximization of Mutual Information

The maximization of mutual information can be used to obtain a capacity achieving solution [56] (see also Appendix in Chapter 4)

$$\max_Q I = \log \left| \mathbf{I} + \mathbf{R}_n^{-1} \mathbf{H} \mathbf{Q} \mathbf{H}^H \right| \qquad (9.93)$$

where Q is the precoder covariance matrix. Using the fact that $|\mathbf{I} + \mathbf{XY}| = |\mathbf{I} + \mathbf{YX}|$ and that $\mathbf{Q} = \mathbf{BB}^H$ [from Equation (9.68)], the mutual information can be expressed as [59]:

$$I = -\log |\mathbf{E}| \qquad (9.94)$$

and therefore, the maximization of I is equivalent to the minimization of $|\mathbf{E}|$ treated in Section 9.3.3.3. Hence, the minimization of the unweighted product of the MSEs, the minimization of the determinant of the MSE matrix, and the maximization of the mutual information are all equivalent criteria with the solution given by a channel-diagonalizing structure and the classical capacity-achieving water-filling for the power allocation:

$$z_{k,i} = \left(\mu^{-1} - \lambda_{k,i}^{-1} \right)^+ \qquad (9.95)$$

9.3.3.5 Minimization of the MAX-MSE

In general, average BER is dominated by the substream with highest MSE. It makes sense then to minimize the maximum of the MSEs (MAX-MSE). Then the objective function is

$$f_0(\{\mathrm{MSE}_{k,i}\}) = \max_{k,i} \{\mathrm{MSE}_{k,i}\} \qquad (9.96)$$

The function $f_0(\{x_i\}) = \max_i \{x_i\}$ is a Schur-convex function [49]. So, the objective function in Equation (9.96) is Schur-convex on each carrier k. Therefore, Equations (9.78)–(9.80), the optimal solution has a nondiagonal MSE matrix $\mathbf{E}_k$ with equal diagonal elements given by Equation (9.84), which have to be minimized (scalarized problem). After minimizing the MSEs, we must still obtain the optimal rotation matrices so that the diagonal elements of the MSE matrices $\mathbf{E}_k$s are identical. The scalarized problem in convex form (the objective is linear and the constraints are all is:

$$\min_{t,\{z_{k,i}\}} \quad t$$

$$\text{s.t.} \quad t \ge \frac{1}{L_k} \left(\left(L_k - \breve{L}_k \right) + \sum_{i=1}^{\breve{L}_k} \frac{1}{1 + \lambda_{k,i} z_{k,i}} \right) \qquad (9.97)$$

$$1 \le k \le N$$

$$\sum_{k,i} z_{k,i} \le P_T$$

$$z_{k,i} \ge 0, \quad 1 \le k \le N, \ 1 \le i \le \breve{L}_k$$

This problem has a multilevel water-filling solution (from the KKT optimality conditions):

$$z_{k,i} = \left(\bar{\mu}_k^{1/2} \lambda_{k,i}^{-1/2} - \lambda_{k,i}^{-1} \right)^+ \tag{9.98}$$

where $\{\bar{\mu}_k^{1/2}\}$ are multiple water levels chosen to satisfy the constraints on t and the power constraint all with equality. For the case of single beamforming (i.e., $L_k = 1$), the solution simplifies to:

$$z_k = \lambda_k^{-1} \frac{P_T}{\sum_l \lambda_l^{-1}} \tag{9.99}$$

For the single-carrier case (or multicarrier cooperative approach), problem (9.97) simplifies to the minimization of the unweighted ARITH-MSE considered in Section 9.3.3.1 with solution $z_i = (\mu^{-1/2}\lambda_i^{-1/2} - \lambda_i^{-1})^+$.

9.3.3.6 Maximization of the ARITH-SINR

Since, in Equations (9.78)–(9.80), we assumed $\text{MSE}_{k,i} \geq \text{MSE}_{k,i+1}$, the SINRs are in increasing order $\text{SINR}_{k,i} \leq \text{SINR}_{k,i+1}$. The objective function to be minimized for the maximization of the (weighted) arithmetic mean of the SINRs (ARITH-SINR) is

$$\tilde{f}_0(\{\text{SINR}_{k,i}\}) = - \sum_{k,i} (w_{k,i} \text{SINR}_{k,i}) \tag{9.100}$$

which can be expressed as a function of the MSEs using Equation (9.77) as:

$$f_0(\{\text{MSE}_{k,i}\}) = \tilde{f}_0(\{\text{MSE}_{k,i}^{-1} - 1\}) = - \sum_{k,i} w_{k,i}(\text{MSE}_{k,i}^{-1} - 1) \tag{9.101}$$

The function $f_0(x_i) = - \sum_i w_i (x_i^{-1} - 1)$ (assuming $x_i \geq x_{i+1} > 0$) is minimized when the weights are in increasing order $w_i \leq w_{i+1}$, and it is then a Schur-concave function [49]. So, the objective function of Equation (9.101) is Schur-concave on each carrier k. Therefore, by Equation (9.78)–(9.80), the diagonal structure is optimal and the SINRs are given by Equation (9.83). The problem expressed in convex form (it is actually an LP since the objective and the constraints are all linear) is:

$$\max_{\{z_{k,i}\}} \sum_{k,i} w_{k,i} \lambda_{k,i} z_{k,i}$$

$$\text{s.t.} \quad \sum_{k,i} z_{k,i} \leq P_T \tag{9.102}$$

$$z_{k,i} \geq 0, \quad 1 \leq k \leq N, \ 1 \leq i \leq \breve{L}_k$$

The optimal solution is to allocate all the available power to the substream with maximum weighted gain ($w_{k,i} \lambda_{k,i}$) (otherwise, the objective value could be increased by transferring power from other substreams to this substream). Although this solution indeed maximizes the weighted sum of the SINRs, it has poor spectral efficiency.

9.3.3.7 Maximization of the GEOM-SINR

The objective function to be minimized for the maximization of the (weighted) geometric mean of the SINRs (GEOM-SINR) is:

$$\tilde{f}_0(\{\text{SINR}_{k,i}\}) = -\prod_{k,i} (\text{SINR}_{k,i})^{w_{k,i}} \tag{9.103}$$

which can be expressed as a function of the MSEs using Equation (9.77) as

$$\tilde{f}_0(\{\text{MSE}_{k,i}\}) = \tilde{f}_0(\{\text{MSE}_{k,i}^{-1} - 1\}) = -\prod_{k,i} (\text{MSE}_{k,i}^{-1} - 1)^{w_{k,i}} \tag{9.104}$$

The maximization of the product of the SINRs is equivalent to the maximization of the sum of the SINRs expressed in decibels. The function $f_0(\{x_i\}) = -\prod_i (x_i^{-1} - 1)^{w_i}$ (assuming $0.5 \geq x_i \geq x_{i+1} > 0$) is minimized when the weights are in increasing order $w_i \leq w_{i+1}$, and it is then a Schur-concave function [49]. So, the objective function (9.104) is Schur-concave on each carrier k, provided that $\text{MSE}_{k,i} \leq 0.5 \forall k$, i (this is a mild assumption since a MSE greater than 0.5 is unreasonable for a practical communication system). Therefore, by Equations (9.78)–(9.80), the diagonal structure is optimal, and the SINRs are given by Equation (9.83). The problem expressed in convex form (the weighted geometric mean is a concave function [64,65]) is:

$$\max_{\{z_{k,i}\}} \prod_{k,i} (\lambda_{k,i} z_{k,i})^{\tilde{w}_{k,i}}$$

$$\text{s.t.} \quad \sum_{k,i} z_{k,i} \leq P_T \tag{9.105}$$

$$z_{k,i} = 0, \quad 1 \leq k \leq N, \ 1 \leq i \leq \breve{L}_k$$

where $\tilde{w}_{k,i} = w_{k,i}/(\sum_{l,j} w_{l,j})$, and it is assumed that $\lambda_{k,i} > 0 \forall k, i$ (otherwise, the problem has the trivial solution $z_{k,i} = 0 \forall k, i$). The solution is easily obtained from the KKT optimality conditions as

$$z_{k,i} = \tilde{w}_{k,i} P_T \tag{9.106}$$

For a uniform weighting $w_{k,i} = 1$, the problem reduces to the maximization of the geometric mean subject to the arithmetic mean:

$$\max_{\{z_{k,i}\}} \prod_{k,i} z_{k,i}^{1/\breve{L}_T}$$

$$\text{s.t.} \quad 1/\breve{L}_T \sum_{k,i} z_{k,i} \leq P_T/\breve{L}_T \tag{9.107}$$

$$z_{k,i} \geq 0$$

where $\breve{L}_T \triangleq \sum_{k=1}^{N} \breve{L}_k$. From the arithmetic–geometric mean inequality ($\prod_k x_k^{1/N} \leq \frac{1}{N} \sum_k x_k$ (with equality if and only if $x_k = x_l \forall k, l$) [60], it follows that the optimal solution is a uniform power allocation:

$$z_{k,i} = P_T/\breve{L}_T \tag{9.108}$$

The uniform power distribution is commonly used due to its simplicity.

9.3.3.8 Maximization of the HARM-SINR

For maximization of the harmonic mean of the SINRs (HARM-SINR) in the case of single beamforming, the objective function to be minimized is:

$$\tilde{f}_0(\{\text{SINR}_{k,i}\}) = \sum_{k,i} \frac{1}{\text{SINR}_{k,i}} \tag{9.109}$$

which can be expressed as a function of the MSEs using Equation (9.77) as:

$$f_0(\{\text{MSE}_{k,i}\}) \sum_{k,i} \frac{\text{MSE}_{k,i}}{1 - \text{MSE}_{k,i}} \tag{9.110}$$

The function f_0 ($\{x_i\} = \Sigma_i$ ($x_i/(1 - x_i)$) (for $0 \leq x_i < 1$) is a Schur-convex function [49].

So, the objective function (9.110) is Schur-convex on each carrier k. Therefore, by Equations (9.78)–(9.80), the optimal solution has a nondiagonal MSE matrix $\mathbf{E}_k$ with equal diagonal elements given by Equation (9.84), which have to be minimized. The scalarized problem in convex form is:

$$\min_{\{t_k\}\{z_{k,i}\}} \sum_k \frac{t_k}{1 - t_k}$$

$$\text{s.t.} \quad 1 > t_k \geq \frac{1}{L_k} \left(\left(L_k - \breve{L}_k \right) + \sum_{i=1}^{\breve{L}_k} \frac{1}{1 + \lambda_{k,i} z_{k,i}} \right)$$

$$\sum_{k,i} z_{k,i} \leq P_T$$

$$z_{k,i} = 0, \quad 1 \leq k \leq N, \quad 1 \leq i \leq \breve{L}_k \tag{9.111}$$

The problem has a multilevel water filling solution

$$z_{k,i} = \left(\bar{\mu}_k^{1/2} \lambda_{k,i}^{-1/2} - \lambda_{k,i}^{-1} \right)^+ \tag{9.112}$$

where $\{\bar{\mu}_k^{1/2}\}$ are multiple water levels chosen to satisfy the lower constraints on the t_ks and the power constraint, all with equality, and also the constraint $\bar{\mu}_k^{1/2} = \upsilon \frac{L_k^{-1/2}}{1-t_k}$, where υ is a positive parameter. For the case of single beamforming (i.e., $L_k = 1$), the solution reduces to $z_k = \lambda_k^{-1/2} (P_T / \sum_l \lambda_l^{-1/2})$. For the single-carrier case (or multicarrier cooperative approach), the problem simplifies to that considered in Section 9.3.3.1.

9.3.3.9 Maximization of the MIN-SINR

The objective function to be minimized for the maximization of the minimum of the SINRs (MIN-SINR) is

$$\tilde{f}_0(\{\text{SINR}_{k,i}\}) = - \min_{k,i} \{\text{SINR}_{k,i}\} \tag{9.113}$$

This design criterion is equivalent to the minimization of the maximum MSE treated with detail in Section 9.3.3.5. In [54], the same criterion was used, imposing a channel diagonal structure.

9.3.3.10   Maximization of the PROD-(1 + SINR)

Let us start with the following maximization:

$$\max \prod_{k,i} (1 + \mathrm{SINR}_{k,i}) \tag{9.114}$$

Using Equation (9.77), this maximization can be equivalently expressed as the minimization of $\prod_{k,i} \mathrm{MSE}_{k,i}$ as in Equation (9.87) with $w_{k,i} = 1$, as the minimization of the determinant of the MSE matrix (Section 9.3.3.3), and as the maximization of the mutual information (Section 9.3.3.4) with the solution given by the capacity-achieving expression Equation (9.95). This result is natural since maximizing the logarithm of (9.114) is tantamount to maximizing the mutual information $I = \sum_{k,i} \log (1 + \mathrm{SINR}_{k,i})$.

9.3.3.11   Minimization of the ARITH-BER

The minimization of the average BER or of the arithmetic mean of the BERs (ARITH-BER) can be considered as the best criterion (assuming that after the linear processing at the receiver, each substream is detected independently). The objective function is:

$$\tilde{f}_0(\{\mathrm{BER}_{k,i}\}) = \sum_{k,i} \mathrm{BER}_{k,i} \tag{9.115}$$

which can be expressed as a function of the MSEs using Equation (9.77) as:

$$f_0(\{\mathrm{MSE}_{k,i}\}) = \sum_{k,i} \mathrm{BER}(\mathrm{MSE}_{k,i}^{-1} - 1) \tag{9.116}$$

The function $f_0(\{x_i\}) = \sum_i \mathrm{BER}\left(x_i^{-1} - 1\right)$ (assuming $\theta \geq x_i > 0$, for sufficiently small θ such that $\mathrm{BER}\left(x_i^{-1} - 1\right) \leq 2 \times 10^{-2}, \forall i$) is a Schur-convex function [49].

The objective function (9.116) is Schur-convex on each carrier k (assuming the same \hbox{constellation}/coding on all substreams of the kth carrier), provided that $\mathrm{BER}_{k,i} \leq 2 \times 10^{-2}$ (interestingly, for BPSK and QPSK constellations, this is true for any value of the BER). Therefore, by Equations (9.78)–(9.80), the optimal solution has a nondiagonal MSE matrix $\mathbf{E}_k$ with equal diagonal elements given by Equation (9.84), which have to be minimized. The scalarized problem in convex form is

$$\min_{\{t_k\},\{z_{k,i}\}} \sum_k \alpha_k Q\left(\sqrt{\beta_k(t_k^{-1} - 1)}\right)$$

$$\mathrm{s.t.} \quad \theta \geq t_k \geq \frac{1}{L_k}\left(\left(L_k - \breve{L}_k\right) + \sum_{i=1}^{\breve{L}_k} \frac{1}{1 + \lambda_{k,i} z_{k,i}}\right)$$

$$1 \leq k \leq N$$

$$\sum_{k,i} z_{k,i} \leq P_T$$

$$z_{k,i} = 0, \quad 1 \leq k \leq N, \ 1 \leq i \leq \breve{L}_k \tag{9.117}$$

The upper bound θ on the MSEs is explicitly included to guarantee the convexity of the BER function and, therefore, of the whole problem. For a general case with $L_k > 1$ and $N > 1$, problem (51) does not have a simple closed-form solution, and one has to resort to general-purpose iterative methods such as interior-point methods (see Section 9.3). For the single-carrier case (or multicarrier cooperative approach), the problem simplifies to the ARITH-MSE criterion considered in Section 9.3.3.1, plus the rotation matrix to make the diagonal elements of the MSE matrix equal.

9.3.4   Constraints in Multicarrier Systems

We can, for example, add constraints on the dynamic range of the power amplifier at each transmit antenna element, as was done in [68]. Consider a Schur-concave objective function and assume for simplicity $L_k = L \forall k$. From the optimal structure in Equation (9.79) $\mathbf{B}_k = \mathbf{U}_{H_k,1} \sum_{B_k,1}$, the total average transmitted power (in units of energy per symbol period) by the ith antenna is

$$P_i = \frac{1}{N} \sum_{k=1}^{N} \sum_{i=1}^{L} |[\mathbf{B}_k]_{i,l}^2| = \frac{1}{N} \sum_{k=1}^{N} \sum_{i=1}^{L} \sigma_{B_k,l}^2 \left|[\mathbf{U}_{H_k,1}]_{i,l}\right|^2 \tag{9.118}$$

which is linear in the variables $\{\sigma_{B_k,i}^2\}$ (for the carrier-cooperative scheme, $P_i = (1/N) \sum_{k=1}^{N} \sum_{l=1}^{NL} |[\mathbf{B}]_{i+(k+1)n_T,l}|^2$. Therefore, the following constraints are linear as $\alpha_i^L \leq P_i \leq \alpha_i^U 1 \leq i \leq n_T$ where α_i^L and α_i^U are the lower and upper bounds for the ith antenna. Similarly, it is straightforward to set limits on the relative dynamic range of a single element in comparison with the total power for the whole array [68] as $\rho_i^L P_{array} \leq P_i \leq \rho_i^U P_{array} 1 \leq i \leq n_T$, where ρ_i^L and ρ_i^U are the relative bounds, and $P_{array} = \sum_{i=1}^{n_T} P_i$ is the total power that is also linear in $\{\sigma_{B_k,i}^2\}$.

We also show how the PAPR (see Chapter 7) can be taken into account into the design of the beam vectors using a convex optimization framework. The PAR is defined as:

$$\text{PAR} \triangleq \max_{0 \leq t \leq T_s} \frac{A^2(t)}{\sigma^2} \tag{9.119}$$

where T_s is the symbol period, $A(t)$ is the zero-mean transmitted signal, and $\sigma^2 = E[A^2(t)]$. Since the number of carriers is usually large ($N \geq 64$), $A(t)$ can be accurately modelled as a Gaussian random process (central-limit theorem) with zero mean and variance σ^2. Using this assumption, the probability that the PAR exceeds certain threshold or, equivalently, the probability that the instantaneous amplitude exceeds a clipping value A_{clip} is

$$\Pr\{|A(t)| > A_{clip}\} = 2Q\left(\frac{A_{clip}}{\sigma}\right) \tag{9.120}$$

The clipping probability of an OFDM symbol is then

$$P_{clip}(\sigma) = 1 - \left(1 - 2Q\left(\frac{A_{clip}}{\sigma}\right)\right)^{2N} \tag{9.121}$$

In other words, in order to have a clipping probability lower than P with respect to the maximum instantaneous amplitude A_{clip}, the average signal power must satisfy

$$\sigma \leq \sigma_{clip}(P) = \frac{A_{clip}}{Q^{-1}\left(\frac{1 - (1 - P)^{1/2N}}{2}\right)} \tag{9.122}$$

When using multiple antennas for transmission, the previous equation has to be satisfied for all transmit antennas. Those constraints can be easily incorporated in any of the convex designs derived in

Section 9.3.3 with a Schur-concave objective function. Using Equation (9.118), the constraint is

$$\frac{1}{N} \sum_{k=1}^{N} \sum_{l=1}^{L} \sigma_{B_k,l}^2 \left| \left[\mathbf{U}_{H_k,1} \right]_{i,l} \right|^2 \leq \sigma_{\text{clip}}^2 \; 1 \leq i \leq n_T \tag{9.123}$$

which is linear in the optimization variables $\{\sigma_{B_k,i}^2\}$. Such a constraint has two effects in the solution: (i) The power distribution over the carriers changes with respect to the distribution without the constraint, and (ii) the total transmitted power drops as necessary.

9.3.5 Performance Examples

For the numerical results, we use the multicarrier modulation OFDM (64 carriers). We consider the multicarrier MIMO model used throughout the section. Perfect CSI is assumed at both sides of the communication link. The frequency selectivity of the channel is modeled using the power delay profile as specified in [70] [see Figure 9.11(a)], which corresponds to a typical large open space indoor environment

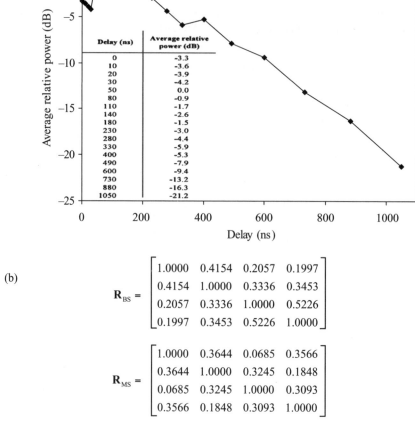

Figure 9.11 (a) Power delay profile type C for HIPERLAN/2. (b) Envelope correlation matrices at the base station (BS) and at the mobile station (MS) © 2000, IEEE.

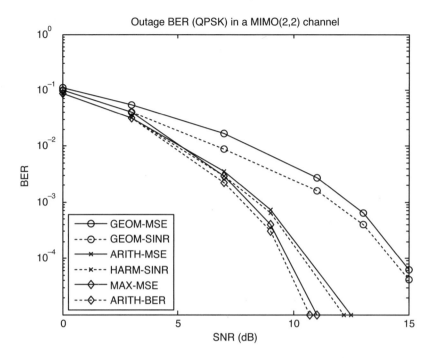

Figure 9.12   BER (at an outage probability of 5%) versus SNR when using QPSK in a 2 × 2 MIMO channel with $L = 1$ for the specified design criteria (without carrier cooperation).

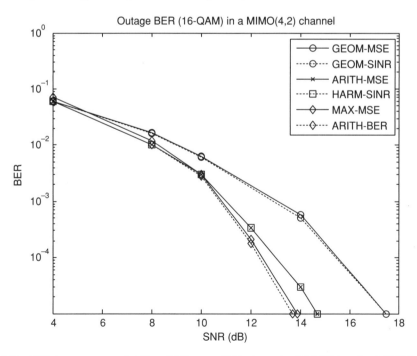

Figure 9.13   BER (at an outage probability of 5%) versus SNR when using 16-QAM in a 4 × 2 MIMO channel (two transmit and four receive antennas) with $L = 1$ for the specified design criteria (without carrier cooperation).

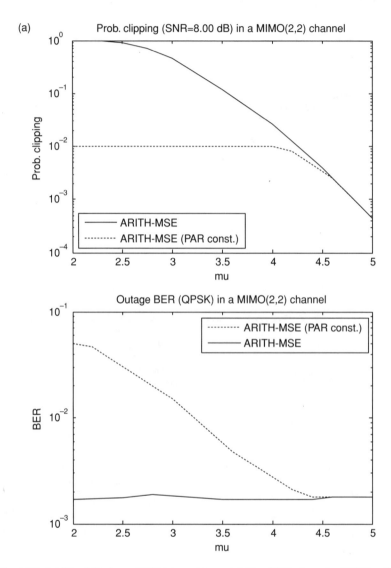

Figure 9.14   (a) Probability of clipping and BER (at an outage probability of 5%) when using QPSK in a 2×2 MIMO channel with $L = 1$ for the ARITH-MSE criterion with and without PAR constraints (without carrier cooperation) as a function of μ (for SNR = 8 dB and $P_{\text{clip}} \leq 10^{-2}$). $A_{\text{clip}} = \mu \sqrt{P_T / n_T}$.

for non-line-of-sight (NLOS) conditions with 150 ns average r.m.s. delay spread and 1050 ns maximum delay (the sampling period is 50 ns) [69]. The spatial correlation of the MIMO channel is modeled according to [71] (which corresponds to a reception hall) specified by the correlation matrices of the envelope of the channel fading at the transmit and receive side given in Figure 9.11(b), where the base station is the receiver (uplink) (see [71] for details of the model). It provides a large open indoor environment with two floors, which could easily illustrate a conference hall or a shopping galleria scenario. The matrix channel generated was normalized so that $\sum_n \mathrm{E}[|\mathbf{H}_{ij}(n)|^2] = 1$. The SNR is defined as the transmitted power normalized with the noise variance. The following design criteria have been compared: ARITH-MSE, GEOM-MSE, MAX-MSE (equivalently, MIN-SINR or MAX-BER),

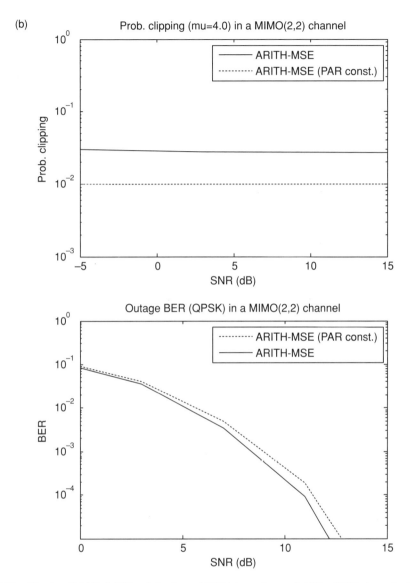

Figure 9.14 (*Continued*) (b) Probability of clipping and BER (at an outage probability of 5%) when using QPSK in a 2×2 MIMO channel with $L = 1$ for the ARITH-MSE criterion with and without PAR constraints (without carrier cooperation) as a function of the SNR (for $\mu = 4$ and $P_{\text{clip}} \leq 10^{-2}$). $A_{\text{clip}} = \mu \sqrt{P_T / n_T}$.

GEOM-SINR, HARM-SINR, and ARITH-BER (benchmark). The utilization of the Chernoff upper bound instead of the exact BER function gives indistinguishable results and is therefore not presented in the simulation results. Unless otherwise specified, carrier-noncooperative approaches are considered. The performance in terms of outage BER (averaged over the channel substreams), i.e. the BER that can be guaranteed with some probability or, equivalently, the BER that is not achieved with some small outage probability. In particular, we consider the BER with an outage probability of 5%. The results are presented in Figures 9.12–9.16.

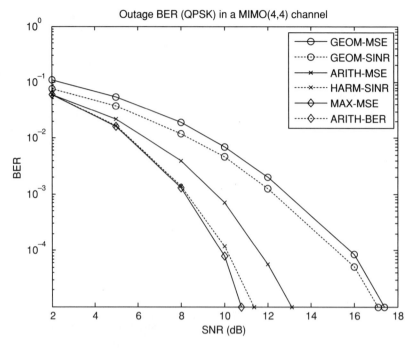

Figure 9.15   BER (at an outage probability of 5%) versus SNR when using QPSK in a 4 × 4 MIMO channel with $L = 2$ for the specified design criteria (without carrier cooperation).

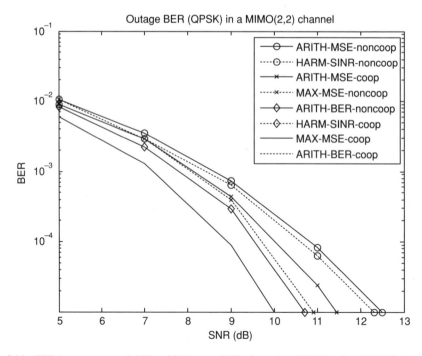

Figure 9.16   BER (at an outage probability of 5%) versus SNR when using QPSK in a 2 × 2 MIMO channel with $L = 1$ for the specified design criteria with (coop) and without (noncoop) carrier cooperation.

References

[1] Ding, Z. and Li, Y. (2001) Blind Equalization and Identification, Signal Processing and Communications Series. New York: Marcel Dekker.

[2] Kailath, T. (1980) *Linear Systems*. Englewood Cliffs, NJ: Prentice-Hall.

[3] Forney, G. D. (1970) Convolution codes I: Algebraic structure, *IEEE Transactions on Information Theory*, **IT-16**, 720–738.

[4] Kung, S.-Y., Wu, Y. and Zhang, X. (2002) Bezout space-time precoders and equalizers for MIMO channels, *IEEE Transactions on Signal Processing*, **50**, 2499–2514.

[5] Xia, X.-G. (1997) New precoding for intersymbol interference cancellation using nonmaximally decimated multirate filterbanks with ideal FIR equalizers, *IEEE Transactions on Signal Processing*, **45**, 2431–2440.

[6] Xia, X.-G., Su, W. and Liu, H. (2001) New precoders for blind equalization: polynomial ambiguity resistant precoders (PARP), *IEEE Transactions on Cirtuits Systems I*, **48**, 193–209.

[7] Yang, J., and Roy, S. (1994) On joint transmitter and receiver optimization for multiple-input–multiple-output (MIMO) transmission systems, *IEEE Transactions on Communication*, **42**, 3221–3231.

[8] Scaglione, A., Giannakis, G. B. and Barbarossa, S. (1999) Redundant filterbank precoders and equalizers, Part I: Identification and optimal designs, *IEEE Transactions on Signal Processing*, **47**, 1988–2006.

[9] Vaidyanathan, P. P. and Mitra, S. K. (1988) Polyphase networks, block digital filtering, LPTV systems, and alias-free QMF banks: a unified approach based on pseudocirculants, *IEEE Transactions on Signal Processing*, **36**, 381–391.

[10] Wiesel, A., Eldar, Y. C. and Shamai (Shitz), S. (2006) Linear precoding via conic optimization for fixed MIMO receivers, *IEEE Transactions on Signal Processing*, **54**, 161–176.

[11] Vojcic, B. R. and Jang, W. M. (1998) Transmitter precoding in synchronous multiuser communication, *IEEE Transactions on Communication*, **46**, 1346–1355.

[12] Joham, M., Kusume, K., Gzara, M. H. and Utschick, W. (2002) Transmit wiener filter for the downlink of TDD DS-CDMA systems, in *Proceedings IEEE 7th International Symposium on Spread Spectrum Techniques Applications (ISSSTA)*, 1, Sep. 2002, pp. 9–13.

[13] Choi, R. L. U. and Murch, R. D. (2003) New transmit schemes and simplified receivers for MIMO wireless communications systems, *IEEE Transactions on Wireless Communication*, **2**, 1217–1230.

[14] Choi, L. U. and Murch, R. D. (2004) Transmit-preprocessing techniques with simplified receivers for the downlink of MISO TDD-CDMA systems, *IEEE Transactions on Vehicular Technology*, **53**, 285–295.

[15] Esmailzadeh, R., Sourour, E. and Nakagawa, M. (1999) Pre-rake diversity combining in time division duplex CDMA mobile communications, *IEEE Transactions on Vehicular Technology*, **48**, 795–801.

[16] Fischer, R. F. H. (2002) Precoding and Signal Shaping for Digital Transmission. Wiley, New York.

[17] Brandt-Pearce, M. and Dharap, A. (2000) Transmitter based multiuser interference rejection for the down-link of a wireless CDMA system in a multipath environment, *IEEE Journal of Selected Areas of Communication*, **18**, 407–417.

[18] Meurer, M., Baier, P. W., Weber, T., Lu, Y. and Papathanassiou, A. (2000) Joint transmission: Advantageous downlink concept for CDMA mobile radio systems using time division duplexing, *Electronics Letters*, **36**, 900–901.

[19] Georgoulis, S. and Cruickshank, D. (2001) Pre-equalization, transmitter precoding, and joint transmission techniques for time division duplex CDMA, in *Proceedings IEE 3G Mob. Communications Technology*, London, Mar. pp. 257–261.

[20] Barreto, A. N. and Fettweis, G. (2003) Joint signal precoding in the downlink of spread spectrum systems, *IEEE Transactions on Wireless Communications*, **2**, 511–518.

[21] Georgoulis, S. L. and Cruickshank, D. G. M. (2004) Transmitter-based inverse filters for reducing MAI and ISI in CDMA-TDD downlink, *IEEE Transactions on Wireless Communications*, **3**, 353–358.

[22] Choi, J. and Perreau, S. (2004) MMSE multiuser downlink multiple antenna transmission for CDMA channels, *IEEE Transactions on Signal Processing*, **52**, 1564–1573.

[23] Wang, S. and Caffery, J. (2003) Linear multiuser precoding for synchronous CDMA, in *Proceedings IEEE Wireless Communications Networking Conference (WCNC2003)*, 1, Mar. 2003, pp. 403–407.

[24] Choi, L. U. and Murch, R. D. (2004) A transmit preprocessing technique for multiuser MIMO systems using a decomposition approach, *IEEE Transactions on Wireless Communications*, **3**, 20–24.

[25] Collin, L., Berder, O., Rostaing, P. and Burel, G. (2004) Optimal minimum distance based precoder for MIMO spatial multiplexing systems, *IEEE Transactions on Signal Processing*, **52**, 617–627.

[26] Fischer, R., Windpassinger, C., Lampe, A. and Huber, J. (2002) Space time transmission using Tomlinson-Harashima precoding, in *Proceedings 4th ITG Conference on Source Channel Coding*, Berlin, January 2002, pp. 139–147.

[27] Schubert, M., and Boche, H. (2002) Joint 'dirty paper' pre-coding and downlink beamforming, in *Proceedings IEEE International Symposium on Spread Spectrum Techniques Applications (ISSSTA 2002)*, **2**, Sep. 2002, pp. 536–540.

[28] Hons, E. S., Khandani, A. K. and Tong, W. (2002) An optimized transmitter precoding scheme for synchronous DS-CDMA, in *Proceedings IEEE International Communications Conference*, **3**, Apr. 2002, pp. 1818–1822.

[29] Ding, Y., Davidson, T. N., Luo, Z.-Q. and Wong, K. M. (2003) Minimum BER block precoders for zero-forcing equalization, *IEEE Transactions on Signal Processing*, **51**, pp. 2410–2423.

[30] Ban, K., Katayama, M., Yamazato, T. and Ogawa, A. (1999) Transmitter precoding with multiple transmit/receive antennas for high data rate communication in bandwidth-limited channels, in *Proceedings of Vehicular Technology Conference (VTC 1999–Fall)*, **3**, 1946–1950.

[31] Yang, J. and Roy, S. (1994) Joint transmitter receiver optimization for multi-input–multi-output systems with decision feedback, *IEEE Transactions on Information Theory*, **40**, 1334–1347.

[32] Scaglione, A., Stoica, P., Barbarossa, S., Giannakis, G. B. and Sampath, H. S. (2002) Optimal designs for space time linear precoders and decoders, *IEEE Transactions on Signal Processing*, **50**, 1051–1064.

[33] Palomar, D. P., Cioffi, J. M. and Lagunas, M. A. (2003) Joint T_x-R_x beamforming design for multicarrier MIMO channels: a unified framework for convex optimization, *IEEE Transactions on Signal Processing*, **51**, 2381–2401.

[34] Palomar, D. P., Lagunas, M. A. and Cioffi, J. M. (2004) Optimum linear joint transmit–receive processing for MIMO channels with QoS constraints, *IEEE Transactions on Signal Processing*, **52**, 1179–1197.

[35] Luo, Z. Q., Davidson, T. N., Giannakis, G. B. and Wong, K. M. (2004) Transceiver optimization for block-based multiple access through ISI channels, *IEEE Transactions on Signal Processing*, **52**, 1037–1052.

[36] Serbetli, S. and Yener, A. (2004) Transceiver optimization for multiuser MIMO systems, *IEEE Transactions on Signal Processing*, **52**, 214–226.

[37] Sampath, H., Stoica, P. and Paulraj, A. (2001) Generalized linear precoder and decoder design for MIMO channels using the weighted MMSE criterion, *IEEE Transactions on Communications*, **49**, 2198–2206.

[38] Chang, J. H., Tassiulas, L. and Farrokhi, F. R. (2002) Joint transmitter receiver diversity for efficient space division multiple access, *IEEE Transactions on Wireless Communications*, **1**, 16–27.

[39] Visotsky, E. and Madhow, U. (1999) Optimum beamforming using transmit antenna arrays, in *Proceedings IEEE Vehicular Technology Conference*, **1**, May 1999, pp. 851–856.

[40] Bengtsson, M. and Ottersten, B. (1999) Optimal downlink beamforming using semidefinite optimization, in *Proceedings 37th Annual Allerton*, Sep. 1999, pp. 987–996.

[41] Boyd, S. and Vandenberghe, L. (2003) *Introduction to Convex Optimization with Engineering Applications*. Cambridge University Press, Stanford, CA.

[42] Sturm, J. F. (1999) Using SEDUMI 1.02, a Matlab toolbox for optimizations over symmetric cones, *Optimization Methodology. Software*, **11–12**, 625–653.

[43] Lobo, M. S., Vandenberghe, L., Boyd, S. and Lebret, H. (1998) Applications of second order cone programming, *Linear Algebra Application*, **284**, 193–228.

[44] Vandenberghe, L. and Boyd, S. (1996) Semidefinite programming, *SIAM Review*, **38**, 49–95.

[45] Boyd, S. and Ghaoui, L. E. (1993) Method of centers for minimizing generalized eigenvalues, *Linear Algebra and Its Applications*, **188–189**, 63–111.

[46] Schubert, M. and Boche, H. (2004) Solution of the multiuser downlink beamforming problem with individual SINR constraints, *IEEE Transactions on Vehicular Technology*, **53**, 18–28.

[47] Tung, T.-L. and Yao, K. (2002) Optimal downlink power-control design methodology for a mobile radio DS-CMDA system, in *Proceedings IEEE Workshop Signal Processing Systems (SIPS-2002)*, pp. 165–170.

[48] Verdu, S. (1998) *Multiuser Detection*. Cambridge University Press, Cambridge.

[49] Palomar, D. P., Cioffi, J. M. and Lagunas, M. A. (2003) Joint T_x-R_x beamforming design for multicarrier MIMO channels: a unified framework for convex optimization, *IEEE Transactions on Signal Processing*, **51**, 2381–2400.

[50] Raleigh, G. G. and Cioffi, J. M. (1998) Spatio-temporal coding for wireless communication, *IEEE Transactions on Communications*, **46**, 357–366.

[51] Bölcskei, H. and Paulraj, A. J. (2002) Multiple-input multiple-output (MIMO) wireless systems, in *The Communications Handbook*, 2nd ed, J. Gibson (Ed.) CRC, Boca Raton, FL, pp. 90.1–90.14.

[52] Yang, J. and Roy, S. (1994) On joint transmitter and receiver optimization for multiple-input-multiple-output (MIMO) transmission systems, *IEEE Transactions on Communications*, **42**, 3221–3231.

[53] Scaglione, A., Giannakis, G. B. and Barbarossa, S. (1999) Redundant filterbank precoders and equalizers, Part I: Unification and optimal designs, *IEEE Transactions on Signal Processing*, **47**, 1988–2006.

[54] Sampath, H., Stoica, P. and Paulraj, A. (2001) Generalized linear precoder and decoder design for MIMO channels using the weighted MMSE criterion, *IEEE Transactions on Communications*, **49**, 2198–2206.

[55] Yang, J. and Roy, S. (1994) Joint transmitter–receiver optimization for multi-input–multi-output systems with decision feedback, *IEEE Transactions on Information Theory*, **40**, 1334–1347.

[56] Cover, T. M. and Thomas, J. A. (1991) *Elements of Information Theory*. Wiley, New York.

[57] Scaglione, A., Barbarossa, S. and Giannakis, G. B. (1999) Filterbank transceivers optimizing information rate in block transmissions over dispersive channels, *IEEE Transactions on Information Theory*, **45**, 1019–1032.

[58] Scaglione, A., Stoica, P., Barbarossa, S., Giannakis, G. B. and Sampath, H. (2002) Optimal designs for space–time linear precoders and decoders, *IEEE Transactions on Signal Processing*, **50**, 1051–1064.

[59] Cioffi, J. M. and Forney, G. D. (1997) Generalized decision-feedback equalization for packet transmission with ISI and Gaussian noise, Ch. 4 in *Communications, Computation, Control and Signal Processing*, A. Paulraj, V. Roychowdhury, and C. D. Schaper (Eds.) Kluwer, Boston, MA.

[60] Horn, R. A. and Johnson, C. R. (1985) *Matrix Analysis*. Cambridge University Press, New York.

[61] Verdú, S. (1998) *Multiuser Detection*. Cambridge University Press, New York.

[62] Viswanath, P. and Anantharam, V. (1999) Optimal sequences and sum capacity of syn\-chro\-nous CDMA systems, *IEEE Transactions on Information Theory*, **45**, 1984–1991.

[63] Marshall, A. W. and Olkin, I. (1979) *Inequalities: Theory of Majorization and Its Applications*. Academic Press, New York.

[64] Boyd, S. and Vandenberghe, L. (2000) Introduction to convex optimization with engineering applications. Stanford University, Stanford, CA, Course Notes. [Online]. Available: http://www.stanford.edu/class/ee364.

[65] Rockafellar, R. T. (1970) *Convex Analysis*, 2nd ed. Princeton University Press, Princeton, NJ.

[66] Wong, K. K., Cheng, R. S. K., Letaief, K. B. and Murch, R. D. (2001) Adaptive antennas at the mobile and base stations in an OFDM/TDMA system, *IEEE Transactions on Communications*, **49**, 195–206.

[67] Luenberger, D. G. (1969) *Optimization by Vector Space Methods*. Wiley, New York.

[68] Bengtsson, M. and Ottersten, B. (2001) Optimal and suboptimal transmit beamforming, in *Handbook of Antennas in Wireless Communications*, L. C. Godara (Ed.) CRC, Boca Raton, FL.

[69] ETSI (2001) Broadband radio access networks (BRAN); HIPERLAN type 2; physical (PHY) layer, ETSI TS 101 475 V1.2.2, pp. 1–41.

[70] ETSI (1998) Channel models for HIPERLAN/2 in different indoor scenarios, ETSI EP BRAN 3ERI085B, pp. 1–8.

[71] Schumacher, L., Kermoal, J. P., Frederiksen, F., Pedersen, K. I., Algans, A. and Mogensen, P. E. (2001) MIMO channel characterization. [Online]. Available: http://www.ist-metra.org, Deliverable D2 V1.1 of IST 1999–11 729 METRA project, pp. 1–57.

[72] Nesterov, Y. and Nemirovsky, A. (1994) Interior-point polynomial methods for convex programming, *SIAM Studies Applied Mathematics*, **13**.

10

Cognitive Networks

Efficient spectrum utilization has attracted significant attention from researchers because most of the allocated spectrum is severely under-utilized. In order to improve spectrum utilization, a new spectrum allocation method, called cognitive radio, is proposed. While in general the terms cognitive radio and cognitive networks include a much broader scale of techniques, based on cognition, in this section we limit our discussion to the network where users are classified into two groups: primary users (PUs) and secondary users (SUs). PUs are licensed users for a given frequency band and have the highest priority to access the allocated band, while the SUs share the bandwidth opportunistically with the PUs only when the bandwidth is not being used currently by the PUs. Therefore, in order to avoid severe interference with the transmission from the PUs, the SUs need to sense channel availability first and then carry out data transmission over idle channels.

In the following we will also use the terms primary network PN for the network of primary users and secondary network SN for the network of secondary users.

10.1 Optimal Channel Sensing in Cognitive Wireless Networks

In this section we will consider a PN consisting of K channels. Without loss of generality, each channel here represents a separate band in the frequency domain and the bandwidth for each channel is assumed to be equal. The time is divided into equal-length slots and all PUs transmit slot by slot in a synchronized manner. The length of each time slot, T_s, is set such that the operation of sensing one channel by the SUs can be completed. For each channel, the occupancy from the PUs is characterized by a two-state (0-idle or 1-busy) discrete-time Markov process.

In general, the sojourn time at the busy state is assumed to be phase type (PH) distributed with parameters $(\alpha, \mathbf{T})$ of order n, where α and $\mathbf{T}$ are a *row vector* and a *square matrix* of order n, respectively. The reason for selecting PH distribution is that most distributions can be represented or approximated by a PH distribution [1]. Every element of $\mathbf{T}$ has the following property $0 \le T_{ij} \le 1$ and $\mathbf{T1} \le \mathbf{1}$ with at least one of the rows strictly less than 1, where $\mathbf{1}$ is a *column vector* of ones of appropriate dimension. Also, the vector α is stochastic, i.e., $\alpha \mathbf{1} = \mathbf{1}$. If $p_b(k)$ is the probability that the busy time will last k time slots, then, according to the PH distribution, we have $p_b(k) = \alpha T^{k-1}\mathbf{t}$ where $\mathbf{t} = \mathbf{1} - \mathbf{T1}$. Similarly, the sojourn time at the idle state is also modeled as a PH distributed random variable with $(\beta, \mathbf{S})$ of order m. β and $\mathbf{S}$ have the same definitions of α and $\mathbf{T}$, respectively. If $p_i(k)$ is defined as the probability that the channel will be idle for k consecutive time slots, then $p_i(k) = \beta S^{k-1}\mathbf{s}$ where $\mathbf{s} = \mathbf{1} - \mathbf{S1}$.

Based on the definitions above, each channel's overall behavior is modeled as a discrete-time semi-Markov process with the state space $\{(0, v) \cup (1, u)\}$, $v = 1, 2, \ldots, m$, $u = 1, 2, \ldots, n$, and the transition matrix

$$\mathbf{P} = \left\| \begin{matrix} \mathbf{S} & \mathbf{s}\alpha \\ \mathbf{t}\beta & \mathbf{T} \end{matrix} \right\| \tag{10.1}$$

If $\pi_{0,i}^t$ and $\pi_{1,j}^t$ are the probabilities that at time t a channel is in state 0 and phase i or state 1 and phase j, respectively, for $i = 1, 2, \ldots, m$ and $j = 1, 2, \ldots, n$ and $\boldsymbol{\pi}_0^t = [\pi_{0,1}^t, \pi_{0,2}^t, \ldots, \pi_{0,m}^t]$, $\boldsymbol{\pi}_1^t = [\pi_{1,1}^t, \pi_{1,2}^t, \ldots, \pi_{1,n}^t], \boldsymbol{\pi}^t = [\boldsymbol{\pi}_0^t, \boldsymbol{\pi}_1^t]$, then we have $\boldsymbol{\pi}^{t+1} = \boldsymbol{\pi}^t \mathbf{P}$.

The stationary distribution of the defined semi-Markov process, $\boldsymbol{\pi} = \boldsymbol{\pi}^t|_{t \to \infty}$, can be calculated as $\boldsymbol{\pi} = \boldsymbol{\pi} P, \quad \boldsymbol{\pi} \mathbf{1} = \mathbf{1}$.

In a cognitive network SUs share the channels opportunistically with the PUs. In general the PUs and the SUs operate asynchronously. We will assume that transmission among the SUs is synchronized with the PUs by following the same time slot structure. However, if the channel is sensed to be idle, the SU will transmit packets over $M > 1$ time slots before sensing the channel again. Since the behavior of the PUs varies on a smaller time scale compared to the sensing process, asynchronism becomes possible, i.e. the channel may become busy during the transmission of the SUs even the channel was sensed to be idle. Obviously, such a scenario is accurate when M is large.

At the beginning of each time slot, the SUs waiting for transmissions need to sense the status of the channels first. In order to guarantee sensing accuracy, each channel sensing will cost one time slot. In addition, owing to the constraints on hardware and power consumption, at any sensing instant, each SU can only sense and ultimately use a portion of all available channels. For the purpose of compatibility, each secondary user is equipped with one single radio, i.e. in each time slot, the SU can only either sense one channel or transmit packets. In this section, each SU will do the sensing independently and no sensing error will be considered. Modeling sensing errors will be discussed later in this chapter.

Channel sensing policies will be discussed for different application scenarios. Since the cooperation of SUs is not always available due to the lack of control channels and may increase network complexity significantly, in this section, we will focus our discussion first on one SU only. Cooperative sensing will be discussed later in this chapter. We first consider the situation where only one channel is sensed for each possible transmission, i.e., $L = 1$, and then extend our discussion to sequential sensing with $L > 1$.

If r_b and r_i represent the probabilities of the channel in busy and idle states, respectively then, based on the Markov model defined earlier, we have $r_b = \boldsymbol{\pi}_1 \mathbf{1}$ and $r_i = \boldsymbol{\pi}_0 \mathbf{1}$.

Since the PUs and the SU operations are asynchronous, it is possible that the channel is idle at the sensing instant but becomes busy during the transmission of the SUs. Such an event will cause a collision with the PUs. In order to model such events, we define two additional probabilities. First the probability that the channel is idle when it was sensed but becomes busy before the end of M slots denoted as $r_{i,b}$, and the probability that the channel was idle when it was sensed and it remained idle for next $M - 1$ time slots denoted as $r_{i,i}$. At the same time $r_{i,i}$ is the probability that the SU makes a successful connection without interference to the PUs. If ξ denotes the channel state ($\xi = 0$ for idle and $\xi = 1$ for busy), and η denotes the number of consecutive idle slots, then we have

$$r_{i,i}(M) = P(\xi = 0, \text{ no} \geq M) = \boldsymbol{\pi}_0 \mathbf{S}^{M-1}(\mathbf{I} - \mathbf{S})^{-1}\mathbf{s} \tag{10.2}$$

and

$$r_{i,b}(M) = \boldsymbol{\pi}_0(\mathbf{I} - \mathbf{S}^{M-1})(\mathbf{I} - \mathbf{S})^{-1}\mathbf{s} \tag{10.3}$$

with $r_b + r_{i,b}(M) + r_{i,i}(M) = 1$. Similarly, we can also derive the probability that an SU finds the channel busy and the channel becomes free after w units as

$$r_{b,v} = \boldsymbol{\pi}_1(\mathbf{I} - \mathbf{T}^w)(\mathbf{I} - \mathbf{T})^{-1}\mathbf{t}, \quad v = 1, 2, \ldots \tag{10.4}$$

The homogeneous scenario refers to the case when all channels' behavior follow same statistics. For multiple identical and independent channels, we can obtain the following probabilities:

a) The probability of at least one channel idle at an arbitrary time with a possibility of no interference during transmission.

$$a_s(K) = \sum_{k=1}^{K} \binom{K}{k} (r_{i,i})^k (r_b + r_{i,b})^{K-k} \tag{10.5}$$

b) The probability of at least one channel idle but with an interference occurring during transmission.

$$a_I(K) = \sum_{k=1}^{K} \binom{K}{k} (r_{i,b})^k (r_b)^{K-k} \tag{10.6}$$

c) The probability of no channel idle during transmission.

$$a_f(K) = (r_b)^K \tag{10.7}$$

For homogeneous channels, random channel selection is obviously the optimal sensing policy. The normalized throughput of the channel is defined as

$$A = \frac{number\ of\ slots\ for\ data\ transmission}{number\ of\ slots\ for\ one\ transmission\ trial} r_{i,i}(M) = \frac{M-1}{M} r_{i,i}(M) \tag{10.8}$$

where $r_{ii}(M)$ is the successful assess probability for any one of the channels.

The *heterogeneous channels* have different statistics and we assume that the busy and idle periods of a channel with index k, $k = 1, 2, \ldots, K$ are governed by PH distributions represented by $(\alpha^{(k)}, \mathbf{T}^{(k)})$ and $(\beta^{(k)}, \mathbf{S}^{(k)})$ of the order of $n^{(k)}$ and $m^{(k)}$, respectively. If channel k is selected, then the achievable normalized throughput is defined as $A^{(k)}$, which can be calculated from (10.8).

The channel sensing policies depend on the optimization criteria.

If we want to maximize the SU's throughput, then the optimal channel, k^*, is $k^* = \arg\max_k A^{(k)}$. This is the most aggressive approach which ignores the possible interference with the PUs.

If we want to minimize the interference introduced from the SU to the PUs then we choose channel k^* such that $k^* = \arg\max_k r_{i,b}^{(k)}(M)$. This is the most conservative approach which does not take into account the throughput achieved by the SU.

The previous polices are either too aggressive or too conservative. For a practical system, a good channel sensing policy should consider the following two aspects jointly: (a) Some tolerance of the interference from the SU is allowed in the PUs; and (b) the SU tries to access any possible opportunities to enlarge its transmission throughput. For this reason we define the objective function as

$$F^{(k)}(a) = aA^{(k)} - (1-a)r_{i,b}^{(k)}(M) \tag{10.9}$$

where a is a parameter taking values between 0 and 1. Now, in order to balance the tradeoff between throughput and interference, we choose channel k^* as $k^* = \arg\max_k F^{(k)}(a)$.

Multiple Channel Sensing for Heterogeneous Channels $(L > 1)$ are used when the SU can exploit a number of channels simultaneously by applying technologies such as orthogonal frequency division multiplexing (OFDM) even though it is equipped with only one radio.

For heterogeneous cases, channel sensing policies need to determine not only the set of channels, but also the order amongst them. If X denotes the number of time slots used for sensing, then, if X is given, the optimal channel sensing policy is to find a ordered set $\mathbf{a} = \{a_1, a_2, \ldots, a_X\}$, where a_j,

$j = 1, 2, \ldots, X$, denotes the channel index. If the overall normalized throughput is defined as

$$F(\mathbf{a}) = \sum_{j=1}^{X} \frac{M-j}{M} r_{i,i}^{(a_j)}(M - j + 1) \tag{10.10}$$

then, the optimal channel sensing can be formulated by the following optimal assignment problem.

$$\begin{cases} \max_{\mathbf{a}} F(\mathbf{a}) \\ s.t.\ a_i \neq a_j,\ \forall i \neq j \end{cases} \tag{10.11}$$

In (10.11), the condition means that each channel can only be sensed once. The problem defined in (10.11) can be solved by exclusive searching. However, the computational complexity of exclusive searching includes the calculation of overall throughput for $K!/X!$ times, which is huge for a large value of K. In order to find the solution simply, an algorithm, based on Munkres algorithm [3], was introduced in [2].

10.2 Optimal Sequential Channel Sensing

In this chapter, we discuss the multi-channel cognitive medium access control problem, in which multiple potential channels (i.e., frequency bands) are available. We consider a cognitive radio network with *opportunistic transmissions* where the secondary user not only senses a channel to decide whether it is free, but also estimates the channel coefficient to decide the transmission rate. If a channel is sensed to be free, but the channel quality between the secondary transceiver pair is not satisfactory, the secondary user may still skip this channel and keep sensing other channels. The opportunistic transmission creates a new degree of freedom for the secondary user. The goal of the secondary user is to find a free and good channel as quickly as possible. In this case, the channel sensing problem can be formulated as an optimal stopping rule problem [4] if the channel sensing order is determined in advance. In this section we discuss an optimal channel sensing order such that the user achieves the maximal gain.

The *Network Model* assumes a number, N, of potential channels having indices from 1 to N. The secondary user is operating in a time-slotted fashion, where the length of each time slot is T. The secondary user also has a sensing order $(s_1, s_2, \ldots, s_N)$, which is a permutation of the set $\{1, 2, \ldots, N\}$. In a given time slot, the secondary user senses the channels sequentially according to the sensing order, until it stops at a channel based on a specific criterion (e.g. the channel is sensed to be free and has acceptable channel quality), and transmits its information in that channel during the remainder of that time slot. It is assumed that accurate channel sensing is achieved, and there is no sensing error. The time needed for sensing a channel and estimating the channel gain, which is assumed to be the same in all the channels, equals τ seconds with $\tau \ll T$.

In each time slot, channel $i (1 \leq i \leq N)$ is free (i.e. no primary user activity) with probability $\theta_i (\in (0, 1))$, which is the availability probability of that channel. For each channel, the *busy/free* status in each slot is assumed to be independent of the status in other slots, and also to be independent of the status in other channels. For each channel, including fading, the signal-to-noise ratio (SNR) is fixed within a time slot, and changes randomly at the beginning of the next time slot. The channel SNR γ is assumed to be independent and identically distributed across time slots and across different channels, with a common probability density function (pdf) denoted by $h_{SNR}(\gamma)$. Opportunistic transmissions are used. If the secondary user decides to transmit in a free channel i, the achievable transmission rate is $f(SNR_i)$ where SNR_i is the SNR of the secondary user in channel i, and $f(\cdot)$ is a non-descending function mapping SNR to the transmission rate. With this model, the channel sensing problem becomes an optimal stopping problem [4–6] as described below.

If channel s_i is sensed to be busy, then the secondary user proceeds to sense channel s_{i+1}. When channel s_i is sensed to be free, the secondary user will transmit at channel s_i as long as the achieved transmission rate at channel s_i is greater than the expected rate if the user proceeds to the next channel (i.e., channel s_{i+1}). Hence, the *utility* (throughput) at channel s_i is given by

$$
\begin{aligned}
u_i &= c_i f(SNR_{s_i}), \text{ if } i = N \text{ (stop at channel } s_i) \\
&= c_i f(SNR_{s_i}), \text{ if } i \neq N \text{ and } c_i f(SNR_{s_x}) > U_{i+1} \text{ (stop at channel } s_i) \\
&= U_{i+1}, \text{ otherwise (proceed to channel } s_{i+1})
\end{aligned}
\tag{10.12}
$$

where c_i is the *effectiveness* of transmitting at channel s_i, given by $c_i = 1 - i\tau/T$, and $U_{i+1}(i \leq N - 1)$ is the expected utility at channel s_{i+1} if the user proceeds to that channel, given by

$$
\begin{aligned}
U_{i+1} &= \theta_{s_{i+1}} E[u_{i+1}] + (1 - \theta_{s_{i+1}}) U_{i+2}, \text{ if } i < N - 1 \\
&= \theta_{i+1} E[u_{i+1}], \qquad\qquad\qquad\quad \text{ if } i = N - 1
\end{aligned}
\tag{10.13}
$$

where $E[\cdot]$ denotes expectation. The sequence $\{U_1, U_2, \ldots, U_N\}$ can be obtained recursively from U_N to U_1 using (10.12) and (10.13) [5].

10.3 Optimal Parallel Multiband Channel Sensing

In this section, multiband joint detection is discussed, which detects the primary signals jointly over multiple frequency bands rather than over one band at a time, as was the case in the previous section.

A primary communication system is assumed to operate over a wideband channel that is divided into K non-overlapping narrowband subbands (e.g. multicarrier modulation based). In a particular geographical region and within a particular time interval, some of the K subbands might not be used by the primary users and are available for opportunistic spectrum access.

The detection problem on subband k is that of choosing between a hypothesis $H_{0,k}(0)$, which represents the absence of primary signals, and $H_{1,k}(1)$, which represents the presence of primary signals. The underlying hypothesis vector is a binary representation of the subbands that are allowed or prohibited from opportunistic spectrum access.

The crucial task of spectrum sensing is to sense the K subbands and identify spectral holes for opportunistic use. For simplicity, we assume that the upper-layer protocols, e.g. the medium access control (MAC) layer, can guarantee that all cognitive radios stay silent during the detection interval such that the only spectral power remaining in the air is emitted by the primary users. In this section, instead of considering one band at a time, as in the previous section, we use a multiband joint detection technique, which takes into account jointly the detection of primary users across multiple frequency bands.

Consider a multipath fading environment, where $h(l), l = 0, 1, \ldots, L - 1$, represents the discrete-time channel impulse response between the primary transmitter and a CR receiver with L denoting the number of resolvable paths. The received baseband signal at the RF front-end can be written as $r(n) = \sum_{l=0}^{L-1} h(l)s(n - l) + v(n)$ where $s(n)$ represents the primary transmitted signal (with cyclic prefix) at time n and $v(n)$ is additive complex white Gaussian noise with zero mean and variance σ_v^2, i.e., $v(n) \sim CN(0, \sigma_v^2)$. In a multipath fading environment, the wideband channel exhibits frequency-selective features and its discrete frequency response can be obtained through a K-point fast Fourier transform (FFT) $(K \geq L)$: $H_k = \left(1/\sqrt{K}\right) \sum_{n=0}^{L-1} h(n)e^{-j2\pi nk/K}, \ k = 1, 2, \ldots K$. In the frequency domain, the received signal at each subchannel can be represented by its discrete Fourier transform (DFT):

$$
R_k = \left(1/\sqrt{K}\right) \sum_{n=0}^{K-1} r(n)e^{-j2\pi nk/K} = H_k S_k + V_k, \ k = 1, 2, \ldots, K
$$

where S_k is the primary transmitted signal at subchannel k and $V_k = \left(1/\sqrt{K}\right) \sum_{n=0}^{K-1} v(n)e^{-j2\pi nk/K}, \ k = 1, 2, \ldots, K$ is the received noise represented in the frequency domain. The random variables $\{V_k\}$ are independent and distributed normally with zero means and variances σ_v^2, i.e., $V_k \sim CN(O, \sigma_v^2)$, since $v(n) \sim CN(O, \sigma_v^2)$ and the DFT is a linear unitary operation. Without loss of generality, we assume that the transmitted signal S_k, channel gain H_k, and additive noise V_k are independent of each other.

The sensing algorithm described in the sequel needs to know only the noise power σ_v^2 and the squared values of the channel frequency responses $\{|H_k|^2\}$, which can be estimated in practice.

We start from signal detection in a single narrowband subband, which will constitute a building block for the multiband joint detection procedure. To decide whether the k th subband is occupied or not, we test the following binary hypotheses: $H_{0,k} : R_k = V_k$ and $H_{1,k} : R_k = H_k S_k + V_k$, where R_k is the secondary received signal, S_k is the primary transmitted signal, and H_k is the channel gain between the primary transmitter and the secondary receiver.

For each subband k, we compute the summary statistic as the sum of received signal energy over an interval of M samples, i.e.

$$Y_k \overset{\Delta}{=} \sum_{n=1}^{M} |R_k(n)|^2, \; k = 1, 2, \ldots, K$$

and the decision rule is chosen as

$$H_{0,k} \text{ if } Y_k < \gamma_k \text{ and } H_{1,k} \text{ if } Y_k > \gamma_k, \quad k = 1, 2, \ldots, K$$

where γ_k is the decision threshold of subband k. For simplicity, we assume that the transmitted signal in each subband has unit power, i.e. $E[|S_k|^2] = 1$.

For large M, the statistics $\{Y_k\}_{k=1}^{K}$ are approximately normally distributed with means

$$E[Y_k] = M\sigma_v^2, \; H_{0,k} = M(\sigma_v^2 + |H_k|^2), \; H_{1,k}$$

and variances

$$Var(Y_k) = 2M\sigma_v^4, \; H_{0,k} = 2M(\sigma_v^2 + 2|H_k|^2)\sigma_v^2, \; H_{1,k}$$

Using the decision rule as defined before, the probabilities of false alarm and detection in the kth subband can be approximately expressed as

$$P_f^{(k)}(\gamma_k) = Pr(Y_k > \gamma_k | H_{0,k}) = Q\left(\frac{\gamma_k - M\sigma_v^2}{\sigma_v^2 \sqrt{2M}}\right) \tag{10.14}$$

and

$$P_d^{(k)}(\gamma_k) = Pr(Y_k > \gamma_k | H_{1,k}) = Q\left(\frac{\gamma_k - M(\sigma_v^2 + |H_k|^2)}{\sigma_v \sqrt{2M(\sigma_v^2 + 2|H_k|^2)}}\right) \tag{10.15}$$

Note that the SNR of such an energy detector is defined as $SNR = |H_k|^2/\sigma_v^2$, which plays an important role in determining detection performance. The choice of the threshold γ_k leads to a tradeoff between the probability of false alarm and the probability of missed detection, $P_m^{(k)}(\gamma_k) = 1 - P_d^{(k)}(\gamma_k)$. Specifically, a higher threshold will result in a smaller probability of false alarm, but a larger probability of miss, and vice versa.

A typical assumption in a cognitive radio network is that if primary signals are detected, the secondary users will not use the corresponding channel, and if no primary signals are detected, then the corresponding frequency band will be used by secondary users.

The probabilities of false alarm and miss have unique implications for such CR networks. Low probabilities of false alarm are necessary to maintain high spectral utilization in CR systems, since a false alarm would prevent the unused spectral segments from being accessed by secondary users. On the other hand, the probability of missed detection measures the interference of secondary users to the primary users, which should be limited in opportunistic spectrum access.

The *multiband joint detection* framework for wideband spectrum sensing is designed to find the optimal threshold vector $\boldsymbol{\gamma} = [\gamma_1, \gamma_2, \ldots, \gamma_K]^T$ so that the cognitive radio system can make efficient use of the unused spectral segments without causing harmful interference to the primary users. For a given threshold vector $\boldsymbol{\gamma}$, the probabilities of false alarm and detection can be represented compactly as

$$\mathbf{P}_f(\boldsymbol{\gamma}) = [P_f^{(1)}(\gamma_1), \ P_f^{(2)}(\gamma_2), \ldots P_f^{(K)}(\gamma_K)]^T$$

and

$$\mathbf{P}_d(\boldsymbol{\gamma}) = [P_d^{(1)}(\gamma_1), \ P_d^{(2)}(\gamma_2), \ldots P_d^{(K)}(\gamma_K)]^T$$

Similarly, the probabilities of missed detection can be written in a vector as

$$\mathbf{P}_m(\boldsymbol{\gamma}) = [P_m^{(1)}(\gamma_1), \ P_m^{(2)}(\gamma_2), \ldots P_m^{(K)}(\gamma_K)]^T$$

The vector $\mathbf{P}_m$ can be expressed as

$$\mathbf{P}_m(\gamma) = 1 - \mathbf{P}_d(\gamma)$$

where **1** denotes the all-one vector.

Consider a CR device sensing the K narrowband subbands in order to make use of the unused ones for opportunistic transmission. Let r_k denote the throughput achievable over the kth subband if used by secondary users, and $\mathbf{r} = [r_1, r_2, \ldots, r_K]^T$. If the transmit power and the channel gains between secondary users are known, $\mathbf{r}$ can be estimated using the Shannon capacity formula. Since $1 - P_f^{(k)}$ measures the opportunistic spectral utilization of subband k, the aggregate opportunistic throughput of the CR system can be defined as

$$R(\boldsymbol{\gamma}) = \mathbf{r}^T[1 - \mathbf{P}_f(\boldsymbol{\gamma})] \tag{10.16}$$

which is a function of the threshold vector $\boldsymbol{\gamma}$. Due to the inherent tradeoff between $P_f^{(k)}(\gamma_k)$ and $P_m^{(k)}(\gamma_k)$, maximizing the sum rate $R(\gamma)$ will result in large $P_m(\gamma)$, thus causing harmful interference for primary users.

The interference for primary users should be limited in a CR network. For a wideband primary communication system, the effect of interference induced by CR devices can be characterized by a cost for each primary user transmitting over the corresponding subbands, i.e., $\mathbf{c} = [c_1, c_2, \ldots, c_K]^T$, where c_k indicates the cost incurred if the primary user in subband k is interfered with. In a special case where the jth primary user is equally important, we may have $\mathbf{c} = \mathbf{1}$. Suppose that J primary users share a portion of the K subbands and each primary user occupies a subset S_j of subbands. The aggregate interference to primary user j can be expressed as

$$\sum_{i \in S} c_i P_m^{(i)}(\gamma_i) \tag{10.17}$$

This expression models, for example, the situation arising in a multiuser orthogonal frequency division multiplexing (OFDM) system, where various primary users have different levels of priority. Alternatively, c_k can be defined as a function of the bandwidth of subband k since in some applications each particular subband does not have to occupy an equal amount of bandwidth.

The objective is to find the optimal thresholds $\{\gamma_k\}_{k=1}^K$ for the K subbands in order to maximize collectively the aggregate opportunistic throughput subject to some interference constraints for each primary user. As such, the opportunistic rate optimization problem in the context of a multiuser primary system can be formulated as

$$\max_{\gamma} R(\gamma) \tag{10.18a}$$

$$s.t. \sum_{i \in S_j} c_i P_m^{(i)}(\gamma_i) \le \varepsilon_j, \ j = 1, 2, \dots J \tag{10.18b}$$

$$\mathbf{P}_m(\gamma) \preceq \alpha \tag{10.18c}$$

$$\mathbf{P}_f(\gamma) \preceq \beta \tag{10.18d}$$

The constraint (10.18c) limits the interference in each subband with $\alpha = [\alpha_1, \alpha_2, \dots, \alpha_K]^T$, and the constraint (10.18d) dictates that each subband should be able to achieve a minimum opportunistic spectral utilization given by $[1 - \beta_1, 1 - \beta_2, \dots, 1 - \beta_K]^T$. For a single-user primary system where all the subbands are used by one primary user, we have $J = 1$.

Some additional factors need to be considered in multi-band joint detection. The subband with a higher opportunistic rate r_k should have a higher threshold γ_k (i.e., a smaller probability of false alarm) such that it can be best used by CRs. The subband that carries a higher priority primary user should have a lower threshold γ_k (i.e., a smaller probability of missed detection) in order to prevent opportunistic access by secondary users. A little compromise on those subbands carrying less important primary users might boost the opportunistic rate considerably. Thus, in the determination of the optimal threshold vector, it is necessary to strike a balance among the channel conditions, the opportunistic throughput, and the relative priority of each subband.

The objective and constraint functions in (10.18) are generally nonconvex, making it difficult to efficiently solve for the global optimum. In most cases, suboptimal solutions or heuristics have to be used. However, this seemingly nonconvex problem can be made convex by exploiting hidden convexity properties and reformulating the problem.

The fact that the Q-function is monotonically non-increasing allows us to transform the constraints in (10.18c) and (10.18d) into linear constraints. Specifically, from (10.18c), we obtain

$$1 - P_d^{(k)}(\gamma_k) \le \alpha_k \ k = 1, 2, \dots, K$$

Substituting expression for $P_d^{(k)}(\gamma_k)$ into this equation gives $\gamma_k \le \gamma_{\max,k}, \ k = 1, 2, \dots, K$ where

$$\gamma_{\max,k} \triangleq M(\sigma_v^2 + |H_k|^2) + \sigma_v \sqrt{2M(\sigma_v^2 + 2|H_k|^2)} Q^{-1}(1 - \alpha_k).$$

Similarly, from (10.18d) we have

$$\gamma_k \ge \gamma_{\min,k}, \ k = 1, 2, \dots, K$$

where $\gamma_{\min,k} = \sigma_v^2[M + \sqrt{2M} Q^{-1}(\beta_k)]$. Consequently, the original problem (10.18) has the following equivalent form

$$\min_{\gamma} \sum_{k=1}^{K} r_k P_f^{(k)}(\gamma_k)$$

$$s.t. \sum_{i \in S_j} c_i P_m^{(i)}(\gamma_i) \le \varepsilon_j, \ j = 1, 2, \dots, J \tag{10.19}$$

$$\gamma_{\min,k} \le \gamma_k \le \gamma_{\max,k}, k = 1, 2, \dots K$$

Although the last constraint is linear, the problem is still non-convex. Fortunately, it can be transformed into a tractable convex optimization problem for low probabilities of false alarm and miss. To establish the transformation, we need the following results which are derived in [7].

1. The function $P_f^{(k)}(\gamma_k)$ is convex in γ_k if $P_f^{(k)}(\gamma_k) \leq 1/2$.
2. The function $P_m^{(k)}(\gamma_k)$ is convex in γ_k if $P_m^{(k)}(\gamma_k) \leq 1/2$.

Since the non-negative weighted sum of a set of convex functions is also convex [8] the problem (10.19) then becomes a convex program for: $0 < \alpha_k \leq 1/2$ and $0 < \beta_k < 1/2, k = 1, 2, \ldots, K$. This regime of probabilities of false alarm and missed detection is of practical interest for achieving rational opportunistic throughput and interference levels in CR networks.

Alternatively, we can formulate the multiband problem into another optimization problem that minimizes the interference from CRs to the primary communication system subject to some constraints on the aggregate opportunistic throughput, i.e.

$$
\begin{aligned}
&\min_{\gamma} \; \mathbf{c}^T \mathbf{P}_m(\gamma) \\
&s.t. \; \mathbf{r}^T[\mathbf{1} - \mathbf{P}_f(\gamma)] \geq \delta \\
&\quad\; \mathbf{P}_m(\gamma) \preceq \alpha \\
&\quad\; \mathbf{P}_f(\gamma) \preceq \beta
\end{aligned}
\tag{10.20}
$$

where β is the minimum required aggregate opportunistic throughput. Like (10.18a), this problem can be transformed into a convex optimization problem by enforcing the conditions discussed earlier.

10.4   Collaborative Spectrum Sensing

In this section, we present a cooperation framework for wideband spectrum sensing, within which CRs can exploit spatial diversity by exchanging local sensing results in order for the secondary network to obtain a more accurate estimate of the unused frequency bands.

Suppose that N spatially distributed CRs sense a wide frequency band collaboratively, in order to find unused spectral segments for opportunistic communication. By combining the summary statistics from individual CRs, a fusion center, which could be one of the CRs, makes the final decision on the presence or absence of primary signals in each of the K subbands. A linear weighting fusion scheme can be used for the final decision. It is assumed that there is a control channel, through which the summary statistics of individual secondary users are transmitted to the fusion center.

Let $Y_k(n)$ denote the summary statistic of the nth secondary user in the kth subband. For each subband $\mathbf{Y}_k = [Y_k(1), Y_k(2), \ldots, Y_k(N)]^T$ are the statistics from individual secondary users. The statistics across the K subbands can be combined compactly in matrix form as follows:

$$
\mathbf{Y} = [\mathbf{Y}_1 \, \mathbf{Y}_2 \ldots \mathbf{Y}_K] = \begin{Vmatrix} Y_1(1) & Y_2(1) & \cdots & Y_K(1) \\ Y_1(2) & Y_2(2) & \cdots & Y_K(2) \\ | & | & \ddots & | \\ Y_1(N) & Y_2(N) & \cdots & Y_K(N) \end{Vmatrix}
\tag{10.21}
$$

The global test statistic is obtained by combining linearly the summary statistics from spatially distributed CRs in each subband k

$$
z_k = \sum_{n=1}^{N} w_k(n) Y_k(n) = \mathbf{w}_k^T \mathbf{Y}_k
\tag{10.22}
$$

where $\mathbf{w}_k = [w_k(1), w_k(2), \ldots, w_k(N)]^T$ are the combining coefficients for subband k. These coefficients can be written compactly as

$$W = [\mathbf{w}_1 \; \mathbf{w}_2 \ldots \mathbf{w}_K] = \left\|\begin{matrix} w_1(1) & w_2(1) & \cdots & w_K(1) \\ w_1(2) & w_2(2) & \cdots & w_K(2) \\ | & | & \ddots & | \\ w_1(N) & w_2(N) & \cdots & w_K(N) \end{matrix}\right\| \tag{10.23}$$

where $w_k(n) \geq 0$, for all k, n.

Since the elements in $\mathbf{Y}_k$ are normally distributed, the test statistics z_k, $k = 1, 2, \ldots, K$, are also normally distributed with means

$$E[z_k] = \begin{cases} M\sigma_v^2 \mathbf{w}_k^T \mathbf{1}, & \text{for } H_{0,k} \\ M\mathbf{w}_k^T(\sigma_v^2 \mathbf{1} + \mathbf{G}_k), & \text{for } H_{1,k} \end{cases} \tag{10.24}$$

and variances

$$Var(z_k) = \begin{cases} 2M\sigma_v^4 \mathbf{w}_k^T \mathbf{w}_k, & \text{for } H_{0,k} \\ 2M\sigma_v^2 \mathbf{w}_k^T \Sigma_k \mathbf{w}_k, & \text{for } H_{1,k} \end{cases} \tag{10.25}$$

where

$$\Sigma_k = E[\mathbf{Y}_k \mathbf{Y}_k^T] = \sigma_v^2 \mathbf{I} + 2diag(\mathbf{G}_k)$$

is a diagonal matrix assuming that the elements of Y_k are independent and

$$G_k = [|H_k(1)|^2, \; |H_k(2)|^2, \ldots, |H_k(N)|^2]^T$$

are the squared magnitudes of the channel gains between the primary transmitter and secondary receivers on each subband. Even though Σ_k becomes a nondiagonal matrix if the elements of $\mathbf{Y}_k$ are not independent, the following derivation is still valid since Σ_k is a positive semidefinite matrix.

In order to decide the presence or absence of the primary signal in subband k, we use the following binary test:

$$Decision = \begin{cases} H_{1,k} & \text{if } z_k \geq \gamma_k, \\ H_{0,k} & \text{if } z_k \leq \gamma_k, \end{cases} \quad k = 1, 2, \ldots, K \tag{10.26}$$

The probabilities of false alarm and detection are now given by

$$P_f^{(k)}(\mathbf{w}_k, \gamma_k) = Q\left(\frac{\gamma_k - M\sigma_v^2 \mathbf{w}_k^T \mathbf{1}}{\sigma_v^2 \sqrt{2M\mathbf{w}_k^T \mathbf{w}_k}}\right) \tag{10.27a}$$

and

$$P_d^{(k)}(\mathbf{w}_k, \gamma_k) = Q\left(\frac{\gamma_k - M\mathbf{w}_k^T(\sigma_v^2 \mathbf{1} + \mathbf{G}_k)}{\sigma_v \sqrt{2M\mathbf{w}_k^T \Sigma_k \mathbf{w}_k}}\right) \tag{10.27b}$$

The goal is to maximize the aggregate opportunistic throughput of the K subbands which is now a function of both the threshold vector γ and the weight coefficient matrix W, i.e. $R(\mathbf{W}, \gamma) = \mathbf{r}^T[1 - \mathbf{P}_f(\mathbf{W}, \gamma)]$.

The *collaborative joint detection problem* is formulated as

$$\max_{\mathbf{W},\boldsymbol{\gamma}} R(\mathbf{W},\boldsymbol{\gamma})$$
$$s.t. \quad \mathbf{c}^T \mathbf{P}_m(\mathbf{W},\boldsymbol{\gamma}) \leq \varepsilon$$
$$\mathbf{P}_m(\mathbf{W},\boldsymbol{\gamma}) \preceq \boldsymbol{\alpha} \tag{10.28}$$
$$\mathbf{P}_f(\mathbf{W},\boldsymbol{\gamma}) \preceq \boldsymbol{\beta}$$

Finding the exact optimal solution for the problem above is difficult, since for any subband the probabilities of false alarm and miss are neither convex nor concave functions of the weight coefficients w_k and the test threshold γ_k according to (10.27).

In the following, we will present two efficient methods for solving the weight coefficients $\mathbf{W}$ and the thresholds $\boldsymbol{\gamma}$, which lead to near-optimal solutions for (10.28) [7]. We still assume the practical conditions $0 < \alpha_k \leq 1/2$ and $0 < \beta_k < 1/2, k = 1, 2, \ldots, K$ unless stated explicitly otherwise.

Joint optimization of $\mathbf{W}$ and $\boldsymbol{\gamma}$, is based on reformulation of (10.28) into an equivalent form with a convex feasible set and an objective function lower bounded by a concave function. Through maximizing the lower bound of the objective function, we are able to obtain a good approximation to the optimal solution of the original problem.

Substituting (10.27a) into the last constraint of (10.28), we have

$$Q^{-1}(\beta_k)\sqrt{2M\mathbf{w}_k^T\mathbf{w}_k} \leq \frac{\gamma_k}{\sigma_v^2} - M\mathbf{w}_k^T\mathbf{1}, k = 1, 2, \ldots, K \tag{10.29}$$

where $Q^{-1}(\beta_k) \geq 0$ since $\beta_k \leq 1/2$. From (10.27b), the second constraint in (10.28) can be expressed as

$$\sqrt{2M\mathbf{w}_k^T\boldsymbol{\Sigma}_k\mathbf{w}_k} \leq \frac{\gamma_k - M\mathbf{w}_k^T(\sigma_v^2\mathbf{1} + \mathbf{G}_k)}{\sigma_v Q^{-1}(1 - \alpha_k)} \tag{10.30}$$

for $k = 1, 2, \ldots, K$, since $\alpha_k \leq 1/2$ and $Q^{-1}(1 - \alpha_k) \leq 0$. From (10.29) and (10.30) we have

$$M\sigma_v^2\mathbf{w}_k^T\mathbf{1} \preceq \gamma_k \preceq M\mathbf{w}_k^T(\sigma_v^2\mathbf{1} + \mathbf{G}_k) \tag{10.31}$$

Since the left-hand side on the constraint (10.29) is a convex function and the right-hand side is a linear function, (10.29) defines a convex set for $\{\gamma_k, w_k\}$. Similarly, (10.30) is also a convex constraint.

We now reformulate problem (10.28) by introducing a new variable

$$\mu_k = \sigma_v \sqrt{2M\mathbf{w}_k^T\boldsymbol{\Sigma}_k\mathbf{w}_k}$$

Define $\gamma' = \gamma_k/\mu_k$ and $\mathbf{w}_k' = \mathbf{w}_k/\mu_k$. The constraints (10.29) and (10.30) can be written as

$$Q^{-1}(\beta_k)\sqrt{2M\mathbf{w}_k'^T\mathbf{w}_k'} \leq \frac{\gamma_k'}{\sigma_v^2} - M\mathbf{1}^T\mathbf{w}_k' \tag{10.32}$$

and

$$\gamma_k' - M(\sigma_v^2\mathbf{1} + \mathbf{G}_k)^T\mathbf{w}_k' \leq \sigma_v Q^{-1}(1 - \alpha_k) \tag{10.33}$$

where (10.33) is actually a linear constraint. The first constraint in (10.28) now becomes

$$\mathbf{c}^T\mathbf{P}_m(\mathbf{W}, \boldsymbol{\gamma}) = \mathbf{c}^T[\mathbf{1} - \mathbf{P}_d(\mathbf{W}, \boldsymbol{\gamma})] \leq \varepsilon$$
$$\Leftrightarrow \mathbf{1}^T\mathbf{c} - \sum_{k=1}^{K} c_k Q[\gamma_k' - M(\sigma_v^2\mathbf{1} + \mathbf{G}_k)^T\mathbf{w}_k'] \leq \varepsilon \tag{10.34}$$

which can be shown to be convex [7]. By changing the variables $\mathbf{W}' = [\mathbf{w}_1', \mathbf{w}_2', \ldots, \mathbf{w}_K']^T$ and $\boldsymbol{\gamma}' = [\gamma_1', \gamma_2', \ldots, \gamma_K']^T$, the objective function in (10.28) becomes

$$
\begin{aligned}
R(\mathbf{W}, \boldsymbol{\gamma}) = \mathbf{r}^T[\mathbf{1} - \mathbf{P}_f(\mathbf{W}, \boldsymbol{\gamma})] &= \sum_{k=1}^{K} r_k \left(1 - Q\left[\frac{\gamma_k - M\sigma_v^2 \mathbf{w}_k^T \mathbf{1}}{\sigma_v^2 \sqrt{2M \mathbf{w}_k^T \mathbf{w}_k}} \right] \right) \\
&= \mathbf{1}^T \mathbf{r} - \sum_{k=1}^{K} r_k Q\left[\left(\frac{\gamma_k'}{\sigma_v^2} - M\mathbf{1}^T \mathbf{w}_k' \right) \times \sqrt{\sigma_v^2 + \frac{2\mathbf{w}_k'^T diag(\mathbf{G}_k)\mathbf{w}_k'}{\mathbf{w}_k'^T \mathbf{w}_k'}} \right]
\end{aligned}
\tag{10.35}
$$

From the Rayleigh-Ritz theorem [9], we have

$$
\min_{1 \leq n \leq N} |H_k(n)|^2 \leq \frac{\mathbf{w}_k'^T diag(\mathbf{G}_k)\mathbf{w}_k'}{\mathbf{w}_k'^T \mathbf{w}_k'} \leq \max_{1 \leq n \leq N} |H_k(n)|^2
\tag{10.36}
$$

Now define a new function

$$
g_k(\gamma_k', \mathbf{w}_k') \triangleq Q\left[\left(\frac{\gamma_k'}{\sigma_v} - M\mathbf{1}^T \mathbf{w}_k' \right) \times \sqrt{\sigma_v^2 + 2\min_{1 \leq n \leq N} |H_k(n)|^2} \right]
\tag{10.37}
$$

for $k = 1, 2, \ldots, K$, which is convex [7]. Now, the aggregate rate can be lower bounded as $R(\mathbf{W}, \boldsymbol{\gamma}) \geq \sum_{k=1}^{N} r_k[1 - g_k(\gamma_k', \mathbf{w}_k')]$.

An efficient suboptimal method to solve (10.28) is to maximize the lower bound of its objective function, i.e.

$$
\max_{\mathbf{W}', \boldsymbol{\gamma}} \sum_{k=1}^{K} r_k[1 - g_k(\gamma_k', \mathbf{w}_k')]
\tag{10.38}
$$

$$
\text{s.t.} -\sum_{k=1}^{K} c_k Q[\gamma_k' - M(\sigma_v \mathbf{1} + \mathbf{G}_k)^T \mathbf{w}_k'] \leq \varepsilon - \mathbf{1}^T \mathbf{c}
$$

$$
Q^{-1}(\beta_k)\sqrt{2M\mathbf{w}_k'^T \mathbf{w}_k'} \leq \frac{\gamma_k'}{\sigma_v^2} - M\mathbf{1}^T \mathbf{w}_k', \, k = 1, 2, \ldots K
$$

$$
\gamma_k' - M(\sigma_v^2 \mathbf{1} + \mathbf{G}_k)^T \mathbf{w}_k' \leq \sigma_v Q^{-1}(1 - \alpha_k), \, k = 1, 2, \ldots K
$$

Due to the conditions $0 < \alpha_k \leq 1/2$ and $0 < \beta_k < 1/2, k = 1, 2, \ldots, K$ this problem is a convex optimization problem and can be efficiently solved.

Sequential optimization divides the optimization of the original problem (10.28) into two stages. In the first stage, referred to as *collaborative optimization*, the weight coefficients $\mathbf{W}$ are chosen in order to maximize a performance measure for signal detection. In the second phase, called *spectral optimization*, the values of $\mathbf{W}$ obtained from *collaborative optimization* are fixed and then the thresholds $\boldsymbol{\gamma}$ is optimized across all the subbands.

Collaborative Optimization is based on *modified detection coefficients* [10, 11], defined as

$$
d_k^2(\mathbf{w}_k) = \frac{(E[z_k|H_{1,k}] - E[z_k|H_{0,k}])^2}{Var(z_k|H_{1,k})} = \frac{M(\mathbf{w}_k^T \mathbf{G}_k)^2}{2\sigma_v^2 \mathbf{w}_k^T \boldsymbol{\Sigma}_k \mathbf{w}_k}
\tag{10.39}
$$

which can be interpreted as a signal-to-noise ratio. For any given probability of false alarm, a larger value of $d_k^2(w_k)$ will result in a larger probability of detection if z_k is normally distributed under both hypotheses $H_{0,k}$ and $H_{1,k}$.

In cooperative optimization, we would like to choose the weight coefficients $\mathbf{w}_k$ in order to achieve the maximum *modified detection coefficient* for each subband where the feasible set for maximizing $d_k^2(w_k)$ is unbounded. To obtain a unique solution, we confine the weight vector to be on the unit-norm ball and pose

$$\max_{\mathbf{w}_k} \ d_k^2(\mathbf{w}_k) \tag{10.40}$$

$$s.t. \ ||\mathbf{w}_k||_2 = 1 \text{ for } k = 1, 2, \dots K$$

This problem can be solved following [7]. First, apply the linear transform $\mathbf{w}_k' = \mathbf{\Sigma}_k^{1/2}\mathbf{w}_k$ where $\mathbf{\Sigma}_k^{1/2}$ is the square root of the matrix $\mathbf{\Sigma}$,

$$\mathbf{\Sigma}_k^{1/2} = diag \begin{bmatrix} \sqrt{\sigma_v^2 + 2|H_k(1)|^2} \\ \sqrt{\sigma_v^2 + 2|H_k(2)|^2} \\ | \\ \sqrt{\sigma_v^2 + 2|H_k(N)|} \end{bmatrix} \tag{10.41}$$

Again, from the Rayleigh-Ritz theorem $d_k^2(w_k)$ is upper bounded by

$$d_k^2(\mathbf{w}_k) = \frac{M\mathbf{w}_k'^T \mathbf{\Sigma}_k^{-T/2}\mathbf{G}_k\mathbf{G}_k^T \mathbf{\Sigma}_k^{-1/2}\mathbf{w}_k'}{2\sigma_v^2 \mathbf{w}_k^T \mathbf{w}_k} \leq \frac{M}{2\sigma_v^2}\lambda_{\max}(\mathbf{\Sigma}_k^{-T/2}\mathbf{G}_k\mathbf{G}_k^T \mathbf{\Sigma}_k^{-1/2}) \tag{10.42}$$

where $\lambda_{\max}(\cdot)$ is the maximum eigenvalue of a matrix. Note that the equality in (10.42) is achieved if $\mathbf{w}_k' = \mathbf{\Sigma}_k^{-T/2}\mathbf{G}_k$ which is the eigenvector corresponding to the maximal eigenvalue of the positive semidefinite matrix $\mathbf{\Sigma}_k^{-T/2}\mathbf{G}_k\mathbf{G}_k^T \mathbf{\Sigma}_k^{-1/2}$ Therefore, the optimal solution of (10.40) is given by $\mathbf{w}_k^o = \mathbf{\Sigma}_k^{-1}\mathbf{G}_k/||\mathbf{\Sigma}_k^{-1}\mathbf{G}_k||_2$.

Spectral Optimization is based on substituting the weight vectors $\{\mathbf{w}_k^o\}_{k=1}^K$ obtained in the first subproblem (10.40) into (10.27), so that the probabilities of false alarm and detection become functions of only the threshold γ_k. Following the procedure in (10.18), we can solve the following subproblem for the threshold vector $\boldsymbol{\gamma}$:

$$\min_{\boldsymbol{\gamma}} \sum_{k=1}^K r_k P_f^{(k)}(\mathbf{w}_k^o, \gamma_k) \tag{10.43}$$

$$s.t. \ \sum_{k=1}^K c_k P_m^{(k)}(\mathbf{w}_k^o, \gamma_k) \leq \varepsilon$$

$$\gamma_{\min,k}' \leq \gamma_k \leq \gamma_{\max,k}', \ k = 1, 2, \dots K$$

where

$$\gamma_{\min,k}' = M\sigma_v^2 \mathbf{1}^T \mathbf{w}_k^o + \sigma_v^2 Q^{-1}(\beta_k)\sqrt{2M\mathbf{w}_k^{oT} \mathbf{w}_k^o}$$

and

$$\gamma_{\max,k}' = M(\sigma_v^2 \mathbf{1} + \mathbf{G}_k)^T \mathbf{w}_k^o + \sigma_v Q^{-1}(1 - \alpha_k)\sqrt{2M\mathbf{w}_k^{oT} \mathbf{\Sigma}_k \mathbf{w}_k^o}.$$

As before, the problem is convex and can be solved efficiently.

The collaborative-spectral joint detection problem can be reformulated to minimize the interference subject to some constraint on the aggregate opportunistic throughput δ, i.e.

$$\min_{\mathbf{W}, \boldsymbol{\gamma}} \mathbf{c}^T \mathbf{P}_m(\mathbf{W}, \boldsymbol{\gamma}) \tag{10.44}$$
$$\text{s.t. } R(\mathbf{W}, \boldsymbol{\gamma}) \geq \delta$$
$$\mathbf{P}_m(\mathbf{W}, \boldsymbol{\gamma}) \preceq \boldsymbol{\alpha}$$
$$\mathbf{P}_f(\mathbf{W}, \boldsymbol{\gamma}) \preceq \boldsymbol{\beta}$$

Near-optimal solutions can be obtained using the same techniques as in solving (10.28).

10.5   Multichannel Cognitive MAC

In this section, we discuss the problem of efficient medium access control (MAC) among cognitive radio devices that are equipped with multiple radios and thus are capable of transmitting simultaneously at different frequencies (channels).

The focus is exclusively on the selfish game between cognitive radio devices (secondary users). It is assumed that the available frequency band is split into orthogonal channels and that channel access is based on the CSMA/CA protocol. A two-tier *medium access control (MAC) game* is defined where in the first step selfish cognitive radio devices first coordinate to achieve a channel allocation of their radios over the available channels. It is assumed that the bandwidth of each channel will be allocated fairly later. Then, in the second step, they optimize the channel access parameters of the CSMA/CA protocol of their radios on each channel to achieve this fair allocation. Selfish users have two choices in order to obtain more throughput on any channel c.

They either use several radios on c to exploit the per radio fairness property of the CSMA/CA backoff algorithm or they can subvert this backoff algorithm by cheating with the backoff parameter. We discuss both options in the MAC game and accordingly we split it into two sub-games: the *channel allocation (CA)* game and the *multiple access (CSMA/CA)* game. The discussion is based on [12].

10.5.1   System Model

It is assumed that in using FDMA the available frequency band is divided into the set C of orthogonal channels of the same bandwidth. The *wireless devices* use multiple channels to communicate at the same time. The *Carrier Sense Multiple Access with Collision Avoidance (CSMA/CA)* protocol is used to resolve conflict at the Medium Access Control (MAC) layer on each channel. *CSMA/CA* protocols rely on the random deferment of packet transmissions for the efficient use of a shared wireless channel amongst many devices in a network. Each radio of each device has a unique MAC layer identifier (the MAC address) which is achieved by the appropriate MAC layer authentication.

Pairs of users communicate with each other over a single hop. Each user participates in only one such communication session and we denote the finite set of communicating pairs by N. Each user owns a device equipped with k radio adapters, all having the same communication capabilities with $k \leq |C|$.

The communication links between two devices are bidirectional and devices always have some packets to exchange. Owing to the bidirectional communication links, the sender and the receiver are able to coordinate and thus to select the same channels to communicate. Each device can hear the transmissions of every other device if they are using the same channel (they reside in a *single collision domain*).

From the network topology and the set of communication pairs, we can construct a *contention graph*. The notion of contention graph was introduced for single channel systems and here it is adapted for the multi-channel scenario.

The medium access problem in a two-tier *medium access control (MAC) game* is modeled as follows. Any communicating pair of devices will be referred to as a *selfish player i*, whose objective is to maximize his total throughput r_i in the network. The term selfish player is used to denote both the communicating pair of devices and the communication link between them. The players first coordinate so as to achieve a channel allocation of their radios over the available channels in C assuming a fair access on a single channel, and then they optimize the channel access parameters of the $CSMA/CA$ protocol of their radios on each channel. So the MAC game is split into two sub-games: the *channel allocation (CA) game* and the *multiple access (CSMA/CA) game*.

The *strategy* s_i of player i defines his decisions in the game, taking the decisions of other players into account. In the CA game, the strategy of player i defines the number of radios player i uses in each channel. In the CSMA/CA game, the strategy defines how persistent player i is in contending for the available bandwidth on a specific channel by adjusting the backoff window parameter in the CSMA/CA protocol used in his radios. We denote the strategy space of player i by S_i. The strategies of the players define the strategy profile $s = \{s_1, \ldots, s_{|N|}\}$. All players other than player i are denoted by '-i' and their strategy profile by $s_{-i} = \{s_1, \ldots, s_{i-1}, s_{i+1} \ldots, s_{|N|}\}$.

The players are assumed to be rational and their objective is to maximize their *payoffs* or utility function in the network. We denote the payoff of player i by u_i. We assume that each player i wants to maximize his total throughput or bit rate (r_i) in the system and thus his payoff function is written as follows:

$$u_i = r_i = \sum_{c \in C} r_{i,c} \tag{10.45}$$

where $r_{i,c}$ is the throughput achieved by player i on channel c.

The total available throughput, defined as the sum of the achieved throughput of all of the players on channel c, $r_c = \sum_i r_{i,c}$ is a non-increasing function of the number of radios deployed on c. If the channel sensing is perfect, r_c is independent of the number of radios on c for the $CSMA/CA$ protocol. In practice, the backoff window values used in the $CSMA/CA$ protocol implementation are not optimal; and owing to packet collisions, r_c becomes a decreasing function of the number of radios on c.

To characterize stability in the MAC game, we introduce the concept of the Nash Equilibrium. The strategy profile $s^* = \{s_1^*, \ldots, s_{|N|}^*\}$ defines a Nash Equilibrium (NE), if for each player i and its strategy $s_i' \in S_i$ we have $u_i(s_i^*, s_{-i}^*) \geq u_i(s_i, s_{-i}^*)$. This means that in a Nash Equilibrium, none of the users can unilaterally change his strategy to increase his payoff.

Since the Nash Equilibrium is often inefficient from the system's point of view the efficiency of the solution can be characterized by the concept of Pareto-optimality. The strategy profile s^{po} is Pareto-optimal if there is no s' such that $u_i(s') \geq u_i(s^{po})$, $\forall i$ with strict inequality for at least one player. This means that in a Pareto-optimal strategy profile s^{po} one cannot improve the payoff of any player i without decreasing the payoff of at least one other player j.

The efficiency of a wireless system can be characterized by using the concept of the *price of anarchy* (*POA*) which is the ratio between the sum of the payoffs achieved by the players using the system-(Pareto-) optimal strategy profile and in the worst-case Nash Equilibrium.

10.5.2 Channel Allocation Game

In the CA game, we define the *strategy* of player i as his channel allocation vector which defines the number of his radios on each of the channels. We denote the number of radios used by player i on channel c by $k_{i,c}$. Since the available number of radio adapters is assumed to be smaller than the total number of channels, i.e., $k \leq |C|$, we can express the strategy of i as $s_i = \{k_{i,1}, \ldots, k_{i,|C|}\}$.

In general players are expected not to use more than one radio on any channel c. On the other hand, the backoff algorithm provides per radio fairness. Thus, any user who uses multiple radios obtains a larger proportion of the total channel throughput without changing protocol parameters. In addition, sophisticated selfish users can manipulate with the backoff mechanism as well.

The set of channels used by player i will be denoted by C_i. The total number of radios employed by player i can be written as $k_i = \sum_{c \in C} k_{i,c}$. Similarly, we can obtain the number of radios using a particular channel $k_c = \sum_{i \in N} k_{i,c}$. The strategy vectors of all of the players define the strategy matrix S (i.e. the strategy profile).

The players are rational and their objective is to maximize their *utility* $u_i = r_i$ in the network where total available throughput r_c on a channel c is a non-increasing function of the number of radios k_c deployed on this channel. If we assume that the channels have the same characteristics, we can characterize the achievable throughput by the function $r(k_c)$. In the CA game, the players assume that the total throughput on channel c will be shared equally among the radio adapters using that channel. This fair throughput allocation is achieved, as shown later in the *CSMA/CA* game. We can then express the utility u_i of player i as:

$$u_i = r_i = \sum_{c \in C} r_{i,c} = \sum_{c \in C} \frac{k_{i,c}}{k_c} \cdot r(k_c) \qquad (10.46)$$

The game above can be implemented as a *round-based* distributed algorithm that works as follows. First, we assume that there exists a random radio assignment of the players over the channels. For simplicity, we assume that no player allocates more than one device on any channel. After the initial channel assignment, each player evaluates the number of radios (which defines the approximate length of the round) on *each of the channels* on the channels he knows (recall that we denote this set by C_i) and decides with probability ρ to improve his total throughput by reorganizing his radios. This probability is necessary to avoid continuous reallocation of the radios if the players change in the same round. We denote the average number of devices on the channels in C_i by K_i (note that K_i is not necessarily an integer). For each channel $b \in C_i$ with $k_b - K_i \geq 1$, player i moves his radio to another channel $c \notin C_i$. The probability to chose a channel $c \notin C_i$ is $1 - |C_i| / |C|$.

10.5.3 CSMA/CA Game

The same *CSMA/CA* game is played on each channel in C between the players who have a radio on that channel. Therefore, it is enough to focus on a single channel $c \in C$ occupied by multiple radio adapters owned by two or more players. All conclusions drawn from this single channel model are applicable to other occupied channels.

We define the set κ of players that have at least one radio on the observed channel $c \in C$ with $N_c \subseteq N$. There can be a number of ways in which players misbehave with their radio adapters. For example, in violation of the standard protocol, player $i \in N_c$ can initialize the contention window size on his radio adapters to a lower value in order to obtain a higher throughput. We will call this lower value $W_{i,\ell}$, the contention window of player $i's$ l-th radio. This mode of cheating is the easiest (and yet highly rewarding), since it does not require changes to be made in the operation of the protocol.

The main conclusions of this study are applicable to any other cheating technique. Those players who tamper selfishly with their adapters, as described above, will be referred to as *cheaters* against the standard protocol and refer to their set as $I_c \subseteq N_c$. In the model, every cheater $i \in I_c$ seeks to maximize the average throughput $r_{i,c}$ on channel c. For cheating player $i \in I_c$, we define the pure-strategy set S_i as $S_i = \{1, 2, \ldots, W_{\max}, W_\infty\} \times \ldots \times \{1, 2, \ldots, W_{\max}, W_\infty\}(k_{i,c}$ times), where $\times$ is the Cartesian product, $k_{i,c}$ is the number of radio adapters player i uses on channel c, $W_{\max} < \infty$ is a positive integer and the symbol W_∞ means that the player i does not transmit at all on channel c using the corresponding adapter, which is equivalent to $W_i = \infty$ for that radio adapter. Note that the set S_i is finite. Strategy $s_i \in S_i$ of each player $i \in I_c$ in the *CSMA/CA* game consist in choosing the contention window values $s_i = (W_{i,1}, W_{i,2}, \ldots, W_{i,k_{lC}})$ such that player $i's$ throughput $r_{i,c}$ is maximized on each channel c.

For easier mathematical treatment, the channel access probabilities $\tau_{i,\ell} = 1/(W_{i\ell} + 1)$ *for* $\ell \in \{1, \ldots, k_{i,c}\}$ is used instead of working directly with contention windows $W_{i,\ell}(\ell \in \{1, \ldots, k_{i,c}\})$. Let τ denote the channel access probabilities of non-cheating nodes. Let us also denote with $K_{i,c}$ the set of radio adapters that player i uses on channel c, that is, $K_{i,c} = \{1, 2, \ldots, k_{i,c}\}$. Then, we can express the (average) throughput $r_{i,c}$ of a cheater $i \in I_c$ on channel c as

$$r_{i,c} = \sum_{\ell \in K_{i,c}} r_{i,c,\ell}$$

where $r_{i,c,\ell}$ is the average throughput of player $i's$ l-th radio adapter used on channel c and it is given by [12]

$$r_{i,c,\ell} = \frac{\tau_{i,\ell} p_{-i,\ell} L}{\tau_{i,\ell} \cdot (p_{-i,\ell}(T^s - T^{id}) - q_{-i,\ell}(T^s - T^c)) + (1 - p_{-i,\ell} - q_{-i,\ell})T^c + q_{-i,\ell}T^s + p_{-i,\ell}T^{id}} \tag{10.47}$$

where L is the average packet length, T^s is the average time needed to transmit a packet of size L (including the inter-frame spacing periods), T^{id} is the duration of the idle period (a single slot) and T^c is the average time spent in the collision and p_{-i} and q_{-i} are given as

$$p_{-i,\ell} = (1 - \tau)^{|N_c| - |I_c|} \times \prod_{\substack{j \in I_c \\ k \in \hat{K}_{j,c}}} (1 - \tau_{j,k}) \tag{10.48}$$

where

$$\hat{K}_{j,c} = \begin{cases} K_{j,c}, & j \neq i; \\ K_{j,c} \backslash \{\ell\}, & j = i. \end{cases}$$

$$q_{-i,\ell} = (1 - \tau)^{|N_c| - |I_c|} \times \sum_{\substack{j \in I_c \\ k \in \hat{K}_{j,c}}} \tau_{j,k} \prod_{\substack{m \in I_c \\ n \in \hat{K}_{m,c}}} (1 - \tau_{m,n}) \tag{10.49}$$

where

$$\hat{K}^{de}_{m,c} = \begin{cases} K_{m,c}, & m \neq j, \ m \neq i; \\ K_{m,c} \backslash \{k\}, & m = j, \ m \neq i; \\ K_{m,c} \backslash \{l\}, & m \neq j, \ m \neq i; \\ K_{m,c} \backslash \{k, l\}, & m = j, \ m \neq i; \end{cases}$$

Note here that the only parameter that a cheating node i has a control over is its own $\tau_{i,\ell}$ by manipulating $W_{i,\ell}$. We denote the game as defined in this subsection by $G_{CSMA/CA} = \langle I_c, (S_i)_{i \in I_c}, (r_{i,c})_{i \in I_c} \rangle$ and call it a *static CSMA/CA game* played on channel $c \in C$.

A desirable solution for the *CSMA/CA* game should exhibit uniqueness, per-radio fairness and Pareto optimality. In order to derive such a solution, the Nash Bargaining Framework (NBF) from the theory of cooperative games is used in [12]. The NBF obtains a unique, fair (per-radio) and Pareto-optimal solution by solving the following maximization problem:

$$\underset{\tau_{i,l}}{maximize} \prod_{i \in I_c} \prod_{\ell=1}^{k_{i,c}} r_{i,c,\ell}(s_1, s_2, \ldots, s_{|I_c|}) \tag{10.50}$$

subject to $s_i \in S_i, \forall i \in I_c$ where $r_{i,c,\ell}(s_1, s_2, \ldots, s_{|I_c|})$ represents the throughput of player $i's$ l-th radio adapter used on channel c. Note that the above optimization problem is discrete and that $\tau_{i,\ell} = \tau^*, \forall i \in I_c$ and $\ell = 1, \ldots, k_{i,c}$, implies $W_{i,\ell} = 1/\tau^* - 1, \forall i \in I_c$ and $\ell = 1, \ldots, k_{i,c}$.

At the core of the strategy that allows the players to converge to this point is the *penalty mechanism*, by which one player can penalize the severe deviations of another player.

Let us consider two arbitrary players i and j from the set I_c and their arbitrary radio adapters m and n used on the common channel c. Let us assume that player i calculates the penalty $p_{j,c,n}$ to be inflicted on player j's n-th radio adapter as

$$p_{j,c,n}(s) = \begin{cases} r_{j,c,n}(s) - r_{i,c,m}(s), & \text{if } r_{j,c,n}(s) > r_{i,c,m}(s) \\ 0, \text{ otherwise} \end{cases} \quad (10.51)$$

and the throughput of player j's n-th adapter is $r_{j,c,n}(s) - p_{j,c,n}(s) = r_{i,c,m}(s)$.

Thus, the two adapters m and n receive the same throughputs. The penalty functions (10.51) is a base of a simple penalizing scheme, in which the packets of the non-cooperative player's adapter that deviate from the given equilibrium point are jammed selectively for a short duration of time, T^{jam}, by the other players using the channel.

Suppose that a player $i \in I_c$ detects the presence of a non-cooperative player $j \in I_c$. Then, if the player i listens to a transmitted packet corresponding to the player j's non-cooperative adapter, it switches to transmission mode and jams enough bits so that the packet cannot be recovered properly at the receiver.

Let the throughput obtained by the two adapters m and n of players i and j over the last observation window, T^{obs}, be $r_{i,c,m}$ and $r_{j,c,n}$, respectively, where $r_{i,c,m} > r_{j,c,n}$. The penalty function (10.51) aims at making the throughputs received by adapters m and n equal. We denote with $r_{i,c,\ell}(t)$ the instantaneous throughput of adapter ℓ. The average throughput received by the adapter m and n should be the same over the total time duration of $T^{obs} + T^{jam}$, that is,

$$\frac{1}{T} \int_t^{t+T} r_{i,c,m}(t)dt = \frac{1}{T} \int_t^{t+T^{obs}} r_{j,c,n}(t)dt$$
$$T = T^{obs} + T^{jam} \quad (10.52)$$

where we have used the fact that the player i jams player j's adapter during the period T^{jam}. If the average throughput over a time period P is $\bar{r}(t, P) = (1/P) \int_t^{t+P} r(t)dt$. Then, from (10.52) we have

$$T^{jam} = T^{obs} \frac{\bar{r}_{j,c,n}(t, T^{obs}) - \bar{r}_{i,c,m}(t, T^{obs})}{\bar{r}_{i,c,m}(t + T^{obs}, T^{jam})} \quad (10.53)$$

In the following, we discribe a simple algorithm that is based on the penalty function (10.51) that leads the players (their adapters on the observed channel) to a unique Pareto-optimal Nash Equilibrium. The main idea is that one player i (his adapter m) acts as a *coordinator* on the observed channel by inflicting penalties on the radios of other players who receive a higher throughput. The equilibrium coordination procedure is an interplay between two simple algorithms.

First, the coordinating adapter m chooses some initial contention window value $W_{i,m} = W_0 > 1$ and begins to penalize the other players' adapters (on the same channel) with $r_{j,c,n}(s) > r_{i,c,m}(s)$, $j \neq i$. The other players (affected adapters) act selfishly and run the adaptation algorithm that maximizes $r_{i,c,m}(s)$ (due to the penalty function (10.51). This algorithm simply attempts to equalize the other adapters' contention windows with the coordinator's contention window $W_{i,m} = W_0$; this value maximizes the throughput of each adapter. This stage lasts until all the adapters that belong to the cheating players on the observed channel stabilize at a *stage Nash equilibrium point* $W_{i',\ell} = W_0$, $i' \in I_c$ and $\ell = 1, \ldots, k_{i,c}$.

After spending some finite time on this point (to let all the adapters to learn that the system has reached an equilibrium point), the coordinator further updates his contention window to value $W_1 > W_0$ and again begins to penalize the other players' adapters (on the same channel) with $r_{j,c,n}(s) > r_{i,c,m}(s)$, $j \neq i$. In turn, the other players selfishly adjust the contention windows of their adapter in order to maximize their throughputs. At this stage, the coordinator compares the throughput $r_{i,c,m}(W_0)$ received at the previous equilibrium point W_0 and the throughput $r_{i,c,m}(W_1)$ at the current equilibrium point W_1. If the difference is $|r_{i,c,m}(W_0) - r_{i,c,m}(W)|$ is smaller than some small value ε, the coordinator stops increasing his contention window.

References

[1] Neuts, M. F. (1981) *Matrix-Geometric Solutions in Stochastic Models.* Johns Hopins University Press, Baltimore.

[2] Cai, J. and Alfa, A. S. (2009) Optimal Channel Sensing in Wireless Communication Networks with Cognitive Radio, *IEEE ICC 2009.*

[3] Munkres, J. (1957) Algorithm for assignment and transportation problems, *Journal of the Society for Industrial and Applied Mathematics*, **5**, 32–38.

[4] Chow, Y. S., Robbins, H. and Siegmund, D. (1971) *Great Expectations: The Theory of Optimal Stopping.* Boston, MA: Houghton Mifflin.

[5] Sabharwal, A., Khoshnevis, A. and Knightly, E. (2007) Opportunistic spectral usage: Bounds and a multi-band *CSMA/CA* Protocol, *IEEE/ACM Trans. Networking*, **15**(3), 533–545.

[6] Hill, T. P. and Hordijk, A. (1985) 'Selection of order of observation in optimal stopping problems,'. *J. Applied Probability*, **22**(1), 177–184.

[7] Quan, Z., Cui, S., Sayed, A. H. and Poor, H. V. (2009) Optimal Multiband Joint Detection for Spectrum Sensing in Cognitive Radio Networks, *IEEE Transactions on Signal Processing*, **57**(3), 1128–1140.

[8] Boyd, S. and Vandenberghe, L. (2003) *Convex Optimization.* Cambridge, U.K.: Cambridge University Press.

[9] Sayed, A. H. (2003) *Fundamentals of Adaptive Filtering.* New York: Wiley.

[10] Quan, Z., Cui, S. and Sayed, A. H. (2008) Optimal linear cooperation for spectrum sensing in cognitive radio networks, *IEEE J. Sel. Topics Signal Process.*, **2**(1), 28–40, Feb. 2008.

[11] Quan, Z., Cui, S. and Sayed, A. H. (2007) An optimal strategy for cooperative spectrum sensing in cognitive radio networks, in *Proc. IEEE GLOBECOM*, Washington DC, Nov., 2947–2951.

[12] Félegyházi, M., Cagalj, M. and Hubaux, J.-P. (2009) Efficient MAC in Cognitive Radio Systems: A Game-Theoretic Approach, *IEEE Transactions on Wireless Communications*, **8**(4), April, 1984–1995.

11

Relay-Assisted Wireless Networks

A. Agustin, J. Vidal, O. Muñoz, and S. Glisic

11.1 Introduction

The growth of wireless networks in recent decades is motivated by their ability to provide communications *anywhere* and *anytime*, encompassing a high proliferation of wireless services and devices such as mobile communications, WiFi (Wireless Fidelity), or cordless phones. However, wireless networks present two important drawbacks: *scarcity of radio spectrum* and *channel impairments*. In this respect, wireless networks must be designed to exhibit a high spectral efficiency and must combat channel impairments (i.e. multipath fading, shadowing, interference and pathloss).

Wireless networks are built around a number of nodes communicating one to each other over a wireless channel, some having a wired backbone with only the last hop being wireless, such as cellular voice and data networks. But the provision of high capacity and reliable wireless multimedia communications to carry bursty packet traffic as well as voice and delay constrained traffic continues to be a challenging aspect of modern and future wireless communication networks. Recent advances in radio transceiver techniques such as Multiple Input Multiple Output (MIMO) architectures have shown an enhancement in the capacity of the current systems by alleviating the effect of channel multipath fading. Similarly, *cooperative communication* is based on collaboration amongst several distributed terminals so as to transmit/receive their intended signals (Laneman, Tse, Wornell, Erkip, Sendonaris, Aazhang, Host-Madsen). This type of communication is based on the seminal work published in the 1970s by van der Meulen, Cover and El Gamal, where a new element is introduced in conventional point-to-point communication, the relay. The new network architecture exhibits some of the properties of MIMO systems, but in contrast to those systems, relay-assisted transmission is able to combat channel impairments due to the shadowing and path-loss provided by the source-destination and relay-destination links.

The application of relay-assisted transmission for practical systems is constrained by current radio technology, which cannot transmit and receive simultaneously in the same band because of the dynamic range of the incoming and outgoing signals through the same antenna element. As a result, the relays must operate in half-duplex mode. In spite of this drawback, relay-assisted multi-hop networks are expected to play a significant role in 4G wireless communications systems, because they can extend coverage cost-effectively, increase spectral efficiency and drive the cost of deploying 3G+ and 4G systems lower.

This chapter aims to unearth the properties of relay-assisted communications and provide some examples of how this kind of transmission can be incorporated in wireless cellular networks.

Advanced Wireless Communications & Internet: Future Evolving Technologies, Third Edition. Savo Glisic.
© 2011 John Wiley & Sons, Ltd. Published 2011 by John Wiley & Sons, Ltd.

The chapter is organized as follows. First, Section 11.2 presents the background of relay-assisted transmission reviewing:

- Relay-assisted versus cooperative communication.
- Definition of transmission protocols for half-duplex relays.
- Strategies of half-duplex relay-assisted transmission (decoding mode, static or dynamic resource allocation, one-way or two-way relays).

Section 11.3 analyzes cooperative communication when terminals collaborate in their transmissions. Its benefits are explored in uplink transmission of cellular networks and in general wireless networks:

- Evaluating the *diversity gain* of cooperative communication.
- Defining the distributed space-time codes to be applied in collaboration of terminals.

Section 11.4 reviews relay-assisted communication:

- Achievable rates for the different strategies (*decode-and-forward*, *amplify-and-forward*) for all half-duplex protocols and performance comparison between protocols.
- Impact of static and dynamic resource allocation.
- Effect of erroneous decoding at the relay when distributed-space time codes are employed.
- Transmission protocols where there are multiple terminals (half-duplex diamond relay channel).

Section 11.5 presents the two-way relay channel (TWRC). In contrast to the conventional approach where a relay is helping a given source to transmit to a destination, the TWRC assumes that the relay helps two terminals exchange their messages. This channel is interesting in that it addresses simultaneously the downlink and uplink transmissions of cellular networks. The following topics are studied:

- Transmission protocols for the TWRC.
- Benefits of the TWRC over the one-way relay channel.

Finally, Section 11.6 explains how the spectral efficiency of relay-assisted transmission with a static resource allocation can be improved. This method is based on the reuse of resources when applied in a cellular system in the downlink. Loosely speaking, in a scenario where idle terminals become relays for active terminals, all transmissions done by relays to different users are allocated in the same temporal slot. The required resources for the transmissions from the relays are reduced at the expense of generating interference to all users. Two solutions are presented to deal with that drawback:

- Distributed power control (at the relay) based on game theory.
- After modeling the statistics of the received interfering power, the transmitted bitrate is adjusted either to maintain certain outage probability or to maximize the sum-throughput of the system.

11.2   Background and Related Work

Communications through wireless channels have generated new problems for transmissions over mobile users owing mainly to time-varying fading channels, interference, and the shadowing effect. The appropriate method to combat these effects is the use of diversity (*diversity gain*). Typically, time and frequency diversity have been considered. In recent years space diversity [3, 4], by means of multiple antenna system (MIMO – Multiple Input Multiple Output) has received a great deal of attention from the research community because it can be combined with the previous forms of diversity and additionally, offers an increase of the total information theoretic capacity (*multiplexing gain*) of the system (see also

Chapter 4). The diversity-multiplexing tradeoff exhibited by multiple antenna channels is shown in [5]. MIMO systems have been proven to increase channel information theoretic capacity linearly with a minimum number of transmitting and receiving antennas. Recently and motivated by the work of [6] and [7], there has been a growing interest in a new space-time method that adopts antennas belonging to multiple terminals, which has been named *cooperative diversity*. The cooperative terminals create a virtual array through distributed transmissions [8–10].

Nevertheless, *cooperative diversity* is based in the classical relay channel, defined originally in [11,12]. Therefore, in order to gain more insight into the benefits of employing an additional terminal (or its antennas) for improving transmission, this section dwells more on the most important aspects to be considered. To this end Section 11.2.1 presents an overview of the relay channel, while the differences with cooperative communication are tackled in Section 11.2.2. In general, the relay terminal, or the terminal that is assisting another, cannot receive and transmit data at the same time, thus obliging that transmission and reception must be done at different instances of time (half-duplex constraint on terminals). Section 11.2.3 presents several protocols that can be defined as a result of that constraint. Finally, Section 11.2.4 delves into the strategies of relay-assisted transmission, taking into account the decoding mode at the relay, the type of the resource allocation (*dynamic* or *static*), and the traffic addressed by the relay (*one-way* or *two-way* relaying).

11.2.1   The Relay Channel

The relay channel is introduced by van der Meulen in [11]. It assumes that there is a source that wants to transmit information to a single destination. However, there is a relay terminal that is able to help the destination (*relay-assisted transmission*). Based on past received symbols, the relay terminal can transmit an additional message to the destination, but connected with the message transmitted by the source. Figure 11.1 illustrates the channel model for that kind of transmission. In general it is assumed that the relay works in *full-duplex mode*, i.e. receiving and transmitting simultaneously. The inner bounds of the information theoretic capacity (capacity in the following) of a discrete memoryless channel are given by [11] based on a timesharing approach. Achievable rates are derived in [12] via three structurally different random coding schemes: *facilitation, cooperation* and *observation*. In the *facilitation* mode, the relay does not help the source actively, but rather, facilitates the source transmission by inducing as little interference as possible. Likewise, in *cooperation* mode, the relay fully decodes the source message and retransmits, jointly with the source using a block Markov encoding. Finally, in *observation* mode the relay encodes a quantized version of its received signal, using ideas from source coding with side information [13–16] (see Chapter 2, Section 2.5). In general, *cooperation* yields the highest achievable rates when the source-relay channel quality is very high, while *observation* yields the highest achievable rates when the relay-destination channel quality is very high. However, for the physically degraded relay

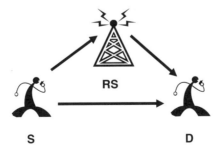

Figure 11.1   The relay channel. Source (S), Relay (RS) and Destination (D).

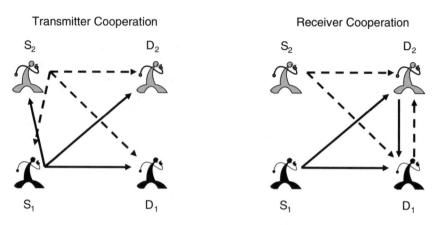

Figure 11.2   Cooperative diversity network. 2 sources (S_1, S_2) and 2 destinations (D_1, D_2).

channel, where the destination receives a corrupted version of what the relay receives, the lower bound attained with the cooperation mode turns out to be the capacity.

The capacity upper bounds for the relay channel (*max-flow-min-cut* theorem [16]) are obtained by the minimum of mutual information obtained by the broadcast channel (transmission from the source to relay and destination, [16]) and the multiple access channel (independent and simultaneous transmission from the source and relay to the destination [16]). The capacity bounds and power allocation for wireless relay channels are given in [17] for *half-duplex* relay and single-antenna terminals. Additionally, algorithms to compute the capacity bounds for the multi-antenna terminals are presented in [18]. In some conditions the upper bound meets the lower bound characterizing the capacity of the MIMO relay channel.

11.2.2   Cooperative Communication

Cooperative communication is based on the relay channel model, Figure 11.1, however, the principal difference between the *relay-assisted* and the *cooperative transmission* is the type of terminal involved in the communication. In *relay-assisted transmission* the relay terminal is an additional terminal which helps the source (it does not have its own information to transmit), while in the *cooperative communication* there are two sources which have to help one another and both have information to transmit.

The cooperative diversity,[1] introduced in [6] and described in section 11.3 considers the scenario depicted in Figure 11.1 but with the relay terminal being another source. Both sources (associated partners) are also responsible for transmitting the information of their partners. The sources are working in full-duplex mode, so that both sources are transmitting to the destination and receiving a noisy version of the partner's transmission. This approach can be seen as the generalization of multiple-access channels with generalized feedback [19–23], to the multipath fading model. The generalized feedback allows the sources to act essentially as relays for one another.

The cooperative diversity for the *ad hoc* four-node network presented in Figure 11.2 and made up of two sources and two destinations is analyzed in [24–26], providing some of the upper and lower bounds of the capacity offered by that network. Each source wants to send a message to a different receiver. Basically, two types of cooperation can be found: *receiver* and *transmitter* cooperation, depending on which nodes are relaying each other's messages. It is assumed that all the nodes work on the full-duplex mode. Although this four-node network is quite similar to a MIMO system (it will be equal if both sources/destinations have prior information about the transmitted/received data by the other node) there

---

[1] This term is defined as user cooperation in [6].

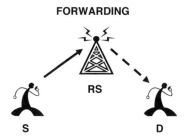

Figure 11.3 Half-duplex forwarding protocol. S, RS and D refer to the source, relay and destination, respectively. Solid lines correspond to the transmission during the *relay-receive phase* and dashed line for the transmission during the *relay-transmit phase*.

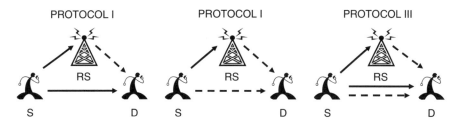

Figure 11.4 Half-duplex relay protocols. S, RS and D refer to the source, relay and destination, respectively. Solid lines correspond to the transmission during the *relay-receive phase* and dashed line for the transmission during the *relay-transmit phase*.

is no *multiplexing gain* for either transmitter or receiver cooperation, but there is an *additive gain* over the direct transmission, [24].

11.2.3   Half-Duplex Relay Protocol Overview

Full duplex terminals are currently unrealistic in practical systems, hence relays are forced to work under half-duplex mode, i.e. there will be an orthogonal duplexing (in time or frequency) between the relay receiving (*relay-receive phase*) and the relay transmitting (*relay-transmit phase*). This phase separation allows for the defininition of several half-duplex relay protocols with various degrees of broadcasting and receiving collision in each *relay-receive* and *relay-transmit phase* among the three terminals (source, destination, and relay). The number of options leads to the four protocol definition[2] [40,41] presented in Figure 11.4 and Figure 11.3, which will be referred to as the forwarding protocol, protocol I, II, and III.

The traditional forwarding protocol consists of a transmission from the source to the relay during the *relay-receive phase* and a transmission from the relay to the destination in the *relay-transmit phase*, see Figure 11.3. In contrast, the relay-assisted protocols depicted in Figure 11.4 are designed to cope also with the source-destination link. Obviously, if that link is of very bad quality as compared with the others, the performance obtained by protocols I, II and III approaches the performance of the forwarding protocol.

Protocol I assumes that the source communicates with the relay and destination during the *relay-receive phase* (solid lines in Figure 11.4). Afterwards, in the *relay-transmit phase*, the relay terminal communicates with the destination (dashed line in Figure 11.4). This protocol shows the same structure as the Broadcast Channel (BC), [16], during the *relay-receive phase*. For example, protocol I is used by

---

[2] The current enumeration differs from the one employed in [40].

the cooperative transmission defined in [7] and [32] and also in the relay-assisted transmission described in [30]. On the other hand, protocol II only allows a source-relay transmission in the *relay-receive phase* (solid line in Figure 11.4), assuming that the destination is not active in that phase. Later on, in the *relay-transmit phase*, source and relay transmit simultaneously to the destination (dashed lines in Figure 11.4), describing a multiple access channel (MAC), [16], in this second phase. Finally, protocol III can be seen as a combination of protocols I and II. The source transmits to the relay and the destination (solid lines in Figure 11.4) in the *relay-receive phase*. Then, in the *relay-transmit phase*, the source and the relay transmit to the destination (dashed lines in Figure 11.4). Notice that the relay is transmitting during the second phase, so that it cannot be aware of the signal being transmitted by the source. This protocol achieves a better spectral efficiency than the previous ones. It is considered for the obtaining of the achievable rates of the relay channel in [17].

11.2.4   Strategies of Relay-Assisted Transmission

The paradigm of conventional source-destination communication is now changed to *source-relay-destination*, and the role played by the relay can be selected over different modes of operation influencing the total achievable rate of the system. In addition to this new element, the half-duplex constraint on the relay imposes the wise design of resources allocated for each phase in order to obtain the highest achievable rate. Therefore, the strategies of relay-assisted transmission have to consider the decoding mode at the relay, resource allocation, and the type of traffic addressed by the relay. In Figure 11.5 we present a classification of the possible strategies for relay-assisted transmission which will be explained in the following.

11.2.4.1   Decoding at the Relay

Basically the three decoding modes analyzed in the literature are: *amplify-and-forward* (AF), *decode-and-forward* (DF), and *compress-and-forward* (CF).

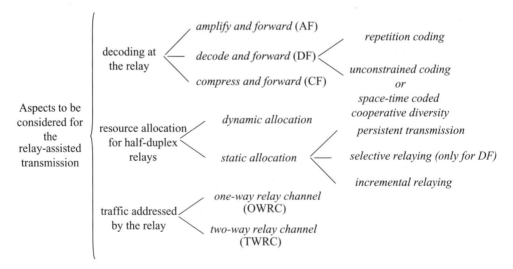

Figure 11.5   Aspects to be considered for the relay-assisted transmission.

Amplify-and-Forward (AF)

This is the simplest strategy that can be used at the relay because it acts as a *dummy* with a constraint on maximum power. The relay amplifies the received signal from the source and transmits it to the destination without making any decision. The main drawback of this strategy is that the relay terminal is also amplifying the received noise. When this strategy is applied to cooperative communication, it is able to obtain a better uncoded bit error rate (BER) than direct transmission [7]. Additionally, the outage probability of cooperative communication is also derived, showing that a diversity order of two is obtained for two cooperative users, as will be explained in Section 11.3.2. When the relay is equipped with multiple antennas and there is channel state information (CSI) of the source-relay and relay-destination links, the AF strategy can attain significant gains over the direct transmission by means of optimum linear filtering the data to be forwarded by the relay [28].

Decode-and-Forward (DF)

The complexity at the receiver increases in comparison with the AF strategy. Now the relay terminal has to estimate the message received from the source and total performance will depend on success in decoding the message correctly. Depending on the type of symbols retransmitted the strategy at the relay is *repetition coding* (RC) or *unconstrained coding* (UC) (named *space-time-coded cooperative diversity* in [29]). In the RC, the relay retransmits the same symbols estimated previously, while in the UC the symbols transmitted are not the same as the received ones, but are related to the same information sent by the source (source and relay are using different codebooks). An illustrative example for UC is the following. Let us assume that the source is going to transmit a coded codeword using a forward error correction (FEC) code, where there are *systematic* and *parity* bits. Once the relay has decoded the signal received in the *relay-receive phase*, it re-encodes the message, selecting new *parity* bits, and transmits them to the destination in *the relay-transmit phase*.

Examples of the *decode-and-forward* scheme under RC are given in [6,7,27]. Basically, the transmission is done over three intervals of time. The first is devoted to transmitting directly to the destination (without cooperation) and the remaining two are used for cooperation. In the second period each source transmits new information to the destination which is estimated by the other source (partner). Afterwards, both sources transmit to the destination a linear combination of their own bits, transmitted in the previous interval of time, and the estimated ones belonging to the other source.

The upper and lower bounds for *unconstrained coding* (UC) (or *space-time-coded cooperative diversity*) in a cooperative scenario where each source can cooperate with *m* terminals in orthogonal mode are shown in [29]. The destination must combine the signals received in two orthogonal periods of time which come from the source and from those cooperative users who were able to decode the message transmitted by the source in the first period of time (protocol I). The total mutual information is the same as two parallel channels [16]. Unconstrained coding is superior to repetition coding in terms of diversity for larger spectral efficiencies.

A practical implementation of *unconstrained coding* (UC) for protocol I can be found in [30] and [31], where the relay terminal, instead of transmitting the same symbols, uses a different part of a rate compatible turbo code, (TC), as in the example given at the beginning of this section. In that case it is assumed that the source transmits a message to the relay and destination in one time interval. Afterwards, the relay transmits to the destination in an orthogonal period of time. The destination must combine both transmissions in order to decode the message and employs a total coded codeword with more protection (*incremental redundancy*). This idea is applied to *cooperative transmission* in [32–34] and is named *coded cooperation*.

Compress-and-Forward (CF)

In this strategy the relay does not decode the data and uses Wyner-Ziv lossy source coding [15] for the estimated symbols of the received signal. Then, the compressed signal is transmitted to the destination by the relay. Depending on the channel gains of the different links CF can be superior to the DF strategy.

This strategy is suggested in Theorem 6 in [12] where the relay works under the *observation* mode (see also Section 11.2.1). In the literature you will find that this decoding strategy is also referred to as Quantize-and-Forward or Estimate-and-Forward. Nevertheless, this approach is not analyzed in this chapter, and more detail can be found in in [17, 18, 35, 36].

11.2.4.2   Resource Allocation in Half-Duplex Relays

The half duplex protocols presented in Figure 11.3 and Figure 11.4 assume a relay-assisted transmission done over orthogonal phases, the *relay-receive* and *relay-transmit phase*. The achievable rate of the system depends on the duration of each phase. Two strategies are possible for defining the duration of each phase: *static* or *dynamic resource allocation relaying*. The adoption of one or another of those strategies may depend on the kind of channel state information at the source and/or the type of system where the relay-assisted transmission is implemented.

Static resource allocation relaying is defined when the duration of each phase is assigned previously. This can be found when the transmission is implemented in a centralized cellular system based on Time Division Multiple Access (TDMA) where transmissions are done over time slots with fixed duration. In that case the relay-assisted transmission needs two time slots (*relay-receive* and *relay-transmit phase*). Despite not employing optimal resource allocation, there are several retransmission strategies at the relay to enhance the spectral efficiency of the relay-assisted transmission: *persistent transmission, incremental relaying, selective relaying*. The previous strategies assume that retransmission is always adopted (*persistent transmission*) or adapt it to the current channel state by exploiting limited feedback from the destination and/or the relay. The source and relay can have knowledge of the success of the transmission at the relay and destination thanks to the existing feedback logical channels in modern communication systems. Two different options are analyzed in [7]: *selective* and *incremental* relaying. *Selective relaying* is applied to the DF because performance is constrained by the success of the transmissions from the source to the relay. When the relay receives wrongly or the source measures that the current channel state of the source-relay link is below a threshold, the relay remains silent and the source transmits again (using repetition or more powerful codes). For *incremental relaying* the relay should be aware of the success or not of the direct transmission during the *relay-receive phase* at the destination. The relay only transmits when the direct transmission (source-destination) is decoded wrongly by the destination. Using this option it is possible to obtain an improvement of spectral efficiency over the *persistent* and *selective* transmission.

Table 11.1 depicts the available types of relay transmission (*persistent, selective relaying* or *incremental relaying*) for the different half-duplex relay protocols. For the forwarding protocol it is required that relays always transmit because the destination is considering only the data received during the *relay-transmit phase*. For protocol I all types of transmission are allowed. For protocol II *incremental relaying* cannot be implemented because the relay is transmitting simultaneously with the source in the *relay-transmit phase* and the destination is considering only the data received in that phase. When the relay is working under *selective relaying* and the relay remains silent because it has not been able to decode the message from the source, the equivalent system is the same as the direct transmission using only the *relay-transmit phase*. For protocol III under *selective relaying* the source is transmitting in both phases and the destination considers the data received in both phases. Additionally, *incremental relaying*

Table 11.1   Types of retransmission at the relay for different protocols

Protocols	Persistent transmission	Selective relaying	Incremental relaying
forwarding	√	×	×
protocol I	√	√	√
protocol II	√	√	×
protocol III	√	√	√

is also possible because the message transmitted by the relay in the *relay-transmit phase* is connected with the transmission done by the source in the *relay-receive phase* (as in protocol I), so the destination might inform the relay about success in decoding the message in that phase and the relay terminal will remain silent in the *relay-transmit phase*. It is worth noticing that under *selective relaying* in protocol I when the relay terminal remains silent, the source transmits during the *relay-transmit phase*, while for protocols II and III the source, by definition, is already transmitting during that phase.

Dynamic resource allocation relaying is assumed when the source has knowledge of the quality of all links (source-destination, source-relay and relay-destination) and designs the resource allocation in order to maximize the spectral efficiency. The impact of the type of resource allocation is analyzed in Section 11.4.

11.2.4.3   Type of Traffic Addressed by the Relay Terminal

Initially the relay terminal was conceived for helping a given source in the transmission of a message to a given destination, as it is sketched in Figure 11.1. Hence, we might envisage such an approach as the one-way relay channel[3] (OWRC), because the relay only works with the messages from the source. However, when relay terminals are applied to cellular networks they have to deal with two types of traffic: the *downlink*, i.e. from a base station to a mobile station, and the *uplink*, from a mobile station to the base station. Using OWRC transmission the relay only works with one type of communication at a time, which reduces the spectral efficiency of the system. In this regard, the two-way relay channel (TWRC) is better suited for such a type of communication. The TWRC model is based on a three terminal network, where there are two terminals that want to exchange their messages with the help of a relay terminal. Those terminals in cellular networks could be the base station and mobile station. Consequently, the two-way relay is working with *downlink* and *uplink* traffic simultaneously. The benefits obtained by the TWRC are shown in [37–39] and are investigated in Section 11.5.

11.3   Cooperative Communications

This section reviews the concepts and properties of cooperative communications which have been investigated in [6, 7, 27, 29]. Let us recall that the terminology of *cooperative communications* (see Section 11.2.2) is reserved for describing the collaboration between terminals in order to help one another in their intended transmissions. Such collaboration is based on the same principles as the relay channel, but taking into account that the terminal that is assisting other users has also its own messages to be transmitted.

In this regard, Section 11.3.1 describes how the cooperative transmission is applied to the wireless cellular networks in the uplink, where the mobile stations have information to be transmitted to the base station, i.e. the destination. Terminals collaborate in order to improve the performance of their transmissions. This collaboration assumes that each terminal becomes a decode-and-forward relay for another terminal, and all can operate in full-duplex mode, i.e. receiving and transmitting simultaneously. Likewise, Section 11.3.2 explores cooperative communication over wireless networks with a half-duplex constraint on terminals, i.e. the periods of time where a terminal is receiving and transmitting are different. Protocol I (see Figure 11.4) is adapted for the investigated scenario, which assumes a static resource allocation. The different transmission strategies presented in Section 11.2.4 are considered when a terminal acts as a relay, such as amplify-and-forward, decode-and-forward (under repetition coding or unconstrained coding), selective or incremental relaying. Finally, Section 11.3.3 analyzes the same scenario as in Section 11.3.2 but assumes that a given terminal can be helped by multiple terminals (relays or virtual antennas), which transmit under a distributed space-time code.

---

[3] All relays operate in one-way mode if it is not specified.

11.3.1   User Cooperation in the Uplink

It was shown in Chapter 4 that multiple transmit antennas provide spatial diversity. Unfortunately, this is not easy to implement in the uplink of a cellular system, due to the size of the mobile unit. In order to overcome this limitation, and yet still emulate transmit antenna diversity, an alternative form of spatial diversity is considered, where diversity gains are achieved via the cooperation of in-cell users. That is, in each cell, each user may have a 'partner'. Each of the two partners would be responsible for transmitting not only their own information, but also the information of their partner, which they receive and detect. Spatial diversity is achieved through the use of the partner's antenna. This is complicated by the fact that the interuser channel is noisy and by the fact that both partners have information of their own to send.

The basic premise of this concept is that both users have information of their own to send, denoted by U_i for $i = 1, 2$, and would like to cooperate in order to send this information to the receiver at the highest rate possible. To distinguish this main/final receiver from the receiving units of the mobiles, we will refer to it as the BS, even though the user cooperation idea is equally applicable to ad hoc networks. The channel model we use is depicted in Figure 11.6. Each mobile receives an attenuated and noisy version of the partner's transmitted signal and uses that, in conjunction with its own data, to construct its transmit signal. The BS receives a noisy version of the sum of the attenuated signals of both users. The mathematical formulation of the model is [6, 27]:

$$Y_0(t) = h_{10}X_1(t) + h_{20}X_2(t) + n_0(t)$$
$$Y_1(t) = h_{21}X_2(t) + n_1(t) \tag{11.1}$$
$$Y_2(t) = h_{12}X_1(t) + n_2(t)$$

where $Y_0(t)$, $Y_1(t)$, and $Y_2(t)$ are the baseband models of the received signals at the BS, user 1 and user 2, respectively, during one symbol period. Also, $X_i(i)$ is the signal transmitted by user i, for $i = 1, 2$, and $n_i(t)$ are the additive channel noise terms at the BS, user 1, and user 2, for $i = 0, 1, 2$, respectively. The fading coefficients, $\{h_{ij}\}$, remain constant over at least one symbol period, and observed over time form independent stationary ergodic stochastic processes, resulting in frequency-nonselective fading.

A critical assumption here is that there is no contribution from $X_2(t)$ in $Y_2(t)$, even though they are actually both present at the terminal belonging to user 2. Since $X_2(t)$ does not go through any fading before it reaches the antenna of user 2, unlike $Y_2(t)$, it may appear that it will have a detrimental effect on

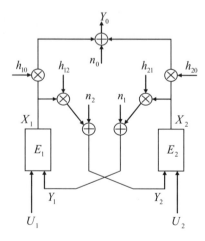

Figure 11.6   Channel model.

the reception of $Y_2(t)$. Provided that user 2 knows the relevant antenna gains, it is assumed that canceling the effects of $X_2(t)$ on $Y_2(t)$ is possible, and thus the model gives an accurate representation. A similar argument can be made in the case of user 1, regarding the effects of $X_1(t)$ on the reception of $Y_1(t)$. In practice, to isolate the transmitted signal from the received one, it may be necessary to use two separate channels, two collocated antennas, or some other means. For example, the CDMA implementations of this concept make use of spreading codes to create two separate channels, thus eliminating the need for echo cancellation. Also, time division among the mobiles has been investigated in [7, 22] where it was shown that cooperation continues to provide full diversity.

In the following we also use additional assumptions: (i) the transmitted signals $X_i(t)$ have an average power constraint of P_i for $i = 1, 2$; (ii) the noise terms $n_i(t)$ are white zero-mean complex Gaussian random processes with spectral height $N_i/2$ for $i = 0, 1, 2$; and (iii) the fading coefficients h_{ij} are zero-mean complex Gaussian random variables with variance ξ_{ij}^2 (Rayleigh fading). It is also assumed that the BS can track the variations in h_{10} and h_{20}, user 1 can track h_{21}, and user 2 can track h_{12}, implying that all the decoding is done with the knowledge of the fading parameters. Due to the reciprocity of the channel, it is also assumed that h_{21} and h_{12} are equal. Finally, for simplicity of analysis and exposition, though with no loss in generality, a synchronous system is assumed.

11.3.1.1 System Capacity

The mathematical model we use is a discrete time version of the model described in (11.1) and is given by:

$$Y_0 = h_{10}X_1 + h_{20}X_2 + n_0$$
$$Y_1 = h_{21}X_2 + n_1 \tag{11.2}$$
$$Y_2 = h_{12}X_1 + n_2$$

with $n_0 \sim N(0, \sigma_0^2)$, $n_1 \sim N(0, \sigma_1^2)$ and $n_2 \sim N(0, \sigma_2^2)$. In general, we assume that $\sigma_1^2 = \sigma_2^2$. The system is causal and transmission is done through blocks of length N, therefore the signal of user 1 at time j, $j = 1, \ldots, N$, can be expressed as $X_1(U_1, Y_1(j - l), Y_1(j - 2), 1/4, Y_1(1))$, where U_1 is the message that user 1 wants to transmit to the BS at that particular block. Similarly, for user 2, we have $X_2(U_2, Y_2(j - l), Y_2(j - 2), 1/4, Y_2(1))$.

The transmission is done for B blocks of length N, where both B and N are large. The mobiles will cooperate based on the signals they receive in the previous block.

We assume mobile 1 divides its information U_1 into two parts: U_{10}, to be sent directly to the BS, and U_{12}, to be sent to the BS via mobile 2. Mobile 1 then structures its transmit signal so that it is able to send the above information as well as some additional cooperative information to the BS. This is done according to $X_1 = X_{10} + X_{12} + e_1$ and divides its total power accordingly $P_1 = P_{10} + P_{12} + P_{e_1}$ where e_1 refers to the part of the signal exchanged to transmit the cooperative information. Thus, X_{10} is allocated power P_{10} for sending U_{10} at rate R_{10} directly to the BS, X_{12} is allocated power P_{12} for sending U_{12} to user 2 at rate R_{12}, and e_1 is allocated power P_{e_1} for sending cooperative information to the BS. It should be noted that the transmission rate of U_{12}, i.e. P_{12}, and the power allocated to U_{12}, i.e. P_{12}, should be such that U_{12} can be perfectly decoded by mobile 2. This perfect reconstruction at the partner forms the basis for cooperation. Mobile 2 structures its transmit signal X_2 and divides its total power P_2 in a similar fashion.

Recall that we transmit for B blocks of length N. Cooperation in block i is achieved by constructing signals e_1 and e_2 based on $(U_{12}(i - 1), U_{21}(i-1))$, both of which are now known at mobile 1 and mobile 2. The receiver waits until all the B blocks have been received and starts decoding from the last block. An achievable rate region with user cooperation is obtained by first considering the above cooperative strategy with constant attenuation factors, and then incorporating the randomness. It is assumed that each block of length N is long enough to observe the ergodicity of the fading distributions.

An achievable rate region for the system given in (11.2) is the closure of the convex hull of all rate pairs (R_1, R_2) such that $R_1 = R_{10} + R_{12}$ and $R_2 = R_{20} + R_{21}$ with

$$R_{12} < E\left\{ C\left(\frac{h_{12}^2 P_{12}}{h_{12}^2 P_{10} + \sigma_1^2} \right) \right\}$$

$$R_{21} < E\left\{ C\left(\frac{h_{21}^2 P_{21}}{h_{21}^2 P_{20} + \sigma_2^2} \right) \right\}$$

$$R_{10} < E\left\{ C\left(\frac{h_{10}^2 P_{10}}{\sigma_0^2} \right) \right\}$$

$$R_{20} < E\left\{ C\left(\frac{h_{20}^2 P_{20}}{\sigma_0^2} \right) \right\}$$

$$R_{10} + R_{20} < E\left\{ C\left(\frac{h_{10}^2 P_{10} + h_{20}^2 P_{20}}{\sigma_0^2} \right) \right\}$$

$$R_{10} + R_{20} + R_{12} + R_{21} < E\left\{ C\left(\frac{h_{10}^2 P_1 + h_{20}^2 P_2 + 2h_{10}h_{20}\sqrt{P_{e1}P_{e2}}}{\sigma_0^2} \right) \right\} \qquad (11.3)$$

for some power assignment satisfying $P_1 = P_{10} + P_{12} + P_{e1}$, $P_2 = P_{20} + P_{21} + P_{e2}$. The function $C(y) = (1/2)\log(1 + y)$ is the capacity of an additive white Gaussian noise (AWGN) channel with signal-to-noise ratio (SNR) y and E denoted expectation with respect to the fading parameters h_{ij}. The proof of Equation (11.3) is based on [6, 21].

Equation (11.3) is shown in Figure 11.7 and Figure 11.8, for different scenarios of channel quality. For the no-cooperation case, the users ignore the signals Y_2 and Y_1, which are equivalent to the

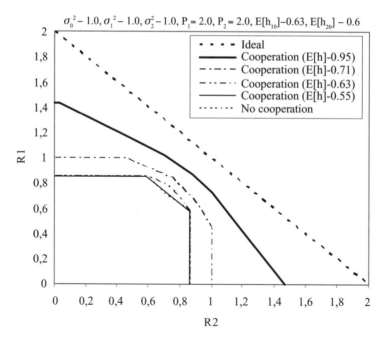

Figure 11.7 Capacity region when the two users face statistically equivalent channels toward the BS.

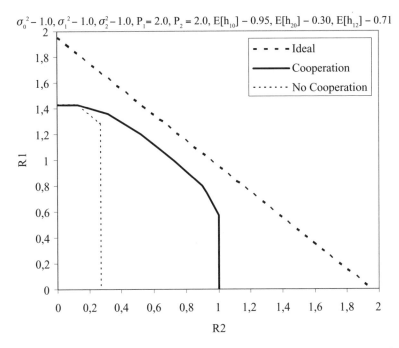

$\sigma_0^2 - 1.0, \sigma_1^2 - 1.0, \sigma_2^2 - 1.0, P_1 = 2.0, P_2 = 2.0, E[h_{10}] - 0.95, E[h_{20}] - 0.30, E[h_{12}] - 0.71$

Figure 11.8 Capacity region when the two users face statistically dissimilar channels toward the BS.

multiple-access channel capacity region. The noiseless interuser channel ($\sigma_1^2 = \sigma_2^2 = 0$) is referred to as ideal cooperation, and is used mostly as an upper bound for the performance of any cooperation scheme.

From Figure 11.7, we can see that when the channels from the users to the BS have similar quality (h_{10} and h_{20} have the same mean) and the channel between the users is better (h_{12} has larger mean), the cooperation scheme greatly improves the achievable rate region. As the interuser channel degrades and the severity of the interuser fading increases, performance approaches that of no cooperation.

When the user–BS links of the two users experience fading with different means, cooperation again improves the achievable rate region, as shown in Figure 11.8. In this case, the user with more fading benefits most from the cooperation. The equal rate point ($R_1 = R_2$) or the maximum rate sum point ($R_1 + R_2$) is increased considerably with cooperation.

In Figure 11.7 and Figure 11.8, the point where any achievable rate curve intersects the Y axis corresponds to user 2 becoming a relay for user 1, and the point where the curve intersects the X axis corresponds to user 1 becoming a relay for user 2.

11.3.1.2 Probability of Outage

If the attenuation factors vary slowly and can be approximated as constants over the B blocks of length N, then over these B blocks we can achieve rates dictated by the current values of h_{10}, h_{20}, h_{12}, and h_{21}. We assume N (the block length) and B (the number of blocks) are large enough to achieve capacity in the case of constant attenuation factors. However, in order to talk about the achievable rate region in Equation (11.3), we need to have even longer block lengths and observe different realizations of our fading amplitudes. When the delay requirements prevent us from having these longer block lengths, the rates achieved will be random quantities based on the current realizations of the fading amplitudes. Some wireless services have minimum requirements on the supported data rates, below which the service is unsustainable. Therefore, we observe an *outage* if the random rates that we can achieve fall below a

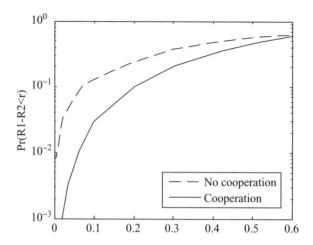

Figure 11.9   Probability of outage $\sigma_0^2 = 1.0, \sigma_1^2 - 1.0, \sigma_2^2 - 1.0, P_1 = 2.0, P_2 = 2.0, \mathrm{E}[h_{10}] - 0.63, \mathrm{E}[h_{20}] - 0.63, \mathrm{E}[h_{12}] - 0.63$.

certain level, which we will call the service sustainability rate, and consider the probability of outage as a performance criterion [43].

In particular, we consider the equal rate point $(R_1 = R_2 = R)$ and calculate the probability of outage versus the service sustainability rate r for the cooperation and the no-cooperation schemes. The probability of outage, $P_{\mathrm{out}} = \Pr(R < r)$, provides us with the probability that the current realization of our slowly fading parameters h_{10}, h_{20}, h_{12}, and h_{21} will not be able to support an equal transmission rate of r for the particular scheme under consideration. The results are given in Figure 11.9. We observe that for all service sustainability rates, the probability of outage for the cooperation scheme is smaller than the probability of outage under no cooperation. This is true despite the fact that the increase in achievable rate due to cooperation is moderate for the scenario depicted in Figure 11.9, as can be seen from Figure 11.7 ($\mathrm{E}[h_{12}] = 0.63$). This demonstrates that even in cases when it does not significantly increase achievable rates, user cooperation is still able to increase robustness against channel variations. We, of course, expect the robustness to improve even more as the interuser channel quality ($\mathrm{E}[h_{12}]$) improves.

11.3.1.3   Cellular Coverage

Our expectation of increased cell coverage is based on a reduction in the required average received power. Assume the mobile is at a distance d km from the BS. Let P_{tx} denote the average power transmitted by the mobile and P_{rx} denote the average power received at the BS (in decibels). Then, in a typical wireless channel, we have (in decibels) $P_{\mathrm{rx}} = P_{\mathrm{tx}} - PL(d)$ where $PL(d)$ is the mean path loss at distance d km. A common model for $PL(d)$ is the Hata model with $PL(d) = B_1 + B_2 \log d$, where B_1 and B_2 are functions of transmitter and receiver antenna height, carrier frequency, and type of environment. Therefore $\log d = (P_{\mathrm{TX}} - P_{\mathrm{RX}} - B_1)/B_2$. For a fixed P_{rx}, the maximum d is achieved when P_{tx} is at its maximum, denoted by $P_{TX}^{\max}$. This is the maximum power that the mobile can transmit. So $d_{\max}\,(P_{RX}) = (P_{TX}^{\max} - P_{\mathrm{RX}} - B_1)/B_2$. The coverage of the cell is thus given by $d_{\max}$. The only quantity of the right-hand side that is variable is P_{rx}.

Assume that two different transmission schemes require different received average powers in order to operate successfully. That is, let scheme 1 require $P_{\mathrm{RX}}^{(1)}$ and scheme 2 require $P_{\mathrm{RX}}^{(2)}$. Define α to be their ratio, that is in decibels $10 \log \alpha = P_{\mathrm{RX}}^{(1)} - P_{\mathrm{RX}}^{(2)}$. From the above, we have $\log \log d_{\max}^{(1)} - \log d_{\max}^{(2)} = (P_{\mathrm{RX}}^{(2)} - P_{\mathrm{RX}}^{(1)})/B_2 = -(10 \log \alpha)/B_2$ resulting in $d_{\max}^{(1)}/d_{\max}^{(2)} = \alpha^{-10/B_2}$. A typical set of values for B_1 and

B_2 is $B_1 = 17.3$ and $B_2 = 33.8$ (obtained for a medium-sized city environment, with carrier frequency of 900 MHz, transmit antenna height of 50 m, receive antenna height of 1 m, and a net antenna gain of 6 dB). Thus, the ratio of the coverage of scheme 1 versus scheme 2, as a function of the ratio of the required received average powers, is $d_{max}^{(1)}/d_{max}^{(2)} = \alpha^{-1/3.38}$.

11.3.2   Cooperative Diversity in Wireless Networks

11.3.2.1   System and Channel Models

In this section we divide the available bandwidth into orthogonal channels and allocate these channels to the transmitting terminals owing to the half-duplex constraint, allowing the protocols to be integrated readily into existing networks. Removing the interference between the terminals at the destination radio simplifies substantially the receiver algorithms and the outage analysis for the purposes of exposition. Figure 11.10 illustrates our channel allocation for an example of the time-division approach with two terminals.

The critical assumptions in this section are the different levels of synchronization between the terminals in order for cooperative diversity to be effective. As suggested by Figure 11.10 and the modeling discussion to follow, we consider the scenario in which the terminals are block, carrier, and symbol synchronous. Given some form of network block synchronization, carrier and symbol synchronization for the network can build upon the same between the individual transmitters and receivers. For the detail of network synchronization see www.wiley.com/go/glisic1.

In characterizing the channel models, we modify notation (11.2) and focus on the message of the 'source' terminal T_s, which potentially employs terminal T_r as a 'relay,' in transmitting to the 'destination' terminal T_d. We utilize a baseband-equivalent, discrete-time channel model for the continuous-time channel, and we consider N consecutive uses of the channel, where N is large.

For direct transmission, we model the channel as:

$$y_d[n] = h_{s,d}x_s[n] + n_d[n]; \quad n = 1.\ldots, N/2, \tag{11.4}$$

where $x_s[n]$ is the source transmitted signal, and $y_d[n]$ is the destination received signal. The other terminal transmits for $n = N/2 + 1, \ldots, N$ as depicted in Figure 11.10(b). Thus, in the baseline system, each terminal utilizes only half of the available degrees of freedom of the channel.

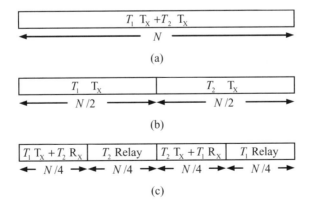

(a)

(b)

(c)

Figure 11.10   Example time-division channel allocations for (a) direct transmission with interference, (b) orthogonal direct transmission, and (c) orthogonal cooperative diversity. Throughout this section we focus on orthogonal transmissions of the form (b) and (c).

For cooperative diversity, we model the channel during the first half of the block as:

$$y_r[n] = h_{s,r}x_s[n] + n_r[n]$$
$$y_d[n] = h_{s,d}x_s[n] + n_d[n]; \quad n = 1, \ldots, N/4 \quad (11.5)$$

where $x_s[n]$ is the source transmitted signal and $y_r[n]$ and $y_d[n]$ are the relay and destination received signals, respectively. For the second half of the block, we have:

$$y_d[n] = h_{r,d}x_r[n] + n_d[n]; \quad n = N/4 + 1, \ldots, N/2 \quad (11.6)$$

where $x_r[n]$ is the relay transmitted signal and $y_d[n]$ is the destination received signal. The representation in the second half of the block is similar, with the roles of the source and relay reversed, as depicted in Figure 11.10(c). Note that half the degrees of freedom are allocated to each source terminal for transmission to its destination, and a quarter of the degrees of freedom are available for communication to its relay.

Channel coefficients $h_{i,j}$ capture the effects of path-loss, shadowing, and frequency nonselective fading, and $n_j[n]$ captures the effects of receiver noise and other forms of interference in the system, where $i \in \{s, r\}$ and $j \in \{r, d\}$. Statistically, we model $h_{i,j}$ as zero-mean, independent, circularly symmetric complex Gaussian random variables with variances $\sigma_{i,j}^2$. Furthermore, we model $n_j[n]$ as zero-mean mutually independent, circularly symmetric, complex Gaussian random sequences with variance N_0.

Two important parameters of the above system are the SNR without fading and the spectral efficiency. If the transmitting terminals have an average power constraint in the continuous-time channel model of P_c joules per second, this translates into a discrete-time power constraint of $P = 2P_c/W$ since each terminal transmits in half of the available degrees of freedom, under both direct transmission and cooperative diversity. Thus, the channel model is parameterized by the SNR variables $SNR |h_{i,j}|^2$, where $SNR := 2P_c/(N_0W) = P/N_0 = y$ is the common SNR without fading.

In addition to SNR, transmission schemes are further parameterized by the rate r bits per second, or spectral efficiency $R := 2r/W$ b/s/Hz attempted by the transmitting terminals. Nominally, one could parameterize the system by the pair (y, R); however, our results lend more insight, and are substantially more compact, when we use either of the pairs (Y, R) or $(y, \Re)$, where

$$Y := \frac{y}{2^R - 1}, \quad \Re = \frac{R}{\log(1 + y\sigma_{s,d}^2)} \quad (11.7)$$

11.3.2.2 Transmission Strategy at the Relay

Among the different decoding modes allowed in a relay, detailed in 11.2.4.1, we look into the amplify-and-forward and decode-and-forward modes. Likewise, since we are working with a *static resource allocation* the cooperative transmission will be improved by means of *selection relaying* or *incremental relaying*, see Section 11.2.4.2.

Amplify-and-Forward

For amplify-and-forward transmission, the appropriate channel model is (11.5)–(11.6). The source terminal transmits its information as $x_s[n]$, say, for $n = 1, \ldots, N/4$. During this interval, the relay processes $y_r[n]$, and relays the information by transmitting $x_r[n] = \beta y_r[n - N/4]$ for $n = N/4+1, 1/4, N/2$. To remain within its power constraint (with high probability), an amplifying relay must use gain $\beta \leq \sqrt{P/(|h_{s,r}|^2 P + N_0)}$ where we allow the amplifier gain to depend upon the fading coefficient $h_{s,r}$ between the source and relay, which the relay estimates to high accuracy. This scheme can be viewed as repetition coding from two separate transmitters, except that the relay transmitter amplifies its own receiver noise. The destination can decode its received signal $y_d[n]$ for $n = 1, \ldots, N/2$ by first appropriately combining the signals from the two subblocks using one of a variety of combining techniques; in the sequel, we focus on a suitably designed matched filter, or maximum-ratio combiner.

Decode-and-Forward

For decode-and-forward transmission, the appropriate channel model is again (11.5)–(11.6). The source terminal transmits its information as $x_s[n]$, for $n = 0, \ldots, N/4$. During this interval, the relay processes $y_r[n]$ by decoding an estimate $x_s[n]$ of the source transmitted signal.

Under a repetition-coded scheme, the relay transmits the signal:

$$x_r[n] = \hat{x}_s[n - N/4] \quad \text{for} \quad n = N/4 + 1 \ldots, N/2$$

Decoding at the relay can take on a variety of forms. For example, the relay might fully decode, i.e. estimate without error, the entire source codeword, or it might employ symbol-by-symbol decoding and allow the destination to perform full decoding. These options allow for trading off performance and complexity at the relay terminal. Note that we focus on full decoding in the sequel. Again, the destination can employ a variety of combining techniques; we focus in the sequel on a suitably modified matched filter.

Selective Relaying

Fixed decode-and-forward is limited by direct transmission between the source and relay. However, since the fading coefficients are known to the appropriate receivers, $h_{s,r}$ can be measured to high accuracy by the cooperating terminals; thus, they can adapt their transmission format according to the realized value of $h_{s,r}$.

If the measured $|h_{s,r}|^2$ falls below a certain threshold, the source simply continues its transmission to the destination, in the form of repetition or more powerful codes. If the measured $|h_{s,r}|^2$ lies above the threshold, the relay forwards what it received from the source, using either amplify-and-forward or decode-and-forward, in an attempt to achieve diversity gain.

Incremental Relaying

Fixed and selection relaying can make inefficient use of the degrees of freedom of the channel, especially for high rates, because the relays repeat all the time. As an alternative, incremental relaying protocols can be used that exploit limited feedback from the destination terminal, e.g. a single bit indicating the success or failure of the direct transmission. We nominally allocate the channels according to Figure 11.10(b). First, the source transmits its information to the destination at spectral efficiency R. The destination indicates success or failure by broadcasting a single bit of feedback to the source and relay, which we assume is detected reliably by at least the relay. If the source-destination SNR is sufficiently high, the feedback indicates success of the direct transmission, and the relay does nothing. If the source-destination SNR is not sufficiently high for successful direct transmission, the feedback requests that the relay amplify-and-forward what it received from the source. In the latter case, the destination tries to combine the two transmissions.

11.3.2.3 Outage Probabilities

As a function of the fading coefficients viewed as random variables, the mutual information for a protocol is a random variable denoted by I; in turn, for a target rate R, $I < R$ denotes the outage event, and $\Pr[I < R]$ denotes the outage probability.

Direct Transmission

For direct transmission, the source terminal transmits over the channel (11.4). The maximum average mutual information between input and output in this case, achieved by independent and identically distributed (i.i.d.) zero-mean, circularly symmetric complex Gaussian inputs, is given by $I_D = \log(1 + y|h_{s,d}|^2)$ as a function of the fading coefficient $h_{s,d}$. The outage event for spectral efficiency R is given by $I_d < R$ and is equivalent to the event $|h_{s,d}|^2 < 1/Y$. For Rayleigh fading, i.e., $|h_{s,d}|^2$ exponentially distributed with parameter $\sigma_{s,d}^{-2}$, the outage probability satisfies

$$p_D^{\text{out}}(y, R) := \Pr[I_D < R] = \Pr\left[|h_{s,d}|^2 < \frac{1}{Y}\right] = 1 - \exp\left(-\frac{1}{Y\sigma_{s,d}^2}\right) \sim \frac{1}{\sigma_{s,d}^2} \cdot \frac{1}{Y} \tag{11.8}$$

Amplify-and-Forward

The amplify-and-forward protocol produces an equivalent one-input, two-output complex Gaussian noise channel with different noise levels in the outputs. The maximum average mutual information between the input and the two outputs, achieved by i.i.d. complex Gaussian inputs, is given by (see the Appendix of Section 11.6.5):

$$I_{AF} = \frac{1}{2}\log(1 + y|h_{s,d}|^2 + f(y|h_{s,r}|^2, y|h_{r,d}|^2)) \tag{11.9}$$

as a function of the fading coefficients, where $f(u,v) := uv/(u + v + 1)$.

The outage event for spectral efficiency R is given by $I_{AF} < R$ and is equivalent to the event:

$$|h_{s,d}|^2 + \frac{1}{y}f(y|h_{s,r}|^2, y|h_{r,d}|^2) < \frac{2^{2R} - 1}{y} \tag{11.10}$$

For Rayleigh fading, i.e., $|h_{i,j}|^2$ independent and exponentially distributed with parameters $\sigma_{i,j}^{-2}$, analytic calculation of the outage probability becomes involved, but we can approximate its high-SNR behavior as:

$$p_{AF}^{\text{out}}(y, R) := \Pr[I_{AF} < R] \sim \left(\frac{1}{2\sigma_{s,d}^2}\frac{\sigma_{s,r}^2 + \sigma_{r,d}^2}{\sigma_{s,r}^2\sigma_{r,d}^2}\right)\left(\frac{2^{2R} - 1}{y}\right)^2 \tag{11.11}$$

where we have utilized the result of Equality 1 in the Appendix of Section 11.6.4, with:

$$u = |h_{s,d}|^2, \ v = |h_{s,r}|^2, \ w = |h_{r,d}|^2, \ \lambda_u = \sigma_{s,d}^{-2}, \ \lambda_v = \sigma_{s,r}^{-2}, \ \lambda_w = \sigma_{r,d}^{-2}$$
$$g(\varepsilon) = (2^{2R} - 1)\varepsilon, \ t = y, h(t) = 1/t$$

Decode-and-Forward

In this case we examine a particular decoding structure at the relay. Specifically, we require the relay to decode fully the source message; examination of symbol-by-symbol decoding at the relay becomes involved because it depends upon the particular coding and modulation choices. The maximum average mutual information for repetition-coded decode-and-forward can be readily shown to be:

$$I_{DF} = \frac{1}{2}\min\left\{\log(1 + y|h_{s,r}|^2), \log(1 + y|h_{s,d}|^2 + y|h_{r,d}|^2)\right\} \tag{11.12}$$

as a function of the fading random variables. The first term represents the maximum rate at which the relay can reliably decode the source message, while the second term represents the maximum rate at which the destination can reliably decode the source message given repeated transmissions from the source and destination. Requiring both the relay and destination to decode the entire codeword without error results in the minimum of the two mutual pieces of information in (11.12).

The outage event for spectral efficiency R is given by $I_{df} < R$ and is equivalent to the event

$$\min \left\{ |h_{s,r}|^2 , |h_{s,d}|^2 + |h_{r,d}|^2 \right\} < \frac{2^{2R} - 1}{y}. \tag{11.13}$$

For Rayleigh fading, the outage probability for repetition-coded decode-and-forward can be computed according to

$$p_{DF}^{\text{out}}(y, R) := \Pr[I_{DF} < R] = \Pr\left[|h_{s,r}|^2 < g(y) \right] +$$
$$+ \Pr\left[|h_{s,r}|^2 \geq g(y) \right] \Pr\left[|h_{s,d}|^2 + |h_{r,d}|^2 < g(y) \right] \tag{11.14}$$

where $g(y) = [2^{2R} - 1]/y$. We examine the large y behavior of Equation (11.14) by computing the limit

$$\frac{1}{g(y)} p_{DF}^{\text{out}}(y, R) = \underbrace{\frac{1}{g(y)} \Pr\left[|h_{s,r}|^2 < g(y) \right]}_{\rightarrow 1/\sigma_{s,r}^2}$$

$$= \underbrace{\Pr\left[|h_{s,r}|^2 \geq g(y) \right]}_{\rightarrow 1} \underbrace{\frac{1}{g(y)} \Pr\left[|h_{s,d}|^2 + |h_{r,d}|^2 < g(y) \right]}_{\rightarrow 0} \rightarrow 1/\sigma_{s,r}^2 \tag{11.15}$$

as $y \rightarrow \infty$ using the results of Results 1 and 2 in Appendix of Section 11.6.4. So,

$$p_{DF}^{\text{out}}(y, R) \sim \frac{1}{2\sigma_{s,r}^2} \frac{2^{2R} - 1}{y} \tag{11.16}$$

The $1/y$ behavior in (11.16) indicates that fixed decode-and-forward does not offer diversity gains for large SNR, because requiring the relay to decode fully the source information limits the performance of decode-and-forward to that of direct transmission between the source and relay.

Selective Relaying
As an example analysis, we determine the performance of selection decode-and-forward. In the case of repetition coding at the relay, we have

$$I_{SDF} = \begin{cases} \dfrac{1}{2} \log(1 + 2y \, |h_{s,d}|^2), & |h_{s,r}|^2 < g(y) \\[2mm] \dfrac{1}{2} \log(1 + y[\,|h_{s,d}|^2 + |h_{r,d}|^2]), & |h_{s,r}|^2 \geq g(y) \end{cases} \tag{11.17}$$

where $g(y) = [2^{2R} - 1]/y$. This threshold is motivated by our discussion of direct transmission.

The first case in Equation (11.17) corresponds to the relay's not being able to decode and the source's repeating its transmission; here, the maximum average mutual information is that of repetition coding from the source to the destination, hence the factor $2y$. The second case in Equation (11.17) corresponds to the relay's ability to decode and repeat the source transmission; here, the maximum average mutual information is that of repetition coding from the source and relay to the destination.

The outage event for spectral efficiency R is given by $I_{SDF} < R$ and is equivalent to the event:

$$\left(\left\{ |h_{s,r}|^2 < g(y) \right\} \cap \left\{ 2 \, |h_{s,d}|^2 < g(y) \right\} \right)$$
$$\cup \left(\left\{ |h_{s,r}|^2 \geq g(y) \right\} \cap \left\{ |h_{s,d}|^2 + |h_{r,d}|^2 < g(y) \right\} \right) \tag{11.18}$$

The first (second) event in (11.18) corresponds to the first (second) case in Equation (11.17). Because the events in Equation (11.18) are mutually exclusive, the outage probability becomes a sum:

$$
\begin{aligned}
p_{SDF}^{out}(y, R) : & = \Pr[I_{SDF} < R] \\
& = \Pr\left[\left|h_{s,r}\right|^2 < g(y)\right] \Pr\left[2\left|h_{s,d}\right|^2 < g(y)\right] \\
& \quad + \Pr\left[\left|h_{s,r}\right|^2 \geq g(y)\right] \Pr\left[\left|h_{s,d}\right|^2 + \left|h_{r,d}\right|^2 < g(y)\right]
\end{aligned}
\tag{11.19}
$$

and we may readily compute a closed-form expression for Equation (11.19). For comparison with our other protocols, we examine the large SNR behavior of Equation (11.19) by computing the limit:

$$
\begin{aligned}
\frac{1}{g^2(y)} p_{SDF}^{out}(y, R) = & \underbrace{\frac{1}{g(y)}\Pr\left[\left|h_{s,r}\right|^2 < g(y)\right]}_{\to 1/\sigma_{s,r}^2} \underbrace{\frac{1}{g(y)}\Pr\left[\left|h_{s,d}\right|^2 < g(y)\right]}_{\to 1/(2\sigma_{s,d}^2)} \\
& + \underbrace{\Pr\left[\left|h_{s,r}\right|^2 \geq g(y)\right]}_{\to 1} \underbrace{\frac{1}{g^2(y)}\Pr\left[\left|h_{s,d}\right|^2 + \left|h_{r,d}\right|^2 < g(y)\right]}_{\to 1/(2\sigma_{s,d}^2 \sigma_{r,d}^2)} \\
& \to \left(\frac{1}{2\sigma_{s,d}^2} \frac{\sigma_{s,r}^2 + \sigma_{r,d}^2}{\sigma_{s,r}^2 \sigma_{r,d}^2}\right)
\end{aligned}
\tag{11.20}
$$

as $y \to \infty$, using the results 1 and 2 of Appendix in Section 11.6.4. Thus, we conclude that the large SNR performance of selection decode-and-forward is identical to that of fixed amplify-and-forward.

11.3.2.4 Performance Bounds for Cooperative Diversity

If we suppose that the source and relay know each other's messages *a priori*, then instead of direct transmission, each would benefit from using a space-time code for two transmit antennas. In this sense, the outage probability of conventional transmit diversity (see Chapter 4) represents an optimistic lower bound on the outage probability of cooperative diversity.

Transmit Diversity Bound
To utilize a space-time code for each terminal, we allocate the channel as in Figure 11.10(b). Both terminals transmit in all the degrees of freedom of the channel, so their transmitted power is $P/2$, half that of direct transmission. The spectral efficiency for each terminal remains R.

For transmit diversity, we model the channel as:

$$
y_d[n] = [h_{s,d} h_{r,d}]\begin{bmatrix} x_s[n] \\ x_r[n] \end{bmatrix} + n_d[n]; \qquad n = 0, \ldots, N/2
\tag{11.21}
$$

An optimal signaling strategy, in terms of minimizing outage probability in the large SNR regime, is to encode information using $[x_s \ x_r]^T$ i.i.d. complex Gaussian, each with power $P/2$ (see Appendix of Section 11.6.6). Using this result, the maximum average mutual information as a function of the fading coefficients is given by:

$$
I_T = \log\left(1 + \frac{y}{2}\left[\left|h_{s,d}\right|^2 + \left|h_{r,d}\right|^2\right]\right)
\tag{11.22}
$$

. The outage event $I_T < R$ is defined by $\left|h_{s,d}\right|^2 + \left|h_{r,d}\right|^2 < (2^R - 1)/(y/2)$.

By using the result of Fact 2 in the Appendix of Section 11.6.4, for $|h_{i,j}|^2$ exponentially distributed with parameters $\sigma_{i,j}^{-2}$, the outage probability becomes:

$$p_T^{out}(y, R) := \Pr[I_T < R] \sim \frac{2}{\sigma_{s,d}^2 \sigma_{r,d}^2} \cdot \left(\frac{2^R - 1}{y}\right)^2 \tag{11.23}$$

Orthogonal Transmit Diversity Bound

The transmit diversity bound (11.23) does not take into account the half-duplex constraint. To capture this effect, we constrain the transmit diversity scheme to be orthogonal. When the source and relay can cooperate perfectly, an equivalent model to Equation (11.21), incorporating the relay orthogonality constraint, consists of parallel channels:

$$\begin{aligned} y_d[n] &= h_{s,d} x_s[n] + n_d[n], \quad n = 0, \ldots, N/4 \\ y_d[n] &= h_{r,d} x_r[n] + n_d[n], \quad n = N/4 + 1, \ldots, N/2 \end{aligned} \tag{11.24}$$

This pair of parallel channels is utilized half as many times as the corresponding direct transmission channel, so the source must transmit at twice the spectral efficiency in order to achieve the same spectral efficiency as direct transmission.

For each fading realization, the maximum average mutual information can be obtained using independent complex Gaussian inputs. Allocating a fraction α of the power to x_s, and the remaining fraction $\bar{\alpha} := (1 - \alpha)$ of the power to x_r, the average mutual information is given by:

$$I_P = \frac{1}{2} \log \left[\left(1 + 2\alpha y \left|h_{s,d}\right|^2\right) \left(1 + 2\bar{\alpha} y \left|h_{r,d}\right|^2\right) \right] \tag{11.25}$$

The outage event $I_P < R$ is equivalent to the outage region $\alpha |a_{s,d}|^2 + \bar{\alpha} |a_{r,d}|^2 + 2\alpha\bar{\alpha} y |h_{s,d}|^2 |h_{r,d}|^2 < (2^{2R} - 1)/2y$, and for high-SNR in Rayleigh fading, using the result of Equality 2 in Appendix of Section 11.6.4, with $u = \alpha \left|h_{s,d}\right|^2$, $v = \bar{\alpha} \left|h_{r,d}\right|^2$, $\lambda_u = 1/(\alpha\sigma_{s,d}^{-2})$, $\lambda_v = 1/(\bar{\alpha}\sigma_{r,d}^{-2})$, $\varepsilon = (2^{2R} - 1)/(2y)$, $t = 2^{2R} - 1$, we have:

$$p_P^{out}(y, R) := \Pr[I_P < R] \sim \frac{1}{4\alpha\bar{\alpha}\sigma_{s,d}^2 \sigma_{r,d}^2} \cdot \frac{2^{2R}[2R \ln(2) - 1] + 1}{y^2} \tag{11.26}$$

which is minimized for $\alpha = 1/2$, yielding:

$$p_P^{out}(y, R) \sim \frac{1}{\sigma_{s,d}^2 \sigma_{r,d}^2} \cdot \frac{2^{2R}[2R\ln(2) - 1] + 1}{y^2} \tag{11.27}$$

Incremental Relaying

The protocols operate at spectral efficiency R when the source-destination transmission is successful, and spectral efficiency $R/2$ when the relay repeats the source transmission. Thus, we examine outage probability as a function of SNR and the *expected* spectral efficiency $\bar{R}$.

For incremental amplify-and-forward, the outage probability as a function of SNR and R is given by:

$$\begin{aligned} p_{IAF}^{out}(y, R) &= \Pr[I_D < R] \Pr[I_{AF} < R/2 \,|\, I_D < R] = \Pr[I_{AF} < R/2] \\ &= \Pr\left[\left|h_{s,d}\right|^2 + \frac{1}{y} f\left(y \left|h_{s,r}\right|^2, y \left|h_{r,d}\right|^2\right) < g(y)\right] \end{aligned} \tag{11.28}$$

where I_D, I_{AF} and $f(\bullet,\bullet)$ are given in Section 11.3.2.3 and $g(y) = [2^R - 1]/y = 1/Y$. The second equality follows from the fact that the intersection of the direct and amplify-and-forward outage events is exactly the amplify-and-forward outage event at half the rate. Furthermore, the expected spectral efficiency can be computed as:

$$
\begin{aligned}
\bar{R} &= R\mathrm{Pr}\left[|h_{s,d}|^2 \geq \frac{1}{Y}\right] + \frac{R}{2}\mathrm{Pr}\left[|h_{s,d}|^2 < \frac{1}{Y}\right] \\
&= R\exp\left(-\frac{1}{Y}\right) + \frac{R}{2}\left[1 - \exp\left(-\frac{1}{Y}\right)\right] \\
&= \frac{R}{2}\left[1 + \exp\left(-\frac{1}{Y}\right)\right] := h_Y(R)
\end{aligned}
\tag{11.29}
$$

where the second equality follows from substituting standard exponential results for $|h_{s,d}|^2$.

A fixed value of $\bar{R}$ can arise from several possible R, depending upon the value of SNR. We define a function $\tilde{h}_Y^{-1}(\bar{R}) := \min h_Y^{-1}(\bar{R})$ to capture a useful mapping from $\bar{R}$ to R.

For fair comparison with protocols without feedback, we characterize a modified outage expression in the large-SNR regime. Specifically, we compare outage of fixed and selection relaying protocols to the modified outage $p_{IAF}^{\mathrm{out}}(y, \tilde{h}_Y^{-1}(\bar{R}))$. For large SNR, we have:

$$
p_{IAF}^{\mathrm{out}}(y, \tilde{h}_Y^{-1}(\bar{R})) \sim \left(\frac{1}{2\sigma_{s,d}^2}\frac{\sigma_{s,r}^2 + \sigma_{r,d}^2}{\sigma_{s,r}^2\sigma_{r,d}^2}\right)\cdot\left(\frac{2^{\bar{R}} - 1}{y}\right)^2
\tag{11.30}
$$

where we have combined the results of Equality 1 and 3 in Appendix of Section 11.6.4.

11.3.3   Distributed Space–Time-Coding Based on Cooperative Diversity

11.3.3.1   System and Channel Model

In this section we focus on a wireless network with a set $M = 1, 2, \ldots, m$ of transmitting terminals. Each transmitting source terminal $s \in M$ has information to transmit to a single destination terminal, denoted $d(s) \notin M$, potentially using terminals $M - \{s\}$ as relays. Thus, there are m cooperating terminals communicating to $d(s)$. If algorithms require the relays to decode fully the source message, the *decoding set $D(s)$* is the set of relays that can decode the message of source s. In the case of amplify-and-forward cooperative diversity, $D(s) = M - \{s\}$.

Both classes of algorithms consist of two transmission phases, as illustrated in Figure 11.11. In the first phase, the source broadcasts to its destination and all potential relays. During the second phase of the algorithms, the other terminals relay to the destination, either on orthogonal subchannels in the

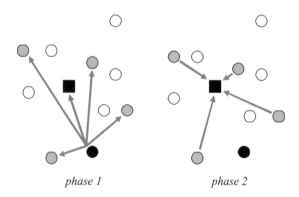

phase 1   *phase 2*

Figure 11.11   The two phases of cooperative diversity transmission.

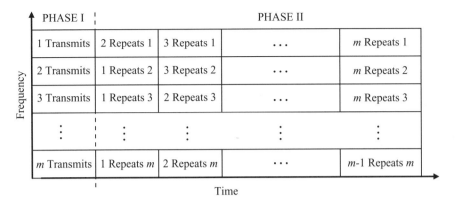

Figure 11.12 Repetition-based medium-access control.

case of repetition-based cooperative diversity, or simultaneously on the same subchannel in the case of space–time-coded cooperative diversity.

An example of channel and subchannel allocations for repetition-based cooperative diversity, in which relays either amplify what they receive or fully decode and repeat the source signal, is shown in Figure 11.12.

Channel and subchannel allocations for space-time-coded cooperative diversity, in which relays utilize a suitable space–time code in the second phase and can therefore transmit simultaneously on the same subchannel, are illustrated in Figure 11.13. Again, transmission between source s and destination $d(s)$ utilizes $1/m$ of the total degrees of freedom in the channel. In contrast to noncooperative transmission and repetition-based cooperative diversity transmission, each terminal, employing space–time-coded cooperative diversity transmits in $\frac{1}{2}$ the total degrees of freedom in the channel. It is important to keep track of these ratios when normalizing power and bandwidth in the sequel.

During the first phase, each potential relay $r \in M - \{s\}$ receives

$$y_r[n] = a_{s,r}x_s[n] + z_r[n] \tag{11.31}$$

in the corresponding subchannel, where $x_s[n]$ is the source transmitted signal and $y_r[n]$ is the received signal at relay r. For decode-and-forward transmission, if the SNR is sufficiently large for r to decode

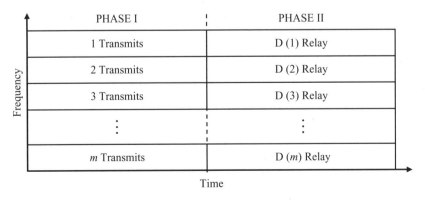

Figure 11.13 Space–time-coded channel allocations across frequency and time for m transmitting terminals. For source s, $D(s)$ denotes the set of decoding relays participating in a space–time code during the second phase.

the source transmission, then r serves as a decoding relay for the source s, so that $r \in D(s)$. For amplify-and-forward transmission, $D(s)$ is the entire set of relays for source s, i.e., $D(s) = M - \{s\}$.

The destination receives signals during both phases. During the first phase, the received signal at $d(s)$ is

$$y_d(s)[n] = a_{s,d(s)} x_s[n] + z_{d(s)}[n] \tag{11.32}$$

in the corresponding subchannel. During the second phase, the equivalent channel models are different for repetition-based and space–time-coded cooperative diversity. For repetition-based cooperative diversity, the destination receives separate retransmissions from each of the relays $r \in M - \{s\}$ and the received signal at $d(s)$ is

$$y_{d_{(s)}}[n] = a_{r,d(s)} x_r[n] + z_{d(s)}[n] \tag{11.33}$$

in the corresponding subchannel, where $x_r[n]$ is the transmitted signal of relay r. For space–time-coded cooperative diversity, all of the relay transmissions occur in the same subchannel and superimpose at the destination, so that:

$$y_{d(s)}[n] = \sum_{r \in D(s)} a_{r,d(s)} x_r[n] + z_{d(s)} \tag{11.34}$$

in the corresponding subchannel.

Parameter $a_{i,j}$ captures the effects of path loss, shadowing, and frequency nonselective fading, and $z_j[n]$ represents the effects of receiver noise and other forms of interference in the system. Statistically, we model $a_{i,j}$ as zero-mean, independent, circularly symmetric complex Gaussian random variables with variances $1/\lambda_{i,j}$, so that the magnitudes $|a_{i,j}|$ are Rayleigh distributed ($|a_{i,j}|^2$ are exponentially distributed with parameter λ_i, j), and the phases of $a_{i,j}$ are uniformly distributed on $[0, 2\pi)$. $z_j[n]$ is modeled as zero-mean mutually independent, circularly symmetric, complex Gaussian random sequences with variance N_0.

For a continuous-time channel with total bandwidth W hertz available for transmission, the discrete-time model contains W two-dimensional symbols per second (2D/s). If the transmitting terminals have an average power constraint in the continuous-time channel model of P_c joules per second (J/s), this translates into a discrete-time power constraint of $P = mP_c/W$ J/2D, since each terminal transmits in a fraction $1/m$ of the available degrees of freedom for noncooperative transmission and repetition-based cooperative diversity. The channel model is parameterized by the SNR random variables $\text{SNR}|a_{i,j}|^2$, where $\text{SNR} = mP_c N_0 W = P/N_0$ is the SNR without fading. For space–time-coded cooperative diversity, the terminals transmit in half the available degrees of freedom, so the discrete-time power constraint becomes $2P/m$.

In addition to SNR, transmission schemes are further parameterized by the spectral efficiency R bits per second per hertz (b/s/Hz) attempted by the transmitting terminals. Throughout this section, R is the transmission rate normalized by the number of degrees of freedom utilized by each terminal under noncooperative transmission, not by the total number of degrees of freedom in the channel. In addition a normalized rate R_{norm} defined as

$$R_{\text{norm}} = \frac{R}{\log(1 + \text{SNR}\sigma_{s,d(s)}^2)} \tag{11.35}$$

will be used. For an AWGN channel with bandwidth (W/m) and SNR given by $\text{SNR}\sigma_{s,d(s)}^2$, $R_{\text{norm}} < 1$ is the spectral efficiency normalized by the *maximum* achievable spectral efficiency, i.e. channel capacity. Results under (*SNR, R*) exhibit a tradeoff between the normalized SNR gain and spectral efficiency of a protocol, while results under (*SNR, R*$_{\text{norm}}$) exhibit a tradeoff between the diversity order and normalized spectral efficiency of a protocol. The latter tradeoff, called the *diversity-multiplexing* tradeoff, was developed originally in the context of multiple-antenna systems.

11.3.3.2 Cooperative Diversity Based on Repetition

Since the channel average mutual information I_{rep} is a function of, the coding scheme, the rule for including potential relays into the decoding set $D(s)$, and the fading coefficients of the channel, it too is a random variable. The event $I_{rep} < R$ when this mutual information random variable falls below some fixed spectral efficiency R is referred to as an outage event, and the probability of an outage event, $Pr[I_{rep} < R]$, is referred to as the outage probability of the channel. Since $D(s)$ is a random set, we have

$$Pr[I_{rep} < R] = \sum_{D(s)} Pr[D(s)] Pr[I_{rep} < R | D(s)] \tag{11.36}$$

For repetition coding, the random codebook at the source is generated independent and identically distributed (i.i.d.) circularly symmetric, complex Gaussian; each of the relays employs the exact same codebook as the source. Conditioned on $D(s)$ being the decoding set, the mutual information between s and $d(s)$ is:

$$I_{rep} = \frac{1}{m} \log \left(1 + SNR \, |a_{s,d(s)}|^2 + SNR \sum_{r \in D(s)} |a_{s,d(s)}|^2 \right) \tag{11.37}$$

$Pr[I_{rep} < R|D(s)]$ involves $|D(s)| + 1$ independent fading coefficients, so we expect it to decay asymptotically in proportion to $1/SNR^{|D(s)|+1}$. In the sequel, we develop the following high-SNR approximation:

$$Pr[I_{rep} < R|D(s)] \sim \left[\frac{2^{mR} - 1}{SNR} \right]^{|D(s)|+1} \times \lambda_{s,d(s)} \prod_{r \subset D(s)} \lambda_{r,d(s)} \times \frac{1}{(|D(s)| + 1)!} \tag{11.38}$$

(11.37) is expressed in such a way that the first term captures the dependence upon SNR and the second term captures the dependence upon $\{\lambda_{i,j}\}$.

To prove (0.37) we start with the following claim;

Claim 1. *If* $u_{k,}$ $k = 1, 2, \ldots, m$, *are positive, independent random variables with* $\liminf_{\varepsilon \to 0} p_{u_k}(\varepsilon u) \geq \lambda_k$ *and* $p_{u_k}(\varepsilon u) \leq \lambda_k$ *then*

$$\lim_{\varepsilon \to 0} \frac{1}{\varepsilon^m} Pr\left[\sum_{k=1}^{m} u_k < \varepsilon \right] = \frac{1}{m!} \prod_{k=1}^{m} \lambda_k \tag{11.39}$$

As an example the exponential distribution satisfies both above requirements. More generally, however, this result suggests that many of our results hold for a much larger class of PDFs, and, in particular, depend mainly upon properties of the PDFs near the origin.

Proof: If $s_n = \sum_{k=1}^{m} u_k, n \leq m$ then

$$Pr\left[\sum_{k=1}^{m} u_k < \varepsilon \right] = Pr[s_m < \varepsilon] = \int_0^\varepsilon p_{s_m}(s) ds = \varepsilon \int_0^1 p_{s_m}(\varepsilon w) dw \tag{11.40}$$

where the last equality results from the change of variables $w = s/\varepsilon$. Thus, it is sufficient for us to compute the limit

$$\lim_{\varepsilon \to 0} \frac{1}{\varepsilon^{(m-1)}} \int_0^1 p_{sm}(\varepsilon w) dw \qquad (11.41)$$

By exploiting Fatou's lemma we get for lower-bound of liminf:

$$\liminf_{\varepsilon \to 0} \frac{1}{\varepsilon^{(m-1)}} \int_0^1 p_{sm}(\varepsilon w) dw \geq \int_0^1 \left\{ \liminf_{\varepsilon \to 0} \frac{1}{\varepsilon^{(m-1)}} p_{sm}(\varepsilon w) \right\} dw \qquad (11.42)$$

$s_m = s_{m-1} + u_m$, and by independence the PDF of s_m is the convolution of the PDFs of s_{m-1} and u_m. Specifically, since u_m is positive, we have:

$$p_{s_m}(s) = \int_0^s p_{s_{m-1}}(s-r) p_{u_m}(r) dr = s \int_0^1 p_{s_{m-1}}(s(1-y)) p_{u_m}(sy) dy \qquad (11.43)$$

where the last equality results from the change of variables $y = r/s$.

Letting

$$A_m(w) = \liminf_{\varepsilon \to 0} \frac{1}{\varepsilon^{(m-1)}} p_{s_m}(\varepsilon w) \qquad (11.44)$$

and substituting into Equation (11.43), again exploiting Fatou's lemma, we obtain the recursion

$$A_m(w) = \liminf_{\varepsilon \to 0} \frac{1}{\varepsilon^{(m-1)}} p_{s_m}(\varepsilon w) \geq w \int_0^1 \left\{ \liminf_{\varepsilon \to 0} \frac{1}{\varepsilon^{(m-2)}} p_{s_{m-1}}(\varepsilon w(1-y)) \right\} \cdot$$
$$\left\{ \liminf_{\varepsilon \to 0} p_{u_m}(\varepsilon wy) \right\} dy \geq \lambda_m w \int_0^1 A_{(m-1)}(w(1-y)) dy \qquad (11.45)$$

where the last inequality follows from the initial assumption $\liminf_{\varepsilon \to 0} p_{u_k}(\varepsilon u) \geq \lambda_k$ and substitution of $A_{(m-1)}(w(1-y))$. Beginning with $A_1(w) \geq \lambda_1$, the recursion (11.45) yields:

$$A_m(w) \geq \frac{1}{(m-1)!} w^{(m-1)} \prod_{k=1}^m \lambda_k \qquad (11.46)$$

As a result, Equation (11.42) with (11.44) and (11.46) yields

$$\liminf_{\varepsilon \to 0} \frac{1}{\varepsilon^m} \Pr[s_m < \varepsilon] \geq \frac{1}{m!} \prod_{k=1}^m \lambda_k \qquad (11.47)$$

To upper-bound the *limsup*, we obtain a recursive upper bound for the PDF of s_m. Letting $B_m(w, \varepsilon) = p_{s_m}(\varepsilon w)$, we have

$$B_m(w, \varepsilon) = \varepsilon w \int_0^1 p_{s_{m-1}}(\varepsilon w(1-y)) p_{u_m}(\varepsilon wy) dy \leq \varepsilon \lambda_m w \int_0^1 B_{m-1}(w(1-y), \varepsilon) dy \qquad (11.48)$$

where the equality comes from the convolution (11.43), and the inequality follows from the initial condition $p_{u_k}(\varepsilon u) \leq \lambda_k$ and substitution of $B_{m-1}(w(1-y))$. Beginning with $B_1(w, \varepsilon) \leq \lambda_1$, Equation (11.48) yields an upper bound very similar to the lower bound in (11.60) (11.460, namely

$$B_m(w, \varepsilon) \leq \varepsilon^{(m-1)} w^{(m-1)} \frac{1}{(m-1)!} \prod_{k=1}^{m} \lambda_k \tag{11.49}$$

Then

$$\limsup_{\varepsilon \to 0} \frac{1}{\varepsilon^{(m-1)}} \int_0^1 p_{S_m}(\varepsilon w) dw \leq \limsup_{\varepsilon \to 0} \frac{1}{\varepsilon^{(m-1)}} \int_0^1 B_m(w, \varepsilon) dw \leq \frac{1}{(m-1)!} \prod_{k=1}^{m} \lambda_k \tag{11.50}$$

Together with the fact that, in general, $\lim \inf \leq \lim \sup$, (11.47), and (11.50) yield the desired result (11.39).

Now we use the result of *Claim 1* to obtain a large SNR approximation for $\Pr[I_{\text{rep}} < R|D(s)]$, the conditional outage probability for repetition decode-and-forward cooperative diversity for source s given a set of decoding relays $D(s)$. As in Equation (11.37), I_{rep} is of the form:

$$I_{\text{rep}} = \frac{1}{m} \log \left(1 + \text{SNR} \sum_{k=1}^{m} u_k \right) \tag{11.51}$$

where u_k are independent exponential random variables with parameters $\lambda_k, k = 1, 2, \ldots, m$. After some algebraic manipulations, the outage probability reduces to exactly the same form as in *Claim 1*:

$$\Pr[I_{\text{rep}} < R|D(s)] = \Pr \left[\sum_{k=1}^{m} u_k < \varepsilon \right] \tag{11.52}$$

with $\varepsilon = (2^{mR} - 1)/SNR \to 0$ as $SNR \to \infty$. Thus, *Claim 1* and continuity yield, for large SNR, the approximation:

$$\Pr[I_{\text{rep}} < R|D(s)] \sim \left[\frac{2^{mR} - 1}{\text{SNR}} \right]^m \frac{1}{m!} \prod_{k=1}^{m} \lambda_k \tag{11.53}$$

If, with the channel allocation illustrated in Figure 11.12, the relays employ independently generated codebooks, corresponding to utilizing parallel channels, the mutual information would become a sum of logarithmic terms:

$$I_{\text{rep}} = \frac{1}{m} \sum_{r \in \{s\} \cup D(s)} \log \left(1 + SNR \left| a_{r,d(s)} \right|^2 \right) \tag{11.54}$$

instead of the log-sum in Equation (11.37), which is larger than (11.37). This means that parallel channel coding is more bandwidth efficient than repetition coding, as we might expect.

In order to calculate the outage probability defined by Equation (11.36) we now consider the term $\Pr[D(s)]$, the probability of a particular decoding set. As one rule for selecting from the potential relays, we can require that a potential relay fully decodes the source message in order to participate in the second phase. Since the realized mutual information between s and r for i.i.d. complex Gaussian codebooks is given by $\log \left(1 + SNR \left| a_{s,r} \right|^2 \right) /m$, under this rule we have:

$$\Pr[r \in D(s)] = \Pr[\left| a_{s,r} \right|^2 > (2^{mR} - 1)/SNR] = \exp \left[-\lambda_{s,r}(2^{mR} - 1)/SNR \right]$$

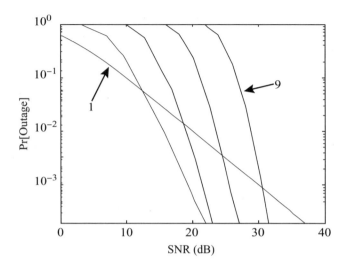

Figure 11.14 Outage probabilities for repetition-based cooperative diversity (numeric integration of the outage probability (solid lines) and calculation of the outage probability approximation (11.71) (dashed lines) versus SNR for different network sizes $m = 1, 3, 5, 7, 9$.

Moreover, since each potential relay makes its decision independently under the above restrictions, and the fading coefficients are independent in our model, we have

$$
\Pr[D(s)] = \prod_{r \in D(s)} \exp\left[-\lambda_{s,r}(2^{mR} - 1)/SNR\right]
$$

$$
\times \prod_{r \in D(s)} (1 - \exp\left[-\lambda_{s,r}(2^{mR} - 1)/SNR\right]) \sim \left[\frac{2^{mR} - 1}{SNR}\right]^{m - |D(s)| - 1} \times \prod_{r \notin D(s)} \lambda_{s,r} \quad (11.55)
$$

Note that any selection means by which $\Pr[r \in D(s)] \sim 1$ and $(1 - \Pr[r \in D(s)])\mu 1/SNR$, for SNR large, independently for each r, will result in similar asymptotic behavior for $\Pr[D(s)]$. Substituting (11.38) and (11.55) into (11.36) results in

$$
\Pr[I_{\text{rep}} < R] \sim \left[\frac{2^{mR} - 1}{SNR}\right]^{m} \times \prod_{D(s)} \lambda_{s,d(s)} \times \prod_{r \in D(s)} \lambda_{r,d(s)} \prod_{r \notin D(s)} \lambda_{s,r} \times \frac{1}{(|D(s)| + 1)!} \quad (11.56)
$$

Figure 11.14 compares the results of numeric integration of the actual outage probability to computing the approximation (11.56), for an increasing number of terminals with $\lambda_{i,j} = 1$. As result (11.56) and Figure 11.14 indicate, repetition decode-and-forward cooperative diversity achieves full spatial diversity of order m, the number of cooperating terminals, for sufficiently large SNR. However, the SNR loss due to bandwidth inefficiency is exponential in m.

In order further to simplify the summation in (11.56), we note that for a given decoding set $D(s)$, either $r \in D(s)$, in which case $\lambda_{r,d(s)}$ appears in the corresponding term in (11.56), or $r \notin D(s)$, in which case $\lambda_{s,r}$ appears in the corresponding term in (11.56). We, therefore, define

$$
\underline{\lambda}_r = \min\{\lambda_{r,d(s)}, \lambda_{s,r}\}, \quad \bar{\lambda}_r = \max\{\lambda_{r,d(s)}, \lambda_{s,r}\} \quad (11.57)
$$

and $\bar{\lambda}_s = \underline{\lambda}_s = \lambda_{s,d(s)}$. Then the product dependent upon $\{\lambda_{i,j}\}$ is bounded by

$$\lambda^m \leq \lambda_{s,d(s)} \prod_{r \in D(s)} \lambda_{r,d(s)} \prod_{r \notin D(s)} \lambda_{s,r} \leq \bar{\lambda}^m \tag{11.58}$$

where $\underline{\lambda}$ is the geometric mean of the $\underline{\lambda}_i$ and $\bar{\lambda}$ is the geometric mean of the $\bar{\lambda}_i$, for $i \in M$. One can see that the upper and lower bounds in (11.58) are independent of $D(s)$. We also note that the bounds in (11.58) coincide, i.e., $\bar{\lambda} = \underline{\lambda}$, if, though not only if, $\bar{\lambda}_i = \underline{\lambda}_i$ for all $i \in M$. Viewing $\lambda_{i,j}$ as a measure of distance between terminals i and j, the class of planar network geometries that satisfy this condition are those in which all the relays lie with arbitrary spacing along the perpendicular bisector between the source and destination.

Substituting (11.58) into (11.60), gives the following simplified asymptotic bounds for outage probability:

$$\Pr[I_{\text{rep}} < R] \geq \left[\frac{2^{mR} - 1}{SNR/\underline{\lambda}} \right]^m \sum_{D(s)} \frac{1}{(|D(s)| + 1)!} \tag{11.59}$$

$$\Pr[I_{\text{rep}} < R] \leq \left[\frac{2^{mR} - 1}{SNR/\bar{\lambda}} \right]^m \sum_{D(s)} \frac{1}{(|D(s)| + 1)!} \tag{11.60}$$

11.3.3.3 Cooperative Diversity Using Space–Time Coding

In this case (11.36) becomes

$$\Pr[I_{\text{stc}} < R] = \sum_{D(s)} \Pr[D(s)] \Pr[I_{\text{stc}} < R \,|D(s)|] \tag{11.61}$$

Conditional on $D(s)$ being the decoding set, the mutual information between s and $d(s)$ for random codebook-generated i.i.d. circularly symmetric, complex Gaussian at the source and all potential relays can be shown to be

$$I_{\text{stc}} = \frac{1}{2} \log \left(1 + \frac{2}{m} SNR \, |a_{s,d(s)}|^2 \right) + \frac{1}{2} \log \left(1 + \frac{2}{m} SNR \sum_{r \in D(s)} |a_{r,d(s)}|^2 \right) \tag{11.62}$$

Equation (11.62) represents the sum of the mutual information for two 'parallel' channels, one from the source to the destination, and one from the set of decoding relays to the destination. Again, $\Pr[I_{\text{stc}} < R|D(s)]$ involves $|D(s)| + 1$ independent fading coefficients, so we expect it to decay asymptotically in proportion to $1/SNR^{|D(s)| + 1}$. In the sequel we develop the following high-SNR approximation

$$\Pr[I_{\text{stc}} < R \,|D(s)|] \sim \left[\frac{2^{2R} - 1}{2SNR/m} \right]^{|D(s)|+1} \times \lambda_{s,d(s)} \prod_{r \in D(s)} \lambda_{r,d(s)} \times A_{|D(s)|}(2^{2R} - 1) \tag{11.63}$$

where

$$A_n(t) = \frac{1}{(n-1)!} \int_0^1 \frac{w^{(n-1)}(1-w)}{(1+tw)} dw, \quad n > 0 \tag{11.64}$$

and $A_0(t) = 1$. Equation (11.63) is expressed in such a way that the first term captures the dependence upon SNR and the second term captures the dependence upon $\{\lambda_{i,j}\}$.

To prove Equation (11.63) we first notice that in (11.62), I_{stc} is of the form

$$I_{stc} = \frac{1}{2} \log \left(1 + \frac{2}{m} SNR u_m \right) + \frac{1}{2} \log \left(1 + \frac{2}{m} SNR \sum_{k=1}^{m-1} u_k \right) \tag{11.65}$$

where again u_k are independent exponential random variables with parameters $\lambda_k, k = 1, 2, \ldots, m$. If we use notation:

$$s_{m-1} = \sum_{k-1}^{m-1} u_k, \, t = (2^{2R} - 1), \, and \, \varepsilon = (2^{2R} - 1)/(2SNR/m)$$

then

$$\Pr[I_{stc} < R|D(s)] = \Pr \left[u_m + s_{m-1} + \frac{2}{m} SNR \, u_m s_{m-1} < \varepsilon \right]$$

$$= \int_0^\varepsilon \Pr \left[u_m < \frac{\varepsilon - s}{1 + (2SNR/m)s} \right] p_{s_{m-1}}(s) ds = \varepsilon \int_0^1 \Pr \left[u_m < \frac{\varepsilon(1-w)}{1+tw} \right] p_{s_{m-1}}(\varepsilon w) dw$$

$$= \varepsilon \int_0^1 \left[1 - \exp \left(-\lambda_m \frac{\varepsilon(1-w)}{1+tw} \right) \right] p_{s_{m-1}}(\varepsilon w) dw \tag{11.66}$$

Equation (11.66) follows from the change of variables $w = s/\varepsilon$, and the last equality follows from substituting the CDF for u_m. We now compute

$$\lim_{\varepsilon \to 0} \frac{1}{\varepsilon^m} \Pr[I_{stc} < R|D(s)] = \frac{1}{(m-2)!} \prod_{k=1}^m \lambda_m \int_0^1 \left[\frac{1-w}{1+tw} \right] w^{(m-2)} dw \tag{11.67}$$

which, provides the large-SNR approximation

$$\Pr[I_{stc} < R|D(s)] \sim \left[\frac{2^{2R} - 1}{2SNR/m} \right]^m \frac{1}{(m-2)!} \prod_{k=1}^m \lambda_m \times \int_0^1 \left[\frac{1-w}{1+(2^{2R}-1)w} \right] w^{(m-2)} dw \tag{11.68}$$

To lower-bound the *lim inf*, we use Fatou's lemma in Equation (11.66) to obtain:

$$\liminf_{\varepsilon \to 0} \frac{1}{\varepsilon^m} \Pr[I_{stc} < R|D(s)] \geq \int_0^1 \left\{ \liminf_{\varepsilon \to 0} \frac{1}{\varepsilon} \left[1 - \exp \left(-\lambda_m \frac{\varepsilon(1-w)}{1+tw} \right) \right] \right\}$$

$$\cdot \left\{ \liminf_{\varepsilon \to 0} \frac{1}{\varepsilon^{(m-2)}} p_{s_{m-1}}(\varepsilon w) \right\} dw = \int_0^1 \lambda_m \left[\frac{1-w}{1+tw} \right] A_{m-1}(w) dw$$

$$\geq \frac{1}{(m-2)!} \prod_{k=1}^m \lambda_k \int_0^1 \left[\frac{1-w}{1+tw} \right] w^{(m-2)} dw \tag{11.69}$$

where the first equality follows from properties of exponentials and substitution of $A_{m-1}(w)$ from Equation (11.43), and the second equality follows from the result (11.46) in the proof of *Claim 1*. To upper-bound the *lim sup*, we derive

$$\limsup_{\varepsilon \to 0} \frac{1}{\varepsilon^m} \Pr[I_{stc} < R|D(s)] \leq \limsup_{\varepsilon \to 0} \int_0^1 \left\{ \frac{1}{\varepsilon} \left[1 - \exp \left(-\lambda_m \frac{\varepsilon(1-w)}{1+tw} \right) \right] \right\}$$

$$\cdot \left\{ \frac{1}{\varepsilon^{(m-2)}} B_{m-1}(w, \varepsilon) \right\} dw < \limsup_{\varepsilon \to 0} \int_0^1 \left\{ \lambda_m \left[\frac{1-w}{1+tw} \right] \right\} \cdot \left\{ \frac{1}{\varepsilon^{(m-2)}} B_{m-1}(w, \varepsilon) \right\} dw$$

$$\leq \limsup_{\varepsilon \to 0} \int_0^1 \left\{ \lambda_m \left[\frac{1-w}{1+tw} \right] \right\} \cdot \left\{ \frac{1}{(m-2)!} w^{(m-2)} \prod_{k=1}^{m-1} \lambda_k \right\} dw$$

$$= \frac{1}{(m-2)!} \prod_{k=1}^{m-1} \lambda_k \int_0^1 \left[\frac{1-w}{1+tw} \right] w^{(m-2)} dw \qquad (11.70)$$

where the first inequality follows from substitution of $B_{m-1}(w, \varepsilon)$ from (11.48), the second inequality follows from he fact that $1 - \exp(-x) \leq x$ for all $x \geq 0$, and the third inequality follows from the result (11.49) in *Claim 1*. Taken together with the fact that *lim inf < lim sup,* Equation (11.69) and (11.70) yield the desired result, Equation (11.67).

The remaining term in Equation (11.61) to be considered is $\Pr[D(s)]$, the probability of a particular decoding set. As before, we require a potential relay to decode the source message fully in order to participate in the second phase, a necessary condition for the mutual information expression (11.62) to be correct. Since the realized mutual information between s and r for i.i.d. complex Gaussian codebooks is given by

$$\frac{1}{2} \log \left(1 + \frac{2}{m} SNR |a_{s,r}|^2 \right) \qquad (11.71)$$

under this condition we have

$$\Pr[r \in D(s)] = \Pr \left[|a_{s,r}|^2 > \frac{2^{2R} - 1}{2SNR/m} \right] = \exp \left[-\lambda_{s,r} \frac{2^{2R} - 1}{2SNR/m} \right] \qquad (11.72)$$

Now, since each potential relay makes its decision independently, and the fading coefficients are independent, we have

$$\Pr[D(s)] = \prod_{\in D(s)} \exp \left[-\lambda_{s,r} \frac{2^{2R} - 1}{2SNR/m} \right] \times \prod_{r \notin D(s)} \left(1 - \exp \left[-\lambda_{s,r} \frac{2^{2R} - 1}{2SNR/m} \right] \right)$$

$$\sim \left[\frac{2^{2R} - 1}{2SNR/m} \right]^{m - |D(s)| - 1} \times \prod_{r \notin D(s)} \lambda_{s,r} \qquad (11.73)$$

Combining (11.63) and (11.73) into (11.61), we obtain

$$\Pr[I_{stc} < R] \sim \left[\frac{2^{2R} - 1}{2SNR/m} \right]^m \times \sum_{D(s)} \lambda_{s,d(s)} \times \prod_{r \in D(s)} \lambda_{r,d(s)} \prod_{r \notin D(s)} \lambda_{s,r} \times A_{|D(s)|}(2^{2R} - 1) \qquad (11.74)$$

Figure 11.15 presents the results of numeric integration of the actual outage probability and the approximation (11.74), for an increasing number of terminals with $\lambda_{i,j} = 1$. As the result (11.74) and Figure 11.15 indicate, space–time-coded cooperative diversity achieves full spatial diversity of order m, the number of cooperating terminals, for sufficiently large SNR. In contrast to repetition-based algorithms, the SNR loss for space–time-coded cooperative diversity is only linear in m.

As in Equations (11.59) and (11.60), we want further to simplify the summation in (11.74). The product dependent upon $\{\lambda_{i,j}\}$ can again be bounded as in Equation (11.58). To avoid dealing with (11.64), we exploit the bounds

$$\frac{1}{(n+1)!(1+t)} \leq A_n(t) \leq \frac{1}{n!} \qquad (11.75)$$

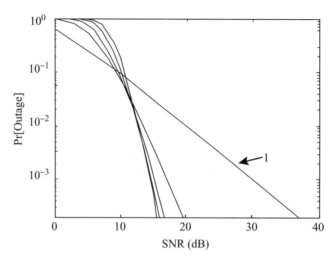

Figure 11.15   Outage probability of space-time coded cooperative diversity.

Combining Equations (11.58) and (11.75) into (11.74), we get the following simplified asymptotic bounds for outage probability:

$$\Pr[I_{\text{stc}} < R] \geq \left[\frac{2^{2R} - 1}{2SNR/(m\underline{\lambda})} \right]^m \sum_{D(s)} \frac{1}{(|D(s)| + 1)!} \tag{11.76}$$

$$\Pr[I_{\text{stc}} < R] \leq \left[\frac{2^{2R} - 1}{2SNR/(m\bar{\lambda})} \right]^m \sum_{D(s)} \frac{1}{|D(s)|!} \tag{11.77}$$

11.4   Relay-Assisted Communications

This section will examine relay-assisted communications with multi-antenna configuration. In contrast to the schemes analyzed in Section 11.3, here we consider that there is one transmitter (source), one half-duplex relay which is not transmitting any information and which aims to help the source in transmitting the messages to a single destination (see Figure 11.1). The topics addressed in this section are grouped into four sections. First, Section 11.4.1 analyzes the performance of the different relay protocols under a static resource allocation. The achievable rates for forwarding protocols and protocols I, II and III under a decode-and-forward and amplify-and-forward relay are reviewed. Continuing with a static resource allocation, Section 11.4.2 explains BER performance when the particular case of protocol III is implemented by means of distributed space-time coding. On the other hand, Section 11.4.3 illustrates the benefits of having a dynamic resource allocation for accommodating all relay-assisted protocols, i.e. optimizing the duration of the different phases of communication. Finally, Section 11.4.4 analyzes a scenario consisting of a source, a destination and two relays, referred to in the literature as the half-duplex relay diamond channel. Several transmission schemes based on protocol III are examined.

11.4.1   Static Resource Allocation Relaying

Half-duplex relay protocols are reviewed under a *static resource allocation* constraint. First, Section 11.4.1.1 looks at protocol I with a *decode-and-forward* relay, analyzing the type of message transmitted by the relay (*repetition* or *unconstrained coding*) and the benefits of *selective relaying*. As a performance metric the *additive capacity gain* [17] obtained by protocol I is evaluated for different antenna configurations. Section 11.4.1.2 presents the achievable rate performance of all half-duplex protocols with r*epetition coding decode-and-forward* or *amplify-and-forward* relays.

11.4.1.1 MIMO Decode-and-Forward Protocol I

This section examines the half-duplex relay protocol I (see Section 11.2.3), under the *decode-and-forward* transmission, the same as that considered in [7] or in Section 11.3.2. The relay-assisted transmission is duplexed in time (TDD – Time Division Duplexing) and extended to a multi-antenna scenario. Such a communication system can be interpreted as a compound channel [49, 50]. We will give expressions for the mutual information[4] of protocol I when the channel state information at the transmitter (CSIT) is assumed at all terminals and for the no CSIT case. That mutual information is obtained by means of Gaussian codebooks. Moreover, *selective relaying* is also analyzed. The following has been presented in [48]. Finally, the *additive capacity gain* obtained by this protocol is evaluated.

In the three-terminal network depicted in Figure 11.4 consisting of a source, a TDD half-duplex relay and a single destination, the relay-assisted transmission under protocol I is done during N symbols in the time domain. It is assumed that *relay-receive* and *relay-transmit phase* present the same size, i.e. $N/2$ symbols. Moreover, the source, relay and destination are equipped with n_s, n_r and n_d antennas, respectively.

During the *relay-receive phase*, the source transmits the signal $\mathbf{x}_s$. Afterwards during the *relay-transmit phase* the relay terminal transmits the signal $\mathbf{x}_r$. The signals received by the relay $\mathbf{y}_r$ and by the destination $\mathbf{y}_d$ during all the transmission are given by,

$$
\begin{aligned}
\mathbf{y}_r(t) &= \mathbf{H}_1\mathbf{x}_s(t) + \mathbf{w}_r(t) && 1 \le t \le N/2 \\
\mathbf{y}_d(t) &= \begin{cases} \mathbf{H}_0\mathbf{x}_s(t) + \mathbf{w}_d(t) & 1 \le t \le N/2 \\ \mathbf{H}_2\mathbf{x}_r(t) + \mathbf{w}_d(t) & N/2 \le t \le N \end{cases}
\end{aligned}
\tag{11.78}
$$

where $\mathbf{H}_1$, $\mathbf{H}_2$ and $\mathbf{H}_0$ denotes the $n_r \times n_s$, $n_d \times n_r$ and $n_d \times n_s$ channel matrices between the source-relay, relay-destination and the source-destination, respectively. The terms $\mathbf{w}_r$ and $\mathbf{w}_d$ denote Gaussian noise with zero mean and covariance matrices $\mathbf{R}_{w_r}$ and $\mathbf{R}_{w_d}$. We assume that $\mathbf{x}_s$ and $\mathbf{x}_r$ are circularly symmetric complex Gaussian vectors with zero mean and transmit covariance matrix $\mathbf{Q}_s = E\left\{\mathbf{x}_s(t)\mathbf{x}_s^H(t)\right\}$ and $\mathbf{Q}_r = E\left\{\mathbf{x}_r(t)\mathbf{x}_r^H(t)\right\}$, respectively. The source and the relay are constrained in their average transmit power by P_s and P_r, respectively. These matrices are variables to be designed under the CSIT case. On the contrary, for no CSIT case we assume $\mathbf{Q}_s = P_s/n_s\mathbf{I}_{n_s}$ and $\mathbf{Q}_r = P_r/n_r\mathbf{I}_{n_r}$.

In the following the signal received by the destination will be rewritten in a more compact way for the direct transmission and for two implementations of the *decode-and-forward* (DF): *unconstrained coding* (UC) and *repetition coding* (RC).

Signal Model
The unconstrained coding (UC) has been defined for the case when the signal transmitted by the relay during the *relay-transmit phase* $\mathbf{x}_r$ is not correlated with the signal transmitted by the source $\mathbf{x}_s$. The signal model at the destination during the *relay-transmit phase* is given by,

$$
\begin{bmatrix} \mathbf{y}_d(t) \\ \mathbf{y}_d(t+N/2) \end{bmatrix} = \begin{bmatrix} \mathbf{H}_0 & \mathbf{0} \\ \mathbf{0} & \mathbf{H}_2 \end{bmatrix} \begin{bmatrix} \mathbf{x}_s(t) \\ \mathbf{x}_r(t+N/2) \end{bmatrix} + \begin{bmatrix} \mathbf{w}_d(t) \\ \mathbf{w}_d(t+N/2) \end{bmatrix} \quad 1 \le t \le N/2
\tag{11.79}
$$

---

[4] In general the capacity of the relay channel is not known (only the degraded case with full-duplex relay [12]). Therefore the different protocols with different decoding strategies provide lower bounds of the capacity of that channel. The capacity of a channel is defined as the maximum mutual information between the transmitted and received signals. However, the maximum mutual information of each protocol with different decoding strategies will not be capacity of the relay channel because we are enforcing a predefined signaling which might not be optimal.

$$\Pr(error) = \Pr(\hat{W}_1 \neq W \text{ or } \hat{W}_2 \neq W)$$

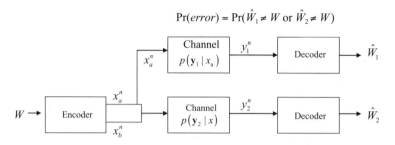

Figure 11.16   Scheme of the compound channel with two receivers.

where the transmitted signals are constrained by $Tr(\mathbf{Q}_s) \leq P_s$ and $Tr(\mathbf{Q}_r) \leq P_r$ subject to $P_s/2 + P_r/2 = P_{direct}$ to keep the total average transmitted power equal to the direct transmission scheme for a fair comparison.

A particular case arises when the relay retransmits the same signal transmitted by the source (repetition-coding, RC) during the *relay-receive phase*, in a nutshell when $\mathbf{x}_r(n + N/2) = \varphi \mathbf{x}_s(n)$ as in [7]. This technique can be used when the number of antennas at the source and the relay are equal ($n_r = n_s$). In such a case, the signal model at the destination during the *relay-receive phase* reduces to,

$$\begin{bmatrix} \mathbf{y}_d(t) \\ \mathbf{y}_d(t+N/2) \end{bmatrix} = \begin{bmatrix} \mathbf{H}_0 \\ \varphi \mathbf{H}_2 \end{bmatrix} \mathbf{x}_s(t) + \begin{bmatrix} \mathbf{w}_d(t) \\ \mathbf{w}_d(t+N/2) \end{bmatrix} \quad \varphi = \sqrt{\frac{P_r}{P_s}} \quad 1 \leq t \leq N/2 \quad (11.80)$$

where the transmitted signal is constrained by $Tr(\mathbf{Q}_s) \leq P_s$.

Mutual Information
The transmission scheme based on protocol I can be seen as a specific case of the compound channel [49] with two receivers presented in Figure 11.16, where a source transmits a message W in N symbols that must be successfully received by two decoders. However, the first one only receives a part of the total transmitted signal, while the other one receives the entire signal. Let us define the transmitted signal by,

$$x(t) = \begin{cases} x_a(t) & 1 \leq t \leq \nu \\ x_b(t) & 1 \leq t \leq N \end{cases}$$

where ν defines the number of symbols where the first receiver is listening. The received symbols at each receiver $\mathbf{y}_1$ and $\mathbf{y}_2$ drawn according to the conditional probability distribution $p_1(\mathbf{y}_1|\mathbf{x}_a)$ and $p_2(\mathbf{y}_2|\mathbf{x})$, where the effect of the noise has been considered.

Proposition 1. *Consider a channel where the transmitted signal* $\mathbf{x}$*, which can be structured as* $\mathbf{x} = (\mathbf{x}_a, \mathbf{x}_b)$*, is received as* $\mathbf{y}_1$ *and* $\mathbf{y}_2$ *by two receivers with channel transition probabilities given by* $p_1(\mathbf{y}_1|\mathbf{x}_a)$ *and* $p_2(\mathbf{y}_2|\mathbf{x})$*, respectively (see Figure 11.16). Then the capacity[5] of the compound channel is given by*

$$C = \sup_{p(x)} \min\left(I(\mathbf{x}_a, \mathbf{y}_1), I(\mathbf{x}, \mathbf{y}_2)\right) \quad (11.81)$$

where the supremum is over all $p(x)$ *that satisfy the system specifications. It has been assumed that no error happens only when both receivers decode the transmitted message without error.*

---

[5] It should be pointed out that protocol I under DF describes a compound channel, where the capacity for that channel is known. However, that value is not the capacity of the relay channel and for that reason it will be referenced as maximum mutual information of protocol I in the following.

Table 11.2 Mutual Information of protocol I under unconstrained coding (UC) or repetition coding (RC) using *Corollary 1*, (0.82)

Coding at the relay	$\mathbf{R}_{w_a}$	$\mathbf{H}_a$	$\mathbf{Q}_a$	$\mathbf{R}_w$	$\mathbf{H}$	$\mathbf{Q}$
Unconstrained coding	$\mathbf{R}_{w_r}$	$\mathbf{H}_1$	$E\left\{\mathbf{x}_s\mathbf{x}_s^T\right\}$	$\begin{bmatrix} \mathbf{R}_{w_d} & \mathbf{0} \\ \mathbf{0} & \mathbf{R}_{w_d} \end{bmatrix}$	$\begin{bmatrix} \mathbf{H}_0 & \mathbf{0} \\ \mathbf{0} & \mathbf{H}_2 \end{bmatrix}$	$E\left\{\begin{bmatrix} \mathbf{x}_s\mathbf{x}_s^T & \mathbf{x}_s\mathbf{x}_r^T \\ \mathbf{x}_r\mathbf{x}_s^T & \mathbf{x}_r\mathbf{x}_r^T \end{bmatrix}\right\}$
Repetition coding	$\mathbf{R}_{w_r}$	$\mathbf{H}_1$	$E\left\{\mathbf{x}_s\mathbf{x}_s^T\right\}$	$\begin{bmatrix} \mathbf{R}_{w_d} & \mathbf{0} \\ \mathbf{0} & \mathbf{R}_{w_d} \end{bmatrix}$	$\begin{bmatrix} \mathbf{H}_0 \\ \varphi\mathbf{H}_2 \end{bmatrix}$	$E\left\{\mathbf{x}_s\mathbf{x}_s^T\right\}$

Proof. It suffices to realize that this particular channel is in fact a compound channel [49, 50] where the set of possible probabilities is by $p_1\left(\mathbf{y}_1|\mathbf{x}_a\right)$ and $p_2\left(\mathbf{y}_2|\mathbf{x}\right)$, as it is depicted in Figure 11.16. We consider that no error happens only when both receivers decode the transmitted message W without error,

$$\Pr\left(\text{no error}\right) = \Pr\left(\hat{W}_1 = W \text{ and } \hat{W}_2 = W\right)$$

where $\hat{W}_1$ and $\hat{W}_2$ are the messages decode by both receivers. Hence, the corresponding model is a compound channel with an uninformed transmitter of the channel transition probability and the capacity expression given by follows (11.82). This result can be alternatively proved in a direct way without referring to compound channels, following the ideas in [16] for the case of a known transition probability

We now particularize the previous general result to the Gaussian case in the following corollary.

Corollary 1. *Consider a vector Gaussian channel where the transmitted signal vector $\mathbf{x}$, which can be structured as $\mathbf{x}^T = \left[\mathbf{x}_a^T, \mathbf{x}_b^T\right]$. When one receiver listens during one half of the time ($v = N/2$), the received signal at both receivers becomes,*

$$\mathbf{y}_a = \mathbf{H}_a\mathbf{x}_a + \mathbf{w}_a$$
$$\mathbf{y} = \mathbf{H}\mathbf{x} + \mathbf{w}$$

where $\mathbf{H}_a$ and $\mathbf{H}$ represent a linear transformation on the transmitted signal and both $\mathbf{w}_a$ and $\mathbf{w}$ are Gaussian distributed with zero mean and covariance matrices $\mathbf{R}_{w_a}$ and $\mathbf{R}_w$. Then the capacity of the compound channel under CSI is given by[6] ,

$$C = \frac{1}{2}\sup_{\mathbf{Q},\mathbf{Q}_a}\min\left(I_a, I_{ab}\right)$$
$$I_a = \log\det\left(\mathbf{I} + \mathbf{R}_{w_a}^{-1}\mathbf{H}_a\mathbf{Q}_a\mathbf{H}_a^H\right) \quad I_{ab} = \log\det\left(\mathbf{I} + \mathbf{R}_w^{-1}\mathbf{H}\mathbf{Q}\mathbf{H}^H\right) \quad (11.82)$$

with $\mathbf{Q} = E\left\{\mathbf{x}\mathbf{x}^H\right\}$ and $\mathbf{Q}_a = E\left\{\mathbf{x}_a\mathbf{x}_a^H\right\}$ being the transmit covariance matrix of $\mathbf{x}$ and $\mathbf{x}_a$. The supremum is over all $\mathbf{Q}$ and $\mathbf{Q}_a$ that satisfy the system specifications. The capacity is achieved when $\mathbf{x}$ is Gaussian distributed with zero mean and covariance matrix $\mathbf{Q}$.

Proof. The proof is straightforward from *Proposition 1* and taking into account that $\mathbf{w}_a$ and $\mathbf{w}$ are Gaussian distributed, [16].

Consequently, *Corollary 1* in (11.82) allow us to provide the mutual information expressions sketched in Table 11.2. Notice that the principal difference between RC and UC lies in the definition of matrix $\mathbf{H}$.

---

[6] All the log functions considered in this work without specify the base are in base 2.

Retransmission Strategies

Apart from the transmission scheme at the relay under decode-and-forward (unconstrained or repetition coding), the relay can operate always transmitting (persistent transmission) or just when it is able to decode the message from the source (selective transmission) as it was described in Section 11.2.4 with Figure 11.5. Hence, following the guidelines of [7] for the repetition coding scheme, we will describe a protocol that avoids bottlenecks in the source-relay link by switching to a direct transmission during the *relay-transmit phase*. In this case, the source is transmitting in both phases and it plays the same role as the relay (*repetition* or *unconstrained coding*). In a nutshell, in the RC the source transmits the same signal in both phases, while in UC, they are uncorrelated. Additionally, it will be assumed equal average power transmission by the source and relay because there is no CSIT. Notice that under no CSIT (i.e. transmit covariance matrices become diagonal in (11.82) and) Table 11.2, the maximum mutual information of protocol I can be written as a function of the capacity of the individual links, also defined under no CSIT.

The capacity of source-relay link under no CSIT with a fixed channel is given by

$$C_1 = \log \det \left(\mathbf{I}_{n_r} + \frac{P_s}{n_s} \mathbf{R}_{w_r}^{-1} \mathbf{H}_1 \mathbf{H}_1^H \right) \tag{11.83}$$

The source transmits in the *relay-transmit phase* (remaining the assisting relay terminal silent) if it has been able to decode the message,

$$C_1 \in S = \{x \in \Re \quad | \quad x \le I_0 \le I_{ab}\} \tag{11.84}$$

where I_{ab} and I_0 are defined by,

$$I_{ab} = \begin{cases} \frac{1}{2} \log \det \left(\mathbf{I}_{n_d} + \frac{P_s}{n_s} \mathbf{R}_{w_d}^{-1} \mathbf{H}_0 \mathbf{H}_0^H \right) + \frac{1}{2} \log \det \left(\mathbf{I}_{n_d} + \frac{P_r}{n_r} \mathbf{R}_{w_d}^{-1} \mathbf{H}_2 \mathbf{H}_2^H \right) & \text{if } UC \\ \frac{1}{2} \log \det \left(\mathbf{I}_{n_s} + \frac{P_s}{n_s} \mathbf{H}_0^H \mathbf{R}_{w_d}^{-1} \mathbf{H}_0 + \frac{P_r}{n_r} \mathbf{H}_2^H \mathbf{R}_{w_d}^{-1} \mathbf{H}_2 \right) & \text{if } RC \end{cases}$$

$$I_0 = \begin{cases} C_{DT} & \text{if } UC \\ \frac{1}{2} \log \det \left(\mathbf{I}_{n_s} + \frac{2P_s}{n_s} \mathbf{H}_0^H \mathbf{R}_{w_d}^{-1} \mathbf{H}_0 \right) & \text{if } RC \end{cases} \tag{11.85}$$

where I_{ab} stands for the mutual information due to the transmissions from the source and relay to destination for the UC and RC, (11.82), and I_0 represents the mutual information obtained through the source-destination link when the source is transmitting in both phases and it is using the UC (in such a case it is the capacity of the source-destination link under no CSIT) or RC mode during the *relay-transmit phase*.

The final expressions for the mutual information of the UC and RC are given by,

$$I_{UC-A} = \begin{cases} \min \left(\frac{1}{2}C_1, I_{ab} \right) & \text{if } C_1 \in S \\ C_{DT} & \text{otherwise} \end{cases}$$

$$I_{RC-A} = \begin{cases} \min \left(\frac{1}{2}C_1, I_{ab} \right) & \text{if } C_1 \in S \\ \frac{1}{2} \log \det \left(\mathbf{I}_{n_d} + \frac{2P_s}{n_s} \mathbf{R}_{w_d}^{-1} \mathbf{H}_0 \mathbf{H}_0^H \right) & \text{otherwise} \end{cases} \tag{11.86}$$

with C_1 defined in (11.83), I_{ab} introduced in (11.85), the mutual information of the direct link and S given in (11.84). When the source transmits during the *relay-receive* and *relay-transmit phase* using unconstrained coding, the capacity of the source-destination link under no CSIT is obtained. In contrast, when the source transmits employs repetition coding, the mutual information obtained is lower than the direct transmission because of the repeated message in both phases. The received signal power increases twice at the expenses of using more resources, motivating the 1/2 factor.

In order to evaluate the performance of the different protocols we assume a scenario multi-antenna terminals and non-frequency selective Rayleigh channels. The transmission is done over intervals of time named frames (with relay-receive and relay-transmit phase) where it is assume that the channel coefficients remain constant, but changing from frame to frame. Different average SNR values are considered to account for the quality of the links including pathloss and slow fading.

We define the outage mutual information at p_{out} [3] as the code rate that is below the mutual information wit a probability $(1-p_{out})$,

$$I^{out} = \arg\max_{R} \ \Pr(I \geq R) \qquad \text{subject to} \Pr(I \geq R) \geq (1 - p_{out})$$

In this regard, we also define the *additive capacity gain* [17] of the outage mutual information over the direct transmission as,

$$a = I^{out} - I_{DT}^{out} \qquad bits/s/Hz \qquad (11.87)$$

where I_{DT}^{out} denotes the outage mutual information of the direct transmission and I^{out} stands for the outage mutual information of the {UC, UC-A, RC, RC-A}.

The *additive capacity gain* at $p_{out} = 5\%$ is explored in a scenario where the $SNR_0 = SNR_2 = SNR$ and $SNR_1 = 2 \times SNR$ (dB) and $n_s = n_r = n_d = 1$ in Figure 11.17 and $n_s = n_r = n_d = \{1,2,3,4\}$ in

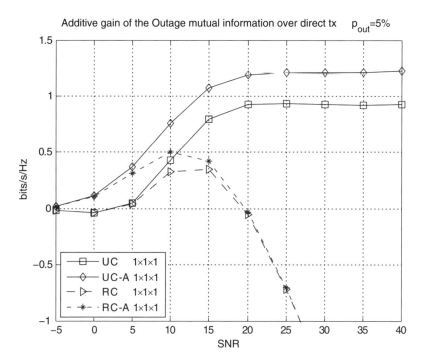

Figure 11.17 Additive gain of the outage mutual information ($p_{out} = 5\%$) over the direct transmission for the different relay schemes $SNR_0 = SNR_2 = SNR$ and $SNR_1 = 2 \times SNR$ (dB). $n_s = n_r = n_d = 1$.

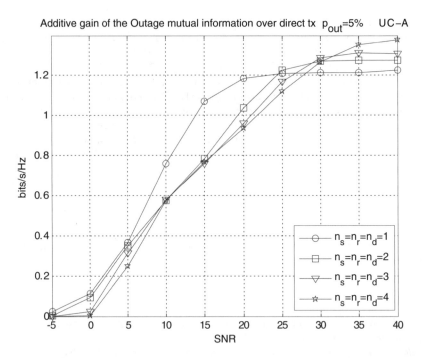

Figure 11.18 Additive gain of the outage mutual information ($p_{out} = 5\%$) over the direct transmission for UC-A $SNR_0 = SNR_2 = SNR$ and $SNR_1 = 2 \times SNR$ (dB). $n_s = n_r = n_d = \{1,2,3,4\}$.

Figure 11.18, just for UC-A. Figure 11.17 shows that RC and RC-A can only get a positive *additive capacity gain* for SNR $\leq$ 20 dB. In contrast, UC and UC-A always get a positive *additive capacity gain*, but for SNR $\geq$ 20 the *additive capacity gain* becomes steady in respect of the SNR. This result shows that there is no multiplexing gain for the relay-assisted transmission because this gain does not increase logarithmically with SNR. Adopting the UC-A scheme, the antennas are increased in Figure 11.18. It can be seen that the additive gain is always positive and gets similar values for the different antenna configuration, around 1.3 bits/s/Hz. The difference is based at which SNR the performance becomes steady, for example for $n_s = n_r = n_d = 1$ it is at SNR $\geq$ 20, $n_s = n_r = n_d = 2$ it is at SNR $\geq$ 30, $n_s = n_r = n_d = 3$ it is at SNR $\geq$ 35 and $n_s = n_r = n_d = 4$ it is at SNR $\geq$ 40. It is important to remark that although there is an *additive capacity gain* by the relay-assisted transmission, depending on the antenna configuration and the SNR, this gain can be neglected when compared with the outage mutual information of the direct transmission, for example at high SNR the increase of 1 bits/s/Hz will not have a great impact on the total performance.

11.4.1.2 Achievable Rates Under Decode-and-Forward Repetition-Coding (DF-RC) and Amplify-and-Forward (AF)

Having discussed protocol I for a *decode-and-forward* transmission scheme in the previous section, this section reviews the achievable rates of the remaining protocols when a relay operates under *decode-and-forward* repetition coding (RC). Additionally, the achievable rates of all protocols when the relay works in *amplify-and-forward* mode are analyzed. In both cases, no CSIT is assumed. The transmissions are done during N symbols in the time domain. Due to the static resource allocation and the repetition coding strategy, the *relay-receive* and *relay-transmit phase* are defined by the following equal-length time intervals, respectively,

$$S_1 = \{ t \mid 1 \le t \le N/2 \}$$
$$S_2 = \{ t \mid N/2 + 1 \le t \le N \}$$

(11.88)

Likewise, we define the variables[7] x_s^1 and x_s^2 for denoting the Gaussian signals transmitted by the source during the *relay-receive* (in S_1) and *relay-transmit phase* (in S_2) respectively, variable x_r stands for the Gaussian signal transmitted by the relay terminal in the *relay-transmit phase*,

$$Q_s^1 = E\{x_s^1 x_s^{1H}\} = \frac{P_s}{n_s} I_{n_s} \quad Q_s^2 = E\{x_s^2 x_s^{2H}\} = \frac{P_s}{n_s} I_{n_s} \quad Q_r = E\{x_r x_r^H\} = \frac{P_r}{n_r} I_{n_r}$$
$$Tr(Q_s^1) = P_s \qquad\qquad Tr(Q_s^2) = P_s \qquad\qquad Tr(Q_r) = P_r$$

(11.89)

Decode-and-Forward

In *decode-and-forward* mode under repetition coding it is assumed that the relay and/or the source are equipped with the same number of antennas, $n_s = n_r$, and the same code words are used by both. It turns out that the achievable rates obtained by the different protocols can be formulated as a function of the mutual information of the different links. Hence, let us define the those expressions by,

$$C_1 = \log\det\left(I_{n_r} + \frac{P_s}{n_s \sigma_{w_r}^2} H_1 H_1^H\right)$$

$$C_0 = \log\det\left(I_{n_d} + \frac{P_s}{n_s \sigma_{w_s}^2} H_0 H_0^H\right), \quad C_b = \log\det\left(I + \frac{P_s}{n_r \sigma_{w_d}^2} H_{eq} H_{eq}^H\right)$$

(11.90)

with n_d, n_r and n_s the number of antennas at destination, relay, and source, respectively. Likewise, H_1, H_0 and H_{eq} denote the channel matrix of the source-relay link, source-destination link and the equivalent channel dependent on the employed protocol.

Table 11.3 presents the signal model and achievable rates attained by the different protocols taking into account the definitions of (11.88)–(11.90). We would like to emphasize protocol III, defined originally in [40] as protocol I. The main idea is to combine the signals at the physical layer thanks to both phases of the relay-assisted transmission being equal. The source transmits a Gaussian signal during the *relay-receive phase* x_s^1 of rate R_r which is received at the relay and destination. The relay decodes the data received and transmits the signal x_r to the destination (*repetition coding*) while the source transmits a Gaussian signal related to another message, x_s^2 of rate R_s. But the definition of repetition coding establishes that $x_r = \varphi x_s^1$. This protocol obtains good performance results in terms of diversity and ergodic capacity gains [40]. The relay must be able to decode the first message, x_s^1, while the destination has to decode the messages x_r and x_s^2 (*successive decoding*). During the *relay-transmit phase* the equivalent channel can be modeled as the conventional MAC channel, [16]. Therefore, the achievable rate region G for this protocol can be written as,

$$G = \begin{cases} R_r \le \dfrac{1}{2}C_1 \\[2mm] R_r \le \dfrac{1}{2}\log\det\left(I_{n_d} + \dfrac{P_r}{n_r \sigma_{w_d}^2} H_2 H_2^H + \dfrac{P_s}{n_s \sigma_{w_d}^2} H_0 H_0^H\right) \\[2mm] R_s \le \dfrac{1}{2}C_0 \\[2mm] R_r + R_s \le \dfrac{1}{2}\log\det\left(I_{2n_d} + \dfrac{P_s}{n_r \sigma_{w_d}^2}\begin{bmatrix} H_0 & 0 \\ \varphi H_2 & H_0 \end{bmatrix}\begin{bmatrix} H_0 & 0 \\ \varphi H_2 & H_0 \end{bmatrix}^H\right) \end{cases}$$

(11.91)

---

[7] A variable that is used in both phases of the relay-assisted transmission will use the super-index 1 for denoting the relay-receive phase and super-index 2 for the relay-transmit phase.

Table 11.3   Signal model and achievable rates of the different static DF-RC protocols $\varphi = \sqrt{\dfrac{P_r}{P_s}}$

Protocol DF-RC	Signal model	Heq	Achievable rate
forwarding	$\mathbf{y}_r(t) = \mathbf{H}_1 \mathbf{x}_s^1(t) + \mathbf{w}_r(t) \qquad t \in S_1$ $\mathbf{y}_d(t) = \begin{cases} 0 & t \in S_1 \\ \mathbf{H}_2 \mathbf{x}_r(t - N/2) + \mathbf{w}_d(t) & t \in S_2 \end{cases}$	$\mathbf{H}_2$	$\dfrac{1}{2}\min(C_1, C_b)$
protocol I	$\mathbf{y}_r(t) = \mathbf{H}_1 \mathbf{x}_s^1(t) + \mathbf{w}_r(t) \qquad t \in S_1$ $\mathbf{y}_d(t) = \begin{cases} \mathbf{H}_0 \mathbf{x}_s^1(t) + \mathbf{w}_d(t) & t \in S_1 \\ \mathbf{H}_2 \mathbf{x}_r(t - N/2) + \mathbf{w}_d(t) & t \in S_2 \end{cases}$	$\begin{bmatrix} \mathbf{H}_0 \\ \varphi \mathbf{H}_2 \end{bmatrix}$	$\dfrac{1}{2}\min(C_1, C_b)$
protocol II	$\mathbf{y}_r(t) = \mathbf{H}_1 \mathbf{x}_s^1(t) + \mathbf{w}_r(t) \qquad t \in S_1$ $\mathbf{y}_d(t) = \begin{cases} 0 & t \in S_1 \\ [\mathbf{H}_2 \ \mathbf{H}_0] \begin{bmatrix} \mathbf{x}_r(t - N/2) \\ \mathbf{x}_s^2(t - N/2) \end{bmatrix} + \mathbf{w}_d(t) & t \in S_2 \end{cases}$	$[\mathbf{H}_0 \ \varphi \mathbf{H}_2]$	$\dfrac{1}{2}\min(C_1, C_b)$
protocol III	$\mathbf{y}_r(t) = \mathbf{H}_1 \mathbf{x}_s^1(t) + \mathbf{w}_r(t) \qquad t \in S_1$ $\mathbf{y}_d(t) = \begin{cases} \mathbf{H}_0 \mathbf{x}_s^1(t) + \mathbf{w}_d(t) & t \in S_1 \\ [\mathbf{H}_2 \ \mathbf{H}_0] \begin{bmatrix} \mathbf{x}_r(t - N/2) \\ \mathbf{x}_s^2(t - N/2) \end{bmatrix} + \mathbf{w}_d(t) & t \in S_2 \end{cases}$	$\begin{bmatrix} \mathbf{H}_0 & 0 \\ \varphi \mathbf{H}_2 & \mathbf{H}_0 \end{bmatrix}$	$\begin{cases} \dfrac{C_b}{2} & C_1 \geq C_b - C_0 \\ \dfrac{C_1 + C_0}{2} & otherwise \end{cases}$

All protocols shown Table 11.3 in are affected by factor 1/2, due to working with an equally sized *relay-receive* and *relay-transmit phase*, the source-relay capacity C_1 and C_b, the definition of which depends on matrix $\mathbf{H}_{eq}$ describing an equivalent system. Those equivalent systems are $(n_d \times n_r)$ MIMO, $(2n_d \times n_r)$ MIMO, similar to $(n_d \times (n_s + n_r))$ MIMO and akin to $(2n_d \times (n_s + n_r))$ MIMO systems for the forwarding, protocol I, protocol II and protocol III, respectively. The difference between protocols II and III and MIMO schemes lies in the power used through the distributed antennas and the equivalent channel $\mathbf{H}_{eq}$ considered for protocol III. For instance in protocol II there are $n_s + n_r$ antennas, but the antennas at the source and the relays transmits with an average power equal to P_s/n_s and P_r/n_r, while in a MIMO system all the antennas should use $P/(n_s + n_r)$, in the non CSIT case.

When the relay-assisted transmission is not limited by the source-relay link, then the performance of DF-RC protocol I and II depends on the antenna configuration (see Appendix in Section 11.6.7).

$$\begin{cases} I_{p-II}^{RC} \leq I_{p-I}^{RC} & if \quad n_s = n_r > n_d \\ I_{p-II}^{RC} = I_{p-I}^{RC} & if \quad n_s = n_r = n_d \\ I_{p-II}^{RC} \geq I_{p-I}^{RC} & if \quad n_s = n_r < n_d \end{cases} \tag{11.92}$$

Hence, the relation between the repetition coding protocols defined by,

$$I_{FW}^{RC} \leq \begin{cases} I_{p-II}^{RC} \leq I_{p-I}^{RC} \\ or \\ I_{p-I}^{RC} \leq I_{p-II}^{RC} \end{cases} \leq I_{p-III}^{RC} \tag{11.93}$$

where the dependence protocol II and I is defined in (11.92).

The multiple antenna case will be evaluated under Rayleigh fading channels whose channel coefficients are constant over a frame where the relay-assisted transmission is carried out. The scenario considered is presented in Figure 11.19, where source and destination are separated at normalized distance equal

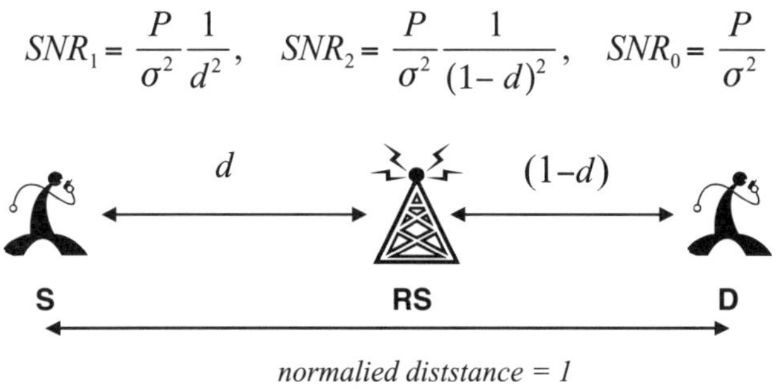

$$SNR_1 = \frac{P}{\sigma^2}\frac{1}{d^2}, \quad SNR_2 = \frac{P}{\sigma^2}\frac{1}{(1-d)^2}, \quad SNR_0 = \frac{P}{\sigma^2}$$

Figure 11.19 Scenario for the evaluation of the half-duplex protocols. SNR_0, SNR_1 and SNR_2 stand for the SNR in the source-destination, source-relay and relay-destination link.

to 1 and the relay terminal is placed at normalized distance d from the source. The signal to noise ratio in the source-relay and relay-destination links depends on the inverse of d^2 and $(1-d)^2$, respectively. The noise power has been assumed equal at all terminals. Results will consider the following combination of antennas (n_s, n_r, n_d): (1,1,1), (1,1,2), (2,1,1), and (2,2,2) at *low SNR* ($SNR_0 = 0$ dB).

Figure 11.20 (top-left) depicts the performance of the RC protocols for $n_s = n_r = n_d = 1$ in terms of average achievable rate. It can be seen that protocol III is the only one which outperforms the direct transmission, obtaining a maximum gain of 29% over the direct transmission when the relay is at $d = 0.4$. Additionally, the performance of protocol I and II is the same due to $n_r = n_s$, (11.92). When the antennas at the destination are increased, $n_s = n_r = 1$, $n_d = 2$ in Figure 11.20 (top-right), the maximums of the average achievable rate are attained in places where the relay is close to the source. On the other hand, when the number of antennas at the source and relay are larger than in the destination, $n_s = n_r = 2$, $n_d = 1$ in Figure 11.20 (bottom-left), the position of the relay where the average achievable rate is maximum is moved to the destination. In this case, all protocols can get a better performance than the direct transmission. It is worth noticing that all protocols get the maximum gains over the direct transmission for this antenna configuration, with protocol III getting a gain of 78% over the 2×1 MIMO system with a single antenna at the destination. Moreover, the relation between protocol I and II when $(n_s = n_r = 2, n_d = 1)$ and $(n_s = n_r = 1, n_d = 2)$ follows (11.93). Finally, for $n_s = n_r = n_d = 2$ in Figure 11.20 (bottom-right) the maximum average achievable rates are obtained with the relay placed in similar positions as in the case $n_s = n_r = n_d = 1$, although the maximum gains of the relay-assisted transmission are better.

Amplify-and-Forward
Under *amplify and forward* mode, the relay does not transmit any codeword, in contrast to the *decode-and-forward* case (signal $\mathbf{x}_r$ in (11.89), it just amplifies the received signal. Hence, the resources allocated to both phases of the *relay-assisted transmission* must be the same, and the definition done in (11.88) is still valid. It is assumed following amplifying gain factor taking when there is not CSIT,

$$\mathbf{G} = \mathbf{I}_{n_r} g \qquad g = \sqrt{\frac{P_r}{Tr\left(\mathbf{H}_1\mathbf{H}_1^H\right)\frac{P_s}{n_s} + n_r\sigma_{w_r}^2}} \tag{11.94}$$

where P_r and P_s are the power at the relay and the source, respectively, n_r and n_s denote the number of antennas at the relay and the source, $\sigma_{w_r}^2$ is the noise power at the relay and $\mathbf{H}_1$ stands for the channel between the source and the relay, with dimensions $n_r \times n_s$.

The source transmits the zero-mean Gaussian signals $\mathbf{x}_s^1$ and $\mathbf{x}_s^2$ during the *relay-receive* and *relay-transmit phase*. The transmitters do not have CSIT (fixed channel realization), so then the definitions

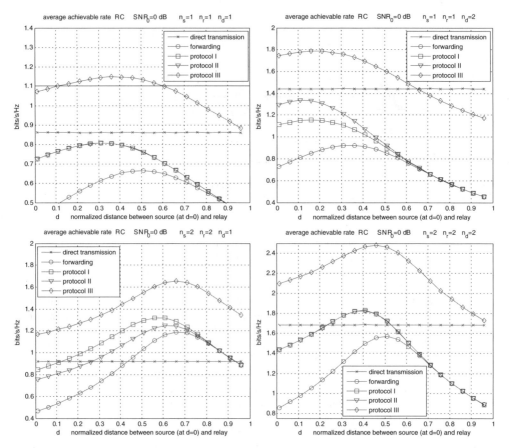

Figure 11.20 Average achievable rate vs. distance between the source and relay. Rayleigh fading. Destination with $SNR_0 = 0$ dB. (top-left) $n_s = 1$, $n_r = 1$, $n_d = 1$, (top-right) $n_s = 1$, $n_r = 1$, $n_d = 2$, (bottom-left) $n_s = 2$, $n_r = 2$, $n_d = 1$, (bottom-right) $n_s = 2$, $n_r = 2$, $n_d = 2$. Repetition-and-Coding (RC).

done in (11.89) are applicable. The mutual information can be written as,

$$I^{AF} = \frac{1}{2} \log \det \left(\mathbf{I} + \frac{P}{n_s} \mathbf{R}_w^{-1} \mathbf{H}_{AF} \mathbf{H}_{AF}^H \right) \qquad (11.95)$$

where $\mathbf{I}$ denotes the identity matrix with size depending on variables $\mathbf{R}_w$. Matrix $\mathbf{R}_w$ and $\mathbf{H}_{AF}$ stands for the noise covariance matrix and the equivalent channel matrix, both depend on the employed protocol. Finally, n_s is the number of transmitting antennas.

Table 11.4 presents the signal model and the values of variables $\mathbf{R}_w$ and $\mathbf{H}_{AF}$ in order to get the mutual information of the different protocols, (11.95). The spectral efficiency of AF protocols is affected by the factor 1/2 and the noise retransmitted by the relays $\sigma_{w_r}^2 \mathbf{H}_2 \mathbf{G} \mathbf{G}^H \mathbf{H}_2^H$. The equivalent system is similar to $(n_d \times n_s)$, $(2n_d \times n_s)$, $(n_d \times 2n_s)$ and $(2n_d \times 2n_s)$ MIMO systems for the forwarding, protocol I, protocol II and protocol III. The number of virtual antennas at destination is increased in protocol I, while protocol II increases the virtual antennas at the source and protocol III increases both.

Table 11.4 Signal model of the different static AF protocols

Protocol AF	Signal model	$\mathbf{H}_{AF}$, $\mathbf{R}_w$
forwarding	$\mathbf{y}_r(t) = \mathbf{H}_1\mathbf{x}_s^1(t) + \mathbf{w}_r(t) \qquad t \in S_1$ $\mathbf{y}_d(t) = \begin{cases} 0 & t \in S_1 \\ \mathbf{H}_2\mathbf{G}\mathbf{H}_1\mathbf{x}_s^1(t-N/2) + \mathbf{w}_d(t) & t \in S_2 \end{cases}$	$\mathbf{H}_{AF} = [\mathbf{H}_2\mathbf{G}\mathbf{H}_1]$ $\mathbf{R}_w = \sigma_{w_d}^2\mathbf{I}_{n_d} + \sigma_{w_r}^2\mathbf{H}_2\mathbf{G}\mathbf{G}^H\mathbf{H}_2^H$
protocol I	$\mathbf{y}_r(t) = \mathbf{H}_1\mathbf{x}_s^1(t) + \mathbf{w}_r(t) \qquad t \in S_1$ $\mathbf{y}_d(t) = \begin{cases} \mathbf{H}_0\mathbf{x}_s^1(t) + \mathbf{w}_d(t) & t \in S_1 \\ \mathbf{H}_2\mathbf{G}\mathbf{H}_1\mathbf{x}_s^1(t-N/2) + \mathbf{w}_d(t) & t \in S_2 \end{cases}$	$\mathbf{H}_{AF} = \begin{bmatrix} \mathbf{H}_0 \\ \mathbf{H}_2\mathbf{G}\mathbf{H}_1 \end{bmatrix}$ $\mathbf{R}_w = \begin{bmatrix} \sigma_{w_d}^2\mathbf{I}_{n_d} & 0 \\ 0 & \sigma_{w_d}^2\mathbf{I}_{n_s} + \sigma_{w_r}^2\mathbf{H}_2\mathbf{G}\mathbf{G}^H\mathbf{H}_2^H \end{bmatrix}$
protocol II	$\mathbf{y}_r(t) = \mathbf{H}_1\mathbf{x}_s^1(t) + \mathbf{w}_r(t) \qquad t \in S_1$ $\mathbf{y}_d(t) = \begin{cases} 0 & t \in S_1 \\ [\mathbf{H}_2\mathbf{G}\mathbf{H}_1 \;\; \mathbf{H}_0] \begin{bmatrix} \mathbf{x}_s^1; t-N/2 \\ \mathbf{x}_s^2; t-N/2 \end{bmatrix} + \mathbf{w}_d(t) & t \in S_2 \end{cases}$	$\mathbf{H}_{AF} = [\mathbf{H}_0 \;\; \mathbf{H}_2\mathbf{G}\mathbf{H}_1]$ $\mathbf{R}_w = \mathbf{I}_{n_d}\sigma_{w_s}^2 + \sigma_{w_r}^2\mathbf{H}_2\mathbf{G}\mathbf{G}^H\mathbf{H}_2^H$
protocol III	$\mathbf{y}_r(t) = \mathbf{H}_1\mathbf{x}_s^1(t) + \mathbf{w}_r(t) \qquad t \in S_1$ $\mathbf{y}_d(t) = \begin{cases} \mathbf{H}_0\mathbf{x}_s^1(t) + \mathbf{w}_d(t) & t \in S_1 \\ [\mathbf{H}_2\mathbf{G}\mathbf{H}_1 \;\; \mathbf{H}_0] \begin{bmatrix} \mathbf{x}_s^1(t-N/2) \\ \mathbf{x}_s^2(t-N/2) \end{bmatrix} + \mathbf{w}_d(t) & t \in S_2 \end{cases}$	$\mathbf{H}_{AF} = \begin{bmatrix} \mathbf{H}_0 & 0 \\ \mathbf{H}_2\mathbf{G}\mathbf{H}_1 & \mathbf{H}_0 \end{bmatrix}$ $\mathbf{R}_w = \begin{bmatrix} \sigma_{w_d}^2\mathbf{I}_{n_d} & 0 \\ 0 & \sigma_{w_d}^2\mathbf{I}_{n_d} + \sigma_{w_r}^2\mathbf{H}_2\mathbf{G}\mathbf{G}^H\mathbf{H}_2^H \end{bmatrix}$

Likewise, the mutual information for AF protocol I can be written in a simple way when the antenna configuration satisfies, $n_d = n_r = 1$ and $n_s \geq 1$, [77], see also Section 11.6.32.

$$I_{p-I}^{AF} = \frac{1}{2}C_0 + \frac{1}{2}\log\left(1 + \frac{\rho_1\rho_2\Theta}{\rho_2 + \rho_1 + 1}\right)$$

$$\Theta = \frac{1 + \rho_0(1 - \xi)}{1 + \rho_0} \qquad \xi = \frac{|\mathbf{h}_1^H\mathbf{h}_0|^2}{\mathbf{h}_0^H\mathbf{h}_0\mathbf{h}_1^H\mathbf{h}_1} = \cos^2(\phi) \qquad C_0 = \log(1 + \rho_0) \qquad (11.96)$$

$$\rho_0 = \frac{P_s\mathbf{h}_0^H\mathbf{h}_0}{\sigma_{w_d}^2 L_0 n_s} \qquad \rho_2 = \frac{P_r h_2 h_2^*}{\sigma_{w_d}^2 L_2 n_r} \qquad \rho_1 = \frac{P_s\mathbf{h}_1^H\mathbf{h}_1}{\sigma_{w_r}^2 L_1 n_s}$$

Moreover, for the single-antenna case the mutual information attained by the amplify-and-forward relaying becomes,

$$I^{AF} = \begin{cases} \frac{1}{2}\log\left(1 + \frac{\rho_2\rho_1}{1 + \rho_1 + \rho_2}\right) & \textit{forwarding} \\[3mm] \frac{1}{2}C_0 + \frac{1}{2}\log_2\left(1 + \frac{\rho_1\rho_2}{(1 + \rho_0)(\rho_2 + \rho_1 + 1)}\right) & \textit{protocol I} \\[3mm] \frac{1}{2}\log_2\left(1 + \frac{\rho_0(1 + \rho_1) + \rho_2\rho_1}{1 + \rho_1 + \rho_2}\right) & \textit{protocol II} \\[3mm] \frac{1}{2}\log_2\left((1 + \rho_0)\left(1 + \frac{\rho_0(1 + \rho_1) + \rho_2\rho_1}{1 + \rho_1 + \rho_2}\right) - \frac{\rho_0\rho_1\rho_2}{1 + \rho_1 + \rho_2}\right) & \textit{protocol III} \end{cases} \qquad (11.97)$$

where

$$\rho_0 = \frac{|h_0|^2 P_s}{\sigma_{w_d}^2} \qquad \rho_1 = \frac{|h_1|^2 P_s}{\sigma_{w_r}^2} \qquad \rho_2 = \frac{|h_2|^2 P_r}{\sigma_{w_d}^2}$$

The AF protocols are evaluated under non-frequency selective Rayleigh channels and the terminals will be equipped with multiple antennas. The channel coefficients are constant during the relay-assisted transmission. Because in the AF mode, the relay only retransmits the signal received in the *relay-received phase* and additionally it is not considered any signal processing at the relay, only the source and destination will feature more than one antenna. Therefore, assuming the scenario considered in Figure 11.19, results are presented in terms of average achievable rate with the following combination of antennas (n_s, n_r, n_d): (1,1,1), (2,1,1), (4,1,1), and (4,1,2) at *low SNR* (SNR$_0$ = 0 dB). Figure 11.21 (top-left) depicts the average achievable rate of the AF protocols for $n_s = 1, n_r = 1, n_d = 1$ where only protocol III can get better performance than the direct transmission for $d \leq 0.75$. In Figure 11.21 (top-right) and Figure 11.21 (bottom-left) the number of antennas at the source is $n_s = 2$ and $n_s = 4$, respectively. It should be remarked that protocol I experiments an important improvement with n_s and $n_r = n_d = 1$ as was predicted by (11.96). For example, the maximum gains of the relay-assisted transmission over the direct transmission for protocol I get a gain up to 4% for $n_s = 4$, while the gain for $n_s = 1$ is around -14%. However, for the configuration $n_s = 4, n_r = 1, n_d = 2$ presented in Figure 11.21(bottom-right), protocol I gets a worse performance than direct transmission (maximum gain of -10%). Protocol III obtains the best performance with a steady maximum gain over the direct transmission for $n_s \geq 2, n_r = 1, n_d \geq 1$ (around 30%).

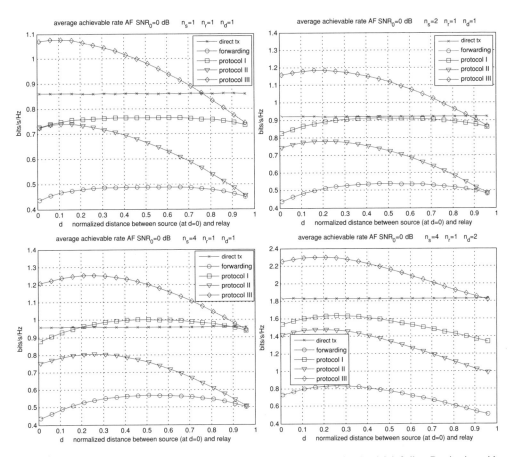

Figure 11.21 Average achievable rate vs. distance between the source and relay. Rayleigh fading. Destination with $SNR_0 = 0$ dB. (top-left) $n_s = 1$, $n_r = 1$, $n_d = 1$, (top-right) $n_s = 2$, $n_r = 1$, $n_d = 1$, (bottom-left) $n_s = 4$, $n_r = 1$, $n_d = 1$, (bottom-right) $n_s = 4$, $n_r = 1$, $n_d = 2$. Amplify-and-Forward.

11.4.2 Distributed Space–Time Coding

Since the introduction of the relay terminal in the system can be seen as new distributed virtual antennas available for the source, the next unavoidable question is how the concept of space–time coding can be implemented. In this regard, this section explores how to adopt the space-time codes for the relay-assisted transmission in a frequency selective channel, dwelling on the impact of an erroneous decoding at the relay on the performance of the communication. The particular case of protocol III, defined in Section 11.2.3, with a static resource allocation is considered here. Although the source is active in both phases of the communication (see Section 11.2.3) the source does not transmit a new message in the second phase as it happens in protocol III presented in Table 11.3. In contrast, the source and relay agree on transmitting the same message using a space-time code.

11.4.2.1 System Description

We illustrate a transmission protocol for such a system by again using the *TDD* scheme, although the same concept could also be implemented with the frequency-division duplexing (*FDD*) mode. In a *TDD* system, each frame is subdivided into consecutive time slots, as shown in Figure 11.22, which is obtained

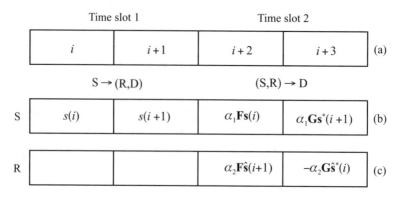

Figure 11.22 Structure of transmit slots: (a) time slots and block indices; (b) information blocks sent by S; and (c) information blocks sent by R.

by further elaboration of Figure 11.10. In the first slot, S (source) transmits and R (relay) receives. In the second slot, S and R transmit simultaneously. D (destination) can be in a listening mode in both time slots (protocol III defined in Section 11.2.3) or only in the second slot (protocol II defined in Section 11.2.3). Here we assume that in the second time slot, S and R transmit using a block Alamouti scheme (see, for example, Chapter 4), as if they were the two antennas of a single node. The extension to other space–time coding techniques is almost straightforward. Of course there are differences between conventional space time coding (*STC*) and the distributed (*DSTC*) concept. First of all, regenerative relays might make decision errors, so that the symbols transmitted from R could be affected by errors. Then, the links between S and D and between R and D do not have the same statistical properties. Also, even if S and R are synchronous, their packets might arrive at D at different times, as S and R are not collocated.

As before, we denote with $h_{sd}^{k_n}(i)$, $h_{sr}^{k_n}(i)$, $and\ h_{rd}^{k_n}(i)$ the channel impulse responses between S and D, S and R, and R and D, respectively; during the time slot indexed with k_n; i is the block index. Each block of symbols $s(i)$ has size M and is linearly encoded, so as to generate the N-size vector $x_s(n) := Fs\ (n)$, where F is the $N \times M$ precoding matrix. All channels are a finite-impulse response (*FIR*) of (maximum) order L_h and time invariant over at least a pair of consecutive blocks. So, a CP (cyclic prefix) of length $L \geq L_h$ is inserted at the beginning of each block, to facilitate elimination of interblock interference (IBI), synchronization, and channel equalization at the receiver.

As shown in Figure 11.22, during the first time slot, S sends, consecutively, the two N-size information symbol blocks $s(i)$ and $s(i + 1)$. The blocks are linearly encoded using the precoding matrix F, so that the corresponding transmitted blocks are $x_s(n) := F_s\ s(n)$, with $n = i,\ i + 1$. Within this time slot, the destination may be in a listening mode or not. After removing the guard interval at the receiver, the N-size vectors $y_d(n)$ and $y_r(n)$ received from D and R are, respectively

$$y_d(n) = \mathbf{H}_{sd}^{k_n}\mathbf{F}_s^n s(n) + \mathbf{n}_d(n), \quad n = i, i + 1$$
$$y_r(n) = \mathbf{H}_{sr}^{k_n}\mathbf{F}_s^n s(n) + \mathbf{n}_r(n), \quad n = i, i + 1 \tag{11.98}$$

where $\mathbf{n}_r(n)$ and $\mathbf{n}_d(n)$ are the additive-noise vectors in R and D, respectively. The channel matrices $\mathbf{H}_{sd}^{k_n}$ and $\mathbf{H}_{sr}^{k_n}$, thanks to the insertion of the CP, are $N \times N$ circulant Toeplitz matrices, with entries $\mathbf{H}_{sd}^{k_n}(i, j) = h_{sd}^{k_n}((i - j)\mathrm{mod}N)\ and\ \mathbf{H}_{sr}^{k_n}(i, j) = h_{sr}^{k_n}((i - j)\mathrm{mod}N)$, respectively. Because of their circulant and Toeplitz structure, $\mathbf{H}_{sd}^{k_n}$ and $\mathbf{H}_{sr}^{k_n}$ are diagonalized as follows: $\mathbf{H}_{sd}^{k_n} = \mathbf{W}\Lambda_{sd}^{k_n}\mathbf{W}^H$ and $\mathbf{H}_{sr}^{k_n} = \mathbf{W}\Lambda_{sr}^{k_n}\mathbf{W}^H$, where $\mathbf{W}$ is the $N \times N$ IFFT matrix with $\{W\}_{kl} = e^{j2\pi kl/N}/\sqrt{N}$, whereas $\Lambda_{sr}^{k_n}$ and $\Lambda_{sd}^{k_n}$ are the $N \times N$ diagonal matrices, whose entries are $\Lambda_{sr}^{k_n}(k, k) = \sum_{l=0}^{L_h-1} h_{sr}^{k_n}(l)e^{-j2\pi lk/N}$ and $\Lambda_{sd}^{k_n}(k, k) = \sum_{l=0}^{L_h-1} h_{sd}^{k_n}(l)e^{-j2\pi lk/N}$, respectively.

The relay node decodes the received vectors and provides the estimated vectors $\hat{s}(i)$ and $\hat{s}(i + 1)$.

During the successive time slot, S and R transmit simultaneously, using the Alamouti block code. Specifically, with reference to Figure 11.22, in the first half of the second time slot, S transmits $x_s(i + 2) = \alpha_1 \, Fs \, (i)$ and R transmits $x_r(i + 2) = \alpha_2 \, F\hat{s} \, (i + 1)$. In the second half, S transmits $x_s(i + 3) = \alpha_1 \, Gs^* \, (i + 1)$ while R transmits $x_r(i + 3) = -\alpha_2 \, G\hat{s}^* \, (i)$. To guarantee maximum spatial diversity, the two matrices G and F are related to each other by $G = JF^*$, as in [44], where J is a time-reversal (plus a one-chip cyclic shift) matrix. If N is even, J has all null entries except the elements of position $(1,1)$ and $(k, N - k + 2)$, with $k = 2, \ldots, N$, which are equal to 1. If N is odd, J is the antidiagonal matrix. The two real coefficients α_1 and α_2 are related to each other by $\alpha_1^2 + \alpha_2^2 = 1$. They are introduced in order to have a degree of freedom in the power distribution between S and R, under a given total transmit power in order to minimize the final average BER.

After discarding the CP, the blocks received by D in the two consecutive time slots $i + 2$ and $i + 3$ are

$$\mathbf{y}_d(i + 2) = \alpha_1 \mathbf{H}_{sd}^{k_i+2} \mathbf{Fs}(i) + \alpha_2 H_{rd}^{k_i+2} \mathbf{F\hat{s}}(i) + \mathbf{n}_d(i + 2)$$
$$\mathbf{y}_d(i + 3) = \alpha_1 \mathbf{H}_{sd}^{k_i+2} \mathbf{Gs}^*(i) - \alpha_2 \mathbf{H}_{rd}^{k_i+2} \mathbf{G\hat{s}}^*(i) + \mathbf{n}_d(i + 3) \qquad (11.99)$$

where $\mathbf{H}_{sd}^{k_n}$ and $\mathbf{H}_{rd}^{k_n}$ represent the channels between S and D, and between R and D, respectively. Exploiting again the diagonalizations $\mathbf{H}_{sd}^{k_n} = \mathbf{W} \mathbf{\Lambda}_{sd}^{k_n} \mathbf{W}^{\mathrm{H}}$ and $\mathbf{H}_{rd}^{k_n} = \mathbf{W} \mathbf{\Lambda}_{rd}^{k_n} \mathbf{W}^{\mathrm{H}}$, if we premultiply in Equation (11.99) $\mathbf{y}_d(i + 2)$ by $\mathbf{W}^{\mathrm{H}}$ and $\mathbf{y}_d^*(i + 3)$ by $\mathbf{W}^T$, we get:

$$\mathbf{W}^H \mathbf{y}_d(i + 2) = \alpha_1 \mathbf{\Lambda}_{sd}^{k_i+2} \tilde{\mathbf{F}}\mathbf{s}(i) + \alpha_2 \mathbf{\Lambda}_{rd}^{k_i+2} \tilde{\mathbf{F}}\hat{\mathbf{s}}(i) + \mathbf{W}^H \mathbf{n}_d(i + 2)$$
$$\mathbf{W}^T \mathbf{y}_d^*(i + 3) = \alpha_1 \mathbf{\Lambda}_{sd}^{k_i+2} \tilde{\mathbf{G}}^* \mathbf{s}(i) - \alpha_2 \mathbf{\Lambda}_{rd}^{k_i+2*} \tilde{\mathbf{G}}^* \hat{\mathbf{s}}(i) + \mathbf{W}^T \mathbf{n}_d^*(i + 3) \qquad (11.100)$$

where $\tilde{\mathbf{F}} := \mathbf{W}^H \mathbf{F}$ and $\tilde{\mathbf{G}} := \mathbf{W}^T \mathbf{G}$. For the sake of simplicity, we assume that orthogonal frequency-division multiplexing (OFDM) is performed at both S and R nodes, so that $N = M$, $F = W$, and thus, $\tilde{\mathbf{F}} = \mathbf{I}_n$ and $\mathbf{G} = \mathbf{W}$. We also introduce the orthogonal matrix:

$$\mathbf{\Lambda}_{k_n} := \begin{pmatrix} \alpha_1 \mathbf{\Lambda}_{sd}^{k_n} & \alpha_2 \mathbf{\Lambda}_{rd}^{k_n} \\ -\alpha_2 \mathbf{\Lambda}_{sd}^{k_n} & \alpha_1 \mathbf{\Lambda}_{sd}^{k_n*} \end{pmatrix} \qquad (11.101)$$

such that $\mathbf{\Lambda}_{k_n}^H \mathbf{\Lambda}_{k_n} := \mathbf{I}_2 \otimes \bar{\mathbf{\Lambda}}_{k_n}^2$, where $\bar{\mathbf{\Lambda}}_{k_n}^2 := \alpha_1^2 \left| \mathbf{\Lambda}_{sd}^{k_n} \right|^2 + \alpha_2^2 \left| \mathbf{\Lambda}_{rd}^{k_n} \right|^2$, whereas $\otimes$ denotes the Kronecker product. We introduce also the unitary matrix $\mathbf{Q}^{Kn} := \mathbf{\Lambda}_{k_n}(\mathbf{I}_2 \otimes \bar{\mathbf{\Lambda}}_{k_n}^{-1})$, satisfying the relationships $\mathbf{Q}^{KnH} \mathbf{Q}^{Kn} = \mathbf{I}_{2n}$ and $\mathbf{Q}^{KnH} \mathbf{\Lambda}^{kn} = \mathbf{I}_2 \otimes \bar{\mathbf{\Lambda}}_{kn}$. Exploiting the above equalities and multiplying the vector $\mathbf{u} := [(\mathbf{W}^H \mathbf{y}_d(i + 2))^T, (\mathbf{W}^T \mathbf{y}_d^*(i + 3))^T]^T$, by the matrix $\mathbf{Q}^{k_i+2H}$, we get:

$$\begin{bmatrix} \mathbf{r}(i) \\ \mathbf{r}(i + 1) \end{bmatrix} := \mathbf{Q}^{k_i+2H} \mathbf{u} = \begin{bmatrix} \left| \tilde{\mathbf{\Lambda}}_{sd}^{k_i+2} \right|^2 & -\tilde{\mathbf{\Lambda}}_{sd}^{k_i+2*} \tilde{\mathbf{\Lambda}}_{rd}^{k_i+2} \\ \tilde{\mathbf{\Lambda}}_{sd}^{k_i+2} \tilde{\mathbf{\Lambda}}_{rd}^{k_i+2*} & \left| \tilde{\mathbf{\Lambda}}_{sd}^{k_i+2} \right|^2 \end{bmatrix} s + \begin{bmatrix} \left| \tilde{\mathbf{\Lambda}}_{rd}^{k_i+2} \right|^2 & \tilde{\mathbf{\Lambda}}_{sd}^{k_i+2*} \tilde{\mathbf{\Lambda}}_{rd}^{k_i+2} \\ -\tilde{\mathbf{\Lambda}}_{sd}^{k_i+2} \tilde{\mathbf{\Lambda}}_{rd}^{k_i+2*} & \left| \tilde{\mathbf{\Lambda}}_{rd}^{k_i+2} \right|^2 \end{bmatrix} \hat{\mathbf{s}} + \mathbf{n}$$

$$(11.102)$$

where

$$\mathbf{s} := [s(i)^T, s(i + 1)^T]^T,$$
$$\hat{\mathbf{s}} := [\hat{s}(i)^T, \hat{s}(i + 1)^T]^T,$$
$$\tilde{\mathbf{\Lambda}}_{sd}^{k_i+2} := \alpha_1 \mathbf{\Lambda}_{sd}^{k_i+2} \bar{\mathbf{\Lambda}}_{kn}^{-1/2}, \qquad (11.103)$$
$$\tilde{\mathbf{\Lambda}}_{rd}^{k_i+2} := \alpha_2 \mathbf{\Lambda}_{rd}^{k_i+2} \bar{\mathbf{\Lambda}}_{kn}^{-1/2},$$
$$\bar{\mathbf{n}} := [\bar{n}^T(i), \bar{n}^T(i + 1)]^T = \mathbf{Q}^{k_i+2H} [\mathbf{n}^T(i + 2), \mathbf{n}^H(i + 3)]^T.$$

If the two transmit antennas use the same power, i.e. $\alpha_1 = \alpha_2$, and there are no decision errors at the relay node, i.e. $\hat{s}(n) = s(n)$, $n = i, i + 1$, the previous equations reduce to the classical block Alamouti equations discussed in Chapter 4.

Since $\mathbf{Q}^{k_i+2\mathrm{H}}$ is unitary, $\mathbf{n}$ is white, $\bar{\mathbf{n}}$ is also white, with covariance matrix $C_n = \sigma_n^2 \mathbf{I}_{2N}$. Furthermore, since all matrices Λ appearing in Equation (11.102) are diagonal, the system Equation (11.102) of $2N$ equations can be decoupled into N independent systems of two equations in two unknowns, each equation representing a single subcarrier. Introducing the vectors $r_k := [r_k(i), r_k(i + 1)]^{\mathrm{T}} s_k := [s_k(i), s_k(i + 1)]^{\mathrm{T}}$, $\hat{s}_k := [\hat{s}_k(i), \hat{s}_k(i + 1)]^{\mathrm{T}}$, and $\bar{\mathbf{n}}_k := [\bar{n}_k(i), \bar{n}_k(i + 1)]^{\mathrm{T}}$, representing the kth subcarrier, with $k = 0, \ldots, N - 1$ [for simplicity of notation, we drop the block index and set $\tilde{\Lambda}_{sd} = \tilde{\Lambda}_{sd}^{k_i+2}(k, k) \, and \, \tilde{\Lambda}_{rd} = \tilde{\Lambda}_{rd}^{k_i+2}(k, k)$] Equation (0.102) is equivalent to

$$\mathbf{r}_k = \begin{bmatrix} |\tilde{\Lambda}_{sd}|^2 & -\tilde{\Lambda}_{sd}^* \tilde{\Lambda}_{rd} \\ \tilde{\Lambda}_{sd} \tilde{\Lambda}_{rd}^* & |\tilde{\Lambda}_{sd}|^2 \end{bmatrix} \mathbf{s}_k + \begin{bmatrix} |\tilde{\Lambda}_{rd}|^2 & \tilde{\Lambda}_{sd}^* \tilde{\Lambda}_{rd} \\ -\tilde{\Lambda}_{sd} \tilde{\Lambda}_{rd}^* & |\tilde{\Lambda}_{rd}|^2 \end{bmatrix} \hat{\mathbf{s}}_k + \bar{\mathbf{n}}_k \tag{11.104}$$

where $\mathbf{r}_k$ represents a sufficient statistic for the decision on the transmitted symbol vector s_k.

Due to synchronization error, in general, the blocks transmitted from S and R arrive at D at different times. However, if the difference in arrival times τ_d is incorporated in the CP used from both S and R, D is still able to get N samples from each received block, without IBI.

If due to synchronization error, the block coming from S arrives with a delay of L_d samples, with respect to arrival instant from R, the only difference, with respect to the case of perfect synchronization, is that the transfer function $\tilde{\Lambda}_{sd}(k)$ in Equation (11.104) will be substituted by $\tilde{\Lambda}_{sd}(k)e^{-j2\pi L_d k/N}$. Such a substitution does not affect the useful term, but only affects the interfering term. The price paid for this robustness is the increase of the CP length L, which, in its turn, reflects into a rate loss. However, this loss can be made small by choosing a block length N much greater than L, or by selecting only relays that are relatively close to the source, so as to make the relative delay small.

The ML detector would assume the knowledge, at the destination node, of the set of error probabilities $p_{e1}(k)$ and $p_{e2}(k)$ at the relay, with $k = 0, \ldots, N - 1$. If this knowledge is not available, a suboptimum scalar detector can be implemented, instead of the ML detector. In the case of BPSK modulation the decision on the transmitted symbol $sk(n)$ can be simply obtained as:

$$\hat{s}(n) = \mathrm{sign}\{\mathrm{Re}[\mathbf{r}(n)]\}, \quad n = i, i + 1 \tag{11.105}$$

For high signal-to-noise ratio (SNR) at the relay (i.e., when R makes no decision errors), the symbol-by-symbol decision in D becomes optimal and, thus, the decoding rule (11.105) provides the same performance as the optimal receiver. When the decision errors at the relay side cannot be neglected, the suboptimal receiver introduces a floor in the BER curve, because the symbol-by-symbol decision treats the wrong received symbols as interference. A cognitive radio should make the choice between the decoding rules as a tradeoff between performance and computational complexity (power consumption), taking into account the need for the ML detector to make available, at the destination node, the error probabilities of the relay node.

Whereas in conventional STC, the transmit antennas typically use the same power over all the transmit antennas, with $DSTC$, it is useful to distribute the available power between source and relay as a function of their relative position with respect to the final destination, since they are not collocated.

11.4.2.2   BER Analysis in DSTC

For *error-free Source to Relay* link, using the same derivations introduced in Section 11.4.2.1, a symbol-by-symbol detector is the optimal detector, and the signal-to-noise ratio on the kth symbol in the nth block is:

$$\text{SNR}_k(n) = y_k(n) = \frac{A^2}{\sigma_n^2} \left(\left| \Lambda_{sd}^{k_i}(k, k) \right|^2 + \alpha \left| \Lambda_{sd}^{k_{i+2}}(k, k) \right|^2 + (1 - \alpha) \left| \Lambda_{rd}^{k_{i+2}}(k, k) \right|^2 \right) \tag{11.106}$$

with $k = 0, 1, \dots, N - 1$; $n = i, i + 1$, and $\alpha = \alpha_1^2$. We assume that the variance of the channel-impulse-response coefficients is proportional to $1/d^{2r}$, with $r \geq 1$, where d is the distance of the link and calculate the BER for fast and slow fading channel.

In fast-fading $\Lambda_{sd}^{k_i}(k, k)$, $\Lambda_{sd}^{k_{i+2}}(k, k)$, and $\Lambda_{rd}^{k_{i+2}}(k, k)$ are independent. The error probability for BPSK, conditioned to a given channel realization, is given by:

$$P_{e/h}(k) = \frac{1}{2} \text{erfc}(\sqrt{0.5 \, y_k}) \tag{11.107}$$

where y_k is given by Equation (11.106). For each subcarrier k, y_k is given by the sum of three statistically independent random variables, each one distributed according to a χ^2 pdf with two degrees of freedom. Thus, the BER P_b averaged over the channel realizations is [45]:

$$P_b = \frac{1}{2} \sum_{k=1}^{Q} \pi_k \left[1 - \sqrt{\frac{\gamma_k}{1 + \gamma_k}} \right] \tag{11.108}$$

where $Q = 3$, and

$$\pi_k := \prod_{i \neq k=1}^{Q} \frac{\gamma_i}{\gamma_i - \gamma_k}; \quad \gamma_1 := \frac{A^2}{\sigma_n^2} \frac{\tilde{\sigma}_h^2}{d_{sd}^r}; \quad \gamma_2 := \frac{A^2}{\sigma_n^2} \frac{\alpha \tilde{\sigma}_h^2}{d_{sd}^r} \quad \gamma_3 := \frac{A^2}{\sigma_n^2} \frac{(1 - \alpha) \tilde{\sigma}_h^2}{d_{sd}^r}; \quad \tilde{\sigma}_h^2 = \sigma_n^2 (L_h + 1)$$

The optimal value of α can be found by minimizing Equation (11.108).

In slow-fading $\Lambda_{sd}^{k_i}(k, k) = \Lambda_{sd}^{k_{i+2}}(k, k)$ *and* $\Lambda_{rd}^{k_i}(k, k) = \Lambda_{rd}^{k_{i+2}}(k, k)$, *the* y_k *on the kth bit of the received block is:*

$$y_k(n) = \frac{A^2}{\sigma_n^2} \left((1 + \alpha) \left| \Lambda_{sd}^{k_i}(k, k) \right|^2 + (1 - \alpha) \left| \Lambda_{rd}^{k_{i+2}}(k, k) \right|^2 \right) \tag{11.109}$$

and the BER averaged over the channel realizations is given by (11.108), with $Q = 2$ and

$$\gamma_1 := \frac{A^2}{\sigma_n^2} \frac{(1 + \alpha) \tilde{\sigma}_h^2}{d_{sd}^r}, \quad \gamma_2 := \frac{A^2}{\sigma_n^2} \frac{(1 - \alpha) \tilde{\sigma}_h^2}{d_{rd}^r}.$$

For *source to relay link with errors* the decision on the transmitted symbol $s_k(i)$ is performed by taking a hard decision on

$$r_k(i) = \left| \tilde{\Lambda}_{sd}^{k_{i+2}} \right|^2 s_k(i) + \left| \tilde{\Lambda}_{rd}^{k_{i+2}} \right|^2 \hat{s}_k(i) + \tilde{\Lambda}_{sd}^{k_{i+2}*} \tilde{\Lambda}_{rd}^{k_{i+2}} \hat{s}_k(i + 1) - \tilde{\Lambda}_{sd}^{k_{i+2}*} \tilde{\Lambda}_{rd}^{k_{i+2}} s_k(i + 1) + \tilde{n}_k(i) \tag{11.110}$$

where $\tilde{\Lambda}_{sd}^{k_{i+2}} := \tilde{\Lambda}_{sd}^{k_{i+2}}(k, k)$; $\tilde{\Lambda}_{rd}^{k_{i+2}} := \tilde{\Lambda}_{rd}^{k_{i+2}}(k, k)$. In the sequel we use the following notation for the possible events: $H_1 = \{\hat{s}_k(i) = s_k(i); \hat{s}_k(i + 1) = s_k(i + 1)\}$; $H_2 = \{\hat{s}_k(i) = -s_k(i); \hat{s}_k(i + 1) = s_k(i + 1)\}$; $H_3 = \{\hat{s}_k(i) = s_k(i); \hat{s}_k(i + 1) = -s_k(i + 1)\}$; $H_4 = \{\hat{s}_k(i) = -s_k(i); \hat{s}_k(i + 1) = -s_k(i + 1)\}$; and $\varepsilon_{1,1} = \{s_k(i) = A; s_k(i + 1) = A\}$, $\varepsilon_{1,-1} = \{s_k(i) = A; s_k(i + 1) = -A\}$, $\varepsilon_{-1,1} = \{s_k(i) = -A; s_k(i + 1) = A\}$, $\varepsilon_{-1,-1} = \{s_k(i) = -A; s_k(i + 1) = -A\}$;. With $p_{e1}(k)$ and $p_{e2}(k)$ the conditional (to a given channel realization) error probabilities, at the relay node, on $s_k(l)$ and s_k (2), respectively, we have $P(H_1) = (1 - p_{e1}(k))(1 - p_{e2}(k))$, $P(H_2) = p_{e1}(k)(1 - p_{e2}(k))$, $P(H_3) = (1 - p_{e1}(k))p_{e2}(k)$, and $P(H_2) = p_{e1}(k)p_{e2}(k)$. Denoting, for simplicity, $u_i = Rr_k(i)$, and assuming that the information symbols are i.i.d. binary phase-shift keying (*BPSK*) symbols that may assume the values A or $-A$ with equal probability, the error probability $Pe/h\ (k)$ on $s_k\ (i)$, conditioned to a given channel realization, is:

$$P_{e/h}(k) := \frac{1}{4} \sum_{j=1}^{4} P(H_j)$$
$$\times [P\{u_i > 0/\varepsilon_{-1,1}, H_j\} + P\{u_i > 0/\varepsilon_{-1,-1}, H_j\}$$
$$+ P\{u_i < 0/\varepsilon_{1,1}, H_j\} + P\{u_i < 0/\varepsilon_{1,-1}, H_j\}] \tag{11.111}$$

Setting $\tilde{n}_k := Re\{\tilde{n}_k(i)\}$, we have:

$$P\{u_i > 0/\varepsilon_{-1,1}, H_j\} = P\left\{-\left(\left|\tilde{\Lambda}_{sd}^{k_{i+2}}\right|^2 - \left|\tilde{\Lambda}_{rd}^{k_{i+2}}\right|^2\right) A + \tilde{n}_r > 0\right\} := \frac{1}{2}\text{erfc}\left(\frac{A\beta_k}{\sigma_n}\right)$$

where β_k is given by:

$$\beta_k = \sqrt{\alpha_1^2 \left|\Lambda_{sd}^{k_n}(k, k)\right|^2 + \alpha_2^2 \left|\Lambda_{rd}^{k_n}(k, k)\right|^2}$$

Using this approach, we find:

$$P\{u_i > 0/\varepsilon_{-1,1}, H_1\} = \frac{1}{2}\text{erfc}\left(\frac{A\beta_k}{\sigma_n}\right); \quad P\{u_i > 0/\varepsilon_{-1,-1}, H_1\} = \frac{1}{2}\text{erfc}\left(\frac{A\beta_k}{\sigma_n}\right)$$

$$P\{u_i > 0/\varepsilon_{-1,1}, H_2\} = \frac{1}{2}\text{erfc}\left(\frac{A\delta_k}{\sigma_n}\right); \quad P\{u_i > 0/\varepsilon_{-1,-1}, H_2\} = \frac{1}{2}\text{erfc}\left(\frac{A\delta_k}{\sigma_n}\right)$$

$$P\{u_i > 0/\varepsilon_{-1,1}, H_3\} = \frac{1}{2}\text{erfc}\left(\frac{A(\beta_k + \gamma_k)}{\sigma_n}\right); \quad P\{u_i > 0/\varepsilon_{-1,-1}, H_3\} = \frac{1}{2}\text{erfc}\left(\frac{A(\beta_k - \gamma_k)}{\sigma_n}\right)$$

$$P\{u_i > 0/\varepsilon_{-1,1}, H_4\} = \frac{1}{2}\text{erfc}\left(\frac{A(\delta_k + \gamma_k)}{\sigma_n}\right); \quad P\{u_i > 0/\varepsilon_{-1,-1}, H_4\} = \frac{1}{2}\text{erfc}\left(\frac{A(\delta_k - \gamma_k)}{\sigma_n}\right)$$

where δ_k and γ_k are defined as:

$$\gamma_k = \frac{2\alpha_1\alpha_2 \text{Re}\left\{\Lambda_{sd}^{k_n*}(k, k)\Lambda_{sd}^{k_n*}(k, k)\right\}}{\sqrt{\alpha_1^2 \left|\Lambda_{sd}^{k_n}(k, k)\right|^2 + \alpha_2^2 \left|\Lambda_{rd}^{k_n}(k, k)\right|^2}}$$

$$\delta_k = \frac{\alpha_1^2 \left|\Lambda_{sd}^{k_n}(k, k)\right|^2 - \alpha_2^2 \left|\Lambda_{rd}^{k_n}(k, k)\right|^2}{\sqrt{\alpha_1^2 \left|\Lambda_{sd}^{k_n}(k, k)\right|^2 + \alpha_2^2 \left|\Lambda_{rd}^{k_n}(k, k)\right|^2}}$$

Repeating this same approach, we are able to find

$$
P_{e/h}(k) = \left[\frac{1}{2}\text{erfc}\left(\frac{A\beta_k}{\sqrt{\sigma_n^2}}\right)\right](1 - p_{e1}(k))(1 - p_{e2}(k)) + \left[\frac{1}{2}\text{erfc}\left(\frac{A\delta_k}{\sqrt{\sigma_n^2}}\right)\right]
$$

$$
\times\, p_{e1}(k)(1 - p_{e2}(k)) + \left[\frac{1}{4}\text{erfc}\left(\frac{A(\beta_k + \gamma_k)}{\sqrt{\sigma_n^2}}\right) + \frac{1}{4}\text{erfc}\left(\frac{A(\beta_k - \gamma_k)}{\sqrt{\sigma_n^2}}\right)\right] \qquad (11.112)
$$

$$
\times\, p_{e2}(k)(1 - p_{e1}(k)) + \left[\frac{1}{4}\text{erfc}\left(\frac{A(\delta_k + \gamma_k)}{\sqrt{\sigma_n^2}}\right) + \frac{1}{4}\text{erfc}\left(\frac{A(\delta_k - \gamma_k)}{\sqrt{\sigma_n^2}}\right)\right] p_{e1}(k)p_{e2}(k)
$$

where, for $n = i + 2,\, i + 3$

$$
p_{e1}(k) = p_{e2}(k) = \frac{1}{2}\text{erfc}\left(\frac{A}{\sqrt{\sigma_w^2(k)}}\right)
$$

with $\sigma_w^2(k) := \sigma_n^2 / \left|\Lambda_{sr}^{k_n}(k,k)\right|^2$, $k_{i+2} = k_{i+3}$.

The average BER can then be obtained by averaging Equation (11.112) over the channel realizations.

As an example, Figure 11.23 shows average BER versus α (setting $\alpha_1 = \alpha_1 = 1 - \alpha_2$ for different values of the distance d_{rd} (and thus, of y_R) between R and D (all distances are normalized with respect to the distance d_{sd} between S and D). Figure 11.23(a), shows the ideal case where there are no errors at the relay node, and Figure 11.23(b), shows average *BER* as a function of α, but for different values of the y_R at the relay node. We can observe that, as y_R decreases, the system tends to allocate less power to the relay node (the optimal value of α is greater than 0.5), as the relay node becomes less and less reliable.

11.4.3 *Dynamic Resource Allocation Relaying*

The literature of relay and cooperative transmission has proposed different schemes based on the relay protocols (forwarding, protocol I, II, and III) where the *relay-receive* and *relay-transmit phase* present the same size (in time or frequency domain), as in [40, 41]. Nevertheless, when there is some CSI about the different links it is possible to optimize the duration of each phase (*resource optimization* or *dynamic resource allocation*) based on that knowledge and thus improve the spectral efficiency of the relay-assisted transmission. In such a case, the resources are allocated to each phase depending on the quality of the links. For example, consider the protocol described in Section 11.2.3, where the source-relay link becomes a bottleneck when it presents poor quality. However, by increasing the *relay-receive phase* duration, the mutual information (in terms of number of bits) in the source-relay link also increases and the bottleneck may be alleviated, at the expense of a reduction of the *relay-transmit phase* duration. The forwarding protocol is considered in [42] and protocol III is examined in [17] and [51] (called the dynamic DF protocol and analyzed from the diversity-multiplexing tradeoff point of view). Regarding protocol I, [52] and [53] investigate resource optimization. Additionally, resource allocation optimization based on the outage probability in each link for protocol I is tackled in [54].

This section investigates the achievable rate of the half-duplex protocols when source and relay terminals transmit with *maximum power* during each phase. This situation is typically found when the relay-assisted transmission is duplexed in the *time domain* (TDD) with *power limited terminals*. Other average power constraints, such as *individual average power* constraint or a *sum average power* constraint are tackled in [2].

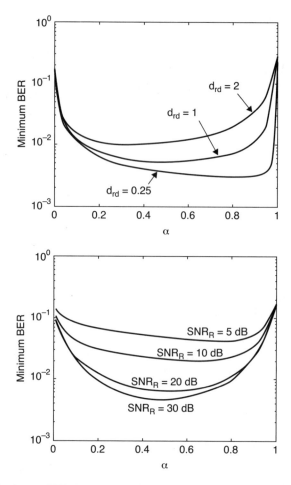

Figure 11.23 Average BER versus α: (a) without errors at the relay; (b) with errors at the relay.

The relay-assisted transmission in the time domain will be carried out during a frame of N symbols. The *relay-receive* and *relay-transmit phase* present a size of αN and $(1 - \alpha) N$, respectively. Let us define those time intervals by,

$$\begin{aligned} S_1 &= \{ t \quad | \quad \quad 1 \leq t \leq \lfloor \alpha N \rfloor \} \\ S_2 &= \{ t \quad | \quad \lfloor \alpha N \rfloor + 1 \leq t \leq N \} \end{aligned} \tag{11.113}$$

where the α parameter will be optimized for the different protocols in order to provide the maximum mutual information. The protocol considered in Section 11.4.1 assume a *static resource allocation* relaying with $\alpha = 0.5$. The assumptions considered in this section are:

- Variables n_s, n_r and n_d denote the number of antennas at the source, relay, and destination, P_s and P_r are the power used by the source and the relay respectively.
- Channel matrices are defined by $\mathbf{H}_1(n_r \times n_s)$, $\mathbf{H}_2(n_d \times n_r)$ and $\mathbf{H}_0(n_d \times n_s)$ for the source-relay, relay-destination, and source-destination link, respectively.
- The noise at the destination and the relay terminals is assumed to be Gaussian with zero mean and covariance matrices $\mathbf{R}_{w_d} = \sigma_{w_d}^2 \mathbf{I}$ and $\mathbf{R}_{w_r} = \sigma_{w_r}^2 \mathbf{I}$ when the transmitted bandwidth is W.

- The source may transmit the signals $\mathbf{x}_s^1$ and $\mathbf{x}_s^2$ during the *relay-receive* and *relay-transmit phase* respectively, depending on the selected protocol. The relay terminal transmits the signal $\mathbf{x}_r$ during the *relay-transmit phase*. All these signals are Gaussian with zero mean and diagonal covariance matrices.

The mutual information of the different protocols can be written as a function of the capacities of the different links under no CSIT and equal average power transmission per antenna defined by,

$$C_1 = \log\det\left(\mathbf{I}_{n_r} + \frac{P_s}{n_s\,\sigma_{w_r}^2}\mathbf{H}_1\mathbf{H}_1^H\right) \qquad C_b = \log\det\left(\mathbf{I}_{n_d} + \frac{P_r}{n_r\,\sigma_{w_d}^2}\mathbf{H}_2\mathbf{H}_2^H + \Delta\right)$$

$$C_0 = \log\det\left(\mathbf{I}_{n_d} + \frac{P_s}{n_s\,\sigma_{w_s}^2}\mathbf{H}_0\mathbf{H}_0^H\right) \tag{11.114}$$

where C_0, C_1 denote the capacities of the source-destination and source-relay link, while C_b is the capacity dependent on variable Δ. If such variable is equal to zero, C_b is equal to the capacity of the relay-destination link, but if it considers the source-destination channel, then C_b denotes the capacity of the (source-relay)-destination link, i.e. source and relay transmitting to destination.

Table 11.5 sketches the signal model and the mutual information attained by the different protocols. It should be remarked that we have defined two types of protocol II (A and B). That protocol (Figure 11.4 in Section 11.2.3) assumes that the source transmits to the relay during the *relay-receive phase*, $\mathbf{x}_s^1$ and both source and relay transmit simultaneously to the destination in the *relay-transmit phase*, $\mathbf{x}_s^2$ and $\mathbf{x}_r$. The destination only considers the signal received in the second phase for decoding the message. However, depending on the type of messages transmitted by the source and relay, two different implementations are possible, named in this section as protocol IIA and protocol IIB. The former assumes that the relay is able to decode all the data transmitted by the relay network. All signals transmitted by the source and relay are related to a common message. For this reason the relay must be able to decode the whole message from the signal received in the *relay-receive phase*. Note that the equivalent MIMO system during the *relay-transmit phase* can use space-time codes with distributed antennas, [9]. Space-time codes transmit linear combinations of symbols in space (antennas) and time, but all the antennas must have access to all the symbols to be transmitted, [55], as happens in this protocol. On the contrary, protocol IIB assumes that the relay can decode only part of the total data delivered to the destination (*partial decoding*). Since this protocol is less restrictive than protocol IIA, larger mutual information values are to be expected. The signal model is the same as protocol IIA, but now since the signals transmitted by the terminals are related to two independent messages, the *relay-transmit phase* resembles the traditional Multiple Access Channel (MAC) [16] with two virtual users. The complexity of the receiver increases because two simultaneous and independent messages are received, requiring a *successive cancellation* scheme. One virtual user is made up of the relay during the *relay-transmit phase*, which transmits a message of rate R_r (using the signals $\mathbf{x}_s^1$ and $\mathbf{x}_r$). The other virtual user is the source during the *relay-transmit phase* transmitting a new message of rate R_s through the signal $\mathbf{x}_s^2$.

The optimal phase duration and attained mutual information can be derived by solving the equations presented in Table 11.5, which are easily formulated from the signal model. More details can be found in [2,56]. The outcome of such derivation is shown in Table 11.6.

It should be remarked that the mutual information of the different protocols shown in Table 11.6 has been derived assuming that the relay is active and must decode the transmitted signal sent by the source. In this regard, the forwarding protocol, protocol I and protocol IIA can achieve a worse performance than the direct transmission. Clearly, in such a case, we avoid the use of the relay protocol and adopt the direct transmission. However, protocols IIB and III can tend to direct transmission thanks to the new message transmitted in the *relay-transmitted phase* by the source, which is not required to be decoded by the relay. In fact, Figure 11.24 illustrates the connections among protocols as a function of parameter α and of the terminal's activity in the relay phase.

Table 11.5 Signal model and achievable rates of dynamic DF protocols using (11.113) and (11.114)

Protocol DF	Signal model	Achievable rate	Δ
forwarding	$\mathbf{y}_r(t) = \mathbf{H}_1\mathbf{x}_s^1(t) + \mathbf{w}_r(t) \quad t \in S_1$ $\mathbf{y}_d(t) = \begin{cases} 0 & t \in S_1 \\ \mathbf{H}_2\mathbf{x}_r(t-N/2) + \mathbf{w}_d(t) & t \in S_2 \end{cases}$	$\displaystyle\max_{0 \le \alpha \le 1} \min\left(\alpha C_1, \; (1-\alpha)C_b\right)$	0
protocol I	$\mathbf{y}_r(t) = \mathbf{H}_1\mathbf{x}_s^1(t) + \mathbf{w}_r(t) \quad t \in S_1$ $\mathbf{y}_d(t) = \begin{cases} \mathbf{H}_0\mathbf{x}_s^1(t) + \mathbf{w}_d(t) & t \in S_1 \\ \mathbf{H}_2\mathbf{x}_r(t-N/2) + \mathbf{w}_d(t) & t \in S_2 \end{cases}$	$\displaystyle\max_{0 \le \alpha \le 1} \min\left(\alpha C_1, \; (1-\alpha)C_b + \alpha C_0\right)$	0
protocol IIA	$\mathbf{y}_r(t) = \mathbf{H}_1\mathbf{x}_s^1(t) + \mathbf{w}_r(t) \quad t \in S_1$ $\mathbf{y}_d(t) = \begin{cases} 0 & t \in S_1 \\ [\mathbf{H}_2 \; \mathbf{H}_0]\begin{bmatrix} \mathbf{x}_r(t-N/2) \\ \mathbf{x}_s^2(t-N/2) \end{bmatrix} + \mathbf{w}_d(t) & t \in S_2 \end{cases}$	$\displaystyle\max_{0 \le \alpha \le 1} \min\left(\alpha C_1, \; (1-\alpha)C_b\right)$	$\dfrac{P_s}{n_s\sigma_{w_d}^2}\mathbf{H}_0\mathbf{H}_0^H$
protocol IIB	The same as in protocol IIA, but the message transmitted by the source in the relay-transmit phase ($\mathbf{x}_s^2$) is a new message independent of the one transmitted in the relay-receive phase. Partial decoding	$\max(\lambda_1, C_0)$ $\lambda_1 = \displaystyle\max_{0 \le \alpha \le 1} (1-\alpha)C_b$ $s.t. \;\; \alpha C_1 \ge (1-\alpha)(C_b - C_0)$	$\dfrac{P_s}{n_s\sigma_{w_d}^2}\mathbf{H}_0\mathbf{H}_0^H$
protocol III	$\mathbf{y}_r(t) = \mathbf{H}_1\mathbf{x}_s^1(t) + \mathbf{w}_r(t) \quad t \in S_1$ $\mathbf{y}_d(t) = \begin{cases} \mathbf{H}_0\mathbf{x}_s^1(t) + \mathbf{w}_d(t) & t \in S_1 \\ [\mathbf{H}_2 \; \mathbf{H}_0]\begin{bmatrix} \mathbf{x}_r(t-N/2) \\ \mathbf{x}_s^2(t-N/2) \end{bmatrix} + \mathbf{w}_d(t) & t \in S_2 \end{cases}$	$\displaystyle\max_{0 \le \alpha \le 1} \min\left(\alpha C_1 + (1-\alpha)C_0, \; \alpha C_0 + (1-\alpha)C_b\right)$	$\dfrac{P_s}{n_s\sigma_{w_d}^2}\mathbf{H}_0\mathbf{H}_0^H$

Table 11.6 Optimum resource allocation of the relay-receive and relay-transmit phase and achievable rates of the different dynamic DF protocols considering (11.114)

Protocol	α^*		Achievable rate		Δ(defined in C_b)
forwarding	$\dfrac{C_b}{C_b + C_1}$		$\dfrac{C_b C_1}{C_b + C_1}$		$\mathbf{0}$
protocol I	$\begin{cases} \dfrac{C_b}{C_1 + C_b - C_0} & if \quad C_1 \geq C_0 \\ 1 & otherwise \end{cases}$		$\begin{cases} \dfrac{C_b C_1}{C_1 + C_b - C_0} & if \quad C_1 \geq C_0 \\ C_1 & otherwise \end{cases}$		$\mathbf{0}$
protocol IIA	$\dfrac{C_b}{C_b + C_1}$		$\dfrac{C_b C_1}{C_b + C_1}$		$\dfrac{P_s}{n_s \, \sigma_{w_d}^2} \mathbf{H_0 H_0^H}$
protocol IIB	$\begin{cases} \dfrac{C_b - C_0}{C_b + C_1 - C_0} & if \, C_1 \geq C_0 \\ 0 & otherwise \end{cases}$		$\begin{cases} \dfrac{C_b C_1}{C_1 + C_b - C_0} & if \quad C_1 \geq C_0 \\ C_0 & otherwise \end{cases}$		$\dfrac{P_s}{n_s \, \sigma_{w_d}^2} \mathbf{H_0 H_0^H}$
protocol III	$\begin{cases} \dfrac{C_b - C_0}{C_1 + C_b - 2C_0} & if \, C_a \geq C_0 \\ 0 & otherwise \end{cases}$		$\begin{cases} \dfrac{C_b C_1 - (C_0)^2}{C_1 + C_b - 2C_0} & if \, C_1 \geq C_0 \\ C_0 & otherwise \end{cases}$		$\dfrac{P_s}{n_s \, \sigma_{w_d}^2} \mathbf{H_0 H_0^H}$

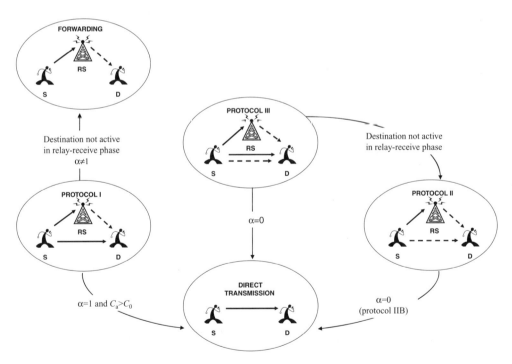

Figure 11.24 Transition between protocols. Solid lines stands for the transmission in the relay-receive phase and dashed lines for the transmissions in the relay-transmit phase.

The mutual information attained by the protocols satisfies the following order,

$$I_{FW} \leq \left\{ \begin{array}{c} I_{p-IIA} \leq I_{p-I} \\ or \\ I_{p-I} \leq I_{p-IIA} \end{array} \right\} \leq I_{p-IIB} \leq I_{p-III} \tag{11.115}$$

where protocols III and IIB get the best achievable rate at the cost of a more complex receiver at the destination. In those cases *successive decoding* is required because two independent messages are received simultaneously. The comparison between protocol I and IIA depends on,

$$\left\{ \begin{array}{ll} I_{p-IIA} \geq I_{p-I} & C_1 > \dfrac{C_b^{IIA} C_0}{C_b^{IIA} - C_b^{I}} \\ I_{p-IIA} < I_{p-I} & otherwise \end{array} \right. \tag{11.116}$$

where C_b^{IIA} and C_b^{I} follow the definition of C_b (11.114) (see Table 11.6 for protocols IIA and I, respectively). When the source-relay link (C_I) is very good, protocol IIA obtains a better achievable rate than using protocol I.

In order to evaluate the half-duplex protocols in the time domain with multiple antennas we have assumed that the channel coefficients involved in the relay-assisted transmission are *Rayleigh* random variables which remain constant over the transmission interval of time (one frame). The relay-assisted transmission is optimized. Because the achievable rate will vary from frame to frame, results will be presented in terms of average achievable rate. The scenario considered will be the same as in Figure 11.19, with source and relay separated by a normalized distance equal to one and the relay placed at distance d from the source (and (1-d) from relay. Results assume an average SNR in the source-destination link equal to $SNR_0 = 0$ dB. Figure 11.25 (top-left) presents the average achievable rate of the different protocols under a Rayleigh channel with $n_s = 1$, $n_r = 1$ and $n_d = 1$. Now, the Rayleigh fading produces that the forwarding protocol gets worse values than direct transmission. Additionally, the performance of protocols IIB and III when the relay is close to destination is better than average achievable rate of the direct transmission. Moreover, if the receiving antennas are increased up $n_d = 2$ (with $n_s = n_r = 1$), Figure 11.25 (top-right) shows that the maximums of the average achievable rates are obtained in positions d where the relay is close to the source. On the other hand, if we increase the number of antennas at the source and relay, $n_s = n_r = 2$ (with $n_d = 1$), Figure 11.25 (bottom-left) the maximums of the average achievable rates move to positions d where the relay is close to destination. In that case, protocol IIA offers a steady average achievable rate independent of the position of the relay. Finally, increasing all the antennas of the terminals, $n_s = n_r = n_d = 2$, Figure 11.25 (bottom-right) the maximum of the average achievable rates are obtained when the relay is placed in similar positions as the case $n_s = n_r = n_d = 1$ (except for protocol I). Moreover, the forwarding protocol when all terminals feature two antennas obtains an average achievable rate better than the direct transmission when the relay is placed in $0.2 \leq d \leq 0.8$.

11.4.4   Dynamic Resource Allocation Relaying with Multiple Relays: Half-Duplex Diamond Relay Channel

In this section we investigate the achievable rates when a multiple *half-duplex* relay assists a single destination (see Figure 11.26). This scenario has been considered in [58] and [59] when the relays operate in *full-duplex* mode. In contrast, we concentrate on *half-duplex relays* and the proper resource allocation for the two phases of the transmission. Moreover, the relays only assist the destination, i.e. they cannot assist amongst themselves. The *half-duplex* constraint at the relays generates a few difficulties when trying to adopt the schemes defined in [58] as how the transmission time between the two phases is selected, how the power is assigned, the optimum messaging or the coordination between relays. One possibility of using two assisting relays is described in [60, 61] where both relays receive and

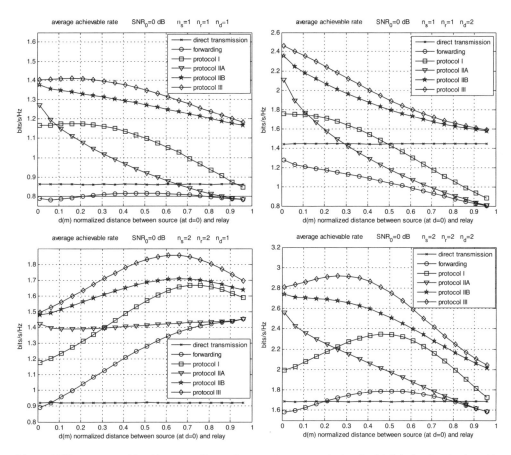

Figure 11.25 Average achievable rate vs. distance between the source and relay. Rayleigh fading. Destination with $SNR_0 = 0$ dB. (top-left) $n_s = 1$, $n_r = 1$, $n_d = 1$, (top-right) $n_s = 1$, $n_r = 1$, $n_d = 2$, (bottom-left) $n_s = 2$, $n_r = 2$, $n_d = 1$, (bottom-right) $n_s = 2$, $n_r = 2$, $n_d = 2$. Dynamic decode-and-forward.

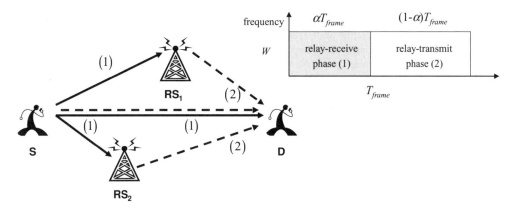

Figure 11.26 Multiple half-duplex relay-assisted transmission. Half-duplex diamond relay channel.

transmit alternatively. On the contrary, here we assume that both relays are receiving or transmitting simultaneously, as is considered in [62] where the present scenario is called the diamond half-duplex relay channel.

The relay protocol is defined as follows: the source sends a signal to the relays and destination during the *relay-receive phase* (solid lines in Figure 11.26). In a subsequent phase, and after the relays have decoded and re-encoded the received messages, the source and relays transmit simultaneously during the *relay-transmit phase*, (dashed lines in Figure 11.26). This protocol is similar to protocol III defined in Section 11.2.3 for a single assisting relay and is different from the approach followed in [62], where the source-destination link is not considered. Two source transmission strategies are considered: either the relays receive *independent messages*, or both receive a *single common message*. In both cases we consider that either the transmitting terminals have a complete (and perfect) CSI of all links (*synchronous case*[8]) so distributed *eigenvector precoding* techniques can be considered, or they do not have it (*asynchronous case*). This perspective was presented in [63].

The optimum duration of each phase of the transmission and power allocation at the source and relays is obtained by maximizing the achievable rate. We formulate the problem for the case where the relays are receiving independent messages with CSI. Some other cases of practical interest (including the single relay-assisted transmission and the no-CSI case) can be seen as particularizations of that problem.

11.4.4.1   Signal Model

All terminals are equipped with a single antenna. In the following ρ_0 denotes the signal to noise (SNR) ratio in the source-destination link, $\rho_{1,i}$ and $\rho_{2,i}$ the SNR in the link between the source and the i-th relay (RS_i) and i-th relay and destination, respectively. All these values have been measured using power P at each terminal. The transmission is carried out in frames of length T_{frame} (s) and bandwidth BW (Hz) (both parameters will be normalized to one). The relays operate in *half-duplex* mode with a duration αT_{frame} and $(1-\alpha)T_{frame}$ for the *relay-receive* and *relay-transmit phase*, respectively. We will assume an *average sum power constraint* for all the terminals. An extension to the MIMO case should consider the achievable rate of the BC (Broadcast channel) with common information introduced in [64].

The signals received at the different terminals are defined by,

$$\left.\begin{aligned} y_{RS_1}(t) &= \sqrt{\rho_{1,1}}e^{j\theta_{1,1}}x_0(t) + n_{RS_1}(t) \\ y_{RS_2}(t) &= \sqrt{\rho_{1,2}}e^{j\theta_{1,2}}x_0(t) + n_{RS_2}(t) \\ y_D^{(1)}(t) &= \sqrt{\rho_0}e^{j\theta_0}x_0(t) + n_D(t) \end{aligned}\right\} \quad if \quad 0 \le t \le \alpha T_{frame}$$

$$y_D^{(2)}(t) = \sqrt{\rho_0}e^{j\theta_0}x_3(t) + \sqrt{\rho_{2,1}}e^{j\theta_{2,1}}x_1(t) + \sqrt{\rho_{2,2}}e^{j\theta_{2,2}}x_2(t) + n_D(t) \quad if \, \alpha T_{frame} < t \le T_{frame}$$

$$\tag{11.117}$$

where n_D, n_{RS_1} and n_{RS_2} are the white Gaussian noise of unitary power at the different terminals. The channel gains are expressed in terms of amplitude (square root of the SNR of each link when the noise power is unitary) and the complex phase: $\theta_0, \theta_{1,1}, \theta_{1,2}, \theta_{2,1}$ and $\theta_{2,}$ for the source-destination, source-RS_1, source-RS_2 links, RS_1-destination and RS_2-destination links respectively. Finally, the sources transmits x_0 in the *relay-receive phase*, while source and relays transmit the signals x_3 (source), x_1(RS_1) and x_2 (RS_2).

Initially, we assume that the relay-assisted transmission sends independent messages to the relays. It consists of three messages W_1, W_2 and W_3, at rates R_1, R_2 and R_3 respectively. The first two messages use the help of the relays, while the third one is sent by the source only in the *relay-transmit phase*. The messages have associated different Gaussian codebooks. Table 11.7 depicts how the messages, terminals and codewords are connected along with the fractions of total power P used by each terminal. There, $\gamma_{0,i}$ with $i = \{1,2,3\}$ stands for the fraction of total power used by the source for transmitting the codewords $m_1^{(1)}, m_2^{(1)}, m_3^{(2)}$. The assisting relays which have decoded the messages W_1 and W_2, only

---

[8] We adopt the definition from [17], distributed terminals have CSIT and they have the carrier phase aligned (synchronized).

Table 11.7 Fraction of power P allocated by each terminal to each codeword used in each phase

Messages	W_1		W_2		W_3	
	Relay-	*Relay-*	*Relay-*	*Relay-*	*Relay-*	*Relay-*
Phase	*Receive*	*Transmit*	*Receive*	*Transmit*	*Receive*	*Transmit*
Codewords	$m_1^{(1)}$	$m_1^{(2)}$	$m_2^{(1)}$	$m_2^{(2)}$	—	$m_3^{(2)}$
S	$\gamma_{0,1}$	$\varphi_{0,1}$	$\gamma_{0,2}$	$\varphi_{0,2}$	$\times$	$\gamma_{0,3}$
RS_1	$\times$	$\varphi_{1,1}$	$\times$	$\varphi_{1,2}$	$\times$	$\times$
RS_2	$\times$	$\varphi_{2,1}$	$\times$	$\varphi_{2,2}$	$\times$	$\times$

transmit $m_1^{(2)}$, $m_2^{(2)}$ along with the source. The variable $\varphi_{j,i}$ with $j = \{0,1,2\}$ and $i = \{1,2\}$ defines the fraction of power allocated by the j-th terminal ($j = 0$ corresponds to the source) to the i-th codeword. This is a general description, where some of the variables could be set directly to zero depending on the scenario definition, i.e. if RS_1 is able to decode both messages and RS_2 just the second one, then $\varphi_{2,1}$ will be zero.

During the *relay-receive phase* only the source is active and is broadcasting the messages W_1 and W_2 to the relays RS_1 and RS_2 using Gaussian signals of power $\gamma_{0,1}P$ and $\gamma_{0,2}P$, respectively, (see Table 11.7). Therefore the signal transmitted by the source x_0 according to (11.117) is,

$$x_0 = \sqrt{\gamma_{0,1}}m_1^{(1)} + \sqrt{\gamma_{0,2}}m_2^{(1)} \tag{11.118}$$

where $m_1^{(1)}$, $m_2^{(1)}$ are the Gaussian codewords of powers P associated with messages W_1 and W_2. Since our scenario is single-antenna, we may assume that the *relay-receive phase* is a degraded channel [16], so one of the relays is able to decode both messages, while the other relay only decodes one. Likewise, during the *relay-transmit phase* the former relay can transmit to destination signals connected to both messages. For example, if RS_1 decodes both messages, then it might transmit two Gaussian signals with power $\varphi_{1,1}P$ and $\varphi_{1,2}P$ conveying message W_1 and W_2 respectively. In contrast, RS_2 will transmit only one signal of power $\varphi_{2,2}P$ connected to the message W_2 ($\varphi_{2,1}$ in Table 11.7 set to zero). In the limiting case where the channel state from the source to both relays is equal, only one message is transmitted.

The source is also active in the *relay-transmit phase* by transmitting two signals to help the relays (with power $\varphi_{0,1}P$ and $\varphi_{0,2}P$) as well as a new signal of power $\gamma_{0,3}P$ connected to message W_3. The notation of the transmitted signals in the general case is described by,

$$\begin{aligned}
x_1 &= \sqrt{\varphi_{1,1}}m_1^{(2)} + \sqrt{\varphi_{1,2}}m_2^{(2)} \\
x_2 &= \sqrt{\varphi_{2,1}}m_1^{(2)} + \sqrt{\varphi_{2,2}}m_2^{(2)} \\
x_3 &= \sqrt{\varphi_{0,1}}m_1^{(2)} + \sqrt{\varphi_{0,2}}m_2^{(2)} + \sqrt{\gamma_{0,3}}m_3^{(2)}
\end{aligned} \tag{11.119}$$

where $m_1^{(2)}$, $m_2^{(2)}$, $m_3^{(2)}$ denote the Gaussian codewords of power P associated to W_1, W_2 and W_3. For the degraded channel where RS_1 is better than RS_2, then $\varphi_{2,1} = 0$ because it cannot decode the message W_1. If RS_2 is better than RS_1, then $\varphi_{1,2} = 0$ because it cannot decode the message W_2.

In order to obtain the spectral efficiency per unit total power, the fraction of power allocated to each one of the different Gaussian codewords must satisfy the *average sum power constraint* considered here,

$$\sum_{j=1}^{3} \gamma_{0,j} + \sum_{j=1}^{2} (\varphi_{0,j} + \varphi_{1,j} + \varphi_{2,j}) = 1 \tag{11.120}$$

in which we are considering that at the source (relay) terminal the transmitted power is inversely proportional to the fraction of time allocated to the first (second) phase.

11.4.4.2 Achievable Rate Region

The achievable rate region described here depends on the maximum decodable rates of messages W_1 and W_2 at the relays during the *relay-receive phase* (denoted by the rate region B_{RS}), as well as the decodable rates of messages W_1, W_2 and W_3 at the destination during both phases of the transmission (denoted by the rate region B_D). In the first case, the rates of the messages W_1 and W_2 are defined by the degraded broadcast channel (BC) [16] to RS_1 and RS_2 transmitting the Gaussian codewords $m_1^{(1)}$ and $m_2^{(1)}$. Assuming that $\rho_{1,1} \geq \rho_{1,2}$ the data rates are given by

$$
B_{ind} = \begin{cases}
B_{RS} = \begin{cases}
R_1 \leq \alpha \log_2 \left(1 + \dfrac{\rho_{1,1}\gamma_{0,1}}{\alpha}\right) \\[2mm]
R_2 \leq \alpha \log_2 \left(1 + \dfrac{\rho_{1,2}\gamma_{0,2}}{\alpha \left(1 + \rho_{1,2}\gamma_{0,1}\right)}\right)
\end{cases} \\[10mm]
B_D = \begin{cases}
R_1 \leq f\left(\gamma_{0,1}\rho_0\right) + g\left(\kappa_1\right) \\
R_2 \leq f\left(\gamma_{0,2}\rho_0\right) + g\left(\kappa_2\right) \\
R_3 \leq g\left(\gamma_{0,3}\rho_0\right) \\
R_1 + R_2 \leq f\left(\left(\gamma_{0,1} + \gamma_{0,2}\right)\rho_0\right) + g\left(\kappa_1 + \kappa_2\right) \\
R_1 + R_3 \leq f\left(\gamma_{0,1}\rho_0\right) + g\left(\kappa_1 + \gamma_{0,3}\rho_0\right) \\
R_2 + R_3 \leq f\left(\gamma_{0,2}\rho_0\right) + g\left(\kappa_2 + \gamma_{0,3}\rho_0\right) \\
\displaystyle\sum_{j=1}^{3} R_j \leq f\left(\left(\gamma_{0,1} + \gamma_{0,2}\right)\rho_0\right) + g\left(\kappa_1 + \kappa_2 + \gamma_{0,3}\rho_0\right)
\end{cases}
\end{cases}
\tag{11.121}
$$

where α denotes the duration of the *relay-receive phase*. Note that in this case, RS_1 is able to decode both messages while RS_2 only can decode message W_2. Likewise, the decodable rates of the different messages (using the codewords $m_1^{(2)}, m_2^{(2)}$ and $m_3^{(2)}$) can be seen as Multiple Access Channel (MAC) with three terminals [16] taking into consideration the signal received during the *relay-receive phase* at the destination (B_D) with the following definitions

$$
\begin{cases}
f(x) = \alpha \log_2\left(1 + \dfrac{x}{\alpha}\right) \qquad g(x) = (1-\alpha)\log_2\left(1 + \dfrac{x}{1-\alpha}\right) \\[3mm]
\kappa_1 = \left(\sqrt{\varphi_{1,1}\rho_{2,1}} + \sqrt{\varphi_{0,1}\rho_0}\right)^2 \quad \kappa_2 = \left(\sqrt{\varphi_{2,2}\rho_{2,2}} + \sqrt{\varphi_{1,2}\rho_{2,1}} + \sqrt{\varphi_{0,2}\rho_0}\right)^2
\end{cases}
\tag{11.122}
$$

where κ_1 and κ_2 are the signal power of the Gaussian codewords $m_1^{(2)}$ and $m_2^{(2)}$ obtained from the assumption of *synchronous* transmission of the terminals (distributed *eigenvector precoding*). Moreover, it has been considered that $\varphi_{2,1} = 0$ because RS_2 is not able to decode message W_1. Otherwise, the roles of RS_1 and RS_2 are exchanged.

It should be emphasized that the achievable rate region of other transmissions strategies, such as asynchronous transmission, relays receiving a common message or the single relay case, can be seen as particular cases of (11.121), as it is sketched in Table 11.8.

11.4.4.3 Resource Allocation Formulation

The resource allocation (which involves the duration of each phase α, the power of the different codewords at the different terminals and bitrate of transmitted messages) can be obtained as the result of the following optimization problem,

$$
\begin{aligned}
&\max_{\mathbf{z}, R_1, R_2, R_3} \; R_1 + R_2 + R_3 \quad s.t. \quad (R_1, R_2, R_3) \in B_{ind}\,(\mathbf{z}) = \{B_{RS} \cap B_D\} \\
&\mathbf{z} = \left[\alpha, \gamma_{0,1}, \gamma_{0,2}, \gamma_{0,3}, \varphi_{0,1}, \varphi_{0,2}, \varphi_{1,1}, \varphi_{1,2}, \varphi_{2,1}, \varphi_{2,2}\right] \\
&\alpha \in [0,1] \quad \sum_{j=1}^{3} \gamma_{0,j} + \sum_{j=1}^{2} \left(\varphi_{0,j} + \varphi_{1,j} + \varphi_{2,j}\right) = 1
\end{aligned}
\tag{11.123}
$$

Table 11.8 Changes in the signal model (11.117) and achievable rate region (11.121) for accommodating other transmission strategies

Transmission strategy	Synchronous case	Asynchronous case
2 relays receiving independent messages	No changes	$\varphi_{0,1} = \varphi_{0,2} = 0,$ $\kappa_2 = \varphi_{2,2}\rho_{2,2} + \varphi_{1,2}\rho_{1,2}$
2 relays receiving a common message	$B_{RS} = \begin{cases} R_2 \le f\left(\rho_{1,1}\gamma_{0,2}\right) \\ R_2 \le f\left(\rho_{1,2}\gamma_{0,2}\right) \end{cases}$ $\gamma_{0,1} = 0,\ \kappa_1 = 0,\ R_1 = 0,\ \varphi_{0,1} = \varphi_{2,1}$ $= \varphi_{1,1} = 0$ $\kappa_2 = \left(\sqrt{\varphi_{0,2}\rho_0} + \sqrt{\varphi_{1,2}\rho_{2,1}} + \sqrt{\varphi_{2,2}\rho_{2,2}}\right)^2$	$B_{RS} = \begin{cases} R_2 \le f\left(\rho_{1,1}\gamma_{0,2}\right) \\ R_2 \le f\left(\rho_{1,2}\gamma_{0,2}\right) \end{cases}$ $\gamma_{0,1} = 0,\ \kappa_1 = 0,\ R_1 = 0,$ $\varphi_{0,1} = \varphi_{0,2} = \varphi_{2,1} = \varphi_{1,1} = 0,$ $\kappa_2 = \varphi_{1,2}\rho_{2,1} + \varphi_{2,2}\rho_{2,2}$
Single relay case	$B_{RS} = R_1 \le f\left(\rho_{1,1}\gamma_{0,1}\right)$ $\gamma_{0,2} = 0,\ \kappa_2 = 0,\ R_2 = 0,\ \varphi_{0,2} = \varphi_{2,1}$ $= \varphi_{2,2} = 0$ $\varphi_{1,2} = 0,\quad \kappa_1 = \left(\sqrt{\varphi_{0,1}\rho_0} + \sqrt{\varphi_{1,1}\rho_{2,1}}\right)^2$	$B_{RS} = R_1 \le f\left(\rho_{1,1}\gamma_{0,1}\right)$ $\gamma_{0,2} = 0,\ \kappa_2 = 0,\ R_2 = 0,\ \varphi_{1,2} = 0$ $\varphi_{0,1} = \varphi_{0,2} = \varphi_{2,1} = \varphi_{2,2} = 0,\ \kappa_1 = \varphi_{1,1}\rho_{2,1}$

where B_{ind} is the rate region defined by the inequalities (11.121). This problem is not convex [57] due to the inequalities of B_{RS} which correspond to the achievable rate region of a degraded broadcast channel (in that case the capacity region). However, the whole problem can be transformed into a convex one by using the duality BC-MAC [65] to define the achievable rate region at the relays as,

$$B_{RS} = \begin{cases} R_1 + R_2 \le \alpha \log_2\left(1 + \dfrac{\lambda_1\rho_{1,1}}{\alpha} + \dfrac{\lambda_2\rho_{1,2}}{\alpha}\right) \\ s.t. \quad \lambda_1 + \lambda_2 = \gamma_{0,1} + \gamma_{0,2} \end{cases} \tag{11.124}$$

where the optimization is done over the variables λ_1 and λ_2. Those variables are related to $\gamma_{0,1}$ and $\gamma_{0,2}$ in the following way

$$\gamma_{0,1} = \lambda_1 \frac{1}{1 + \rho_{1,2}\lambda_2} \qquad \gamma_{0,2} = \lambda_2\left(1 + \frac{\rho_{1,2}\lambda_1}{1 + \rho_{1,2}\lambda_2}\right) \tag{11.125}$$

Since the resource allocation is defined as a convex problem we can use efficient techniques to find the optimal solution of the problem [57].

11.4.4.4 Examples

The relay-assisted protocols defined previously are evaluated in a Gaussian scenario where the source-destination distance is normalized to 1. The RS_1 and RS_2 are placed at distance d_1 and d_2 from the source, and $(1-d_1)$ and $(1-d_2)$ from the destination. The path-loss of each link is proportional to the square of the distance between the terminals. The SNR of the different links is given by defining the ρ_0 (source-destination link). The complex phases of the channels shown in (11.117) are equal to zero, (θ_0, $\theta_{1,1}$, $\theta_{1,2}$, $\theta_{2,1}$, and $\theta_{2,2}$) but these value only are known at the transmitters under CSI.

Figure 11.27 presents the achievable rate for the transmission using a single relay placed at different positions when CSI is available (square-line) or not (circle-line). The CSI is helpful to deal with the case where the relay-destination link limits the communication. Those rates are compared with *synchronous* and *asynchronous* transmission of *independent* or *common messaging* with two relays, being RS_2 placed at $d_2 = 0.2$. The best performance is obtained by the *synchronous* transmission of *independent messages*.

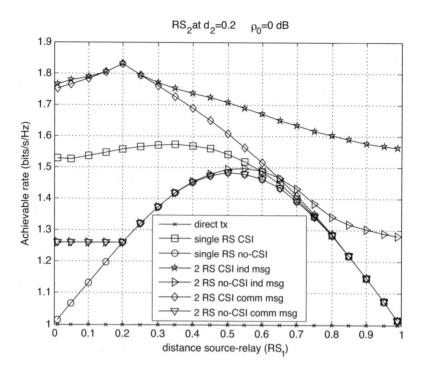

Figure 11.27 Achievable rate for the single and 2-relays with/without CSI vs distance between source-RS$_1$, RS$_2$. $\rho_0 = 0$ dB. Reproduced by permission of IEEE 2008.

When both relays are at the same distance $d_1 = d_2 = 0.2$, *independent* and *common messaging* coincide. However, in certain configurations the adoption of a *common messaging* strategy provides no rate gains compared to the use of a single relay case, even when CSI is considered. The main drawback of *common messaging* is that the achievable rate is dominated by the relay with the worst channel quality. For the CSI case, there is a rate gain over a single relay case for $0 \le d_1 \le 0.65$. In that region, the relay-destination link limits the communication but thanks to the *synchronous* transmission of the two relays the equivalent link is improved. The source-relay channel limits the performance when RS$_1$ is close to the source and then RS$_2$ is not helpful. For the *asynchronous* transmission, *common messaging* does not offer any rate gain for $0.2 \le d_1 \le 1$. However, for those positions of RS$_1$, *independent messaging* under *asynchronous* transmission improves the single relay case because the position of RS$_1$ does not limit the total transmission.

11.5   Two-Way Relay-Assisted Communications

In order to enhance the spectral efficiency of half-duplex relay communications the two-way relay channel (TWRC) has been introduced in [60, 66, 67]. In this type of channel, also named *bidirectional relaying*, two terminals exchange simultaneously their messages through the same half-duplex relay. Note that both terminals work as sources and destinations, but not simultaneously.

In contrast to the protocols defined for OWRC (see Section 11.2.3) only two protocols are devised for the TWRC, *forwarding* and *protocol I*. Other protocols like protocol III in Section 11.2.3, [17] require simultaneous transmission from terminals and relay, an impossible scenario with half-duplex terminals.

The TWRC-*forwarding* protocol is sketched in Figure 11.28 at the top (see for example [60, 68, 69] (two-step scheme). The transmission consists of two phases. In the first phase both terminals (T$_1$ and

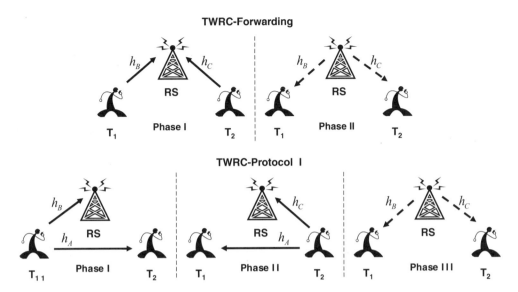

Figure 11.28   Forwarding and Protocol I for the TWRC. Terminals T_1 and T_2 exchange messages and the relay (RS) help both terminals.

T_2) transmit simultaneously to the relay (RS), while in the second phase the relay transmits to both terminals. On the other hand, TWRC-*protocol I* presented in Figure 11.28 at the bottom consist of three phases, where in the first two each terminal transmit to the relay and the other terminal. In the last one, the relay transmits to both terminals. Note that the consideration of the direct link between terminals by TWRC-protocol I is done at the cost of designing the communication in three orthogonal phases. This protocol has been considered in [66] and [69].

The work presented in this section assumes that the transmission strategy at the relay is based on the *network coding relaying* (NCR), where the relay generates a common signal intended for both terminals from the messages received in the first phase employing the random binning technique. The achievable rate region is shown in [68, 70, 71] for static resource allocation (equal phase duration).

This section analyzes which protocol (TWRC-*forwarding* or TWRC-*protocol I*) gets the best performance based on individual power constraint (*max-power*). We investigate the optimal resource allocation in terms of duration of phases, transmitted power, and data rate. Resource optimization based on an average sum power constraint or assuming another transmission strategy at the relay can be found in [72].

11.5.1   Scenario and Signal Model

The transmission is carried out in frames of length T_{frame} with N channel uses and normalized bandwidth equal to 1 Hz. For the TWRC-*forwarding* protocol (see Figure 11.28-top) there are two phases where both terminals are transmitting or receiving simultaneously; the duration of each phase is proportional to α_1 and α_2, respectively. For TWRC-*protocol I*, three phases are defined (see Figure 11.28-bottom). The first two are proportional to $\alpha_{1,1}$ and $\alpha_{1,2}$ (*phase I, phase II*) and the last phase is proportional to α_2 (*phase III*).

Terminals T_1, T_2 and RS are equipped with a single antenna. The channel gains of the direct (T_1-T_2), T_1-RS and RS-T_2 links are referenced by the variables h_A, h_B and h_C in Figure 11.28-bottom and remain constant over the whole transmission. In general, the perfect channel state information (CSI) at the transmitters is not known, but there is knowledge about the signal-to-noise ratio (SNR) of each link. The transmission is carried out over an Additive White Gaussian Noise (AWGN) channel. The mutual

Table 11.9 Received signals at the terminals for the two analyzed TWRC protocols depicted in Figure 11.28 and considering the definitions done in (11.127)

TWRC protocols	Phase I ($k \in S_1$)	Phase II ($k \in S_2$)	Phase III ($k \in S_3$)
forwarding	$y_{RS}(k) = h_B x_1(k) +$ $+ h_C x_2(k) + n_r(k)$	$y_{RS}(k) = 0$	×
	$y_{T_1}(k) = 0$	$y_{T_1}(k) = h_B x_{RS}(k) + n_1(k)$	
	$y_{T_2}(k) = 0$	$y_{T_2}(k) = h_C x_{RS}(k) + n_2(k)$	
protocol I	$y_{RS}(k) = h_B x_1(k) + n_r(k)$	$y_{RS}(k) = h_C x_2(k) + n_r(k)$	$y_{RS}(k) = 0$
	$y_{T_1}(k) = 0$	$y_{T_1}(k) = h_A x_2(k) + n_1(k)$	$y_{T_1}(k) = h_B x_{RS}(k) + n_1(k)$
	$y_{T_2}(k) = h_A x_1(k) + n_2(k)$	$y_{T_2}(k) = 0$	$y_{T_2}(k) = h_C x_{RS}(k) + n_2(k)$

information of the different links assuming an equal noise power at all terminals is given by,

$$ I_X = N \log_2 \left(1 + \frac{P}{\sigma^2} |h_X|^2 \right), \quad I_{BC} = N \log_2 \left(1 + \frac{P}{\sigma^2} |h_B|^2 + \frac{P}{\sigma^2} |h_C|^2 \right) \quad (11.126) $$

where $X \in \{A, B, C\}$, i.e. I_A, I_B, I_C, stand for the mutual information of T_1-T_2, T_1-RS and T_2-RS links, and I_{BC} is the mutual information at the relay when both terminals transmit simultaneously, (T_2, T_1)-RS link. Finally, N stands for the channel uses employed. Let us define the following time intervals for the different phases of the TWRC protocols,

$$ \begin{cases} S_1^{fw} = \{k \mid 0 \le k \le \alpha_1 N\} \\ S_2^{fw} = \{k \mid \alpha_1 N < k \le \alpha_2 N\} \end{cases} \quad \begin{cases} S_1^{pI} = \{k \mid 0 \le k \le \alpha_{1,1} N\} \\ S_2^{pI} = \{k \mid \alpha_{1,1} N < k \le (\alpha_{1,1} + \alpha_{1,2}) N\} \\ S_3^{pI} = \{k \mid (\alpha_{1,1} + \alpha_{1,2}) N < k \le \alpha_2 N\} \end{cases} \quad (11.127) $$

The signals received by the terminals in the different phases are summarized in Table 11.9, where x_1, x_2 and x_{RS} denote the transmitted signals by T_1, T_2 and RS, respectively. This latter signal, x_{RS}, is transmitted from the RS to both terminals. However, such signal carries information from terminals T_1, T_2 (received in Phase I for the forwarding protocol and in phases I and II for protocol I). Hence, when it is transmitted to both terminals, they already know part of the message, that can be seen as *back-propagating self-interference*, as it is defined in [67]. Consequently, it can be removed completely if the channel is known without errors at the receiver.

The achievable rate regions attained by the TWRC-forwarding protocol (see also [68]) and TWRC-protocol I according to the signal model presented in Table 11.9 are defined by

$$ B^{FW}(\alpha_1, \alpha_2) = \begin{cases} R_1 \le \alpha_1 I_B, \quad R_1 \le \alpha_2 I_C \\ R_2 \le \alpha_1 I_C, \quad R_2 \le \alpha_2 I_B \\ R_1 + R_2 \le \alpha_1 I_{BC} \end{cases}, \quad B^I(\alpha_{1,1}, \alpha_{1,2}, \alpha_2) = \begin{cases} R_1 \le \alpha_{1,1} I_B \\ R_2 \le \alpha_{1,2} I_C \\ R_1 \le \alpha_{1,1} I_A + \alpha_2 I_C \\ R_2 \le \alpha_{1,2} I_A + \alpha_2 I_B \end{cases} $$

$$ (11.128) $$

11.5.2 Resource Allocation

The optimal selection of the phase duration and data rate are found as the maximization of,

$$ \max_{\alpha, R_1, R_2} R_1 + R_2 \quad s.t. \begin{cases} (R_1, R_2) \in B(\alpha) \\ 0 \le f(\alpha) \le 1 \end{cases} \quad (11.129) $$

where R_1, R_2 stand for the rate transmitted by terminal T_1 and T_2, $\boldsymbol{\alpha}$ is a vector that contains the duration of the different phases of the communication, f defines the linear connection between duration of phases, $B(\boldsymbol{\alpha})$ denotes the achievable rate region for a given $\boldsymbol{\alpha}$.

11.5.2.1 TWRC-Forwarding Protocol

The phase duration for TWRC-forwarding protocol is given by,

$$
\alpha_1^{FW} = \begin{cases} \dfrac{I_C}{I_{BC}} & \text{if } \dfrac{I_C}{I_B}(I_C + I_B) \le I_{BC}, \ \dfrac{I_B}{I_C} \ge \dfrac{I_C}{I_B} \\[2ex] \dfrac{I_C + I_B}{I_{BC} + I_C + I_B} & \text{if } \dfrac{I_C}{I_B}(I_C + I_B) > I_{BC}, \ \dfrac{I_B}{I_C} \ge \dfrac{I_C}{I_B} \\[2ex] \dfrac{I_C + I_B}{I_{BC} + I_C + I_B} & \text{if } \dfrac{I_C}{I_B}(I_C + I_B) > I_{BC}, \ \dfrac{I_B}{I_C} < \dfrac{I_C}{I_B} \\[2ex] \dfrac{I_B}{I_{BC}} & \text{if } \dfrac{I_C}{I_B}(I_C + I_B) \le I_{BC}, \ \dfrac{I_B}{I_C} < \dfrac{I_C}{I_B} \end{cases}, \quad \alpha_2^{FW} = 1 - \alpha_1^{FW} \tag{11.130}
$$

The previous result can be obtained by observing that the constraints depicted in (11.128) describe a pentagonal shape for the achievable rate region as a function of R_1 and R_2. Let us define the following piecewise functions which account for the first two pairs of constraints of (11.128),

$$
\varphi_1(\alpha) = \begin{cases} \alpha I_B & \text{if } \alpha \le \alpha_c \\ (1-\alpha) I_C & \text{if } \alpha > \alpha_c \end{cases}, \quad \varphi_2(\alpha) = \begin{cases} \alpha I_C & \text{if } \alpha \le \alpha_d \\ (1-\alpha) I_B & \text{if } \alpha > \alpha_d \end{cases} \tag{11.131}
$$

with

$$
\alpha_c = \frac{I_C}{I_B + I_C}, \quad \alpha_d = \frac{I_B}{I_C + I_B} \tag{11.132}
$$

Since we are interested in maximizing the sum-rate, we look for a solution satisfying the sum-rate constraint, so we have,

$$
\alpha I_{BC} - \varphi_1(\alpha) \le \varphi_2(\alpha) \quad \rightarrow \quad \alpha I_{BC} \le \varphi_1(\alpha) + \varphi_2(\alpha) \tag{11.133}
$$

Therefore, we have to analyze the different cases due to the functions φ_1 and φ_2, (11.131). For example, solving $\alpha_c < \alpha_d$, $\alpha_c \le \alpha \le \alpha_d$, the solution to (11.133) is given by,

$$
\alpha = \frac{I_C}{I_{BC}} \tag{11.134}
$$

This solution is valid for a certain values of α, obtaining the following conditions,

$$
I_{BC} \ge \frac{I_C}{I_B}(I_C + I_B), \quad \frac{I_B}{I_C} \ge \frac{I_C}{I_B} \tag{11.135}
$$

which corresponds to $\alpha \le \alpha_d$, $\alpha_c < \alpha_d$ respectively. Notice that (11.134) and (11.135) are present in the first row of the solution of phase duration α_1^{FW} in (11.130). By evaluating the other cases of (11.133) and following the same procedure, we obtain the other solutions depicted in (11.130).

Finally, the sum-rate attained by the TWRC-forwarding protocol is given by,

$$
R_1 + R_2 = \alpha_1^{FW} I_{BC} \tag{11.136}
$$

11.5.2.2   TWRC-Protocol I

The phase duration for TWRC-protocol I is given by,

$$
\begin{cases}
\alpha_{1,1}^I = \dfrac{(I_C - I_A)\, I_C}{(I_C - I_A)\, I_C + I_B\,(I_B - I_A) + (I_B - I_A)\,(I_C - I_A)} \\[4mm]
\alpha_{1,2}^I = \alpha_{1,1}^I \dfrac{(I_B - I_A)\, I_B}{(I_C - I_A)\, I_C}, \\[4mm]
\alpha_2^I = \alpha_{1,2}^I \dfrac{(I_C - I_A)}{I_B}
\end{cases}
\tag{11.137}
$$

The result shown in (11.137) is obtained by getting the conditions to be satisfied by α_2 from the achievable rate region depicted in (11.128), i.e.,

$$
\alpha_2^I = \alpha_{1,2}^I \frac{(I_C - I_A)}{I_B}, \qquad \alpha_2^I = \alpha_{1,1}^I \frac{(I_B - I_A)}{I_C}
$$

After equating them and taking into account that $\alpha_{1,1} + \alpha_{1,2} + \alpha_2 = 1$ we get (0.137). Finally, the sum-rate attained by TWRC-protocol I is given by,

$$
R_1 + R_2 = \alpha_{1,1}^I I_B + \alpha_{1,2}^I I_C
\tag{11.138}
$$

11.5.2.3   Examples

The evaluation of the two-way relay-assisted transmission is carried out in a scenario where both terminals (T_1 and T_2) are separated by a normalized distance equal to 1. The relay is placed at a distance d from terminal T_1 and (1-d) from terminal T_2. The propagation exponent of each link is equal to 3. The noise power at the different terminals will be the same. We assume an additive white Gaussian noise channel. Hence the SNR in the different links is given by,

$$
\rho_A = \frac{P}{\sigma^2} \qquad \rho_B = \frac{1}{d^3}\frac{P}{\sigma^2} = \frac{1}{d^3}\rho_A \qquad \rho_C = \frac{1}{(1-d)^3}\rho_A
$$

The OWRC transmission considered in this section, [17] or protocol III in Section 11.2.3, assumes a four orthogonal phase transmission to account for the transmission of T_1 to T_2 and then from T_2 to T_1. All phases are optimized accordingly.

Figure 11.29 presents the sum-rate obtained by both terminals when direct transmission, OWRC or TWRC strategies are considered and $\rho_A = 10$ dB. With NCR we note that the transmission strategy introduced in this section and by XOR is a simpler technique than NCR where the relay performs a bit-wise XOR operation of the received bits in order to get the message transmitted, see [72]. We can observe that the OWRC is able to get better results than TWRC-*protocol I* and TWRC-*forwarding* for a wide range of positions in the relay. Let us stress that T_1 and T_2 transmit simultaneously with RS, hence those terminals expend more energy in the OWRC than in the TWRC. Moreover, the position of the relay determines which protocol is better. This performance can be explained by observing the performance of the forwarding protocol. According to (11.130) there are four different zones for calculating the sum-rate. The first describes the situation where the relay moves from T_1 up to $d = 0.33$ while $I_C\,(I_C + I_B) \le I_{BC}\,I_B$. In such a case the sum-rate is equal to I_C (using (11.130) and (11.1360). The last condition in (11.130) describes the relay approaching to T_2 and the sum-rate becomes I_B. In those cases the sum-rate obtained by the *forwarding* protocol is always superior to the one obtained by *protocol I,* given by (11.138) with (11.137). Nevertheless, for the remaining zones defined in (11.130), the best

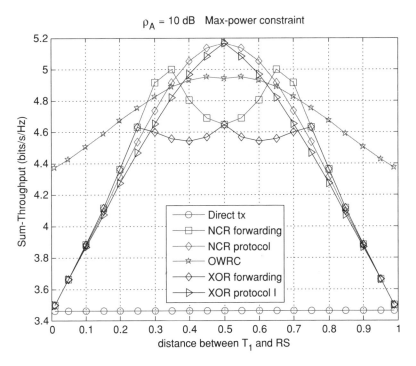

Figure 11.29   Sum-rate as a function of the position of the RS from T_1 for different TWRC and OWRC transmissions when $\rho_A = 10$ dB. Max-power constraint. Reproduced by permission of IEEE © 2009.

protocol depends on the channel configuration, as Figure 11.29 depicts. Finally, the XOR approach can get similar results that the NCR for the TWRC-*protocol* I, but it presents significant losses when it is implemented under the TWRC-*forwarding* protocol.

11.6   Relay-Assisted Communications With Reuse of Resources

Relay-assisted transmission duplexed in time (TDD) is adapted to a cellular system based on TDMA in the *downlink* by using protocol I as defined in Section 11.2.3 and examined in Sections 11.4.1 and 11.4.3. For an easy adaptation of the protocol without requiring large modifications of the current wireless cellular standards, it is assumed that there will be a static resource allocation based on a fixed-duration slot TDMA. Despite the fact that such an approach may be spectrally inefficient (see Section 11.4.1) this section will show that by means of a cell-wide spatial reuse of the *relay-transmit phase* by multiple relays associated with different destinations, the spectral efficiency of communication is improved significantly. Basically, a single time slot is used to allocate all *relay-transmit phases* of the destinations served in the same frame, which is referred to as a *relay slot*. Therefore, having a frame of N_u+1 time slots, we can allocate N_u destinations. If the number of allocated destinations grows, the increased utilization of resources is negligible. Figure 11.30 depicts an example of the spatial reuse of a relay slot using protocol I. In this example there are two active destinations with two assisting relays (one per destination) and the communications will be carried out in a frame with three time slots. During the first time slot (*relay-receive phase* for destination-relay pair 1), the source transmits a signal which is received by the destination and the assisting relay. Afterwards, the source transmits a new signal that is received by the second destination and its associated relay (*relay-receive phase* for destination-relay pair 2). Finally, in

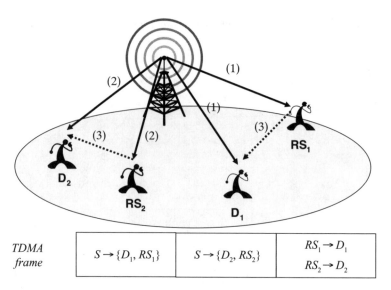

TDMA frame	$S \rightarrow \{D_1, RS_1\}$	$S \rightarrow \{D_2, RS_2\}$	$RS_1 \rightarrow D_1$ $RS_2 \rightarrow D_2$

Figure 11.30 Relay-assisted transmission for two destinations. Solid lines: transmissions during the relay-receive phase (in two orthogonal slots). Dashed lines: transmissions in the common relay-transmit phase.

the third time slot, *relay slot*, both assisting relays transmit to its associated destinations (dashed lines in Figure 11.30). Each destination is received in two time slots which can be seen as an increment of the *virtual receiving antennas* [9].

However, this gain arises at the expense of transforming the *relay-transmit phase* (third time slot in Figure 11.30) into an interference channel (IFC). Two possible solutions are envisaged in order to deal with the interference:

1. Distributed *power control* introduced in Section 11.6.2. Each relay must adjust the transmitted power.
2. Centralized *rate control* presented in Section 11.6.3. It is based on the knowledge of the statistics of the interfering power when the transmitted power is kept constant. In such a case, the source tunes the data rate in order to combat the outage events due to interference.

The interfering power during the *relay-transmit phase* can be reduced by controlling those terminals which produce the larger interference with other destinations by a distributed power algorithm. For example, in [75] a game theory framework is proposed for maximizing the rate between source-relay pairs under simultaneous transmissions. It formulates a non-cooperative game in order to obtain the suitable power allocation strategy for each pair under OFDMA. Section 11.6.2 presents an analysis of the cell-wide spatial reuse of the relay slot in the *downlink* which is also based on game theory [76]. The distributed power algorithm is applied to relay terminals which are selected from the idle terminals deployed around the cell (*user relaying*).

On the other hand, when the interfering power statistics are known, the considered way for dealing with the interference is by tuning the rate of each transmission (*rate control*), [77, 78]. In that case, the destinations served in the same frame are selected randomly over the cell in order to guarantee that the interference is a random process. For certain cases it is possible to model the interfering power statistics, not the actual value of the interfering power. In such a case, the source may select a conservative rate to maximize the throughput of the system or keep a maximum outage probability.

11.6.1 Scenario and Signal Model Definition

The centralized cellular system (of area A m^2) is built upon a single source, the base station (BS), equipped with n_s antennas and N_{tot} terminals distributed spatially according to a *Poisson* distribution with parameter λ (average number of terminals per unit area) and having n_d antennas. Because the assisting relays will be selected from those idle terminals (*user relaying*) the number of antennas at the relay will be the same as the active destinations, $n_r = n_d$.

Each destination interested in relay-assisted transmission selects the nearest idle terminal as its assisting relay, which has a probability p_s to be selected successfully (success probability). Since the terminal deployment has a *Poisson* distribution, the distance, named r, between a destination and its assisting relay becomes a random variable modeled by a Rayleigh distribution with *pdf* given by [79],

$$f_r(r) = 2\pi \lambda_{relay} p_s r \exp\left(-\pi \lambda_{relay} p_s r^2\right) \qquad (11.139)$$

where λ_{relay} denotes the idle terminal density of the cell, p_s is the success probability which accounts for a reduction in the density on the number of candidate relays. If $p_s = 1$, the nearest terminal will always be the assisting relay. The success probability, p_s, may depend on several factors common to all terminals, for example the shadowing at the relay, the probability of having bad channel conditions in the source-relay link or the probability of this terminal being an active destination and hence not available as a relay. This last situation can be considered very unlikely if the number of terminals in the cell is much higher than the number of active terminals at each frame.

The source serves N_u destinations selected among all terminals in TDMA-based transmission using a time frame length T$_{frame}$. The process described in Figure 11.30 for the relay-assisted transmission is extended for N_u destinations, where the frame is made up of N_u+1 time slots of length T$_{slot}$. Notice that with that frame definition it is possible to serve up to N_u+1 destinations using direct transmission. However, the number of served destinations (N_u) and the frame length (T$_{frame}$) must be the same in the relay-assisted and direct transmission for a fair comparison. Towards this end, the time slots considered in direct transmission must be larger than in the relay-assisted transmission case, leading to define the following ratio

$$\frac{T_{slot}}{T_{frame}} = \begin{cases} \dfrac{1}{N_u} & \text{direct tx} \\[2mm] \dfrac{1}{N_u + 1} & \text{relay tx} \end{cases} \qquad (11.140)$$

Taking into account the assumptions previously presented, for a given N_{tot} terminal in a region with area A, the different densities considered here are defined as,

$$\lambda = \frac{N_{tot}}{A} \quad \lambda_{active} = \frac{N_u}{A} \quad \lambda_{relay} = \frac{N_{tot} - N_u}{A} \qquad (11.141)$$

with λ the terminal density, λ_{active} the density of the active destinations in the TDMA frame and λ_{relay} the idle terminal density.

The channel matrices of the different relay links are denoted by,

$$\begin{array}{llll} \mathbf{H}_0 & (n_d \times n_s) & \text{source-destination link} \\ \mathbf{H}_1 & (n_r \times n_s) & \text{source-relay link} & (11.142) \\ \mathbf{H}_2 & (n_d \times n_r) & \text{relay-destination link} \end{array}$$

Additional sub-indexes can be considered in order to specify the channel matrix for the different destinations. Moreover, it has been assumed that there is no channel state information at the transmitter (CSIT), so equal average transmitted power per antenna is adopted.

In direct transmission the source transmits orthogonally to each destination, so the normalized mutual information for each destination is defined as [3],

$$I^{DT} = \frac{T_{slot}}{T_{frame}} \log \det \left(\mathbf{I}_{n_d} + \frac{P_s}{n_s \sigma_{n_d}^2} \mathbf{H}_0 \mathbf{H}_0^H \right) \qquad (11.143)$$

where $\sigma_{n_d}^2$ is the noise power at the destination, n_d and n_s denote the number of antennas at the destination and source, p_s is the power used by the source and T_{slot} and T_{frame} are defined by (11.140).

Protocol I is adopted for the relay-assisted transmission, and relays work under *amplify-and-forward* (AF) or *decode-and-forward* (DF) mode. For the AF approach, the signal received at the destination during both phases is modeled by,

$$\begin{bmatrix} \mathbf{y}^{(I)} \\ \mathbf{y}^{(II)} \end{bmatrix} = \begin{bmatrix} \mathbf{H}_0 \\ \mathbf{H}_2 \mathbf{G} \ \mathbf{H}_1 \end{bmatrix} \mathbf{x} + \begin{bmatrix} \mathbf{I}_{n_d} & \mathbf{0}_{n_d} & \mathbf{0}_{n_d} \\ \mathbf{0}_{n_d} & \mathbf{H}_2 \mathbf{G} & \mathbf{I}_{n_d} \end{bmatrix} \begin{bmatrix} \mathbf{n}_{n_d} \\ \mathbf{n}_{n_d} \\ \mathbf{n}_{n_r} \end{bmatrix} + \begin{bmatrix} \mathbf{0}_{n_d} \\ \boldsymbol{\upsilon} \end{bmatrix} = \mathbf{H}_{AF} \mathbf{x} + \mathbf{n}_b \qquad (11.144)$$

where $\mathbf{y}^{(I)}$ and $\mathbf{y}^{(II)}$ are the received signal at the destination in each phase of the relay-assisted transmission, $\mathbf{x}$ denotes the signal transmitted by the source, $\mathbf{n}_{n_d}$ and $\mathbf{n}_{n_r}$ stand for the white Gaussian noise of zero-mean and covariance $\sigma_{n_d}^2$ and $\sigma_{n_r}^2$, $\boldsymbol{\upsilon}$ accounts for the interference received in the *relay-transmit phase* with interfering power σ_i^2. The interference is Gaussian when all the relays transmit Gaussian codewords. Finally, $\mathbf{G}$ is a linear combining matrix at the relay defined by,

$$\mathbf{G} = g\mathbf{I}_{n_r} \qquad g = \sqrt{\frac{P_r}{n_r} \left/ \left(trace \left(\mathbf{H}_1^H \mathbf{H}_1 \right) \frac{P_s}{n_s} + n_r \sigma_{n_r}^2 \right) \right.} \qquad (11.145)$$

with P_r the power transmitted by the relay. The noise plus interference covariance matrix is given by,

$$\mathbf{R}_b = \begin{bmatrix} \sigma_{n_d}^2 \mathbf{I}_{n_d} & \mathbf{0} \\ \mathbf{0} & \sigma_{n_d}^2 \mathbf{H}_2 \mathbf{G} \mathbf{G}^H \mathbf{H}_2^H + \sigma_{n_d}^2 \mathbf{I}_{n_d} + \boldsymbol{\upsilon} \boldsymbol{\upsilon}^H \end{bmatrix} \qquad (11.146)$$

The normalized achievable rate of the AF relay-assisted transmission without CSIT is,

$$I^{AF} = \frac{T_{slot}}{T_{frame}} \log \det \left(\mathbf{I}_{2n_d} + \frac{P_s}{n_s} \mathbf{R}_b^{-1} \mathbf{H}_{AF} \mathbf{H}_{AF}^H \right) = \frac{1}{N_u + 1} \log \det \left(\mathbf{I}_{2n_d} + \frac{P_s}{n_s} \mathbf{R}_b^{-1} \mathbf{H}_{AF} \mathbf{H}_{AF}^H \right) \qquad (11.147)$$

Note that the relay-assisted system resembles a MIMO $2n_d \times n_s$ system for each destination in a TDMA system, except for the extra channel resources.

On the other hand, for the decode-and-forward approach the relay will be working under *unconstrained coding* (UC) (see Sections 11.2.4.1 and 11.4.1.1) hence the signal received at the destination is modeled as,

$$\begin{bmatrix} \mathbf{y}^{(I)} \\ \mathbf{y}^{(II)} \end{bmatrix} = \begin{bmatrix} \mathbf{H}_0 & \mathbf{0}_{n_d} \\ \mathbf{0}_{n_d} & \mathbf{H}_2 \end{bmatrix} \begin{bmatrix} \mathbf{x}_s \\ \mathbf{x}_r \end{bmatrix} + \begin{bmatrix} \mathbf{n}_{n_d} \\ \mathbf{n}_{n_d} \end{bmatrix} + \begin{bmatrix} \mathbf{0}_{n_d} \\ \boldsymbol{\upsilon} \end{bmatrix} = \mathbf{H}_{DF} \mathbf{x} + \mathbf{n}_b \qquad (11.148)$$

where $\mathbf{x}_s$ and $\mathbf{x}_r$ are the signals transmitted from the source and relay, respectively. The covariance matrices of the signal transmitted and the noise plus interference are,

$$\mathbf{Q}_x = \begin{bmatrix} \frac{P_s}{n_s} \mathbf{I}_{n_s} & \mathbf{0} \\ \mathbf{0} & \frac{P_r}{n_r} \mathbf{I}_{n_r} \end{bmatrix} \qquad \mathbf{R}_b = \begin{bmatrix} \sigma_{n_d}^2 \mathbf{I}_{n_d} & \mathbf{0} \\ \mathbf{0} & \sigma_{n_d}^2 \mathbf{I}_{n_d} + \boldsymbol{\upsilon} \boldsymbol{\upsilon}^H \end{bmatrix} \qquad (11.149)$$

The achievable rate of the DF approach taking into account the signal model of (11.148) and the source-relay transmission without CSIT, Section 11.4.1.1, is defined as

$$I^{DF} = \frac{1}{N_u + 1} \min \left(\log \det \left(\mathbf{I}_{n_r} + \frac{P_s}{n_s \sigma_{n_r}^2} \mathbf{H}_1 \mathbf{H}_1^H \right), \log \det \left(\mathbf{I}_{2n_d} + \mathbf{R}_b^{-1} \mathbf{H}_{DF} \mathbf{Q}_x \mathbf{H}_{DF}^H \right) \right) \quad (11.150)$$

with $\mathbf{Q}_x$, $\mathbf{R}_b$ defined in (11.149) and $\mathbf{H}_{DF}$ defined in (11.149).

11.6.2 Distributed Power Control

Game theory is a tool for analyzing multi-person decision making. It has been used by economists, but in recent years it has also been employed to analyze different problems in different areas of communications, such as power control in CDMA wireless systems [82]. Here, the objective is to apply the game theory framework to the new scenario obtained due to the spatial reuse of the relay slot, illustrated by Figure 11.30. There are N_u assisting relays transmitting simultaneously to their associated destinations generating interference with the remaining terminals which will depend on the power transmitted by each relay. Game theory offers a framework to control the interference in a distributed way.

11.6.2.1 Game Theory Overview

In *non-cooperative*[9] games each player involved in the game pursues its own interests which may be partly conflicting with others [80, 81], see also Chapter 16. As the players do not always have complete control over the outcome of the game, conflicts with other players arise. The components of game are:

- The set of players named Ω with Q players. $\Omega = \{1, 2, 3, \ldots, Q\}$.
- The set of pure strategies $X \subseteq P^Q$ with $\mathbf{p} = [p_1, \ldots p_Q]^T \in X$ where p_q represents the strategy of the q-th player. Additionally, $\mathbf{p}_{-q}$ will define the strategy of all the remaining Q-1 players.
- The set of utility (or payoff) functions that map the strategies of each user to real numbers. The function u_q will be the utility function for the q-th player which also depends on the strategy of all the players.

The *non-cooperative* game is defined by

$$G = \left\{ \Omega, X, \{u_q\}_{q \in \Omega} \right\} \quad (11.151)$$

where all players always maximize their utility function rationally, i.e. they do not try to cheat. Each player modifies its strategy in order to maximize its utility function in each iteration of the game. The *readjustment scheme* is defined as how the players update their actions. For example in this work, all players update their actions simultaneously.

Two concepts are important in game theory, the *Nash equilibrium* and the *Pareto solution*. The Pareto solution is obtained when it is impossible to improve the utility of any player without degrading the utility of other player. We can get an optimal set $\mathbf{p}^*$ such as fulfills the following equation,

$$u_q \left(p_q^*, \mathbf{p}_{-q}^* \right) \geq u_q \left(p_q, \mathbf{p}_{-q} \right) \qquad \forall \, p_q \in x_q, \, q \in \Omega \quad (11.152)$$

The *Nash equilibrium* (NE) is a strategy profile where no player may get better utility by unilaterally deviating, providing a stable operating point. Each player finds a solution p_q^* such as,

$$u_q \left(p_q^*, \mathbf{p}_{-q} \right) \geq u_q \left(p_q, \mathbf{p}_{-q} \right) \qquad \forall \, p_q \in x_q, \, q \in \Omega \quad (11.153)$$

---

[9] It is a type of game and is not connected with cooperative/relay-assisted transmission.

The number of NE in a game, if it exists, can be more than one. Therefore if it desired that the game always converge to the same point for each possible strategy, we must prove its existence and uniqueness. The Pareto solution might not coincide with any of the NE points.

In [83] the existence and uniqueness of the *Nash equilibrium* is proved when the utility functions used by the players are concave. Moreover, [84] defines the concept of *standard* utility functions which describe games with a unique *Nash equilibrium* point. A utility function is *standard* if for all $\mathbf{p} \geq 0$ the following properties are satisfied:

- *Positivity*

$$u_q(\mathbf{p}) > 0 \qquad \forall q \in \Omega, \quad \mathbf{p} \in X$$

- *Monotonicity*[10]

$$if \quad \mathbf{p} \geq \mathbf{p}' \quad u_q(\mathbf{p}) \geq u_q(\mathbf{p}') \quad \forall q \in \Omega, \quad \mathbf{p}, \mathbf{p}' \in X \tag{11.154}$$

- *Scalability*

$$\text{For all} \quad \alpha > 1, \quad \alpha u_q(\mathbf{p}) > u_q(\alpha \mathbf{p}) \qquad \forall q \in \Omega, \quad \mathbf{p} \in X$$

Another way for deriving sufficient conditions for existence and uniqueness is the use of the called *potential games*, [85]. This type of game is produced when the incentive of all players to change their strategy can be expressed by a global utility function V. There is an *exact potential game* if it exist the function V such that,

$$u_q\left(p_q, \mathbf{p}_{-q}\right) - u_q\left(p_q^+, \mathbf{p}_{-q}\right) = V\left(p_q, \mathbf{p}_{-q}\right) - V\left(p_q^+, \mathbf{p}_{-q}\right) \qquad \forall \, p_q, p_q^+ \in x_q, \, q \in \Omega \tag{11.155}$$

An ordinal potential game is defined by,

$$u_q\left(p_q, \mathbf{p}_{-q}\right) - u_q\left(p_q^+, \mathbf{p}_{-q}\right) > 0 \Rightarrow V\left(p_q, \mathbf{p}_{-q}\right) - V\left(p_q^+, \mathbf{p}_{-q}\right) > 0 \quad \forall \, p_q, p_q^+ \in x_q, \, q \in \Omega \tag{11.156}$$

In such a case, the game defined by the utility function V presents the same NE points as the original game (11.151). See for example [86], where the different utility functions of the players can be reduced to a common utility function.

Likewise, several methodologies are reviewed in [87] in order to characterize the equilibrium in terms of existence, uniqueness, selection, and efficiency. The efficiency of a NE point can be improved by *pricing* [88]. It consists in modifying the utility function of each player, sometimes incorporating some common value known by all players.

11.6.2.2 Spatial Reuse of the Relay Slot

The power control in the relay transmissions with spatial reuse can be defined as game

$$G = \left\{ \Omega, \, X, \, \{u_q\}_{q \in \Omega} \right\} \tag{11.157}$$

where Ω is the set of players, in this case the destination-relay pairs active in the *relay-transmit phase* (or relay slot), X denotes the set of pure strategies, where the set of the values of power transmitted by

---

[10] When two vector are defined as p ≥ p', the condition is satisfied in each element of the vector.

the q-th relay is in the set X_q and u_q the utility of each player,

$$\Omega = \{1, 2, 3, \ldots, N_u\}$$
$$X_q = [0, P_{max}] \qquad \forall q \in \Omega \qquad (11.158)$$
$$X = X_1 \times X_2 \times \ldots X_Q$$

The proposed utility function for each user considered here is,

$$u_q(\mathbf{p}) = \frac{I(p_q, \mathbf{p}_{-q})}{P_s + p_q} \quad bits/Joule \qquad p_q \in \mathbf{x}_q \quad \forall q \in \Omega \qquad (11.159)$$

where $I()$ is the mutual information of the relay-assisted transmission (in AF, equation (11.147) or in DF, equation (11.150)), p_s is the power used at the source which transmits with fixed power, p_q is the selected transmission power of the q-th assisting relay and $\mathbf{p}_{-q}$ defines the power vector of the remaining relays. The proposed utility function gives the theoretical maximum number of bits/s/Hz without error we can transmit per unit of employed energy. The consideration of the total required power, including the source power, p_s, allows a fair comparison of mutual information with respect to the direct transmission scheme in terms of equal transmitted power. Hence, it is possible to identify when the use of relay-assisted transmission does not improve mutual information, in that case the utility function presents a maximum for p_q equal to zero, and therefore that relay is switched off automatically.

The game is played during several iterations by the destination-relay pairs in order to adjust the powers at the relays. This implies that the channel must remain constant if expressions defined in (11.147) for the AF and (11.150) for the DF are used. In order to avoid this limitation, the players will use approximations of the previous expressions assuming uncorrelated channels, large number of antennas and $n_s \geq n_r \geq n_d$,

$$\tilde{I}^{AF}(\mathbf{p}) = \frac{1}{N_u + 1} \left(n_d \log(1 + \gamma_0) + n_d \log\left(1 + \frac{\gamma_1 \gamma_2}{1 + \gamma_1 + \gamma_2}\right) \right)$$
$$\tilde{I}^{DF}(\mathbf{p}) = \frac{1}{N_u + 1} \min\left(n_r \log(1 + \gamma_1), n_d \log(1 + \gamma_0) + n_d \log(1 + \gamma_2)\right) \qquad (11.160)$$

with $\gamma_0, \gamma_1, \gamma_2$ denoting the signal to noise ratio (SNR) in the source-destination, source-relay and relay-destination link due to the path-loss. The definition for the q-th player is given by,

$$\gamma_{0,q} = \frac{P_s}{L_{0,q}\sigma_{n_d}^2} \qquad \gamma_{1,q} = \frac{P_s}{L_{1,q}\sigma_{n_r}^2} \qquad \gamma_{2,q} = \frac{p_q/L_{2,q}}{\sigma_{n_d}^2 + \sum\limits_{k=1}^{Q} \frac{p_k}{L_{2,q-k}}} \qquad (11.161)$$

where $L_{0,q}, L_{1,q}, L_{2,q}$ stand for the path-loss in the source-destination, source-relay and relay-destination link. Additionally, $L_{2,q-k}$ defines the path-loss between the q-th relay and the k-th destination (which is interfered by the q-th relay).

11.6.2.3 Existence and Uniqueness of the Nash Equilibrium

The proposed utility function in (11.159) is not concave. It is not a *standard* function (11.154), because it can not satisfy the monotonicity property. Additionally it can not be formulated as a *potential game* because there is not a function V valid for all the users satisfying the properties of (11.155) or (11.156). Therefore some additional results are needed to prove the existence and uniqueness of the Nash equilibrium points of a game with the proposed utility function.

Existence of the Nash Equilibrium

Proposition 1. *The game* $G = \left\{\Omega, X, \{u_q\}_{q\in\Omega}\right\}$ *has a least one Nash equilibrium point if for all* $q\in\Omega$:

- x_q *is a non-empty, convex, and compact subset of a Euclidean space.*
- $u_q(p)$ *is continuous in* p *and quasi-concave in* p_q.

Proof. See [89].

By definition the action sets x_q are non-empty and convex (11.158). Each x_q is closed so it is also compact, satisfying the first condition. Moreover, the proposed utility function in (11.159) is quasi-concave.

A function $y(x)$ is quasi-concave if and only if either [90]

- y *is non-increasing.*
- y *is non-decreasing.*
- There is a x^* such that $y(x)$ is non-decreasing for $x < x^*$ and non-increasing for $x > x^*$.

Therefore the first and second derivatives of the function u_q over the variable p_q have to be studied in order to know their properties. The function u_q (11.159) can be redefined as,

$$u(p) = f(p)\,g(p) \qquad f(p) = I(p) \quad g(p) = \frac{1}{P_s + p} \tag{11.162}$$

where $I()$ denotes the mutual information for the AF or DF defined in (11.160). The set of p values which set to zero the first derivative of u is given by,

$$\Psi = \left\{p \mid f'(p)\,g(p) + f(p)\,g'(p) = 0\right\} \tag{11.163}$$

The second derivative is defined by

$$\frac{u''(p)}{u(p)} = \frac{f''(p)}{f(p)} - 2\left(\frac{g'(p)}{g(p)}\right)^2 + \frac{g''(p)}{g(p)} \qquad p \in \Psi \tag{11.164}$$

Taking into account the current value of the $g()$ function (11.162) and its derivatives, the second derivative of the utility simplifies to,

$$\frac{u''(p)}{u(p)} = \frac{f''(p)}{f(p)} \quad \rightarrow \quad u''(p) = f''(p)\,g(p) \qquad p \in \Psi \tag{11.165}$$

For the AF approach the equation (11.165) is always negative because the mutual information increases logarithmically with p. Hence, there is no minimum, or equivalently, the utility function is monotonic (*non-increasing* or *non-decreasing*) or it has a single maximum (conditions for quasi-concavity). On the other hand, for the DF the mutual information (11.160) depends on the minimum of two functions, so the evaluation of (11.165) is more elaborated. Assume that p_0 is the power which bounds that the mutual information is limited by the source-relay or the (source,relay)-destination link. When the power of the relay p goes from 0 to p_0 the mutual information is given by the expression of the (source, relay)-destination link and increases logarithmically with p. As a consequence for that interval of values the utility function is monotonic or it has a single maximum. When $p > p_0$ the mutual information is limited by the source-relay link which is constant and the utility function $u(p) = g(p)$ decreases with p (see Equation (11.162)). Therefore, the utility function is also quasi concave.

Uniqueness of the Nash Equilibrium

Unfortunately, the uniqueness of the Nash equilibrium in the analyzed problem only can be assessed by means of simulations. In this regard *definition 4.5* of [80] is adopted,

Definition 11.1 In a game G (11.157), a Nash equilibrium p_q^* with $q \in \Omega$, is stable with respect to an *readjustment scheme S* if it can be obtained as the limit of the iteration,

$$p_q^* = \lim_{\tau \to \infty} p_q(\tau) \tag{11.166}$$

$$p_q(\tau + 1) = \arg \max_{p_q \in \mathbf{x}_q} u_q\left(p_q, \mathbf{p}_{-q}^{S_\tau}\right), \qquad p_q(0) \in \mathbf{x}_q, \quad q \in \Omega \tag{11.167}$$

where equation (11.167) is also known as the *best response* of the q-th player, the superscript S_τ indicates that the precise choice of $\mathbf{p}_{-q}^{S_\tau}$ depends on the *readjustment scheme* selected, that is, how the players update their actions. If the iteration of *Definition 1* converges, the Nash equilibrium is unique with respect to the scheme S. For example the scheme considered here is

$$\mathbf{p}_{-q}^{S_\tau} = \mathbf{p}_{-q}(\tau) \tag{11.168}$$

which describes the situation where the players update their actions simultaneously in response to the most recent actions of the other players

11.6.2.4 Distributed Power Computation

The game will be played on each TDMA frame. The following procedure summarizes the distributed power computation,

1. Initialization ($\tau = 0$). Firstly, all the relays will transmit with some initial power $p_q \in \mathbf{x}_q$ (see equation (11.158)).
2. During the *relay-transmit phase* (relay slot) of the τ-th frame, each destination measures the interfering power received (which will be the denominator of $\gamma_{2,q}$ in (11.161)), and $\gamma_{1,q}, \gamma_{0,q}$ of the q-th player.
3. The decision of the power used at each relay for the next frame ($\tau + 1$). Each player independently obtains the *best response*, (11.167), i.e. selects that power which maximizes its own utility function u_q, (11.159) given the measured interfering power. Note that the value of the interfering power depends on the actions taken by the other players in the previous frame.
4. The iterative procedure continues until all players find that the change in their power levels is less than a pre-defined bound, or an upper limit on the number of iterations is reached.

The following examples will show that the game converges to the same power values for different power initializations.

11.6.2.5 Examples

For the simulations, the following scenario has been considered: a square area of 900×900 m^2, one source in the center of the cell using omni-directional antennas, terminals are uniformly distributed in the cell with Rayleigh flat fading channels. The distance loss model is the usual inverse law with propagation exponent to be 4. The source transmission power is 30 dBm ($p_s = 1000$ mW) and the terminals have maximum transmission power of 23 dBm ($P_{max} = 200$ mW). The thermal noise at the terminals is -115.2 dBm, [73]. We consider that each user has chosen a relay terminal around which is at a distance according to (11.139), (*user relaying*). Figure 11.31 shows the evolution in the transmitter power for the relays assigned to three destinations named 2, 3, and 6 for a specific scenario realization where the assisting relays create interference with the associated destinations to those relays. It has considered three different

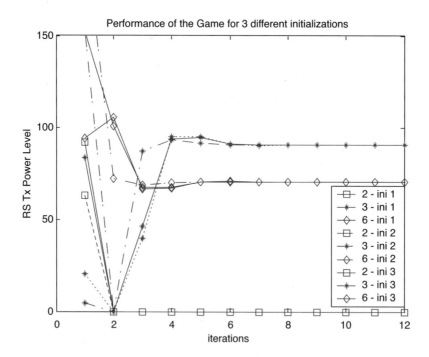

Figure 11.31 Convergence of the algorithm. The destination-relay pairs $\{2, 3, 6\}$ describe a scenario where those destinations are receiving interference from other relays. Reproduced by permission of IEEE © 2004.

initializations of the iterative algorithm, considering the AF scheme. It is worth noting that the power of the assisting relay 2 converges to zero, that it is destination 2 decides to not use the relay-assisted transmission. The algorithm converges, after a few iterations, to the same stationary point for the relay's power with independence of the initial power values. The same behavior, not plotted in the figure for clarity, is observed over the remaining destinations served in the same frame. Figure 11.32 shows the cumulative function of the cell capacity or sum-throughput[11] for n_s equal to two antennas at the source (BS) with and without relay-assisted transmission (cross solid line). Two different terminal densities are considered which will impact on the terminal selection as assisting relay, $\lambda = 10^{-3}$ terminals/m^2 (circles) and $\lambda = 10^{-4}$ terminals/m^2 (diamonds). The circle and diamond solid lines correspond to the AF approach with $n_r = 1$ antenna at the assisting relay while the dotted lines corresponds to the DF approach with $n_r = 2$ antennas. As in the *user relaying* the assisting relay will be near the destination, we will assume that the assisting relays are equipped with $n_r = 2$ antennas in order to avoid the limitation of the source-relay link in the DF. In such a case, the source-relay links present a 2×2 MIMO configuration. It can be seen that the relay-assisted transmission improves the direct transmission (2×1 MISO). When DF is assumed, gains of 66% ($\lambda = 10^{-4}$) and 90% ($\lambda = 10^{-3}$) over the direct transmission are possible. On the other hand, using AF the gains are 45% ($\lambda = 10^{-4}$) and 66% ($\lambda = 10^{-3}$). Although the AF is inferior to the DF, it only requires of a single antenna and simple operation (*amplify-and-forward*) at the relay-assisting terminal. Basically, these gains of the relay-assisting transmission depend on:

- The reuse of the relay slot (number of terminal using relay-assisted transmission in the same frame).
- Interference at the relay-transmit phase (relay slot).
- The position of the relay terminal.

---

[11] Total throughput delivered by the source in each frame.

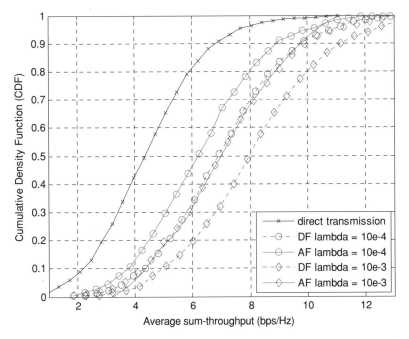

Figure 11.32 CDF of the average sum-throughput for direct and relay-assisted transmission (AF-solid lines, DF – dotted lines) with different terminal densities ($\lambda = 10$-3 – circles, $\lambda = 10$-4 – diamonds). $n_s = 2$, $n_d = 1$, $n_r = 1$ (AF) and $n_r = 2$ (DF).

11.6.3 Rate Control Interference Management in Relay-Assisted Users

This section addresses *amplify-and-forward* relay-assisted transmission under interference limitation principles. It presents a statistical model for the interfering power, by adapting some previous results from [91]. The interfering power received at the destinations during the *relay-transmit phase* (see Figure 11.30) is modeled by an α-stable distribution, [92], which depends on several aspects: the terminal density, the transmitted power at the relay, the distance between assisting relay-destination pairs, and the spatial traffic pattern. The main difference with respect to the approach considered in Section 11.6.2 is that the relays are transmitting with a constant power rather than adjusting the transmitted power. Hence, for combating the outage events produced by the interference the source adapts its data rate in each transmission (*rate control*). Two strategies can be followed: selecting the rate which maximizes the throughput or selecting a conservative rate to keep the outage probability below a certain threshold. While the first option is found commonly in *best effort* traffic services where retransmissions are allowed, the second could be adopted for *delay sensitive* traffic services where retransmissions are not permitted. The scenario defined in Section 11.6.1 applies here with the following system assumptions:

1. The assisting relays will be selected from those terminals in idle mode. All the terminals present the same opportunity to be assisting relays. Each destination is assisted by one relay.
2. All terminals are equipped with a single antenna ($n_r = n_d = 1$) and simple receivers, while the source has n_s antennas.
3. The source has perfect knowledge of the current SNR of the source-relay, source-destination and relay-destination (without interference) links, but not of the exact complex channel gain, i.e. there is no *perfect* CSI. The channel coefficients remain fixed during all of the relay-assisted transmission.
4. As a result of the unknown interference, outage events appear at the relay slot.

5. Each assisting relay adjusts the power level in order to set a *target* SNR (measured without interference) in the relay-destination link, SNR_t. Its value is the same for all the relay-assisted users in the cell, becoming a parameter of the system. The transmitted power depends on the quality of the relay-destination link.

6. The source selects randomly N_u active destinations to be served in each TDMA frame (*round robin packet scheduling*). Consequently, the transmitting relays change in each frame, and each destination will observe a random interfering power.

A target signal to noise ratio (SNR_t) in the relay-destination link (without interference and only due to the path-loss in the relay link) is established to be the same for all active relay-destination pairs around the cell. This implies that the power transmitted by a relay, although kept fixed per terminal, is a random variable because of the distance between each assisting relay-destination (named r) is also a random variable (11.139). It is defined by,

$$P_r = \min\left(P_{MAX}, \frac{\sigma^2 \cdot SNR_t \cdot r^\gamma}{K_r}\right), \quad SNR_t = \frac{K_r P_r}{r^\gamma \sigma^2} \tag{11.169}$$

where P_{MAX} is the maximum power allowed at the relay, σ^2 denotes the noise power, K_r considers the effect of shadowing, r is the distance between the pair relay-destination, (11.139) and γ stands for the propagation exponent.

11.6.3.1 Interfering Power Modeling

Since all the assisting relays (associated to different destinations) transmit simultaneously during the relay slot (*relay-transmit phase*), they induce interference with all active destinations. Taking into account the assumptions of previous sections, the interfering power received at each destination during the relay slot becomes a random variable. A similar scenario is investigated in [91], where the statistics of the interference power are obtained in an ad hoc network of *Poisson* distributed terminals, all transmitting the same power in a channel affected only by path-loss. Now we need to incorporate to that model the shadowing and the multipath fading, taking into account that in our case each relay transmits with power according to the *target* SNR (*assumption 5*).

Following a similar procedure as in [91], the interfering power received from a transmitter placed at distance r_i, $g(r_i)$, satisfies the following two conditions:

1. $-\quad \lim_{r_i \to 0} g(r_i) = \infty, \quad \lim_{r \to \infty} g(r_i) = 0, \quad g(r_i)$ is monotonically decreasing

2. $-\quad \lim_{r_i \to \infty} r_i^2 g(r_i) = 0 \quad$ (finite interfering power in a network) $\tag{11.170}$

Note that the interfering power when r_i tends to zero will be very large but not necessarily ∞. The interfering power received at a destination from a single interfering relay due to the multipath and/or fading is,

$$\eta(K_i, r_i) = K_i g(r_i) = K_i \frac{P_r}{r_i^\gamma} \tag{11.171}$$

where P_r is the power transmitted by the interfering relay, r_i denotes the distance between the destination and the interfering relays, γ is the propagation exponent and K_i is a constant that takes into account the shadowing and fast fading effects in the interfering relay-destination link (with the same distribution for all the users). The power transmitted by each assisting relay is variable according to equation (11.169), so the interference power observed at each destination depends on r_i and also on r, distance between the interfering relay and its associated destination. Hence, the received interfering power can be described by,

$$
g(r_i, r) = \begin{cases} \dfrac{1}{K_r} N_0 \cdot SNR_t \left(\dfrac{r}{r_i}\right)^{\gamma} & r \le \Delta \\ \dfrac{P_{MAX}}{r_i^{\gamma}} & otherwise \end{cases} \qquad \Delta = \left(\dfrac{K_r P_{MAX}}{N_0 SNR_t}\right)^{\frac{1}{\gamma}} \quad \gamma > 2 \qquad (11.172)
$$

Let Y_a be the interfering power received from the active assisting relays in D_a, a disk of radius a,

$$
Y_a = \sum_{r_i \le a} \eta(K_i, r_i, r) = \sum_{r_i \le a} K_i g(r_i, r) \qquad (11.173)
$$

After some calculation, see Appendix in Section 11.6.8 for details, we obtain the interference power Y (Y_a when $a \to \infty$) as an α-stable variable [92] whose characteristic function depends on the α-th moment of the variable K_i:

$$
\phi_Y(\omega) = \exp\left(-K \cos(\alpha\pi/2)\, \omega^{\alpha}\left(1 - j\tan(\alpha\pi/2)\right)\right) \qquad (11.174)
$$

where α and K are given by,

$$
\alpha = \gamma/2 \qquad K = \Upsilon \mu_{K_i, \alpha} \qquad (11.175)
$$

$$
\Upsilon = \Gamma(1-\alpha)\lambda_{active}\pi\left(\left(\dfrac{N_0 SNR}{K_r}\right)^{\alpha} \dfrac{\left(1 - \left(1 + \pi\lambda_{relay}\, p_s \Delta^2\right)\Omega\right)}{\pi\lambda_{relay}\, p_s} + \Omega P_{MAX}^{\alpha}\right) \qquad (11.176)
$$

$$
\Omega = \exp\left(-\pi\lambda_{relay}\, p_s \Delta^2\right) \qquad \mu_{K_i, \alpha} = \int K_i^{\alpha} f_{K_i}(K_i)\, dK_i
$$

where $\Gamma(\cdot)$ the Gamma function, λ_{active} depends on the number of active destinations in each frame, λ_{relays} depends on the probability of a nearby terminal to become a relay, λ is the terminal density, (11.141), Δ is given by equation (11.172) and $\mu_{K_i, \alpha}$ is the αth-moment of the random variable K_i. When $\gamma = 4$ it is possible to obtain closed-form expressions for its probability and cumulative density function (*pdf* and *cdf*) of the interfering power as,

$$
f_y(y) = \dfrac{K}{2\sqrt{\pi}} y^{-3/2} \exp\left(-\dfrac{K^2}{4y}\right), \qquad F_y(y < x) = erfc\left(\dfrac{K}{2\sqrt{x}}\right) \qquad (11.177)
$$

11.6.3.2 Properties of AF Relay-Assisted Transmission

This section analyzes the properties of the TDD *amplify and forward* (AF) transmission of a destination served in the TDMA frame (11.140). First, it will be derived a simple expression for the mutual information when the source is equipped with possible multiple antennas and the relays and destinations feature a single antenna, $n_s \ge 1$, $n_r = n_d = 1$. The mutual information (11.147) is given by,

$$
I^{AF} = \dfrac{T_{slot}}{T_{frame}} \log \det\left(\mathbf{I}_2 + \dfrac{P_s}{n_s}\mathbf{R}_b^{-1}\mathbf{H}_{AF}\mathbf{H}_{AF}^H\right) = \dfrac{1}{N_u + 1}\log_2(\Psi) \qquad (11.178)
$$

Assuming that the noise power at the different links is the same, i.e. $\sigma_{n_d}^2 = \sigma_{n_r}^2 = \sigma^2$ in (11.146) and the channel matrices are given by column vectors due to the antenna configuration, then the Ψ variable is defined by,

$$\Psi = \left(1 + \frac{\mathbf{h}_0^H \mathbf{h}_0 \frac{P_s}{n_s L_0}}{\sigma^2}\right)\left(1 + \frac{\frac{P_s}{n_s L_2 L_1} h_2 g \mathbf{h}_1^H \mathbf{h}_1 g^* h_2^*}{\frac{h_2 g g^* h_2^*}{L_2}\sigma^2 + \sigma^2 + Y}\right) - \frac{\frac{P_s \mathbf{h}_1^H \mathbf{h}_0}{n_s \sqrt{L_0 L_1}} \frac{g^* h_2^* h_2 g}{L_2} \frac{\mathbf{h}_0^H \mathbf{h}_1 P_s}{n_s \sqrt{L_0 L_1}}}{\sigma^2\left(\frac{h_2 g g^* h_2^*}{L_2}\sigma^2 + \sigma^2 + Y\right)} \qquad (11.179)$$

where Y denotes the interfering power received in the relay link, L_0, L_1 and L_2 account for the path-loss terms in each link and $\mathbf{h}_0$, $\mathbf{h}_1$ and h_2 are the channel complex gains of dimensions $n_s \times 1$, $n_s \times 1$ and 1×1, respectively. After a tedious transformation the mutual information becomes,

$$I^{AF} = \frac{1}{(N_u + 1)}(m_{DT} + m_{AF})$$

$$m_{DT} = \log(1 + SNR_0) \quad m_{AF} = \log\left(1 + \frac{SNR_1 SNR_2 \Theta}{SNR_2 + (1 + Y/\sigma^2)(SNR_1 + 1)}\right) \qquad (11.180)$$

with the following variable definition,

$$\Theta = \frac{1 + SNR_0(1 - \xi)}{1 + SNR_0} \qquad \xi = \frac{\left|\mathbf{h}_1^H \mathbf{h}_0\right|^2}{\mathbf{h}_0^H \mathbf{h}_0 \mathbf{h}_1^H \mathbf{h}_1} = \cos^2(\phi)$$

$$SNR_0 = \frac{P_s \mathbf{h}_0^H \mathbf{h}_0}{\sigma^2 L_0 n_s} \qquad SNR_2 = \frac{P_{TX} h_2 h_2^*}{\sigma^2 L_2} \qquad SNR_1 = \frac{P_s \mathbf{h}_1^H \mathbf{h}_1}{\sigma^2 L_1 n_s} \qquad (11.181)$$

where SNR_0, SNR_1 and SNR_2 are the SNR in the source-destination, source-relay and relay-destination links, respectively. Finally, m_{DT} and m_{AF} stand for the mutual information in the source-destination and source-relay-destination link. Note that the benefits of the relay-assisted transmission depend on the m_{AF} value.

It is relevant to point out the importance of ξ and Θ in the mutual information for the *amplify and forward* scheme, (0.180). When $n_s = 1$, then $\xi = 1$ (channels $\mathbf{h}_0$ and $\mathbf{h}_1$ are vectors of dimension 1) and $\Theta = 1/(1 + SNR_0)$, penalizing the mutual information of m_{AF} (11.180). However, allowing $n_s > 1$ in Rayleigh fading channels, ξ becomes a random variable distributed as a Beta function [93] with parameters $(1, n_s-1)$. As n_s increases the distribution of ξ concentrates around zero and Θ around 1. Therefore, m_{AF} is defined in the following interval which also limits the mutual information of the AF scheme, I^{AF},

$$\log\left(1 + \frac{SNR_1 SNR_2/(1 + SNR_0)}{SNR_2 + (1 + Y/\sigma^2)(SNR_1 + 1)}\right) \le m_{AF} \le \log\left(1 + \frac{SNR_1 SNR_2}{SNR_2 + (1 + Y/\sigma^2)(SNR_1 + 1)}\right) \qquad (11.182)$$

where its actual value depends on the variable ξ.

When comparing the direct transmission and the relay-assisted AF both systems must use the same average transmitted power in order to evaluate the efficiency of the communication. Because both transmission schemes use a different resource allocation, as defined by equation (11.140), the direct transmission should transmit with the following *cooperative power* (P_{coop}),

$$P_{coop} = \frac{N_u}{N_u + 1}(P_s + P_r) \qquad (11.183)$$

with P_r denoting the current power transmitted by the assisting relay (11.169).

11.6.3.3   Rate Selection in Relay-Assisted Transmission

It is assumed that the outage events are due to the interference generated by the simultaneous transmissions of all assisting relays, which statistic is modeled by (11.174). The source has two criteria to select a conservative rate (*rate control*), R^{AF}: fixing the probability that the transmission is not in outage (ζ) or maximizing the throughput.

The probability of the selected rate, R^{AF} is not in outage is defined as,

$$\zeta = 1 - P_{out}\left(T_{frame} I^{AF} < T_{slot} R^{AF}\right) = F_y\left(Y < \left(\sigma^2\left(\frac{SNR_2}{1 + SNR_1}\left(\frac{SNR_1 \Theta}{2^{R^{AF} - m_{DT}} - 1} - 1\right) - 1\right)\right)\right)$$

(11.184)

where m_{DT} is defined in (11.181), Y is the interfering power received and F_y denotes the *cdf* of the interference which is the erfc function when $\alpha = 1/2$, (11.177). Assuming that the following parameter ε_0 is connected with ζ through,

$$\zeta = erfc(\varepsilon_0) \qquad \varepsilon_0 = erfc^{-1}(\zeta) = \frac{K}{2\sqrt{\left(\sigma^2\left(\frac{SNR_2}{1 + SNR_1}\left(\frac{SNR_1 \Theta}{2^{R^{AF} - m_{DT}} - 1} - 1\right) - 1\right)\right)}}$$

(11.185)

The selected rate R^{AF} for a given probability ζ of not being in outage (using the variable ε_0 (11.185)) is,

$$R^{AF} = \log\left(1 + \frac{SNR_1 \Theta}{1 + \left(1 + \left(\frac{K}{2\sigma\varepsilon_0}\right)^2\right)\left(\frac{1 + SNR_1}{SNR_2}\right)}\right) + \log(1 + SNR_0)$$

(11.186)

with K modeling the interfering power in the relay slot defined in (11.175), which will vary with the number of destinations server in a TDMA frame (N_u). The throughput for a TDMA destination using relay-assisted transmission for a given probability of not being in outage (ζ) is,

$$T_{AF} = \frac{1}{N_u + 1} R^{AF} \zeta$$

(11.187)

Additionally, if the relay-assisted rate R^{AF} is selected to maximize the throughput then

$$T_{AF}^* = \max_{\zeta}\left(\frac{1}{N_u + 1} R^{AF} \zeta\right)$$

(11.188)

where R^{AF} is connected with ζ as equations (11.186) and (11.185) show.

11.6.3.4   Examples

This section present the results obtained by relay-assisted and direct transmission, showing under what configurations relay-assisted transmission is beneficial. Our scenario is defined by a circular cell with radius 700m and the source placed at the center with the following terminal densities $\lambda = 5 \times 10^{-4}$ terminals/m^2. These values may describe different population densities in rural and suburban regions, see Table 1 in [94]. The multipath fading is modeled by a complex Rayleigh random variable of zero mean and unitary variance and the path-loss at a distance d in the source-destination (L_0), source-relay (L_1) and relay-destination (L_2) is given by,

$$\begin{cases} L_0 = L_1 = 43.65 + 3 \times 10 \times \log_{10}(d) & (dB) \\ L_2 = 43.65 + \gamma \times 10 \times \log_{10}(d) & (dB) \quad \gamma = 4 \end{cases} \quad (11.189)$$

The power transmitted by the source in the direct transmissions is $p_s = 30$ dBm and the maximum power transmitted by an assisting relay is $P_{MAX} = 20$ dBm. The noise power is set to $\sigma^2 = -102$ dBm. Considering only the path-loss effect, the received SNR at the border of the cell is 3 dB, while the received SNR is -1.65 dB when an assisting relay is at 100 meters and transmitting with P_{MAX} (relay-destination link). Finally, the interfering power in the relay slot is modeled by an α-stable random variable according to equations (11.175) and (11.176).

Results are obtained with Monte-Carlo simulations, measuring average *sum-throughput*, i.e. the total throughput delivered by the source to all the destinations served in each frame. The destinations are selected randomly each time (*assumption 6*). The AF *sum-throughput* is compared with the *sum-throughput* obtained by the direct transmission using cooperative power, (11.183) in the following way,

$$\Delta T (\%) = \frac{E\left\{\sum_{n=1}^{N_u} T_{AF}(n)\right\} - E\left\{\sum_{n=1}^{N_u} T_{DT}(n)\big|_{P_{coop}}\right\}}{E\left\{\sum_{n=1}^{N_u} T_{DT}(n)\big|_{P_{coop}}\right\}} \times 100 \quad (11.190)$$

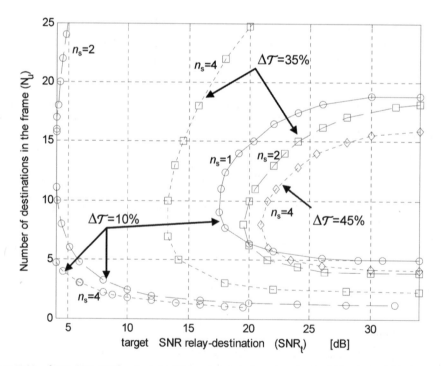

Figure 11.33 $\{10\%, 35\%, 45\%\}$ gain plots of the average sum-throughput for the AF (under individual throughput maximization, (11.188)) over the direct transmission with cooperative power for different values as a function of the target SNR (11.169) and N_u. $n_s = \{1, 2, 4\}$. Propagation scenario defined by (11.169). Reproduced by permission of IEEE © 2008.

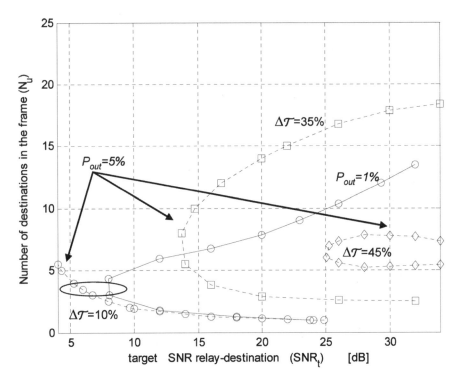

Figure 11.34 {10%, 35%, 45%} gain plots of the average sum-throughput for the AF over the direct transmission with cooperative power when outage probability is fixed as a function of the target SNR and N_u. Terminal density $\lambda = 5 \times 10^{-4}$ users/m^2, $n_s = 4$, $P_{out} = \{1\%,5\%\}$. Propagation scenario defined by (11.189). Reproduced by permission of IEEE © 2008.

when $N_u = 1$ the conventional AF (under *user relaying*) protocol is compared with the direct transmission. Results are presented as contour plots for different throughput gains (11.190) as a function of the number of destinations (N_u) and the target SNR (SNR_t) selected in the relay-destination link (11.169).

Figure 11.33 depicts the {10%, 35% and 45%} gains of the average *sum-throughput* AF over the average *sum-throughput* of the direct transmission with cooperative power. The source maximizes the throughput for each destination according to equation (11.180). By increasing the number of antennas at the source it is minimized the constraints in terms of target SNR at the relay-destination link and N_u in order to achieve some throughput gains. For example with $n_s = 1$ it is required a target SNR ≥ 17 dB and $6 \leq N_u \leq 19$ to obtain $\Delta T = 10\%$ of throughput gain. The conventional AF ($N_u = 1$) gets a lower throughput gain because of the position of the relay. Whereas with $n_s = 2$ and $n_s = 4$ that throughput gain (ΔT) can be obtained for a lower SNR_t. In such a case, AF without reuse gets $\Delta T = 10\%$ because in some cases it is required $N_u = 1$.

Figure 11.34 shows that the relay-assisted transmission with $n_s = 4$ antennas at the source also achieves some throughput gains over the direct transmission when the source selects a rate to have an error probability (due to the interfering power at the relay slot) of 1% ($\zeta = 0.99$) and 5% ($\zeta = 0.95$), (11.187). When the outage probability grows the throughput gains are improved, for example the throughput gain does not get the $\Delta T = 35\%$ for $P_{out} = 1\%$. That gain is achieved when $P_{out} = 5\%$ and even $\Delta T = 45\%$ is possible for that configuration. Additionally, we can see that working with $P_{out} = 5\%$ needs more SNR_t and/or N_u to get the same gains as when we are working to maximize the throughput (see Figure 11.33 with $n_s = 4$).

Appendix 11.1

Asymptotic CDF Approximations

All of the results from Section 11.2 are of the form

$$\lim_{t \to t_0} \frac{P_{u(t)}(g_1(t))}{g_2(t)} = c \tag{A11.1}$$

where t is a parameter of interest; $P_{u(t)}$ ($g_1(t)$) is the CDF of a certain random variable u(t) that can, in general, depend upon t; $g_1(t)$ and $g_2(t)$ are two (continuous) functions; and t_0 and c are constants. (A11.1) implies the approximation $P_{u(t)}(g_1(t)) \sim c g_2(t)$ for t close to t_0.

Results 1. Let u be an exponential random variable with parameter λ_u. Then, for a function $g(t)$ continuous about $t = t_0$ and satisfying $g(t) \to 0$ as $t \to t_0$

$$\lim_{t \to t_0} \frac{1}{g(t)} P_u(g(t)) = \lambda_u \tag{A11.2}$$

Results 2. Let $w = u + v$, where u and v are independent exponential random variables with parameters λ_u and λ_v, respectively. Then the CDF

$$P_w(w) = \begin{cases} 1 - \left[\left(\dfrac{\lambda_v}{\lambda_v - \lambda_u} \right) e^{-\lambda_u w} + \left(\dfrac{\lambda_u}{\lambda_u - \lambda_v} \right) e^{-\lambda_v w} \right], \lambda_u \neq \lambda_v \\ 1 - (1 + \lambda w)e^{-\lambda w}, \lambda_u = \lambda_v = \lambda \end{cases} \tag{A11.3}$$

$$\lim_{\varepsilon \to 0} \frac{1}{\varepsilon^2} P_w(\varepsilon) = \frac{\lambda_u \lambda_v}{2} \tag{A11.4}$$

and

$$\lim_{t \to t_0} \frac{1}{g^2(t)} P_w(g(t)) = \frac{\lambda_u \lambda_v}{2} \tag{A11.5}$$

if a function $g(t)$ is continuous about $t = t_0$ and satisfies $g(t) \to 0$ as $t \to_0$,

Equality 1. Let u, v, and w be independent exponential random variables with parameters λ_u, λ_v, and λ_w, respectively. Let ε be positive, and let $g(\varepsilon) > 0$ be continuous with $g(\varepsilon) \to 0$ and $\varepsilon / g(\varepsilon) \to c <$ as $\varepsilon \to 0$. Then for $f(x,y) = (xy)/(x + y + l)$

$$\lim_{t \to t_0} \frac{1}{g^2(t)} \Pr[u + \varepsilon f(v/\varepsilon, \; w/\varepsilon) < g(\varepsilon)] = \frac{\lambda_u(\lambda_v + \lambda_w)}{2} \tag{A11.6}$$

and

$$\lim_{t \to t_0} \frac{1}{g^2(h(t))} \Pr[u + h(t) f(v/h(t), w/h(t)) < g(h(t))] = \frac{\lambda_u(\lambda_v + \lambda_w)}{2} \tag{A11.7}$$

if $h(t)$ is continuous about $t = t_0$ and satisfies $h(t) \to 0$ as $t \to t_0$.

The following lemma will be useful in the proof of *Equality 1*.

Lemma 1. *For the following set of parameters: positive δ; $r_\delta := \delta f(v/\delta, w/\delta)$, where v and w are independent exponential random variables with parameters λ_v and λ_w, respectively; continuous $h(\delta) > 0$ with $h(\delta) \to 0$ and $\delta/h(\delta) \to d < \infty$ as $\delta \to 0$; probability $Pr[r_\delta < \delta]$ satisfies*

$$\lim_{\delta \to 0} \frac{1}{h(\delta)} \Pr[r_\delta < h(\delta)] = \lambda_v + \lambda_w \tag{A11.8}$$

Proof of Lemma 1: First we look at the lower bound

$$
\begin{aligned}
\Pr[r_\delta < h(\delta)] &= \Pr[1/v + 1/w + \delta/(vw) > 1/h(\delta)] \\
&> \Pr[1/v + 1/w > 1/h(\delta)] \geq \Pr[\max(1/v, 1/w) > 1/h(\delta)] \\
&= 1 - \Pr[v \geq h(\delta)] \Pr[w \geq h(\delta)] = 1 - \exp[-(\lambda_v + \lambda_w)h(\delta)]
\end{aligned}
\tag{A11.9}
$$

so, by using Result *1* we have

$$\lim_{\delta \to 0} \frac{1}{h(\delta)} \Pr[r_\delta < h(\delta)] \geq \lambda_v + \lambda_w \tag{A11.10}$$

To prove the other direction, let $l > 1$ be a fixed constant:

$$
\begin{aligned}
\Pr[r_\delta < h(\delta)] &= \Pr[1/v + 1/w + \delta/(vw) > 1/h(\delta)] \\
&= \int_0^\infty \Pr\left[1/v > \frac{1/h(\delta) - 1/w}{1 + \delta/w}\right] p_w(w)dw \\
&\leq \Pr[w < lh(\delta)] + \int_{lh(\delta)}^\infty \Pr\left[1/v > \frac{1/h(\delta) - 1/w}{1 + \delta/w}\right] p_w(w)dw
\end{aligned}
\tag{A11.11}
$$

But

$$\Pr[w < lh(\delta)]/h(\delta) \leq \lambda_w l \tag{A11.12}$$

which takes care of the first term of Equation (A11.11). To bound the second term of (A11.11), let $k > l$ be another fixed constant, and note that

$$
\begin{aligned}
\int_{lh(\delta)}^\infty &\Pr\left[1/v > (1/h(\delta) - 1/w)/(1 + \delta/w)\right] p_w(w)dw = \\
\int_{lh(\delta)}^\infty &\Pr\left[\frac{1}{v} > \frac{1/h(\delta) - 1/w}{1 + \delta/w}\right] p_w(w)dw + \int_{lh(\delta)}^{kh(\delta)} \Pr\left[\frac{1}{v} > \frac{1/h(\delta) - 1/w}{1 + \delta/w}\right] p_w(w)dw \\
&\leq \Pr\left[\frac{1}{v} > \frac{1 - 1/k}{h(\delta) + \delta/k}\right] + \lambda_w \int_{lh(\delta)}^{kh(\delta)} \Pr\left[\frac{1}{v} > \frac{1/h(\delta) - 1/w}{1 + \delta/w}\right] p_w(w)dw
\end{aligned}
\tag{A11.13}
$$

where the first term in the bound of Equation (A11.13) follows from the fact that $\Pr[1/v > (1/h(\delta) - 1/w)/(1 + \delta/w)]$ is nonincreasing in w, and the second term in the bound of (A11.13) follows from the fact that $p_w(w) = \lambda_w \exp(-\lambda_w w) \leq \lambda_w$ Now, the first term of (A11.13) satisfies

$$\Pr\left[1/v > \frac{1 - 1/k}{h(\delta) + \delta/k}\right]/h(\delta) \leq \frac{1 + \delta/(kh(\delta))}{1 - 1/k} \tag{A11.14}$$

and, by a change of variable $w' = w/h(\delta)$, the second term of (A11.13) satisfies

$$\frac{1}{h(\delta)} \int_{lh(\delta)}^{kh(\delta)} \Pr\left[1/v > \frac{1/h(\delta) - 1/w}{1 + \delta/w}\right] dw =$$

$$= h(\delta) \int_{l}^{k} \frac{1}{h(\delta)} \left(1 - \exp\left[-\frac{\lambda_v(h(\delta) + \delta/w')}{(1 - 1/w')}\right]\right) dw' \tag{A11.15}$$

$$\leq h(\delta) \int_{l}^{k} \lambda_v \left(\frac{1 + \delta/(w'h(\delta))}{(1 - 1/w')}\right) dw'$$
$$\underbrace{\qquad\qquad\qquad\qquad\qquad\qquad\qquad}_{B(\delta, h(\delta), k, l)}$$

where $B(\delta, h(\delta), k, l)$ remains finite for any $k > l > 1$ as $\delta \to 0$.
Combining Equations (A11.12), (A11.14), and (A11.15), gives

$$\frac{1}{h(\delta)} \Pr[r_\delta < h(\delta)] \leq \lambda_w l + \lambda_v \left(\frac{1 + \delta/(kh(\delta))}{1 - 1/k}\right) + h(\delta) B(\delta, h(\delta), k, l) \tag{A11.16}$$

and

$$\limsup_{\delta \to 0} \frac{1}{h(\delta)} \Pr[r_\delta < h(\delta)] \leq \lambda_w l + \lambda_v \left(\frac{1 + d/k}{1 - 1/k}\right)$$

since $\lim_{\delta \to 0} B(\delta, h(\delta), k, l) < \infty$ and, by assumption, $h(\delta) \to 0$ and $\delta/h(\delta) \to d$ as $\delta \to 0$. The constants $k > l > 1$ are arbitrary. In particular, k can be chosen arbitrarily large, and l arbitrarily close to 1. Hence,

$$\limsup_{\delta \to 0} \frac{1}{h(\delta)} \Pr[r_\delta < h(\delta)] \leq \lambda_w + \lambda_v \tag{A11.17}$$

Combining Equation (A11.10) with (A11.17), the lemma is proved.
Proof of Equality 1:

$$\Pr[u + \varepsilon f(v/\varepsilon, w/\varepsilon) < g(\varepsilon)] = \Pr[u + r_\varepsilon < g(\varepsilon)] = \int_{0}^{g(\varepsilon)} \Pr[r_\varepsilon < g(\varepsilon) - u] p_u(u) \, du$$

$$= g(\varepsilon) \int_{0}^{g(\varepsilon)} \Pr[r_\varepsilon < g(\varepsilon)(1 - u')] \lambda_u \, e^{-\lambda_u g(\varepsilon)u'} \, du' \tag{A11.18}$$

$$= g^2(\varepsilon) \int_{0}^{1} (1 - u') \frac{\Pr[r_\varepsilon < g(\varepsilon)(1 - u')]}{g(\varepsilon)(1 - u')} \times \lambda_u \, e^{-\lambda_u g(\varepsilon)u'} \, du'$$

where in the second equality we have used the change of variables $u' = u/g(\varepsilon)$. But by *Lemma 1* with $\delta = \varepsilon$ and $h(\delta) = g(\delta)(1 - u')$, the quantity in brackets approaches $\lambda_v + \lambda_w$ as $\varepsilon \to 0$, so we expect

$$\lim_{\varepsilon \to 0} \frac{1}{g^2(\varepsilon)} \Pr[u + r_\varepsilon < g(\varepsilon)] = \lambda_u(\lambda_v + \lambda_w) \int_{0}^{1} (1 - u) \, du = \frac{\lambda_u(\lambda_v + \lambda_w)}{2} \tag{A11.19}$$

To verify Equation (A11.19), fully we must utilize the lower and upper bounds developed in *Lemma 1*. Using the lower bound (A11.10), (A11.18) satisfies

$$\liminf_{\varepsilon \to 0} \frac{1}{g^2(\varepsilon)} \Pr[u + r_\varepsilon < g(\varepsilon)] \geq \lim_{\varepsilon \to 0} \int_0^1 \frac{1 - \exp[-(\lambda_v + \lambda_w)g(\varepsilon)(1 - u')]}{g(\varepsilon)}$$

$$\times \lambda_u \, e^{-\lambda_u g(\varepsilon)u'} \, du' = \lambda_u(\lambda_v + \lambda_w) \int_0^1 (1 - u') du' = \frac{\lambda_u(\lambda_v + \lambda_w)}{2}$$

(A11.20)

where the first equality results from the Dominated Convergence theorem [46] after noting that the integrand is both bounded by and converges to the function $\lambda_u (\lambda_v + \lambda_w)(1 - u')$. Using the upper bound (A11.17), Equation (A11.18) satisfies

$$\limsup_{\varepsilon \to 0} \frac{1}{g^2(\varepsilon)} \Pr[u + r_\varepsilon < g(\varepsilon)]$$

$$\leq \limsup_{\varepsilon \to 0} (\lambda_v/(1 - 1/k) + \lambda_w l) \int_0^1 (1 - u')\lambda_u \, e^{-\lambda_u g(\varepsilon)u'} \, du'$$

$$+ \limsup_{\varepsilon \to 0} \varepsilon/g(\varepsilon) \int_0^1 \lambda_v \lambda_u \, e^{-\lambda_u g(\varepsilon)u'}/(k - 1) \, du'$$

$$+ \limsup_{\varepsilon \to 0} g(\varepsilon)D(\varepsilon, g(\varepsilon), k, l)$$

(A11.21)

where the last equality results from the fact $\varepsilon/g(\varepsilon) \to c$ and the fact that $D(\varepsilon, g(\varepsilon), k, l) := \int_0^1 (1 - u')^2 B(\varepsilon, g(\varepsilon)(1 - u'), k, l)\lambda_u e^{-\lambda_u g(\varepsilon)u'} du'$ remains finite for all $k > l > 1$ even as $\varepsilon \to 0$. Again, the constants $k > l > 1$ are arbitrary. In particular, k can be chosen arbitrarily large, and l arbitrarily close to 1. Hence,

$$\limsup_{\varepsilon \to 0} \frac{1}{g^2(\varepsilon)} \Pr[u + r_\varepsilon < g(\varepsilon)] \leq \frac{\lambda_u(\lambda_v + \lambda_w)}{2}$$

(A11.22)

Combining Equation (A11.21) and (A11.22) completes the proof.

Equality 2. Let u and v be independent exponential random variables with parameters λ_u and λ_v, respectively. Let ε be positive and let $g(\varepsilon) > 0$ be continuous with $g(\varepsilon) \to 0$ as $\varepsilon \to 0$. Then for

$$h(\varepsilon) := \varepsilon^2[(g(\varepsilon)/\varepsilon + 1)\ln(g(\varepsilon)/\varepsilon + 1) - g(\varepsilon)/\varepsilon]$$

(A11.23)

we have

$$\lim_{\varepsilon \to 0} \frac{1}{h(\varepsilon)} \Pr[u + v + uv/\varepsilon < g(\varepsilon)] = \lambda_u \lambda_v$$

(A11.24)

and

$$\lim_{t \to t_0} \frac{1}{h(\varepsilon(t))} \Pr[u + v + uv/\varepsilon(t) < g(\varepsilon(t))] = \lambda_u \lambda_v$$

(A11.25)

if $\varepsilon(t)$ is continuous about $t = t_0$ with $\varepsilon(t) \to 0$ as $t \to t_0$.

Proof. First, we write CDF in the form

$$
\Pr[u + v + uv/\varepsilon < g(\varepsilon)]
$$

$$
= \int_0^\infty \Pr[u + v + uv/\varepsilon < g(\varepsilon) \,|\, v = v] p_v(v) \, dv
$$

$$
= \int_0^{g(\varepsilon)} \Pr\left[u < \frac{g(\varepsilon) - v}{1 + v/\varepsilon} \,\Big|\, v = v\right] \lambda_v \, e^{-\lambda_v v} \, dv \qquad\qquad \text{(A11.26)}
$$

$$
= \int_0^{g(\varepsilon)} \left[1 - \exp\left(-\lambda_u \left[\frac{g(\varepsilon) - v}{1 + v/\varepsilon}\right]\right)\right] \lambda_v \, e^{-\lambda_v v} \, dv
$$

$$
= g(\varepsilon) \int_0^1 \left[1 - \exp\left(-\lambda_u \left[\frac{g(\varepsilon)(1 - w)}{1 + g(\varepsilon)w/\varepsilon}\right]\right)\right] \lambda_v \, e^{-\lambda_v g(\varepsilon)w} \, dw
$$

where the last equality follows from the change of variables $w = v/g(\varepsilon)$. To upper-bound Equation (A11.26), we use the identities $1 - e^{-x} \le x$ for all $x \ge 0$ and $e^{-y} \le 1$ for all $y \ge 0$, so that (A11.26)

$$
\Pr[u + v + uv/\varepsilon < g(\varepsilon)] \le g^2(\varepsilon) \lambda_u \lambda_v \int_0^1 \frac{1 - w}{1 + g(\varepsilon)w/\varepsilon} \, dw
$$

$$
= \lambda_u \lambda_v g^2(\varepsilon) \frac{(g(\varepsilon)/\varepsilon + 1)\ln(g(\varepsilon)/\varepsilon + 1) - g(\varepsilon)/\varepsilon}{(g(\varepsilon)/\varepsilon)^2} = \lambda_u \lambda_v h(\varepsilon)
$$

and

$$
\limsup_{\varepsilon \to 0} \frac{1}{h(\varepsilon)} \Pr[u + v + uv/\varepsilon < g(\varepsilon)] \le \lambda_u \lambda_v. \qquad\qquad \text{(A11.27)}
$$

To lower-bound Equation (A11.26), we use the concavity of $1 - e^{-x}$, i.e. for any $t > 0$, $1\, 1 - e^{-x} \ge ((1 - e^{-t})/t)x$, for all $x \le t$ and the identity $e^{-y} \ge 1 - y$ for all $y \ge 0$, so that Equation (A11.26) becomes

$$
\Pr[u + v + uv/\varepsilon < g(\varepsilon)]
$$

$$
\ge g(\varepsilon) \int_0^1 \left[\left(\frac{1 - e^{-\lambda_u g(\varepsilon)}}{\lambda_u g(\varepsilon)}\right) \frac{\lambda_u g(\varepsilon)(1 - w)}{1 + wg(\varepsilon)/\varepsilon}\right] \lambda_v (1 - \lambda_v g(\varepsilon)w) \, dw
$$

$$
= \lambda_u \lambda_v g^2(\varepsilon) \left(\frac{1 - e^{-\lambda_u g(\varepsilon)}}{\lambda_u g(\varepsilon)}\right) \times \int_0^1 \left[\frac{1 - w}{1 + wg(\varepsilon)/\varepsilon}\right](1 - \lambda_v g(\varepsilon)w) \, dw
$$

$$
\ge \lambda_u \lambda_v g^2(\varepsilon) \left(\frac{1 - e^{-\lambda_u g(\varepsilon)}}{\lambda_u g(\varepsilon)}\right)(1 - \lambda_v g(\varepsilon)) \times \int_0^1 \frac{1 - w}{1 + wg(\varepsilon)/\varepsilon} \, dw
$$

$$
= \lambda_u \lambda_v g^2(\varepsilon) \left(\frac{1 - e^{-\lambda_u g(\varepsilon)}}{\lambda_u g(\varepsilon)}\right)(1 - \lambda_v g(\varepsilon)) \times \frac{(g(\varepsilon)/\varepsilon + 1)\ln(g(\varepsilon)/\varepsilon + 1) - g(\varepsilon)/\varepsilon}{(g(\varepsilon)/\varepsilon)^2}
$$

$$
= \lambda_u \lambda_v \left(\frac{1 - e^{-\lambda_u g(\varepsilon)}}{\lambda_u g(\varepsilon)}\right)(1 - \lambda_v g(\varepsilon)) h(\varepsilon)
$$

and

$$
\liminf_{\varepsilon \to 0} \frac{1}{h(\varepsilon)} \Pr[u + v + uv/\varepsilon < g(\varepsilon)]
$$

$$
\ge \lambda_u \lambda_v \lim_{\varepsilon \to 0} \left(\frac{1 - e^{-\lambda_u g(\varepsilon)}}{\lambda_u g(\varepsilon)}\right)(1 - \lambda_v g(\varepsilon)) = \lambda_u \lambda_v \qquad\qquad \text{(A11.28)}
$$

Since the bounds in Equations (A11.27) and (A11.28) are equal, *Equality 2* is proved.

Equality 3. For $f_t(s) \to g\{s\}$ pointwise as t$\to t_0$, and $f_t(s)$ monotone increasing in s for each t let $h_t(s)$ be such that $h_t(s) \le s$, $h_t(s) \to s$ pointwise as $t \to t_0$, and $h_t(s)/s$ is monotone decreasing in s for each t. For $\tilde{h}_t^{-1}(r) := \min h_t^{-1}(r)$ we have

$$\lim_{t \to t_0} f_t(\tilde{h}_t^{-1}(r)) = g(r) \qquad (A11.29)$$

Proof. Since $h_t(s) \le s$ for all t, we have $r \le \tilde{h}_t^{-}1(r)$, and consequently $f_t(r) \le f_t(\tilde{h}_t^{-1}(r))$ *because* $f_t(*)$) is monotone increasing. Thus,

$$\liminf_{t \to t_0} f_t(\tilde{h}_t^{-1}(r)) \ge g(r) \qquad (A11.30)$$

For upper bound fix $\delta > 0$. *Lemma 2* shows that for each r there exists t^* such that $\tilde{h}_t^{-1}(r) \le r/(1-\delta)$ *for all t such that* $|t - t_0| < |t^* - t_0|$. So

$$f_t(\tilde{h}_t^{-1}(r)) \le f_t(r/(1-\delta)) \quad \text{and} \quad \limsup_{t \to t_0} f_t(\tilde{h}_t^{-1}(r)) \le g(r/(1-\delta))$$

Since δ can be made arbitrarily small we have

$$\limsup_{t \to t_0} f_t(\tilde{h}_t^{-1}(r)) \le g(r) \qquad (A11.31)$$

Combining (A11.30) with (A11.31), we obtain the desired result.

The following *Lemma* is used in the proof of the upper bound of *Equality 3.*

Lemma 2. *For $h_{t(s)}$ such that $h_{t(s)} \le s$, $h_t(s) \le s$ pointwise as* t $\to t_0$, *and $h_t(s)/s$ is monotone decreasing in s for each* t, *define* $\tilde{h}_t^{-1}(r) := \min h_t^{-1}(r)$. *For each* $r_0 > 0$ *and any $\delta > 0$, there exists* t^* *such that $\tilde{h}_t^{-1}(r) < r_0/(1-\delta)$ for all t such that* $|t - t_0| < |t * -t_0|$.

Proof. Fix $r_0 > 0$ and $\delta > 0$, and select s_0 such that $s_0 > r_0/(1-\delta)$.

Because $h_t(s)/s \to 1$ point-wise as $t \to t_0$, for each $s > 0$ and any $\delta > 0$, there exists a t^* such that $h_t(s) \ge s(1-\delta)$, all $t : |t - t_0| < |t * -t_0|$.

Also, since $h_t(s)/s$ is monotone decreasing in s, if t^* is sufficient for convergence at s_0, then it is sufficient for convergence at all $s \le s_0$. Thus, for any, $s_0 > 0$ and $\delta > 0$ there exists a t^* such that $h_t(s) \ge s(1-\delta)$, all $s \le s_0, t : |t - t_0| < |t^* - t_0|$.

In the rest of the proof, we only consider $s \le s_0$ and t such that $|t - t_0| < |t * - t_0|$. Consider the interval $I = [r_0, r_0/(1-\delta)]$, and note that $s \in I$ implies $s < s_0$. Since $h_t(s) < s$, we have $h_{t(r_0)} < r_0$. Also, since $h_{t(s)} > s(1-\delta)$ by the above construction, we have $h_t(r_0/(1-\delta)) > r_0$. By continuity, $h_{t(s)}$ assumes all intermediate values between $h_t(r_0)$ and $h_t(r_0/(1-\delta))$ on the interval $(r_0, r_0/(1-\delta))$ (Theorem 4.23 in [47]), in particular, there exists an $s_1 \in (r_0, r_0/(1-\delta))$ such that $h_t(x_1) = r_0$. The result follows from $\tilde{h}_t^{-}1(r) \le x \le r_0/(1-\delta)$, where the first inequality follows from the definition of $\tilde{h}_t^{-}1(\cdot)$ and the second inequality follows from the fact that $x_1 \in I$.

Amplify-and-Forward Mutual Information

We write the equivalent channel (11.5), with relay processing, in vector form as

$$\mathbf{y}_d[n] = \mathbf{h}x_s[n] + \mathbf{B}\mathbf{n}[n]$$

where

$$\mathbf{y}_d[n] = \begin{bmatrix} y_d[n] \\ y_d[n + N/4] \end{bmatrix}; \qquad \mathbf{h} = \underbrace{\begin{bmatrix} hs, d \\ h_{r,d}\beta h_{s,r} \end{bmatrix}}_{H}$$

$$\mathbf{B} = \underbrace{\begin{bmatrix} 0 & 1 & 0 \\ h_{r,d}\beta & 0 & 1 \end{bmatrix}}_{B}; \qquad n[n] = \underbrace{\begin{bmatrix} n_r[n] \\ n_d[n] \\ n_d[n + N/4] \end{bmatrix}}_{n[n]}$$

The source signal has power constraint $E[x_s] \le P_s$, and relay amplifier has constraint

$$\beta \le \sqrt{\frac{P_r}{|h_{s,r}|^2 P_s + N_r}} \tag{A11.32}$$

and the noise has covariance $E[\mathbf{nn}^T] = \mathrm{diag}(N_r, N_d, N_d)$. Since the channel is memoryless, the average mutual information satisfies

$$I_{AF} \le I(\mathbf{x}_s; \mathbf{y}_d) \le \log \det \left(\mathbf{I} + (P_s \mathbf{hh}^T)(\mathbf{B}E[\mathbf{nn}^T]\mathbf{B}^T)^{-1} \right)$$

with equality for $\mathbf{x}_s$, zero-mean, circularly symmetric complex Gaussian. Noting that

$$\mathbf{hh}^T = \begin{bmatrix} |h_{s,d}|^2 & h_{s,d}(h_{r,d}\beta h_{s,r})^* \\ h_{s,d}^* h_{r,d}\beta h_{s,r} & |h_{r,d}\beta h_{s,r}|^2 \end{bmatrix}$$

$$\mathbf{B}E[\mathbf{nn}^T]\mathbf{B}^T = \begin{bmatrix} N_d & 0 \\ 0 & |h_{r,d}\beta|^2 N_r + N_d \end{bmatrix}$$

we have

$$\det \left(\mathbf{I}_2 + (P_s \mathbf{hh}^T)(\mathbf{B}E[\mathbf{nn}^T]\mathbf{B}^T)^{-1} \right)$$

$$= 1 + \frac{P_s |h_{s,d}|^2}{N_d} + \frac{P_s |h_{r,d}\beta h_{s,r}|^2}{\left(|h_{r,d}\beta|^2 N_r + N_d \right)} \tag{A11.33}$$

Because (A11.33) is increasing in β, the amplifier power constraint (A11.32) should apply, yielding,

$$I_{AF} = \log(1 + |h_{s,d}|^2 y_{s,d} + f\left(|h_{s,r}|^2 y_{s,r}, |h_{r,d}|^2 y_{s,d} \right)$$

where $y_{a,b} = SNR_{a,b}$ and is $f(\cdot, \cdot)$ given in Section 11.2.3

Input Distributions for Transmit Diversity Bound

An equivalent channel model for the two-antenna case can be summarized as

$$y[n] = \underbrace{\begin{bmatrix} h_1 & h_2 \end{bmatrix}}_{h} \underbrace{\begin{bmatrix} x_1[n] \\ x_2[n] \end{bmatrix}}_{x[n]} + n[n] \tag{A11.34}$$

where $\mathbf{h}$ represents the fading coefficients and $\mathbf{x}[n]$ the transmit signals from the two transmit antennas, and $\mathbf{n}[n]$ is a zero-mean, white complex Gaussian process with variance N_0 that captures the effects of noise and interference. If $\mathbf{Q} = E\,[\mathbf{x}\mathbf{x}^T]$ is the covariance matrix for the transmit signals, then the power constraint on the inputs may be written in the form $tr(\mathbf{Q}) \leq P$.

We want to find a distribution on the input vector $\mathbf{x}$, subject to the power constraint, that minimizes outage probability, i.e.,

$$\min_{p_X:\,tr(Q)\leq P} \Pr[I(x;y\,|h=a\,) < R] \tag{A11.35}$$

The optimization (A11.35) can be restricted to optimization over zero-mean, circularly symmetric complex Gaussian inputs, because Gaussian codebooks maximize the mutual information for each value of the fading coefficients a, i.e.

$$\min_{Q:\,tr(Q)\leq P} \Pr\left[\log\left(1 + \frac{\mathbf{h}\mathbf{Q}\mathbf{h}^T}{N_0}\right) < R\right] \tag{A11.36}$$

We now argue that $\mathbf{Q}$ diagonal is sufficient, even if the components of a are independent but not identically distributed. We write $\mathbf{h} = \tilde{h}\mathbf{\Sigma}$, where $\tilde{h}$ is a zero-mean, i.i.d. complex Gaussian vector with unit variances and $\mathbf{\Sigma} = \mathrm{diag}(\sigma_1, \sigma_2)$. Thus, the outage probability in (A11.36) may be written as:

$$\Pr\left[\log\left(1 + \frac{\tilde{\mathbf{h}}\mathbf{\Sigma}\,\mathbf{Q}\mathbf{\Sigma}^T\tilde{\mathbf{h}}^T}{N_0}\right) < R\right]$$

Now consider an eigendecomposition of the matrix $\mathbf{\Sigma}Q\mathbf{\Sigma}^T = \mathbf{U}\mathbf{D}\mathbf{U}^T$, where $\mathbf{U}$ is unitary and $\mathbf{D}$ is diagonal. Using the fact that the distribution of $\tilde{h}$ is rotationally invariant, i.e., $\mathbf{h}\mathbf{U}$ has the same distribution as $\tilde{h}$ for any unitary $\mathbf{U}$, we observe that the outage probability for covariance matrix $\mathbf{\Sigma}Q\mathbf{\Sigma}^T$ is the same as the outage probability for the diagonal matrix $\mathbf{D}$.

For $\mathbf{D} = \mathrm{diag}(d_1, d_2)$, the outage probability can be written in the form

$$\Pr\left[d_1\,|h_1|^2 + d_2\,|h_2|^2 < \frac{2^R - 1}{\mathrm{SNR}}\right]$$

which, using *Result 2*, decays in proportion to $1/(\mathrm{SNR}^2 \det D)$ for large SNR if $d_1, d_2 \neq 0$. Thus, minimizing the outage probability for large SNR is equivalent to maximizing

$$\det\quad \mathbf{D} = \det\mathbf{\Sigma}\mathbf{Q}\mathbf{\Sigma}^T = \sigma_1^2\sigma_2^2(Q_{1,1}Q_{2,2} - |Q_{1,2}|^2) \tag{A11.37}$$

such that $Q_{1,1} + Q_{2,2} \leq P$. Clearly, (A11.37) is maximized for $Q_{1,1} = Q_{2,2} = P/2$ and $Q_{1,2} = Q_{2,1} = 0$. Thus, zero-mean, i.i.d. complex Gaussian inputs minimize the outage probability in the high-SNR regime.

Protocol I and II Under Repetition-Coding

The expressions for the mutual information of protocol I and protocol II (assuming that source-relay link is not limiting the total expressions) depend on,

$$C_b^I = \log\det\left(\mathbf{I}_{2n_d} + \frac{P_s}{\sigma^2 n_s}\begin{bmatrix}\mathbf{H}_0 \\ \varphi\mathbf{H}_2\end{bmatrix}\begin{bmatrix}\mathbf{H}_0^H & \varphi\mathbf{H}_2^H\end{bmatrix}\right) \tag{A11.38}$$

$$C_b^{II} = \log\det\left(\mathbf{I}_{n_d} + \frac{P_s}{\sigma^2 n_s}\begin{bmatrix}\mathbf{H}_0 & \varphi\mathbf{H}_2\end{bmatrix}\begin{bmatrix}\mathbf{H}_0^H \\ \varphi\mathbf{H}_2^H\end{bmatrix}\right) = \log\det\left(\mathbf{I}_{n_d} + \frac{P_s}{\sigma^2 n_s}\left(\mathbf{H}_0\mathbf{H}_0^H + \varphi^2\mathbf{H}_2\mathbf{H}_2^H\right)\right) \tag{A11.39}$$

with $\mathbf{H}_0$ and $\mathbf{H}_2$ of dimensions $n_d \times n_s$ and $n_d \times n_r$, respectively, and σ^2 the noise power. Because we are considering repetition coding, $n_s = n_r$ by definition in Section 11.4.1.2. The motivation of this section is obtain the conditions when protocol I is better than protocol II. The following determinant property will be considered,

$$\det\left(\mathbf{I}_n + \mathbf{AB}\right) = \det\left(\mathbf{I}_m + \mathbf{BA}\right) \tag{A11.40}$$

where $\mathbf{A}$ and $\mathbf{B}$ are matrices of dimensions $n \times m$ and $m \times n$, respectively.

Applying the determinant property (A11.40) to the expression for protocol I, (A11.38), it can be rewritten as,

$$C_b^I = \log \det\left(\mathbf{I}_{n_s} + \frac{P_s}{\sigma^2 n_s}\left[\mathbf{H}_0^H \varphi \mathbf{H}_2^H\right]\begin{bmatrix}\mathbf{H}_0 \\ \varphi \mathbf{H}_2\end{bmatrix}\right) = \log \det\left(\mathbf{I}_{n_s} + \frac{P_s}{\sigma^2 n_s}\left(\mathbf{H}_0^H \mathbf{H}_0 + \varphi^2 \mathbf{H}_2^H \mathbf{H}_2\right)\right) \tag{A11.41}$$

- If $n_s = n_d$,
  For this configuration, the eigenvalues of $\mathbf{H}_0 \mathbf{H}_0^H$ and $\mathbf{H}_2 \mathbf{H}_2^H$ are the same as the eigenvalues of $\mathbf{H}_0^H \mathbf{H}_0$ and $\mathbf{H}_2^H \mathbf{H}_2$, respectively. Therefore, equations (A11.39) and (A11.41) are equal, $C_b^I = C_b^{II}$
- If $n_s < n_d$
  In this case, the mutual information for protocol II (A11.39) can be written in the following way by using the property (A11.40),

$$C_b^{II} = \log \det\left(\mathbf{I}_{2n_s} + \frac{P_s}{\sigma^2 n_s}\begin{bmatrix}\mathbf{H}_0^H \mathbf{H}_0 & \varphi \mathbf{H}_0^H \mathbf{H}_2 \\ \varphi \mathbf{H}_2^H \mathbf{H}_0 & \varphi^2 \mathbf{H}_2^H \mathbf{H}_2\end{bmatrix}\right) \tag{A11.42}$$

For $n_s = 1$ and $n_d > 1$, the previous equation is given by,

$$C_b^{II} = \log \det\left(\mathbf{I}_2 + \frac{P_s}{\sigma^2 n_s}\begin{bmatrix}\mathbf{h}_0^H \mathbf{h}_0 & \varphi \mathbf{h}_0^H \mathbf{h}_2 \\ \varphi \mathbf{h}_2^H \mathbf{h}_0 & \varphi^2 \mathbf{h}_2^H \mathbf{h}_2\end{bmatrix}\right) = \log\left(1 + \frac{P_s}{\sigma^2}\left(\mathbf{h}_0^H \mathbf{h}_0 + \varphi^2 \mathbf{h}_2^H \mathbf{h}_2 + \varphi^2 \Delta_1\right)\right)$$

with $\mathbf{h}_0$ and $\mathbf{h}_2$ column vectors that contains the channel coefficients of the source-destination and relay-destination link, and Δ_1 given by,

$$\Delta_1 = \mathbf{h}_0^H \mathbf{h}_0 \mathbf{h}_2^H \mathbf{h}_2 - \mathbf{h}_0^H \mathbf{h}_2 \mathbf{h}_2^H \mathbf{h}_0 \tag{A11.43}$$

However, the previous equation can also been defined in the following way,

$$\Delta_1 = \det\left(\begin{bmatrix}\mathbf{h}_0^H \mathbf{h}_0 & \mathbf{h}_0^H \mathbf{h}_2 \\ \mathbf{h}_2^H \mathbf{h}_0 & \mathbf{h}_2^H \mathbf{h}_2\end{bmatrix}\right) = \det\left(\Psi\right) \tag{A11.44}$$

where the matrix Ψ is semi-definite positive by construction, so Δ_1 is always greater or equal to 0. Likewise, that means that protocol II (A11.42) provides larger mutual information values than protocol I (A11.41), $C_b^I \leq C_b^{II}$, for such antenna configuration.

When $n_d > 1$, it is more difficult to prove analytically the previous proposition. For that reason, in the following we will show experimental results that confirms the previous assert, investigating the values of

$$\phi = C_b^{II} - C_b^I \tag{A11.45}$$

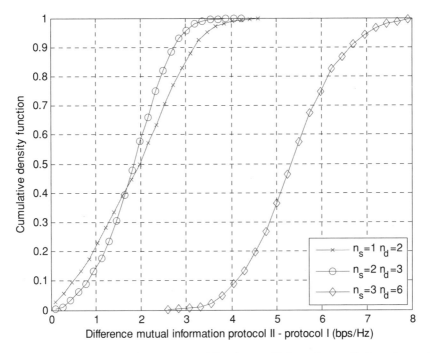

Figure 11.35 CDF of the difference between protocol II and protocol I. $\{n_s = 1, n_d = 2\}$, $\{n_s = 2, n_d = 3\}$ and $\{n_s = 3, n_d = 6\}$. $P_s/ = 10$.

Figure 11.35 presents the cumulative density function (CDF) of ϕ (Pr($\phi < x_0$) with x_0 the value of the x-axis defined in (A11.45). It has assumed a large number of independent Rayleigh fading realizations for the channel, $P_s/\sigma^2 = 10$ and different antenna configurations $\{n_s = 1, n_d = 2\}$ (crosses), $\{n_s = 2, n_d = 3\}$ (circles) and $\{n_s = 3, n_d = 6\}$ (diamonds). For all those configurations where $n_s < n_d$, ϕ is always greater than 0. For example for $\{n_s = 2, n_d = 3\}$ the 90% of different channel realizations the variable $\phi > 1$ bps/Hz, while for $\{n_s = 3, n_d = 6\}$ $\phi > 4$ bps/Hz.

- If $n_s > n_d$

  For this antenna configuration, similar steps can be done as in the previous section, but now assuming that the mutual information of protocol I can be written as,

$$C_b^I = \log \det \left(I_{2n_d} + \frac{P_s}{\sigma^2 n_s} \right) \begin{bmatrix} H_0 H_0^H & \varphi H_0 H_2^H \\ \varphi H_2 H_0^H & \varphi^2 H_2 H_2^{bf\,H} \end{bmatrix} \tag{A11.46}$$

and this equation has to be compared with protocol II, C_b^{II} defined in (A11.39).
For $n_d = 1$ and $n_s > 1$, the previous equation is given by,

$$C_b^I = \log \det \left(I_2 + \frac{P_s}{\sigma^2 n_s} \begin{bmatrix} h_0^H h_0 & \varphi h_0^H h_2 \\ \varphi h_2^H h_0 & \varphi^2 h_2^H h_2 \end{bmatrix} \right) = \log \left(1 + \frac{P_s}{\sigma^2 n_s} \left(h_0^H h_0 + \varphi^2 h_2^H h_2 + \varphi^2 \Delta_1 \right) \right)$$

where h_0 and h_2 stand for column vectors that contains the channel coefficients of the source-destination and relay-destination link. In fact, that channel coefficients are in the row of matrix H_0 and H_2 of dimensions ($n_d \times n_s$) of (A11.46). Additionally, Δ_1 given by,

$$\Delta_1 = \mathbf{h}_0^H \mathbf{h}_0 \mathbf{h}_2^H \mathbf{h}_2 - \mathbf{h}_0^H \mathbf{h}_2 \mathbf{h}_2^H \mathbf{h}_0 \tag{A11.47}$$

which also can be obtained as a result of the determinant of a semi-definite positive matrix, as in (A11.44). The previous equation has to be compared with C_b^{II},

$$C_b^{II} = \log \det \left(1 + \frac{P_s}{\sigma^2 n_s} \left(\mathbf{h}_0^H \mathbf{h}_0 + \varphi^2 \mathbf{h}_2^H \mathbf{h}_2 \right) \right)$$

Therefore, the comparison between C_b^I and C_b^{II} follow similar guidelines as in the previous case, although now for $n_s > n_d$, $C_b^I \geq C_b^{II}$.

Interference Power Model

The complete proof of the interference power model can be found in [77], but here we sketch the basic aspects to be considered. Assuming the scenario defined in Section 11.6.1 and the assumptions of Section 11.6.3 let Y_a be the interference power received from the active relays in D_a (a disk of radius a),

$$Y_a = \sum_{r_i \leq a} \eta(K_i, r_i, r) = K_i \sum_{r_i \leq a} g(r_i, r) \tag{A11.48}$$

where K_i is a random variable modeling the multipath fading or shadowing channel gain that affects the interference and $g(.)$ (see Equation (11.172)) is a function which models the relation between the interfering power and the distance between the destination and the interfering relay, r_i. The function η will be a piecewise function defined as,

$$\eta(K_i, r_i, r) = \begin{cases} \eta_1(K_i, r_i, r) = K_i g(r_i, r) = K_i \dfrac{N_0 SNR}{K_r r_i^\gamma} r^\gamma & r \leq \Delta \\[2ex] \eta_2(K_i, r_i) = K_i g(r_i) = K_i \dfrac{P_{MAX}}{r_i^\gamma} & otherwise \end{cases} \tag{A11.49}$$

where Δ is defined in (11.172).

Taking into account the distribution of the users and their contribution of the interference, the characteristic function of Y may be evaluated as,

$$\phi_Y(\omega) = \lim_{a \to \infty} \left(\mathbf{E} \left\{ \mathbf{E} \left\{ e^{j\omega Y_a} | k \text{ in } D_a \right\} \right\} \right) =$$

$$\lim_{a \to \infty} \left(\sum_{k=0}^{\infty} \frac{(\lambda_{active} \pi a^2)^k}{k!} \exp\left(-\lambda_{active} \pi a^2\right) \mathbf{E} \left\{ e^{j\omega Y_a} | k \text{ in } D_a \right\} \right)$$

$$\lim_{a \to \infty} \left(\exp \left(\int_0^\infty \lambda_{active} \pi a^2 \left(\int_0^\infty \int_0^\infty \exp(j\omega \eta(K_i, r_i, r)) f_{r_i, r}(r_i, r) \, dr_i dr - 1 \right) f_{K_i}(K_i) dK_i \right) \right) \tag{A11.50}$$

The previous function can be integrated by parts due to the piece-wise linear function η, (A11.49), where it is exploited the independence between random variables r_i (distance of interfering relay to a destination) and r (distance of each assisting relay to its associated destination). Some of the obtained results can be neglected when we evaluate $a \to \infty$.

$$\phi_Y(\omega) = \lim_{a \to \infty} \exp\left(\int_0^\infty \lambda_{active} \pi a^2 (A(K_i) + B(K_i) - 1) f_{K_i}(K_i) dK_i\right)$$

$$A(K_i) = \int_0^\Delta f_r(r) dr \int_0^\infty \exp(j\omega\eta_1(K_i, r_i, r)) f_{ri}(r_i) dr_i \qquad (A11.51)$$

$$B(K_i) = \int_\Delta^\infty f_r(r) dr \int_0^\infty \exp(j\omega\eta_2(K_i, r_i)) f_{ri}(r_i) dr_i$$

$$f_{ri}(r_i) = \begin{cases} 2r_i/a^2 & r_i \le a \\ 0 & \text{otherwise} \end{cases}$$

References

[1] Glisic, S. G. (2007) *Advanced Wireless Communications. 4G Cognitive and Cooperative Broadband Technology.* 2e John Wiley & Sons, Ltd.

[2] Agustin, A. (2008) Relay-assisted transmission and Radio Resource Management for wireless Networks, PhD Thesis. Universitat Politecnica de Catalunya (UPC) (Technical University of Catalonia). May. Barcelona. Spain. Available on: http://www.tdx.cat/TDX-0701108-121341/.

[3] Telatar, I. E. (1999) Capacity of multi-antenna Gaussian channels, *European Transactions on Telecommunications*, **10**, 585–595.

[4] Foschini, G. and Gans, M. J. (1998) On limits of wireless communications in a fading environment when using multiple antennas, *Wireless Personal Communications*, **6**(3), 311–335.

[5] Zheng, L. and Tse, D. N. C. (2003) Diversity and multiplexing: A fundamental tradeoff in multiple antenna channels, *IEEE Trans. on Information Theory*, **49**(5), 1073–1096.

[6] Sendonaris, A., Erkip, E. and Aazhang, B. (2003) User cooperation diversity-part I: System description, *IEEE Trans. on Communications*, **51**(11), 1927–1938.

[7] Laneman, J., Tse, D. N. C. and Wornell, G. W. (2004) Cooperative diversity in wireless networks: Efficient protocols and outage behavior, *IEEE Trans. Information Theory*, **50**(12), 3062–3080.

[8] Wittneben, A. and Rankov, B. (2003) Impact of cooperative relays on the capacity of rank deficient MIMO channels, in *Proc. IST Mobile & Wireless Communications Summit (IST-2003)*, Aveiro (Portugal), June.

[9] Dohler, M. (2003) Virtual Antenna Arrays, PhD Thesis, King's College London, London, UK.

[10] Dohler, M., Lefranc, E. and Aghvami, A. H. (2002) Virtual Antenna Arrays for Future Wireless Mobile Communication Systems, *ICT 2002, Conference CD-ROM*, Beijing, China, June.

[11] van der Meulen, E. C. (1971) Three-terminal communication channels, *Adv. Appl. Prob.*, **3**, 120–154.

[12] TCover, T. M. and El Gamal, A. A. (1979) Capacity theorems for the relay channel, *IEEE Trans. on Information Theory*, **25**(5), 474–584, Sept.

[13] Slepian, D. S. and Wolf, J. K. (1973) Noseless coding of correlated information sources, *IEEE Trans. on Information Theory*, **19**, 471–489.

[14] Wyner, A. D. (1975) On source coding with side information at the decoder, *IEEE Trans. on Information Theory*, **21**, 294–300.

[15] Wyner, A. D. and Ziv, J. (1976) The rate-distortion function for source coding with side information at the decoder, *IEEE Trans. on Information Theory*, **22**(6), 1–10.

[16] Cover, T. M. and Thomas, J. A. (1991) *Elements of Information Theory.* John Wiley & Sons.

[17] Host-Madsen, A. and Zhang, J. (2005) Capacity bounds and power allocation for wireless relay channels, *IEEE Trans. on Information Theory*, **51**(6), 2020–2040.

[18] Wang, B., Zhang, J. and Host-Madsen, A. (2005) On the capacity of MIMO relay channels, *IEEE Trans. on Information Theory*, **51**(1), 29–43.

[19] King, R. C. (1978) Multiple Access Channels with Generalized feedback, PhD dissertation, Stanford University, Palo Alto.

[20] Carleial, A. B. (1982) Multiple-access channels with different generalized feedback signals, *IEEE Trans. on Information Theory*, **28**, 841–850.

[21] Willems, F. M. J., van der Meulen, E. C. and Schalkwijk, J. P. M. (1983) An achievable rate region for the multiple access channel with generalized feedback, in *Proc. Allerton Conference Communications, Control and Computing*, Monticello.

[22] Willems, F. M. J. (1983) The discrete memoryless multiple access channel with partially cooperating encoders, *IEEE Trans. on Information Theory*, **29**(3), 441–445.

[23] Willems, F. M. J. and van der Meulen, E. C. (1985) The discrete memoryless multiple access channel with cribbing encoders, *IEEE Trans. on Information Theory*, **31**, 313–327.

[24] Host-Madsen, A. (2006) Capacity bounds for cooperative diversity, *IEEE Trans. on Information Theory*, **52**(4), 1522–1544.

[25] Jindal, N., Mitra, U. and Goldsmith, A. (2004) 'Capacity of ad-hoc networks with node cooperation', in *Proc. IEEE International Symposium of Information Theory (ISIT-2004)*, Chicago (USA), June.

[26] Host-Madsen, A. and Nosratinia, A. (2005) The multiplexing gain of wireless networks, in *Proc. IEEE International Symposium of Information Theory (ISIT-2005)*, Adelaide, Australia, 2310–2314.

[27] Sendonaris, A., Erkip, E. and Aazhang, B. (2003) User cooperation diversity-part II: Implementation aspects and performance analysis, *IEEE Trans. on Communications*, **51**(11), 1939–1948.

[28] Muñoz, O., Vidal, J. and Agustin, A. (2007) Linear transceiver design in nonregenerative relays with channel state information, *IEEE Trans. on Signal Processing*, **44**(6), 2593–2604.

[29] Laneman, J. N. and Wornell, G. W. (2003) Distributed Space-Time Coded protocols for exploiting Cooperative Diversity in Wireless Networks, *IEEE Trans. on Information Theory*, **49**(10), 2415–2425.

[30] Valenti, M. C. and Zhao, B. (2003) Distributed turbo codes: Towards the capacity of the relay channel, in *Proc. IEEE Vehicular Technology Conf. Fall (VTC-Fall 2003)*, Orlando, FL.

[31] Agustin, A., Vidal, J. and Muñoz, O. (2005) Hybrid turbo FEC/ARQ systems and distributed space-time coding for cooperative transmission, *International Journal of Wireless Information Networks (IJWIN)*, **12**(4), 263–280.

[32] Hunter, T. E. and Nosratinia, A. (2002) Cooperation diversity through coding, in *Proc. IEEE International Symposium of Information Theory, (ISIT-2002)*, Laussane, Switzerland.

[33] Hunter, T. E., Nosratinia, A., Hedayat, A. and Janani, M. (2004) Coded cooperation in wireless communications: space-time transmission and iterative decoding, *IEEE Trans. on Signal Processing*, **52**(2), 362–371.

[34] Nosratinia, A., Hunter, T. E., Hedayat, A. and Janani, M. (2004) Cooperative communications in wireless networks, *IEEE Communications Magazine*, **42**(10), 74–80.

[35] Kramer, G., Gastpar, M. and Gupta, P. (2005) Cooperative Strategies and Capacity Theorems for Relay Networks, *IEEE Trans. on Information Theory*, **51**(9).

[36] Simoens, S., Muñoz, O., Vidal, J. and Del Coso, A. (2009) On the Gaussian MIMO Relay channel with full Channel State Information, *IEEE Trans. on Signal Processing*, **57**(9), 3588–3599.

[37] Larsson, P., Johansson, N. and Sunell, K. E. (2006) Coded bi-directional relaying, in *Proc. IEEE Vehicular Technology Conf. (VTC-FALL)*, May, Melbourne, Australia.

[38] Rankov, B. and Wittneben, A. (2007) Spectral efficient protocols for half-duplex fading relay channels, *IEEE Journal Selec. Areas in Communications*, **25**(2).

[39] Knopp, R. (2006) Two-Way Radio Networks with a Star Topology, in *Proc. IEEE Intl. Zurich Seminar on Communications (IZS)*, Feb.

[40] Nabar, R., Bölcskei, H. and Kneubühler, F. (2004) Fading relay channels: performance limits and space-time signal design, *IEEE Journal Selected Areas Communications (JSAC)*, **22**(6), 1099–1109.

[41] Ochiani, H., Mitran, P. and Tarokh, V. (2006) Variable rate two phase collaborative communications protocols for wireless networks, *IEEE Trans. on Information Theory*, **52**(9), 4299–4313.

[42] Dohler, M., Gkelias, A. and Aghvami, A. H. (2004) Resource allocation for FDMA-based regenerative multihop links, *IEEE Trans. on Wireless Communications*, **3**(6), 1989–1993.

[43] Ozarow, L., Shamai, S. and Wyner, A. (1994) Information theoretic considerations for cellular mobile radio, *IEEE Transaction on Vehicular Technology*, **43**, 359–377.

[44] Barbarossa, S. and Cerquetti, F. (2001) Simple space–time coded SS-CDMA systems capable of perfect MUI/ISI elimination, *IEEE Communications Letters*, **5**, 471–473.

[45] Proakis, J. (1989) *Digital Communications*, 3rd ed. McGraw-Hill, New York, p. 783.

[46] Adams, M. and Guillemin, V. (1996) *Measure Theory and Probability*. Birkhäuser, Boston, MA.

[47] Rudin, W. (1964) *Principles of Mathematical Analysis*, 2nd ed. McGraw-Hill, New York.

[48] Palomar, D. P., Agustin, A., Muñoz, O. and Vidal, J. (2004) Decode and Forward Protocol for Cooperative Diversity in Multi-Antenna Wireless Networks, in *Proc. IEEE Annual Conference on Information Sciences and Systems (CISS-2004)*, Princeton, NJ (USA), Feb. 17–19.

[49] Wolfowitz, J. (1978) *Coding Theorems of Information Theory*, 3rd ed. Berlin, Germany. Springer-Verlag.

[50] Lapidoth, A. and Narayan, P. (1998) Reliable communication under channel uncertainty, *IEEE Trans. on Information Theory*, **44**(6), 2148–2177.

[51] Azarian, K., El Gamal, H. and Schniter, P. (2005) On the achievable diversity-multiplexing tradeoff in half-duplex cooperative channels, *IEEE Trans. on Information Theory*, **51**(12), 4152–4172.

[52] Zhao, B. and Valenti, M. C. (2003) Some new adaptive protocols for the wireless relay channel, in *Proc. Allerton Conference on Communications, Control and Computing*, Monticello, IL.

[53] Liang, Y. and Veeravalli, V. V. (2005) Gaussian orthogonal relay channels: optimal resource allocation and capacity, *IEEE Trans. on Information Theory*, **51**(9), 3284–3289.

[54] Khormuji, M. N. and Larsson, E. G. (2007) Analytical results on block length optimization for decode-and-forward relaying with CSI feedback, in *Proc. 8th IEEE Workshop on Signal Processing Advances in Wireless Communications (SPAWC-2007)*, Helsinki, Finland.

[55] Tarokh, V., Jafarkhani, H. and Calderbank, A. R. (1999) Space-Time block codes from orthogonal designs, *IEEE Trans. on Information Theory*, **45**, 1456–1467.

[56] Vivier, E. (2009) *Radio Resource Management in WiMAX. From Theoretical Capacity to System Simulations.* ISTE Ltd and John Wiley & Sons, Inc., Chapter 8.

[57] Boyd, S. and Vandenberghe, L. (2004) *Convex Optimization*. Cambridge University Press.

[58] Kramer, G., Gastpar, M. and Gupta, P. (2005) Cooperative strategies and capacity theorems for relay networks, *IEEE Trans. on Information Theory*, **51**(9), 3037–3063.

[59] Reznik, A., Kulkarni, S. R. and Verdu, S. (2004) Degraded Gaussian Multirelay channel: Capacity and Optimal Power allocation, *IEEE Trans. on Information Theory*, **50**(12), 3037–3046.

[60] Rankov, A. and Wittneben, A. (2007) Spectral efficient protocols for half-duplex fading relay channels, *IEEE Journal on Sel. Areas in Communications*, **25**(2), 379–389.

[61] Fan, Y., Wang, C., Thompson, J. and Vincent Poor, H. (2007) Recovering multiplexing loss through successive relaying using repetition coding, *IEEE Trans. on Wireless Communications*, **6**(12), 4484–4493.

[62] Xue, F. and Sandhu, S. (2007) Cooperation in a Half-Duplex Gaussian Diamond Relay channel, *IEEE Trans. on Information Theory*, **53**(10).

[63] Agustin, A. and Vidal, J. (2008) Resource optimization in the decode-and-forward multiple relay-assisted channel, *Proc. 9th IEEE Workshop on Signal Processing Advances in Wireless Communications (SPAWC-2008)*, Recife-Pernambuco, Brazil.

[64] Weingarten, H., Steinberg, Y. and Shamai, S. (2006) On the Capacity Region of the Multi-Antenna Broadcast Channel with Common Messages, in *Proc. IEEE International Symposium on Information Theory (ISIT-2006)*, Seattle, USA, July.

[65] Vishwanath, S., Jindal, N. and Goldsmith, A. (2004) On the duality of Gaussian Multiple-Access and Broadcast channels, *IEEE Trans. on Information Theory*, **50**(5), 768–783.

[66] Larsson, P., Johansson, N. and Sunell, K. E. (2006) Coded bi-directional relaying, in *Proc. IEEE Vehicular Technology Conf. (VTC-Fall)*, May, Melbourne, Australia.

[67] Rankov, B. and Wittneben, A. (2006) Achievable rate regions for the two-way relay channel, in *Proc. IEEE International Symposium on Information Theory (ISIT)*, July, Seattle, USA.

[68] Knopp R. (2006) Two-way radio networks with a star topology, in *Proc. IEEE Intl. Zurich Seminar on Communications (IZS)*, Feb., Zurich.

[69] Popovski, P. and Yomo, H. (2007) Physical Network Coding in Two-way wireless relay channels, *IEEE Proc. of Intl. Conf. on Communications (ICC)*, Jun.

[70] Xie, L. L. (2007) Network coding and random binning for multi-user channels, in *Proc. IEEE Canadian Workshop on Information Theory (CWIT)*, Edmonton, Canada, June.

[71] Oechtering, T. J, Schnurr, C., Bjelakovic, I. and Boche, (2008) Broadcast capacity region of two-phase bidirectional relaying, *IEEE Trans. on Information Theory*, **54**(1).

[72] Agustin, A., Vidal, J. and Muñoz, O. (2009) Protocols and resource allocation for the two-way relay channel with half-duplex terminals, in *Proc. IEEE Intl. Conf. on Communications (ICC)*, Dresden, Germany, June.

[73] Holma, H. and Toskala, A. (2004) *WCDMA for UMTS : Radio Access for Third Generation Mobile Communications*, John Wiley & Sons.

[74] Dohler, M. (2003) Virtual Antenna Arrays, PhD Thesis, King's College London, London, UK.

[75] Scutari, G., Barbarossa, S. and Ludovici, D. (2003) Cooperation diversity in multihop wireless networks using opportunistic driven multiple access, in *Proc. IEEE 3rd Workshop on Signal Processing Advances in Wireless Communications (SPAWC-2003)*, Rome, Italy, June.

[76] Agustin, A., Muñoz, O. and Vidal, J. (2004) A game theoretic approach for cooperative MIMO systems with cellular reuse of the relay slot, in *Proc. IEEE International Conference on Acoustics, Speech and Signal Processing (ICASSP-2004)*, Montreal, Canada, May.

[77] Agustin, A. and Vidal, J. (2008) Amplify-and-forward cooperation under interference-limited spatial reuse of the relay slot, *IEEE Transactions on Wireless Communications*, 7(5), part 2, May.

[78] Agustin, A. and Vidal, J. (2007) High spectral efficiency of multiple-relays assisted amplify-and-forward cooperative transmissions under spatial reuse, in *Proc. IEEE Vehicular Technology Conference Fall (VTC-Fall-2007)*, Baltimore, USA, Sept.

[79] Haining, R. (2003) *Spatial Data Analysis: Theory and Practice*, Cambridge University Press.

[80] Basar, T. and Olsder, G. J. (1999) *Dynamic Noncooperative Game Theory. SIAMS Classics in Applied Mathematics*. 2nd ed.

[81] Fudenberg, D. and Tirole, J. (1992) *Game Theory*, MIT Press, Cambridge, MA.

[82] MacKenzie, A. B. and Wicker, S. B. (2001) Game theory in communications: Motivation, explanation and application to power control, in *Proc. IEEE Global Communications Conference (GLOBECOM-2001)*, San Antonio, TX, USA, Nov.

[83] Rosen, J. B. (1965) Existence and uniqueness of equilibrium points for concave n-person games, *Econometrica*, 33(3), 520–534.

[84] Yates, R. D. (1995) A framework for uplink power control in cellular radio systems, *IEEE Journal on Selected Areas in Communications*, 13(7), 1341–1348.

[85] Monderer, D. and Shapley, L. S. (1996) Potential games, *Games and Economics Behaviour*, 14, 124–143.

[86] Scutari, G., Barbarossa, S. and Palomar, D. P. (2006) Potential games: A framework for vector power control problems with coupled constraints, in *Proc. IEEE International Conference on Acoustics, Speech and Signal Processing (ICASSP-2006)*, Toulouse, France, May.

[87] Lasaulce, S., MDebbah, M. and Altman, E. (2009) Methodologies for analyzing equilibria in wireless games, *IEEE Signal Processing Magazine*, 26(6), 41–52.

[88] Saraydar, C. U., Mandayam, N. B. and Goodman, D. J (2002) Efficient power control via pricing in wireless data networks, *IEEE Transactions on Communications*, 50(2), 291–303.

[89] Debreu, G. (1952) A Social Equilibrium Existence Theorem, in Proceedings of the National Academy of Sciences, 886–893.

[90] Osborne, M. J. (2007) Mathematical methods for economic theory: a tutorial, 1997-2007. Available online: http://www.economics.utoronto.ca/osborne/MathTutorial/.

[91] Sousa, E. and Silvester, J. (1990) Optimum transmission ranges in a direct sequence spread spectrum multihop packet radio network, *IEEE Journal on Selected Areas in Communications*, 8(5), 762–771.

[92] Shao, M. and Nikias, C. L. (1993) Signal Processing with Fractional Lower Order Moments: stable processes and their applications, *in Proc. of the IEEE*, 81(7), 986–1010.

[93] Burke and Zeidler, CDMA reverse link spatial combining gains: optimal vs. MRC in a faded voice-data system having a single dominant high data user, in *Proc. IEEE Global Communications Conference (GLOBECOM-2001)*, San Antonio, TX, USA, Nov.

[94] A comparative analysis of Mobile WIMAX Deployment alternative in the access networks, Wimax Forum website May 2007.

12

Biologically Inspired Paradigms in Wireless Networks

In this chapter we will discuss new paradigms in wireless networks inspired by existing biological concepts in the human body or other living organisms.

12.1 Biologically Inspired Model for Securing Hybrid Mobile Ad Hoc Networks

Research in networking turned recently to the immune system as a wealthy source of inspiration. Network security is one of the most interesting fields that represent fertile ground for applying immunity. This approach can also be used to resolve in part hybrid network problems. A hybrid network may consist of wireless and non-wireless components, hybrid protocols and nodes. Each (the cabled and the wireless network) has its own architecture and protocols, which are sometimes the same and sometimes different with regard to different types of network components. This makes the network sensitive to different types of threats. In the text that follows we will present a solution for such network security following the same concept as the body's innate immune system.

12.1.1 Immune System and Terminology

The immune system is a comprehensive, integrated system protecting the body against various diseases and deadly microbes. It consists of a set of elements (cells and molecules) and complex processes which help to perform this function. The difficulty comes from the fact that the immune system, in order to fulfill this function, has to discriminate between the *self* and the *non-self*, which are the elements (proteins) inside the body and which are not, respectively. This complexity results from the huge disproportion in the number of patterns where the non-self pattern ratio is about 10^{16} and that of the self is about 10^6[1], so it may be visualized just how difficult the process is. The immune system has some unique features which enable it to bypass both the difficulties and the complexity. These features are classified in [2] as multilayered protection, uniqueness, imperfect detection, adaptability, distributed detection, and multi layered protocol. The immune system, which performs a distributed detection, consists mainly of white blood cells which are known as lymphocytes. Recognizing the non-self patterns is known as binding. The elimination of the foreign pattern occurs when receptors, which cover the surface of the lymphocytes, bond to the pathogen successfully. Binding becomes stronger between the pathogens and the receptors owing to their complementary surface shape, which is known as '*affinity*'. In order to

Advanced Wireless Communications & Internet: Future Evolving Technologies, Third Edition. Savo Glisic.
© 2011 John Wiley & Sons, Ltd. Published 2011 by John Wiley & Sons, Ltd.

detect most pathogens and to solve the problem the immune system must have an important source of robustness: diversity. Diversity assumes that the immune system must contain many different types of genetic matter. Receptors are constructed from inherited genes or libraries and made by combining elements from different libraries that result in a number of randomly possible combinations, which provides the diversity of the immune system. The immune system learns to recognize specific pathogens through a process known as '*maturation*' through which a kind of lymphocytes '*B-cell*', when activated, generate receptors called '*antibodies*'. Upon binding, these antibodies either inactivate the pathogens or identify them for phagocytes or other immune system defenses to be eliminated. In general, the immune system is tolerant to self. This tolerance is enabled by another class of lymphocytes known as *T-cells*. While a protein is circulated, T-cells are exposed to most self protein, so if a maturing T-cell binds to any of the proteins it will be removed, while the survivors will be tolerant to most proteins. The T-cells represent distributed sensors for B cells through a mechanism known as co-stimulation.

A T –cell will only send a signal to B-cell if it recognizes the pathogen that the B-cell has captured. For more detail on modeling an immune system see [4].

12.1.2   Security Issues in MANETs

Mobile Ad hoc Networks (MANETs) are stand-alone, autonomous networks which support multihop communication. They do not rely on any existing infrastructure. However, the integration of MANETs and infrastructure networks such as the Internet extends network coverage and increases the application domain of ad hoc networks. It offers convenient Infrastructure-free communication over the shared wireless medium. Ad-hoc networks are created on demand without the support of fixed infrastructure such as central servers. The organization of this network is based on groups of nodes. Each MANET router may not know its neighborhoods a priori, but should determine its neighborhoods dynamically and track changes as the network evolves. Similarly for MANET network membership, MANET routers may leave or join a MANET, and the MANET may partition or merge with others. Attacks against MANETs are generally categorized as follows [3]:

- Attacks against the routing protocol, where the attacker advertises wrong routing information, sends false route error packets, performs selective routing, drops or does not forward routing protocol packets etc.
- Attacks that aim at exhausting the resources of other nodes in the MANET. The attacker might achieve this by sending many RREQs for bogus destinations, dropping packets selectively and thus causing affected nodes to send a large number of control packets and RREQs.
- Cooperative attacks, where malicious nodes cooperate strategically with each other to cause harm.
- Routing fabrication, attackers tempers with the normal routing procedures, it's achieved through alteration of the routing message 'fields', or by insertion of false routing message.

12.1.3   A Bio-Inspired Security Model

Biology and computing have merged together, yielding two central disciplines [4], computational biology and the artificial immune system. Simulating and applying some features of the natural immune system in the body may help to reduce risks arising from numerous threats of attack in hybrid mobile ad hoc networks.

The hybrid mobile ad hoc networks may contain both wireless and wired networks merged together, each with its own protocols and means of achieving communication. The striking feature of an immune system that is designed to apply in mobile ad hoc networks is the processes by which it generates detectors, identifies and eliminates foreign material and remembers the patterns of previous infections. These are all highly parallel and distributed. To build a system with the same features, it must at least have the following components [2]: a stable definition of self, the ability to prevent or detect and subsequently eliminate dangerous foreign activities (infections), memory of previous infections, a method of recognizing new infections, autonomy in managing responses, and a method of protecting the operating system itself from self attacks.

Figure 12.1 illustrates the model of a bio-inspired protocol. Each process leads to the next, starting with the random generation of what are labeled as agents. The generated agent has to be applied to a mechanism that discriminates between self and non-self behavior (event). The difficulty of protecting computer systems can be considered as the dilemma of learning to distinguish self from the other.

While self can be represented as corrupted data, from legitimate users, non-self can be an unauthorized user, foreign code in the form of computer viruses or an unanticipated code in the form of a Trojan horse. If the agent passes the self test it is then activated and becomes a mature item of the immunity system. This phase is equivalent to negative selection in the immune system. If the agent fails the self test it is then eliminated. Following the activation mode, the agent will be broadcast to every node. Each node generates and broadcasts a set of agents randomly. Each node will have a copy of other nodes' agents; consequently, building a security database in each node, applying diversity (in terms of immunology) and distributability to make sure that any non self will be treated correctly. Broadcasting a copy of the foreign event (simulating the memory, proliferation in the immunity) guaranties a much faster response to an event in the future.

To avoid a multitude of agents within the network a timer is set to control the lifetime of each agent (TTL – time to live). This timer is inherent in each agent. Survival of the agent depends either on a timer or foreign event (non-self event detected) and subsequently the next procedure is initiated. After detecting the event, the next stage for the agent is approximate matching where a mechanism is applied to decide the strength of the matching (affinity) between the agent and the event (binding). Applicable matching schemes can be Hamming distance, edit distance, and r-contiguous bits [6].

The stronger the matching, the faster the event elimination. If the matching is not proficient, a request to recall another agent takes place. On the other hand, upon detecting the non-self event, a scanning for the source address of the sender node is initiated. If the same source address had been encountered before, the node is treated as an untrusted source and a mechanism will be applied to isolate the sender node

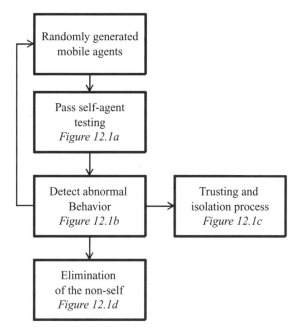

Figure 12.1 The model of a bio-inspired protocol.

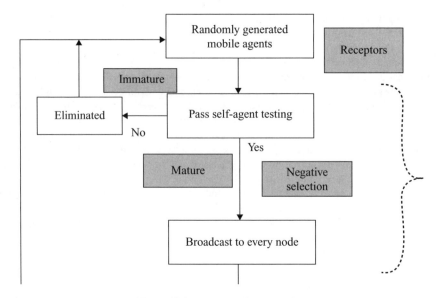

Figure 12.1   (a) Pass self-agent testing.

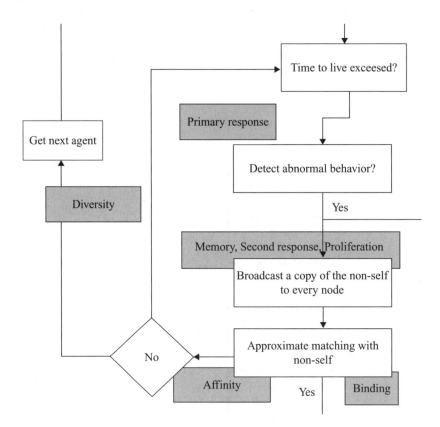

Figure 12.1   (b) Detect abnormal behavior.

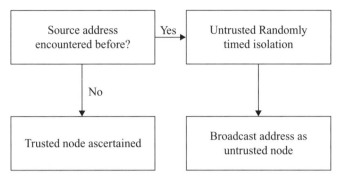

Figure 12.1   (c) Trusting and isolation process.

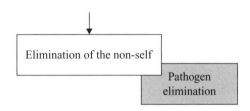

Figure 12.1   (d) Elimination of the non-self.

gradually. An exponential algorithm may be used for this function. The malicious address is broadcast to all nodes. If the source address never came across, the node will be treated as trusty.

Consequently, the node's database is automatically and continuously updated. For more information see [5–8].

12.2   Biologically Inspired Routing in Ad Hoc Networks

A biologically inspired algorithm, swarm intelligence, is presented to route messages in mobile wireless ad-hoc networks. This adaptive algorithm, most often referred to as *Termite*, uses stigmergy to reduce the amount of control traffic needed to maintain a high data goodput. Stigmergy is a process by which information is communicated indirectly between individuals through their environment. The Termite environment is the contents of all routing tables. The movement of packets is influenced at each node, and communicating nodes observe this influence to update their own tables. Strong routing robustness is achieved through the use of multiple paths; each packet is routed randomly and independently.

12.2.1 ·Routing in MANET

Conventional routing schemes in MANET are used to adapt only to the network topology, such as finding a shortest hop path or perhaps a minimum energy path. However, a MANET environment is affected by many more factors than simply changes in topology. Additional factors may include traffic congestion, link quality and variability, relative node mobility and local topological stability, or the effects of a specific medium access (MAC) layer protocol.

A biologically inspired approach is proposed to adapt to the aggregate effects of each of these phenomena by finding paths of maximum throughput [9]. Furthermore, the proposal maintains strong routing robustness and low control traffic overhead.

A social insect paradigm suggests a probabilistic routing algorithm. Information about the network environment, including topology, link quality, traffic congestion, etc., is determined from the arrival rate of packets at each node. Packets are considered to route themselves and are able to influence the paths of others by updating routing parameters at each node. The collection of these parameters from all nodes across the network constitutes the environment in which the packets exist. This Termite environment is a representation of the network environment. The interaction between packets and their environment spreads information implicitly about net- work conditions and thus reduces the need to generate explicit control traffic. The method of communicating information indirectly through the environment is known as stigmergy.

There exists relatively little work with regard to biologically inspired algorithms for routing in MANETs. A number of examples exist for wired networks including Ant Based Control [10] and *Ant Net* [11, 12]. In general, mobile agents travel the network while updating each visited node with routing information. Another major feature is the use of stigmergy.

Mobile Ants Based Routing (MABR) is introduced as the first routing algorithm for MANETs inspired by social insects [13]. The approach presented in *Ant Net* is extended to ad-hoc networks by abstracting the network into logical links and nodes based on relative node location. Location data is assumed from positioning devices. An optimized greedy routing algorithm is used to forward messages between logical nodes.

The *Ant-Colony Based Routing Algorithm* (ARA) presents a detailed routing scheme for MANETs, including route discovery and maintenance mechanisms [14,15]. Route discovery is achieved by flooding forward ants to the destination while establishing reverse links to the source. A similar mechanism is employed in other algorithms such as AODV. Routes are maintained primarily by data packets as they flow through the network. In the case of a route failure, an attempt is made to send the packet over an alternate link. Otherwise, it is returned to the previous hop for similar processing. A new route discovery sequence is launched if the packet is eventually returned to the source.

12.2.2   Swarm Intelligence Based Routing

We will start with a brief introduction to swarm intelligence. For more details see [16]. This framework is used to define rules for each packet to follow, which result in routing behavior. Reduced control traffic and quick route discovery and repair are additional benefits.

When packets are sent from a source to a destination, each follows a bias towards its destination while biasing its path in reverse towards its source. This bias is known as *pheromone*. Pheromone is laid on the communications links between nodes. Packets are attracted towards strong pheromone gradients but the next hop is always randomly decided. Following a destination pheromone trail while laying source pheromone along the same trail, increases the likelihood of packets following that reverse path to the source. This is positive feedback. In order to prevent old routing solutions from remaining in the collective network memory, exponential pheromone decay is introduced as negative feed-back. Pheromone increases linearly per packet, but decreases exponentially over time.

Each node maintains a table tracking the amount of pheromone on each neighbor link. Each node has a distinct pheromone scent. The table may be visualized as a matrix with neighbor nodes listed along the side and destination nodes listed across the top. Rows correspond to neighbors and columns to destinations.

An entry in the pheromone table is referenced by $P_{n,d}$ where n is the neighbor index and d denotes the destination index. In other words, $P_{n,d}$ is the amount of pheromone from node d on the link with neighbor n.

The pheromone table is analogous to a routing table. The table can have up to N rows and D columns. The number of neighbors of a node is N, and D is the number of destinations in the network except for the node itself. Each cell in the table contains the amount of pheromone on the link to neighbor n from destination d. The table contains at most $N \cdot D$ entries, but its size ultimately depends on which destinations have been heard from, as well as the number of neighbors.

When a packet arrives at a node, the pheromone for the source of the packet is incremented by a constant, γ. The nominal value of γ is one. Only packets addressed to a node must be processed. A node is said to be addressed if it is the intended next hop recipient of the packet. Latter on we will introduce also an optional promiscuous mode. The pheromone update procedure when a packet from source s is delivered from previous hop p is given as

$$P'_{p,s} = P_{p,s} + \gamma \tag{12.1}$$

where a prime indicates the updated value.

To account for pheromone decay, each value in the pheromone table is multiplied periodically by the decay factor, $e^{-\tau}$, where $\tau \geq 0$ is the decay rate. The nominal pheromone decay interval is one second; this is called the decay period giving

$$P'_{n,d} = P_{n,d} \cdot e^{-\tau} \tag{12.2}$$

If all of the pheromone for a particular node decays, then the corresponding row *and/or* column is removed from the pheromone table. Removal of an entry from the pheromone table indicates that no packet has been received from that node in quite some time. It has likely become irrelevant and no route information must be maintained. A column (destination listing) is considered decayed if all of the pheromone in that column is equal to a minimum value. If that particular destination is also a neighbor then it cannot be removed unless all entries in the neighbor row are also decayed. A row is considered decayed if all of the pheromone values on the row are equal to the pheromone floor.

Neighbor nodes must be handled specially because they can forward packets as well as originate packets. A decayed column indicates that no traffic has been seen which was sourced by that node. However, a neighbor's role as traffic source may be secondary to its role as a traffic relay. The neighbor row must be declared decayed before the neighbor node can be removed from the pheromone table.

If a neighbor is determined to be lost by means of communications failure (the neighbor has left communications range), the neighbor row is simply removed from the pheromone table.

There are three values governing the bounds on pheromone in the table *pheromone ceiling*, the *pheromone floor*, and the *initial pheromone*. When a packet is received from an unknown source, a new entry for that node is created in the pheromone table. In the case of a neighbor node, a new column and row will be created (neighbor nodes are also potential destinations). If the source is not a neighbor, only a column is entered into the table. Each pheromone value in the new cells will be assigned the initial pheromone value. During the course of pheromone decay, no value is allowed to fall below the pheromone floor. This allows unused nodes to be easily detected. Likewise, no pheromone value is allowed to exceed the pheromone ceiling. These bounds prevent extreme differences in pheromone from upsetting the calculation of next hop probabilities.

Upon arrival to a node i, an incoming packet with destination d is routed randomly based on the amount of $d's$ pheromone present on the neighbor links of i. A packet is never forwarded to the same neighbor from which it was received. If i has only one neighbor, i.e. the node that the packet was just received from, the packet is dropped. The forwarding function is the probability that the packet will be forwarded to n

$$p_{n,d} = \frac{(P_{n,d} + K)^F}{\sum_{i=1}^{N}(P_{i,d} + K)^F} \tag{12.3}$$

The constants F and K are used to tune the routing behavior of the algorithm.

There are five types of packets used by the algorithm *data, hello, seed, route request (RREQ)*, and *route reply (RREP)*. The latter four types are control messages. Each packet type consists of at least six fields, including *source address, destination address, previous hop address, next hop address, message identification*, and *Time-To-Live (TTL)*. Data packets may contain additional fields such as *data length* and *bulk data*.

Data Packets are routed normally through the network. If a node does not know how to forward a packet, which is the case when the node's pheromone table does not contain the packet's destination, the packet is stored and a route request is issued. If a reply is not received within a given time period, *rreq timeout*, the data packet is dropped and considered lost.

Route Request Packets (RREQ) are sent when a node needs to find a path to an unknown destination. Route requests perform a random walk over the network until a node is found which contains some pheromone for the requested destination. In a random walk, a packet uniformly randomly chooses its next hop, except for the link it arrived on. If a route request cannot be forwarded, it is dropped. Any number of RREQ packets may be sent for each route request, the exact number of which may be tuned for a particular environment.

A *Route Reply Packet* (RREP) is returned to the requestor once a route request packet is received by a node containing pheromone to the requested destination. The RREP message is created such that the source of the packet appears to be the requested destination and the destination of the packet is the requestor. The reply packet extends pheromone for the requested destination back to the requestor without any need to change the way in which pheromone is recorded at each node. The reply packet is routed normally through the network by probabilistically following a pheromone trail to the requestor. Intermediate nodes on the return path automatically discover the requested node.

Hello Packets are used to search for neighbors when a node has become isolated. Hello packets are broadcast at a regular interval until a reply is received. A reply is sent by all nodes who hear the original hello. The node will stop sending hello packets until the pheromone table is again empty. Hello broadcasts may be avoided at the routing layer by an analogous mechanism at the MAC layer.

Seed Packets are used to actively spread a node's pheromone throughout the network. Seeds make a random walk through the network and serve to advertise a node's existence. They can be useful for reducing the necessary number of explicit route request transactions. This is equivalent to conventional proactive routing.

Figure 12.2 shows system performance for the simulation scenario defined by the following parameters: simulation area 50×50 [sq. meters], transmission range 10 [meters], channel bit rate 1 [Mbps], initial

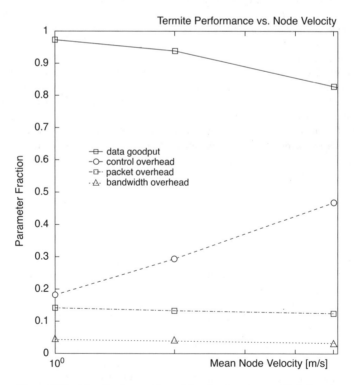

Figure 12.2   Data Goodput vs. Control Overhead vs. Average Node Speed.

pheromone 1, pheromone ceiling 10000, pheromone floor 0.1, rreq timeout 2 [seconds], τ (decay rate) 0.105, decay period 1 [second], Data TTL 32 [hops], RREQ TTL 32 [hops], RREP TTL 32 [hops], Seed TTL 4 [hops], Seed period 30 [seconds], Hello period 1 [second], RREQs per Route Request 2.

12.3 Analytical Modeling of AntNet as Adaptive Mobile Agent Based Routing

By retaining the core ideas of the artificial ant colony paradigm from the previous section, we now discuss an analytical model for the adaptive routing problem in datagram networks. We will refer to this model as *the AntNet algorithm*. Based on the discussion from the previous section its main characteristics can be summarized as follows:

1. At regular intervals, and concurrently with the data traffic, from each network node mobile agents [18–20] (in this section also referred to interchangeably as ants) are asynchronously launched towards randomly selected destination nodes.
2. Agents act concurrently and independently, and communicate in an indirect way, through the information they read and write locally to the nodes.
3. Each agent searches for a minimum cost path joining its source and destination nodes.
4. Each agent moves step-by-step towards its destination node. At each intermediate node a greedy stochastic policy is applied to choose the next node to move to. The policy makes use of (a) local agent-generated and maintained information, (b) local problem-dependent heuristic information, and (c) agent-private information.
5. While moving, the agents collect information about the time length, the congestion status and the node identifiers of the followed path.
6. Once they have arrived at the destination, the agents go back to their source nodes by moving along the same path as before but in the opposite direction.
7. During this backward travel, local models of the network status and the local routing table of each visited node are modified by the agents as a function of the path they followed and of its goodness.
8. Once they have returned to their source node, the agents die.

AntNet is conveniently described in terms of two sets of *homogeneous mobile agents* called in the following *forward* and *backward* ants. Agents in each set possess the same structure, but they can sense different inputs and they can produce different, independent outputs. They can be broadly classified as *deliberative* agents, because they behave *reactively* retrieving a pre-compiled set of behaviors, and at the same time they maintain a complete internal state description. Agents communicate in an indirect way, according to the stigmergy paradigm, through the information they concurrently read and write in two data structures stored in each network node k as indicated in Figure 12.2.

First through the routing table T_k, organized as in vector-distance algorithms [17], but with probabilistic entries where T_k defines the probabilistic routing policy currently adopted at node k. For each possible destination d and for each neighbor node n, T_k stores a probability value P_{nd} expressing the goodness (desirability), under the current network-wide routing policy, of choosing n as next node when the destination node is d:

$$\sum_{n \in N_k} P_{nd} - 1, \ d \in [1, N], \ N_k = \{neighbors(k)\} \tag{12.4}$$

Secondly, through an array $M_k(\mu_d, \sigma_d^2, W_d)$, of data structures defining a simple parametric statistical model for the traffic distribution over the network as seen by the local node k. The model is adaptive and described by sample means and variances computed over the trip times experienced by the mobile agents, and by a moving observation window W_d used to store the best value W_{best_d} of the agents' trip time.

$$
\begin{array}{c}
\textit{network node} \\
\textit{outgoing links} \left\|
\begin{array}{cccc}
P_{11} & P_{12} & \cdots & P_{1N} \\
P_{21} & P_{22} & \cdots & P_{2N} \\
\vdots & & & \\
P_{L1} & P_{L1} & \cdots & P_{LN}
\end{array}
\right\|
\end{array}
$$

For each destination d in the network, an estimated mean and variance, μ_d and σ_d^2, give a representation of the expected time to go and of its stability. Arithmetic, exponential and windowed strategies are used to compute the statistics. Changing strategy does not affect performance much, and exponential model [21] with updates

$$
\mu_d \leftarrow \mu_d + \eta(0_{k \to d} - \mu_d)
$$
$$
\sigma_d^2 \leftarrow \sigma_d^2 + \eta((0_{k \to d} - \mu_d)^2 - \sigma_d^2)
$$

(12.5)

where $0_{k \to d}$ is the new observed agent's trip time from node k to destination d. The factor η weights the number of most recent samples that will really affect the average. The weight of the $t_i - th$ sample used to estimate the value of μ_d after j samplings, with $j > i$, is: $\eta(1 - \eta)^{j-i}$. In this way, for example, if $\eta = 0.1$, approximately only the latest 50 observations will really influence the estimate, for $\eta = 0.05$, the latest 100, and so on. Therefore, the number of effective observations is $\approx 5(1/\eta)$.

The moving observation window W_d is used to compute the value W_{best_d} of the best agents' trip time towards destination d as observed in the last w samples. After each new sample, w is incremented modulus $|W|_{max}$, and $|W|_{max}$ is the maximum allowed size of the observation window. The value W_{best_d} represents a short-term memory expressing a moving empirical lower bound of the estimate of the time to go to node from the current node.

T and M can be seen as memories local to nodes capturing different aspects of the network dynamics. The model M maintains absolute *distance/time* estimates to all the nodes, while the routing table gives relative probabilistic goodness measures for each link- destination pair under the current routing policy implemented over all the network.

AntNet *Algorithm*
1. At regular intervals Δt from every network node s, a mobile agent (forward ant) $F_{s \to d}$ is launched toward a destination node d to discover a feasible, low-cost path to that node and to investigate the load status of the network. Forward ants share the same queues as data packets, so that they experience the same traffic loads. Destinations are locally selected according to the data traffic patterns generated by the local workload:
   if f_{sd} is a measure (in bits or in number of packets) of the data flow $s \to d$, then the probability of creating at node s a forward ant with node d as destination is

$$
p_d = \frac{f_{sd}}{\sum_{d=1}^N f_{sd'}}
$$

(12.6)

In this way, ants adapt their exploration activity to the varying data traffic distribution.
2. While traveling toward their destination nodes, the agents keep memory of their paths and of the traffic conditions found. The identifier of every visited node k and the time elapsed since the launching time to arrive at this k-th node are pushed onto a memory stack $S_{s \to d}(k)$.
3. At each node k, each traveling agent headed towards its destination d selects the node n to move to choosing among the neighbors it did not already visit, or over all the neighbors in case all of them had been previously visited. The neighbor n is selected with a probability, referred to as goodness,

P'_{nd} computed as the normalized sum

$$P'_{nd} = \frac{P_{nd} + \alpha l_n}{1 + \alpha(|N_k| - 1)} \tag{12.7}$$

where P_{nd} is given by (12.3) and a heuristic correction factor l_n takes into account the state (the length) of the n-th link queue of the current node k. The heuristic correction l_n is a [0, 1] normalized value proportional to the length q_n (in bits waiting to be sent) of the queue of the link connecting the node k with its neighbor n.

$$l_n = 1 - \frac{q_n}{\sum\limits_{n'=1}^{|N_k|} q_{n'}} \tag{12.8}$$

The value of α weights the importance of the heuristic correction with respect to the probability values stored in the routing table. Parameter l_n effects the instantaneous state of the node's queues, and assuming that the queue's consuming process is approximately stationary or slowly varying, l_n gives a quantitative measure associated with the queue waiting time. The routing table's values, on the other hand, are the outcome of a continual learning process and capture both the current and the past status of the whole network as seen by the local node. Correcting these values with the values of l allows the system to be simultaneously more 'reactive' and at the same not to sensitive to all the network fluctuations. Agent's decisions are taken on the basis of a combination of a long-term learning process and an instantaneous heuristic prediction. Depending on the characteristics of the problem, the best value to assign to the weight α can vary, but if α ranges between 0.2 and 0.5, performance doesn't change much. For lower values, the effect of l is vanishing, while for higher values the resulting routing tables oscillate and, in both cases, performance degrades.

4. In the case that a cycle is detected, that is, if an ant is forced to return to an already visited node, the cycle's nodes are popped from the ant's stack and all the memory about them is destroyed. If the cycle lasted longer than the lifetime of the ant before entering the cycle, (that is, if the cycle is greater than half the ant's age) the ant is destroyed. In fact, in this case the agent wasted a lot of time probably because of a wrong sequence of decisions and not because of congestion states. Therefore, the agent is carrying an old and misleading memory of the network state and it is counterproductive to use it to update the routing tables.

5. When the destination node d is reached, the agent $F_{s \to d}$ generates another agent (backward ant) $B_{d \to s}$, transfers to it all of its memory, and dies.

6. The backward ant takes the same path as that of its corresponding forward ant, but in the opposite direction. At each node k along the path it pops its stack $S_{s \to d}(k)$ to find the next hop node. Backward ants do not share the same link queues as data packets. They use higher priority queues, because their task is to quickly propagate to the routing tables the information accumulated by the forward ants.

7. When it arrives at a node k coming from a neighbor node f, the backward ant updates the two main data structures of the node, the local model of the traffic M_k and the routing table T_k, for all the entries corresponding to the (forward ant) destination node d. To some extend the updates are performed also on the entries corresponding to every node $k' \in S_{k \to d}, k' \neq d$ on the 'sub-paths' followed by ant $F_{s \to d}$ after visiting the current node k. If the elapsed trip time of a sub-path is statistically 'good' (i.e., it is less than $\mu + I(\mu, \sigma)$, where I is an estimate of a confidence interval for μ), then the time value is used to update the corresponding statistics and the routing table. At the same time, trip times of sub-paths not considered good, in the same statistical sense as defined above, are not used because they don't give a correct idea of the time to go toward the sub-destination node. One should keep in mind that, all the forward ant routing decisions have been made only as a function of the destination node so that sub-paths are side effects, and they are intrinsically sub-optimal because of the local variations in the traffic load. In case of a good sub-path we can use it since the ant discovered an additional good route at zero cost.

With respect to a generic destination node $d' \in S_{k \to d}$ M and T are updated as follows:

a) M_k is updated with the values stored in the stack memory $S_{s \to d}(k)$. The time elapsed to arrive (for the forward ant) to the destination node d' starting from the current node is used to update the mean and variance estimates, $\mu_{d'}$ and $\sigma_{d'}^2$, and the best value over the observation window $W_{d'}$. In this way, a parametric model of the traveling time to destination d' is maintained. The mean value of this time and its dispersion can vary strongly, depending on the traffic conditions: a poor time (path) under low traffic load can be a very good one under heavy traffic load. The statistical model has to be able to capture this variability and to follow in a robust way the fluctuations of the traffic. This model plays a critical role in the routing table updating process which is elaborated in more detail in the sequel.

b) The routing table T_k is changed by incrementing the probability $P_{fd'}$ (i.e., the probability of choosing neighbor f when destination is d') and decrementing, by normalization, the other probabilities $P_{nd'}$. The amount of the variation in the probabilities depends on a measure of goodness we associate with the trip time $T_{k \to d'}$ experienced by the forward ant. This time represents the only available explicit feedback signal to score paths. It provides an indication about the goodness r of the followed route because it is proportional to its length from a physical point of view (number of hops, transmission capacity of the used links, processing speed of the crossed nodes) and from a traffic congestion point of view since the forward ants share the same queues as data packets.

The time measure T, composed by all the sub-paths' elapsed times, is not an exact error measure, since we don't know the optimal trip times, which depend on the whole network load status. When the network is in a congested state, all the trip times will be poor with respect to the times observed in low load situations. Still, a path with a high trip time should be scored as a good one if its trip time is significantly lower than the other trip times observed in the same congested situation.

For this reason, T can only be used as a reinforcement signal in a credit assignment defined problem typical of the reinforcement learning field [22].

In this case the reinforcement $r = r(T, M_k)$ is a function of the goodness of the observed trip time as estimated on the basis of the local traffic model. r is a dimensionless value, $r \in (0, 1]$, used by the current node k as a positive reinforcement for the node f the backward ant $B_{d \to s}$ comes from. Parameter r takes into account some average of the so far observed values and of their dispersion to evaluate the goodness of the trip time T, such that the smaller T is, the higher r is, which is an issue to be further discussed latter. Based on the previous discussion we have

$$P_{fd'} \leftarrow P_{fd'} + r(1 - P_{fd'}) \tag{12.9}$$

In this way, small probability values are increased proportionally more than big probability values, favoring in this way a quick exploitation of new, and good, discovered paths.

Probabilities $P_{nd'}$ for destination d' of the other neighboring nodes n implicitly receive a negative reinforcement by normalization. That is, their values are reduced so that the sum of probabilities will still be 1:

$$P_{nd'} \leftarrow P_{nd'} - r P_{nd'}, \ n \in N_k, \ n \neq f \tag{12.10}$$

Every discovered path receives a positive reinforcement in its selection probability, and the reinforcement is (in general) a non-linear function of the goodness of the path, as estimated using the associated trip time.

In this way, not only the (explicit) assigned value r plays a role, but also the (implicit) ant's arrival rate. This strategy is based on trusting paths that receive either high reinforcements, independent of their frequency, or low and frequent reinforcements. For any traffic load condition, a path receives one or more high reinforcements only if it is much better than previously explored paths. During a transient phase after a sudden increase in network load all paths will probably have high traversing times with respect to those learned by the model M in the preceding, low congestion, situation. So, in this case good paths can only be differentiated by the frequency of ants' arrivals. As could be seen from above, the reinforcement r is a critical quantity that should reflect at least three following aspects: (a) paths should receive an increment in their selection probability proportional to their goodness, (b) the goodness is a relative measure, which depends on the traffic conditions, that can be estimated by means of the model M, and (c) it is important not to follow all the traffic fluctuations. This last aspect is particularly important. Uncontrolled oscillations in the routing tables are one of the main problems in shortest paths routing [21]. It is very important to be able to set the best trade-off between stability and adaptability.

In the following we discuss several ways to assign the r values trying to take into account the above three requirements:

a) The simplest way is to use $r = const.$ independently of the ant's experiment outcomes, the discovered paths are all rewarded in the same way. In this case, what is at work is the implicit reinforcement mechanism due to the differentiation in the ant arrival rates. Ants traveling along faster paths will arrive at a higher rate than other ants, hence their paths will receive a higher cumulative reward. In this case, the core of the algorithm is based on the capability of 'real' ants to discover shortest paths communicating by means of pheromone trails. The problem of this approach lies in the fact that, although ants following longer paths arrive delayed, they will nevertheless have the same effect on the routing tables as the ants who followed shorter paths.

   With this strategy, the algorithm shows moderately good performance. This suggests that the implicit' component of the algorithm, based on the ant arrival rate, plays a very important role. To compete with state-of-the-art algorithms, the available information about path costs has to be used.

b) More elaborate approaches define r as a function of the ant's trip time T, and of the parameters of the local statistical model M. Different linear, quadratic and hyperbolic combinations of the T and M values can be used. The best results can be obtained for

$$r = c_1(W_{best}/T) + c_2((I_{sup} - I_{inf})/[(I_{sup} - I_{inf}) + (T - I_{inf})]) \qquad (12.11)$$

where W_{best} is the best trip time experienced by the ants traveling toward the destination d, over the last observation window W. The maximum size of the window (the maximum number of considered samples before resetting the W_{best} value) is assigned on the basis of the coefficient η of (12.5). Parameter η weights the number of samples effectively giving a contribution to the value of the μ estimate, defining a sort of moving exponential window. Following the expression for the number of effective samples as discussed below (12.7), we can set $|W|_{max} = 5(c/\eta)$, with $c < 1$. In this way, the long- form exponential mean and the short-term windowing are referring to a comparable set of observations, with the short-term mean evaluated over a fraction c of the samples used for the long-term one. I_{sup} and I_{inf} are convenient estimates of the limits of an approximate confidence interval for μ. I_{inf} is set to W_{best}, while $I_{sup} = \mu + z(\sigma/\sqrt{|W|})$, with $z = 1/\sqrt{(1 - \gamma)}$ where $gamma$ gives the selected confidence level. The expression is obtained in [22] by using the Tchebycheff inequality that allows the definition of a confidence interval for a random variable following any distribution. Usually, for specific probability densities the Tchebycheff bound is too high, but here it is used to avoid to make assumptions on the distribution of μ since only a raw estimate of the confidence interval is needed.

The first term in (12.11) evaluates the ratio between the current trip time and the best trip time observed over the current observation window. This term is corrected by the second, that evaluates how far the value T is from I_{inf} in relation to the extension of the confidence interval, that is, considering the stability in the latest trip times. The coefficients c_1 and c_2 weight the importance of each term. The first term is the most important one, while the second term plays the role of a correction. In [22] $c_1 = 0.7$ and $c_2 = 0.3$ was used. It was observed that c_2 shouldn't be too big (0.35 is an upper limit), otherwise performance starts to degrade significantly. The behavior of the algorithm is quite stable for c_2 values in the range 0.15 to 0.35 but setting c_2 below 0.15 slightly degrades performance. The algorithm is very robust to changes in γ, which defines the confidence level: varying the confidence level in the range from 75% to 95% changes performance little. The best results have been obtained for values around 75%–80%.

The value r obtained from (12.12) is finally transformed by function $s(x)$:

$$s(x) = \left(1 + exp\left(\frac{a}{x|N_k|}\right)\right)^{-1}, \quad x \in (0, 1], \quad a \in R^+$$

$$r \leftarrow \frac{s(r)}{s(1)} \tag{12.12}$$

This transformation of r values allows the system to be more sensitive in rewarding good (high) values of r, while having the tendency to saturate the rewards for bad (near to zero) r values. The scale is compressed for lower values and expanded in the upper part. In such a way an emphasis is put on good results, while bad results play a minor role.

The ratio $a/|N_k|$ determines a parametric dependence of the transformed reinforcement value on the number $|N_k|$ of neighbors of the reinforced node k. The greater $|N_k|$, the higher the reinforcement. The reason to do this is that we want to have a similar, strong, effect of good results on the probabilistic routing tables, independent of the number of neighbor nodes.

12.3.1 Performance Example

The sample network topology is shown in Figure 12.3. The topological properties are summarized by a triple of numbers $(\mu, \sigma, N) = (2.2, 0.8, 14)$ indicating respectively the mean shortest path distance, in terms of hops, between all pairs of nodes, the variance of this average, and the total number of nodes. Sessions over the network are characterized by their inter-arrival time distribution (Poison) and by their geographical distribution (uniform). The latter is controlled by the probability assigned to each node to be selected as a session start or end-point.

Figure 12.4 represents the sensitivity of *AntNet* with respect to the ant launching rate. For the data traffic model above, the interval Δg between two consecutive ant generations is decreasing progressively (g is the same for all nodes). g values are sampled at constant intervals over a logarithmic scale ranging from 0.06 to 25 seconds. The dashed curve represents the interpolated generated routing overhead expressed, as the fraction of the available network bandwidth used by routing packet. The solid curve represents utility normalized to its highest value, where utility is defined as the ratio between the delivered Throughput and the packet delay. The value used for delivered Throughput is the Throughput value at time 1000 averaged over ten trials, while for packet delay we used the 90-th percentile of the empirical distribution.

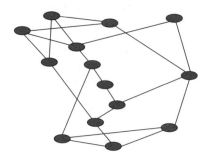

Figure 12.3 Example network topology, each edge in the graph represents a pair of directed links. Link band-width is 1.5 Mbit/sec, propagation delays range from 4 to 20 msec.

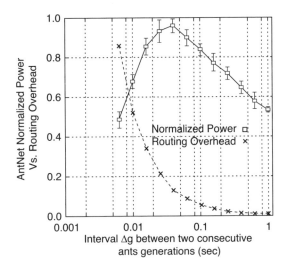

Figure 12.4 AntNet utility vs. routing overhead. Utility is defined as the ratio between delivered throughput and packet delay.

In the figure, we can see how an excessively small Δg causes an excessive growth of the routing overhead, with consequent reduction of the utility. Similarly, when Δg is too big, the utility diminishes slowly and tends toward a plateau because the number of ants is not enough to generate and maintain up-to-date statistics of the network status. In the middle of these two extreme regions a wide range of g intervals gives rise to similar, very good values of utility, while, at the same time, the routing overhead falls down quickly toward negligible values. This figure confirms strongly the robustness of *AntNet's* internal parameter settings.

12.4 Biologically Inspired Algorithm for Optimum Multicasting

In this section the problem of minimum power multicasting in wireless networks with sectored antennas is considered. For omnidirectional antenna systems, a transmission from node i to node j will also reach all nodes which are closer to i than j. Depending on the network geometry, this strategy can be highly power efficient. In ideal sectored antenna systems, however, this phenomenon is sector specific since only those nodes which are located in the same sector as j will receive the transmission implicitly. Though this might seem an apparent disadvantage, the higher gains associated with directional antennas (as opposed to omnidirectional antennas) and more flexibility in intercell interference management allow for reduced transmission powers without sacrificing the signal-to-interference ratio at the receiver. In this section, we discuss a biologically inspired algorithm for solving the problem to near-optimality at a very reasonable computation time. The performance of the algorithm is compared with a more conventional approach based on a mixed integer linear programming model.

We assume a fixed N-node network with a specified source node which has to broadcast (multicast) a message to all (some) other nodes in the network. We assume static broadcasting (multicasting) which means that the same tree is used for the entire broadcast (multicast) duration. Any node can be used as a relay node to reach other nodes in the network.

All nodes are assumed to have S-sector antennas. The number of sectors, is $S = 360/\theta$, where θ (in degrees) is the beamwidth.

Each sector is assumed to span the angular region $[(s-1)360/\theta, (s)360/\theta]$ in the 2-D plane, where $1 \leq s \leq S$ is the sector index. We ignore sidelobe effects and assume that when sector s is switched on,

100% of the radiated power is confined within that sector, providing an uniform gain within the angular region spanned by the sector. We consider antennas with 100% efficiency. That is, we ignore any antenna power losses.

Several operating modes are possible in networks (a) Directional transmit, directional receive (*DTxDRx*) (b) Directional transmit, omni-directional receive (*DTxODRx*) (c) Omni-directional transmit, directional receive (ODTxDRx) and (d) Omni-directional transmit, omni- directional receive (ODTxO-DRx). We focus on the directional transmit, omni-directional receive (*DTxODRx*) operating mode.

For this operating mode, the transmitter power at i necessary to support the link $(i \rightarrow j)$, P_{ij}, is proportional (accounting for *link/antenna* gains and other factors) to d_{ij}^{α}/S, where d_{ij} is the Euclidean distance between nodes i and j. If (x_i, y_i) are the coordinates of node i and α (typically in the range $2 \leq \alpha \leq 4$) is the channel loss exponent, d_{ij} is given by $d_{ij} = [(x_i - x_j)^2 + (y_i - y_j)^2]^{1/2}$. For simplicity, we set the proportionality constant to be equal to 1 and therefore $P_{ij} = d_{ij}^{\alpha}/S$.

In addition we assume that power expenditures due to signal reception and processing are negligible compared to signal transmission and hence the *cost* of a multicast tree is determined solely by the choice of transmitter powers.

Let Y be a vector of node transmission powers, the element Y_i representing the total transmission power cost of node i. For an S-sector antenna, Y_i can be written as the sum of $Y_{i,s}$, where $Y_{i,s}$ is the transmission power cost corresponding to sector s of node i. We assume that each node has a constraint on the maximum transmitter power it can use per sector, denoted by $Y_{i,s}^{max}$ so that $0 \leq Y_{i,s} \leq Y_{i,s}^{max} : \forall i \in N$ where N is the set of all nodes in the network and $|N| = N$.

Also, let E the set of all directed edges and D the set of destination nodes, $D \subseteq \{N \backslash source\}$. The notation $(i \rightarrow j)$ will be used to denote a directed edge from node i to j. The notation (i, j) will be used to refer to the node pair. Let the cardinality of these sets be E and D respectively; i.e., $E = |E|$ and $D = |D|$. Using the transmitter power constraint, the set of all edges, E, is given by

$$E = \{(i \rightarrow j) : (i, j) \in N, i \neq j, P_{ij} \leq Y_{i,s}^{max}\}$$

The third condition in the right hand side of this relation defines the set of nodes which are reachable by a direct transmission from any transmitting node depending on its maximum sector power constraint.

Note that a transmission from node i to node j, referred to as *actual transmission,* would also be received by all nodes which are located within the same sector as j and are geometrically closer to i than j which is referred to as *implicit transmission.*

Let $\{X_{ij} : (i \rightarrow j) \in E\}$ be a set of binary variables such that $X_{ij} = 1$ if the transmission $i \rightarrow j$ is used in the optimum tree and 0 otherwise. Also, let $ne(i, s)$ be the set of neighbours of node i which are within radio range of i and are located within the same sector, s, w.r.t node i. For example, in Figure 12.5,

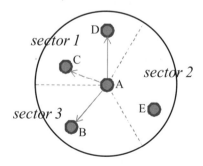

Figure 12.5   Multicast illustration with a 3-sector antenna.

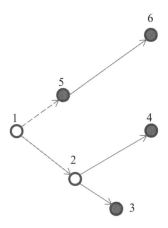

Figure 12.6   Example multicast tree.

$ne\,(A, 1) = \{C, D\}, ne\,(A, 2) = \{E\}$ and $ne\,(A, 3) = \{B\}$ so

$$Y_{i,s} = \max_{j}\{X_{ij}P_{ij} : j \in ne(i, s)\} \qquad (12.13)$$

where $X_{ij} = 1$ if node j is reached from node i (actually or implicitly) and 0 otherwise. For minimum power multicasting with sectored antennas, our objective function is:

$$minimize\,\left(\sum_{i=1}^{N} Y_i\right) = minimize\,\left(\sum_{i=1}^{N}\sum_{s=1}^{S} Y_{i,s}\right) \qquad (12.14)$$

where $Y_{i,s}$ is the transmission power cost corresponding to sector s of node i. The objective function, (12.14), has to be solved subject to the following constraints: (a) all destination nodes must be reached, either actually or implicitly, (b) the source node must reach at least one other node, and (c) the tree must be *connected*; i.e., there must be directed paths from the source to all destination nodes, possibly involving other intermediate nodes. Figure 12.6 presents an example of multicast tree. Node 1 is the source. The shaded nodes are the destination nodes and the dashed line represents implicit transmissions. Assume that all nodes are equipped with a 3-sector antenna. Suppose that (a) nodes 2 and 5 are located in sector 3 w.r.t node 1 (b) node 3 is located in sector 1 w.r.t node 2 (c) node 4 is located in sector 3 w.r.t node 2 and (c) node 6 is located in sector 1 w.r.t node 5. The cost of the multicast tree is therefore: $Y_{1,3} + Y_{2,1} + Y_{2,3} + Y_{5,1} = P_{12} + P_{23} + P_{24} + P_{56}$.

In the next section, we first develop a mixed integer programming (MILP) model for optimal solution of the above optimization problem.

12.4.1   Mixed Integer Programming Model

Let $\{F_{ij} : \forall(i \rightarrow j) \in E\}$ be a set of flow variables (F_{ij} represents the flow from node i to node j). The general multicast problem can be interpreted as a single-origin multiple-destination uncapacitated flow problem, with the source having D units of supply (no demand) and the destination nodes having one unit of demand each. For other nodes, the net in-flow must equal the net out-flow, since they serve only as relay nodes.

For example, for the multicast tree in Figure 12.6, the flow variables are: $F_{12} = F_{15} = 2, F_{23} = F_{24} = 1, F_{56} = 1$. All other flow variables are 0.

The supply and demand constraints discussed above can be expressed as the following *flow conservation equations*

$$\sum_{j=1}^{N} F_{ij} = D; i = source, \ (i \rightarrow j) \in E$$

$$\sum_{j=1}^{N} F_{ji} = 0; i = source, \ (i \rightarrow j) \in E$$

$$\sum_{j=1}^{N} F_{ji} - \sum_{j=1}^{N} F_{ij} = 1; \ \forall i \in D, \ (i \rightarrow j) \in E \qquad (12.15)$$

$$\sum_{j=1}^{N} F_{ji} - \sum_{j=1}^{N} F_{ij} = 0; \forall i \in R, \ (i \rightarrow j) \in E$$

$$R \overset{\Delta}{=} N \backslash \{D \cup source\}$$

The set of constraints which couple the flow variables and the X_{ij} variables is:

$$D \cdot X_{ij} - F_{ij} \geq 0; \forall (i \rightarrow j) \in E \qquad (12.16)$$

which ensures that $X_{ij} = 1$ if $F_{ij} > 0$. For the multicast tree in Figure 12.2, the status of the X_{ij} variables are $X_{12} = X_{15} = X_{23} = X_{24} = X_{56} = 1$, the rest being equal to 0. The coefficient of X_{ij} in (12.16) is due to the fact that the maximum flow out of any node on a single link is equal to the number of destination nodes. The same equation, however, leaves open the possibility of X_{ij} being equal to 1 for $F_{ij} = 0$. We show later that this possibility can be discounted since it would unnecessarily increase the cost of the optimum solution. It should also be noted that the smallest integer value of X_{ij} which satisfies (12.16) for any nonzero flow out of node i, if $\sum_j X_{ij} \geq 1$ is 1. Consequently, we can simply define the X_{ij}s to be integers, instead of explicitly declaring them to be binary variables.

Next, we write down constraints linking the X_{ij} variables and the power variables. As discussed earlier the cost of spanning in multiple nodes, located within the same sector, from node i is simply the cost incurred in reaching the farthest node. This condition is expressed as:

$$Y_{i,s} - P_{ij} X_{ij} \geq 0; \ \ \forall i \in N, \ \ \ \forall j \in ne(i, s) \qquad (12.17)$$

It is now clear that, if there is no flow out of node i (i.e., $\sum_j F_{ij} = 0$), setting $X_{ij} = 1$ would result in a positive value for Y_i and thereby unnecessarily increase the cost of the optimal solution. In the case when there are multiple flows out of node i, i.e., $\sum_j F_{ij} > 1$ and $j^* \in ne(i, s)$ is such that $\hat{Y}_{i,s} = P_{ij^*} X_{ij^*} = \max_j(P_{ij} X_{ij} : j \in ne(i, s))$ is part of the optimal solution, setting $X_{ij} = 1, j \neq j^*$, would not affect the cost of the optimal solution if $P_{ij} X_{ij} \leq P_{ij^*} X_{ij^*}$. If, however, $P_{ij} X_{ij} > P_{ij^*} X_{ij^*}$, this solution cannot be optimal since it can easily be improved by setting $X_{ij} = 0$.

The final set of constraints express the integrality of the X_{ij} variables and non-negativity of the F_{ij} and $Y_{i,s}$ variables.

$$X_{ij} \geq 0, \ integer; \ \forall (i \rightarrow j) \in E$$

$$F_{ij} \geq 0; \forall (i \rightarrow j) \in E \qquad (12.18)$$

$$0 \leq Y_{i,s} \leq Y_{i,s}^{max} ; \ \forall i \in N, s = 1, 2, \cdots S$$

Solving the objective function (12.14), subject to constraints (12.15) to (12.18) solves the minimum power multicast problem in wireless networks with sectored antennas. The number of integer variables in the MILP model is equal to E while the number of continuous variables is equal to $E + SN$. The number of constraints is approximately on the order of $E + N(1 + S)$.

12.4.2   Ant Colony System (ACS) Model

In this section we further elaborate principles discussed in Section 12.2 by introducing the following notation:

$t = $ time index, $t^{max} = $ *maximum* time index, $N_A = $ *number* of Type-A ants, $N_B = $ *number* of Type-B ants, $\tau^{min} = $ *minimum* pheromone level, $\tau^{max} = $ *maximum* pheromone level

$\tau_{ij}(t) = $ *pheromone level on edge* $(i \to j)$ *at time* t, $\tau^{min} \leq \tau_{ij}(t) \leq \tau^{max}$

$\eta_{ij} = $ *local* visibility of node j from node $i \triangleq 1/P_{ij}$, $\beta_A = $ *tunable* parameter to control η_{ij} for Type-A ants with $0 < \beta_A \leq 1$, $\beta_B = $ *tunable* parameter to control η_{ij} for Type-B ants with $0 < \beta_B < \beta_A \leq 1$

$T_m(t) = $ *tree* developed by ant m at time t, $C_m(t) = $ *the* cost of $T_m(t)$

$\rho = $ *pheromone* decay coefficient $\rho \in (0, 1]$, $q = $ *uniformly* distributed random variable over the interval $[0, 1]$, $q_0 = $ *tunable* parameter with $q_0 \in [0, 1]$, initial pheromone level on all edges at time $t = 0$ is

τ^{min}; $\tau_{ij}(0) = \tau^{min}$: $\forall(i \to j) \in E$, $k = $ *transmission* step number,

$NR^k = $ *set* of new nodes reached at transmission step k

$(NR^0 = [source])$, $NR^{0:k} = $ *set* of all nodes reached till transmission step k equal by definition to $\cup_{x=0}^{k} NR^x$, $NNR^k = $ *set* of all nodes *not reached* till transmission step k equal to $N \backslash NR^{0:k}$.

Node i is said to be newly reached at step k if $i \in NR^k$ but $i \notin NR^{0:k-1}$.

Tree building by an ant is an iterative process which starts with a transmission from the source and continues till all the intended destination nodes are reached (*i.e.*, when $D \subseteq NR^{0:k}$). The iteration converges in at most $N - 1$ iterations (*i.e.*, $k \leq N - 1$).

Due to the inherently broadcast nature of the wireless medium, the number of iterations can range from 1 (if all destination nodes are located within the same sector and are within direct radio range of the source) to $N - 1$ (if exactly one new node is reached during each iteration and the last edge chosen reaches a destination node).

In general, at any transmission step $k(k \geq 1)$, an ant m can travel from any node which has been reached till step $k - 1$, to any node which has not yet been reached till step $k - 1$. The set of possible edges to choose from (edge list) el_m^k, is given by:

$$el_m^k = \{(i \to j) : i \in NR^{0:k-1}, j \in NNR^{k-1}, (i \to j) \in E\} \tag{12.19}$$

The decision rule governing which edge is chosen at step k of the tree building process, initially defined in Section 12.2 is farther elaborated as

Algorithm 1: Edge Selection at Iteration k by Ant m [24]

1. Generate el_m^k.
2. Randomly choose a transmitting node from el_m^k say f^k.

3. Let $A^{k,m} = \{a_{ij} : i = f_k, (i \to j) \in el_m^k\}$ be the decision matrix based on which ant m makes its decision for selecting an edge at step k. The probabilities $\{a_{ij}\}$ are computed as follows:

$$a_{ij} = \begin{cases} \dfrac{[\tau_{ij}(t)][\eta_{ij}]^{\beta_A}}{\Sigma_\zeta [\tau_{x\zeta}(t)][\eta_{i\zeta}]^{\beta_A}}, & i = f^k, \quad \text{for A ants} \\[4mm] \dfrac{[\tau_{ij}(t)][\eta_{ij}]^{\beta_B}}{\Sigma_\zeta [\tau_{i\zeta}(t)][\eta_{i\zeta}]^{\beta_B}}, & i = f^k, \quad \text{for B ants} \end{cases} \quad ; (f^k \to x) \in el_m^k.$$

4. Sample q from a uniform distribution over [0, 1].
5. if $(q < q_0)/ * \textit{Deterministic decision making} */$
  Choose the strongest edge, (f^k, t^k), from $A^{k,m}$.
  $(f^k, t^k) = \arg \max_{i,j}\{a_{ij}\}$

  else /* *Explore. Probabilistic decision making.* */
  Choose an edge (f^k, t^k) from $A^{k,m}$ probabilistically
  end if
6. Set $NR^k = \{t^k\}$. If other nodes, located within the same sector as t^k, are reached implicitly, include them in NR^k.
7. Update the sets:
  $NR^{0:k} \leftarrow NR^{0:k-1} \cup NR^k$
  $NNR^k \leftarrow N \setminus NR^{0:k}$
8. Stop if $D \subseteq NR^{0k}$
  Otherwise, increment $k(k \leftarrow k + 1)$ and repeat steps 1–7.

Starting with $k = 0$, this decision rule is executed till all the intended desti-nation nodes are reached. It can be easily seen from Figure 12.3 that the worst case complexity of the tree-building procedure is on the order of $O(N^2)$.

It can be seen from *Algorithm 1* that edges are chosen either deterministically or probabilistically. The extent to which probabilistic decisions are made is controlled by the tunable parameter q_0.

The factors which determine the desirability of choosing an edge (i, j) at iteration k and time t are:

1. Local visibility of node j from i, scaled exponentially by the parameter β_A or β_B, depending on the type of ant. The higher the local visibility, the higher the desirability of choosing that edge. The degree of desirability can be varied by properly selecting β_A and β_B.
2. Pheromone level, $\tau_{ij}(t)$, on the edge at time t. Since edges which are part of better solutions are positively reinforced presence of a high pheromone level on an edge is used to boost the desirability of choosing that edge. A very high pheromone level on any edge, therefore, makes it much more probable for that edge to be included in the final tree.

A pseudo code representation of the ACO algorithm is given in *Algorithm 2*.

Algorithm 2: Pseudo Code Representation of ACS Algorithm [24]

1. Set $t = 0$.
2. Set $\tau_{ij}(0) = \tau^{\min} : \forall (i \to j) \in E$.
3. $T^{best} \to$ tree grown by the global best ant and Y^{best} its cost.
4. $T^{best}(t) \to$ the best tree grown by any ant during iteration t
         and $Y^{best}(t)$ its cost.
5. while $(t < t^{\max})$
    for $(m = 1 : N_A + N_B)/ *$ ant number $*/$
    /* *Tree building depends on whether the ant is Type A or B* */

•*Build the tree $T_m(t)$*; / * see Algorithm 1*/
/*prune (T,D) is a function which takes as input a tree T and
the destination set, D, and outputs the pruned tree * /
•$T_m(t) \leftarrow prune(T_m(t), D)$;
•*Compute the cost $C_m(t)$ of $T_m(t)$*
/ * Pheromone update */
• for all $(i \rightarrow j) \in T_m(t)$
    $\tau_{ij}(t) \leftarrow \rho/C_m(t) + (1 - \rho)\tau_{ij}(t)$
end for
/* Pheromone thresholding
• for all $(i \rightarrow j) \in T_m(t)$
$$\tau_{ij}(t) = \begin{cases} \tau^{\min} & \tau_{ij}(t) < \tau^{\min} \\ \tau^{\max} & \tau_{ij}(t) > \tau^{\max} \end{cases}$$
    end for
end for
if $(t == 0)$
    $T^{best} \leftarrow T^{best}(t), \ Y^{best} \leftarrow Y^{best}(t)$;
else
  if $(Y^{best}(t) < Y^{best})$
      $T^{best} \leftarrow T^{best}(t), \ Y^{best} \leftarrow Y^{best}(t)$;
      end if

  end if
  / * Increment t*/
  •$t \leftarrow t + 1$;
end while
6. Print T^{best} and Y^{best}

Performance of the algorithm above has been compared to optimum algorithm presented in Section 12.4.1 in [24]. Performance degradation of less than 10% was obtained while the computation time was reduced significantly.

12.5   Biologically Inspired (BI) Distributed Topology Control

By adjusting the transmission power of mobile nodes, topology control aims to reduce interlink interference, reduce overall energy consumption, and increase effective network capacity, subject to connectivity constraints.

Topology control could be considered as a power assignment problem in a network with the set of $V = \{v_1, v_2, \ldots, v_n\}$ nodes. Each node is equipped with the same finite set P_s of available power discrete levels, where $P_s = \{p_1, p_2, \ldots, p_k\}$, $p_1 < p_2 < \cdots < p_k$, and $p_k \equiv P_{full}$. So, the size of the search space of the problem is $O(k^n)$. The network is assumed to be connected when every node is assigned P_{full}, which is the case when no topology control is in effect. The goal of topology control is to assign transmission power to each node (generate power assignment $A = \{P_v, |\forall v_i \in V\}$) so as to achieve optimization objectives subject to given connectivity constraints.

In this section, we discuss the possibility of using the biological paradigm of *Swarm Intelligence*, introduced in Section 12.2, to design a distributed topology control algorithm referred to as *Swarm Intelligence*-based Topology Control (SITC) for mobile ad hoc networks. SITC is scalable, so that each node determines its own power assignment based on local information collected from neighbor nodes located within its full power transmission range.

SITC attempts to either

a) minimize the maximum power used by any node in the network, $P_{max} = \max_{i=1}^{n} P_v$, (*minimax* objective); or

b) minimize the total power used by all of the nodes in the network, $P_{tot} = \sum_{i=1}^{n} P_v$, ' (*minitotal* objective), subject to connectivity constraints. In addition, the protocol may achieve a common power optimization, defined by $\forall i, 1 \leq i \leq n, P_v, = P_{compow}$, as a by-product while minimizing maximum power, and may properly assign power to nodes with non-uniform distribution while minimizing total power.

In SITC, the (intensity of) pheromone in swarm intelligence is used to denote the *goodness* of power assignments. Initially, power is searched and assigned randomly. Gradually, better power assignments, those that achieve either the *MinMax* or the *MinTotal* objective subject to connectivity constraints, are preferred in the search and their associated pheromone intensities increase, which at the same time further reinforces these power assignments. Through such a positive feedback process, SITC quickly converges to (locally) optimal power assignments. In contrast, negative feedback via pheromone *evaporation* renders stale old power assignments that are no longer effective. In addition, amplification of fluctuation explores and discovers new or better power assignments to accommodate changing topology due to mobility. In the end, nodes decide power assignments locally by collecting local information, via sending and receiving ant packets, without relying on any central entity (multiple interaction).

12.5.1   SITC Protocol Description

The protocol described in the previous section applies also to an Ant Colony System (ACS), to the power control problem of ad hoc networks. Specifically, each source (root) node applies the ACS algorithm directly to 'compute' a minimum-power broadcast tree with global location information so as to update the pheromone trails in a centralized manner. Therefore, this might not be applied directly to mobile ad hoc networks. In contrast, SITC is a distributed protocol that uses only local information, and adapts well to mobility. Furthermore, SITC deals with the topology control problem in general with power related optimization objectives, which is different from that described in the previous section.

In the SITC protocol every node v_i executes a neighbor discovery protocol periodically using the full power P_{full} to obtain its (current) neighbor set, $N(v_i)$ (the set of nodes that are located within $v_i's$ full power transmission range). In addition, node v_i (periodically) broadcasts ant packets using different transmission power levels, and the values of the transmission power are also carried inside the ant packets. An ant packet will then be *relayed* (via broadcast) by using the *same* (original) transmission power. In the meantime, upon receiving ant packets, node v_i evaluates the condition under which node v_i can receive ant packets originated from *all*, and relayed *only* by, the nodes in $N(v_i)$ by using power p. If this condition holds, SITC assigns p to P_{v_i}. It has been proved [25, T.8.8] that when this condition is satisfied at every node, the resulting power assignment guarantees network connectivity.

Figures 12.7 and 12.8 illustrate how such a mechanism guarantees network connectivity. Figure 12.7(a) shows the transmission power ranges of three discrete power levels in $P_s : p_1 < p_2 < p_3 \equiv P_{full}$, and the topology when every node is assigned p_3. Figure 12.7(b) shows the routes via which ant packets originated from node l's neighbors are relayed with power p_2. Although, for each neighbor, multiple routes could exist, the route with the lowest hop count is shown. Similar to Figure 12.7(b), Figure 12.7(c) shows the routes when ant packets are relayed with power p_1. Since ant packets originated from *all* of node l's neighbors can be relayed with either p_1, p_2, or p_3 (the trivial case) and $p_1 < p_2 < p_3$, node 1 is assigned p_1. Similarly, Figure 12.8(a) and Figure 12.8(b) show the cases for node 3. Since node 7 is not node $3's$ neighbor, ant packets originated from node 7 are ignored. For the same reason, even an ant packet originated from node 2 can be relayed with p_1 to node 3 via node 7, that packet will also be ignored by node 3. Therefore, node 3 is assigned p_2, which could have been p_1 if node 7 were node $3's$ neighbor. After every node applying the same algorithm, we obtain the final power assignment and the

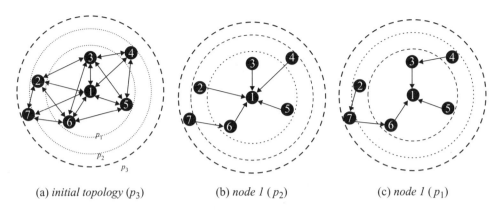

| (a) *initial topology* (p_3) | (b) *node 1* (p_2) | (c) *node 1* (p_1) |

Figure 12.7   Examples of SITC's connectivity guarantee.

corresponding topology is shown in Figure 12.8(c). The resulting topology is connected with P_{max} equal to p_2. If node 7 were node $3's$ neighbor, SITC could have resulted in P_{max} equal to p_2. As mentioned early, SITC is a distributed algorithm where nodes only collect local information from nearby nodes (neighbor sets), and thus does not ensure global optimality with respect to both the *minimax* and the *mintotal* objectives.

When there is no mobility, every node v_i *only* needs to originate ant packets using each one of the power levels *once*. In the meantime it also needs to collect ant packets from all of its neighbors to determine $P_{i_{min}}$ which is the *minimal* power of all those power levels (for instance, p) that satisfy the aforementioned condition '*node v_i can receive ant packets originated from all, and relayed only by, the nodes in $N(v_i)$ by using power p'*'.

In the presence of mobility, the neighbor set of each node will keep changing. Therefore, a node may not have enough time to collect ant packets transmitted with all the different power levels originated from all its neighbors before its neighbor set changes, unless the broadcast period is short enough, which imposes high control overhead. In the following we discuss how swarm intelligence can be used as a heuristic search mechanism to discover a minimized power level that best meets the condition aforementioned.

In SITC, each node has a pheromone table which maintains a time-varying pheromone value for each of its neighbors for every power level, as shown in Table 12.1. The 'goodness' of a power assignment for a node is defined as the weighted sum of pheromone and heuristics values of the node's neighbors,

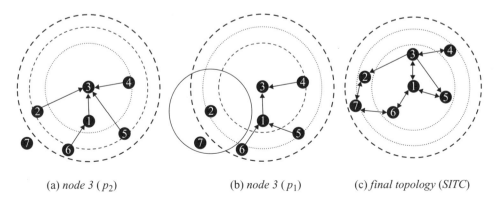

| (a) *node 3* (p_2) | (b) *node 3* (p_1) | (c) *final topology* (*SITC*) |

Figure 12.8   Examples of SITC's connectivity guarantee.

Table 12.1　Pheromone table at node v_1 in Figure 12.7 and Figure 12.8

Node $v_1(t_0)$	v_2	v_3	v_4	v_5	v_6	v_7
$p1$	$\tau_{v_1,v_2,p1}(t)$	$\tau_{v_1,v_3,p1}(t)$	$\tau_{v_1,v_4,p1}(t)$	$\tau_{v_1,v_5,p1}(t)$	$\tau_{v_1,v_6,p}1(t)$	$\tau_{v_1,v_7,p1}(t)$
$p2$	$\tau_{v_1,v_2,p2}(t)$	$\tau_{v_1,v_3,p2}(t)$	$\tau_{v_1,v_4,p2}(t)$	$\tau_{v_1,v_5,p2}(t)$	$\tau_{v_1,v_6,p}2(t)$	$\tau_{v_1,v_7,p2}(t)$
$p3$	$\tau_{v_1,v_2,p3}(t)$	$\tau_{v_1,v_3,p3}(t)$	$\tau_{v_1,v_4,p3}(t)$	$\tau_{v_1,v_5,p3}(t)$	$\tau_{v_1,v_6,p}3(t)$	$\tau_{v_1,v_7,p3}(t)$

which is proportional to the number of neighbors from which ant packets originated are received, and the inverse of power level of these ant packets. It is so defined to serve as a heuristics for the optimal local power assignment, in which case a node receives ant packets with the lowest common power from the maximum number of its neighbors. In addition, pheromone values degrade (evaporate) over time such that the previously favored power levels that are not reinforced for some time, say, due to mobility, will be gradually unfavored. SITC balances the positive feedback which reinforces pheromone intensities and expedites the convergence to good power assignments, and the amplification of fluctuation which searches for new or better power assignments. At the same time, pheromone values are evaporated over time to stale old power assignments that are no longer effective. In such way SITC converges quickly to (local) optimal power assignments while adapts well to mobility.

Algorithm 3 presents the SITC protocol, and the notations used in the protocol are listed in the following table

---
Notation used in the SITC protocol

P_s- the power set common to all nodes

k -the number of power levels in P_s

P_{full}- the maximum power in P_s (full power)

P_{v_j}- the selected power for node v_j

seq (v_j)- the sequence number for ant packets originated from node v_j
　　totalPower -the total power left for an ant packet to be forwarded

relay Set -the set of nodes that have originated or forwarded an ant
　　packet

objective -the objective of the algorithm {*minmax, mintotal*}

Each node, say v_j, executes this protocol asynchronously and independently from other nodes, and SITC protocols running on different nodes interact via sending and receiving ant packets. SITC is an *event-driven* protocol which reacts to three types of events: (a) ticks of periodic timer to originate ant packets; (b) receptions of ant packets to update the local pheromone table and decide whether to forward ant packets; and (c) ticks periodic timer to evaporate pheromone values. The field *totalPower* or *totalP* in an ant packet and the condition using it (i.e., $totalP \geq tx\,P$ in line (22) of *Algorithm 3* are not required for the correctness of the algorithm. Instead, the use of *totalPower* or *totalP* is to limit how far an ant packets could be forwarded. Line (22) discards ant packets that are forwarded by too many neighbors. Condition, $totalP \geq tx\,P$ is checked up each time an ant packet is forwarded. By doing so, an ant packet does not use more total power for delivery from a node to its neighbors than the case of no topology control.

Algorithm 3: SITC Algorithm

1. **INPUT:** $P_s = \{p_1, p_2 \ldots, p_k\}$, $p_1 < p_2 < \cdots < p_k$, $p_k \equiv P_{full}$,
　　objective $\in \{minmax, mintotal\}$.
2. **OUTPUT:** $P_{v_i} \in P_s$ is the power assignments for nodes v_j.
3. **START**
4. $P_{v_j} \leftarrow P_{full}$ entries in pheromone table are set to 0;
5. **DO**

6. **upon event** periodic timer for *ant packet origination* ticks
7. select power P_{best} according to Equation (12.20);
8. $P_{v_j} \leftarrow P_{best}$
9. **if** *objective==minmax* **then**
10. $txPower \leftarrow P_{best}$;
11. **end if**
12. **if** *objective ==mintotl* **then**
13. choose power *txPower* to the next power level in a round-robin order;
14. **end if**
15. $seq(v_j) + +$; *totalPower* $\leftarrow P_{full}$; *relaySet* $\leftarrow \{v_j\}$;
16. With power *txPower*,
    broadcast ant packet $< v_j, seq(v_j), txPower, totalPower,$
    relaySet $>$;
17. **end upon event**
18. **upon event** receiving an ant packet $< origin, seq, txP\ totalP, relay\ S >$
19. **if** the ant packet, identified by $< origin, seq >$, has not been received
    recently **then**
20. Update local pheromone table according to Equation (12.3) with
21. $org \leftarrow origin$ and $p \leftarrow txP$;
22. **if** $(\forall v_i \in relayS, v_i$ is $v_j's$ neighbor) $\wedge(totalP \geq txP)$ **then**
23. broadcast ant packet $< origin, seq, txPtotalP - txP\ relay$
    $S \cup \{v.\} >$;
24. **end if**
25. **end if**
26. **end upon event**
27. **upon event** periodic timer *for pheromone evaporation* ticks
28. Evaporate local pheromone values according to Equation (12.4);
29. **end upon event**
30. **end DO**
31. **END**

Each node v_j has a pheromone table which maintains a time-varying pheromone value $\tau_{v_j org,p}(t)$ for different originators of ants (org) with different power levels (p). When the periodic timer to originate ant packets ticks, in line (7) SITC selects power level P_{best} as

$$P_{best}(t) = \begin{cases} \arg\max_{p \in P_s} \left(\Phi_{v_j p}(t) + \beta \cdot \eta_p\right); & \text{if } q \leq q_0 \\ P; & \text{if } q \geq q_0 \end{cases} \qquad (12.20)$$

where $\Phi_{v_j,p}(t) \equiv \sum_{org \in NbrSet(v_j)(t)} \tau_{v_j,org,p}(t)$ and $NbrSet(v_j)(t)$ is $v_j's$ neighbor set at time t. q is a random variable uniformly distributed over $[0, 1]$, q_0 is a tunable parameter in $[0, 1]$, and $P \in P_s$ is the power level randomly selected with probability

$$prob_{v_j P}(t) = \frac{\Phi_{v_l,P(t)+\beta \cdot \eta_P}}{\sum_{p \in P_s} \Phi_{v_j p(t)+\beta \cdot \eta_P}} \qquad (12.21)$$

η_P is a heuristics defined as $(\gamma + 1/p)$ that contributes inversely with power level, and thus lower power levels are favored, where γ is used to adjust the relative weight of different power levels. β balances the relative weights between τ and η. q_0 sets the preference between exploration of new solutions and exploitation of knowledge collected from old solutions.

Upon receiving an ant packet originated from node org with power p, SITC increases the corresponding pheromone value as

$$\tau_{v_j,org,p}(t) \leftarrow \tau_{v_j,org,p}(t) + \tau_0 \tag{12.22}$$

When the pheromone evaporation timer with time interval T_{ph} ticks, SITC degrades pheromone values in the pheromone table as,

$$\tau_{v_j nbr,p}(t) \leftarrow \tau_{v_j nbr,p}(t) \cdot (1/2)^{Re} \tag{12.23}$$
$$\forall nbr \in NbrSet(v_j)(t) \text{ and } \forall p \in P_s$$

where Re is evaporation rate given as $R_e = T_{ph}/T_{ph/2}$. Parameter $T_{ph/2}$ is the time it takes for the pheromone value to be halved referred to as pheromone half life. In addition we have $\tau_0 = \gamma + 1/P_{full}$.

12.6   Optimization of Mobile Agent Routing in Sensor Networks

In conventional fusion architectures, all the sensor data is sent to a central location where it is fused [17, Chapter 14]. But, the transmission of data in military distributed sensor network (DSN) increases the risk of being intercepted in addition to consuming the limited resources such as battery power and network bandwidth. To meet these challenges, the concept of *mobile agent-based distributed sensor networks* (MADSNs) has been proposed [26] wherein a mobile agent visits the sensors selectively and fuses the appropriate measurement data incrementally. Mobile agents are special programs that can be sent from a source node to be executed at remote nodes. Upon arrival at a remote node, a mobile agent presents its credentials, obtains access to local services and data to collect needed information or perform certain actions, and then departs with the results.

The order and number of nodes on the route traversed by a mobile agent determine the energy consumption, path loss, and detection accuracy and, hence, have a significant impact on the overall performance of a MADSN. The computation of a suitable route, in practice, involves a tradeoff between the cost (energy consumption and path loss) and the benefit (detection accuracy). Visiting more sensors improves the quality of fused data, but also raises the communication and computing overheads. Alternative algorithms based on the *local closest first* (LCF) and *global closest first* (GCF) heuristics have been used to compute routes for mobile agents in MADSNs [26]. Their performance has been quite satisfactory for small DSNs that are systematically deployed in simple environments. However, their performance deteriorates as the network size grows and the sensor distributions become more complicated.

In this section we consider DSNs with geographically distributed sensors that are deployed for the purposes of target classification and tracking, both of which require the fusion of measurements from a number of sensors. The measurements from the sensors that receive strong signals from the target are often most useful in the fusion process and, thus, the fusion step is preceded by identifying a subset of sensor nodes that strongly indicate the presence of a target. By identifying such sensor nodes, the complexity and amount of sensor data for fusion operation can be significantly reduced. Thus, we focus on identifying a route for a mobile agent through such sensor nodes by utilizing the *signal strengths of the sensor nodes*. The mobile agent then visits these nodes and performs the required fusion of the sensor information available at these nodes. Each sensor can transmit messages with certain energy costs, and the transmissions are subject to wireless propagation losses. We formulate the *mobile agent routing problem* (MARP) in an MADSN as a combinatorial optimization problem involving the cost of communication, path loss due to wireless propagation, and signal energy received by the sensors. The overall routing objective is to maximize the sum of the signal energy received at the visited nodes while minimizing the power needed for communication and the path losses. Therefore, the objective function is directly proportional to the signal energy and inversely proportional to the path loss and cost of communication. We show that MARP is NP-complete by using a reduction from a variation of the known solution for $3D$ traveling salesman problem. We then discuss in the subsequent section an approximate solution based on a *genetic algorithm* (GA) that employs a two-level encoding scheme and genetic operators tailored to the objective function, which aims to achieve the total signal energy equal to or above a given level.

An MADSN typically consists of three types of components: *processing elements* (PE), *sensor nodes*, and *communication network*, where the first two elements are interconnected via a wireless communication network. A group of neighboring sensor nodes that are commanded by a single PE forms a *cluster*. The amount of signal energy that reaches an individual sensor is an effective indicator of how close the node is to a potential target. In target detection applications, a leaf node with higher signal energy carries more information and should have higher priority of being visited. To simplify computation, we use a quantitative value to represent the level of signal energy detected by a local sensor node.

Wireless communication links need to be established between neighboring nodes as the mobile agent migrates along a route.

A mobile agent is sent from a *processing element* and is expected to visit a subset of sensors within the *cluster* to fuse data collected in the coverage area. Generally, the more sensors visited, the higher the accuracy achieved using any reasonable data fusion algorithm will be [27]. At the same time, it is important to select an appropriate route so that the required signal energy level can be achieved with a low cost in terms of total energy consumption and path loss.

For simplicity we will consider a sensor network that contains one PE, labeled as S_0, and N leaf nodes, labeled S_i, $i = 1, 2, \ldots, N$, some of which might be down. The sensor nodes are spatially distributed in a surveillance region of interest, each of which is responsible for collecting measurements in the environment. The signal energy detected by sensor node S_i is denoted by s_i, $i = 1, 2, \ldots, N$. Sensor node S_i takes time $t_{i,acq}$ for data acquisition and time $t_{i,proc}$ for data processing. The wireless communication link with physical distance d_{ij} between sensor node S_i and S_j has channel width W bits and operates at frequency B Hz. Some sensor nodes may be down temporarily due to intermittent failures.

The routing objective is to find a path for a mobile agent that satisfies the desired detection accuracy while minimizing the energy consumption and path loss. The required path is computed based on the current state of DSN and the mobile agent traverses the nodes along the path while performing the fusion operation. The energy consumption depends on the processor operational power and computation time and the path loss is directly related to the physical length of the selected path. In the following we define these quantities.

12.6.1 Objective Function for System Optimization

The objective function for the mobile agent routing problem is based on three aspects of a routing path: energy consumption, path loss, and detected signal energy.

The energy consumption at a sensor node is determined by the processing speed and the computation time. We assume that the message includes the mobile agent code of size M bits and measured data of size D bits. For a given desired resolution, a fixed data size D is used to store the partially integrated data at each sensor. The time for the message to be transmitted over a wireless channel of bandwidth BW bps is calculated as $t_{msg} = \lceil (M + D)/BW \rceil$.

The energy consumption EC of path P, consisting of nodes $P[0], P[1], \ldots, P[H - 1]$, is defined as:

$$EC(P) = a \cdot (t_{0,setup} + t_{0,proc}) \cdot F_0^2 + P_{0,t} \cdot t_{msg}$$
$$+ \sum_{k=1}^{H1} \{b \cdot \{(t_{P[k],acq} + t_{P[k],proc}) \cdot F_{P[k]}^2\} + P_{P[k],t} \cdot t_{msg}\}) \tag{12.24}$$

where the kth leaf node $S_{P[k]}$ on path P has data acquisition time $t_{P[k],acq}$, data processing time $t_{P[k],proc}$, operational level $F_{P[k]}$, transmitting power $P_{P[k],t}$ and node $P[0] = 0$ corresponds to the PE. Here, the operational level refers to the processor operating frequency, the square of which determines its corresponding operating power level. Coefficients a and b are chosen suitably to 'normalize' the operational level to its power level.

Path loss calculation is based on the fact that the power received by sensor S_j has the following relation with the power transmitted by sensor S_i

$$P_{j,r}(d_{l,j}) = P_{x,t} \cdot \frac{G_{x,t} G_{Jr}, \lambda^2}{(4\pi)^2 d_{l,j}^2 \beta} \tag{12.25}$$

where $G_{i,t}$ is the gain of sensor S_l as a transmitter and $G_{j,r}$ is the gain of sensor S_j as a receiver, λ is the signal wavelength and β is the system loss factor. The physical distance $d_{i,j}$ between S_i and S_j is computed from their spatial locations. *Path loss (PL)* is given as

$$PL(d_{l,j}) = 10 \log \frac{P_{i,t}}{P_{j,r}} = 10 \log \left[\frac{(4\pi)^2 \beta}{G_{i,t} G_{j,r} \lambda^2} \cdot d_{i,j}^2 \right] \tag{12.26}$$

So, the total path loss along path P can be calculated as:

$$PL(P) = \sum_{k=0}^{H-1} [10 \log \left\{ \frac{(4\pi)^2 \beta}{G_{P[k],t} G_{P[(k+1) \bmod H],r} \lambda^2} \times d_{P[k], P[(k+1) \bmod H]}^2 \right\} \tag{12.27}$$

Signal energy collection is based on the fact that a sensor node detects a certain amount of energy emitted by the potential target, which may be used by a mobile agent for integration. A mobile agent always tries to accumulate as much signal energy as possible for accurate decision in target detection. The sum of the detected signal energy SE along path P is defined as

$$SE(P) = \sum_{k=1}^{H1} s_{P[k]} \tag{12.28}$$

where $s_{P[k]}$ is the signal energy detected by the kth sensor node on path P. By combining the above three components of a routing path we define the objective function as

$$O(P) = SE(P) \left\{ \frac{1}{EC(P)} + \frac{1}{PL(P)} \right\} \tag{12.29}$$

where the terms *SE(P)*, *EC(P)*, and *PL(P)* are assumed to be 'normalized' to appropriately reflect the contribution by various loss terms. Intuitively, this objective function prefers paths with higher signal energies by penalizing those with high path losses and energy consumption. For the detection applications, we are interested in concluding the presence of a target in the monitoring area, which is determined by a threshold level of detected energy. For example, this threshold can be determined to maximize the probability of detection while keeping the false alarm rate below a specified quantity. A path providing high signal energy at the expense of a considerable amount of energy consumption and path loss may not be preferable.

Alternative objective functions may be also used as long as they correctly reflect the tradeoff between detected signal energy, energy consumption, and path loss.

12.7 Epidemic Routing

Epidemic routing [28] is an approach for routing in sparse *and/or* highly mobile networks in which there may not be a contemporaneous path from source to destination. Epidemic routing is based on a so-called *store-carry-forward* paradigm where a node receiving a packet buffers and carries that packet as it moves. Later when the occasion permits, it passes the packet on to new nodes that it encounters on its way. Analogous to the spread of infectious diseases, each time a packet-carrying node encounters a new node that does not have a copy of that packet, the carrier is said to *infect* this new node by passing on a packet copy. The newly infected nodes, pass the infection, in the same way. The destination receives the packet when it first meets an infected node. The minimum delivery delay is achieved at the expense of increased use of resources such as buffer space, bandwidth, and transmission power. Variations of epidemic routing have also been used that exploit this trade-off between delivery delay and resource consumption, including K-hop schemes, probabilistic forwarding, and spray-and-wait [29, 30].

In this section, we present unified framework, based on Differential Equations (DE), to study epidemic routing and its variations. The starting point is [31,32] where the authors consider common node mobility models (e.g., random waypoint and random direction mobility) and show that nodal inter-meeting times are nearly exponentially distributed when transmission ranges are small compared to the network's area, and node velocity is sufficiently high. This observation suggests that Markovian models of epidemic routing can lead to quite accurate performance predictions. [31] develops Markov chain models for epidemic routing and 2-hop forwarding, deriving the average source-to-destination delivery delay and the number of copies of a packet at the time of delivery. An analytical study of such Markov chain models is quite complex for even simple epidemic models.

*DE*s as a fluid limit of Markovian models such as [31] is developed in [32], under an appropriate scaling as the number of nodes increases. This approach enables derivation of closed-form formulas for the performance metrics considered in [31], obtaining matching results.

The *network model* assumes $N + 1$ mobile nodes moving in a closed area according to a random mobility model. When two nodes come within transmission range of each other, they can forward packets to each other.

We assume the inter-meeting time of any pair of nodes is an exponential random variable with rate β. As the node density is low and the transmission range is short, we ignore interference among nodes. When two nodes meet, the transmission between them succeeds instantaneously. There are $N + 1$ source-destination pairs, with each node being the source of one flow, and the destination of another flow. Each source generates packets according to Poisson process with rate λ. Each data packet includes a sequence number in its header.

The *Epidemic Routing Protocol* works in such a way that each node stores and forwards packets destined for other nodes [28]. Along with the data packet, each node maintains a *summary vector* that indicates the set of packets that are stored in its buffer.

When two nodes come within transmission range of each other, they first exchange their summary vectors. Next, based on this information, each node requests packets that are not in its buffer. Finally, they transmit the requested packets to each other.

Under epidemic routing, packets can arrive at the destination out of order. The sequence number allows the destination node to reorder packets and discard duplicates.

Performance Metrics is based on the analogy between epidemic routing and disease spreading. The specific packet is considered as a disease, and a node that has a copy of a packet is called an *infected* node. A node that does not have a copy of a packet, but can potentially store and forward a copy, a *susceptible* node. Once a node carrying a copy meets the destination, it deletes the copy and keeps 'packet-delivered' information so that it will not be forwarded the packet again. We call such information *anti-packet* and say that the nodes are *recovered*. The average *lifetime*, L, of a packet is the time from the moment when the packet is generated at the source node to the time instant when all copies of the packets are removed (i.e. no more infected nodes for this packet in the network).

Three performance metrics, *delivery delay, loss probability* and *power consumption*, are of interest. The delivery delay of a packet, T_d, is the duration of the time from when the packet is generated at the source to the time the packet is first delivered to the destination. For the case where nodes have a limited amount of buffer, a packet might be dropped from the network before it is delivered. The loss probability is the probability of a packet being dropped from the network before delivery.

Two metrics related to the power consumption are considered: the number of times a packet is copied in its entire lifetime, G, and the number of times a packet is copied at the time of delivery, C.

It was shown in [31] that the pairwise meeting time between nodes can be approximated as distributed exponentially, if nodes move in a limited region (of area, A) according to common mobility models [32] with transmission range (d) small compared to A, and sufficiently high speed. In this case the pairwise meeting rate β can be approximated as $\beta \approx 2wdE[V]/A$ where w is a constant specific to the mobility models, and $E[V]$ is the average relative speed between two nodes. Under this approximation, [31] showed that the evolution of the number of infected nodes can be modeled as a Markov chain.

The *Differential equation (DE)* based modeling approach starts from the Markov model for simple epidemic routing. Given $n_I(t)$, the number of infected nodes at time t, the transition rate from state n_I to state $n_I + 1$ is $r_N(n_I) = \beta n_I(N - n_I)$, where N is the total number of nodes in the network (excluding the destination). If we rewrite the rates as $r_N(n_I) = N\lambda(n_I/N)(1 - n_I/N)$ and assume that $\lambda = N\beta$ is constant, we can prove [33] that, as N increases, the fraction of infected nodes (n_I/N) converges asymptotically to the solution of the following equation

$$i'(t) = \lambda i(t)(1 - i(t)), \quad \text{for} \quad t \geq 0 \tag{12.30}$$

with initial condition $i(0) = \lim_{N \to \infty} n_I(0)/N$.

The average number of infected nodes then converges to $I(t) = Ni(t)$ as $\forall \varepsilon > 0$, $\lim_{N \to \infty}$ $Prob\{| \sup_{s \leq t}\{n_I(s)/N - i(s)\}| > \varepsilon\} = 0$. The following equation can be derived for $I(t)$,

$$I'(t) = \beta I(N - I) \tag{12.30a}$$

with initial condition $I(0) = Ni(0)$. Such differential equation DE, which results as a fluid limit of a Markov model as N increases, has been commonly used in epidemiology studies, and was first applied to epidemic routing in [34] as an approximation.

The *delay under epidemic routing* T_d will be characterized by its Cumulative Distribution Function (CDF) $P(t) = Pr(T_d < t)$ which can be derived by starting from $P_N(t)$ the CDF of T_d when the number of nodes in the system is $N + 1$, i.e., there are N nodes plus one destination node.

$$
\begin{aligned}
P_N(t + dt) - N(t) &= Prob\{t \leq T_d < t + dt\} \\
&= Prob \{\text{destination meets an infected node in } [t, t + dt]|T_d > t\} \\
&= Prob \{\text{destination meets an infected node in } [t, t + dt]\}(1 - P_N(t)) \\
&= E\{Prob\{\text{destination meets one of the } n_I(t) \text{ infected nodes in } [t, \ t + dt]|n_I(t)\}\} \times (1 - P_N(t)) \\
&\approx E\{\beta n_I(t)dt\}(1 - P_N(t)) \\
&= \beta E\{n_I(t)\}(1 - P_N(t))dt = \lambda E\left\{\frac{n_I(t)}{N}\right\}(1 - P_N(t))dt
\end{aligned}
$$

which results in

$$\frac{dP_N}{dt} = \lambda E\left\{\frac{n_I(t)}{N}\right\}(1 - P_N(t))$$

As $N \to \infty$ we have $E\{n_I(t)/N\} \to i(t)$, and $P_N(t)$ converges to the solution of the following equation:

$$P'(t) = \lambda i(t)(1 - P(t)) \tag{12.31}$$

For a finite population of size N we can consider:

$$P'(t) = \beta I(t)(1 - P(t)) \tag{12.31a}$$

Equation (12.31a) was proposed in [34], based on an analogy with a Markov process. Solving (12.30a) and Equation (12.31a) with $I(0) = 1$ *and* $P(0) = 0$, gives

$$I(t) = \frac{N}{1 + (N-1)e^{-\beta Nt}}$$

$$P(t) = 1 - \frac{N}{N - 1 + e^{\beta Nt}}$$

(12.32)

and the average delivery delay as:

$$E[T_d] = \int_0^\infty (1 - P(t))dt = lnN/(\beta(N-1))$$

(12.33)

The average number of copies of a packet in the system when the packet is delivered to the destination, $E[C_{ep},]$ will be derived later in this section.

When a node delivers a packet to the destination, it should delete the copy from its buffer both to save storage space, and to prevent the node from infecting other nodes. But if the node does not store any information to keep itself from receiving the packet again (i.e. becomes susceptible to the packet), a packet would generally be copied, and the infection would never die out. In order to prevent a node from being infected by a packet multiple times, an anti-packet can be stored in the node when the node delivers a packet to the destination. This scheme is referred to as IMMUNE scheme. A more aggressive approach to delete obsolete copies is to propagate the anti-packets among the nodes. The anti-packet can be propagated only to those infected nodes (IMMUNE scheme) or also to susceptible nodes (VACCINE scheme).

The infection and recovery process can be modeled by Markov model. In order to derive the limiting equation the number of destinations, n_D, need to be normalized to the number of nodes N. We first consider IMMUNE scheme. Let $n_R(t)$ denote the number of recovered nodes at time t, then the state can be denoted as $(n_I(t), n_R(t))$ and transition rate

$$r_N((n_I(t), n_R(t)), (n_I(t) + 1, n_R(t))) = \beta n_I(t)(N - n_I(t) - n_R(t))$$

(12.34)

and

$$r_N((n_I(t), n_R(t)), (n_I(t) - 1, n_R(t) + 1)) = \beta n_I(t)n_D$$

(12.35)

The transition rates can be similarly written in a 'density dependent' form, given that the number of destinations n_D scales in a manner similar to the scaling of the number of initially infected nodes, i.e., $\lim_{N\to\infty} n_D/N = d$. Therefore by Theorem 3.1 in [33], we get as N increases, the fraction of infected nodes (n_I/N) and recovered nodes (n_R/N) converge asymptotically to the solution of the following equations:

$$i'(t) = \lambda i(t)(1 - i(t) - r(t)) - \lambda i(t)d, \; for \; t \geq 0$$

$$r'(t) = \lambda i(t)d, \; for \; t \geq 0$$

(12.36)

where $d = n_D/N$, and $i(0) = \lim_{N\to\infty} n_I(0)/N, r(0) = 0$.

The number of infected and recovered nodes then converges to $I(t) = Ni(t), \; R(t) = Nr(t)$. From Eq.(12.36) we have

$$I'(t) = \beta I(N - I - R) - \beta I n_D$$

$$R'(t) = \beta I n_D$$

(12.37)

with $I(0) = Ni(0), R(0) = 0$. We consider $I(0) = 1, \; R(0) = 0, \; D = 1$.

DE models for IMMUNE and VACCINE scheme can be also derived from Markov model. If for simplicity the dependence on time is omitted for *IMMUNE* : $r_N((n_I, n_R), (n_I + 1, n_R)) = \beta n_I(N - n_I - n_R)$, and $r_N((n_I, n_R), (n_I - 1, n_R + 1)) = \beta n_I(n_R + n_D)$. The limiting equations are:

$$
\begin{aligned}
i'(t) &= \lambda i(t)(1 - i(t) - r(t)) - \lambda i(t)(r(t) + d), \ \text{for } t \geq 0 \\
r'(t) &= \lambda i(t)(r(t) + d), \ \text{for } t \geq 0
\end{aligned}
\tag{12.38}
$$

and consequently

$$
\begin{aligned}
I'(t) &= \beta I(N - I - R) - \beta I(1 + R) \\
R'(t) &= \beta I(1 + R)
\end{aligned}
\tag{12.39}
$$

For *VACCINE* we need to specify how many destination nodes have received the packet. Let n_{DR} denotes this number. For the previous schemes there is no such a need because a destination can recover only an infected node. Hence even if the destination has not received the packet, the destination receives it when it meets the infected node.

We assume that all the destinations have to receive the packets from an infected node. Here different assumptions can be made, for example a destination could receive the packet from another destination, or a destination could receive the antipacket from a recovered node and propagate it without having received the packet. The latter case is meaningful when we deal with an anycast communication (the packet has to reach at least one of the destinations) or if we can rely on the fact all the destinations will receive a copy of the packet from the destination that started the recovery process. These different assumptions lead to minor differences in the final equations. The state transition rates are:

$$
\begin{aligned}
r_N((n_I, n_R, n_{DR}), (n_I + 1, n_R, n_{DR})) &= \beta n_I(N - n_I - n_R) \\
r_N((n_I, n_R, n_{DR}), (n_I - 1, n_R + 1, n_{DR})) &= \beta n_I(n_R + n_{DR}) \\
r_N((n_I, n_R, n_{DR}), (n_I - 1, n_R + 1, n_{DR} + 1)) &= \beta n_I(n_D - n_{DR}) \\
r_N((n_I, n_R, n_{DR}), (n_I, n_R + 1, n_{DR})) &= \beta(N - n_I - n_R)(n_R + n_{DR})
\end{aligned}
\tag{12.40}
$$

By using notation $d_r(t) = \lim_{N \to \infty} (n_{DR}/N)$ we have the limit

$$
\begin{aligned}
i'(t) &= \lambda i(t)(1 - i(t) - r(t)) - \lambda i(t)(r(t) + d), \ \text{for } t \geq 0 \\
r'(t) &= \lambda i(t)(r(t) + d) + \lambda(1 - i(t) - r(t))(r(t) + d_r(t)), \ \text{for } t \geq 0 \\
d_r'(t) &= \lambda i(t)(d - d_r(t)), \ \text{for } t \geq 0
\end{aligned}
\tag{12.41}
$$

For the average values ($Ni(t)$, $Nr(t)$ and $Nd_r(t)$), with $N_D = 1$, we observe that $Nd_r(t)$ satisfies the same *DE* as $P(t)$, and

$$
\begin{aligned}
I'(t) &= \beta I(t)(N - I(t) - R(t)) - \beta I(t)(R(t) + 1) \\
R'(t) &= \beta I(t)(1 + R(t)) + \beta(N - I(t) - R(t))(R(t) + P(t)) \\
P'(t) &= \beta I(t)(1 + P(t))
\end{aligned}
\tag{12.42}
$$

These DE models enable us to evaluate the number of times a packet is copied during its lifetime, and the average storage requirement.

The *number of times a packet is transmitted* $G_{ep}(N)$ is a random variable taking value between $[0, \infty.]$ The power consumption grows linearly with $G_{ep}(N)$[35], [36]. For IMMUNE scheme (15.37) models the infection and recovery process. Note that as $R(t)$ is a strictly increasing function of t, $I(R)$ is well defined. Dividing the two Equation (15.37) yields:

$$dI/dR = N - I - R - 1$$

The solution to this DE with initial condition $I(0) = 1$ is $I(R) = (-N + 1)e^{-R} - R + N$.
As $\lim_{t \to \infty} I(t) = 0$, we can solve $I(R) = 0$ for R to find $\lim_{t \to \infty} R(t)$. For N large enough ($N > 10$), the
solution gives $\lim_{t \to \infty} R(t) \approx N$. Since $I(t) + R(t) - (I(0) + R(0)) = I(t) + R(t) - 1$ is the number of
times a packet is copied in the system by time t, we have $E[G_{ep}(N)] = \lim_{t \to \infty} I(t) + R(t) - 1 \approx N - 1$.
Similarly, for IMMUNE scheme, from (15.39), we can solve $I(R)$ as

$$I(R) = \left(-R^2 + (N - 1)R + 1\right)/(R + 1)$$

As $\lim_{t \to \infty} I(t) = 0$, we find $\lim_{t \to \infty} R(t)$ by solving $I(R) = 0$ for R. $I(R) = 0$ has two roots $(N - 1 \pm \sqrt{N^2 - 2N + 5})/2$. Discarding the negative root, we have $\lim_{t \to \infty} R(t) = (N - 1 + \sqrt{N^2 - 2N + 5})/2$.
Therefore, for IMMUNE scheme, we have

$$E[G_{ep}(N)] = \lim_{t \to \infty} (I(t) + R(t) - 1) = \left(N - 3 + \sqrt{N^2 - 2N + 5}\right)/2 \qquad (12.43)$$

For VACCINE scheme, the DEs are solved numerically to get the total number of nodes that ever get
infected by a packet.

12.8 Nano-Networks

Probably the most obvious need for biologically inspired paradigms in networking is in the area of
nano-networks. In general nano-networks refer to the architectures interconnecting nano-machines [38–
46]. A nano-machine is a device, consisting of nano-scale components, able to perform specific tasks
at nano-level. These tasks would include communicating, computing, data storing, sensing, and perhaps
actuating.

Nano-machines such as chemical sensors, nano-valves, nano-switches, or molecular elevators [40],
cannot execute complex tasks by themselves. The exchange of information and commands between
networked nano-machines will allow them to work in a cooperative and synchronous manner to perform
more complex tasks such as in-body drug delivery or disease treatments. The work space of a single
nano-machine is extremely limited. Nano-networks will allow dense deployments of interconnected
nano-machines. Thus, larger application scenarios will be enabled, such as monitoring and control of
chemical agents in ambient air. In some application scenarios, nano-machines will be deployed over
large areas, ranging from meters to kilometers. In these scenarios, the control of a specific nano-machine
is extremely difficult due to its small size. Nano-networks will enable the interaction with remote
nano-machines by means of broadcasting and multihop communication mechanisms.

Communication between nano-machines can be realized through nano-mechanical, acoustic, electromagnetic, and chemical or molecular communication means.

Although applied in a very specific scenarios all these principles except molecular communications are
intuitively easily understood. *Molecular communications* can be defined formally as the use of molecules
as messages between transmitters and receivers.

Due to the size and principles of traditional acoustic transducers and radiofrequency transceivers, their
integration at molecular or nano-scale is not feasible. By contrast, molecular transceivers are conceived
intrinsically at nano-scale. These are nano-machines which are able to emit and receive molecules.

In nano-mechanical communication, transmitters and receivers need to be in direct contact. This is
not a restriction for molecular communication over large areas, where transmitters and receivers can be
remotely located as long as the transmitted molecules reach the intended receiver.

Several biological structures found in living organisms can be considered as nano-machines. Most of these biological nano-machines can be found in cells. Biological nano-machines in cells include: nano-biosensors, nano-actuators, biological data storing components, tools, and control units. The expected features of future nano-machines are already present in a living cell, which can be defined as a self-replicating collection of nano-machines [63]. Several biological nano-machines are interconnected in order to perform more complex tasks such as cell division. The resulting nano-network is based on molecular signaling. This communication technique is also used for inter-cell communication allowing multiple cells to cooperate to achieve a common objective such as the control of hormonal activities or immune system responses in humans.

The bio-hybrid approach uses these biological nano-machines as models to develop new nano-machines or to use them as building blocks integrating them into more complex systems such as nano-robots. Following this approach, the use of a biological nano-motor to power a nano-device has been reported in [48]. Another example in this direction is the use of bacteria as controlled propulsion mechanisms for the transport of micro-scale objects [49].

The most important and expected features of future nano-machines can be described as follows:

- Nano-machines will be intrinsically self-contained which means that each nano-machine will contain a set of instructions or code to realize the intended task. These instructions or sequence of operations can be embedded in the molecular structure of nano-machines, or can be read from another molecular structure in which the instruction set is stored.
- Self-assembly is defined as the process in which several disordered elements form an organized structure without external intervention, as a result of local interactions between them. At nano-level, self-assembly is naturally driven by molecular affinities between two different elements. Self-assembly will leverage the development of nano-machines and will allow them to interact with external molecules in an autonomous way.
- Self-replication is defined as the process in which a device makes a copy of itself using external elements. This potential process will enable the creation of large number of nano-machines to realize macroscopic tasks in an inexpensive way [49]. Similar to the first feature, self-replication implies that the nano-machine contains the instructions to create a copy of itself.
- Locomotion is the ability to move from one place to another. Nano-machines are aimed to accomplish specific tasks, which are usually described by a spatial-temporal actuation. This means that a nano-machine should be located in the right place at the right time to accomplish the task. However, no single nano-machine is able to move towards a previously identified target. More complex systems could use embedded nano-sensors and nano-propellers to detect and follow specific traces of the target. Locomotion will enable the use of nano-machines in applications where mobile actors are needed, e.g., nano-robots for disease treatments [50].
- Communication between nano-machines is needed to allow them to realize more complex tasks in a cooperative manner. At this level, as explained earlier in this section, the most promising technique is based on molecular communication. Further advances in nano-sensors and nano-actuators are expected to enable the integration of molecular transceivers into nano-machines.

A nano-machine could consist of one or more components, resulting in different levels of complexity, which could be from simple molecular switches to nano-robots [50]. The most complete nano-machines will include the following architecture components:

1. *Control unit* which is aimed at executing the instructions to perform the intended tasks. To achieve this goal, it can control all the other components of the nano-machine. The control unit could include a storage unit, in which the information of the nano-machine is saved.
2. *Communication unit* which consists of a transceiver able to transmit and receive messages at nano-level, e.g. molecules.

3. *Reproduction unit* with function designed to fabricate each component of the nano-machine using external elements, and then assemble them to replicate the nano-machine. This unit is provided with all the instructions needed to realize this task.
4. *Power unit* which is aimed at powering all the components of the nano-machine. The unit will be able to scavenge energy from external sources such as light, temperature and store it for a later distribution and consumption.
5. *Sensor and actuators* which act as an interface between the environment and the nano-machine. Several sensors and/or actuators can be included in a nano-machine, like temperature sensors, chemical sensors, clamps, pumps, motor or locomotion mechanisms.

Currently such complex nano-machines cannot be built. However, there exist systems found in the nature, such as living cells, with similar architectures. According to the bio-hybrid approach these biological models, i.e. the cells, can be used to learn and understand the principles governing the operation of nano-machines and their interactions.

Similar to the architecture of a nano-machine, a cell contains the following components:

1. *Control unit.* The nucleus can be considered as the control unit of the cell. It contains all the instructions to realize the intended cell functions.
2. *Communication.* The gap junctions and hormonal and pheromonal receptors, located on the cell membrane, act as molecular transceivers for inter-cell communication.
3. *Reproduction.* Several nano-machines are involved in the reproduction process of the cell such as the centrosome and some molecular motors. The code of the nano-machine is stored in molecular sequences, which are duplicated before the cell division. Each resulting cell will contain a copy of the original DNA sequence.
4. *Power unit.* Cells can include different nano-machines for power generation. One of them is the *mitochondrion* that generates most of the chemical substances, which are used as energy in many cellular processes. Another interesting nano-machine is the *chloroplast*, which converts sunlight into chemical fuel.
5. *Sensors and actuators.* Cells can include several sensors and actuators such as the Transient Receptor Potential channels for tastes and the flagellum of the bacteria for locomotion. The chloroplast of the plants can also be considered as an actuator since it transforms water to oxygen that is later released to the environment.

A bio-hybrid approach is not only used to develop novel nano-machines, but also to understand their interactions in larger systems such as cells. These interactions, enabled exclusively by molecular communication techniques, are essential since this is the only way to explore their capabilities so as to achieve more complex tasks in a cooperative manner.

Among all of the expected features of future nano-machines, our interest in this section is focused on communication capabilities. This is the only feature that enables them to work in a synchronous, supervised, and cooperative manner to pursuit a common objective.

Nano-machines communication can include the two following bidirectional scenarios: (1) Communication between a nano-machine and a larger system such as electronic micro-devices, and (2) Communication between two or more nano-machines.

Different communication technologies, such as electromagnetic, acoustic, nano-mechanical or molecular; have been proposed for each scenario in [51]. In this section we will limit our interest on molecular communications and short range communications using calcium signaling.

Molecular communication is defined as the transmission and reception of information encoded in molecules. Molecular communication is a new and interdisciplinary field that spans nano, bio, and communication technologies [52–55]. Unlike the other communication techniques, the integration process of molecular transceivers in nano-machines is more feasible due to the size and natural domain of molecular transceivers, i.e. nano-scale framework. These transceivers are nano-machines which are able to react to specific molecules, and to release others as a response to an internal command. Molecular communication can be used to interconnect multiple nano-machines, resulting in nano-networks as defined earlier in this section. Nano-networks expand the capabilities of single nano-machines in the following terms:

- More complex objectives can be achieved if multiple nano-machines cooperate. Nano-networks enable this cooperation by providing mechanisms to exchange information between different nano-machines such as molecular motors or nano-switches.
- Single nano-machines can only perform tasks at nano-level, and therefore their workspace is very limited. However, if a large number of nano-machines are interconnected, they can pursuit macro-scale objectives, and work over larger areas, such as treatment of cancer tumors or air pollution monitoring.
- If multiple nano-machines are deployed over large areas, the interaction with a specific nano-machine is extremely difficult due to its small size. This interaction includes procedures such as nano-machines activation/deactivation, configuration of parameters, data acquisition or actuation commands. Nano-networks will enable this interaction by providing the infrastructure and mechanism to broadcast the information over these areas. In addition, two nano-machines could interact indirectly by using other nano-machines as repeaters. Besides expanding capabilities of single nano-machines, nano-networks represent a potential solution for some applications where available communication networks and micro-devices are not suitable. Compared to current communication network technologies, nano-networks have the following advantages:

  (a) The reduced size of nano-machines and the resulting nano-network components can be an advantage in many applications where the dimension of the involved systems is critic. For instance, in the biomedical field, nano-machines can be used for intra-body applications allowing nano-networks to lead to nano-invasive and more selective treatments.

  (b) Biocompatibility is defined as the quality of a device to operate accordingly in biological environments without affecting them negatively. In some biomedical applications, many electronic devices have to cope with hostile environments as well as many organisms which reject implants and drugs. Nano-technologies can be used to enhance the compatibility between nano-machines and natural organs or tissues by means of more friendly materials and interfaces. For instance, bio-hybrid nano-machines compounded by biological elements can interact with natural processes without any side effects. In addition, nano-machines and molecular messages may also be programmed to deactivate after completing the nano-network task preventing removal procedures. Nano-technologies, allow the control of materials at molecular level. Using these materials, we can design nano-networks nodes according to specific environmental conditions improving the biocompatibility of the system.

  (c) Chemical reactions are highly efficient in terms of energy consumption [53]. These reactions will power the nano-networks nodes and processes. Chemical reactions can also represent complex computation and decision processes, which in traditional communication could mean multiple operations.

Details on short range communications using molecular motors can be found in [54–59] and short range communications using calcium signaling in [60–61].

12.9 Genetic Algorithm Based Dynamic Topology Reconfiguration in Cellular Multihop Wireless Networks

The genetic algorithm represents probably the most widely used biologically inspired optimization tool. In this section we present a genetically based algorithm for dynamic topology reconfiguration in cellular multihop wireless networks. When considering a multicell scenario with nonuniform traffic distribution in multihop wireless networks, the search for the optimum topology becomes an NP-hard problem. For such problems, exact algorithms based on exhaustive search are only useful for small toy models, so heuristic algorithms such as genetic algorithms (GA) must be used in practice. For this purpose, we present here a specific sequential genetic algorithm (SGA) [62] to optimize the relaying topology in multihop cellular networks aware of intercell interference and spatial traffic distribution. The topologies will be encoded as a set of chromosomes and special crossover and mutation operations are used to search for the optimum topology. Improvement in the fitness function is controlled sequentially as newer generations evolve and whenever the improvement is increased sufficiently the current topology is updated by the new one having higher fitness. Illustrations will show that SGA provides both high performance improvements in the system and fast convergence (one order of magnitude faster than exhaustive search) in a dynamic network environment.

12.9.1 System Model

Network and Intercell Interference Model-Up link: we consider a cellular network with a set of $I=\{i\}$ base stations. Let us assume that in a reference cell with index $i=r$, there is a reference user m_r connected to the reference access point AP_r (base station) with channel gain $G_{m_r,r}$. At the same time a cochannel interfering user m_i is connected to access point AP_i in cell i, $i \in \mathcal{I}_{-r} = \{i \neq r\}$ with channel gain $G_{m_i,i}$. Assuming that all the users are expected to reach their respective access points with the same required received power $S_{m_i,i}$, we have for the transmitted powers $P_{m_i,i}$ of the useful and interfering signal

$$S_{m_i,i} = G_{m_i,i} P_{m_i,i} ; i \in \mathcal{I} \tag{12.44}$$

Let $I_{m_i,r}$ represents the interference power at the position of the reference receiver AP_r due to the interfering cochannel signal transmitted in cell *i*. This can be presented as

$$I_{m_i,r} = G_{m_i,r} P_{m_i,i} ; i \in \mathcal{I}_{-r} \tag{12.45}$$

where $G_{m_i,r}$ is the gain of the channel between the interfering user m_i and AP_r. The signal to interference plus noise ratio $SINR_{m_r,r}$ at AP_r in the presence of all interfering users m_i is

$$SINR_{m_r,r} = \frac{S_{m_r,r}}{n_r + \sum_{i\neq r} I_{m_i,r}} = \left(N_r^{-1} + \sum_{i\neq r} \frac{G_{m_i,r}}{G_{m_i,i}} \right)^{-1} \tag{12.46}$$

where n_r is the background noise power, $N_r = P/n_r = SNR$, and the channel capacity per unit spectra can be represented as

$$\mathbf{c} = (c_{m_r,r}, c_{m_1,1}, \ldots, c_{m_i,i}, \ldots, c_{m_{N_c-1},N_c-1}); i \in \mathcal{I}_{-r}$$
$$c_{m_r,r} = \log(1 + SINR_{m_r,r}) \tag{12.47}$$

where N_c is the number of cells. The network capacity is defined as

$$C = \sum_r c_{m_r,r} ; r \in \mathcal{I} \tag{12.48}$$

If the radio resource management is defined as channel assignment function $A(m_r)$ responsible to allocate to each user m_r proper channel (time slot, frequency bin) then the optimum assignment is defined as

$$A(m_r) = \max_{A(m_r)} C ; r \in \mathcal{I} \tag{12.49}$$

Equations (12.44)–(12.49) can be easily modified to include multiple interfering signals in the same channel from the reference cell as well as from each interfering cell. Such example is used in section represnting the simulation results.

Downlink: In this scenario the reference access point AP_r is providing power S_{r,m_r} to the reference user m_r. At the same time, interfering AP_i providing the same signal level S_{i,m_i} to the user m_i, is producing interference to the useful signal of user m_r. So, we have

$$
\begin{aligned}
S_{i,m_i} &= G_{i,m_i} P_{i,m_i} ; i \in \mathcal{I} \\
I_{i,m_r} &= G_{i,m_r} P_{i,m_i} = S_{i,m_i} G_{i,m_r}/G_{i,m_i} ; i \in \mathcal{I}_{-r} \\
SINR_{r,m_r} &= \left(N_r^{-1} + \sum_{i \neq r} G_{i,m_r}/G_{i,m_i} \right)^{-1}
\end{aligned}
\tag{12.50}
$$

where G_{i,m_i} is the channel gain between AP_i and user m_i, P_{i,m_i} is the power needed at AP_i to provide power S_{i,m_i} for user m_i, I_{i,m_r} is the interference power at m_r produced by AP_i and $SINR_{r,m_r}$ is SINR at m_r for the signal transmitted by AP_r. Equations (12.47) and (12.48) now become

$$
\mathbf{c} = (c_{r,m_r}, c_{1,m_1}, \ldots, c_{i,m_i}, \ldots, c_{N_c-1,m_{N_c}-1}); i \in \mathcal{I}_{-r}
\tag{12.51}
$$

$$
c_{r,m_r} = \log(1 + SINR_{r,m_r})
$$

$$
C = \sum_r c_{r,m_r}
\tag{12.52}
$$

And the optimum radio resource management is again defined by (12.49).

Relaying and Scheduling: We will use notation $r(m_{r_2}, m_{r_1}, \mathbf{m}_{i_2}, \mathbf{m}_{i_1})$, to denote simultaneous transmission (relaying) on reference route from user m_{r_1} to $m_{r_2}, r \in \mathcal{I}$ and interfering users from m_{i_1} to $m_{i_2}, i \in \mathcal{I}_{-r}$ position in all interfering cells. Under these conditions the corresponding link capacity will be denoted as $c_r(m_{r_2}, m_{r_1}, \mathbf{m}_{i_2}, \mathbf{m}_{i_1})$. This capacity can be calculated by the following set of equations

$$
\begin{aligned}
S_{m_{r_1},m_{r_2}} &= P_{m_{r_1},m_{r_2}} G_{m_{r_1},m_{r_2}}, ; S_{m_{i_1},m_{i_2}} = P_{m_{i_1},m_{i_2}} G_{m_{i_1},m_{i_2}} \\
I_{m_{i_1},m_{r_2}} &= P_{m_{i_1},m_{i_2}} G_{m_{i_1},m_{r_2}} = S_{m_{i_1},m_{i_2}} G_{m_{i_1},m_{r_2}}/G_{m_{i_1},m_{i_2}} \\
SINR_{m_{r_1},m_{r_2}} (\mathbf{m}_{i_2}, \mathbf{m}_{i_1}) &= \frac{S_{m_{r_1},m_{r_2}}}{n_r + \sum_{i \neq r} I_{m_{i_1},m_{r_2}}} = \\
&= \left(N_r^{-1} + \sum_{i \neq r} G_{m_{i_1},m_{r_2}}/G_{m_{i_1},m_{i_2}} \right)^{-1}
\end{aligned}
\tag{12.53}
$$

$$
\mathbf{m}_{i1} = (m_{11}, m_{21}, \ldots, m_{N_c1}); \mathbf{m}_{i2} = (m_{12}, m_{22}, \ldots, m_{N_c2}); i \in \mathcal{I}_{-r}
$$

$$
c_r(m_{r_2}, m_{r_1}, \mathbf{m}_{i_2}, \mathbf{m}_{i_1}) = \log\left(1 + SINR_{m_{r_1},m_{r_2}}(\mathbf{m}_{i_2}, \mathbf{m}_{i_1})\right); i \in \mathcal{I}_{-r}
$$

We define now the multihop (H hops) route as a series of relaying transmissions

$$
\mathfrak{R}_r(m_{rH}, m_{rH-1}, \ldots, m_{r_2}, m_{r_1}, \mathbf{m}_{iH}, \mathbf{m}_{iH-1}, \ldots, \mathbf{m}_{i_2}, \mathbf{m}_{i_1})
\tag{12.54}
$$

The capacity of the route is then defined as

$$
\begin{aligned}
c_{\mathfrak{R}_r} &= \min_h c_{\mathfrak{R}_r,r}(m_{rh}, m_{rh-1}, \mathbf{m}_{ih}, \mathbf{m}_{ih-1}); h = 2, \ldots, H \\
c_{\mathfrak{R}_r,r} &= \log\left(1 + SINR_{m_{r_1},m_{r_2}}(m_{rh}, m_{rh-1}, \mathbf{m}_{ih}, \mathbf{m}_{ih-1})\right)
\end{aligned}
\tag{12.55}
$$

which is equal to the minimum link capacity on the route. The optimum set of relaying routes is defined as

$$\{\Re_r\} = \max_{H,\Re_r} C_\Re; \quad \text{where } C_\Re = \sum_r c_{\Re_r,r}; r \in \mathcal{I} \tag{12.56}$$

Due to the fact that a node cannot receive and transmit the signal simultaneously on the same channel, only a subset of transmissions can be active simultaneously. For that reason scheduling in different time slots will be introduced.

Two Dimensional Relaying Topology: The cochannel interference can be further reduced by scheduling different transmissions in different subchannels (time slots). All necessary transmissions between all users and their respective access points should be completed in B slots (scheduling cycle) in both directions: uplink and downlink.

As an illustration, for the two cells scenario and notation shown in Figure 12.9(a), a possible (feasible) topology is shown in Figure 12.9(b). For a systematic presentation of the problem the cell area is divided into concentric rings (e.g. three rings for each cell in Figure 12.9(a)). It is assumed that one cochannel user from each ring has unidirectional connection with the corresponding access point. For the uplink, the topology consists of four partial topologies representing transmissions in four consecutive time slots ($B = 4$). In the first time slot (the first partial topology) there are two simultaneous transmissions: packet originating from ring 3 in reference cell r is transmitted from ring 3 to ring 2 and at the same time packet originating from ring 1 of interfering cell i is transmitted from ring 1 to access point AP_i. In the second time slot (the second partial topology), packet originating from ring 3 in cell r is transmitted from ring 2 to AP_r and at the same time packet originating from ring 2 of cell i is transmitted from ring 2 to access point AP_i. Similarly the same notation is used for transmission in time slot 3 and 4. The topology for the downlink is presented in the lower part of Figure 12.9(b).

From the figure we can see that, with frequency reuse factor one, only seven time slots are required for all six users to transmit on the up and down link. Classical TDMA scheme would require $6 + 6 = 12$ slots (separate slot for each transmission).

These seven partial topologies together are referred to as a *possible* or *feasible two dimensional (time and space) topology* and will be represented in the sequel by a given topology index t. For this concept (12.54) becomes

$$\{\Re_r^{(2)}\} = \max_{B, H, \Re_r^{(2)}} C_{\Re^{(2)}}; \quad \text{where } C_{\Re^{(2)}} = \sum_r c_{\Re^{(2)}}; r \in \mathcal{I} \tag{12.57}$$

and $\Re^{(2)}$ is two dimensional relaying topology to be elaborated in more detail in the next section. For each slot $b = 1, \ldots, B$, parameter $c_{\Re_r^{(2)}}$ is given by corresponding (12.55).

To generalize this modeling and further explain the meaning of scheduling interval B, we use analogy with standard contention graph modeling of the MAC layer operation by representing users in the relaying rings from the previous examples as nodes in the network. Designing a MAC protocol can be modeled as a bandwidth allocation problem at the link layer. When considering link layer flows, contention relations between the links can be represented by a *link conflict graph*. In such a graph, vertices represent link flows and edges between vertices denote contention between links, which is the situation where there is interference between either the sender or the receiver of one link with either the sender or the receiver of the other link. A fully connected subgraph in a conflict graph is referred to as a *clique* and *maximal clique*, is a clique not contained in any larger clique. Therefore, a maximal clique represents a 'channel resource', which has a given fixed capacity. The basic requirement for feasibility of a schedule or bandwidth assignment is that the total flow rate in each clique does not exceed the clique's capacity, subject to the conflict constraints. In addition, the bandwidth allocation should satisfy some performance requirement such as fairness.

Assuming all nodes use omnidirectional antennas to transmit packets in the same shared wireless channel, a link conflict graph can be used to describe the contention relations between link flows, and

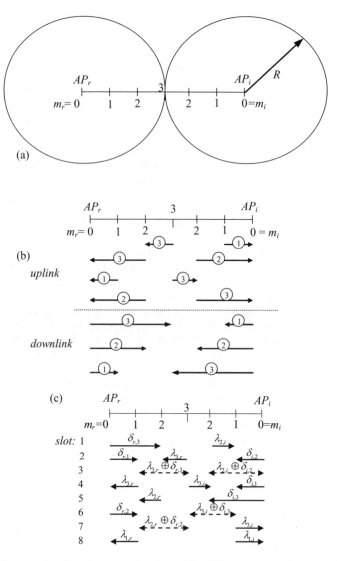

Figure 12.9 (a) Modeling interfering users positions for 2-cells; (b) Possible transmission schedule. (c) Possible schedule by using network coding.

each maximal clique is treated as a 'channel resource' with a given fixed capacity. The capacity of a clique depends on the topology of the network, and the fairness principle under consideration. In the previous example transmissions within the same clique are shared on the time scale by scheduling transmissions in different time slots. After B slots (scheduling cycle) the transmissions can be repeated in the same order.

Traffic Modeling: If we denote by n_i the number of rings in cell i and $N = \sum_{i=1}^{N_c} n_i$ the total number of rings in the network, the vector. $\boldsymbol{\lambda}_i = (\lambda_{1,i}, \ldots, \lambda_{n_i,i})$, $\boldsymbol{\lambda}_i \in \mathbf{R}^{n_i}$ defines the amount of generated source traffic by the users situated in the different rings in cell i to be transmitted to the access point AP_i on the uplink, and $\boldsymbol{\delta}_i = (\delta_{i,1}, \ldots, \delta_{i,n_i})$, $\boldsymbol{\delta}_i \in \mathbf{R}^{n_i}$ the traffic that the access point AP_i is transmitting to the users on the downlink. For the same traffic vectors λ_i, δ_i the base station can schedule the transmission through different channels (time slots, frequency bins) resulting in *temporal and spatial MAC protocol*. The overall network traffic on the uplink and downlink is defined as $\lambda = (\lambda_1, \lambda_2, \ldots, \lambda_{N_c}) = (\lambda_1, \ldots, \lambda_N)$ and $\delta = (\delta_1, \delta_2, \ldots, \delta_{N_c}) = (\delta_1, \ldots, \delta_N)$ respectively.

The base stations jointly assign an access vector $\mathbf{a} = (\mathbf{a}_1, \mathbf{a}_2, \ldots, \mathbf{a}_{N_c}) = (a_1, \ldots, a_N)$, where $N = \sum_{i=1}^{N_c} n_i$, to the different rings to give them permission to transmit, where each component P_s with $a_{\lambda_n}, a_{\delta_n} \in (0, 1)$. With $a_{\lambda_n}=1$ the users from ring n are allowed to transmit uplink otherwise not, and with $a_{\delta_n}=1$ the users from ring n are allowed to transmit downlink otherwise not. In the two cell case $\mathbf{a} = (\mathbf{a}_1, \mathbf{a}_2)$, the first half of the coefficients represents the permissions to transmit for the rings in referent cell r and the second half for rings in interfering cell i, $i \in \mathcal{I}_{-r}$.

We consider symmetric bidirectional transmission (duplex connection) in the sense that the access point will only transmit to the users situated in the rings activated by $\mathbf{a}$ where both components of $a_n = (a_{\lambda_n}, a_{\delta_n})$ are active, or not active, simultaneously.

Bidirectional Relaying and Networking Coding: In this section we introduce network coding and combine it with the previous results on optimum relaying to reduce the number of slots needed for the users to complete their transmissions and achieve further improvements in the system performance.

Let us assume that the hops are indexed in increasing order for uplink as $h^{(up)}$ and for downlink as $h^{(down)}$. By combining the uplink and downlink traffic from the previous hop at hop h as $y_h^{(down, up)} = y_{h-1}^{(down)} \oplus y_{h-1}^{(up)}$ the number of overall time slots needed for transmission in B cycles can be reduced. The optimization process defined by (12.57) now becomes

$$\{\mathfrak{R}_r^{(2)}\} = \max_{B, H, \oplus, \mathfrak{R}_r^{(2)}} C_{\mathfrak{R}^{(2)}}; \text{ where } C_{\mathfrak{R}^{(2)}} = \sum_r c_{\mathfrak{R}_r^{(2)}}$$
$$\mathfrak{R} = \mathfrak{R}^{(up)} \cup \mathfrak{R}^{(down)}$$
(12.58)

To elaborate this concept in more detail an example of possible topology that includes network coding is shown in Figure 12.9(c) for the two cell scenario from Figure 12.9(a). The traffic between users and access point is bidirectional, so given a schedule that alternates the transmissions between the different rings, after a certain number of time slots all intermediate users ($m_i, i \in \mathcal{I}$) have information frames buffered for transmission in both directions. Whenever an opportunity arises, the intermediate users combine two information frames, one for each direction, with a simple *XOR* operation and send it to its neighbors in a single omnidirectional transmission. Both receiving nodes already know one of the frames combined (have it stored from the previous transmission), while the other frame is new. Thus, one transmission allows two users to decode a new packet, effectively doubling the capacity of the path, reducing the power consumption of the transmitter node and reducing the number of time slots required to complete the transmission.

The transmission schedule presented in Figure 12.9(c) defines a possible topology for two cell scenario and access vector $\mathbf{a} = \mathbf{1}$. In this case all rings have duplex connection and the topology consists of eight partial topologies representing transmissions in eight consecutive time slots. In the first time slot (the first partial topology) there are two simultaneous transmissions; packet originating from the access point AP_r (addressed to user in ring 3 is transmitted to ring 2 in cell r and at the same time packet originating from ring 2 (addressed to access point AP_i) of cell i is transmitted from ring 2 to ring 1. In the second time slot (the second partial topology), packet originating from access point AP_r (addressed to user in ring 1) is transmitted to ring 1, at the same time packet originating from ring 3 (addressed to access point AP_r) is transmitted from ring 3 to ring 2 and, packet originating at AP_i (addressed to user in ring 2) is transmitted to ring 1 in the adjacent cell. Similarly, the same notation is then used for transmissions in time slots from 3 to 8. As already discussed earlier these eight partial topologies together are referred to as a *possible two dimensional* (*time and space*) *topology* and will be represented in the sequel by a given topology index (t).

So there will be limited interference transmission for three users per cell in eight channels (eight slots in Figure 12.9(c)) giving the intercell throughput $6/8 = 3/4$, as opposed to the $6/12 = 1/2$ in a conventional TDMA system where each cell uses a half of available channels (slots). Although scheduling in Figure 12.9(b) requires seven time slots it also assumes transmissions over three rings which requires higher power.

12.9.2   System Optimization

The examples of topologies presented in the previous section are based on intuition and we need a systematic approach to system optimization. The optimization of the relaying topology in multihop cellular networks should provide the answer to the questions 'who is transmitting to whom' and 'when', so as to ensure the best system performance. In the next step, we first introduce a set of definitions used in the analysis. For simplicity, in the following we will describe the optimization only for one direction of the traffic and then at the end of the section make some additional comments on how we extend it for bidirectional case. For $i \in \mathcal{I}$,

- We define the topology matrix $\mathbf{T}_i = \left[T(m_{i2}, m_{i1}) \right]$ with $T(m_{i2}, m_{i1}) = 1$, if node m_{i1} is transmitting to m_{i2} and 0 otherwise, with indexes $m_{i1}, m_{i2} = 0, 1, 2, \ldots, n_i$, and $m_{i2} < m_{i1}$. Each (m_{i2}, m_{i1}) pair is represented by a specific link index l as shown in Figure 12.10 for the case of two cells. With this notation the column vector of equivalent (source + relayed) rates in cell i becomes

$$\mathbf{x}_i = \left[x_{m_{i2}} \right] \text{ where } x_{m_{i2}} = x_{sm_{i2}} + \sum_{m_{i1}} T(m_{i2}, m_{i1}) x_{m_{i1}} \qquad (12.59)$$

  for each direction of the traffic, where $x_{sm_{i2}}$ is the source rate of user m_{i2}, $x_{m_{i2}}$ and $x_{m_{i1}}$ are the overall rate of the traffic relayed through user m_{i2} and m_{i1} respectively. The overall topology matrix will be formally defined as $\mathbf{T} = diag\,[\mathbf{T}_i]$ and $\mathbf{x} = \left[\mathbf{x}_i \right]$ is the concatenated column vector of the overall aggregate rates.
- The routing matrix $\mathbf{R} = [r_{ln}]$ has entries $r_{ln} = 1$ if source n (n = 1,2,..,N) is using link l (l=1,2,..,L) and 0 otherwise. Recall that, parameter N is the number of overall rings in the network. Parameters $r_{ln} = r_{lm_{i1}}$ of the routing matrix $\mathbf{R}$ are calculated as $r_{lm_{i1}} = \bigcup_{m_{i2}} r_{lm_{i2}} T(m_{i2}, m_{i1})$, and the number of links is given by $L = \sum_{m_{i1}} T(m_{i2}, m_{i1})$.
- The scheduling set Π will be combined with the routing matrix $\mathbf{R}$ resulting into two dimensional routing protocol characterized by *extended routing matrix* $\mathbf{R}^{(2)} \in \mathfrak{R}^{(2)}$. By assuming that the scheduling cycle within the maximum clique has B steps, the optimization process will include:
  (a) Utility function

$$U = (1/B) \sum_n a_n \log(x_n)/P_n \qquad (12.60)$$

  where x_n and P_n are the aggregate rate (source + relays) and aggregate power respectively needed for transmission of information from the source n to the access point, a_n is the access parameter ($a_n = 1$ indicates that source n is active and $a_n = 0$ otherwise). We assume that the buffers of the users are infinite, and we maximize the sum of the transmission rates and leave to the internal node fairness policy as a parameter to control the share of the channel between the local and relayed traffic.
  (b) Constraint $\mathbf{R}^{(2)} \mathbf{x}^{(2)} \leq \mathbf{c}^{(2)} (\mathbf{R}^{(2)})$ with the following definitions of *extended system parameters*

$$\mathbf{x}^{(2)T} = \left(\mathbf{x}^T(1), \mathbf{x}^T(2), \ldots, \mathbf{x}^T(B) \right) ; \mathbf{c}^{(2)T} = \left(\mathbf{c}^T(1), \mathbf{c}^T(2), \ldots, \mathbf{c}^T(B) \right)$$
$$\mathbf{R}^{(2)} = diag \, \|\mathbf{R}(b)\| \ , \ b = 1, 2, .., B \qquad (12.60a)$$
$$\mathbf{R} \in \mathfrak{R}, \ \mathbf{x} \in \Pi \ \Rightarrow \mathbf{R}^{(2)} \in \mathfrak{R}^{(2)}$$

  where $\mathbf{c}$ are the logical link capacities calculated as discussed in Section II which capture the functional dependency of power control and interference level in the network.

(c) Each component of the set of feasible routes in $\mathfrak{R}^{(2)}$ should provide directional connection for each terminal to the corresponding access point. This means that the sequence of links generated in a clique cycle must provide connection for all terminals to the corresponding access point. To define this constraint explicitly we introduce the link hopping distance h_l and the vector $\mathbf{h} = (h_1, \ldots, h_L)$. h_l represents the number of rings that link l is hoping over, from its transmitter/receiver to the corresponding receiver/transmitter. Similarly the source hopping distance is denoted as $\mathbf{d} = (d_1, \ldots, d_N)$. The sum of link hopping distances on the route from source n to the access point should be equal to the source hopping distance

$$\sum_b \mathbf{R}^T(b)\mathbf{h}(b) = \mathbf{d} \qquad (12.60b)$$

(d) The overall transmission rate is defined as

$$(\mathbf{I} - \mathbf{T})\mathbf{x} = \mathbf{x}_n \quad \leftarrow \quad x_{m_{i2}} = x_{sm_{i2}} + \sum_{m_{i1}} T(m_{i2}, m_{i1}) x_{m_{i1}} \qquad (12.60c)$$

The formulation of the problem obtained by Equations (12.60a)–(12.60c) can be summarized as:

$$\begin{aligned} \mathbf{P}: \underset{\mathbf{T}, \mathbf{x}}{\text{maximize}} \quad & U \\ \text{subject to} \quad & \mathbf{R}^{(2)}\mathbf{x}^{(2)} \le \mathbf{c}^{(2)}(\mathbf{R}^{(2)}); \quad \sum_b \mathbf{R}^T(b)\mathbf{h}(b) = \mathbf{d} \qquad (12.61) \\ & (\mathbf{I} - \mathbf{T})\mathbf{x} = \mathbf{x}_n ; \quad \mathbf{R} \in \mathfrak{R}, \ \mathbf{x} \in \Pi \ \Rightarrow \mathbf{R}^{(2)} \in \mathfrak{R}^{(2)} \end{aligned}$$

In the case of bidirectional traffic an independent set of Equations [17] should be written for both directions and (12.61) should be modified to include overall utility function $U = U^{(up)} + U^{(down)}$ with separate set of constrains for both directions.

The optimum topology search defined by (12.61) will be referred as TSL algorithm. The algorithm may result in high complexity and for this reason in the next section we define an evolutionary sequential genetic algorithm (SGA) that can be used for readjustment of the topology due to traffic variation.

12.9.3 Ga-Tsl Algorithm Design

Genetic Algorithms (GAs) are adaptive methods based on the mechanics of natural selection [63]. The basic principles of GAs are described in many texts [63–67]. They are very efficient in directing the search toward the relatively prospective regions of the search space. Empirical studies have shown that genetic algorithms do converge on global optima for a large class of NP-hard problems [64].

Encoding Scheme: As in the previous section, for simplicity of presentation we start this section by considering only uplink transmission, which will be further extended to the bidirectional case.

We encode the topologies as a set of chromosomes, where each chromosome defines a partial topology. A chromosome consists of a number of gene-instances $\boldsymbol{\gamma} = (\gamma_1, \ldots, \gamma_N)$ that in our design corresponds to users that are transmitting from specific rings. We use binary coding scheme and the value of the gene will be 1 if the corresponding user is transmitting in that time slot or 0 otherwise.

On the other hand, the phenotype information for a genotype instance is represented by the set of active links that the corresponding users are activating in each time slot. For the two cell example, using the notation of the links shown in Figure 12.10 for uplink transmission, the phenotype of gene γ_1 is 1 (link l_1 is used), for gene γ_2 can be 2 or 4 (link l_2 or l_4 can be used), and so on.

With this scheme the topology is given by a set of chromosomes that define the partial topologies generated in B time slots and are denoted by $PT_\gamma^b(t)$, where t is the index of the topology, b is the index of the time slot $b = 1, \ldots, B$ and γ the index of the gene.

To illustrate the encoding scheme we consider a simple example of a possible topology ($t = 1$), for two cell case and $\mathbf{a}=1$, that consists of the set of links $T(1) = \{PT_\gamma^b(1)\} = \{\{l_1\}, \{l_2\}, \{l_3\}, \{l_7\}, \{l_8\}, \{l_9\}\}$

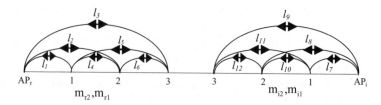

Figure 12.10   Link notation.

as shown in Figure 12.11. In this case every user transmits in a separate time slot directly to its access point. The partial topologies are:

$$PT_\gamma^{b=1}(1) = \{l_1\} \rightarrow genotype = (100000) \rightarrow phenotype = 1 - 0 - 0 - 0 - 0 - 0$$
$$PT_\gamma^{b=2}(1) = \{l_2\} \rightarrow genotype = (010000) \rightarrow phenotype = 0 - 2 - 0 - 0 - 0 - 0$$
$$PT_\gamma^{b=3}(1) = \{l_3\} \rightarrow genotype = (001000) \rightarrow phenotype = 0 - 0 - 3 - 0 - 0 - 0$$
$$PT_\gamma^{b=4}(1) = \{l_7\} \rightarrow genotype = (000100) \rightarrow phenotype = 0 - 0 - 0 - 7 - 0 - 0$$
$$PT_\gamma^{b=5}(1) = \{l_8\} \rightarrow genotype = (000010) \rightarrow phenotype = 0 - 0 - 0 - 0 - 8 - 0$$
$$PT_\gamma^{b=6}(1) = \{l_9\} \rightarrow genotype = (000001) \rightarrow phenotype = 0 - 0 - 0 - 0 - 0 - 9$$

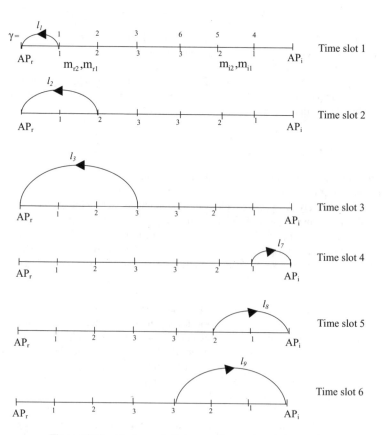

Figure 12.11   The transmission pattern for topology index t = 1.

As we can see, the previous topology consists of a set of six chromosomes (partial topologies) that give the information as to which user is transmitting in each time slot and which link is being used.

To extend the previous notation to bidirectional links, we duplicate the genes. So, every partial topology is defined by two genotypes (uplink/downlink) and two phenotypes. The vector of gene-instances now should include also the access point $\gamma = (\gamma_0, \gamma_1, \ldots, \gamma_{N+N_c-1})$. With this notation transmission pattern presented in Figure 12.6 is defined by the following partial topologies:

$$PT_\gamma^{b=1}(t) = \left\{ l_1^{(down)} \right\}$$
$$\rightarrow \begin{cases} genotype^{(up)} = \mathbf{0}; \ phenotype^{(up)} = \mathbf{0} \\ genotype^{(down)} = (10000000); \ phenotype^{(down)} = 1 - 0 - 0 - 0 - 0 - 0 - 0 - 0 \end{cases}$$

$$PT_\gamma^{b=2}(t) = \left\{ l_{10}^{(up)} \right\}$$
$$\rightarrow \begin{cases} genotype^{(up)} = (00000010); \ phenotype^{(up)} = 0 - 0 - 0 - 0 - 0 - 0 - 10 - 0 \\ genotype^{(down)} = \mathbf{0}; \ phenotype^{(down)} = \mathbf{0} \end{cases}$$

$$PT_\gamma^{b=3}(t) = \left\{ l_4^{(up)} \right\}$$
$$\rightarrow \begin{cases} genotype^{(up)} = (00100000); \ phenotype^{(up)} = 0 - 0 - 4 - 0 - 0 - 0 - 0 - 0 \\ genotype^{(down)} = \mathbf{0}; \ phenotype^{(down)} = \mathbf{0} \end{cases}$$

$$PT_\gamma^{b=4}(t) = \left\{ l_7^{(down)} \right\}$$
$$\rightarrow \begin{cases} genotype^{(up)} = \mathbf{0}; \ phenotype^{(up)} = \mathbf{0} \\ genotype^{(down)} = (00001000); \ phenotype^{(down)} = 0 - 0 - 0 - 0 - 7 - 0 - 0 - 0 \end{cases}$$

$$PT_\gamma^{b=5}(t) = \left\{ l_1^{(up)}, l_4^{(down)}, l_7^{(up)}, l_{10}^{(down)} \right\}$$
$$\rightarrow \begin{cases} genotype^{(up)} = (01000100); \ phenotype^{(up)} = 0 - 1 - 0 - 0 - 0 - 7 - 0 - 0 \\ genotype^{(down)} = (01000100); \ phenotype^{(down)} = 0 - 4 - 0 - 0 - 0 - 10 - 0 - 0 \end{cases}$$

where upper index (up) corresponds to uplink transmission and $(down)$ to downlink.

Population Structure and Initialization: The initial population consists of the topologies (sets of chromosomes) that define all possible routes to the access point. In this way the reproduction operators will produce as result all possible feasible topologies through the pass of the generations. For example, for the two cell case and $\mathbf{a} = \mathbf{1}$, the initial population is formed by:

- $T(1) = \{PT_\gamma^b(1)\} = \{\{l_1\}, \{l_2\}, \{l_3\}, \{l_7\}, \{l_8\}, \{l_9\}\}$ as defined in the previous section, with the genotypes and phenotypes for uplink/downlink transmission and $l_j = \left\{ \{l_j^{(up)}\}, \{l_j^{(down)}\} \right\}$.
- $T(2) = \{\{l_6\}, \{l_4\}, \{l_1\}, \{l_{12}\}, \{l_{10}\}, \{l_7\}\}$ where the genotypes and phenotypes are defined in the same way as before.
- $T(3) = \{\{l_5\}, \{l_1\}, \{l_2\}, \{l_{11}\}, \{l_7\}, \{l_8\}\}$.

Fitness Function and Sequential GA: The quality of the solution is judged by the fitness function (f) defined as in [17] and is included in the global optimization defined by [18] with f = U for a 'given **T**'. The topologies that correspond to the best values of the fitness are kept in the pool for possible future reconfiguration of the network, and the optimum topology which corresponds to the highest fitness will be assigned to the network.

In this chapter we compare convergence of GA and exhaustive search method. For these purpose we use as '*given* **T**' an optimum topology found by exhaustive search. In Section VI we discuss other options to calculate **T**. In a practical implementation of GA this variable is not known so we run GA as long as we find a topology that provides $\Delta f \geq threshold$ where Δf is the difference between the previously accepted solution and the new one. This algorithm will be referred to as sequential GA (SGA) since the usable topology T_{new} is produced as a sequence of solutions satisfying $\Delta f \geq threshold$.

Crossover Operator: In our algorithm, we adopt arithmetic crossover [67] that defines new chromosomes as a result of the following arithmetic operation:

$$\{T_{new}\} = \left\{PT_\gamma^b(\mathbf{t})\right\} = \left\{PT_{\gamma_w}^{b_i}(t)\right\} OR \left\{PT_{\gamma_w}^{b_j}(t)\right\}$$ (12.62)

where $b = 1, \ldots, B - 1; b_i, b_j = 1, \ldots, B$ with $b_i \neq b_j$ and $w = 0, \ldots, N + N_c - 1$.

From each initial topology, we generate a new set of topologies $\{T_{new}\}$ that consists of $\binom{B}{2}$ topologies as a result of OR operation between all pairs of chromosomes that define certain topology $T(t)$. If the new topology is not feasible it will be excluded from the population.

To guarantee that all feasible routes are available through the passage of the generations, we allow a certain level of elitism and we let the chromosomes of the initial population survive to the next generation.

Mutation Operation: Mutation probability is normally set to be small. This is because though mutation can enable the algorithm to search a new search space, it actually destroys the current pattern. If the optimum topology is not obtained after successive crossover operations, we mutate the chromosome choosing for a given gene γ a different phenotype, which means that for a certain user (gene) we select another link from the set of possible links represented in Figure 12.2 for the two cell scenario.

As the epitasis [66], defined as the level of dependency between different genes, is very high due to the relay, this is the only mutation that we allow.

12.9.4   Traffic Cognitive Topology Control (TC^2)

To model the traffic in a Traffic Cognitive Network (TCN), we use the traffic vectors λ and δ described in Section II. For a given spatial traffic distribution the access vector should be varied in time to provide

$$E(\mathbf{a}_\lambda(t)) = \lambda(t) \text{ and } E(\mathbf{a}_\delta(t)) = \delta(t)$$ (12.63)

The variation of the traffic in the network is defined by the vector $\mathbf{\Delta}=(\Delta_1, \ldots, \Delta_N)$. If the traffic in the network changes owing to users that became inactive then the corresponding component of $\mathbf{\Delta}$ where the change occurs is negative; on the other hand, if a new source appears in the network this component is positive.

The differential access vector corresponding to the traffic variation is given by $\mathbf{a}' = \mathbf{a}_I XOR \, \mathbf{a}_F$, where $\mathbf{a}_I$ is the access vector corresponding to the initial traffic and $\mathbf{a}_F$ is the access vector after the traffic has changed (final). We assume that the traffic distribution is observed in time intervals short enough to detect each change in the traffic so that traffic change only in one ring is assumed in a given observation instant.

TC^2 algorithm described below uses an exhaustive search (TSL) algorithm to find initial optimal topology and SGA-TSL for tracking traffic variations. Different options to initialize SGA-TSL are discussed in the following section.

The TC^2 algorithm works as follows:

1. Calculate the access vector $\mathbf{a}_I$ for a given traffic distribution λ and δ, by using (12.63).
2. Use TSL to generate the set of feasible candidate topologies T for $\mathbf{a}_I$.
3. For each topology in the set T:

    Calculate the aggregate powers needed for users to deliver the information.

    Calculate the utility using (12.60).

    Use CVX [68, 69] to optimize the source rates in (12.61).

4. Go to 3. until the optimum topology $\mathbf{T}_{opt}$ is obtained or if the traffic distribution in the network has changed ($\mathbf{a}' \neq \mathbf{0}$), use SGA-TSL to find the new optimum topology $\mathbf{T}_{opt}$ and go to 3.

The operation of *SGA-TSL* program can be summarized as

1. B_1: number of time slots of the initial optimum topology $\mathbf{T}^{B_1}$ for $\mathbf{a}_I$ (initial traffic)
2. f_1: value of the fitness function (utility) associated with the initial topology $\mathbf{T}^{B_1}$
3. B': number of time slots of the optimum topology $\mathbf{T}^{B'}$ for $\mathbf{a}'$ (differential access vector)
4. f_0: value of the fitness function (utility) associated with the initialization of the new topology $\mathbf{T}^{B_0}$ (after traffic variation)
5. N_A: number of active rings in the network after the traffic variation $N_A = \sum_n \mathbf{a}_{Fn}$
6. **procedure** Check_traffic_variation
7. $E(\mathbf{a}_{F\lambda}(t)) = \lambda(t); E(\mathbf{a}_{F\delta}(t)) = \delta(t)$ {Assign access vector $\mathbf{a}_F$ based on the existing traffic}
8. $\mathbf{a}' = \mathbf{a}_I$ XOR $\mathbf{a}_F$
9. If $\mathbf{a}' \neq \mathbf{0}$
10. Apply TSL($\mathbf{a}'$) to obtain the optimum topology for $\mathbf{a}'$.
11. Check the value of Δ to initialize the new topology:
12. If $\Delta_n > 0$, initialize the new topology as the set of chromosomes $\mathbf{T}^{B_0} = \{PT^{B_1}, PT^{B'}\}$.
13. If $\Delta_n < 0$, the new topology is initialized as the set of the different chromosomes in both
14.   sets $\mathbf{T}^{B_0} = \{PT^{B_1} - PT^{B'}\}$
15.   Calculate the fitness function (f_0) by using (12.60) and (18) with $f_0 \rightarrow U$
16.   Initialize f = f_0; n_m = 0; $\mathbf{a}_I = \mathbf{a}_F$
17.   end
18. **end**
19. **procedure** Calculate_fitness
20.   Calculate f by using (12.60) and (12.61) with $f \rightarrow U$
21.   If (f-f_0) > threshold
22.   Reconfigure the system with $\mathbf{T}_{opt} = \mathbf{T}_{new}$ that corresponds to the fitness $f_{opt} = f$
23. f_0 = f;
24.   end
25. **end**
- - - - - - - - - - - - - SGA-TSL algorithm - - - - - - - - - -
26. While (1)
27.   Check_traffic_variation
28.   for b = 1 to B_0
29. $\mathbf{T}_{new}$ = crossover(PT^b, $PT^{(b+1) \bmod B_0}$);
30.   Calculate_fitness
31.   end
32.   If (n_m<N_A) {number of mutations<number of rings actived}
33. $\mathbf{T}_{new}$ = mutation($\mathbf{T}^{B_0}, \gamma_{n\_m}$);
34.   n_m++;
35.   Calculate_fitness
36.   end
37. end

Lines 1 to 5 define the variables that will be used in the program. From line 6 to 18, the procedure *Check_traffic_variation* is defined. In line 6, the access vector $\mathbf{a}_F$ is assigned depending on the traffic variation. Line 8 calculates the differential access vector $\mathbf{a}'$. Line 9 examines whether the traffic has changed. In line 10 TSL program is used to obtain the optimum topology associated with the traffic variation in the network, $\mathbf{T}^{B'}$. This topology is needed to initialize the algorithm. This topology has only one active source and TSL program should complete the search in few iterations. From lines 11 to 14 the topology depending on the traffic variation is initialized to start the program. If a new source has appeared, the new topology $\mathbf{T}^{B_0}$ is initialized as the set (union) of partial topologies (chromosomes) of the initial topology $\mathbf{T}^{B_1}$ and the set of partial topologies corresponding to $\mathbf{T}^{B'}$. On the other hand, if a source has become inactive then the new topology $\mathbf{T}^{B_0}$ is initialized as the difference between the two sets of partial topologies. In line 15, the fitness function f_0 for the new topology is calculated. In line 16 the instant fitness value f, the number of mutations n_m and access vector $\mathbf{a}_I$ are initialized.

From lines 19 to 25 the procedure *Calculate_fitness* is defined. In line 20, the fitness value f is calculated as result of the optimization described by (18). Line 21 checks if the new fitness value f is higher than the previous one f_0 plus certain threshold. The threshold can be zero or a positive value depending on how often the traffic changes in the network. If the previous condition holds, line 22 reconfigures the system with the new topology $\mathbf{T}_{new}$.

From lines 26 to 37, SGA-TSL algorithm is described. Line 27 checks if the traffic in the network has changed. In line 28, the time slots of the new topology $\mathbf{T}^{B_0}$ are assigned to index b. The crossover of the partial topologies corresponding to $\mathbf{T}^{B_0}$ is performed in line 29 to obtain the new topology $\mathbf{T}_{new}$. In line 30, procedure *Calculate_fitness* is called.

Lines 32 to 36 perform the mutation operation if f did not follow the previous requirements for the topology to be updated. As the mutation is performed gene by gene, line 32 checks if the number of mutations is less than the number of active rings in the network. Line 33 realizes the mutation operation over gene n_m. Line 34 updates the number of mutations. In line 35 procedure *Calculate_fitness* is called again to check if the new fitness value f has improved compared to the previous one f_0 to reconfigure the network with the topology associated to f. Finally, the algorithm goes back to 26 to check again the traffic or to continue with the crossover and mutation operations to find a better fitness.

The algorithm may be implemented in one of the base stations, where cooperating base stations must exchange information about the traffic distribution (λ_i, δ_i) in their respective cells i. The base station or an equivalent coordinating unit should pass the information about the resulting access vector $\mathbf{a}_F$ back to the cooperating base stations.

12.9.5   Performance Examples

In this section, we provide several examples to illustrate the performance of the algorithms. We calculate the link capacities $c_r(m_{r2}, m_{r1}, \mathbf{m}_{i2}, \mathbf{m}_{i1})$ as specified in Section 12.9.1. While the analysis is general, for the sake of simplicity in this section, the channel gains used to calculate $SINR_{m_{r1}, m_{r2}}(\mathbf{m}_{i2}, \mathbf{m}_{i1})$ are $G_{m_{i1}, m_{r2}} \sim 1/d^{\alpha}_{m_{i1}, m_{r2}}$ and $G_{m_{i1}, m_{i2}} \sim 1/d^{\alpha}_{m_{i1}, m_{i2}}$, where $d^{\alpha}_{m_{i1}, m_{r2}}$ is the distance between the interfering transmitter in ring m_{i1} and reference receiver in ring m_{r2}, analogous for $d^{\alpha}_{m_{i1}, m_{i2}}$ and, α is the propagation constant. In the simulations we use $\alpha = 4$, and SNR = 10. The calculation of the distances is straightforward from the geometry presented in Figure 12.9(a).

In Figures 12.12 and 12.17 we present the utility for different access vectors a versus the topology index (t) for the scenario presented in Figure 12.9(a). The resulting topologies, indexed by t, represent a certain combination of the active links and are defined as collection of partial topologies (links) generated in B slots and will be represented formally as $\Im^{(2)} = \bigcup_b T^{(b)} = \bigcup_b L^{(b)}(l)$, where $L^{(b)}$ indicates the set of active links.

Figure 12.12, presents the utility function versus the topology index for a = [010010]. With this access vector user in ring 2, in both, cell r and i have permission to transmit. As the number of possible topologies obtained for this access vector is very high, we plot the segment of topologies close to the optimum topology. The maximum utility is $u = 0.6991$ by using network coding, and with no coding the maximum utility is $u = 0.5826$.

In Figure 12.13 the transmission pattern is shown for one of the optimum topologies (topology index $t = 7$) in the case with no coding, for the previous access vector, defined by the set of links $T_7 = \left\{ \{l_1^{(down)}\}, \{l_{10}^{(up)}\}, \{l_4^{(down)}, l_7^{(up)}\}, \{l_4^{(up)}\}, \{l_1^{(up)}\}, \{l_7^{(down)}\}, \{l_{10}^{(down)}\} \right\}$. We can see that isolated short range transmissions are favored which can simultaneously reduce the intercell interference and power consumption.

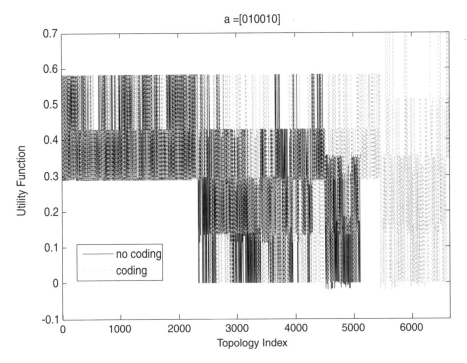

Figure 12.12   Utility function for access vector a = [010010].

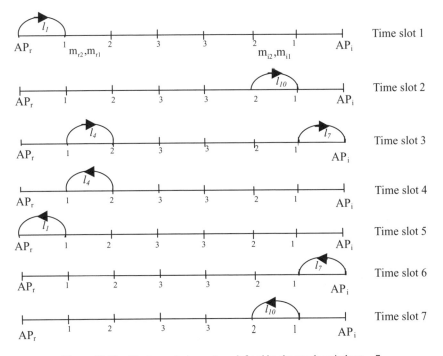

Figure 12.13   The transmission pattern defined by the topology index t = 7.

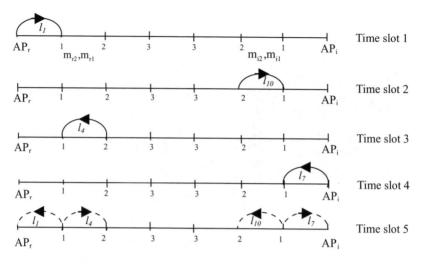

Figure 12.14   The transmission pattern defined by the topology index t = 5571.

In Figures 12.14 and 12.15 the transmission patterns for two topology indices that correspond to the maximum utility with coding ($t = 5571$ and $t = 5621$) for the previous access vector are plotted. The optimum topologies for the two cases are given by $T_{5571} = \left\{ \{l_1^{(down)}\}, \{l_{10}^{(up)}\}, \{l_4^{(up)}\}, \{l_7^{(down)}\}, \{l_1^{(up)} \oplus l_4^{(down)}, l_7^{(up)} \oplus l_{10}^{(down)}\} \right\}$ and $T_{5621} = \left\{ \{l_1^{(down)}\}, \{l_4^{(up)}\}, \{l_7^{(down)}\}, \{l_1^{(up)} \oplus l_4^{(down)}, l_{10}^{(up)}\}, \{l_7^{(up)} \oplus l_{10}^{(down)}\} \right\}$.

We can see an improvement in the number of slots needed with coding (five slots) compared to seven slots in the case with no coding. So for the same type of isolated and short range transmissions the utility function is improved by reducing the number of slots.

In Figure 12.16 the overall capacity for the previous access vector **a** is presented. For the optimum topologies with coding we can see that the overall capacity of the system improves by a factor of 4 compared with the case without coding.

In Figure 12.17, the utility function is shown for $\mathbf{a} = [010100]$. With this access vector user from ring 2 in cell r and user from ring 1 in cell i have permission to transmit. The maximum utility is obtained for topology index $t = 478$ ($u_{478} = 0.8739$) by using network coding. We see a significant improvement

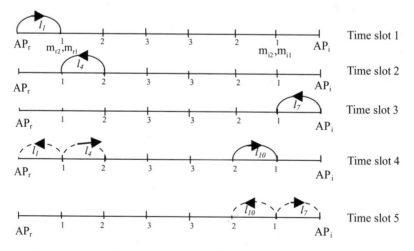

Figure 12.15   The transmission pattern defined by the topology index t = 5621.

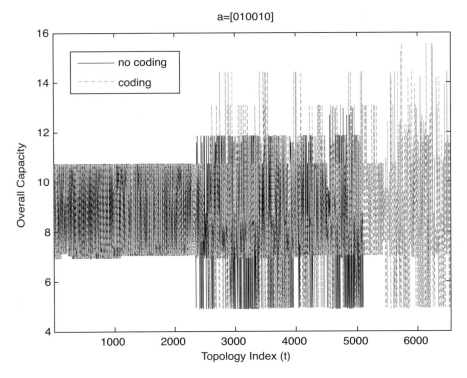

Figure 12.16 Overall capacity for a = [010010].

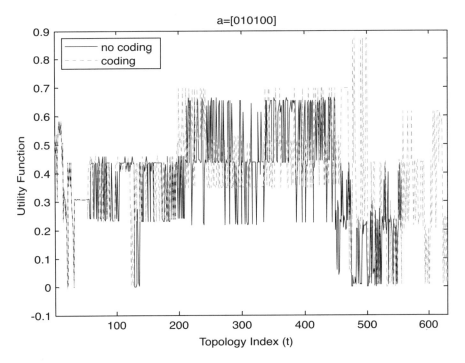

Figure 12.17 Utility function for access vector a = [010100].

compared with the maximum utility with no coding, obtained for topology index $t = 215$ ($u_{215} = 0.6640$). Both utilities are higher than in the previous case due to lower interference level.

The optimum topologies for both cases are given by

$$T_{478} = \left\{ \{l_1^{(down)}\}, \{l_7^{(down)}\}, \{l_4^{(up)}\}, \{l_1^{(up)} \oplus l_4^{(down)}, l_7^{(up)}\} \right\}$$

and

$$T_{215} = \left\{ \{l_1^{(down)}\}, \{l_4^{(up)}\}, \{l_4^{(down)}, l_7^{(up)}\}, \{l_7^{(down)}\}, \{l_1^{(up)}\} \right\}$$

In Table 12.1 the reconfiguration results obtained with the SGA-TSL algorithm are presented for uplink transmission and localized variations in traffic Δ when the initial state in the network was defined by the optimum topology associated with access vector $\mathbf{a}_I$. The new optimum topology, as a result of the reconfiguration of the initial optimum topology, is associated with the access vector $\mathbf{a}_F$.

As mentioned in Section 12.9.4, we consider that the traffic distribution is observed in time intervals short enough to detect each change in the traffic so that traffic changed only in one ring is assumed in a given observation instant. The traffic change is given by access vector $\mathbf{a}'$.

As the first step, we show the number of generated topologies needed to find the optimum one by exhaustive search N_{es} and by GA N_{ga}, and the fitness value associated to the new optimum topology.

- Entries from t = 1 to t = 4 in Table 12.2 represent the reconfiguration results for $\mathbf{a}_I = (100001)$ when a new source appears in ring 2 ($\mathbf{a}' = (010000)$), 3 ($\mathbf{a}' = (001000)$), 4 ($\mathbf{a}' = (000100)$), or 5 ($\mathbf{a}' = (000010)$), respectively. In the first case, topology t = 1, $N_{es} = 209$ compared to $N_{ga} = 8$ is obtained when a new source appears in ring 2. For t = 2, as a new source appears in ring 3, we have more routes than in the previous case so $N_{es} = 412$, and $N_{ga} = 11$. For t = 3, as the new source just introduces one more route we obtain $N_{es} = 59$ and $N_{ga} = 1$. For t = 4, we have $N_{es} = 163$ and $N_{ga} = 9$. The optimum topologies are presented for $\mathbf{a}_I$, $\mathbf{a}'$ and $\mathbf{a}_F$. From t = 5 to 10, we show more examples for different traffic patterns and traffic variations for $\Delta > 0$.

- From t = 11 to 13, the reconfiguration results for $\mathbf{a}_I = (100111)$ are shown when a source becomes inactive in ring 4 ($\mathbf{a}' = (000100)$), 5 ($\mathbf{a}' = (000010)$), or 6 ($\mathbf{a}' = (000001)$), respectively. For t = 11, we eliminate one route from the initial topology and we have $N_{es} = 163$ compared to $N_{ga} = 1$. For t = 12, more routes are eliminated than in the previous case, so we need to generate less topologies than before to obtain the optimum topology resulting in $N_{es} = 59$ and $N_{ga} = 5$. For t = 13, the source in the border of the cell became inactive so the number of routes is significantly reduced and we have $N_{es} = 22$ and $N_{ga} = 1$. From t = 14 to 20, other examples are shown for different traffic patterns and traffic variations for $\Delta < 0$.

We can see that the number of generated topologies by using GA in the search for the optimum one is significantly reduced. In some of the examples the mutation operation is needed to obtain the optimum topology, and is indicated by arrow '$\downarrow$' over the mutated link l_k.

Due to the symmetry of the scenario presented in Figure 12.9(a), $\mathbf{a}_{F_1} = (\mathbf{a}_1, \mathbf{a}_2)$ and $\mathbf{a}_{F_2} = (\mathbf{a}_2, \mathbf{a}_1)$ will produce symmetrical topologies. The previous examples represent some illustrative cases to show the performance of the algorithm for different traffic variations. In order to calculate the total number of possible combinations, we denote by $|\mathbf{a}_1|$ and $|\mathbf{a}_2|$ the number of rings active in cell 1 and 2 respectively. For $\Delta > 0$, taking into account the symmetry of the scenario considered, the number of combinations where the changes can occur is $\sum_{|\mathbf{a}_1|=0}^{n_1} \sum_{|\mathbf{a}_2|=|\mathbf{a}_1|}^{n_2} \binom{n_1}{|\mathbf{a}_1|} \binom{n_2}{|\mathbf{a}_2|} \binom{n_1 + n_2 - |\mathbf{a}_1| - |\mathbf{a}_2|}{1}$ where $n_1 = n_2 = 3$ (3 rings per cell). For $\Delta < 0$, the number of combinations is $\sum_{|\mathbf{a}_1|=1}^{n_1} \sum_{|\mathbf{a}_2|=|\mathbf{a}_1|}^{n_2} \binom{n_1}{|\mathbf{a}_1|} \binom{n_2}{|\mathbf{a}_2|} \binom{|\mathbf{a}_1| + |\mathbf{a}_2|}{1}$.

The performance of the algorithm is evaluated using two performance parameters: the success ratio $SR = 1/N_t$ where N_t is the number of generated topologies in the search for the optimum solution and, the

Table 12.2   Different topologies after the traffic variation Δ

Δ_n a_I	t	a'	a_F
>0 (100001) $T^{B_1} = \{\{l_{12}\}, \{l_1\}, \{l_8\}\}$	1	(010000) $T^{B'} = \{l_4, l_1\}$	(110001) $T_{new} = \{\{l_{12}\}, \{l_4\}, \{l_1\}, \{l_1, l_8\}\}$ $N_{es} = 209,\ N_{ga} = 8$ $f_{opt} = 0.6554$
	2	(001000) $T^{B'} = \{l_6, l_2\}$	(101001) $T_{new} = \{\{l_{12}\}, \{l_6\}, \{l_1\}, \{l_2, l_8\}\}$ $N_{es} = 412,\ N_{ga} = 11$ $f_{opt} = 0.6554$
	3	(000100) $T^{B'} = \{l_7\}$	(100101) $T_{new} = \{\{l_{12}\}, \{l_1\}, \{l_8\}, \{l_7\}\}$ $N_{es} = 59,\ N_{ga} = 1$ $f_{opt} = 0.6554$
	4	(000010) $T^{B'} = \{l_{10}, l_7\}$	(100011) $T_{new} = \{\{l_{12}\}, \{l_1\}, \{l_{10}\}, \{l_8\}, \{l_7\}\}$ $N_{es} = 163,\ N_{ga} = 9$ $f_{opt} = 0.5243$
>0 (010100) $T^{B_1} = \left\{\left\{\overset{\downarrow}{l}_4\right\}, \{l_1, l_7\}\right\}$	5	(000010) $T^{B'} = \{l_{10}, l_7\}$	(010110) $T_{new} = \left\{\{l_{10}\}, \{l_7\}, \left\{\overset{\downarrow}{l}_2, l_7\right\}\right\}$ $N_{es} = 70,\ N_{ga} = 8$ $f_{opt} = 0.5888$
>0 (010001) $T^{B_1} = \left\{\{l_4\}, \left\{l_1, \overset{\downarrow}{l}_9\right\}\right\}$	6	(000100) $T^{B'} = \{l_7\}$	(010101) $T_{new} = \left\{\left\{\overset{\downarrow}{l}_{12}\right\}, \{l_1, l_8\}, \{l_7\}, \{l_4\}\right\}$ $N_{es} = 205,\ N_{ga} = 13$ $f_{opt} = 0.5243$
>0 (001001) $T^{B_1} = \left\{\{l_6\}, \{l_2\}, \left\{\overset{\downarrow}{l}_{12}\right\}, \{l_8\}\right\}$	7	(010000) $T^{B'} = \{l_4, l_1\}$	(011001) $T_{new} =$ $\left\{\{l_6\}, \{l_4\}, \left\{\overset{\downarrow}{l}_{11}, l_1\right\}, \{l_7, l_2\}\right\}$ $N_{es} = 1579,\ N_{ga} = 27$ $f_{opt} = 0.4406$
>0 (010010) $T^{B_1} = \left\{\{l_4\}, \left\{l_1, \overset{\downarrow}{l}_8\right\}\right\}$	8	(001000) $T^{B'} = \{l_6, l_2\}$	(011010) $T_{new} =$ $\left\{\{l_6\}, \{l_4\}, \left\{l_1, \overset{\downarrow}{l}_{10}\right\}, \{l_2, l_7\}\right\}$ $N_{es} = 560,\ N_{ga} = 4$ $f_{opt} = 0.4957$
>0 (001100) $T^{B_1} = \{\{l_6\}, \{l_2\}, \{l_7\}\}$	9	(010000) $T^{B'} = \{l_4, l_1\}$	(011100) $T_{new} = \{\{l_6\}, \{l_4\}, \{l_2\}, \{l_1\}, \{l_7\}\}$ $N_{es} = 154,\ N_{ga} = 9$ $f_{opt} = 0.5243$
>0 (110010) $T^{B_1} = \{\{l_{10}\}, \{l_2, l_7\}, \{l_1\}\}$	10	(000100) $T^{B'} = \{l_7\}$	(110110) $T_{new} = \{\{l_7\}, \{l_{10}\}, \{l_2, l_7\}, \{l_1\}\}$ $N_{es} = 257,\ N_{ga} = 1$ $f_{opt} = 0.6600$

Table 12.2   (*Continued*)

Δ_n	a_I	t	a'	a_F
		11	(000100) $T^{B'} = \{l_7\}$	(100011) $T_{new} = \{\{l_{12}\}, \{l_{10}\}, \{l_1\}, \{l_7\}, \{l_8\}\}$ $N_{es} = 163, N_{ga} = 2$ $f_{opt} = 0.5243$
<0	(100111) $T^{B_1} =$ $\{\{l_{12}\}, \{l_{10}\}, \{l_8\}, \{l_7\}, \{l_1, l_7\}\}$	12	(000010) $T^{B'} = \{\{l_{10}\}, \{l_7\}\}$	(100101) $T_{new} = \{\{l_1\}, \{l_7\}, \{l_{12}\}, \{l_8\}\}$ $N_{es} = 59, N_{ga} = 5$ $f_{opt} = 0.6554$
		13	(000001) $T^{B'} = \{\{l_{12}\}, \{l_8\}\}$	(100110) $T_{new} = \{\{l_7\}, \{l_{10}\}, \{l_1, l_7\}\}$ $N_{es} = 22, N_{ga} = 1$ $f_{opt} = 0.6815$
<0	(100011) $T^{B_1} = \{\{l_{12}\}, \{l_{10}\}, \{l_1\}, \{l_7\}, \{l_8\}\}$	14	(000001) $T^{B'} = \{\{l_{12}\}, \{l_8\}\}$	(100010) $T_{new} = \{\{l_{10}\}, \{l_1, l_7\}\}$ $N_{es} = 8, N_{ga} = 3$ $f_{opt} = 0.5823$
<0	(001110) $T^{B_1} = \left\{\{l_7\}, \{l_{10}\}, \left\{\overset{\downarrow}{l}_3, l_7\right\}\right\}$	15	(000010) $T^{B'} = \{\{l_{10}\}, \{l_7\}\}$	(001100) $T_{new} = \left\{\left\{\overset{\downarrow}{l}_6\right\}, \{l_2\}, \{l_7\}\right\}$ $N_{es} = 20, N_{ga} = 2$ $f_{opt} = 0.5826$
<0	(101110) $T^{B_1} = \left\{\{l_1\}, \{l_6\}, \{l_2\}, \{l_7\}, \left\{\overset{\downarrow}{l}_8\right\}\right\}$	16	(001000) $T^{B'} = \{l_6, l_2\}$	(100110) $T_{new} = \left\{\{l_7\}, \left\{\overset{\downarrow}{l}_{10}\right\}, \{l_1, l_7\}\right\}$ $N_{es} = 23, N_{ga} = 11$ $f_{opt} = 0.6815$
<0	(110110) $T^{B_1} = \{\{l_7\}, \{l_{10}\}, \{l_2, l_7\}, \{l_1\}\}$	17	(100000) $T^{B'} = \{l_1\}$	(010110) $T_{new} = \{\{l_7\}, \{l_{10}\}, \{l_2, l_7\}\}$ $N_{es} = 71, N_{ga} = 1$ $f_{opt} = 0.5888$
<0	(011001) $T^{B_1} =$ $\left\{\left\{\overset{\downarrow}{l}_6\right\}, \{l_4\}, \left\{l_1, \overset{\downarrow}{l}_{11}\right\}, \{l_2, l_7\}\right\}$	18	(010000) $T^{B'} = \{l_4, l_1\}$	(001001) $T_{new} = \left\{\{l_{12}\}, \left\{\overset{\downarrow}{l}_3, \overset{\downarrow}{l}_8\right\}\right\}$ $N_{es} = 105, N_{ga} = 4$ $f_{opt} = 0.4369$
<0	(111010) $T^{B_1} =$ $\left\{\{l_6\}, \{l_4\}, \{l_1\}, \left\{\overset{\downarrow}{l}_{10}, l_1\right\}, \{l_2, l_7\}\right\}$	19	(010000) $T^{B'} = \left\{\overset{\downarrow}{l}_4, l_1\right\}$	(101010) $T_{new} = \left\{\{l_6\}, \left\{\overset{\downarrow}{l}_2, \overset{\downarrow}{l}_8\right\}, \{l_1\}\right\}$ $N_{es} = 166, N_{ga} = 9$ $f_{opt} = 0.5816$
<0	(011010) $T^{B_1} =$ $\left\{\{l_6\}, \{l_4\}, \left\{\overset{\downarrow}{l}_{10}, l_1\right\}, \{l_2, l_7\}\right\}$	20	(001000) $T^{B'} = \{l_6, l_2\}$	(010010) $T_{new} = \left\{\{l_4\}, \left\{l_1, \overset{\downarrow}{l}_8\right\}\right\}$ $N_{es} = 22, N_{ga} = 4$ $f_{opt} = 0.4462$

Table 12.3 Performance evaluation of GA-TSL

t	a_F	SR_{ga}	SR_{es}	F
1	(110001)	1/8	1/209	209/8
2	(101001)	1/11	1/412	412/11
3	(100101)	1/1	1/59	59/1
4	(100011)	1/9	1/163	163/9
5	(010110)	1/8	1/70	70/8
6	(010101)	1/13	1/205	205/13
7	(011001)	1/27	1/1579	1579/27
8	(011010)	1/4	1/560	560/4
9	(011100)	1/9	1/154	154/9
10	(110110)	1/1	1/257	257/1
11	(100011)	1/2	1/163	163/2
12	(100101)	1/5	1/59	59/5
13	(100110)	1/1	1/22	22/1
14	(100010)	1/3	1/8	8/3
15	(100010)	1/2	1/20	20/2
16	(100110)	1/11	1/23	23/11
17	(010110)	1/1	1/71	71/1
18	(001001)	1/4	1/105	105/4
19	(101010)	1/9	1/166	166/9
20	(010010)	1/4	1/22	22/4

improvement factor F given by $F = N_{es}/N_{ga}$. The performance parameters for the previous examples are shown in Table 12.2.

- Entry $t = 1$, in Table 12.3, shows $SR_{ga} = 1/8$ compared to $SR_{es} = 1/209$ to obtain the optimum topology for $a_F = (110001)$ which gives an improvement factor $F = 209/8$.
- Entry $t = 2$, in Table 12.3, shows $SR_{ga} = 1/11$ compared to $SR_{es} = 1/412$ to obtain the optimum topology for $a_F = (101001)$. In this case the number of new routes introduced by a' is higher than in the previous case, so N_t is increased. And it gives an improvement factor $F = 412/11$.
- Entry $t = 3$, F is significantly increased ($F = 59/1$) because the new optimum topology is obtained by concatenating the active links from the initial topology and $T^{B'}$, which is the first operation of SGA-TSL algorithm for $\Delta_n > 0$. Equivalently, for $\Delta_n < 0$ the first operation to obtain the new optimum topology is eliminating the active links of $T^{B'}$ from the initial topology. This is the case of $t = 13$ ($F = 22/1$), and $t = 17$ ($F = 71/1$).

In all cases, independently of the location of the traffic variation in the network, N_{ga} is at least one order of magnitude less than N_{es}.

So far, we have initialized the SGA-TSL algorithm by the optimum topology calculated by exhaustive search. In more complicated scenarios exhaustive search is not practical. In the following we calculate the optimum topology for the cases $t = 1,2,3,4$ from Tables 12.2 and 12.3, when the initial topology is any of the feasible topologies for $a_l = [100001]$. In Figure 12.18 we represent the utility versus the topology index for access vector a_l. We can see that there are 20 feasible topologies for that access vector, and the optimum topologies correspond to indexes 16 to 18 in the figure. To show the robustness of our algorithm we initialized GA with any feasible topology (SGA-AFT), to calculate the optimum topology in the cases $t = 1,2,3,4$. In Figure 12.19 we show that for obtaining the optimum topology for $a_F = [101001]$ starting for any feasible topology, in the worst case we have $N'_{ga} = 15$, compared to $N_{ga} = 8$ needed for SGA-TSL . For $a_F = [110010]$, in the worst case we need $N'_{ga} = 31$ to obtain the optimum topology, compared to $N_{ga} = 10$ need by SGA-TSL. For $a_F = [100011]$, SGA-AFT, in the worst case needs $N'_{ga} = 17$, compared to $N_{ga} = 9$. And for $a_F = [100101]$ we obtain $N'_{ga} = 13$, compared to N_{ga}

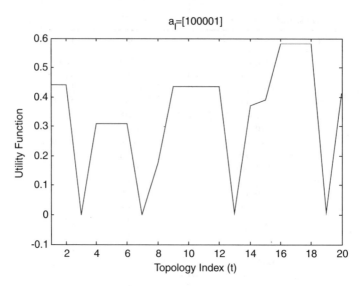

Figure 12.18   Utility function for access vector $a_I = [100001]$.

= 1. We can see that our proposed algorithm outperforms an exhaustive search even when the initial topology is not the optimum one.

In summary, in this section we have presented a sequential GA algorithm for the efficient dynamic reconfiguration of relaying topology in multihop cellular networks due to traffic variations in the network. Depending on the traffic load, there may be situations where searching for the new optimum topology will be NP-hard. Through numerical simulations we have shown that by using SGA the number of operations required to reconfigure the optimum topology is significantly reduced independently of the

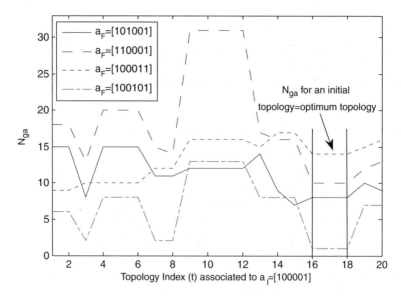

Figure 12.19 N_{ga} to obtain the optimum topology for access vectors $a_F = [101001], [110010], [100011], [100101]$ when $a_I = [100001]$ and GA is initialized by any feasible topology t.

initial topology of the network. The utility function used in the optimization process drives the solution towards the topology favoring simultaneously isolated and short range transmissions. As expected, within these solutions further improvements are obtained by using network coding to reduce the number of slots needed for transmission.

The performance of the algorithm is characterized using two performance parameters: the success ratio SR which is inversely proportional to the number of generated topologies in the search process, and the improvement factor F defined as the ratio of the number of generated topologies by exhaustive search and number of generated topologies by GA algorithm. In addition to optimum performance in terms of network utility, numerical results demonstrate also significant improvements in the convergence rate of the new algorithm.

SGA may be implemented in one of the base stations. Cooperating base stations must exchange information about the traffic distribution, and the coordinating base station should pass the information about the resulting access vector $\mathbf{a}_F$ back to the cooperating base stations. This level of coordination between the base stations seems to be already considered in practice, i.e. coordinated multipoint transmission, where a cluster of base stations jointly perform beamforming in order to reduce intercell interference.

References

[1] Somayaji, A., Hofmeyr, S. and Forrest, S. (1997) *Principles of a Computer Immune System*, New Security Paradigms Workshop Langdale, Cumbria UK.

[2] Forrest, S., Hofmeyr, S. and Somayaji, A. (1997) Computer immunology. *Communications of the ACM*, **40**(10), 88–96.

[3] Hofmeyr, S. A. and Forrest, S. (2000) Architecture for an artificial immune system. *Evolutionary Computation Journal*, **8**(4), 443–473.

[4] Monroy, R., Saab, R. and Godínez, F. (2004) Sobre un Modelo Computacional del Sistema Inmune *Computación y Sistemas*, **7**(4) 249–259.

[5] Forrest, S., Balthrop, J., Glickman, M. and Ackley, D. (2002) *Computation in the Wild*.

[6] Mohamed, Y. A. and Abdullah, A. B. (2009) Immune inspired framework for ad hoc network security Control and Automation, *IEEE International Conference on Digital Object Identifier: 10.1109/ICCA.2009.5410147*, 297–302.

[7] Nian, L., Diangang, W., Xuemei, H., Sunjun, L. and Kui, Z. (2009) Research on network security situation awareness technology based on artificial immunity system, *International Forum on Information Technology and Applications, IFITA '09*, Vol, **1**, 10.1109/IFITA. 487, pp. 472–475.

[8] Zhang, X., Liang, Z. and Ding, Y. (2009) A distributed collaborative control scheme based on multi-immune agent, *International Joint Conference on Computational Sciences and Optimization, CSO 2009*. Vol. 1, 10.1109/CSO.2009.323, 651–655.

[9] The Second Mediterranean Workshop on Ad-Hoc Networks, 2003.

[10] Schoonderwoerd, R., Holland, O., Bruten, J. and Rothkrantz, L. (1996) Ant-based load balancing in telecommunications networks, *Adaptive Behavior*.

[11] Di Caro and Dorigo, G. (1997) Mobile agents for adaptive routing, *Technical Report, IRIDIA/97-12*, Universit Libre de Bruxelles, Belgium.

[12] Baran, B. and Sosa, R. (2000) A new approach for AntNet routing, *Proceedings of the Ninth International Conference on Computer Communications and Networks*.

[13] Heissenbuttel, M. and Braun, T. (2003) Ants-based routing in large scale mobile ad-hoc networks, *Kommunikation in Verteilten Systemen (KiVS)*.

[14] Gunes, M., Sorges, U. and Bouazizi, I. (2002) ARA – The ant-colony based routing algorithm for MANETs, *Proceedings of the ICPP Workshop on Ad Hoc Networks (IWAHN 2002)*, IEEE Computer Society Press, 79–85.

[15] Gunes, M., Kahmer, M. and Bouazizi, I. (2003) Ant routing algorithm (ARA) for mobile multi-hop ad-hoc networks – new features and results, *The Second Mediterranean Workshop on Ad-Hoc Networks*.

[16] Bonabeau, E., Dorigo, M. and Theraulaz (1999) *Swarm Intelligence: From Natural to Artificial Systems*, Oxford University Press.

[17] Glisic, S. and Lorenzo, B. (2009) *Advanced Wireless Networks, Cognitive, Cooperative and Opportunistic 4G Technology*, 2e, John Wiley and Sons, London.

[18] Arsie, A., Savla, K. and Frazzoli, E. (2009) Efficient routing algorithms for multiple vehicles with no explicit communications, *IEEE Transactions on Automatic Control*, **54**(10), 2302–2317.

[19] An, B. and Lesser, V. (2009) Characterizing contract-based multiagent resource allocation in networks, *IEEE Transactions on Systems, Man, and Cybernetics, Part B: Cybernetics*, **PP**(99), 1–12.

[20] Wedde, H. F., Lehnhoff, S., Senge, S., and Lazarescu, A. M. (2009) Bee inspired bottom-up self-organization in vehicular traffic management, *Third IEEE International Conference on Self-Adaptive and Self-Organizing Systems, 2009. SASO '09*, 278–279.

[21] Peterson, L. L. and Davie, B. (1996) *Computer Networks: A System Approach*, Morgan Kaufmann.

[22] Bertsekas, D. and Tsitsiklis, J. (1996) *Neuro-Dynamic Programming*. Athena Scientific.

[23] Di Caro, G. and Dorigo, M. (1998) AntNet: distributed stigmergetic control for communications networks, *Journal of Artificial Intelligence Research* **9**, 317–365.

[24] Das, A. K., Marks, R. J.. El-Sharkawi, M., Arabshahi, P. and Gray, A. (2004) Optimization Methods for Minimum power multicasting in wireless networks with sectored antennas, Proceedings of IEEE Wireless Communications and Networking Conference 2004, 1299–1304.

[25] Shen, C.-C., Huang, Z. and Jaikaeo, C. (2005) Ant-based distributed topology control algorithms for mobile ad hoc networks, *Wireless Networks* **11**, 299–317.

[26] Qi, H., Iyengar, S. S. and K. Chakrabarty (2001) Multi-resolution data integration using mobile agents in distributed sensor networks, *IEEE Trans. Systems, Man, and Cybernetics Part C: Applications and Rev.*, **31**(3), 383–391.

[27] Rao, N. S. V. (2002) Multisensor fusion under unknown distributions: finite sample performance guarantees, in *Multisensor Fusion*, A.K. Hyder, E. Shahbazian and E. Waltz, eds.

[28] Small, T. and Haas, Z. J. (2005) Resource and performance tradeoffs in delay-tolerant wireless networks. In *ACM Workshop on Delay Tolerant Networking*.

[29] Spyropoulos, T., Psounis, K. and Raghavendra, C. S. (2005) Spray and wait: an efficient routing scheme for intermittently connected mobile networks. In *Workshop on Delay Tolerant Networking*.

[30] Vahdat, A. and Becker, D. (2000) Epidemic routing for partially connected ad hoc networks. *Technical Report CS-200006*, Duke University.

[31] Groenevelt, R., Nain, P. and Koole, G. (2005) The message delay in mobile ad hoc networks. In *Performance*, October.

[32] Bettstetter, C. (2001) Mobility modeling in wireless networks: categorization, smooth movement, and border effects. In *ACM SIGMOBILE Mobile Computing and Communications Review*, **5**(3), July.

[33] Kurtz, T. G. (1970) Solutions of ordinary differential equations as limits of pure jump markov processes. *Journal of Applied Probabilities*, 49–58.

[34] Small, T and Haas, Z. J. (2003) The shared wireless infostation model – a new ad hoc networking paradigm. In *Mobihoc*.

[35] Kermack, W. O. and McKendrick, A. G. (1927) A contribution to the mathematical theory of epidemics. In *Proc. Royal Society of London*, vol. A **115**, 700–721.

[36] Daley, D. J. and Gani, J. (1999) *Epidemic Modelling*. Cambridge University Press.

[37] Zhang, E., Neglia, G., Kurose, J. and Towsley, D. (2005) Performance modeling of epidemic routing, *U Mass Computer Science Technical Report 2005-44*.

[38] Alfano, G. and Miorandi, D. (2006) On information transmission among nanomachines, in *Proceedings of the First International Conference on Nano-Networks (NanoNet'06)*, September.

[39] Aylott, J. W. (2003) Optical nanosensors – an enabling technology for intracelular measurements, *Analyst* **128**, 309–312.

[40] Badjic, J. D., Balzani, V., Credi, A., Silvi, S. and Stoddar, J. F. (2004) A molecular elevator, *Science* **303**, 1845–1849.

[41] Ballardini, R., Balzani, V., Credi, A., Gandolfi, M. and Venturi, M. (2001) Artificial molecular-level machines: which energy to make them work? *Accounts of Chemical Research* **34**, 445–455.

[42] Balzani, V., Gómez-López, M. and Stoddart, J. F. (1998) Molecular machines, *Accounts of Chemical Research* **31**, 405–414.

[43] Balzani, V., Credi, A., Silvi, S. and Venturi, M. (2006) Artificial nanomachines based on interlocked molecular species: recent advances, *Chemical Society Reviews* **35**, 1135–1149.

[44] Batteas, J., Chidsey, C., Kagan, C. and Seideman, T. (2007) Building electronic functions into nanoscale molecular architectures, *National Science Foundation, Technical Report*.

[45] Bauschlicher, C., Ricca, A. and Merkle, R. (1997) Chemical storage of data, *Nanotechnology* **8**, 1–5.

[46] Behkam, B. and Sitti M. (2007) Bacterial flagella-based propulsion and on/off motion control of microscale objects, *Applied Physics Letters* **90**.

[47] Whitesides, G. M. (2001) The once and future nanomachine, Scientific American (September), 78–83.

[48] Soong, R. K., Bachand, G. D., Neves, H. P., Olkhovets, A. G., Craighead, H. G. and Montemagno, C. D. (2000) Powering an inorganic nanodevice with a biomolecular motor, *Science* **24**, 1555–1558.

[49] Behkam, B. and Sitti, M. (2007) Bacterial flagella-based propulsion and on/off motion control of microscale objects, *Applied Physics Letters* **90**.

[50] Cavalcanti, A., Shirinzadeh, B. Freitas, R. A. and Hogg, T. (2008) Nanorobot architecture for medical target identification, *Nanotechnology* **19**(1), 1–12.

[51] Freitas, R. A. (1999) *Nanomedicine, Volume I: Basic Capabilities*. Landes Biosience.

[52] Moore, M., Enomoto, A., Nakano, T., Egashira, R., Suda, T., Kayasuga, A., Kojima, H., Sakakibara, H. and Oiwa, K. (2006) A design of a molecular communication system for nanomachines using molecular motors, in *Proceedings of the Fourth Annual IEEE International Conference on Pervasive Computing and Communications (PerCom'06)*, March.

[53] Moore, M., Enomoto, A., Nakano, T., Okaie, Y. and Suda, T. (2007) Interfacing with nanomachines through molecular communication, in Proceedings of the IEEE International Conference on Systems, Man and Cybernetics, October, 18–23.

[54] Moritani, Y., Hiyama, S. and Suda, T. (2006) Molecular communication among nanomachines using vesicles, in *Proceedings of NSTI Nanotechnology Conference*, May.

[55] Moritani, Y., Hiyama, S. and Suda, T. (2006) Molecular communication for health care applications, in *Proceedings of the Fourth Annual IEEE International Conference on Pervasive Computing and Communications (PerCom'06)*, March.

[56] Serreli, V., Lee, C., Kay, E. R. and Leigh, D. A. (2007) A molecular information ratchet, *Nature* **445**, 523–527.

[57] Tulu, U., Fagerstrom, C., Ferenz, N. and Wadswort, P. (2006) Molecular requirements for kinetochore-associated microtubule formation in mammalian cells, *Current Biology* **16**, 536–541.

[58] Hiratsuka, Y., Tada, T., Oiwa, K., Kanayama, T. and Uyeda, T. Q. P. (2001) Controlling the direction of kinesin-driven microtubule movements along microlithographic tracks, *Biophysical Journal* **81**, 1555–1561.

[59] Koltz, I., and Hunston, D. (1984) Mathematical models for ligand-receptor binding. real sites, ghost sites, *Journal of Biological Chemistry* **259**, 10060–10062.

[60] Berridge, M. J. (1997) The AM and FM of calcium signalling, *Nature* **386**, 759–780.

[61] Nakano, T., Suda, T., Moore, M., Egashira, R., Enomoto, A. and Arima, K. (2005) Molecular communication for nanomachines using intercellular calcium signaling, in *Proceedings of the Fifth IEEE Conference on Nanotechnology*, 478–481.

[62] Lorenzo, B. and Glisic, S. (2010) Genetic Algorithm Based Dynamic Topology Reconfiguration in Cellular Multihop Wireless Networks, submitted to *IEEE Transactions on Mobile Computing*, July 2010.

[63] Holland, J. H. (1975) *Adaptation in Natural and Artificial Systems*, MIT Press.

[64] Davis, L. (1991) *Handbook of Genetic Algorithms*. Van Nostrand Reinhold.

[65] Grefenstette, J. J. (1986) Optimization of control parameters for genetic algorithms, *IEEE Trans. Systems, Man and Cybernetics*, **16**(1), 122–128.

[66] Goldberg, D. E. (1989) *Genetic Algorithms in search, optimization and machine learning*, Addison-Wesley.

[67] Michalewicz, Z. (1992) *Genetic Algorithms + Data Structures = Evolution Programs*. Springer-Verlag.

[68] Grant, M. and Boyd, S. (2009) CVX: Matlab software for disciplined convex programming (web page and software). http://stanford.edu/~boyd/cvx, June.

[69] Grant, M. and Boyd, S. (2008) Graph implementations for nonsmooth convex programs, *Recent Advances in Learning and Control (atribute to M. Vidyasagar)*, in V. Blondel, S. Boyd, and H. Kimura, eds, 95–110, *Lecture Notes in Control and Information Sciences*, Springer, 2008. http://stanford.edu/~boyd/graph_dcp.html.

13

Positioning in Wireless Networks

13.1 Mobile Station Location in Cellular Networks

13.1.1 Introduction

In this section we present a general mathematical framework that covers all the positioning techniques for processing absolute and/or relative distance measurements between the mobile station and multiple base transceiver stations. Based on this scheme, a general measure of positioning accuracy is introduced and then analyzed in the special cases defined by the three most feasible techniques. A geometrical interpretation of the formulas that define the location accuracy and a comparison between the techniques from the geometric conditioning point of view are also given.

The Federal Communications Commission (FCC) mandated US cellular operators, in 1996, to locate mobile phones calling the emergency number 911 by October 2001 [1, 2]. The European Union has been taking steps towards similar regulation [3]. Emergency services, beside commercial applications such as vehicle fleet management, intelligent transport systems, and location-based billing [4, 5], are also examples of applications of mobile station (MS) positioning. The first studies on positioning of mobile phones were published in the 1970s [6]. Since that time, several techniques for locating the MS by measuring attenuation, direction of arrival, and delay of the radio signals exchanged between the MS and multiple base transceiver stations (BTSs) have been proposed (see [7] for an overview). Simulation results of phase-ranging and pulse-ranging techniques are presented in [8] and [9]. Methods for locating mobiles by processing attenuation measurements are presented in [10–14]. In [15], methods involving time-of-arrival (TOA) and angle-of-arrival (AOA) measurements in a code-division, multiple-access, network are analyzed. Positioning through AOA measurements is discussed in [16]. Performance of tracking algorithms that process absolute and/or relative propagation delay measurements are presented using simulations in [17–19] and verified experimentally in [20] and [21–23].

Standardization activities for MS positioning are presented in [24–26]. Along the years, the standardization group T1P1.5 has considered four alternatives:

  (i) the *network-assisted GPS* method [27], which calculates the MS location by using GPS technology;
 (ii) *time advance* [28];
(iii) enhanced observed time difference (E-OTD) [29];
 (iv) *uplink TOA* [30].

The *network-assisted GPS* method is beyond the scope of this section and the remaining methods will be the focus of this discussion. The more general goal of this section is to analyze the accuracy of the multilateration techniques when used in cellular networks. Such analysis is carried out in a general manner

Advanced Wireless Communications & Internet: Future Evolving Technologies, Third Edition. Savo Glisic.
© 2011 John Wiley & Sons, Ltd. Published 2011 by John Wiley & Sons, Ltd.

so that it applies to the third and incoming fourth generation of universal mobile telecommunications systems [31].

The *time advance, enhanced OTD,* and *uplink TOA* methods estimate the MS coordinates by processing with multilateration techniques absolute distances (ADs) and/or relative distances (RDs) between the MS and multiple BTSs. A general analysis of the multilateration techniques that process only ADs (*circular multilateration*) or only RDs (*hyperbolic multilateration*) can be found in [32–35]. The technique that combines ADs and RDs (referred to as *mixed multilateration*) is of major interest for cellular applications.

In this section we present the basics of the *time advance, enhanced OTD,* and *uplink TOA* methods as a background for later analysis. The mathematical formulation of the generic multilateration problem and its linear weighted least squares (WLS) solution are also presented. The WLS location estimate is then used to derive an explicit expression for the positioning accuracy measure. The accuracy of circular, hyperbolic, and mixed multilateration is geometrically interpreted.

13.1.2   MS Location Estimation Using AD and RD Measurements

The *time advance* method uses the existing timing advance (TA) parameter, which is introduced to avoid overlapping of bursts transmitted by the MS during a call in TDMA systems [28, 38, 39]. For positioning purposes, the TA is considered to be an estimate of the absolute distance between MS and serving BTS, and is used to implement the *circular multilateration* technique. In a two-dimensional scenario, as assumed throughout this section, ADs from $N \geq 3$ different stations are needed to find the MS coordinates at the intersection of N circumferences centered at the BTSs with radii equal to the AD measurements.

In a TDMA-based network, the TA is estimated by the serving BTS only when the MS is in *connected mode* (i.e., the MS is communicating with the serving BTS using a dedicated channel) [39]. As a consequence, ADs from multiple BTSs can only be measured by sequentially forcing the communication to be handed over from one BTS to another until all the N BTSs have been accessed. In practice, when a positioning handover occurs, the TA is estimated by the new serving BTS, but the handover request is not responded to and the connection returns to the previous serving BTS [28].

The E-OTD method is based on three parameters: observed time difference (OTD), real time difference (RTD), and geometric time difference GTD = RTD – OTD. If a burst is transmitted by BTS_1 and BTS_2 at the instants t_{Tx1} and t_{Tx2} respectively, and received by the MS at instants t_{Rx1} and t_{Rx2} respectively, then the RTD is $t_{Tx2} - t_{Tx1}$ and the OTD is $t_{Rx2} - t_{Rx1}$. The GTD is a scaled measure of the relative distance between the MS and the pair BTS_1, BTS_2; in fact, GTD = RTD – OTD = $(t_{Rx1} - t_{Tx1}) - (t_{Rx2} - t_{Tx2}) = (d_1 - d_2)/c$ = RD/c, with c being the speed of light and $d_1 = c(t_{Rx1} - t_{Tx1})$, $d_2 = c(t_{Rx2} - t_{Tx2})$ being the lengths of the propagation path between the MS and BTS_1, BTS_2, respectively [29]. The possible positions of an MS observing a constant GTD value are located on a hyperbola having foci at BTS_1 and BTS_2. In a two-dimensional scenario, the MS position is calculated via *hyperbolic multilateration* at the intersection of at least two hyperbolas; thus $N \geq 3$ BTSs are needed to implement this technique. Because of the implementation [39], the *enhanced OTD* method takes one of the N available BTSs as *reference BTS* and uses it to calculate *all* the $N - 1$ RDs. As a consequence, linear dependence between multiple equations is avoided.

13.1.3   The Circular, Hyperbolic, and Mixed Multilateration

Let $\{BTS_1, \ldots, BTS_N\}$ be the N base stations available for locating the MS. BTS_1 is the reference station eventually used to evaluate RDs. Let $S = \{1, \ldots, N\}$ be an *ordered* set of indexes identifying the BTSs. Within S, the following *ordered* subsets are defined as in Figure 13.1, where:

$S_c = \{$indexes identifying the BTSs involved in AD measurements$\}$;

$S_h = \{$indexes identifying the BTSs involved in RD measurements$\}$;

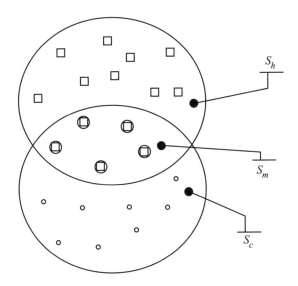

Figure 13.1   Graphical representation of the subsets S_c, S_m, and S_h.

$S_m = S_c \cap S_h = \{\text{indexes identifying the BTSs involved } both \text{ in AD and RD measurements}\}$;

$\bar{S}_c = S_c - S_m = \{\text{indexes identifying the BTSs involved } only \text{ in AD measurements}\}$;

$\bar{S}_h = S_h - S_m = \{\text{indexes identifying the BTSs involved } only \text{ in RD measurements.}\}$.

The number of elements in each subset is $N_c = size \{S_c\}$, $N_h = size \{S_h\}$, $N_m = size \{S_m\}$, $\bar{N}_c = size$ $\{\bar{S}_c\}$ and $\bar{N}_h = size \{\bar{S}_h\}$. Moreover $S = S_c \cup S_h = \bar{S}_c \cup \bar{S}_h \cup S_m$ and $N = N_c + N_h - N_m = \bar{N}_c + \bar{N}_h + \bar{N}_m$.

The general multilateration problem is described by a set of equations in which each AD defines a circumference and each RD defines a hyperbola. The known BTS coordinates are $(x_1, y_1), \ldots, (x_n, y_n)$, and the unknown MS coordinates are (x, y). The AD measured between the MS and the BTS$_i$ $(i \in \bar{S}_c)$ is d_i; the RD measured between the MS and the pair of stations $\{\text{BTS}_1, \text{BTS}_l\}(l \in \bar{S}_h - \{1\})$ is r_{1l}; the AD and RD involving the BTS identified by S_m are d_p $(p \in S_m)$ and $r_{1q}(q \in S_m - \{1\})$, respectively.

$$
\begin{cases}
\sqrt{(x-x_i)^2 + (y-y_i)^2} = d_i, & i \in \bar{S}_c \\
\sqrt{(x-x_1)^2 + (y-y_1)^2} - \sqrt{(x-x_l)^2 + (y-y_l)^2} = r_{1l}, & l \in \bar{S}_h - \{1\} \\
\sqrt{(x-x_p)^2 + (y-y_p)^2} = d_p, & p \in S_m \\
\sqrt{(x-x_1)^2 + (y-y_1)^2} - \sqrt{(x-x_q)^2 + (y-y_q)^2} = r_{1q}, & q \in S_m - \{1\}
\end{cases}
\tag{13.1}
$$

The circular, hyperbolic, and mixed multilateration can be derived from Equation (13.1) by properly defining S_c, S_h, and S_m. In *circular multilateration* all the available BTSs are used only to measure ADs, and no RD measurements are available; thus S_h and S_m are empty and $\bar{S}_c$ coincides with S_c: $S_m = S_h = \bar{S}_h = 0$; $\bar{S}_c = S_c = S = \{1, \ldots, N\}$. In *hyperbolic multilateration* all the available BTSs are used to measure only RDs. No ADs are used. BTS$_1$ is only used to measure RDs; thus S_m is empty and BTS$_1$ belongs to subset $\bar{S}_h$, which in turn coincides with S_h: $S_m = S_c = \bar{S}_c = 0$; $\bar{S}_h = S_h = S = \{1, \ldots N\}$. Finally, in *mixed multilateration* all the available BTSs are used to measure RDs. BTS$_1$ is also used to measure ADs; thus BTS$_1$ belongs to S_h and is the only element in S_m, which coincides with S_c (BTS$_1 \notin \bar{S}_h$). Since no BTSs are used only to measure ADs, $\bar{S}_c$ is empty: $S_h = \{1, \ldots, N\}$; $S_m = S_c = \{1\}$; $\bar{S}_h = \{2, \ldots, N\}$; $\bar{S}_c = 0$.

13.1.4   WLS Solution of the Location Problem

In real applications, Equation (13.1) should be further modified to include the presence of noise. In this section, x and y are estimated by using a linear WLS algorithm [41, 48], following the same approach of [35] and [37]. To apply the WLS estimation, Equation (13.1) is linearized by assuming that an *a priori* estimate of the MS position, $P^{(0)} = (x^{(0)}, y^{(0)})$, is available. $P^{(0)}$ could be determined, for instance, from a previous iteration of the WLS algorithm or by classical calculus to find the intersection of a number of circumferences defined by Equation (13.1). If $P^{(0)}$ is sufficiently close to $P = (x, y)$, the equations resulting from the ith AD measurement and the lth RD measurement can be accurately represented by their linear approximation in a neighborhood of $P^{(0)}$:

$$d_i \simeq d_i^{(0)} - u_{i,x}\Delta x - u_{i,y}\Delta y$$
$$r_{1l} \simeq d_{1l}^{(0)} - w_{1l,x}\Delta x - w_{1l,y}\Delta y \tag{13.2}$$

where

$$\Delta x = x - x^{(0)}; \qquad \Delta y = y - y^{(0)};$$
$$u_{i,x} = (x_i - x^{(0)})/(d_i^{(0)}); \qquad u_{i,y} = (y_i - y^{(0)})/(d_i^{(0)});$$
$$d_i^{(0)} = \sqrt{(x_i - x^{(0)})^2 + (y_i - y^{(0)})^2}; \qquad r_{1l}^{(0)} = d_1^{(0)} - d_l^{(0)};$$
$$w_{1l,x} = u_{1,x} - u_{l,x}; \qquad w_{1l,y} = u_{i,y} - u_{l,y}.$$

A graphical representation is given in Figure 13.2 where $\hat{\mathbf{u}}_i = u_{i,x}\hat{\mathbf{x}} + u_{i,y}\hat{\mathbf{y}}$ is a unit vector originated at the MS and directed toward BTS_i ($\hat{\mathbf{x}}$ and $\hat{\mathbf{y}}$ are the unit vectors in the x and y directions) and $w_{1l,x}$ and $w_{1l,y}$ are the x–y components of the difference vector $\mathbf{w}_{1l} = \hat{\mathbf{u}}_1 - \hat{\mathbf{u}}_l$.

Stacking the linearized Equations (13.2) as in Equation (13.1) results in a linear system $\mathbf{Ax} = \mathbf{b}$ defined by the 2×1 unknown vector $\mathbf{x}$, the $(\bar{N}_c + \bar{N}_h + 2N_m - 1) \times 2$ design matrix $\mathbf{A}$, and the

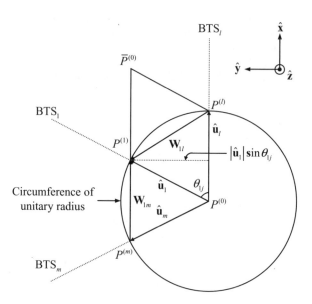

Figure 13.2   Definition of the basic vectors: $\mathbf{w}_{1l}, \hat{\mathbf{u}}_1, \hat{\mathbf{u}}_l$. The area of the parallelogram of vertices $P^{(0)}, P^{(l)}, \bar{P}^{(0)}, P^{(1)}$ is A_{1j}. The area of the parallelogram of vertices $P^{(0)}, P^{(l)}, P^{(1)}, P^{(m)}$ is B_{1m}.

$(\bar{N}_c + \bar{N}_h + 2N_m - 1) \times 1$ observation vector $\mathbf{b}$ defined as follows:

$$\mathbf{A} = \begin{bmatrix} \mathbf{U}_{\bar{c}} \\ \mathbf{W}_{\bar{h}} \\ \mathbf{U}_m \\ \mathbf{W}_m \end{bmatrix}; \mathbf{x} = \begin{bmatrix} \Delta x \\ \Delta y \end{bmatrix}; \mathbf{b} = \begin{bmatrix} \mathbf{D}_{\bar{c}} \\ \mathbf{R}_{\bar{h}} \\ \mathbf{D}_m \\ \mathbf{R}_m \end{bmatrix} \qquad (13.3)$$

$\mathbf{U}_{\bar{c}} = |u_{i,x}\, u_{i,y}|_{i \in \bar{S}_c}$ is a $\bar{N}_c \times 2$ matrix with the x–y components of $\widehat{\mathbf{u}}_i (i \in \bar{S}_c)$ on its columns. Analogously, $\mathbf{W}_{\bar{h}} = [w_{1l,x} w_{1l,y}]_{l \in \bar{S}_h - \{1\}}$, $\mathbf{U}_m = [u_{p,x} u_{p,y}]_{p \in S_m}$ and $\mathbf{W}_m = [w_{1q,x} w_{1q,y}]_{q \in S_m - \{1\}}$. The vectors $\mathbf{D}_{\bar{c}} = [d_i^{(0)} - d_i]_{i \in \bar{S}_c}$, $\mathbf{R}_{\bar{h}} = [r_{1l}^{(0)} - r_{1l}]_{l \in \bar{S}_h - \{1\}}$, $\mathbf{D}_m = [d_p^{(0)} - d_p]_{p \in S_m}$ and $\mathbf{R}_m = [r_{1q}^{(0)} - r_{1q}]_{q \in S_m - \{1\}}$ are obtained by stacking the measurements in a single column. The WLS solution of the linear problem $\mathbf{Ax} = \mathbf{b}$ that minimizes the scalar cost function $J(\mathbf{x}) = (\mathbf{Ax} - \mathbf{b})^T \mathbf{Q_b}^{-1} (\mathbf{Ax} - \mathbf{b})$ is $\widehat{\mathbf{x}} = \mathbf{G}^{-1}\mathbf{g}$, where $\mathbf{G} = \mathbf{A}^T \mathbf{Q_b}^{-1} \mathbf{A}$, $\mathbf{g} = \mathbf{A}^T \mathbf{Q_b}^{-1} \mathbf{b}$, and $\mathbf{Q}_b$ is the covariance matrix of $\mathbf{b}$ [41, 48].

13.1.5 Accuracy Measure

The covariance matrix of $\widehat{\mathbf{x}}$ is $\mathbf{Q}_{\widehat{\mathbf{x}}} = \mathrm{Cov}\{\widehat{\mathbf{x}}\} = \mathbf{G}^{-1}$. The diagonal entries of $\mathbf{Q}_{\widehat{\mathbf{x}}}$ are the variances of the location error along the directions x and y. The square root of their sum, corresponding to the square root of the trace of $\mathbf{Q}_{\widehat{\mathbf{x}}}$, is the accuracy measure considered hereafter: $M = \sqrt{\mathrm{Tr}\{\mathbf{Q}_{\widehat{\mathbf{x}}}\}}$. In order to derive an explicit expression of M, $\mathbf{Q}_{\widehat{\mathbf{x}}} = \mathbf{G}^{-1} = (\mathbf{A}^T \mathbf{Q_b}^{-1} \mathbf{A})^{-1}$ must be evaluated; $\mathbf{A}$ is defined in Equation (13.3), and $\mathbf{Q}_b$ is derived below.

Let c_i and f_l be the measurement errors affecting the ith AD measurement d_i and the lth RD measurement r_{1l}, respectively. In TDMA-based systems, such errors can be assumed uncorrelated. ADs and RDs are averages of 'raw' propagation delays estimated during periods longer than the typical coherence time of the mobile radio channel. Under this hypothesis, $\mathbf{Q}_b$ has the following block-diagonal structure (matrices $\mathbf{0}$ are properly sized matrices with all zero entries):

$$\mathbf{Q_b} = \begin{bmatrix} \mathbf{S}_{e,\bar{c}} & \mathbf{0} & \mathbf{0} & \mathbf{0} \\ \mathbf{0} & \mathbf{S}_{f,\bar{h}} & \mathbf{0} & \mathbf{0} \\ \mathbf{0} & \mathbf{0} & \mathbf{S}_{e,m} & \mathbf{0} \\ \mathbf{0} & \mathbf{0} & \mathbf{0} & \mathbf{S}_{f,m} \end{bmatrix} \qquad (13.4)$$

where

$$\mathbf{S}_{e,\bar{c}} = \mathrm{diag}\{\sigma_{e,i}^2\}_{i \in \bar{S}_c};$$
$$\mathbf{S}_{f,\bar{h}} = \mathrm{diag}\{\sigma_{f,l}^2\}_{l \in \bar{S}_h - \{1\}};$$
$$\mathbf{S}_{e,m} = \mathrm{diag}\{\sigma_{e,p}^2\}_{p \in \bar{S}_m};$$
$$\mathbf{S}_{f,m} = \mathrm{diag}\{\sigma_{f,q}^2\}_{q \in \bar{S}_m - \{1\}};$$
$$\sigma_{e,i}^2 = \mathrm{Var}\{c_i\} \text{ are variances of the AD and RD measurement errors.}$$

$\mathbf{A}$ and $\mathbf{Q}_b$ defined in Equation (13.3) and (13.4) can be used to derive $\mathbf{G} = \mathbf{A}^T \mathbf{Q_b}^{-1} \mathbf{A}$

$$\mathbf{G} = \begin{bmatrix} G_{11} & G_{12} \\ G_{12} & G_{22} \end{bmatrix} \qquad (13.5)$$

with

$$G_{11} = \sum_{i \in \bar{S}_c} \sigma_{e,i}^{-2} u_{i,x}^2 + \sum_{p \in S_m} \sigma_{e,p}^{-2} u_{p,x}^2 + \sum_{l \in \bar{S}_h - \{1\}} \sigma_{f,l}^{-2} w_{1l,x}^2 + \sum_{q \in \bar{S}_m - \{1\}} \sigma_{f,q}^{-2} w_{1q,x}^2$$

$$G_{12} = \sum_{i \in \bar{S}_c} \sigma_{e,i}^{-2} u_{i,x} u_{i,y} + \sum_{p \in S_m} \sigma_{e,p}^{-2} u_{p,x} u_{p,y} + \sum_{l \in \bar{S}_h - \{1\}} \sigma_{f,l}^{-2} w_{1l,x} w_{1l,y} + \sum_{q \in \bar{S}_m - \{1\}} \sigma_{f,q}^{-2} w_{1q,x} w_{1q,y}$$

$$G_{22} = \sum_{i \in \bar{S}_c} \sigma_{e,i}^{-2} u_{i,y}^2 + \sum_{p \in S_m} \sigma_{e,p}^{-2} u_{p,y}^2 + \sum_{l \in \bar{S}_h - \{1\}} \sigma_{f,l}^{-2} w_{1l,y}^2 + \sum_{q \in \bar{S}_m - \{1\}} \sigma_{f,q}^{-2} w_{1q,y}^2 \tag{13.6}$$

Inversion of **G** leads to the following general expression of the accuracy measure:

$$M = \sqrt{\text{Tr}\{\mathbf{Q}_{\hat{x}}\}} = \sqrt{\frac{G_{11} + G_{22}}{\text{Det}\{\mathbf{G}\}}} \tag{13.7}$$

where

$$G_{11} + G_{22} = \sum_{i \in \bar{S}_c} \sigma_{e,i}^{-2} + \sum_{p \in S_m} \sigma_{e,p}^{-2} + \sum_{l \in \bar{S}_h - \{1\}} \sigma_{f,l}^{-2} |w_{1l}|^2 + \sum_{q \in \bar{S}_m - \{1\}} \sigma_{f,q}^{-2} |w_{1q}|^2 \tag{13.8}$$

and

$$
\begin{aligned}
\text{Det}\{\mathbf{G}\} = & \left[\sum_{i \in \bar{S}_c} \sigma_{e,i}^{-2} u_{i,x}^2 + \sum_{p \in S_m} \sigma_{e,p}^{-2} u_{p,x}^2 + \sum_{l \in \bar{S}_h - \{1\}} \sigma_{f,l}^{-2} w_{1l,x}^2 + \sum_{q \in \bar{S}_m - \{1\}} \sigma_{f,q}^{-2} w_{1q,x}^2 \right] \\
& \cdot \left[\sum_{i \in \bar{S}_c} \sigma_{e,i}^{-2} u_{i,y}^2 + \sum_{p \in S_m} \sigma_{e,p}^{-2} u_{p,y}^2 + \sum_{l \in \bar{S}_h - \{1\}} \sigma_{f,l}^{-2} w_{1l,y}^2 + \sum_{q \in \bar{S}_m - \{1\}} \sigma_{f,q}^{-2} w_{1q,y}^2 \right] \\
& - \left[\sum_{i \in \bar{S}_c} \sigma_{e,i}^{-2} u_{i,x} u_{i,y} + \sum_{p \in S_m} \sigma_{e,p}^{-2} u_{p,x} u_{p,y} + \sum_{l \in \bar{S}_h - \{1\}} \sigma_{f,l}^{-2} w_{1l,x} w_{1l,y} \right. \\
& + \left. \sum_{q \in \bar{S}_m - \{1\}} \sigma_{f,q}^{-2} w_{1q,x} w_{1q,y} \right] \cdot \left[\sum_{j \in \bar{S}_c} \sigma_{e,j}^{-2} u_{j,x} u_{j,y} + \sum_{r \in S_m} \sigma_{e,r}^{-2} u_{r,x} u_{r,y} \right. \\
& + \left. \sum_{m \in \bar{S}_h - \{1\}} \sigma_{f,m}^{-2} w_{1m,x} w_{1m,y} + \sum_{s \in \bar{S}_m - \{1\}} \sigma_{f,s}^{-2} w_{1s,x} w_{1s,y} \right]
\end{aligned}
\tag{13.9}
$$

The structure of M shows that the location accuracy is affected by the measurement accuracy ($\sigma_{e,i}, \sigma_{f,l}$) and the reciprocal position of the MS and the BTSs (represented by the x–y components of vectors $\hat{\mathbf{u}}_i$ and $\mathbf{w}_{ij}$). These two contributions can be separated only if all the (uncorrelated) measurement errors have the same standard deviation.

13.1.6   Circular Multilateration

In this case we have $S_h = S_m = \bar{S}_h = 0$ and $\bar{S}_c = S_c$. Using this in Equation (13.7) gives:

$$M_c^2 = \frac{\sum_{i \in S_c} \sigma_{e,i}^{-2}}{\text{Det}\{\mathbf{G}_c\}} \tag{13.10}$$

Det $\{\mathbf{G}_c\}$ can be obtained from Equation (13.9) by retaining only the sums over indexes i and j:

$$\text{Det}\{\mathbf{G}_c\} = \sum_{i,j \in S_c} \sigma_{e,i}^{-2} \sigma_{e,j}^{-2} (u_{i,x}^2 u_{j,y}^2 - u_{i,x} u_{i,y} u_{j,x} u_{j,y}) \tag{13.11}$$

The cross product between two vectors $\mathbf{a} = a_x \hat{\mathbf{x}} + a_y \hat{\mathbf{y}}$ and $\mathbf{b} = b_x \hat{\mathbf{x}} + b_y \hat{\mathbf{y}}$ can be written as $\mathbf{a} \times \mathbf{b} = (a_x b_y - a_y b_x) \hat{\mathbf{z}}$ ($\hat{\mathbf{z}} = \hat{\mathbf{x}} \times \hat{\mathbf{y}}$ is the unit vector in the z direction). With this observation in mind, the quantity between parentheses in Equation (13.11) becomes $\text{Det}\{\mathbf{G}_c\} = u_{i,x} u_{j,y} \hat{\mathbf{z}} \cdot (\mathbf{u}_i \times \mathbf{u}_j)$, where $\cdot$ represents the scalar product. $\text{Det}\{\mathbf{G}_c\}$ can be further modified by excluding from the double sum the terms occurring when $j = i$ since $\mathbf{u}_i \times \mathbf{u}_j = \mathbf{0}$. S_c (as well as S_h and S_m) is an ordered set of indexes; thus the formula $\sum_i \sum_{j \neq i} a_i b_j = \sum_i \sum_{j > i} (a_i b_j + a_j b_i)$, valid for i and j spanning the same set of *ordered* values, can be used and $\text{Det}\{\mathbf{G}_c\}$ becomes:

$$\text{Det}\{\mathbf{G}_c\} = \sum_{i \in S_c} \sum_{\substack{j \in S_c \\ j > i}} \sigma_{e,i}^{-2} \sigma_{e,j}^{-2} |\mathbf{u}_i \times \mathbf{u}_j|^2 \tag{13.12}$$

An interpretation of $\text{Det}\{\mathbf{G}_c\}$ in Equation (13.12) comes from the geometric definition of cross product (see Figure 13.2). In a polar reference system (ρ, θ) centered at the MS position, each unit vector $\hat{\mathbf{u}}_i$ can be expressed as $\hat{\mathbf{u}}_i = \cos \theta_i \hat{\mathbf{x}} + \sin \theta_i \hat{\mathbf{y}}$, with θ_i being the angle of $\hat{\mathbf{u}}_i$ measured counterclockwise from $\hat{\mathbf{x}}$. By definition, $|\hat{\mathbf{u}}_i \times \hat{\mathbf{u}}_j|^2 = (|\hat{\mathbf{u}}_i||\hat{\mathbf{u}}_j|| \sin \theta_{i,j}|)^2$, where $\theta_{ij} = \theta_i - \theta_j$; thus $|\hat{\mathbf{u}}_i \times \hat{\mathbf{u}}_j|^2$ is the area of the parallelogram determined by $\hat{\mathbf{u}}_i$ and $\hat{\mathbf{u}}_j$, $A_{ij} = |\hat{\mathbf{u}}_i||\hat{\mathbf{u}}_j|| \sin \theta_{i,j}|$, raised to the second power. Introducing, A_{ij} in Equation (13.12), we have

$$M_c^2 = \frac{\sum_{i \in S_c} \sigma_{e,i}^{-2}}{\sum_{i \in S_c} \sum_{\substack{j \in S_c \\ j > i}} \sigma_{e,i}^{-2} \sigma_{e,j}^{-2} |\mathbf{u}_i \times \mathbf{u}_j|^2} = \frac{\sum_{i \in S_c} \sigma_{e,i}^{-2}}{\sum_{i \in S_c} \sum_{\substack{j \in S_c \\ j > i}} \sigma_{e,i}^{-2} \sigma_{e,j}^{-2} A_{ij}^2} \tag{13.13}$$

Equation (13.13) shows the dependence of the positioning accuracy on the AD measurements' accuracy $(\sigma_{e,j})$ and on the geometric conditioning (A_{ij}). Only if $\sigma_{e,i} = \sigma_{e,j} \triangleq \sigma_e$ the two contributions can be separated and $M_c = \sigma_e \text{ GDOP}_c$ can be written as a product of σ_e and the geometric dilution of precision GDOP_c:

$$\text{GDOP}_c = \sqrt{N_c \Big/ \sum_{i \in S_c} \sum_{\substack{j \in S_c \\ j > i}} A_{ij}^2} \tag{13.14}$$

For a given N_c, the location accuracy improves if GDOP_c is small, or equivalently if the areas A_{ij}s in the denominator of Equation (13.14) are large. Geometrically, this means that the unit vectors $\hat{\mathbf{u}}_i$, $\hat{\mathbf{u}}_j$, and then the BTSs, must be widely (angularly) separated with respect to the MS. Notice that since $|\hat{\mathbf{u}}_i| = 1$, $A_{ij} = |\sin(\theta_i - \theta_j)|$, thus GDOP_c depends only on the *angular* distribution of the BTSs around the MS and not on the distance between MS and BTSs, which might affect σ_e, for instance.

13.1.7 Hyperbolic Multilateration

In this case we have $S_m = S_c = \bar{S}_c = 0$ and $\bar{S}_h = S_h$ so that Equation (13.7) gives

$$M_h^2 = \frac{\sum_{l \in S_h - \{1\}} \sigma_{f,l}^{-2} |w_{1l}|^2}{\text{Det}\{\mathbf{G}_h\}} \tag{13.15}$$

Det$\{\mathbf{G}_h\}$ can be obtained from Equation (13.9) by retaining only the sums over indexes l and m. Introducing the area of the parallelogram determined by w_{1i} and w_{1m}, B_{lm} (see Figure 13.2), we have

$$M_h^2 = \frac{\sum_{l \in S_h - \{1\}} \sigma_{f,l}^{-2} |w_{1l}|^2}{\sum_{l \in S_h - \{1\}} \sum_{\substack{m \in S_h - \{1\} \\ m > l}} \sigma_{f,l}^{-2} \sigma_{f,m}^{-2} |\mathbf{w}_{1l} \times \mathbf{w}_{1m}|^2}$$

$$= \frac{\sum_{l \in S_h - \{1\}} \sigma_{f,l}^{-2} |w_{1l}|^2}{\sum_{l \in S_h - \{1\}} \sum_{\substack{m \in S_h - \{1\} \\ m > l}} \sigma_{f,l}^{-2} \sigma_{f,m}^{-2} B_{lm}^2} \tag{13.16}$$

If $(\sigma_{f,l} = \sigma_{f,m} \overset{\Delta}{=} \sigma_f)$, we have $M_h = \sigma_f \, \text{GDOP}_h$ where

$$\text{GDOP}_h = \sqrt{\frac{\sum_{l \in S_h - \{1\}} \sigma_{f,l}^{-2} |w_{1l}|^2}{\sum_{l \in S_h - \{1\}} \sum_{\substack{m \in S_h - \{1\} \\ m > l}} B_{lm}^2}} \tag{13.17}$$

is the GDOP for the hyperbolic multilateration.

13.1.8   Mixed Multilateration

In this case Equation (13.7) gives

$$M_m^2 = \frac{\sigma_{e,1}^{-2} + \sum_{l \in S_h - \{1\}} \sigma_{f,l}^{-2} |w_{1l}|^2}{\text{Det}\{\mathbf{G}_m\}} \tag{13.18}$$

because $S_m = S_c = \{1\}$, $\bar{S}_h = S_h - \{1\}$ and $\bar{S}_c = S_c - S_m = 0$. Det$\{\mathbf{G}_h\}$ can be derived from Equation (13.9) by retaining the sums over indexes l, m and by substituting $p = r = 1$. Introducing the definitions of A_{1m} and B_{lm} as before, the expression for M_m^2, becomes

$$M_m^2 = \frac{\sigma_{e,1}^{-2} + \sum_{l \in S_h - \{1\}} \sigma_{f,l}^{-2} |w_{1l}|^2}{\sum_{l \in S_h - \{1\}} \sum_{\substack{m \in S_h - \{1\} \\ m > l}} \sigma_{f,l}^{-2} \sigma_{f,m}^{-2} |\mathbf{w}_{1l} \times \mathbf{w}_{1m}|^2 + \sum_{m \in S_h - \{1\}} \sigma_{e,1}^{-2} \sigma_{f,m}^{-2} |\hat{\mathbf{u}}_1 \times \hat{\mathbf{u}}_m|^2}$$

$$= \frac{\sigma_{e,1}^{-2} + \sum_{l \in S_h - \{1\}} \sigma_{f,l}^{-2} |w_{1l}|^2}{\sum_{l \in S_h - \{1\}} \sum_{\substack{m \in S_h - \{1\} \\ m > l}} \sigma_{f,l}^{-2} \sigma_{f,m}^{-2} B_{lm}^2 + \sum_{m \in S_h - \{1\}} \sigma_{e,1}^{-2} \sigma_{f,m}^{-2} A_{1m}^2} \tag{13.19}$$

A GDOP for the mixed multilateration can be defined only if AD and RD measurement errors have the same standard deviation $(\sigma_{e,1} = \sigma_{f,l} \overset{\Delta}{=} \sigma_m)$.

$$\text{GDOP}_m = \sqrt{\frac{1 + \sum_{l \in S_h - \{1\}} |w_{1l}|^2}{\sum_{l \in S_h - \{1\}} \sum_{\substack{m \in S_h - \{1\} \\ m > l}} B_{lm}^2 + \sum_{m \in S_h - \{1\}} A_{1m}^2}} \tag{13.20}$$

In the denominator of Equation (13.20), the areas B_{lm}s represent the RDs' contributions, while the area A_{1m} represents the contribution of the AD from BTS_1.

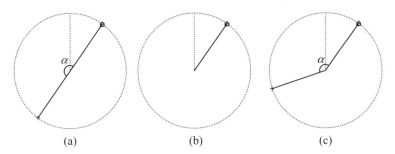

Figure 13.3 Infinite GDOP configurations when AT = 3, x, o, and Δ identify the BTSs and the center of the circumference the MS. (a) and (b) Circular, mixed, and hyperbolic trilateration (GDOP→ ∞ if MS and BTSs are aligned; α can be either zero or π. (c) Hyperbolic trilateration (GDOP→ ∞ if at least two BTSs are aligned with the MS; α can assume any value).

13.1.9   Performance Results for Three Stations

When the minimum number of BTSs is available ($N = 3$), the location techniques are named 'trilaterations'. GDOPc, GDOP$_h$ and GDOP$_m$ can be expressed as follows:

$$GDOP_c^2 = \frac{3}{|\hat{\mathbf{u}}_1 \times \hat{\mathbf{u}}_2|^2 + |\hat{\mathbf{u}}_1 \times \hat{\mathbf{u}}_3|^2 + |\hat{\mathbf{u}}_2 \times \hat{\mathbf{u}}_3|^2} = \frac{3}{A_{12}^2 + A_{13}^2 + A_{23}^2} \tag{13.14a}$$

$$GDOP_h^2 = \frac{|\mathbf{w}_{12}|^2 + |\mathbf{w}_{13}|^2}{|\mathbf{w}_{12} \times \mathbf{w}_{13}|^2} = \frac{|\mathbf{w}_{12}|^2 + |\mathbf{w}_{13}|^2}{B_{23}^2} \tag{13.17a}$$

$$GDOP_m^2 = \frac{1 + |\mathbf{w}_{12}|^2 + |\mathbf{w}_{13}|^2}{|\mathbf{w}_{12} \times \mathbf{w}_{13}|^2 + |\hat{\mathbf{u}}_1 \times \hat{\mathbf{u}}_2|^2 + |\hat{\mathbf{u}}_1 \times \hat{\mathbf{u}}_3|^2} = \frac{1 + |\mathbf{w}_{12}|^2 + |\mathbf{w}_{13}|^2}{B_{23}^2 + A_{12}^2 + A_{13}^2} \tag{13.20a}$$

and analyzed in the polar reference system (ρ, θ) introduced earlier. Assuming $\theta_1 = 0$, the GDOPs become functions of $\theta_2, \theta_3 \in [0, 2\pi]$ and can be plotted graphically. These plots can be used to determine under what conditions GDOP→ ∞ (e.g., even with noiseless measurements, a unique solution of the problem does not exist) or GDOP = GDOP$_{min}$ (e.g., the problem is optimally conditioned).

The results generalized to any θ_1 value are summarized in Figures 13.3 and 13.4 and Table 13.1 and can be interpreted as follows:

In the case of *circular trilateration:* (i) GDOP$_c$ → ∞ when $\theta_2 \in \theta_1 + \{0, \pi\}$, and $\theta_3 \in \{\theta_2 - \pi, \theta_2, \theta_2 + \pi\}$. These conditions are met when the three BTSs lie along a straight line passing through the MS

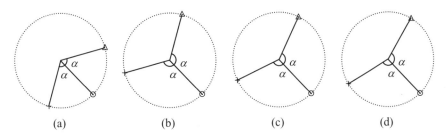

Figure 13.4 Minimum GDOP configurations when N = 3. x, o, and Δ identify the BTSs and the center of the circumference the MS. (a) and (b) Circular trilateration (GDOP$_c$ = GDOP$_{c,min}$ if a is either $\pi/3$ or $2\pi/3$ rad). (c) Hyperbolic trilateration (GDOP$_h$ = GDOP$_{h,min}$ if $\alpha = 0.61\pi$). (d) Mixed trilateration (GDOP$_m$ = GDOP$_{m,min}$ if $\alpha = 0.58\pi$).

Table 13.1   Comparison of the GDOP for the trilateration techniques

	Conditions of GDOP→ ∞	$\text{GDOP}_{\min}$ value	Conditions of GDOP = $\text{GDOP}_{\min}$
Circular trilateration	$\theta_2 \in \theta_1 + \{0, \pi\}$ and $\theta_3 \in \{\theta_2 - \pi, \theta_2, \theta_2 + \pi\}$	1.15	$\theta_2 \in \theta_1 + \{\pi/3, 2\pi/3, 4\pi/3, 5\pi/3\}$ and $\theta_3 \in \{\pi - \theta_2, 2\pi - \theta_2, 3\pi - \theta_2\}$
Hyperbolic trilateration	$\theta_1 = \theta_2$ or $\theta_1 = \theta_3$ or $\theta_2 = \theta_3$ or $\theta_1 = \theta_2 = \theta_3$	0.92	$\theta_2 \in \theta_1 + \{0.61\pi, 1.39\pi\}$ and $\theta_3 = 2\pi - \theta_2$
Mixed trilateration	$\theta_2 \in \theta_1 + \{0, \pi\}$ and $\theta_3 \in \{\theta_2 - \pi, \theta_2, \theta_2 + \pi\}$	0.88	$\theta_2 \in \theta_1 + \{0.58\pi, 1.42\pi\}$ and $\theta_3 = 2\pi - \theta_2$

(in fact, $\hat{\mathbf{u}}_1$, $\hat{\mathbf{u}}_2$ and $\hat{\mathbf{u}}_3$ are parallel and all the cross products in the denominator of GDOP_c are null). It can be easily shown, moreover, that the alignment between all the BTSs and the MS is the condition giving $\text{GDOP}_c \to \infty$ in the general case with $N \geq 3$.

(ii) $\text{GDOP}_c = \text{GDOP}_{c,\min} \simeq 1.15$ when $\theta_2 \in \theta_1 + \{\pi/3, 2\pi/3, 4\pi/3, 5\pi/3\}$ and $\theta_3 \in \{\pi - \theta_2, 2\pi - \theta_2, 3\pi - \theta_2\}$, meaning that the BTSs are angularly separated from the MS viewpoint by either $\pm 2\pi/3$ (for example, $\theta_1 = 0$, $\theta_2 = 2\pi/3$, $\theta_3 = -2\pi/3$) or $\pm\pi/3$ (for example, $\theta_1 = 0$, $\theta_2 = \pi/3$, $\theta_3 = -\pi/3$). In the first case, the BTSs are maximally separated; in the second case, they define the same parallelograms as if they were separated by $\pm 2\pi/3$.

In the case of *hyperbolic trilateration:* (i) $\text{GDOPh} \to \infty$ when at least two out of the three angles $\theta_1, \theta_2, \theta_3$ are equal. In these conditions, at least one vector among $\mathbf{w}_{12}$ and $\mathbf{w}_{13}$ is null and $B_{23} = 0$. Geometrically, the MS is the origin of half a straight line passing through at least two of the three BTSs (see Figure 13.3). In the general scheme with $N \geq 3$, $\text{GDOP}_h \to \infty$ when the MS is at the origin of half a straight line passing through at least N BTSs. (ii) Numerical minimization of GDOP_h results in two minima ($\text{GDOP}_{h,\min} \simeq 0.92$) verified when $\theta_2 \in \theta_1 + \{0.61\pi, 1.39\pi\}$ and $\theta_3 = 2\pi - \theta_2$).

Finally in the case of *mixed trilateration:* (i) $\text{GDOP}_m \to \infty$ in the same conditions as GDOP_c. (ii) Numerical minimization of GDOP_m results in two minima ($\text{GDOP}_{m,\min} \simeq 0.88$) verified when $\theta_2 \in \theta_1 + \{0.58\pi, 1.42\pi\}$ and $\theta_3 = 2\pi - \theta_2$.

13.1.10   Performance Results for N Stations

The combined effect of number and geographical distribution of the BTSs on the GDOP is analyzed in the regular network of hexagonal omnidirectional cells. Two relevant cases are considered. In the first one, the MS is located in a central cell; thus no particular restriction on the geometric conditioning is expected to be introduced by the BTS distribution. In the second case, the MS is located in a border cell of the network where the distribution of the available BTSs introduces serious limitations on the geometric conditioning of the problem. For both cases, the $N = 3, \ldots, 10$ BTSs nearest the MS are considered, and the closest one is used as a reference station for RDs calculation.

Table 13.2 reports maximum, minimum, and mean values of GDOP_c, GDOP_h, and GDOP_m when the MS is located in the central cell. Analogous values for the MS located in the border cell are listed in Table 13.3 [43].

Results in Table 13.1 prove that for $N = 3$ and a given value of θ_1, $\text{GDOP}_h \to \infty$ in an uncountable number of cases while GDOP_c and GDOP_m go to infinity in only four cases. Table 13.3 shows that when the MS is in the border cell, GDOP_h has maximum and mean values always greater than 1 and well above the corresponding values of GDOP_c and GDOP_m, which in turn are comparable ($\text{GDOP}_{h,\text{ mean}} > \text{GDOP}_{m,\text{ mean}} > \text{GDOP}_{c,\text{ mean}}$; $\text{GDOP}_{h,\text{ max}} \gg \text{GDOP}_{m,\text{ max}} \gg \text{GDOP}_{c,\text{ max}}$). From the robustness point of view (of particular interest when $N = 3$ or when the geometry of the problem is poor), the hyperbolic multilateration is thus the least reliable technique because a low number of BTSs and a poor geometric

Table 13.2 GDOP$_c$, GDOP$_h$, AND GDOP$_m$ when the MS is located in the central cell [43] © IEEE 2001

N	GDOP$_c$			GDOP$_h$			GDOP$_m$		
	min.	max.	mean	min.	max.	mean	min.	max.	mean
3	1.15	1.53	1.24	0.94	1.50	1.14	0.92	1.45	1.12
4	1.00	1.03	1.01	0.81	0.92	0.83	0.79	0.90	0.81
5	0.89	0.91	0.91	0.71	0.76	0.73	0.70	0.74	0.71
6	0.82	0.83	0.82	0.67	0.72	0.69	0.66	0.70	0.68
7	0.76	0.77	0.76	0.65	0.70	0.67	0.64	0.69	0.66
8	0.71	0.73	0.72	0.59	0.65	0.63	0.58	0.64	0.63
9	0.67	0.68	0.68	0.58	0.59	0.59	0.58	0.59	0.58
10	0.63	0.66	0.66	0.53	0.57	0.55	0.52	0.57	0.55

conditioning offer the highest GDOP values. Moreover, when $N = 3$, it is enough that two BTSs out of three are aligned with the MS to make GDOP$_h$ infinite. The circular technique, on the other hand, is the most robust technique as its GDOP has the lowest values when the geometry of the problem is poor. When $N = 3$, GDOP$_c \to \infty$ only when the three BTSs are aligned. In this sense, circular and mixed techniques are equivalent GDOP$_m \to \infty$ under the same conditions as GDOP$_c$).

For $N = 3$ and a fixed value of θ_1, GDOP$_c$ has eight minima, while GDOP$_h$ and GDOP$_m$ have only two minima; however, GDOP$_{m, \min}$ < GDOP$_{h, \min}$ < 1 < GDOP$_{c, \min}$, meaning that when $N = 3$, the accuracy for the circular trilateration is always worse than the AD measurements' accuracy $M_c = \sigma_c \cdot$ GDOP$_c$ > σ_c). Analogous conclusions can be drawn from Table 13.2: given N, minimum, mean, and maximum values of GDOP calculated when the MS in the central cell satisfy the inequality GDOP$_m$ < GDOP$_h$ < GDOP$_c$. This simply means that when the geometric conditioning is not problematic, the mixed and hyperbolic techniques are the most accurate while the circular technique has the highest GDOP values.

Based on the GDOP analysis above, the mixed multilateration turns out to be the location estimation technique that, among those considered in this section, offers the best compromise between robustness and accuracy.

13.2 Relative Positioning in Wireless Sensor Networks

In this section we discuss self-configuration in wireless sensor networks as a general class of estimation problem via the Cramér–Rao bound (CRB). Specifically, we consider sensor location estimation when sensors measure received signal strength (RSS) or time-of-arrival (TOA) between themselves and neighboring sensors. A small fraction of sensors in the network have a known location, whereas the remaining locations must be estimated. CRBs and maximum-likelihood estimators (MLEs) under Gaussian and log-normal models for the TOA and RSS measurements, are derived respectively.

Table 13.3 GDOP$_c$, GDOP$_h$, and GDOP$_m$ when the MS is located in the border cell [43] © IEEE 2001

N	GDOP$_c$			GDOP$_h$			GDOP$_m$		
	min.	max.	mean	min.	max.	mean	min.	max.	mean
3	1.15	1.51	1.23	0.95	8.81	2.44	0.93	1.58	1.34
4	1.00	1.35	1.07	0.81	5.23	2.11	0.79	0.57	1.21
5	0.89	1.22	0.95	0.71	4.63	1.99	0.70	1.44	1.13
6	0.82	1.13	0.86	0.66	4.54	1.76	0.65	1.42	1.05
7	0.76	0.98	0.80	0.60	4.16	1.54	0.59	1.30	0.99
8	0.71	0.95	0.75	0.56	3.36	1.44	0.56	1.28	0.94
9	0.67	0.86	0.70	0.55	3.11	1.38	0.55	1.23	0.91
10	0.64	0.83	0.67	0.52	3.03	1.33	0.52	1.2	0.89

We assume a network in which a small proportion of devices, called reference devices, have *a priori* information about their coordinates. All devices, regardless of their absolute coordinate knowledge, estimate the range between themselves and their neighboring devices. Such location estimation is called 'relative location' because the range estimates collected are predominantly between pairs of devices of which neither has absolute coordinate knowledge. These devices without *a priori* information we call blindfolded devices. In cellular location estimation systems, described in the previous section, location estimates are made using only ranges between a blindfolded device and reference devices. Relative location estimation requires simultaneous estimation of multiple device coordinates. Greater location estimation accuracy can be achieved as devices are added into the network, even when new devices have no *a priori* coordinate information and range to just a few neighbors.

Emerging applications for wireless sensor networks will depend on automatic and accurate location of thousands of sensors. In environmental sensing applications 'sensing data without knowing the sensor location is meaningless'. In addition, relative location estimation may enable applications such as inventory management, intrusion detection, traffic monitoring, and locating emergency workers in buildings.

To design a relative location system that meets the needs of these applications, several capabilities are necessary. The system requires a network of devices capable of peer-to-peer range measurement, an *ad-hoc* networking protocol, and a distributed or centralized location estimation algorithm. For range measurement, using received signal strength (RSS) is attractive from the point of view of device complexity and cost but is traditionally seen as a coarse measure of range. Instead, time-of-arrival (TOA) range measurement can be used. In this section, we will show that both RSS and TOA measurements can lead to accurate location estimates in dense sensor networks.

13.2.1   Performance Bounds

In network self-calibration problems, parameters of all devices in a network must be determined. Information comes both from measurements made between pairs of devices and a subset of devices that know *a priori* their parameters. A network self-calibration estimator estimates the unknown device parameters. For example, distributed clock synchronization in a network could be achieved by devices observing pair-wise timing offsets when just a small number of devices are synchronous. For details of mutual node synchronization in wireless networks see Chapter 13 in the previous edition of this book [45].

Specifically, consider a vector of device parameters $\gamma = [\gamma_1, \ldots, \gamma_{n+m}]$. Each device has one parameter. Devices $1 \ldots n$ are blindfolded devices, and devices $n + 1 \ldots n + m$ are reference devices. The unknown parameter vector is $\theta = [\theta_1, \ldots, \theta_n]$, where $\theta_i = \gamma_i$, for $i = 1 \ldots n$. Note that $\{\gamma_i, i = n + 1 \ldots n + m\}$ are known. Devices i and j make pair-wise observations $X_{i,j}$ with density $f_{X/\gamma}(X_{i,j}/\gamma_i, \gamma_j)$. We include also the case when devices make incomplete observations since two devices may be out of range or have limited link capacity. Let $H(i) = \{j$: device j makes pair-wise observations with device $i\}$. By convention, a device cannot make a pair-wise observation with itself, so that $i \notin H(i)$. By symmetry, if $j \in H(i)$, then $i \in H(j)$.

We assume by reciprocity that $X_{i,j} = X_{j,i}$; thus, it is sufficient to consider only the lower triangle of the observation matrix $\mathbf{X} = ((X_{i,j}))_{i,j}$ when formulating the joint likelihood function. In practice, if it is possible to make independent observations on the links from i to j and from j to i, then we assume that a scalar sufficient statistic can be found. Finally, we assume that $\{X_{i,j}\}$ are statistically independent for $j < i$. This assumption can be somewhat oversimplified for the RSS case but necessary for analysis. The log of the joint conditional pdf is

$$l(\mathbf{X}/\gamma) = \sum_{i=1}^{m+n} \sum_{\substack{j \in H(i) \\ j < i}} l_{i,j}$$

$$l_{i,j} = \log f_{\mathbf{X}/\gamma}(X_{i,j}/\gamma_i, \gamma_j) \tag{13.21}$$

The MLE will produce the vector $\hat{\theta} = \max_{\theta} l(\mathbf{X}/\gamma)$. The CRB on the covariance matrix of any unbiased estimator $\hat{\theta}$, including MLE estimator, is cov $(\hat{\theta}) \geq \mathbf{F}_{\theta}^{-1}$, where the Fisher information matrix (FDM) $\mathbf{F}_{\theta}$ is defined as

$$
\mathbf{F}_{\theta} = -\mathrm{E}\nabla_{\theta}(\nabla_{\theta}l(\mathbf{X}/\gamma))^T =
\begin{bmatrix}
f_{1,1} & \cdots & f_{1,n} \\
\vdots & \ddots & \vdots \\
f_{n,1} & \cdots & f_{n,n}
\end{bmatrix}
\tag{13.22}
$$

The diagonal elements $f_{k,k}$ of $\mathbf{F}$ given in Equation (13.22) are:

$$
f_{k,k} = E\left(\frac{\partial}{\partial\theta_k}l(\mathbf{X}/\theta)\right)^2 = E\left(\sum_{j\in H(k)}\frac{\partial}{\partial\theta_k}l_{k,j}\right)^2
$$

$$
f_{k,k} = \sum_{j\in H(k)}\sum_{p\in H(k)} E\left(\frac{\partial}{\partial\theta_k}l_{k,j}\right)\left(\frac{\partial}{\partial\theta_k}l_{k,p}\right)
$$

Since $X_{k,j}$ and $X_{k,p}$ are independent random variables, and $E\left((\partial/\partial\theta_k)l_{k,j}\right) = 0$, the expectation of the product is only nonzero for $p = j$. Thus, $f_{k,k}$ simplifies to the $k = l$ result in Equation (13.23). The off-diagonal elements similarly simplify:

$$
f_{k,k} = \sum_{j\in H(k)}\sum_{p\in H(l)} E\left(\frac{\partial}{\partial\theta_k}l_{k,j}\right)\left(\frac{\partial}{\partial\theta_k}l_{l,p}\right)
$$

Here, due to independence and zero mean of the two terms, the expectation of the product will be zero unless both $p = k$ and $j = l$. Thus, the $k \neq l$ result in Equation (13.23):

$$
f_{k,l} =
\begin{cases}
-\sum_{j\in H(k)} \mathrm{E}\left[\dfrac{\partial^2}{\partial\theta_k^2}l_{k,j}\right], & k = l \\[2mm]
-\mathrm{I}_{H(k)}(l)\mathrm{E}\left[\dfrac{\partial^2}{\partial\theta_k\partial\theta_l}l_{k,l}\right], & k \neq l
\end{cases}
\tag{13.23}
$$

where $\mathrm{I}_{h(k)}(l)$ is an indicator function: 1 if $l \in H(k)$ or 0 otherwise.

Intuitively, as more devices are used for location estimation, the accuracy increases for all the devices in the network. For an n-device network, there are $O(n)$ parameters but $O(n^2)$ variables $\{X_{i,j}\}$ used for their estimation. The remaining question is, what are sufficient conditions to ensure the CRB decreases as devices are added to the network? For a network of n blindfolded devices and m reference devices, consider adding one additional blindfolded device. For the n and $(n + 1)$ blindfolded device cases, let $\mathbf{F}$ and $\mathbf{G}$ be the FIMs defined in Equation (13.22), respectively. If $[\mathbf{G}^{-1}]_{ul}$ is the upper left $n \times n$ block of $\mathbf{G}^{-1}$ and if for the $(n + 1)$ blindfolded device case

Condition (i): $(\partial/\partial\theta_{n+1})l_{k,n+1} = \pm(\partial/\partial\theta_k)l_{k,n+1}, \forall k = 1 \ldots n$ and
Condition (ii): device $n + 1$ makes pair-wise observations between itself and at least one blindfolded device and at least two devices in total; then it can be shown that two properties hold:
Property (i): $\mathbf{F}^{-1} - [\mathbf{G}^{-1}]_{ul} \geq 0$ in the positive semi-definite sense, and
Property (ii): $\mathrm{tr}\mathbf{F}^{-1} > \mathrm{tr}[\mathbf{G}^{-1}]_{ul}$.

The Gaussian and log-normal distributions used in the next section meet condition (i). Property (i) implies that the additional unknown parameter introduced by the $(n + 1)$st blindfolded device does not impair the estimation of the original n unknown parameters. Furthermore, property (ii) implies that the sum of the CRB variance bounds for the n unknown parameters strictly decreases. Thus, when a blindfolded device enters a network and makes pair-wise observations with at least one blindfolded device and at least two devices in total, the bound on the average variance of the original n coordinate estimates is reduced.

To prove the above properties, compare $\mathbf{F}$, which is the FIM for the n blindfolded device problem, to $\mathbf{G}$, which is the FIM for the $n + 1$ blindfolded device case. Partition $\mathbf{G}$ into blocks:

$$\mathbf{G} = \begin{bmatrix} \mathbf{G}_{ul} & \mathbf{g}_{ur} \\ \mathbf{g}_{ll} & \mathbf{g}_{lr} \end{bmatrix}$$

where $\mathbf{G}_{ul}$ is an $n \times n$ matrix, g_{lr} is the scalar Fisher information for θ_{n+1}, and $\mathbf{g}_{ur} = \mathbf{g}_{ll}^T$ are $n \times 1$ vectors with kth element:

$$g_{ur}(k) = I_{H(n+1)}(k) E \left(\frac{\partial}{\partial \theta_k} l_{k,n+1}^{n+1} \right) \left(\frac{\partial}{\partial \theta_{n+1}} l_{k,n+1}^{n+1} \right)$$

$$g_{lr} = \sum_{j \in H(n+1)} E \left(\frac{\partial}{\partial \theta_k} l_{n+1,j}^{n+1} \right)^2$$

Here, we denote the log-likelihood of the observation between devices i and j in Equation (13.21) as $l_{i,j}^n$ and $l_{i,j}^{n+1}$ for the n and $(n + 1)$ blindfolded device cases, respectively. If $l^n(\mathbf{X}|\gamma_n)$ and $l^{n+1}(\mathbf{X}|\gamma_{n+1})$ are the joint log-likelihood function in Equation (13.21) for the n and $n + 1$ blindfolded device cases, respectively, then we have:

$$l^{n+1}(\mathbf{X}/\gamma_{n+1}) = \sum_{i=1}^{m+n+1} \sum_{\substack{j \in H(i) \\ j < i}} l_{i,j}^{n+1} = l^n(\mathbf{X}/\gamma_n) + \sum_{j \in H(n+1)} l_{n+1,j}^{n+1}$$

Since $l_{n+1,j}^{n+1}$ is a function only of parameters $\gamma_{n+1} = \theta_{n+1}$ and γ_j, then

$$\frac{\partial^2}{\partial \theta_k \partial \theta_l} \sum_{j \in H(n+1)} l_{n+1,j}^{n+1} = \begin{cases} I_{H(n+1)}(k) \dfrac{\partial^2}{\partial \theta_k^2} l_{n+1,k}^{n+1}, & l = k \\ 0, & l \neq k. \end{cases}$$

Thus, $\mathbf{G}_{ul} = \mathbf{F} + \text{diag}(\mathbf{h})$, where $\mathbf{h} = \{h_1, \ldots, h_n\}$, and $h_k = I_{H(n+1)}(k) E((\partial/\partial\theta_k) l_{n+1,k}^{n+1})^2$. Compare the CRB for the covariance matrix of the first n devices in the n and $n + 1$ device cases, given by $\mathbf{F}^{-1}$ and $[\mathbf{G}^{-1}]_{ul}$, respectively. Here, $[\mathbf{G}^{-1}]_{ul}$ is the upper left $n \times n$ submatrix of $\mathbf{G}^{-1}$:

$$\left[\mathbf{G}^{-1} \right]_{ul} = \left\{ \mathbf{G}_{ul} - \mathbf{g}_{ur} g_{lr}^{-1} \mathbf{g}_{ll} \right\}^{-1} = \{ \mathbf{F} + \mathbf{J} \}^{-1}$$

where $\mathbf{J} = \text{diag}(\mathbf{h}) - \mathbf{g}_{ur} \mathbf{g}_{ur}^T / g_{lr}$. Both $\mathbf{F}$ and $\mathbf{J}$ are Hermitian. We know that $\mathbf{F}$ is positive semidefinite. Let $\lambda_k(\mathbf{F})$, $k = 1 \ldots n$ be the eigenvalues of $\mathbf{F}$ and $\lambda_k(\mathbf{F} + \mathbf{J})$, $k = 1 \ldots n$ be the eigenvalues of the sum, both listed in increasing order. Then, if we can show that $\mathbf{J}$ is positive semidefinite, then it is known [46] that

$$0 \leq \lambda_k(\mathbf{F}) \leq \lambda_k(\mathbf{F} + \mathbf{J}), \qquad \forall k = 1, \ldots, n \tag{13.24}$$

Since the eigenvalues of a matrix inverse are the inverses of the eigenvalues of the matrix

$$\lambda_k \left(\{ \mathbf{F} + \mathbf{J} \}^{-1} \right) \leq \lambda_k(\mathbf{F}^{-1}), \qquad \forall k = 1, \ldots, n. \tag{13.25}$$

which proves Property (i). If we can show that $\text{tr}(\mathbf{J}) > 0$, then $\text{tr}(\mathbf{F} + \mathbf{J}) > \text{tr}(\mathbf{F})$, and therefore, $\sum_{k=1}^{n} \lambda_k(\mathbf{F} + \mathbf{J}) > \sum_{k=1}^{n} \lambda_k(\mathbf{F})$. This, with (13.24), implies that $\lambda_j(\mathbf{F} + \mathbf{J}) > \lambda_j(\mathbf{F})$ for at least one

$j \in 1 \ldots n$. So, in addition to (13.25) $\lambda_j \left(\{ \mathbf{F} + \mathbf{J} \}^{-1} \right) < \lambda_j(\mathbf{F}^{-1})$, for some $j \in 1, \ldots n$, that implies $\mathrm{tr}(\{ \mathbf{F} + \mathbf{J} \}^{-1}) < \mathrm{tr}(\mathbf{F}^{-1})$ which proves Property (ii).

To show positive semidefiniteness and positive trace of $\mathbf{J}$ we notice that the diagonal elements of $\mathbf{J}$, $[\mathbf{J}]_{k,k}$ are

$$[\mathbf{J}]_{k,k} = h_k - \mathbf{g}_{ur}^2(k)/g_{lr} \tag{13.26}$$

If $k \notin H(n+1)$, then $h_k = 0$ and $\mathbf{g}_{ur}(k) = 0$; thus, $[\mathbf{J}]_{k,k} = 0$. Otherwise, if $k \in H(n+1)$

$$[\mathbf{J}]_{k,k} = E \left(\frac{\partial l_{n+1,k}^{n+1}}{\partial \theta_k} \right)^2 - \frac{\left[E \left(\frac{\partial l_{n+1,k}^{n+1}}{\partial \theta_k} \right) \left(\frac{\partial l_{n+1,k}^{n+1}}{\partial \theta_{n+1}} \right) \right]^2}{\sum_{j \in H(n+1)} E \left(\frac{\partial l_{n+1,j}^{n+1}}{\partial \theta_{n+1}} \right)^2}$$

Because of reciprocity, the numerator is equal to the square of the $j = k$ term in the sum in the denominator. Thus:

$$[\mathbf{J}]_{k,k} \geq E \left(\frac{\partial l_{n+1,k}^{n+1}}{\partial \theta_k} \right)^2 - E \left(\frac{\partial l_{n+1,k}^{n+1}}{\partial \theta_k} \right) \left(\frac{\partial l_{n+1,k}^{n+1}}{\partial \theta_{n+1}} \right) = 0$$

The equality will hold if k is the only member of the set $H(n+1)$. When Condition (ii) holds, $[\mathbf{J}]_{k,k}$ will be strictly greater than zero. Thus, $\mathrm{tr}\mathbf{J} > 0$. Next, we show that $\mathbf{J}$ is diagonally dominant [46], i.e.

$$[\mathbf{J}]_{k,k} \geq \sum_{\substack{j=1 \\ j \neq k}}^{n} |[\mathbf{J}]_{k,j}| = \sum_{\substack{j=1 \\ j \neq k}}^{n} \frac{|\mathbf{g}_{ll}(k)\mathbf{g}_{ll}(j)|}{g_{lr}}$$

where $[\mathbf{J}]_{k,k}$ is given in Equation (13.26). Since $H(n+1) \neq 0$, thus, $g_{lr} > 0$, and an equivalent condition is

$$g_{lr} h_k \geq |\mathbf{g}_{ur}(k)| \sum_{j=1}^{n} |\mathbf{g}_{ur}(j)| \tag{13.27}$$

If $k \notin H(n+1)$, then $h_k = 0$ and $\mathbf{g}_{ur}(k) = 0$ and equality holds. If $k \in H(n+1)$, then

$$9_{lr} h_k = E \left(\frac{\partial l_{k,n+1}^{n+1}}{\partial \theta_k} \right)^2 \sum_{j \in H(n+1)} E \left(\frac{\partial l_{n+1,j}^{n+1}}{\partial \theta_{n+1}} \right)^2$$

Due to Condition (i)

$$E \left(\frac{\partial l_{k,n+1}^{n+1}}{\partial \theta_k} \right)^2 = \left| E \left(\frac{\partial l_{k,n+1}^{n+1}}{\partial \theta_{n+1}} \right) E \left(\frac{\partial l_{k,n+1}^{n+1}}{\partial \theta_k} \right) \right|$$

Thus

$$g_{lr} h_k = |\mathbf{g}_{ur}(k)| \left[\sum_{\substack{j \geq 1 \\ j \in H(n+1)}} |\mathbf{g}_{ur}(j)| + \sum_{\substack{j < 0 \\ j \in H(n+1)}} \left| E \left(\frac{\partial l_{j,n+1}^{n+1}}{\partial \theta_{n+1}} \frac{\partial l_{j,n+1}^{n+1}}{\partial \theta_j} \right) \right| \right]$$

Since $g_{ur}(j) = 0$ if $j \in H(n + 1)$, we can include in the first sum all $j \in 1 \ldots n$. Since the second sum is ≥ 0, (13.27) is true.

Diagonal dominance implies that $\mathbf{J}$ is positive semidefinite, which proves (13.25). Note that if $H(n+1)$ includes ≥ 1 reference device, the second sum is > 0, and the inequality in (13.27) is strictly > 0, which implies positive definiteness of $\mathbf{J}$ and assures that the CRB will strictly decrease.

13.2.2   Relative Location Estimation

Here we use the general framework from the previous section for device location estimation using pair-wise RSS or TOA measurements in a wireless network. For a network of m reference and n blindfolded devices, the device parameters are $\gamma = [\mathbf{z}_1, \ldots, \mathbf{z}_{m+n}]$, where, for a two-dimensional (2D) system, $\mathbf{z}_i = [x_i, y_i]^T$. An extension of these results to 3D is also possible. The relative location problem corresponds to the estimation of blindfolded device coordinates $\theta = [\theta_x, \theta_y]$ (where $\theta_x = [x_1, \ldots, x_n], \theta_y = [y_1, \ldots, y_n]$) given the known reference coordinates $[x_{n+1}, \ldots, x_{n+m}, y_{n+1}, \ldots, y_{n+m}]$. In the TOA case, $X_{i,j} = T_{i,j}$ is the measured TOA between devices i and j (in seconds), and in the RSS case, $X_{i,j} = P_{i,j}$ is the measured received power at device i transmitted by device j (in milliwatts). As discussed in the previous section, only a subset $H(k)$ of devices make pair-wise measurements with device k, $((T_{i,j}))_{i,j}$ and $((P_{i,j}))_{i,j}$ are taken to be upper triangular matrices, and these measurements are assumed statistically independent. In addition, assume that $T_{i,j}$ is Gaussian distributed with mean $d_{i,j}/c$ and variance σ_T^2, which is denoted:

$$T_{i,j} \sim N(d_{i,j}/c, \sigma_T^2), \quad d_{i,j} = d(\mathbf{z}_i, \mathbf{z}_j) = \|\mathbf{z}_i - \mathbf{z}_j\|^{1/2} \tag{13.28}$$

where c is the speed of propagation, and σ_T^2 is not a function of $d_{i,j}$. We assume that $P_{i,j}$ is log-normal; thus, the random variable $P_{i,j}$ (dBm) $= 10\log_{10} P_{i,j}$ is Gaussian:

$$P_{i,j}(\text{dBm}) \sim N(\bar{P}_{i,j}(\text{dBm}), \sigma_{\text{dB}}^2)$$
$$\bar{P}_{i,j}(\text{dBm}) = P_0(\text{dBm}) - 10n_p \log_{10}(d_{i,j}/d_0) \tag{13.29}$$

where $\bar{P}_{i,j}$ (dBm) is the mean power in decibel milliwatts, σ_{dB}^2 is the variance of the shadowing, and P_0 (dBm) is the received power in decibel milliwatts at a reference distance d_0. Typically, $d_0 = 1$ m, and P_0 is calculated from the free space path loss formula. The path loss exponent n_p is a function of the environment. For particular environments, n_p may be known from prior measurements. Although we derive the CRB assuming n_p is known, it could have been handled as an unknown 'nuisance' parameter. Given (13.28), the density of $P_{i,j}$ is

$$f_{P/\gamma}(P_{i,j}/\gamma) = \frac{10/\log 10}{\sqrt{2\pi\sigma_{\text{dB}}^2}} \frac{1}{P_{i,j}} \exp\left[-\frac{b}{8}\left(\log\frac{d_{i,j}^2}{\tilde{d}_{i,j}^2}\right)\right]$$
$$b = \left(\frac{10n_p}{\sigma_{\text{dB}}\log 10}\right)^2; \quad \tilde{d}_{i,j} = d_0\left(\frac{P_0}{P_{i,j}}\right)^{1/n_p} \tag{13.30}$$

Here, $\tilde{d}_{i,j}$ is the MLE of range $d_{i,j}$ given received power $P_{i,j}$.

For simplicity let us first consider *one-dimensional TOA example* that could, be applied to location estimation on an assembly line. Consider n blindfolded devices and m reference devices with combined parameter vector $\gamma = [x_1, \ldots, x_{n+m}]$. The unknown coordinate vector is $\theta = [x_1, \ldots, x_n]$. Assume all devices make pair-wise measurements with every other device, i.e. $H(k) = \{1, \ldots, k-1, k+1, \ldots, m+n\}$. The distribution of the observations is given by (13.28) with $d_{i,j} = |x_j - x_i|$ and $l_{i,j} = (\partial^2/\partial x_j^2)l_{i,j} = -(\partial^2/\partial x_j \partial x_i)l_{i,j} = -1/\sigma_T^2 c^2, \forall i \neq j$, which are constant with respect to the random variables $T_{i,j}$. Thus, the FIM, which is calculated using Equation (13.23), is $\mathbf{F}_T = [(n+m)\mathbf{I}_n - \underline{1}\underline{1}^T]/(\sigma_T c^2$, where $\mathbf{I}_n$ is the $n \times n$ identity matrix, and $\underline{1}$ is an n by 1 vector of ones. For $m \geq 1$, $\mathbf{F}_T^{-1} = \sigma_T^2 c^2\left[m\mathbf{I}_n + \underline{1}\underline{1}^T\right]/(m(n+m))$.

The CRB on the variance of an unbiased estimator for x_i is:

$$\sigma_{x_i}^2 \geq \sigma_T^2 c^2 (m+1) / (m(n+m)) \tag{13.31}$$

Equation (13.31) implies that the variance $\sigma_{x_i}^2$ is reduced more quickly by adding reference (m) than blindfolded (n) devices. However, if m, is large, the difference between increasing m and n is negligible.

In the case of *two-dimensional location estimation* we denote by $\mathbf{F}_R$ and $\mathbf{F}_t$ the FIMs for the RSS and TOA measurements, respectively. Each device has two parameters, and we can see that the FIM will have a similar form to Equation (13.22) if partitioned into blocks

$$\mathbf{F}_R = \begin{bmatrix} \mathbf{F}_{Rxx} & \mathbf{F}_{Rxy} \\ \mathbf{F}_{Rxy}^T & \mathbf{F}_{Ryy} \end{bmatrix}, \quad \mathbf{F}_T = \begin{bmatrix} \mathbf{F}_{Txx} & \mathbf{F}_{Txy} \\ \mathbf{F}_{Txy}^T & \mathbf{F}_{Tyy} \end{bmatrix} \tag{13.32}$$

where $\mathbf{F}_{Rxx}$ and $\mathbf{F}_{Txx}$ are given by Equation (13.22) using only the x parameter vector $\theta = \theta_x$, and $\mathbf{F}_{Ryy}$ and $\mathbf{F}_{Tyy}$ are given by Equation (13.22) using only $\theta = \theta_y$. The off-diagonal blocks $\mathbf{F}_{Rxy}$ and $\mathbf{F}_{Txy}$ are similarly derived.

To derive the elements of the submatrices of Equation (13.32) for $\mathbf{F}_R$, we use Equation (13.30) and (13.21) to get

$$l_{i,j} = \log\left(\frac{10 \log 10}{\sqrt{2\pi\sigma_{dB}^2}} \frac{1}{P_{i,j}}\right) - \frac{b}{8}\left(\log\frac{d_{i,j}^2}{\hat{d}_{i,j}^2}\right)^2 \tag{13.33}$$

Since $d_{i,j} = \sqrt{(x_i - x_j)^2 + (y_i - y_j)^2}$ we have

$$\frac{\partial}{\partial x_j} l_{i,j} = -\frac{b}{2}\left(\log\frac{d_{i,j}^2}{\hat{d}_{i,j}^2}\right)\frac{x_j - x_i}{d_{i,j}^2} \tag{13.34}$$

Since $(\partial/\partial x_j)l_{i,j} = -(\partial/\partial x_j)l_{i,j}$, the log-normal distribution of RSS measurements meets Condition (i). The second partials differ based on whether or not $i = j$ and if the partial is taken w.r.t. y_i or x_i. For example:

$$\frac{\partial^2 l_{i,j}}{\partial x_j \partial y_j} = -b\frac{(x_i - x_j)(y_i - y_j)}{d_{i,j}^4}\left[-\log\left(\frac{d_{i,j}^2}{\hat{d}_{i,j}^2}\right) + 1\right]$$

$$\frac{\partial^2 l_{i,j}}{\partial x_j \partial y_i} = -b\frac{(x_i - x_j)(y_i - y_j)}{d_{i,j}^4}\left[-\log\left(\frac{d_{i,j}^2}{\hat{d}_{i,j}^2}\right) - 1\right] \tag{13.35}$$

Since $E[\log(d_{i,j}^2/\hat{d}_{i,j}^2)] = 0$, the FIM simplifies to take the form:

$$[\mathbf{F}_{Rxx}]_{k,l} = \begin{cases} b\displaystyle\sum_{i\in H(k)}\frac{(x_k - x_i)^2}{\|\mathbf{z}_k - \mathbf{z}_i\|^4}, & k = l \\ -b\mathbf{I}_{H(k)}(l)\dfrac{(x_k - x_l)^2}{\|\mathbf{z}_k - \mathbf{z}_l\|^4}, & k \neq l \end{cases}$$

$$[\mathbf{F}_{Rxy}]_{k,l} = \begin{cases} b\displaystyle\sum_{i\in H(k)}\frac{(x_k - x_i)(y_k - y_i)}{\|\mathbf{z}_k - \mathbf{z}_i\|^4}, & k = l \\ -b\mathbf{I}_{H(k)}(l)\dfrac{(x_k - x_l)(y_k - y_l)}{\|\mathbf{z}_k - \mathbf{z}_l\|^4}, & k \neq l \end{cases}$$

$$
\left[\mathbf{F}_{Ryy} \right]_{k,l} =
\begin{cases}
b \displaystyle\sum_{i \in H(k)} \frac{(y_k - y_i)^2}{\|\mathbf{z}_k - \mathbf{z}_i\|^4}, & k = l \\[2ex]
-b\mathrm{I}_{H(k)}(l) \dfrac{(y_k - y_l)^2}{\|\mathbf{z}_k - \mathbf{z}_l\|^4}, & k \neq l
\end{cases}
\tag{13.36}
$$

For the TOA case, the derivation is very similar, and the details are omitted for brevity:

$$
\left[\mathbf{F}_{Txx} \right]_{k,l} =
\begin{cases}
\dfrac{1}{c^2 \sigma_T^2} \displaystyle\sum_{i \in H(k)} \frac{(x_k - x_i)^2}{\|\mathbf{z}_k - \mathbf{z}_i\|^2}, & k = l \\[2ex]
-\dfrac{1}{c^2 \sigma_T^2} \mathrm{I}_{H(k)}(l) \dfrac{(x_k - x_l)^2}{\|\mathbf{z}_k - \mathbf{z}_l\|^2}, & k \neq l
\end{cases}
$$

$$
\left[\mathbf{F}_{Txy} \right]_{k,l} =
\begin{cases}
\dfrac{1}{c^2 \sigma_T^2} \displaystyle\sum_{i \in H(k)} \frac{(x_k - x_i)(y_k - y_i)}{\|\mathbf{z}_k - \mathbf{z}_i\|^2}, & k = l \\[2ex]
-\dfrac{1}{c^2 \sigma_T^2} \mathrm{I}_{H(k)}(l) \dfrac{(x_k - x_l)(y_k - y_l)}{\|\mathbf{z}_k - \mathbf{z}_l\|^2}, & k \neq l
\end{cases}
$$

$$
\left[\mathbf{F}_{Tyy} \right]_{k,l} =
\begin{cases}
= \dfrac{1}{c^2 \sigma_T^2} \displaystyle\sum_{i \in H(k)} \frac{(y_k - y_i)^2}{\|\mathbf{z}_k - \mathbf{z}_i\|^2}, & k = l \\[2ex]
-\dfrac{1}{c^2 \sigma_T^2} \mathrm{I}_{H(k)}(l) \dfrac{(y_k - y_l)^2}{\|\mathbf{z}_k - \mathbf{z}_l\|^2}, & k \neq l
\end{cases}
\tag{13.37}
$$

Note that $\mathbf{F}_R \propto n_p^2 / \sigma_{dB}^2$ while $\mathbf{F}_T \propto 1/(c^2 \sigma_T^2)$. These SNR quantities directly affect the CRB. For TOA measurements, the dependence on the device coordinates is in unitless distance ratios, indicating that the size of the system can be scaled without changing the CRB as long as the geometry is kept the same. However, in the case of RSS measurements, the variance bound scales with the size of the system even if the geometry is kept the same due to the d^4 terms in the denominator of each term of $\mathbf{F}_R$. These scaling characteristics indicate that TOA measurements would be preferred for sparse networks, but for sufficiently high density, RSS can perform as well as TOA.

If $\hat{x}_i$ and $\hat{y}_i$ are unbiased estimates of x_i and y_i, for the case of TOA measurements, the trace of the covariance of these estimates satisfies:

$$
\sigma_i^2 \geq \mathrm{tr}\{\mathrm{cov}_\theta(\hat{x}_i, \hat{y}_i)\} = \mathrm{Var}_\theta(\hat{x}_i) + \mathrm{Var}_\theta(\hat{y}_i) \geq
$$
$$
\geq \left(\left[\mathbf{F}_{Txx} - \mathbf{F}_{Txy} \mathbf{F}_{Tyy}^{-1} \mathbf{F}_{Txy}^{-1} \right]^{-1} \right)_{i,i} + \left(\left[\mathbf{F}_{Tyy} - \mathbf{F}_{Txy} \mathbf{F}_{Txx}^{-1} \mathbf{F}_{Txy}^{-1} \right]^{-1} \right)_{i,i}
\tag{13.38}
$$

For RSS measurements, $\mathbf{F}_y$ in Equation (13.38) should be replaced with $\mathbf{F}_R$. For the case of one blindfolded device, a simple expression can be derived for both RSS and TOA measurements.

Consider the network having blindfolded device 1 and reference devices $2 \ldots m+1$. This example, with a single pair of unknowns x_1 and y_1, is equivalent to many existing location systems, and a bound for the variance of the location estimator has already been derived in for TOA measurements in Section 13.1. In the case of RSS measurements:

$$
\sigma_1^2 \triangleq E\left[(\hat{x}_1 - x_1)^2 + (\hat{y}_1 - y_1)^2\right] \geq \frac{F_{Rxx} + F_{Ryy}}{F_{Rxx} F_{Ryy} - F_{Rxy}^2}
$$

from which we obtain

$$
\sigma_1^2 \triangleq \frac{1}{b} \frac{\displaystyle\sum_{i=2}^{m+1} d_{1,i}^{-2}}{\displaystyle\sum_{i=2}^{m} \sum_{j=i+1}^{m+1} \left(\frac{d_{1\perp i,j}, d_{i,j}}{d_{1,i}^2 d_{1,j}^2} \right)^2}
$$

where the distance $d_{1\perp i,j}$ is the shortest distance from the point (x_1, y_1) to the line segment connecting device i and device j. For the case of TOA measurements, we obtain:

$$\sigma_1^2 \triangleq c^2 \sigma_T^2 m \left[\sum_{i=2}^{m} \sum_{j=i+1}^{m+1} \left(\frac{d_{1\perp i,j}, d_{i,j}}{d_{1,i} d_{1,j}} \right)^2 \right]^{-1} \tag{13.39}$$

In Section 13.1, the ratio $d_{1\perp i,j}, d_{i,j}/(d_{1,i} d_{1,j})$ has been called the geometric conditioning $A_{i'j}$ of device 1 w.r.t. references i and j. $A_{i,j}$ is the area of the parallelogram specified by the vectors from device 1 to i and from device 1 to j, normalized by the lengths of the two vectors. The geometric dilution of precision (GDOP), which is defined as $\sigma_1/(cs_T)$, is

$$\text{GODP} = \sqrt{\frac{m}{\sum_{i=2}^{m} \sum_{j=i+1}^{m+1} A_{i,j}^2}}$$

which matches the result in (13.14). The CRBs are shown in Figure 13.5 when there are four reference devices located in the corners of a 1 by 1 m square. The minimum of Figure 13.5(a) is 0.27. Since the CRB scales with size in the RSS case, the standard deviation of unbiased location estimates in a traditional RSS system operating in a channel with $\sigma_{dB}/n = 1.7$ is limited to about 27% of the distance between reference devices. This performance has prevented use of RSS in many existing location systems and is the motive for having many blindfolded devices in the network. Note that in the TOA case, σ_1 is proportional to $c\sigma_T$ and thus, $c\sigma_T = 1$ was chosen in Figure 13.5(b).

For general n and m, we calculate the MLE of θ. In the case of TOA measurements, the MLE is:

$$\hat{\boldsymbol{\theta}}_T = \arg \min_{\{\mathbf{z}_i\}} \sum_{i=1}^{m+n} \sum_{\substack{j \in H(i) \\ j < i}} (cT_{i,j} - d(\mathbf{z}_i, \mathbf{z}_j))^2 \tag{13.40}$$

where $\mathbf{z}_i = [x_i, y_i]^T$. The MLE for the RSS case is:

$$\hat{\boldsymbol{\theta}}_{\tilde{R}} = \arg \min_{\{\mathbf{z}_i\}} \sum_{i=1}^{m+n} \sum_{\substack{j \in H(i) \\ j < i}} \left(\ln \frac{\tilde{d}_{i,j}^2}{d^2(\mathbf{z}_i, \mathbf{z}_j)} \right)^2 \tag{13.41}$$

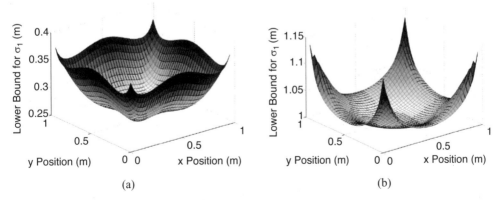

(a) (b)

Figure 13.5 σ_1 (in meters) for the example system versus the coordinates of the single blindfolded node for (a) RSS with $\sigma_{dB}/n = 1.7$ or (b) TOA with $c\sigma_T = 1$ m © , IEEE 2003.

Unlike the MLE based on TOA measurements, the RSS MLE is readily shown to be biased. Specifically, for a single reference and single blindfolded device, the range estimate between the two devices is $\tilde{d}_{1,2}$. Using Equation (13.30), the mean of $\tilde{d}_{1,2}$ is given by:

$$E[\tilde{d}_{1,2}] = Cd_{1,2}, \text{ where, } C = \exp\left[\frac{1}{2}\left(\frac{\ln(10)}{10}\frac{\sigma_{\text{dB}}}{n_p}\right)^2\right] \tag{13.42}$$

For typical channels, $C \approx 1.2$, adding 20% bias to the range. Motivated by Equation (13.42), a bias-reduced MLE can be defined as:

$$\hat{\boldsymbol{\theta}}_R = \arg\min_{\{z_i\}} \sum_{i=1}^{m+n} \sum_{\substack{j\in H(i) \\ j<i}} \left(\ln\frac{\tilde{d}_{i,j}^2/C^2}{d^2(z_i, z_j)}\right)^2 \tag{13.43}$$

However, there remains residual bias. Estimation results using methodology presented in this section and measurements from [44] are shown in Figures 13.6 and 13.7.

13.3 Average Performance of Circular and Hyperbolic Geolocation

In this section we provide more details on performance analysis of geolocation methods in terms of their theoretical positioning errors. Comparison is established in two different ways: strict and average. As in Section 13.1, in the strict type, methods are examined for a particular geometric configuration of base stations (BSs) with respect to mobile position, which determines a given noise profile affecting the estimates. In the average type, methods are evaluated in terms of the expected covariance matrix of the position error over an ensemble of random geometries, so that comparison is geometry independent.

In this section, the accuracy of four geolocation methods is defined in terms of their position error covariance matrices: (i) C_{st}–*circular with known clock* (ii) $C_{s\tilde{i}}$–*circular with unknown clock* (iii) C_{hc}–*hyperbolic correlated, and* (iv) $C_{h\tilde{c}}$–*hyperbolic uncorrelated*. In scenario (i), the available measurements consist of the absolute TOAs of the signals transmitted between the mobile and the beacons. A circular technique is used to compute the position estimate. In scenario (ii), the available measurements consist of the pseudo-TOAs of the signals transmitted between the mobile and the beacons. A circular technique is used to compute the position estimate along with the clock offset. In scenario (iii), the available measurements consist of the TDOAs obtained from differences between pairs of pseudo-TOAs. A hyperbolic technique is used to compute the position estimate. In scenario (iv), the available measurements consist of uncorrelated TDOAs obtained at different times. A hyperbolic technique is used to compute the position estimate.

13.3.1 *Signal Models and Performance Limits*

For *circular methods*, the TOA/pseudo-TOA measurements obtained from N beacons can be expressed as a function of the mobile position $x = [x_1, x_2]^T$ as:

$$\hat{t}_n = \tilde{t}_n(x, t_0) + u_n = t_n(x) + t_o + u_n \tag{13.44}$$

where $t_n(x) = \frac{1}{c}\|x - x_n\|$, $x_n = \left[x_{1,n}, x_{2,n}\right]^T$ is the nth beacon position, for $1 \le n \le N$, t_o is the clock offset, which has the same value for all n, u_n are the measurement errors, and c is the speed of light. Note that $t_o = 0$ in the case of absolute TOAs, while it becomes an unknown parameter in the case of pseudo-TOAs.

If the measurements be stacked into vector $\hat{\mathbf{t}}$, we have:

$$\hat{\mathbf{t}} = \tilde{\mathbf{t}}(x, t_0) + \mathbf{u} = \tilde{\mathbf{t}}(x) + t_0\mathbf{1} + \mathbf{u} \tag{13.45}$$

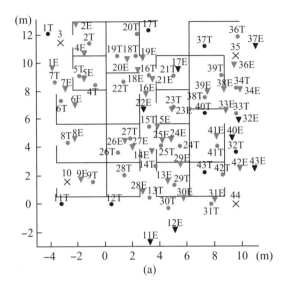

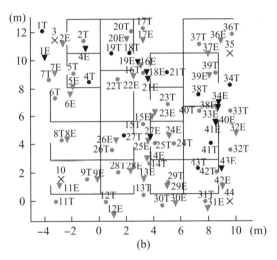

Figure 13.6 True (•#T) and estimated (▼ #E) location using (a) RSS, and (b) TOA data for measured network with four reference devices (X#). Higher errors are indicated by darker text [44] ©, IEEE 2003.

where **1** is the all-ones column vector. The $N \times N$ covariance matrix of the noise vector **u** is $\mathbf{R} = E[\mathbf{uu}^T]$. It is assumed throughout the section that matrix **R** is known. In practice, **R** is to be estimated, with a consequent degradation in the performance.

For hyperbolic methods the $(N-1)$ TDOA measurements are given by:

$$\hat{\mathbf{d}} = \mathbf{H}\tilde{\mathbf{t}}(x, \; t_0) + v = \mathbf{H}\mathbf{t}(x) + \mathbf{v} \tag{13.46}$$

with **H** representing any full-rank $(N-1) \times N$ matrix such that $\mathbf{H1} = \mathbf{0}$, where **0** is the all-zeros column vector. Note that this condition implies that the measurement vector $\hat{\mathbf{d}}$ is not affected by the presence of the nonzero clock offset t_o in Equation (13.46). The common example of **H** is $\mathbf{H} = [\mathbf{1}, -\mathbf{I}]$, where all TDOAs are obtained with respect to a common pseudo-TOA. This first (or reference) pseudo-TOA

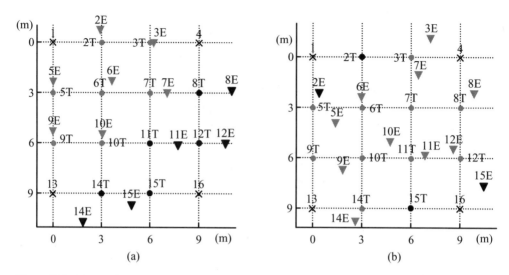

Figure 13.7 True (• #T) and estimated (▼ #E) location for the (a) parking lot, and (b) residential home tests using four reference devices (X#). Higher errors are indicated by darker text [44] © , IEEE 2003.

corresponds to the all-ones column of $\mathbf{H}$, and will henceforth be called the pivot BS. The analysis in this section is general for any $\mathbf{H}$, unless otherwise stated. The $(N-1) \times (N-1)$ covariance matrix of the noise vector $\mathbf{v}$ is $\mathbf{R}_h = E[\mathbf{v}\,\mathbf{v}^T]$. Only in the 'hyperbolic-correlated' scenario, are the TDOAs are computed as $\hat{\mathbf{d}} = \mathbf{H}\hat{\mathbf{t}}$, and then it holds from Equation (13.45) that $\mathbf{v} = \mathbf{H}\mathbf{u}$, and $\mathbf{R}_h = \mathbf{R}_{hc} = \mathbf{H}\mathbf{R}\mathbf{H}^T$. However, in the 'hyperbolic-uncorrelated' scenario, the structures of $\mathbf{R}_h = \mathbf{R}_{h\bar{c}}$ and $\mathbf{R}$ are not necessarily related in a simple way. Indeed, in the 'hyperbolic-uncorrelated' scenario, matrix $\mathbf{H}$ in Equation (13.46) only shows how the pairs of BS are selected to obtain the $N-1$ TDOA measurements, but measurements remain independent. In that scenario, the relationship between $\mathbf{R}_{h\bar{c}}$ and $\mathbf{R}$ for any $\mathbf{H}$ can be expressed as $\mathbf{R}_h = \mathbf{R}_{h\bar{c}} = (\mathbf{H}\mathbf{R}\mathbf{H}^T)_\circ \mathbf{I}$ where '∘' stands for the component-wise Schur–Hadamard product between two matrices. $\mathbf{R}_{hc}$ and $\mathbf{R}_{h\bar{c}}$ are the specific versions of $\mathbf{R}_h$ for the correlated and uncorrelated version, respectively, of the hyperbolic approach.

The performance limits are computed as the performance of the linearized weighted least squares (WLS) estimator, which coincides with the Cramer–Rao bound in the case of Gaussian noise.

As in Equation (13.2) for circular approaches, the linearization of Equation (13.45) at the true position yields

$$\Delta\hat{\mathbf{t}} = \mathbf{F}\Delta\mathbf{x} + \Delta t_0 \mathbf{1} + \mathbf{u} \tag{13.47}$$

where the $N \times 2$ matrix of partial derivatives $\mathbf{F}$ can be calculated as:

$$[\mathbf{F}]_{n,i} = \frac{\partial t_n(x)}{\partial x_i} = \frac{1}{c}\frac{x_i - x_{i,n}}{\|x - x_n\|} = \frac{1}{c}\frac{x_i - x_{i,n}}{\sqrt{\left(x_1 - x_{1,n}\right)^2 + \left(x_2 - x_{2,n}\right)^2}} \tag{13.48}$$

with $i = 1, 2$, and where $[\mathbf{M}]_{l,l'}$ denotes the lth row l'th column element of a matrix $\mathbf{M}$. The WLS estimator [48] of Δx and Δt_o is the minimizer of

$$\left(\Delta\hat{\mathbf{t}} - \mathbf{F}\Delta x - \Delta t_0 \mathbf{1}\right)^T \mathbf{R}^{-1} \left(\Delta\hat{\mathbf{t}} - \mathbf{F}\Delta x - \Delta t_0 \mathbf{1}\right) \tag{13.49}$$

which, using standard differential matrix calculus, can be expressed as:

$$\begin{bmatrix} \Delta \mathbf{x} \\ \Delta t_0 \end{bmatrix} = \left(\mathbf{F}_1^T \mathbf{R}^{-1} \mathbf{F}_1 \right)^{-1} \mathbf{F}_1^T \mathbf{R}^{-1} \Delta \hat{\mathbf{t}}$$

(13.50)

where $\mathbf{F}_1 = [\mathbf{F}, \mathbf{1}]$, and the covariance matrix of $\Delta \mathbf{x}$ is given by $\mathbf{C}_{s\tilde{t}} = \mathbf{T}^T \left(\mathbf{F}_1^T \mathbf{R}^{-1} \mathbf{F}_1 \right)^{-1} \mathbf{T}$ with $\mathbf{T}^T = [\mathbf{I}, \mathbf{0}]$. Following the same procedure, the covariance matrix of $\Delta \mathbf{x}$ in the case of known clock t_o s

$$\mathbf{C}_{st} = \left(\mathbf{F}_1^T \mathbf{R}^{-1} \mathbf{F}_1 \right)^{-1}$$

For hyperbolic approaches, we have the linearization of Equation (13.46) that can be written as:

$$\Delta \hat{\mathbf{d}} = \mathbf{HF} \Delta \mathbf{x} + \mathbf{v}$$

(13.51)

and the WLS estimator of $\Delta \mathbf{x}$ becomes the minimizer of

$$\left(\Delta \hat{\mathbf{d}} - \mathbf{HF} \Delta \mathbf{x} \right)^T \mathbf{R}_h^{-1} \left(\Delta \hat{\mathbf{d}} - \mathbf{HF} \Delta \mathbf{x} \right)$$

(13.52)

given by

$$\Delta \mathbf{x} = \left(\mathbf{F}^T \mathbf{H}^T \mathbf{R}_h^{-1} \mathbf{HF} \right)^{-1} \mathbf{F}^T \mathbf{H}^T \mathbf{R}_h^{-1} \Delta \hat{\mathbf{d}}$$

(13.53)

with covariance matrix $\mathbf{C}_h = \left(\mathbf{F}^T \mathbf{H}^T \mathbf{R}_h^{-1} \mathbf{HF} \right)^{-1}$ where the subindex in $\mathbf{C}_h$ and $\mathbf{R}_h$ stands for a generic hyperbolic approach. For the uncorrelated hyperbolic approach, this covariance matrix becomes:

$$\mathbf{C}_{h\tilde{c}} = \left(\mathbf{F}^T \mathbf{H}^T \mathbf{R}_{h\tilde{c}}^{-1} \mathbf{HF} \right)^{-1} = \left(\mathbf{F}^T \mathbf{H}^T \left[(\mathbf{HRH}^T)^{\circ} I \right]^{-1} \mathbf{HF} \right)^{-1}$$

(13.54)

and, in the specific case that the TDOAs are computed as differences between pseudo-TOAs (correlated hyperbolic approach), we have $\mathbf{R}_h = \mathbf{R}_{hc} = \mathbf{HRH}^T$, and the corresponding covariance matrix is given by

$$\mathbf{C}_{hc} = \left(\mathbf{F}^T \mathbf{H}^T \mathbf{R}_{h\tilde{c}}^{-1} \mathbf{HF} \right)^{-1}$$
$$\mathbf{C}_{hc} = \left(\mathbf{F}^T \mathbf{H}^T \left(\mathbf{HRH}^T \right)^{-1} \mathbf{HF} \right)^{-1}$$

(13.55)

13.3.2   *Performance of Location Techniques*

From the previous section we have for $\mathbf{C}_{s\tilde{t}}^{-1}$ and $\mathbf{C}_{hc}^{-1}$:

$$\mathbf{C}_{s\tilde{t}}^{-1} = \left(\mathbf{T}^T \left(\mathbf{F}_1^T \mathbf{R}^{-1} \mathbf{F}_1 \right)^{-1} \mathbf{T} \right)^{-1}$$
$$\mathbf{C}_{hc}^{-1} = \mathbf{F}^T \left(\mathbf{H}^T \left(\mathbf{HRH}^T \right)^{-1} \mathbf{H} \right) \mathbf{F}^T$$

(13.56)

By forming the product $\mathbf{F}_1^T \mathbf{R}^{-1} \mathbf{F}_1$ in $\mathbf{C}_{s\tilde{t}}^{-1}$, as:

$$\mathbf{F}_1^T \mathbf{R}^{-1} \mathbf{F}_1 = \begin{bmatrix} \mathbf{F}^T \\ \mathbf{1}^T \end{bmatrix} \mathbf{R}^{-1} [\mathbf{F} \ \ \mathbf{1}] = \begin{bmatrix} \mathbf{F}^T \mathbf{R}^{-1} \mathbf{F} & \mathbf{F}^T \mathbf{R}^{-1} \mathbf{1} \\ \mathbf{1}^T \mathbf{R}^{-1} \mathbf{F} & \mathbf{1}^T \mathbf{R}^{-1} \mathbf{1} \end{bmatrix}$$

(13.57)

and using block matrix inversion in Equation (13.57), we have:

$$\mathbf{F}_1^T \mathbf{R}^{-1} \mathbf{F}_1 = \begin{bmatrix} \mathbf{B}_{11} & \mathbf{b}_{12} \\ \mathbf{b}_{21}^H & \beta \end{bmatrix}$$

$$\mathbf{B}_{11} = \left[\mathbf{F}^T \mathbf{R}^{-1} \mathbf{F} - \frac{\mathbf{F}^T \mathbf{R}^{-1} \mathbf{1} \mathbf{1}^T \mathbf{R}^{-1} \mathbf{F}}{\mathbf{1}^T \mathbf{R}^{-1} \mathbf{1}} \right]^{-1}$$

We also have from (13.56) that:

$$\mathbf{C}_{s\bar{\imath}}^{-1} = \left[[\mathbf{I}\ \mathbf{0}] \left(\mathbf{F}_1^T \mathbf{R}^{-1} \mathbf{F}_1 \right)^{-1} \begin{bmatrix} \mathbf{I} \\ \mathbf{0}^T \end{bmatrix} \right] = \mathbf{B}_{11}^{-1}$$

$$\mathbf{B}_{11}^{-1} = \mathbf{F}^T \mathbf{R}^{-1} \mathbf{F} - \frac{\mathbf{F}^T \mathbf{R}^{-1} \mathbf{1} \mathbf{1}^T \mathbf{R}^{-1} \mathbf{F}}{\mathbf{1}^T \mathbf{R}^{-1} \mathbf{1}} = \mathbf{F}^T \left(\mathbf{R}^{-1} - \frac{\mathbf{R}^{-1} \mathbf{1} \mathbf{1}^T \mathbf{R}^{-1}}{\mathbf{1}^T \mathbf{R}^{-1} \mathbf{1}} \right) \mathbf{F} \qquad (13.58)$$

By using the fact that

$$\mathbf{R}^{-1} - \frac{\mathbf{R}^{-1} \mathbf{1} \mathbf{1}^T \mathbf{R}^{-1}}{\mathbf{1}^T \mathbf{R}^{-1} \mathbf{1}} = \mathbf{H}^T (\mathbf{H} \mathbf{R} \mathbf{H}^T)^{-1} \mathbf{H} \qquad (13.59)$$

is true for any matrix $\mathbf{H}$ that fulfills $\mathbf{H} \mathbf{1} = \mathbf{0}$, we have

---

result 1 $\mathbf{C}_{s\bar{\imath}} = \mathbf{C}_{hc}$ $\qquad\qquad$ (13.60)

Proof of Equation (13.59): If Ξ is a matrix and $\mathbf{G}$ is any Gram matrix, the matrix $\mathbf{G}_\Xi^\perp$ is a projection matrix

$$\mathbf{G}_\Xi^\perp = \mathbf{G} - \mathbf{G} \Xi [\Xi^T \mathbf{G} \Xi]^{-1} \Xi^T \mathbf{G}$$

that fulfills $\mathbf{G}_\Xi^\perp \Xi = \mathbf{G} \Xi - \mathbf{G} \Xi [\Xi^T \mathbf{G} \Xi]^{-1} \Xi^T \mathbf{G} \Xi = 0$.
 Hence, for $\mathbf{G} = \mathbf{R}$ and $\Xi = \mathbf{H}^T$, we have $\mathbf{R}_{\mathbf{H}^T}^\perp = \mathbf{R} - \mathbf{R} \mathbf{H}^T (\mathbf{H} \mathbf{R} \mathbf{H}^T)^{-1} \mathbf{H} \mathbf{R}$ and $\mathbf{R}_{\mathbf{H}^T}^\perp \mathbf{H}^T = 0$. Consequently, $\mathbf{R}_{\mathbf{H}^T}^\perp$ has rank one (because $\mathbf{H}$ is almost full rank) and $\mathbf{R}_{\mathbf{H}^T}^\perp = \lambda \mathbf{a} \mathbf{a}^T$, with $\mathbf{H}\, \mathbf{a} = \mathbf{0}$. Therefore, $\mathbf{a} = \mathbf{1}$ fulfills the required orthogonality with the columns of $\mathbf{H}$ and $\mathbf{R} - \mathbf{R} \mathbf{H}^T (\mathbf{H} \mathbf{R} \mathbf{H}^T)^{-1} \mathbf{H} \mathbf{R} = \lambda \mathbf{1} \mathbf{1}^T$.
  Multiplying by $\mathbf{R}^{-1}$ on both sides $\mathbf{R}^{-1} - \mathbf{H}^T (\mathbf{H} \mathbf{R} \mathbf{H}^T)^{-1} \mathbf{H} = \lambda \mathbf{R}^{-1} \mathbf{1} \mathbf{1}^T \mathbf{R}^{-1}$.
  Now, as $\mathbf{H} \mathbf{1} = \mathbf{0}$, multiplication by $\mathbf{1}^T$ and $\mathbf{1}$ on both sides yields $\mathbf{1}^T\, \mathbf{R}^{-1}\, \mathbf{1} = \lambda \mathbf{1}^T \mathbf{R}^{-1} \mathbf{1} \mathbf{1}^T \mathbf{R}^{-1} \mathbf{1}$.
  Thus, $\lambda = (\mathbf{1}^T \mathbf{R}^{-1} \mathbf{1})^{-1}$, and $\mathbf{R}^{-1} - \mathbf{H}^T (\mathbf{H} \mathbf{R} \mathbf{H}^T)^{-1} \mathbf{H} = \dfrac{\mathbf{R}^{-1} \mathbf{1} \mathbf{1}^T \mathbf{R}^{-1}}{\mathbf{1}^T \mathbf{R}^{-1} \mathbf{1}}$ which is Equation (13.59).

Next we compare $\mathbf{C}_{s\bar{\imath}} = \mathbf{T}^T \left(\mathbf{F}_1^T \mathbf{R}^{-1} \mathbf{F}_1 \right)^{-1} \mathbf{T}$ and $\mathbf{C}_{st} = \left(\mathbf{F}_1^T \mathbf{R}^{-1} \mathbf{F}_1 \right)^{-1}$ to prove that $\mathbf{C}_{s\bar{\imath}} > \mathbf{C}_{st}$ for any geometry or noise profile. We know from Equation (13.58) that

$$\mathbf{C}_{s\bar{\imath}}^{-1} = \mathbf{F}^T \left(\mathbf{R}^{-1} - \frac{\mathbf{R}^{-1} \mathbf{1} \mathbf{1}^T \mathbf{R}^{-1}}{\mathbf{1}^T \mathbf{R}^{-1} \mathbf{1}} \right) \mathbf{F} \qquad (13.61)$$

Hence

$$\mathbf{C}_{s\bar{\imath}}^{-1} = \mathbf{C}_{st}^{-1} - \mathbf{q} \mathbf{q}^T \qquad (13.62)$$

where

$$q = \frac{F^T R^{-1} 1}{\sqrt{1^T R^{-1} 1}} \qquad (13.63)$$

Using the matrix inversion lemma in Equation (13.62), we see that $C_{s\bar{\imath}}$ is more positive definite than C_{st} (i.e., $C_{s\bar{\imath}} - C_{st}$ is a positive definite matrix), giving

$$\textit{result 2} \quad C_{s\bar{\imath}} = C_{st} + \frac{qq^T}{1 + q^T C_{st} q} \geq C_{st} \qquad (13.64)$$

and we conclude from Equation (13.64) that absolute TOA measurements generate a C_{st} smaller than $C_{s\bar{\imath}}$ for any geometry or noise correlation. However, C_{st} and $C_{s\bar{\imath}}$ are not simply related by a scaling factor.

Strict comparison between C_{hc} and $C_{h\bar{c}}$ is not possible, as some scenarios will deliver better performance for $C_{h\bar{c}}$ and others will not. The judgment as to which method is better, and to what extent, will require the average performance analysis that is presented in the next section.

13.3.3 Average Performance of Location Techniques

This section presents an efficient procedure for comparing algorithms in terms of their average covariance matrix, as well as an analytical lower bound in the positive definite sense. The placement of BSs is defined according to a uniform random distribution on a circular area centered at the true mobile position, and depending on a density parameter ρ (BS/km^2).

From Section 13.3.1 and Equations (13.54), and (13.55), the covariance matrices of the four geolocation algorithms can be expressed as:

$$C = \left(F^T A F\right)^{-1} \qquad (13.65)$$

where A is referred to as the information matrix. As defined in Section 13.3.1, we have for each geolocation algorithm:

$$\begin{aligned}
A_{st} &= R^{-1} \\
A_{s\bar{\imath}} &= H^T (HRH^T)^{-1} H \\
A_{hc} &= A_{s\bar{\imath}} \\
A_{h\bar{c}} &= H^T [(HRH^T) \circ I]^{-1} H
\end{aligned} \qquad (13.66)$$

In Equation (13.65), geometrical information is included in matrix F and in the noise power distribution defined by the measurement covariance matrix R. From Equation (13.48), the rows of F constitute the directive cosines of the vectors aligned with the paths from the mobile to the base stations (or generic references) with respect to the coordinate axes. Then:

$$F = F(\alpha) = \frac{1}{c} \begin{bmatrix} \cos(\alpha_1) & \sin(\alpha_1) \\ \cos(\alpha_2) & \sin(\alpha_2) \\ & \\ \cos(\alpha_N) & \sin(\alpha_N) \end{bmatrix} \qquad (13.67)$$

with $\alpha = [\alpha_1, \alpha_2, \ldots, \alpha_N]^T$, and α_n the angle with respect to the horizontal coordinate axis of the vector from the mobile to the nth BS. The diagonal of $\mathbf{R}$ contains the noise power of all individual pseudo-TOA measurements, which depends mainly on the distance between the mobile and each BS. So, the specific position of the BSs affects the covariance matrix $\mathbf{C}$ in two different ways: in their angular distribution implicit in $\mathbf{F}$, and in their distance distribution implicit in $\mathbf{R}$. Hence, the geometry-independent average of the position error covariance matrix can be expressed as:

$$\bar{\mathbf{C}} = E_G \left[\left(\mathbf{F}^T \mathbf{AF} \right)^{-1} \right] = E_N E_\alpha \left[\left(\mathbf{F}^T \mathbf{AF} \right)^{-1} \right] \tag{13.68}$$

with E_G the expectation operator with respect to all possible geometry scenarios, which can be split into the expected value with respect to the angles α, (E_α), and the expected value with respect to the power distribution (E_N) (related to the distance distribution). Its analytical computation is too involved due to the particular way in which geometry affects the covariance structure. So, the expression derived in the sequel exploits a rotational property of the 2D location case in such a way that the structure of Equation (13.68) can be studied without computing its expected value analytically.

For convenience, let us define the expectation with respect to the angles in (13.68) as $\overline{\mathbf{C}_N}$:

$$\bar{\mathbf{C}} = E_N \overline{\mathbf{C}_N} \tag{13.69}$$

where

$$\overline{\mathbf{C}_N} = E_\alpha \left[\left(\mathbf{F}^T \mathbf{AF} \right)^{-1} \right] \tag{13.70}$$

The angles α are independent, uniformly distributed random variables in $[0, 2\pi]$ due to the uniform random placement of BSs in the coverage area. Thus, the infinite set of all possible angular configurations of the BSs can be divided into an infinite number of subsets within which angular configurations differ in a rotation. That is:

$$[\alpha_1, \alpha_2, \ldots, \alpha_N] = \alpha_1 \cdot \mathbf{1}^T + [0, \alpha_2 - \alpha_1, \ldots, \alpha_N - \alpha_1] \tag{13.71}$$

with the rotated angles $\alpha_i - \alpha_1$, $2 \le i \le N$ a set of mutually independent random variables uniformly distributed in $[0, 2\pi]$ and also independent of the equally uniformly distributed leading angle α_1. Then, the expectation (E_α) in Equation (13.70) can be expressed as the composition of two expectations: the expectation over the rotation α_1 applied to the expectation over the subset of rotated angles $SS = \{\alpha_i -\alpha_1, 2 \le i \le N\}$. From this division into subsets of the geometry, Equation (13.70) becomes:

$$\overline{\mathbf{C}_N} = E_{SS} E_\beta \left[\left(\mathbf{F}_{SS,\beta}^T \mathbf{AF}_{SS,\beta} \right)^{-1} \right] \tag{13.72}$$

where $\mathbf{F}_{SS,\beta}$ denotes $\mathbf{F}$ for a particular rotation angle $\beta = \alpha_1$ and subset configuration. From Equation (13.71), we can express the generic subset $\mathbf{F}_{SS,\beta}^T$ matrix as $\mathbf{F}_{SS,\beta}^T = \mathbf{G}_\beta \mathbf{F}_{SS}^T$, where $\mathbf{F}_{SS}^T = \mathbf{F}_{SS,0}^T$ denotes the canonic $\mathbf{F}$ matrix of the subset in the specific case $\beta = 0$, and $\mathbf{G}_\beta$ is a unitary-rotation matrix defined as:

$$\mathbf{G}_\beta = \begin{bmatrix} \cos(\beta) & \sin(\beta) \\ -\sin(\beta) & \cos(\beta) \end{bmatrix} \tag{13.73}$$

Now, Equation (13.72) can be expressed as:

$$\overline{\mathbf{C}_N} = E_{SS} E_\beta \left[\left(\mathbf{G}_\beta \mathbf{F}_{SS}^T \mathbf{AF}_{SS} \mathbf{G}_\beta^T \right)^{-1} \right] = E_{SS} E_\beta \left[\mathbf{G}_\beta \left(\mathbf{F}_{SS}^T \mathbf{AF}_{SS} \right)^{-1} \mathbf{G}_\beta^T \right] \tag{13.74}$$

since $\mathbf{G}_\beta$ is unitary. The expectation over β gives

$$\overline{\mathbf{C}_N} = \frac{1}{2} E_{SS} \text{tr} \left\{ \left(\mathbf{F}_{SS}^T \mathbf{A} \mathbf{F}_{SS} \right)^{-1} \right\} \cdot \mathbf{I} \tag{13.75}$$

Proof of (13.75) : Here we prove that for symmetric matrix A holds

$$\mathbf{B} = E_\beta [\mathbf{G}_\beta \mathbf{A} \mathbf{G}_\beta^T] = \frac{1}{2} \text{tr}\{\mathbf{A}\} \cdot \mathbf{I}$$

where E_β denotes the expectation operation over β. A is a 2×2 matrix:

$$\mathbf{A} = \begin{bmatrix} a_{11} & a_{12} \\ a_{21} & a_{22} \end{bmatrix}$$

and $\mathbf{G}_\beta$ is the previously defined unitary rotation matrix:

$$\mathbf{G}_\beta = \begin{bmatrix} \cos(\beta) & \sin(\beta) \\ -\sin(\beta) & \cos(\beta) \end{bmatrix}$$

Then, the expectation over the uniformly distributed random variable β can be expressed as

$$E_\beta [\mathbf{G}_\beta \mathbf{A} \mathbf{G}_\beta^T] = \begin{bmatrix} b_{11} & b_{12} \\ b_{21} & b_{22} \end{bmatrix}$$

Now, taking the expectation over the uniform distribution of β,

$$b_{11} = \frac{a_{11} + a_{22}}{2}, \quad b_{12} = \frac{a_{12} - a_{21}}{2}, \quad b_{21} = \frac{a_{21} - a_{12}}{2}, \quad b_{22} = \frac{a_{11} + a_{22}}{2}$$

For A a symmetric matrix, we finally obtain:

$$E_\beta [\mathbf{G}_\beta \mathbf{A} \mathbf{G}_\beta^T] = \begin{bmatrix} \frac{a_{11} + a_{22}}{2} & 0 \\ 0 & \frac{a_{11} + a_{22}}{2} \end{bmatrix} = \frac{1}{2} \text{tr}\{\mathbf{A}\} \cdot \mathbf{I}$$

Applying Equation (13.75) in (13.69) to the four algorithms associated with the information matrices $\mathbf{A}$ defined in Equation (13.66), we obtain:

$$\bar{\mathbf{C}}_{st} = \frac{1}{2} E_{N,SS} \text{tr} \left\{ \left(\mathbf{F}_{SS}^T \mathbf{R}^{-1} \mathbf{F}_{SS} \right)^{-1} \right\} \cdot \mathbf{I}$$

$$\bar{\mathbf{C}}_{s\bar{\imath}} = \frac{1}{2} E_{N,SS} \text{tr} \left\{ \left(\mathbf{F}_{SS}^T \mathbf{H}^T (\mathbf{H} \mathbf{R} \mathbf{H}^T)^{-1} \mathbf{F}_{SS} \right)^{-1} \right\} \cdot \mathbf{I}$$

$$\bar{\mathbf{C}}_{hc} = \bar{\mathbf{C}}_{s\bar{\imath}}$$

$$\mathbf{C}_{h\bar{c}} = \frac{1}{2} E_{N,SS} \text{tr} \left\{ \left(\mathbf{F}_{SS}^T \mathbf{H}^T \left[(\mathbf{H} \mathbf{R} \mathbf{H}^T) \circ \mathbf{I} \right]^{-1} \mathbf{H} \mathbf{F}_{SS} \right)^{-1} \right\} \cdot \mathbf{I} \tag{13.76}$$

As an illustration, numerical results for (13.76) are presented in Figures 13.8 and 13.9 with the same parameters as in [47]. A completely random uniform distribution of the BSs in the visibility circle of the

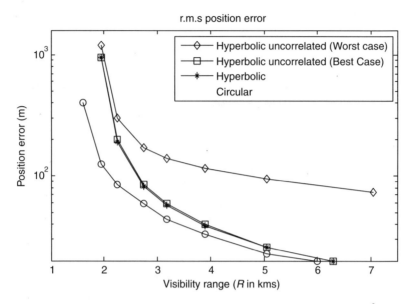

Figure 13.8   Mean performance with a constant BS density of $\rho = 0.25$ BS/km^2.

mobile is assumed. It is also assumed that the mobile can observe BSs within a radius of R m where BSs are uniformly distributed with a certain density of ρ BS/km^2.

If d is the distance from the mobile to a randomly selected beacon, then the probability density function (pdf) of d is given by $p_d(d) = 2d/R^2$ for $0 \leq d \leq R$ [49]. On the other hand, the delay spread, which constitutes the principal phenomenon that disturbs the TOA measurements, is given by [49] as $\sigma = T_1\, d^{\varepsilon}\, y$ where T_1 is the mean value of σ at $d = 1$ km, ε is an exponent between 0.5 and 1.0, and y is a

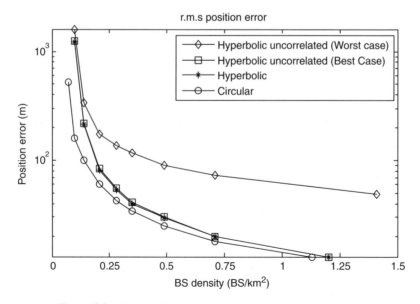

Figure 13.9   Mean performance with a constant visibility range of 3 km.

log-normal random variable generated by $y = 10^{Y/10}$, where Y is Gaussian with standard deviation σ_Y in the interval between 2 and 6 dB. The performance results shown hereafter assume that the TOA covariance matrix $\mathbf{R}$ is diagonal, i.e. $\mathbf{R} = \text{diag}\left(\left[\sigma_1^2, \sigma_2^2, \ldots, \sigma_N^2\right]\right)$ with σ_n samples of the random variable σ for the nth beacon. In all the simulations presented here, T_1, ε, and σ_Y have been chosen for the suburban environment following [49]. That is, $T_1 = 0.4~\mu$, $\varepsilon = 0.5$, and $\sigma_Y = 4$ dB. The results are shown in Figures 13.8 and 13.9.

References

[1] Revision of the commissions rules to ensure compatibility with enhanced 911 emergency calling systems RM-8143, CC Docket 94–102, FCC, Washington, DC, July 26, 1996. FCC.

[2] Woerner, D., Rappaport, T. S., Reed, J. H. and Krizman, K. J. (1998) An overview of the challenges and progress in meeting the E-911 requirement for location system, *IEEE Communications Magazine*, **36**, 30–37.

[3] Koolen, L. and Rantalainen, T. (1999) Mobile standardization and regulatory issues–A European perspective, presented at the IBC Conference on Mobile Location Services, London, Nov. 1999.

[4] Collier, W. and Weiland, R. (1994) Smart cars, smart highways, *IEEE Spectrum*, **31**, 27–33.

[5] Jurgen, R. (1991) Smart cars and highways go global, *IEEE Spectrum*, **28**, 26–36.

[6] Ott, G. D. (1977) Vehicle location in cellular mobile radio system, *IEEE Transactions on Vehicular Technology*, **26**, 43–46.

[7] Rappaport, T. S., Reed, J. H. and Woerner, D. (1996) Position location using wireless communications on highways of the future, *IEEE Communications Magazine*, **34**, 33–41.

[8] Turin, G. L., Jewell, W. S. and Johnston, T. L. (1972) Simulation of urban vehicle-monitoring systems, *IEEE Transactions on Vehicular Technology*, **21**, 9–16.

[9] Staras, H. and Honickman, S. N. (1972) The accuracy of vehicle location by trilateration in a dense urban environment, *IEEE Transactions on Vehicular Technology*, **21**, 38–43.

[10] Hata, M. and Nagatsu, T. (1980) Mobile location using signal strength measurements in a cellular system, *IEEE Transactions on Vehicular Technology*, **29**, 245–252.

[11] Song, H. L. (1994) Automatic vehicle location incellular communications systems, *IEEE Transactions on Vehicular Technology*, **43**, 902–908.

[12] Kennemann, O. (1994) Continuous location of moving GSM mobile stations by pattern recognition techniques, in Proceedings 5th IEEE International Symposium on Personal, Indoor and Mobile Radio Communications (PIMRC'94) and ICCC Regional Meeting on Wireless Computer Networks (WCN), 2, 630–634.

[13] Hellembrandt, M., Mathar, R. and Scheibenbogen, M. (1997) Estimating position and velocity of mobiles in a cellular radio network, *IEEE Transactions on Vehicular Technology*, **46**, 65–71.

[14] Hellembrandt, M. and Mathar, R. (1999) Location tracking of mobiles in cellular radio networks, *IEEE Transactions on Vehicular Technology*, **48**, 1558–1562.

[15] Caffery Jr., J. and Stüber, G. L. (1998) Subscriber location in CDMA cellular networks, *IEEE Transactions on Vehicular Technology*, **47**, 406–416.

[16] Cesbron, F. and Arnott, R. (1998) Locating GSM mobiles using antenna arrays, *Electronics Letters*, **34**, 1539–1540.

[17] Pent, M., Spirito, M. A. and Turco, E. (1997) Method for positioning GSM mobile stations using absolute time delay measurements, *Electronics Letters*, **33**, 2019–2020.

[18] Spirito, M. A. and Mattioli, A. G. (1998) On the hyperbolic positioning of GSM mobile stations, in *Proceedings 1998 URSI International Symposium Signals, Systems, and Electronics (ISSSE'98)*, Pisa, Italy, 1998, pp. 173–177.

[19] Spirito, M. A. (1999) Further results on GSM mobile station location, *Electronics Letters*, **35**, 867–869.

[20] Spirito, M. A. and Guidarelli Mattioli, A. (1999) Preliminary experimental results of a GSM mobile phones positioning system based on timing advance, in *Proceedings 1999 IEEE Vehicular Technology Conference (VTC'99—Fall)*, Amsterdam, the Netherlands, September 1999, pp. 2072–2076.

[21] Ruutu, V., Alanen, M., Gunnarsson, G., Rantalainen, T. and Teit-tinen, V.-M. (1998) Mobile phone location in dedicated and idle modes, in *Proceedings 9th IEEE International Symposium on Personal, Indoor and Mobile Radio Communications (PIMRC'98)*, **1**, 1998, pp. 456–460.

[22] Spirito, M. A. and Wylie-Green, M. P. (1999) Time of arrival (TOA) estimation in the GSM and EDGE mobile communication systems, in *Proceedings 1999 Finnish Signal Processing Symposium (FINSIG'99)*, Oulu, Finland, May 1999, pp. 64–68.

[23] Spirito, M. A. and Wylie-Green, M. P. (1999) Mobile stations location in future TDMA mobile communication systems, in *Proceedings 1999 IEEE Vehicular Technology Conference (VTC'99—Fall)*, Amsterdam, the
Netherlands, September 1999, pp. 790–794.

[24] *Report of Location Services*, Nokia Telecommunications, Doc. TDoc SMG1 418/95. ETSI STC-SMG1.

[25] *Service requirements for a mobile location service*, Doc. SMG1 Tdoc.158/97, Mar. 1997. ETSI-SMG1.

[26] All T1P1.5 standardization documents can be found in the T1P1.5 documents—PCS directory [Online]. Available: http://www.t1.org

[27] *Evaluation Worksheet for Assisted GPS*, Doc. T1P1.5/98–132r3, June 1998. T1P1.

[28] Evaluation Sheet for the Time Advance Positioning Method, Doc. T1P1.5/98–033, Jan. 1998. T1P1.

[29] Evaluation Sheet for the Enhanced Observed Time Difference (E-OTD) Method, Doc. T1P1.5/98–021r7, June
1998. T1P1.

[30] Evaluation Sheet for the Uplink TOA Positioning Method, Doc. T1P1.5/98–034r6, Aug. 1998. T1P1.

[31] Functional Stage 2 Specification of Location Services in UTRAN, Doc. 3G TR 25.305. 3GPP.

[32] Friedlander, B. (1987) A passive localization algorithm and its accuracy analysis, *IEEE Journal of Oceanic
Engineering*, **12**, 234–244.

[33] Fenwick, A. J. (1999) Algorithms for position fixing using pulse arrival times, in Proceedings of Institute of
Electrical Engineers Radar, Sonar and Navigation, **146**, 208–212.

[34] Foy, W. (1976) Position-location solutions by Taylor-series estimation, *IEEE Transactions on Aerospace Electronics Systems*, **12**, 187–193.

[35] Torrieri, D. (1984) Statistical theory of passive location systems, *IEEE Transactions on Aerospace Electronics
Systems*, **20**, 183–197.

[36] Massatt, P. and Rudnick, K. (1991) Geometric formulas for dilution of precision calculations, *Journal of the
Institute of Navigation*, **37**, 379–391.

[37] Lee, H. (1975) A novel procedure for assessing the accuracy of hyperbolic multilateration systems, *IEEE
Transactions on Aerospace Electronics Systems*, **11**, 2–15.

[38] European Digital Telecommunications System (Phase 2); Mobile Radio Interface Layer 3 Specification (GSM
04.08), Aug. 1997. ETSI TC-SMG.

[39] European Digital Telecommunications System (Phase 2); Radio Subsystem Synchronization (GSM 05.10),
May 1996. ETSI TC-SMG.

[40] Collins, J., Hoffmann-Wellenhof, B. and Lichtenegger, H. (1992) *Global Positioning System–Theory and
Practice*. Springer Verlag, New York.

[41] Scharf, L. L. (1990) Statistical Signal Processing–Detection, Estimation, and Time Series Analysis. Wiley, New
York.

[42] Steele, R. (1992) *Mobile Radio Communications*, Pentech, London.

[43] Spirito, M. A. (2001) On the accuracy of cellular mobile station location estimation. *IEEE Transactions on
Vehicular Technology*, **50**, 674–685.

[44] Patwari, N., Hero, A. O., Perkins, M., Correal, N. S. and O'Deam, R. J. (2003) Relative location estimation in
wireless sensor networks, *IEEE Transactions on Signal Processing*, **51**, 2137–2148.

[45] Glisic, S. (2004) *Advanced Wireless Communications*. Wiley, Chichester.

[46] Horn, R. A. and Johnson, C. R. (1990) *Matrix Analysis*. Cambridge University Press, New York.

[47] Urruela, A., Sala, J. and Riba, J. (2006) Average performance analysis of circular and hyperbolic geolocation,
IEEE Transactions on Vehicular Technology, **55**, 52–65.

[48] Scharf, L. L. (1991) Statistical Signal Processing, Detection, Estimation, and Time Series Analysis. Addison-
Wesley, Reading, MA, Chapter 9.

[49] Greenstein, L. J., Erceg, V., Yeh, Y. S. and Clark, M. V. (1997) A new path-gain/delay-spread propagation
model for digital cellular channels, *IEEE Transactions on Vehicular Technology*, **46**, 477–485.

14

Wireless Networks Connectivity

14.1 Survivable Wireless Networks Design

Given a set of nodes and a set of commodities, the *capacitated survivable network design problem* involves designing the topology and dimensioning the links so that the network can carry all traffic demands and assure full recovery from a range of link failures. In this context a *commodity* is defined by an origin node, a destination node, and a traffic demand expressing, for instance, a specific required bandwidth. We focus on the so-called *single- facility capacitated survivable network design problem*, denoted as the SFCSND problem.

A *facility* is a constant given capacity resource.

Various traffic rerouting strategies exist to withstand single link failures. The fastest and safest method, but obviously the most expensive, is *dedicated rerouting*. This strategy requires the simultaneous routing of traffic demands along *working paths* (i.e., used for the nominal state) and dedicated *rerouting paths* (i.e., used for the failure state). In order to reduce cost, the spare capacity devoted to protection is usually shared among several rerouting paths. In this section we look at such *shared rerouting* modes. Fast rerouting mechanisms have already been developed for IP networks (based on Multi Protocol Label Switching solutions), and for automatically switched optical networks. Most of them are based on a *link protection* scheme. This first shared rerouting strategy, also called *local rerouting*, tries to reroute traffic locally between the extremities of the failed link. It generally offers the fastest failure detection and recovery. A more flexible and economical shared protection, called *path protection* or *end-to-end rerouting*, propagates failure information to the destination nodes of traffic demands, in order to set up rerouting paths between them. This flexibility is particularly appropriate for the upper layers of networks such as IP or ATM. Several variants of path protection may be considered, depending on the working paths and the resources available during the rerouting process. Here we focus only on *partial end-to-end rerouting without recovery*. This means that only the interrupted working paths have to be rerouted (in contrast to global rerouting which stipulates the rerouting of all working paths), and network resources will be devoted exclusively either to the nominal or to the restoration state (i.e., *without recovery*). Once the failure is over, the working paths can thus be simply reused. From here on, the term *end-to-end rerouting* will refer to partial end-to-end rerouting without recovery. This approach does not take into account the case of partial end-to-end rerouting with recovery where network resources used for traffic routing, that are released after a failure, could also be used for traffic rerouting. The decomposition approach presented in this section could not be directly applied for such a restoration strategy.

Advanced Wireless Communications & Internet: Future Evolving Technologies, Third Edition. Savo Glisic.
© 2011 John Wiley & Sons, Ltd. Published 2011 by John Wiley & Sons, Ltd.

14.1.1   Analytical Model

As indicated earlier in this section we concentrate only on the case of a single facility with a constant capacity, that is, the SFCSND problem which will be referred to as *network design (ND)* problem in the sequel for brevity. The graphs we consider in the model are finite, simple and undirected. In a graph $G = (V, E)$ the elements of V are called the *vertices* or *nodes* and those of E the *edges or links*. If $e \in E$ is an edge with extremities u and v, then we write $e = (u, v)$. For an edge $e \in E$, let E_e be the set of edges different to e, that is, $E_e = E \backslash \{e\}$, and G_e denote the graph (V, E_e).

In the network design, the input network is represented as a graph $G = (V, E)$ where the vertices correspond to the known nodes of the network and the problem is to select the links that will be inserted to connect certain nodes. The edge set E represents the potential links among which we are going to make a selection. To model this topological aspect of the problem, we associate with each edge $e \in E$ a *topological variable* x_e so that $x_e = 1$ if the corresponding link is part of the network, and 0 otherwise. Given a installation cost vector $a \in R_+^{|E|}$ associated with the edges of G, one part of the problem is to find a minimum installation cost subgraph of G, according to whether the commodity and survivability requirements are met, where the cost of a subgraph is the sum of the cost of its edges.

The selected links have to be capable of carrying traffic between some pairs of nodes. Traffic data is represented by a set of commodities $K = \{1, 2, \ldots, |K|\}$ where each index $k \in K$ is associated with an origin vertex $0_k \in V$, a destination vertex $d_k \in V$ and a specific traffic demand (e.g., a required bandwidth) $B_k \in R_+$ to be satisfied. The second purpose of the ND problem is to load capacity onto edges in multiples of a given facility whose capacity is equal to $\lambda \in Z_+$, in order to allow the routing of all the commodities in K. Let us denote by $y_e \in Z_+$ the *dimensioning variable* corresponding to the number of units λ loaded onto edge $e \in E$. If a cost $c_e \in R_+$ corresponds to the loading of a single facility onto edge $e \in E$, the aim is to determine the minimum loading cost capacity to be installed onto the selected edges.

In addition, some links of the network, if fitted in, may fail. So we also consider a set of link failure indexes $L = \{1, 2, \ldots, |L|\}$ where $|L| \leq |E|$ and index $l \in L$ represents the failure of the link corresponding to the edge $e_l \in E$. Including this survivability issue means that the overall problem is to find a subgraph of G (related to the topological variables $x_e, e \in E$) and the capacities we have to load onto its edges (related to the dimensioning variables $y_e, e \in E$) so that all the commodities in K may be carried in the nominal state as well as in the failure state (i.e., when the link corresponding to any index $l \in L$ fails), and the global cost (i.e., installation and loading costs) is minimum.

As mentioned earlier, the survivability requirement is implemented by traffic rerouting (local and partial end-to-end rerouting without recovery). In both cases, the overall design problem can be formulated as an optimal synthesis problem with non simultaneous multi-commodity requirements. Before defining the program analytically, we first introduce the following notations. Let $P(k)$ be the finite set of elementary paths (without cycles) of G between the extremities o_k and d_k of the commodity $k \in K$. The quantity of traffic related to a commodity $k \in K$, carried through the path $p \in P(k)$ is denoted by the *nominal flow variable* $f_p^k \in R_+$. These variables allow us to deal with the routing of all traffic demands in the nominal state. In the case of rerouting we have to differentiate both rerouting strategies in terms of the rerouting paths.

In the case of the end-to-end rerouting strategy, we have to reroute the commodities interrupted by a link failure from their origin to their destination nodes. For a commodity $k \in K$ and an edge failure index $l \in L$, $Q(l, k)$ denotes the set of elementary paths of G_{e_l} between 0_k and d_k. Given a commodity $k \in K$, the *end-to-end rerouting flow variable* $h_q^{l,k} \in R_+$ represents the quantity of rerouted traffic carried on path $q \in Q(l, k)$ when link $e_l \in E$ fails, $l \in L$.

End-to-end rerouting strategy: With this notation, the network design problem based on the end-to-end rerouting strategy can be formulated as

$$\text{Minimize} \sum_{e \in E} (a_e x_e + c_e y_e) \tag{14.1}$$

subject to

$$\sum_{p \in P(k)} f_p^k = B_k \ \forall k \in K, \tag{14.1a}$$

$$\sum_{q \in Q(l,k)} h_q^{l,k} - \sum_{\substack{p \in P(k) \\ p \ni e_l}} f_p^k = 0 \quad \forall l \in L, \ k \in K, \tag{14.1b}$$

$$\sum_{k \in K} \left(\sum_{\substack{p \in P(k) \\ p \ni e}} f_p^k + \sum_{\substack{q \in Q(l,k) \\ q \ni e}} h_q^{l,k} \right) \le \lambda y_e \qquad \forall e \in E_{e_l}, \ l \in L, \tag{14.1c}$$

$$y_e \le M x_e \ \forall e \in E, \tag{14.1d}$$

$$x_e \in \{0, 1\} \ \forall e \in E, \tag{14.1e}$$

$$y_e \in Z_+ \ \forall e \in E, \tag{14.1f}$$

$$f_p^k \ge 0 \ \forall p \in P(k), \ k \in K, \tag{14.1g}$$

$$h_q^{l,k} \ge 0 \ \forall q \in Q(l, k), \ l \in L, \ k \in K \tag{14.1h}$$

Equation (14.1) is obviously based on an edge-path multi flow formulation. The overall objective function is to minimize the global facility installation cost, that is, the sum of the installation and loading costs. Constraints (14.1a) ensure that all the commodities are routed in the nominal state. When a link failure occurs, constraints (14.1b) guarantee that all the interrupted traffic, given by the second term of the left-hand side, is rerouted between the extremity nodes of the failed link. Constraints (14.1c) mean that the capacity loaded onto edge $e \in E$ must be higher than the total traffic carried in both nominal and rerouting situations by all paths containing this edge. Constraints (14.1d) force the capacity to be loaded onto existing links. The value M gives the maximum number of capacity units that can be loaded onto a link. The integrity of the topological and dimensioning variables is given by constraints (14.1e) and (14.1f), respectively. The last two constraints (14.1g) and (14.1h) express the nonnegativity of nominal and end-to-end rerouting flow variables.

Local rerouting strategy: In this case, the interrupted traffic must be rerouted between the extremity nodes of the failed link. For this reason we have to rewrite some constraints in (14.1). Given an index $l \in L$ and its associated link $e_l = (u_l, v_l) \in E$, let $Q(l)$ be the finite set of elementary paths of G_{e_l} between the vertices u_l and v_l. To each edge failure index $l \in L$ and path $q \in Q(l)$ we associate a *local rerouting flow variable* $g_q^l \in R_+$ corresponding to the amount of traffic flowing through the path q when the link e_l is down. The mathematical formulation of the problem based on the local rerouting strategy is similar to (14.1), except for constraints (14.1b), (14.1c), and (14.1h), which are replaced by the same type of constraints as

$$\sum_{q \in Q(l)} g_q^l - \sum_{k \in K} \sum_{\substack{p \in P(k) \\ p \ni e_l}} f_p^k = 0 \qquad \forall l \in L, \tag{14.2.a}$$

$$\sum_{k \in K} \sum_{\substack{p \in P(k) \\ p \ni e}} f_p^k + \sum_{\substack{q \in Q(l) \\ q \ni e}} g_q^l \le \lambda y_e \ \forall e \in E_{e_l}, \ l \in L, \tag{14.2.b}$$

$$g_q^l \ge 0 \ \forall q \in Q(l), \ l \in L. \tag{14.2.c}$$

involving this time the local rerouting flow variables. The size of both mixed-integer linear programs may be very large because of the huge number of paths in $P(k)$, $Q(l)$ and $Q(l, k)$. However, the working paths in $P(k)$ and the rerouting paths in $Q(l, k)$ or $Q(l)$) are independent. Decomposition methods such as Benders' decomposition [1] can therefore be used to obtain optimal solutions in a reasonable time for both variants of the problem. For more detail see also [2].

14.2   Survivability of Wireless Ad Hoc Networks

Compared with traditional networks, wireless ad hoc networks are more vulnerable to malicious attacks as well as random failures due to their unique features, such as constrained node energy, error-prone communication media, and dynamic network topology. Therefore, it is the first major goal for a survivable wireless ad hoc network to establish and maintain a connected topology, whenever it is practical. Based on this observation, as a fundamental topology property and prerequisite for all networking operations, topology connectivity is a critical index for the survivability of wireless ad hoc networks, especially in the presence of malicious attacks and random failures.

For the analytical modeling of the problem we assume that N mobile nodes in a wireless ad hoc network are randomly and uniformly distributed over a 2-D square with area A. The node transmission radius r, is assumed to be identical for all nodes. Thus, the underlying communication graph of a wireless ad hoc network is modeled by a geometric random graph $G(N, r)$ [4] where N denotes the vertex set with $|N| = N$ and an edge exists between two vertices only if their distance is no greater than r. Analogous to $G(N, r)$, a wireless ad hoc network is denoted by $M(N, r)$ (or only M for brevity).

In a wireless ad hoc network, node cooperation in routing process is an essential requirement to maintain protocol operations and network connectivity [5], [6]. However, since in a future cognitive network every node is an autonomous system, it may decide how to act in the network by itself. To consider the potential impacts of various misbehaviors, the geometric random graph model aforementioned should be extended by introducing an additional assumption that all nodes operate independently in the following *four* states:

> *Cooperative (C) state* when the node complies with all routing and forwarding rules, which includes being able to initiate and response route discoveries correctly and forward control and data packets for others at the best effort.

> *Selfish (S) state* when a node can initiate and response route discoveries for its own purposes but may not forward control or data packets for others for the sake of power saving.

> *Malicious (M) state* when a node launches denial of service *DoS* attacks on the network layer, which means being cooperative in the routing stage but reluctant in forwarding data packets, or disrupting legitimate path selections by broadcasting fake route replies.

> In this section we focus on two types of malicious behaviors: *Jellyfish* and *Blackhole* attacks [7]. The *Jellyfish* attack is referred to as to any malicious behavior that complies with all routing rules but reorders, delays, or drops data packets; while the *Blackhole* attack is referred to as to a brutal behavior that always claims the shortest path to the destination so that path selections and all data traffics can be trapped later on.

> *Failed (F) state* when a node is unable to initiate or response route discoveries.

Various survivability definitions have been proposed in different disciplines [8]–[13].

In this section, we consider the survivability of a wireless ad hoc network to be a basic network capability designed to maintain a connected topology in the presence of malicious adversaries and random failures and use *connectivity* as the metric to define it as follows:

> *Given a wireless ad hoc network M, let κ(M) denote the (vertex)-connectivity of M. The network survivability of M, denoted by NS_k(M), is defined as the probability that all active (un-failed) nodes*

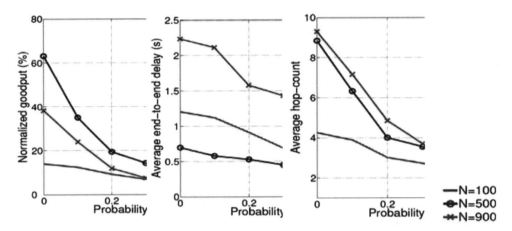

Figure 14.1 Impacts of misbehaving nodes on network performance vs probability of maliciousness.

are k-connected, i.e.,

$$NS_k(M) = Pr(\kappa(M_a) = k) \tag{14.3}$$

where M_a is the network induced by all active nodes of M.

In the presence of node misbehavior this definition is further modified as: *Given a wireless ad hoc network M with four node behavior states described previously, find out the quantitative relationship between four behaviors and the network survivability, i.e., $NS_k(M)$, in closed forms.*

For a wireless ad hoc network M in the presence of node misbehaviors and failures, when the number of nodes N is sufficiently large, the network survivability defined in (14.3) is upper bounded asymptotically by

$$NS_k(M) \leq \left(e^{-\mu_a P_B} \left(1 - \frac{\Gamma(k, \mu_a P_c)}{\Gamma(k)} \right) \right)^{N/\lambda \pi r^2} \tag{14.4}$$

and lower bounded asymptotically by

$$NS_k(M) \geq 1 - N(1 - P_f) \left(1 - e^{-\mu_a P_B} \left(1 - \frac{\Gamma(k, \mu_a P_c)}{\Gamma(k)} \right) \right) \tag{14.5}$$

where $\mu_a = N(1 - P_f)/(\lambda \pi r^2)$ and λ is the node density, and $\Gamma(h) = (h - 1)!$ and $\Gamma(h, x) = (h - 1)!e^{-x} \sum_{l=0}^{h-1} x^l/l!$ are the complete and incomplete Gamma functions, respectively [3]. Parameters P_x are the probabilities that the node is in the one of four states x, defined earlier in the section.

An additional insight into the negative effect of node misbehaviors can be obtained from the simulation results presented in Figure 14.1 where misbehaving nodes simply drop all data packets to be forwarded once paths are established.

14.3 Network Dimensioning

This section focuses on a specific network design task called *network dimensioning* in which the network structure is given and the task is to find values of link capacities that are optimal according to a certain criterion.

The problem will be defined in the framework of two layers: the traffic layer (*TL*), which defines the demands between each pair of nodes in this layer, and the link resources layer (*LL*), which consists of resources for implementing the demands defined in the *TL*. In general, one link in the *TL* may be realized by several paths in the *LL*.

This approach is generic, which means that it can be applied to any two layers telecommunication networks. It follows that the design tasks for different networks and different resource layers can be solved by the same optimization procedures.

As an example, in a core network we can consider a SDH network and assume that the LL is the STM-1 sublayer of the SDH layer, whereas the TL is the VC-12 sublayer. These two sublayers constitute the synchronous layer of the SDH network. The demand volume in TL is expressed as the number of VC-12 containers (carrying 2 *Mb/s* streams). The capacity of a TL link is expressed in STM-1 modules (one STM-1 module can transport up to 63 VC-12 containers). Analog examples can also be found in wireless networks.

The multicommodity flow networks analyzed in this section are modeled as multigraphs. Such a multigraph $G = (V, E, P)$ is composed of a set of nodes V, a set links E, and an incidence relation P defined over the product $V \times E \times V$ that specifies how the nodes are connected by the links. In general there might exist more than a single distinct link between a pair of nodes. Our interest is restricted to undirected multigraphs without loops.

Simple path p between nodes v, w is a sequence of links without cycles, joining v and w designated as $p = (e_1, e_2, \ldots, e_{L(p)})$.

We define an integer function $y: E \to N$, where $N = \{0, 1, \ldots\}$, that determines capacities of links.

In the network design, we distinguish between the TL and the LL network. For an illustration see Figure 14.2.

We assume that the set of nodes V of the TL is contained by the set of nodes of the LL. There may exist nodes from LL that are not in TL. LL links are treated as resources that are used to implement the TL network. In the sequel, we use the term *link* for the LL link and the TL links are called *demands*. D denotes the set of demands. The demand d is characterized by the following attributes.

1. The pair of end nodes $v(d)$ and $t(d)$.
2. The demand volume $n(d)$, expressed in the demand capacity units (DCU).
3. The set of *admissible paths* $P(d)$; each path $p \in P(d)$ is a simple path in the LL connecting nodes $v(d)$ and $t(d)$ and can be used to allocate the DCU of d.

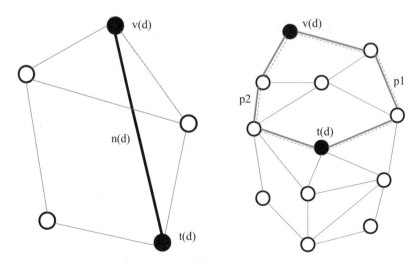

Figure 14.2   Definition of the demand in TL and possible realization in LL.

Link capacities $y(c)$ are expressed in link capacity units (LCUs). We assume that one LCU is equal to m times DCU; therefore, if the capacity of a link in LL is equal to an integer multiple of $mDCU$, it cannot be further increased by one DCU unit but rather by a whole module of $mDCU$. This feature is called the *modularity* of links and it reflects the modularity of real telecommunication networks. This makes the design difficult. With no modularity, i.e. $m = 1$, the design is a simple linear problem; with $m > 1$, it becomes NP-complete.

In the following we will specify a set S of states. In each state, LL and TL structures may differ from the nominal state structure, i.e. different demand volumes and link capacities. These states can be used to model LL failures and are *failure/demand(f/d)* states, which the network should be robust enough to withstand. Each state $s \in S$ is characterized by a set of the link failure coefficients $\alpha(s, e)$. Parameter $\alpha(s, e)$ is a fraction of the nominal capacity of the link $y(e)$ available in the f/d state s. In the nominal state $s_0 \in S$, $\alpha(s_0, e) = 1$ for all links. For $s \neq s_0$, $0 \leq \alpha(s, e) \leq 1$.

We focus on a simplified f/d state model by assuming that in each state $s \in S$:

1. there is exactly one link $e^* \in E$ for which $\alpha(s, e^*) = 0$; for all other links, $\alpha(s_J c) = 1$;
2. demands are unchanged and are the same as in the nominal state s_0.

Obviously, the model above does not cover demand variations and all possible failures.

An *allocation function* (flow function) $\mathbf{x}^s$ specifies the way the demands are allocated to their admissible paths in f/d state s. Hence, $\mathbf{x}^s : \mathbf{Q} \rightarrow \mathbf{N}$ is a nonnegative integer-valued function (whose values are referred to as *flows*) defined over the set of pairs

$$\mathbf{Q} = \{(d, p) : d \in D, \ p \in P(d)\}$$

Each individual flow x_{dp}^s (expressed in DCU) is a portion of the capacity $m \cdot \alpha(s, e) \cdot y(e)$ traversing link e in situation s, allocated to path $p \in P(d)$. Flows in the nominal state (for $s = s_0$) are denoted by x_{dp} and the allocation function in nominal state s_0 by $\mathbf{x}$.

Each allocation function $\mathbf{x}^s$ implies the load $l(\mathbf{x}^s, e)$ of link e

$$l(\mathbf{x}^s, e) = \sum_{d \in D} \sum_{p \in P(d), e \in p} x_{dp}^s \tag{14.6}$$

In addition, $f_e(l)$ denotes the minimal capacity of link e expressed in LCUs necessary to accommodate load l. In general, the dimensioning function $f_e(\cdot)$ can be nonlinear. We restrict ourselves to the dimensioning function $f_e(l) = \lceil l/m \rceil$.

In the nominal state $s_0 \in S$, we compute link capacities $y(e)$ according to $y(e) = f(l(x, e))$.

Taking into account the set of failures S that the network should be able to withstand, we compute the capacity of link e as

$$y(e) = \max_{s \in S} f(l(\mathbf{x}^s, e))$$

The total cost C of the network is defined as a function of link capacities $y(e)$ as

$$C = \sum_{e \in E} \xi(e) \cdot y(e) \tag{14.7}$$

where $\xi : E \rightarrow R^+$ is the cost of one LCU on link e. In general, C may be given by a more complicated nonlinear function.

In practice, there may be a need for aggregation of the demand allocation vector where each demand is allocated to at most ω admissible paths. The whole flow allocated to each path is treated as an indivisible module and restored together. In the sequel, $\omega = 1$ denotes complete aggregation [each demand d is allocated as a single module of $n(d)$ DCU]; $\omega = \infty$ implies complete disaggregation, i.e. flow $n(d)$ can be distributed over all admissible paths.

14.3.1   Network Design Objectives

In general network design objective (*GNDO*) consists of finding allocation patterns $\mathbf{x}^s$ for all $s \in S$, including allocation pattern $\mathbf{x}$, such that the network cost (14.7) is minimized with the following constraints:

$$\sum_{p \in P(d)} x_{dp}^s \geq n(d), \quad \forall d \in D, \ s \in S,$$

$$\sum_{d \in D} \sum_{p \in P(d)} x_{dp}^s \leq m \cdot \alpha(s, e) \cdot y(e), \quad \forall e \in E, \ s \in S \quad (14.8)$$

The set of demands D, the set of f/d states S, and coefficients $\alpha(s, c)$ are given. The first line of constraint (14.8) guarantees complete allocation of demands and the second line ensures sufficient link capacity to realize flows in all f/d states.

In the following, we describe design objectives that are special cases of the above problem.

1. *Design D0 – Nominal state design* is derived from *GNDO* by setting $S = \{s_0\}$.
2. *Design D1–design with restricted reallocation* is the *GNDO* with an additional constraint preserving nominal flows not affected: $x_{dp}^s \geq x_{dp} \cdot \prod_{e \in p} \alpha(s, e)$, $\forall d \in D$, $p \in P(d)$, $s \in S$. The broken flow is distributed randomly over all admissible paths that are disjoint with the failed links (see Table 14.1).
3. *Design D2–design with state-dependent backup path*: Let denote the set of all the f/d states breaking path p as $T(p) = \{s \in S : \exists e \in p \ \alpha(s, e) = 0\}$. We assume that for each f/d state each nominal flow x_{dp} is assigned a protection path $q \in B_{dp}^s$, where $B_{dp}^s = \{p \in P(d) : \forall_{e \in p} \alpha(s, e) = 1\}$ is a set of admissible paths (operating links not affected by failure). The predefined protection path q is used to restore the entire flow x_{dp} in the state s.

The following constraints have to be satisfied: $\forall s \in S \backslash T(p)$ $x_{dp}^s \geq x_{dp}$ (the flow is not reallocated from an operational path, but can be increased by the reallocation of flow from broken paths) $\forall s \in T(p)$ $x_{dq}^s \geq x_{dp}$ *where* $q \in B_{dp}^s$ (the entire flow from broken paths is reallocated to the protection paths). See Table 14.2 for an example.

Table 14.1   Example allocation in nominal state s_0 and reconfiguration in the failure state s_1; Design D1. Failure of a link causes breakdown of certain paths, flow traversing broken paths is reallocated randomly over the remaining admissible paths

		state s_0			state s_1	
	n(d)	Admissible paths	Allocation	Broken paths	Remaining paths	Allocation
d_1	24	(p_1, p_2, p_3)	$(10, 8, 6)$	p_2	(p_1, p_3)	$(12, 0, 12)$
d_2	30	(p_1, p_2)	$(10, 20)$	—	(p_1, p_2)	$(10, 20)$
d_3	38	(p_1, p_2, p_3, p_4)	$(8, 6, 14, 10)$	p_1	(p_2, p_3, p_4)	$(0, 8, 16, 14)$
d_4	46	(p_1, p_2, p_3)	$(20, 20, 6)$	p_3	(p_1, p_2)	$(24, 22, 0)$

Table 14.2 Reconfiguration in the failure state s_1 when the allocation in nominal state s_0 is as in Table 14.1; Design D2. Failure of a link causes breakdown of certain paths, flow traversing broken paths is reallocated on the protection paths

		state s_1		state s_2		
	n(d)	Broken/protection path	Allocation	Broken paths	Protection paths	Allocation
d_1	24	p_2/p_1	(18, 0, 6)	p_3	p_2	(10, 14, 0)
d_2	30	$-/-$	(10, 20)	p_1	p_2	(0, 30)
d_3	38	$(p_1, p_2)/p_3$	(0, 0, 28, 10)	—	—	(8, 6, 14, 10)
d_4	46	p_3/p_2	(20, 26, 0)	(p_2, p_3)	p_1	(46, 0, 0)

14.3.2   Genetic Algorithm

The method is based on the selection scheme from the $(\mu + \lambda)$ evolution strategy [15] as summarized bellow

Genetic Algorithm

begin
$t = 0$
initialize P_0
evaluate P_0
while (not *stop_criterion*) do
begin
$O_t = $ *randomly reproduce* P_t
crossover (x)
state (x)
$P_{t+1} = $ select the best from $(P_t \cup O_t)$
$t = t + 1$
   end
end

The algorithm maintains a population P_t of *chromosomes*, each being a tabulated allocation function as in Tables 14.1 and 14.2. Initially, chromosomes are generated randomly, according to a special procedure that aims to preserve diversity in the population. In the main loop of the algorithm, an offspring population O_t of λ chromosomes is generated by reproducing randomly selected elements from P_t. Then, chromosomes are randomly mated into pairs, crossed over with a certain probability p_c, and mutated. A new base population P_{t+1} is generated by selecting μ best chromosomes from the sum of the old base population and the offspring population.

A chromosome is identical to the allocation function in the nominal state **x**. It is represented in a tabulated form, so a chromosome is a vector containing nonnegative demand allocation vectors $\mathbf{x}_d$. An illustration is shown in Figure 14.3.

Any feasible chromosome satisfies the following constraints:

$$x_{dp} \geq 0 \;\; and \;\; \sum_{p \in P(d)} x_{dp} = n(d) \tag{14.9}$$

In the sequel we define *demand aggregation* $\omega(x_d)$ as the number of admissible paths to be used for allocation of the demand d

$$\omega(\mathbf{x}_d) = \sum_{p \in P(d)} z(x_{dp})$$

$$z(x_{dp}) = \begin{cases} 0, & if \;\; x_{pd} = 0 \\ 1, & if \;\; x_{pd} > 0 \end{cases} \tag{14.10}$$

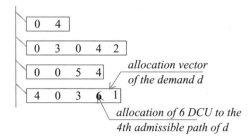

Figure 14.3   Chromosome-allocation function.

and the chromosome aggregation is

$$\omega(\mathbf{x}) = \max_{d \in D} \omega(\mathbf{x}_d) \tag{14.11}$$

The chromosome is an allocation function $\mathbf{x}$ describing allocation in the *nominal state only*. In the design D0, the chromosome's fitness is computed according to (14.7). In the robust design, we need allocation patterns $\mathbf{x}^s$ for all f/d states $s \in S$. $\mathbf{x}^s$ is derived from $\mathbf{x}$ in the following way.

For *Design D1* the broken flow is split randomly over all admissible paths that are disjoint with the failed links while for *D2* a flow traversing the broken path is reallocated to one of the predefined protection paths.

An initial population P_0 is generated randomly. Each chromosome consists of demand allocation vectors $\mathbf{x}_d$ satisfying (14.9) and additionally $\omega(\mathbf{x}_d) \leq \omega_i$, where ω_i is a parameter called the *initial aggregation*. The number of different allocation vectors for the demand d that can be obtained during the initialization process is given by

$$n_i(\mathbf{x}_d) = \sum_{\omega=1}^{\omega_i} \binom{h(d)}{\omega} \cdot \begin{cases} 0, & if\ n(d) < \omega \\ 1, & if\ n(d) = \omega \\ F(n(d) - \omega, \omega), & otherwise \end{cases} \tag{14.12}$$

where $h(d)$ is the number of admissible paths for the demand d and $F(n, \omega)$ is given by

$$F(n, \omega) = 1 + \sum_{c=1}^{n} F(c, \omega - 1), \quad F(1, \omega) = \omega, \quad F(n, 1) = 1 \tag{14.13}$$

If $\mathbf{p}(d) = \{p_1, p_2, \ldots, p_\omega\}$ is the set of admissible paths, which can be used to allocate the demand d, then (14.13) is designed in the following way: (a) The flow of n DCUs can be allocated to path p_1 in one way only ($p = \{n, 0, \ldots, 0\}$); (b) If the flow of c DCUs is allocated to p_1, then $n - c$ DCUs still must be allocated to $\omega - 1$ remaining paths; (c) The flow of one DCU can be allocated to the vector $\mathbf{p}(d)$ in ω various ways; (d) Finally, for the vector that contains only one path, n DCUs can be allocated in one way only. Since set $p(d)$ containing ω paths should be chosen from $h(d)$ possibilities, we have (14.12).

Genetic operators take advantage of the assumption that parental chromosomes are feasible. In the sequel, a parameter called *working aggregation* ω_w is used. This value is used by genetic operators to put an upper bound on aggregation of the resulting chromosomes; ω_w may differ from ω_i and $\omega_w \geq \omega_i$.

Mutation is performed for each chromosome. Each demand allocation vector $\mathbf{x}_d$ is mutated separately with probability p_m. For each mutated demand allocation vector $\mathbf{x}_d$, a new vector $\mathbf{x}'_d$ is generated randomly, irrespectively of $\mathbf{x}_d$, and in such a way that $\omega(\mathbf{x}'_d) \leq \omega_w$.

Then, the number different allocation vectors that can be generated during the mutation is given by (14.12) with $n_i(\mathbf{x}_d) \to n_w(\mathbf{x}_d)$ and $\omega_i \to \omega_d$.

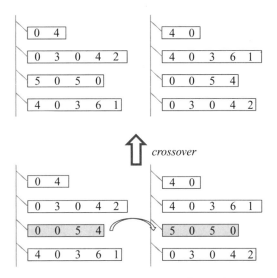

Figure 14.4 Example crossover with demand allocation vectors x_2^1 and x_2^2 (allocation vectors for the demand $d = 2$) undergo interchange.

A *crossover* method based on uniform crossover [16, 17] is used, where genes are equal to demand allocation vectors $\mathbf{x}_d$. Offspring chromosomes are created as a concatenation of demand allocation vectors picked up randomly from both parents

$$\mathbf{x}'_{d^1} = \mathbf{x}^1_d \quad and \quad \mathbf{x}'_{d^2} = \mathbf{x}^2_d \; with \; probability \; 0.5$$

or

$$\mathbf{x}'_{d^2} = \mathbf{x}_{d^1} \quad and \quad \mathbf{x}'_{d^1} = \mathbf{x}_{d^2} \; with \; probability \; 0.5$$

where $\mathbf{x}^1$, $\mathbf{x}^2$, and $\mathbf{x}'^1$, $\mathbf{x}'^2$ denote the parental chromosomes and their offspring with $1 \le \omega(\mathbf{x}'^1)$, $\omega(\mathbf{x}'^2) \le \max(\omega(\mathbf{x}^1), \omega(\mathbf{x}^2))$. Figure 14.4 illustrates an example crossover.

14.3.3 Integer Programming Method

A common approach to the network design problems is to formulate them as IP or mixed-integer programming (MIP) tasks. In the first case, all of the problem variables are integers, whereas in the second case, some of them may be real numbers. Here we treat design problems as linear IP tasks where we assume that both the objective function and the constraints take linear form. An implementation of the IP method provided in the existing optimization packages can be used. In this framework:

Design D0 consists of minimizing (14.7) subject to the constraints

$$\sum_{P \in P(d)} x_{dp} = n(d), \quad for \quad \forall d \in D$$

$$\sum_{d \in D} \sum_{p \in P(d)} a_{dp}(e) \cdot x_{dp} \le m \cdot y(e), \quad \forall e \in E$$

$$x_{dp} \ge 0, \quad for \quad \forall d \in D, \; p \in P \tag{14.14}$$

$$y(e) \ge 0, \quad for \quad \forall e \in E$$

where $a_{dp}(e) = 1$, *if* $e \in p$ & $p \in P(d)$ and 0 otherwise. For complete disaggregation ($\omega = \infty$)), we assume that $x_{dp} \in \{0, 1, \ldots, n(d)\}$ and for full aggregation ($\omega = 1$), we assume that $x_{dp} \in \{0, n(d)\}$.

Design D1 consists of minimizing (14.7) subject to the constraints

$$\sum_{p \in P(d)} x_{dp}^s = n(d), \; for \; \forall \; d \in D$$

$$\sum_{d \in D} \sum_{p \in P(d)} a_{dp}(e) \cdot x_{dp}^s \le m \cdot \alpha(s, e) \cdot y(e), \quad for \quad \forall e \in E, \; s \in S$$

$$x_{dp}^s \ge x_{dp} \cdot \prod_{\{e \in E : a_{dp}(e)=1\}} \alpha(s, e), \quad for \quad \forall d \in D, \; p \in P, \; s \in S \qquad (14.15)$$

$$x_{dp}^s \ge 0, \; for \quad \forall d \in D, \; p \in P, \; s \in S$$

$$y(e) \ge 0, \quad for \quad \forall e \in E$$

For full disaggregation ($\omega = \infty$)), we assume that $x_{dp}^s \in \{0, 1, \ldots, n(d)\}$ and for aggregation ($\omega = 1$), we assume that $x_{dp}^s \in \{0, n(d)\}$. If the dimensioning function $f_e(l)$ is linear and the variables are continuous or integer, then we arrive at the MIP formulation. For the case of all the continuous variables, the problem reduces to LP formulation.

14.3.4 Simulated Annealing

This method maintains a single working solution x^t. One phase of the method consists of performing a random perturbation x'' of the working solution. If this makes an improvement in terms of the evaluation function, the perturbation is accepted as a working solution for the next step $x^{t+1} := x''$. Otherwise x'' is not always rejected; there is still a chance that it will be accepted with probability p_a. The key issue in SA is to control properly the acceptance probability. The most common method is to define $p_a = \exp\left(-\left[f(x'') - f(x^t)\right]/T\right)$ where T is a parameter called *temperature*. In general, the worse the new solution is, the less probable is its acceptance and $f(x^t)$ denotes the evaluation function value of x^t. The value of T is used to balance exploration and exploitation. The general idea is to decrease the temperature in successive steps.

In [18] the temperature is kept constant through $L = L_0 \cdot |D|$ successive steps, where L_0 is a parameter. Then, the new value of T is set according to the following formula

$$T := T(1 + T\ln(1 + \delta)/3\sigma(T))$$

where $\sigma(T)$ is the standard deviation of the objective function values generated at temperature T, and δ is a tuning parameter.

In general for $\omega_w = \infty$, it should be expected that for smaller networks GA yields results that are slightly worse than the exact solution obtained by LP, but if $\omega_w = 1$, it outperforms LP. Moreover, LP might be unable to find an exact solution in reasonable time without exceeding the time limit.

For larger networks when $\omega_w = 1$, GA performance should be outstanding, whereas LP might fail completely for both *D0* and *D1* design. In some cases, LP is unable to reach any integer solution in reasonable computation time. For $\omega_w = \infty$, LP usually yields results only slightly better than the GA with much longer computation times. In some cases, LP might be still unable to converge, and its intermediate result would be in general poorer than that yielded by the GA. In addition to this it is interesting to compare the convergence speed of GA, SA, and a kind of local search (LS) approach (which is, in our case, a stochastic hill-climbing method). An LS algorithm is defined as SA at zero temperature (which implies no likelihood of accepting a worse solution).

GA can outperform other methods. Computation time would be comparable for GA, SA, and LS methods. LS yields its best results in a slightly shorter period of time. For more details see [14].

14.4   Survivable Network Under General Traffic

The conventional approach for designing a survivable network, presented in the previous section, is to start with one static traffic matrix and compute the required working and spare capacities for the links. In this section we call the resulting network planned with this approach the Static Traffic Survivable Network (STSN). Different from the conventional paradigm, General Traffic Survivable Network GTSN considers all possible traffic patterns under the input and output constraints at the nodes. Thus, GTSN requires higher link capacities than the STSN. One key issue that concerns the planning and deployment of GTSN in practice is how GTSN compares with the traditional STSN in terms of the capacity requirement. This is the extra cost for dynamic demand provisioning.

A simple way to understand the GTSN is that it incorporates the nonblocking network property into a survivable network. Nonblocking network was first studied in a classical paper by Clos [20] for electronic circuit switching networks. A network is said to be nonblocking if an egress node with an allowable outbound traffic capacity of τ can be connected to any ingress node with an allowable inbound traffic capacity $\geq \tau$ and vice versa. In this generalized definition, the constraints on the number of intermediate stages become the constraints on the edge (or link) capacities, assuming the traffic can be switched in any arbitrary manner at each network node or interconnection. The illustration of the network is given in Figure 14.5.

In a real network, the capacity constraints at the input and output interfaces of each network node (e.g. the bandwidth of the network interface cards) limit the traffic flowing into or out of a network. More generally, for any real network, the following deterministic constraints must hold:

$$\sum_{s\in N, s\neq d} r_{sd} \leq O_d \;\forall d \in N \quad \text{and} \quad \sum_{d\in N, s\neq d} r_{sd} \leq I_s \;\forall s \in N \qquad (14.16)$$

where I_s is the maximum input capacity of node s and O_d is the maximum output capacity of node d, r_{sd} represents the demand from node s to node d and N is the set of network nodes.

If we plan a network in such a way that there is sufficient capacity for every valid traffic pattern that is consistent with the I/O constraints (14.16), then we can achieve a nonblocking network which guarantees a routing path can always be found for a connection request when resources at both the source and destination are available.

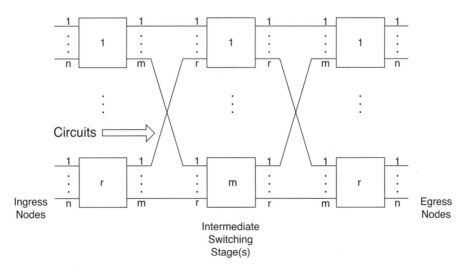

Figure 14.5   Symmetric switching network.

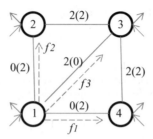

Figure 14.6 An example of four nodes GTSN.

If in addition we take survivability into consideration, then the resulting network can guarantee the full survivability under the dynamic change of traffic. We refer to such network the GTSN where the term 'generalized' means the generalization to the set of all allowable traffic demands under the deterministic constraints of the network – namely, the maximum input or output traffic flow that a node can handle.

The four-node network shown in Figure 14.6 is a GTSN. In the figure, the number on the arrows going *into/out* of each network node represents the node's *I/O* bandwidth capacity. The number *without/with* bracket on each network link represents the *working/spare* capacity of that link. This network can satisfy any possible demand matrix satisfying the *I/O* constraint with single-path routing while it is single-fault tolerant under link restoration and single-path rerouting. Six possible demand matrices that it can satisfy are shown in Figure 14.7. The routing and rerouting information are shown in Table 14.3.

This GTSN networking concept represents design methodology that results in a network that fulfills the requirement of both dynamic bandwidth provisioning and full survivability.

14.4.1   Analytical Model for GTSN

In this section, we present a mathematical framework using the mixed integer linear programming MILP formulations to evaluate the capacity requirement for a GTSN.

We represent a network with an undirected graph $G(N, E)$, where N represents the set of nodes, and E represents the set of edges. Each edge in E has two extremities of nodes $\{i, j\}$, $i \in N$ and $j \in N$. For each edge $\{i, j\} \in E$, c_{ij} represents the cost of per-unit capacity reservation. For simplicity, all

(a)

$$
\begin{array}{c}
\quad 1\ 2\ 3\ 4 \\
\begin{array}{c}1\\2\\3\\4\end{array}
\begin{bmatrix}
0 & 1 & 0 & 0 \\
1 & 0 & 0 & 0 \\
0 & 0 & 0 & 1 \\
0 & 0 & 0 & 0
\end{bmatrix}
\end{array}
$$

(b)

$$
\begin{array}{c}
\quad 1\ 2\ 3\ 4 \\
\begin{bmatrix}
0 & 0 & 1 & 0 \\
1 & 0 & 0 & 0 \\
0 & 0 & 0 & 1 \\
0 & 1 & 0 & 0
\end{bmatrix}
\end{array}
$$

(c)

$$
\begin{array}{c}
\quad 1\ 2\ 3\ 4 \\
\begin{bmatrix}
0 & 0 & 0 & 1 \\
0 & 0 & 1 & 0 \\
0 & 1 & 0 & 0 \\
1 & 0 & 0 & 0
\end{bmatrix}
\end{array}
$$

(d)

$$
\begin{array}{c}
\quad 1\ \ 2\ \ 3\ \ 4 \\
\begin{bmatrix}
0 & 0 & \tfrac{1}{2} & \tfrac{1}{2} \\
0 & 0 & \tfrac{1}{2} & \tfrac{1}{2} \\
\tfrac{1}{2} & \tfrac{1}{2} & 0 & 0 \\
\tfrac{1}{2} & \tfrac{1}{2} & 0 & 0
\end{bmatrix}
\end{array}
$$

(e)

$$
\begin{array}{c}
\quad 1\ \ 2\ \ 3\ \ 4 \\
\begin{bmatrix}
0 & \tfrac{1}{2} & 0 & \tfrac{1}{2} \\
0 & 0 & \tfrac{1}{2} & \tfrac{1}{2} \\
\tfrac{1}{2} & \tfrac{1}{2} & 0 & 0 \\
\tfrac{1}{2} & 0 & \tfrac{1}{2} & 0
\end{bmatrix}
\end{array}
$$

(f)

$$
\begin{array}{c}
\quad 1\ \ 2\ \ 3\ \ 4 \\
\begin{bmatrix}
0 & \tfrac{1}{2} & \tfrac{1}{2} & 0 \\
\tfrac{1}{2} & 0 & 0 & \tfrac{1}{2} \\
\tfrac{1}{2} & 0 & 0 & \tfrac{1}{2} \\
0 & \tfrac{1}{2} & \tfrac{1}{2} & 0
\end{bmatrix}
\end{array}
$$

Figure 14.7 Six traffic matrices supported by the four-node GTSN.

Table 14.3   Routing and Rerouting Information of the GTSN from Figure 14.6

(s,d) pair	Working path	Failed link	Rerouting path
(1,2)	$1 \rightarrow 3 \rightarrow 2$	(1,3)	$1 \leftrightarrow 2 \leftrightarrow 3$
(1,3)	$1 \rightarrow 3$	(2,3)	
(1,4)	$1 \rightarrow 3 \rightarrow 4$	$2 \leftrightarrow 1 \leftrightarrow 4 \leftrightarrow 3$	
(2,1)	$2 \rightarrow 3 \rightarrow 1$	(3,4)	
(2,3)	$2 \rightarrow 3$	$3 \leftrightarrow 2 \leftrightarrow 1 \leftrightarrow 4$	
(2,4)	$2 \rightarrow 3 \rightarrow 4$		
(3,1)	$3 \rightarrow 1$		
(3,2)	$3 \rightarrow 2$		
(3,4)	$3 \rightarrow 4$		
(4,1)	$4 \rightarrow 3 \rightarrow 1$		
(4,2)	$4 \rightarrow 3 \rightarrow 2$		
(4,3)	$4 \rightarrow 3$		

edges are assumed to have the same cost metric $c_{ij} = 1$ and the same length. The capacity of each edge represents the maximum traffic that can flow through the edge from both directions simultaneously. For each node $n \in N$, the input and output capacity bounds I_n and O_n are given. I_n and O_n are assumed to be nonnegative integers.

The traffic demands among all source-destination (sd) pairs are represented by the traffic matrix R in which each element r_{sd} represents a specific traffic demand from node $s \in N$ to node $d \in N$. We assume $r_{nn} = 0, \forall n \in N$. A traffic matrix R is said to be allowable if all its elements simultaneously satisfy I/O constraints (14.16) . The set of allowable traffic matrices R will be denoted by $\mathfrak{R}$. It is obvious that if the value of each element r_{sd} in R is a real number, the cardinality of $\mathfrak{R}$ will be infinite. Even if the value of each element r_{sd} in R is restricted to be an integer, the cardinality of $\mathfrak{R}$ is at least of $O(|N|!)$, which is an extremely large number.

A network is referred to as nonblocking if it can satisfy all the allowable traffic matrices R in $\mathfrak{R}$ with a specific routing model. The definitions of *wide-sense nonblocking* and *rearrangeably nonblocking* in circuit switching are generalized as follows. If there exists a routing algorithm that can meet any allowable traffic demand without having to rearrange any existing connections, the network is said to be wide-sense nonblocking. If there exists a routing algorithm that can meet any allowable traffic demand by rearranging some existing connections, the network is said to be rearrangeably nonblocking. A wide-sense nonblocking network is a rearrangeably nonblocking network but not vice versa. For practical reasons wide-sense nonblocking networks are more interesting because existing network connections need not be torn down, rerouted and reconnected whenever there is a change in the traffic demand. This simplifies the network management and operations and improves the restoration speed when there is a link failure.

In this section, we assume that the routing model is fixed single-path routing, i.e. each traffic demand r_{sd} is routed along one single path which will not change with time no matter what the current traffic matrix R is. A formulation based on fixed single- path routing result in a network with sufficient capacity such that any given working demand can be carried by the fixed single route from a source to a destination. Therefore, the resulting network planned out by the model in this section is a wide-sense nonblocking network and it can meet any allowable traffic demand without having to rearrange any existing connections.

Any link failure is assumed to cut off the bandwidth capacities in both directions. A network is said to be survivable if it has sufficient diversity (multi-edge-connected) and spare capacities for rerouting a certain portion of traffic demands under a set of failure scenarios. The process of rerouting the broken traffic is called restoration. If the restoration portion is 100%, the network is a fully survivable network. For a survivable network, there are two general restoration schemes: *link restoration* and *path restoration*. In link restoration, when an edge fails, only the working flow on this edge will be rerouted with alternative routes that connect the two ends of the failed edge. In path restoration, all traffic demands r_{sd} passing through the faulty edge will be rerouted with alternative routes. The number of alternative routes can be single or multiple. In case of link failures, some route rearrangements are required by the restoration process and these rearrangements are not considered as traffic rearrangements under the nonblocking network definition.

A GTSN has two properties: it can satisfy dynamic traffic demands which vary with time under the deterministic I/O capacity constraints at each node, and it is fully survivable under link failures. A GTSN is α-survivable (α is an integer) if the network is fully protected up to α-link failures anywhere in the network.

In this section, for simplicity, we only consider the capacity planning of 1-survivable GTSN under link restoration with single or multiple rerouting paths. The problem is stated as follows:

Given a network topology that is at least two-edge connected with a set of nodes N and a set of edges E, the I/O constraints I_n, O_n for each node $n \in V$, how to compute the working and spare capacity of each edge $\{i, j\} \in E$ in order to obtain a 1-survivable GTSN with minimum cost? The resulting GTSN should be able to carry any allowable traffic demand with fixed single-path routing while guarantee full survivability under single link failure using link restoration.

In the following we will use the following variables: w_{ij} working capacity on edge $\{i, j\}$, x_{ij}^{sd} portion of flow for the traffic demand r_{sd} routed on edge $\{i, j\}$ from node i to node j, s_{ij} spare capacity on edge $\{i, j\}$,

z_{ij}^{fg} restoration flow on edge $\{i, j\}$ for rerouting the demand volume through edge $\{f, g\}$ (that is equal to w_{fg}) from node i to node j when link $\{f, g\}$ is down.

The capacity planning problem of GTSN can be written down as

$$\text{Min} \sum_{\{i,j\}\in E} (w_{ij} + s_{ij})$$

subject to

a) $\quad \displaystyle\sum_{j:\{i,j\}\in E} (x_{ij}^{sd} - x_{ji}^{sd}) = \begin{cases} 1, & \text{if } i = s \\ -1, & \text{if } i = d \ \forall i \in N \\ 0, & \text{if } i \neq s, d \ \forall s, d \in N, \ s \neq d \end{cases}$

b) $\quad \displaystyle\sum_{s,d\in N, s\neq d} r_{sd}(x_{ij}^{sd} + x_{ji}^{sd}) \leq w_{ij} \forall R \in \Re, \{i, j\} \in E$ $\hspace{2cm}$ (14.17)

c) $\quad \displaystyle\sum_{j:\{i,j\}\in E} (z_{ij}^{fg} - z_{ji}^{fg}) = \begin{cases} \omega_{fg}, & \text{if } i = f \\ -\omega_{fg}, & \text{if } i = g \ \forall i \in N \\ 0, & \text{if } i \neq f, g \ \forall \{f, g\} \in E \end{cases}$

d) $\quad z_{ij}^{fg} + z_{ji}^{fg} \leq s_{ij}, \ \forall\{i, j\} \in E, \{f, g\} \in E$

e) $\quad z_{ij}^{ij} = z_{ji}^{ij} = 0, \ \forall\{i, j\} \in E$

f) $\quad z_{ij}^{fg}, z_{ji}^{fg} \geq 0, \ \forall\{i, j\} \in E, \{f, g\} \in E$

g) $\quad w_{ij}, s_{ij}$ are nonnegative integers $\quad \forall\{i, j\} \in E$

h) $\quad x_{ij}^{sd}, x_{ji}^{sd}$ are 0 or 1 $\forall\{i, j\} \in E, s, d \in N, \quad s \neq d$

The objective function, defined by the first line of (14.17) minimizes the total capacity installed in the network. We assume that once the capacities are installed, there will not be any additional routing cost. The formulation (14.17) assumes single-path routing and multiple-path rerouting in the case of single-link failure using link restoration. For single-path routing and single-path rerouting using link restoration (14.17) becomes

$$\text{Min} \sum_{\{i,j\}\in E} (w_{ij} + s_{ij})$$

subject to

$$\sum_{j:\{i,j\}\in E} (x_{ij}^{sd} - x_{ji}^{sd}) = \begin{cases} 1, & if\ i = s \\ -1, & if\ i = d\ \forall i \in N \\ 0, & if\ i \neq s, d\ \forall s, d \in N,\ s \neq d \end{cases}$$

$$x_{ij}^{sd} + x_{ji}^{sd} \leq \omega_{ij}^s + \varpi_{ij}^d \forall \{i, j\} \in E,\ s,\ d \in N,\ s \neq d$$

$$\sum_{n \in N} (I_n \omega_{ij}^n + O_n \varpi_{ij}^n) \leq \omega_{ij} \forall \{i, j\} \in E$$

$$\sum_{j:\{i,j\}\in E} (z_{ij}^{fg} - z_{ji}^{fg}) = \begin{cases} \omega_{fg}, & if\ i = f \\ -\omega_{fg}, & if\ i = g\ \forall i \in N \\ 0, & if\ i \neq f, g\ \forall \{f, g\} \in E \end{cases} \qquad (14.17a)$$

$$\omega_{fg} + M(z_{ij}^{fg} + z_{ji}^{fg} - 1) \leq s_{ij} \forall \{i, j\} \in E, \{f, g\} \in E$$

$$z_{ij}^{ij} = z_{ji}^{ij} = 0 \forall \{i, j\} \in E$$

$$z_{ij}^{fg}, z_{ji}^{fg}\ are\ 0\ or\ 1\ \forall \{i, j\} \in E, \{f, g\} \in E$$

s_{ij} are nonnegative integers $\forall \{i, j\} \in E$

ω_{ij} are nonnegative integers $\forall \{i, j\} \in E$

x_{ij}^{sd}, x_{ji}^{sd} are 0 or 1, $\forall \{i, j\} \in E, s, d \in N, s \neq d$

$\omega_{ij}^n, \varpi_{ij}^n \geq 0 \forall \{i, j\} \in E, n \in N$

where $M =$ the upper bound of $\omega_{ij} = \min\{\sum_{n \in N} I_n, \sum_{n \in N} O_n\}$. Similar formulations can also be given for path restoration.

The constraints *a)* in (14.17) are the classical flow balance constraints that ensure the routing between each *sd* pair [21]. Constraints *b)* ensure that enough working capacity is installed on each edge $\{i, j\}$ so that the total edge flow is satisfied for each allowable traffic matrix $R \in \mathfrak{R}$. Constraints *c)* are the flow balance constraints that ensure the rerouting of the broken working flow passing through the failed edge $\{f, g\}$ through multiple paths. Constraints *d)* ensure that enough spare capacity is assigned on each edge $\{i, j\}$ so that the total rerouted flow is satisfied for any affected working flow upon any single link failure. Constraints *e)* ensure the rerouted flow variables to be zero on the failed edge. Constraints *f)* restrict the rerouted flow variables to be nonnegative. Constraints *g)* restrict the working and spare capacity to be nonnegative integers. Constraints *h)* restrict the flow variables to be either 0 or 1 so that single-path routing is ensured. Once the capacity allocation for the links is solved, some additional post processing is required to find the routing and rerouting paths but the problem can be solved explicitly in polynomial time [21].

The spare capacity planning problem alone is known to be NP-hard [22]. If we restrict our problem to only one specific traffic demand that satisfies the I/O constraints, the problem reduces to the spare capacity planning problem which is NP-hard. What makes the problem even harder is the exponentially increasing constraints *b)*. The number of constraints in *b)* depends on the total number of allowable traffic matrices. As discussed earlier, the total number of allowable traffic matrices will be at least of order $O(|N|!)$ and can even be infinite and so is the number of constraints in *b)*. Therefore, we cannot construct the whole integer linear program explicitly. In order to find a lower complexity algorithm, we have to simplify the formulation first.

The capacity planning problem in GTSN is large in size because of the at least exponentially increasing ($|N|!$ or more) constraints *b)*. In order to tighten the formulation and obtain a form suitable for optimization, we want to eliminate constraints *b)*. First, if we fix the flow variables with feasible values for all sd pairs in constraints b), it can be seen that the optimal working capacity of link $\{i, j\}$ is just equal to the maximum traffic through that link over all allowable traffic matrices R. The maximum traffic through the link $\{i, j\}$ with fixed flow variables can be calculated from the following linear program denoted as *maximum traffic via link* $\{i, j\}(MT_{ij})$

$$\text{Max} \sum_{s,d \in N, s \neq d} r_{sd}(x_{ij}^{sd} + x_{ji}^{sd})$$

subject to

$$\sum_{s \in N, s \neq d} r_{sd} \leq O_d \ \forall d \in N$$

$$\sum_{d \in N, s \neq d} r_{sd} \leq I_s \ \forall s \in N \tag{14.18}$$

$$r_{sd} \geq 0 \ \forall s, \ d \in N, \ s \neq d$$

The constraints of this linear program are actually the I/O constraints (14.16) and characterize all valid traffic matrices. The objective function maximizes the traffic through $\{i, j\}$.

It may appear that if we replace constraints b) of the normal form with the above linear program, we can represent all allowable traffic matrices implicitly and obtain a compact form. Unfortunately, it is not so straightforward. If we simply eliminate constraints b) and insert the linear program into the normal form, the objective function in (14.18) will become nonlinear and thus the whole formulation becomes a nonlinear program.

In order to avoid the nonlinear constraints and use the linear program above to obtain a compact formulation, we can apply the duality transformation [23] to the linear program (14.18) as follows:

Dual of

$$MT_{ij}(DMT_{ij}) \ : \ \text{Min} \sum_{n \in N}(I_n \omega_{ij}^n + O_n \varpi_{ij}^n) \tag{14.19}$$

subject to

$$x_{ij}^{sd} + x_{ji}^{sd} \leq \omega_{ij}^s + \varpi_{ij}^d \ \forall \{i, j\} \in E \ s, \ d \in N, \ s \neq d$$
$$\omega_{ij}^n, \ \varpi_{ij}^n \geq 0 \ \forall \{i, j\} \in E, \ n \in N.$$

The objective value of MT_{ij} is equal to that of its dual. Since the objective of DMT_{ij} is a linear function, we can obtain a compact formulation simply by putting the linear program of DMT_{ij} into the normal form and eliminating constraints b). This dual form will not contain any nonlinear constraints and is given as follows.

$$\text{Min} \sum_{\{i,j\} \in E}(w_{ij} + s_{ij}) \tag{14.20}$$

subject to

a) $\quad \displaystyle\sum_{j:\{i,j\} \in E}(x_{ij}^{sd} - x_{ji}^{sd}) = \begin{cases} 1, & if \ i = s \\ -1, & if \ i = d \ \forall i \in N \\ 0, & if \ i \neq s, d \ \forall s, d \in N, \ s \neq d \end{cases}$

b) $\quad x_{ij}^{sd} + x_{ji}^{sd} \leq \omega_{ij}^s + \varpi_{ij}^d \forall \{i, j\} \in E, \ s, d \ \in N, \ s \neq d$

c) $\quad \displaystyle\sum_{n \in N}(I_n \omega_{ij}^n + O_n \varpi_{ij}^n) \leq \omega_{ij} \forall \{i, j\} \in E$

d) $\quad \displaystyle\sum_{j:\{i,j\}\in E} (z_{ij}^{fg} - z_{ji}^{fg}) = \begin{cases} \omega_{fg}, & if\ i = f \\ -\omega_{fg}, & if\ i = g\ \forall i \in N \\ 0, & if\ i \neq f, g\ \forall \{f, g\} \in E \end{cases}$

e) $\quad z_{ij}^{fg} + z_{ji}^{fg} \leq s_{ij} \forall \{i, j\} \in E, \{f, g\} \in E$

f) $\quad z_{ij}^{fg}, z_{ji}^{fg} \geq 0 \forall \{i, j\} \in E, \{f, g\} \in E$

g) $\quad s_{ij}$ are nonnegative integers $\quad \forall \{i, j\} \in E$

h) $\quad \omega_{ij}$ are nonnegative integers $\forall \{i, j\} \in E$

i) $\quad x_{ij}^{sd}, x_{ji}^{sd}$ are 0 or 1 $\forall \{i, j\} \in E,\ s,\ d \in N,\ s \neq d$

j) $\quad \omega_{ij}^{n}, \varpi_{ij}^{n} \geq 0 \forall \{i, j\} \in E,\ n \in N$

This compact form of the problem is limited in size and can be constructed explicitly.

By relaxing the constraints on the spare capacity variables s_{ij}, the working capacity variables ω_{ij} and the flow variables x_{ij}^{sd} to any linear value, $s_{ij} \geq 0 \forall \{i, j\} \in E,\ \omega_{ij} \geq 0 \forall \{i, j\} \in E\ 0 \leq x_{ij}^{sd}, x_{ji}^{sd} \leq 1 \forall \{i, j\} \in E, s, d \in N, s \neq d$, the above MILP can be transformed into a linear program which can be solved in polynomial time. The optimal solution of the linear program represents the optimal cost for deploying a GTSN with fractional capacities, multiple-path routing and link restoration with multiple-path rerouting. Therefore, it can serve as a lower bound for the cost of the GTSN and can be used to calibrate the performance of various optimization schemes.

Further simplification is obtained by *the* following simple two-phase sequential approach for solving the problem. The approximation involves decomposing (14.20) into two sub-problems: 1) working capacity planning –the planning of the nonblocking part of GTSN, and 2) spare capacity planning –the planning of the fully survivable part of GTSN. In the first phase, we optimize the working capacity requirement for building a nonblocking network. In the second phase, we optimize the spare capacity requirement to make the nonblocking network fully survivable and form a GTSN. Therefore, (14.20) is decomposed a working capacity part

$$\text{Min} \sum_{\{i,j\}\in E} \omega_{ij} \qquad\qquad (14.21)$$

subject to constraints *a)–c)* and *h)–j)*.

Spare capacity part

$$\text{Min} \sum_{\{i,j\}\in E} s_{ij} \qquad\qquad (14.22)$$

subject to constraints *d)–g)*. w_{ij} for all links $\{i, j\} \in E$ are variables that need to be optimized in (14.21) during the first phase. They will be the input parameters for optimizing the spare capacity in (14.22) during the second phase.

14.5   Stochastic Geometry and Random Graphs Theory

Wireless networks are basically limited by the intensity of the received useful signals and by their interference. Since both of these signals depend on the spatial location of the nodes, mathematical techniques have been developed to provide communication-theoretic results accounting for the network's geometrical configuration. Most of the time, the location of the nodes in the network can be modeled as random, following for example a Poisson point process. In this case, different techniques based on *stochastic geometry* and the *theory of random geometric graphs* –including point process theory, percolation theory, and probabilistic combinatorics, have enabled the results on the connectivity, the capacity, the outage probability, and other fundamental limits of wireless networks. In this section we present some of these techniques, discuss their application to model wireless networks, and present some of the main results as illustration.

In a wireless network with many concurrent transmissions the SINR is the relevant figure of merit for the system. In this case the received signal power is random due to the random spatial distribution of the users and the interference power is governed by a number of stochastic processes such as the random spatial distribution of the nodes, shadowing, and fading. The SINR for a receiver placed at the origin 0 in the 2 or 3-dimensional Euclidean space can be written as $SINR = S/(W + I)$ with $I = \sum_{i \in T} P_i h_i \ell(||x_i||)$, where S, W and I are the desired signal, noise, and interference powers, respectively. The summation for I is taken over the set of all interfering transmitters T, P_i is the transmit power, h_i is a random variable that characterizes the cumulative effect of shadowing and fading, and ℓ is the path loss function, assumed to depend only on the distance $||x_i||$ from the origin of the interferer situated at position x_i in space. Parameter ℓ can be modeled as a power law, $\ell(||x_i||) = k_0 ||x_i||^{-\alpha}$, or in environments where absorption is dominant, as an exponential law, $\ell(||x_i||) = k_0 \exp(-\gamma ||x_i||)$. While in a large system, the unknowns are T, h_i, and x_i, and P_i, it is the locations of the interfering nodes that most influence the SINR levels, and hence, the performance of the network. Wireless network performance is mostly interference-limited, and a large number of users contribute to the interference in vastly varying magnitudes, as described by the interference function $I = \sum_{i \in T} P_i h_i \ell(||x_i||)$, which is a *shot noise process*. In one dimension, a shot noise process is defined as $I(x) = \sum_{i=-\infty}^{\infty} g(x - X_i)$ where $\{X_i\}$ is a stationary Poisson process on $\mathbb{R}$ and $g(x)$ is the impulse response of a linear system. In its two-dimensional generalization, x represents a point on the plane and the Poisson point process is on $\mathbb{R}^2$. When $g(x)$ depends only on the Euclidean norm $||x||$, it can be identified with the path loss function $\ell(||x||)$, and $I(x)$ is the aggregate interference received at point x in a wireless network, without fading.

14.5.1 Stochastic Geometry

Stochastic geometry is a branch of applied probability which allows the study of random phenomena on the plane or in higher dimensions. It is intrinsically related to the theory of point processes which are the most basic objects studied in stochastic geometry. Visually, a point process can be depicted as a random collection of points in space. More formally, a point process (PP) is a measurable mapping Φ from some probability space to the space of point measures (a point measure is a measure which is locally finite and which takes only integer values) on some space E. Each such measure can be represented as a discrete sum of Dirac measures on E or $\Phi = \sum_i \delta_{X_i}$.

The random variables $\{X_i\}$, which take their values in E are the points of Φ. Most often, the space E is the Euclidean space $\mathbb{R}^d$ of dimension $d \geq 1$. The *intensity measure* Λ of Φ is defined as $\Lambda(B) = E\Phi(B)$ for Borel B, where $\Phi(B)$ denotes the number of points in $\Phi \cap B$. In mathematics, a Borel set is any set in a topological space that can be formed from open sets (or, equivalently, from closed sets) through the operations of countable union, countable intersection, and relative complement.

Poisson point processes: If Λ is a locally finite measure on some metric space E a point processes Φ is Poisson on E if a) For all disjoint subsets $A_1, \cdots A_n$ of E, the random variables $\Phi(A_i)$ are independent; b) For all sets A of E, the random variables $\Phi(A)$ are Poisson.

Conditionally on the fact that $\Phi(A) = n$, these n points are independently (for homogeneous PPP also uniformly) located in A. Based on this if $E = R^d$, the Laplace functional is defined for general point processes Φ by

$$L_\Phi(f) = E[e^{-\int_{R^d} f(x)\Phi(dx)}] = E[e^{-\Sigma_{X \in \Phi} f(X)}] \qquad (14.23)$$

where f is a non-negative function on R^d. In the Poisson case,

$$L_\Phi(f) = \exp(-\int_{R^d} (1 - e^{-f(x)})\Lambda(dx)) \qquad (14.24)$$

The appealing features of PPPs are their invariance to a large number of key operations. For example: (a) The superposition of two or more independent PPPs (which is defined as the sum of the associated point measures) is again a PPP; this can be extended to denumerable sums under some conditions; (b) The independent thinning of a PPP is again a PPP; this can be extended to the case of location-dependent thinning, where a point is retained or not depending on its location; (c) The point process obtained by displacing point X_i independently of everything else according to some Markov kernel $K(X_i, \cdot)$ that defines the distribution of the displaced position of the point X_i yields another PPP; this result is often referred to as the displacement theorem. In each case, the intensity of the resulting PPP can be obtained in closed form from that of the initial PPP and the involved transformations (e.g., the thinning probability or the kernel K).

If $\rho(x, \cdot)$ is the probability density pertaining to the Markov kernel applied to a PPP of intensity $\lambda(x)$ on R^d the displaced points form a PPP of intensity

$$\lambda'(y) = \int_{R^d} \lambda(x)\rho(x, y)\, dx \tag{14.25}$$

In the case of $\lambda(x){=}\lambda$ and $\rho(x, y)$ is a function of $y - X$ only, then $\lambda'(y) = \lambda$ for all y.

Slivnyak's theorem for PPPs states that the law of $\Phi - \delta_x$ conditional on the fact that Φ has a point at x is the same as the law of Φ. In other words, the reduced Palm probability $P^{x!}$ of a PPP is the distribution of this Poisson point process itself. This is usually expressed as $P^{x!} = P$, for all points $x \in \Phi$. This means that the properties seen from a point $x \in R^d$ are the same whether we condition on having a point $x \in \Phi$ or not, if the point at x is not considered. Based on this, we have for the mean number of points within distance r of x:

$$E\Phi(B(x;r)\backslash\{x\}) = E(\Phi(B(x;r)\backslash\{x\})|x \in \Phi) = \Lambda(B(x;r)) \tag{14.26}$$

where $B(x;r)$ is the ball of radius r centered at x.

Stationary point processes: The theory of stationary point processes SPP is based on the concept of marks and Palm probability [24–28]. A *mark* of some point of a SPP is a quantity that remains unchanged when the collection of points is transported by a global translation operation. As an example, the local configuration of neighbors of point X, which is defined as the collection of points in a ball of radius R centered at X, is a mark of this point. If R is infinity, this mark is the universal mark of X, *the point process seen from X*.

The Palm probability P^o of a stationary point process is the law of this universal mark, which can be shown to be the same for all points. It can be understood as the law of the point process given that it has a point at the origin. As defined here P^o is a probability on the space of point measures.

If $\Phi - X$ is the global translation of all points of Φ by the vector X, then

$$E[\sum_{X \in \Phi} f(X, \Phi - X)] = \int_N \int_{R^d} f(x, \phi)\lambda dx\, P^o(d\phi) \tag{14.27}$$

where N is the space of simple point sequences, for all positive functions $f(x, \phi)$ of $x \in R^d$ parameter ϕ a point measure on R^d and λ is the intensity of Φ, that is the mean number of points per unit space. The last formula is Campbell's formula used for computing the mean values of sums on the points of a stationary point process. If Φ is a stationary PPP of intensity λ on R^d then for non-negative f,

$$E(\sum_{X \in \Phi} f(X)) = \lambda \int_{R^d} f(x)dx$$

$$var\left(\sum_{X \in \Phi} f(X)\right) = \lambda \int_{R^d} f^2(x)dx \tag{14.28}$$

These expressions can be used, for example, to calculate the mean and variance of the interference in a network or to determine the mean node degree.

Boolean models and random geometric graphs: The basic model of stochastic geometry and the continuum percolation is the Boolean or *germ-grain model*. The model is based on a Poisson point process Φ, with points $\{X_i\}$, which are also called *germs*, and on an independent sequence of *iid* compact sets $\{K_i\}$ called the *grains*. The Boolean model is defined formally as

$$\Theta = \bigcup_i (X_i + K_i); \; X_i + K_i = \{X_i + y, y \in K_i\} \tag{14.29}$$

The Boolean model is analytically tractable and is used in the analysis of some key parameters in wireless networks. An example is coverage as the distribution of the number of grains that intersect a given compact set, for instance a given location of the space. This distribution is Poisson.

Gilbert's random disk graph: This is a model for wireless networks that is a special case of the Boolean model described above. Assume that the compact sets described in the last subsection are all balls of radius $r/2$. We define the random disk graph of a Poisson point Φ process of intensity λ with range r, denoted as $G_{\lambda, r}$ as the graph with nodes the points of Φ and with edges between X and Y if the two grains touch, i.e., if $||X - Y|| \le r$. This is the basic random geometric graph and a central object in random graph theory where the following questions are of particular interest: (a) Does this random graph has an infinite component? This is equivalent to Θ having an infinite component. This property is referred to as *percolation*. An important result is that there is a deterministic critical value $0 < r_c < \infty$ such that when $r < r_c$, there is no such infinite component with probability 1 (there is no percolation for all realizations of Φ), whereas when $r > r_c$, there is an infinite component with probability 1 (there is percolation for all realizations of Φ). (b) In case percolation occurs, what is the fraction of nodes included in the infinite component? In case it does not occur, what is the typical size of a component? [29, 30].

Voronoi tessellation: A tessellation is a collection of open, pairwise disjoint polyhedra (polygons in the case of R^2) whose closures cover the space, and which is locally finite (i.e., the number of polyhedra intersecting any given compact set is finite). Given a simple point measure (or point sequence) ϕ on R^d and a point $x \in R^d$, the *Voronoi cell* $C_x(\phi)$ of the point $x \in R^d$ w.r.t. ϕ to is defined as the set

$$C_x(\phi) = \{y \in R^d : ||y - x|| < \inf_{x_i \in \phi, x_i \ne x} ||y - x_i||\}) \tag{14.30}$$

For a simple point process $\Phi = \sum_i \delta_{X_i}$ on R^d the *Voronoi tessellation* or *mosaic* generated by Φ is defined as the marked point process $V = \sum_i \delta_{(X_i, C_{X_i}(\Phi) - X_i)}$, whose marks are the Voronoi cells shifted to the origin.

The Delaunay triangulation generated by a simple point measure ϕ is a graph with the set of vertices ϕ and edges connecting each $y \in \phi$ to any of its Voronoi neighbors. The Delaunay triangulation of a Poisson point process is an object of central importance in communications. In a regular periodic (say hexagonal or triangular) grid it is obvious how to define the neighbors of a given vertex. However, for irregular patterns of points like a realization of a PPP, which is often used to model set of nodes in mobile ad hoc networks, this notion is much less evident. The Delaunay triangulation offers some purely geometric definition of neighborhood in such patterns. For more on Poisson-Voronoi tessellations (mean cell size, mean number of sides of the cell, etc.) and Poisson-Delaunay graphs (mean degree, mean length of a typical edge, mean size of a typical triangle) see [31–33].

14.5.2   Link Outage Probability

In this section we apply some of the techniques introduced above to study the interference in large ad hoc networks and the outage probability of any given link.

We discuss first the total interference power measured at a point x in the network, given by $I(x) = \sum_{Y \in \Phi_t} \ell(||x - Y||)$ where Φ_t is a point process of transmitters (assumed to be interferers) on R^2. Φ_t is typically a subset of a larger point process Φ since it constitutes the nodes selected by the MAC scheme to transmit concurrently. For example, if nodes in a homogeneous PPP of intensity 1 transmit independently and randomly with probability p (slotted ALOHA), Φ_t is a PPP with intensity p. The Laplace transform of the interference is derived as follows. Let $\Phi - \{R_i = ||X_i||\}$ be the distances of the points of a d-dimensional uniform PPP of intensity μ from an arbitrary origin 0. Then Φ is an inhomogeneous PPP with intensity function $\lambda(r) = \mu c_d d r^{d-1}$, where $c_d = |B(0, 1)|$ is the volume of the d-dimensional unit ball. Considering the interference as a shot noise process $I(x) = \sum_{i=-\infty}^{\infty} g(x - X_i)$, and also accounting for the fading terms, we can identify $h_r \ell(r) = h_r r^{-\alpha}$ for iid fading h with the impulse response of the shot noise process. The Laplace transform of the interference is $L_I(s) = E[e^{-sI}] = E[\prod_{R \in \Phi} \exp(-sh_R R^{-\alpha})]$. This is a Laplace functional with $f(r) = s\ell(r) = sh_r r^{-\alpha}$. The expectation is over both Φ and h, but since the fading is assumed independent of the point process, the expectation over h can be moved inside the product, so we have from (14.24)

$$L_I(s) = \exp\{-\int_0^\infty E_h[1 - e^{-shr^{-\alpha}}]\lambda(r)dr\} = \exp(-\mu c_d E[h^\delta]\Gamma(1 - \delta)s^\delta) \qquad (14.31)$$

where $\delta = d/\alpha < 1$. For Rayleigh fading, $E[h^\delta] = \Gamma(1 + \delta)$, and $L_I(s) = \exp(-\mu c_d s^\delta \pi \delta / \sin(\pi \delta))$ and the interference has a *stable distribution* with characteristic exponent δ and dispersion $\mu c_d E[h^\delta]\Gamma(1 - \delta)$. Since $\delta < 1$, I does not have any finite moments.

A closed-form expression for the interference distribution only exists for $\delta = 1/2$; this is the inverse Gaussian or Lévy distribution [34].

Using the distribution of the distances to the n-th nearest neighbor [35], the distributions of the interference (without fading) from the n-th nearest neighbor are easily found. The tail probabilities do not depend on the presence or type of fading and are given by $P(I_n > x) \sim (1/n!)(\lambda c_d)^n x^{-n\delta}$, $x \to \infty$. This means that moments $E(I_n^p)$ exists for $p < n\delta$. For example, if interference-canceling techniques are used and the interference from the k nearest interferers can be cancelled, we need $k > \alpha$ in two-dimensional networks to have a finite second moment.

If the non-singular path loss model $\ell(x) = (1 + ||x||)^{-\alpha}$ is used, the tail probability of the interference reflects the tail probability of the fading process: If the fading has an exponential or power-law tail, so does the interference. This holds for general motion-invariant processes [36]. Other approaches to the singularity issue are given in [37, 38].

An outage of a wireless link occurs when a packet transmission fails. In many situations, it is justified to equate the outage event to the event that $SINR < T$ for some threshold T that depends on the physical layer parameters such as rate of transmission, modulation, and coding. The success probability is defined as $p_s = P(\text{SINR} > T)$. In the case where the desired signal S is subject to Rayleigh fading, the probability over a link of distance R is

$$p_s = Pr(S > T(W + I)) = \exp(-TR^\alpha W/P)E(e^{-TR^\alpha I}) \qquad (14.32)$$

where P is the transmit power. The first term depends only on the noise (or SNR), while the second depends only on the interference (or SIR) which is the Laplace transform of the interference evaluated at $s = TR^\alpha$. So, in a d-dim. interference-limited network whose nodes are distributed as a uniform PPP of intensity λ with ALOHA channel access with probability p, the success probability is given by (15.31), replacing s by TR^α i.e $p_s = \exp(-\lambda p c_d R^d E[h^\delta]\Gamma(1 - \delta)T^\delta)$. The equivalence of Laplace transforms and success probabilities with some generalizations has been pointed out in [39–41].

Throughput: The transmission success probability in the previous subsection is derived assuming that the desired transmitter transmits while the received listens. To optimize the network parameters, such as the ALOHA transmit permission probability p, the unconditioned success probabilities must be considered. In the case of ALOHA and half-duplex transceivers, the *spatial throughput* is modified as $p(1 - p)p_s(p)$. Finding the optimum p means finding the optimum trade-off between spatial reuse (a larger p results in a higher density of concurrent transmissions) and success probabilities (a larger p results in higher interference and thus a lower success probability). In [41], a related metric, the *spatial density of success* pp_s is optimized in function of p. In some cases, the optimal value, which is known in closed form, does not depend on the intensity λ of the underlying point process.

The same framework can also be used to find the optimum value for the SINR threshold T that maximizes the *area spectral efficiency*. A larger T permits higher transmission rates (or spectral efficiencies if normalized by the bandwidth) but results in lower success probabilities. Similarly, the *probabilistic progress*, usually defined as the product of distance times success probability can be maximized by finding the optimum link distance R [42]. Choosing a larger R may increase the progress but comes with the disadvantage of a lower success probability [43]. In [41], an expression for progress based on extremal shot noise is derived, and in [44], the distribution and the mean value of the throughput of a typical user (as given by Shannon's formula) are given using Fourier transform techniques.

14.5.3   Percolation and Connectivity

Percolation models are used to model the connectivity of wireless multi-hop networks. The main property of percolation models is that they exhibit a *phase transition* in their connectivity behavior. Depending on some (continuous) parameters, the components of the model are either all in finite (*sub-critical* case) or one giant component forms (*super-critical* case). In the context of networking, such a transition affects a great deal the performance of the system. Without a giant component, the network would be completely fragmented and unusable. It is therefore of prime importance to characterize the conditions under which the network is super-critical. In this section, we cover some of the basics of percolation theory, starting with the simplest models and proof techniques, and then moving on to models that are more appropriate for wireless networks. Percolation theory deals mostly with models of infinite size. We start with these classical models and address finite networks briefly.

The bond percolation model is defined as follows: consider the infinite square lattice Z^2 and connect each pair of nearest neighbors independently with probability p. Then define the component (or *cluster*) of the origin C as the set of elements of Z^2 that are connected to the origin by a sequence of adjacent edges. We define the *percolation probability* as $\theta(p) := P(|C| = \infty)$.

It was shown that *there exists a number* $0 < p_c < 1$ *such that* $\theta(p) = 0$ *for* $p < p_c$ *and* $\theta(p) > 0$ *for* $p > p_c$ [45].

In the particular case of bond percolation on Z^2, the exact value of p_c is known to be $1/2$ [46]. Other specific cases and properties of $\theta(p)$ can be found in [30]. To prove the above property, we need first to observe that $\theta(p)$ is an increasing function. It is then enough to prove that there exists some probability $p_1 > 0$ such that $\theta(p_1) = 0$ and some probability $p_2 < 1$ such that $\theta(p_2) > 0$. In the sequel we will find probabilities p_1 and p_2 by using simple arguments.

We start from the observation that if the origin belongs to an infinite cluster, then for any integer n, one can find in the lattice a self-avoiding path of length n starting at the origin. Thus we have

$$\theta(p) \leq P(\exists \text{ a path of length } n \text{ starting at o}) \forall n \tag{14.33}$$

If all direct neighbors of Z^2 were connected by an edge, the number $\kappa(n)$ of such path would be bounded from above by $4 \cdot 3^{n-1}$. Since edges are present with probability p, each of these κ paths exists with probability p^n. Using the union bound, we find that

$$P(\exists \text{ a path of length } n \text{ starting at o}) \leq 4p(3p)^{n-1}$$

If $p < 1/3$, this quantity tends to zero when n increases, so that (14.32) implies $\theta(p) = 0$. So we have established that $p_c \geq 1/3$.

To prove the existence of an infinite cluster for sufficiently large p consider the dual lattice of Z^2, which consists of vertices that are shifted by half a unit in both directions. An edge is placed between two direct neighbors of the dual lattice if it does not intersect an edge of the direct lattice.

The proof is based on the fact that if a component is finite in the original lattice, it is necessarily surrounded by a circuit in the dual lattice. Therefore it is enough to show that the probability that a circuit surrounds the origin in the dual lattice is less than one to prove that a vertex (e.g., the origin) belongs to an infinite cluster with positive probability. Let us estimate the number $\sigma(n)$ of possible circuits of length $2n$ that surround the origin: it is easy to see that it is bounded by $\sigma(n) \leq (n - 1) \cdot 3^{2(n-1)}$. Therefore, the probability that there exists a circuit around the origin with all edges closed upper bounded by

$$P(closed\ circuit) \leq \sum_{n=2}^{\infty} (1 - p)^{2n} \sigma(n) = 9(1 - p)^4 / [1 - 9(1 - p)^2]^2$$

One can verify that when $p > 1 - 1/(2\sqrt{3}) \approx 0.71$, the sum above converges to a number smaller than one. As a consequence, the origin belongs to an infinite cluster with positive probability. Since the existence of an infinite cluster does not depend on the state of a finite number of edges, we can use Kolmogorov's zero-one law to conclude that the probability that such a cluster exists is either zero or one [47]. If it was zero, then the origin would belong to an infinite cluster also with probability zero. Therefore, whenever $\theta(p) > 0$, an infinite cluster exists with probability one.

Continuum percolation: The random geometric graph: The basic random geometric graph or disk graph $G_{\lambda,r}$, relies on two assumptions (a) the nodes' location follows a two-dimensional Poisson point process; (b) each node can communicate directly to any other node within a radius r around it. The latter assumption comes from the following model: assume that nodes emit with a certain power P and that this signal is attenuated over distance according to a deterministic decreasing function $\ell(d)$. If receivers can receive data if the signal is at least β times stronger than the ambient noise of power W, then the transmission radius is defined by $r = max\{d : P\ell(d)/W \geq \beta\}$.

Similarly to the discrete model, we denote by $\theta(\lambda, r)$ the probability that a node located at the origin belongs to an infinite cluster. Due to its simplicity, the graph $G_{\lambda,r}$ can be rescaled while keeping its connectivity properties. If all distances are divided by γ, the underlying PPP is transformed into another Poisson process with intensity $\gamma^2 \lambda$. Thus, the graph $G_{\gamma^2\lambda,r/\gamma}$ has the same connectivity properties as $G_{bda,r}$ and we have $\theta(\gamma^2\lambda, r/\gamma) = \theta(\lambda, r)$. As in the bond percolation model, one can show that a phase transition occurs in $G_{\lambda,r}$.

Nearest-neighbors networks: In this model, each node connects to its k nearest neighbors. This model is for example suitable for a dense wireless network where nodes use power control algorithm in order to be connected only to their k first neighbors.

An important property of this model is that the value of λ does not affect the connectivity of the model (it is called a *scale-free* model) and its only relevant parameter is k. One can show that there exists a critical value of k for which a giant component forms, and below which only finite clusters are observed. In two dimensions, the critical value is $k = 3$ [48].

Random connection model: This model is another generalization of Gilbert's model. For each pair of nodes, we consider the distance x between them. Then an edge between them is added with probability $g(x)$ where g is a function from R^+ to $[0, 1]$ such that $\int_0^{\infty} xg(x)dx < \infty$.

This model takes some randomness of the wireless channel into account so that the nodes connect to each other probabilistically, depending on their distances. The probability of failed connection can for example model a shadow fading effect. Gilbert's original model can be retrieved by setting $g(x) = u(r - x)$. A similar phase transition as in Gilbert's model can be observed at some critical density of nodes λ [49, 50]. One relevant question is how this critical density changes with the shape of the connection function $g(x)$. The work in [51]–[55] has shown that under some spreading transformations on g, the critical density cannot increase. The spread-out limiting case has also been worked out in [53], showing that in the case of very long, unreliable connections, the critical density has a limit corresponding to that of an independent branching process, i.e. the average degree of the corresponding random graph at the percolation threshold tends to 1. Alternative proofs of these spreading results also appear in [54, 55].

Signal-to-interference ratio graph (STIRG) model: The connectivity criterion $r = \max\{d : P\ell(d)/W \geq \beta\}$ compares the received signal to the ambient noise only. However, if several nodes are using the same channel, interference degrades the received signals. In the so-called *STIRG* model [48], the SNR threshold is replaced by an SINR threshold as in (14.32) so that the nodes X_i, $X_j \in \Phi$ are connected by an edge if

$$\frac{P\ell(\|X_i - X_j\|)}{W + \gamma(\max\{I(X_i); I(X_j)\} - P\ell(\|X_i - X_j\|))} \geq T$$

where $I(X_i) = \sum_{X_k \in \Phi \setminus \{X_x\}} P\ell(\|X_i - X_k\|)$. This condition ensures that the two nodes have a sufficiently high SINR to exchange data in both directions despite the interference of all the other nodes. The factor $\gamma \leq 1$ serves as a weight for the interference term and models the gain of the spread spectrum scheme (if any).

Connectivity in finite networks: In finite networks, there is no infinite cluster, and therefore no percolation in the sense as defined at the beginning of this section. However, if one considers a sufficiently large network, one expects to observe a similar phenomenon: if the density of nodes is large enough, a component that contains a large fraction of the nodes should emerge. This intuition in the case of Gilbert's disk graph was confirmed in [56] by the following result:

Consider the restriction of a Boolean model to a square of size $\sqrt{n} \times \sqrt{n}$, whose connectivity graph is denoted by $G_{\lambda,r}(n)$. Denote by $C_{\lambda,r}(\eta, n)$ the event that there exists in $G_{\lambda,r}(n)$ a component that contains at least ηn vertices. Then we have

$$\lim_{n \to \infty} P(C_{\lambda,r}(\eta, n)) = \begin{cases} 1 & \text{if } \eta < \theta(\lambda, r) \\ 0 & \text{if } \eta > \theta(\lambda, r) \end{cases}$$

As a consequence of the above result, since $\theta(\lambda, r)$ is never equal to one, there is always a non-vanishing fraction of disconnected nodes in the network. However, if one lets the connectivity range r increase with n, the fraction of connected nodes can be made to converge to one. If the connectivity range of the nodes is a function $r(n)$ of the number of nodes, the condition for *asymptotic connectivity* (i.e. the condition when the probability that all nodes are connected tends to one when the n increases) is given by

$$r(n) = \sqrt{(\log n)/n} + c(n)$$

where $c(n)$ is any function such that $\lim_{n \to \infty} c(n) = \infty$. This result can be deduced from [70] and has been published in its explicit form in [71]. A similar condition on the rate at which $p \to 1$ to observe full connectivity in a bond percolation model on an $n \times n$ grid has been derived in [72].

Nearest-neighbor model: For the nearest-neighbors model a similar result on asymptotic connectivity has been derived in [59]. The rate at which the number of neighbors k must increase with n is in the range $0.3043 \log n \leq k(n) \leq 0.5139 \log n$.

References

[1] Benders, J. F. (1962) Partitioning procedures for solving mixed variables programming problems, *Numerische Mathematik*, **4**, 238–252.

[2] Kerivin, H., Nace, D. and Pham, T. (2005) Design of capacitated survivable networks with a single facility, *IEEE/ACM Transactions on Networking*, **13**(2), 248–261.

[3] Xing, F. and Wang, W. (2009) On the survivability of wireless ad hoc networks with node misbehaviors and failures, *IEEE, Transactions on Dependable and Secure Computing*, **PP**(99) , 1–15.

[4] Penrose, M. (2003) *Random Geometric Graphs*. Oxford University Press.

[5] Glisic, S. and Lorenzo, B. (2009) *Advanced Wireless Networks, Cognitive, Cooperative and Opportunistic 4G Technology*, 2e, John Wiley and Sons, London.

[6] Buttyan, L. and Hubaux, J.-P. (2003) Stimulating Cooperation in Self-Organizing Mobile Ad Hoc Networks, *Mobile Networks and Applications*, **8**(5), 579–592.

[7] Aad, I., Hubaux, J.-P. and Knightly, E. W. (2004) Denial of service resilience in ad hoc networks, in *Proc. of ACM MobiCom*.

[8] Ellison, R., Fisher, D., Linger, R., Lipson, H. Longstaff, T. and Mead, N. (1997) Survival network systems: an emerging discipline, *SEI, CMU, Tech. Rep. CMU/SEI-97-TR-013*, [Online]. Available at: //www.ceM.org/research/97tr013.pdf.

[9] Telecom Glossary 2000, Institute for Telecommunication Services, NTIA, *DOC, Tech. Rep. ANS T1.523-2001*, Feb. 2001. [Online]. Available at: www.atis.org/tg2k/.

[10] Knight, J. C. and Sullivan, K. J. (2000) On the definition of survivability, Dept. of Computer Science, University of Virginia, Techni*cal Report CS-TR-33-00*, Jan.

[11] Snow, A. P., Varshney, U. and Malloy, A. D. (2000) Reliability and survivability of wireless and mobile networks, *IEEE Computer Magazine*, **33**(7), 449–454.

[12] Chen, D. Garg, S. and Trivedi, K. S. (2002) Network survivability performance evaluation: a quantitative approach with applications in wireless ad-hoc networks, in *Proc. of the ACM International Workshop on Modeling, Analysis, and Simulation of Wireless and Mobile Systems (MSWiM')*, 61–68.

[13] Mannie, E. and Papadimitriou, D. (eds), Recovery (protection and restoration) terminology for generalized multi-protocol label switching (GMPLS), *IETF, RFC 4427*, [Online]. Available at: //www.ietf.org/rfc/rfc4427 .txt

[14] Arabas, J. and Kozdrowski, S. (2001) Applying an evolutionary algorithm to telecommunication network design, *IEEE Transactions on Evolutoinary Computation*, **5**(4), 309–322.

[15] Michalewicz, Z. (1996) *Genetic Algorithms +Data Structures = Evolution Programs*, 3rd ed. Berlm, Germany: Springer-Verlag.

[16] Reed, J., Toombs, R. and Barricelli, N. A. (1967) Simulation biological evolution and machine learning, *J Theor Bio.*, **17**, 319–94.

[17] Syswerda, G. (1989) Uniform crossover in genetic algorithms, In *Proceedings of the International Conference on Genetic Algorithms*, J. D. Schaffer, ed. San Mateo: Morgan Kaufmann, 2–9.

[18] Dekkers, A. and Aarts, E. (1991) Global optimization and simulated annealing, *Mathematical Programming*, **50**, 367–393.

[19] Ho, K. S. and Cheung, K. W. (2007) Generalized survivable network, *IEEE/ACM Transactions on Networking*, **15**(4), 750–760.

[20] Clos, C. (1953) A study of nonblocking networks, *Bell Syst. Tech. J.*, **32**, 406–424.

[21] Ahuja, R. K., Magnanti, T. L. and Orlin, J. B. (1993) *Network Flows: Theory, Algorithms Applications*. Upper Saddle River, NJ: Prentice-Hall.

[22] Liu, Y., Tipper, D. and Siripongwutikom, P. (2005) Approximating optimal spare capacity allocation by successive survivable routing, *IEEE/ACM Trans. Netw.*, **13**(1), 198–211.

[23] Ho, K. S. and Cheung, K. W. (2005) Capacity planning of link restorable optical networks under dynamic change of traffic, in Proc. APOC', Shanghai, China, Nov. 6-10, vol. 6022, 6022–38.

[24] Stoyan, D., Kendall, W. and Mecke, J. (1996) *Stochastic Geometry and Its Applications*, 2nd ed. John Wiley and Sons.

[25] Penrose, M. (2003) *Random Geometric Graphs. Oxford Studies m Probability*, Oxford University Press.

[26] Franceschetti, M. and Meester, R. (2007) *Random Networks for Communication: from Statistical Physics to Information Systems*, Cambridge University Press.

[27] Daley, D. and Jones, D. V. (1988) *An Introduction to the Theory of Point Processes*. New York: Springer.

[28] Baccelli, F. and Bremaud, P. (2003) *Elements of Queueing Theory. Palm Martingale Calculus and Stochastic Recurrences*. Springer.

[29] Meester, R. and Roy, R. (1996) *Continuum percolation*. Cambridge University Press.

[30] Grimmett, G. (1999) *Percolation*. Springer.

[31] Moller, J. *Lectures on Random Voronoi Tessellations*, ser. Lecture Notes in Statistics. New York: Springer-Verlag, vol. 87.

[32] Baccelli, F., Klein, M., Lebourges, M. and Zuyev, S. (1997) Stochastic geometry and architecture of communication networks, *J. Telecommunication Systems*, **7**,–209–227.

[33] Baccelli, F. and Zuyev, S. (1999) Poisson-Voronoi spanning trees with applications to the optimization of communication networks, *Operations Research*, **47**(4), 619–631.

[34] Sousa, E. S. and Silvester, J. A. (1990) Optimum transmission ranges in a direct-sequence spread-spectrum multihop packet radio network, *IEEE J. Sel. Areas Commun.*, **8**(5), 762–771.

[35] Haenggi (2005) On distances in uniformly random networks, *IEEE Trans. on Information Theory*, **51**(10), 3584–3586.

[36] Ganti, R. K. and Haenggi, M. (2007) Interference in ad hoc networks with general motion-invariant node distributions, in *2008 IEEE International Symposium on Information Theory (ISIT'08)*, Toronto, Canada, Jul.

[37] Inaltekin, H., Chiang, M., Poor, H. V. and Wicker, S. (2009) The behavior of unbounded path-loss models and the effect of singularity on computed network characteristics, *IEEE Journal on Sel. Areas in Communications*, **27**(7), 1078–1092.

[38] Gowaikar, R. and Hassibi, B. (2009) Achievable throughput in two-scale wireless networks, *IEEE Journal on Sel. Areas in Communications*, **27**(7), 1169–1179.

[39] Linnartz, J.P. M. G. (1992) Exact analysis of the outage probability in multiple-user radio, *IEEE Trans. Commun.*, **40**(1), 20–23.

[40] Zorzi, M. and Pupolin, S. (1995) Optimum transmission ranges in multihop packet radio networks in the presence of fading, *IEEE Trans.Commun.*, **43**(7), 2201–2205.

[41] Baccelli, F., Blaszczyszyn, B. and Muhlethaler, P. (2006) An ALOHA protocol for multihop mobile wireless networks, *IEEE Transactions on Information Theory*, **52**(2), 421–436.

[42] Haenggi, M. (2009) Outage, Local throughput, and capacity of random wireless networks, *IEEE Transactions on Wireless Communications*, **8**(8) 4350–4359.

[43] Hunter, A., Andrews, J. G. and Weber, S. (2008) The transmission capacity of ad hoc networks with spatial diversity, *IEEE Trans. on Wireless Communications*, **7**(12), Part 1, 5058–5071.

[44] Baccelli, F., Błaszczyszyn, B. and Muhlethaler, P. (2009) Stochastic analysis of spatial and opportunistic aloha, *IEEE Journal on Selected Areas in Communications*, **27**(7), 1105–1119.

[45] Broadbent, S. R. and Hammersley, J. M. (1957) Percolation processes in crystals and mazes, in *Proceedings of the Cambridge Philosophical Society*, 629–641.

[46] Seymour, P. D. and Welsh, D. J. A. (1978) Percolation probabilities on the square lattice, *Ann. Discrete Math.*, **3**, 227–245.

[47] Grimmett, G. R. and Stirzaker, D. (1992) *Probability and Random Processes*. Oxford University Press.

[48] Haggstrom, O. and Meester, R. (1996) Nearest neighbor and hard sphere models in continuum perolcation, *Random Structures and Algorithms*, **9**(3), 295–315.

[49] Dousse, O., Franceschetti, M., Macris, N., Meester, R. and Thiran, P. (2006) Percolation in the signal to interference ratio graph, *Journal of Applied Probability*, **43**(2), 552–562.

[50] Penrose, M. (1991) On a continuum percolation model, *Advances in Applied Probability*, **23**(3), 536–556.

[51] Balister, P., BBollobas, B. and Walters, M. (2005) Continuum percolation with steps in the square of the disc, *Random Structures and Algorithms*, **26**(4), 392–403.

[52] Franceschetti, M., Booth, L., Cook, Meester, M. and Bruck, J. (2005) Continuum percolation with unreliable and spread out connections, *Journal of Statistical Physics*, **118**(3-4), 721–734.

[53] Penrose, M. (1993) On the spread-out limit for bond and continuum percolation. *Annals of Applied Probability*, **3**(1), 253–276.

[54] Bollobas, B., Janson, S. and Riordan, O. (2007) Spread-out percolation in R^d *Random Structures and Algorithms*, **31**(2), 239 –246.

[55] Balister, P., BBollobas, B. and Walters, M. (2004) Continuum percolation with steps in an annulus, *Annals of Applied Probability*, **14**, 1869–1879.

[56] Penrose, M. and Pisztora, A. (1996) Large deviations for discrete and continuous percolation, *Advances in Applied Probability*, **28**, 29–52.

[57] Gupta, P. and Kumar, P. R. (1998) Critical power for asymptotic connectivity in wireless networks, *Stochastic Analysis, Control, Optimization and Applications: A Volume in Honor of W.H. Fleming*, 1998, W. M. McEneany, G. Yin, and Q. Zhang, (eds) Birkhauser.

[58] Franceschetti, M. and Meester, R. (2006) Critical node lifetimes in random networks via the Chen-Stein method, *IEEE Trans. on Information Theory*, **52**(6), 2831–2837.

[59] Balister, P., Bollobas, B., Sarkar, A. and Walters, M. (2005) Connectivity of random k-nearest neighbor graphs, *Adv. Appl. Prob.*, **37**, 1–24.

15

Advanced Routing and Network Coding

15.1 Conventional Routing Versus Network Coding

A conventional router in a network can merely route, or forward, messages. Each message on an output link must be a *copy* of a message that arrived earlier on an input link. On the other hand, network coding allows each node in a network to perform some computation. In other words, in network coding, each message sent on a node's output link can be some *function* or *mixture* of messages that arrived earlier on the node's input links. Thus, network coding is generally the transmission, combining (encoding) and recombining (decoding) of messages arriving at nodes *inside* the network, such that the transmitted messages can be recombined (or decoded) at their final destinations.

Network coding has several advantages over conventional routing. The first is the potential of network coding to improve throughput. Consider the following situation. Two streams of information, both at bit rate b bits per second, arrive at a node, contending for an output link, having capacity b bits per second. With network coding, it may be possible to increase throughput by pushing both streams through the bottleneck link at the same time. The method is simple. Using network coding, the node can mix the two streams together by taking their exclusive-OR (XOR) bit-by-bit and sending the mixed stream through the link. In this case, XOR is the function computed at the node. This increases the throughput of the network if the two streams can be separated before they reach their final destinations. This can be done using side information if it is available downstream.

As an example, Figure 15.1 illustrates an idealized network of routers, links. In the example, all the links are directional and have the same capacity, say $b = 1$ bits per second.

Source situated at node 1, denoted as s_1, wishes to send information to sink t_2, and source s_2, from node 2, wishes to send information to sink t_1. Observe that s_1 can reach t_2 only by the path $s_1 \rightarrow 1 \rightarrow 3 \rightarrow 4 \rightarrow t_2$, and that s_2 can reach t_1 only by the path $s_2 \rightarrow 2 \rightarrow 3 \rightarrow 4 \rightarrow t_1$. These share the bottleneck link $3 \rightarrow 4$.

The question is at what rates, r_1 and r_2, can the two sessions communicate reliably? If the network nodes can only route information, then clearly the bottleneck link $3 \rightarrow 4$ must be timeshared between the two sessions, giving rise to the set of achievable rates $\{(r_1, r_2) : 0 \leq r_1, 0 \leq r_2, r_1 + r_2 \leq b\}$, which is shown in Figure 15.2 (left). However, if the network nodes can perform network coding, then both sessions can communicate reliably at rate b, giving rise to the set of achievable rates $\{(r_1, r_2): 0 \leq r_1 \leq b, 0 \leq r_2 \leq b\}$, which is shown in Figure 15.2 (right). To achieve the pair of rates (b, b), for example, the two streams can be mixed at node 3 using an XOR operation, and they can be purified downstream at nodes t_1 and t_2, again using XOR operations.

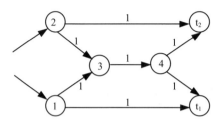

Figure 15.1 Two unicast sessions contending for a bottleneck link.

Clearly, it is not possible for either session to communicate reliably at a rate greater than b, since the sender and receivers in each session are connected to the network through a single link of capacity b. Hence the region in Figure 15.2 (right) is the set of all achievable rates, which is called the *capacity region* for these sessions in this network.

It turns out to be very difficult, in general, to determine the capacity region for an arbitrary set of sessions in an arbitrary network. To be specific, let a network (V,E,c) be represented by a set of nodes (or vertices) V, a set of directed links (or edges) E, and real-valued capacities $c(e)$ on each link $e \in E$, and each session (s, T) be represented by a sender $s \in V$ and a set of receiving terminals $T \subseteq V$. If the set of receivers consists of only a single node t, then the session is said to be a *unicast* session. Otherwise is said to be a *multicast* session. Given a network (V,E, c) and a set of sessions $(s_1, T_1), \ldots, (s_N, T_N)$ with respective communication rates $r_1, \ldots, r_N$, the answer to the question, '*is the vector of communication rates $(r_1, \ldots, r_N)$ achievable, or not?*' remains a difficult problem. In many specific networks and sets of sessions, the answer may be obvious, however, a precise characterization of the capacity region for an arbitrary network and set of sessions has remained elusive. We will have more detail about this later.

In a specific case, when there is only a single session (s, T) in any given network $(V, E, c)_s$ the *capacity* region $[0, C]$, or simply the *capacity* C, is now characterized well, thanks to the recent invention of network coding. Most of the remainder of this section is devoted to this single session case.

In the unicast case, a session (s, t) consists of a single sender s and a single receiver t. Let $r(s, t)$ designate an achievable rate of communication from s to t in a given network (V,E,c). It has been known that an upper bound on $r(s, t)$ is the *value* of any $s - t$ *cut* through the network. An $s - t$ cut is a partition of the network into two subsets, U and $\overline{U}$, the first containing s and the second containing t. The value of this cut is the sum of the capacities of the links from nodes in U to nodes in $\overline{U}$.

The minimum of the values of all such $s - t$ cuts, herein designated $MinCut(s, t)$ is clearly also an upper bound on $r(s, t)$. That is, $r(s, t) \leq MinCut(s, t)$ [1, 2] .

In the broadcast case, a session consists of a single sender s and a set of receivers V consisting of all the nodes in the network (V, E, c). Let $r(s, V)$ stands for an achievable rate at which s can broadcast common information reliably to all the other nodes in the network. Clearly, an upper bound on $r(s, V)$ is the maximum rate at which s can communicate reliably with any single receiver $v \in V$. Thus, $r(s, V) \leq \min_{v \in V} MinCut(s, v)$. The upper bound, $\min_{v \in V} MinCut(s, v)$, is also achievable [3] so that,

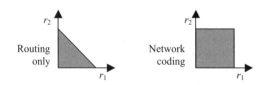

Figure 15.2 Achievable communication rate regions for routing only and network coding.

$C = \min_{v \in V} MinCut(s, v)$ can be considered the broadcast capacity of the network. A maximal set of spanning trees achieving the broadcast capacity can be found in polynomial time.

In the multicast case, a session consists of a single sender s and a set of receivers $T \subseteq V$. Let $r(s, T)$ designate an achievable rate at which s can multicast common information reliably to all the nodes in T. Clearly, $MinCut(s, t)$ remains an upper bound on this rate for any $t \in T$. Thus, $r(s, T) \leq \min_{t \in T} MinCut(s, t)$. Unfortunately, unlike the broadcast case, the upper bound may not be achievable by routing information through a set of edge-disjoint trees. The example shown in Figure 15.3 demonstrates such a case.

Figure 15.3(a) shows a seven-node network with unit-capacity directed links. A sending node s is on the top and two receiving nodes $T = \{t_1, t_2\}$ are at the bottom. Figure 15.3(b) and (c) show that there are two edge-disjoint directed paths from the sender to each of the two receivers. Hence, $\min_{t \in T} MinCut(s, t) = 2$. However, multicast at this rate is not achievable merely by routing information along a set of edge-disjoint trees. Figure 15.3(d) – (h) show the only possible Steiner trees from s to T. (A *Steiner tree*, also known as a multicast tree, from s to T in a graph (V, E) is a tree rooted at s that reaches every node in T through edges in E.) Any two of these Steiner trees share at least one edge. Hence, in this network the maximum number of edge-disjoint Steiner trees from s to T is 1. Thus routing along a maximal set of edge-disjoint Steiner trees cannot in general achieve throughput equal to the upper bound $\min_{t \in T} MinCut(s, t)$. In addition, finding such a maximal set of edge-disjoint Steiner trees turns out to be NP-hard.

Nevertheless, reliable multicast from s to T in Figure 15.3 can occur at the upper bound if network coding is used. Figure 15.3(i) shows how two unit-bandwidth streams, a and b, can be encoded at an interior node to produce a mixed stream, $a + b$, from which stream a can later be subtracted to recover stream b, and vice versa, thus delivering both streams to both receivers. Here, addition and subtraction are operations over a finite field, specifically XOR operations.

Furthermore, reliable multicast at a rate equal to the upper bound, $\min_{t \in T} MinCut(s, t)$, can always be achieved in any network using network coding [4, 5, 6]. Thus, $h = \min_{t \in T} MinCut(s, t)$ can be considered to be the *multicast capacity*, or simply the *capacity* (since the formula works for unicast and broadcast as well) for an arbitrary single session in an arbitrary directed network.

More generally, *linear* network coding is sufficient to achieve the multicast capacity [5, 6]. Linear network coding means that the messages (e.g., y_1, y_2, and y_3 at the input of a node) can be considered vectors of elements from a finite field, and the functions performed at the nodes can be simple linear combinations over this field $f_1(y_1, y_2, y_3) = \alpha_1 y_1 + \alpha_2 y_2 + \alpha_3 y_3$ at one output and $f_2(y_1, y_2, y_3) = \beta_1 y_1 + \beta_2 y_2 + \beta_3 y_3$ at the other possible output link . Furthermore, all the decoding at the receivers can be performed using linear operations. A polynomial time algorithm for finding the encoding and decoding coefficients in directed acyclic networks can be found in [7].

In summary, using linear network coding the multicast capacity in a directed network can always be achieved, and the coding coefficients necessary to achieve the capacity can be computed in polynomial time. On the other hand, if only routing can be used, not only is it generally impossible to achieve the multicast capacity but computing the set of edge-disjoint multicast trees necessary to achieve the best possible routing is a problem that is NP-hard in general.

These basic concepts outlined in this section will be further elaborated in the rest of the chapter.

15.2   A Max-Flow Min-Cut Theorem

In this section, we define formally the theorem which characterizes the admissible coding rate region for the single-source multicast problem. In other words, the information source X_1 is generated at node s and is multicast to nodes $t_1, \ldots, t_L$. We will call s the *source* and $t_1, \ldots, t_L$ the *sinks* of the graph G. For a specific L, the problem will be referred to as the one-source L-sink problem.

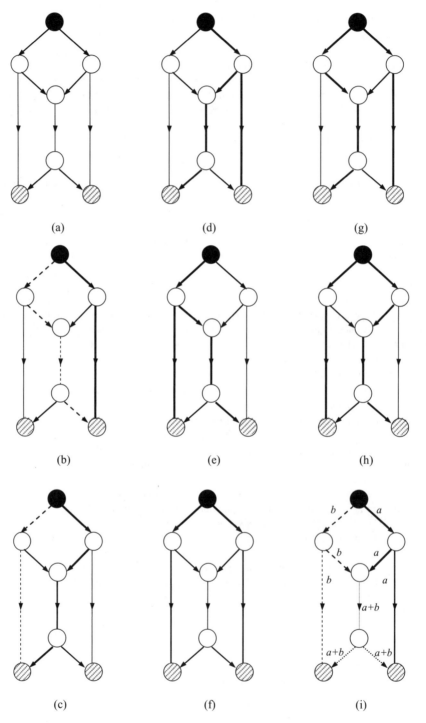

Figure 15.3   A multicast scenario.

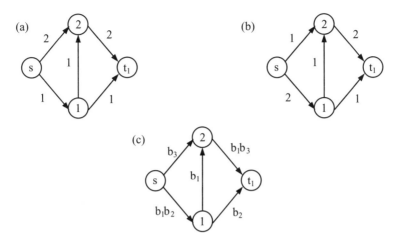

Figure 15.4   A one-source one-sink network.

Let $G = (V, E)$ be a graph with source s and sinks $t_1, \ldots, t_L$. The capacity of an edge $(i, j) \in E$ is given by R_{ij}, and let $R = [R_{ij}, (i, j) \in E]$. The subgraph of G from s to $t_l, l = 1, \ldots, L$, refers to the graph $G_l = (V, E_l)$, where $E_l = \{(i, j) \in E : (i, j)$ is on a directed path from s to $t_l\}$. $F = [F_{ij}, (i, j) \in E]$ is a *flow* in G from s to t_l if for all $(i, j) \in E$ we have $0 \leq F_{ij} \leq R_{ij}$ such that for all $i \in V$ except for s and t_l $\sum\limits_{i':(i',i) \in E} F_{i'i} = \sum\limits_{j:(i,j) \in E} F_{ij}$ i.e., the total flow into node i is equal to the total flow out of node i. F_{ij} is referred to as the value of F in the edge (i, j). The value of F is defined as $\sum\limits_{j:(s,j) \in E} F_{sj} - \sum\limits_{i:(i,s) \in E} F_{is}$ which is equal to $\sum\limits_{i:(i,t_l) \in E} F_{it_l} - \sum\limits_{j:(t_l,j) \in E} F_{t_l j}$. F is a *maxflow* from s to t_l in G if F is a flow from s to t_l whose value is greater than or equal to any other flow from s to t_l. Evidently, a *maxflow* from s to t_l in G is also a *maxflow* from s to t_l in G_l. For a graph with one source and one sink (for example, the graph G_l), the value of a *maxflow* from the source to the sink is called the *capacity* of the graph.

   For a graph $G = (V, E)$ with source s, sinks $t_1,..,t_L$, and the capacity of an edge (i, j) being R_{ij}, we say that (R, h, G) is admissible if and only if the values of a maxflow from s to $t_l, l = 1, \ldots, L$ are greater than or equal to h, the rate of the information source.

   The spirit of this conjecture resembles that of the well known *Max flow Min-cut Theorem* in graph theory [8]. For illustration we use the example in Figure 15.4. Figure 15.4(a) shows the capacity of each edge. By the Max flow Min-cut Theorem [8], the value of a *max* flow from s to t_1 is 3, since the min cut is 3, so the flow in Figure 15.4(b) is a max-flow. In Figure 15.4(c), we show how we can send three bits b_1, b_2, b_3 from s to t_1 based on the max-flow in Figure 15.4(b). The conjecture is trivially seen to be true for $L = 1$, because when there is only one sink, we only need to treat the raw information bits as physical entities. The bits are routed at the intermediate nodes according to any fixed routing scheme, and they will all eventually arrive at the sink. Since the routing scheme is fixed, the sink knows which bit is coming in from which edge, and the information can be recovered accordingly.

   In Figure 15.5 we illustrate that the conjecture is true for $L = 2$.

   Figure 15.5(a) shows the capacity of each edge. It is easy to check that the value of a min cut (max flow) from s to t_1 and to t_2 are 5 and 6, respectively. So the conjecture asserts that we can send 5 bits b_1, b_2, b_3, b_4, b_5 to t_1 and t_2 simultaneously, and Figure 15.5 (b) shows such a scheme. Note that in this scheme, bits only need to be replicated at the nodes to achieve optimality.

   In Figure 15.6 we show another example to illustrate that the conjecture is true for $L = 2$.

   Figure 15.6(a) shows the capacity of each edge. It is easy to check that the value of a min cut (max flow) from s to t_l is 2, $l = 1, 2$. So the conjecture asserts that we can send 2 bits b_1, b_2 to t_1 and t_2

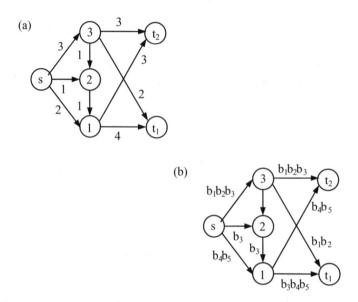

Figure 15.5   A one-source two-sink network without coding.

simultaneously, and Figure 15.6(b) shows such a scheme, where $+$ denotes modulo 2 addition. At t_1, b_2 can be recovered from b_1 and $b_1 + b_2$. Similarly, b_1 can be recovered at t_2. Note that when there is more than one sink, we can no longer think of information as a real entity, because information needs to be replicated or transformed at the nodes. In this example, information is coded at node 3, which is unavoidable. For $L \geq 2$, network coding is in general necessary in an optimal multicast scheme.

Finally, Figure 15.7 illustrates that the conjecture is true for $L = 3$. Figure 15.7(a) shows the capacity of each edge. It is easy to check that the values of a min cut (max flow) from s to all the sinks are 2. Figure 15.7(b), shows how we can multicast 2 bits b_1, b_2 to all the sinks.

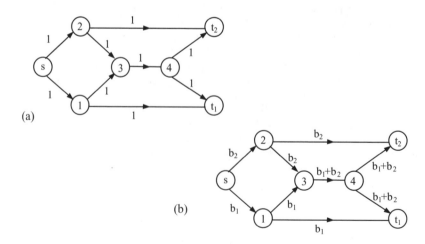

Figure 15.6   A one-source two-sink network with coding.

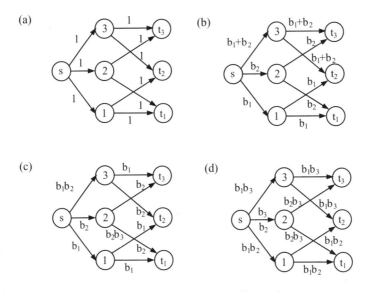

Figure 15.7   A one-source three-sink network.

15.3   Algebraic Formulation of Network Coding

In this section we extend the notation to allow multiple edges between two vertices in the network graph $G = (V, E)$ and hence, E is a subset of $E \subseteq V \times V \times Z_+$, where the last integer enumerates edges between two vertices. Edges (links) are denoted by round brackets $(v_1, v_2, i) \in E$ and assumed to be directed. We also denote edges simply as (v_1, v_2) if no confusion can arise. The head and tail of an edge $e = (v', v, i)$ is denoted by $v = head(e)$ and $v' = $ tail(e).

We define $\Gamma_I(v)$ as the set of edges that end (*index I-input*) at a vertex $v \in V$ and $\Gamma_O(v)$ as the set of edges originating (*index O-output*) at v. Formally, we have $\Gamma_I(v) = \{ e \in E : head(e) = v \}$ and $\Gamma_O(v) = \{ e \in E : tail(e) = v \}$. The *in-degree* $\delta_I(v)$ of **v** is defined as $\delta_I(v) = |\Gamma_I(v)|$, while the *out-degree* $\delta_O(v)$ is defined as $\delta_O(v) = |\Gamma_O(v)|$.

A network is called *cyclic* if it contains directed cycles, i.e., if there exists a sequence of edges $(v_0, v_1), (v_1, v_2), \ldots, (v_n, v_0)$ in G and *acyclic* otherwise. To each link $e \in E$ we associate a nonnegative number $C(e)$, called the capacity of e.

Let $X(v) = \{X(v, 1), X(v, 2), \ldots, X(v, \mu(v))\}$ be a collection of $\mu(v)$ discrete random processes that are observable at node v. We want to allow communication between selected nodes in the network, i.e., we want to replicate, by means of the network, a subset of the random processes in $X(v)$ at some different node v'. We define a connection $c=(v, v', X(v, v')) \in V \times V \times P_{X(v)}$, where $P_{X(v)}$ denotes the power set of $X(v)$. The rate $R(c)$ of a connection c is defined as $R(c) = \sum\limits_{i:X(v,i)\in X(v,v')} H(X(v, i))$, where $H(X)$ is the entropy rate of a random process X.

Given a connection $c = (v, v', X(v, v'))$, we call $v = source(c)$ and $v' = sink(c)$. For notational convenience, we will always assume that $source (c) \neq sink(c)$.

A node v can send information through a link $e = (v, u)$ originating at v at a rate of at most $C(e)$ bits per time unit. The random process transmitted through link e is denoted by $Y(e)$. In addition to the random processes in $X(v)$, node v can observe random processes $Y(e')$ for all e' in $\Gamma_I(v)$. In general, the random process $Y(e)$ transmitted through link $e = (v, u) \in \Gamma_O(v)$ will be a function of both $X(v)$ and $Y(e')$ if e' is in $\Gamma_I(v)$.

If v is the sink of any connection c, the collection of $u(v)$ random processes $Z(v) = \{Z(v, 1), Z(v, 2), \ldots, Z(v, u(v))\}$ denotes the output at $v = sink (c)$. A connection $c = (v, v', X(v, v'))$ is established successfully if a (possibly delayed) copy of $X(v, v')$ is a subset of $Z(v')$.

For a network G with a set C of desired connections we will make a number of simplifying assumptions:

1. The capacity of any link in G is a constant,
2. Each link in the communication network has the same delay.
3. Random processes $X(v, l), l \in \{1, 2, \ldots, \mu(v)\}$ are independent and have a constant and integral entropy rate.
4. The random processes $X(v, l)$ are independent for different v.

In addition to the above constraints, we assume that communication in the network is performed by transmission of vectors (symbols) of bits. The length of the vectors is equal in all transmissions and we assume that all links are synchronized with respect to the symbol timing.

Any binary vector of length m can be interpreted as an element in F_{2^m}, the finite field with 2^m elements. The random processes $X(v, l)$, $Y(e)$, and $Z(v, l)$ can, hence, be modeled as discrete processes $X(v, l) = \{X_0(v, l), X_1(v, l), \ldots\}$, $Y(e) = \{Y_0(e), Y_1(e), \ldots\}$, and $Z(v, l) = \{Z_0(v, l), Z_1(v, l), \ldots\}$ that consist of a sequence of symbols from F_{2^m}.

We say that a delay-free communication network $G = (V, E)$ is a F_{2^m}-linear network, if for all links the random process $Y(e)$ on a link $e = (v, u, i) \in E$ satisfies

$$Y(e) = \sum_{l=1}^{\mu(v)} \alpha_{l,e} X(v, l) + \sum_{e':head(e')=tail(e)} \beta_{e',e} Y(e') \tag{15.1}$$

where the coefficients $\alpha_{l,e}$ and $\beta_{e',e}$ are elements of F_{2^m}.

The above definition is concerned with the formation of random processes that are transmitted on the links of the network. We call the network *time-varying* or *time-invariant*, depending on whether or not the coefficients $\alpha_{l,e}$ and $\beta_{e',e}$ are time varying or not respectively.

The output $Z(v, l)$ at any node v is formed from the random processes $Y(e)$ for $e \in \Gamma_I(v)$. We will restrict ourselves to the case where $Z(v, l)$ are also linear combinations of the $Y(e)$, i.e.,

$$Z(v, j) = \sum_{e':head(e')=v} \varepsilon_{e',j} Y(e') \tag{15.2}$$

where the coefficients $\varepsilon_{e',j}$ are elements of F_{2^m}.

For a given network G and a given set of connections C, we formally define a network coding problem as a pair (G, C). The problem is to give succinct algebraic conditions under which a set of desired connections is feasible. This is equivalent to finding elements $\alpha_{l,e}, \beta_{e',e}$, and $\varepsilon_{e',j}$ in a suitably chosen field F_{2^m} such that all desired connections can be established successfully by the network. Such a set of numbers $\alpha_{l,e}, \beta_{e',e}$, and $\varepsilon_{e',j}$ is called a *solution* to the network coding problem (G, C). If a solution exists, the network coding problem is called *solvable*.

The network problem is solvable if and only if the rate of the connection $R(c)$ is less than or equal to the minimum value of all cuts between v and v'[1, 2].

The Ford-Fulkerson labeling algorithm [1] gives a way for finding a solution for point-to-point connections provided a network problem is solvable. The algorithm is graph theoretic by design and finds, a solution such that all parameters $\alpha_{l,e}$ and $\beta_{e',e}$ are either zero or one.

In the remainder of this section, we develop some theory and notation necessary for more complex setups. We first consider a point-to-point setup. Let node v be the only source in the network. We let $\mathbf{x} = (X(v, 1), X(v, 2), \ldots, X(v, \mu(v)))$ denote the vector of input processes observed at v. Similarly, let v' be the only sink node in a network. We let $\mathbf{z} = (Z(v', 1), Z(v', 2), \ldots, Z(v', u(v')))$ be the vector of output processes.

In an $F_{2^m} - linear$ network we can set a *transfer matrix* $\mathbf{M}$ describing the relationship between an input vector $\mathbf{x}$ and an output vector $\mathbf{z}$ as $\mathbf{z} = \mathbf{x}\mathbf{M}$. For a fixed set of coefficients $\alpha_{l,e}, \beta_{e',e}$, and $\varepsilon_{e',j}$, $\mathbf{M}$ is a matrix whose coefficients are elements in the field F_{2^m} We can go a step further and consider the

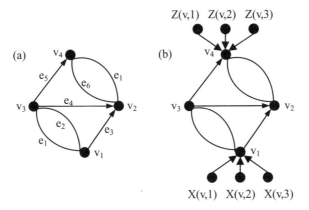

Figure 15.8   (a) Example network. (b) the random processes to be transmitted in the network.

coefficients as indeterminate variables. Hence, we consider the elements of matrix $\mathbf{M}$ as polynomials over the ring $F_2[\ldots, \alpha_{l,e}, \ldots, \beta_{e',e}, \ldots, \varepsilon_{e',j} \ldots]$ of polynomials in the variables $\alpha_{l,e}$, $\beta_{e',e}$, and $\varepsilon_{e',j}$.

For the network example of Figure 15.8, the following set of equations governs the parameters $\alpha_{l,e}$, $\beta_{e',e}$, and $\varepsilon_{e,j}$ and the random processes in the network.

$$Y(e_1) = \alpha_{1,e_1} X(v, 1) + \alpha_{2,e_1} X(v, 2) + \alpha_{3,e_1} X(v, 3)$$
$$Y(e_2) = \alpha_{1,e_2} X(v, 1) + \alpha_{2,e_2} X(v, 2) + \alpha_{3,e_2} X(v, 3)$$
$$Y(e_3) = \alpha_{1,e_3} X(v, 1) + \alpha_{2,e_3} X(v, 2) + \alpha_{3,e_3} X(v, 3)$$
$$Y(e_4) = \beta_{e_1,e_4} Y(e_1) + \beta_{e_2,e_4} Y(e_2) \qquad\qquad (15.3)$$
$$Y(e_5) = \beta_{e_1,e_5} Y(e_1) + \beta_{e_2,e_5} Y(e_2)$$
$$Y(e_6) = \beta_{e_3,e_6} Y(e_3) + \beta_{e_4,e_6} Y(e_4)$$
$$Y(e_7) = \beta_{e_3,e_7} Y(e_3) + \beta_{e_4,e_7} Y(e_4)$$

$$Z(v', 1) = \varepsilon_{e_5,1} Y(e_5) + \varepsilon_{e_6,1} Y(e_6) + \varepsilon_{e_7,1} Y(e_7)$$
$$Z(v', 2) = \varepsilon_{e_5,2} Y(e_5) + \varepsilon_{e_6,2} Y(e_6) + \varepsilon_{e_7,2} Y(e_7) \qquad (15.4)$$
$$Z(v', 3) = \varepsilon_{e_5,3} Y(e_5) + \varepsilon_{e_6,3} Y(e_6) + \varepsilon_{e_7,3} Y(e_7)$$

It can be shown that the transfer matrix describing the relationship between $\mathbf{x}$ and $\mathbf{z}$ is

$$\mathbf{M} = \mathbf{A} \begin{pmatrix} \beta_{e_1,e_5} & e_1, e_4 \beta e_4, e_6 & \beta_{e_1,e_4}\beta_{e_4,e_7} \\ \beta_{e_2,e_5} & \beta_{e_2,e_4}\beta_{e_4,e_6} & \beta_{e_2,e_4}\beta_{e_4,e_7} \\ 0 & \beta_{e_3,e_6} & \beta_{e_3,e_7} \end{pmatrix} \mathbf{B}^T \qquad (15.5)$$

where

$$\mathbf{A} = \begin{pmatrix} \alpha_{1,e} & \alpha_{1,e_2} & \alpha_{1,e_3} \\ \alpha_{2,e_1} & \alpha_{2,e_2} & \alpha_{2,e_3} \\ \alpha_{3,e} & \alpha_{3,e_2} & \alpha_{3,e_3} \end{pmatrix} \qquad \mathbf{B} = \begin{pmatrix} \varepsilon_{e_5} & \varepsilon_{e_5,2} & \varepsilon_{e_5,3} \\ \varepsilon_{e_6,1} & \varepsilon_{e_6,2} & \varepsilon_{e_6,3} \\ \varepsilon_{e_7} & \varepsilon_{e_7,2} & \varepsilon_{e_7,3} \end{pmatrix} \qquad (15.6)$$

The determinant of matrix $\mathbf{M}$ equals

$$det(\mathbf{M}) = det(\mathbf{A})(\beta_{e_1,e_5}\beta_{e_2,e_4} - \beta_{e_2,e_5}\beta_{e_1,e_5})(\beta_{e_4,e_6}\beta_{e_3,e_7} - \beta_{e_4,e_7}\beta_{e_3,e_6})det(\mathbf{B}).$$

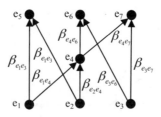

Figure 15.9   Directed labeled line graph G corresponding to the network depicted in Figure 15.8(a).

We can choose parameters in an extension field F_2^m so that the determinant of M is nonzero over F_{2^m}. Hence, we can choose A as the identity matrix and B so that the overall matrix M is also an identity matrix. One such solution (found by the Ford-Fulkerson algorithm) would be to let $\beta_{e_1,e_5} = \beta_{e_2,e_4} = \beta_{e_4,e_6} = \beta_{e_3,e_7} = 1$ while all other parameters of type $\beta_{e',e}$ are chosen to equal zero. Clearly, a point-to-point communication between v and v' is possible at a rate of three bits per unit time. We note that, over the algebraic closure $\overline{F}$ there exists an infinite number of solutions to the posed networking problem, namely, all assignments to parameters $\beta_{e',e}$ which render a nonzero determinant of the transfer matrix M.

In a linear communication network defined by (15.1) and (15.2), any node v_i transmits, on an outgoing edge, a linear combination of the symbols observed on the incoming edges. We say that any edge $e = (u, v)$ *feeds into* edge $e' = (v, u')$ if head (e) is equal to tail (e'). The *directed labeled line graph* of $G = (V, E)$ is defined as $G^*(V, E)$ with vertex set $V = E$ and edge set $E = \{(e, e') \in E^2 :$ head $(e) = tail(e')\}$. Any edge $e = (e, e') \in E$ is labeled with the corresponding label $\beta_{e',e}$. Figure 15.9 shows the directed labeled line graph of the network in Figure 15.8.

The adjacency matrix F of the graph G with elements $F_{i,j}$ is given as

$$F_{i,j} = \begin{cases} \beta_{e_i,e_j} & head(e_i) = tail(e_j) \\ 0, & otherwise \end{cases} \tag{15.7}$$

In order to consider the case that a network contains multiple sources and sinks, we consider

$$\mathbf{x} = (x_1, x_2, \ldots, x_\mu) = (X(v_1, 1), X(v_1, 2), \ldots, X(v_1, \mu(v_1)), X(v_2, 1), \ldots, X(v_{v_{|V|}}, \mu(v_{|V|})))$$

as the vector of input processes on all vertices in V. If a vertex v in a network is not a source node, we set the corresponding parameter $\mu(v)$ equal to zero. $\mathbf{x} = (x_1, x_2, \ldots, x_\mu)$ is a vector of length $\mu = \sum_i \mu(v_i)$. We define the entries of a $\mu \times |E|$ matrix A as

$$A_{i,j} = \begin{cases} \alpha_{l,e_j} & x_i = X(tail(e_j), l) \\ 0, & otherwise \end{cases} \tag{15.8}$$

Similarly, we use the notation $\mathbf{z} = (z_1, z_2, \ldots, z_\mu) = (Z(v_1, 1), Z(v_1, 2), \ldots, Z(v_1, u(v_1)), Z(v_2, 1), \ldots, Z(v_{|V|}, u(v_{|V|})))$ be the vector of output processes. If v_j is not a sink node of any connection, we let $u(v_j)$ be equal to zero. $\mathbf{z}$ is a vector of length $u = \sum_i u(v_i)$. Let the entries of a $u \times |E|$ matrix $\mathbf{B}$ be

$$B_{i,j} = \begin{cases} \varepsilon_{e_j,l} & z_i = Z(head(e_j), l) \\ 0, & otherwise \end{cases} \tag{15.9}$$

From the definition of matrices $\mathbf{F}$, $\mathbf{A}$ and $\mathbf{B}$, we can easily find the transfer matrix of the overall network as

$$\mathbf{M} = \mathbf{A}(\mathbf{I} - \mathbf{F})^{-1}\mathbf{B}^T \qquad (15.10)$$

where $\mathbf{I}$ is the $|E| \times |E|$ identity matrix [5].

15.4 Random Network Coding

In this section we present a distributed random linear network coding approach for transmission and compression of information in general multisource multicast networks. Network nodes select linear mappings independently and randomly from inputs onto output links over some field. This achieves capacity with probability exponentially approaching 1 with the code length. The random linear coding performs compression when necessary in a network, generalizing error exponents for linear Slepian-Wolf coding (see Chapter 1) in a natural way. Benefits of this approach are decentralized operation and robustness to network changes or link failures.

The basic network coding model, used in the previous section, is further extended to include networks with cycles, link delays and correlated sources. As in the previous section, a network is represented as a directed graph $G = (V, E)$, where V is the set of network nodes and E is the set links, such that information can be sent noiselessly from node i to j for all $(i, j) \in E$. Each link $l \in E$ is associated with a nonnegative real number c_l representing its transmission capacity in bits per unit time.

Nodes i and j are called the *origin* and *destination*, respectively, of link (i, j). The origin and destination of a link $l \in E$ are denoted $o(l)$ and $d(l)$, respectively. We assume $o(l) \neq d(l) \forall l \in$ E. The information transmission on a link $l \in E$ is obtained as a coding function of information previously received at $o(l)$. We assume that r discrete memoryless information source processes $X_1, X_2, \ldots, X_r$ are random binary sequences. We denote the Slepian-Wolf region of the sources $R_{SW} = \{(R_1, R_2, \ldots, R_r) : \sum_{i \in S} R_i >$ $H(X_S | X_{S^c}) \forall S \subseteq \{1, 2, \ldots, r\}\}$ where $X_S = (X_{i_1}, X_{i_2}, \ldots, X_{i_{|S|}})$, $i_k \in S, k = 1, \ldots, |S|$. Source process X_i is generated at node $a(i)$, and multicast to all nodes $j \in b(i)$, where $a : \{1, \ldots, r\} \to V$ and $b : \{1, \ldots, r\} \to 2^V$ are arbitrary mappings. In this section, we consider the (multisource) multicast case where $b(i) = \{\beta_1, \ldots, \beta_d\}$ for all $i \in [1, r]$. The nodes $a(1), \ldots, a(r)$ are called *source nodes* and the nodes $\beta_1, \ldots, \beta_d$ are called *receiver nodes*, or receivers. For simplicity, we assume subsequently that $a(i) \neq \beta_j \forall i \in [1, r], j \in [1, d]$. The mapping a, the set $\{\beta_1, \ldots, \beta_d\}$ and the Slepian-Wolf region R_{SW} specify a set of multicast *connection requirements*. The connection requirements are satisfied if each receiver is able to reproduce, from its received information, the complete source information. A graph $G = (V, E)$, a set of link capacities $\{c_l | l \in E\}$, and a set of multicast connection requirements C specify a multicast connection problem.

In the previous section, the source information processes, the receiver output processes, and the information processes transmitted on each link, are sequences of length-u of a finite field F_q, $q = 2^u$. The information process Y_j transmitted on a link j is formed as a linear combination, in F_q, of link j's *inputs*, i.e., source processes X_i for which $a(i) = o(j)$ and random processes Y_l for which $d(l) = o(j)$, if any. For the delay-free case, this is represented by the equation

$$Y_j = \sum_{\{i : a(i) = o(j)\}} a_{i,j} X_i + \sum_{\{l : d(l) = o(j)\}} f_{l,j} Y_l \qquad (15.11)$$

$Z_{\beta,i} = \sum_{\{l : d(l) = \beta\}} b_{\beta,i,l} Y_l$. The i-th output process $Z_{\beta,i}$ at receiver node β is a linear combination of the information processes on its terminal links, represented as

$$Z_{\beta,i} = \sum_{\{l : d(l) = \beta\}} b_{\beta,i,l} Y_l \qquad (15.12)$$

Unit delay links are characterized by the following linear coding equations

$$Y_j(t+1) = \sum_{\{i:a(i)=o(j)\}} a_{i,j} X_i(t) + \sum_{\{l:d(l)=o(j)\}} f_{l,j} Y_l(t)$$

$$Z_{\beta,i}(t+1) = \sum_{u=0}^{\mu} b'_{\beta,i}(u) Z_{\beta,i}(t-u) + \sum_{\{l:d(l)=\beta\}} \sum_{u=0}^{\mu} b''_{\beta,i,l}(u) Y_l(t-u) \tag{15.13}$$

where $X_i(t)$, $Y_j(t)$, $Z_{\beta,i}(t)$, $b'_{\beta,i}(t)$, and $b''_{\beta,i,l}(t)$ are the values of the corresponding variables at time t, respectively, and μ represents the memory required. Links with longer delay are modeled as links in series. These equations, as with the random processes in the network, can be represented algebraically in terms of a delay variable D

$$Y_j(D) = \sum_{\{i:a(i)=o(j)\}} D a_{i,j} X_i(D) + \sum_{\{l:d(l)=o(j)\}} D f_{l,j} Y_l(D)$$

$$Z_{\beta,i}(D) = \sum_{\{l:d(l)=\beta\}} b_{\beta,i,l}(D) Y_l(D) \tag{15.14}$$

where

$$b_{\beta,i,l}(D) = \frac{\sum_{u=0}^{\mu} D^{u+1} b''_{\beta,i,l}(u)}{1 - \sum_{u=0}^{\mu} D^{u+1} b'_{\beta,i}(u)}$$

and $X_i(D) = \sum_{t=0}^{\infty} X_i(t) D^t$, $Y_j(D) = \sum_{t=0}^{\infty} Y_t(j) D^t$ with $Y \cdot (0) = 0$ and $Z_{\beta,i}(D) = \sum_{t=0}^{\infty} Z_{\beta,i}(t) D^t$ with $Z_{\beta,i}(0) = 0$.

The coefficients $\{a_{i,j}, f_{l,j}, b_{\beta,i,l}\}$ can be collected into $r \times |E|$ matrices

$$\mathbf{A} = \begin{cases} (a_{i,j}) & \text{in the acyclic delay free case} \\ (D a_{i,j}) & \text{in the general case with delays} \end{cases}$$

and

$\mathbf{B}_{\beta} = (b_{\beta,i,l})$, and the $|E| \times |E|$ matrix

$$\mathbf{F} = \begin{cases} (f_{l,j}) & \text{in the acyclic delay free case} \\ (D f_{l,j}) & \text{in the general case with delays} \end{cases}$$

whose structure is constrained by the network. A pair $(\mathbf{A}, \mathbf{F})$ or tuple $(\mathbf{A}, \mathbf{F}, \mathbf{B}_{\beta_1}, \ldots, \mathbf{B}_{\beta_d})$ can be called a linear network code.

Linearly correlated sources are modeled as given linear combinations of underlying independent processes, each with an entropy and bit rate of one bit per unit time. To simplify the notation in our subsequent development, we work with these underlying independent processes in a similar manner as for the case of independent sources: the jth column of the $\mathbf{A}$ matrix is a linear function $\sum_k \alpha_{k,j} \mathbf{x}_j^k$ of given column vectors $\mathbf{x}_j^k \in F_2^r$, where $\mathbf{x}_j^k$ specifies the mapping from r underlying independent processes to the kth source process at $o(j)$. We can also consider the case where $\mathbf{x}_j^k \in F_{2^m}^r$ by restricting network coding to occur in $F_q, q = 2^{mn}$. A receiver that decodes these underlying independent processes is able to reconstruct the linearly correlated source processes.

For acyclic graphs, we assume such indexing of links in E, so that if $d(l_1) = o(l_2)$ for any links l_1, l_2, then l_1 has a lower index than l_2. Such indexing always exists for acyclic networks. It then follows that matrix $\mathbf{F}$ is upper triangular with zeros on the diagonal.

From (15.10) we use notation $\mathbf{G} = (\mathbf{I} - \mathbf{F})^{-1}$. For the acyclic delay-free case, the sequence $(\mathbf{I} - \mathbf{F})^{-1} = \mathbf{I} + \mathbf{F} + \mathbf{F}^2 + \cdots$ converges since $\mathbf{F}$ is nilpotent for an acyclic network. For the case with delays, $\mathbf{G} = (\mathbf{I} - \mathbf{F})^{-1}$ exists since the determinant of $\mathbf{I} - \mathbf{F}$ is nonzero in its field of definition $F_2(D, \dots, f_{l,j}, \dots)$, as seen by letting $D = 0$ [5] .

From (15.19), the mapping from source processes $[X_1, \dots, X_r]$ to output processes $[Z_{\beta,1}, \dots, Z_{\beta,r}]$ at a receiver β is given by the transfer matrix $\mathbf{AGB}_\beta^T$. For a given multicast connection problem, if some network code $(\mathbf{A}, \mathbf{G}, \mathbf{B}_{\beta_1}, \dots, \mathbf{B}_{\beta_d})$ in a field F_q (or $F_q(D)$) satisfies the condition that $\mathbf{AGB}_{\beta_k}^T$ has full rank r for each receiver $\beta_k, k = 1, \dots, d$, then $\tilde{\mathbf{B}}_{\beta_k} = (\mathbf{B}_{\beta_k} \mathbf{G}^T \mathbf{A}^T)^{-1} \mathbf{B}_{\beta_k}$ satisfies $\mathbf{AG}\tilde{\mathbf{B}}_{\beta_k}^T = \mathbf{I}$, and $(\mathbf{A}, \mathbf{G}, \tilde{\mathbf{B}}_{\beta_1}, \dots, \tilde{\mathbf{B}}_{\beta_d})$ is a *solution* to the multicast connection problem in the same field. A multicast connection problem for which there exists a solution in some field F_q or $F_q(D)$ is called *feasible*, and the corresponding connection requirements are said to be feasible for the network.

In the sequel, where we consider choosing the value of $(\mathbf{A}, \mathbf{G})$ by distributed random coding, we say that if for a receiver β_k there exists some value of $\mathbf{B}_{\beta_k}$ such that $\mathbf{AGB}_{\beta_k}^T$ has full rank r, then $(\mathbf{A}, \mathbf{G})$ is a *valid* network code for β_k. A network code $(\mathbf{A}, \mathbf{G})$ is *valid* for a multicast connection problem if it is valid for all receivers.

Consider now a multicast connection problem on an arbitrary network with independent or linearly correlated sources, and a network code in which some or all network code coefficients $\{a_{i,j}, \alpha_{k,j}, f_{l,j}\}$ are chosen uniformly at random from a finite field F_q where $q > d$ (d is the number of receivers), and the remaining code coefficients, if any, are fixed. If there exists a solution to the network connection problem with the same values for the fixed code coefficients, then the probability that the random network code is valid for the problem is at least $(1 - d/q)^\eta$, where η is the number of links j with associated random coefficients $\{a_{i,j}, \alpha_{k,j}, f_{l,j}\}$[9].

15.5 Gossip Based Protocol and Network Coding

In *gossip-based* protocols [11], nodes communicate with each other in communication steps called *rounds*, and the amount of information exchanged in each round between two communicating nodes is limited. Further, there is no centralized controller and every node in the network acts only based on the state or information of the node, and not that of the overall network. Thus, *gossip-based* protocols are inherently distributed and easily implementable, and provide powerful alternatives to *flooding-* and *broadcast-based* protocols for dissemination of messages. There is well documented literature on the practical aspects of gossip-based message dissemination [12, 13]. A common performance measure in most of the work is the time required to disseminate a single message to all the nodes in the network. In [14, 15] the scenario where there are multiple messages, each with a unique destination was considered. The authors considered *spatial gossip* where the nodes are embedded in a metric-space with some distance metric between the nodes. In [16], the authors have studied gossip-like distributed algorithms for computing averages at the nodes. In the framework they consider, each node starts with a possibly distinct value of a certain parameter, and the goal is to find the time required for every node to compute the average of the parameter values at the nodes, using gossip-like algorithms.

In this section, we discuss a problem, in which the network seeks to disseminate multiple messages simultaneously to all the nodes in a distributed and decentralized manner. Thus, each node wants to compute the exact messages at every other node.

Any *gossip* protocol is designed in two steps [14]. One is the design of gossip algorithm by which, in every round, every node decides its communication partner, either in a deterministic, or in a randomized manner. The other important aspect is the design of *gossip-based protocol* by which any node, upon deciding the communication partner according to the *gossip* algorithm, decides the content of the message to send to the communication partner. The main difference of this section lies in discussing a *gossip-based protocol* using the idea of random linear coding (RLC) [10], which was discussed in Section 15.4.

In a network, with a given number of nodes, each of the nodes has a distinct message to start with, and the goal is to disseminate all the messages among all the nodes. In the first step, each node picks up a node at random (*gossip-based* communication) and transmits the message it has. In the second step, some of the nodes already have two distinct messages. For transmitting a message in the second step, these nodes have to decide which message to transmit without the knowledge of the contents of its communication partner. Obviously, the constraint is imposed by the fact that only one message can be transmitted and nodes only have local knowledge.

The key intuition behind the idea of a coding-based approach is in the following. Suppose there are k distinct elements in a finite field of size q. Consider two approaches to store the elements in a database with k slots. Suppose each slot chooses an element at random without the knowledge of what the other slots are choosing. Then, the probability that all the elements are there in the database is very low. Now, consider a second approach in which each slot in the database stores a random linear combination of the messages. All the messages can be recovered from the database only if the linear combination chosen by the slots are linearly independent. Since there are $((q^k - 1)(q^k - q)(q^k - q^2)\ldots(q^k - q^{k-1}))$ ways of choosing k linearly independent combination of the k elements in a finite field of size q, the probability that the elements can be recovered from the database is much higher in the latter scenario. This is the key idea which is at the heart of the RLC-based protocol [10] we present in this section.

The model of the protocol assumes that there are n nodes and k messages. Initially, each node has one of the k messages indexed by the elements in the set $M = \{m_1, m_2, \ldots m_k\}$. The nodes are indexed by elements of the set $[n]$. Initially, every node has only one out of the k messages. If V_{m_i} is the set of nodes that start out with the message $m_i \in M$, we also assume $|V_{m_1}| = |V_{m_2}| = \cdots = |V_{m_k}|, \forall m \in M$, i.e., at the beginning each message is equally spread in the network.

Suppose there are only k distinct messages at k nodes initially and no messages with the other $n - k$ nodes. There are two phases of message dissemination if all the nodes require all the messages. The first phase ends when every node has at least one message and the second phase starts once every node has at least one message and ends with all the nodes having all the messages.

Gossip Algorithm advances in *rounds* indexed by $t \in Z^+$. The communication graph in round t, G_t, is obtained in a randomized manner as follows. In the beginning of each round, each node $u \in [n]$ calls a communication partner v chosen uniformly from $[n]$. As proposed in [11], we consider two versions of the *rumor mongering* model for message exchange.

1. *Pull model*, where a message is transmitted from a *called* node to the *caller* node according to a suitable protocol described in the sequel. Thus, the communication process is initiated by the receiving node.
2. *Push model*, where the message is transmitted from a *caller* node to the *called* node. Thus, the communication process is initiated by the transmitting node.

There can be other variants of these basic models, for instance, by combining *push* and *pull* as considered in [13].

Gossip-Based Protocols will be adopted by the *caller* node in the *push* model and the *called* node in the *pull* model. Below, we describe two protocols for message transmission we consider in this section [10].

a) *Random Message Selection (RMS)* is a simple strategy, where the transmitting node simply looks at the messages it has received and picks any of the messages with equal probability to transmit to the receiving node. So, if R is the set of messages at node v, then v transmits a 'random' message e to its communicating partner, where $Pr(e = m) = I_{(m \in M_v)}/|M_v|$, where $I_{(m \in M_v)}$ is the indicator variable of the event $(m \in M_v)$.
b) *Random Linear Coding (RLC)* considers the messages as vectors over the finite field F_q of size $q \geq k$. If the message size is m bits, this can be done by viewing each message as an $r = \lceil m/\log_2(q) \rceil$ -dimensional vector over F_q (instead of viewing each message as an m-dimensional vector over the binary field). Therefore, let $m_i \in F_q^r (m_i, i = 1, 2, \ldots K$, are the messages) for some integer r. Thus, the messages are over a vector space with the scalars in F_q. All the additions and the multiplications in the following description are assumed to be over F_q. In the RLC protocol, the nodes start collecting

several linear combinations of the messages in M. Once there are k independent linear combinations with a node, it can recover all the messages successfully. Let S_v denote the set of all the coded messages (each coded message is a linear combination of the messages in M) with node v at the beginning of a round. In other words, if $f_l \in S_v$, where $l = 1, 2 \ldots |S_v|$, then $f_l \in F_q^r$ has the form

$$f_l = \sum_{i=1}^{k} a_{li} m_i, \quad a_{li} \in F_q \tag{15.15}$$

In addition, the protocol ensures that $a_{li}s$ are known to v. This can be done with minimal overhead with each packet in a manner to be described soon.

Therefore, if the node v has to transmit a message to u, then v transmits a 'random' coded message with payload $e \in F_q^r$ to u, where

$$e = \sum_{f_l \in S_v} \beta_l f_l, \quad \beta_l \in F_q \tag{15.16}$$

and

$$Pr(\beta_l = \beta) = \frac{1}{q} \; \forall \beta \in F_q \tag{15.17}$$

For decoding purposes, the transmitting nodes also send the 'random coding vectors' as overhead with each packet. This can be achieved by padding an additional $k \log_2 q$ bits with each message. The structure of the overhead associated with a packet can be seen better from the following representation of the payload part of the transmitted message e in (15.16)

$$e = \sum_{f_l \in S_v} \beta_l f_l = \sum_{f_i \in S_v} \beta_l \sum_{i=1}^{k} a_{li} m_i \text{ where } a_{li} \in F_q^{:})$$

$$= \sum_{i=1}^{k} \theta_i m_i \quad \left(\text{with } \theta_i = \sum_{f_l \in S_v} \beta_l a_{li} \in F_q \right) \tag{15.18}$$

The $\theta_i s$ are sent as overhead with the transmitted messages. So, once the $\beta_l's$ are selected in randomized manner according to (15.17), the transmitting nodes can obtain the values of $\theta_i's$ $(i = 1, 2 \ldots k)$ and send as overhead. This overhead would clearly require a padding of additional $k \log_2(q)$ bits. We call the overhead $(\theta_1, \theta_2, \ldots, \theta_k) \in F_q^k$, the transmitted 'code vector'.

For decoding of the messages in the RLC approach, the nodes start collecting the 'code vectors' as the protocol progresses. Once the dimension of the subspace spanned by the received 'code vectors' at a node becomes k, the node can recover all the messages.

The expected time (rounds) required for all the nodes to receive (decode) all the messages, and also the time required to receive all the messages with high probability for four cases: RLC with *pull*, RLC with *push*, RMS with *pull*, and RMS with *push* are derived in [10].

These result can be summarized as follows:

RLC with pull: Suppose $q \geq k$. Let $\overline{T}_{RLC}^{pull}$ be the random variable denoting the time required by all the nodes to get all the messages using an RLC approach with *pull mechanism*. Then, under assumption made earlier we have

$$\overline{T}_{RLC}^{pull} \leq 3.46k + O(\sqrt{k \ln(k)} \ln(n)), \; w.p. \; 1 - O(1/n) \tag{15.19}$$

where *w.p.* stands for *with probability* and O for *'order of'*. If T_{RLC}^{pull} is the time required for a particular node to get all the messages, then

$$E\ [T_{RLC}^{pull}] \leq 3.46k + O(\sqrt{k \ln(k)} \ln(n)) \tag{15.20}$$

RLC with a push: Suppose $q \geq \max(k, \ln(n))$. If $\overline{T}_{RLC}^{push}$ is the random variable representing the time required for all the nodes to get all the messages using an RLC protocol with push mechanism, then we have

$$\overline{T}_{RLC}^{push} \leq 5.96k + O(\sqrt{k} \ln(k) \ln(n)), \ \ \text{w.p. } 1 - O(1/n) \tag{15.21}$$

If T_{RLC}^{push} is the time required for a particular node to get all the messages with RLC based *push*, then

$$E\ [T_{RLC}^{push}] \leq 5.96k\ + O(\sqrt{k} \ln(k) \ln(n)) \tag{15.22}$$

If the nodes do not manipulate the packets and simply *store* and *forward* the packets which we refer to as RMS or 'random message selection,' we use the notation $T = \Omega(n \ln(n))$ to imply $T \geq cn \ln(n)$ for a suitable constant $c > 0$.

RMS with pull: Suppose $k = \alpha n$ for some $\alpha \leq 1$, and let $T_k^{RMSpull}$ be the time required for all the nodes to get all the k messages using an RMS protocol with *pull* mechanism. Then, we have [10]

$$ET_k^{RMSpull} = \Omega(n \ln n) $$

and

$$\lim_{k \to \infty} \Pr(T_k^{RMSpull} = \Omega(n \ln(n))) = 1 \tag{15.23}$$

RMS with a push: Let $k = \alpha n$ for some $\alpha \leq 1$, and let $T_k^{RMSpush}$ be the time required for all the nodes to get all the k messages using an RMS protocol with *push* mechanism. Then, we have

$$ET_k^{RMSpush} = \Omega(n \ln(n)) \tag{15.24}$$

and

$$\lim_{k \to \infty} Pr(T_k^{RMSpush} = \Omega(n \ln(n))) = 1$$

15.6   Network Coding With Reduced Complexity

In order to simplify the system and still achieve the multicast capacity, we can restrict the network coding only to the relaying nodes. Before we discuss this problem in more detail we will summarize the main results in network coding presented so far.

In the following, the network will be still represented by a directed graph $G = (V, E)$, where the vertex set V and the edge set E denote the nodes and links, respectively. It will be assumed that each edge, say from v to w, can be used to transfer information from its *tail* v to its *head* w at *unit-rate*. Multiple edges with the same tail and head are generally allowed, so as to represent different link capacities. Given G, the source set s and the set of destination nodes T, the *multicast capacity* refers to the maximum multicast rate from s to T.

An upper bound of the multicast capacity was established by examining the *cuts* that separate s from any destination $t \in T$. Given two nodes $s, t \in V$, an $s - t$ cut $(U, \overline{U})$ refers to a partition of the nodes $V = U + \overline{U}$ with $s \in U, t \in \overline{U}$. Denote the set of edges going from U to $\overline{U}$ by $\delta^{out}(U) \triangleq \{ e \in E | tail(e) \in U,\ head(e) \in \overline{U} \}$.

The *capacity* of the cut refers to the sum capacity of edges going from U to $\overline{U}$, which is equal to the cardinality of the cut since each edge has unit capacity. Denote the minimum capacity of an $s - t$ cut in (V, E) by $\rho_{s,t} \overset{\Delta}{=} \min_{U:s\in U, t\in\overline{U}} |\delta^{out}(U)|$. The significance of an $s - t$ cut comes from the fact that it exhibits a bottleneck for communication from s to t. It is intuitively clear that all information t can get from s must be derived from the information flowing across the cut. Since the capacity of any $s - t$ cut is an upper bound on the rate at which information can be transferred from s to t, $C_{s,T} \overset{\Delta}{=} \min_{t\in T} \rho_{s,t}$ provides an upper bound of the multicast capacity and is referred to as the cut bound of multicast capacity. The next question is whether s can multicast information to T at rate $\min_{t\in T} \rho_{s,t}$ and whether it is achievable by *conventional routing*, where nodes in the network store and forward information.

In the case where there is only a single destination node (unicast), i.e., $T = \{t\}$, a fundamental theorem in graph theory by Menger [18, 19] shows that the cut bound $\rho_{s,t}$ is achievable by conventional routing.

Menger's Theorem states that in a directed graph G, the maximum number of edge-disjoint $s - t$ paths (paths from s to t) is $\rho_{s,t}$. In other words, the *unicast capacity* can be achieved by routing information along $\rho_{s,t}$ edge-disjoint $s - t$ paths [18, 20].

In the case, where all nodes other than the source s are destinations (broadcast), i.e., $T = V - \{s\}$, another fundamental theorem in graph theory, by Edmonds [21], shows that the cut bound $C_{s,T}$ is again achievable by conventional routing.

Edmonds' Theorem states that in a directed graph $G = (V, E)$, the maximum number of edge disjoint directed spanning trees at $s \in V$ is $C_{s, V-\{s\}}$[21].

In other words, *broadcast capacity* can be achieved by routing information with $C_{s, V-\{s\}}$ edge-disjoint trees, each connecting s to all other nodes.

According to [19], in the graph theory literature some conjectures were made regarding possible generalizations of Edmonds' theorem to the general case, where there exist *Steiner nodes*. (The Steiner nodes are nodes other that the source s and the destinations T. They are used as relays.) However these conjectures have been disproved by concrete examples. One conjecture is about the existence of $C_{s,T}$ edge-disjoint *Steiner trees*, each connecting the source s with the destination nodes in T. Figure 15.10 shows an example graph G_1 where there are no two edge-disjoint Steiner trees connecting s with T although $C_{s,T} = 2$ [22]. *The source node s is shown with a filled square, the destinations t_1, t_2, t_3 are shown with unfilled squares, and the only Steiner node u is shown with circle. It can be checked that $\rho_{s,t_1} = \rho_{s,t_2} = \rho_{s,t_3} = 2$ and $\rho_{s,u} = 1$. Therefore, $C_{s,T} = 2$ and $C_{s,V-\{s\}} = 1$.*

It was shown in Sections 15.1 and 15.2 that routing in general cannot achieve the cut bound $C_{s,T}$. This result is stronger than saying $C_{s,T}$ edge-disjoint Steiner trees do not generally exist. In these sections we have shown that in a directed graph $G = (V, E)$ with unit capacity edges, the multicast capacity from source s to destination node T is $C_{s,T} = \min_{t\in T} \rho_{s,t}$ and it can be achieved by performing (linear time-invariant) network coding.

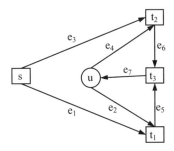

Figure 15.10 An example graph G_1 to show that there are no two edge-disjoint Steiner trees connecting s with T although $C_{s,T} = 2$.

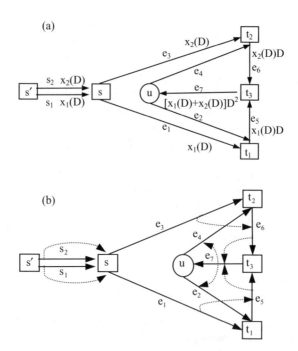

Figure 15.11 (a) An example linear time-invariant network coding solution achieving the multicast capacity. The graph shown is $G[s, 2] \triangleq (V \cup \{s'\}, E \cup \{s_1, s_2\})$. (b) Coding relations among the edges, illustrated by dashed arrows.

We now use the example graph G_1, in Figure 15.10, to introduce some additional details into the concept of linear time-invariant (LTI) network coding. The network is assumed to operates as a synchronous system with a discrete time index n running from 0 to $+\infty$. The source node s generates two information symbols (e.g., two bits), x_{1n} and x_{2n}, at time n. We use polynomial representations (in terms of the delay operator D) of two symbol streams $\{x_{1n}\}$ and $\{x_{2n}\}$, as

$$x_1(D) \triangleq \sum_{n=0}^{+\infty} x_{1n} D^n \;\; and \;\; x_2(D) \triangleq \sum_{n=0}^{+\infty} x_{2n} D^n \tag{15.25}$$

Assume that it takes one time unit for a symbol to propagate through an edge. Let $G[s, h]$ be the graph obtained by adding to G a new vertex s' and h *source edges*, $s_1, \ldots, s_h$. Figure 15.11(a) shows $G[s, 2]$ and illustrates an LTI network coding solution achieving the multicast capacity. The information carried by each edge is shown next to each edge. The i-th source edge s_i carries the i-th stream of symbols $x_i(D)$. This represents how information is originally injected into the network. As shown in Figure 15.11(a), $x_1(D)$ flows on e_1 and is then forwarded to e_5 after a unit delay, resulting in $x_1(D)D$ flowing on e_5. Similarly, $x_2(D)D$ flows on e_6. The received information from e_5 and e_6 is combined at t_3 to produce $[x_1(D) + x_2(D)]D^2$, which flows on e_7. The stream $[x_1(D) + x_2(D)]D^2$ is further forwarded to e_2 and e_4, after a unit delay. With this coding solution, t_1 receives $x_1(D)D$ from e_1 and $[x_1(D) + x_2(D)]D^4$ from e_2. It can then recover both source streams $x_1(D)$ and $x_2(D)$ after certain delay. Similarly, it can be checked that t_2 and t_3 can recover the source information. This confirms that this LTI coding solution achieves the multicast capacity.

Let us take now a more abstract look at how network coding is applied at each edge in this simple example. Figure 15.11(b) shows the *coding relations* among the edges, determined by the LTI network coding solution of Figure 15.11(a). We draw a dashed arrow from edge f to edge e if $head(f) = tail(e)$ and f is involved in the linear combination producing the information flowing on e. Given an LTI coding solution, for each edge $e \in E$, the edges with dashed arrows pointing to e are said to be the *coding predecessors* of e. For example, the information flowing on e_5 is produced as a delayed version of the information flowing on e_1, thus e_5 has only one coding predecessor e_1. If an edge e has more than one coding predecessor, then the information flowing on e is produced by mixing information on the coding predecessors.

It is seen from this example that mixing is in general required for optimality. Meanwhile, by Edmonds' theorem, when all nodes other than the source node are destinations, no mixing is necessary for achieving the multicast capacity. Since network coding is a more complex operation than routing, Ahlswede et al.'s theorem does not imply Edmonds' theorem.

While network coding promises the highest rate, the use of mixing presumably incurs a higher processing complexity at the nodes, which is less desirable from a practical standpoint. For example, when $C_{s,V-\{s\}} \approx C_{s,T}$, there may be incentive to use a routing-based approach that achieves rate $C_{s,V-\{s\}}$.

In this section we discuss how is possible to achieve the multicast capacity without paying the full complexity price, by using mixing only at some 'critical places' and routing at others. Where are the critical places? In Figure 15.11(b), information mixing is only performed at e_7. What distinguishes e_7 from other edges? In this example, only e_7 is entering a Steiner node and all other edges are entering destinations. We are thus motivated to classify the edges into two categories. (a) edges entering Steiner nodes, which we call *Steiner edges*; (b) edges entering destination nodes, which we call *terminal edges*. The main result of this section is the following.

In a directed graph $G = (V, E)$ with unit capacity edges, the multicast capacity from source node s to destinations T, $C_{s,T}$, is still achievable even if information mixing is only allowed on Steiner edges. In other words, there exists a capacity-achieving LTI network coding solution in which each terminal edge carries a delayed version of the information stream carried by one of its predecessor edges, or zero-hence each terminal edge performs routing only. This is proved as the *Unifying Theorem* in [17].

The main idea in establishing the unifying theorem is to force edges entering destinations to perform routing only. For acyclic graphs, it is possible to force routing via an algebraic approach. However, it is difficult to extend the algebraic approach to cyclic graphs. For cyclic graphs, it is possible to force routing via a graph theoretic approach.

To specify these procedures we let for a graph G, $V(G)$ and $E(G)$ denote its vertex set and edge set, respectively. For a cut $(U, \overline{U})$, let the edges going from $\overline{U}$ to U be $\delta^{in}(U)$ with $\delta^{in}(U) = \delta^{out}(\overline{U})$. We do not distinguish one-element vertex sets from its single element, e.g., $\delta^{in}(v) \triangleq \delta^{in}(\{v\})$.

Let $h \triangleq C_{s,T}$. When necessary, we write $\rho_{s,t}(G)$ instead of $\rho_{s,t}$, and $C_{s,T}(G)$ instead of $C_{s,T}$, to specify the graph G. Denote the set of terminal edges and Steiner edges by $E_T \triangleq \{e : \text{head}(e) \in T\}$ and $E_S \triangleq \{e : \text{head}(e) \in V(G[s, h]) - T\}$, respectively. The source edges $s_1, \dots, s_h$ are treated as Steiner edges as a notational convention.

For acyclic graphs, we can assume that edges have no delay without essential loss of generality and focus at one time slot, instead of discussing streams of data. Thus each edge $e \in E$ will carry one symbol, which we denote by y_e. The source edges $s_1, \dots, s_h$ will carry the h source symbols $x_1, \dots, x_h$, respectively.

Since the graph is acyclic, we can assume that each node waits for all symbols carried by its incoming edges to arrive before it sends out symbols on its outgoing edges. In a linear network code assignment, the symbol on edge e is a linear combination of the symbols on the edges entering tail (e), namely $y_e = \sum_{e':head(e')=tail(e)} w_{e,e'} y_{e'}$ We call the coefficients $\{w_{e,e'}\}$ the *mixing coefficients*. By induction, y_e on any edge e is a linear combination of the source symbols, namely $y_e = \sum_{i=1}^{h} q_{e,i} x_i$. The vector $\mathbf{q}_e \triangleq [q_{e,1}, \dots, q_{e,h}]$ is known as the global coding vector along edge e. It can be determined recursively as $\mathbf{q}_e = \sum_{e':head(e')=tail(e)} w_{e,e'} \mathbf{q}_{e'}$ where q_{s_i} is the ith unit vector ε_i. Since each y_e is a linear combination of the source symbols, any destination t receiving h symbols with linearly independent global coding vectors can recover the source symbols.

A capacity-achieving linear network code assignment is an assignment of mixing coefficients $\{w_{e,e'}\}$ such that all destinations $t \in T$ can recover all source symbols. Various algorithms for finding a capacity-achieving linear network code assignment has been proposed in the literature [22–25].

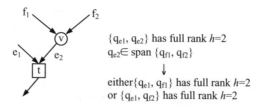

Figure 15.12   An illustration of the algebraic approach [17]. Reproduced by permission of IEEE © 2006.

Note that a capacity-achieving linear network code assignment for an acyclic graph $G = (V, E)$ exists if and only if there is a set of global coding vectors $\{q_e\}_{e \in E}$ that satisfy:

$$\forall e \in E, \quad \mathbf{q}_e \in span\{\mathbf{q}_f, head(f) = tail(e)\} \tag{15.26}$$
$$\forall t \in T, \{\mathbf{q}_e, head(e) = t\} \text{ has full rank } h$$

Given an acyclic graph G, s, T, let $\{w_{e,e'}\}$ be a capacity-achieving linear network coding assignment. This determines the global encoding vectors $\{\mathbf{q}_e\}_{e \in E}$, which satisfy (15.26).

We now show how to force terminal edges to perform routing only, starting with a capacity-achieving linear network coding solution. Visit the terminal edges in E_T in a topological order. When $e \in E_T$ is visited, we force e to perform routing only, by setting $\mathbf{q}_e := \mathbf{q}_f$ for some predecessor edge f with $head(f) = tail(e)$, while ensuring (15.26) are satisfied. We now show why it is always possible to do so.

Suppose the terminal edge e (with $head(e) = t$) is currently being visited. For simplicity, we label the edges entering $tail(e)$ as $f_1, \dots, f_n$. We label the edges entering $head(e) = t$ as $e_1, \dots, e_{m+1}$ with $e_{m+1} = e$. This is illustrated in Figure 15.12. From (15.26), $\mathbf{q}_e \in span\{\mathbf{q}_{f_1}, \dots, \mathbf{q}_{fn}\}$. From (15.26), $\{\mathbf{q}_{e_1}, \dots, \mathbf{q}_{e_m}, \mathbf{q}_e\}$ has full rank h. Therefore, there must exist an edge $f_i, 1 \leq i \leq n$ such that $\{\mathbf{q}_{e_1}, \dots \mathbf{q}_{e_m}, \mathbf{q}_{f_i}\}$ has full rank h. We then set $\mathbf{q}_e := \mathbf{q}_{f_i}$. After this change, it is easy to see that (15.26) is still satisfied, and (15.26) is satisfied for e and all edges e' with tail $(e') \neq t$. In addition, (15.26) is satisfied for all outgoing edges of t, because $\{\mathbf{q}_{e_1}, \dots, \mathbf{q}_{e_m}, \mathbf{q}_e\}$ has full rank after setting $\mathbf{q}_e := \mathbf{q}_{f_i}$.

In addition, since the terminal edges are considered in a topological order, after $e \in E_T$ is visited and $\mathbf{q}_e$ is set to $\mathbf{q}_f$ for some predecessor edge f, $\mathbf{q}_f$ will not be updated in subsequent steps. Hence each step forces one more terminal edge to perform routing only. After $|E_T|$ steps, we will have a capacity-achieving linear coding solution where all terminal edges perform routing only.

The algebraic argument presented above does not readily carry over to cyclic graphs. The complication due to cycles can be handled by resorting to a graph theoretic approach [17].

15.7   Multisource Multicast Network Switching

In the section, by considering network switching as a special form of network coding, we discuss the achievable information rate region for *multi* source multicast network switching. Network switching is essentially a problem of multicast-route packing in the multicast networks. Based on this, the theory of games was used in [26] to formulate the network switching as a *matrix game* between the first 'player' of links and the second 'player' of multicast routes in the multicast networks. It was proven that the maximum achievable information rate at each probabilistic direction in the information rate region is the reciprocal of the value of the corresponding game. Consequently, the maximum achievable information rate can be computed in a simple way by applying the existing theory and algorithms for the computation of the value of a matrix game, especially for such multicast networks with links all having unit capacity as the multicast networks discussed so far in this chapter. For multicast networks with links having arbitrary positive real-valued capacity, by using convex optimization, [26] develops a simple and efficient iterative algorithm to find the maximum achievable information rates for multisource multicast network switching.

We start with some preliminaries in the directed graphs and multicast networks and introduce the relevant notion. The network will be represented again by a directed graph (or digraph) $G = (V, E)$, where V is the finite set of nodes (or vertexes) in the network and $E \subseteq G \times G^d = \{\,(\,u,\ v)\,|u \in V\ and\ v\ \in V\}$ is the set of links (arcs, or directed edges) representing the communication channels. We denote by $|V|$ and $|E|$ the number of nodes and the number of links in the network, respectively. The digraph $G = (V, E)$ is said to be a digraph of order V and size $|E|$.

For a given digraph $G = (V,\ E)$ of *order m* with $V = \{v_1, v_2, \ldots, v_m\}$, a set (or sequence) of l links in E of the form $\{(u_1, u_2), (u_2, u_3), \ldots, (u_l, u_{l+1})\}$ is said to be a *path* of length ℓ if the nodes $u_1, u_2, \ldots, u_\ell$ are distinct. We also denote the above path by $\{u_1 \to u_2, u_2 \to u_3, \ldots, u_\ell \to u_{\ell+1}\}$ or $u_1 \to u_2 \to u_3 \to \cdots \to u_\ell \to u_{\ell+1}$. If $u_1 = u_{\ell+1}$, we say that the path is *closed* and it is a *directed cycle* of length ℓ. Otherwise, we say that the path is *open* and the node u_{l+1} is *reachable* by the node u_1. If the digraph $G = (V, E)$ has a directed cycle of some length ℓ as $u_1 \to u_2 \to\to u_\ell \to u_1$, then G is said to be *cyclic*. Otherwise, G is said to be *acyclic*.

For the underlying digraph $G = (V, E)$ of the network, we define a *capacity function* $\Theta\ :\ G \times G \to R^+ = [0, +\infty)$ such that $\Theta(u, v)$ is positive for all $(u, v) \in E$ and zero otherwise. In the following, we use the triplet $N = (V, E, \Theta)$ to denote the point-to-point communication network. Let $V = \{v_1, v_2, \ldots, v_m\}$ with $m = |V|$. We can define the matrix $C = (c_{ij})$ with $c_{ij} = \Theta(v_i, v_j)$ for all $i, j = 1, 2, \ldots, m$. If $(v_i, v_j) \in E$, then $c_{ij} > 0$ is the capacity of the link (v_i, v_j) in bits per channel use. The aforementioned matrix C is said to be the *capacity matrix* of the network, which is a real-valued $m \times m$ matrix with zero diagonal entries and nonnegative off-diagonal entries. Conversely, for such an $m \times m$ matrix $C = (c_{ij})$, we can also define a corresponding *network* $N = (V, E, \Theta)$ of *order m* with the set of nodes $V = \{v_1, v_2, \ldots, v_m\}$, the set of links $E = \{(v_i, v_j)|c_{ij} > 0, i, j = 1, 2, \ldots, m\}$, and the capacity function $\Theta: G \times G \to R^+$ as $\Theta(v_i, v_j) = c_{ij}$ for all $i, j = 1, 2, \ldots, m$. A network $N = (V, E, \Theta)$ is called *cyclic* (respectively, *acyclic*) if the underlying digraph $G = (V, E)$ is *cyclic* (respectively, *acyclic*).

Suppose that the information transmission over each link is essentially error-free if and only if the transmission rate over the link is not greater than the capacity of the link. Let X be an information source of bit sequences generated at a specific *source node* $s \in V$. The information for source X is transported over the network N from the source node s to other specific and distinct *sink* (or *destination*) *nodes* $t_1, t_2, \ldots, t_L$ at each of which we can reconstruct the information for source X sent from s. We say that the information for source X is *multicast* to the set of L distinct sink nodes $\{t_1, t_2, \ldots, t_L\} \subseteq V$ from the source node $s \notin \{t_1, t_2, \ldots, t_L\}$. The network N becomes functionally a *multicast network*.

In the model we allow the capacity of each link to be arbitrarily real-valued. Therefore, we can update the capacity matrix C of the network N so that we can assume that every link has one channel use per unit time without loss of generality. For example, for some link $(v_i, v_j) \in E$ with capacity c_{ij} in bits per channel use and bandwidth of $\rho(\rho \in (0, +\infty))$ channel uses per unit time and, consequently, the bit rate of ρc_{ij} in bits per unit time, we can update the capacity c_{ij} to ρc_{ij}. The link can now be regarded as having capacity ρc_{ij} in bits per channel use and bandwidth of one channel use per unit time and, consequently, the same bit rate of ρc_{ij} in bits per unit time. In what follows, for the convenience of mathematical formulation and analysis, we consider the *transmission rate* of each link measured in terms of bits per channel use instead of bits per unit time, under the assumption that every link has one channel use per unit time.

We employ the network $N = (V, E, \Theta)$ of order m to multicast the information for source X from the source node s to the sink nodes $t_1, t_2, \ldots, t_L$. Suppose that k bits of information for source X is multicast from s to $t_1, t_2, \ldots, t_L$ and each link has an appropriate number of channel uses in the multicast procedure. Let n be the largest number of channel uses among all links in the network. We define the ratio $R = k/n$ as the *information rate* of the multicast network in terms of bits per network use.

A rate R in bits per network use is said to be *asymptotically achievable* or simply *achievable* if for any arbitrarily small number $\varepsilon > 0$, there exist two positive integers n and k with $k/n > R - \varepsilon$ such that k bits of information for source X can be multicast from the source node s to the L sink nodes $t_1, t_2, \ldots, t_L$ and each link has at most n channel uses in the multicast procedure.

A basic problem in the multicast networks is the determination of the maximum achievable information rate.

In the previous sections of this chapter we demonstrated that a multicast network with network coding can achieve a multicast information rate that is beyond the reach of the multicast approach with network switching.

The aforementioned discussion is focused on the *single*-source multicast networks. The information transportation problem can be generalized to *multi* source multicast networks. Let $N = (V, E, \Theta)$ be a given network with the underlying digraph $G = (V, E)$ of order m. Let $X_1, X_2, \ldots, X_r$ be r mutually independent information sources, which are assumed to be generated at specific *source nodes* $s_1, s_2, \ldots, s_r \in V$, respectively. These source nodes $s_1, s_2, \ldots, s_r$ are allowed to be repeated. For example, $s_1 \in V$ and $s_2 \in V$ may be or may not be the same source node. Now, for each source index $d = 1, 2, \ldots, r$, we employ the network $N = (V, E, \Theta)$ of order m to multicast information for source X_d from the source node s_d to the corresponding L_d distinct sink nodes $t_1^{(d)}, t_2^{(d)}, \ldots, t_{L_d}^{(d)} \in V$ with $s_d \notin \{t_1^{(d)}, t_2^{(d)}, \ldots, t_{L_d}^{(d)}\} \subseteq V$. Therefore, the network N becomes functionally a multisource multicast network. If $r = 1$, the network N is said to be a single-source multicast network. We continue to assume that the information transmission over each link is essentially error-free if and only if the transmission rate over the link is not greater than its capacity.

For each $d = 1, 2, \ldots, r$, suppose that $k^{(d)}$ bits of information for source X_d is multicast from the source node s_d to the corresponding sink nodes $t_1^{(d)}, t_2^{(d)}, \ldots, t_{L_d}^{(d)}$. In the multicast procedure, each link has a certain number of channel uses. Let n be the largest number of channel uses among all links in the multicast network. We define the vector of ratios

$$\omega = (\omega_1, \omega_2, \ldots, \omega_r) = \left(\frac{k^{(1)}}{n}, \frac{k^{(2)}}{n}, \ldots, \frac{k^{(r)}}{n} \right) \tag{15.27}$$

as the information rate tuple of the multicast network, each rate component of which is in terms of bits per network use. A rate tuple $\omega = (\omega_1, \omega_2, \ldots, \omega_r)$ is said to be *asymptotically achievable* or simply *achievable* if for any arbitrarily small number $\varepsilon > 0$, there exist $r + 1$ positive integers $k^{(1)}, k^{(2)}, \ldots, k^{(r)}$ and n with $k^{(d)}/n > \omega_d - \varepsilon$ such that $k^{(d)}$ bits of information for source X_d can be *multicast* from the source node s_d to the corresponding L_d sink nodes $t_1^{(d)}, t_2^{(d)}, \ldots, t_{L_d}^{(d)}$ for each $d = 1, 2, \ldots, r$ and each link in the network has at most n channel uses in the multicast procedure. The set of all achievable rate tuples, denoted by Ω, is said to be the achievable information rate region of the multisource multicast network.

A fundamental problem in a multisource multicast network is the determination of the achievable information rate region Ω. It is clear that the achievable information rate region for multisource multicast network with *network coding* will contain that for the same multisource multicast network with *network switching*. However, unlike the general single-source multicast network coding, a *theoretical* or *computational* determination of the achievable information rate region for the general *multi* source multicast network coding remains unsolved.

In this section, it will be shown that, for network switching as a special form of network coding, we can make a complete determination theoretically and computationally of the maximum achievable information rate region for *multi* source multicast networks whose underlying digraphs may be *acyclic* or *cyclic* and whose links may have arbitrary positive *integer* or *real-valued* capacity.

15.7.1 Conventional Route Packing Problem

A necessary and sufficient condition for a set of links in E, denoted by T, to be used for a successful multicast of one bit of information for source X from s to $t_1, t_2, \ldots, t_L$ with network switching is that T includes all the links of some open path from s to t_ℓ for each $p = 1, 2, \ldots, L$. For the *maximum efficiency* of the network N, it is clear that the aforementioned set of links T for multicast network switching must be *minimal* in the sense that T with deletion of its any link would not include all the links of any given open path from s to t_l in the underlying digraph G for *some* $l \in \{1, 2, \ldots, L\}$.

Let the set of links $T \subseteq E$ be a *minimal* set in the above sense which can be used to multicast one bit of information for source X from s to $t_1, t_2, \ldots, t_L$ with network switching. We denote by N' the set of nodes that are a sender or a receiver of some link in T. We call (N', T) as a digraph *induced* by T, which is a subgraph of $G = (V, E)$. Because T is minimal in the above sense, its induced digraph (N', T) must be a rooted tree with the unique root node s. A *rooted tree* is defined as an acyclic digraph with a unique node, called its *root node*, which has the property that there exists a *unique* path from the root node to each other node. The nodes in a rooted tree that have no output link are called *leaves*. It is seen that (N', T) is a rooted tree with the root node s and all leaves being sink nodes (but possibly a sink node $t_l \in N'$ not being a leaf).

In the following, we call the set of links $T \subseteq E$ or its induced digraph (N', T) as a *multicast route* of the underlying digraph $G = (V, E)$ from the source node $s \in V$ to the sink nodes $t_1, t_2, \ldots, t_L \in V$ if the digraph (N', T) induced by T is a rooted tree of G with $t_1, t_2, \ldots, t_L \in N$, whose root node is s and whose leaves are all sink nodes. If $L = 1$, a multicast route T is the set of links of an open path from the source node s to the sink node t_1.

For the underlying digraph $G = (V, E)$ of order m, we denote the set of links by $E = \{e_1, e_2, \ldots, e_I\}$ with $I = |E|$ and the set multicast routes by $\{T_1, T_2, \ldots, T_J\}$ with J being the number of distinct multicast routes of G from the source node $s \in V$ to the sink nodes $t_1, t_2, \ldots, t_L \in V$. For each multicast route T_j, we define an *indicator function* over E as $\chi_{T_j}(e_i) = 1$ if $e_i \in T_j$ and 0 if $e_i \notin T_j$, for $i = 1, 2, \ldots, I$ and $j = 1, 2, \ldots, J$.

In the general multisource case, let $X_1, X_2, \ldots, X_r$ be r mutually independent information sources and $N = (V, E, \Theta)$ be a given network with the underlying digraph $G = (V, E)$ of order m. For each $d = 1, 2, \ldots, r$, we want to employ the network N with network switching to multicast $k^{(d)}$ bits of information for source X_d from the source node $s_d \in V$ to the other L_d distinct reachable sink nodes $t_1^{(d)}, t_2^{(d)}, \ldots, t_{L_d}^{(d)} \in V$. It is allowed for two source nodes to be identical. In this case, however, we require that the two sets of their corresponding sink nodes cannot be identical.

Let N' denote the set of positive integers. For each $d = 1, 2, \ldots, r$, we denote the set of multicast routes of the underlying digraph $G = (V, E)$ from the source node s_d to the sink nodes $t_1^{(d)}, t_2^{(d)}, \ldots, t_{L_d}^{(d)}$ by $\{T_1^{(d)}, T_2^{(d)}, \ldots, T_{J_d}^{(d)}\}$ with $J_d \in N'$. From the preceding discussion, we know that each bit of the $k^{(d)}$ bits of information for source X_d will be multicast from s_d to $t_1^{(d)}, t_2^{(d)}, \ldots, t_{L_d}^{(d)}$. Let $k_j^{(d)} \in N \cup \{0\}$ be the number of bits that are multicast from s_d to $t_1^{(d)}, t_2^{(d)}, \ldots, t_{L_d}^{(d)})$ with the jth multicast route $T_j^{(d)}$ for $j = 1, 2, \ldots, J_d$. The overall number of multicast bits from source d is $k^{(d)} = k_1^{(d)} + k_2^{(d)} + \cdots + k_{J_d}^{(d)}$, for $d = 1, 2, \ldots, r$. We define the following subset

$$\Phi = \bigcup_{\substack{k_j^{(d)} \in N' \cup \{0\} \\ 1 \leq d \leq r, 1 \leq j \leq J_d}} \left\{ \omega = (\omega_1, \omega_2, \ldots, \omega_r) | 0 \leq \omega_d \leq \frac{k^{(d)}}{n} \quad \text{for} \quad d = 1, 2, \ldots, r \right\} \subseteq R^r$$

where n is given by

$$n = \max_{1 \leq i \leq I} \frac{1}{\Theta(e_i)} \sum_{d=1}^{r} \sum_{j=1}^{J_d} k_j^{(d)} \chi_{T_j^{(d)}}(e_i) \tag{15.28}$$

The closure of the subset Φ is defined as

$$\Omega = cl(\Phi) = cl \left(\bigcup_{\substack{k_j^{(d)} \in N' \cup \{0\} \\ 1 \leq d \leq r, 1 \leq j \leq J_d}} \left\{ \omega = (\omega_1, \omega_2, \ldots, \omega_r) | 0 \leq \omega_d \leq \frac{k^{(d)}}{n} \quad \text{for} \quad d = 1, 2, \ldots, r \right\} \right) \subseteq R^r \tag{15.29}$$

For any given network $N = (V, E, \Theta)$ with links having arbitrary positive real valued capacity, the achievable information rate region for multisource multicast network switching from the source node s_d to the other L_d distinct reachable sink nodes $t_1^{(d)}, t_2^{(d)}, \ldots, t_{L_d}^{(d)}$ for $d = 1, 2, \ldots, r$, denoted by $\Omega \subseteq R^r$, is given by (15.29), namely, the closure of the subset Φ with n given by (15.28) and with $\{e_1, e_2, \ldots, e_I\}$ being the set of links E and $\{T_1^{(d)}, T_2^{(d)}, \ldots, T_{J_d}^{(d)}\}$ being the set of multicast routes from s_d to $t_1^{(d)}, t_2^{(d)}, \ldots, t_{L_d}^{(d)}$ for $d = 1, 2, \ldots, r$ [26].

For *single-source* multicast network switching, it follows from above that the maximum achievable information rate is given by

$$R^* = \sup_{\substack{k_j \in N' \cup \{0\} \\ 1 \le j \le J}} \frac{k_1 + k_2 + \cdots + k_J}{n} \tag{15.30}$$

where (15.28) reduces to

$$n = n(k_1, k_2, \ldots k_J) = \max_{1 \le i \le I} \frac{1}{\Theta(e_i)} \sum_{j=1}^{J} k_j \chi_{T_j}(e_i) \tag{15.28a}$$

with $\{e_1, e_2, \ldots, e_I\}$ being the set of links E and $\{T_1, T_2, \ldots, T_J\}$ being the set of multicast routes from s to $t_1, t_2, \ldots, t_L$. In the *finite* case, it can also be interpreted as a route packing problem as follows. Assume that the capacity of each link in the network is one. Given any finite number n, find the maximum number of multicast routes with multiplicity which have the property that each link in E can belong to at most n multicast routes. More precisely, this route packing problem can be generally formulated as

$$\text{maximize } k = k_1 + k_2 + \cdots + k_J$$
$$\text{subjected to } \sum_{j=1}^{J} k_j \chi_{T_j}(e_i) \le n\Theta(e_i) \quad \text{for } i = 1, 2, \ldots, I \tag{15.31}$$

where $k_j \in N' \cup \{0\}$ for $j = 1, 2, \ldots, J$. Therefore, the maximum achievable information rate R^* can be thought of as the supremum of k/n for all $n \in N$, where $k = k(n)$ denotes the maximum value in (15.30) for any given $n \in N$. This becomes the maximum fractional Steiner tree packing problem [27–29]. The problem of finding the Steiner packing number, namely, the maximum achievable information rate R^* for single-source *network switching*, of any given network $N = (V, E, \Theta)$ is NP-hard [28]. To simplify the problem, in the next section, a connection between network switching and matrix game will be established and, consequently, the maximum achievable information rate R^* can often be calculated with only some, but not all, of the multicast routes by using the dominance relation and successive-elimination technique in the game theory.

15.7.2 Multicast Network Switching as Matrix Game

This section establishes the connection of the network switching with the theory of games. As a result, the complexity of finding the maximum achievable information rate region for network switching can be reduced to solving an equivalent problem in the theory of games. Then, we can apply the existing concepts, results, methods, and algorithms in the field of game theory to provide a satisfactory theoretical and computational solution for maximum information flow in the multisource multicast networks with network switching.

For the basics of matrix game theory the reader is referred to [30–32]. We denote the set of all I-dimensional probability distribution vectors (or mixed strategies) by

$$\Delta^I = \{(p_1, p_2, \ldots, p_I) \in R^I \mid \sum_{i=1}^{I} p_i = 1 \}$$

and

$$p_i \geq 0 \quad \text{for } i = 1, 2, \ldots, I\} \tag{15.32}$$

For a matrix game with the $I \times J$ payoff matrix A, a necessary and sufficient condition for a mixed-strategy pair (p^*, q^*) with $p^* \in \Delta^I$ and $q^* \in \Delta^J$ to be a Nash equilibrium point and for a real number $v \in R$ to be the value of the game is that every component of the vector $(p^*)^T A \in R^J$ is greater than or equal to v and every component of the vector $A q^* \in R^I$ is less than or equal to v [32].

For the multisource multicast network switching, the achievable information rate region is a subset in a high-dimensional Euclidean space. For the connection of multisource network switching with a matrix game whose value is a one-dimensional scalar, we should provide a *scalar-valued* representation for the high-dimensional achievable information rate region.

The region $\Omega \subseteq R^r$ is *star-like* with respect to the origin in the sense that for any point $\omega = (\omega_1, \omega_2, \ldots, \omega_r) \in \Omega$, we have $\rho \omega = (\rho \omega_1, \rho \omega_2, \ldots, \rho \omega_r) \in \Omega$ for all $0 \leq \rho < 1$. Furthermore, the region $\Omega \subseteq R^r$ is a *closed* subset in R^r.

For any nonorigin point $\omega = (\omega_1, \omega_2, \ldots, \omega_r) \in \Omega \subseteq R^r$, we say that the probability distribution vector

$$D = D(\omega) \triangleq \left(\frac{\omega_1}{\varpi}, \frac{\omega_2}{\varpi}, \ldots, \frac{\omega_r}{\varpi} \right) \in \Delta^r \; ; \; \varpi = \sum_{d=1}^{r} \omega_d \tag{15.33}$$

is the *probabilistic direction* of ω. It is clear that, for each nonorigin point $\omega \in \Omega$, we have $\omega = \varpi D(\omega)$.

Based on the star-like and closed properties of Ω, we can define the following function.

For the achievable information rate region Ω and any given probabilistic direction $D \in \Delta^r$, we say that

$$\Re = \Re(D) \triangleq \sup\{\rho \in R^+ | \rho D \in \Omega\} \tag{15.34}$$

is the maximum achievable directional information rate in Ω with respect to D. The function $\Re = \Re(D)$ for $D \in \Delta^r$ is said to be the *rate-direction function* of the achievable information rate region Ω.

For the achievable information rate region Ω, its rate-direction function $\Re(D)$ plays a role of an indicator. For any given nonorigin point $\omega = (\omega_1, \omega_2, \ldots, \omega_r) \in R^r$ with $\omega_d \geq 0$ for $d = 1, 2, \ldots, r$, $\omega \in \Omega$ if and only if $\varpi \leq \Re(D(\omega))$.

In this sense, the determination of the achievable information rate region Ω for multisource network switching can reduce to that of its rate-direction function $\Re(D)$. As expected, for the single-source network switching (i.e., $r = 1$ implying $\Delta^r = \{(1)\}$), the maximum achievable *directional* information rate $\Re(D)$ for $D \in \Delta^r$ is reduced to the maximum achievable information rate R^* of the single-source multicast network.

For the multicast network switching, we are interested in finding the rate-direction function $\Re(D)$ for $D \in \Delta^r$. It can be shown that for any given network $N = (V, E, \Theta)$ with links having arbitrary positive real-valued capacity, the rate-direction function $\Re(D)$ for multisource multicast network switching from the source node s_d to the other L_d distinct reachable sink nodes $t_1^{(d)}, t_2^{(d)}, \ldots, t_{L_d}^{(d)}$ for $d = 1, 2, \ldots, r$ has the two properties [26]

$$\Re(D) > 0, \text{ for all } a = (\delta_1, \delta_2, \ldots, \delta_r) \in \Delta^r \tag{15.35}$$

and

$$\frac{1}{\Re(D)} = \min_{\substack{q^{(d)} \in \Delta^{J_d} \\ 1 \leq d \leq r}} \max_{p \in \Delta^I} \left(\sum_{d=1}^{r} p^T A_D^{(d)} q^{(d)} \right)$$

$$\tag{15.36}$$

$$= \min_{\substack{q^{(d)} \in \Delta^{J_d} \\ 1 \leq d \leq r}} \max_{p \in \Delta^I} \sum_{i=1}^{I} \sum_{d=1}^{r} \sum_{j=1}^{J_d} p_i \left(\frac{\delta_d}{\Theta(e_i)} \chi_{T_j^{(d)}}(e_i) \right) q_j^{(d)}$$

where $A_D^{(d)} = (a_{ij}^{(d)})$ is an $I \times J_d$ matrix given by

$$a_{ij}^{(d)} = \frac{\delta_d}{\Theta(e_i)} \chi_{T_j^{(d)}}(e_i) \qquad (15.37)$$

for $i = 1, 2, \ldots, I$ and $j = 1, 2, \ldots, J_d$ with $\{e_1, e_2, \ldots, e_I\}$ being the set of links E and $\{T_1^{(d)}, T_2^{(d)}, \ldots, T_{J_d}^{(d)}\}$ being the set of multicast routes from s_d to $t_1^{(d)}, t_2^{(d)}, \ldots, t_{L_d}^{(d)}$ for $d = 1, 2, \ldots r$.

The optimization problem in (15.36) is a general *minimax* problem and applying the general *minimax* theory [33, p. 218, Theorem 5.2]), we know that

$$\min_{\substack{q^{(d)} \in \Delta^{J_d} \\ 1 \le d \le r}} \max_{p \in \Delta^I} \left(\sum_{d=1}^{r} p^T A_D^{(d)} q^{(d)} \right) = \max_{p \in \Delta^I} \min_{\substack{q^{(d)} \in \Delta^{J_d} \\ 1 \le d \le r}} \left(\sum_{d=1}^{r} p^T A_D^{(d)} q^{(d)} \right) \qquad (15.38)$$

Moreover, there exist $p^* \in \Delta^I$ and $q^{(d)*} \in \Delta^{J_d}$ for $d = 1, 2, \ldots, r$ such that

$$\min_{\substack{q^{(d)} \in \Delta^{J_d} \\ 1 \le d \le r}} \max_{p \in \Delta^I} \left(\sum_{d=1}^{r} p^T A_D^{(d)} q^{(d)} \right) = \max_{p \in \Delta^I} \left(\sum_{d=1}^{r} p^T A_D^{(d)} q^{(d)*} \right) = \sum_{d=1}^{r} (p^*)^T A_D^{(d)} q^{(d)*} \qquad (15.39)$$

and

$$\max_{p \in \Delta^I} \min_{\substack{q^{(d)} \in \Delta^{J_d} \\ 1 \le d \le r}} \left(\sum_{d=1}^{r} p^T A_D^{(d)} q^{(d)} \right) = \min_{\substack{q^{(d)} \in \Delta^{J_d} \\ 1 \le d \le r}} \left(\sum_{d=1}^{r} (p^*)^T A_D^{(d)} q^{(d)} \right)$$

$$= \sum_{d=1}^{r} (p^*)^T A_D^{(d)} q^{(d)*} \qquad (15.40)$$

From (15.36), it is clear to see that the rate-direction function $\Re(D)$ is a bounded and continuous function for $D \in \Delta^r$. So, from (15.34) we know that the achievable information rate region Ω is a bounded star-like region. Because Ω is also closed, it is a nonempty *star-like* and *compact* subset in R^r.

Now, we can provide the following game-theoretic formulation for multisource multicast network switching [26].

Rate-direction function game: For any given network $N = (V, E, \Theta)$ with links having arbitrary positive real-valued capacity, the rate-direction function $\Re(D)$ with $D = (\delta_1, \delta_2, \ldots, \delta_r) \in \Delta \Delta^r$ for multisource multicast network switching from the source node s_d to the other L_d distinct reachable sink nodes $t_1^{(d)}, t_2^{(d)}, \ldots, t_{L_d}^{(d)}$ for $d = 1, 2, \ldots, r$ equals the reciprocal of the value of the corresponding matrix game with the $I \times \prod_{d=1}^{r} J_d$ payoff matrix $A_D = (a_{ij})$ given by

$$a_{ij} = \frac{\delta_1}{\Theta(e_i)} \chi_{T_{j_1}^{(1)}}(e_i) + \frac{\delta_2}{\Theta(e_i)} \chi_{T_{j_2}^{(2)}}(e_i) + \cdots + \frac{\delta_r}{\Theta(e_i)} \chi_{T_{j_r}^{(r)}}(e_i) \qquad (15.41)$$

for $i = 1, 2, \ldots, I$ and

$$j \in I_r \overset{\Delta}{=} \{(j_1, j_2, \ldots, j_r) | 1 \le j_d \le J_d \text{ for } d = 1, 2, \ldots, r\}$$

with $\{e_1, e_2, \ldots, e_I\}$ being the set of links E and $\{T_1^{(d)}, T_2^{(d)}, \ldots, T_{J_d}^{(d)}\}$ being the set of multicast routes from s_d to $t_1^{(d)}, t_2^{(d)}, \ldots, t_{L_d}^{(d)}$ for $d = 1, 2, \ldots, r$.

So, network switching for multisource multicast networks can be interpreted as a matrix game. Player I has the pure strategies as the set of links $\{e_1, e_2, \ldots, e_I\}$. Player II has the pure strategies as the

Cartesian product of the r sets of multicast routes $\{T_1^{(d)}, T_2^{(d)}, \ldots, T_{J_d}^{(d)}\}$ from s_d to $t_1^{(d)}, t_2^{(d)}, \ldots, t_{L_d}^{(d)}$ for $d = 1, 2, \ldots, r$. The payoff matrix is given by (15.41), where a_{ij} is the needed number of channel uses in the link e_i per bit of information with its proportion δ_d for source X_d transported over the multicast route $T_{j_d}^{(d)}$ for $i = 1, 2, \ldots, I, d = 1, 2, \ldots, r$ and $j_d = 1, 2, \ldots, J_d$. Player I chooses a mixed strategy $p \in \Delta^I$ and player II chooses a mixed strategy $q \in \Delta^{\prod_{d=1}^{r} J_d}$. In a play of the game, player II pays the amount a_{ij} to player I if player I chooses the ith link e_i and player II chooses the *set* of multicast routes $\{T_{j_1}^{(1)}, T_{j_2}^{(2)}, \ldots, T_{j_r}^{(r)}\}$ for $i = 1, 2, \ldots, I$ and $j = (j_1, j_2, \ldots, j_r)$ with $1 \leq j_d \leq J_d$ for $d = 1, 2, \ldots, r$. In the sense of mixed strategies, the *value* of the game is the expected payoff $(p^*)^T A_D q^*$ when both players I and II choose their optimal mixed strategies $p^* \in \Delta^I$ and $q^* \in \Delta^{\prod_{d=1}^{r} J_d}$ respectively.

The optimal mixed strategy $q^* \in \Delta^{\prod_{d=1}^{r} J_d}$ for player II provides an optimal network switching scheme for multisource multicast network switching. As a result, the multicast network achieves a maximum achievable *directional* information rate with respect to $D \in \Delta\Delta^r$ and the directional information rate $\Re(D)$ is the *reciprocal* of the value of the corresponding game with payoff matrix A_D.

Sum Rate Game: For the achievable information rate region Ω of the multicast network switching, we are also interested in finding the maximum possible sum rate for any point $\omega \in \Omega$ as the sum of its components. More precisely, for the achievable information rate region Ω, we say that

$$\Re^* = \Re^*(\Omega) \overset{\Delta}{=} \sup\{\omega_1 + \omega_2 + \cdots + \omega_r | \omega = (\omega_1, \omega_2, \ldots, \omega_r) \in \Omega\} \quad (15.42)$$

is the maximum achievable information sum-rate in Ω.

Obviously, for the single-source network switching (i.e., $r = 1$), the maximum achievable information *sum-rate* $\Re^*$ is reduced to the maximum achievable information rate R^* of the single-source multicast network.

Moreover, $\Re^*$ can be viewed as the maximum value among all maximum achievable *directional* information rates $\Re(D)$ in Ω, as [26]

$$\Re^* = \max\{\omega_1 + \omega_2 + \cdots + \omega_r | \omega = (\omega_1, \omega_2, \ldots, \omega_r) \in \Omega\}$$
$$= \max_{D \in \Delta^r} \Re(D) \quad (15.42)$$

The maximum achievable information sum-rate $\Re^*$ for multisource multicast network switching has a game-theoretic formulation as follows. For any given network $N = (V, E, \Theta)$ with links having arbitrary positive real-valued capacity, the maximum achievable information sum-rate $\Re^*$ for multisource multicast network switching from the source node s_d to the other L_d distinct reachable sink nodes $t_1^{(d)}, t_2^{(d)}, \ldots, t_{L_d}^{(d)}$ for $d = 1, 2, \ldots, r$ equals the reciprocal of the value of the matrix game with the $I \times \sum_{d=1}^{r} J_d$ payoff matrix $U = (a_{ij})$ given by

$$a_{ij} = \frac{1}{\Theta(e_i)} \chi_{T_{j_d}^{(d)}}(e_i) \quad (15.43)$$

for $i = 1, 2, \ldots, I$ and $j \in \{j\}_r \overset{\Delta}{=} \{(d, j_d) | d = 1, 2, \ldots, r$ and $j_d = 1, 2, \ldots, J_d\}$ with $\{e_1, e_2, \ldots, e_I\}$ being the set of links E and $\{T_1^{(d)}, T_2^{(d)}, \ldots, T_{J_d}^{(d)}\}$ being the set of multicast routes from s_d to $t_1^{(d)}, t_2^{(d)}, \ldots, t_{L_d}^{(d)}$ for $d = 1, 2, \ldots, r$ [26].

Now, network switching for multisource multicast networks can be interpreted as another matrix game as follows. Player I has the pure strategies as the set of links $\{e_1, e_2, \ldots, e_I\}$. Player II has the pure strategies as the *union* of the r sets of multicast routes $\{T_1^{(d)}, T_2^{(d)}, \ldots, T_{J_d}^{(d)}\}$ from s_d to $t_1^{(d)}, t_2^{(d)}, \ldots, t_{L_d}^{(d)}$ for $d = 1, 2, \ldots, r$. The payoff matrix is given by (15.43), where a_{ij} is the needed number of channel uses in the link e_i for one bit of information for source X_d transported over the multicast route $T_{j_d}^{(d)}$ for $i = 1, 2, \ldots, I, d = 1, 2, \ldots, r$ and $j_d = 1, 2, \ldots, J_d$. Player I chooses a mixed strategy $p \in \Delta^I$ and player II chooses a mixed strategy $q \in \Delta^{\sum_{d=1}^{r} J_d}$. In a play of the game, player II pays the amount a_{ij} to player I if player I chooses the ith link e_i and player II chooses the multicast route $T_{j_d}^{(d)}$ for $i = 1, 2, \ldots, I$ and $j = (d, j_d)$ with $d = 1, 2, \ldots, r$ and $j_d = 1, 2, \ldots, J_d$. In the sense of mixed strategies, the *value* of the game is the expected payoff $(p^*)^T U q^*$ when both player I and player II choose their optimal mixed strategies $p^* \in \Delta^I$ and $q^* \in \Delta^{\sum_{d=1}^{r} J_d}$ respectively. The optimal mixed strategy $q^* \in \Delta^{\sum_{d=1}^{r} J_d}$ for player II provides an optimal network switching scheme for multisource multicast network switching. Consequently, the multicast network achieves a maximum achievable information *sum-rate* and the information sum-rate $\Re^*$ is the *reciprocal* of the value of the game with payoff matrix U given by (15.43).

It should be noted that in the *Rate-direction function game* formulations for multisource multicast network switching, player II in the game for maximum achievable *directional* information rate has a mixed strategy of the *product* dimension $\prod_{d=1}^{r} J_d$ over the *Cartesian product* of the sets of the multicast routes for r sources, while in the *sum-rate game* has the *sum* dimension $\sum_{d=1}^{r} J_d$ over the union of the sets of the multicast routes for r sources. For the *single-source* multicast network switching where $r = 1$, they both reduce to a mixed strategy of the dimension $J = J_1$ over the set of the multicast routes for the single source. In the sequel, the maximum achievable information rate for multisource multicast network switching denotes both its maximum achievable directional information rate and its maximum achievable information sum-rate.

15.7.3 Computation of Maximum Achievable Information Rate for Single-Source Multicast Network Switching

As an illustration, we will calculate first R^* with network switching for multicast network in Figure 15.13. For this purpose, in general, we need to find the set of all multicast routes $\{T_1, T_2, \ldots, T_J\}$ from s to $t_1, t_2, \ldots, t_L$ so that we can determine the indicator function $\chi_{T_j}(e_i)$ for any link in the set of links $E = \{e_1, e_2, \ldots, e_I\}$.

In the sequel, we describe a method to find the set of all multicast routes $\{T_1, T_2, \ldots, T_J\}$ from s to $t_1, t_2, \ldots, t_L$. Recall that a multicast route from s to $t_1, t_2, \ldots, t_L$ is a set of links in E such that 1) it has the property that it includes all the links of an open path from s to t_l for each $l = 1, 2, \ldots, L$, and 2) the resultant set with deletion of its any link would not have the property in 1).

For the case of a single sink node when $L = 1$, a multicast route reduces to an open path from s to t_1.

First, we find the set of all open paths from s to t_l for each $l = 1, 2, \ldots, L$. There is a rooted-tree method to solve this *path enumeration* problem [34]. For example from Figure 15.13, we enumerate all open paths from source node s to sink node t_1. Beginning at s, we may first go to node i_1 or i_2. If we have come to i_1, we may next visit node t_1 or i_3, and so on. A rooted tree with the root node s showing all possibilities is given in the figure and there are three open paths in total from s to t_1. It is seen that the sink node t_1 is a leaf node in the rooted tree and every leaf node in the rooted tree is the sink node t_1. Then, all the paths from s to t_1 in the rooted tree are those in the original multicast network.

Now, we can address the *multicast-route enumeration* problem from s to $t_1, t_2, \ldots, t_L$. If $L = 1$, it reduces to the *path enumeration* problem from s to t_1. If $L \geq 2$, then we can use the following *induction* method.

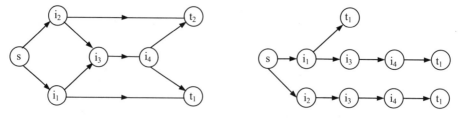

Figure 15.13   A rooted tree for the path enumeration from s to t in example network.

Let $\{\phi_1^{(\ell)}, \phi_2^{(l)}, \ldots, \phi_{K_l}^{(l)}\}$ with $K_l \in N$ be the set of all the K_l open paths from source node s to the sink node t_l for $l = 1, 2, \ldots, L$, which can be found by using the above rooted-tree method.

Suppose that $\{T_1^{(l)}, T_2^{(l)}, \ldots, T_{J_l}^{(p)}\}$ is the set of all the J_l multicast routes from source node s to the sink nodes $t_1, t_2, \ldots, t_l$ for $l = 1, 2, \ldots, L$. They can be found by the following procedure.

If $l = 1$, then $J_1 = K_1$ and $T_j^{(1)} = \phi_j^{(l)}$ for $j = 1, 2, \ldots, J_1$.

If $l \geq 2$, we suppose that the set of all the J_{l-1} multicast routes from the source node s to the sink nodes $t_1, t_2, \ldots, t_{l-1}$, namely, $\{T_1^{(l-1)}, T_2^{(l-1)}, \ldots, T_{J_{l-1}}^{(l-1)}\}$, has been found.

For any given $j = 1, 2, \ldots, J_{l-1}$, we establish the concatenation of the multicast route $T_j^{(l-1)}$ with each open path in $\{\phi_1^{(l)}, \phi_2^{(l)}, \ldots, \phi_{K_l}^{(l)}\}$ from s to t_l as follows.

> For $i = 1, 2, \ldots, K_l$, let
>
> $\phi_i^{(l)} = \{s \to u_1, u_1 \to u_2, \ldots, u_{n-2} \to u_{n-1}, u_{n-1} \to t_l.\}$
>
> Let $u_0 = s$ and $u_n = t_l$ and define
>
> $l^* = \max\{0 \leq k \leq n | u_k$ is a trans. or rec. node of some link in $T_j^{(l-1)}.\}$
>
> It is clear that $0 \leq \ell^* \leq n$

We denote $T_{ji}^{(l)} = C(T_j^{(l-1)}, \phi_i^{(l)})$ as the *concatenation* of $T_j^{(l-1)}$ with $\phi_i^{(l)}$, which can be defined as follows.

> If $l^* = n$, then $T_{ji}^{(l)} = C(T_j^{(l-1)}, \phi_i^{(l)}) = T_j^{(\ell-1)}$
>
> If $0 \leq l^* \langle n$, then
>
> $T_{ji}^{(\ell)} = C(T_j^{(\ell-1)}, \phi_i^{(\ell)}) = T_j^{(l-1)} \cup \{u_{l^*} \to u_{\ell^*+1}, \ldots, u_{n-1} \to t_l\}$

We say that $T_{ji}^{(l)} = C(T_j^{(l-1)}, \phi_i^{(l)})$ is *new*, if $i = 1$ or $1 < i \leq K_l$ but $T_{ji}^{(l)}$ is a different set of links from each of $T_{j1}^{(l)}, T_{j2}^{(l)}, \ldots, T_{j(i-1)}^{(l)}$, and *old* otherwise.

If $\{T_{ji_1}^{(l)}, T_{ji_2}^{(l)}, \ldots, T_{ji_{L_j}}^{(l)}\}$ is the set of all $L_j (1 \leq L_j \leq K_l)$ new ones among $T_{ji}^{(l)}$ for $i = 1, 2, \ldots, K_l$, then, $\{T_1^{(l)}, T_2^{(l)}, \ldots, T_{J_l}^{(l)}\} = \bigcup_{j=1}^{l_e-1} \{T_{ji_1}^{(\ell)}, T_{ji_2}^{(\ell)}, \ldots, T_{ji_{L_j}}^{(l)}\}$ is the set of all multicast routes from s to $t_1, t_2, \ldots t_e$.

When $l = L$, we find the set of all multicast routes from s to $t_1, t_2, \ldots, t_L$ given by $\{T_1, T_2, \ldots, T_J\} = \{T_1^{(L)}, T_2^{(L)}, \ldots, T_{J_L}^{(L)}\}$ where $J = J_L$ and $T_j = T_j^{(L)}$ for $j = 1, 2, \ldots, J$. Moreover, we have

$$\max_{1 \leq l \leq L} K_l \leq J_L \leq \prod_{l=1}^{L} K$$

As illustration, from the rooted tree shown in Figure 15.13(b), for the path enumeration from s to t_1, we see that there are three open paths from s to t_1 as follows:

> 1) $\phi_1^{(1)} = \{s \to i_1, i_1 \to t_1\}$;
> 2) $\phi_2^{(1)} = \{s \to i_1, i_1 \to i_3, i_3 \to i_4, i_4 \to t_1\}$;
> 3) $\phi_3^{(1)} = \{s \to i_2, i_2 \to i_3, i_3 \to i_4, i_4 \to t_1\}$.

There also are three open paths from s to t_2 as follows.

> 1) $\phi_1^2 = \{s \to i_2, i_2 \to t_2\}$;
> 2) $\phi_2^2 = \{s \to i_1, i_1 \to i_3, i_3 \to i_4, i_4 \to t_2\}$;
> 3) $\phi_3^2 = \{s \to i_2, i_2 \to i_3, i_3 \to i_4, i_4 \to t_2\}$.

For $l = 1$, the set of all multicast route from s to t_1 is given by $\{T_1^{(1)}, T_2^{(1)}, T_3^{(1)}\}$, where $T_j^{(1)} = \phi_j^{(1)}$ for $j = 1, 2, 3$. For $l = 2$, we make the concatenation operations as follows.

a) $T_{11}^{(2)} = C(T_1^{(1)}, \phi_1^{(2)}) = \{s \rightarrow i_1, i_1 \rightarrow t_1, s \rightarrow i_2, i_2 \rightarrow t_2\}$ *(new)*;

b) $T_{12}^{(2)} = C(T_1^{(1)}, \phi_2^{(2)}) = \{s \rightarrow i_1, i_1 \rightarrow t_1, i_1 \rightarrow i_3, i_3 \rightarrow i_4, i_4 \rightarrow t_2\}$ *(new)*;

c) $T_{13}^{(2)} = C(T_1^{(1)}, \phi_3^{(2)}) = \{s \rightarrow i_1, i_1 \rightarrow t_1, s \rightarrow i_2, i_2 \rightarrow i_3, i_3 \rightarrow i_4, i_4 \rightarrow t_2\}$ *(new)*,

d) $T_{21}^{(2)} = C(T_2^{(1)}, \phi_1^{(2)}) = \{s \rightarrow i_1, i_1 \rightarrow i_3, i_3 \rightarrow i_4, i_4 \rightarrow t_1, s \rightarrow i_2, i_2 \rightarrow t_2\}$ *(new)*;

e) $T_{22}^{(2)} = C(T_2^{(1)}, \phi_2^{(2)}) = \{s \rightarrow i_1, i_1 \rightarrow i_3, i_3 \rightarrow i_4, i_4 \rightarrow t_1, i_4 \rightarrow t_2\}$ *(new)*;

f) $T_{23}^{(2)} = C(T_2^{(1)}, \phi_3^{(2)}) = \{s \rightarrow i_1, i_1 \rightarrow i_3, i_3 \rightarrow i_4, i_4 \rightarrow t_1, i_4 \rightarrow t_2\}$ *(old)*;

g) $T_{31}^{(2)} = C(T_3^{(1)}, \phi_1^{(2)}) = \{s \rightarrow i_2, i_2 \rightarrow i_3, i_3 \rightarrow i_4, i_4 \rightarrow t_1, i_2 \rightarrow t_2\}$ *(new)*;

h) $T_{32}^{(2)} = C(T_3^{(1)}, \phi_2^{(2)}) = \{s \rightarrow i_2, i_2 \rightarrow i_3, i_3 \rightarrow i_4, i_4 \rightarrow t_1, i_4 \rightarrow t_2\}$ *(new)*

i) $T_{33}^{(2)} = C(T_3^{(1)}, \phi_3^{(2)}) = \{s \rightarrow i_2, i_2 \rightarrow i_3, i_3 \rightarrow i_4, i_4 \rightarrow t_1, i_4 \rightarrow t_2\}$ *(old)*.

Since, $L = 2$, we find the set of seven multicast routes from s to t_1, t_2 given by $\{T_j | 1 \leq j \leq 7\} = \{T_j^{(2)} | 1 \leq j \leq 7\} = \{T_{11}^{(2)}, T_{12}^{(2)}, T_{13}^{(2)}, T_{21}^{(2)}, T_{22}^{(2)}, T_{31}^{(2)}, T_{32}^{(2)}\}$.

According to (15.36), (15.37), (15.43), the maximum achievable information rate $R^* = 1/val(A)$, where $A = (a_{ij})$ is given by $a_{ij} = \chi_{T_j}(e_i)/\Theta(e_i)$ for $i = 1, 2, \ldots, I$ and $j = 1, 2, \ldots, J$ with $\{e_1, e_2, \ldots, e_I\}$ being the set of links E and $\{T_1, T_2, \ldots, T_J\}$ being the set of multicast routes from s to $t_1, t_2, \ldots, t_L$. The optimal mixed strategy for player II provides an optimal network switching scheme to achieve the information rate R^*.

It is known that the *dominance relation* in matrix game [30, 32] can be used for reducing the number of pure strategies without changing the value of the game.

A vector $(a_1, a_2, \ldots, a_I) \in R^I$ is said to be *upper-bounded* by the vector $(b_1, b_2, \ldots, b_I) \in R^I$ if $a_i \leq b_i$ for $i = 1, 2, \ldots, I$. A row vector of the payoff matrix A is said to be dominated, if it is upper-bounded by another row vector of A, and softly dominated, if it is upper-bounded by a convex linear combination of the remaining row vectors of A. In contrast, a column vector of the payoff matrix A is said to be dominated if another column vector of A is upper-bounded by it, and softly dominated if a convex linear combination of the remaining column vectors of A are upper-bounded by it. A well-known result in matrix game is that a resultant payoff matrix from a successive elimination of row and column vectors that are dominated (and even softly dominated) ones of the previous payoff matrix will have the same value of the game.

For the example network from Figure 15.31, it follows from the preceding multicast-route enumeration example that the payoff matrix is given by

$$A = (\chi_{T_j}(e_i))_{9 \times 7} = \begin{array}{c} \\ e_1 \\ e_2 \\ e_3 \\ e_4 \\ e_5 \\ e_6 \\ e_7 \\ e_8 \\ e_9 \end{array} \begin{array}{c} T_1\ T_2\ T_3\ T_4\ T_5\ T_6\ T_7 \\ \left\| \begin{array}{ccccccc} 1 & 1 & 1 & 1 & 1 & 0 & 0 \\ 1 & 0 & 1 & 1 & 0 & 1 & 1 \\ 1 & 1 & 1 & 0 & 0 & 0 & 0 \\ 0 & 1 & 0 & 1 & 1 & 0 & 0 \\ 0 & 0 & 1 & 0 & 0 & 1 & 1 \\ 1 & 0 & 0 & 1 & 0 & 1 & 0 \\ 0 & 1 & 1 & 1 & 1 & 1 & 1 \\ 0 & 0 & 0 & 1 & 1 & 1 & 1 \\ 0 & 1 & 1 & 0 & 1 & 0 & 1 \end{array} \right\| \end{array} \qquad (15.44)$$

where $E = \{e_i | 1 \leq i \leq I\} = \{s \rightarrow i_1, s \rightarrow i_2, t_1 \rightarrow t_1, i_1 \rightarrow i_3, \quad i_2 \rightarrow i_3, i_2 \rightarrow t_2, i_3 \rightarrow i_4, i_4 \rightarrow t_1, i_4 \rightarrow t_2\}$ with the ith entries in both sets being identical for $i = 1, 2, \ldots, I$ with $I = 9$.

One can see that the two row vectors for e_3 and e_4 are *dominated* by the row vector for e_1, the two row vectors for e_5 and e_6 are *dominated* by the row vector for e_2, and the two row vectors for e_8 and e_9 are *dominated* by the row vector for e_7. By eliminating of dominated row vectors, we obtain the following approximation for the payoff matrix:

$$A_1 = \begin{matrix} & \begin{matrix} T_1 & T_2 & T_3 & T_4 & T_5 & T_6 & T_7 \end{matrix} \\ \begin{matrix} e_1 \\ e_2 \\ e_7 \end{matrix} & \left\| \begin{matrix} 1 & 1 & 1 & 1 & 1 & 0 & 0 \\ 1 & 0 & 1 & 1 & 0 & 1 & 1 \\ 0 & 1 & 1 & 1 & 1 & 1 & 1 \end{matrix} \right\| \end{matrix} \tag{15.44a}$$

The row vectors are for the links e_1, e_2, e_7 and the column vectors for the multicast routes $T_1, T_2, \ldots, T_7$.

Now, in the payoff matrix A_1, the column vectors for T_3 and T_4 are *dominated* by the column vector for T_1, the column vectors for T_5 is *dominated* by the column vector for T_2, and the column vector for T_7 are *dominated* by the column vector for T_6. The resultant payoff matrix with elimination of dominated column vectors is given by

$$A_2 = \begin{matrix} & \begin{matrix} T_1 & T_2 & T_6 \end{matrix} \\ \begin{matrix} e_1 \\ e_2 \\ e_7 \end{matrix} & \left\| \begin{matrix} 1 & 1 & 0 \\ 1 & 0 & 1 \\ 0 & 1 & 1 \end{matrix} \right\| \end{matrix}$$

where row vectors are for the links e_1, e_2, e_7 and column vectors for the multicast routes T_1, T_2, T_6.

According to conditions discussed bellow (15.32) it is seen that $p^* = q^* = (1/3, 1/3, 1/3)^T \in \Delta^3$ are optimal mixed strategies of players I and II and the value of the game is $val(A_2) = 2/3$. From this, we know that the maximum achievable information rate R^* in bits per network use is given by $R^* = 1/val(A) = 1/val(A_1) = 1/val(A_2) = 3/2$ and an optimal switching scheme is to use the three multicast routes T_1, T_2, T_6 with the same relative frequency $1/3$.

15.8 Optimization of Wireless Multicast Ad-Hoc Networks

Now, the matrix games framework, discussed in the previous section, will be used for joint optimization of routing, multicast diversity, link scheduling, and network coding in multicast wireless ad-hoc networks. The impact of multicast diversity on scheduling is controlled by using the topology compression concept quantified through the compressed multicast topology matrix. To define the topology matrix, first a set of all possible multicast paths, including network coded paths, is identified using the method underlined in Section 15.7. After that this set is compressed, and then by switching between some of remaining paths with appropriate usage rates (frequencies), achievable throughput or throughput per sum of transmitting powers on the route are optimized. A link conflict free environment is designed by appropriate conflict free network partitioning (realizations) using network graph coloring. For a *pre-designed coloring scheme*, link scheduling partitions topology matrix into multiple sub-matrices, one for each color (network realization), referred to as partial topologies. Each sub matrix, with links weighted by the inverse of the link capacities, is used as a payoff matrix for an asymmetrical matrix game where against any single move of one player, its opponent has multiple (K) moves, corresponding to different partial topologies or their equivalent colors. The strategy sets for the players are links and paths. Such a two player (dimension) game will be formally referred to as Asymmetrical Matrix Game with notation $AMG(2,K,1)$ and the value of this game is inverse of the throughput. In this notation 2 indicates number of players, and K and 1 are the number of decisions taken by players at each step. At the equilibrium, the mixed strategy vector of the first player indicates the optimum percentage of time or optimum number of time slots dedicated to the selected network realizations for a given partitioning of the network graph while the mixed strategy vectors of the second player indicate the optimum usage rates of the paths. By adding a constraint to the game, instead of throughput, throughput per power can be optimized, which is a more reasonable criterion in a wireless ad-hoc network. The AMG game optimizes network coding and routing under a given scheduling strategy. The AMG game is solved by extending fictitious playing method. The dominancy theory, discussed in the previous section, is also used in this game to simplify game for large networks.

15.8.1   The System Model

In this section, the problem of the definition of scheduling and routing in wireless multicast *ad-hoc* network is presented. Assume a multi-source multicast wireless *ad-hoc* network including N nodes. This network is defined as $G(\mathbf{V}, \mathbf{E}, \zeta, \mathbf{s}_1, \mathbf{s}_2, \ldots, \mathbf{s}_M)$ where V is a $2 \times N$ matrix which presents location of nodes with two rows corresponding to x and y positions of N nodes, E is set of all L virtual wireless links defined by topology control, ζ is set of M sources and s_i is set of sinks corresponding to the ith source. Wireless propagation for this network assumes the following: (a) Omni-directional transmission; (b) Presence of interference due to simultaneous transmission; (c) TDMA as multiple access scheme for different hops without inter time slot interference; (d) Network partitioning, based on network graph coloring is used to avoid interference from relatively close interferes; (e) Only one single packet can be transmitted or received by any node at its assigned time slot.

The following assumptions are used for network coding: (a) The network coded paths have already been identified by one of the methods available in the literature [33]; (b) Coded blocks must be decodable at all sinks. This assumes that a required number of independent routs to the sinks are supported by the network; (c) If a relay node receives signal from source via two or more different paths, this node is eligible to perform network coding.

For an example of 15 nodes shown in Figure 15.14(a), where the first node is source and last two nodes are sinks, a possible network topology including 28 links is shown in Figure 15.14(b).

Conflict Free Operation assumes that, S_{ij}, in dB, is the power required at node j, for transmitting node i to reach the receiving node j at distance d_{ij}. It is assumed that $S_{ij} \propto s_i d_{ij}^{-\alpha}$ where α is attenuation factor and s_i transmit power at node i. Accordingly, by definition of the conflict free scheduling, any node k $\neq$i, j, receiving the signal from node m, will be interfered by link l_{ij} if and only if $S_{mk} \leq S_{ik} + \beta$, where β, in dB, is accepted interference margin between two links. In other word, link l_{ij} is adjacent to l_{mk} for any m and any k $\neq$ i, j if

$$S_{mk} \leq S_{ik} + \beta \tag{15.45}$$

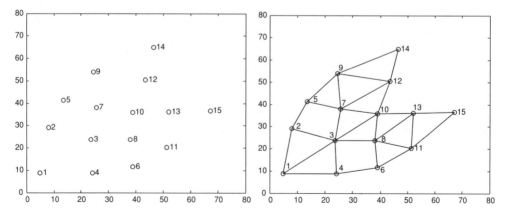

Figure 15.14   (a) A typical 15-node wireless ad-hoc network with first node as source and last two nodes as multicast sinks.(b) Topology defined for network in Figure 15.14(a).

Alternatively, the two links are adjacent if

$$S_{ij} \leq S_{mj} + \beta \tag{15.46}$$

As a consequence of (15.45) and (15.46), these two links are adjacent if they are physically adjacent i.e. they have a common node. Whenever l_{ij} and l_{mk} are physically adjacent or either (15.45) or (15.46) hold, they cannot be painted by the same color in the network graph.

In general conflict free scheduling, for a given interference the margin is based on conventional graph coloring techniques. A simple and low complexity algorithm for minimal coloring, i.e. painting all links with minimum required number of colors is defined as follows:

Step 1. The first link is assigned to the first (i=1) partial topology (color), i.e. $T_1 = \{l_1 = l_{1,2}\}$.

Step 2. The next unassigned link l_j is chosen.

Step. 3. Using (1), if for any value of i, l_j can be simultaneously activated with all members of T_i, it is added to T_i as $T_i \leftarrow T_i \cup l_j$.

Step 4. If for all i, where $i \leq i_{max}$, last step cannot be run, l_j is considered as first element of T_{imax+1} and $i_{max} \leftarrow i_{max} + 1$.

Step 5. If all links are allocated to T_i, we have a complete set of partial topologies and if not go to Step 2 and run last two steps for all remaining links.

Although the outcome of this algorithm is the minimal set of partial topologies, it is not unique and depends on order of links selection. When links usage rates are different, links with the highest usage rates should be grouped together whenever possible and such partial topology should have as high a usage rate as possible. So the optimal method, in order to generate a complete set of network realizations, needs to know the rates r_{links} of link usage and with an optimal method at each iteration, we must sort links versus their usage rates to make a new set of network realizations.

If for illustration we use our suboptimal scheduling algorithm, for interference margin 0 dB we will have the nine following network realizations (*partial topologies*).

$$T_1 = \{l_{1,2}, l_{4,6}, l_{9,14}, l_{10,13}\}, \quad T_2 = \{l_{1,4}, l_{2,5}, l_{6,11}, l_{8,10}, l_{12,14}, l_{3,7}\},$$
$$T_3 = \{l_{5,9}, l_{11,15}, l_{10,12}, l_{2,3}\}, \quad T_4 = \{l_{5,7}, l_{6,8}, l_{13,15}, l_{4,3}\}, \quad T_5 = \{l_{7,10}, l_{9,12}, l_{11,13}, l_{1,3}\},$$
$$T_6 = \{l_{3,8}, l_{7,9}\}, \quad T_7 = \{l_{7,12}, l_{8,13}\}, \quad T_8 = \{l_{3,10}\}, T_9 = \{l_{8,11}\}.$$

where links from each set can be activated at the same time slot. This partitioning set is minimal in the sense that number of required time slots to activate all links at least once, referred to as *clique cycle*, is minimal. In other word, chromatic number for this topology when $\beta = 0dB$, is 9. The optimization of this process will be discussed later.

Figure 15.15 shows these network realizations. Any network realization needs some specific time slots. For any path usage rate vector r_{paths} there are optimum fractions for the number of time slots that should be dedicated to each network realization. Consequently, for a given optimization criterion (throughput or throughput per power) the system is searching for the equilibrium in which the source is choosing certain parts for the multicast and the network is searching for a proper switching between these parts, in different network realizations, to achieve their optimum usage rate. The problem is summarized as solving a game with payoff matrix A, considering the constraint on link scheduling imposed by network realizations. The construction of matrix A is discussed in the following.

Multicast Paths Identification procedure, described in the previous section, is used to find the set of all possible multicast paths from the set of sources to the set of sinks. Assume $P_i^{m,n}$ as ith non-cyclic

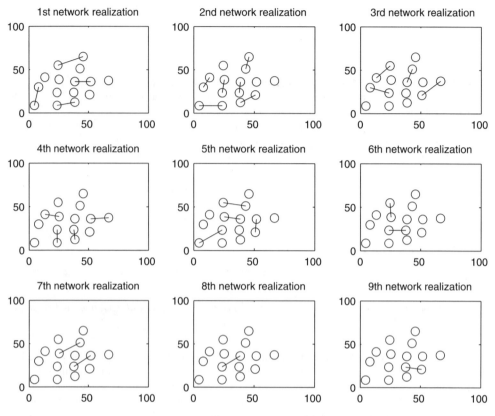

Figure 15.15   Complete set of conflict free network realizations for $\beta = 0$.

directed path from the mth source to its nth corresponding sink. The set of all multicast paths from source m is calculated as follows,

$$P^m_{i_1,i_2,\dots i_{|s_m|-1},i_{|s_m|}} = \bigcup_{n=1}^{|s_m|} P^{m,n}_{i_n} \tag{15.47}$$

Were $\cup$ is union and $|\,.\,|$ is cardinality operator. Accordingly multicast gain (G) is defined as follows,

$$G = \sum_{m,\langle i_s\rangle} y^m_{\langle i_s\rangle} |P^m_{\langle i_s\rangle}| - \sum_{m,n,\langle i_s\rangle} y^m_{\langle i_s\rangle} |P^{m,n}_{\langle i_s\rangle}| \tag{15.48}$$

where $\langle i_s\rangle = i_1, i_2, \dots, i_{|s_m||1}, i_{|s_m|}$ and $y^m_{\langle i_s\rangle}$ is normalized rate of the path $P^m_{\langle i_s\rangle}$ with, $\sum_{m,\langle i_s\rangle} y^m_{\langle i_s\rangle} = 1$.

In our sample topology, from Figure 15.14 the unicast paths from source $(m = 1)$ to the first sink $(s = 1)$ are

$$P^{(1,1)}_1 = \{l_{1,2}, l_{2,5}, l_{5,9}, l_{9,14}\},$$
$$P^{(1,1)}_2 = \{l_{1,2}, l_{2,5}, l_{5,7}, l_{7,10}, l_{10,12}, l_{12,14}\},$$
$$P^{(1,1)}_3 = \{l_{1,2}, l_{2,5}, l_{5,9}, l_{9,12}, l_{12,14}\},$$
$$P^{(1,1)}_4 = \{l_{1,2}, l_{2,3}, l_{3,7}, l_{7,10}, l_{10,12}, l_{12,14}\},$$

$$P_5^{(1,1)} = \{l_{1,2}, l_{2,3}, l_{3,7}, l_{7,12}, l_{12,14}\},$$

$$P_6^{(1,1)} = \{l_{1,2}, l_{2,3}, l_{3,7}, l_{7,10}, l_{10,12}, l_{12,14}\},$$

$$P_7^{(1,1)} = \{l_{1,3}, l_{3,10}, l_{10,12}, l_{12,14}\},$$

$$P_8^{(1,1)} = \{l_{1,4}, l_{4,3}, l_{3,7}, l_{7,9}, l_{9,14}\},$$

$$P_9^{(1,1)} = \{l_{1,4}, l_{4,6}, l_{6,8}, l_{8,10}, l_{10,12}, l_{12,14}\},$$

$$P_{10}^{(1,1)} = \{l_{1,4}, l_{4,3}, l_{3,7}, l_{7,10}, l_{10,12}, l_{12,14}\}.$$

And the unicast paths from source *(m = 1)* to the second sink *(s = 2)* are

$$P_1^{(1,2)} = \{l_{1,4}, l_{4,6}, l_{6,11}, l_{11,15}\},$$

$$P_2^{(1,2)} = \{l_{1,4}, l_{4,6}, l_{6,8}, l_{8,10}, l_{10,13}, l_{13,15}\},$$

$$P_3^{(1,2)} = \{l_{1,4}, l_{4,6}, l_{6,11}, l_{11,13}, l_{13,15}\},$$

$$P_4^{(1,2)} = \{l_{1,4}, l_{4,3}, l_{3,8}, l_{8,10}, l_{10,13}, l_{13,15}\},$$

$$P_5^{(1,2)} = \{l_{1,4}, l_{4,3}, l_{3,8}, l_{8,13}, l_{13,15}\},$$

$$P_6^{(1,2)} = \{l_{1,4}, l_{4,3}, l_{3,8}, l_{8,10}, l_{10,13}, l_{13,15}\},$$

$$P_7^{(1,2)} = \{l_{1,3}, l_{3,10}, l_{10,13}, l_{13,15}\},$$

$$P_8^{(1,2)} = \{l_{1,2}, l_{2,3}, l_{3,8}, l_{8,11}, l_{11,15}\},$$

$$P_9^{(1,2)} = \{l_{1,2}, l_{2,5}, l_{5,7}, l_{7,10}, l_{10,13}, l_{13,15}\},$$

$$P_{10}^{(1,2)} = \{l_{1,2}, l_{2,3}, l_{3,8}, l_{8,10}, l_{10,13}, l_{13,15}\}.$$

Therefore, we have 10 unicast paths to each sink and to form a multicast path set we just need to concatenate the paths above using (15.47).

As a result we will have one hundred store and forward multicast routing paths. In general, we could reduce this number by eliminating in advance those paths requiring excessive power or in accordance with any other criteria. For instance, only those paths requiring maximum 3 dB or less power than the minimum power path are selected, which would result in a negligible deviation from optimality. Owing to the low complexity of this technique, such a reduction in complexity is normally not necessary.

Network Coded Paths are identified in the next step. In the optimization process combining n input signals will be represented as having an equivalent capacity $C_c = nC_{min}$ where C_{min} is the minimal input link capacity and n is the number of combined input signals. The code selection techniques for guaranteed signal recoverability are described in the previous sections and the literature [33]. For an illustration of this and without loss of generality, we assume that network coding is performed at just two relay nodes, but this is not a restriction and any other network coded realization can be involved. From the example in Figure 15.14(b), we can see that although all intermediate nodes 2, 3, . . . , 13 receive the signal from the source through two independent paths, only nodes 3 and 10 meet all of the assumptions specified at the beginning of this section for network coding. By considering these nodes for performing XOR operation, the following network coded paths can be used.

$$P_1^{NC} = \{l_{1,2}, l_{1,3}, l_{1,4}, l_{2,5}, l_{3,10}, l_{4,6}, l_{5,9}, l_{6,11}, l_{9,14}, l_{10,12}, l_{10,13}, l_{11,15}, l_{12,14}, l_{13,15}\},$$

$$P_2^{NC} = \{l_{1,2}, l_{1,4}, l_{2,3}, l_{2,5}, l_{3,10}, l_{4,3}, l_{4,6}, l_{5,9}, l_{6,11}, l_{9,14}, l_{10,12}, l_{10,13}, l_{11,15}, l_{12,14}, l_{13,15}\},$$

$$P_3^{NC} = \{l_{1,2}, l_{1,4}, l_{2,3}, l_{2,5}, l_{3,7}, l_{3,8}, l_{3,10}, l_{4,3}, l_{4,6}, l_{5,9}, l_{6,11}, l_{7,10}, l_{8,10}, l_{9,14}, l_{10,12}, l_{10,13}, l_{11,15}, l_{12,14}, l_{13,15}\},$$

$$P_4^{NC} = \{l_{1,2}, l_{1,4}, l_{2,5}, l_{4,6}, l_{5,7}, l_{5,9}, l_{6,8}, l_{6,11}, l_{7,10}, l_{8,10}, l_{9,14}, l_{10,12}, l_{10,13}, l_{11,15}, l_{12,14}, l_{13,15}\},$$

$$P_5^{NC} = \{l_{1,2}, l_{1,4}, l_{2,5}, l_{4,6}, l_{5,7}, l_{6,8}, l_{7,9}, l_{7,10}, l_{8,10}, l_{8,11}, l_{9,14}, l_{10,12}, l_{10,13}, l_{11,15}, l_{12,14}, l_{13,15}\}.$$

Table 15.1   Compressed multicast topology matrix

	$P_1 = P_{1,1}$	$P_2 = P_{1,2}$	$\cdots$	$P_{100} = P_{10,10}$	$P_{101} = P_1^{NC}$	$P_{102} = P_2^{NC}$	$\cdots$	$P_{105} = P_5^{NC}$
$l_1 = l_{1,2}$	1	1	$\cdots$	1	$\frac{1}{2}$	$\frac{1}{2}$	$\cdots$	$\frac{1}{2}$
$l_2 = l_{1,3}$	0	0	$\cdots$	0	$\frac{1}{2}$	0	$\cdots$	0
$l_3 = l_{1,4}$	0	0	$\cdots$	0	$\frac{1}{2}$	$\frac{1}{2}$	$\cdots$	$\frac{1}{2}$
$l_{21} = l_{11,15}$	1	0	$\cdots$	0	$\frac{1}{2}$	$\frac{1}{2}$	$\cdots$	$\frac{1}{2}$
$l_{28} = l_{13,15}$	0	1	$\cdots$	1	$\frac{1}{2}$	$\frac{1}{2}$	$\cdots$	$\frac{1}{2}$

Topology Compression of the selected paths due to the omnidirectional nature of the propagation and inherent multicast gain is performed next. If one multicast path consists of two links carrying the same information and originating from the same source, a link with lower power can be removed from the link set of that path. We refer to this process as topology compression which reduces the network interference graph.

For instance, by using (15.46) and (15.47) $P_{1,1}^{(1)} = P_{1,1} = P_1^{(1,1)} \cup P_1^{(1,2)}$ will be $P_{1,1} = \{l_{1,2}, l_{2,5}, l_{5,9},$ $l_{9,14}, l_{1,4}, l_{4,6}, l_{6,11}, l_{11,15}\}$.

where upper index *(1)* is dropped for simplicity. After this compression $l_{1,4}$ is eliminated because $S_{1,4} \le S_{1,2}$ and whenever $l_{1,2}$ is fired, $l_{1,4}$ receives signal too. Consequently $P_{1,1}$ is simplified as, $P_{1,1} = \{l_{1,2}, l_{2,5}, l_{5,9}, l_{9,14}, l_{4,6}, l_{6,11}, l_{11,15}\}$.

Finally compressed topology matrix A is formed representing all multicast and network coded compressed paths as its columns and links set as its rows. The elements of this matrix are one, if links (of unit capacity) corresponding to row number is member of compressed multicast paths corresponding to the column number. This element equals $1/C_c$, if links corresponding to row number is member of compressed network coded paths corresponding to the column number and C_c is the number of combined signals.

In summary we have

$$a_{ij} = \begin{cases} 1, & \text{if } l_i \in P_j \text{ and } P_j \text{ is a compressed multicast path.} \\ 1/C_c, & \text{if } l_i \in P_j \text{ and } P_j \text{ is a network coded path.} \\ 0, & \text{Otherwise} \end{cases}$$

The columns of this matrix present the required link capacities for each path to transfer one packet. Table 15.1 shows some elements of this 28×105 matrix for topology shown in Figure 15.14(b).

This example matrix A has elements defined as follows: $a_{i,j} = 1$ when $j \le 100$ and $l_i \in P_j$, $a_{i,j} = 0.5$ when $j > 100$ (due to network coding with $C_c = 2$) and $l_i \in P_j$, otherwise $a_{i,j} = 0$. When multiple paths usage rate vector is r_{paths}, the link usage rate r_{links}, can be expressed as,

$$r_{links} = A r_{paths} \qquad (15.49)$$

15.8.2   *Matrix Game Formulation of Joint Routing and Scheduling*

The model of switching (routing) based on matrix game theory for wired networks, where no scheduling is necessary and all links are activated simultaneously, has been presented in Section 15.7. The simultaneous activation of the links in wired networks makes the problem easy to solve because just a single A matrix is the payoff matrix of the game. In this game, the strategy set of the first player is a set of links and the strategy set of the second player is a set of all multicast paths. The decision of the first player potentially creates the bottleneck link of the partial topology. In other words, first player decides which links in each topology has to be used more than others.

In a scheduled routing scheme, based on topology partitioning, against any decision of the second player, i.e. any values of the r_{paths}, the first player, i.e. the network, will make multiple decisions, each corresponding to each partial topology. Consequently, for a wireless ad-hoc network with given scheduling sets T_k, $for k = 1, 2, \ldots, K$, A must be broken into $i_{max} = K$ smaller matrices (sub matrices). Assume A_ks for $k = 1, 2, \ldots, K$ as K payoff matrices, defined corresponding to K partial topologies with size I_k x J. We denote the set of all I_k dimensional probability distribution vectors for the second player and J dimensional probability distribution vector of the first player by

$$\Delta^{I_k} = \{ \left(x_1^k, x_2^k, \ldots, x_{I_k}^k \right) \in R^{I_k} \mid \sum_{i=1}^{I_k} x_i^k = 1 \} \tag{15.50}$$

$$\Delta^J = \{ \left(y_1^k, y_2^k, \ldots, y_J^k \right) \in R^J \mid \sum_{i=1}^{J} y_i^k = 1 \} \tag{15.51}$$

where $x_i^{(k)}$s and y_js are elements of the probability strategy vectors x_k and y. The vector y is normalized as

$$y = r_{paths} / \left| r_{paths} \right| \tag{15.52}$$

For a given routing vector y, the minimum number of time slots to deliver a single packet from source to multicast sinks which is inverse of the multicast throughput is

$$t_{tot} = \sum_{k=1}^{K} \max_{x_k} \left(x_k^T A_k y \right) \tag{15.53}$$

In a given network $G (V, E, A, T)$, where T is a complete set of partial topologies, the right hand side of (15.53) is maximized with a sequence of pure strategies defined by,

$$i_{k,opt} = \arg \max(A_k y) \tag{15.54}$$

where maximization is taken over elements of the vector $A_t y$. To prove it we assume $i_{k,max} = \arg \max (A_k y)$, and then we have

$$x_k^T A_k y = \sum_{i=1}^{I_k} x_i^{k_i} (A_k) y \le \sum_{i=1}^{I_k} x_i^k i_{k \, max} (A_k) y =_{i_{k \, max}} (A_k) y \sum_{i=1}^{I_k} x_i^k = i_{k \, max} (A_k) y$$

Comparing (11) and (12), it is easy to see $i_{k,opt} = i_{k,max}$ and

$$t_{tot} = \sum_{k=1}^{K} i_{k \, max} (A_k) y \tag{15.55}$$

In Section 15.7 we have shown that for wired networks, the value of A is the inverse of the throughput. In other words, the value of A is the minimum number of time slots required to deliver one single packet from one of the sources involved in A to its corresponding multicast sinks. In this game the model for the wireless networks, the value of each A_k is the minimum required number of time slots for each partial topology to deliver one packet and consequently the total number of required time slots for whole topology to deliver one packet which is the inverse of the maximum throughput is

$$t_{tot} = \min_{y} \max_{x_k} \sum_{k=1}^{K} \left(x_k^T A_k y \right) \tag{15.56}$$

Table 15.2 Payoff matrix corresponding to the first network realization (A_1)

	$P_1 = P_{1,1}$	$P_2 = P_{1,2}$	...	$P_{100} = P_{10,10}$	$P_{101} = P_1^{NC}$	$P_{102} = P_2^{NC}$	...	$P_{105} = P_5^{NC}$
$l_1 = l_{1,2}$	1	1	...	1	½	½	...	½
$l_2 = l_{4,6}$	1	1	...	0	½	½	...	½
$l_3 = l_{9,14}$	1	0	...	0	½	½	...	½
$l_4 = l_{10,13}$	0	0	...	0	0	0	...	0

Table 15.3 Payoff matrix corresponding to the second network realization (A_2)

	$P_1 = P_{1,1}$	$P_2 = P_{1,2}$	...	$P_{100} = P_{10,10}$	$P_{101} = P_1^{NC}$	$P_{102} = P_2^{NC}$	...	$P_{105} = P_5^{NC}$
$l_1 = l_{1,4}$	0	0	...	0	½	½	...	½
$l_2 = l_{2,5}$	1	1	...	0	½	0	...	½
$l_3 = l_{6,11}$	1	1	...	1	½	½	...	0
$l_4 = l_{8,10}$	0	0	...	1	0	0	...	0
$l_5 = l_{12,14}$	0	1	...	1	½	½	...	½
$l_6 = l_{3,7}$	0	0	...	1	0	0	...	0

The right hand side of (15.56) is defined as the value of our *AMG(2,K,1)* game. For topology shown in Figure 15.14(b), assuming 0dB interference margin, the first three partial topology matrices of our nine payoff matrices are given by Tables 15.2–4.

Payoff matrices $A_4, A_5, \ldots, A_9$ are calculated in the same way as the three matrices above. Therefore we have $K = 9$ payoff matrices corresponding to nine network realizations resulting in *AMG(2,9,1)* game. Assume x_k, $k = 1,2,\ldots,9$ as probability vectors of the network's multiple mixed strategies and y as probability vector of the source's single mixed strategy. All 9 x_ks and y are probability vectors and therefore their elements have unit sum. When paths are selected with probabilities y, payoff received by source at kth network realization $A_k y$ and when link are chosen by probabilities x_k, $k = 0,1,2,\ldots,9$ total cost paid by network will be $\sum_{k=1}^{K} x_k^T A_k$ which is total amount of network resource consumption. For any given set of x_ks and y the value of game is

$$\eta(x_1, x_2, \ldots, x_8, x_9, y) = \sum_{k=1}^{K} x_k^T A_k y = 1/R \tag{15.57}$$

where R is the multicast throughput of the game. The average throughput or throughput per average number of time slots, is inverse of the value of the game as discussed in Section 15.7. When players are supposed to choose their best pure strategy source looks for y_j that gives, lowest $\sum_{k=1}^{K} x_k^T A_k$, i.e. $j_{opt} = \arg\min(\sum_{k=1}^{K} x_k^T A_k)$, and network at kth network realization, chooses $x_{k,i}$ that gets the highest value of $A_k y$ i.e. $i_{k,opt} = \arg\max(A_k y)$.

If the payoff matrices set give a single and unique equilibrium solution, these two relations are self-contained, but in the practical solutions there are multiple equilibriums and consequently mix strategy gives the optimum solution. The optimal solution for this problem is achieved by the min-max

Table 15.4 Payoff matrix corresponding to the third network realization (A_3)

	$P_1 = P_{1,1}$	$P_2 = P_{1,2}$	...	$P_{100} = P_{10,10}$	$P_{101} = P_1^{NC}$	$P_{102} = P_2^{NC}$	...	$P_{105} = P_5^{NC}$
$l_1 = l_{5,9}$	1	1	...	0	½	½	...	0
$l_2 = l_{11,15}$	1	1	...	0	½	½	...	½
$l_3 = l_{10,12}$	0	1	...	1	½	½	...	½
$l_4 = l_{2,3}$	0	0	...	1	0	½	...	0

theory as follows,

$$y = \arg\ \min_y \sum_{k=1}^{K} x_k^T A_k y$$

$$x_k = \arg\ \max_{x_k} \sum_{k=1}^{K} x_k^T A_k y \qquad (15.58)$$

The final step is calculation of r_{paths} i.e. path rate per time slot from y. First we define $\theta_j = 1$ if P_j is a compressed multicast path and C_c if P_j is a network coded path. Throughput is related to r_{paths} by the following equation

$$R = \sum_{j=1}^{J} r_j \theta_j \qquad (15.59)$$

where r_j is jth element of the r_{paths}. Consequently using (15.52) and (15.59), r_{paths} is calculated as,

$$r_{paths} = \frac{R\, y}{\sum_{j=1}^{J} y_j \theta_j} \qquad (15.60)$$

15.8.3 Interference Controlled Scheduling

In this section, we discuss an optimization criterion which controls the amount of interference at links and maintains the fixed value of the signal to interference ratio (capacity). The conflict free network realizations are defined by introducing the interference margin between links, so that even the links of the same network realization still interfere with each other and sometimes the total received interference at one link may exceed the threshold and corrupt the performance of all of the paths that use such link. Controlling interference not only helps the paths to avoid this problem but also can improve performance and increase total achieved throughput. To control the interference we need to calculate the interference received from other active links in the same network realization. We call the interference matrix for kth network realization as I_k where its elements are calculated as follows,

$$i_{i,j}^k = \left| \frac{d_{j_1,j_2}^k}{d_{j_1,i_2}^k} \right|^\alpha \qquad (15.61)$$

where d_{j_1,j_2}^k is the distance between the transmitter and receiver of the jth link of the kth network realization, d_{j_1,i_2}^k is the distance between the transmitter of the jth link and the receiver of the ith link of the same network realization and α is the attenuation factor. It is easy to see that the numerator in (15.61) is proportional to the minimum transmission power of the jth link of the kth network realization. To control interference per link by a predefined upper bound we just need to modify A_ks as follows,

$$A_k^{(new)} = \frac{I_k A_k}{1 + ISR_{max}} \qquad (15.62)$$

where ISR_{max} is the maximum received interference to signal ratio for active links. One interesting property of this method is introducing interference even when there is no interference source by increasing the transmission rate to above pre-defined capacity. The modification of A_k forces all of the nodes to have the same interference to signal ratios, resulting in the same link capacities. This scheme is optimal in terms of capacity because when all links have the same capacity, the capacity of each path will be the same as the capacity of each individual link.

15.8.4   Extended Fictitious Playing and Dominancy Theory for AMG Games

For a given $AMG(2,K,1)$ with positive payoff matrices, mixed strategy vectors x_k for $k \in [1, K]$ and y present a Nash equilibrium if and only if for some value γ, and any arbitrary $i_k \in [1, I_k]$ and $j \in [1, J]$,

$$\sum_{k=1}^{K} x_k^T (A_k)_j \, \gamma \quad \text{and} \quad \sum_{k=1}^{K} {}_{i_k} (A_k) y \gamma \tag{15.63}$$

where $(A_k)_j$ and ${}_{i_k}(A_k)$ are jth column and i_kth row of A_k respectively. To prove it we first prove the necessary conditions. At the equilibrium, the first player chooses y to get minimum payoff (i.e. chooses the path with the lowest cost). Consequently, for any arbitrary $y^{(0)}$, $\sum_{k=1}^{K} x_k^T A_k y^{(0)} \geq \sum_{k=1}^{K} x_k^T A_k y$. One possible selection for $y^{(0)}$ is a vector with 1 at its jth element and zero at other locations. Such $y^{(0)}$ by multiplying $\sum_{k=1}^{K} x_k^T A_k$ from the right hand side, selects its jth element, i.e. $\sum_{k=1}^{K} x_k^T (A_k)_j$. In other words, at the equilibrium, for any value of $j \in [1, J]$, $\sum_{k=1}^{K} x_k^T (A_k)_j$ is greater or equal to $\sum_{k=1}^{K} x_k^T A_k y$. Similarly, at the equilibrium, the second player chooses its best set of x_k strategy vectors to get maximum payoff and therefore for any arbitrary set of $x_k^{(0)}$ vectors, $\sum_{k=1}^{K} x_k^{(0)} {}^T A_k y \leq \sum_{k=1}^{K} x_k^T A_k y = \gamma$. Consequently, for the special case when only the i_kth element of $x_k^{(0)}$s has value 1 and 0 otherwise, $x_k^{(0)} {}^T A_k y$ chooses only the i_kth elements of $A_k y$ namely ${}_{i_k}(A_k) y$ so that for any set of of i_k, $\sum_{k=1}^{K} {}_{i_k}(A_k) y$ is less or equal to γ.

To prove the sufficiency condition, we can use necessity conditions as follows. If for any $j \in [1, J]$, $\sum_{k=1}^{K} x_k^T (A_k)_j$ is greater or equal to γ then $\sum_{k=1}^{K} x_k^T (A_k)_j \, y_j^{(0)} \geq y_j^{(0)} \gamma$ resulting in $\sum_{j=1}^{J} \sum_{k=1}^{K} x_k^T (A_k)_j \, y_j^{(0)} \geq \sum_{j=1}^{J} y_j^{(0)} \gamma$ and consequently $\sum_{k=1}^{K} x_k^T A_k y^{(0)} \geq \gamma$. Similarly, if for any set of $[i_1, i_2, \ldots, i_K]$, $\sum_{k=1}^{K} {}_{i_k}(A_k) y \leq \gamma$, then $\sum_{i_k=1}^{I_k} \sum_{k=1}^{K} {}_{i_k} x_{i_k}^{(0)} (A_k) y \leq \sum_{i_k=1}^{I_k} x_{i_k}^{(0)} \gamma$ and considering $\sum_{i_k=1}^{I_k} x_{i_k}^{(0)} = 1$, we will have $\sum_{k=1}^{K} x_k^{(0)} {}^T A_k y \leq \gamma$.

Algorithm 1: Extending the Fictitious Playing Algorithm for Non-Constrained AMG Games

In the theory of matrix games, an iterative algorithm called the fictitious play method is available in order to find the optimum mixed strategies. This algorithm was proposed initially in [34] and its convergence was demonstrated in [35]. In the fictitious playing algorithm, each player chooses its best pure strategy against its opponent's mixed strategy and then updates its own mixed strategy vector based on its new pure strategy. Apparently, an AMG game has one major difference as compared to a conventional matrix game, namely against the multiple decisions of the first player, the second player takes only one decision, consequently the total payoff received by the second player is the accumulated payoff expected from multiple decisions taken by the first player. Therefore, the FP algorithm is modified as follows,

Step 1. Initialization of x_k and y as : $y^{(0)} = [A_n]_m^T$ and $x_{k,i}^{(0)} = 0$ for $\forall$ k and i and any arbitrary n and m where $[A_n]_m$ means mth row of A_n.

Step 2. For $\forall$ k, $x_{k,m}^{(t)} \leftarrow x_{k,m}^{(t-1)} + 1$ where m=arg max $(A_k \, y)$, max is element-wise maximization operator and superscript (t) means the t-th iteration of the algorithm.

$$z^{(t)} = \sum_{k=1}^{K} x_k^{(t)} {}^T A_k / \sum_{i=1}^{I_k} x_{k,i}^{(t)}$$

Step 3. In this algorithm, unlike conventional FP, normalization by $\sum_{i=1}^{I_k} x_{k,i}^{(t)}$ is necessary.

Step 4. $y_q^{(t+1)} \leftarrow y_q^{(t)} + 1$ where $q = $ arg min $z^{(t)}$.

Step 5. Run the last two steps to get convergence. After convergence, probability vectors x_k and y are normalized as $\mathbf{x}_k = \mathbf{x}_k^{(t_{\max})} / \sum_{i=1}^{l_k} x_{k,i}^{(t_{\max})}$ and $\mathbf{y} = \mathbf{y}^{(t_{\max})} / \sum_{j=1}^{J} y_j^{(t_{\max})}$.

Steps 2-5 must be run to reach equilibrium where payoff received by both players, are almost the same.

Algorithm 2: Extending the Fictitious Playing Algorithm for Constrained AMG Games

In wireless networks maximization of throughput per power is also important. Any restriction on this power makes the game non-linear. Assuming s_i as minimum transmission power of ith links and $s = [s_1, s_2, \ldots, s_{28}]^T$ as link power vector, the total transmission power required for path probability vector y is calculated as follows, $S_{tot} = s^T A y$. Consequently, considering the restriction on total required transmission power, (15.58) is modified by following normalization.

$$y = \arg\min_y \sum_{k=1}^{K} x_k^T A_k y \left(s^T A_k y \right) \tag{15.64}$$

Normalization by power does not affect (15.58) because maximization is taken over x_k and power normalization term is independent of it. According to (15.57), $\sum_{k=1}^{K} x_k^T A_k y$ is the inverse of throughput and consequently to normalize throughput to power this term must be multiplied by total power $S_{tot} = s^T A y$. Therefore for any given x_k values, payoff received by the second player after choosing its jth strategy is $\sum_{k=1}^{K} x_k^T [A_k]_j s^T [A_k]_j$ and consequently only change in *Algorithm 1* is in Step 3. In other words, for the power constrained *AMG* game model, Steps 1, 2, 4, and 5 are the same as in *Algorithm 1*, and Step 3 is modified as follows,

$$\text{Step 3. } \mathbf{z}^{(t)} = \sum_{k=1}^{K} \left(\mathbf{x}_k^{(t)\,T} \mathbf{A}_k \right) \cdot \left(\mathbf{A}_k s^T \right) / \sum_{i=1}^{l_k} x_{k,i}^{(t)},$$

where $(\cdot)$ is element wise vector multiplication.

Dominancy Theory for AMG(2,K,1)

Dominancy theory is one of the salient properties of the matrix game theory which reduces the complexity of the problems by removing unnecessary parameters resulting in reduced size of matrix game. For instance, in our application, many paths, links, or components are considered in the model whereas most of these parameters (players' strategies) can be ignored without solving the game because these dominant strategies never contribute to the mixed equilibrium. Dominancy theory for *AMG(2,K,1)* has a simple difference compared to conventional theory, expressed by the following theorem. In a *AMG(2,K,1)* jth columns of all payoff sub-matrices are dominated by qth columns if and only if,

$$a_{i_k,j}^k \geq a_{i_k,q}^k \; for \; \forall \, k \in [1, K] \; and \; i_k \in [1, I_k] \tag{15.65}$$

Correspondingly, inside each sub-matrix, row i_k is dominated by row i_p if and only if

$$a_{i_k,j}^k \leq a_{i_p,i}^k \; for \; j \in [1, J] \tag{15.66}$$

Figures 15.16–18 illustrate the system performance. Mix strategy with interference control offers the best performance.

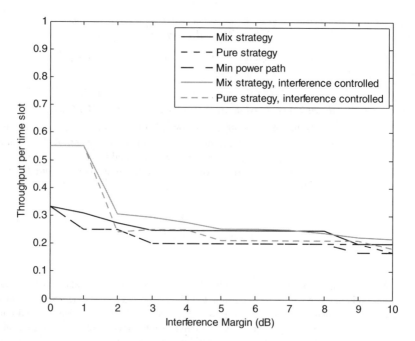

Figure 15.16   Throughput w.r.t. β.

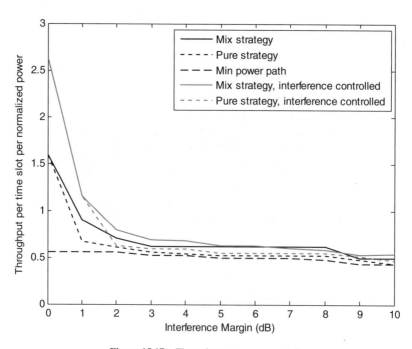

Figure 15.17   Throughput per power w.r.t. β.

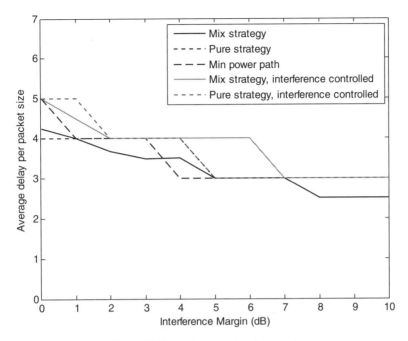

Figure 15.18   Delay per packet size w.r.t. β.

15.9   Optimization of Multicast Wireless Ad-Hoc Network Using Soft Graph Coloring and Non-Linear Cubic Games

In this section we present game theoretic models for joint routing, network coding, and the scheduling problem. As discussed in the previous section, modeling of this problem by matrix games is convenient because large matrix games can be simplified by the properties of the games such as the dominancy theory. To define such a game model, first routing and network coding are modeled by using the approach discussed in the previous section, based on a compressed topology matrix that takes into account the inherent multicast gain of the network. The topology matrix includes the set of all possible paths, including network coded paths, from sources to their corresponding sinks. These paths are identified and compressed, and then by switching between some of them with appropriate usage rates (frequencies), achievable throughput or throughput per sum of transmitting powers on the route is optimized.

The scheduling is optimized by so-called *network graph soft coloring*. Conventional scheduling algorithms are based on graph coloring where each link is painted with only one color, which we refer to as *hard coloring*. A conventional minimal graph coloring scheme is optimum if all links have the same quality and are also used with the same rate. In a real wireless *ad-hoc* network, wireless links usually neither have the same quality nor are used with the same rate. Therefore this problem should be solved with a different concept, which we refer to as soft graph coloring. Soft graph coloring is designed by switching between different *components* of a wireless network graph, which we refer to as graph fractals, with appropriate usage rates. Therefore, each link can be painted with more than one color selected with appropriate probabilities. The network components, represented by graph fractals, are a new paradigm in network graph partitioning that enables modeling of the network optimization problem by using the matrix game framework. Although topology components are types of partial topology, unlike conventional network realizations characterized by a specific color in the graph, elements (links) of a *topology component* set can have intersections. In the game discussed in this section, which is a *nonlinear cubic* game, the strategy sets of the players are links, path, and network components. The outputs of

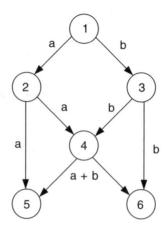

Figure 15.19   Typical butterfly network with network coding.

this game model are the mixed strategy vectors of the second and the third players at equilibrium. The strategy vector of the second player specifies optimum multi-path routing and network coding solution while the mixed strategy vector of the third player indicates the optimum switching rate among different network components or membership probabilities for an optimal soft scheduling approach. The optimum throughput is the value of the *nonlinear cubic* game at equilibrium. When throughput per power is an optimization criterion one additional constraint must be added in order to normalize the value of the game by power. The *nonlinear cubic* game is solved by extending the fictitious playing method. Numerical and simulation results prove the superior performance of this technique as compared to other conventional schemes.

15.9.1   System Model

Assumptions about the network model and the conflict-free concept are the same as in the previous section. Although the paths identification procedures are the same as in the previous section here we introduce some additional detail and use a different network example for the purposes of illustration. Unicast path identification can be done using any algorithm available in the literature. For instance, the Dijkstra algorithm finds the shortest path and can be extended easily to find $I^{m,n}$ shortest path from mth source to its nth corresponding sink. Assume $P_i^{m,n}$ as ith non-cyclic directed path from the mth source to its nth corresponding sink. The set of all multicast paths from source m to the set of sinks $\mathbf{s}_m$ is calculated again by using (15.47). We also define network coded paths as sets of links to carry the minimum required systematic and parity data from sources to sinks, whereas received packets at sinks are decodable. For the simple butterfly network in Figure 15.19 where the first node is source and the last two nodes are multicast sinks we can define nine multicast paths as follows,

$$
\begin{aligned}
P_1 &= P_{1,1}^1 = \{l_{12}, l_{25}, l_{13}, l_{36}\}, \quad & P_2 &= P_{1,2}^1 = \{l_{12}, l_{25}, l_{24}, l_{46}\}, \\
P_3 &= P_{1,3}^1 = \{l_{12}, l_{25}, l_{13}, l_{34}, l_{46}\}, \quad & P_4 &= P_{2,1}^1 = \{l_{12}, l_{24}, l_{45}, l_{13}, l_{36}\}, \\
P_5 &= P_{2,2}^1 = \{l_{12}, l_{24}, l_{45}, l_{46}\}, \quad & P_6 &= P_{2,3}^1 = P_{3,2}^1 = \{l_{12}, l_{24}, l_{45}, l_{13}, l_{34}, l_{46}\}, \\
P_7 &= P_{3,1}^1 = \{l_{13}, l_{34}, l_{45}, l_{36}\}, \quad & P_8 &= P_{3,3}^1 = \{l_{13}, l_{34}, l_{45}, l_{46}\}, \\
P_9 &= P_{NC} = \{l_{12}, l_{13}, l_{24}, l_{25}, l_{34}, l_{36}, l_{45}, l_{46}\}
\end{aligned}
$$

where the last path is the network coded path and contains all of the links of the butterfly network.

The next step is compression of the selected set of paths owing to the omnidirectional nature of the propagation and inherent multicast gain. If one multicast path consists of two or more links carrying the same information and originating from the same source, they can be represented with only one equivalent link depending on the definition of adjacency, as discussed in the previous section. If link adjacency is defined based on the interference margin, those links are replaced with the one with higher power. If the k-hops adjacency definition is used, those links are substituted with one new link. We refer to this process as topology compression that reduces the network interference graph and simplifies the scheduling process. For instance, for topology in Figure 15.19 and using the k-hops adjacency definition, $P_1, P_2, \ldots, P_9$ are compressed as:

$$P_1 = \left\{ l_{1(2,3)}, l_{25}, l_{36} \right\}, \quad P_2 = \left\{ l_{12}, l_{2(4,5)}, l_{46} \right\}, \quad P_3 = \left\{ l_{1(2,3)}, l_{25}, l_{34}, l_{46} \right\},$$
$$P_4 = \left\{ l_{1(2,3)}, l_{24}, l_{45}, l_{36} \right\}, \quad P_5 = \left\{ l_{12}, l_{24}, l_{4(5,6)} \right\}, \quad P_6 = \left\{ l_{1(2,3)}, l_{24}, l_{4(5,6)}, l_{34} \right\},$$
$$P_7 = \left\{ l_{13}, l_{3(4,6)}, l_{45} \right\}, \quad P_8 = \left\{ l_{13}, l_{34}, l_{4(5,6)} \right\},$$
$$P_9 = \left\{ l_{12}, l_{13}, l_{2(4,5)}, l_{3(4,6)}, l_{4(5,6)} \right\}$$

where, for instance, new link $l_{1(2,3)}$ means a link originating from node 1 and ending at both nodes 3 and 5 realized with only one transmission. With this notation, the problem of optimum multicast routing is finding optimum usage rates for the identified paths.

15.9.2   *Conventional Scheduling: Motivating Example*

In general, conflict free scheduling is based on conventional network graph coloring techniques. When conventional graph coloring is used, any link is painted by just one color and no two adjacent links are painted with the same color. When all links have the same capacity and are used with the same rate, any minimal conventional coloring is the optimum solution for links scheduling. The following simple example shows the inefficiency of coloring when links must be activated with different usage rates. Assume that we have three links l_1, l_2, l_3 with usage rates $r_1 = 3, r_2 = 1, r_3 = 2$. This means that within a given time frame, referred to as clique cycle to be minimized, link l_1 is used during three time slots ($r_1 = 3$), link l_2 during one slot only ($r_2 = 1$) and link l_3 during 2 slots ($r_3 = 2$). In addition, due to a given link adjacency, link 1 can be activated with the other two simultaneously and the other two cannot be activated together. The minimal coloring schemes for these three links with their assigned rates are as follows

1. $T_1 = \{l_1, l_2\}$ *and* $T_2 = \{l_3\}$ required number of time slots (clique cycle) is 5.
2. $T_1 = \{l_1, l_3\}$ *and* $T_2 = \{l_2\}$ required number of time slots is 4.
   Obviously none of these coloring schemes is optimum and the optimum scheduling for these links is,
3. $T_1 = \{l_1, l_2\}$ during the first time slot and $T_2 = \{l_1, l_3\}$ during the next two slots; required number of time slots is 3.

Therefore in the optimum solution T_1 and T_2 have non-empty intersection and we refer to the non-overlapped partial topologies components of the topology as represented by *graph fractals*. A topology or network component set is a complete set of links that can be activated at the same time. Any single link is also a component and we refer to them as first generation network components and τth generation of network components refers to the set of all components with τ members. The network components of each generation are the parents of the next generation. For our three-links example, we have five components as follows, $T_1 = \{l_1\}, T_2 = \{l_2\}, T_3 = \{l_3\}, T_4 = \{l_1, l_2\}$ *and* $T_5 = \{l_1, l_3\}$, where the first three components are the parents of the last two. Consequently in a general case in order to optimize the scheduling the appropriate network components and their optimal usage rates must be found.

15.9.3   Matrix Game Modeling for Optimum Scheduling

Formulation of the Game: Given link activation rate vector r and link capacity vector c with I elements each and network component set ζ with J elements, we define a payoff matrix H as follows,

$$h_{ij} = \begin{cases} \dfrac{c_i}{r_i}, & \text{if } l_i \in \zeta_j \\ 0, & \text{otherwise} \end{cases} \tag{15.67}$$

where r_i and c_i are the activation rate and capacity respectively for the ith link and ζ_j is the jth network component. Assume H as payoff matrix of the min-max zero sum game between two players where their strategy sets are the links and network components sets respectively. In the following we prove that the mixed strategy vector of the second player gives the optimum network component rate.

Theorem 15.1   mixed equilibrium of the zero sum game with payoff matrix (15.67) gives the optimum scheduling for usage rate vector r.

Proof. Assume x and y as mixed strategies of the players at equilibrium. Since y is normalized to 1 (sum of its component equals 1), activation of network components in average needs one time slot. Parameter $\tilde{r}_i$ as activation rate of ith link supported by components of vector y is computed as

$$\frac{\tilde{r}_i}{r_i} =_i [H] y \tag{15.68}$$

where $_i [H]$ is the ith row of the H. Therefore, for a given component rate vector y, the link with minimum supported rate, which is the bottleneck of the game is,

$$i_{minimum} = \arg \min_i (_i [H] y) \tag{15.69}$$

Consequently optimum component rate vector must maximize $\min_i (_i [H]\, y)$ as

$$y = \arg \max_y \min_i (_i [H]\, y) \tag{15.70}$$

And theoretically (15.70) is equivalent to

$$x = \arg \min_x \left(x^T H y \right) \tag{15.71}$$
$$y = \arg \max_y \left(x^T H y \right) \tag{15.72}$$

where (15.71) and (15.72) define equilibrium for game defined by payoff matrix H and value

$$Throughput = Value = max_y\ it min_x \left(\mathbf{x}^T \mathbf{H} \mathbf{y} \right) \tag{15.73}$$

Formulation of the problem as a game gives us the chance to simplify the model using some of the special properties of the games, such as the dominancy theory. For instance, in a general case while solving this game, because of dominancy, components tagged as a parent can be ignored because apparently any parent is dominated by its child. Therefore we just need to consider the last generation generated (born) from any link. In addition, the calculation of the mix strategy for linear games is easy and can be performed by the *fictitious playing* method (FP) that was discussed in the previous section.

FP Algorithm to Solve the Game Model: the FP algorithm for a min-max zero-sum game has following steps,

Algorithm 1. FP algorithm for min-max game

Step 1. Initialization of $\hat{x}$ as,

$$\hat{x}_0 = {}_i[H]^T \qquad (15.74)$$

where $_i[H]$ is an arbitrary row of H ($\hat{x}$ actually will be Hy). Then set iteration number $k = 1$.

Step 2. Finding best strategy for the first player at kth iteration as,

$$i_k = \arg\min_m \hat{x}_{k-1,m} \qquad (15.75)$$

where $\hat{x}_{k,m}$ is mth element of the $\hat{x}_k$.

Step 3. Updating $\hat{y}_k$ as, $\hat{y}_k = \hat{y}_{k-1} + [H]_{i_k}$.

Step 4. Finding best strategy for the second player at kth iteration as,

$$j_k = \arg\max_n \hat{y}_{k,n} \qquad (15.76)$$

where $\hat{y}_{k,n}$ is nth element of the $\hat{y}_k$.

Step 5. If the algorithm has not δ converged to equilibrium where δ is an small threshold set $kk + 1$ then update $\hat{x}_k$ as,

$$\hat{x}_k = \hat{x}_{k-1} + {}_{j_k}[H] \qquad (15.77)$$

and then return to Step 2.

Algorithm is δ converged if,

$$\hat{y}_{k,j_k} - \hat{x}_{k,i_k} \leq \delta \qquad (15.78)$$

When the algorithm approaches equilibrium the mixed strategies for both players are calculated as probabilities of the strategies taken at all iterations, i.e. averaging over i_ks and j_ks.

As an illustration we present the numerical results for the network coded butterfly network shown in Figure 15.19. For this network path P_9 has six links $L = \begin{bmatrix} l_{12}, & l_{13}, & l_{1(2,3)}, & l_{2(4,5)}, & l_{3(4,6)}, & l_{4(5,6)} \end{bmatrix}^T$ involved in the path. When these links have the same capacity, the conventional and optimum scheduling are the same because they also have the same link usage rate. Therefore, we present the result for the case when links have different capacities. Assume C as vector of link capacity rates as,

$$C = [1.2844, 0.7916, 0.7916, 1.1035, 1.4644, 0.6325]^T$$

where the capacity of $l_{1(2,3)}$ is the minimum capacity of l_{12} and l_{13}. The adjacency matrix for these six links is,

$$A = \begin{bmatrix} 1 & 1 & 1 & 1 & 0 & 0 \\ 1 & 1 & 1 & 1 & 0 & 0 \\ 1 & 1 & 1 & 1 & 0 & 1 \\ 0 & 1 & 1 & 1 & 1 & 1 \\ 0 & 0 & 0 & 1 & 1 & 1 \\ 1 & 0 & 1 & 1 & 1 & 1 \end{bmatrix} \qquad (15.79)$$

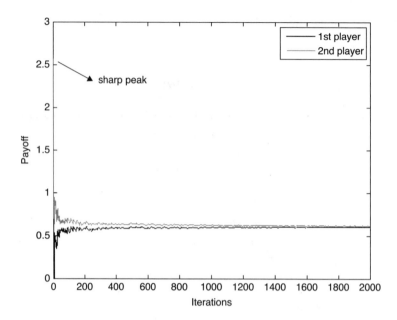

Figure 15.20 Payoff received by two players w.r.t. iteration number when scheduling in single path butterfly is optimized.

Based on adjacency matrix A, eleven network components are defined where the first six are single link parents and dominated by the last five. The dominating components are as follows, $C_1 = \{l_{12}, l_{3(4,6)}\}$, $C_2 = \{l_{12}, l_{4(5,6)}\}$, $C_3 = \{l_{13}, l_{2(4,5)}\}$, $C_4 = \{l_{13}, l_{4(5,6)}\}$, $C_5 = \{l_{1(2,3)}, l_{4(5,6)}\}$.

After solving the game defined by (15.71), (15.72) and running FP with 2000 iterations, x and y at equilibrium have values as follows,

$$x = [0.2245, 0.3671, 0.3671, 0.0121, 0.0099, 0.0193]^T$$
$$y = [0.2256, 0.0104, 0.2942, 0.0875, 0.3823]^T$$

(15.80)

Figure 15.20 presents the transient behaviour of the FP where the lower and upper curves are payoff received by the first and second players respectively. We can see that the payoff received by the second player faces a sharp peak at initial iteration but this peak disappears after few iterations and both payoffs converge to the same value. On the other hand, as we expect payoff received by the second player is always higher than that received by the first player. At equilibrium both payoffs approach the value of the game which is throughput equal to a 0.61 packet/time slot. At the same time for conventional scheduling with the following three partial topologies, $T_1 = \{l_{12}, l_{3(4,6)}\}$, $T_2 = \{l_{13}, l_{2(4,5)}\}$, $T_3 = \{l_{4(5,6)}\}$, the achieved throughput is a 0.55 packet/time slot. The throughput gain of the optimum scheduling as compared to the conventional one is about 11%.

15.9.4   Cubic Matrix Game Modeling for Joint Optimum Routing, Network Coding and Scheduling

Formulation of the Game: Assume z as a vector of path usage rates in the problem of multipath routing. In this case, we modify the game model from the previous subsection by adding a third player with mix strategy vector z. Clearly, the link usage rates are linear combinations of the path usage rates and therefore according to (15.67), the proposed three-player game will be defined by modifying parameter r_i in (15.67) by a linear combination of the elements of vector z. Consequently, this game in general form becomes a *nonlinear cubic* game where the strategy sets of its players are links, paths, and network components respectively.

The actual payoff and cost received by the first and third players respectively is a nonlinear function of the mix strategy chosen by the second player. According to (15.67), for a given z, if the first and the second players choose their ith and jth pure strategy, their payoff and cost will be $1/\sum_{k=1}^{K} a_{ijk} z_k$, where a_{ijk} as an element of the three dimensional (cubic) matrix is defined as follows,

$$a_{ijq} = \begin{cases} 1, & \text{if } l_i \in P_j, \, l_i \in \kappa_q \text{ and } P_j \text{ is a compressed multicast path,} \\ \frac{1}{C_c}, & \text{if } l_i \in P_j, \, l_i \in \kappa_q \text{ and } P_j \text{ is a compressed network coded path,} \\ \infty, & \text{if } l_i \notin P_j, \\ 0, & \text{Otherwise.} \end{cases} \tag{15.81}$$

and C_c is the number of XOR combined links at the node with network coding. For our general problem of joint optimal modeling of the routing, scheduling, and network coding using game theory, the set of payoff matrices introduced in Section 15.9.3 is replaced by the same number of matrices (this time three dimensional) with elements a_{ijk}. The mixed strategies and value of this game are calculated as follows,

$$x = \arg\min{}_x \sum_{i=1}^{L} \sum_{j=1}^{J} \frac{x_i y_j}{\sum_{k=1}^{K} a_{ijk} z_k} \tag{15.82}$$

$$y = \arg\max{}_y \sum_{i=1}^{L} \sum_{j=1}^{J} \frac{x_i y_j}{\sum_{k=1}^{K} a_{ijk} z_k} \tag{15.83}$$

$$z = \arg\max{}_z \sum_{i=1}^{L} \sum_{j=1}^{J} \frac{x_i y_j}{\sum_{k=1}^{K} a_{ijk} z_k} \tag{15.84}$$

$$\text{Throughput} = \text{Value} = \min{}_x \max{}_y \max{}_z \sum_{i=1}^{L} \sum_{j=1}^{J} \frac{x_i y_j}{\sum_{j=1}^{J} a_{ijk} z_k} \tag{15.85}$$

Since the value of this game is a linear function of the mix strategy vector of the first and second player, the dominancy theory is still applicable for their strategies but this property of linear games is not valid for the third player's strategies.

Theorem 15.2 *Dominancy theory for the first two players, in the proposed nonlinear cubic game:* Strategy i_1 of the first player is dominated by its strategy i_2 if for any j and k, $a_{i_1 jk} \leq a_{i_2 jk}$ and in the same way, strategy j_1 of the second player is dominated by its strategy j_2 if for any i and k, $a_{ij_1 k} \geq a_{ij_2 k}$.

Proof. We only need to prove the first part of the theorem and the second part is proved in the same way. Assume x, y, and z as strategy vectors at equilibrium. If $a_{i_1 jk} a_{i_2 jk}$ for any j and k, then $\sum_{k=1}^{K} a_{i_1 jk} z_k \leq \sum_{k=1}^{K} a_{i_2 jk} z_k$ and consequently,

$$val = \sum_{i=1}^{L} \sum_{j=1}^{J} \frac{x_i y_j}{\sum_{k=1}^{K} a_{ijk} z_k} = \sum_{\substack{i=1, \\ i \neq i_1, \\ i=i_2}}^{L} \sum_{j=1}^{J} \frac{x_i y_j}{\sum_{k=1}^{K} a_{ijk} z_k} + \sum_{j=1}^{J} \frac{x_{i_1} y_j}{\sum_{k=1}^{K} a_{i_1 jk} z_k} + \sum_{j=1}^{J} \frac{x_{i_2} y_j}{\sum_{k=1}^{K} a_{i_2 jk} z_k} \tag{15.86}$$

and,

$$val \sum_{\substack{i=1, \\ i=i_1, \\ i=i_2}}^{L} \sum_{j=1}^{J} \frac{x_i y_j}{\sum_{k=1}^{K} a_{ijk} z_k} + \sum_{j=1}^{J} \frac{(x_{i_1} + x_{i_2}) y_j}{\sum_{k=1}^{K} a_{i_2 jk} z_k} \tag{15.87}$$

Therefore the payoff received by the first player is less than or equal to the value of the game if the player chooses strategy i_2 instead of i_1 and consequently i_1 strategy is dominated by i_2. Inequality is possible if $x_{i_1} = 0$.

Extended Factitious Playing for Nonlinear Cubic Game: To solve (15.82)–(15.84), we first prove that against any mixed strategies of the second and third players, the best option of the first player is a pure strategy.

Theorem 15.3 For given mixed strategy vectors y and z, the first player receives minimum payoff by a pure strategy and for the given mixed strategy vectors x and z, the second player receives maximum payoff by a pure strategy.

Proof. Assume, $i_{min} = \arg\min_i \sum_{j=1}^{J} \frac{y_j}{\sum_{k=1}^{K} a_{ijk} z_k}$. Consequently we have

$$\sum_{i=1}^{L} \sum_{j=1}^{J} \frac{x_i y_j}{\sum_{k=1}^{K} a_{ijk} z_k} \geq \sum_{i=1}^{L} \sum_{j=1}^{J} \frac{x_i y_j}{\sum_{k=1}^{K} a_{imax jk} z_k} = \sum_{i=1}^{L} x_i \sum_{j=1}^{J} \frac{y_j}{\sum_{k=1}^{K} a_{imax jk} z_k} \tag{15.88}$$

Considering $\sum_{i=1}^{l} x_i = 1$, right hand side of (15.88) is simplified as,

$$\sum_{i=1}^{L} \sum_{j=1}^{J} \frac{x_i y_j}{\sum_{k=1}^{K} a_{ijk} z_k} \sum_{j=1}^{J} \frac{y_j}{\sum_{k=1}^{K} a_{imax jk} z_k} \tag{15.89}$$

i.e. the payoff received by the first player when it chooses i_{min}th strategy is minimum and proof is complete. The second part of the theorem for the second player is proved in the same way but there is no such a property for the third player and consequently the conventional FP algorithm cannot be used for the third player.

Algorithm 2: Extending the Fictitious Playing Algorithm for nonlinear cubic game
   Step 1. Initialization of y, z as follows,

$$q = 0, \quad y_j^{(q)} = 0, \text{ for } j \in [2, J] \text{ and } y_1^{(q)} = 1, \text{ and}$$
$$z_k^{(q)} = 0, \text{ for } k \in [2, K] \text{ and } z_1^{(q)} = 1$$

Step 2. Setting iteration number $q \leftarrow q + 1$ and calculation of the payoff received by different strategies of the first player as,

$$V_{x,i}^{(q)} = \sum_{j=1}^{J} \frac{y_j^{(q-1)}}{\sum_{k=1}^{K} a_{ijk} z_k^{(q-1)}} \tag{15.90}$$

Step 3. Choose best strategy for the first player against $y^{(q)}$ and $z^{(q)}$ as,

$$i_{min}^{(q)} = \arg \, \min_i V_{x,i}^{(q)} \tag{15.91}$$

Step 4. Updating $x^{(q)}$ as $x_{i_{min}^{(q)}}^{(q)} \leftarrow x_{i_{min}^{(q)}}^{(q-1)} + 1$ and $x_i^{(q)} \leftarrow x_i^{(q-1)}$ for $\forall i \neq i_{min}^{(q)}$.

Step 5. Calculate the payoff received by different strategies of the second player as,

$$V_{y,j}^{(q)} = \sum_{i=1}^{I} \frac{x_i^{(q)}}{\sum_{k=1}^{K} a_{ijk} z_k^{(q-1)}} \tag{15.92}$$

Step 6. Choosing the best strategy for the second player against $x^{(q)}$ and $z^{(q)}$ as,

$$j_{max}^{(q)} = \arg \, \max_j V_{y,j}^{(q)} \tag{15.93}$$

Step 7. Updating $y^{(q)}$ as $y_{j_{max}^{(q)}}^{(q)} \leftarrow y_{j_{max}^{(q)}}^{(q-1)} + 1$ and $y_j^{(q)} \leftarrow y_j^{(q-1)}$ for $\forall j \neq j_{max}^{(q)}$.

Step 8. Updating $z^{(q)}$ using the following iterative algorithm which is based on steepest descent algorithm. (Note, the value of the game is a non-linear function of z and consequently the best strategy of the third player at iteration q cannot be found by just maximization over the strategy set of this player.)

8.2. Initialize $z_0^{(q)}$ as $z_0^{(q)} = \frac{z^{(q-1)}}{|z^{(q-1)}|}$ and go to the first iteration $t = 1$.

8.2a. Updating z_t as

$$\tilde{z}_{t,k}^{(q)} = z_{t-1,k} - \mu \sum_{i=1}^{I} \sum_{j=1}^{J} \frac{x_i^{(q)} y_j^{(q)} \left(a_{ijk} - a_{ijK} \right)}{\left(\sum_{k=1}^{K} a_{ijk} z_{t-1,k}^{(q)} \right)^2 \sum_{i=1}^{I} x_i^{(q)} \sum_{j=1}^{J} y_j^{(q)}}, \quad \text{for } k \neq K \tag{15.94}$$

$$\tilde{z}_{t,k}^{(q)} = z_{t-1,k} + \mu \sum_{k=1}^{K-1} \sum_{i=1}^{I} \sum_{j=1}^{J} \frac{x_i^{(q)} y_j^{(q)} \left(a_{ijk} - a_{ijK} \right)}{\left(\sum_{l=1}^{K} a_{ijk} z_{t-1,l}^{(q)} \right)^2 \sum_{i=1}^{I} x_i^{(q)} \sum_{j=1}^{J} y_j^{(q)}}, \quad \text{for } k = K \tag{15.95}$$

where μ is the step size.

8.3. Assuming ρ as minimum of $\tilde{z}_{t,k}$ versus k, $z_{t,k}$ is updated as,

$$z_{t,k} = \begin{cases} \dfrac{\tilde{z}_{t,k}}{|\tilde{z}_{t,k}|}, & \text{if } \rho \geq 0 \\[4mm] \dfrac{\tilde{z}_{t,k} - \rho}{|\tilde{z}_{t,k} - \rho|} & \text{if } \rho < 0 \end{cases} \tag{15.96}$$

8.4. $t + 1$, and return to 8.2 for the next iteration. This sub-algorithm is repeated until have convergence.

Step 9. Updating $z^{(q)}$ as $z^{(q)} = z^{(q-1)} + z_t^{(q)}$.

Step 10. Going to step 2, and repeating steps 2–9 to have convergence.

Step 11. After Q iterations when convergence condition holds calculated mixed strategy vector are normalized to one as $x^{(eq)} = \frac{x^{(Q)}}{|x^{(Q)}|}$, $y^{(eq)} = \frac{y^{(Q)}}{|y^{(Q)}|}$, and $z^{(eq)} = \frac{z^{(Q)}}{|z^{(Q)}|}$. This normalization during iterations before convergence is not necessary and can be performed only once after convergence.

Illustration for Nonlinear Cubic Game: for illustration, the butterfly network shown in Figure 15.19 is used again, with the eight paths defined in Section 15.9.2 and the link capacities and network components set defined in Section 15.9.3. This joint optimization and routing problem is solved by using the FP algorithm for *non-linear cubic* game. For 1-hop adjacency, paths 4, 5, 6, and 8 are dominated by other five paths and therefore we actually need to find rates for the four remaining paths. Figure 15.21 shows the transient behavior of the payoffs received by three players. It can be seen that, for the lower iteration index, the payoff received by the third player has a sharp peak. Although this peak disappears for higher iteration indices, it can be avoided by better initialization in the steepest descent algorithm used to update mix strategy of the third player. As expected, although the payoff received by the first player is always lower than that received by its opponents, all of the payoff converges asymptotically to the same value. The mixed strategies of the three players at equilibrium are,

$$x = [0.0025, 0.0055, 0.0015, 0.2928, 0.2245, 0.4733]^T \tag{15.97}$$
$$y = [0.1025, 0.0510, 0, 0.0625, 0.7839]^T \tag{15.98}$$
$$z = [0.2284, 0.0030, 0.3003, 0.2289, 0.2394]^T \tag{15.99}$$

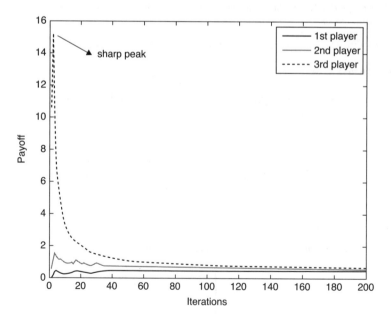

Figure 15.21 Payoff received by 3 players w.r.t. iteration number from non-linear cubic game when joint routing and scheduling in multipath butterfly network is optimized.

The value of the game which is equivalent to the throughput is 0.76 which shows about a 25% gain compared to single path routing with optimum scheduling and a 36% gain compared to single path routing with conventional scheduling.

Simulation results are presented for the case where network graphs are generated randomly with nodes distributed uniformly over unit squares and the Monte Carlo simulation technique is used.

Multiple Single-Path Unicast Sessions: The game model discussed in Section 15.9.3 is applied to the scenario obtained by random selection of sources and their corresponding sinks where for each source-sink pair the shortest path (physically the shortest part that requires minimum power) is calculated using the Dijkstra algorithm [36]. The attenuation factor is assumed to be 4 and the number of packets to be delivered from each source to its corresponding sink is chosen randomly by Poisson distribution.

The optimal algorithm is compared to the conventional scheduling based on network graph hard coloring. The average number of required time slots to carry each packet from source to its corresponding sink is considered as the performance criterion. This parameter is reciprocal of the value of the game calculated by (15.73). Results are averaged over 1000 independent runs. At each run, link usage rate is calculated from the load on each path and then using (15.45) the adjacency matrix is defined. Finally, using (15.67) the game model is defined and is solved using the FP algorithm presented in Section 15.9.3. The results are summarized in Figures 15.22–23.

Figure 15.22 presents the average number of required time slots versus the number of parallel source-sink pairs for different numbers of source-sink sessions when $N = 10$ nodes. We can see from the figure that the optimum scheduling outperforms the conventional scheme. In this case the proposed optimum scheduling needs between 11 and 25% fewer time slots on average.

Figure 15.23 presents the same results for $N = 20$. In this case, we can see that the scheduling gain compared to the conventional scheduler is even higher than in Figure 15.22.

Multipath Multicast Routing and Scheduling: here, the *nonlinear cubic* game model presented in Section 15.9.4 is simulated over randomly generated graphs with uniformly distributed nodes for cases $N = 10$ and 20 nodes, and the number of sinks between one to five and the results are averaged over 100

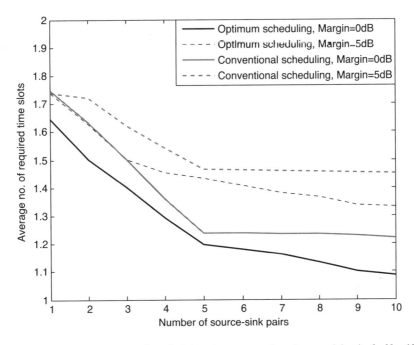

Figure 15.22   Average number. of required time slots w.r.t. number of source-sink pairs for $N = 10$.

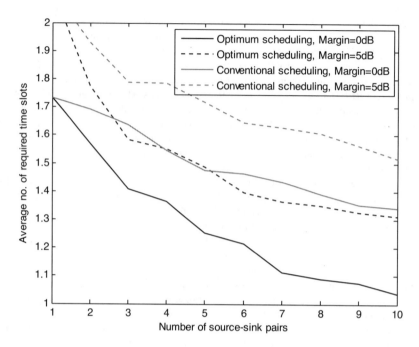

Figure 15.23   Average number of required time slots w.r.t. number of source-sink pairs for N = 20.

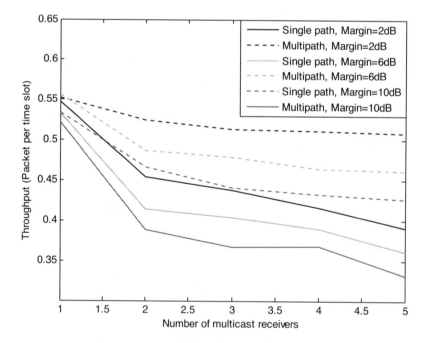

Figure 15.24   Multicast throughput of the optimized multipath and single-path routing w.r.t. number of multicast sinks for different values of interference margin when N = 10 nodes.

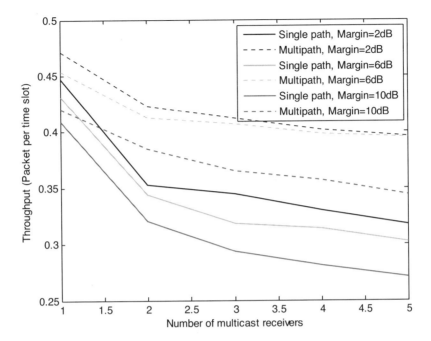

Figure 15.25   Multicast throughput of the optimized multipath and single-path routing w.r.t. number of multicast sinks for different values of interference margin when N = 20 nodes.

independent runs. In each run, one node is selected randomly as the source and some of the other nodes are selected as its corresponding sinks. Then, using a modified version of the Dijkstra algorithm up to five of the shortest paths from source to each sink are identified and then by combining the identified unicast path using (15.47) the set of multicast paths is constructed. In the next step, the wireless links involved in the set of multicast paths are identified and their adjacency matrix is defined using (15.45). Finally, the non-*linear cubic game* is defined using (15.81) and solved by using the modified FP algorithm and the throughput is calculated using (15.85).

Figure 15.24 presents the simulation results for N = 10 nodes. It can be seen that in all of the range of interference margin, optimized multipath routing outperforms the single path case and this improvement, when there are more multicast sinks, is higher. One additional interesting result is the throughput floor observed in the curves of the multipath routing which proves that after settling to these floors, the multicast throughput does not decrease further with the increase of multicast sinks; and this floor can be considered as throughput for broadcasting.

Figure 15.25 presents the simulation results for N = 20 nodes. In this case we can see a lower throughput compared to Figure 15.24 which is due to the greater number of required hops between source and sinks; however, in this case the performance gain compared to single-path routing is still slightly higher than in Figure 15.24.

15.10   Joint Optimization of Routing and Medium Contention in Multihop Multicast Wireless Network

In the previous section we discussed optimization of the multicast networks by using contention free links obtained by proper scheduling. In this section we present a general framework for the joint

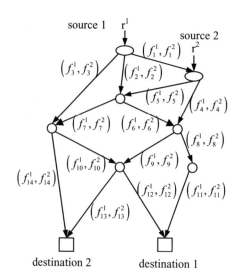

Figure 15.26   Routing graph at the network layer: $(r, f) \in N$.

optimization problem, which involves variables from both the network layer (routing) and the physical layer (interference/power control). A given level of interference will determine the link capacities. So the joint optimization of routing and scheduling considered in the previous section is now replaced by joint optimization of routing and power (interference) control. We show how dual problem (Lagrange relaxation) and subgradient optimization can be applied to decompose the overall optimization problem into a sequence of smaller subproblems, each of which only involves variables from either the network layer or the physical layer and after that we discuss the interactions between the two subproblems.

The formulation of the throughput maximization problem in wireless mesh networks is based on the following facts: (a) Throughput is realized by routing flows from sources to destinations; (b) At each transmission link, the aggregated flow rate cannot exceed the link capacity; (c) The link capacity is a function of signal-to-interference-and-noise ratio (SINR), which in turn is determined by the power levels at all the transmitters.

In what follows we will use the following terminology

$G = (V, E)$ - *network topology.*

S - *set of multiple data sessions supported in the network.*

$\mathbf{r} = \{r^i\}$ *be the set of multicast throughput for each session $i \in S$.*

$\mathbf{f}$ - *flow rate vector* $\{f_l^i\}$, *where i is the session index $i \in S$ and l is the link index $l \in E$.*

N - *routing region, a set of (r, f) such that flow rate f can support multicast throughput r.*

Figure 15.26 illustrates the routing (flow and the throughput) variables in a network.

At the physical layer we will be interested in the *capacity region* C. Let $\mathbf{p}$ be the set of power consumption on each link in E, and $\mathbf{c}$ be the set of achievable link capacities. The capacity region defines a set of $(\mathbf{c}, \mathbf{p})$ such that the link power $\mathbf{p}$ can support link capacity $\mathbf{c}$. The capacity region characterizes the tradeoff between link capacity and power allocation due to the shared nature of wireless mesh networks. Figure 15.27 illustrates power and capacity variables in a network.

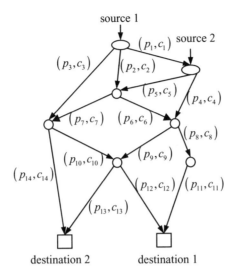

source 1

source 2

(p_1, c_1)

(p_3, c_3)

(p_2, c_2)

(p_5, c_5)

(p_4, c_4)

(p_7, c_7) (p_6, c_6)

(p_8, c_8)

(p_{10}, c_{10}) (p_9, c_9)

(p_{14}, c_{14})

(p_{12}, c_{12}) (p_{11}, c_{11})

(p_{13}, c_{13})

destination 2   destination 1

Figure 15.27   Capacity/interference graph at the physical layer (c,p) ∈C.

The physical layer can support the network traffic only if the aggregated flow of different multicast sessions on each link is less than the link capacity. In this section, we adopt the following utility

$$U(\mathbf{r}) = \sum_i U_i(r^i) = \sum_i \log(1 + r^i)$$

The throughput optimization problem can now be formulated as

$$\begin{aligned}
\max \ & U(\mathbf{r}) \\
\text{s.t. } & (\mathbf{r}, \mathbf{f}) \in N \\
& (\mathbf{c}, \mathbf{p}) \in C \\
& \sum_{i \in S} f_l^i \le c_l, \ \forall l \in E
\end{aligned}$$ (15.100)

where the constraint $(\mathbf{r}, \mathbf{f}) \in N$ models the interdependence between the achievable multicast throughput $\mathbf{r}$ and the data flow routing scheme $\mathbf{f}$. The constraint $(\mathbf{c}, \mathbf{p}) \in C$ models the interdependence between the link capacity vector $\mathbf{c}$ and the link power consumption $\mathbf{p}$. The constraint $\sum_i f_l^i \le c_l$ reflects the fact that the aggregated flow rate at each link is upper bounded by the link capacity where i is the index of data sessions, and l is the index of links. The detailed characterization of the regions N and C are independent of this general formulation and will be discussed in the next section.

In the following we will present a dual decomposition approach for a distributed solution of the overall network optimization problem (15.100). In general, when the utility function $U(\mathbf{r})$ is concave and when both N and C are convex regions, generic convex optimization methods can be used to solve the overall optimization problem (15.100). In a wireless mesh network, decentralized and scalable implementations are preferred. In this section we present an optimization solution framework, where the original problem is decomposed into smaller subproblems. Each of these subproblems can be solved efficiently in a distributed fashion.

As the first step we relax link capacity constraint $\sum_i f_l^i \leq c_l$ by introducing price into the objective function

$$L = U(\mathbf{r}) + \sum_l \lambda_l \left[c_l - \sum_i f_l^i \right] \tag{15.101}$$

The maximization of the Lagrangian above now consists of two sets of variables: network-layer variables $(\mathbf{r}, \mathbf{f})$, and physical-layer variables $(\mathbf{c}, \mathbf{p})$. In other words, the Lagrangian optimization problem is now decoupled into two disjoint parts. The network-layer part is a routing subproblem max $U(\mathbf{r}) - \sum_l \lambda_l \sum_i f_l^i$, s.t. $(r, f) \in N$ and the physical-layer part is a power control subproblem max $\sum_l \lambda_l c_l$ s.t. $(\mathbf{c}, \mathbf{p}) \in C$. Thus, the global maximization problem decomposes into two parts: routing at the network layer and power control at the physical layer. The power control subproblem ensures that the maximal capacity is provided in individual network links, while the routing subproblem ensures that the link capacity is efficiently utilized to maximize the multicast throughput.

The key requirement that allows the decoupling of the network optimization problem into routing and power control is the underlying convexity structure of the problem. Further, as strong duality holds, the optimization problem (15.100) can be solved efficiently via its dual. Therefore the following primal-dual algorithm can solve the entire network optimization problem.

Primal-Dual Algorithm
1. *Set $t = 0$. Initialize $\lambda^{(0)}$.*
2. *In primal domain, solve the following subproblems:*

$$\max_{\mathbf{r}, \mathbf{f}} U(\mathbf{r}) - \sum_l \lambda_l \sum_i f_l^i, \ s.t. \ (\mathbf{r}, \mathbf{f}) \in N \tag{15.102}$$

$$\max_{\mathbf{c}, \mathbf{p}} \sum_l \lambda_l c_l, \ s.t. \ (\mathbf{c}, \mathbf{p}) \in C \tag{15.103}$$

3. *In dual domain, update the dual variables*

$$\lambda_l^{(t+1)} = [\lambda_l^{(t)} + v_l^{(t)} (c_l - \sum_i f_l^i)]^+; \ \text{where } [\cdot]^+ \text{ denotes } \max(0, \cdot) \tag{15.104}$$

4. *Set $t = t + 1$. Return to step 2) until convergence.*

The flow routing subproblem (15.102) with network coding can be stated as follows:

$$\max U(\mathbf{r}) - \sum_l \lambda_l \sum_i f_l^i$$
$$s.t. \ r^i \leq \sum_{l \in I(T_j^i)} e_l^{i,j}, \ \forall i, \forall j, \forall T_j^i \in V$$
$$e_l^{i,j} \leq f_l^i, \ \forall i, \forall j, \forall l \in E \tag{15.105}$$
$$\sum_{l \in O(n)} e_l^{i,j} = \sum_{l' \in I(n)} e_{l'}^{i,j}, \ \forall i, \forall j, \forall n \in V \backslash \{s^i, T_j^i\}$$
$$f_l^i \geq 0, \ e_l^{i,j} \geq 0, \ r^i \geq 0$$

The first inequality represents the constraint that the ith session multicast throughput r^i is less than or equal to the sum of all the conceptual flow rates from source s^i to each of its jth destination T_j^i. The second inequality represents the fact that the actual flow rate f_l^i of session i on link l is the maximum of all the conceptual flows from source to destinations in that session. The third equality constraint represents the law of flow conservation for conceptual flows, where $I(n)$ is defined as the set of links that are incoming to node n, and $O(n)$ is the set of links that are outgoing from node n.

Consider a network with G_{ll}, p_l, and σ_l^2 as the link gain, power, and noise variance, respectively. If G_{lj} as the interference coefficient from link j to link l and each node has a power budget $P_{n,max}$, the power control subproblem (15.103) with a physical-layer interference model may be formulated as follows:

$$
\begin{aligned}
&\max \sum_l \lambda_l c_l \\
&\text{s.t. } c_l = \log(1 + SINR_l), \ \forall l \in E \\
&\quad SINR_l = \frac{G_{ll} p_l}{\sum_{j \neq l} G_{lj} p_j + \sigma_l^2} \ \forall l \in E \\
&\quad \sum_{l \in O(n)} p_l \leq P_{n,max}, \ p_l \geq 0 \ \forall n \in V, \ \forall l \in E
\end{aligned}
\tag{15.106}
$$

where c_l is the capacity of link l, $SINR_l$ is the signal-to-interference-and-noise ratio of link l, and n is the node index.

Because of interference, the power control subproblem (15.106) is a nonconvex optimization problem that is inherently difficult to solve.

Geometric Programming can be used in high SINR scenarios, to solve the problem (15.106). The idea is to first approximate the link capacity $c_l = \log(1 + SINR_l) \approx \log(SINR_l)$ assuming that the SINR is much larger than 1. Then, through a logarithmic transformation of power vector, the transformed problem becomes a convex optimization problem.

Game Theoretic Approach, in a power control game, considers each link as a player with an aim of maximizing its payoff function. In such game each link player l maximizes its payoff function as follows:

$$
Q_l = \lambda_l \log\left(1 + \frac{G_{ll} p_l}{\sum_{j \neq k} G_{lj} p_j + \sigma_l^2}\right) - m_l p_l - \mu_n p_l
\tag{15.107}
$$

where Q_l is the payoff for link player l, p_l is the player's action, m_l is the dual variable summarizing the effect of interference to all other links, and μ_n is the dual variable that indicates the price of transmitter power at node n. A sensible choice for m_l is $-\partial \sum_{s \neq l} c_s / \partial p_l$. In other words, m_l is the rate at which other users' achievable data rates decrease with an additional amount of power. The power price μ_n reflects how tight the resource at node n is being utilized by its outgoing links under the constraint $\sum_{l \in O(n)} p_l \leq P_{n,max}$. The algorithm can be summarized as follows

Power Control Game Algorithm

1. *start $p^{(0)}, m^{(0)}, \mu^{(0)}, t = 0$.*
2. *Set $\tilde{p}^{(0)} = p^{(t)}$ and $i = 0$, iteratively update*

$$
\tilde{p}_l^{(i+1)} = [\lambda_l / \left(m_l^{(t)} + \mu_n^{(t)}\right) - \sum_{j \neq l} G_{lj} \tilde{p}_l^{(i)} / G_{ll} - \sigma_l^2 / G_{ll}]^+
$$

$i = i + 1$ *(repeat until $\tilde{p}^{(i)}$ converges), set $p^{(t+1)}(t) = \tilde{p}^{(i)}$.*

3. *Update price μ_n with stepsize γ_n*

$$
\mu_n^{(t+1)} = \left[\mu_n^{(t)} + \gamma_n^{(t)} \left(\sum_{l \in O(n)} p_l^{(t)} - P_{n,max}\right)\right]^+
$$

4. *Update the message m_l*

$$m_l^{(t+1)} = \sum_{s \neq l} G_{ls} \lambda_s \frac{SINR_s^{(t+1)}}{G_{ss} p_s^{(t+1)}} \frac{SINR_s^{(t+1)}}{1 + SINR_s^{(t+1)}}$$

5. *Set $t = t + 1$. Return to step 2. until convergence.*

The expression for optimal p_l is obtained by setting the derivative of Q_l, with respect to p_l, to zero. Such a locally optimal p_l strikes a balance between maximizing its own rate and minimizing its interference to other links (which is taken into account via m_l). A large value for m_l indicates that link l is producing severe interference to other links. This is reflected in the power update as a larger m_l leads to a lower p_l. Similarly, the value of the pricing variable μ_n indicates the tightness of the per-node power constraint. A high value for μ_n signals that the supply for power is tight and it instructs link l to reduce its power.

The Distributed Solution can be obtained as

Distributed Primal-Dual Algorithm
  1. Set $t = 0$. Initialize $\lambda^{(0)}$.

  2. In primal domain,

  2.1. Solve the routing subproblem by network coding technique in a distributed manner

$$\max_{\mathbf{r},\mathbf{f}} U(\mathbf{r}) - \sum_l \lambda_l \sum_i f_l^i, \text{ s.t. } (\mathbf{r}, \mathbf{f}) \in N$$

  2.2. Solve the power control subproblem by Power Control Game Algorithm, in a distributed manner

$$\max_{\mathbf{c},\mathbf{p}} \sum_l \lambda_l c_l, \quad s.t. \ (\mathbf{c}, \mathbf{p}) \in C$$

  with the optimal solution denoted as $(\mathbf{c}^*, \mathbf{p}^*)$.

  3. In dual domain,

$$\lambda_l^{(t+1)} = [\lambda_l^{(t)} + v_l^{(t)}(c_l^* - \sum_i f_l^{i,*})]^+$$

  4. Set $t = t + 1$. Return to step 2) until convergence.

The network-layer multicast flow routing step (2.1) is solved in a three-phase solution: (a) session separation; (b) distributed min cast flow computation; and (c) utility maximization at the source.

In phase (a), the joint multisession optimization is separated into smaller intrasession optimization problems. A critical assumption of the optimization in (2.1) is that the objective function $\max(U(\mathbf{r}) - \sum_l \lambda_l \sum_i f_l^i) = \sum_i \max(U_i(r^i) - \sum_l \lambda_l f_l^i)$ is separable.

As a result, we can solve the joint optimization in step (2.1) by solving a sequence of projections within each multicast session.

The separated intrasession optimizations are then solved in phase (b) and phase (c). Phase (b) makes necessary preparations for the final optimization in phase (c) by computing the minimum weighted (with given weight vector λ) bandwidth consumption needed to achieve a unit end-to-end multicast throughput.

Once the minimum weighted bandwidth consumption ($b^* = \sum_l \lambda_l f_l^i$) for one unit multicast through-put is determined, the source s_i makes the final optimization decision in phase (c). The source first collects the value of b^* as the output from phase (b), then transforms the objective function from $\max(U_i(r^i) - \sum_l \lambda_l f_l^i)$ into $\max(U_i(r^i) - b^* r^i)$, and performs a local single-variable maximization based on the function curve of U and the value of b^*. Since each of phases (a)–(c) can be implemented in a distributed way, we have a distributed algorithm for the flow routing module 2.1). Step 2.2 which is a power control game can be also implemented in a distributed way. For this purposes we decompose the message update in step 4 of *power control game algorithm* as

$$m_l^{(t+1)} = \sum_{s \neq l} G_{ls} bcm_s$$

$$bcm_s = \lambda_s \frac{SINR_s^{(t+1)}}{G_{ss} p_s^{(t+1)}} \frac{INR_s^{(t+1)}}{1 + SINR_s^{(t+1)}}$$

In other words, we use a two-step message-passing mechanism. In the first step (a), each link calculates its broadcast message bcm_s, based on local information (λ_s, $SINR_s$, G_{ss}, p_s), and broadcasts to the network. In the second step (b), each link collects broadcast messages from others, and computes the message m_l locally, where the interference term G_{ls} can be estimated, for example, by pilots. Since the update of μ_n only requires the outgoing link power allocation from node n and its budget $P_{n,\max}$, it can be updated locally. Hence, given the dual variables m_l and μ_n, the power update in step 2 of the power control game can be achieved locally.

In step 3 of *Distributed Primal-Dual Algorithm*, the update of the dual variable λ_l in the lth link only requires the local capacity c_l and the rates of local flows $\sum_i f_l^i$. Therefore, the shadow prices λ can be updated locally. For the reasons given above, the overall throughput maximization problem in mesh networks has a distributed implementation.

15.11 Routing and Network Stability

In this section we present a formulation of a general power control problem for time-varying wireless networks, the characterization of the network layer capacity region, and the development of capacity achieving routing and power allocation algorithms that offer delay guarantees and control of queue overflows. These algorithms are valid for systems with general arrival and channel processes, including ad hoc networks with mobility. This approach unifies notions of *network capacity, network optimization*, and *network control*.

15.11.1 Time varying network with queuing

We assume a network with N nodes and time-varying channels. Let $\underline{S}(t) = (S_{ab}(t))$ denote the matrix process of channel states, where $S_{ab}(t)$ represents the current state of channel (a, b) (attenuation values *and/or* noise levels). Time is slotted with slots normalized to integral units $t \in \{0, 1, 2, \ldots\}$. We assume that channels do not change their state for the duration of a time slot, and are known to the network controller at the beginning of each slot. Such information can be obtained either through direct mea-surement (where time slots are assumed to be long in comparison to the required measurement time) or through a combination of measurement and channel prediction. The channel process $\underline{S}(t)$ takes values on a finite-state space, and is ergodic with time average probabilities $\pi_{\underline{S}}$ for each state $\underline{S}$.

Every time slot, a controller determines transmission rates on each link by allocating a power matrix $\underline{P}(t) = (P_{ab}(t))$ subject to a total power constraint $\sum_{b \neq i} P_{ib}(t) \leq P_i^{tot}$ for all nodes i represented in the form $\underline{P}(t) \in \Pi$, where Π is a compact set of acceptable power allocations that includes the power limits for each node.

Link rates are determined by a corresponding rate-power curve $\underline{\mu}(\underline{P}, \underline{S}) = (\mu_{ab}(\underline{P}, \underline{S}))$. It is assumed that data can be split continuously, so that each time slot the transmission rate $\mu_{ab}(\cdot)$ determines the number of bits that can be transferred over the wireless link (a, b). The transmission rate over a link (a, b) of the network depends on the full matrix of power allocation decisions because communication rates over the link may be influenced by interference from other channels. For example, achievable data rates could be approximated

$$\mu_{ab}(\underline{P}, \underline{S}) = \log \left(1 + \frac{\alpha_{ab} P_{ab}}{N_b + \alpha_{ab} \sum_{j \neq b} P_{aj} + \sum_{i \neq a} \alpha_{ib} \sum_j P_{ij}} \right) \tag{15.108}$$

where N_b and α_{ab} represent noise and fading coefficients associated with the particular channel state $\underline{S}$.

Each network node i maintains a set of output queues for storing data according to its destination. For convenience, we classify all data flowing through the network as belonging to a particular *commodity* $c \in \{1, \ldots, N\}$, representing the destination node for the data. A network control algorithm makes decisions about power allocation, routing, and scheduling. As a general algorithm might schedule multiple commodities to flow over the same link on a given slot, we define $\mu_{ab}^{(c)}(t)$ as the rate offered to commodity c traffic along link (a, b) during time slot t. The controller, makes the following decisions:

Choose $\underline{P}(t)$ such that $\underline{P}(t) \in \Pi$ (*power allocation decision*) and choose $\mu_{ab}^{(c)}(t)$ such that $\sum_c \mu_{ab}^{(c)}(t) \leq \mu_{ab}(t) = \Delta \mu_{ab}(\underline{P}(t), \underline{S}(t))$ which is a *routing/scheduling* decision. In the special case where there is no power allocation, the $\mu_{ab}(t)$ process is determined purely by the dynamic channel states of the network, and any network control algorithm reduces to pure routing and scheduling.

If $A_i^{(c)}(t)$ is the amount of commodity c bits that arrive exogenously to the network at node i during slot t and $U_i^{(c)}(t)$ is the current backlog of bits in node i destined for node c then the $U_i^{(c)}(t)$ processes evolve according to the following queuing dynamics:

$$U_i^{(c)}(t + 1) \leq \max[U_i^{(c)}(t) - \sum_b \mu_{ib}^{(c)}(t), \ 0] + \sum_a \mu_{ai}^{(c)}(t) + A_i^{(c)}(t) \tag{15.109}$$

For a single queue in isolation, with an input process $A(t)$, a time varying server process $\mu(t)$ and the unfinished work function $U(t)$ representing the amount of unprocessed bits remaining in the queue at time t, we define the following 'overflow' function $g(V)$:

$$g(V) = \lim_{t \to \infty} \sup (1/t) \sum_{\tau=0}^{t-1} Pr[U(\tau) > V] \tag{15.110}$$

as a measure of the fraction of time the unfinished work is above a certain value V.

A single server queuing system is stable if $g(V) \to 0$ as $V \to \infty$.

A network of queues is stable if all individual queues are stable.

Assume input processes $A_i^{(c)}(t)$ are stationary and ergodic with rates λ_{ic}, so that $\lim_{t \to \infty} \frac{1}{t} \sum_{\tau=0}^{t-1} A_i^{(c)}(\tau) = \lambda_{ic}$ with probability 1. Let (λ_{ic}) represent the corresponding $N \times N$ *rate matrix*, having all diagonal entries (λ_{ii}) equal to zero. Assume the channel process $\underline{S}(t)$ is stationary and ergodic with steady-state channel probabilities $\pi_{\underline{S}}$.

The network capacity region Λ for such a network, is the closure of the set of all rate matrices (λ_{ic}) that can be stably supported over the network considering all possible algorithms.

We show that stabilizing policies do not require knowledge of future events and, hence, such knowledge does not expand the region of stabilizable rates. To build intuition about the set Λ, we first consider the capacity region of a traditional wire line network with no time variation, defined on a weighted graph with N nodes, E edges, and edge capacities given by a link rate matrix (R_{ab}).

The network capacity region is described implicitly as the set of all arrival rate matrices (λ_{ic}) for which there exist multicommodity flow variables $f_{ab}^{(c)}$ (for $a, b, c \in \{1, \ldots, N\}$) that satisfy a set of flow conservation equations, and which additionally satisfy the link constraint $\sum_c f_{ab}^{(c)} \leq R_{ab}$ for all links (a, b).

This constraint ensures that the total flow over any link does not exceed the link capacity.

The time-varying channel conditions of a wireless network require link capacities to be defined in a time average sense. Even the resulting time average link rates are not fixed, but depend on the (potentially non ergodic) power allocation policy. Thus, instead of describing the network as a single weighted graph (R_{ab}) of link rates, the network is described by a *collection* of graphs, or a *graph family* Γ. We define the graph family Γ as the following set of node-to-node transmission rate matrices:

$$\Gamma = \sum_S \pi_S \, ch \, \{\underline{\mu}(\underline{P}, S) | \underline{P} \in \Pi\} \tag{15.111}$$

where *ch()* stands for convex hull and addition and scalar multiplication of sets is used.

For sets A, B and scalars α, β, the set $\alpha A + \beta B$ is defined as $\{\gamma | \gamma = \alpha a + \beta b \text{ for some } a \in A, b \in B\}$. The convex hull of a set A is defined as the set of all convex combinations $p_1 a_1 + p_2 a_2 + \cdots + p_k a_k$ of elements $a_i \in A$ (where $\{p_i\}$ are probabilities summing to 1).

Therefore, a transmission rate matrix $\underline{R} = (R_{ab})$ is in graph family Γ if and only if $\underline{R}$ can be represented as $\underline{R} = \sum_S \pi_S \underline{R}_S$ for some set of matrices $\underline{R}_S$, each one being inside the convex hull of the set of node-to-node transmission rates achievable by power allocation under channel state $\underline{S}$.

The capacity region Λ is the set of all input rate matrices (λ_{ic}) for which there exist multicommodity flow variables $\{f_{ab}^{(c)}\}$ satisfying

$$f_{ab}^{(c)} \geq 0, \; f_{aa}^{(c)} = f_{ab}^{(a)} = 0 \; \forall a, \; b, \; c$$

$$\lambda_{ic} \leq \sum_b f_{ib}^{(c)} - \sum_a f_{ai}^{(c)} \forall i, c \quad \text{where } i \neq c \tag{15.112}$$

$$\sum f_{ab}^{(c)} \leq R_{ab} \; \forall a, \; b, \; \text{and for some } (R_{ab}) \in \Gamma$$

Every time slot the network controller observes the channel state $S(t)$ and the matrix of queue backlogs $\underline{U}(t) = (U_i^{(c)}(t))$ and performs routing and power control as follows [37]:

Dynamic Routing Algorithm (DRA)
1. *For all links (a, b), find commodity $c_{ab}^*(t)$ such that*

$$\boxed{c_{ab}^*(t) = \arg \max_{c \in \{1,...,N\}} \{U_a^{(c)}(t) - U_b^{(c)}(t)\}}$$

and define

$$W_{ab}^*(t) = \max[U_a^{(c_{ab}^*(t))}(t) - U_b^{(c_{ab}^*(t))}(t), \; 0]$$

2. *Power Allocation: Choose a matrix $\underline{P}(t)$ such that*

$$\underline{P}(t) = \arg \max_{\underline{P} \in \Pi} \sum_{a,b} \mu_{ab}(\underline{P}, \underline{S}(t)) W_{ab}^*(t)$$

3. *Routing: Define transmission rates as follows:*

$$\mu_{ab}^{(c)}(t) = \begin{cases} \mu_{ab}(\underline{P}(t), \underline{S}(t)), & c = c_{ab}^*(t) \ and \ W_{ab}^*(t) > 0 \\ 0, & otherwise \end{cases}$$

15.11.2   Network Delay

Let μ_{max}^{out} and μ_{max}^{in} be the maximum transmission rate out of any node and into any node, respectively, under the best channel conditions

$$\mu_{max}^{out} = \max_{\{i, \underline{S}, \underline{P} \in \Pi\}} \sum_{b} \mu_{ib}(\underline{P}, \underline{S})$$

$$\mu_{max}^{in} = \max_{\{i, \underline{S}, \underline{P} \in \Pi\}} \sum_{a} \mu_{ai}(\underline{P}, \underline{S})$$

In addition assume that the second moment of exogenous arrivals to any node is bounded every time slot by some finite maximum value A_{max}^2, regardless of past history. If $H(t) = (\underline{S}(\tau); \underline{A}(\tau))|_{\tau=0}^{t-1}$ then, for all nodes i and all t, we have $E\{(\sum_c A_i^{(c)}(t))^2 | H(t)\} \le A_{max}^2$. Since $E\{A\} \le \sqrt{E\{A^2\}}$ for any random variable A, we also have $E\{\sum_c A_i^{(c)}(t) | H(t)\} \le A_{max}$.

If $T_{\underline{S}}(t_0, K)$ is the set of time slots at which $\underline{S}(t) = \underline{S}$ during the interval $t_0 \le \tau \le t_0 + K - 1$, and $||T_{\underline{S}}(t_0, K)||$ is the total number of such slots then for a given value $\delta > 0$, the *convergence interval* K is defined as the smallest number of time slots such that for any t_0, any (i, c), and regardless of past history, we have

$$|\lambda_{ic} - \frac{1}{K} \sum_{\tau=t_0}^{t_0+K-1} E\{A_i^{(c)}(\tau) | H(t_0)\}| \le \delta \tag{15.113}$$

$$\sum_{\underline{s}} |\pi_{\underline{S}} - E\{||T_{\underline{S}}(t_0, K)|| | H(t_0)\} / K| \le \delta / \max\{\mu_{max}^{out}, \mu_{max}^{in}\}$$

Such a value K must exist for any stationary and ergodic channel and arrival processes with arrival rates (λ_{ic}) and channel probabilities $\pi_{\underline{S}}$, respectively. This convergence interval represents the time period over which the network is expected to reach steady-state, regardless of past history. If a finite interval size K exists for any given $\delta > 0$, then we say that the arrival processes and channel processes for which this assumption holds are *rate convergent* and *channel convergent*, respectively. We note that for systems with independent identically distributed arrivals and channel states, steady-state is exactly achieved every time slot, so that $K = 1$ even when $\delta = 0$.

For N-node wireless network with capacity region Λ and rate matrix (λ_{ic}) such that ($\lambda_{ic} + \varepsilon$) $\in \Lambda$ for some $\varepsilon > 0$ DRA stabilizes the system and guarantees bounded average congestion satisfying [37]

$$\overline{\sum_{i,c} U_i^{(c)DRA}} \le \frac{KBN}{\varepsilon} + \frac{(K-1)\tilde{B}N}{\varepsilon} \tag{15.114}$$

for K corresponding to $\delta = \varepsilon/6$ in (15.113),

$$B = (A_{max} + \mu_{max}^{in})^2 + (\mu_{max}^{out})^2, \quad \tilde{B} = 2 \max[\mu_{max}^{out}, \mu_{max}^{in}](\mu_{max}^{out} + \mu_{max}^{in} + A_{max})$$

and $\overline{\sum_{i,c} U_i^{(c)}} = \limsup_{t \to \infty} \frac{1}{t} \sum_{\tau=0}^{t-1} \left[\sum_{i,c} E\{U_i^{(c)}(\tau)\} \right]$ where ε can be viewed as the 'distance' measure of the rate matrix to the boundary of the capacity region. Equation (15.114) is proven in the next section.

For an input matrix (λ_{ic}), where each user i *sends at total rate* λ, so that $\sum_c \lambda_{ic} = \lambda$ where the matrix satisfies $(\lambda_{ic} + \varepsilon) \in \Lambda$, $R = \lambda + N\varepsilon$ is defined as the row sum of the $(\lambda_{ic} + \varepsilon)$ matrix. If $\rho = \lambda/R$ represent an effective loading on each user (assumed independent of N) and having in mind that $\lambda = \rho R$, and $\varepsilon N = R(1 - \rho)$, from Little's theorem, the average bit delay satisfies $\overline{D}_{bit} = (1/N\lambda) \sum_{i,c} \overline{U}_i^{(c)}$ or

$$\overline{D}_{bit} \le \frac{KBN + (K-1)\tilde{B}N}{\varepsilon N\lambda} = \frac{KBN + (K-1)\tilde{B}N}{\rho(1-\rho)R^2} \tag{15.115}$$

Note that when the network is lightly loaded, there is very little information contained in the differential backlog values. Hence, packets might take many false turns, which could lead to significant delay for large networks (consider the above delay bound for ρ small). Performance can often be improved by using the DRA with a restricted set of desirable routes for each commodity. However, restricting the routes may reduce network capacity, and may be harmful in time-varying situations, where networks change and links fail. Alternatively, we can keep the full set of routes, but incorporate a *bias* into the DRA so that in low loading situations, nodes are encouraged to route packets in the direction of their destinations. This idea was suggested in the following enhanced *DRA (EDRA)* algorithm [37], defined in terms of constants $\theta_a^c > 0$ and $V_a^c \ge 0$.

For all links (a, b), and all commodities c, EDRA defines $W_{ab}^{(c)} = \theta_a^c(U_a^c(t) + V_a^c) - \theta_b^c(U_b^c(t) + V_b^c)$, and c_{ab}^* as the maximizer of $W_{ab}^{(c)}$ over all $c \in \{1, \ldots, N\}$. Power allocation and routing is done as before, solving the optimization in DRA with respect to $W_{ab}^* = W_{ab}^{(c_{ab}^*)}$.

The V_a^c parameters can be chosen as scaled hop count estimates between nodes a and c, so that, in the absence of backlog information, data is routed to reduce the remaining distance to the destination. The θ_a^c values are any weights for prioritizing commodity c service in node a. This EDRA algorithm can be shown to stabilize the network for any constants $\theta_i^c > 0$ and $V_i^c \ge 0$. We note that the weight $\theta_a^c(V_a^c + U_a^c)$ can be used in the same manner as a routing table, and the unfinished work quantities can be updated each time slot by having neighboring nodes transmiting their backlog changes over a low bandwidth control channel. As each wireless link transmits only a single commodity every time slot, the number of such backlog increments required to be transmitted over the control channel by any user is on the order of the number of neighboring nodes.

15.11.3   Lyapunov Drift and Network Stability

The performance of the *DRA* can be analyzed by comparing it to a stationary algorithm that makes scheduling decisions according to the multicommodity flow variables $\{f_a^{(a)}\}$ of (15.112). Suppose the rate matrix (λ_{ic}) and the channel probabilities $\pi_{\underline{S}}$ were known in advance, with $\varepsilon > 0$ such that $(\lambda_{ic} + \varepsilon) \in \Lambda$. Then, a set of multicommodity flow variables $\{f_{ab}^{(c)}\}$ and a link rate matrix $(R_{ab}) \in \Gamma$ must exist that satisfy the constraints (15.112) with respect to the rate matrix $(\lambda_{ic} + \varepsilon)$ that can be rewritten as $(\lambda_{ic} + \varepsilon) \le \sum_b f_{ib}^{(c)} - \sum_a f_{ai}^{(c)}$, for $i \ne c$ and $(\sum_c f_{ab}^{(c)}) \le (R_{ab})$.

A *stationary randomized* power allocation algorithm $\underline{P}^{STAT}(\tau)$ can be implemented so that the resulting $\mu_{ab}^{STAT}(t)$ satisfies $\lim_{t\to\infty}(1/t)\sum_{\tau=0}^{t-1} \mu_{ab}^{STAT}(\tau) = R_{ab}$ with probability 1 for all (a, b) resulting in graph family acheviability [37]. Every time slot in which the channel state $S(t) = \underline{S}$ is observed, the power matrix $\underline{P}^{STAT}(t)$ is chosen randomly from a finite set of allocations $\{\underline{P}_{\underline{S}}^1, \ldots, \underline{P}_{\underline{S}}^m\}$ according to a set of probabilities $\{q_{\underline{S}}^1, \ldots, q_{\underline{S}}^m\}$. This policy requires complete knowledge of the input rates λ_{ic}, channel probabilities $\pi_{\underline{S}}$, and flow variables $f_{ab}^{(c)}$.

In this protocol scheduling/routing is implement in such a way that for every link (a, b) such that $\sum_c f_{ab}^{(c)} > 0$, algorithm transmits the single commodity $\hat{c}_{ab}$, where $\hat{c}_{ab}$ is chosen randomly among $c \in \{1, \ldots, N\}$ with probability $f_{ab}^{(c)}/\sum_d f_{ab}^{(d)}$. Only a fraction $\sum_d f_{ab}^{(d)}/R_{ab}$ of the instantaneous link rate is used, so that

$$
\mu_{ab}^{(c)STAT}(t) = \begin{cases} \mu_{ab}^{STAT}(t) \dfrac{\sum_d f_{ab}^{(d)}}{R_{ab}} & if \ c = \hat{c}_{ab} \\ 0, & otherwise \end{cases}
\tag{15.116}
$$

If a node does not have enough (or any) bits of a certain commodity to send over its output links, *null bits* are delivered. Since $E\{\mu_{ab}^{(c)STAT}(t) \,|\mu_{ab}^{STAT}(t)\,\} = \mu_{ab}^{STAT}(t)\, f_{ab}^{(c)}/R_{ab}$ the processes $\mu_{ab}^{(c)STAT}(t)$ are rate convergent with time average rates $f_{ab}^{(c)}$.

In order to analyze delay, we consider the rates $\mu_{ab}(t)$ and the input processes $A_i^{(c)}(t)$ averaged over K slot intervals. In particular, for any time t_0, we define $\tilde{\mu}_{ab}^{(c)}(t_0) = (1/K)\sum_{\tau=t_0}^{t_0+K-1} \mu_{ab}^{(c)}(\tau)$ and $\tilde{A}_i^{(c)}(t_0) = (1/K)\sum_{\tau=t_0}^{t_0+K-1} A_i^{(c)}(\tau)$. Since we have that $\sum_b f_{ib}^{(c)} - \sum_a f_{ai}^{(c)} - \lambda_{ic} \geq \varepsilon \ for \ i \neq c$ it follows that for a suitably large value of K, the expectations of $\tilde{\mu}_{ab}^{(c)}(t_0)$ and $\tilde{A}_i^{(c)}(t_0)$ with fixed $\delta = \varepsilon/6$ satisfy [37]

$$
E\left\{\sum_b \tilde{\mu}_{ib}^{(c)STAT}(t_0) - \sum_a \tilde{\mu}_{ai}^{(c)STAT}(t_0)|H(t_0)\right\} - E\{\tilde{A}_i^{(c)}(t_0)|H(t_0)\} \geq \frac{\varepsilon}{2}
\tag{15.117}
$$

For the stability and delay analysis we start by defining $L(\underline{U}) = \sum_{i,c} [U_i^{(c)}]^2$ as a *Lyapunov function* of unfinished work, representing a scalar measure of network congestion. For a given control policy X and a given unfinished work $\underline{U}(t)$ at time t, define the K-*step Lypapunov drift* $\Delta_K^X(\underline{U}(t))$, as $\Delta_K^X(\underline{U}(t)) = E\{L(\underline{U}(t+K)) - L(\underline{U}(t))|\underline{U}(t)\}$. If there exists a positive integer K such that for all time slots t, the K-step Lyapunov drift satisfies $\Delta_K^X(\underline{U}(t)) \leq C - \sum_{i,c} \theta_{ic} U_i^{(c)}(t)$ for some positive constants C, $\{\theta_{ic}\}$, and if $E\{L(\underline{U}(t_0))\} < \infty$ for $t_0 \in \{0, 1, \ldots, K-1\}$, then the network is stable [37] and

$$
\overline{\sum_{i,c} \theta_{ic} U_i^{(c)}} = \limsup_{t\to\infty} \frac{1}{t} \sum_{\tau=0}^{t-1} \left[\sum_{i,c} \theta_{ic} E\{U_i^{(c)}\}(\tau)\right] \leq C
\tag{15.118}
$$

The intuition behind (15.118) is that drift is negative whenever backlog is sufficiently large, leading to negative feedback and stability. The fact that Lyapunov drift is compared after K slots (rather than after a single slot) is required for systems with non i.i.d. dynamics. Similar K-slot analysis of Lyapunov drift has been used in [38] to address stability, and similar drift statements for i.i. $d.$ systems where $K = 1$ are found in [39–42].

For any control policy X resulting in decision variables $\mu_{ab}^{(c)}(t)$, the K-step Lyapunov drift at any slot t_0 satisfies [37]

$$
\Delta_K^X(\underline{U}(t_0)) \leq K^2 BN - 2K[\Phi^X(\underline{U}(t_0)) - \beta(\underline{U}(t_0))]
\tag{15.119}
$$

where $B = (A_{max} + \mu_{max}^{in})^2 + (\mu_{max}^{out})^2$ and

$$
\Phi^X(\underline{U}(t_0)) = \sum_{i,c} U_i^{(c)}(t_0) E\{[\sum_b \tilde{\mu}_{ib}^{(c)}(t_0) - \sum_a \tilde{\mu}_{ai}^{(c)}(t_0)]|\underline{U}(t_0)\}
$$

$$
\beta(\underline{U}(t_0)) = \sum_{i,c} U_i^{(c)}(t_0) E\{\tilde{A}_i^{(c)}(t_0)|\underline{U}(t_0)\}
\tag{15.120}
$$

To compare performance *DRA* and *STAT* we use the definitions of $\Phi^X(\underline{U}(t_0))$ and $\beta(\underline{U}(t_0))$ given in (15.120) together with inequality (15.117) resulting into

$$\Phi^{STAT}(\underline{U}(t_0)) - \beta(\underline{U}(t_0)) \geq \frac{\varepsilon}{2} \sum_{i,c} U_i^{(c)}(t_0) \tag{15.121}$$

Plugging this inequality directly into the Lyapunov drift expression (15.119) and using (15.118) yields the following bound on network congestion under STAT:

$$\overline{\sum_{i,c} U_i^{(c)STAT}} \leq \frac{K^2 BN}{K\varepsilon} = \frac{KBN}{\varepsilon} \tag{15.122}$$

We now consider a frame-based modification of the *DRA (DRAm)*. Scheduling, power allocation, and routing under *DRAm* are done every time slot exactly as in the *DRA* algorithm, with the exception that backlog updates are performed only every K slots. Specifically, for any time slot τ with $\{t_0, t_0 + 1, \ldots, t_0 + K - 1\}$, power is allocated to maximize $\sum_{ab} \mu_{ab}(\underline{P}, \underline{S}(\tau)) W_{ab}^*(t_0)$ subject to $\in \Pi$. Thus channel state information but out of data backlog information is used every slot. One can show that [37]

$$\Phi^{DRAm}(\underline{U}(t_0)) \geq \Phi^{STAT}(\underline{U}(t_0))$$
$$\Phi^{DRA}(\underline{U}(t_0)) \geq \Phi^{DRAm}(\underline{U}(t_0)) - (K-1)N\tilde{B}/2 \tag{15.123}$$

Using (15.123) with (15.121) gives

$$\Phi^{DRA}(\underline{U}(t_0)) - \beta(\underline{U}(t_0)) \geq \frac{\varepsilon}{2} \sum_{i,c} U_i^{(c)}(t_0) - \frac{(K-1)\tilde{B}N}{2}$$

Using this bound in the Lyapunov drift expression (15.119) yields

$$\Delta_K^{DRA}(\underline{U}(t_0)) \leq K^2 BN + K(K-1)\tilde{B}N - K\varepsilon \sum_{i,c} U_i^{(c)}(t_0)$$

Applying the drift theorem and (15.118) proves that DRA is stable with the congestion bound as given in (15.114).

15.12 Lagrangian Decomposition of the Multicomodity Flow Optimization Problem

In this section, we consider the flow problem for a wireless network and compare the dynamic Lyapunov drift approach, presented in the previous section, to a more traditional static optimization technique that assumes known rate matrix (λ_{ij}). A convex optimization problem corresponding to a multicommodity flow in the wireless network is formulated, and it is shown that a classical subgradient search method for solving the problem via convex duality theory corresponds exactly to a deterministic network simulation of the DRA from the previous section. Notions of duality are used to consider static network optimization, where dual variables play the role of prices charged by the network to multiple users competing for shared network resources in order to maximize their own utility. In our context, the dual variables correspond to *queue backlog* work prices. This illustrates a relationship between static optimization and the DRA and contributes to a growing theory of *dynamic optimization*, suggesting that static algorithms can be modified and applied in dynamic settings while preserving analytical optimality.

We restrict attention to time invariant systems, so that the rate- power curve is only a function of power: $\mu(\underline{P}, \underline{S}) = \mu(\underline{P})$. Given a rate matrix (λ_{ij}), the problem of finding a multicommodity flow corresponds to the following convex optimization problem

$$Maximize : 1$$
$$Subject\ to : \lambda_{ic} + \sum_a f_{ai}^{(c)} \le \sum_b f_{ib}^{(c)}\ \forall i,\ c,\ such\ that\ i \ne c \tag{15.124}$$
$$(\{f_{ab}^{(c)}\}, \{\mu_{ab}\}) \in \Theta$$

where Θ is the set of all variable

$$f_{ab}^{(c)} \ge 0,\ for\ all\ a,\ b,\ c \in \{1, \ldots, N\}$$
$$f_{aa}^{(c)} = f_{ab}^{a)} = 0,\ for\ all\ a, b, c \in \{1, \ldots, N\} \tag{15.125}$$
$$(\sum_c f_{ab}^{(c)}) - (\mu_{ab}),\ for\ some\ (\mu_{ab}) \in \Gamma$$

The artificial utility function '1' is used to pose this multicommodity flow problem in the framework of an optimization problem. The optimization problem is convex, and has a dual formulation, where the optimal solution of the dual problem exactly corresponds to an optimal solution of the original 'primal' problem (15.124). To form the dual problem we introduce nonnegative Lagrange multipliers $\{U_i^{(c)}\}$ for each of the inequality constraints in (15.124), and define the dual function

$$L(\{U_i^{(c)}\}) = \max_{(\{f_{ab}^{(c)}\}, \{\mu_{ab}\}) \in \Theta} \left[1 + \sum_{i \ne c} U_i^{(c)} \left(\sum_b f_{ib}^{(c)} - \sum_a f_{ai}^{(c)} - \lambda_{ic} \right) \right] \tag{15.126}$$

The dual problem to (15.124) is

$$Minimize : L(\{U_i^{(c)}\})$$
$$Subject\ to : U_i^{(c)} \ge 0,\ for\ all\ i,\ c \in \{1, \ldots, N\} \tag{15.127}$$

The dual problem is always convex, and the minimizing solution can be obtained using classical subgradient search methods where the function $-L(\{U_i^{(c)}\})$ is maximized.

15.13   Flow Optimization in Heterogeneous Networks

In this section, we discuss a set of decoupled algorithms for resource allocation, routing, and flow control for general networks with both wireless and wireline data links and time varying channels. In a network with N nodes and L links the condition of each link is described by a time varying *link state vector* $\vec{S}(t) = (S_1(t), \ldots, S_L(t))$, where $S_l(t)$ is a parameter characterizing the communication channel for link l. If l is a wireless link, $S_l(t)$ may represent the current attenuation factor or noise level. In an unreliable wired link, $S_l(t)$ may take values in the two-element set $\{0, 1\}$, indicating whether link l is available (1) or not (0) for communication. Channels hold their state for the duration of a timeslot, and potentially change states on slot boundaries. The channel state vectors are discretized, and the vectors $\vec{S}(t)$ are independent and identically distributed (i.i. d.) over timeslots.

For each channel state $\vec{S}$, let $\Gamma_{\vec{S}}$ denote the set of link transmission rates available for resource allocation decisions when $\vec{S}(t) = \vec{S}$. In particular, every timeslot t the network controllers are constrained to choosing a transmission rate vector $\vec{\mu}(t) = (\mu_1(t), \ldots, \mu_L(t))$ such that $\vec{\mu}(t) \in \Gamma_{\vec{S}(t)}$.

As an example, consider the heterogeneous network consisting of three separate groups of links A, B, and C. Set A may represent a wireless sensor system that connects to a wired infrastructure through some uplink access points, set B represents the wired data links, and set C represents the downlink channels of a basestation that transmits to different users. For a given channel state $\vec{S}$, the set of feasible transmission rates $\Gamma_{\vec{S}}$ reduces to a product of rates corresponding to the three independent groups $\Gamma_{\vec{S}} = \Gamma_{\vec{S}_A}^A \times \Gamma^B \times \Gamma_{\vec{S}_C}^C$.

Set $\Gamma_{\vec{S}_A}^A$ might contain a continuum of link rates associated with the channel interference properties and power allocation options of the sensor nodes, and depends only on the link states S_A of these nodes. Set Γ^B might contain a single vector $(C_1, \ldots, C_k)$ representing the fixed capacities of the k wired links. For an example of two downlink channels, set $\Gamma_{\vec{S}_C}^C$ might represent a set of two vectors $\{(\phi_{S_1}, 0), (0, \phi_{S_2})\}$, where ϕ_{S_x} is the rate available over link i if this link is selected to transmit on the given timeslot t when $S_i(t) = S_i$. The ground network from the previous example can be augmented to include a collection of *satellite nodes*, increasing route diversity and enabling connectivity to other sub-networks.

Data is transmitted from node to node over potentially multi-hop paths to reach its destination. Data arrives to the network according to random processes with well-defined time average rates, with $\lambda_n^{(c)}$ representing the long-term average rate of new exogenous arrivals to source node n intended for destination node c (*bits/slot*) and $(\lambda_n^{(c)})$ representing the matrix of exogenous arrival rates. As in the previous section the *network layer capacity region* Λ is defined as the closure of the set of all arrival matrices that are stably supportable by the network, considering all possible multi-hop routing and resource allocation policies (possibly those with perfect knowledge of future events). In the same section, a routing and power allocation policy was developed to stabilize a general wireless network whenever the rate matrix $(\lambda_n^{(c)})$ is within the capacity region Λ. In this section we treat heterogeneous networks and develop distributed algorithms for flow control, routing, and resource allocation that provide optimal fairness in cases when arrival rates are *either inside or outside the network capacity region*.

We use *utility functions* $g_n^{(c)}(r)$, representing the 'satisfaction' received by sending data from node n to node c at a time average rate of r*bits/slot*. The utility functions are assumed to be non-decreasing and concave. The goal is to support a fraction of the traffic demand matrix $(\lambda_n^{(c)})$ to achieve long term throughputs $(\overline{r}_n^{(c)})$ that maximize the sum of user utilities within the following optimization problem:

$$
\begin{aligned}
Maximize &: \sum_{n,c} g_n^{(c)}(\overline{r_n^{(c)}}) \\
Subject\ to &: (\overline{r}_n^{(c)}) \in \Lambda \\
& 0 \leq \overline{r}_n^{(c)} \leq \lambda_n^{(c)} \text{ for all } (n, c)
\end{aligned}
\tag{15.128}
$$

The second line of (15.128) is the *stability constraint* which ensures that the long term admitted rates are stabilizable by the network. The third line of (15.128) is the *demand constraint* that ensures the rate provided to session (n, c) is no more than the incoming traffic rate of this session.

Because the functions $g_n^{(c)}(r)$ are non-decreasing, it is clear that if $(\lambda_n^{(c)}) \in \Lambda$, then the above optimization is solved by the matrix $(r_n^{*(c)}) = (\lambda_n^{(c)})$. If $(\lambda_n^{(c)}) \notin \Lambda$, then the solution $(r_n^{*(c)})$ will lie somewhere on the capacity region boundary. The above optimization could in principle be solved if the arrival rates $(\lambda_n^{(c)})$ and the capacity region Λ were known in advance, and all users could coordinate by sending data according to the optimal solution. However, the capacity region depends on the channel probabilities, which are unknown to the network controllers and to the individual users. Furthermore, the individual users do not know the data rates or utility functions of other users. In this section, we present a practical dynamic control strategy [43] that yields a resulting set of throughputs $(\overline{r}_n^{(c)})$ that are arbitrarily close to the optimal solution of (15.128). The distance to the optimal solution is shown to decrease as $1/V$, where V is a control parameter affecting a tradeoff in average delay for data that is served by the network.

To further elaborate this problem consider a heterogeneous network with N nodes, L links, and time varying channels $\vec{S}(t)$. For each link $l \in \{1, \ldots, L\}$ we define $t(l)$ and $r(l)$ as the corresponding transmitting and receiving nodes, respectively. Each node of the network maintains a set of output queues for storing data according to its destination. All data (from any source node) that is destined for a particular node $c \in \{1, \ldots, N\}$ is classified as *commodity c data*, and we let $U_n^{(c)}(t)$ represent the backlog of commodity c data currently stored in the network layer at node n. A *network layer control algorithm* makes decisions about routing, scheduling, and resource allocation in reaction to current channel state and queue backlog information. The objective is to deliver all data to its proper destination, potentially by routing over multi-hop paths.

In order to avoid loops and excessive delays we restrict routing options so that data follows a particular path or set of paths to the destination. To enforce this constraint, for each commodity $c \in \{1, \ldots, N\}$ we define L_c as the set of all data links l that are acceptable for commodity c data to traverse. As a general algorithm might schedule multiple commodities to flow over the same link on a given timeslot, we define $\mu_l^{(c)}(t)$ as the rate offered to commodity c traffic along link l during timeslot t. The transmission rates and routing variables are chosen by a *dynamic scheduling and routing algorithm DSRA* defined as

Resource (Rate) Allocation
Choose a transmission rate vector $\vec{\mu}(t) = (\mu_1(t), \ldots, \mu_L(t))$ such that $\vec{\mu}(t) \in \Gamma_{\vec{S}(t)}$

Routing/Scheduling: For each link l and each commodity c choose $\mu_l^{(c)}(t)$ to satisfy

$$\sum_c \mu_l^{(c)}(t) \leq \mu_l(t)$$
$$\mu_l^{(c)}(t) = 0 \text{ if } l \notin L_c$$

(15.129)

A set of *flow controllers* acts at every node to limit the new data admitted into the network layer. In conventional IP/TCP protocol this is equivalent to the congestion window control. Specifically, new data of commodity c that arrives to source node n is first placed in a *transport layer storage reservoir* (n, c). A control value determines the amount of data $R_n^{(c)}(t)$ released from this reservoir on each timeslot. The $R_n^{(c)}(t)$ process acts as the exogenous arrival process to the queue $U_n^{(c)}(t)$. Endogenous arrivals consist of commodity c data transmitted to node n from other network nodes. Define Ω_n as the set of all links l such that $t(l) = n$, and define Θ_n as the set of all links such that $r(l) = n$. Every timeslot the backlog $U_n^{(c)}(t)$ changes as:

$$U_n^{(c)}(t+1) \leq \max \left[U_n^{(c)}(t) - \sum_{l \in \Omega_n} \mu_l^{(c)}(t), 0 \right] + \sum_{l \in \Theta_n} \mu_l^{(c)}(t) + R_n^{(c)}(t)$$

(15.130)

Data leaves the network when it reaches its destination, and so we define $U_n^{(n)}(t) = 0$ for all n and all t.

If $A_n^{(c)}(t)$ represent the new commodity c data arriving to the system at source node n during slot t, and $L_n^{(c)}(t)$ the current backlog in the flow control reservoir at time t the control decision variables $R_n^{(c)}(t)$ are chosen every timeslot according to the following restrictions:
Flow Control: Choose $R_n^{(c)}(t)$ such that

$$R_n^{(c)}(t) \leq L_n^{(c)}(t) + A_n^{(c)}(t) \text{ for all } t$$
$$\sum_c R_n^{(c)}(t) \leq R_n^{max} \text{ for all } t$$

(15.131)

where the constants R_n^{max} are chosen to be positive and suitably large. The first flow control constraint ensures that admitted data is less than or equal to the actual data available, and the second is important for limiting the burstiness of the admitted arrivals.

15.13.1 Protocol Optimization with Infinite Buffers

If K_n is the number of internal queues kept by node n, where $K_n \in \{0, 1, \ldots, N - 1\}$ and D is the set of all node-commodity pairs (n, c) associated with internal queues in the network, then $D = \sum_{n=1}^N K_n$ represents the number of such queues. The integer D defines the *relative dimension* of the network.

We assume that all active traffic sessions are within the set $\{D\}$, so that $R_n^{(c)}(t) = 0$, $g_n^{(c)}(r) = 0$ for all $(n, c) \notin \{D\}$. We further define $U_n^{(c)}(t) = 0$ for all t when $(n, c) \notin \{D\}$.

Suppose the admitted traffic from each flow control value (n, c) has a well-defined time average

$$\bar{r}_n^{(c)} = \lim_{t \to \infty} (1/t) \sum_{\tau=0}^{t-1} E\{R_n^{(c)}(\tau)\}.$$

The network layer capacity region Λ is the set of all time average rate matrices $(\bar{r}_n^{(c)})$ that the network can stably support, considering all possible routing and resource allocation algorithms that satisfy the constraints (15.129) and the queueing dynamics (15.130) described above.

A queue $U(t)$ with general stochastic arrival and transmission rate processes is *strongly stable* if $\lim \sup (1/t) \sum_{\tau=0}^{t-1} E\{U(\tau)\} < \infty$.

That is, we define strong stability (hereafter called 'stability') in terms of a finite time average backlog.

If channels $\vec{S}(t)$ are i.i.d. over slots, a fixed input rate matrix $(r_n^{(c)})$ is in the capacity region Λ if and only if $r_n^{(c)} = 0$ for all $(n, c) \notin D$ and there exists a stationary randomized resource allocation algorithm that chooses particular transmission rates $\mu_l^{*(c)}(t)$ based only on the current channel state $\vec{S}(t)$, and satisfies for all slots t

$$E\left\{ \sum_{l \in \Omega_n} \mu_l^{*(c)}(t) - \sum_{l \in \Theta_n} \mu_l^{*(c)}(t) | \underline{U}(t) \right\} = r_n^{(c)}, \forall n \neq c \tag{15.132}$$

$$\mu_l^{*(c)}(t) = 0, \text{ if } t(l) = c \text{ or if } l \notin L_c$$

where $\underline{U}(t) = (U_n^{(c)}(t))$ represents the matrix of current queue backlogs.

In the sequel we will also use parameter μ_{sym} to be the largest rate that is simultaneously supportable by all sessions $(n, c) \in D$, so that $(\mu_{sym} 1_n^{(c)}) \in \Lambda$ (where $1_n^{(c)}$ is equal to 1 if $(n, c) \in D$, and zero else).

In what follows we present a practical control algorithm that stabilizes the network and ensures that utility is arbitrarily close to optimal, with a corresponding tradeoff in network delay [43]. By definition functions $g_n^{(c)}(r)$ in (15.128) represent the utility of supporting rate r communication from node n to node c with $g_n^{(c)}(r) = 0$ if there is no active session of traffic originating at node n and destined for node c. To highlight the fundamental issues of routing, resource allocation, and flow control, in this section we assume that all active sessions (n, c) have *infinite backlog* in their corresponding reservoirs. A modified algorithm is developed in section 15.3.3 for the general case of finite demand matrices $(\lambda_n^{(c)})$ and finite buffer reservoirs.

The following control strategy is *decoupled* into separate algorithms for resource allocation, routing, and flow control. The strategy combines a *flow control technique* together with a generalization of the Dynamic Routing Algorithm (DRA) described in the previous section and will be referred to as DFRA.

Dynamic Flow/Routing Control Algorithm (DFRA)

F algorithm (flow control): Every timeslot, the flow controller at each node n observes the current level of queues $U_n^{(c)}(t)$ for each commodity $c \in \{1, \ldots, N\}$. It then chooses $R_n^{(c)}(t) = x_n^{(c)}$, where the $x_n^{(c)}$ values solve

$$\textit{Maximize}: \sum_{c=1}^{N} [V g_n^{(c)}(x_n^{(c)}) - 2 x_n^{(c)} U_n^{(c)}(t)] \tag{15.133}$$

$$\textit{Subject to}: (x_n^{(c)}) \geq 0, \sum_{c=1}^{N} x_n^{(c)} \leq R_n^{\max}$$

where $V > 0$ is a constant that controls the extent to which utility optimization is emphasized.

R&S algorithm (Routing and Scheduling) – Each node n observes the backlog in all neighboring nodes j to which it is connected by a link l (where $t^{(l)} = n$, $r(l) = j$).

For $W_l^{(c)}(t) = U_{t(l)}^{(c)}(t) - U_{r(l)}^{(c)}(t)$ representing the differential backlog of commodity c data, define $W_l^*(t) = \max_{[c|l \in L_c]} \{W_l^{(c)}(t), 0\}$ as the maximum differential backlog over link l, and let $c_l^*(t)$ represent the maximizing commodity. Data of commodity $c_l^*(t)$ is selected for (potential) routing over link l whenever $W_l^*(t) > 0$.

Resource Allocation – The current channel state $\vec{S}(t)$ is observed, and a transmission rate vector $\vec{\mu}(t)$ is selected by maximizing $\sum_l W_l^*(t)\mu_l(t)$ subject to the constraint $\vec{\mu}(t) \in \Gamma_{\vec{S}(t)}$. The resulting transmission rate of $\mu_l(t)$ is offered to commodity $c_l^*(t)$ data on link l (provided that $W_l^*(t) > 0$). If any node does not have enough bits of a particular commodity to send over all outgoing links requesting that commodity, null bits are delivered.

To analyze the performance of DFRA algorithm, we define the maximum transmission rates out of and into a given node n as $\mu_{\max,n}^{out} = \max_{(\vec{S},\vec{\mu} \in \Gamma_{\vec{S}})} \sum_{l \in \Omega_n} \mu_l$, $\mu_{\max,n}^{in} = \max_{(\vec{S},\vec{\mu} \in \Gamma_{\vec{S}})} \sum_{l \in \Theta_n} \mu_l$. We assume that each node n knows its own value of $\mu_{\max,n}^{out}$, since nodes would typically have a prespecified set of modulation and coding strategies to choose from, with a well-defined maximum. Assume that the flow control constants $R_n^{\max}$ are selected to satisfy $R_n^{\max} \geq \mu_{\max,n}^{out}$ for all n. As $R_n^{\max}$ is only used at node n, it can easily be set to satisfy this inequality. Assume utilities $g_n^{(c)}(r)$ are non-negative, non-decreasing, and concave, and define $G_{\max} = \max_{\sum_c r_n^{(c)} \leq R_n^{\max} \forall n} \sum_{n,c} g_n^{(c)}(r_n^{(c)})$. Define the constant B as

$$B = (1/N) \sum_{n=1}^{N} [(R_n^{\max} + \mu_{\max,n}^{in})^2 + (\mu_{\max,n}^{out})^2] . \tag{15.134}$$

If channel states are i.i.d. over timeslots and all active reservoirs have infinite backlog, then for any flow parameter $V > 0$ the *DFRA* algorithm stabilizes the network with

$$\sum_{n,c} \overline{U_n^{(c)}} \leq (2\mu_{sym})^{-1} (BN + VG_{\max})$$

$$\liminf_{t \to \infty} \sum_{n,c} g_n^{(c)}(\overline{r}_n^{(c)}(t)) \geq \sum_{n,c} g_n^{(c)}(r_n^{*(c)}) - BN/V \tag{15.135}$$

where $(r_n^{*(c)})$ is the optimal solution of (15.128), and

$$\sum_{n,c} \overline{U_n^{(c)}} = \limsup_{t \to \infty} \frac{1}{t} \sum_{\tau=0}^{t-1} \left[\sum_{n,c} E\{U_n^{(c)}(\tau)\} \right]$$

$$\overline{r}_n^{(c)}(t) = \frac{1}{t} \sum_{\tau=0}^{t-1} E\{R_n^{(c)}(\tau)\} \tag{15.136}$$

Since the above result holds for all $V > 0$, the value of V can be chosen so that BN/V is arbitrarily small, resulting in achieved utility that is arbitrarily close to optimal. This comes at the cost of a linear increase in network congestion with the parameter V. By Little's theorem, average queue backlog is proportional to average bit delay, and hence performance can be pushed towards optimality with a corresponding tradeoff in end-to-end network delay. The proof of (15.135) will be elaborated in the next section.

If the resource allocation policy of DFRA is replaced with any (potentially randomized) resource allocation policy $\vec{\mu}(t)$ that satisfies $\sum_l W_l^*(t)E\{\mu_l(t)|\underline{U}(t)\} \geq \theta \left(\max_{\vec{\mu} \in \Gamma_{\vec{S}(t)}} \sum_l W_l^*(t)\mu_l \right) - C$ for some fixed constants θ and C such that $0 < \theta \leq 1$ and $C \geq 0$, then

$$\overline{\sum_{n,c} U_n^{(c)}} \leq (2C + BN + VG_{\max}) / 2\mu_{sym}\theta$$

$$\liminf_{t \to \infty} \sum_{n,c} g_n^{(c)}(\overline{r}_n^{(c)}(t)) \geq \sum_{n,c} g_n^{(c)}(\tilde{r}_n^{*(c)}) - (2C + BN)/V \qquad (15.137)$$

where $(\tilde{r}_n^{*(c)})$ is the optimal solution to the following optimization:

$$\begin{aligned} Maximize \ &: \ \sum_{n,c} g_n^{(c)}(r_n^{(c)}) \\ Subject\ to \ &: \ (r_n^{(c)}) \in \theta \Lambda \\ & \quad 0 \leq r_n^{(c)} \leq \lambda_n^{(c)} \end{aligned} \qquad (15.138)$$

In words, allocating resources to come within a factor θ of the optimal solution of the *DFRA* resource allocation yields a utility that is close to the optimal utility with respect to a θ scaled version of the capacity region.

In a special case when utilities are linear, so that $g_n^{(c)}(r) = \alpha_n^{(c)} r$ for some nonnegative weights $\alpha_n^{(c)}$, the resulting objective is to maximize the weighted sum of throughput, and the resulting F segment of *DFRA* algorithm has a simple threshold form, where some commodities receive as much of the $R_n^{\max}$ delivery rate as possible, while others receive none. In the special case where the user at node n desires communication with a single destination node c_n (so that $g_n^{(c)}(r) = 0$ for all $c \neq c_n$), the flow control algorithm (15.133) reduces to maximizing $V\alpha_n^{(c_n)}r - 2U_n^{(c_n)}r$ subject to $0 \leq r \leq R_n^{\max}$, and the solution is the following threshold rule:

$$R_n^{(c_n)}(t) = \begin{cases} R_n^{\max}, & U_n^{(c_n)}(t) \leq \left(V\alpha_n^{(c\Omega)}\right)/2 \\ 0, & otherwise \end{cases}$$

In other words when backlog in the source queue is large, we should refrain from sending new data.

Utility functions of the form $g_n^{(c)}(r) = \log(1 + \beta r_n^{(c)})$ and maximizing a sum of such utilities over any convex set Λ leads to *proportional fairness*. In the special case when there is only one destination c_n for each user n, the flow control algorithm reduces to maximizing $V \log(1 + \beta r) - 2U_n^{(c_n)}r$ subject to $0 \leq r \leq R_n^{\max}$, which leads to so called '$1/U$' flow control function

$$R_n^{(c_n)}(t) = \min\left(\max\left(\frac{V}{2U_n^{(c_n)}} - \frac{1}{\beta}, 0\right), R_n^{\max}\right)$$

where the flow control value restricts flow according to a continuous function of the backlog level at the source queue, being less conservative in its admission decisions when backlog is low and more conservative when backlog is high.

The drawback of this $1/U$ policy is that the resulting flow control variables $R_n^{(c)}(t)$ are real numbers (not necessarily integers or integer multiples of a given packet length), and hence it is implicitly assumed that packets can be fragmented for admission to the network. The joint flow and routing DFRA (*JDFRA*) algorithm presented in Section 15.13.3 overcomes this issue.

15.13.2 Lyapunov Drift Analysis of the DFRA Algorithm

Here we elaborate more detail leading to (15.135) by using the Lyapunov drift analysis. For $\underline{U}(t) = (U_n^{(c)}(t))$ representing a process of queue backlogs that evolves according to some probability law, the *Lyapunov function* $L(\underline{U})$ is defined in Section 15.11.3. If $R_n^{(c)}(t)$ represent an input process affecting the system, and the values are bounded so that $\sum_{n,c} g_n^{(c)}(R_n^{(c)}(t)) \leq G_{\max}$, utility functions $g_n^{(c)}(r)$ are non-negative and concave, and g^* represent a 'target utility' value, then a given queue backlog vector $\underline{U}(t)$, we define the one step (K=1) *conditional Lyapunov drift* $\Delta(\underline{U}(t))$ as $\Delta(\underline{U}(t)) = E\{L(\underline{U}(t+1)) - L(\underline{U}(t))|\underline{U}(t)\}$ where the conditional expectation is with respect to the random one step queuing dynamics given the current backlog $\underline{U}(t)$.

If there are positive constants V, ε, B such that for all timeslots t and all unfinished work matrices $\underline{U}(t)$, the Lyapunov drift satisfies

$$\Delta(\underline{U}(t)) - V \sum_{n,c} E\{g_n^{(c)}(R_n^{(c)}(t))|\underline{U}(t)\} \leq B - \varepsilon \sum_{n,c} U_n^{(c)}(t) - Vg^* \tag{15.139}$$

then time average utility and congestion satisfy

$$\lim_{t \to \infty} \sup \frac{1}{t} \sum_{\tau=0}^{t-1} \sum_{n,c} E\{U_n^{(c)}(\tau)\} \leq (B + VG_{\max})/\varepsilon \quad \lim_{t \to \infty} \inf \sum_{n,c} g_n^{(c)}(\bar{r}_n^{(c)}(t)) \geq g^* - B/V \tag{15.140}$$

where $\bar{r}_n^{(c)}(t)$ is defined in (15.136).

To prove (15.140) we assume that (15.139) holds and take expectations over the distribution of $\underline{U}(t)$. By using the definition of $\Delta((t))$ together with the law of iterated expectations yields

$$E\{L(\underline{U}(t+1)) - L(\underline{U}(t))\} - V \sum_{n,c} E\{g_n^{(c)}(R_n^{(c)}(t))\}$$

$$\leq B - \varepsilon \sum_{n,c} E\{U_n^{(c)}(t)\} - Vg^*$$

Summing over $t \in \{0, 1, \ldots, M-1\}$ yields

$$E\{L(\underline{U}(M))\} - E\{L(\underline{U}(0))\} - V \sum_{\tau=0}^{M-1} \sum_{n,c} E\{g_n^{(c)}(R_n^{(c)}(\tau))\}$$

$$\leq BM - \varepsilon \sum_{\tau=0}^{M-1} \sum_{n,c} E\{U_n^{(c)}(\tau)\} - VMg^* \tag{15.141}$$

Using non-negativity of the Lyapunov function and the utility functions as well as the fact that $\sum_{n,c} g_n^{(c)}(R_n^{(c)}(\tau)) \leq G_{\max}$, we rearrange the terms of (22) and divide by $M\varepsilon$ to yield

$$(1/M) \sum_{\tau=0}^{M-1} \sum_{n,c} E\{U_n^{(c)}(\tau)\} - E\{L(\underline{U}(0))\}/M\varepsilon \leq (B + V.G_{\max})/\varepsilon \tag{15.142}$$

Taking the limsup as $M \to \infty$ yields the backlog bound (15.140).

The utility bound (15.140) is proved similarly. Indeed, we again rearrange (15.141) and divide by MV to yield

$$\sum_{n,c} \frac{1}{M} \sum_{\tau=0}^{M-1} E\{g_n^{(c)}(R_n^{(c)}(\tau))\} \geq g^* - \frac{B + E\{L(\underline{U}(0))\}/M}{V} \tag{15.143}$$

By concavity of $g_n^{(c)}(r)$ together with Jensen's inequality, we have

$$(1/M) \sum_{\tau=0}^{M-1} E\{g_n^{(c)}(R_n^{(c)}(\tau))\} \leq g_n^{(c)}((1/M) \sum_{\tau=0}^{M-1} E\{R_n^{(c)}(\tau)\}) \tag{15.144}$$

Using this fact in the left hand side of (15.143) and taking the *liminf* as $M \to \infty$ yields (15.141). This result suggests that a good control strategy is to greedily minimize the following drift metric every timeslot: $\Delta(\underline{U}(t)) - V \sum_{n,c} E\{g_n^{(c)}(R_n^{(c)}(t))|\underline{U}(t)\}$.

This is the principle behind the control algorithm presented in the sequel. We first develop an expression for Lyapunov drift from the queuing dynamics (15.130). Any general queue with backlog $U(t)$ and queuing law $U(t + 1) = \max[U(t) - \mu(t), 0] + A(t)$ has a Lyapunov drift given by

$$E\{U^2(t + 1) - U^2(t)|U(t)\} \le \mu_{\max}^2 + A_{\max}^2 - 2U(t)E\{\mu(t) - A(t)|U(t)\}$$

with $A_{\max}$ and $\mu_{\max}$ being upper bounds on the arrival and server variables $A(t)$ and $\mu(t)$. This well-known fact follows simply by squaring the queuing equation and taking expectations. Applying this to the specific queuing law (15.130) for queue (n, c) and summing the result over all (n, c) pairs yields the following expression for Lyapunov drift [37]

$$\Delta(\underline{U}(t)) \le NB - 2 \sum_{n,c} U_n^{(c)}(t) \times E \left\{ \sum_{l \in \Omega_n} \mu_l^{(c)}(t) - \sum_{l \in \Theta_n} \mu_l^{(c)}(t) - R_n^{(c)}(t) \, |\underline{U}(t) \right\} \qquad (15.145)$$

where B is defined in (15.134). In the sequel we define the *flow function* $\Psi\left(\underline{U}(t)\right)$ and the *network function* $\Phi(\underline{U}(t))$ as

$$\Psi(\underline{U}(t)) = \sum_{n,c} E\{V g_n^{(c)}(R_n^{(c)}) - 2U_n^{(c)} R_n^{(c)}|\underline{U}\}$$

$$\Phi\left(\underline{U}(t)\right) = 2 \sum_{n,c} U_n^{(c)} E \left\{ \sum_{l \in \Omega_n} \mu_l^{(c)} - \sum_{l \in \Theta_n} \mu_l^{(c)}|\underline{U} \right\} \qquad (15.146)$$

where we have simplified the notation for $\underline{U}(t)$, $\mu_l^{(c)}(t)$, and $R_n^{(c)}(t)$ by using $\underline{U}$, $\mu_l^{(c)}$, and $R_n^{(c)}$. Subtracting the utility component $V \sum_{n,c} E\{g_n^{(c)}(R_n^{(c)})|\underline{U}\}$ from both sides of (15.145) yields

$$\Delta((t)) - V \sum_{n,c} E\{g_n^{(c)}(R_n^{(c)}(t))|\underline{U}\} \le NB - \Phi\left(\underline{U}(t)\right) - \Psi\left(\underline{U}(t)\right) \qquad (15.147)$$

Given a $\underline{U}(t)$ matrix at time t, the *DFRA* policy is designed to *greedily minimize the right hand side of* (15.147) *over all possible routing, resource allocation and flow control options* resulting into:

$$\Psi^{DFRA}(\underline{U}(t)) \ge \sum_{n,c} [V g_n^{(c)}(r_n^{*(c)}) - 2U_n^{(c)}(t) r_n^{*(c)}]$$

$$\Phi^{DFRA}(\underline{U}(t)) \ge 2 \sum_{n,c} U_n^{(c)}(t) E \left\{ \sum_{l \in \Omega_n} \mu_l^{*(c)}(t) - \sum_{l \in \Theta_n} \mu_l^{*(c)}(t)|\underline{U}(t) \right\} \qquad (15.148)$$

where $(r_n^{*(c)})$ are any particular values that satisfy constraints in (15.133) for all n, and $(\mu_l^{*(c)}(t))$ are any particular (potentially randomized) routing and resource allocation decisions at time t.

That the *DFRA* flow control strategy (15.133) maximizes $\Psi(\underline{U}(t))$ over all feasible choices of the $R_n^{(c)}(t)$ values follows immediately by comparing (15.133) and (15.146). That the routing and resource allocation policy of *DFRA* maximizes $\Phi(\underline{U}(t))$ is proven in [37], and can be understood by switching the sums in the definition of $\Phi(\underline{U}(t))$ i.e. $\Phi(\underline{U}(t)) = 2 \sum_l \sum_c E\{\mu_l^{(c)}(t)|\underline{U}\}[U_{t(l)}^{(c)} - U_{r(l)}^{(c)}]$.

If $\Phi^{DFRA}(\underline{U}(t))$ represents the above function when the rates $\mu_l^{(c)}(t)$ are chosen at time t according to DFRA, and if $\Phi^*(\underline{U}(t))$ represents the above $\Phi(\underline{U}(t))$ function when any alternate feasible rates $\mu_l^{*(c)}(t)$ are chosen at time t (possibly via randomization), then we have

$$\Phi^*(\underline{U}(t)) \le 2 \sum_l \sum_c E\left\{\mu_l^{*(c)}(t)|\underline{U}(t)\right\}W_l^*(t) \le \Phi^{DFRA}(\underline{U}(t))$$

where

$$W_l^*(t) = \max[U_{tran(l)}^{(c)}(t) - U_{rec(l)}^{(c)}(t), 0]$$

In order to use the Lyapunov drift result to study the performance of the *DFRA* algorithm, it is important to first compare performance to the utility of a *near-optimal* solution to the optimization problem (15.128). For this purposes, for any $\varepsilon > 0$, the set Λ_ε will be define as $\Lambda_\varepsilon = \{(r_n^{(c)})|(r_n^{(c)} + \varepsilon 1_n^{(c)}) \in \Lambda, r_n^{(c)} \ge 0$ for all $(n, c)\}$ where $1_n^{(c)}$ is equal to 1 whenever $(n, c) \in D$, and zero else. So, the set Λ_ε can be viewed as the resulting set of rate matrices within the network capacity region when an ε-wide layer of the boundary is stripped away from the D effective dimensions. Note that this set is compact and nonempty whenever $\varepsilon \le \mu_{sym}$.

The *near-optimal operating point* $(r_n^{*(c)}(\varepsilon))$ is defined as a solution to the following optimization problem:

$$Maximize : \sum_{n,c} g_n^{(c)}(r_n^{(c)})$$

$$Subject\ to : (r_n^{(c)}) \in \Lambda_\varepsilon \qquad\qquad (15.149)$$

$$(r_n^{(c)}) \le \lambda_n^{(c)} \text{ for all } (n, c)$$

This optimization differs from the optimization in (15.128) in that the set Λ is replaced by the set Λ_ε. If utility functions $g_n^{(c)}(r)$ are non-negative and concave, and if there is a positive scalar μ_{sym} such that $(\mu_{sym}) \in \Lambda$, then

$$\sum_{n,c} g_n^{(c)}(r_n^{*(c)}(\varepsilon)) \to \sum_{n,c} g_n^{(c)}(r_n^{*(c)}) \text{ as } \varepsilon \to 0 \qquad\qquad (15.150)$$

The proof uses convexity of the capacity region Λ, and is given in [43, 44 Chapter 5.5.2]. The proof of (15.135) relies on the following two inequalities:

$$\Psi^{DFRA}(\underline{U}(t)) \ge \sum_{n,c} [Vg_n^{(c)}(r_n^{*(c)}(\varepsilon)) - 2U_n^{(c)}(t)r_n^{*(c)}(\varepsilon)]$$

$$\Phi^{DFRA}(\underline{U}(t)) \ge 2 \sum_{n,c} U_n^{(c)}(t)(r_n^{*(c)}(\varepsilon) + \varepsilon) \qquad\qquad (15.151)$$

where $(r_n^{*(c)}(\varepsilon))$ is the optimal solution of problem (15.149). The first inequality follows from (15.148) by using $r_n^{*(c)} = r_n^{*(c)}(\varepsilon)$. This is based on the fact that these are valid flow control choices because all reservoirs are infinitely backlogged and so it is always possible to choose $R_n^{(c)}(t) = r_n^{*(c)}(\varepsilon)$. Also, the matrix $(r_n^{*(c)}(\varepsilon))$ is within Λ and hence $\sum_c r_n^{*(c)}(\varepsilon) \le R_n^{\max}$ for all n. The second inequality in (15.151) follows by plugging the particular $\mu_l^{*(c)}(t)$ policy related to (15.132) (for $(r_n^{*(c)}(\varepsilon) + \varepsilon 1_n^{(c)}) \in \Lambda$) into

(15.148). Plugging the above two inequalities directly into (15.147) yields for *DFRA*:

$$\Delta(\underline{U}(t)) - V \sum_{n,c} E\{g_n^{(c)}(R_n^{(c)}(t))|\underline{U}\} \le NB - 2\sum_{n,c} U_n^{(c)}(t)\,(r_n^{*(c)}(\varepsilon) + \varepsilon)$$

$$-V \sum_{n,c} g_n^{(c)}(r_n^{*(c)}(\varepsilon)) + 2\sum_{n,c} U_n^{(c)}(t) r_n^{*(c)}(\varepsilon) =$$

$$= NB - 2\varepsilon \sum_{n,c} U_n^{(c)}(t) - V\sum_{n,c} g_n^{(c)}(r_n^{*(c)}(\varepsilon))$$

which has the exact form as (15.139). Thus,

$$\overline{\sum U_n^{(c)}} \le (NB + VG_{\max})/(2\varepsilon)$$
$$\liminf_{t\to\infty} \sum_{n,c} g_n^{(c)}(\overline{r}_n^{(c)}(t)) \ge \sum_{n,c} g_n^{(c)}(r_n^{*(c)}(\varepsilon)) - NB/V \tag{15.152}$$

The performance bounds in (15.152) hold for any value ε such that $0 < \varepsilon \le \mu_{sym}$. However, the particular choice of ε only affects the bound calculation and does not affect the *DFRA* control policy or change any sample path of system dynamics.

We can thus optimize the bounds separately over all possible ε values. The second bound in (15.152) is clearly maximized by taking a limit as $\varepsilon \to 0$, so that the right hand side converges to $\sum_{n,c} g_n^{(c)}(r_n^{*(c)}) - NB/V$ (using (15.150)). The first bound in (15.152) is minimized as $\varepsilon \to \mu_{sym}$, yielding: $\sum_{n,c} \overline{U_n^{(c)}} \le (NB + VG_{\max})/(2\mu_{sym})$. This proves (15.135). The proof of (15.137) follows by noting that the suboptimal allocation bound changes the second inequality of (15.148) into

$$\Phi^{DFRA}(\underline{U}(t)) \ge -2C + 2\sum_{n,c} U_n^{(c)}(t)$$

$$\times E\left\{ \sum_{l\in\Omega_n} \theta\mu_l^{*(c)}(t) - \sum_{l\in\Theta_n} \theta\mu_l^{*(c)}(t) \,|\underline{U}(t) \right\}$$

This modifies inequality (15.151) into

$$\Phi^{DFRA}(\underline{U}(t)) \ge -2C + 2\theta\sum_{n,c} U_n^{(c)}(t)(r_n^{*(c)}(\varepsilon) + \varepsilon)$$

and the proof then proceeds exactly as before.

15.13.3   *Optimization of DFRA with Finite Buffers*

The algorithm from the previous section assumes there is always an amount of data $R_n^{(c)}(t)$ available in reservoir (n, c), and the flow variable $R_n^{(c)}(t)$ has only $R_n^{\max}$ constraint. In practice all transport layer storage reservoirs have a finite (sometimes zero) buffer. If $L_n^{(c)}(t)$ represent the current backlog in the reservoir buffer the flow control decisions are now subject to the additional constraint $R_n^{(c)}(t) \le L_n^{(c)}(t) + A_n^{(c)}(t)$ (where $A_n^{(c)}(t)$ is the amount of new commodity c data exogenously arriving to node n at slot t).

Any arriving data that is not immediately admitted to the network is stored in the reservoir, or dropped if the reservoir has no extra space.

Assume the $A_n^{(c)}(t)$ arrivals are i.i.d. over timeslots with arrival rates $\lambda_n^{(c)} = E\{A_n^{(c)}(t)\}$. It can be shown that for any matrix $(\lambda_n^{(c)})$ (possibly outside of the capacity region), modifying the *DFRA* flow algorithm to maximize (15.133) subject to the additional reservoir backlog constraint yields the same performance guarantees (15.135) when utility functions are linear [44]. For nonlinear utilities, such a strategy can be shown to maximize the time average of $\sum_{n,c} E\{g_n^{(c)}(R_n^{(c)}(t))\}$ over all strategies that make immediate admission/rejection decisions upon arrival, but may not necessarily maximize $\sum_{n,c} g_n^{(c)}(E\{R_n^{(c)}(t)\})$, which is the utility metric of interest. In [43] this problem is solved by defining additional *flow state variables* $Z_n^{(c)}(t)$ for each reservoir (n, c), and fixing $Z_n^{(c)}(0) = V R_n^{\max}/2$ for all (n, c) (any initial condition can be used this one works well experimentally). For each flow control process $R_n^{(c)}(t)$, a new process $Y_n^{(c)}(t)$ is defined as

$$Y_n^{(c)}(t) = R_n^{\max} - R_n^{(c)}(t) \tag{15.153}$$

where $Y_n^{(c)}(t) \geq 0$ for all t. The $Y_n^{(c)}(t)$ variables represent the difference between the maximum value and the actual value of admitted data on session (n, c). The $Z_n^{(c)}(t)$ state variables are updated every slot as

$$Z_n^{(c)}(t + 1) = \max[Z_n^{(c)}(t) - \gamma_n^{(c)}(t), \ 0] + Y_n^{(c)}(t) \tag{15.154}$$

where $\{\gamma_n^{(c)}(t)\}$ are additional flow control decision variables. The cost function is now defined as

$$h_n^{(c)}(\gamma) = g_n^{(c)}(R_n^{\max}) - g_n^{(c)}(R_n^{\max} - \gamma) \tag{15.155}$$

For $\overline{\gamma}_n^{(c)}$ representing the time average value of the decision variables $\gamma_n^{(c)}(t)$ we design a policy to stabilize the network queues $U_n^{(c)}(t)$ and the flow state 'queues' $Z_n^{(c)}(t)$ while minimizing the cost $\sum_{n,c} h_n^{(c)}(\overline{\gamma}_n^{(c)})$.

In other words, if the $Z_n^{(c)}(t)$ queues are stabilized, it must be the case that the time average of the 'server process' $\gamma_n^{(c)}(t)$ is greater than or equal to the time average of the 'arrival process' $Y_n^{(c)}(t)$: $\overline{Y}_n^{(c)} \leq \overline{\gamma}_n^{(c)}$. From (15.153), this implies $\overline{r}_n^{(c)} \geq R_n^{\max} - \overline{\gamma}_n^{(c)}$, and hence

$$\sum_{n,c} h_n^{(c)}(\overline{\gamma}_n^{(c)}) = \sum_{n,c} g_n^{(c)}(R_n^{\max}) - \sum_{n,c} g_n^{(c)}(R_n^{\max} - \overline{\gamma}_n^{(c)})$$

$$\geq \sum_{n,c} g_n^{(c)}(R_n^{\max}) - \sum_{n,c} g_n^{(c)}(\overline{r}_n^{(c)}) \tag{15.156}$$

From this relation we can see that minimizing $\sum_{n,c} h_n^{(c)}(\overline{\gamma}_n^{(c)})$ over all feasible $\overline{\gamma}_n^{(c)}$ values is related to maximizing $\sum_{n,c} g_n^{(c)}(\overline{r}_n^{(c)})$ over all feasible $\overline{r}_n^{(c)}$ values. Based on the above observations a modified *DFRA* referred to as *mDFRA* algorithm was proposed in [43] as follows:

mDFA Algorithm

For η being any fixed constant such that $0 < \eta \leq 1$, every timeslot and for each node n, choose $R_n^{(c)}(t) = x_n^{(c)}$ to solve

$$\begin{aligned} \textit{Maximize} : & \sum_c [\eta Z_n^{(c)}(t) - U_n^{(c)}(t)]x_n^{(c)} \\ \textit{Subject to} : & \sum_c x_n^{(c)} \leq R_n^{\max} \\ & x_n^{(c)} \leq L_n^{(c)}(t) + A_n^{(c)}(t) \end{aligned} \tag{15.157}$$

Additionally, the flow controllers at each node n choose $\gamma_n^{(c)}(t)$ for each session (n, c) to solve

$$
\begin{aligned}
&Maximize: \ Vg_n^{(c)}(R_n^{\max} - \gamma_n^{(c)}) + 2\eta Z_n^{(c)}(t)\gamma_n^{(c)} \\
&Subject\ to: \ 0 \leq \gamma_n^{(c)} \leq R_n^{\max}
\end{aligned}
\tag{15.158}
$$

The flow states $Z_n^{(c)}(t)$ are then updated according to (15.154). Routing and resource allocation within the network is the same as in DFRA. The optimization of $R_n^{(c)}(t)$ in (15.157) is solved by a simple control policy where no data is admitted from reservoir (n,c) if $U_n^{(c)}(t) > \eta Z_n^{(c)}(t)$, and otherwise as much data as possible is delivered from the commodities of node n with the largest non-negative values of $\eta Z_n^{(c)}(t) - U_n^{(c)}(t)$ subject to the $R_n^{\max}$ constraint. These decisions also enable the strategy to be implemented optimally in systems where admitted data is constrained to integral units, a feature that *DFRA* does not have. As in *DFRA*, the constraint $\sum_{(c)} x_n^{(c)} \leq R_n^{\max}$ can be replaced by the simpler constraint $x_n^{(c)} \leq R_n^{\max}$ at the cost of increasing the system delay bound.

The $\gamma_n^{(c)}(t)$ variable assignment in (15.158) involves maximizing a concave function of a single variable, and can be solved easily by finding the critical points over $0 \leq \gamma_n^{(c)} \leq R_n^{\max}$.

For arbitrary rate matrices $(\lambda_n^{(c)})$ (possibly outside of the capacity region), for any $V > 0$, and for any reservoir buffer size (possibly zero), the *mDFRA* algorithm stabilizes the network and yields a congestion bound:

$$
\overline{\sum_{n,c} U_n^{(c)}} \leq \frac{NB + 2\eta N \sum_n (R_n^{\max})^2 + VG_{\max}}{2\mu_{sym}} f
$$

In addition, the time average utility satisfies [43]

$$
\liminf_{t \to \infty} \sum_{n,c} g_n^{(c)}(\overline{r}_n^{(c)}(t)) \geq \sum_{n,c} g_n^{(c)}(r_n^{*(c)}) - \frac{NB + 2\eta N \sum_n (R_n^{\max})^2}{V}
$$

15.14 Dynamic Resource Allocation in Computing Clouds

Providing an efficient communication network in distributed computing clouds environment represent a new challenge.

To understand the nature of the problem in this section we consider a virtualized data center with M servers that host a set of N applications. The set of servers is denoted by S and the set of applications by A. Each server $j \in S$ hosts a subset of the applications. It does so by providing a virtual machine (VM) for every application hosted on it. An application may have multiple instances running across different VMs in the data center. We define the following indicator variables for $i \in \{1, 2, \ldots, N\}$, $j \in \{1, 2, \ldots, M\}$: $a_{ij} = 1$ if application i is hosted on server j and $a_{ij} = 0$ otherwise.

If $a_{ij} = 1 \forall i, j$, then each server can host all applications. In general, applications may be multi-tiered and the different tiers corresponding to an instance of an application may be located on different servers and VMs. In a time-slotted system, every slot, new requests arrive for each application i according to a random arrival process $A_i(t)$ that has a time average rate λ_i *requests/slot*.

The control architecture for the virtualized data center consists of three components as shown in Figure 15.28. Every slot, for each application $i \in A$, an *Admission Controller A/C(i)* determines whether to admit or decline the new requests. The requests that are admitted are stored in the Router buffer W_i before being routed to one of the servers hosting that application by the *Router*. Each server $j \in S$ has a set of resources (such as CPU, disk, memory, network resources, etc.) that are allocated to the VMs hosted on it by its *Resource Controller*.

All servers in the data center are assumed to be resource constrained. Specifically, in the basic model, we focus on CPU frequency and power constraints.

Modern CPUs can be operated at different speeds at runtime by using techniques such as Dynamic Frequency Scaling (DFS), Dynamic Voltage Scaling (DVS), or a combination Dynamic Voltage and

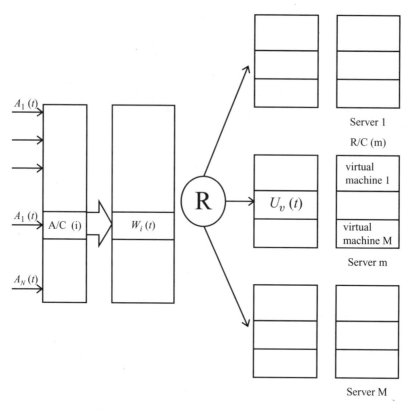

Figure 15.28   Virtualized Data Center Architecture.

Frequency Scaling (DVFS). These techniques result in a *non-linear* power-frequency relationship. As an example, the power-frequency relationship for Dell PowerEdge R610 server is well-approximated by a quadratic model, i.e., $P(f) = P_{min} + \alpha(f - f_{min})^2$ [45]. The CPUs can run at a finite number of operating frequencies in an interval $[f_{min}, f_{max}]$ with an associated power consumption $[P_{min}, P_{max}]$. This allows a tradeoff between performance and power costs. All servers in our model are assumed to have identical CPU resources.

Additionally, the servers may be turned OFF in order to further save energy costs. This can be advantageous if the workload is low. While an inactive server does not consume any power, it also cannot provide any service to the applications hosted on it. We thus assume that, in any slot, new requests can only be routed to active servers. Inactive servers can be turned active to handle increases in workload.

Let $A_i(t)$ denote the number of new request arrivals for application i in slot t. Let $R_i(t)$ be the number of requests out of $A_i(t)$ that are admitted into the Router buffer $W_i(t)$ for application i by the Admission Controller. Thus, for all i, t, we have $0 \le R_i(t) \le A_i(t)$.

Let $R_{ij}(t)$ be the number of requests for application i that are routed from its Router buffer to server j in slot t. Then the queueing dynamics for $W_i(t)$ is given by $W_i(t + 1) = W_i(t) - \sum_j R_{ij}(t) + R_i(t)$.

For each application i, the admitted requests can only be routed to those servers that host application i and are active in slot t. Thus, the routing decisions $R_{ij}(t)$, given the set of active servers $S(t)$ in slot t, must satisfy the following constraints every slot:

$$R_{ij}(t) = 0 \text{ if } j \notin S(t) \text{ or } a_{ij} = 0$$
$$0 \le \sum_{j \in S(t)} a_{ij} R_{ij}(t) \le W_i(t) \tag{15.159}$$

Every slot, the Resource Controller allocates the resources of each server among the VMs that host the applications running on that server. This allocation is subject to the available control options. For example, the Resource Controller may allocate different fractions of the CPU (or different number of cores in case of multi-core processors) to the VMs in that slot. The Resource Controller may also use available techniques such as DFS, DVS, DVFS, etc. to modulate the current CPU speed which affects the CPU power consumption. We use I_j to denote the set of all such control options available at server j. This includes the option of making server j. If $I_j(t) \in I_j$ is the particular control decision at server j in slot t under any policy and $P_j(t)$ is the corresponding power consumption, then, the queueing dynamics for the requests of application i at server j is

$$U_{ij}(t+1) = \max[U_{ij}(t) - \mu_{ij}(I_j(t)), 0] + R_{ij}(t) \tag{15.160}$$

where $\mu_{ij}(I_j(t))$ denotes the service rate (in units of requests/slot) provided to application i on server j in slot t by taking control action $I_j(t)$. We assume that, for each application, the expected value of this service rate as a function of the control action $I_j(t)$ is known for all $I_j(t) \in I_j$. This can be obtained by application profiling and application modeling techniques [47, 48]. Thus, in every slot t, a control policy needs to make the following decisions:

1. *If $t = nT$ (i.e., beginning of a new frame), determine the new set of active servers $S(t)$. Else, continue using the active set already computed for the current frame.*
2. *Admission control decisions $R_i(t)$ for all applications i.*
3. *Routing decisions $R_{ij}(t)$ for the admitted requests.*
4. *Resource allocation decision $I_j(t)$ at each active server.*

In the following we will present an online control policy that maximizes a joint utility of the sum throughput of the applications and the energy costs of the servers subject to the available control options and the structural constraints imposed by this model [46]. It is desirable to develop a flexible and robust resource allocation algorithm that *automatically* adapts to time varying workloads. The technique of Lyapunov optimization, discussed in the previous sections, is used to design such an algorithm.

Consider any policy η for this model that takes control decisions $S^\eta(t)$, $R_i^\eta(t)$, $R_{ij}^\eta(t)$, $I_j^\eta(t) \in I_j$, $P_j^\eta(t)$ for all i, j in slot t. Under any feasible policy η, these control decisions must satisfy the admission control constraint $0 \le R_i(t) \le A_i(t)$, routing constraints (15.159), and the resource allocation constraint $I_j(t) \in I_j$ every slot for all i, j.

For the time average expected rate of admitted requests for application i under policy η, $r_i^\eta = \lim_{t\to\infty} (1/t) \sum_{\tau=0}^{t-1} E\{R_i^\eta(\tau)\}$ let $r = (r_1, \ldots, r_N)$ denote the vector of these time average rates. Similarly, let $e_j^\eta = \lim_{t\to\infty} (1/t) \sum_{\tau=0}^{t-1} E\{P_j^\eta(\tau)\}$ denote the time average expected power consumption of server j under policy η. The expectations above are with respect to the possibly randomized control actions that policy η might take.

For non-negative weights α_i and β our objective is to design a policy η that solves the following *stochastic optimization problem*:

$$Maximize : \sum_{i \in A} \alpha_i r_i^\eta - \beta \sum_{j \in S} e_j^\eta$$

$$Subject\ to : 0 \le r_i^\eta \le \lambda_i \forall i \in A \tag{15.160a}$$
$$I_j^\eta(t) \in I_j \forall j \in S, \forall t,$$
$$\mathbf{r} \in \mathbf{\Lambda}$$

where, $\mathbf{\Lambda}$ is the *capacity region* of the data center model as described above. It is defined as the set of all possible long-term throughput values that can be achieved under *any* feasible resource allocation strategy.

Suppose (15.160) is a feasible and let r_i^* and $e_j^* \forall i$, j denote the optimal value of the objective function, potentially achieved by some arbitrary policy. Using the techniques developed in [49], it can be shown that to solve (15.160) and achieve the optimal value of the objective function, it is sufficient to consider only the class of stationary, randomized policies that take control decisions independent of the current queue backlog every slot. This policy chooses an active set of servers, at the beginning of each frame, according to a stationary distribution in an i.i.d. fashion. Once chosen, other control decisions are likewise taken in an i.i.d. fashion according to stationary distributions. For the basic model with homogeneous application hosting and identical CPU resources, in choosing an active server set, we do not need to consider all possible subsets of S. We define the following collection O of subsets of S: $O = \{\emptyset, \{1\}, \{1,2\}, \{1,2,3\}, \ldots, \{1, 2, 3, \ldots, M\}\}$.

Then for any arrival rate vector $(\lambda_1, \ldots, \lambda_N)$, there exists a frame-based stationary randomized control policy that chooses active sets from O every frame, makes admission control, routing and resource allocation decisions every slot independent of the queue backlog and yields the following steady state values:

$$\lim_{t \to \infty} \frac{1}{t} \sum_{\tau=0}^{t-1} \left[\sum_{i \in A} \alpha_i E\{R_i(\tau)\} - \beta \sum_{j \in S} E\{P_j(\tau)\} \right] = \sum_{i \in A} \alpha_i r_i^* - \beta \sum_{j \in S} e_j^* \qquad (15.161)$$

Computing the optimal stationary, randomized policy requires knowledge of all system parameters (like workload statistics) as well as the capacity region in advance. This policy would not be adaptive to unpredictable changes in the workload and must be recomputed.

For this reason, we use the framework of Lyapunov Optimization to develop an optimal control algorithm for this model.

Let $W_i(t)$, $U_{ij}(t) \forall i$, j be the queue backlog values in slot t. These are initialized to 0. Every slot, the *Dynamic Control Algorithm* (DCA) uses the backlog values in that slot to make joint Admission Control, Routing and Resource Allocation decisions. As the backlog values evolve over time,

$$W_i(t + 1) = W_i(t) - \sum_j R_{ij}(t) + R_i(t)$$
$$U_{ij}(t + 1) = \max[U_{ij}(t) - \mu_{ij}(I_j(t)), 0] + R_{ij}(t) \qquad (15.162)$$

the control decisions made by DCA adapt to these changes. However, we note that this is implemented using knowledge of current backlog values only and does not rely on knowledge of statistics governing future arrivals. Thus, DCA solves for the objective in (15.160) by implementing a sequence of optimization problems over time.

The DCA algorithm operates as follows.

Admission Control
For each application i, choose the number of new requests to admit $R_i(t)$ as the solution to the following problem:

$$\textit{Maximize} : \ R_i(t)[V\alpha_i - W_i(t)]$$
$$\textit{Subject to} : \ 0 \leq R_i(t) \leq A_i(t) \qquad (15.163)$$

where $V \geq 0$ is a control parameter that is input to the algorithm. This problem has a simple threshold-based solution. In particular, if the current Router buffer backlog for application i, $W_i(t) > V\alpha_i$, then $R_i(t) = 0$ and no new requests are admitted. Else, if $W_i(t) \leq V\alpha_i$, then $R_i(t) = A_i(t)$ and all new requests are admitted.

Routing: Given an active server set $S(t)$, routing follows a simple Join the Shortest Queue policy. For any application i, let $j' \in S(t)$ be the active server with the smallest queue backlog $U_{ij'}(t)$. If $W_i(t) > U_{ij'}(t)$, then $R_{ij'}(t) = W_i(t)$, i.e., all requests in the Router buffer for application i are routed to server j'. Else, $R_{ij}(t) = 0 \forall j$ and no requests are routed to any server for application i. In order to make *these decisions, the Router requires the queue backlog information $U_{ij}(t) \forall i, j$. Given this information, we note that this routing decision can be performed separately for each application.*

Resource Allocation: At each active server $j \in S(t)$, choose a resource allocation $I_j(t)$ that solves the following problem:

$$\begin{aligned} Maximize : &\sum_i U_{ij}(t) E\{\mu_{ij}(I_j(t))\} - V\beta P_j(t) \\ Subject\ to : &\ I_j(t) \in I_j, P_j(t) \geq P_{\min} \end{aligned} \tag{15.164}$$

The optimal solution would allocate resources so as to maximize the service rate of the most backlogged application.

If $t = nT$, then a new active set $S^*(t)$ for the current frame is determined by solving the following:

$$\begin{aligned} S^*(t) = \underset{S(t) \in O}{argmax} &\left[\sum_{ij} U_{ij}(t) E\{\mu_i \cdot (I_j(t))\} - V\beta \sum_j P_J(t) \right. \\ &\left. + \sum_{ij} R_{ij}(t)(W_i(t) - U_{ij}(t)) \right] \end{aligned}$$

subject to : $j \in S(t), I_j(t) \in I_j, P_j(t) \geq P_{\min}$

constraints $0 \leq R_i(t) \leq A_i(t)$ and (15.159) $\tag{15.165}$

For the case where (a) all queues are initialized to 0; (b) all arrivals in a slot $A_i(t)$ are i.i.d. and are upper bounded by finite constants so that $A_i(t) \leq A^{\max}$ for all i, t; (c) $\mu_{\max}$ is the maximum service rate (in *requests/slot*) over all applications in any slot, the implementing the DCA algorithm every slot for any fixed control parameter $V \geq 0$ and frame size T yields the following performance bounds [46]:

$$\begin{aligned} W_i(t) &\leq W_l^{\max} = V\alpha_i + A_i^{\max} \\ U_{ij}(t) &\leq 2W_i^{\max} = 2(V\alpha_i + A_i^{\max}) \\ \liminf_{t \to \infty} \frac{1}{t} \sum_{\tau=0}^{t-1} &\left[\sum_{i \in A} \alpha_i E\{R_i(\tau)\} - \beta \sum_{j \in S} E\{P_j(\tau)\} \right] \geq \\ \sum_{i \in A} \alpha_i r_i^* - \beta &\sum_{j \in S} e_j^* - \frac{BT}{V} \end{aligned} \tag{15.166}$$

where $B = \left[\Sigma_i (A_l^{\max})^2 + NM\mu_{\max}^2 \right]/2.$

References

[1] Bertsekas, D. P. (1998) *Network Optimization: Continuous and Discrete Models*, Belmont, MA: Athena Scientific.

[2] Elias, P. A. Feinstein, A. and Shannon, C.E. (1956) A note on the maximum flow through a network, *IEEE Trans. Info. Theory*, **IT-2**, 117–119.

[3] Edmonds, J. (1973) Edge-Disjoint Branchings in (ed.) R. Rustin, *Combinatorial Algorithms*, New York: Academic Press.

[4] Alswede, R., et al. (2000) Network information flow, *IEEE Trans. Information Theory*, **46**(4), 1204–1216.

[5] Koettar R. and Medard M (2003) An algebraic approach to network coding, *IEEE/ACM Trans. Networking*, **11**(5), 782–795.

[6] Li S.-Y. R., Yeung, R. W. and Cai, N. (2003) Linear network coding, *IEEE, Trans. Information Theory*, **49**(6), 371–381.

[7] Jaggi, S., et al. (2005) Polynomial time algorithms for multicast network code construction, *IEEE Trans. Information Theory*, **51**, 1973–1982.

[8] Bollobas, B. (1979) *Graph Theory, An Introductory Course*, New York: Springer-Verlag.

[9] Ho, T., Médard, M., Koetter, R., Karger, D., Effros, M., Shi, J. and B. Leong (2006) A random linear network coding approach to multicast, *IEEE Transactions on Information Theory*, **52**(10), 4413–4430.

[10] Deb, S., Médard, M. and Choute, C. (2006) Algebraic gossip: a network coding approach to optimal multiple rumor mongering, *IEEE Transactions on Information Theory*, **52**(6), June 2006, 2486–2507.

[11] Dermers, A. et al. (1987) Epidemic algorithms for replicated database maintenance, in *Proc. ACM Sympo. Principles of Distributed Computing*, Vancouver, BC, Canada.

[12] Minski, Y. (2002) Spreading Rumors Cheaply, Quickly, and Reliably, Ph.D. dissertation, Cornell Univ., Ithaca, NY.

[13] Karp, R., Schindelhauer, C., Shenker, S. and Vocking B. (2000) Randomized rumor spreading, in *Proc. Foundations of Computer Science*, Redondo Beach, CA.

[14] Kempe, D., Kleinberg, J. M. and Demers, A. J. (2001) Spatial gossip and resource location protocols, in *Proc. ACM Symp. Theory of Computing*, Heraklion, Crete, Greece.

[15] Kempe, D. and Kleinberg, J. (2002) Protocols and impossibility results for gossip-based communication mechanisms, in *Proc. 43rd IEEE Symp. Foundations of Computer Science*.

[16] Boyd, S., Ghosh, A., Prabhakar, B. and Shah Gossip and mixing times of random walks on random graphs. [Online]. Available at: http://www.stanford.*edu/~oyd/gossip_gnr.html*.

[17] Wu, Y., Jain, K. and Kung, S.-Y. (2006) A unification of network coding and tree-packing (routing) theorems, *IEEE Transactions on Information Theory*, **52**(6), 2398–2409.

[18] Menger, K. (1927) Zur allgemeinen Kurventheorie, Fund. *Math.*, **10**, 95–115.

[19] Schrijver, A. (2003) *Combinatorial Optimization: Polyhedra and Efficiency*, New York, Springer.

[20] Ford, L. R. and Fulkerson, D. R. (1956) On the max-flow min-cut theorem of networks, in *Linear Inequalities and Related Systems*, H. Kuhn and A. Tucker (eds) Princeton, NJ: Princeton University Press, Annals of Mathematics Studies, **38**, 215–221.

[21] Edmonds, J. (1973) Edge-disjoint branchings, in *Combinatorial Algorithms*, R. Rustin, (ed.) New York: Academic, 91–96.

[22] Ho, T., Koetter, R., Medard, M., Karger, D. R. and Effios (2003) The benefits of coding over routing in a randomized setting, in *Proc. Int. Symp. Inf. Theory*, Yokohama, Japan.

[23] Jaggi, S., Chou, P. A. and Jain, K. (2005) Low complexity optimal algebraic multicast codes, in *Proc. Int. Symp. Inf Theory*, Yokohama, Japan.

[24] Sander, P., Egner, S. and Tolhuizen, L. (2003) Polynomial time algorithms for network information flow, in *Proc. Symp. Parallel Algorithms Architect*. (SPAA), San Diego, CA, 286–294.

[25] Harvey, N. J. A., Karger, D. R. and Murota, K. (2005) Deterministic network coding by matrix completion, in *Proc. ACM-SIAM Symp. Discrete Algorithms (SODA 05)*, Vancouver, Canada.

[26] Liang, X.-B. (2006) Matrix games in the multicast networks: maximum information flows with network switching, *IEEE Transactions on Information Theory*, **52**(6), 2433–2466.

[27] Agarwa, A. and Charikar, M. (2004) On the advantage of network coding for improving network throughput, in *Proc. 2004 IEEE Information Theory Workshop*, San Antonio, TX, 247–249.

[28] Jain, K., Mahdian, M. and Salavatipour, M. (2003) Packing Steiner trees, in. *Proc 14th Annu. Annu. –ACM SIAM Symp. Discrete Algorithms*, Baltimore, MD, 266–274.

[29] Wu, Y., Chou, P. A. and Jain, K. (2004) A comparison of network coding and tree packing, in *Proc, 2004 IEEE Int. Symp. Information Theory*, Chicago, IL, 145.

[30] McKinsey, J. C. C. (1952, 2003) *Introduction to the Theory of Games*. New York: McGraw-Hill. Also: Mineaole, NY: Dover.

[31] Owen, G. (1995) *Game Theory*, 3rd ed. San Diego, CA: Academic.

[32] Szép, J. and Forgo, F. (1985), *Introduction to the Theory of Games*. Dordrecht, The Netherlands: Reidel.

[33] Chi, K., Jiang, X., Horiguchi, S. and Guo, M. (2008) Topology design of network coding-based multicast networks, *IEEE Transactions on Parallel and Distributed Systems*, **19**(5), 627–664.

[34] Brown, G. W. (1951) Iterative solution of games by fictitious play, in *Activity Analysis of Production and Allocation*, T. C. Koopmans (ed.) NewYork, Wiley, 374–376.

[35] Robinson, J. (1951) An iterative method of solving a game, *Ann. Math.*, **54**(2), 296–301.

[36] Dijkstra, E. W. (1959) A note on two problems in connection with graphs, in *Numerishe Mathematik*, **1**, 267–271.

[37] Neely, M. J., Modiano, E. and Rohrs, C. E. (2005) Dynamic power allocation and routing for time-varying wireless networks, *IEEE Journal on Selected Areas in Communications*, **23**(1), 89–103.

[38] Tassiulas, L. (1997) Scheduling and performance limits of network with constantly changing topology, *IEEE Trans. Inf Theory*, **43**(3), 1067–1073.

[39] Leonardi, E., Melia, M., Neri, F. and Ajmone Marson, M. (2001) Bounds on average Delay queued cell-based switches, in *Proc. IEEE INFOCOM*, 1095–1103.

[40] McKeown, N., Anantharam, V. and Walrand, J. (1996) Achieving 100% throughput in an input-queued switch, in *Proc. INFOCOM*, 296–302.

[41] Kumar, P. R. and Meyn, S. P. (1995) Stability of queueing networks and sched- uling policies, *IEEE Trans. Autom. Control*, **40**(2), 251–260.

[42] Meyn, S. and Tweedie, R. Markov *Chains and Stochastic Stability*. New York: Springer-Verlag.

[43] Neely, M. J., Modiano, E. and Li, C.-P. (2008) Fairness and optimal stochastic control for heterogeneous networks, *IEEE/ACM Transactions On Networking*, **16**(2), 396–409.

[44] Neely, M. J. (2003) Dynamic power allocation and routing for satellite and wireless network, PhD dissertation, LIDS, Massachusetts Inst. Technol., Cambridge, MA.

[45] Gandhi, A., Harchol-Balter, M., Das, R. and Lefurgy, C. (2009) Optimal power allocation in server farms. In *Proceedings of SIGMETRICS*, June.

[46] Urgaonkar, R., Kozat, U. C., Igarashi, K. and M. J. Neely (2010), Dynamic resource allocation and power management in virtualized data centers, *IEEE/IFIP Network Operations and Management Symposium – NOMS 2010*, 479–486.

[47] Wang, X., Lan, D., Fang, X., Ye, M. and Chen, Y. (2008) A resource management framework for multi-tier service delivery in autonomic virtualized environments. In *Proceedings of NOMS*.

[48] Urgaonkar, B., Shenoy, P. and Roscoe, T. (2002) Resource overbooking and application profiling in shared hosting platforms. In *Proceedings of OSDI*.

[49] Georgiadis, L., Neely, M. J. and Tassiulas, L. (2006) Resource allocation and crosslayer control in wireless networks. *Foundations and Trends in Networking*, **1**(1), 1–149.

16

Network Formation Games

Network formation games (NFG) describe the interaction between a collection of nodes that wish to form a network. Such models have been introduced and studied in the economics literature [1–3]. Each of the nodes in the network is a decision maker and a network is formed through interaction between the decision makers. We are interested in understanding and characterizing the networks that result when decision makers interact to choose their connections. In particular, we will focus on the role of bilateral contracting and the dynamic process of network formation in shaping the network structure. Following standard game theory terminology, we refer to the decision makers as *players*. These games have the following properties: (1) Nodes in the network are strategic agents, (2) Links in the network represent bilateral agreements between their endpoints, (3) Nodes choose whom to accept connections from and whom to connect to, (4) Given a network topology, nodes' payoffs are obtained from two contributions: (a) intrinsic utility extracted from participating in the network; and (b) transfers of utility, or 'payments', between nodes participating in the same link.

16.1 General Model of Network Formation Games

In this contest the players are the set of nodes of the network. Nodes receive a reward that depends on the network topology that arises from formation. The game models a scenario where each link in the network is the result of a bilateral 'contract' between its end nodes. Each contract has a node seeking the agreement and a node accepting it, and therefore there is some utility transfer from the seeking node to the accepting one. In addition to the utility transfers, nodes obtain utility that depends on the topology created.

In the following we will use the following notations and assumptions:

- $G = (V, E)$ - *network topology graph*, consisting of a set V of n nodes and edges E. The nodes will be the players in the NFG.
- All edges in G are *undirected*.
- ij - an undirected edge between i and j.
- $ij \in G$ when the edge ij is present in E.
- $G + ij$ and $G - ij$ denote, respectively, adding and subtracting the link ij to graph G.
- $C_i(G)$ - the monetary *cost* to node i, $i \in V$ of network topology G. P_{ij} - payment from i to j; if no undirected link ij exists, or if $i = j$, then $P_{ij} = 0$.
- $\mathbf{P} = (P_{ij}, i, j \in V)$ - the *payment matrix*.

Given a payment matrix P, the total transfer of utility to node i is the sum of payments received by i minus the sum of payments made by i, that is: $T_i(\mathbf{P}) = \sum_j P_{ji} - P_{ij}$. Thus the total utility of node i in

Advanced Wireless Communications & Internet: Future Evolving Technologies, Third Edition. Savo Glisic.
© 2011 John Wiley & Sons, Ltd. Published 2011 by John Wiley & Sons, Ltd.

graph G is

$$U_i(\mathbf{P}, G) = T_i(\mathbf{P}) - C_i(G) \tag{16.1}$$

We consider a network formation game where each node selects other nodes they wish to connect to, as well as those they are willing to accept connections from. Formally, each node selects simultaneously a subset $F_i \subseteq V$ of nodes i is willing to accept connections from, and a subset $T_i \subseteq V$ of nodes i wishes to connect to. This is a generalization of Myerson's announcement game described in [4]. If $\mathbf{T} = (T_i, i \in V)$ and $\mathbf{F} = (F_i, i \in V)$ denote the composite strategy vectors, then an undirected link is formed between two nodes i and j if i wishes to connect to j (i.e., $j \in T_i$), and j is willing to accept a connection from i (i.e., $i \in B$). All edges that are formed in this way define the network topology $G(T, F)$ realized by the strategy vectors $\mathbf{T}$ and $\mathbf{F}$; i.e., $j \in T_i, i \in F_j$ implies that $ij \in G(\mathbf{T}, \mathbf{F})$.

From (16.1) we see that the utility of all nodes is defined by G and $\mathbf{P}$. The next question is how to determine the entries of the payment matrix $\mathbf{P}$ using the strategy vectors. If $i \in F_j$ and $j \in T_i$, then a contract is formed from i to j; we denote this contract by (i, j), and refer to the *directed* graph $\Gamma(\mathbf{T}, \mathbf{F})$ as the *contracting graph*. The contracting graph captures the directionality link formation: a link is only formed if one node asks for the link, and the target of the request accepts.

The contracting graph Γ and the network topology G define the transfers between the nodes. This is formally, defined by a *contracting function* $Q(i, j; G)$ that gives the payment in a contract from i to j when the network topology is G. If $Q(i, j; G)$ is negative, then j pays i. Given strategy vectors $\mathbf{T}$ and $\mathbf{F}$, the payment matrix $\mathbf{P}(\mathbf{T}, \mathbf{F})$ at the outcome of the game is given by

$$P_{ij}(\mathbf{T}, \mathbf{F}) = \begin{cases} Q(i, j; G(\mathbf{T}, \mathbf{F})) & \text{if } (i, j) \in \Gamma(\mathbf{T}, \mathbf{F}) \\ 0, & \text{otherwise} \end{cases} \tag{16.2}$$

Thus given strategy vectors $\mathbf{T}$ and $\mathbf{F}$, the payoff to node i is $U_i(G(\mathbf{T}, \mathbf{F}), \mathbf{P}(\mathbf{T}, \mathbf{F}))$. For simplicity, we will often use the shorthand $G = G(\mathbf{T}, \mathbf{F})$, $\Gamma = \Gamma(\mathbf{T}, \mathbf{F})$, and $\mathbf{P} = \mathbf{P}(\mathbf{T}, \mathbf{F})$ to represent specific instantiations of the network topology, contracting graph, and payment matrix, respectively, arising from strategy vectors $\mathbf{T}$ and $\mathbf{F}$. A triple $(G, \Gamma, \mathbf{P})$ arising from strategic decisions of the nodes will be referred to as a *feasible outcome* if there are strategy vectors $\mathbf{T}$ and $\mathbf{F}$ that give rise to $(G, \Gamma, \mathbf{P})$.

The contracting function could be defined so that a system designer or an external regulator require that nodes must have pre-negotiated tariffs which are encoded in the contracting function. In this case, the value of the contract is dependent on the surrounding network topology. A second option is that the contracting function does not assume the existence of the regulator; instead, we presume that the value of the contracting function is the outcome of a negotiation process. This negotiation takes place *holding the network topology fixed*; i.e., the negotiation is used to determine the value of the contract, given the topology that is currently in place. One example is simply that $Q(i, j; G)$ is the result of a Rubinstein bargaining game of alternating offers between and j, where i makes the first offer [5, 6]. We will investigate this example in further detail later.

The contracting functions allow design flexibility in a distributed setting. Instead of focusing on a particular contracting function, we will be interested in contracting functions exhibiting two natural properties: *monotonicity* and *anti-symmetry*. Given $j \neq i$, we define the *difference* in cost to node i between graph G and graph $G + ij$ as $\Delta C_i(G, ij) = C_i(G + ij) - C_i(G)$. If $ij \in G$, then $\Delta C_i(G, ij) = 0$. If G is a graph such that $ij \notin G$ and $ik \notin G$ then we say that the contracting function is *monotone* if

$$\Delta C_j(G, ij) > \Delta C_k(G, ik) \quad \text{if and only if}$$
$$Q(i, j; G + ij) > Q(i, k; G + ik)$$

We say that the contracting function Q is *anti-symmetric* if, for all nodes i and j, and for all graphs G, we have $Q(i, j; G) = -Q(j, i; G)$.

A contracting function that is anti-symmetric has the property that at any feasible outcome of the game, the payment for a link ij does not depend on which node asked for the connection.

16.1.1   Stability and Efficiency of the Game

In this segment, we use a game-theoretic notion of equilibrium, called *pairwise stability* [2]. The pairwise stability requires that no unilateral deviations by a single node are profitable, and that no bilateral deviations by any pair of nodes are profitable. From the system perspective we study the *efficiency* of the network as the performance criterion. We measure the efficiency of a network topology via the total value obtained by all nodes using that topology.

The simplest notion of equilibrium is the celebrated *Nash equilibrium*: a strategy profile $(\mathbf{T}, \mathbf{F})$ is a Nash equilibrium if no node i can make a profitable *unilateral deviation*, i.e., strictly improve its payoff by altering either or both of the sets T_i and F_i. However, the Nash equilibrium lacks sufficient predictive power in our model due to the presence of trivial equilibriums. For example, $F_i = T_i = \emptyset$ is a Nash equilibrium regardless of the cost structure or contracting function: no node can affect the outcome through a unilateral deviation, so no unilateral deviation is profitable. The inadequacy of the Nash equilibrium as a solution concept is well known for NFGs [3].

We consider a notion of stability that is robust to *bilateral* deviations, known as *pairwise stability* since link formation is inherently *bilateral*. The consent of two nodes is required to form a single link. The pairwise stability of a strategy vector requires that both no unilateral deviations are profitable and that no two nodes can collude to improve their payoff. Suppose that the current strategy vectors are $\mathbf{T}$ and $\mathbf{F}$, and the current network topology and contract graph are $G = G(\mathbf{T}, \mathbf{F})$ and $\Gamma = \Gamma(\mathbf{T}, \mathbf{F})$ respectively. Suppose that two nodes i and j attempt to bilaterally deviate; this involves changing the pair of strategies (T_i, F_i) and (T_j, F_j) together. Any deviation will of course change both the network topology, as well as the contract graph. However, we assume that any contracts present both before and after the deviation *retain the same payment*. This is consistent with the notion of a contract: unless the deviation by i and j entails either breaking an existing contract or forming a new contract, there is no reason that the payment associated to a contract should change. Based on this the formal definition of pairwise stability is:

Assume Q is a contracting function. Given strategy vectors $\mathbf{T}$ and $\mathbf{F}$, let $G = G(\mathbf{T}, \mathbf{F})$, $\Gamma = \Gamma(\mathbf{T}, \mathbf{F})$ and $\mathbf{P} - \mathbf{P}(\mathbf{T}, \mathbf{F})$. Given strategy vectors T' and F', define $G' = G(\mathbf{T}', \mathbf{F}')$ and $\Gamma' = \Gamma(\mathbf{T}', \mathbf{F}')$. Define P' according to

$$
P'_{k\ell} = \begin{cases} P_{kl}, & \text{if } (k,l) \in \Gamma' \text{and } (k,l) \in \Gamma; \\ Q(k, \ell; G'), & \text{if } (k,l) \in \Gamma' \text{and } (k,l) \notin \Gamma; \\ 0, & \text{otherwise} \end{cases} \tag{16.3}
$$

Then $(\mathbf{T}, \mathbf{F})$ is a *pairwise stable equilibrium* if: (1) No unilateral deviation is profitable, i.e., for all i, and for all **T'** and **F'** that differ from $\mathbf{T}$ and $\mathbf{F}$ (respectively) only in the $i'th$ components

$$
U_i(\mathbf{P}, G) \geq U_i(\mathbf{P}', G')
$$

and (2) no bilateral deviation is profitable, i.e., for all pairs i and j, and for all T' and F' that differ from T and F only in the $i'th$ and $j'th$ components

$$
U_i(\mathbf{P}, G) < U_i(\mathbf{P}', G') \Rightarrow U_j(\mathbf{P}, G) > U_j(\mathbf{P}', G')
$$

Note that (16.3) is a formalization of the discussion above. When nodes i and j deviate to the strategy vectors $\mathbf{T}$ and **F'**, all payments associated to preexisting contracts remain the same. If a contract is formed, the payment becomes the value of the contracting function given the new graph. Finally, if a contract is broken, the payment of course becomes zero. These conditions give rise to the new payment matrix **P'**. Nodes then evaluate their payoffs before and after a deviation. The first condition in the definition ensures no unilateral deviation is profitable, and the second condition ensures that if node i benefits from a bilateral deviation with j, then node j must be strictly worse off.

Regarding the pairwise stability of the network topology and contracting graph, we will say that a feasible outcome $(G, \Gamma, \mathbf{P})$ is a *pairwise stable outcome* if there exists a pair of strategy vectors $\mathbf{T}$ and $\mathbf{F}$ such that (1) $(\mathbf{T}, \mathbf{F})$ is a pairwise stable equilibrium; and (2) $(\mathbf{T}, \mathbf{F})$ give rise to $(G, \Gamma, \mathbf{P})$. Note that for all i and j such that $ij \in G$ we must have $P_{ij} = Q(i, j; G)$ in a pairwise stable outcome.

A useful property of pairwise stable outcomes can be formulated as: Let (G, Γ, P) be a pairwise stable outcome. Then for all nodes i and j, if $(i, j) \in \Gamma$ and $(j, i) \in \Gamma$, then $Q(i, j; G) = 0$ and $Q(j, i; G) = 0$ [8].

Given two feasible outcomes $(G, \Gamma, \mathbf{P})$ and $(G', \Gamma', \mathbf{P}')$, we say that $(G, \Gamma, \mathbf{P})$ Pareto dominates $(G', \Gamma', \mathbf{P}')$ if all players are better off in (G, Γ, P) than in (G', Γ', P'), and at least one is strictly better off. A feasible outcome is *Pareto efficient* if it is not Pareto dominated by any other feasible outcome. Since payoffs to nodes are *quasi linear* in our model, i.e., utility is measured in monetary units, it can be shown that an outcome $(G, \Gamma, \mathbf{P})$ is Pareto efficient if $G \in \arg\min\limits_{G'} S(G')$, where $S(G)$ is the *social cost function* $S(G) = \sum_{i \in V} C_i(G)$.

Such feasible outcomes are called *efficient*. The preceding condition does not involve the contracting function. Contracts induce zero-sum monetary transfers among nodes, and do not affect global efficiency. Given a graph G, we define the *efficiency* of G as the ratio $S(G)/S(G_{eff})$, where G_{eff} is the network topology in an efficient outcome.

16.1.2   Traffic Routing Utility Model

In this section we specify an NFG model where the nodes' utility depends on the traffic that is routed through the network. In particular, nodes extract utility per unit of data they successfully send through the network and experience per-unit routing costs when in the data network, as well as maintenance costs per adjacent link.

We assume that each user i wants to send one unit of traffic to each node in the network. We refer to this as a *uniform all-to-all* traffic matrix. We assume that given a network topology, traffic is routed along shortest paths, where the length of a path is measured by the number of hops. In the case of multiple shortest paths of equal length, traffic is split equally among all available paths. Each node faces three types of costs: *routing costs*, *link maintenance cost*, and *disconnection cost*. Let $f_i(G)$ be the total traffic that transits through i plus the total traffic received by i. We assume that node i faces a positive routing cost of per unit of traffic. Thus given a graph G, *the total routing cost* experienced by node i is $R_i(G) = c_i f_i(G)$. A maintenance cost of $\pi > 0$ is incurred by the endpoints of each link (the effective cost of a single link is 2π). Thus given a graph $G = (V, E)$, *the total link maintenance cost* incurred by node i is $M_i(G) = \pi d_i(G)$, where $d_i(G)$ is the degree of node i in the graph G. We assume that each node experiences a cost of $\lambda > 0$ per unit of traffic not sent because the network is not connected. Thus given a graph G, the cost to a node i from incomplete connectivity, or disconnection cost, is $D_i(G) = \lambda(n - n_i(G))$, where $n_i(G)$ is the number of nodes i can reach in the graph G. Thus the total cost to a node i in a graph G is

$$C_i(G) = R_i(G) + M_i(G) + D_i(G) \tag{16.4}$$

We now characterize pairwise stable outcomes, given the cost model [4]. Let $(G, \Gamma, \mathbf{P})$ be a pairwise stable outcome. Then G is a forest (i.e., all connected components of G are trees) [8].

The preceding relation shows the 'minimality' of pairwise stable graphs: since our payoff model does not include any value for redundant links, any pairwise stable equilibria must be forests. An interesting open direction for our model includes the addition of a utility for redundancy (e.g. for robustness to failures).

Most of the pairwise stable equilibria we discuss are framed under the following assumption on the disconnectivity cost λ.

Assumption 1: Given a contracting function Q, the disconnectivity cost $\lambda > 0$ is such that for all disconnected graphs G and for all pairs i and j that are disconnected in G, there holds $\Delta C_i(G, ij) + Q(i, j; G + ij) < 0$ and $\Delta C_i(G, ij) - Q(j, i; G + ij) < 0$.

This implies that if nodes i and j are not connected in G, then both are better off by forming the link ij using either the contract (i, j) or (j, i). If Q is anti-symmetric the second condition is trivially satisfied.

The preceding assumption is meant to ensure that we can restrict attention to connected graphs in our analysis. From our utility structure, it is easy to see that only the payments and disconnectivity costs act as incentives to nodes to build a connected network topology. But payments alone are not enough to induce connectivity, since of course the node paying for a link feels a negative incentive due to the payment. We emphasize that the preceding assumption is made assuming that *the contracting function and all other model parameters are given,* so that the threshold value of λ necessary to satisfy the preceding assumption may depend on these other parameters. We will see that this assumption has interesting implications for our model. It is clear from our model that if all other model parameters are fixed, then a λ satisfying the preceding assumption must exist.

Suppose that *Assumption 1* holds, and that Q is monotone. Let $(G, \Gamma, \mathbf{P})$ be a feasible outcome where G is a tree. Then $(G, \Gamma, \mathbf{P})$ is pairwise stable if and only if no pair of nodes can profitably deviate by simultaneously breaking one link and forming another [7].

In other words, given nodes i and j and any link $ik \in G$, let $G = G - ik + ij$, $\Gamma' = (\Gamma \setminus \{(i, k), (k, i)\}) \cup \{(i, j)\}$, and define $\mathbf{P}'$ as in (16.3). Then $U_i(\mathbf{P}, G) < U_i(\mathbf{P}', G') \Rightarrow U_j(\mathbf{P}, G) > U_j(\mathbf{P}', G')$.

It is important to note that without specifying all the parameters in the model, it is impossible to characterize further the set of pairwise stable outcomes.

For a full characterization of the set of efficient outcomes let $c_{\min}$ be the minimum per-unit routing cost, i.e., $c_{\min} = \min_{u \in V} c_u$. Let $V_{\min} \subseteq V$ be the set of nodes in V with per-unit routing cost of $c_{\min}$. Let $(G, \Gamma, \mathbf{P})$ be a feasible outcome of the static game. Then a) for all $0 < \pi < c_{\min}$, $(G, \Gamma, \mathbf{P})$ is efficient if and only if $G = K_n$; and b) for all $\pi > c_{\min}$, $(G, \Gamma, \mathbf{P})$ is efficient if and only if $G = T_{\min}$; where K_n is the complete network over n vertices, and $T_{\min}$ is any star centered around a node from $V_{\min}$ [8].

So the previous discussion shows that there exist parameter regimes under which all trees can be pairwise stable. Combined with the preceding arguments, we conclude there exist parameter regimes in which there exist inefficient pairwise stable equilibria: in particular, if π is large enough, then all trees are pairwise stable, but the only efficient graphs are stars centered at nodes of minimum routing cost.

To answer the question whether efficient graphs are pairwise stable, we start with the following definition.

A $c_{\min}$-*centered network* is a tree such that all internal nodes – i.e., all nodes that are not leaves–have minimum routing cost. Formally, if i is not a leaf, then $c_i \leq c_j$. for all $j \neq i$.

Suppose that *Assumption 1* holds and let $(G, \Gamma, \mathbf{P})$ be a feasible outcome such that the topology G is a $c_{\min}$-centered network. Then $(G, \Gamma, \mathbf{P})$ is pairwise stable [7].

16.1.3 Network Formation Game Dynamics

Myopic (short term) dynamics refer to the property of the dynamics that at any given round, nodes update their strategic decisions only to optimize their current payoff. This is in contrast to dynamics that consider some long-term objective.

We consider a discrete-time myopic dynamics that includes two stages at every round. At round k, both a node u_k and an edge $u_k v_k$ are *activated.* At the first stage of the round, with probability $p_d \in [0, 1]$, node u_k can choose to unilaterally break the edge $u_k v_k$ if it is profitable to do so; and, with probability $1 - p_d$, the link (and thus all contracts associated with) $u_k v_k$ is broken, regardless of node $u_k's$ preference. In the second stage, u_k selects a node w and proposes to form the contract (u_k, w) to w, with associated payment given by the contracting function. Node w then decides whether to accept or reject, and the dynamics then continue to the next round given the new triple of network topology, contracting graph, and payment matrix. It is crucial to note that $u_k's$ strategic decisions are made so that its utility is maximized *at the end of the round.* We contrast this with $w's$ strategic decision, which is made to maximize its utility at the end of the second stage given its utility at the end of the first stage.

We consider two variations on our basic model of dynamics: either $p_d = 1$, or $p_d < 1$. When $p_d = 1$, node u_k can choose to break either or both of the contracts associated with $u_k v_k$ (if they exist). When $p_d < 1$, provided all links are activated infinitely often, all links are broken infinitely often *regardless of the activated node's best interest.* For ease of exposition, unless otherwise stated, all the subsequent discussion assumes $p_d = 1$.

We call an *activation process* any discrete-time stochastic process $\{(u_k, v_k)\}_{k \in N}$ where the pairs (u_k, v_k) are i.i.d. random pairs of distinct nodes from V. A realization of an activation process is called an *activation sequence.* The activation process is said to be *uniform* if, for all k, u and v, $u \neq v$, the probability that $(u_k, v_k) = (u, v)$ is uniform over all ordered pairs. Thus $V [(u_k, v_k) = (u, v)] = 1/n(n - 1)$.

If (u_k, v_k) is the pair selected at the beginning of round k and $(G^{(k)}, \Gamma^{(k)}, \mathbf{P}^{(k)})$ is the state at the beginning of the round, in a single round k, our dynamics consist of two sequential stages, as follows: 1) If $u_k v_k \in G^{(k)}$, then node u_k decides whether to break the contract (u_k, v_k) (if it exists), the contract (v_k, u_k) (if it exists), or both. 2) Node u_k decides if it wishes to form a contract with another w_k. If it chooses to do so, then u_k asks to form the contract (u_k, w_k), and w_k can accept or reject. The contract is added to the contracting graph if w_k accepts the contract.

Node u_k takes actions in stages 1 and 2 that maximize its utility *at the end of the round*; in the case that no action can strictly improve node $u'_k s$ utility in a stage, we assume that u_k takes no action at that stage. At stage 1 node u_k only breaks (u_k, v_k) *and/or* (v_k, u_k) if a profitable deviation is anticipated to be possible at stage 2. At stage 2, node w_k accepts $u'_k s$ offer if this yields a higher utility to w_k than the state *at the beginning of stage 2.*

At stage 2, we only consider a very specific bilateral deviation between u_k and w_k. This is consistent with our discussion above. In general, at stage 2, there may be multiple choices w_k that maximize the utility of node u_k. This will be discussed later in this section.

The rules for updating the contracting graph $\Gamma^{(k+1)}$, at the end round k, are summarized in the form *Action(s) by* $u_k \rightarrow \Gamma^{(k+1)}$ as follows:

(a) breaks$(u_k, v_k) \rightarrow \Gamma^{(k)} \setminus \{(u_k, v_k)\}$; (b) breaks$(v_k, u_k) \rightarrow \Gamma^{(k)} \setminus \{(v_k, u_k)\}$

(c) *Adds* $(u_k, w_k) \rightarrow \Gamma^{(k)} \cup \{(u_k, w_k)\}$; (d) breaks (u_k, v_k)
and $(v_k, u_k) \rightarrow \Gamma^{(k)} \setminus \{(u_k, v_k), (v_k, u_k)\}$; (e) breaks (u_k, v_k) and

adds $(u_k, w_k) \rightarrow \Gamma^{(k)} \setminus \{(u_k, v_k)\}) \cup \{(u_k, w_k)\}$

The first three actions (a), (b) and (c) are the basic actions the first node of the selected pair can do during a round. The last two actions (d) and (e) are compositions of two of the basic actions.

We define $G^{(k+1)}$ to be the associated network topology: i.e., $ij \in G^{(k+1)}$ if and only if either $(i, j) \in \Gamma^{(k+1)}$ or $(j, i) \in \Gamma^{(k+1)}$ (or both). In all cases, the payment vector $\mathbf{P}^{(k+1)}$ is updated as in (16.3), first after *stage 1*, and then after *stage 2*. The nodes are, interested in the maximization of the utility after stage 2.

In general the *state* of the dynamics at round k, $(G^{(k)}, \Gamma^{(k)}, \mathbf{P}^{(k)})$, need not be a *feasible outcome* because the payment matrix may not be consistent with the current contracting graph. When contracts are updated, only payments associated to the added or deleted contracts are updated–all other payments remain the same. This motivates the following definition which will be used when discussing feasibility of outcomes.

Let $(G, \Gamma, \mathbf{P})$ *be a triple consisting of a (undirected) network topology, a (directed) contracting graph, and a payment matrix. We say that the edge ij is adapted in $(G, \Gamma, \mathbf{P})$ if the following conditions hold:*

1. *If* $(i, j) \in \Gamma$, *then* $P_{ij} = Q(i, j; G)$; *otherwise* $P_{ij} = 0$.
2. *If* $(j, i) \in \Gamma$, *then* $P_{ji} = Q(j, i; G)$; *otherwise* $P_{ji} = 0$.
3. $ij \in G$ *if and only if either* $(i, j) \in \Gamma$ *or* $(j, i) \in \Gamma$.

If every edge ij is adapted to $(G, \Gamma, \mathbf{P})$, then $(G, \Gamma, \mathbf{P})$ must be a feasible outcome. Further, note that if the initial state of our dynamics was a feasible outcome, then condition 3 of the preceding definition is satisfied in every round.

The following definition captures convergence.

Given any initial feasible outcome $(G^{(0)}, \Gamma^{(0)}, \mathbf{P}^{(0)})$ and an instance of the activation process AP, we say the dynamics *converge* if there exists K such that for $k > K$

$$(G^{(k+1)}, \Gamma^{(k+1)}, \mathbf{P}^{(k+1)}) = (G^{(k)}, \Gamma^{(k)}, \mathbf{P}^{(k)})$$

The dynamics converge uniformly if for every $\varepsilon > 0$ there exist a K such that

$$Pr[(G^{(k+1)}, \Gamma^{(k+1)}, \mathbf{P}^{(k+1)}) = (G^{(k)}, \Gamma^{(k)}, \mathbf{P}^{(k)}), \ \forall k \geq K] \geq 1 - \epsilon$$

where the probability is taken w.r. t. the AP.

16.2 Knowledge Based Network Formation Games

In this section we discuss distributed topology control algorithms that combine network-formation games with machine learning. The algorithms rely on game players to pursue selfish actions through low-complexity greedy algorithms with low or no signaling overhead.

Convergence and stability are ensured through proper mechanism design that eliminates infinite adaptation process. The framework also includes game-theoretic extensions to influence behavior such as fragment merging and preferring links to weakly connected neighbors. Knowledge acquisition (learning) allows adaptations that prevent node starvation, reduce link flapping, and minimize routing disruptions by incorporating network layer feedback in cost/utility tradeoffs.

This framework is particularly suitable to cognitive radio networks because it can be extended to handle heterogeneous users with different utility functions and conflicting objectives [10].

16.2.1 System Model

Topology control consists of two processes: topology creation and topology maintenance. Topology creation includes a process of neighbor discovery. We assume that nodes have the capability to add or delete individual links, e.g., by using directional antennas.

Some of the algorithms to be described in this section rely on having a more global, connected component view of network connectivity. Such a view can for example be constructed by propagating and merging local neighborhoods. This process already occurs as a side function of some routing protocols. Having a global view of the network topology available, nodes can determine which paths to use in order to route *packets/messages*. This map also helps verify properties like k-connectivity. We present several different algorithms, based on game theory using local or global information, and with little overhead.

16.2.2 Network Formation as a Non-Cooperative Game

To formally define our topology control games, suppose we are given a network topology in the form of a graph $G = (V, E)$. A strategic form game $\Gamma(N, A = (C_i)_{i \in N}, U = (u_i)_{i \in N})$ is defined as follows: the set of players is the set of nodes V, $N = |V|$. The action space of the nodes (players) consists of the ability to propose, accept or reject edges incidental to themselves. Let $q_i \in C_i, i = 1, \ldots, N$ refer to the actions of player/node i (in terms of proposing, accepting or rejecting specific edges in G). We will refer to the actions of all other players not including i via the vector $q_{-i}, i = 1, \ldots, N$. In general, the payoff for a player depends on the total realized graph $G = (V, E)$ resulting from the players' combined actions $(q_i C_i, i = 1, \ldots, N)$. In the following, we will often write the utility functions as a function of this graph, rather than as a function of the combined actions. Suppose the utility function for node i is defined as $u_i(G) = g_i(G) - c_i(G)$.

As an example, let gain $g(G) = M$ if the network is connected, $g(G) = 0$ otherwise, and $c_i(G) = \alpha \delta_i$ where δ_i is the degree of node i, or $c_i(G) = \alpha \sum_{e \in E, e \sim i} w_e$, the 'weighted' degree of node i. We require that M is chosen large enough such that $M \geq \max_i c_i(G)$. We can also modify the benefit function to include the number of nodes in the same connected component of G as node i (the number of nodes that can be reached from node i, possibly in multiple hops), denoted $g_i(G) = M \cdot f_i(G)$. This can be helpful if the topology is not (initially) connected.

16.2.3   Network Formation as a Cooperative Game

We can also view the topology control network formation process as a cooperative (coalitional) game, where the connected nodes form the coalitions, which are perhaps the outcome of noncooperative bargaining (negotiations). A game in coalitional form is a set of N players, together with a characteristic (value) function v that assigns a value $v(S)$, or 'worth', to every coalition (subset) S, and satisfies $v(\emptyset) = 0$ and $v(S \cup T) \geq v(S) + v(T)$ whenever S and T are disjoint subsets of $1, \ldots, N$. Since the function v is super-additive, by assumption cooperation can only help, but never hurt. We can furthermore allow side-payments between players, specified in contracts. The formed links then correspond to contracts in effect, as discussed in the previous section. In the setting of topology control discussed in this section, let $G = (V, E)$ be the current topology without link $\{i, j\}$. Then the *incremental utility* of adding link $\{i, j\}$ is $[v(\{i, j\}; G) - v(\{i\}; G) - v(\{j\}; G)]$. If $v(\{i\}; G)$ is lower than $v(\{j\}; g)$, node i has more incentive in establishing link $\{i, j\}$ than node j. A scheme that divides the *incremental utility* evenly transfers some marginal utility of link $\{i, j\}$ from node i to node j. This may entice node j to accept to form link $\{i, j\}$. More details on coalition games will be presented later in this chapter.

16.2.4   Dynamic Network Formation and Topology Control

In the remainder of this section we assume that nodes discover their neighbors dynamically, according to the general update schedule.

Suppose there are N nodes (players). For each node i there is an (infinite) sequence $\{T_k^i, i = 1, \ldots, N, k = 1, 2, 3, \ldots\}$. At time T_k^i, node i uses its best response (e.g. power update) to (the result of) the policies used by all other players up to (just before) time T_k^i.

Here we assume that the times T_k^i are distinct. One natural way in which this condition is (approximately) satisfied in MANETs, is if the nodes use a random access scheme. Another way in which this occurs is through natural node mobility.

Upon discovery of a neighbor, a quality measure for the possible link to that neighbor is available in the form of a single number, e.g. a link utility. If the quality depends on multiple different attributes, then we assume the node has a ranking on the bundles of attributes. Whether nodes establish (additional) links depends on the marginal utility that the new link provides, given the currently established links. We next give a number of examples for defining the (marginal) utility.

16.2.5   Greedy Utility Maximization

We describe the utility functions in terms of the utility increment of an additional link. First, we describe a utility function which models the effect of diminishing returns of additional links: more links add reliability, but are also costly to maintain. To model this effect, we can define the utility increment of the (additional) link when the current degree of node i equals $\delta - 1$ as $\Delta u_i(\delta) = 1 - \alpha \delta$, $\alpha = 1/\delta_{\max}$.

The competing effects of adding reliability and increasing cost determine the value of $\delta_{\max}$. We could model the reliability effect explicitly with formula $\alpha \delta = -C \log(p^\delta) = \delta [E \log(1/p)$, where p is the link failure probability, and C is a weighting parameter. The full utility function is: $u_i(\delta) = \int \Delta u_i(\delta)d\delta = u_i(\delta) = \delta(1 - \alpha\delta/2)$ which takes its maximum at $\delta = \lfloor 1/\alpha \rfloor$ equal $u_i(\delta) = \lfloor 1/\alpha \rfloor (1 - \alpha \lfloor 1/\alpha \rfloor /2)$.

The utility incremental function does not take into account relative quality of links, nor connectivity properties. It treats all other nodes equally, irrespective of link quality or degree of the other node. In fact, if the possible links arrive in any order, then the first δ_{max} links are accepted, and any further ones are rejected until one of the existing links δ_{max} is broken/delited. This has the benefit of allowing participation of weak nodes, but the drawback of perhaps rejecting better link candidates. A choice of higher x allows both weak and strong links to be included, and increases the likelihood of overall network connectivity and reliability, at the cost of greater link maintenance cost.

16.2.6   Non-Cooperative Link Utility Maximization

The utility of (potential) link $\{i, j\}$, denoted as b_{ij} can be related to bandwidth, or to expected traffic load between i and j. We can define a new link utility incremental function, given there are $\delta - 1$ existing links, as $\Delta u_i(\delta) = b_{ij} - \alpha\delta b_{ij} = (1 - \alpha\delta)b_{ij}, \alpha = 1/\delta_{max}$.

This utility incremental function takes into account link quality, but assumes existing links are not deleted unless they fail. Furthermore, it does not take into account the connectivity /degree of neighboring node j. Therefore, weak nodes and new nodes which have no connectivity can get starved (especially when they have weak links when first discovered).

16.2.7   Cooperative Link Utility Maximization Via Utility Transfer

Consider the scenario in Figure 16.1, where Node 1 has *potential* links to Nodes 2, 3, and 4. Suppose that achievable bandwidth as indicated by the neighbor discovery protocol is $b_{14} > b_{13} > b_{12}$. Non-cooperative utility maximization will select link 1–4 to be established first, followed by link 1–3 and then link 1–2. However, it is Node 2 that has the fewest links and, consequently, the greatest utility increment when opening new links. If Node 1 reaches its maximum configured degree after establishing link 1-4 and link 1-3, Node 2 may remain weakly connected. To correct such starvation, a weak new node j may accept to transfer some of its utility for link $\{i, j\}$ (the first links that the new nodes establish are the most valuable).

A fair cost for the link is the Rubinstein bargaining value [7,9,10], which splits the increment in utility of node j relative to node i for link $\{i, j\}$ evenly between nodes i and j.

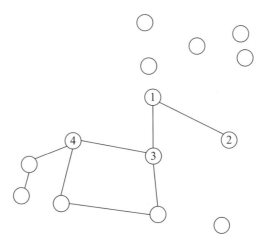

Figure 16.1   Utility transfer scenario.

16.2.8   Preferential Attachment with Knowledge of Component Sizes

If global topology information is available in the form of the graph $G = (V, E)$, then we can assess whether two nodes i and j are in the same connected component or not. Suppose node i and node j are in different components of G, with sizes $f_i(G)$ and $f_j(G)$ respectively. Then the utility increment of adding a link from node i to node j for node i follows from Section 16.2.5 with $g_i(\{i, j\}; g) = Mf_j(G)$ and $c_i(\{i, j\}; G) = \alpha Mf_j(G)\delta_i$ as $\Delta u_i(\{i, j\}; G) = Mf_j(G)(1 - \alpha\delta_i)$.

The utility increment for node j is $\Delta u_j(\{i, j\}; G) = Mf_i(G)(1 - \alpha\delta_j)$. If node i and node j are in the same connected component of G, then the marginal utility of adding a link $\{i, j\}$ between i and j reduces to $\Delta u_i(\{i, j\}; G) = (1 - \alpha\delta_i)$, $\alpha = 1/\delta_{max}$. To take into account specific link (throughput) utility, $\Delta u_i(\{i, j\}; G) = Mf_j(G)(1 - \alpha\delta_i)$ now becomes $\Delta u_i(\{i, j\}; G) = Mf_j(G)(1 - \alpha\delta_i)b_{ij}$ if node i and node j are in different components of G and $\Delta u_i(\{i, j\}; G) = (1 - \alpha\delta_i)b_{ij}$, $\alpha = 1/\delta_{max}$ otherwise.

Under the condition that M is chosen large enough so that increasing the number of reachable nodes is always preferred over choosing higher throughput, for example $M \geq \max_i c_i(G) = (\alpha\delta_i - 1)b_{ij}$, then the greedy, best response selection rule, i.e., the rule which picks $\max_{\exists\{i,j\}_{(t), t \in [t_0, t_0+T_s]}} \{\Delta u_i(\{i, j\}; G)\}$ favors connecting to the largest different component of G.

As an example in Figure 16.2, Node 1 has potential links to nodes within its own fragment as well as to Nodes 2 and 3 in other fragments. The rules above will encourage Node 1 to establish actual links to Node 2 or Node 3 before considering links within its own fragment, thereby healing network fragmentation.

16.2.9   Preferential Attachment with Knowledge of Neighbor Degrees

The utility increment defined in the previous section relies on having knowledge of the global topology G. However, the only information about G needed are the sizes $f_i(G)$ and $f_j(G)$ of the components containing i and j, respectively, as well as the degrees δ_i resp. δ_j of these nodes. If $f_i(G)$ and $f_j(G)$ are unavailable, we may approximate these by $(\delta_i + 1)$ and $(\delta_j + 1)$ respectively, as long as we can still

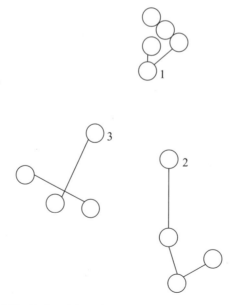

Figure 16.2   Preferential attachment to merge disconnected fragments.

assess if i and j are part of different connected components or not giving $\Delta u_i(\{i, j\}; G) = M(\delta_j + 1)(1 - \alpha\delta_i)b_{ij}$ if i and j are in different components of G, and $\Delta u_i(\{i, j\}; G) = (1 - \alpha\delta_i)b_{ij}$ if i and j are in the same component of G. This is a local preferential attachment scheme that favors connecting nodes from (large) previously disjoint components.

16.2.10   LADA: Link Addition/Deletion Algorithm

In the algorithm that each node uses to add or delete incident links, first node i discovers candidate links during a selection time window of length T_s. To select candidate links to add, it iteratively finds links which maximize $\max\limits_{\exists\{i,j\}_{(t)}, t\in[t_0, t_0+T_s]} \{\Delta u_i(\{i, j\}; G)\}$. If T_s is small, then the links basically arrive and are accepted one-by-one, as long as doing so still increases total utility. If the utility increment of each candidate link is negative, then in principle, we could stop. However, a node may still have an incentive to add a new link at the expense of the weakest current link. The weakest current link is also the link $\{i, j\}$, such that adding the link right back after deleting it would add the smallest utility increment $\min\limits_{\exists\{i,j\}_{(t)}, t\in[t_0, t_0+T_s]} \{\Delta u_i(-\{i, j\}; G)\}$.

16.3   Coalition Games in Wireless Ad Hoc Networks

In this section, we present a framework for the analysis of self organized distributed coalition formation process for spectrum sharing in an interference channel for large scale ad hoc networks. In this approach a concept of coalition clusters within the network is used where mutual interdependency between different clusters is characterized by the concept of spatial network correlation. Then by using stochastic models of the process we give up some details characteristic for coalition game theory in order to be able to include some additional parameters for network scaling. The applications of this model are: (a) Estimation of average time τ to reach grand coalition and its variance σ_τ^2 through closed form equations. These parameters are important in designing the process in a dynamic environment; (b) Dimensioning the coalition cluster within the network; (c) Modeling the network spatial correlation characterizing mutual visibility of the interfering links; (d) Modeling of the effect of the new link activation/inactivation on the coalition forming process; (e) Modeling the effect of link mobility on the coalition forming process. Prior to introducing the specific models for dealing with the problems above we summarize in the next section the applications of coalition games in wireless networks in general.

16.3.1   Applications of Coalition Games in Wireless Networks

Cooperation as a new networking paradigm has been used to improve the performance of the physical layer [11, 12] up to the networking layers [13]. Coalitional games have proved to be a very powerful tool for designing fair, robust, practical, and efficient cooperation strategies in communication networks.

A part of the research in this field is restricted to applying standard coalitional game models and techniques to study very limited aspects of cooperation in networks. This is due mainly to lack of the literature that tackles coalitional games. In fact, most pioneering game theoretical reference works, such as [14–16], focus on noncooperative games, touching slightly on coalitional games in a few chapters.

For most of Section 16.3, following upon Section 16.3.1, we present the modeling of coalition games for spectrum sharing in an interference channel for applications in large scale wireless ad hoc networks.

A coalitional game is defined formally by the pair (N, v), where N is the set of players and v is the coalition value. The most common form of a coalitional game is the characteristic form, whereby the value of a coalition S depends solely on the members of that coalition, with no dependence on how the players in $N \setminus S$ are structured. The characteristic form was introduced, along with a category of coalitional games known as games with transferable utility (TU), by Von Neumann and Morgenstern [17].

Examples of utility transfer were discussed in the previous section. The value of a game in characteristic form with TU is a function over the real line defined as v. This characteristic function associates with every coalition S a real number quantifying the gains of S. The TU property implies that the total utility represented by this real number can be divided in any manner between the coalition members. The values in TU games are thought of as monetary values that the members in a coalition can distribute among themselves using an appropriate fairness rule (one such rule being an equal distribution of the utility). The amount of utility that a player i receives from the division of $v(S)$ constitutes the player's payoff.

In a coalitional game with nontransferable utility (NTU), the payoff that each player in a coalition S receives is dependent on the joint actions that the players of coalition S select. The action space depends on the underlying noncooperative game [18]. The value of a coalition S in an NTU game, $v(S)$, is no longer a function over the real line, but a set of payoff vectors, where each element of a vector represents a payoff that player i can obtain within coalition S given a certain strategy selected by i while being a member of S. Given this definition, a TU game can be seen as a particular case of the NTU framework [14].

The class of canonical coalitional games, is the most popular category of games in coalitional game theory. The main requirements for classifying a game as canonical, are as follows:

1. The coalitional game is in characteristic form (TU or NTU).
2. Cooperation, i.e. the formation of large coalitions, is never detrimental to any of the involved players. Hence, in canonical games no group of players can do worse by cooperating, i.e. by joining a coalition than by acting noncooperatively. This pertains to the mathematical property of superadditivity.
3. The main objectives of a canonical game are to study the properties and stability of the grand coalition, i.e. the coalition of all the players in the game, and to study the gains resulting from cooperation with negligible or no cost, as well as the distribution of these gains in a fair manner to the players.

The use of canonical games for the study of rate allocation in multiple access channels (MAC), within communication networks is presented in [11, 19–21]. The models tackle the problem of how to allocate the transmission rates fairly between a number of users accessing a wireless Gaussian MAC channel. In this model, the users are bargaining for obtaining a fair allocation of the total transmission rate available. Every user, or group of users (coalition), that does not obtain a fair allocation of the rate can threaten to act on its own, which can reduce the rate available for the remaining users. The game is modeled as a coalitional game defined by (N, v) where N is the set of players, i.e., the wireless network users that need to access the channel, and v is the maximum sum-rate that a coalition S can achieve.

In [22], canonical games are used for studying the cooperation possibilities between single antenna receivers and transmitters in an interference channel. The model considered in [22] consists of a set of transmitter-receiver pairs, in a Gaussian interference channel. The authors study the cooperation between the receivers under two coalitional game models: a TU model where the receivers communicate through noise-free channels and jointly decode the received signals, and an NTU model where the receivers cooperate by forming a linear multiuser detector (in this case the interference channel is reduced to a MAC channel). Further, the authors study the transmitters' cooperation problem under perfect cooperation and partial decode and forward cooperation, while considering that the receivers have formed the grand coalition. Since all the considered games are canonical (as we will see later), the main interest is in studying the properties of the grand coalitions for the receivers and the transmitters.

In [22], canonical coalitional games are used to solve an inherent problem in packet forwarding ad hoc networks where the users that are located in the centre of the network, known as backbone nodes, have a mutual benefit to forward each others' packets. In contrast, users located at the boundary of the network, known as boundary nodes, are not helped by the backbone nodes due to the fact that the backbone nodes do not need the help of the boundary nodes at any time. Hence, in such a setting, the boundary nodes end up having no way of sending their packets to other nodes, and this is a problem known as the curse of the boundary nodes. In [22], a canonical coalitional game model is proposed between a player set N which includes all boundary nodes and a single backbone node. In this model, forming a coalition provides the following benefits:

1. By cooperating with a number of boundary nodes and using cooperative transmission, the backbone node can reduce its power consumption.
2. In return, the backbone node agrees to forward the packets of the boundary nodes. For cooperative transmission, in a coalition S, the boundary nodes act as relays while the backbone node acts as a source.

Beyond packet forwarding, many other applications such as in [23, 24], and [25] utilize different techniques for studying the grand coalition in a variety of communications applications.

In a coalition formation game, one must answer the question '*how to form a coalitional structure that is suitable to the studied game*'. In addition, the evolution of this structure is important, notably when changes to the nature of the game can occur due to external or internal factors (e.g., what happens to the coalition structure if one or more players leave the game). In many applications, coalition formation consists of finding a coalitional structure that either maximizes the total utility (social welfare) if the game is TU, or finding a structure with Pareto optimal payoff distribution for the players if the game is NTU. For achieving such a goal, a centralized approach can be used; however, such an approach is generally NP-complete [26–29]. The reason is that finding an optimal partition requires iterating over all the partitions of the player set N. The number of partitions of a set N grows exponentially with the number of players in N and is given by a value known as the Bell number [26].

In fact, the complexity of the centralized approach as well as the need for distributed solutions have sparked a huge growth in the coalition formation literature that aims to find low complexity and distributed algorithms for forming coalitions [26–29]. The approaches used for distributed coalition formation are quite varied and range from heuristic approaches [26], Markov chain-based methods [27], to set theory-based methods [28], as well as approaches that use bargaining theory or other negotiation techniques from economics [29]. Although there are no general rules for distributed coalition formation, a work such as [28] provides generic rules that can be used to derive application-specific coalition formation algorithms. Although [28] does not construct a coalition formation algorithm explicitly, the mathematical framework presented can be used to develop such algorithms.

Reference work [28] presents two main rules for forming or breaking coalitions, referred to as merge and split . The basic idea behind the rules is that, given a set of players N, any collection of disjoint coalitions $(S1, \ldots, Sl)$ can agree to merge into a single coalition G, if this new coalition G is preferred by the players over the previous state depending on the selected comparison order. Similarly, a coalition S splits into smaller coalitions if the resulting collection $(S1, \ldots, Sl)$ is preferred by the players over S.

The formation of virtual MIMO systems through distributed cooperation has received increasing attention recently (see [12] and [30] and the references therein). The problem involves a number of single antenna users that cooperate and share their antennas to benefit from spatial diversity or multiplexing, and hence form a virtual MIMO system. Most of the literature that has studied the problem is either devoted to analyzing the link level information theoretical gains from distributed cooperation, or focused on assessing the stability of the grand coalition, for cooperation with no cost, such as in [12] described previously. However, there is a lack of literature that studies how a network of users can interact to form virtual MIMO systems, notably when there is a cost for cooperation. Hence, a study of the network topology and dynamics that result from the interaction of the users is needed and, for this purpose, coalition formation games are quite an appealing tool. These considerations motivated the work in [30] where a network of single antenna transmitters that send data in the uplink of a TDMA system to a receiver with multiple antennas is considered. In a noncooperative approach, each single antenna transmitter sends its data in an allocated slot. For improving their capacity, the transmitters can interact for forming coalitions, whereby each coalition S is seen as a single user MIMO that transmits in the slots that were held previously by the users of S. After cooperation, the TDMA system schedules one coalition per time slot.

In cognitive radio networks, the unlicensed secondary users (SUs) are supposed to sense the environment to detect the presence of the licensed primary user (PU) and transmit during periods where the PU is inactive. Collaborative spectrum sensing (CSS) has been considered [31] for improving the sensing performance of the SUs, in terms of reducing the probability of missing the detection of the PU (probability of miss), hence decreasing the interference on the PU and simultaneously controlling the probability of detecting falsely that the PU is transmitting . Even though CSS decreases the probability of miss, it also increases the false alarm probability, i.e. the probability of detecting falsely that the PU is transmitting. Therefore, CSS presents an inherent trade off between reducing the probability of miss (reducing interference on the PU) and maintaining good false alarm probability, which corresponds to good spectrum utilization.

The coalition formation games have already been applied in [32] to improve the physical layer security of wireless nodes through cooperation among the transmitters while in [33] coalition formation among a number of autonomous agents, such as unmanned aerial vehicles, is studied in the context of data collection and transmission in wireless networks.

In canonical and coalition formation games, the utility or the value of a coalition does not depend on how the players are interconnected within the coalition. However, it has been shown that, in certain scenarios, the underlying communication structure between the players in a coalitional game can have a major impact on the utility and other characteristics of the game [34, 35]. By the underlying communication structure, we mean the graph representing the connectivity of the players among each other, i.e. which player communicates with which one inside each and every coalition which we refer to as *coalitional graph games*.

The idea of having a value dependent on a graph of communication between the players was first introduced by Myerson in [34], through the graph function for TU games. In [14, Chapter 9.5], using the Myerson framework of [34], an extensive form game is proposed for forming the network graph.

For forming the graph, a broad range of approaches exist, and are grouped into two types: myopic and far sighted (these approaches are sometimes referred to as dynamics of network formation [36]). The main difference between these two types is that, in myopic approaches, the players play their strategies given the current state of the network, while in far sighted algorithms, the players adapt their strategy by learning, and predicting future strategies of the other players [37–39].

The latest researches in wireless networks consider introducing a new node, the relay station (RS) for improving the network's capacity and coverage. The introduction of the RS impacts the network architecture. For example, WiMAX networks such as the mesh network is replaced by a tree architecture that connects the base station (BS) to its subordinate RSs. Reference work [40] models the problem of the uplink tree formation in 802.16j using coalitional graph games, namely network formation games.

In [41] the application of network formation games in 802.16j was extended and the algorithm was adapted to support the tradeoff between improving the effective throughput by relaying and the delay incurred by multi-hop transmission, for voice over IP services in particular.

In [42] a stochastic approach for network formation is provided. In this model, a network of nodes that are interested in forming a graph for routing traffic among them is considered. Each node in this model aims at minimizing its cost function that reflects the various costs that routing traffic can incur (routing cost, link maintenance cost, disconnection cost, etc.).

The usage of the network formation games in routing applications is not solely restricted to forming the network, but also for studying properties of an existing network. For instance in [43] the authors study the stability and the flow of the traffic in a given directed graph. For this purpose, several concepts from network formation games such as pairwise stability are used. In addition, the work in [43] generalizes the concept of pairwise stability making it more suitable for directed graphs.

In the next section we present, in more detail, modelling of coalition games for spectrum sharing in interference channel for applications in large scale wireless ad hoc networks.

16.3.2 Stochastic Model of Coalition Games for Spectrum Sharing in Large Scale Wireless Ad Hoc Networks

The network consists of a number of links whose transmitters and receivers are located randomly within a certain area A. In the case when all links transmit simultaneously, the signal received by a reference receiver r is given as

$$y_r = h_{rr} x_r + \sum_{i \neq r} h_{ri} x_i + n_r \tag{16.5}$$

where the second term in right hand side of (16.5) is interference received from other transmitters. SINR and its equivalent maximum transmission rate are calculated as,

$$SINR_r = \frac{g_{rr} P_r}{\sum_{i \neq r} g_{ri} P_i + N_0} \tag{16.6}$$

$$R_r = \frac{1}{2} \log_2 (1 + SINR_r) \tag{16.7}$$

where P_i is the transmitted power from transmitter of link i, g_{ir} the channel gain between the transmitter of link i and receiver of link r and N_0 is the receiver background noise power. In this scenario a given transmitter can *spread* its signal over entire available bandwidth or *share* the bandwidth with other users in the network in certain proportion. For the convenience of the presentation let us assume at the beginning that the mutual channel gains are known. This assumption has significant impact on the coalition forming process and for this reason will be reconsidered later.

For the above assumption for communication between the users i and j during the process of coalition negotiation some signalling capacity $R_{ij} = R_c$ is required. In the literature the signalling network is referred to as underlying network [23, 24] and the system performance depends significantly on its capacity. In general more sophisticated algorithms, leading faster to optimum operation of the network, would also require more information to be exchanged between the users which consequently require more signalling capacity.

If the interfering level from other users is too high, the reference user r may consider proposing the coalition $S(r, j)$ to a given user j which consists of sharing the spectra in certain proportion.

Again, for simplicity, let us assume that the spectra will be shared evenly (one half for each). The extension to the case of so called fair partitioning of the value of the game where the pay off to the users is proportional to their contribution to the value of the game (Shapley value) [14] is straightforward. Under these conditions we have

$$SINR_{r \in S} = \frac{g_{rr} P_r}{\sum_{i \in S^c} g_{ri} P_i + N_0} \tag{16.8}$$

$$R_{r \in S} = \frac{1}{2 \times 2} \log_2 (1 + SINR_{r \in S}) - R_S \tag{16.9}$$

where $R_S = R_c$ and S^c is the complementary set of users not belonging to coalition S. The signal to interference to noise ratio is now improved because the two users in the coalition do not transmit simultaneously, but the user can use only a half of the available spectra (or time slots). The condition for the coalition formation is that the both users are benefiting from such coalition which can be formally expressed as

$$R_{r \in S} \geq R_r \tag{16.10}$$

This simple concept will be elaborated later in more detail.

16.3.2.1 Distributed Stochastic Interference Channel Model

In this section, we generalize the previous concept in order to be used in large scale networks. Like in any other generalization, we will have to give up some detail in the system modelling, characteristic for coalition game theory, in order to be able to include some additional system parameters relevant to network scaling. We will assume that every user in the network identifies N as the strongest interferers (seen by its receiver) and initiate process to establish coalition with these links in order to reduce the mutual interference. The deviation from this assumption to the case where the user is aware only of the aggregate interference would be to initiate the coalition forming process with randomly chosen interferer.

In *fully correlated network (FCN)*, the same set of interfering signals is observed by all receivers. The strength of the interfering signals observed with different receivers is uncorrelated. This models the case where the cluster size N equals the overall number of terminals in the network M.

In *partially correlated network (PCN)*, with correlation

$$\rho = \left(1 - \frac{a}{N}\right) \tag{16.11}$$

the transmitter of specific user i, whose receiver identifies N interfering links, is not observed by a receiver from the set of the interfering links. The strength of the interfering signals observed with different receivers is again uncorrelated. In a real network, in every particular time instance, parameter a might be different for each link but in average the statistical channel model assumes that parameter a is the same for each link. In a fully connected one hop network with M terminals and cluster size N with $M>N$, probability distribution function $p(a)$ can be expressed as

$$p = \left(1 - \frac{N}{M}\right) \tag{16.12}$$

$$p(a) = \binom{N}{a} p^a (1 - p)^{N-a} \tag{16.13}$$

In (16.12), p represents the probability that for the two links i and j, the receiver of link i observes the interference from transmitter of link j but the receiver of link j does not see the interference from transmitter of link i.

16.3.2.2 Initiation of Coalition Formation

Random Initiation
In this case the order in which users are initiating coalition formation is random and the initiations are mutually independent and uncorrelated. This can be implemented either in centralized or distributed way (by running a Bayesian test). On wining the Bayesian test a specific user may take different action depending on the knowledge of the channel gains that will be discussed further in the next section.

Round Robin Initiation
At the association with the network, users my be allocated an ID so that the right for initiation of the coalition formation process can be granted in round robin fashion to the users currently associated with the network.

16.3.2.3 Coalition Formation Protocol (CFP)

Once getting the right to initiate coalition formation process, and known channel coefficients, the user i proposes the coalition formation to the link j with the interfering level

a) higher than the level of its useful signal;
b) maximum interfering level;

c) interfering level with no more than δ positions below its own received signal level on the ordered list of the received signals;

d) to the arbitrary chosen link (when only the aggregate level of interference is known).

The interfering link j will accept the coalition if the interfering level from the user i is

a) higher than the level of its useful signal;

b) its maximum interfering level;

c) no more than δ positions bellow the level of the received signal level on the ordered list of received signals;

d) the interfering link j accepts the coalition conditionally, initiates the channel sharing mode, checks the effects of the coalition forming and confirm or reject the coalition based on the experienced value obtained by the coalition.

16.3.3   CFP Modeling

The graph presentation of the CFP is shown in Figure 16.3. Assume there are M available links and the process starts with a group of N randomly selected users $G(N)$ in singleton status where $N \ll M$. A given link i creates an ordered list of interferers.

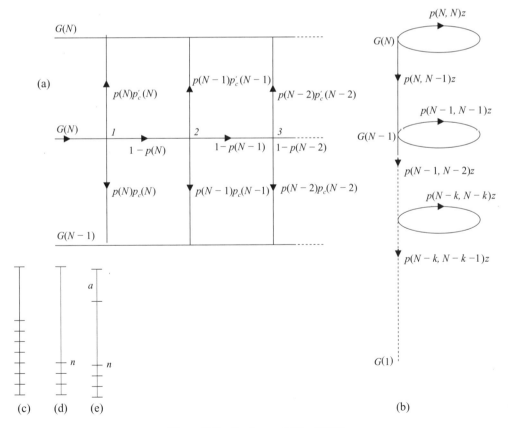

Figure 16.3   Graph presentation of CFP.

$L(i) = \{l(i, j)\} = \{l(i, 1), l(i, 2), \ldots, l(i, N)\}$ including its own signal by allocating the lowest index $j = 1$ to the weakest signal level. Node 1, on graph in Figure 16.3, represents the case when the useful signal level is the lowest on the list i.e. $l(i) = l(i, 1)$. In the statistical interference channel model, this will happen with probability $p_1 = p(N) = 1/N$. Under the condition represented by node 1 and option (a) in Section 16.3.2 the link will propose the coalition formation to the link k with the interfering level $l(i, k)$ higher than the level of its useful signal that would happen with probability one for $N > 1$ since all other links have higher level $l(i) = l(i, 1) \le l(i, k)$; for $\forall k$. Link k will have its own list $L(k) = \{l(k, j)\} = \{l(k, 1), l(k, 2), \ldots, l(k, N)\}$ and its own level $l(k) = l(k, n)$ can be with same probability $\frac{1}{N}$ anywhere on the list $L(k)$. The level of the link i, proposing the coalition, is also uniformly distributed on the list $L(k)$. Extension to the case when these probabilities are arbitrary distribution functions is straightforward. Link k will accept the coalition if the level of the link proposing the coalition is higher than its own level on the list $L(k)$. This will occur with probability

$$p_c^{FCN}(N) = p\{l(k, i) > l(k, n) \text{ for } n = 0, 1, \ldots, i - 1, i + 1, \ldots, N\}$$

$$= \sum_{n=1}^{N} p_n p_{i > n \mid i \ne n} \tag{16.14}$$

and consequently,

$$p_c^{FCN}(N) = \begin{cases} \displaystyle\sum_{n=1}^{N} \dfrac{N - n}{N(N-1)} = \dfrac{1}{2} & \text{for } N \ge 2 \\[4mm] 0 & \text{for } N = 1 \end{cases} \tag{16.15}$$

In partially correlated network (PCN), with correlation factor $\rho = (1 - a/N)$, the transmitter of the specific link i, whose receiver identifies N interfering links, is not seen by a receivers from the set of the interfering links. In this case (16.11) should be modified as

$$p_c^{PCN(a)}(N) = p\{l(k, i) > l(k, n) \ \& \ l(k, i) \in L(k)$$
$$\text{for } n = 0, 1, \ldots, i - 1, i + 1, \ldots, N\} =$$

$$= \sum_{n=1}^{N-\alpha} p_n p_{i > n \mid i \ne n \& l(k,i) \in L(k)} \tag{16.16}$$

In the right hand side of (16.15), the a unobservable interferers are excluded from the list $L(k)$ and consequently,

$$p_c^{PCN(a)}(N) = \begin{cases} \displaystyle\sum_{n=1}^{N-a} \dfrac{N - a - n}{N(N-1)} = \dfrac{(N-a)(N-a-1)}{2N(N-1)} & \text{for } N \ge a + 1 \\[4mm] 0 & \text{for } N \le a \end{cases} \tag{16.17}$$

And by averaging of (16.17) over a using (16.13), we have

$$p_c^{PCN}(M, N) = \begin{cases} \displaystyle\sum_{a=0}^{N} \binom{N}{a} \left(1 - \dfrac{N}{M}\right)^a \left(\dfrac{N}{M}\right)^{N-a} \dfrac{(N-a)(N-a-1)}{2N(N-1)} & \text{for } N \ge 2 \\[4mm] 0 & \text{for } N = 1 \end{cases} \tag{16.18}$$

After some manipulations (16.18) is simplified as

$$p_c^{PCN}(M, N) = \begin{cases} \dfrac{1}{2}\left(\dfrac{N}{M}\right)^2 & \text{for } N \geq 2 \\ 0 & \text{for } N = 1 \end{cases} \tag{16.19}$$

When the coalition request in accepted, the process in Figure 16.3 will move to the level $G(N-1)$ which represents the same cluster of links where two links have created coalition that will be represented by coalition head which will be negotiating the further coalition process on behalf of the coalition. Effectively from that point on there will be $N-1$ negotiators.

Depending on the capacity of the underlying network, that determines the amount of information that can be distributed about the coalition, the level of interference caused by the coalition will be represented either by the level of the coalition head or the sum of the levels of the coalition members. If the proposal for coalition is not accepted, which occurs with probability $p_c'(N) = 1 - p_c(N)$, the process will remain on level $G(N)$ and the proposals for coalition will be initiated from the same level. In this case, coalition cluster is updated by substituting the link that has rejected the coalition request with a new one from $M - N$ remained links.

In the node 1 of the graph in Figure 16.1, the level of the user i on list $L(i)$ will not be the lowest with probability $p'(N) = 1 - p_1 = 1 - p(N)$ and the process will move to node 2, which represents the hypothesis that this level is the second on the list given that it is not the lowest. In general, node k represents the probability that the level of the useful signal is the k-th on the list given that it is not lower that k. Under these assumptions, the transition probabilities are represented in the graph. By summing up all possible ways to go from $G(N)$ back to $G(N)$ we have the transition probability $p(N, N)$ and to move to $G(N-1)$ the transition probability $p(N, N-1)$ as

$$p(N, N) = \sum_{k=0}^{N-1} p(N-k)p_c'(N-k)\prod_{i=0}^{k-1}p'(N-i)$$

$$= \frac{1}{N}\sum_{k=0}^{N-1}p_c'(N-k) \tag{16.20}$$

$$p(N, N-1) = \sum_{k=0}^{N-1}p(N-k)p_c(N-k)\prod_{i=0}^{k-1}p'(N-i)$$

$$= 1 - p(N, N) \tag{16.21}$$

and the graph from Figure 16.3(a) can be replaced by an equivalent graph from Figure 16.3(b), where z represents a fixed time T, in z transform, needed for initiation and decision on coalition forming.

By substituting (16.15) in (16.20) and (16.21), the transient probabilities for a FCN are derived simply as

$$p^{FCN}(N, N) = \frac{N+1}{2N} \tag{16.22}$$

$$p^{FCN}(N, N-1) = \frac{N-1}{2N} \tag{16.23}$$

From (16.18) we can see $p^{FCN}(N, N)$ is bounded as

$$(\text{for } N \gg 1) \quad 1/2 < p^{FCN}(N, N) \leq 1 \quad (\text{for } N = 1) \tag{16.24}$$

And for a PCN which is a more general model, transient probabilities are computed as follows,

$$
p_M^{PCN}(N, N) = \begin{cases} 1 - \dfrac{1}{2N} \displaystyle\sum_{k=0}^{N-2} \left(\dfrac{N-k}{M-k}\right)^2 & \text{for } N \geq 2 \\[4mm] 1 & \text{for } N = 1 \end{cases} \tag{16.25}
$$

$$
p_M^{PCN}(N, N-1) = \begin{cases} \dfrac{1}{2N} \displaystyle\sum_{k=0}^{N-2} \left(\dfrac{N-k}{M-k}\right)^2 & \text{for } N \geq 2 \\[4mm] 0 & \text{for } N = 1 \end{cases} \tag{16.26}
$$

From (16.25), the upper and lower bounds of $p^{PCN}(N, N)$ are as

$$
1 - \frac{N^2}{4M^2} \leq p_M^{PCN}(N, N) \leq 1 - \frac{(N-1)N}{2M^2} \tag{16.27}
$$

In (16.27) the upper bound is hold for $N = 1$ and lower bound is hold for $N = 2$. Both equations (16.22) and (16.24) for $N = 1$, give the same result i.e. $p^{FCN}(1, 1) = p^{PCN}(1, 1) = 1$ which is an obvious result because in this case there are no remaining coalition candidates.

16.3.4   Modeling Coalition Game Dynamics

16.3.4.1   Link Activation/Inactivation

If during the coalition forming a new link becomes active and its level falls within the range of the list $L(i)$, which occurs with probability p^+, the process will have one more singleton which is presented in the graph in Figure 16.4(a) by a transition to the previous state, from state $G(N-k)$ to $G(N-k+1)$. The process may be designed either to extend the size of the cluster to $G(N+1)$ if arrivals occur at state $G(N)$ or to keep the size of the cluster constant. Reordering of the list $L(i)$ is necessary in both cases.

Any inactivation of the link, which happens with probability p^-, may cause additional changes in the process that is presented in graph from Figure 16.4(b). If a singleton becomes inactive, which probability is $p^1(N-k)$, the process will move to the next state, form state $G(N-k)$ to $G(N-k-1)$. If a link, from a coalition of m members, becomes inactive, the coalition will split and the process will move to different states from the graph. The probability of moving i steps back to state $G(N-k+i)$ is equal to probability that the link will become inactive while belonging to coalition of size i after which all members of coalition will become singletons. This will be given as

$$
p_k^{+i} = \frac{\dbinom{k}{i}}{\displaystyle\sum_{j=1}^{k} \dbinom{k}{j}} \tag{16.28}
$$

16.3.4.2   Channel Dynamics

Owing to mobility or fading the interfering signal level will change in time. In order to use the existing system models Figures 16.3–4, we will characterize the channel seen by link k, by a Markov model of

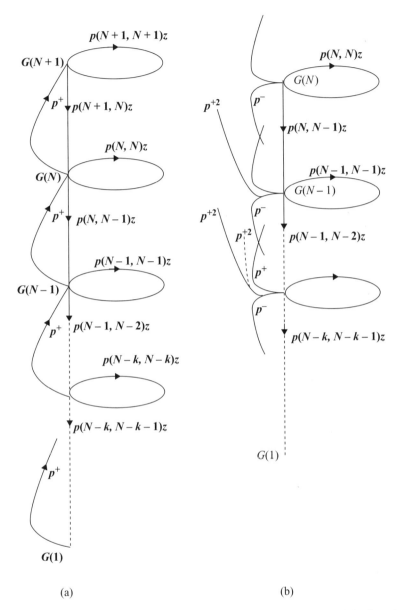

Figure 16.4   Modelling of the a) arrivals of a new link b) departure of a link during the coalition formation process.

its interference list $L(k)$. If due to changes in the level, a signal disappears from the list with probability p^- or join the list with probability p^+ then the same model used in previous section can be used.

16.3.5   *System Performance*

In this section, we evaluate the performance of CFP using two different methods first using process graph transfer function and second using absorbing Markov chain theory.

16.3.5.1 Performance Analysis Using Transfer Function

For a static network, represented by the graph in Figure 16.3, the transfer function between the point $G(N)$ and $G(1)$ representing the grand coalition is

$$U(z) = \prod_{k=2}^{N} U(k, z) = \prod_{k=2}^{N} \frac{p(k, k-1) z}{1 - p(k, k) z} \tag{16.29}$$

$$Ln(U(z)) = \sum_{k=2}^{N} [Ln(p(k, k-1) z) - Ln(1 - p(k, k) z)] \tag{16.30}$$

The average time for the process to reach the grand coalition starting form all singleton states and its variance are given as [44,45]

$$\tau = \left[\frac{\partial}{\partial z} LnU(z) \right]_{z=1} T \tag{16.31}$$

$$\sigma_\tau^2 = \left[\frac{\partial^2}{\partial z^2} U(z) + \frac{\partial}{\partial z} U(z) + \left(\frac{\partial}{\partial z} U(z) \right)^2 \right]_{z=1} T^2 \tag{16.32}$$

And from (16.29)–(16.32) for $N > 1$ we have,

$$\tau = T \sum_{k=2}^{N} \frac{1}{[1 - p(k, k)]} \tag{16.33}$$

$$\sigma_\tau^2 = T^2 \sum_{k=2}^{N} \frac{p(k, k)}{[1 - p(k, k)]^2} \tag{16.34}$$

For a FCN from (16.32), (16.33), and (16.34) we have

$$\tau^{FCN} = T \sum_{k=2}^{N} \left(\frac{2k}{k-1} \right) \tag{16.35}$$

$$\sigma_\tau^{2FCN} = T^2 \sum_{k=2}^{N} \frac{2k(k+1)}{(k-1)^2} \tag{16.36}$$

The upper and lower bounds for τ^{FCN} and σ_τ^{2FCN} are

$$2(N-1)T \le \tau^{FCN} < 2[N + Ln(N-1)]T \tag{16.37}$$

$$2(N-1)T^2 \le \sigma_\tau^{2FCN} < \left(2N + 4 + \frac{2\pi^2}{3} + 6\log(N-1) \right) T^2 \tag{16.38}$$

And for PCN from (16.25), (16.33), and (16.34) we have

$$\tau^{PCN} = T \sum_{k=2}^{N} \frac{2k}{\sum_{m=0}^{k-2} \left(\frac{k-m}{M-N+k-m} \right)^2} \tag{16.39}$$

$$\sigma_\tau^{2PCN} = T^2 \sum_{k=2}^{N} \frac{2k\left(2k - \sum_{m=0}^{k-2}\left(\frac{k-m}{M-N+k-m}\right)^2\right)}{\left(\sum_{m=0}^{k-2}\left(\frac{k-m}{M-N+k-m}\right)^2\right)^2} \qquad (16.40)$$

The upper and lower bounds for τ^{PCN} are

$$\frac{2(M-N+2)^2 \, Ln\,(N)^2}{N-1} < \tau^{PCN} < 4M^2\left(\frac{\pi^2}{6} - 1\right) \qquad (16.41)$$

The bounds in (16.39) are tight ones for $M \gg N$.

16.3.5.2   Performance Analysis Using Markov Theory

Owing to the multiple loops in the graph, *Markov chain theory* may be more convenient for representation of the process from Figures 16.4(a) and 16.4(b). By using standard theory of absorbing Markov chains, we can calculate the mean time τ and its variance σ_τ^2 for the dynamic coalition game starting from the initial state of all singleton coalitions to reach stable coalition structure. Let P represents the state transition probability matrix of an absorbing Markov chain in canonical form [25–28]

$$P = \begin{bmatrix} I & 0 \\ R & Q \end{bmatrix} \qquad (16.42)$$

where I is an identity matrix, 0 is a matrix with all zero entries, R is the matrix of transition probabilities from transient to absorbing states and Q is the matrix of transition probabilities between the transient states. The matrix

$$F = (I - Q)^{-1} \qquad (16.43)$$

is called the *fundamental matrix* of P. By using F, one can calculate the mean time τ for the game before the coalition game process converges to the absorbing state:

$$\tau = FT \qquad (16.44)$$

where τ is a column vector whose ith entry τ_i is the average time before the process reaches a stable CS, i.e., an absorbing state of the Markov chain, given that the process starts in state $i + 1$ and T is a column vector whose components are the respective state 'dwell' times. The last element of τ i.e. τ_{N-1} is the average time to reach G(1) starting at G(N) and therefore $\tau_{N-1} = \tau$. The variance of required time before absorption is calculated as

$$\sigma_\tau^2 = (2F - I) - \tau_{sq} \qquad (16.45)$$

where τ_{sq} is a column vector whose entries are the square of the entries of τ.

For CFP only the last state i.e. $N = 1$ is absorbing state and other states are transient states. Therefore $I = 1$ and elements of Q are defined as

$$q_{ij} = \begin{cases} P(i+1, j+1), & \text{if } i \in [j-1, j] \\ 0 & \text{otherwise, } \text{for } i, j \in [1\ N-1] \end{cases} \qquad (16.46)$$

and since there is only one absorbing state, R is a vector whose elements are defined as

$$r_i = P(i+1, i), \quad if \ i = 1, \ otherwise \ 0, \ for \ i \in [1 \ N-1] \tag{16.47}$$

16.3.6   Cluster Size Optimization

The size of the cluster N is the parameter to be optimized. This optimization is performed through maximization of the throughput received by the cluster head. Using (7) and (8) and definition of the coalition formation process, for the network with M links, available rate for a singleton with index 0 is

$$R_0 = \frac{1}{2} \log_2 \left(1 + \frac{S}{N_0 + \sum_{i \neq 0} I_i} \right) \tag{16.48}$$

If we assume that the same rate R_c, is dedicated for communication between the cluster head with each coalition candidate and that the spectra share is the same for any member of the coalition, the rate per member is

$$R_N = \frac{1}{2N} \log_2 \left(1 + \frac{y}{1 + \sum_{i \notin S^c} y_i} \right) - (N-1) R_c \tag{16.49}$$

where y is *SNR* of useful signal (signal of coalition head) and y_i is SNR of ith interfering signal. At the beginning, S^c is unknown and therefore the interference term $\sum_{i \notin S^c} y_i$ is estimated as follows,

$$\sum_{i \notin S^c} y_i = \sum_{i=1}^{M} y_i - \sum_{i \in S^c} y_i = M \bar{y} - \sum_{i \in S^c} y_i \tag{16.50}$$

Since interference coming from the member of coalition cluster is greater than useful signal term $\sum_{i \in S^c} y_i$ can be estimated as,

$$\sum_{i \in S^c} y_i \cong y + (N-1) E(y_i|y_i > y) = y + (N-1) \frac{\int_y^\infty x P(x) dx}{\int_y^\infty P(x) dx} \tag{16.51}$$

where $P(.)$ is probability distribution function for y_is which is dependent to the propagation environment and distribution of nodes. For getting rid of $P(y)$, we approximate right hand side of (16.51) as

$$E(y_i|y_i > y) = y + \frac{\int_0^\infty x P(x+y) dx}{\int_0^\infty P(x+y) dx} \cong y + \bar{y} \tag{16.52}$$

Consequently

$$\sum_{i \in S^c} y_i \cong Ny + (N-1) \bar{y} \tag{16.53}$$

From (16.49), (16.50), and (16.53) the rate per member as a function of N is expressed as

$$R_N = \frac{1}{2N} \log_2 \left(1 + \frac{y}{1 + (M - N + 1)\bar{y} - Ny} \right) - (N - 1) R_c \qquad (16.54)$$

Equation (16.54) shows that by increasing the N, not only the number of interferers decreases, but also the average of received interference from each interferer decreases and this is the direct result of selection of the candidates with stronger interference.

Extension to the fair distribution of the game value, proportional to the contribution to the gain is straightforward. The optimum N is calculated through the maximization of sum throughput. For illustration purposes as an example assume that R_c is proportional to the available data rate as

$$R_c = \alpha R = \frac{\alpha}{2} \log_2 \left(1 + \frac{y}{1 + (M - N + 1)\bar{y} - Ny} \right) \qquad (16.55)$$

Then R_N is derived as

$$R_N = \frac{1 - \alpha N (N - 1)}{2N} \log_2 \left(1 + \frac{y}{1 + (M - N + 1)\bar{y} - Ny} \right) \qquad (16.56)$$

where $\bar{y}$ is the average signal to noise ratio in S^c. In this case optimum N is the closest integer to the solution of the following equation

$$Ln \left(1 + \frac{y}{1 + M\bar{y} - N(y + \bar{y})} \right) =$$
$$\frac{N(1 - \alpha N) y (y + \bar{y})}{(1 + \alpha N^2)(1 + (M - N + 1)\bar{y} - Ny)(1 + (M - N + 1)\bar{y} - (N - 1)y)} \qquad (16.57)$$

In the case where R_c is constant optimum, N is the closest integer to the solution to the following equation

$$R_c Ln(2) =$$
$$\frac{y(y + \bar{y})}{(1 + (M - N + 1)\bar{y} - Ny)(1 + (M - N + 1)\bar{y} - (N - 1)y)} \qquad (16.58)$$
$$- \frac{1}{2N^2} Ln \left(1 + \frac{y}{1 + (M - N + 1)\bar{y} - Ny} \right)$$

16.3.7   Performance Example

In this Section, we present simulations results for the above models. All nodes are assumed uniformly distributed on a rectangular and results are averaged over 20 000 independent runs. Figures 16.5–8 present throughput and average optimum cluster size w.r.t. SNR for both fix and proportional R_c and compare throughput achieved by them with no coalition case i.e. when $N = 1$.

Figure 16.5 presents optimum cluster size for $M = 10$ and 25. For low SNR values when the performance is rather dependent to SNR than interference, coalition does not much improve the performance and therefore the optimum cluster size is small. On the other hand, for large values of R_c and α, optimum cluster size is small because larger cluster sizes need more dedicated bandwidth for coalition negotiation. Figure 16.6 confirms this result and we can see for large values of R_c and α coalition does not much improve the throughput. In a practical network, when nodes are stationary the coalition negotiation is performed only once and when we have a dynamic network, further negotiations are necessary upon

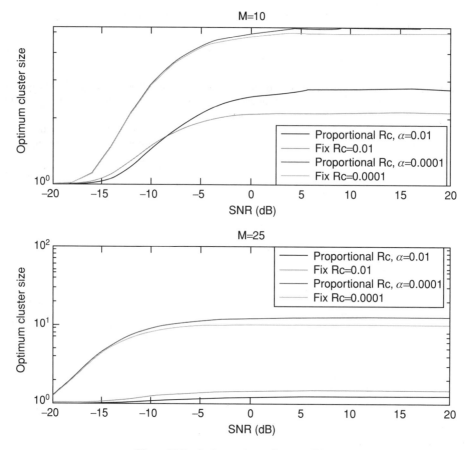

Figure 16.5   Optimum cluster size w.r.t. SNR.

mobility in the networks and actually the average required R_c is low. From Figure 16.5, we can see, for very small values of R_c and α the average optimum cluster size is almost $M/2$. In this case and from Figure 16.7, we can see the throughput gain comparing to no coalition case is large especially for M = 25 where the optimum cluster size is larger and using coalition provides higher possibility to increase the throughput.

Figure 16.8 presents numerical results of τ and σ_τ^2 w.r.t. N for different values of M and normalized $T = 1$. It is easy to see that curves for both τ and σ_τ^2 have the same shape but just they have different scales. In all cases for smaller values of N/M more snapshots are required for coalition negotiations. This is because for small N/M the number of rejected coalition requests is higher.

16.4   HD Game Based TCP Selection

The Internet today can accept the evolution of protocols at various network layers. On the TCP layer the end users have a large amount of freedom in the choice of the protocols they use for congestion control. In the absence of a centralized policy for a global deployment of a unique protocol to perform a given task, we are witnessing today in the Internet a competitive evolution between various versions of protocols. The evolution consists of the upgrading of existing protocols, abandonment of some protocols, and appearance of new ones. In this section we discuss the modeling capabilities of

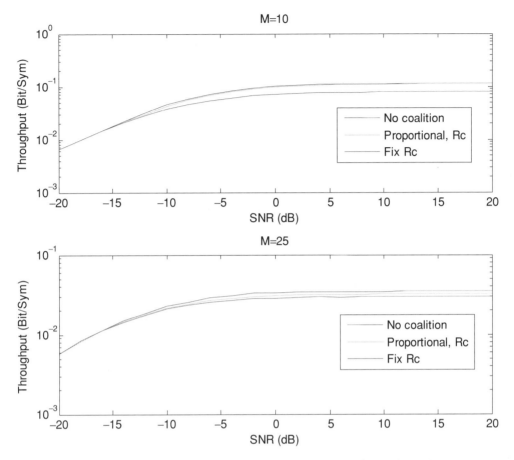

Figure 16.6   Throughput w.r.t. SNR when cluster size is optimized for $\alpha = R_c = 0.01$.

the evolutionary game paradigm for explaining past evolution and predicting the future one [48]. In particular, using this paradigm we derive conditions under which a successful protocol would dominate and suppress the use of other protocols, or various competing protocols could coexist. In the latter case we also predict the share of users that would use each of the protocols. We further use evolutionary games to propose guidelines for upgrading protocols in order to achieve desirable stability behavior of the system.

When transferring data between nodes, flow control protocols are needed to regulate the transmission rates so as to adapt to the available resources. A connection that looses data units has to retransmit them later. In the absence of adaptation to the congestion, the on going transmissions along with the retransmissions can cause cumulative increase of congestion in the network and further retransmissions by this and/or by other connections. This type of phenomenon, motivated the evolution of the Internet transport protocol, TCP, to a protocol that reduces dramatically the connection's throughput upon congestion detection. For the details on variety of adaptive TCP's see [49], Chapter 9. The possibilities to deploy freely new versions of protocols on terminals connected to the Internet creates a competing environment between protocols. There are two main approaches for predicting whether one version of a protocol would dominate another. In *Inter-Population Competition* (*IRPC*) one examines local interactions between connections of different types that interact with each other (by sharing some common bottleneck link). If a connection that corresponds to one version performs better in such an interaction,

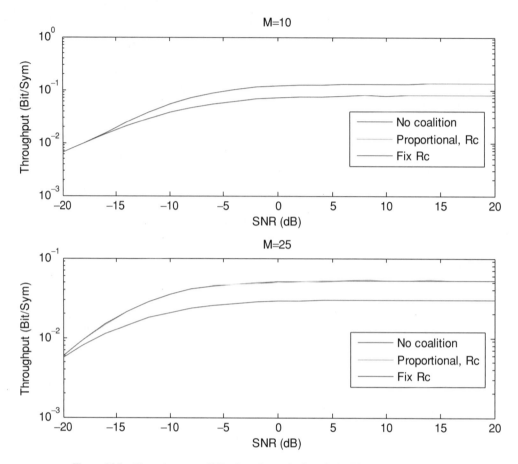

Figure 16.7   Throughput w.r.t. SNR when cluster size is optimized for $\alpha = R_c = 0.0001$.

then the IRPC approach predicts that it would dominate and that the other version would vanish. In *Intra-Population Competition (IAPC)* one studies the performance of a version of a protocol assuming a world where all connections use that version. This is repeated with the other version. One then predicts that the version that gives a better world would dominate.

Reference work [48] addresses the dominance question with the evolutionary game paradigm and provides an alternative answer along with a more detailed analysis of this competition scenario. This approach predicts whether one can expect one protocol to dominate the other or whether the two protocols can be expected to coexist. It provides the tools for computing the share of the population that is expected to use each version in case the versions would coexist.

In addition, it provides a description of the dynamics of the competition, which may result in stable behavior that consists of a convergence to some equilibrium, or it may display instabilities and oscillations. By identifying the conditions for stable behavior, one can provide recommendations for upgrading protocols so as to avoid undesirable oscillating behavior.

16.4.1   Evolutionary Stable Strategy

The *Evolutionary Stable Strategy* (ESS) is characterized by a property of robustness against invaders (mutations). If an ESS is reached, then the proportions of each population do not change in time and

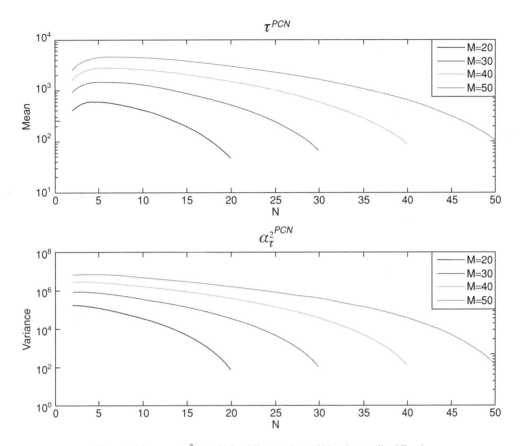

Figure 16.8 τ and σ_τ^2 w.r.t. N for different values of M and normalized T = 1.

the populations are immune from being invaded by other small populations. This notion is stronger than Nash equilibrium in which it is *only* requested that a single user would not benefit by a change (mutation) of its behavior.

Consider now a large population of players. Each individual needs occasionally to take some action. We focus on some (arbitrary) tagged individual. The actions of some M (in general random number) of other individuals may interact with the action of the tagged individual (e.g. some other connections share a common bottleneck). For simplicity, we shall restrict ourselves here to interactions that are limited to pairwise, i.e. to $M = 1$. This will correspond to networks operating at light loads.

We define by $J(p, q)$ the expected payoff for our tagged individual if it uses a strategy p when meeting another individual who adopts the strategy q. This payoff is called 'fitness' and strategies with larger fitness are expected to propagate faster in a population.

We assume that there are N pure strategies. A strategy of an individual is a probability distribution over the pure strategies. An equivalent interpretation of strategies is obtained by assuming that individuals choose pure strategies and then the probability distribution represents the fraction of individuals in the population that choose each strategy.

Suppose that the whole population uses a strategy q and that a small fraction ε (called mutations) adopts another strategy p. Evolutionary forces are expected to select q against p if:

$$J(q, \varepsilon p + (1 - \varepsilon)q) > J(p, \varepsilon p + (1 - \varepsilon)q) \qquad (16.59)$$

A strategy q is said to be ESS if for every $p \neq q$ there exists some $\hat{\varepsilon}_y > 0$ such that (16.59) holds for all $\varepsilon \in (0, \hat{\varepsilon}_y)$. In fact, we expect that if for all $p \neq q$,

$$J(q, q) > J(p, q) \tag{16.60}$$

then the mutations fraction in the population will tend to decrease (as it has a lower reward, meaning a lower growth rate). The strategy q is then immune to mutations. If it does not but if still the following holds,

$$J(q, q) = J(p, q) \text{ and } J(q, p) > J(p, p) \forall p \neq q \tag{16.61}$$

then a population using q are 'weakly' immune against a mutation using p since if the mutant's population grows, then we shall frequently have individuals with strategy q competing with mutants; in such cases, the condition $J(q, p) > J(p, p)$ ensures that the growth rate of the original population exceeds that of the mutants. A strategy is ESS if and only if it satisfies (16.60) or (16.61) [50]. The conditions to be an ESS can be related to and interpreted in terms of Nash equilibrium in a matrix game. The situation in which an individual, say player 1, is faced with a member of a population in which a fraction p chooses strategy A is then translated to playing the matrix game against a second player who uses mixed strategies (randomizes) with probabilities p and $1 - p$, resp. The central model that we shall use to investigate protocol evolution is introduced in the sequel along with its matrix game representation. For more on the relation between ESS and Nash equilibria, see [50].

The Hawk and Dove (HD) *game* that will be used in the sequel is based on the following rational. Consider a large population of animals. Occasionally two animals find themselves in competition on the same piece of food. An animal can adopt an aggressive behavior (*Hawk*) or a peaceful one (*Dove*). The matrix in Table 16.1 presents the fitness of an arbitrary player I associated with the possible outcomes of the game as a function of the actions taken by each one of the two players. We assume a symmetric game so the utilities of any animal (player II) as function of its actions and those of a potential adversary (player I), are the same as those player I depicted in Table 16.1(a).

The utilities (i.e. fitness) represent the following: The entry on the position $D - D$ is a result of a peaceful, equal-sharing of the food which translates to a fitness of 0.5 to each player.

The entry on the position H-H is a result of a fight in which with equal chances, one or the other player obtains the food but also in which there is a positive probability for each one of the animals to be wounded. Then the fitness of each player is $0.5 - d$, where the 0.5 term is as in the $D - D$ position and the $-d$ term represents the expected loss of fitness due to being injured.

The entries on the positions $H - D$ or $D - H$ are results in zero fitness to the D and in one unit of utility for the H that gets all the food without fight.

One can think of other scenarios that are not covered in the original $H - D$ game, such as the possibility of a Hawk to find the Dove, in a $H - D$ entry, more delicious than the food they compete over. A generalized version [51] of the HD game given in Table 16.1(b) is characterized by $A_{11} < A_{22} < A_{12}$ and $A_{21} < A_{22}$. In that case, if $A_{11} > A_{21}$ then the pure strategy H is the unique ESS. If $A_{11} < A_{21}$ then there is a unique ESS $p = (p_L, p_H)$, it is a mixed strategy given by $p_H = u/(u + v)$ where $A_{ij} = J(i, j), i, j \in \{H, D\}, u = A_{12} - A_{22}, v = A_{21} - A_{11}$.

At this stage we define the replicator dynamics which describes the evolution in the population of the various strategies. In the replicator dynamics, the share of a strategy in the population grows at a

Table 16.1 A HD game in matrix form

		Player II					Player II	
		H	D				H	D
a)	H	$0.5 - d$	1	b)		H	A_{11}	A_{12}
Player I	D	0	0.5		Player I	D	A_{21}	A_{22}

rate equal to the difference between the payoff of that strategy and the average payoff of the population. Specifically, consider N strategies and let x be the N-dimensional vector whose ith element x_i is the population share of strategy i (i.e. the fraction of the population that uses strategy i). Thus, we have $\sum_i x_i = 1$ and $x_i \geq 0$. In the sequel we denote by $J(i, k)$ the expected payoff (or the fitness) for a player using strategy i when it encounters a player with strategy k. If we define $J(i, \mathbf{x}) = \sum_j J(i, j)x_j$ the replicator dynamics is defined as

$$\dot{x}_i(t) = x_i K (J(i, \mathbf{x}) - \sum_j x_j J(j, \mathbf{x}))$$

$$= x_i K \left(\sum_j x_j J(i, j) - \sum_j \sum_k x_j J(j, k)x_k \right) \tag{16.62}$$

where K is a positive constant and $\dot{x}_i(t) := dx_i(t)/dt$. Note that the right hand side vanishes when summing over i. This is compatible with the fact that we study here the share of each strategy rather than the size of the population that uses each one of the strategies.

In Eq. (16.62), the fitness of strategy i at time t has an instantaneous impact on the rate of growth of the population size that uses it. A more realistic model for replicator dynamic would have some delay. In other words, the fitness acquired at time t will have impact on the rate of growth τ time later. We then have:

$$\dot{x}_i(t) = x_i K \left(\sum_j x_j(t - \tau) J(i, j) - \sum_j \sum_k x_j J(j, k)x_k(t - \tau) \right) \tag{16.62}$$

where K is some positive constant.

To relate the above discussion with the TCP selection problem we review few basic characteristics of these protocols. For the detail on a variety of adaptive TCPs see [49], Chapter 9. There are various versions of the TCP protocol among which the mostly used one is New-Reno. The degree of 'aggressiveness' varies from version to version. The behavior of New-Reno is approximately AIMD (Additive Increase Multiplicative Decrease): it adapts to the available capacity by increasing the window size in a linear way by adding α packets every round trip time and when it detects congestion it decreases the window size to β (multiply) times its value. In New-Reno the constants α and β are 1 and 1/2, respectively.

More aggressive TCP versions are also available, such as HSTCP (High Speed TCP) [52] and Scalable TCP [53–55]. HSTCP can be modeled by an AIMD behavior where α and β are not constant anymore: α and β have minimum values of 1 and of 1/2, respectively and both increase as the window size increases. In other words, the values of α and β increase as new acknowledgements arrive, as long as no congestion is detected. Scalable TCP is an MIMD (Multiplicative Increase Multiplicative Decrease) protocol, where the window size increases exponentially instead of linearly and is thus more aggressive. Versions of TCP which are less aggressive than the New-Reno also exist, such as Vegas [56].

The performance of networks in which various transport protocols coexist, are analyzed in [57–61]. In all these papers, the population size using each type of protocol is fixed. ˙

Competition between aggressive and well behaved congestion control mechanisms within a game theoretic approach is analyzed in [48, 62, 63]. Their conclusions in a wire line context was that if connections can choose selfishly between a well behaved cooperative behavior and an aggressive one then the Nash equilibrium is obtained by all users being aggressive and thus in a congestion collapse.

Reference work [48] introduces two models of competition between TCP versions, both can be modeled within the framework of the Hawk and Dove game. This allows us to predict whether a given version of TCP is expected to dominate others (ESS in pure strategies, which means that some versions of TCP would disappear) or whether several versions would coexist. The first model is adapted to competition in wireless networks and the second to wireline networks. In the following we will study only the first model for competition in wireless networks.

16.4.2   TCP Protocol Competition in Wireless Networks

TCP performances in terms of energy consumption and average goodput within wireless networks have been studied in [49, 64, 65]. It was shown that the TCP New-Reno can be considered as well performing within wireless environment among all other TCP variants and allows for greater energy savings. Indeed, a less aggressive TCP, as TCP New-Reno, may generate lower packet loss than other aggressive TCP. Thus the advantage of an aggressive TCP in terms of throughput could be compensated with energy efficiency of a more gentle TCP version. In next sections, we shall illustrate another consideration that affects the competition between TCP versions. The goal of this section is to illustrate this point, as well as its possible impact on the evolution of the share of TCP versions, through a simple model of an aggressive TCP.

We consider two populations of connections, all of which use AIMD TCP. A connection of population i is characterized by a linear increase rate α_i and a multiplicative decrease factor β_i. Let $\zeta_i(t)$ be the transmission rate of connection i at time t. We consider the following model for competition:

(a) The RTT (round trip times) are the same for all connections; (b) There is light traffic in the system in the sense that a connection either has all the resources it needs or it shares the resources with one other connection; (c) Losses occur whenever the sum of rates reaches the capacity C : $\zeta_1(t) + \zeta_2(t) = C$; (d) Losses are synchronized: when the combined rates attain C, both connections suffer from a loss. The rate of connection i is reduced by the factor $\beta_i < 1$.e) As long as there are no losses, the rate of connection i increases linearly by a factor α_i.

We say that a TCP connection i is more aggressive than a connection j if $\alpha_i \geq \alpha_j$ and $\beta_i \geq \beta_j$. Let $\bar{\beta_i} = 1 - \beta_i$. Let y_n and z_n be the transmission rates of connection i and j, respectively, just before a loss occurs so $y_n + z_n = C$. Just after the loss, the rates are $\beta_1 y_n$ and $\beta_2 z_n$. It takes $T_n = (C - \beta_1 y_n - \beta_2 z_n) / (\alpha_1 + \alpha_2)$ time to reach again C.

This yields $y_{n+1} = \beta_1 y_n + \alpha_1 T_n = q y_n + (\alpha_1 C \beta_2) / (\alpha_1 + \alpha_2)$, where $q = (\alpha_1 \beta_2 + \alpha_2 \beta_1) / (\alpha_1 + \alpha_2)$. The solution is given by

$$y_n = q^n y_0 + \left(\frac{\alpha_1 C \beta_2}{\alpha_1 + \alpha_2} \right) \frac{1 - q^n}{1 - q} \tag{16.63}$$

Since TCP retransmits lost packets, losses present energy inefficiency and the loss rate is included explicitly in the utility of a user through the term representing energy cost. We thus consider fitness of the form $J_i = Thp_i - \lambda R$ for connection i which is the difference between the throughput Thp_i and the loss rate R weighted by the so called tradeoff parameter, λ. This parameter allows us to model the tradeoff between the valuation of losses and throughput in the fitness. Next we show that our competition model between aggressive and non-aggressive TCP connections can be formulated as a HD game. We study how the fraction of aggressive TCP in the population at (the mixed) ESS depends on the tradeoff parameter λ.

Since $|q| < 1$, we get the following limit y of y_n when $n \to \infty$:

$$\bar{y} = \left(\alpha_1 C \bar{\beta_2} \right) / \left((\alpha_1 + \alpha_2)(1 - q) \right) = \left(\alpha_1 \bar{\beta_2} C \right) / \left(\alpha_1 \bar{\beta_2} + \alpha_2 \bar{\beta_1} \right)$$

It can be seen that the share of the bandwidth (just before losses) of a user is increasing in its aggressiveness. Hence the average throughput of connection 1 is $Thp_1 = ((1 + \beta_1) \alpha_1 \beta_2 C) / (2 (\alpha_1 \beta_2 + \alpha_2 \beta_1))$. The average loss rate of connection 1 is the same as that of connection 2 $R = (1/T) = (\alpha_1/\beta_1 + \alpha_2/\beta_2)/C$ where $T = (\bar{\beta_1} \beta_2 C) / (\alpha_1 \beta_2 + \alpha_2 \beta_1)$ with T being the limit as $n \to \infty$ of T_n.

If H corresponds to (α_H, β_H) and D to (α_D, β_D) with $\alpha_H \geq \alpha_D$ and $\beta_H \geq \beta_D$, then, for $i = 1, 2$, $Thp_i(H, H) = Thp_i(D, D)$. Since the loss rate for any user is increasing in $\alpha_1, \alpha_2, \beta_1, \beta_2$ it then follows that $J(H, H) < J(D, D)$, and $J(H, D) < J(D, D)$. We conclude that the utility that describes a tradeoff between average throughput and the loss rate leads to the HD game structure.

The mixed ESS is given by the following probability of using H[51,64]:
$x^*(\lambda) = (\eta_1 - \eta_2\lambda)/\eta_3$ where

$$\eta_1 = \left(\bar{\mu}\frac{1+\beta_1}{2} - \frac{1+\beta_2}{4}\right)C, \quad \eta_2 = \frac{1}{C}\left(\frac{\alpha_1}{\bar{\beta}_1} - \frac{\alpha_2}{\bar{\beta}_2}\right) \tag{16.64}$$

$$\eta_3 = C\left(\frac{1}{2} - \mu\right)\frac{\beta_1 - \beta_2}{2}, \quad \mu = \frac{\alpha_2(\bar{\beta}_1)}{\alpha_2(\bar{\beta}_1) + \alpha_1(\bar{\beta}_2)} \tag{16.65}$$

where $\bar{\mu} = 1 - \mu$. Note that η_2 and η_3 are positive. Hence, the equilibrium point x^* decrease linearly on λ. We conclude that applications that are more sensitive to losses would be less aggressive at ESS.

References

[1] Bala, V. and Goyal, S. (2000) A noncooperative model of network formation, *Econometrica*, **68**(5), 1181–1230.
[2] Jackson, M. O. and Wolinsky, A. (1996) A strategic model of social and economic networks, *J. Econ. Theory*, **71**(1), 44–74.
[3] Jackson, M. O. (2003). *Group Formation in Economics*, G. Demange and M. Wooders (eds) Cambridge, Cambridge University Press, 11–57.
[4] Myerson, R. B. (1977) Graphs and cooperation in games, *Math. Oper. Res.*, **2**(3), 225–229.
[5] Osborne, M. J. and Rubinstein, A. (1990) *Bargaining and Markets*. San Diego, CA: Academic Press.
[6] Muthoo, A. (1999) *Bargaining Theory with Applications*. Cambridge, Cambridge University Press.
[7] Arcaute, E., Johari, R. and Mannor, S. (2009) Network formation: bilateral contracting and myopic dynamics, *IEEE Transactions on Automatic Control*, **54**(8), 1765–1778.
[8] Arcaute, E., Johari, R. and Mannor, S. (2007) Network formation: bilateral contracting and myopic dynamics, Management Science and Engineering Department, Stanford University, Stanford, CA, Tech. Rep.
[9] Osborne, M. J. and Rubinstein, A. (1990) *Bargaining and Markets*, Academic Press, San Diego, California.
[10] Van Den Berg, E., Fecko, M. A., Samtani, S., Lacatus, C. and Patel, M. (2010) Cognitive topology control based on game theory, *IEEE MILCOM*, San Jose, CA, Oct.–Nov., 2010 Paper #1569299231.
[11] La, R. and Anantharam, V. A game-theoretic look at the Gaussian multi-access channel, *DIMACS Series in Discrete Math. Theoretical Comp. Sci.*, **66**,.87–106.
[12] Mathur, S., Sankaranarayanan, L. and Mandayam, N. (2008) Coalitions in cooperative wireless networks, *IEEE J. Select. Areas Commun.*, **26**, 1104–1115.
[13] Han, Z. and Liu, K. J. (2008) *Resource Allocation for Wireless Networks: Basics, Techniques, and Applications*. Cambridge, Cambridge University Press.
[14] Myerson, R. B. (1991) *Game Theory, Analysis of Conflict*. Cambridge, MA: Harvard University Press.
[15] Basar, T. and Olsder, G. J. (1999) *Dynamic Noncooperative Game Theory* (Series in Classics in Applied Mathematics). Philadelphia, PA, SIAM.
[16] Owen, G. (1995) *Game Theory*, 3rd ed. London, Academic.
[17] von Neumann, J. and Morgenstern, O. (1944) *Theory of Games and Economic Behaviour*. Princeton, NJ, Princeton University Press.
[18] Aumann, R. J. and Peleg, B. (1960) Von neumann-morgenstern solutions to cooperative games without side payments, *Bull. Amer. Math. Soc.*, **6**(3), 173–179.
[19] Adlakha, S., Johari, R. and Goldsmith, A. (2007) Competition in wireless systems via Bayesian interference games – *Arxiv preprint, arXiv.*
[20] Etkin, R., Parekh, A. and Tse, D. (2007) Spectrum sharing for unlicensed bands. *IEEE Journal on Selected Areas in Communications*, **25**(3), 7–528.
[21] Laufer, A. and Leshem, A. (2005) Distributed coordination of spectrum and the prisoner's dilemma, in *Proc. of the First IEEE Symposium on New Frontiers in Dynamic Spectrum Access Networks (DySpan)*, 94–100.
[22] Han, Z. and Poor, V. (2009) Coalition games with cooperative transmission: A cure for the curse of boundary nodes in selfish packet-forwarding wireless networks, *IEEE Trans. Commun.*, **57**, 203–213.
[23] Madiman, M. (2008) Cores of cooperative games in information theory, *EURASIP J. Wireless Commun. Networking*, vol. 2008.
[24] Aram, A., Singh, C. and Sarkar, (2009) Cooperative profit sharing in coalition based resource allocation in wireless networks, in *Proc. IEEE INFOCOM*, Rio de Janeiro, Brazil.

[25] Cai, J. and Pooch, U. (2004) Allocate fair payoff for cooperation in wireless ad hoc networks using shapley value, in *Proc. Int. Parallel and Distributed Processing Symp.*, Santa Fe, NM, April, 219–227.

[26] Sandholm, T., Larson, K., Anderson, M., Shehory, O. and Tohme, F. (1999) Coalition structure generation with worst case guarantees, *Artifi. Intell.*, **10**, 209–238.

[27] Ray, D. (2007) *A Game-Theoretic Perspective on Coalition Formation.* New York, Oxford University Press.

[28] Apt, K. and Witzel, A. (2006) A generic approach to coalition formation, in *Proc.Int. Workshop Computational Social Choice (COMSOC)*, Amsterdam, The Netherlands.

[29] Arnold, T. and Schwalbe, U. (2002) Dynamic coalition formation and the core, *J.Econ. Behavior Org.*, **49**, 363–380.

[30] Saad, W., Han, Z., Debbah, M. and Hjørungnes, A. (2008) A distributed merge and split algorithm for fair cooperation in wireless networks, in *Proc. Int. Conf. Communications, Workshop Cooperative Communications and Networking*, Beijing, China, May, 311–315.

[31] Saad, W., Han, Z., Debbah, M., Hjørungnes, A. and Basar, T. (2009) Coalitional games for distributed collaborative spectrum sensing in cognitive radio networks, in *Proc. IEEE INFOCOM*, Rio de Janeiro, Brazil, Apr.

[32] Saad, W., Han, Z., Basar, T., Debbah, M. and A. Hjørungnes (2009) Physical layer security: Coalitional games for distributed cooperation, in *Proc. 7th Int. Symp. Modeling and Optimization in Mobile, Ad Hoc, and Wireless Networks (WiOpt)*, Seoul, June.

[33] Saad, W., Han, Z., Basar, T., Debbah, M. and A. Hjørungnes (2009) A selfish approach to coalition formation among unmanned aerial vehices in wireless networks, in *Proc. Int. Conf. Game Theory for Networks*, Istanbul, Turkey, May.

[34] Myerson, R. (1977) Graphs and cooperation in games, *Math. Oper. Res.*, **2**, 225–229, June.

[35] Herings, P., van der Laan, G. and Talman, D. (2002) Cooperative games in graph structure, Res. Memoranda, Maastricht Res. Sch. Econ., Technol., Org.,Memo. Maastricht, The Netherlands, no. 11, Aug.

[36] Jackson, M. O. (2003) A survey of models of network formation: Stability and efficiency, California Inst. Technol., Div. Humanities Social Sci., CA, Rep. 1161, Nov.

[37] Demange, G. and Wooders, M. (2006) *Group Formation in Economics.* New York: Cambridge University Press.

[38] Bala, V. and Goyal, S. (2000) Noncooperative model of network formation, *Econometrica*, **68**, 1181–1230.

[39] Derks, J., Kuipers, J., Tennekes, M. and Thuijsman, M. (2008) Local dynamics in network formation, in *Proc. 3rd World Congress Game Theory Soci.*, IL, July.

[40] Saad, W., Han, Z., Debbah, M. and Hjørungnes, A. (2008) Network formation games for distributed uplink tree construction in IEEE 802.16j networks, in *Proc. IEEE Global Communication Conf.*, New Orleans, LA, 1–5.

[41] Saad, W., Han, Z., Debbah, M., Hjørungnes, A. and Basar, T. (2009) A game-based self-organizing uplink tree for VoIP services in IEEE 802.16j networks, in *Proc. Int. Conf. Communications*, Dresden, Germany, June.

[42] Arcaute, E., Johari, R., and Mannor, S. (2007) Network formation: Bilateral contracting and myopic dynamics, *Lecture Notes Comput. Sci.*, vol. **4858**, 191–207, Dec.

[43] Johari, R., Mannor, S. and Tsitsiklis, J. (2006) A contract based model for directed network formation, *Games Econ. Behaviour*, **56**, 201–224.

[44] Holmes, J. and Chen, C. (1977) Acquisition time performance of PN spread-spectrum systems, *IEEE Transactions on Communications*, **25**(8), 778–784.

[45] Glisic, S. and Vucetic, B. (1997) *Spread Spectrum CDMA Systems for Wireless Communications*, Artech House.

[46] Kemeney, J. G. and Snell, J. L. (1976) *Finite Markov Chains*, Springer-Verlang.

[47] Howard, R. A. (1960) *Dynamic Programming and Markov Processes*, John Wiley & Sons, Ltd.

[48] Altman, E., El-Azouzi, R., Hayel, Y. and Tembine, H. (2009) The evolution of transport protocols: An evolutionary game perspective, *Computer Networks* 53 1751–1759.

[49] Glisic, S. and Lorenzo, B. (2009) Advanced wireless networks, cognitive, cooperative and opportunistic 4G technology, John Wiley and Sons, Chichester, London.

[50] Weibull, J. (1995) *Evolutionary Game Theory*, MIT Press.

[51] van Damme, E. (1991) *Stability and Perfection of Nash Equilibria*, Springer, Berlin.

[52] Souza, E., Agarwal, D. A. (2002) A highspeed TCP study: characteristics and deployment issues, *Technical Report LBNL-53215*, Lawrence Berkeley National Laboratory.

[53] Kelly, T. (2003) Scalable TCP: improving performance in highspeed wide area networks, *Comput. Commun. Rev.*, **32**(2)).

[54] El-Khoury, R. and Altman, E. (2004) Analysis of scalable TCP, in *HET-NETs'04 International Working Conference on Performance Modelling and Evaluation of Heterogeneous Networks*, West Yorkshire, UK.

[55] Altman, E., Avratchenkov, K., Barakat, C., Kherani, A. A. and Prabhu, B. J. (2004) Analysis of scalable TCP, in *Proceedings of the Seventh IEEE International Conference on High Speed Networks and Multimedia Communications (HSNMC'04)*, Toulouse, France, June 30–July 2.

[56] Brakmo, L. S., O'Malley, S. and Peterson, L. L. (1994) TCP Vegas: new techniques for congestion detection avoidance, *Comput. Commun. Rev.* **24**(4), 24–35.

[57] Tang, A., Wang, J., Low, S. and Chiang, M. (2007) Equilibrium of heterogeneous congestion control: existence and uniqueness, *IEEE/ACM Trans.Netw.*

[58] Bonald, T. (1999) Comparison of TCP Reno and TCP Vegas: efficiency and fairness, *Perform. Eval.* **36–37**(1–4), 307–332.

[59] Ait-Hellal, O. and Altman, E. (2000) Analysis of TCP Vegas and TCP Reno, *Telecommun. Syst.* **15**, 381–404.

[60] EAltman, E., Avrachenkov, K. and Prabhu, B. (2005) Fairness in MIMD congestion control algorithms, *Telecommun. Syst.* **30**(4), 387–415.

[61] Lopez, L., Fernandez, A. and Cholvi, V. (2005) A game theoretic analysis of protocols based on fountain codes, in *IEEE ISCC'2005*, June.

[62] Lopez, L., Rey, G., Fernandez, A. and Paquelet, S. (2005) A mathematical model for the TCP tragedy of the commons, *Theor. Comput. Sci.*, **343**, 4–26.

[63] Garg, R., Kamra, A. and Khurana, V. (2002) A game-theoretic approach towards congestion control in communication networks, SIGCOMM *Comput. Commun. Rev., Arch.* **32**(3), 47–61.

[64] Tao, Y. and Wang, Z. (1997) Effect of time delay and evolutionarily stable strategy, *J. Theor. Biol.*, **187**, 111–116.

Index